Dov Bahat

Avinoam Rabinovitch

Vladimir Frid

Tensile Fracturing in Rocks

Tectonofractographic and Electromagnetic Radiation Methods

Dov Bahat
Avinoam Rabinovitch
Vladimir Frid

Tensile Fracturing in Rocks

Tectonofractographic and Electromagnetic Radiation Methods

With 302 Figures

Springer

Authors

Prof. Dov Bahat

Department of Geological and Environmental Sciences
Ben Gurion University of the Negev, POB 653
Beer Sheva, 84105 Israel
Head, the Deichmann Rock Mechanics Laboratory of the Negev
e-mail: bahat@bgumail.bgu.ac.il, phone: +972-8-6472623, fax: +972-8-6472997

Prof. Avinoam Rabinovitch

Physics Department
Ben Gurion University of the Negev, POB 653
Beer Sheva, 84105 Israel
The Deichmann Rock Mechanics Laboratory of the Negev
e-mail: avinoam@bgumail.bgu.ac.il, phone: +972-8-6461172

Dr. Vladimir Frid

Department of Geological and Environmental Sciences
Ben Gurion University of the Negev, POB 653
Beer Sheva, 84105 Israel
The Deichmann Rock Mechanics Laboratory of the Negev
e-mail: vfrid@bgumail.bgu.ac.il, phone: +972-8-6461770

Library of Congress Control Number: 2004108439

ISBN 3-540-21456-9 **Springer Berlin Heidelberg New York**

Springer is a part of Springer Science+Business Media
springeronline.com

Printed in Germany

Cover design: Erich Kirchner, Heidelberg
Typesetting: Büro Stasch (stasch@stasch.com), Bayreuth
Production: Luisa Tonarelli
Printing: Mercedes-Druck, Berlin
Binding: Stein & Lehmann, Berlin

Printed on acid-free paper 32/3141/LT – 5 4 3 2 1 0

Preface

'Tensile Fracturing in Rocks' presents field observations on fracturing of sedimentary rocks and granite outcrops from various provinces in three continents. It also combines results of recent experiments conducted at different laboratories around the world with current theories on fracturing. In treating faults, this book limits itself to faults that are associated with joint sets produced by definable causes and occasionally to cases where interaction between the two types of fracture – faults and joints – is not clear. The book's subject matter is divided over six chapters, which are briefly described below.

Chapter 1 summarizes current key concepts in fracture physics. It starts with a presentation of the elastic theory of fracture, and concentrates on the results of linear elastic fracture mechanics. The chapter touches also upon other fracture properties, e.g., crack nucleation, dynamic fracturing and slow fracturing processes. Nucleation is addressed by statistical mechanics methods incorporating modern approaches of thermal and fiber bundle processes. The analyses of dynamic fracturing and slow fracturing focus on the differences, as compared to the linear elastic approach. The controversy in interpreting experimental dynamic results is highlighted, as are the surface morphology patterns that emerge in fracturing and the non-Griffith crack extension criterion in very slow fracturing processes.

Chapter 2 concentrates on basic topics in fracture geology. Fractography is applied to improve the understanding of some intricacies, such as nucleation and fracture growth in rocks. A review of fourteen lab experiments and a numerical experiment is followed by a discussion on the properties of mirror planes and fringes on joint surfaces. These are exemplified on sedimentary and granitic rocks. Concepts of conjugate hybrid joints and longitudinal splitting are re-evaluated in the light of recent experiments. Various aspects of joint classification are reviewed. The chapter's last section (2.4) relates to mutual relationships of primary and secondary fractures, such as wing cracks, pinnate joints and branching fractures.

Chapter 3 presents observations on sedimentary rocks in four fracture provinces: Two in the USA, one in south Britain and one in south Israel. Although not representing all types and styles of fracturing, the combined data allow a synthesis of joint formation in different tectonic settings.

Chapter 4 presents four fracture provinces in granites (in Germany, the Czech Republic and in California) that exhibit a wide spectrum of fracture settings. The syntectonic, uplift and post-uplift joint features resemble their equivalents in sedimentary rocks. However, granites also display unique fracture features that are associated with late cooling stages of the pluton.

Chapter 5 deals with a new method of investigating embryonic stages of rock failure by measuring high-frequency electromagnetic radiation (EMR) from micro-cracks. We review previous EMR investigations, present our new EMR model and show the extracted EMR 'fingerprints' in rocks and other materials when strained. EMR signals from laboratory tests as well as from quarry blasting are analyzed. Underground EMR preceding rock-bursts and gas outburst are studied, and some criteria for their forecasting based on the obtained parameters are derived. The EMR results described here are expected to lead to a novel forecasting method of earthquake nucleation.

Chapter 6 addresses some particular problems in fracture geology, including procedures for the study of joints. A discussion on previous paleostress studies in the Middle East, particularly in the Israeli Negev, leads to a model on regional jointing in the broad context of plate tectonics. This chapter contains a new approach in studying joint length and a brief account of the procedure of estimating the fracture paleostresses on a joint surface by a fractographic method. Also included is a detailed characterization of the unique fracture geometry of a fault termination zone. Additional topics are, the description of a variety of fault-joint geometric and genetic relationships and a characterization of a new joint category, termed 'surface joints'. We also take account of hydro-geological research that relies on fracture distribution data in the Beer Sheva syncline. Finally, we present here major features of a new mechanical model of *en echelon* segmentation that differs from the more conventional ideas.

This book is a sequel to a previous volume in this series, 'Tectonofractography' (1991), and many premises are not re-introduced here. Some terms are introduced in a special section on nomenclature.

Acknowledgements

The authors wish to thank all those who contributed to the writing of this book. Several persons contributed immensely to the field work and data collection. These include Peter and Elfriede Bankwitz who collaborated in the study of fracturing in granites in central Europe, particularly, the pre-uplift fracture during the cooling of the pluton. Jointing in sandstone (Zion National Park, Utah) and in granites (the Sierra Nevada Batholith, California) was studied in cooperation with Ken Grossenbacher and Kenzi Karasaki. Some of the initial observations that were made with them on post-uplift joints and oblique fractures during my (DB) Sabbatical year at LBNL matured during the years to concrete data, and they should share the credit for it. Section 4.12 was mostly contributed by Ken Grossenbacher. Important help was also given by Paul Witherspoon and Sally Benson. This book contains numerous citations from works by Terry Engelder and his associates that contributed greatly to the understanding of jointing. Working with Tsafrir Levi on his thesis has been a delight for one of us (DB). Julia Goldbaum enriched considerably our understanding of EMR phenomena. We also acknowledge the important contributions by Yoav Avni and Menachem Friedman, and instructive field trips with Yehuda Eyal and Ram Weinberger. Charles Howard contributed greatly to the writing of Sect. 6.8.

Special thanks are due to colleagues who critically reviewed chapters in draft: Chapter 2, Ram Weinberger; Chapter 3, Ithamar Perath; Chapter 4, Reginald Shagam; Chapter 6, Meir Abelson and Ronit Nativ.

Figures 2.35e and 3.35c were donated by Ran Kagan and Alon Raviv, respectively, and Fig. 2.69b and 3.10e were photographed during inspections of quarries with Peter and Elfriede Bankwitz.

Most of our experiments were conducted in the Ruth and Horst Deichmann Rock Mechanics Laboratory of the Negev, the Department of Geological and Environmental Sciences, the Ben Gurion University of the Negev.

For important technical help, we owe thanks to Vyacheslav Palchik, Dina, Isaak and Michael Frid, Julia and Boris Goldbaum, Avi Sharon, Leonid Baluashvili, Eli Shimshilashvili, Rivka Eini, Ran Novitsky, Hagit Lev, Ania Arbel, Daniela Vida, Michal Barima, Ronen Shavit from BGU, and Pascual Benito from LBNL. We also owe thanks to H. Hertrich from Solnhofer Bodenplatten-Industrie GmbH and Co KG for help in our experiments.

The help extended by members of the department of the Geological and Environmental Sciences was crucial to the completion of this book. Financial assistance was received from the Earth Sciences Administration (Ministry of Energy and Infrastructure), and thanks are due to Yosef Bartov for his interest and assistance. Support was also given by the Israel Science Foundation, the Ezvonot Committee of the Israeli Government Estate Distributions and by the Vatat and NRCN funds.

Many authors and publishers contributed valuable photographs and diagrams upon our request, enabling us to fundamentally upgrade our book. All readers are invited to write to us about errors that they find in the book.

The ideas presented in this book were beginning to crystallize when a sudden deterioration in my (DB) wife Hanna's health overshadowed all literary thinking. Dan and Merav, Zvi and Jonathan were with us during her last months of suffering. This book is dedicated to her memory.

Contents

Chapter 1

Fracture Physics

In this chapter we give a brief account of the theory of fracture, restricting ourselves to brittle fracture. The structure of this chapter is as follows. The first part describes the Griffith criterion of fracture initiation, since this criterion constituted the turning point in our understanding of the basics of fracture. We next consider the question of nucleation both of flaws and of fractures in a solid. This process has recently had a breakthrough and is related to the origin of fracture initiation. The general elastic theory of fracture is considered next – linear elastic fracture mechanics (LEFM) as the only complete theory and the rational extension of the Griffith approach. The analysis of two-dimensional problems, for some of which analytical solutions can be obtained, is emphasized. The last parts deal with two important topics that do not fall within and can not be dealt with by LEFM, namely, dynamic crack propagation and creep. In both of these topics, new advances have been made in recent years and both are therefore a subject matter for contemporary research.

A note of caution is in order. Although the Griffith criterion and the related linear elastic fracture mechanics methods have helped scientists in dealing with static cracks and engineers in building safer structures, the basic dynamics of the fracture problem are as of yet not well understood. This relative ignorance stems from the fact that on the one hand, fracturing is neither an equilibrium nor a near equilibrium process and is therefore not amenable to perturbation procedures, and on the other hand, fracturing, whose principle part occurs at a very restricted area and is highly nonlinear, influences the whole extension of the body in question, and it is therefore not easy to correlate between its different scales.

A full understanding of crack propagation would entail a quantum mechanical calculation of all atoms in a certain vicinity of the crack. Since each atom contains several electrons, a full "many-body" calculation (see, e.g., Fabrocini et al. 2002) should be made. Obviously, even numerically, this procedure is unachievable in the foreseeable future. A less presumptuous attempt would be to consider each atom as a point entity and to calculate fracture propagation by assuming a certain interaction among these entities. Even a classical (rather than a quantum mechanical) calculation would have to assume an approximation of only two or at most three body forces, and it would have to assume, moreover, that these two body forces (say of a Lennard Jones type; see, e.g., Ashcroft and Mermin 1981) can be extended to inter-atomic distances far greater (or smaller) than equilibrium ones. Even under such approximations, the number of atoms and the time period necessary for a useful calculation are as of yet beyond the scope of available computational means. Thus, the maximum number of atoms for which such numerical calculations have been tried is the boasted 10^9 (Abraham et al. 2002). However, for a meaningful dynamical calculation a number about a thousand times higher is needed. As for the time periods needed, these should be scaled by atomic vibrations, which entail at least around 10^6 time steps even for such a short time as 1 μs, rendering this task unattainable at the present stage.

There are at present several groups of scientists around the world who try to "connect" the different scales that appear in fracture calculations. The idea is to use a quantum mechanical type of calculation in order to obtain the correct two-body atomic force even for large inter-atomic distances. These forces are to be used only for a relatively small number of atoms near the crack tip and the so called "process zone" to obtain an accurate microscopic (classical) calculation. The latter should yield inter alia the energy needed for crack advance, the stresses and strains near the tip and the appearance and motion of dislocations for non-brittle materials. The boundary conditions for this calculation would be the connection to the major part of the material, which is to be treated in a continuous manner (linear or non-linear fracture mechanics). This ambitious plan is still in its infancy but promises to become a real asset in understanding fracture dynamics in the future.

1.1 The Griffith Criterion

1.1.1 Stress and Strain

Consider the bar in Fig. 1.1 whose initial length and cross section are l_0 and $A = b \times b$, respectively. We apply a force in the x-direction, which causes the bar to elongate by x. We define $\sigma_{xx} = f/A$, as the *stress* operating inside the bar in the x-direction. Hence, stress has the units of force per area. It is assumed that the bar elongates with the continued application of the force in such a way as to preserve elasticity, namely that it operates like a spring, maintaining a linear relationship between the applied stress f/A and the elongation, x, $f = kx$ (Hook's law). This relationship together with Newton's law $f = ma$ are the basis of all of elasticity theory (see, e.g., Timoshenko and Goodier 1984). Defining the non-dimensional variable $\varepsilon_{xx} = x/l_0$ as the strain in the bar (the change in length divided by the initial length), one can write Hook's law as

$$\varepsilon_{xx} = \frac{\sigma_{xx}}{E} \tag{1.1}$$

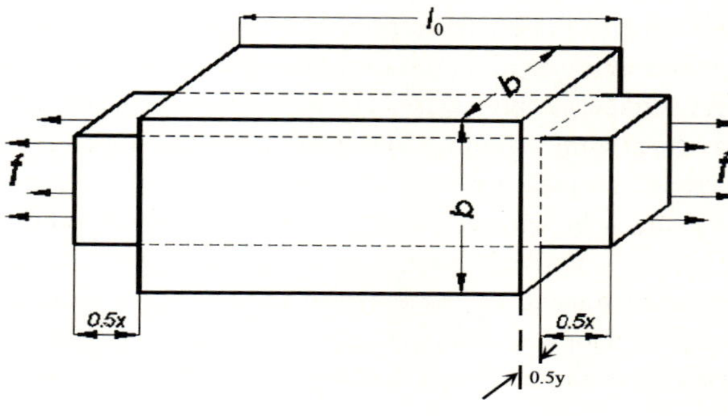

Fig. 1.1. Schematic stress-strain behavior of a bar ($l_0 \times b \times b$) under uniaxial force f

where E is Young modulus, related in the case of the bar, to the Hook's constant k through

$$k = \frac{f}{x} = \frac{\sigma_{xx}A}{\varepsilon_{xx}l_0} = \frac{AE}{l_0} \tag{1.2}$$

Under the same stress, there also occurs a (small) reduction of size of the bar in the lateral directions. Thus, if in the y-direction the decrease of width is denoted by $-y$ (the minus sign comes to denote a *reduction* of size), the strain in this direction is $\varepsilon_{yy} = -y/b$ and for elastic materials is given by

$$\varepsilon_{yy} = -\nu\varepsilon_{xx} = -\frac{\nu\sigma_{xx}}{E} \tag{1.3}$$

where ν is a non-dimensional constant called the Poisson ratio, which for different materials usually assumes values between 0.1 and 0.5. A similar reduction occurs in the z-direction (ε_{zz}).

An important issue to be used frequently in the following is the value of the elastic energy stored in the material. For the bar problem, suppose we apply the force quasi statically to avoid acceleration and kinetic energy and increase its magnitude from zero to a final value F. The elastic energy is given by the work needed to create the final state:

$$W = \int_0^{\Delta l} f(x)\mathrm{d}x = k\int_0^{\Delta l} x\mathrm{d}x = k\frac{\Delta l^2}{2} = \frac{l_0 AE\varepsilon_\mathrm{f}^2}{2} = \frac{VE\varepsilon_\mathrm{f}^2}{2} = V\frac{\varepsilon_\mathrm{f}\sigma_\mathrm{f}}{2} = V\frac{\sigma_\mathrm{f}^2}{2E} \tag{1.4}$$

where Δl is the final elongation of the bar, $\varepsilon_\mathrm{f} = \Delta l / l_0$ is the final strain, σ_f is the final stress ($= E\varepsilon_\mathrm{f}$) and V is the initial bar volume. We have therefore obtained the important result that the elastic energy per unit volume (W/V) stored in the sample is given by

$$\frac{W}{V} = \frac{1}{2}\sigma\varepsilon = \frac{1}{2}E\varepsilon^2 = \frac{1}{2}\frac{\sigma^2}{E} \tag{1.5}$$

Although this result was obtained for the specific case of a bar acted upon by a uniaxial force, this result is quite general and we will use it later, e.g., for the Griffith criterion.

In the general case, i.e., for a body of a general shape acted upon by a general distribution of forces and displacements, elasticity theory considers an infinitesimal cube inside the body and the force components per unit area on each side of the cube constitute the stress *tensor* σ_{ij}, where $i = 1,2,3$ and $j = 1,2,3$ denote the x-, y- and z-coordinates, respectively. Thus, for example $\sigma_{12} = \sigma_{xy}$ denotes the force (per unit area) component in the y-direction acting on the cube face, which is perpendicular to the x-direction. This tensor can in general depend both on the location of the cube within the body, **r**, and on time. Therefore, σ is generally written as $\sigma_{ij}(r,t)$. It can be shown that σ is a symmetric tensor, namely

$$\sigma_{ij} = \sigma_{ji} \tag{1.6}$$

The force per unit area acting on a general surface perpendicular to a unit vector $\hat{n} = n_1\hat{x} + n_2\hat{y} + n_3\hat{z}$ ($\hat{x}, \hat{y}, \hat{z}$ are unit vectors in the x-, y- and z-directions respectively) is called the *traction* **t** and its components are given by

$$t_i = \sum_{j=1}^{3} \sigma_{ij} n_j \qquad (= \sigma_{i1}n_1 + \sigma_{i2}n_2 + \sigma_{i3}n_3) \tag{1.7}$$

Next, displacements from equilibrium position are defined such that under the action of forces, a point within the body initially (with no force) located at the position (x, y, z) is displaced to the position $(x + u_1, y + u_2, z + u_3)$ by the action of the forces. The displacement vector is thus denoted by $\mathbf{u} = u_1\hat{x} + u_2\hat{y} + u_3\hat{z}$. Newton's equation of motion (force = mass × acceleration) is transferred into the infinitesimal cube to yield (per unit volume)

$$\sum_j \frac{\partial \sigma_{ij}}{\partial x_j} + f_i = \rho \frac{\partial^2 u_i}{\partial t^2} \tag{1.8}$$

where the first term on the left denotes the effect of the surface forces (by measuring their difference in the i-direction between the two sides of the cube, while f_i is the ith component of the so-called "body force" if such a force exists (e.g., gravity when it is relatively large compared with surface forces and can not be ignored) and ρ is the density (mass per unit volume) at the position (x, y, z). Equation 1.8 is the basic equation of motion of an elastic medium and all further linear elastic considerations are conducted in order to solve it. An important special case of Eq. 1.8 is static equilibrium, where all movements are zero. In that case,

$$\sum_j \frac{\partial \sigma_{ij}}{\partial x_j} + f_i = 0 \qquad \text{(static equilibrium)} \tag{1.9}$$

or, when no body forces exist,

$$\sum_j \frac{\partial \sigma_{ij}}{\partial x_j} = 0 \qquad \text{(static equilibrium, zero body force)} \tag{1.10}$$

In order to solve Eq. 1.8, one should have a way to connect σ with u. This is done via the generalized Hook's law in the following way. Firstly, a strain tensor ε_{ij} is defined as

$$\varepsilon_{ij} = \frac{1}{2}\left(\frac{\partial u_i}{\partial x_j} + \frac{\partial u_j}{\partial x_i}\right) \tag{1.11}$$

As can be seen, ε is also a symmetric tensor, i.e., $\varepsilon_{ij} = \varepsilon_{ji}$. The most general Hook's law can now be written as a linear relation between σ and ε,

$$\sigma_{ij} = \sum_{k,l} C_{ijkl}\varepsilon_{kl} \tag{1.12}$$

where C_{ijkl} is a constant tensor of the 4th rank replacing the Young modulus of Eq. 1.1. It is a material property. For isotropic homogeneous materials, C_{ijkl} has only two components λ and μ (called the Lamé constants) different from zero and the law can be written as

$$\sigma_{ij} = 2\mu\varepsilon_{ij} + \lambda\delta_{ij}\sum_k \varepsilon_{kk} \tag{1.13}$$

where δ_{ij} is the Kronecker δ,

$$\delta_{ij} = \begin{cases} 1 & \text{if} \quad i = j \\ 0 & \text{if} \quad i \neq j \end{cases} \tag{1.14}$$

For example, $\sigma_{xz} = \sigma_{13}$ equals $2\mu\varepsilon_{13}$ while $\sigma_{zz} = \sigma_{33} = 2\mu\varepsilon_{33} + \lambda(\varepsilon_{11} + \varepsilon_{22} + \varepsilon_{33})$. The Lamé constants are connected to the Young modulus and the Poisson ratio ν through

$$\nu = \frac{\lambda}{2(\mu + \lambda)} \qquad \text{and} \qquad E = \frac{\mu(2\mu + 3\lambda)}{\mu + \lambda} \tag{1.15}$$

Now, the right hand side of Eq. 1.8 can be differentiated with respect to x_i and x_j to relate it through Eq. 1.11 with ε_{ij}, and the left-hand side of Eq. 1.8 in its turn can be transformed by Eq. 1.12 or 1.13 to be given in terms of ε_{ij} (as well). In this way, the ensuing equation will have similar variables on both sides. There is, however, a problem arising from the connection between the displacements and the strain (Eq. 1.11). Although the displacement **u** (a vector with three components) uniquely defines ε_{ij} (a symmetric tensor with six components), the ε_{ij} does not uniquely define the displacements. There is a redundancy. In order to insure a unique **u**, the strain components must satisfy an additional criterion (called the compatibility equations, see also Sect. 1.3.3.1)

$$\frac{\partial^2 \varepsilon_{ij}}{\partial x_k \partial x_l} + \frac{\partial^2 \varepsilon_{kl}}{\partial x_i \partial x_j} = \frac{\partial^2 \varepsilon_{ik}}{\partial x_j \partial x_l} + \frac{\partial^2 \varepsilon_{jl}}{\partial x_i \partial x_k} \tag{1.16}$$

The mathematical complexity of this set of equations (Eqs. 1.8–1.16) both for the dynamic but even for the static cases makes exact analytic solutions for elastic problems almost unobtainable. Exceptions are problems that due to their symmetry can be reduced to simpler forms. The most common solvable problems are "two-dimensional" (2D) ones, where the influence of the third dimension is ignored. Two types of such problems are of importance. In the first, we deal with a thin plate, where it is assumed that the thickness is much smaller than the other dimensions and can be ignored (plane stress); and in the second a very thick specimen is considered where the third dimension is ignored (plane strain) under the condition that *no change* of the applied force occurs across the thickness of the plate and the whole system is homogeneous and therefore solutions do not change with the coordinate in this direction.

1.1.2 Energy Considerations and the Griffith Criterion

Fractures occur in solids at much lower stresses than those which can be deduced from simple estimations. For example, if we assume that the solid breaks when the distance between its atomic planes become elongated by, say, 1 / 10 of the equilibrium distance, this would mean that $\varepsilon \sim 1/10$ and $\sigma = E\varepsilon \sim E/10$. Most solids break under loads that are two or three orders of magnitude lower than this value. This discrepancy was resolved by Griffith (1920). In principle his argument is that the stress is not transferred to the solid homogeneously. Rather, due to the existence of flaws in the body (and almost all solids contain flaws, unless grown in a very special method – controlled whisker procedure), the stress becomes concentrated at the tip of such a flaw. Hence, strains and stresses at this location become much larger than those prevailing in the rest of the material – leading to fracture there. In fact, it is the longest flaw (favorably oriented), which sets the limit for the critical stress, the stress above which fracture would occur, as is shown presently.

Consider (Fig. 1.2), a large plate of thickness b and cross section A, which is operated on by a constant force f or a stress σ_0 ($=f/A$) and in which there exists a flaw (fracture) of length c. The fracture evidently has caused the elastic energy in its vicinity to be relieved. We can estimate the volume wherefrom the elastic energy was relieved to be approximately equal to the half cylinder shown.

This volume is

$$V \sim \frac{1}{2}\pi c^2 b \tag{1.17}$$

And according to Eq. 1.5, the total relieved energy is

$$W_{\mathrm{E}} \sim V\frac{\sigma^2}{2E} = \frac{\pi \sigma_0^2}{4\ E} c^2 b \tag{1.18}$$

Therefore, should the fracture propagate by an additional increment Δc, the change in relieved energy would be

Fig. 1.2.
A through crack in a semi infinite slab and the (schematic) region in which energy is relaxed

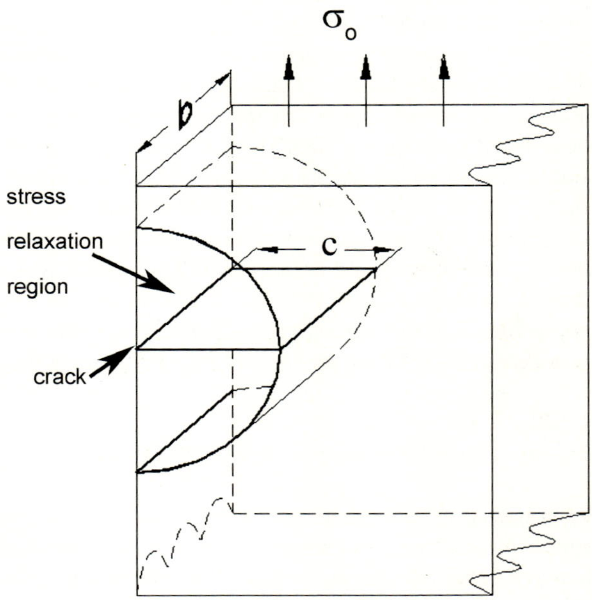

$$\Delta W_E \sim \frac{\pi b}{2E} \sigma_0^2 c \Delta c \qquad (1.19)$$

By the way, the energy per unit length $\Delta W_E / \Delta c$ is usually denoted by the (not very fortunate) term "energy release rate" G (in honor of Griffith).

According to Griffith, the energy thus released from the elasticity stored reservoir should go in its entirety (or almost so) to create the areas of the new addition to the fracture surface. If we assume that the energy needed for the creation of a new unit of fracture area, Γ, is a material constant (it is in fact half of the energy needed to break the chemical bonds between the two newly created surfaces), then the energy required for an increment Δc of the fracture is

$$\Delta W_s = 2\Gamma b \Delta c \qquad (1.20)$$

The factor of two comes because two new surfaces are created above and below the increment. Energy balance entails that the change in the released increment of elastic energy be equal to the increment of surface energy, i.e.,

$$2\Gamma b \Delta c \sim \frac{\pi b}{2E} \sigma_0^2 c \Delta c \qquad \text{or} \qquad \frac{\Delta W_E}{\Delta c} = \frac{\Delta W_s}{\Delta c}$$

yielding

$$\sigma_0^2 c \sim \frac{4}{\pi} \Gamma E \qquad (1.21)$$

A similar calculation applies for an internal flaw (crack). Assume that this crack is circular of radius c. The volume from which the elastic energy was released is

$$V = \frac{4}{3} \pi c^3$$

and therefore in this case

$$W_E = \frac{4}{3} \pi c^3 \frac{\sigma_0^2}{2E}$$

where σ_0 is the remote stress perpendicular to the flaw. The surface energy needed to create this flaw is $W_s = 2pc2\Gamma$. Griffith's incremental energy balance,

$$\frac{\partial W_E}{\partial c} = \frac{\partial W_s}{\partial c}$$

yields

$$\sigma_0^2 c = 2E\Gamma \qquad (1.21')$$

Although the method used here to obtain Eq. 1.21 or 1.21' is somewhat crude, the result, except perhaps for the coefficient, is accurate. This is Griffith's condition. Let us first calculate the value of σ_0 needed to initiate a crack and compare it to the homogeneous estimate of $E/10$ we had before. The decrease with respect to $E/10$ of σ_0 needed for fracture commencement can be seen as a measure of the enhancement of stress at the crack tip relative to σ_0. From Eq. 1.21,

$$\sigma_0 = \frac{E}{q}; \qquad q \approx \sqrt{\frac{Ec}{\Gamma}} \tag{1.22}$$

Therefore, the magnitude of q is a measure of stress increase (relative to σ_0) at the tip of a flaw of length c.

Let us examine soda lime glass as an example. The relevant constants for glass are $E \sim 6.2 \times 10^{10}$ Newton m^{-2} and $\Gamma \sim 0.54$ Joule m^{-2}, hence for glass $q \sim 3.4 \times 10^5 \sqrt{c}$. This relation is plotted in Fig. 1.3. It is seen that our first estimate of q, $q \sim 10$ is obtained only for flaw lengths of the order of atomic distances, while for usual flaw lengths of the order of 0.1 mm, q is of the order of several thousands.

According to the Griffith criterion, Eq. 1.21, therefore, the important experimental variable "responsible" for crack initiation is the product $\sigma_0^2 c$ of $\sigma_0 \sqrt{c}$ and not σ_0 itself. When this variable reaches a certain threshold, which depends on the specific material and on the geometry of the loading system, fracture should commence. This threshold according to Eq. 1.21 depends on the Young modulus and the surface energy of the fracturing sample. As will be shown next (Sect. 1.3), a more elaborate and accurate treatment of linear elasticity (called linear elastic fracture mechanics (LEFM) an established branch of mechanical engineering nowadays) yields the same result as Eq. 1.21, up to a change of coefficient, for uniaxial constant loading. This LEFM treatment has been used for many years now to achieve solutions for engineering static safety problems and has proven extremely useful for that purpose. Basically, it is an extension of the Griffith approach. As we shall point out, however, this approach, so successful for regular static cracking problems, is inadequate when dealing with the following two important issues:

1. Fracture development under low stresses in very long time periods (the so-called "creep" situation).
2. Dynamic crack propagation.

We dedicate a section for each of these problems.

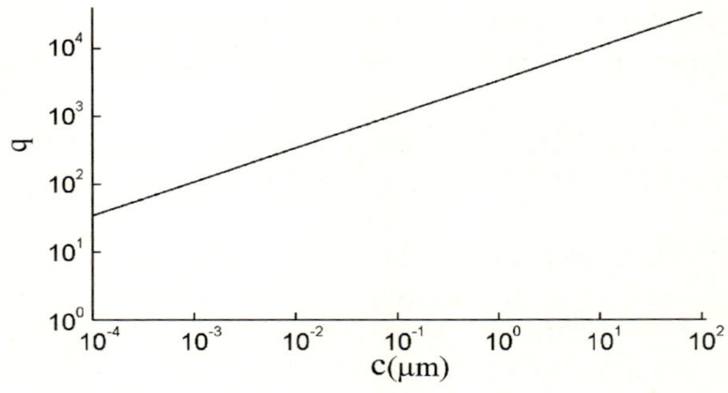

Fig. 1.3. The amount q by which the stress σ_0, needed to cause fracture, is decreased from E as a function of the Griffith-fracture length c

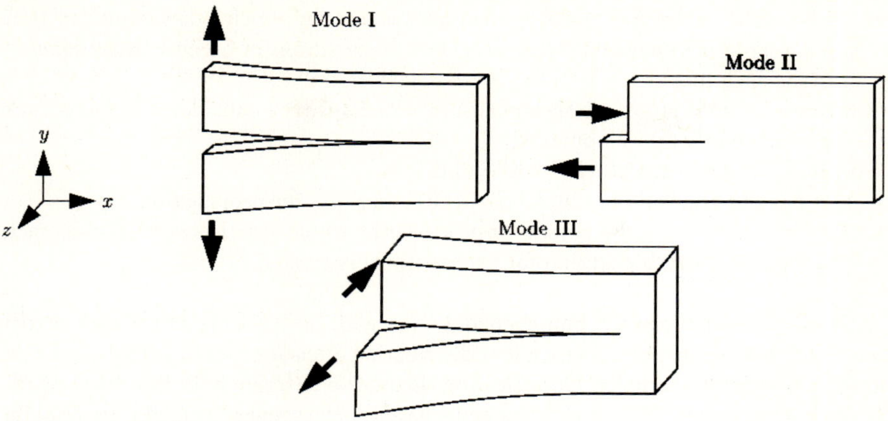

Fig. 1.4. The three fracture modes

1.1.3 Fracture Modes

Cracks can appear in solids under stress only in three different geometrical situations, called modes, or in situations that are combinations of these modes. Figure 1.4 shows these three geometries called mode I, mode II and mode III. Almost all engineering treatments of fracture problems are developed in a mode I geometry, which is the prevalent mode of fracture. The presence of mode II and III in addition to mode I is manifested by changes of crack directions as in the appearance of lancets, striae or plumes, undulations etc. In geological cracks, we discern between "joints" that are created in a mode I geometry and "folds" that are created under the other mode types. In the description of all three modes (Fig. 1.4), the fracture surface is in the x-z-plane and the fracture propagates in the x-direction. In mode I, stress (tension or compression) is applied in the $\pm y$-direction and the displacements **u** also occur in the same direction. Modes II and III are shear modes. In mode II, the shearing stress and the displacements are in the same direction ($\pm x$) as that of the crack propagation. Mode III fracture is also called an "antiplane" mode, since stress and displacements occur in the $\pm z$-directions, out of the plane of the fracture and normal to the plane of the specimen.

1.2 Nucleation

1.2.1 General

Let us first define the three different nucleation processes that are used here (see also Sect. 2.2.2 for additional applications of the term "nucleation"):

1. "Flaw nucleation" is the process or processes by which a seed for the subsequent creation of a propagating fracture is created. The seed is usually a flaw in the otherwise "perfect" or homogeneous solid, which facilitates the appearance of a crack in its continuation (see Sect. 1.1.2).

2. "Microcrack nucleation" is the process or processes by which a flaw (which is itself not a microcrack) develops small cracks at the positions of highest stress concentrations such that
   a the combined structure, flaw plus small cracks, can be considered a microcrack for subsequent developments,
   b this combined structure is stable, and
   c it has not reached the critical (Griffith) condition for propagation.
3. "Crack nucleation" is the process or processes by which a *critical crack* (i.e., a crack fulfilling the Griffith criterion for propagation) is created.

Crack nucleation can be "homogeneous", i.e., with no flaws present (a rare occurrence) or "heterogeneous", where a flaw distribution already exists in a solid. Note that in the literature the term "crack nucleation" is used to describe both flaw- and crack-nucleation processes. The terms homogeneous and heterogeneous are also used to describe materials. Thus a homogeneous material is an ideal material, which is made up of the same chemical substance, is devoid of grains and is completely devoid of flaws. The only existing material that is almost completely homogeneous is a glass wisker, i.e., a glass fiber grown under very strict conditions of purity. Since no flaws (or almost no flaws) are created in the production phase, q of Eq. 1.22 is of the order of ten and this material fractures only under very high stresses. Its fracturing is also "homogeneous", namely all parts of it separate at the same time and the material integrates into dust under a large blast. Heterogeneous brittle materials can be divided into three groups according to the degree of severity of the heterogeneity. The least heterogeneous material is regular glass, containing flaws of all kinds (see below). Next in line are rocks or ceramic materials having the same chemical composition. These materials are grainy materials containing grain boundaries and "triple points" as inherent flaws. This group further divides into isotropic (i.e., having the same properties in any direction) and unisotropic. The most heterogeneous (third) group of materials are ceramics or rocks which consist of substances having different chemical compositions. These are usually both grainy and unisotropic. Homogeneous (heterogeneous) nucleation is a nucleation process in a homogeneous (heterogeneous) material.

We use the term "flaw" here as a general term for defect. This is the usual meaning of the term in the fracture literature, while, e.g., in solid state physics the common term is "defect". A flaw can be two-dimensional in the form of a microcrack, or three-dimensional, such as a pore, an inclusion, a vacancy, a dislocation etc. The term "Griffith flaw" is used in the literature generally to describe a flaw in the form of a microcrack (the common form of flaw in glass). A three-dimensional flaw can be transformed to a microcrack by developing actual microcracks at points where high stress concentrations exist (critical or singular points; see below). Note that regular glass, besides Griffith flaws, can contain other flaws (voids etc.) generated during its formation stages or by external environmental degradation. In general, the appearance of flaws in a solid or on its surfaces can be induced by the activity of two mechanisms, "external agents" operation or internal processes. The first mechanism includes surface abrasions and erosion, corrosion by chemical substances, radiation and impact effects, etc. The operation of each agent is subject matter for extensive research. A short review can be found in Lawn (1993, Chap. 9). We shall not consider here the flaws that develop during the production phase and include pores, vacancies, grain boundaries, inclusions etc. (see *ibid*). In this section, we will concentrate on the internal processes and specifically on microcrack and *crack* nucleations. Note that the use of the term "homogeneous nucleation" in the literature is somewhat controversial. In the words of Frost (2001), "roughly it means the appearance of a crack where there was none present before. But the condition of there being no crack before depends on the precision of our observation." In physics and metallurgy, nucleation means an appearance of a new "phase" in a position "*where absolutely nothing* of the sort existed before." In fracture nomenclature, this term usually means "where absolutely nothing *could be seen (optically)* before." We will have more to say on this subject when discussing thermal nucleation below.

A nucleation of even a microcrack in a perfect (homogeneous) solid would entail stresses of the order of $E/10$ (or $G/10$; Sect. 1.1.2), or corresponding amounts of energy. Following the Griffith model, we look for stress concentration mechanisms that could facilitate this process. In Sect. 1.2.2 we shall review (Frost 2001 and Evans 1974) the possible stress concentration mechanisms capable of reducing the stress (energy) needed for microcrack nucleation, and in Sect. 1.2.3 and 1.2.4 we will present two modern theoretical approaches to crack nucleation. The first one is "thermal nucleation", which has recently been verified experimentally and the second is the so-called "fiber bundle model", which can apparently be used for general grainy materials.

1.2.2
Stress Concentration Avenues to Microcrack Nucleation

The stress concentration mechanisms (see Frost 2001) are elastic stress concentrations resulting from

1. Interior flaws.
2. Elastic anisotropy.
3. Thermal expansion mismatch or thermal expansion coefficient anisotropy, and inelastic mechanisms.
4. Dislocation pile-up.
5. Grain-boundary sliding.

We shall briefly touch upon two of the elastic mechanisms. For the inelastic ones see Frost (2001) and references therein.

1.2.2.1
Elastic Anistropy

We shall treat only heterogeneous anisotropic materials of the second group (Sect. 1.2.1). In heterogeneous materials of the third group (which are made up of grains of *different* substances), there are additional stress concentration possibilities resulting from the mismatch of elastic constants between neighboring grains, which can be treated in a similar way to the one presented here.

Tvergaard and Hutchinson (1988) considered stress concentrations in 2D solids. The treatment was extended to 3D problems by Ghahremani et al. (1990), who treated grains in hexagonal tiling where all boundaries meet at 120° – which is the case we will consider here. Picu and Gupta (1996) extended the treatment to general 2D configurations and Picu (1996, 1997) extended it to several 3D problems.

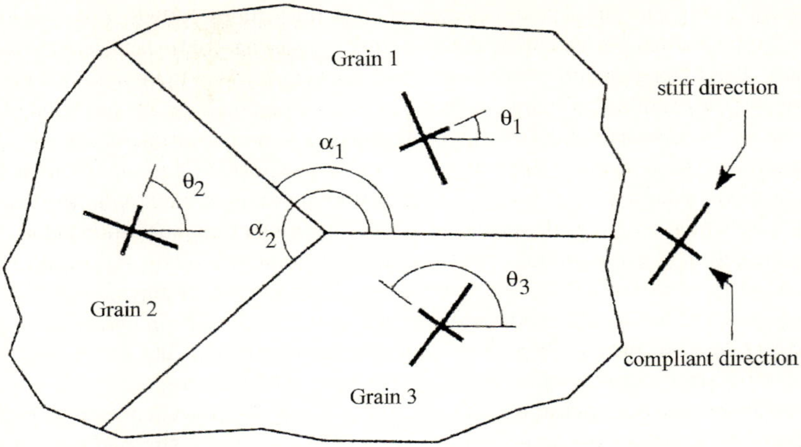

Fig. 1.5. Geometry used for the calculation of stress singularity due to anisotropic elasticity (Frost 2001)

Fig. 1.6.
A scanning electron micrograph of a typical triple point void in sintered Al_2O_3 (Evans 1978)

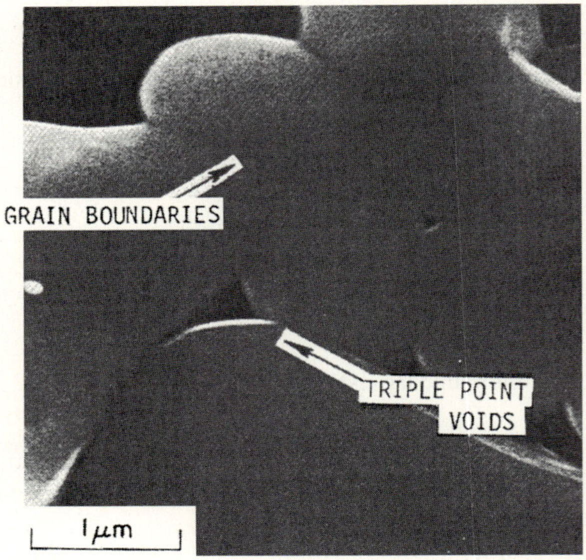

Each crystalline grain of an anisotropic solid has a "stiff" elastic direction and a "compliant" direction, i.e., the strain in the first direction is smaller than the one in the second direction for the same applied stress. When adjacent to each other, grains can have several orientation angles. Consider Fig. 1.5, which shows a rather simple geometry, in a 2D situation. Even here there are several (five) orientations: α_1, α_2 (here 120° and 240°, respectively) specifying the relative orientations of the grain boundaries and three θ measuring the angles of the stiff directions in each grain. Picu and Gupta (1996) calculated the stress singularity, which arises in such a geometry, in polar coordinates (r, θ) around the triple point (Fig. 1.6):

$$\sigma_{ij}(r,\theta) = kr^{\gamma}\phi_{ij}(\theta) \tag{1.23}$$

where k is a constant, γ is an exponent defining the stress singularity and $\phi_{ij}(\theta)$ is a function of θ only with no singularities. The value of γ is crucial for nucleation. In order to achieve stress concentration, γ has to be negative (in which case there exists a real σ singularity, i.e., $\sigma \to \infty$ for $r \to 0$), and for *crack nucleation* to occur, γ has to be less than -0.3 (Frost and Gupta 1993), since only then a σ singularity that is sufficiently high to overcome atomic cohesion extends beyond atomic dimensions. Picu and Gupta (1996) considered random orientations of the crystal axes for the 120–240° triple junctions and for other symmetric ($\theta_1 = -\theta_3$) situations. Specific problems should be checked there. For example, for ice they have shown that no crack nucleation can occur from elastic anisotropy.

1.2.2.2
Thermal Expansion Mismatch or Anisotropy (Evans 1978)

Evans (1978) calculated the stress concentration, which occurs in polycrystalline (grainy) solids due to the thermal expansion anisotropy. As for the elastic anisotropy case (Sect. 1.2.2.1), thermal expansion singularities were found at grain corners (triple points; Fig. 1.6) and were shown to decrease rapidly with distance from these corners. The maximum stress concentration for defects located on grain boundaries was again found for those defects residing at the triple points. Grain arrays were shown (Fig. 1.7)

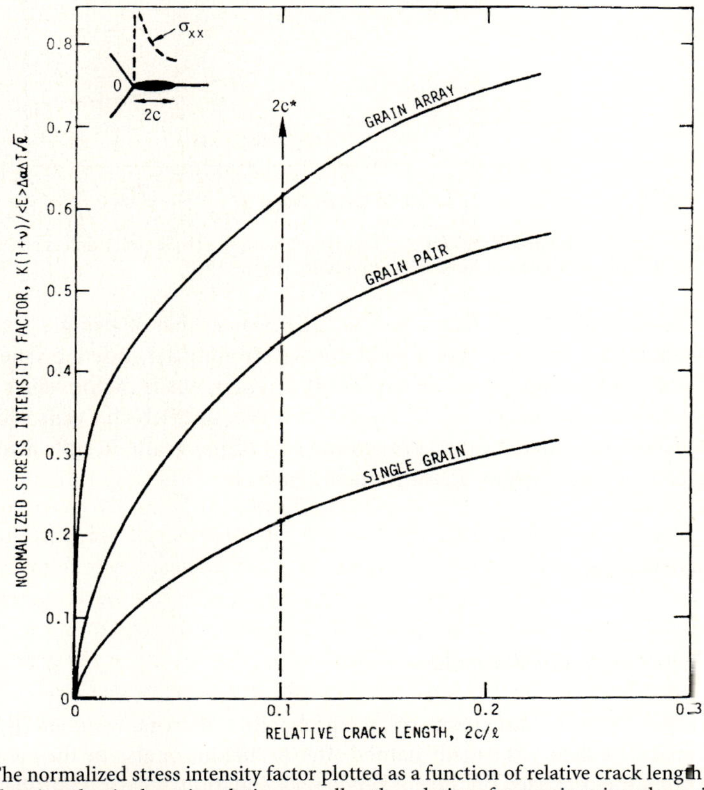

Fig. 1.7. The normalized stress intensity factor plotted as a function of relative crack length for regular hexagonal grains; the single grain solution, as well as the solutions for a grain pair and a grain array are shown (Evans 1978)

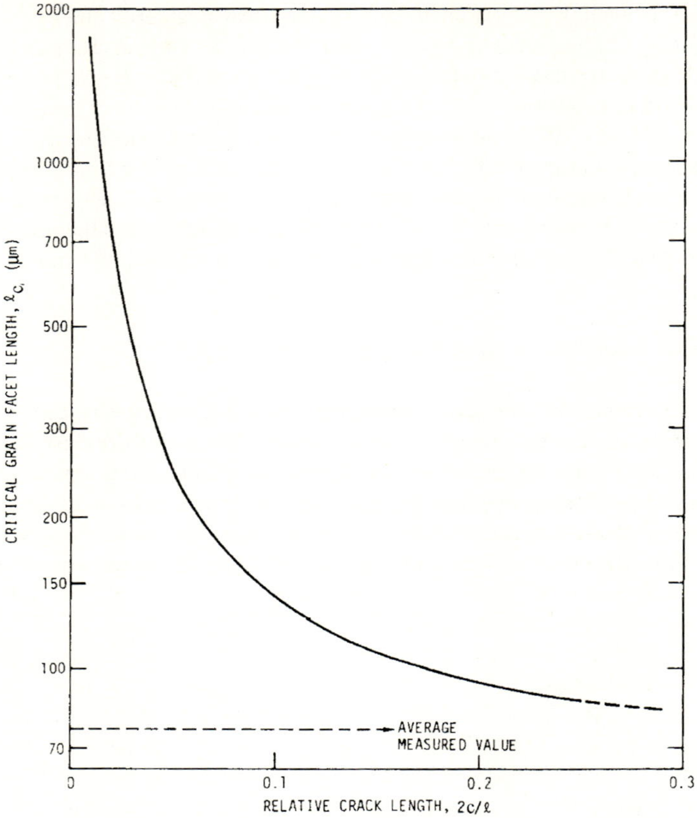

Fig. 1.8. The variation of the predicted critical facet length with the triple point defect size for Al_2O_3. Also shown is the average value obtained experimentally (Evans 1978)

to magnify the stress intensity factor (see Sect. 1.3.3) over that of a single grain by up to ~4. "The incidence of microfracture will thus be strongly dependent on the relative orientation of neighboring grains" (Evans 1978). Stress intensity factor values are crucially dependent on the ratio between the defect (flaw) size (which Evans transforms to an equivalent crack size; Evans 1974) to the size of the grain. Results also yield a critical grain size (facet length) l_c, for inducing microfracture (Fig. 1.8).

1.2.3 Thermal Nucleation

1.2.3.1 *Thermal Fluctuations and Arrhenius Process*

In many areas of physics and chemistry, the probability of crossing an energy barrier is encountered. The theory is usually named after Arrhenius or also by the name "reaction rate". For a theory based on the Fokker-Planck equation, see e.g., Gardiner (1990). A naïve explanation of this approach is as follows. We are dealing with a system (A) that

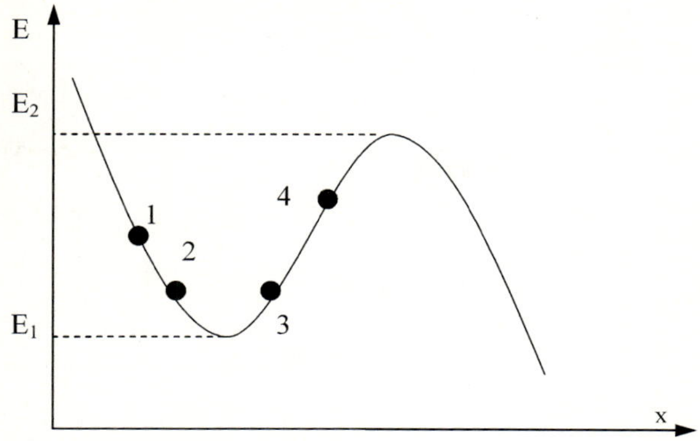

Fig. 1.9.
A schematic "barrier passing". The energy of the system is shown as a function of a parameter x (which can represent different variables, e.g., crack length, stress, strain etc.); x is often called the "reaction parameter"

resides in an equilibrium (minimum energy) situation (Fig. 1.9), at a certain "initial" energy level E_1 and interacts with a big system (such as the atmosphere) called a "thermal bath", which keeps the temperature of A constant. Due to the heat exchange with the bath, the system A constantly changes its energy level around E_1 (say by moving along positions 1, 2, 3, 4, etc.). These changes are called "thermal fluctuations". Now suppose that A has a second equilibrium energy state E_2 ($>E_1$). Note that E_2 of Fig. 1.9 is *not* an equilibrium point, since it is not an energy minimum. If we now start with a large number (an ensemble) of identical systems, each having two energy levels E_1 and E_2, the thermal fluctuations will establish a steady state whereby the ratio of the number of systems residing in E_2 to the number of systems residing in E_1 equals $\exp\{-(E_2 - E_1)/(\mathrm{k}T)\}$, where k is a universal (Boltzmann) constant. This situation is called Maxwell's distribution. On this basis, we assume (to get the Arrhenius process) that the probability of a single system (such as that depicted in Fig. 1.9), having only one equilibrium energy level, E_1, and being in contact with a heat bath that keeps it at a constant temperature T, to reach a higher energy level E_2 ($>E_1$), is proportional to $\exp\{-(E_2 - E_1)/(\mathrm{k}T)\}$. This is evidently an approximation, since E_2 is not an equilibrium state. Thus, according to this theory, the probability of a system overcoming an energy "barrier" of $\Delta E = E_2 - E_1$ (Fig. 1.9), per unit time is

$$\nu = \nu_0 \exp\left(-\frac{\Delta E}{\mathrm{k}T}\right) \tag{1.24}$$

where ν_0 is a "normalization" factor and $\Delta E = E_2 - E_1$. In a more generalized case where pressure or volume changes can occur during the process, ΔE would be replaced by the enthalpy difference of the system between the two states.

1.2.3.2
Fracture and Microfracture Nucleation

Figure 2.1 depicts the pattern observed on a fractured glass sample. Here we relate only to the two smallest forms, i.e., c_i and c_cr. The former is created by the *initial* flaw (or actually the microcrack developed from it), while the latter, c_cr, is the critical fracture length wherefrom the crack starts to propagate (above which the Griffith criterion holds). For a physical understanding consider (Fig. 1.10) the total energy of the system as a function of crack length, c. See also Sect. 2.2.2.1 for characteristic lengths of c_i and c_cr.

Fig. 1.10.
A schematic representation of the difference between "microfracture nucleation" ($E_1 \rightarrow E_3$ via E_2) and "fracture nucleation" ($E_1 \rightarrow E_4 \rightarrow E_5, E_6$ or $E_3 \rightarrow E_4 \rightarrow E_5, E_6$). *A:* System with no flaws (no microfractures); *B:* System with flaws

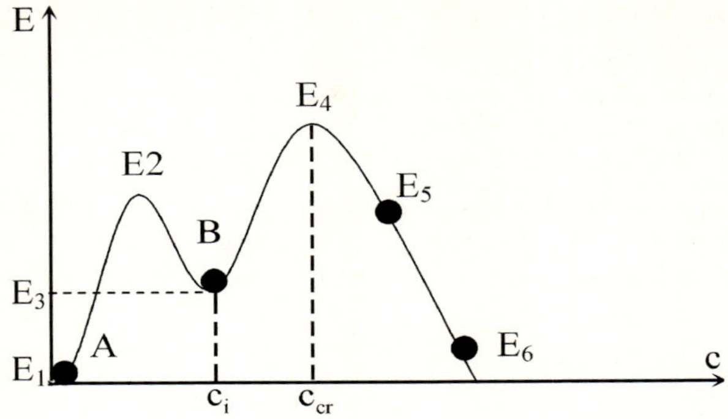

As discussed above (Sect. 1.2.1), the term "nucleation" can be used for "microfracture nucleation", i.e., nucleation of a microfracture c_i (around a void, an inclusion or even a small existing microfracture), whose length is smaller than c_{cr}; and it can be used for "fracture nucleation", namely the appearance of the critical crack c_{cr} that causes material failure. While both of these "nucleations" usually proceed in the same manner, i.e., by thermal fluctuations, there is a major difference in their final stability. Microfractures, once nucleated, can remain stable (and do not heal) because they reach a stable or quasi-stable equilibrium, i.e., an energy minimum (or a local minimum) in which the system resides (such as E_3, Fig. 1.10). To remove it from this equilibrium position back to E_1 (to heal the flaw), an energy barrier ($E_2 - E_3$) has to be overcome. In order to have nucleated this flaw, from the equilibrium position E_1, the thermal fluctuations should obviously have taken the system over the barrier $E_2 - E_1$ (and would eventually transfer the system from E_1 to E_3). On the other hand, "crack nucleation" would bring the system from E_1 (if no flaws exist) or from E_3 (in a system with a microcrack) to E_4. Then the system ceases to be stable. It reaches in fact the Griffith criterion, which means that the crack starts propagating indefinitely, thus gaining energy from the field (going to E_5, E_6, etc).

We shall describe only the second procedure, namely the nucleation of a crack leading to eventual rupture. The procedure of "microcrack nucleation" is principally similar. For microcrack nucleation, however, the derivation of the energy barrier ($E_2 - E_1$) from the physical situation is of course different from the one we show here and depends on the existing flaw (see Sect. 1.2.2).

1.2.3.3
Single versus Multiple Crack Nucleation

Two scenarios are conceived in the literature for the fracture process of brittle materials. The first consists of a nucleation and growth of a single critical crack that after reaching the Griffith criterion totally breaks the material. If the rate of stress increase when loading the material is high enough and a critical crack or an effective critical flaw already exists there, then the nucleation part becomes very short. The second scenario (e.g., Spyropoulos et al. 2002), which is assumed to hold in "grainy" materials, i.e., groups 2 and 3 heterogeneous materials (Sect. 1.2.1) and below the critical stress, comprises three to four stages:

1. Nucleation of "many" microcracks under the combined effects of the existing stress and thermal fluctuations.
2. Initial growth of such microcracks.
3. Coalescence of the microcracks into macrocracks.
4. Rupture of the material. In this section we treat the thermal nucleation process, which is related to the first scenario and also briefly the "fiber bundle" model for grainy materials. Other models such as percolation (Hermann and Roux 1990) and self-organized criticality (P. Bak et al. 1987, 1988) will not be touched upon. The models chosen here were those which (recently) have been shown to agree with experiments.

1.2.3.4
Thermal Crack Nucleation

For many years, the time to failure of a material under a constant remote stress, σ_0, below that which would cause immediate fracture (Griffith) was calculated by the Zhurkov (1965) or the Mogi (1962) assumption

$$\tau = \tau_0 \mathrm{e}^{-\mathrm{b}\sigma_0}$$

where τ_0 and b are constants. Careful measurements (see below) by Ciliberto and his collaborators have shown that this was not the correct $\tau(\sigma_0)$ relation.

We shall follow an adapted version of the model proposed by Pomeau (1992) for fracture of two- and three-dimensional solids. A similar model was proposed by Golubovic and Feng (1991), and was extended to fractal dimensions by Bonn et al. (1998), who carried out an experiment on polymer gels to verify their theory.

Crack nucleation in 2D and 3D (no flaws – homogeneous nucleation). Nucleation of a critical crack for a material under a constant remote stress, σ_0, i.e., the probability of the appearance of a crack large enough as to fulfill the Griffith criterion, is accomplished by thermal fluctuations, via an Arrhenius-type process (Sect. 1.2.3.1). Assuming there are no initial flaws in the material, the rate of appearance of a critical crack is given by (Pomeau 1992)

$$\nu = \nu_0 \exp\left(-\frac{W}{\mathrm{k}T}\right) \tag{1.25}$$

where ν_0 is a normalizing rate, k is the Boltzmann constant, T is the absolute temperature and W is the "liberated" enthalpy. Here we assume that volume and pressure are unchanged in the process and therefore W becomes the energy barrier needed to be overcome by the process. In a method similar to the one described in Sect. 1.1.2, for a 2D solid, the elastic energy that would be liberated by a fracture of length c is (Fig. 1.11):

$$W_1 \cong -\frac{\sigma_0^2}{2E}\pi c^2 b \tag{1.26}$$

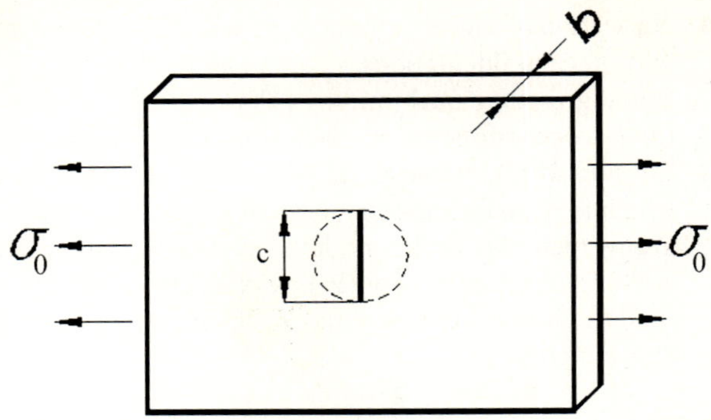

Fig. 1.11.
A schematic "2D" homogeneous nucleation of a crack of length *c* in an infinite plate under remote tensile stress σ_0

where σ_0 is the remote stress field in the medium, E is the Young modulus and b is the thickness (assumed small with respect to the other dimensions) of the sample. The energy needed to create the two new surfaces is

$$W_2 = 2\Gamma cb \tag{1.27}$$

where Γ is the surface energy. The total energy is therefore

$$W = -\frac{\sigma_0^2}{2E}\pi bc^2 + 2\Gamma bc \tag{1.28}$$

Figure 1.12a depicts W as a function of c. Note that for small c values, W increases linearly with c and then the parabolic part becomes dominant leading to a maximum. This maximum is given by $\partial W/\partial c = 0$, resulting in

$$c_{\max} = \frac{2\Gamma E}{\pi\sigma_0^2}; \qquad W_{\mathrm{m}} = \frac{2\Gamma^2 Eb}{\pi\sigma_0^2} \tag{1.29}$$

Now, if no flaws exist in the material, the critical barrier needed to be overcome by the creation of a critical crack is equal to W_{m}, and hence the rate of nucleation of such flaws is

$$\nu = \nu_0 \exp\left\{-\frac{\sigma_1^2}{\sigma_0^2}\right\} \tag{1.30}$$

where

$$\sigma_1^2 \cong \frac{2\Gamma^2 Eb}{\pi kT}$$

In three dimensions, the calculation is similar to that appearing in section 1, Eq. 1.23'. The total energy is given (Fig. 1.12b) by

$$W = -\frac{\sigma_0^2}{2E}\frac{4}{3}\pi c^3 + 2\Gamma\pi c^2 \tag{1.31}$$

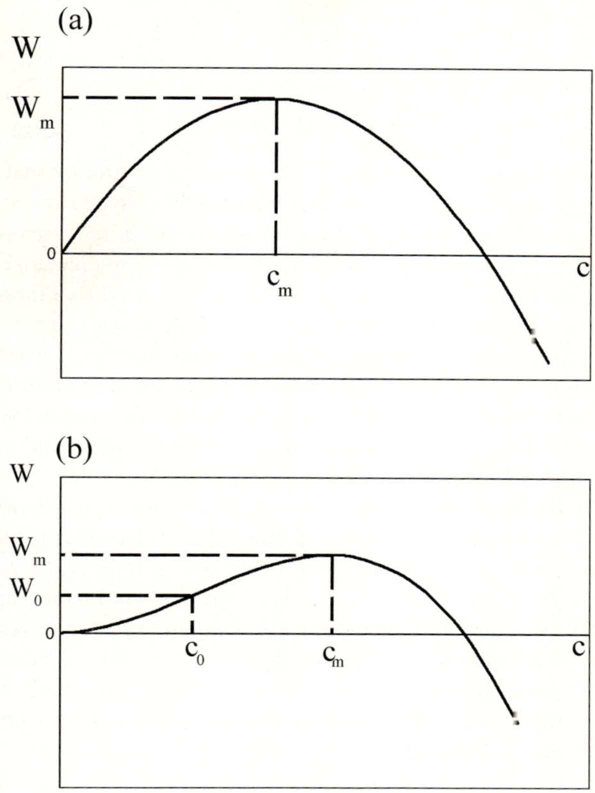

Fig. 1.12. Total energy (relieved elastic and surface energies) as a function of c. **a** 2D case. **b** 3D case. W_m: Maximum energy occurring at crack length c_m; W_0: Energy for existing crack of length c_0. Note that there should be a dip (energy minimum) at W_0 (not shown)

the maximal energy, which appears for $c_m = 2\Gamma E / \sigma_0^2$, is given by

$$W_m = \frac{8}{3}\pi\frac{\Gamma^3 E^2}{\sigma_0^4} \tag{1.32}$$

and the nucleation rate becomes

$$v = v_0 \exp\left(-\frac{\sigma_1^4}{\sigma_0^4}\right), \qquad \sigma_1^4 = \frac{8\pi\Gamma^3 E^2}{3kT} \tag{1.33}$$

The way to experimentally verify these results is to measure the time to fracture under a constant stress load that is below the critical stress, i.e., in a creep condition. Thus, in a sequence of papers (Guarino et al. 1999; Scorretti et al. 2001; Ciliberto et al. 2001; Guarino et al. 2002; Politi et al. 2002) Ciliberto and his group performed such measurements on several materials. According to Eq. 1.30 and Eq. 1.33, the times to failure for 2D and 3D materials under a constant σ_0 are given by the reciprocals of the rates, namely,

$$\tau_2 = \tau_{02} \exp\left\{\frac{\sigma_1^2}{\sigma_0^2}\right\} \tag{1.34}$$

and

$$\tau_3 = \tau_{03} \exp\left\{\frac{\sigma_1^4}{\sigma_0^4}\right\} \tag{1.35}$$

respectively. Here τ_{02} and τ_{03} are characteristic times, equal to the $(1/\nu_0)$, for 2D and 3D, respectively. Note that the dependence on σ_0, especially for the 3D case, is very strong: A small change in σ_0 can lead to a large change in τ. Therefore, the measurements must be done very delicately to avoid a situation where the material either breaks immediately or does not break at all (the time to failure is above that of the experiment). Ciliberto's group succeeded in obtaining very accurate measurements and verifying the $\tau(\sigma_0)$ behavior. For example, Fig. 1.13 (adapted from Fig. 4 of Guarino et al. 2002) depicts their results for wood samples (3D). This result, similar results for Perspex by the same group and 2D (Pauchard and Meunier 1993, 1998) and fractal sample (Bonn et al. 1998) time to fracture results, all lead to the same conclusion, namely that the new nucleation theory embodied in Eq. 1.34 and Eq. 1.35 is the correct one. For example, Fig. 1.13 shows the inadequacy of Mogi's (1962) or Zhurkov's (1965) approach and the apparent applicability of Eq. 1.35.

The result of Ciliberto's group did seem to have problems:

1. The distribution of times to failure was not Poissonian as expected.
2. The measured temperature dependence of τ for a linearly increasing load (see below) seemed to yield a temperature of ~3 000 K, while experiments were carried out near room temperature. To explain this discrepancy, they resorted to the "fiber bundle" model (see below), arguing that the spread in fiber strengths could lead to such high effective temperatures.

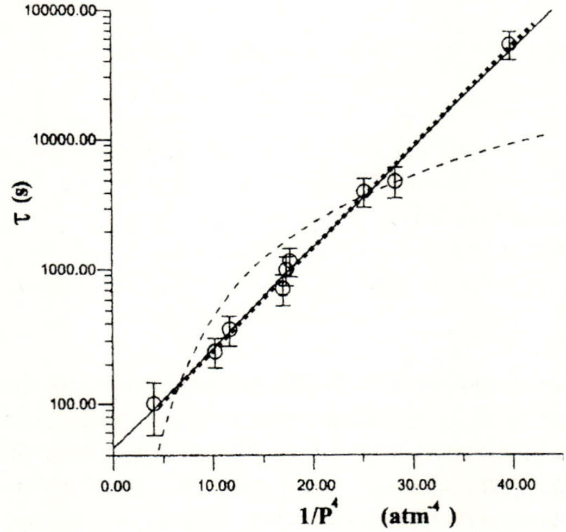

Fig. 1.13. Measurements on wood samples (Guarino et al. 2002). The time τ needed to break the samples under an imposed constant pressure P is here plotted as a function of $1/P^4$ in a semilog scale. The *dashed line* represents the solution proposed by Mogi (1962) ($\tau = ae^{-bP}$). The *continuous line* is the solution proposed by Pomeau for microcrystals ($\tau = \tau_0 \exp[(P_1/P)^4]$, Eq. 1.35 with P instead of σ). In the plot $\tau_0 = 50.5$ s and $P_1 = 0.63$ atm. Every point is the average of ten samples. The *error bar* is the statistical uncertainty. *Dotted line:* Results of our calculation Eq. 1.41 and Eq. 1.42 (with average $c_0 = 1.25 \times 10^{-6}$ cm)

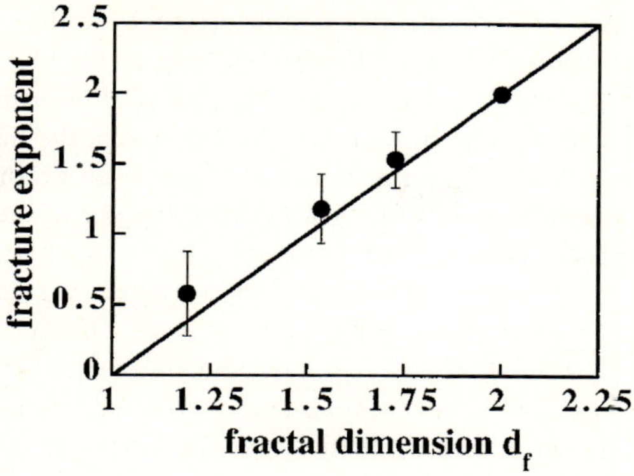

Fig. 1.14.
The absolute value of the measured exponent of activation energy vs. force as a function of the fractal dimension of the gels. The point at $d_f = 2$ is the measurement of Pauchard and Meunier (1993) for a two-dimensional crystal. The line is the prediction from the scaling arguments, Eq. 1.36 (Bonn et al. 1998)

Two-dimensional solids were fractured by Pauchard and Meunier (1993, 1998), and the

$$\exp\left\{\frac{\sigma_1^2}{\sigma_0^2}\right\}$$

relation was shown to hold for the averaged breakage times. As mentioned above, time to failure of polymer gels, whose fractal dimensions were measured by visible light scattering, was measured by Bonn et al. (1998), and the exponent of σ_0 in the delayed time to fracture was shown to agree with these fractal dimensions. Thus, the critical length was predicted to be $c_m = 2(d-1)\Gamma E / d\sigma_0^2$, where d is the fractal dimension of the solid. Hence the energy barrier should be proportional to

$$\frac{\Gamma^d E^{d-1}}{\sigma_0^{2d-2}} \tag{1.36}$$

Figure 1.14 (Fig. 5 of Bonn et al. (1998), adapted) shows the agreement between the (light scattering) measured dimensions, d_f and the dimensions d, obtained from the σ_0 dependence of the delayed times to fracture (derived from Eq. 1.36). These authors also discuss the question of "lattice trapping" (see, e.g., Marder 1996b) and show that the latter cannot be the reason for the delayed fractures since this phenomenon involves delay times that are shorter (or of much lower energy barriers) than those measured.

Thermal nucleation in the presence of flaws (microcracks, heterogeneous nucleation). Thermal nucleation was considered here (Sect. 1.2.3.4) as a process for which the thermal fluctuations were supposed to overcome an energy barrier that started from zero – thus completely disregarding the existing flaw distribution in the material. In this section, we treat nucleation where the stressed sample already contains internal (or surface) flaws, which reduce the energy barrier (Rabinovitch et al. 2004).

The calculation is carried out for the three-dimensional case. Results are easily extended to solids of other dimensions.

It is therefore assumed that at the moment of loading, the sample already possesses a distribution of microcracks each having a different length c_i perpendicular to the direction of the highest principle stress. Note that a flaw can be looked upon as an "effective microfracture" (Evans 1974) in that the highest stress concentration at a point on its perimeter can be equated to that arising from a 2D microcrack with a (different) effective length. According to Fig. 1.12b, the barrier becomes smaller since the thermal fluctuations need only to augment an existing flaw from c_0, say, to c_m, rather than create a crack of length c_m from the start. The energy barrier now is given by (Fig. 1.12b) $W = W_m - W_0$, where

$$W_0 = -\frac{2}{3}\pi\frac{\sigma_0^2}{E}c_0^3 + 2\pi\Gamma c_0^2 \tag{1.37}$$

The energy barrier, W, is larger than zero for $c_0 < c_m$. The time to fracture under such conditions is therefore

$$\tau = \tau_0 \exp\left[\frac{1}{kT}\left(\frac{\gamma}{\sigma_0^4} + \alpha\sigma_0^2 c_0^3 - \beta c_0^2\right)\right] \tag{1.38}$$

where

$$\alpha = \frac{2\pi}{3E}, \qquad \beta = 2\pi\Gamma \qquad \text{and} \qquad \gamma = \frac{8}{3}\pi\Gamma^3 E^2 \tag{1.39}$$

Clearly, Eq. 1.38 shows that flaws facilitate failure. Thus, denoting by

$$\tau_1 = \tau_0 \exp\left(\frac{\gamma}{\sigma_0^4 kT}\right)$$

(the Pomeau result, Eq. 1.33), Fig. 1.15 depicts the decrease in τ with c_0 $(0 \le c_0 \le c_m)$ from τ_1 to τ_0.

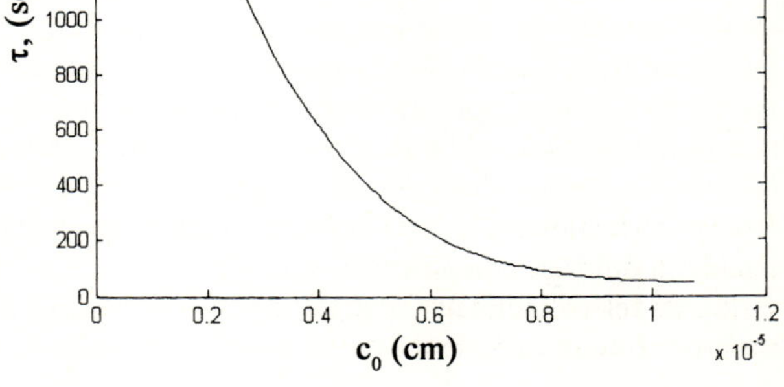

Fig. 1.15. The lowering of the time to failure for an existing crack of length c_0 in a 3D sample due to the decrease of the energy barrier. The value $c_0 = 0$ is for nucleation where no flaws exist. Here $c_m = 1.3 \times 10^{-5}$ cm, above which $\tau = \tau_0$

Now, the most "important" crack, the one for which the barrier is smallest and hence leads to the shortest time to fracture if augmented, is the crack with the *longest c*. Denoting the length of this crack by c_0, Eq. 1.37 and Eq. 1.38 provide the lowest energy barrier and the shortest (hence the measured) time to failure, respectively.

According to the statistics of extreme values (e.g., Castillo 1988; Kotz and Nadarajah 2000) the longest cracks can have only one of three distributions, Gumbel, Frechet or Weibull. Since the "basin of attraction" of the Gumbel distribution includes the exponential and the Gaussian ones, both of which serve as possible distributions of flaws (Lawn 1993), it is conceivable that the longest flaws here assume the Gumbel distribution. The probability density function of c_0 is therefore assumed (Kotz and Nadarajah 2000) to be

$$p(c_0) = \frac{1}{l}\exp\left(-\frac{c_0 - m}{l}\right)\exp\left(-e^{-(c_0 - m)/l}\right) \tag{1.40}$$

where l and m are parameters.

The following procedure was adopted. For specific chosen values of l and m, a c_0 value is chosen randomly from the $p(c_0)$ distribution. The τ for this c_0 is calculated by Eq. 1.38 for a definite remote stress σ_0 and temperature T. Note that the choice mechanism can generate c_0 values that are already above the Griffith criterion. For these cases, τ is chosen to be τ_0 (see Fig. 1.15). In this way, a histogram of τ for these values of σ_0 and T is obtained, from which the average, $\overline{\tau}$, of τ under these conditions is calculated. Changing σ_0 for the same T, these averages are drawn as a function of σ_0, $\overline{\tau}(\sigma_0)$. This procedure is repeated for different l and m values until the $\overline{\tau}(\sigma_0)$ agrees with the experimental results, at which point the values of l and m are considered to be the correct ones for the flaw distribution. Results are shown in Fig. 1.13 (*dotted line*). It is seen that although $\ln(\tau)$ of Eq. 1.38 is no longer directly proportional to $1/\sigma_0^4$, the changes are minimal and the agreement with experiment is good.

A straightforward understanding of the situation can be obtained by considering $\ln(\tau)$ as a function of $x = 1/\sigma_0^4$. By Eq. 1.38, we obtain

$$\ln\tau = \eta + \frac{1}{\mathrm{k}T}\left(\gamma x + \frac{\alpha c_0^3}{\sqrt{x}}\right) \tag{1.41}$$

where

$$\eta = \tau_0 \exp\left(\frac{-\beta c_0^2}{\mathrm{k}T}\right)$$

Hence $\ln(\tau)$ has a minimum at

$$x_1 = \frac{1}{\tilde{\sigma}^4} = \left(\frac{\alpha}{2\gamma}\right)^{2/3} c_0^2 \tag{1.42}$$

below which it diverges. Therefore, if the stress values used in the experiment are much smaller than $\tilde{\sigma}$, then $\ln(\tau)$ would still follow an almost straight line. For Fig. 1.13, the largest experimental σ or (P) is ~0.7 atm, while for our simulations $\overline{c}_0 = 1.2 \times 10^{-6}$ cm and therefore $\tilde{\sigma} \approx 3.5$ atm.

A second experiment, analyzed, e.g., in Guarino et al. (2002) is where a load, which is linearly increasing with time ($\sigma_0 = A_p t$ for a constant A_p), is applied to different samples until they fail. The breaking times are drawn as a function of A_p in a log-log plot yielding a straight line. In the experiment (Guarino et al. 2002) two temperature values were used, 300 K and 380 K, with almost no difference in the obtained straight lines. On the basis of this invariance, the investigators concluded that there was something amiss and invoked (Politi et al. 2002) fiber bundle theory to explain this apparent difficulty.

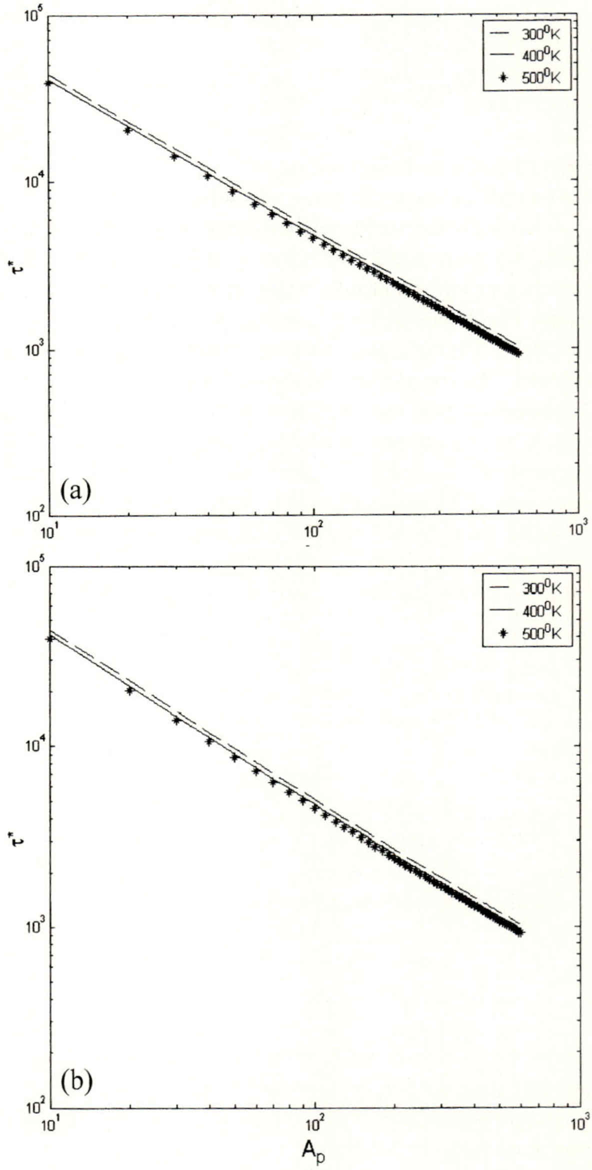

Fig. 1.16. Time to failure for a linearly increasing stress as a function of stress rate A_p. **a** For $c_0 = 0$ simulating Guarino et al. (2002) results. **b** Averaged times for c_0 chosen according to Eq. 1.40

The difficulty is however only a sham difficulty as can be seen from the following "numerical experiment". First we follow the method used by these authors (Guarino et al. 2002) to treat breaking time problems where the stress changes with time. We regard (Guarino et al. 2002) the variable $1/\tau$ (either of Eq. 1.35 for "no flaw" or of Eq. 1.38 for nucleation under existing flaw distribution) as the "density of damage" per unit time. Breaking is certain to occur after a time τ_1 if

$$\int_0^{\tau_1} \frac{1}{\tau(t)} dt = 1 \tag{1.43}$$

We now apply Eq. 1.43 to find τ_1 numerically by integrating Eq. 1.35 with $\sigma_0 = A_p t$, or perform a similar calculation using Eq. 1.38 with an appropriate choice process (of c_0) as described before. We obtain (Fig. 1.16) that in both cases *results are almost independent of temperature*. The results of Guarino et al. (2002, Fig. 8) are therefore completely natural and in line with both the "no flaw" and nucleation under existing flaw distribution theories.

We turn now to the time distribution function. Experimental results (Guarino et al. 2002, Fig. 6) show a distribution, which is definitely neither Poissonian (which would have been expected from the Arrhenius approach (e.g., Gardiner 1990)) nor Gaussian. Using Eq. 1.38 and choosing c_0 values randomly according to Eq. 1.39, a series of τ values is thereby generated. In this way, we calculated τ-distribution functions for several temperatures. Distribution results are shown in Fig. 1.17 for different values of temperature. It is seen that

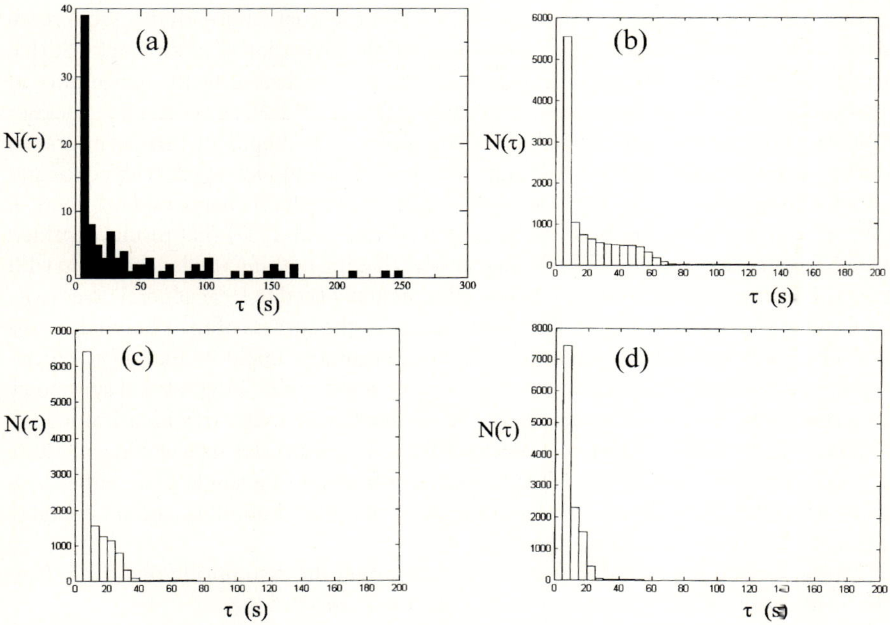

Fig. 1.17. Time distributions for different temperatures. **a** Experimental results (Guarino et al. 2002) $T \approx 300$ K. **b** Results of our calculations (Eqs. 1.38, 1.39 and 1.43) $T = 300$ K. **c** As in **b** but for $T = 400$ K. **d** As in **b** but for $T = 500$ K

1. The distribution function for 300 K is very similar to the experimental results.
2. The change of distribution function with temperature is interesting.

Thus, a relatively small temperature increase causes the "balance" of the distribution to move somewhat towards shorter times, while a large increase (in Fig. 1.17 from 300 K, say, to 500 K) even alters the whole pattern of the distribution, moving it away from a decreasing exponential towards a more "bulky" shape.

It is therefore seen that thermal nucleation is indeed the process behind failure of polycrystalline samples under creep conditions. An improvement of the theory, by the use of nucleation where account is taken of the existing flaw distribution inside the sample, leads

1. To an easier fracture and, more importantly.
2. To an accurate change of lifetime distribution functions with temperature.

1.2.4 Fiber Bundle Model

1.2.4.1 *Introduction*

A brittle heterogeneous solid of the second or third group (Sect. 1.2.1) under an applied stress can break via what we have called the "second scenario" (Sect. 1.2.3.3), namely the appearance of a large number of microcracks. These can either grow or increase in number, leading to their coalescence and eventual failure. In this section, we discuss fiber bundle models which treat the multiple nucleation of microcracks. Such a nucleation is supposed to occur in rock fracture, as is evidenced by the appearance of ever-increasing acoustic emission pulses (see, e.g., Mogi 1962; Hirata et al. 1987; Lockner 1993) and of electromagnetic radiation emissions (see Chap. 5 of this monograph) during such fracture processes. Although the fiber bundle model was developed for and seems most appropriate to describe a solid made up of fibers such as wood or "fibers in matrix" composite materials, it can be shown (Hirata et al. 1987) that results obtained using it agree with "regular" field damage models having nothing whatsoever to do with fibers. The fiber bundle model (FBM) was thus actually used to treat general cases (e.g., Guarino et al. 2002) of fracture. The exact reason for the success of the FBM to describe non-fibrous materials is still unknown. A crude argument might be that most experiments either in tension or compression are conducted under a cylindrical symmetry (uniaxial stress or triaxial stress where the perpendicular axes carry identical loads). For a grainy material, a string of grains parallel to the symmetry axis would constitute a "chain of beads" not unlike a fiber. Moreover, a fracture of a single grain either by a transversal through fracture or a fracture along the grain boundary could be looked upon as a failure of the chain – the fiber.

We shall discuss the FBM only in 1D, namely under the assumption that the lattice-like connection of the fibers is ignored. This restriction is done

1. Since the 1D case can often be analytically or at least asymptotically solved.
2. Since it retains the main features and results of the more complex models.

More realistic models, usually containing entangled numerical calculations, are based on a "chain of bundles" system (see Hidalgo et al. 2002; Gucer and Gurland 1962; Smith and Phoenix 1981; Smith 1982).

Treated here are two cases defined according to how the load is redistributed, following a fiber rupture.

1. The stress is redivided equally among all the remaining fibers (ELS). For this treatment, we will follow mainly the approach of Turcotte et al. (2003).
2. The stress is transferred in decreasing amounts to the "neighbors" of the ruptured fiber (VLS). For this treatment, we will mainly follow the approach of Hidalgo et al. (2002).

Our treatment is meant to be explanatory rather than exhaustive, since other approaches to the unequal redistribution of stress are abundant.

1.2.4.2
Uniform (or Equal) Load Sharing (Turcotte et al. 2003) ELS (Mean-Field Approximation)

Consider (Fig. 1.18) a rod made up of N_0 fibers. Denote by $N(t)$ the number of unbroken fibers at time t. The rate of fibers failure is given by

$$\frac{\mathrm{d}N(t)}{\mathrm{d}t} = -\nu(\sigma_\mathrm{f})N(t) \tag{1.44}$$

where ν is a parameter (called "hazard rate"), which is dependent on the applied fiber stress at time t ("no delay" is usually assumed). If the rod is loaded by a constant force, F_0, the fiber stress in it at time t is

$$\sigma_\mathrm{f}(t) = \frac{F_0}{N(t)s} \tag{1.45}$$

where s is the area of a fiber. This is the uniform load sharing assumption namely that following a failure of a fiber, the load is equally redistributed among the other fibers.

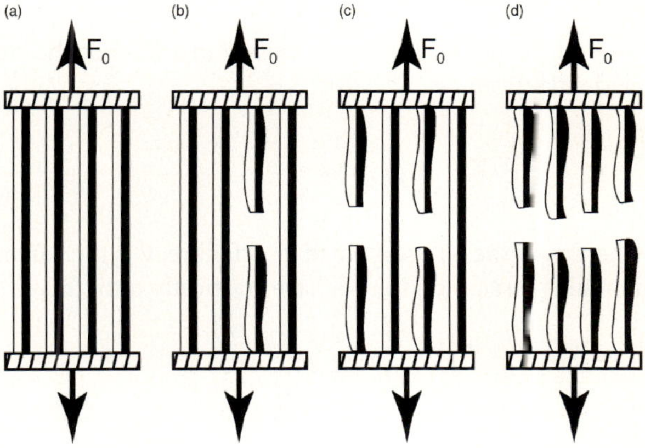

Fig. 1.18. A schematic illustration of the failure of a fiber-bundle with uniform load sharing. **a** Each of the fibers carries one-quarter of the load F_0. **b** One fiber has failed and each remaining fiber carries one-third of the load F_0. **c** Two fibers have failed and each remaining fiber carries one-half of the load F_0. **d** All four fibers have failed and no load is carried (from Turcotte et al. 2003)

In order to complete the presentation of the problem, the functional form of $v(\sigma)$ should be specified. An empirical form for $v(\sigma)$ is

$$v = v_0 \left(\frac{\sigma}{\sigma_0} \right)^{\rho} \tag{1.46}$$

where v_0 is the rate related to the initial stress σ_0 $(= F_0 / (N_0 s))$ and ρ is a constant exponent. In order to agree with experimental results, it is shown by Newman and Phoenix (2001) that the range of ρ values is between 2 to 5. Solving Eqs. 1.44–1.46 yields

$$N(t) = N_0 (1 - \rho v_0 t)^{1/\rho} \tag{1.47}$$

The rod fails when $N = 0$. Hence, the time to failure is given by

$$t_{\mathrm{f}} = \frac{1}{\rho v_0} \tag{1.48}$$

Note that this result does not depend on the number of initial fibers but does depend on σ_0. In view of Eq. 1.46, no agreement with Eq. 1.34 or Eq. 1.35 of the thermal nucleation model (Sect. 1.2.3.4) is to be expected. As noted above, Ciliberto et al. (2001) try to build a combined model of thermal nucleation with fiber bundles to explain their experimental results.

A very nice feature of the fiber bundle model is its ability to provide an explanation for the energy emission rate from a fracture either by acoustic emission or by EMR (see Chap. 5). To calculate the rate of say the acoustic emission (ae) in this model, we assume it to be a fraction, η_{ae}, of the total strain energy release rate, $\mathrm{d}W / \mathrm{d}t$:

$$\frac{\mathrm{d}W_{\mathrm{ae}}}{\mathrm{d}t} = \eta_{\mathrm{ae}} \frac{\mathrm{d}W}{\mathrm{d}t} \tag{1.49}$$

while $\mathrm{d}W / \mathrm{d}t$ is given by

$$\frac{\mathrm{d}W}{\mathrm{d}t} = -w_{\mathrm{f}} \frac{\mathrm{d}N}{\mathrm{d}t} \tag{1.50}$$

where w_{f} is the stored elastic energy in a fiber at the time of its failure. The latter is evidently given by

$$w_{\mathrm{f}} = \frac{1}{2} \frac{\sigma_{\mathrm{ff}}^2}{E} V \tag{1.51}$$

where σ_{ff} is the stress in the fiber at failure, V is the volume of a fiber and E is its Young modulus (assuming they behave elastically until failure). Using Eqs. 1.44–1.51 yields

$$\frac{\mathrm{d}W_{\mathrm{ae}}}{\mathrm{d}t} = \eta_{\mathrm{ae}} \frac{N_0 V}{2E} \upsilon_0 \sigma_0^2 \frac{1}{(1 - t / t_{\mathrm{f}})^{(\rho+1)/\rho}} \tag{1.52}$$

Fig. 1.19.
Cumulative acoustic energy emissions $e_a(t)$ at time t normalized by the total acoustic energy emissions e_{tot} at the time of rupture (t_f) as a function of $1 - t / t_f$. A constant pressure difference was applied at $t = 0$ (Guarino et al. 1999)

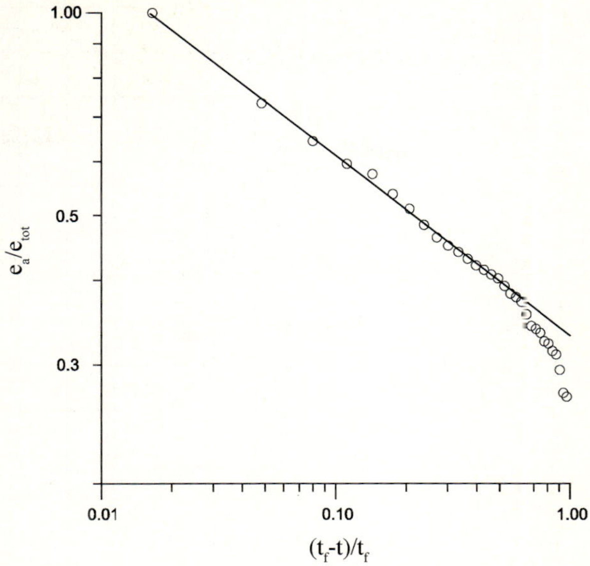

A comparison with experiment (e.g., Guarino et al. 2002) of the acoustic energy release as a function of time from failure of chipboard and fiberglass yields an excellent agreement with Eq. 1.52 for $\rho = 3.7$, or $(\rho + 1) / \rho = 0.27$ (Turcotte et al. 2003). Figure 1.19 shows an example of such an agreement for a cumulative acoustic emission as a function of $(1 - t / t_f)$ (Guarino et al. 1998, 1999).

1.2.4.3
Variable Load Sharing (VLS)

The method of stress sharing following a failure of one fiber that is opposite to the one discussed above is called local load sharing (LLS). Already, Coleman (1957) considered the influence of friction among fibers. This friction would tend to concentrate the load in the vicinity of the broken fiber rather than spread it like in the ESL. Here we follow the numerical approach of Hidalgo et al. (2002) (see the list of references there for an accumulation of papers in the area). The model assumes a power law decrease of stress away from the failed fiber. Denoting the stress in fiber i at time t by $\sigma_i(t)$, the additional stress there at time $t + \tau$, $\sigma_i(t + \tau) - \sigma_i(t)$ is obtained by contributions from all failed fibers $B(\tau)$ in the interval τ in the following manner:

$$\sigma_i(t+\tau) - \sigma_i(t) = \sum_{j \in B(\tau)} \sigma_i(t) F(r_{ij}, \gamma) \tag{1.53}$$

where

$$F(r_{ij}, \gamma) = \frac{r_{ij}^{-\gamma}}{\left(\sum_{i \in I} r_{ij}^{-\gamma} \right)} \tag{1.54}$$

Fig. 1.20.
Illustration of the model construction. X indicates a fiber that is going to break, and O is an intact fiber in the square lattice (Hidalgo et al. 2002)

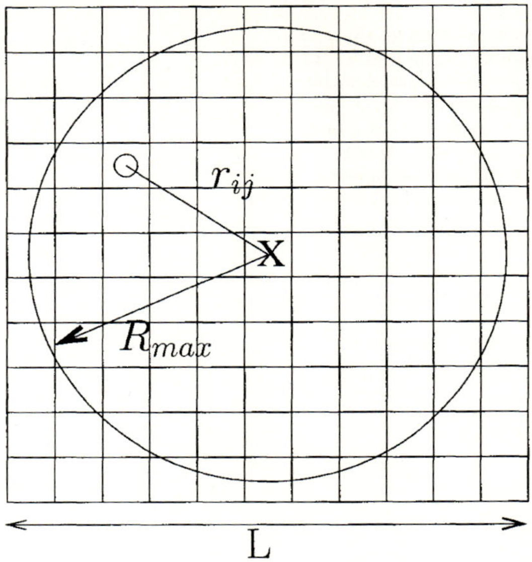

is the normalized contribution fraction from a failed fiber j (at point X in Fig. 1.20) to the fiber i. The sum on i in the normalizing denominator runs over all unfailed fibers at time t, and r_{ij} is the distance from fiber j to fiber i. Hidalgo et al. (2002) have used periodic boundary conditions on an $L \times L$ grid so that the largest value of r used was $(L-1)/\sqrt{2}$. The number of fibers is $N = L^2$. $B(\tau)$, the set of all fibers that break in the increment τ is defined as follows. The system is driven by first applying on it a small value of stress (σ) which is equally distributed among the L^2 fibers. The strengths of the fibers are assumed to be distributed according to a Weibull distribution.

$$P(\sigma_{\text{th}}) = 1 - \exp\left[-\frac{\sigma_{\text{th}}}{\sigma_0}\right]^{\rho} \tag{1.55}$$

where ρ is the so-called Weibull index, which controls the spread in threshold value, and σ_0 is a normalization parameter. Each fiber is ascribed a certain threshold value σ_{th} chosen at random from this distribution. Now the fiber with the lowest value of $(\sigma_{\text{th}} - \sigma/N)$ is sought, and this minimum value is added to all intact fibers. Thus, at least one fiber is going to break and its load is transferred via Eq. 1.53 to the other ones. This may cause more fibers to fail at this stage, defining ultimately the avalanche $B(\tau)$ of all fibers that failed at this stage. When the failures stop, σ is increased again by the same method.

The important parameter is γ (see Eq. 1.53 and Eq. 1.54). It defines the effective range of the influence of one failure. For $\gamma = 0$, the range is infinite and the model is equivalent to the ELS discussed above. For a large γ, this range shrinks and for $\gamma = \infty$ only its nearest neighbors are affected by the failure.

Results of numerical simulations (Monte Carlo type) show some interesting results. Figure 1.21 shows the ultimate strength of the complete bundle as a function of γ for different values of L. Here σ_c is the stress (per fiber) above which all fibers were broken. Three ranges of γ are observed. For $\gamma \leq 2$, the behavior is similar to that of $\gamma = 0$, namely the ELS model. In this range, the ultimate strength depends neither on γ nor on L and can

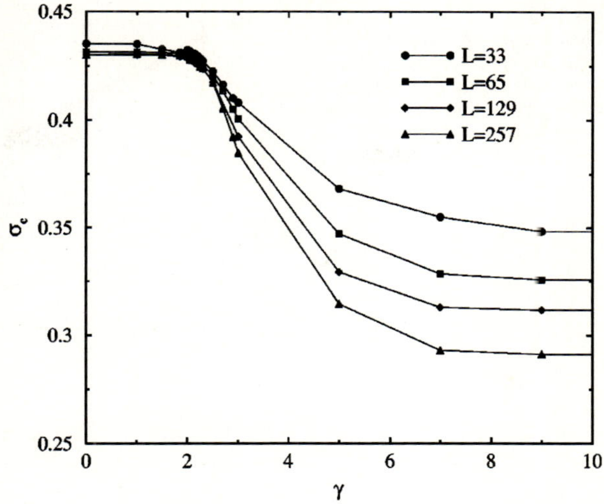

Fig. 1.21. Ultimate strength of the material for different system sizes as a function of the effective range of interaction γ. A cross-over from mean-field to short-range behavior is clearly observed (Hidalgo et al. 2002)

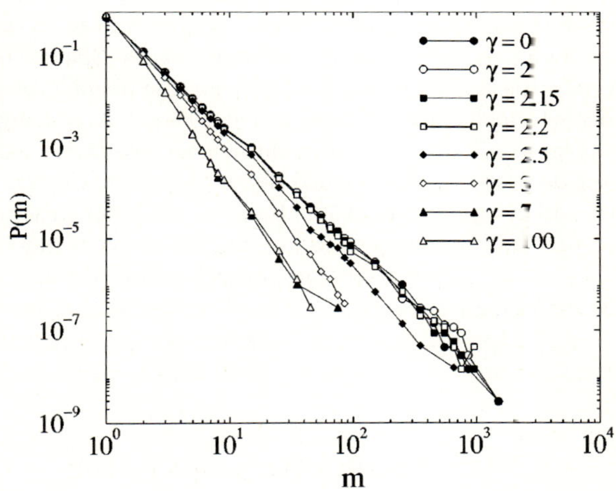

Fig. 1.22. Avalanche size distributions for different values of the exponent of the stress-transfer function γ ($L = 257$) (Hidalgo et al. 2002)

in fact be calculated exactly. For the Weibull distribution, exact calculation yields $\sigma_c = (\rho e)^{-1/\rho}$. For $\gamma \geq 2$ (local load sharing), σ_c depends on L and, as was shown by Hidalgo et al. (2002), approaches zero for $L \to \infty$ as $\sigma_c(N) \sim 1/\ln N$. This behavior is in line with other short range models (see e.g., Harlow 1985; Kloster et al. 1997). However, for $\gamma \leq 7$, σ_c depends on γ, actually decreasing with increasing γ, while for $\gamma \geq 7$ (extreme local load sharing) σ_c is independent of γ due probably to the discrete nature of the model.

The avalanche size-distribution is shown in Fig. 1.22 for $L = 257$ and different values of γ. For small γ values (mean field approximation or ELS), the distribution can be fitted by a power law.

$$P(m) \sim m^{-5/2} \tag{1.56}$$

A similar result was obtained by exact calculations in this limit (Harlow 1985; Kloster et al. 1997). For a more localized interaction ($\gamma > 2$ say), the causality correlation be-

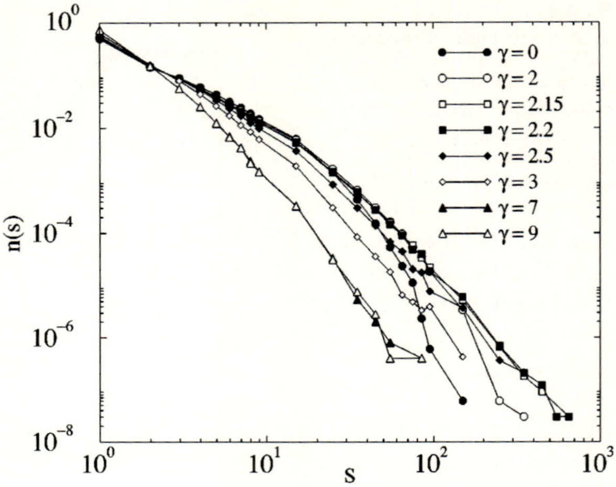

Fig. 1.23. Cluster size distributions for different values of the stress-transfer function exponent γ. Clearly, two different groups of curves can be distinguished ($L = 257$) (Hidalgo et al. 2002)

tween failing fibers seems to decrease and no relation like Eq. 1.56 can be derived. For small γ values, "any given element could be near to its rupture point regardless of its position in the lattice", while for large γ values "spatial correlation (is) important and concentration of stress take(s) place in the fibers located at the perimeter of an already formed cluster" (Hidalgo et al. 2002) thus reducing the avalanche size.

Figure 1.23 depicts the distribution of cluster sizes, namely the spatially connected (via nearest-neighbors) failed fibers. Shown are clusters just before total failure; the transition from ELS to LLS is obvious. The cluster structure for a complete mean field ($\gamma = 0$) "can be mapped to percolation clusters generated with the probability $0 < P(\sigma_c) < 1$" (*ibid*) except for the exact critical percolation probability, which depends on the Weibull index. For more localized interaction, we approach the "first scenario" (Sect. 1.2.3.3), where a single cluster (crack) is formed leading to complete breakdown. Thus, for large γ values, the total number of clusters decreases.

1.3 Linear Elastic Fracture Mechanics (LEFM)

1.3.1 General

Although the basic physical model (Newton's second law, Hook's law and the boundary conditions which assume that the crack sides are traction free) are simple, the ensuing differential equations (Eq. 1.8–1.16) become mathematically complicated, especially those related to 3D problems. Therefore, most of the theoretical approaches were carried out for 2D problems. Even for 2D, the task of solving these equations is complicated. However, as will be shown, the solution here can be divided into two parts. Firstly, due to the sharpness of the crack tip, the stresses in the immediate neighborhood of the tip turn out to be singular with asymptotic forms (at the tip), which depend only on the mode of loading (of which there are only three; Sect. 1.1.3) and on three related constants (one for each mode). Secondly, the influence of the *external* boundary conditions, i.e., the geometry of the problem and the applied stresses or strains on the surface of the specimen, is manifested *only* via these constants (called stress intensity factors and denoted by K_{I}, K_{II} and K_{III}). The energy supplied to the crack tip, which leads to the latter's ability to propagate, is shown to depend only on these stress intensity factors. A balance between this energy and the energy needed to form the new surfaces created by the fracture then yields the extension of the Griffith criterion. Since the surface energy depends on the specific material the specimen is made of, a corresponding material based critical parameter, K_c is defined. Energy balance implies that when K_{I} say, equals the critical parameter K_{Ic}, fracture becomes possible. Since an absolute majority of fractures actually occur in a mode I situation, the usual result of LEFM is the criterion

$$K_{\mathrm{I}} \geq K_{\mathrm{Ic}} \tag{1.57}$$

as a necessary condition of cracking. A second condition, which (together with Eq. 1.57) constitutes a sufficient requirement, is that the equilibrium implied by Eq. 1.21 would not be a stable one (see below). The whole effort of static LEFM is therefore concentrated in calculating K_{I} for specific geometries and loadings.

1.3.2
Equations of Equilibrium and of Motion

For the sake of completeness, we present here the general equations of motion and equilibrium for a homogeneous solid in the elastic approximation, resulting from Eq. 1.8 in three dimensions. They are written in terms of the displacements u in contrast to the presentation in Eqs. 1.8–1.16, which are formulated through stresses (or strains). For actual problems where boundary conditions are usually given in terms of stresses or tractions, the second representation (Eqs. 1.8–1.16) is preferable. Thus, in terms of **u**, the equations become

$$\rho \frac{\partial^2 \mathbf{u}}{\partial t^2} = (\lambda + \mu)\nabla(\nabla \cdot \mathbf{u}) + \mu \nabla^2 \mathbf{u} \qquad \text{(zero body force)} \tag{1.58}$$

or, in terms of the components,

$$\begin{aligned}
\rho \frac{\partial^2 u_x}{\partial t^2} &= (\lambda + \mu)\left[\frac{\partial^2 u_x}{\partial x^2} + \frac{\partial^2 u_y}{\partial x \partial y} + \frac{\partial^2 u_z}{\partial x \partial z}\right] + \mu\left[\frac{\partial^2 u_x}{\partial x^2} + \frac{\partial^2 u_x}{\partial y^2} + \frac{\partial^2 u_x}{\partial z^2}\right] \\
\rho \frac{\partial^2 u_y}{\partial t^2} &= (\lambda + \mu)\left[\frac{\partial^2 u_x}{\partial x \partial y} + \frac{\partial^2 u_y}{\partial y^2} + \frac{\partial^2 u_z}{\partial y \partial z}\right] + \mu\left[\frac{\partial^2 u_y}{\partial x^2} + \frac{\partial^2 u_y}{\partial y^2} + \frac{\partial^2 u_y}{\partial z^2}\right] \\
\rho \frac{\partial^2 u_z}{\partial t^2} &= (\lambda + \mu)\left[\frac{\partial^2 u_x}{\partial x \partial z} + \frac{\partial^2 u_y}{\partial y \partial z} + \frac{\partial^2 u_z}{\partial z^2}\right] + \mu\left[\frac{\partial^2 u_z}{\partial x^2} + \frac{\partial^2 u_z}{\partial y^2} + \frac{\partial^2 u_z}{\partial z^2}\right]
\end{aligned} \tag{1.58'}$$

where ρ is the material density and λ, μ are (see Sect. 1.1.1) the Lamé constants, and it is assumed that body forces are negligible. In the case of equilibrium and static conditions, the left-hand side (or sides) of Eq. 1.58 or 1.58' is zero.

There are several good references in which solutions to Eq. 1.58 are given (e.g., Erdogan 1983; Leibowitz 1968, and see also next section). Several works discuss the steady state solutions while others (see Fineberg and Marder 1999 and references therein) treat the dynamic version. Let us reemphasize that while the steady state (Griffith) results are very useful for basic understanding and for engineering applications, the dynamic solutions are controversial and do not, in general, scrupulously agree with the experimental results (see Sect. 1.4).

Together with Eq. 1.11 and Eq. 1.13, Eq. 1.58 (1.58') constitute a complete set of equations for the linear elastic problem.

1.3.2.1
Antiplane Stress Fracture (Mode III)

A very important 2D case, the equations for which are derived from Eqs. 1.58, is the mode III case. In mathematical terms, this is the simplest case to treat, and can be used as a basis for understanding all types of fracture.

For mode III (see Fig. 1.4), displacements are only in the z-direction; therefore, $u_x = u_y = 0$ and u_z does not depend on z, namely $u_z = u_z(x,y)$. Hence Eqs. 1.58 become

$$\rho\frac{\partial^2 u_z}{\partial t^2} = \mu\left(\frac{\partial^2 u_z}{\partial x^2} + \frac{\partial^2 u_z}{\partial y^2}\right) \tag{1.59}$$

Defining

$$c = \sqrt{\frac{\mu}{\rho}}$$

Eq. 1.59 becomes the so-called "2D wave equation"

$$\frac{1}{c^2}\frac{\partial^2 u_z}{\partial t^2} = \nabla^2 u_z \tag{1.60}$$

where

$$\nabla^2 = \frac{\partial^2}{\partial x^2} + \frac{\partial^2}{\partial y^2}$$

By Eq. 1.11 the only non-zero strain components in this case are

$$\varepsilon_{13} = \frac{1}{2}\frac{\partial u_z}{\partial x} \qquad \text{and} \qquad \varepsilon_{23} = \frac{1}{2}\frac{\partial u_z}{\partial y}$$

leading, by Eq. 1.13, to the only non-zero stress components

$$\sigma_{13} = \mu\frac{\partial u_z}{\partial x} \qquad \text{and} \qquad \sigma_{23} = \mu\frac{\partial u_z}{\partial y}$$

Even in this case, and even for a time-independent solution (where Eq. 1.60 becomes the Laplace equation, $\nabla^2 u_x = 0$), the actual solution may be complicated due to the prevailing boundary conditions.

1.3.3 Two-dimensional (2D) Problems and Stress Intensity Factors

1.3.3.1 *General*

In two dimensions (mode III, Sect. 1.3.2.1, plane stress, plane strain, see above), the equations become much simpler, and it is also easier to understand the problems involved in their solution. We shall treat here only equilibrium problems with no body forces. Extensions to dynamical problems and to problems including body forces can be found in books specializing in those cases (see above and Sect. 1.4). Thus, the 2D equations of equilibrium for stresses in the x- and y-directions stemming from Eq. 1.10 become

$$\begin{cases} \dfrac{\partial \sigma_{xx}}{\partial x} + \dfrac{\partial \sigma_{xy}}{\partial y} = 0 \\ \dfrac{\partial \sigma_{yy}}{\partial y} + \dfrac{\partial \sigma_{xy}}{\partial x} = 0 \end{cases} \tag{1.61}$$

For an elementary derivation of these equations, see e.g., Timoshenko and Goodier (1984) p. 26ff, and for a more mathematical approach see e.g., Landau and Lifshitz (1963) p. 17ff.

It now becomes obvious why the compatibility equations (Eq. 1.16) are needed, since Eq. 1.61 are a set of two equations for three unknowns (σ_{xx}, σ_{yy} and σ_{xy}) and as such are not a complete set. The connections with strains are (Hook's law, Eqs. 1.13 in 2D):

$$\begin{cases} \varepsilon_{xx} = \dfrac{\sigma_{xx} - \nu\sigma_{yy}}{E} \\ \varepsilon_{yy} = \dfrac{\sigma_{yy} - \nu\sigma_{xx}}{E} \\ \varepsilon_{xy} = \dfrac{\sigma_{xy}}{2G} = \dfrac{\sigma_{xy}}{2\mu} \end{cases} \tag{1.62}$$

where $G \equiv \mu$ is the so-called shear modulus. But the three components of strain are obtained from only two displacement components u_x and u_y, which are the basic physical variables (this is actually the reason why Eq. 1.61 consists of only two equations). In 2D, the definition of strains (Eq. 1.11) becomes

$$\varepsilon_{xx} = \frac{\partial u_x}{\partial x}, \qquad \varepsilon_{xy} = \frac{1}{2}\left(\frac{\partial u_x}{\partial y} + \frac{\partial u_y}{\partial x}\right), \qquad \varepsilon_{yy} = \frac{\partial u_y}{\partial y} \tag{1.63}$$

(three components of strain obtained from two displacements). However, Eqs. 1.63 are not independent. For taking the second derivative of the first with respect to y, of the second with respect to x and y and of the third with respect to x, we arrive at the "compatibility equations" in 2D

$$\frac{\partial^2 \varepsilon_{xx}}{\partial y^2} + \frac{\partial^2 \varepsilon_{yy}}{\partial x^2} = 2\frac{\partial^2 \varepsilon_{xy}}{\partial x \partial y} \tag{1.64}$$

which can be written (Timoshenko and Goodier 1984) in terms of stresses as

$$\nabla^2(\sigma_x + \sigma_y) = 0 \tag{1.65}$$

where

$$\nabla^2 = \frac{\partial^2}{\partial x^2} + \frac{\partial^2}{\partial y^2}$$

The complete set to be solved for σ is now Eq. 1.61 and Eq. 1.65. We discuss here two types of solutions to this set. One is accomplished by the so-called Airy or stress-function and the second is known as the Westergaard method.

1.3.3.2
Airy-Function (Stress-Function)

A new function Ψ is defined such that the stress components are given by:

$$\sigma_{xx} = \frac{\partial^2 \Psi}{\partial y^2}, \qquad \sigma_{yy} = \frac{\partial^2 \Psi}{\partial x^2}, \qquad \sigma_{xy} = -\frac{\partial^2 \Psi}{\partial x \partial y} \tag{1.66}$$

In this manner, Eqs. 1.61 are immediately fulfilled and in order to fulfill Eq. 1.65, Ψ has to satisfy the following equation:

$$\nabla^4 \Psi = \frac{\partial^4 \Psi}{\partial x^4} + 2\frac{\partial^4 \Psi}{\partial x^2 \partial y^2} + \frac{\partial^4 \Psi}{\partial y^4} = 0 \tag{1.67}$$

To get a solution for a 2D elastic problem, one has to solve Eq. 1.67 with the appropriate boundary conditions. The latter can be given in terms of displacements (u_i), stresses, or tractions (forces, see Eq. 1.7) on the surface of the medium (including the surface of the fracture itself).

1.3.3.3
Westergaard Method (Westergaard 1939)

This method is based on the properties of analytic functions of a complex variable. A variable $z = x + iy$ is defined where x and y are the coordinates of the 2D problem and $i = \sqrt{-1}$. Analytical functions $Z(z)$ are defined as "complex functions of a complex variable which possess a derivative (Z') wherever the function is defined" (Ahlfors 1979). A well-known property of such functions is that their real and imaginary parts are harmonic, i.e., have the properties

$$\frac{\partial \operatorname{Re} Z}{\partial x} = \frac{\partial \operatorname{Im} Z}{\partial y} = \operatorname{Re} Z', \qquad \frac{\partial \operatorname{Im} Z}{\partial x} = -\frac{\partial \operatorname{Re} Z}{\partial y} = \operatorname{Im} Z' \tag{1.68}$$

and fulfill the Laplace equation

$$\left(\frac{\partial^2}{\partial x^2} + \frac{\partial^2}{\partial y^2}\right)(\operatorname{Re} Z) = \left(\frac{\partial^2}{\partial x^2} + \frac{\partial^2}{\partial y^2}\right)(\operatorname{Im} Z) = 0 \tag{1.69}$$

where ReZ and ImZ are the real and imaginary parts of Z, which can be written as

$$Z = Z(z) = Z(x + iy) = \operatorname{Re} Z + i \operatorname{Im} Z \tag{1.70}$$

$Z', \overline{Z}$ and $\overline{\overline{Z}}$ are defined as the first derivative and the first and second integrals of Z, respectively so that

$$Z' = \frac{dZ}{dz}, \qquad Z = \frac{d\overline{Z}}{dz}, \qquad \overline{Z} = \frac{d\overline{\overline{Z}}}{dz} \tag{1.71}$$

Westergaard then shows that for a restricted group of cases the stress components σ_{ij} can be derived from Z in the form

$$\begin{aligned} \sigma_{xx} &= \operatorname{Re} Z - y \operatorname{Im} Z' \\ \sigma_{yy} &= \operatorname{Re} Z + y \operatorname{Im} Z' \\ \sigma_{xy} &= -y \operatorname{Re} Z' \end{aligned} \tag{1.72}$$

Using Eq. 1.68, it is immediately seen that the stress components defined by Eq. 1.72 fulfill the equilibrium equations (Eq. 1.61) for any analytical function $Z(z)$. The restriction mentioned above is seen from Eqs. 1.72 and amounts to the condition that at $y = 0$, $\sigma_{xx} = \sigma_{yy}$ and $\sigma_{xy} = 0$. The displacements u_x and u_y are given by

$$\begin{aligned} 2\mu u_x &= (1 - 2\nu)\operatorname{Re} \overline{Z} - y \operatorname{Im} Z \\ 2\mu u_y &= 2(1 - \nu)\operatorname{Im} \overline{Z} - y \operatorname{Re} Z \end{aligned} \tag{1.73}$$

where ν is the Poisson ratio, since it can be shown that u_x and u_y of Eq. 1.73 can generate the stress components via Hook's law.

The Westergaard method can be correlated with the Airy or stress function Ψ defined above, for it is easily seen that Ψ can be given by

$$\Psi = \operatorname{Re} \overline{\overline{Z}} + y \operatorname{Im} \overline{Z} \tag{1.74}$$

1.3.3.4
Application to the Internal Crack under Tension

According to Westergaard (1939), if we use for Z the function

$$Z = \frac{\sigma_0}{\sqrt{1 - c^2 / z^2}} \tag{1.75}$$

then Eq. 1.72 and Eq. 1.73 would describe the situation for a crack (of length $2c$) stretching between $x = -c$ and $x = c$ at $y = 0$ in an infinite plane under a remote stress field σ_0 in the y-direction (Fig. 1.24). For $y = 0$ by Eq. 1.72 outside of the crack, the tension is

$$\sigma_{yy} = \frac{\sigma_0}{\sqrt{1 - c^2 / x^2}} \tag{1.76}$$

and inside the length of the crack, by Eq. 1.73, the y displacement is

$$2u_y = \frac{4(1-\nu^2)}{E} \operatorname{Im} \overline{Z} = \frac{4(1-\nu^2)\sigma_0}{E} \sqrt{c^2 - x^2} \tag{1.77}$$

Therefore, the shape of the crack in the vicinity of the tip is parabolic (the complete crack shape is elliptic). At the tips $x = \pm c$, there exists a singularity of σ_{yy}, while u_y is regular.

Irwin (1957) used the solution Eq. 1.75 to calculate the (singular) stress fields near the tip of the crack and to show that these fields can be connected via the energy release rate to the Griffith criterion. Substituting for z near the tip,

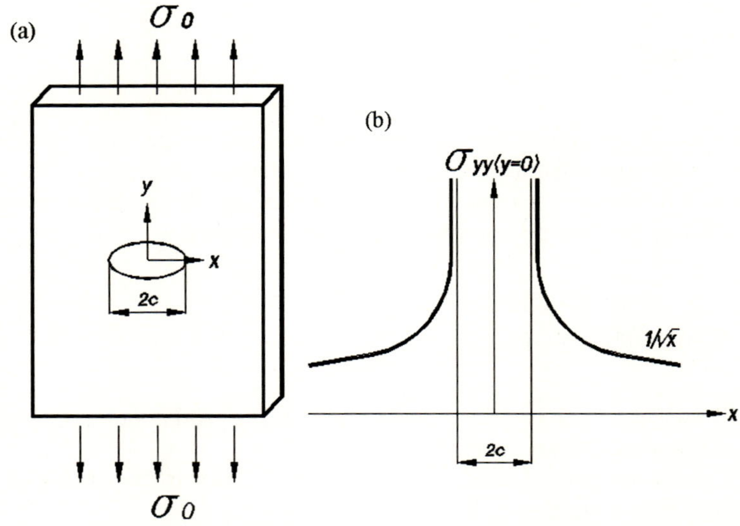

Fig. 1.24. Internal crack (according to Westergaard's paper), which has opened from $z = -c$ to $z = +c$ under the influence of a remote tension σ_0 and the resulting σ_{yy} (at $y = 0$) as a function of x

$$z = c + re^{i\theta} \tag{1.78}$$

where r is the small distance ($r \ll c$) from the tip as an origin

$$r = [(x-c)^2 + y^2]^{1/2} \tag{1.79}$$

and θ is the polar angle, again with the tip as an origin

$$\theta = \arctan\frac{y}{x-c} \tag{1.80}$$

the solution of Eq. 1.75 can be developed in powers of $(r/c)^{1/2}$. The first and dominant term for small r becomes (Irwin 1957)

$$\begin{aligned}\sigma_{xx} &= \frac{K_\mathrm{I}}{\sqrt{2\pi r}}\cos\frac{\theta}{2}\left(1+\sin\frac{\theta}{2}\sin\frac{3\theta}{2}\right)\\ \sigma_{yy} &= \frac{K_\mathrm{I}}{\sqrt{2\pi r}}\cos\frac{\theta}{2}\left(1-\sin\frac{\theta}{2}\sin\frac{3\theta}{2}\right)\end{aligned} \qquad \text{mode I} \quad (1.81)$$

and (see e.g., Lawn 1993)

$$\sigma_{xy} = \frac{K_\mathrm{I}}{\sqrt{2\pi r}}\sin\frac{\theta}{2}\cos\frac{\theta}{2}\cos\frac{3\theta}{2}$$

where $K_\mathrm{I}\,(=\sigma_0\sqrt{\pi c})$ is independent of r and θ and is discussed presently. For mode II, similar calculations lead to a similar result:

$$\begin{aligned}\sigma_{xx} &= -\frac{K_\mathrm{II}}{\sqrt{2\pi r}}\sin\frac{\theta}{2}\left(2+\cos\frac{\theta}{2}\cos\frac{3\theta}{2}\right)\\ \sigma_{yy} &= \frac{K_\mathrm{II}}{\sqrt{2\pi r}}\sin\frac{\theta}{2}\cos\frac{\theta}{2}\cos\frac{3\theta}{2}\\ \sigma_{xy} &= \frac{K_\mathrm{II}}{\sqrt{2\pi r}}\cos\frac{\theta}{2}\left(1-\sin\frac{\theta}{2}\sin\frac{3\theta}{2}\right)\end{aligned} \qquad \text{mode II} \quad (1.82)$$

and for completeness, the mode III result is obtained from Eq. 1.59 above, with the left side equaling zero:

$$\begin{aligned}\sigma_{xx} &= \sigma_{yy} = 0\\ \sigma_{xz} &= -\frac{K_\mathrm{III}}{\sqrt{2\pi r}}\sin\frac{\theta}{2}\\ \sigma_{yz} &= \frac{K_\mathrm{III}}{\sqrt{2\pi r}}\cos\frac{\theta}{2}\end{aligned} \qquad \text{mode III} \quad (1.83)$$

A more elaborate approach for the stress field near the tip of the crack was undertaken by Williams (1957). In his method, the stresses are directly expanded in a series. The leading term agrees with the Westergaard-Irwin results (Eq. 1.81). The other terms in σ come in powers of $r^{1/2}$; thus, the next two terms are a constant (r^0) and a term proportional to $r^{1/2}$. These terms and especially the sign of the constant term are sometimes used for a more elaborate criterion for crack propagation (e.g., Cotterell and Rice 1980).

1.3.3.5
Connection between the Stress Intensity Factors K_I (K_{II}, K_{III}) and the Energy Release Rate

Figure 1.25 shows the (elliptic) crack shape (Eq. 1.77) at its tip (to where the origin of the x-y-coordinate system has been transferred). Irwin (1957) then calculates the energy needed to drive the crack forward, assuming that it can be calculated from the crack vicinity. More comprehensive energy calculations where the energy of the whole medium is accounted for can be found, e.g., in Freund (1972a,b, 1973, 1974). For the static case, however, the outcome of these calculations yields essentially the same results as Irwin's.

Thus, according to Irwin (1957), suppose the crack had moved by a small increment α (see Fig. 1.25). The energy needed to perform this extension is equal to the work needed to close it down, since the theory is linear. Now, the force (per unit area) on the "bulge" for $\theta \cong 0$ and $0 \le x \le \alpha$, $\alpha \ll c$ is, by Eq. 1.81, approximately

$$f = \frac{K_I}{\sqrt{\pi}} \frac{1}{\sqrt{2x}} \beta = \beta f_1(x) \tag{1.84}$$

where β measures the "closure fraction" changing from 0 at the beginning of closure to 1 (end of closure). The y extension u_y for $0 \le x \le \alpha \ll c$ is given by Eq. 1.77 approximately as

$$u_y = \frac{4K_I}{E\sqrt{\pi}} \sqrt{2(\alpha - x)}(1-\beta) = (1-\beta)u_y(x) \tag{1.85}$$

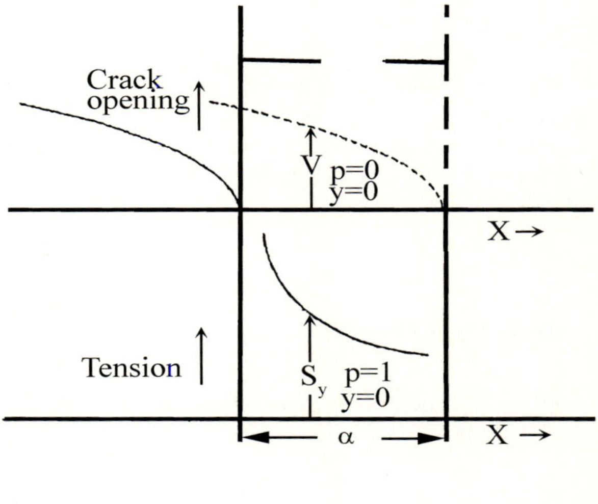

Fig. 1.25. Irwin's (1957) linear-elastic theory crack opening and stresses near the tip of the crack

The work of closure is therefore

$$W = \int f\,\mathrm{d}u_y = \int_0^1 \beta\,\mathrm{d}\beta \int_0^\alpha f_1(x)u_y(x)\mathrm{d}x = \frac{2K_\mathrm{I}^2}{E\pi}\int_0^\alpha \left(\frac{\alpha - x}{x}\right)^{1/2} \mathrm{d}x = \frac{\alpha K_\mathrm{I}^2}{E} \tag{1.86}$$

and per unit crack extension, namely the energy release rate is

$$\frac{W}{\alpha} = G = \frac{K_\mathrm{I}^2}{E} \tag{1.87}$$

As stated above, this is a simple mathematical method to obtain the energy derivative with respect to crack length (G).

These results (the set of Eqs. 1.75, 1.77, 1.81–1.83 and 1.87) constitute the basis of the LEFM for the treatment of 2D equilibrium problems. Thus, by a connection with Griffith's criterion, the necessary condition for a flaw to increase in length or for a crack to develop is $G > G_\mathrm{c}$ or $K_\mathrm{I} > K_\mathrm{Ic}$. Here, however, we have also obtained a method of calculating G (or K). For a crack of length $2c$ in an infinite medium (Fig. 1.26a), the above results yield (as in Griffith 1920)

$$K_\mathrm{I} = \sigma_0\sqrt{\pi c} \tag{1.88}$$

For different geometries, in order to calculate K_I say, the elastic equations should be solved only very near the crack's tip. The value of K_I is then obtained by taking the limit of $\sigma_{yy}\sqrt{2\pi r}$ say as $r \to 0$, for $\theta = 0$. Several methods, both analytical and numerical, were developed for the calculation of K (see e.g., Peterson 1974; Pook 2000; Murakami 1987, 1992, 2001). We present here only a few characteristic results of K_I calculations:

1. Many results are written in the form

   $$K = q\sigma\sqrt{\pi c} \tag{1.89}$$

   where q is a geometric factor and c is a characteristic crack (or flaw) dimension (explained below). For example, for an edge crack (Fig. 1.26b), $q = 1.12$ and

   $$K = 1.12\sigma\sqrt{\pi c} \tag{1.90}$$

2. For an elliptical crack in an infinite plate (Fig. 1.26c) under tensile stress perpendicular to the plane of the crack,

   $$K = \frac{M}{Q}\sigma\sqrt{\pi c} \tag{1.91}$$

   where Q is the elliptic integral of the second kind:

   $$Q = \int_0^{\pi/2} (1 - m^2 \sin^2\theta)^{1/2}\mathrm{d}\theta\,, \qquad m = \sqrt{1 - c^2/a^2} \tag{1.92}$$

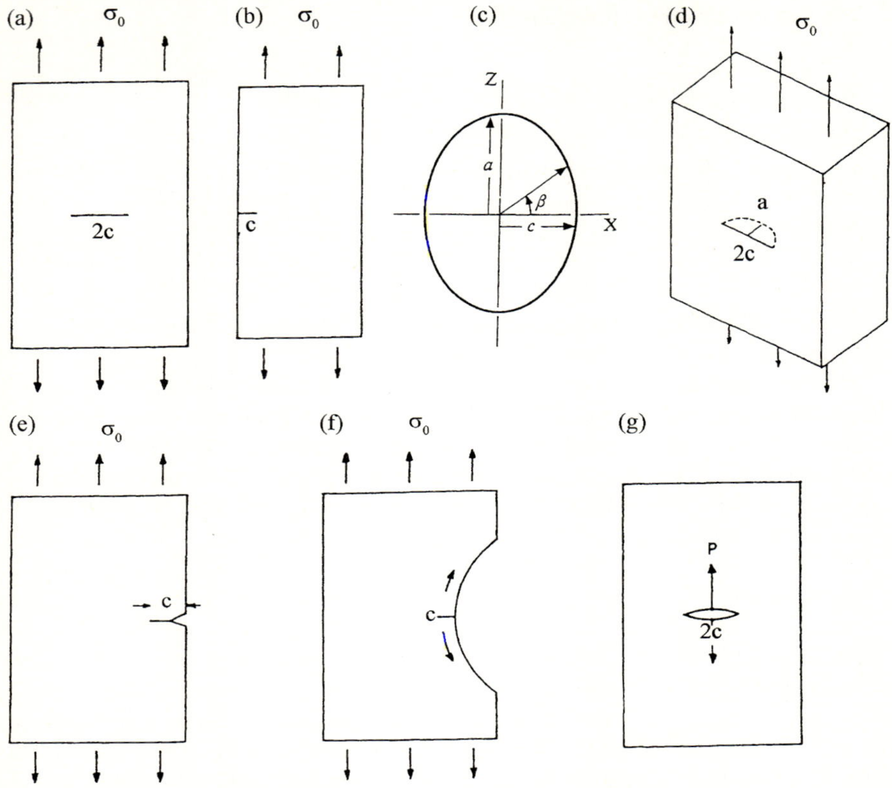

Fig. 1.26. Several cases of K_I values (after Murakami): **a** Internal crack in an infinite plate. **b** Edge crack in an infinite plate. **c** Elliptic crack in the x-z-plane in an infinite body under uniaxial tension σ in the y-direction. **d** Semi elliptical surface crack in a semi infinite body under uniaxial tension σ. **e, f** Edge crack in an infinite plate under uniaxial tension: **e** Sharp notch; **f** blunted notch. **g** Two point loads in middle of crack in an infinite plane (Pook 2000)

where a and c are the large and small elliptic radii, and (Rooke and Cartwright 1976)

$$M = [\cos^2\beta + (c/a)^2 \sin^2\beta]^{1/4} \tag{1.93}$$

describes the change of K_I along the ellipse. Note, e.g., that $M = 1$ for $\beta = 0$ (in the direction of c), while $M = \sqrt{c/a}$ (<1) in the direction of a. Here q of Eq. 1.89 is equal to M/Q.

3. For a penny-shaped crack loaded as in No. 2, $a = c$ and

$$K = 2\sigma\sqrt{c/\pi} \tag{1.94}$$

($q = 2/\pi$ in this case).

4. For a semi-elliptic crack at the edge of a semi-infinite plate (Fig. 1.26d) under remote tensile stress,

$$K = 1.12\sigma\frac{\sqrt{\pi\alpha}}{Q} \tag{1.95}$$

and for a semi penny-shaped crack, etc.

$$K = 0.713\sigma\sqrt{\pi c} \tag{1.96}$$

5. The value of c depends on the sharpness of the flaw. For instance, for a sharp groove followed by a crack (Fig. 1.26e), c is the sum of lengths of the groove and the crack, while for a blunt groove (Fig. 1.26f), only the crack length is taken as c.
6. An important case is a crack in an infinite plate operated on by two point loads as in Fig. 1.26g. This case can be used as a basis for K_{I} calculations using it within a Green's function method, namely considering any load as a superposition of point loads. In this case,

$$K = \frac{P}{b\sqrt{\pi c}} \tag{1.97}$$

where P is the value of the point load and b is the thickness of the plate. Note that here, uncharacteristically, c appears to the power of −½ instead of the usual ½.

1.4 Dynamic Fracture

1.4.1 Mott and Mott-like Approaches

The first attempt to address dynamic fracture was made by Mott (1948). His approach was based only on an energy balance and dimensional considerations. Thus, according to Mott (1948), there are three terms which combine to form the total energy of the system: The kinetic energy T, the liberated elastic (potential) energy U and the dissipative energy S. The only changing variables are the crack length c and its velocity

$$v = \frac{\mathrm{d}c}{\mathrm{d}t} = \dot{c}$$

Therefore by a purely dimensional analysis, the respective energies (for a crack in a slab of constant thickness b under a constant load) should be given by

$$T = \alpha v^2 c^2 b \tag{1.98}$$

$$U = -\beta c^2 b \tag{1.99}$$

$$S = \gamma c b \tag{1.100}$$

where α, β and γ are constants. Note that both T and U are proportional to the volume $c^2 b$ of the material in which elastic energy has been relaxed and which was set in motion by the moving crack with different velocities across it, which are assumed to be proportional to the crack velocity v. The dissipative energy, S, as before, is assumed to be proportional to the newly created area (cb) via a surface energy parameter γ.

The total energy of the system is

$$E_{tot} = T + P = T + U + S = \alpha v^2 c^2 b - \beta c^2 b + \gamma c b \tag{1.101}$$

where P is the total non-kinetic energy.

Mott then went on to assume that the crack develops under the condition $\partial E_{tot} / \partial c = 0$, while assuming that $\partial v / \partial c = 0$. This procedure yields

$$v = V\left(1 - \frac{c_0}{c}\right)^{1/2} \tag{1.102}$$

where the final velocity $V = (\beta / \alpha)^{1/2}$, and $c_0 = \gamma / 2\beta$. Since α and β are unknown from first principles, V and c_0 cannot be obtained in this manner from known properties (e.g., sound velocities). But, since for $c = c_0$, $v = 0$, it is assumed that c_0 is the critical microcrack (or effective flaw) length (c_{cr}) (see Sect. 1.2.3.2 and 1.2.3.4). However, a different interpretation of c_0 is presented below in this section.

Mott's assumption that $\partial v/\partial c = 0$ (which is obviously valid only for large values of v approaching the final velocity) has incurred some heavy criticism. Thus, foremost among others, Berry (1960) and Dulaney and Brace (1960) tried to improve on Mott's derivation. To understand their calculations, consider the non-kinetic energy P as a function of c as obtained by Dulaney and Brace (1960) (Fig. 1.27). It is immediately seen that the (static) Griffith criterion for the initiation of a crack, namely $\partial P / \partial c = 0$, occurs at c_0 and that for $c > c_0$ there is an ever-increasing gain of non-kinetic energy for crack elongation. Assuming now that the (dynamic) condition for crack propagation is $E_{tot} = P_0$, Dulaney and Brace (1960), obtained for the crack velocity

$$v = V\left(1 - \frac{c_0}{c}\right) \tag{1.103}$$

This relation for the crack velocity was deemed to be preferable over Eq. 1.102 (for example Fineberg, Gross, Marder and Swinney (Fineberg et al. 1991) and Marder (1993))

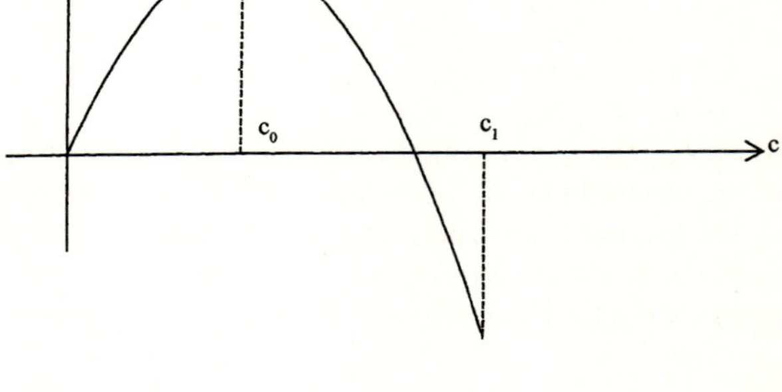

Fig. 1.27. Sum of potential and dissipative energies, P as a function of crack length c; c_0 is the Griffith criterion; c_1 is the beginning of dynamic fracture (Rabinovitch 1994)

although it suffers from two severe drawbacks. Firstly, for velocities that are close to V, even according to Dulaney and Brace (1960), Mott's approximation (Eq. 1.102) should be valid. Secondly, regarding Eq. 1.103 as a differential equation, for $v = dc/dt$, it cannot be integrated to obtain $c(t)$ (the improper integral diverges).

We have assumed (Rabinovitch 1994) that the dynamic crack process does not start when the crack length is at c_0. Rather, a delay of fast propagation occurs and the crack's fast movement is initiated only at $c_1 > c_0$. That means that we identify c_0 with c_i (the initial microcrack or effective flaw) and c_1 with c_{cr} (see Sect. 1.2.3.2, 1.2.3.4 and Fig. 2.1). Note however that this assumption entails that for $K_I > K_{Ic}$ ($K_I = K_{Ic}$ occurs at $c_0 = c_i$) the crack propagates with a very low velocity until its length reaches $c_1 = c_{cr}$. While the origin of this delay might be in dispute (for example Schoeck and Pichl 1990; Bernstein and Ness 2003), for our purpose it is the fact that between c_0 and c_1, energy is dissipated into avenues other than the kinetic or potential energies discussed above, which is important. It is therefore assumed that the actual critical length occurs at c_1 ($\geq c_0$), and only for $c > c_1$ can the energy balance à la Mott be performed. Thus, for $c \geq c_1$, the condition E_{tot} = constant with the provision that $v \cong 0$ at $c = c_1$ yields

$$v = V\left(1 - \frac{2c_0}{c} - \frac{k}{c^2}\right)^{1/2} \tag{1.104}$$

where $V = (\beta/\alpha)^{1/2}$, $c_0 = \gamma/2\beta$ as before and

$$k = c_1(c_1 - 2c_0) \tag{1.105}$$

Several points should be noted.

1. Equation 1.104 transfers over to the Mott relation (Eq. 1.102) for large values of c.
2. Very close to c_1, say for $c = c_1 + \varepsilon$ (where ε is a small quantity), the velocity behaves as $\varepsilon^{1/2}$.
3. Equation 1.104 is easily integrated (Gradshteyn and Ryzhik 1980) to yield

$$V(t - t_0) = g(c) + c_0 \ln\frac{g(c) + c - c_0}{c_1 - c_0} \tag{1.106}$$

   where $g(c) = (c^2 - 2c_0c + b)^{1/2}$ and t_0 is the time where $c = c_1$. Note that $g(c_1) = 0$ and that, for $t \gg t_0$, $c \approx Vt$.
4. Equation 1.103 is a (non-generic) special case of Eq. 1.104 for $c_1 \rightarrow c_0$. In this case, unlike the divergent integral obtained in trying to integrate Eq. 1.103 directly, the limit $c_1 \rightarrow c_0$ in Eq. 1.106 is well defined.
5. Since Eq. 1.104 contains three parameters V, c_0 and c_1, which can be adjusted to experimental results, it is not remarkable that such experiments can be approximated quite well by this equation. For example, the results depicted in Fig. 1.31 (see below) by Kobayashi et al. (1974), which were shown by Marder (1993) not to agree with the constant $G(t)$ approach (see below, Sect. 1.4.3), are in a very good agreement with Eq. 1.106 with $V \approx 575$ m s^{-1}, $c_0 \approx 1.33$ cm and $c_1 \approx 3.10$ cm (not shown in Fig. 1.31).

In Sect. 1.4.2 and 1.4.3, we present the linear elastic approach to dynamic fracture.

1.4.2
2D Dynamic Crack

We first consider here the (asymptotic) constant velocity solution. Therefore, starting from Eq. 1.58 we look for solutions that describe crack propagation at a constant velocity of magnitude (v) in the x-direction (Fineberg and Marder 1999, Sect. 2.4). We therefore consider displacements **u**, which have a "wavelike" behavior or, alternatively, where the origin stays at the crack tip, $x = x' - vt$, thus,

$$\frac{\partial}{\partial x'} = \frac{\partial}{\partial x}; \qquad \frac{\partial}{\partial t} = \frac{\partial}{\partial x}\cdot\frac{\partial x}{\partial t} = -v\frac{\partial}{\partial x}; \qquad \frac{\partial^2}{\partial t^2} = v^2\frac{\partial^2}{\partial x^2}$$

and Eq. 1.58 becomes

$$(\lambda + \mu)\nabla(\nabla \cdot \mathbf{u}) + \mu\nabla^2\mathbf{u} = \rho v^2 \frac{\partial^2 \mathbf{u}}{\partial x^2} \tag{1.107}$$

General solutions can be obtained by defining two auxiliary functions $\Phi(z_\alpha)$ and $\psi(z_\beta)$, where now z_α and z_β depend on the longitudinal and transversal acoustic wave velocities (c_ℓ and c_t respectively, defined in Eq. 1.109)

$$z_\alpha = x + i\alpha y\,, \qquad z_\beta = x + i\beta y \qquad \text{and} \tag{1.108}$$

$$\left.\begin{aligned} \alpha^2 &= 1 - \frac{\rho v^2}{\lambda + 2\mu} = 1 - \frac{v^2}{c_\ell^2} \\ \beta^2 &= 1 - \frac{\rho v^2}{\mu} = 1 - \frac{v^2}{c_t^2} \end{aligned}\right\} \qquad \begin{aligned} c_\ell &= \sqrt{\frac{\lambda + 2\mu}{\rho}} \\ c_t &= \sqrt{\frac{\mu}{\rho}} \end{aligned} \tag{1.109}$$

The displacements are given (as can be checked by inserting Eqs. 1.110 in Eq. 1.107) by

$$\begin{aligned} u_x &= \mathrm{Re}\,\Phi(z_\alpha) - \beta\,\mathrm{Im}\,\psi(z_\beta) \\ u_y &= -\alpha\,\mathrm{Im}\,\Phi(z_\alpha) - \mathrm{Re}\,\psi(z_\beta) \end{aligned} \tag{1.110}$$

The stress fields are given by

$$\begin{aligned} \sigma_{xx} &= \mu(1 + 2\alpha^2 - \beta^2)\,\mathrm{Re}\,\Phi'(z_\alpha) - 2\beta\mu\,\mathrm{Im}\,\psi'(z_\beta) \\ \sigma_{yy} &= -\mu(1 + \beta^2)\,\mathrm{Re}\,\Phi'(z_\alpha) + 2\beta\mu\,\mathrm{Im}\,\psi'(z_\beta) \\ \sigma_{xy} &= \mu[2\alpha\,\mathrm{Im}\,\Phi'(z_\alpha) + (\beta^2 + 1)\,\mathrm{Re}\,\psi'(z_\beta)] \end{aligned} \tag{1.111}$$

where

$$\Phi' = \frac{\partial \Phi(z)}{\partial z} \qquad \text{and} \qquad \psi' = \frac{\mathrm{d}\psi(z)}{\mathrm{d}z}$$

Note that our Φ and ψ are twice the Φ and ψ of Fineberg and Marder (1999).

In order to obtain a solution for a specific problem, "all" that is needed is to find the appropriate Φ and ψ functions. An important example (Fineberg and Marder 1999) is the near-tip behavior of a propagating crack under a mode I loading. Using the static solution (Sect. 1.3.3.4) as a clue and the symmetry properties of the problem, i.e., $u_x(-y) = u_x(y)$ and $u_y(-y) = -u_y(y)$, the appropriate solution is

$$\Phi(z) = \frac{C}{\sqrt{z}}; \qquad \psi(z) = i\frac{D}{\sqrt{z}} \tag{1.112}$$

(where the cuts of the $\sqrt{\ }$ are along the negative x-axis). The boundary condition on the crack surface is that σ_{xy} and σ_{yy} are zero there (relaxed stress). Using these conditions yields

$$\frac{D}{C} = \frac{2\alpha}{\beta^2 + 1} \tag{1.113}$$

As was done for the static case (Sect. 1.3.3.4), near the tip, we can develop the results in powers of $r^{1/2}$, where r is the small distance from the crack tip. We denote

$$\begin{aligned} z_\alpha &= r_\alpha e^{i\theta_\alpha}, \quad \text{so that} \quad r_\alpha = \sqrt{x^2 + \alpha^2 y^2} \quad \text{and} \quad \tan\theta_\alpha = \frac{\alpha y}{x} \\ z_\beta &= r_\beta e^{i\theta_\beta}, \quad \text{so that} \quad r_\beta = \sqrt{x^2 + \beta^2 y^2} \quad \text{and} \quad \tan\theta_\beta = \frac{\beta y}{x} \end{aligned} \tag{1.114}$$

and the leading terms of the stress are again of the order of $r^{-1/2}$. These are (Freund 1990)

$$\begin{aligned} \sigma_{xx} &= \frac{K(t)B}{\sqrt{2\pi r}}\left[-(1+2\alpha^2-\beta^2)\left(\frac{r}{r_\beta}\right)^{1/2}\cos\left(\frac{\theta_\alpha}{2}\right) - \frac{4\alpha\beta}{1+\beta^2}\left(\frac{r}{r_\beta}\right)^{1/2}\cos\left(\frac{\theta_\beta}{2}\right)\right] \\ \sigma_{yy} &= \frac{K(t)B}{\sqrt{2\pi r}}\left[-(1+\beta^2)\left(\frac{r}{r_\alpha}\right)^{1/2}\cos\left(\frac{\theta_\alpha}{2}\right) + \frac{4\alpha\beta}{1+\beta^2}\left(\frac{r}{r_\beta}\right)^{1/2}\cos\left(\frac{\theta_\beta}{2}\right)\right] \\ \sigma_{xy} &= \frac{2K(t)B\alpha}{\sqrt{2\pi r}}\left[\left(\frac{r}{r_\alpha}\right)^{1/2}\sin\left(\frac{\theta_\alpha}{2}\right) - \left(\frac{r}{r_\beta}\right)^{1/2}\sin\left(\frac{\theta_\beta}{2}\right)\right] \end{aligned} \tag{1.115}$$

where the dynamical stress intensity factor (mode I) is

$$K(t) = \lim_{\substack{x\to 0 \\ y\to 0}} \sqrt{2\pi x}\,\sigma_{yy}(t) \quad \text{and} \tag{1.116}$$

$$B = \frac{1+\beta^2}{4\alpha\beta - (1+\beta^2)^2} \tag{1.117}$$

As is shown presently (Sect. 1.4.3), $K(t)$ is related to the energy flux into the crack tip. Eqations 1.115 can be used to calculate the direction of crack propagation (see e.g., Fineberg and Marder 1999). Alternatively, it can be used (Rabinovitch et al. 1993, see Fig. 2.29) to calculate the direction of secondary cracks that develop ahead of the propagating primary crack. The method used there is as follows. A certain radius $r_0 > 0$ is defined. Eqations 1.115 specify the stress field as a function of $\theta = \mathrm{arctg}(y/x)$. At each θ, the two principal directions and principal stresses (the stress tensor can be digonalized) are calculated. θ_1 is calculated for which the major principal stress σ_1 is maximal and the axis of this principal stress is evaluated. The secondary crack is assumed to originate at (r_0, θ_1) and its direction (α_1) to be perpendicular to the axis thus found. Results (Fig. 1.28) show that both θ_1 and α_1 depend on crack velocity. For low velocities ($v \ll C_\ell$), θ_1 is ~60° and secondary cracks emerge almost parallel to the primary crack, while for higher crack velocities, both θ_1 and α_1 increase in value. The possibility of $\theta_1 > 90°$ is unclear. Experimental results (Rabinovitch et al. 2000a) qualitatively confirm these predictions for low crack velocities.

In addition to Eq. 1.115, the leading terms of the velocity components of the movements near the crack tip are obtained as

$$\frac{du_x}{dt} = \dot{u}_x = -\frac{K(t)Bv}{\sqrt{2\pi r\mu}}\left[\left(\frac{r}{r_\alpha}\right)^{1/2}\cos\left(\frac{\theta_\alpha}{2}\right) - \frac{2\alpha\beta}{1+\beta^2}\left(\frac{r}{r_\beta}\right)^{1/2}\cos\left(\frac{\theta_\beta}{2}\right)\right]$$

$$\frac{du_y}{dt} = \dot{u}_y = -\frac{K(t)Bv}{\sqrt{2\pi r\mu}}\left[\left(\frac{r}{r_\alpha}\right)^{1/2}\sin\left(\frac{\theta_\alpha}{2}\right) - \frac{2}{1+\beta^2}\left(\frac{r}{r_\beta}\right)^{1/2}\sin\left(\frac{\theta_\beta}{2}\right)\right] \tag{1.118}$$

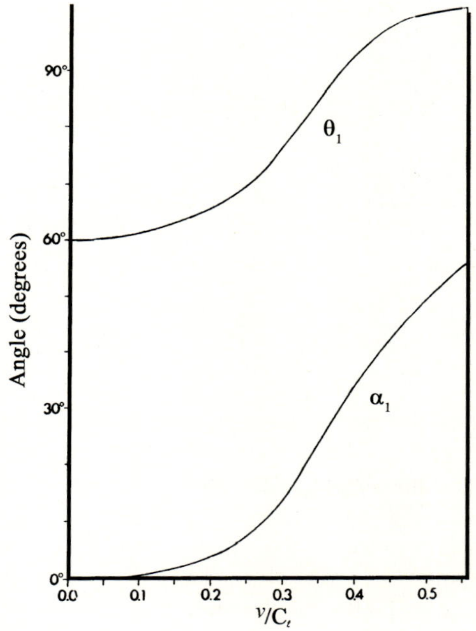

Fig. 1.28. The angle θ_1 at which σ_1 attains its maximum and the secondary crack direction α_1, vs. normalized crack speed v/C_ℓ (C_ℓ: Longitudinal-wave speed) (Rabinovitch et al. 1993)

Note that unlike the situation in the static case, here $r^{-1/2}$ singularities exist for $\dot{u}_x$ and $\dot{u}_y$ besides those for σ_{ij}. Eqations 1.115 and 1.118 demonstrate that the near-tip LEFM dynamic solutions, like the static case ones, can be divided into two parts. The (r, θ) part that does not and the $K(t)$ part that does depend on the geometry of the sample, the crack length and the remote stresses. It is thus the value of $K(t)$ that should dominate crack extension even for dynamic fracturing.

1.4.3
Energy Calculations

The 3D elastic (potential) energy stored in the sample per unit volume is given by (Eq. 1.5, generalized)

$$e = \frac{1}{2}\sum_{i,j} \varepsilon_{ij}\sigma_{ij}\,, \qquad i,j = 1,2,3 \tag{1.119}$$

In two dimensions, the indices of both ε and σ change only between 1 and 2 (x and y). We denote the indices in the 2D problem by α and β, each of which takes the values 1 and 2. The potential energy (per unit area) in 2D is therefore

$$e = \frac{1}{2}\sum_{\alpha,\beta} \varepsilon_{\alpha\beta}\sigma_{\alpha\beta} \tag{1.120}$$

Using the definition Eq. 1.11 of

$$\varepsilon_{\alpha\beta}\left(= \frac{1}{2}\left(\frac{\partial u_\alpha}{\partial x_\beta} + \frac{\partial y_\beta}{\partial x_a}\right)\right)$$

and the symmetry of σ ($\sigma_{\alpha\beta} = \sigma_{\beta\alpha}$), e can be written as

$$e = \sum_{\alpha,\beta} \frac{\partial u_\alpha}{\partial x_\beta}\sigma_{\alpha\beta} \tag{1.121}$$

The kinetic energy (per unit area in 2D) is given by

$$\tau = \frac{1}{2}\rho\sum_{\alpha}\left(\frac{du_\alpha}{dt}\right)^2 \left(= \frac{1}{2}\rho\left[\left(\frac{du_x}{dt}\right)^2 + \left(\frac{du_y}{dt}\right)^2\right]\right) \tag{1.122}$$

where du_α / dt is the αth component of the velocity and ρ is the material density.

Consider now (e.g., Kanninen and Popelar 1985) the cracked body depicted in Fig. 1.29, where S is its outer boundary on which forces (tractions) **t** are applied, $c(t)$ is the crack tip position which changes with time with a velocity **v**, and Γ^* is a small contour through which the energy flux is delivered to the tip; this Γ^* is eventually reduced to zero, and R is the volume (area in 2D) of the sample.

Fig. 1.29.
Dynamic crack propagation in 2D (after Kanninen and Popelar 1985)

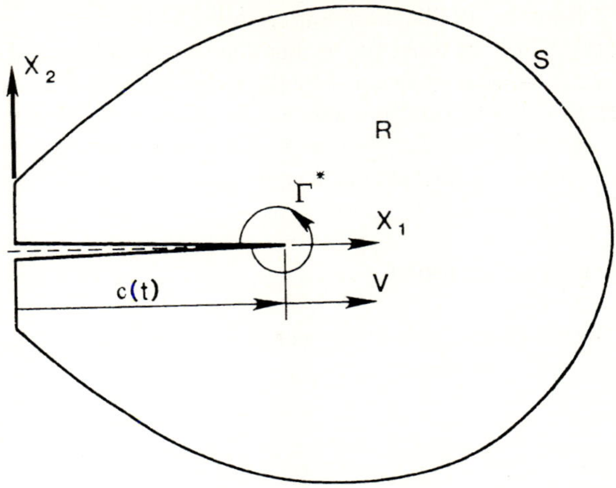

Energy balance is enacted as follows. The rate of work, $\dot{P}$, performed by the outer forces (tractions) on S is equal to the rate of the total potential U and kinetic T energy increase in R plus the energy flux Fl into the tip

$$\dot{P} = \dot{U} + \dot{T} + Fl \tag{1.123}$$

where

$$\dot{P} = \int \sum_{\alpha} \dot{u}_{\alpha} t_{\alpha} \mathrm{d}S \tag{1.124}$$

t_{α} is the αth component of the traction **t**; the total kinetic energy is

$$T = \int \tau \mathrm{d}A = \lim_{\Gamma^* \to 0} \int_R \frac{1}{2} \rho \sum_{\alpha} \left(\frac{\mathrm{d}u_{\alpha}}{\mathrm{d}t} \right)^2 \mathrm{d}A \tag{1.125}$$

and the total potential energy is

$$U = \int e \mathrm{d}A = \lim_{\Gamma^* \to 0} \int_R \sum_{\alpha,\beta} \frac{\partial u_{\alpha}}{\partial x_{\beta}} \sigma_{\alpha\beta} \, \mathrm{d}A \tag{1.126}$$

As always, a dot above a variable means a derivative with respect to time of this variable. We are interested in the flux into the tip, $Fl = \dot{P} - \dot{U} - \dot{T}$. The calculation of the separate terms is quite involved, and only a sketch is given here. A specific complication arises in the calculations due to the fact that the contour Γ^* propagates in time with the crack tip, making R time-dependent. Therefore, for example, $\dot{U}$ includes two terms, one which comes from the rate of change of u_{α} with time and the other is the contribution from the flux of the density of U through Γ^*.

Consider (Fig. 1.30) a small surface (line in 2D), $\mathrm{d}S = \mathrm{AB}$ which moves with a velocity **v**. All through space (area) there exists a density (an amount per unit volume) of some variable, e.g., energy. In the case of the potential energy, the density is

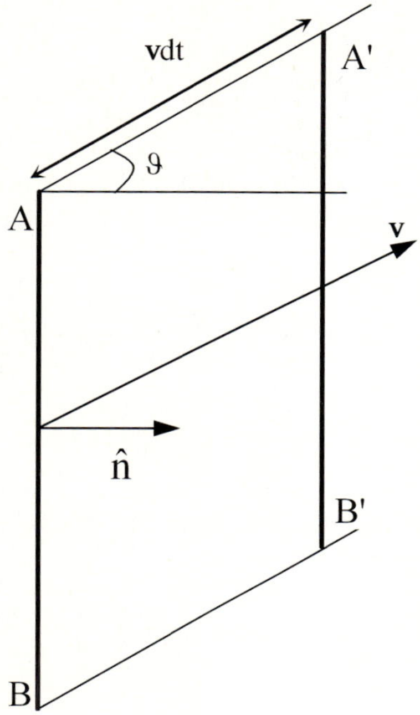

Fig. 1.30.
Schematic representation of an area dS (here the *line AB*) moving through space filled with a density of a specific physical quantity

$$e = \sum_{\alpha,\beta} \frac{\partial u_\alpha}{\partial x_\beta} \sigma_{\alpha\beta}$$

and in the case of the kinetic energy it is

$$\tau = \frac{1}{2} \rho \sum_\alpha \left(\frac{\mathrm{d}u_\alpha}{\mathrm{d}t} \right)^2$$

During the time dt, AB has moved by an amount $\mathbf{v}\mathrm{d}t$ to the position A'B'. The area (volume in 3D) traversed by dS is that of the parallelogram ABB'A', namely

$$\mathrm{d}S\, v\, \mathrm{d}t \cos\theta = \mathrm{d}S\, \mathbf{v} \cdot \hat{n}\, \mathrm{d}t$$

where $\hat{n}$ is a unit vector normal to dS. Therefore the amount of energy which passed through dS in the time dt is $e\mathrm{d}S\mathbf{v} \cdot \hat{n}\mathrm{d}t$ and the contribution of the movement of dS to the time rate of change of energy, $\Delta\dot{U}$, is

$$\Delta\dot{U} = \sum_{\alpha,\beta} \frac{\partial u_\alpha}{\partial x_\beta} \sigma_{\alpha\beta} \mathrm{d}S\mathbf{v} \cdot \hat{n}$$

Hence, the contribution from the movement of the whole Γ^* is

$$\int_{\Gamma^*} \sum_{\alpha,\beta} \frac{\partial u_\alpha}{\partial x_\beta} \sigma_{\alpha\beta} v_n \mathrm{d}S$$

Therefore,

$$\dot{U} = \lim_{\Gamma^* \to 0} \left[\int_R \sum_{\alpha,\beta} \frac{\partial \dot{u}_\alpha}{\partial x_\beta} \sigma_{\alpha\beta} \mathrm{d}A - \int_{\Gamma^*} \sum_{\alpha,\beta} \frac{\partial u_\alpha}{\partial x_\beta} \sigma_{\alpha\beta} v_n \mathrm{d}S \right] \tag{1.127}$$

where $v_n = v\cos\theta$ is the component of v normal to Γ^*, θ being the angle between the normal to Γ^* and the v-direction. A similar term appears for $\dot{T}$. Hence, as can easily be shown,

$$\begin{aligned} Fl = & \int_S t_\alpha \dot{u}_\alpha \mathrm{d}S + \lim_{\Gamma^* \to 0} \int_{\Gamma^*} \left(\sum_{\alpha,\beta} \frac{\partial u_\alpha}{\partial x_\beta} \sigma_{\alpha\beta} + \frac{1}{2} \rho \sum_\alpha u_\alpha^2 \right) v_n \mathrm{d}S \\ & - \lim_{\Gamma^* \to 0} \int_R \left(\sigma_{\alpha\beta} \frac{\partial \dot{u}_\alpha}{\partial x_\beta} + \rho \ddot{u}_\alpha \dot{u}_\alpha \right) \mathrm{d}A \end{aligned} \tag{1.128}$$

The next step (e.g., Kanninen and Popelar 1985) is to use the Gauss theorem after inserting the relation

$$\sigma_{\alpha\beta} \frac{\partial \dot{u}_\alpha}{\partial x_\beta} = \frac{\partial}{\partial x_\beta} (\sigma_{\alpha\beta} \dot{u}_\alpha) - \dot{u}_\alpha \frac{\partial}{\partial x_\beta} \sigma_{\alpha\beta} \tag{1.129}$$

After cancellation, this leads to a simpler equation

$$Fl = \lim_{\Gamma^* \to 0} \int_{\Gamma^*} \left[\left(\sum_{\alpha,\beta} \left(\frac{\partial u_\alpha}{\partial x_\beta} \sigma_{\alpha\beta} + \frac{1}{2} \rho \dot{u}_\alpha^2 \right) \right) v_n + t_\alpha \dot{u}_\alpha \right] \mathrm{d}S \tag{1.130}$$

Three remarks are in order.

1. Equation 1.130 means that the flux of energy into the crack tip can be evaluated by using the values of the stress and displacement fields only in the immediate vicinity of the tip.
2. The flux of energy into the tip is given by the sum of fluxes of the kinetic and potential energies into Γ^*, as the tip is propagating, and the rate of work done on Γ^* by the tractions from the exterior of Γ^*.
3. The flux is only obtained in the limit where $\Gamma^* \to 0$. A non-zero different contour around the tip would yield a different value of the right-hand side of Eq. 1.130. An exception to this general state of affairs is the case of a crack moving at a constant velocity, $\mathbf{v} = v_0 \hat{x}$. Here

$$\dot{u}_\alpha = \frac{\partial u_\alpha}{\partial t} = \frac{\partial u_\alpha}{\partial x} \frac{\partial x}{\partial t} = -v_0 \frac{\partial u_\alpha}{\partial x}$$

and Eq. 1.130 becomes

$$Fl = v_0 \int_{\Gamma^*} \left[\left(\sum_{\alpha\beta} \frac{\partial u_\alpha}{\partial x_\beta} \sigma_{\alpha\beta} + \frac{1}{2} \rho v_0^2 \sum_\alpha \left(\frac{\partial u_\alpha}{\partial x} \right)^2 \right) n_x - t_\alpha \frac{\partial u_\alpha}{\partial x} \right] \mathrm{d}S \tag{1.131}$$

For this case (Atkinson and Eshelby 1968), the value of the integral does not depend on the choice of the contour Γ^*, so there is no need to go to the limit $\Gamma^* \to 0$.

Now, to obtain the energy release rate G, recall that G is the rate of change of energy with crack length. Denoting this length by c,

$$G = \frac{\mathrm{d}E_{\mathrm{tot}}}{\mathrm{d}c} = \frac{\mathrm{d}E_{\mathrm{tot}}}{\mathrm{d}t}\left(\frac{\mathrm{d}t}{\mathrm{d}c}\right) = \frac{\mathrm{d}E_{\mathrm{tot}}}{\mathrm{d}t} \cdot \frac{1}{\left(\frac{\mathrm{d}c}{\mathrm{d}t}\right)} = \frac{1}{v_0}\frac{\mathrm{d}E_{\mathrm{tot}}}{\mathrm{d}t} = \frac{1}{v_0} Fl \tag{1.132}$$

Inserting Eqs. 1.115 and 1.118 for $\sigma_{\alpha\beta}$ and $\dot{u}_\alpha$ in Eqs. 1.130 and 1.132 finally yields the result

$$G = \frac{\alpha v^2}{2\mu C_\mathrm{t}^2} \frac{K^2(t)}{[4\alpha\beta - (1+\beta^2)^2]} \tag{1.133}$$

where α, β and C_t were defined in Eq. 1.109, and v_0 was replaced by v.

Equation 1.133 is sometimes written as

$$G(V) = G(0)g(v) \tag{1.134}$$

where $G(0)$ is the velocity independent energy release rate (e.g., Eq. 1.87) and $g(v)$ (a universal function) contains the v-dependent part of Eq. 1.133. A good approximation for $g(v)$ is (Freund 1972a, see also 1990)

$$g(v) \sim 1 - \frac{v}{C_\mathrm{R}} \tag{1.135}$$

where C_R is the Rayleigh wave velocity – the velocity of a wave propagating on the surface of the solid.

Equation 1.133 is the dynamic LEFM equivalent of the static case. It describes the driving force for a crack to grow. If the dissipative energy per unit crack length, which is needed for opening bonds, creating secondary damage (e.g., secondary cracks) ahead of the crack tip, etc. is denoted by Γ (which can be velocity dependent), then the condition for crack propagation is

$$G = \Gamma \tag{1.136}$$

Eqations 1.134–1.136 are the ones used to analyze dynamic crack propagation. For example, they imply that when V approaches C_R, $G(0)$ should go to infinity to sustain the propagation, namely that C_R is the crack's maximal velocity. In experiments conducted on brittle materials under constant load, the crack velocity invariably increases monotonically in time, up to a certain terminal velocity. It can subsequently stay at this velocity, which is then the maximal velocity or it can fluctuate around this terminal velocity (Fineberg and Marder 1999), in which case we define the maximal velocity as the time average of the fluctuations. In the literature, the terms maximal and terminal velocities are therefore usually interchangeable. The energy release rate is proportional to $K^2(t)$. Thus, according to Eq. 1.134, the crack propagates as long as $K(t)$ is larger than a dynamic critical stress intensity factor called K_{ID}. The latter depends on the material as before but also on the crack velocity (and probably on the temperature). By Eq. 1.133 and Eq. 1.136,

$$K_{ID}(v,T) = \frac{C_t}{v}\sqrt{\frac{2\mu}{\alpha}[4\alpha\beta - (1+\beta^2)^2]\Gamma} \tag{1.137}$$

which is usually called "crack resistance".

If Γ and its dependence on v were known, Eq. 1.136, or alternatively

$$K(t) = K_{ID}(v,T) \tag{1.138}$$

would provide a relation wherefrom the velocity of the propagating crack could be obtained. Therefore, Eqs. 1.134–1.136 or Eq. 1.138 are commonly used by engineers for these calculations. There exist, however, several problems with this approach. As mentioned earlier, the biggest discrepancy with measurement is that the maximal crack velocity is not C_R. Experiments yield values of terminal velocities ranging between ~$0.2C_R$ and $0.7C_R$. But even for small velocities where Eq. 1.136 and Eq. 1.138 are supposed to be a good approximation to the exact situation, there exist several factors (not included in Eq. 1.136 and Eq. 1.138) that influence crack speed. We enumerate here briefly some of these:

1. The rate of loading (Liu et al. 1998; Arakawa and Takahashi 1987).
2. The geometry of the sample (Kobayashi and Mall 1978).
3. The non-uniqueness of the stress intensity factor in defining crack velocities (Dally et al. 1985).
4. Additional effects (Ravi-Chandar and Knauss 1987).

The prevalent measuring methods of dynamic fracture parameters are described, e.g., in Yang and Ravi-Chandar (2001), in Fineberg and Marder (1999) and in Knauss (2000). They include both well-established methods that have been used for a very long time and new developments of the last ten years.

We return now to the question of maximal crack velocity. Table 1.1 (adapted from Fineberg and Marder 1999) shows that for homogeneous brittle materials, the maximal crack speed never exceeds $0.7C_R$. On the other hand, if we assume that $K(t)$ under a constant load σ_0 is proportional to $\sqrt{c(t)}\,\sigma_0$ (as is the case e.g., for an infinite plate) where $c(t)$ is the instantaneous crack length, Eqs. 1.134–1.136, with the help of Eq. 1.133 yield for a constant Γ

$$q\left(1 - \frac{v}{C_R}\right)c\sigma_0^2 = \Gamma \tag{1.139}$$

Table 1.1. Maximal crack velocities in some homogeneous brittle materials

Material	V_{max}/C_R
Glass	0.47 – 0.66
PMMA	0.58 – 0.62
Homalite	0.33 – 0.41

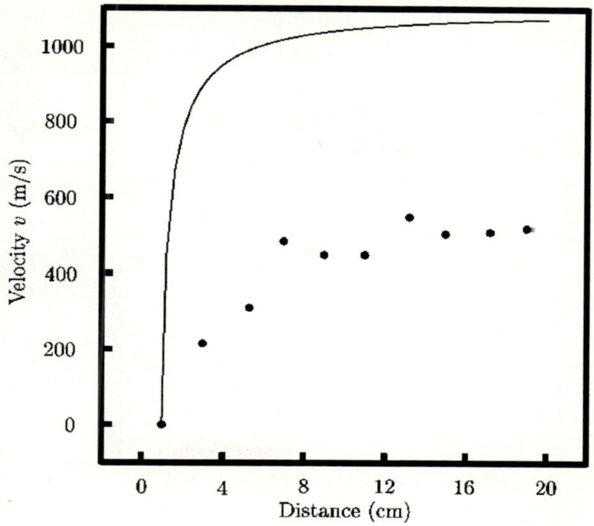

Fig. 1.31. Data of Kobayashi et al. (1974) in PMMA, showing the discrepancy with linear elastic theory (*full line*)

where q is a constant, or

$$1 - \frac{v}{C_R} = \frac{c_0}{c} \tag{1.140}$$

where $c_0 = \Gamma / q\sigma_0^2$ is the length where crack initiation occurred. Equation 1.140 can be used to find the expected LFEM crack velocity

$$v = C_R\left(1 - \frac{c_0}{c}\right) \tag{1.141}$$

which is equivalent to Eq. 1.103, with C_R as the terminal velocity. Figure 1.31 shows the experimental results of Kobayashi et al. (1974) mentioned before, together with v according to Eq. 1.141. This figure dramatically emphasizes the discrepancy between the LEFM theory and the experimental results. This discrepancy obviously arises from the processes occurring in the tip vicinity. These concern atomic nonlinear forces, not included in the linear theory (e.g., i–iv above), which lead to instabilities in the crack propagation and to the collapse of the LEFM theory.

Besides our simplistic semiphenomenological approach (see Sect. 1.4.1 and Rabinovitch 1994), there exist at the moment two lines of thought that try to explain these instabilities. They will be denoted by the "velocity threshold" and the "dissipation process" (Sects. 1.4.3.1 and 1.4.3.2, respectively).

1.4.3.1
Velocity Threshold, Microbranching

In a series of careful experiments, conducted mainly on PMMA and Homalite but also on glass samples, J. Fineberg, M. Marder and their group (Fineberg et al. 1991, 1992; Sharon et al. 1995; Sharon and Fineberg 1998, 1999) measured the accurate changes of

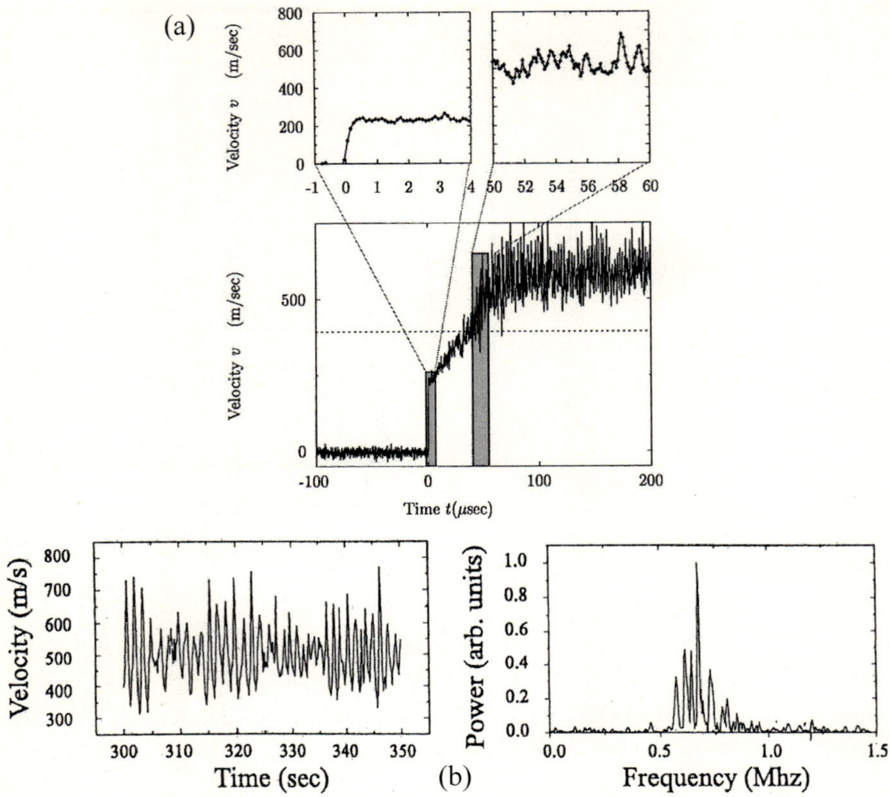

Fig. 1.32. A typical measurement of the velocity of a crack tip in PMMA. **a** as a function of time: After an initial jump to about 150 m s^{-1} the crack accelerates smoothly up to a critical velocity v_c (*dotted line*). Beyond this velocity, strong oscillations in the instantaneous velocity of the crack develop and the mean acceleration of the crack decreases (Fineberg and Marder 1999). **b** The steady-state above v_c and the corresponding power spectrum. Note the existence of a well-defined timescale (from Hauch and Marder 1998)

crack velocities in those materials and the simultaneous appearance of secondary "patterns" around the propagating cracks. It was found that in PMMA at and above a critical crack velocity of $v_c = 0.36C_R$ (which is independent of sample geometry, sample thickness, and the applied stress or acceleration rate of the crack) velocity fluctuations appear (Fig. 1.32) together with microbranching (Fig. 1.33) emanating from the main crack and ending inside the samples. Note that v_c is smaller than the final (or terminal) velocity. They have concluded, therefore, that the fluctuations were due to these microbranching or "branching attempts". They have also identified these microbranching with the "mist" and "hackle" patterns found in brittle materials (Fineberg et al. 1991; Sharon and Fineberg 1996). An additional feature of these velocity measurements in PMMA (not observed in glass) is a velocity jump at the beginning of crack propagation (Fig. 1.34) (see in this connection Rabinovitch and Bahat 1979). In a series of "microscopic" calculations Marder and coworkers (Marder and Gross 1995; Hauch and Marder 1998; Marder and Liu 1993) obtained results that support these experiments. Their models are based on a lattice of points connected by springs whose performance is predetermined. Calculation results show

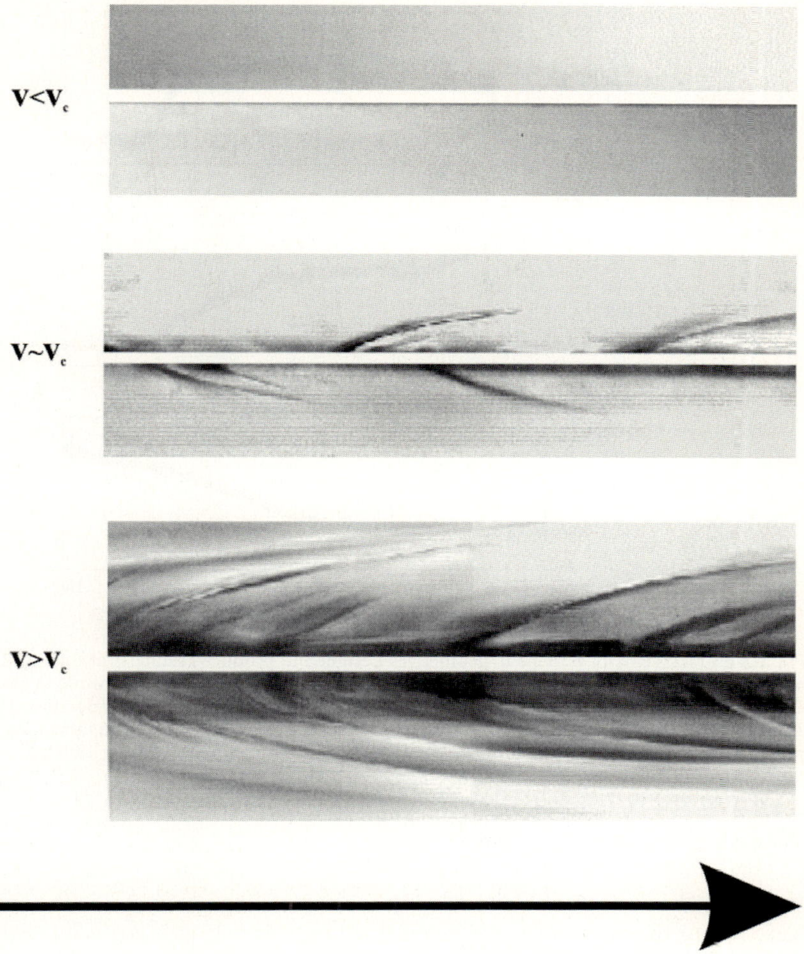

Fig. 1.33. Images of local crack branches in the x-y-plane in PMMA. The *arrow*, of length 250 μm, indicates the direction of propagation. All figures are to scale with the path of the main crack in white. *Top:* $v < v_c$; *center:* $v = 1.18v_c$; *bottom:* $v = 1.45v_c$ (from Sharon et al. 1995)

1. A velocity "gap" – the initial crack velocity is immediately greater than zero.
2. Instabilities arise whenever the crack velocity exceeds a certain threshold velocity – above which the crack direction bounces and surface roughness is thereby created.

These results agree with some atomistic (e.g., Abraham et al. 1997) and finite element (Xu and Needleman 1994) calculations in the identification of a threshold velocity above which crack direction starts to "wiggle". Such a velocity dependent instability was already predicted by Yoffe (1951), who showed that the stress field around the crack tip was velocity dependent: For small crack velocities, the stress field at the tip perpendicular to r, $\sigma_{\theta\theta}$, had a maximum at the crack propagation direction ($\theta = 0$), while above a certain threshold velocity the maximum was shifted to non-zero θ values (in Yoffe's calculation $\theta = \pm 60°$).

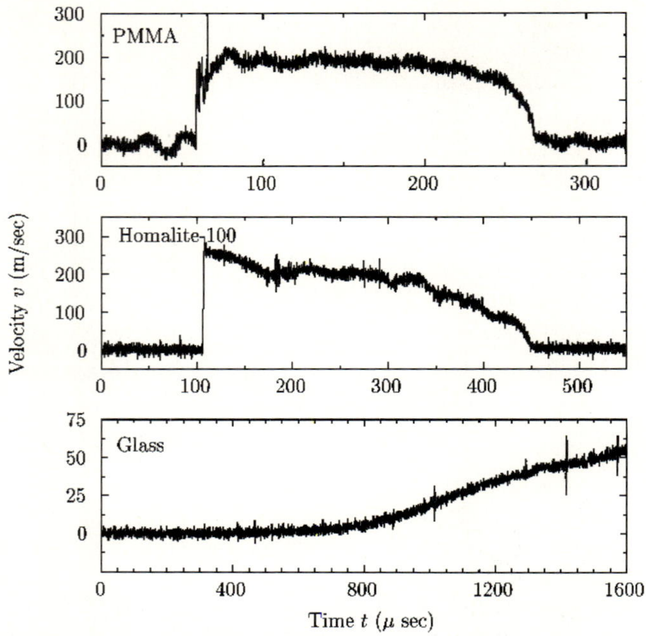

Fig. 1.34. Crack velocity vs. time in geometries designed to produce gradual crack arrest. In PMMA and Homalite, cracks decelerate slowly towards zero velocity on the scale of μs, while in glass, cracks are able to accelerate slowly from zero velocity and travel stably at very low velocities. These data show that jumps in velocity from zero are not intrinsic properties of fracture dynamics in amorphous materials. Data of Hauch and Marder (1998) (from Fineberg and Marder 1999)

1.4.3.2
Dissipation Processes

In this model, instead of crack velocity, the controlling parameter is the energy release rate or the stress intensity factor. However, it is the *static* stress intensity factor and not the *dynamic* one which is experimentally seen to cause the different changes in crack behavior and the corresponding surface patterns (different roughness features). It is assumed that the excess energy released from the elastic field is not "used" solely to increase the kinetic energy of the medium. Rather, it goes into the creation of microcracks or other damage mechanisms ahead of the propagating crack. Several investigators in this school of thought are Kerkhof (1973), Johnson (1992), Ravi-Chandar and Knauss (1984a–d). See also Ravi-Chandar (1998) for a thorough discussion and comparison with the velocity threshold line of thought.

This line of thought is based on the fractographic investigation of fractured surfaces of brittle materials (glass, glass ceramic, ceramics and rocks; see Chap. 2 for a detailed description) in which a universal pattern appears. Figure 2.1b is a representative schematic picture. It is seen that after the crack initiation (at c_i or c_{cr}; see Sect. 1.4.1) in a sequence the following patterns appear: "Mirror", "mist", "hackle" and "bifurcation". The "mirror" is a rather flat surface, reflecting light (hence the name). A careful atomic microscopy measurement, however, shows that "roughness" increases across the whole range of the mirror, although it stays invisible to the naked eye. Crack velocity increases gradually (Fig. 1.31) across the mirror and usually reaches its terminal value there. The "mist" is a region of numerous secondary fractures, while the "hackle" appears as a set of parabolic (or hyperbolic) secondary markings, the planes of their propagations making acute angles with the primary crack plane. Bifurcations (or branching) and attempted bifurcations appear only *after* the mirror-mist-hackle sequence has occurred (there are however some conflicting observations; Sect. 2.2.3.14) and ultimately induce a splitting of the crack into two or more branches.

The demarcation lines between the different patterns, i.e., the mirror-mist (r_m in Fig. 2.1) mist-hackle (r) and hackle-bifurcation (r_b), are sometimes sharp and sometimes gradual. Many investigators (e.g., Bansal 1977; Johnson and Holloway 1966; Shand 1959) have found that these transition lines can empirically be obtained by relations of the following form:

$$r_i \sigma_0^2 = A_i \tag{1.142}$$

where r_i can assume the values of r_m, r, or r_b; σ_0 is the remote stress in the sample and A_i ($= A_m$, A, A_b respectively) are material constants. This relation was used, e.g., by Bahat et al. (1982) to evaluate the shape of the mirror boundary when $\sigma_0(\theta)$ changed with the polar angle and by Bahat and Rabinovitch (1988) to evaluate the paleostress in a rock by the shape of its transition line into the hackle zone (mist was barely discernable there).

The empirical result embodied by Eq. 1.142 is very intriguing. It relates the static stress intensity factor (which is proportional to $\sigma_0\sqrt{r_i}$ for an infinite plate) to the transition in crack behavior. This could mean either that dynamic effects, such as the changes of K_I with velocity (see above) do not count, or that all of these transitions occur for the same (or nearly the same) velocity. In either case, Eq. 1.142 seems to indicate that the controlling parameter is K_I, at least for materials other than PMMA, and not the crack velocity.

1.4.4
Fracture Pattern Analysis

A complete understanding of pattern formations on brittle crack surfaces is still missing, although more than a century has passed since they were detected (see Bahat 1991a). The identification of the mist zone as a zone studded by secondary microcracks is perhaps due to Ravi-Chandar and Knauss (1984a–d) and as mentioned above (Sect. 1.4.3.2), the relations (Sect. 1.142) were obtained by many investigators but without a firm theoretical foundation. In a series of papers (Rabinovitch et al. 1994, 2000a,b), we have found some properties of these patterns. These are presently discussed.

Based on the directions of the secondary cracks (Rabinovitch et al. 1993, see Sect. 1.4.2), we have calculated (Rabinovitch et al. 1994) the lengths of these cracks as a function of distance across the mist zone.

This calculation was based on two assumptions:

1. A flaw (or a microcrack), which is found ahead of the propagating crack, starts to grow when the *static* stress intensity factor at its tip becomes equal to K_{Ic}.
2. The primary crack overtakes the secondary crack since the former moves faster, thereby causing the secondary crack to halt.

The outcome was that the lengths of *all* secondary cracks should increase across the mist zone by a common magnifying factor M which itself is growing monotonically as a function of the primary crack length. Although these results agreed quite nicely with the roughness experiment by Ravi-Chandar and Knauss (1984c), this agreement showed only a proportionality correlation. Indeed, a painstaking experiment (Rabinovitch et al. 2000b) showed a rather surprising result. The experiment was conducted on a glass bottle and the lengths of some 12000 secondary cracks across the mist zone were carefully measured under a microscope. It emerged that

1. The appearance of secondary cracks was abrupt (at the end of the "mirror plane").
2. The length distribution of these cracks consisted of two parts – Gaussian for short lengths (up to ~6 μm) and a negative exponential tail above 6 μm.
3. Length increase across the mist zone did behave like the predicted one with a common increasing magnification M, but only a small fraction (around 9%) of the secondary cracks were really magnified.

These features are not understood as of yet.

A more detailed calculation of the interaction between the primary crack and the secondary cracks in the mist zone was carried out in Rabinovitch et al. 2000a. A penny-shaped flaw (microcrack) was assumed to start its propagation as a secondary crack when the static stress intensity factor at its tip, caused by the primary crack stress field, reached K_{Ic}. It was further assumed that the velocity of the secondary crack increased approximately in an exponential manner, reaching ultimately the "terminal (maximal) velocity" of the material. Each part of the secondary crack overtaken by the primary one ceased to grow, since stress there was relaxed. In this way, the two-dimensional shapes of the secondary cracks could be calculated. Results (Fig. 1.35) showed that these shapes should change along the way from the mirror to the hackle zones from being circular to elliptic and to "teardrop"-like. Since the stress field ahead of the primary crack increases along its way (recall that $K_I \sim \sigma_0\sqrt{2\pi R}$, where R is the primary crack radius), the secondary flaws start to grow as cracks at distances further and further ahead of the primary one. Finally, they acquire such a "forestart" that the primary crack cannot overtake them anymore. This point is assumed mark the beginning of the hackle zone. A secondary crack shape calculation in this zone yields for it a hyperpolic form (Fig. 1.35). Experimental secondary crack shapes (Fig. 1.36) are seen to agree well with the calculated ones.

These results constitute a step in understanding the mirror-mist-hackle-bifurcation sequence of surface patterns.

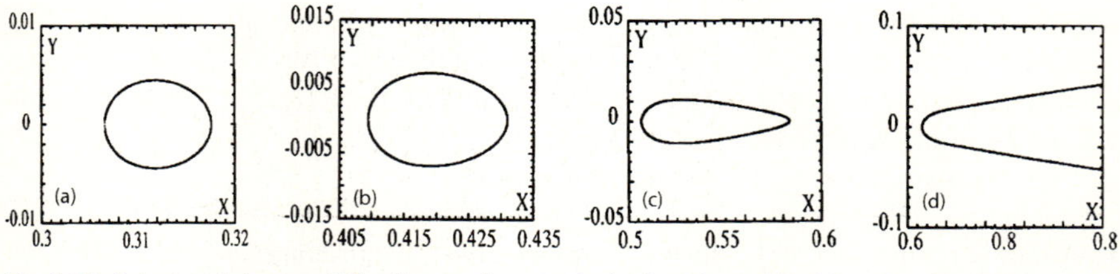

Fig. 1.35. Calculated shapes of "final" secondary cracks in the mist and hackle zones, for different radii R of the primary crack (flaw size $c_0 = 0.002$ mm). x and y scales are in mm. **a** $R = 0.3$ mm, the beginning of the mist zone (*oval*); **b** $R = 0.4$ mm, in the mist zone (*elliptic*); **c** $R = 0.495$ mm, towards the end of the mist zone (*tear drop*); **d** $R = 0.6$ mm, in the hackle zone (*hyperbolic*) (Rabinovitch et al. 2002a)

Fig. 1.36.
Experimental results for soda lime fracture. **a** The beginning of the mist zone (*oval*), scale bar 2.5 μm; **b** towards the end of the mist zone (*teardrop*), scale bar 10.0 μm; **c** at the mist hackle transition (*two semiparallel lines*), scale bar 10.0 μm; **d** in the hackle zone (*neighboring hyperbolic forms*, e.g., A and B), scale bar 20.0 μm (Rabinovitch et al. 2000a)

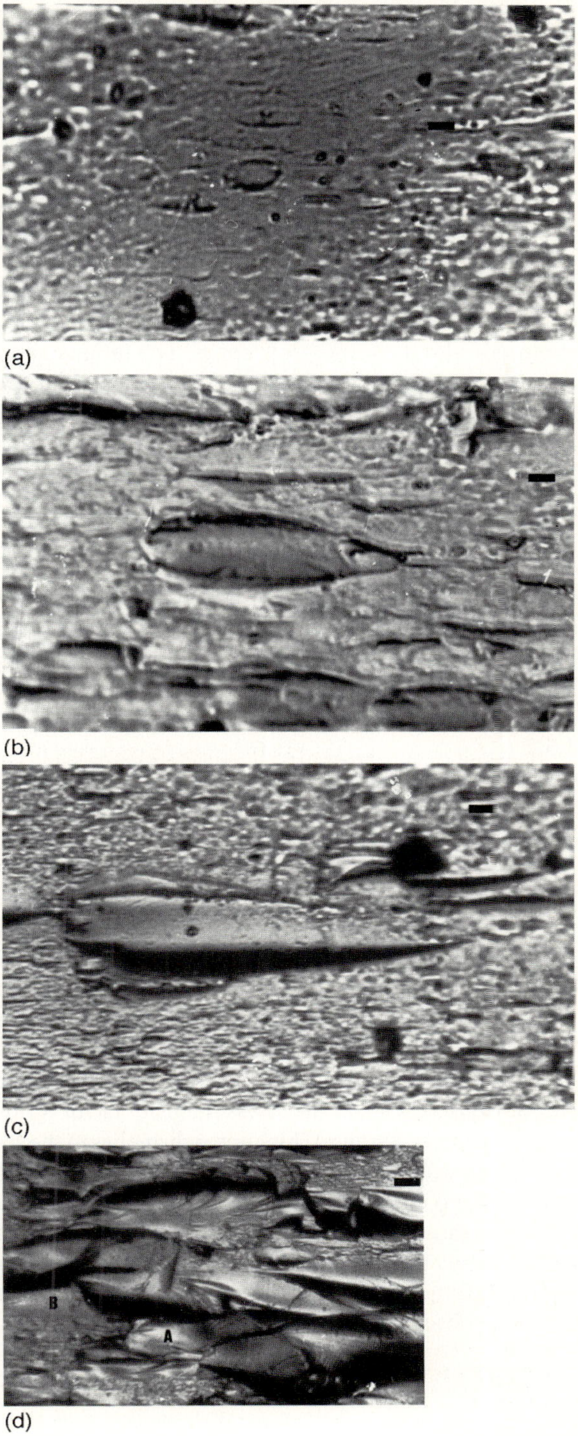

1.5 Subcritical Cracking

1.5.1 General

This section concerns situations for which the Griffith criterion is not fulfilled, i.e., situations where $K < K_c$, hence the name "subcritical". Experimentally this area is divided broadly into two parts called "fatigue" and "creep". The term "fatigue" is nowadays used specifically for a situation where stress (or K) is periodically changed between two values, while the term "creep" is reserved for the description of a gradual rupture under a constant low stress. The fatigue case, which is very important for metallic samples, will not be discussed here except to mention the *empirical*, so-called "Paris law" for the rate of crack length, c, increase with the number of cycles N (Paris and Erdogan 1963)

$$\frac{dc}{dN} = C(\Delta K)^p \tag{1.143}$$

where $\Delta K = K_{max} - K_{min}$, K_{max} and K_{min} being the maximum and minimum K values between which the load is cyclically changed and C and p are material constants. Since this relation does not always agree with experiments, more sophisticated relations have appeared (e.g., Suresh 1998).

1.5.2 Creep

This phenomenon of rupture under low stress conditions is also called by many other names, such as corrosion, stress corrosion, corrosion fatigue, static fatigue, etc. It is directly related to the question of flaw and crack nucleation, see Sect. 1.2. Since experiments can only be carried out within a finite temporal span (of a maximum duration of the order of 1 year), geological results can not be directly verified in the laboratory. As we shall show, these results demonstrate an approach to fracture, which is different than the one governing the behavior of shorter times (higher stresses).

The usual accepted result of creep is that of Wiederhorn (1967) or Evans and Johnson (1975). See also Wiederhorn (1978). In this approach, the governing parameter of rupture is again the stress intensity factor K. Results are given in the form of crack velocity vs. K-value. This approach applies to the scenario (see Sect. 1.2.3.3) where a single flaw grows under the external stress field. The whole range of K values ($0 \le K < K_c$) is customarily divided into four parts. In the lowest K region, $K < K_{Isc}$, no crack increase is detected. A controversy exists in the literature whether $K_{Isc} = 0$ or constitutes a real (>0) lower bound. The other three ranges of K (Fig. 4.34) are termed regions I, II and III, respectively. In region I, a very fast increase of fracture velocity with K_I is observed. It is usually empirically described by

$$v = AK_I^n \tag{1.144}$$

where A and n are material constants. A compilation of such constants for rocks can be found in Atkinson (1984) and Atkinson and Meredith (1987a,b). Region II is called the stress corrosion zone, where crack velocity changes slowly with K_I, the change being strongly dependent on the properties of the environment, i.e., temperature, water content of air, material contaminations, etc. Region III is again a transition range (to the supercritical region), where velocity is initially increased very rapidly and eventually saturates to the "final" constant (maximal) velocity of the material.

However, this behavior of crack velocity can not be used for fracture under extremely low stresses. The latter is important especially for geologic situations, where processes occur very slowly and presumably under extremely low stresses (or strains). One of the most important manifestations of such processes is the abundant single layer jointing that has been extensively studied. In the next section we present in detail our contribution to this study and show that the fracture probability under these extreme conditions is not proportional to K_I but rather to σ itself. This result is in line with a recent experiment by Yoon et al. (2000).

1.5.3 The Dependence of Jointing in Rock Layers Solely on Stress

1.5.3.1 *General*

We show that the criterion for fracture for the jointed layers is really based only on the level of stress (σ_0) they have endured during the elapsed geological times. It is therefore conceivable that such a fracture criterion based on σ_0 alone is the prevailing one for extreme below-threshold-extended-time fractures for all brittle materials. This result is shown to be in line with results obtained in current experiments on "lengthy" laboratory creep. Note that we are dealing only with joints that were formed quasi-statically. It is observed that dynamic joints are abundant in granites, whereas quasi-static ones are prevalent in sedimentary rock layers (Bahat et al. 2003). Joint spacing in such layered rocks is considered to be originated by two main processes and possibly by intermediate ones.

The first process envisages the layer denoted by A, as being "sandwiched" by "upper" and "lower" layers, each of which is made of different material than that of A. The neighboring layers, because of their higher compliance, are usually less jointed than A. But they tend to stretch the latter when they themselves are stretched, thus transferring to it the stress field. This process is referred to as the Cox-Hobbs case according to the two authors who analyzed it (Cox 1952; Hobbs 1967). In this process, the three layers are treated as perfectly connected (glued) to each other and the shear stress transferred at the interlayers is transposed into tensile stress in the middle layer. The Cox-Hobbs model can be divided into two parts. The first part, based on Cox (1952), calculates the stress transfer from the outer layers to A, and the second part treats the process of joint "in filling", i.e., the addition of more and more fractures to the layer A as a function of strain (or stress) increase.

The second process by which a layer is stretched and jointed is when it is completely separated from its neighboring layers and is itself pulled upon, either by geologic (tectonic) occurrences or by pore pressure, or both. This process is called the Pollard-Segall (Pollard and Segall 1987) process after the two authors who analyzed it. These two extreme processes of layer fracturing actually appear in nature, as do intermediate modes in which the "gluing" of the layers is less than absolute and friction exists between layer A and its neighbors, thus transferring to A only part of the stress in them. We emphasize again that we are considering here extremely slow phenomena, which no laboratory measurements can exactly simulate. The experiments that have the closest similarity to them are those related to creep or fatigue (Nomenclature). Although the durations of both experiment types are of course much smaller than geologic ones, we shall use results of creep experiments to glean some information about geologic jointing.

1.5.3.2
The Model by Rives et al. (1992)

We start by analyzing the shapes of the joint spacing distributions obtained experimentally. It such an experiment, a "measuring" line (scan-line) is drawn parallel to the layer; the joints cut it at intersection points; and the distribution of the spacings between these points is analyzed. It was observed that the shape of these distributions is gradually changing with the increase of the density of cracks (number of joints per unit length). The latter is evidently due to the stress magnitude in the layer. The higher this magnitude is, the higher the joint density is and vice versa. The prevailing conjecture regarding this change of shape was that of Rives et al. (1992), who claimed that with increasing joint density the distribution passed from a negative exponential one through a lognormal one to a normal distribution (for very high densities). This conjecture was "supported" by a numerical experiment where a line segment representing the spacing measuring line was cut by randomly applied points (representing the intersection points on the line) with the provision that a new joint could not be closer to an existing one than a certain distance, *d*. The latter was used as a representative of the "shadow" area, namely, the area around an existing crack where it is highly improbable, due to stress release there, that an additional crack would appear. The interpretation of the numerical experiment apparently showed that the joint spacing distribution actually behaved as predicted. However, Rabinovitch et al. (1999a) considered two difficulties in this interpretation of the numerical experiment. They showed that the numerical result of Rives et al. (1992) did not fit the exact analytical solution of their numerical experiment, which can be obtained from a 1D car parking process model (CPP) (Widom 1966). Moreover, the exact CPP results (Fig. 1.37b,c) are shown not to be compatible with existing joint spacing distribution results, as explained below.

1.5.3.3
The Exact Shadow Model

One of the problems with the CPP model is the analysis of the "shadow". It is not true that a new crack is *completely* prevented from occurring in a close neighborhood of an existing crack (Bai and Pollard 2000a, see below). The probability of such an event is only *decreased* as we approach the existing crack. We denote this probability by $q(x)$, namely, $q(x)\mathrm{d}x$ is the probability that a new crack can appear at a distance between x and $x + \mathrm{d}x$ from an existing crack (examples are shown in Fig. 1.38). Note that for the CPP model $q(x) = 0$ for say $|x| < b$, $2b$ denotes the excluded area around an existing crack, while $q(x) = \lambda$ say, for $|x| > b$, where λ measures the crack density as we will show later.

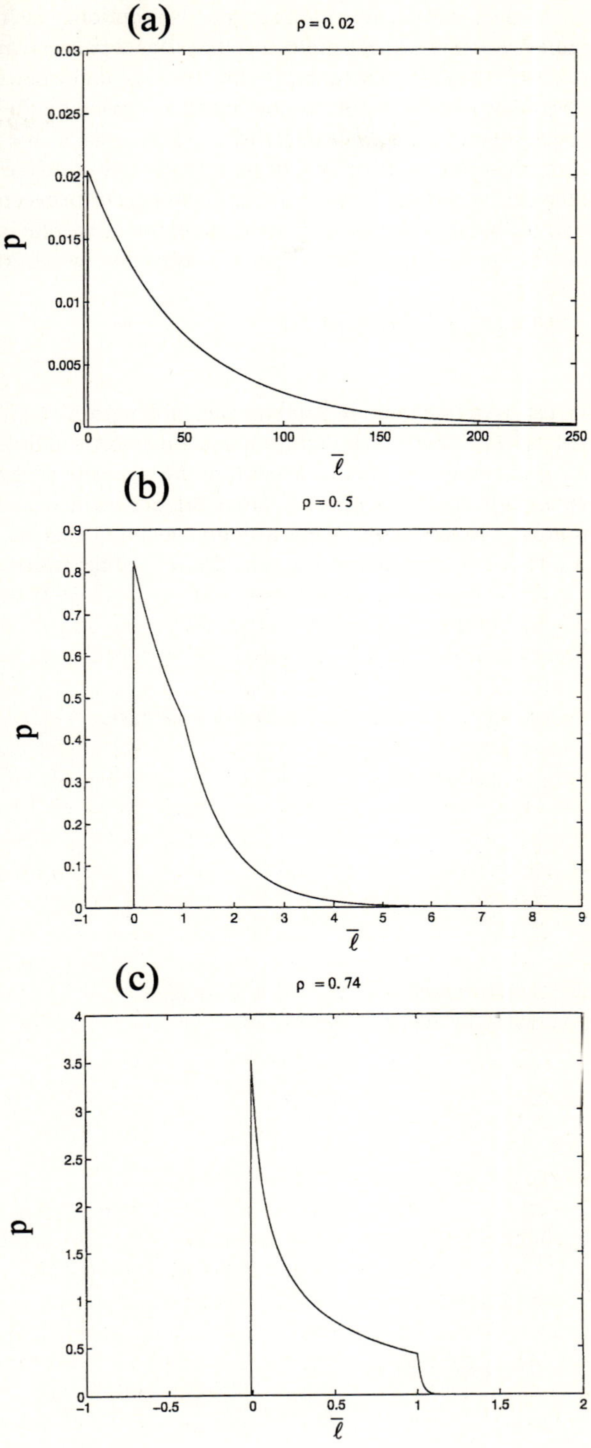

Fig. 1.37.
The calculated probability distribution function p of normalized crack spacing $\bar{\ell}$ obtained under "random sequential addition" of "cracks" onto a scanline (CPP). Three curves are shown for: **a** Very low crack density ($\rho = 0.02$), **b** an intermediate density ($\rho = 0.5$), and, **c** a density very near jamming ($\rho = 0.74$) (Rabinovitch et al. 1999a). Here $\bar{\ell} = (\ell / d) - 1$, where ℓ is the intercrack distance; $\rho = (N/L)d$, where N is the total number of cracks and L is the total length of the scanline. "Jamming" or saturation here means that no additional cracks can be added

We next calculate the joint spacing distribution, which we denote by $p(l)$ (Fig. 1.39), namely $p(l)\mathrm{d}l$ is the probability that the distance between neighboring joints would reside exactly in the interval between the distance l and the distance $l + \mathrm{d}l$ (Rabinovitch and Bahat 1999). The calculation of this distribution is based on the "shadow" probability $q(x)$, in the following way. The probability $p(l)\mathrm{d}l$ is equal to the product of two terms as follows: First, the probability that no fracture has occurred between $x = 0$ and $x = l$, and second, the probability that it actually did occur between $x = l$ and $x = l + \mathrm{d}l$. The second probability is of course $q(l)\mathrm{d}l$, while the first one is somewhat more difficult to obtain. Let us denote this "negative" probability by $y(l)$. We presently show that

$$y(l) = \exp\left[-\int_0^l q(x)\mathrm{d}x\right]$$

Let us divide l into N intervals each of length Δx ($= l/N$). As mentioned, the probability of an additional fracture appearing in Δx at a distance x from the origin (where a crack resides) is $q(x)\Delta x$. Therefore the opposite probability, namely that no crack appears in Δx is $(1 - q(x)\Delta x)$. Since the occurrence probabilities in all intervals are mutually independent, the ensuing probability that no fracture has appeared in all N intervals is the product, denoted by "Π", of the individual ones, namely

$$y(l) = \prod_{i=1}^{N} [1 - q(x_i)\Delta x] \tag{1.145}$$

Taking the logarithm of both sides we have

$$\ln y(l) = \sum_i \ln[1 - q(x_i)\Delta x] \tag{1.146}$$

where

$$\sum_{i=1}^{N}$$

denotes the sum between $i = 1$ and $i = N$.

Passing to the limit as $\Delta x \to 0$, we get

$$\lim_{\substack{\Delta x \to 0 \\ N \to \infty}} \ln y(l) = -\int_0^l q(x)\mathrm{d}x \tag{1.147}$$

where use has been made of the relation

$$\lim_{\varepsilon \to \infty} \ln|1 - \varepsilon| = -\varepsilon \tag{1.148}$$

And therefore

$$y(l) = \exp\left[-\int_0^l q(x)\mathrm{d}x\right]$$

Hence, the product of probabilities mentioned before yields

$$p(l) = q(l)\exp\left[-\int_0^l q(x)\mathrm{d}x\right] \tag{1.149}$$

Equation 1.149 is an interesting result. It states in fact that to obtain $p(l)$ one has only to know the shadow probability $q(x)$. A nice feature of Eq. 1.149 is that it is normalized for any $q(x)$. This feature is proven as follows. We denote

$$u(l) = \int_0^l q(x)\mathrm{d}x \quad \text{then}$$

$$q(l) = \frac{\mathrm{d}u(l)}{\mathrm{d}l}$$

and

$$\int_0^\infty p(l)\mathrm{d}l = \int_0^\infty \frac{\mathrm{d}u(l)}{\mathrm{d}l}\exp(-u)\mathrm{d}l = \int_0^\infty \exp(-u)\mathrm{d}u = 1 \tag{1.150}$$

In Rabinovitch and Bahat (1999), several possible qualitatively different shadow distributions were considered:

$$q(x) = \begin{cases} \lambda x^\alpha & 0 \le x < \mu \\ \lambda[1-(2\mu - x)^\alpha] & \mu \le x < 2\mu \\ \lambda & 2\mu \le x \end{cases} \tag{1.151}$$

where μ is a parameter measuring the shadow range and α is an exponent defining the qualitatively different distributions.

For each such $q(x)$ (Fig. 1.38), the related $p(l)$ is given in Fig. 1.39. These possibilities of $q(x)$ were applied to several pure experimental distributions ("pure" here means that no addition of distributions from different positions was carried out; each distribution originated from a specific layer which went through the same stress history). The results (Rabinovitch and Bahat 1999) showed that only two types of shadow distributions appeared, one (Eq. 1.151) with $\alpha = 1$ and one with $\alpha = 3$. This fact reminds us of the two main processes mentioned above for layer fracturing, namely the Cox-Hobbs process and the Pollard-Segall process. It was shown (Rabinovitch and Bahat 1999) that in fact, the $\alpha = 1$ distribution was related to rocks for which the Cox-Hobbs process was the preferred jointing method, while the $\alpha = 3$ distribution agreed with the Pollard-Segall one. This result led us to the conclusion that $q(x)$ was proportional to the stress field $\sigma(x)$, where x is the distance from an existing crack. This conclusion is discussed in the remainder of this chapter. First we show the agreement of $\alpha = 1$ and 3 with the two $\sigma(x)$ behaviors; in Sect. 1.5.3.4 we discuss the theoretical implications of this conclusion and show their agreement with a field measurement; while in Sect. 1.5.3.5 a recent creep experiment is shown to lend support to this conclusion. (*a*) For the Cox-Hobbs case, $\sigma(x)$ is given by

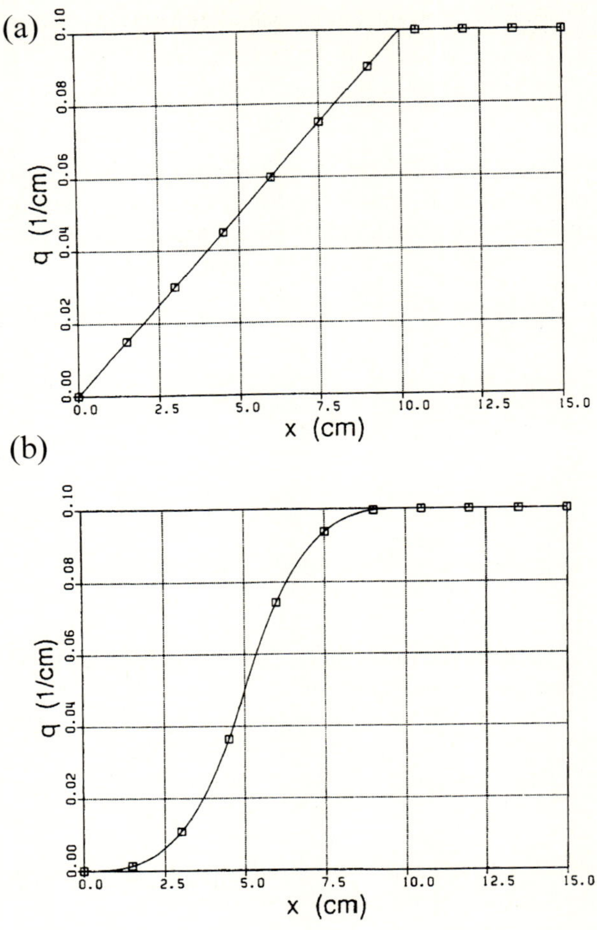

Fig. 1.38.
The complying shadow – the probability $q(x)$ (Eq. 1.151) of a second crack, a distance x away from an existing one, for $\mu = 5$ cm, $\lambda = 0.1\ \text{cm}^{-1}$, for **a** $\alpha = 1$, and **b** $\alpha = 3$

$$\sigma(x) = \sigma_0 \left[1 - \exp\left(\frac{-x}{\delta} \right) \right] \tag{1.152}$$

where (Cox 1952; Hobbs 1967), $\delta = (d/2)(E/G_N)^{1/2}$, d is the layer thickness, E is its Young modulus, G_N is the shear modulus of the neighboring layers, and σ_0 is the remote stress. For short distances, namely $x \ll \delta$, the exponent can be developed in a Taylor series to yield $\sigma(x) \approx \sigma_0 x / \delta$, which is proportional to x. (*b*) For the Pollard-Segall case, $\sigma(x)$ is given (Pollard and Segall 1987) by

$$\sigma(x) = 8\sigma_0 \left(\frac{x}{\delta} \right)^3 \left[4 \left(\frac{x}{d} \right)^2 + 1 \right]^{-3/2} \tag{1.153}$$

and for small values of x, $x \ll d$, $(x/d)^2$ can be neglected with respect to 1 and $\sigma(x) \approx 8\sigma_0 (x/d)^3$ which is proportional to x^3. The powers of x, the α, for the two cases are therefore $\alpha = 1$ and $\alpha = 3$, which are exactly equal to the powers of x in the $q(x)$ relations obtained experimentally.

Fig. 1.39.
Simulated fracture spacing distributions, $p(l)$ (Eq. 1.149, 1.150), for $B = 100$, $\mu = 5$ cm, $\lambda = 0.2$ cm^{-1}, for **a** $\alpha = 1$, **b** $\alpha = 3$, and **c** $\alpha = 10$

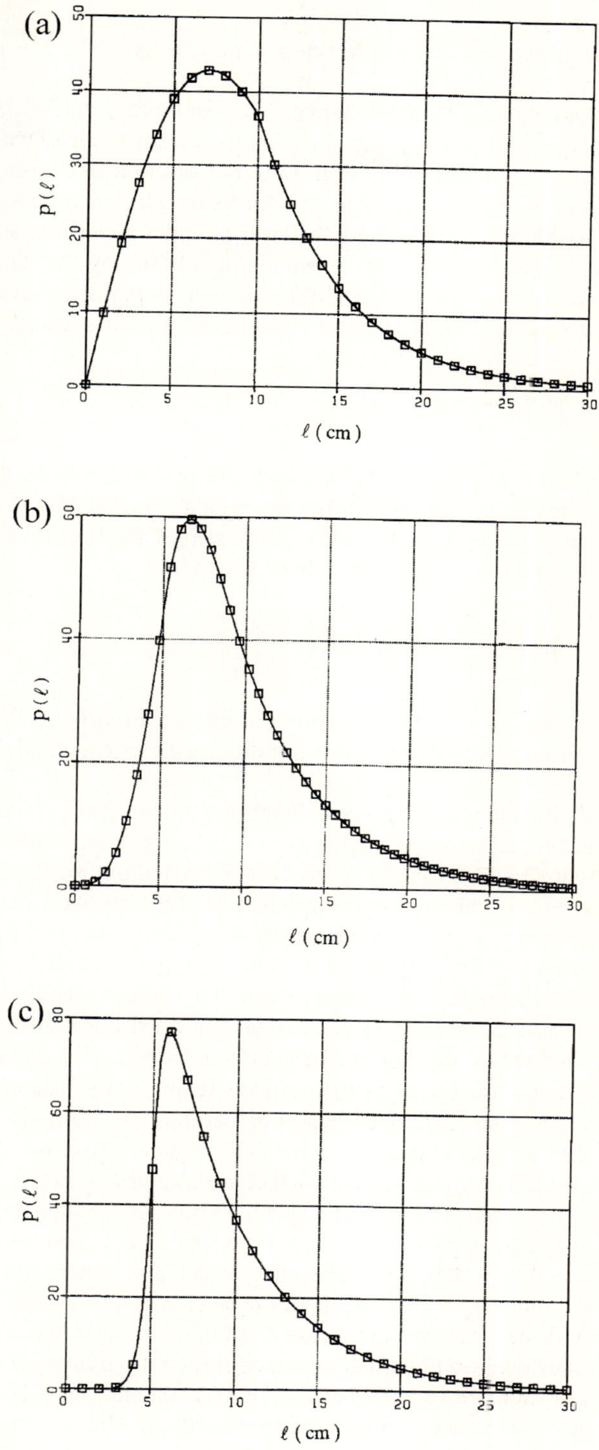

1.5.3.4
The Relation between the Shadow and Stress

Hypothetical K_I dependence. The assumption that $q(x)$, the probability of the appearance of a new crack at a distance x from an already existing one, is proportional to the stress field $\sigma(x)$ itself (and not say, to K_I!) is a very unusual one. To see what happens for "regular" assumptions let us calculate q for a hypothetical case where the Griffith criterion is used, the layer is under a constant stress σ_0 and the flaw length distribution is an inverse exponential. Denoting the flaw length by c, it is thus assumed that $K_I \propto \sigma\sqrt{c}$ and that the flaw length distribution $p(c)$, where $p(c)\mathrm{d}c$ is the probability that the flaw length is between c and $c+\mathrm{d}c$, is given by

$$p(c) = \frac{1}{c_0}\exp\left(-\frac{c}{c_0}\right) \tag{1.154}$$

where c_0 is the average flaw length. The probability that a flaw grows under these stress conditions is the probability that $K_I \geq K_{Ic}$ or that $K_I^2 \geq K_{Ic}^2$. The latter is equivalent to the condition that $\sigma^2 c \geq A$, where $A \propto K_{Ic}^2$. Thus, the probability that a flaw would grow is equal to the probability that $c \geq A/\sigma^2$, or

$$q(x) = \int_{A/\sigma^2}^{\infty} p(c)\mathrm{d}c = \exp\left[-\frac{A}{\sigma^2 c_0^2}\right] \tag{1.155}$$

The ensuing q is evidently not linearly proportional to $\sigma(x)$ and leads to a joint spacing distribution that is not observed experimentally.

Verification of the $q = \gamma\sigma$, Relationship. We establish our proposed relation, $q = \gamma\sigma$, in two ways. In the first one we assume this relation to hold, and calculate the exact $p(l)$ that should arise. This procedure is carried out for the Cox-Hobbs case, and the results are compared with the field data of Becker and Gross (1996) and shown to agree quite well with the experimental distribution obtained there. The second verification of the $q = \gamma\sigma$ relation is derived from a recent extended creep experiment which agrees with this relation for low σ values. The whole procedure will be elaborated here, since several interesting results, besides the verification of the $q(x)$ vs. $\sigma(x)$ relation, were obtained. According to Hobbs (1967), joint spacing should continuously decrease as long as stress (or strain) continues to act on the fractured layer. Experiments, however, display the phenomenon of "saturation", where the density of joints apparently does not grow above a certain critical value. This value is deemed by some investigators (see Bai and Pollard 2000a,b for a review of previous results) to depend on the layer thickness. Bai and Pollard (2000a,b) also challenged the Cox-Hobbs model. This challenge consists of two parts. In the main part, they carry out a finite-element method calculation of a layer containing four joints under strain. In contrast to the Hobbs' assumption that the stress in the layer is always tensile, their numerical simulations indicate that between the inner pair of joints there might appear compressive stresses. The latter are obtained for values of the ratio of inter-fracture spacing to layer thickness that are smaller than a certain threshold, which in turn depends weakly on the elastic parameters of the media and on the overburden stress. Such compressive stresses would obviously prevent the formation of additional joints in the region where they operate. In the second part of their challenge Bai and Pollard (2000a,b) show that if they extend the basic assumptions of Hobbs to the limit, the results do not fulfill the elasticity theory equations.

We take a different approach and consider Hobbs' (1967) calculation of the stress transfer from the outer layers to A to be a valid *approximation*. As such it does not have to strictly fulfill all the elasticity equations. On the other hand, we develop a different approach to the infilling process than that of Hobbs and show that a modified form of saturation does exist within this approach (see below). This would explain the "closely spaced fractures" for which Bai and Pollard (2000b) invoke an additional mechanism (flaw size, length and position distribution, possible pore pressure).

We now return to our statistical model for the Cox-Hobbs case. Assuming that $q(x) = \gamma\sigma(x)$ where γ is a proportionality constant, we have, by Eq. 1.152.

$$q(x) = \gamma\sigma_0\left[1 - \exp\left(-\frac{x}{\delta}\right)\right] \tag{1.156}$$

By Eq. 1.149 we therefore obtain

$$p(l) = \frac{\eta}{l}[1 - \exp(-z)]\exp[-\eta(z - 1 + e^{-z})] \tag{1.157}$$

where $z = l/\delta$ and $\eta = \gamma\sigma_0\delta$.

Equation 1.157 is the joint spacing distribution for the Cox-Hobbs case, according to our assumption that $q(x)$ is proportional to $\sigma(x)$. Some remarks are in order regarding the different parameters that appear in Eq. 1.157 (or Eq. 1.160, see below):

1. Lengths are always to be measured in dimensions of $\delta = (d/2)(E/G_N)^{1/2}$. The latter has dimensions of length and, in a sense, normalizes bed thicknesses. Normalized spacings $z = l/\delta$ can then be compared between different beds.
2. The proportionality constant $\gamma = q/\sigma$) has dimensions of length · (stress)$^{-1}$ or of (surface tension)$^{-1}$. It can probably be deduced from creep experiments (see below) if the latter can be seen as applicable to geologic processes.
3. The parameter $\eta = \gamma\sigma_0\delta$ is dimensionless. A crude interpretation for it is that it provides the approximate probability that a second crack would appear within a distance δ of an existing fracture. Higher η values therefore imply higher joint densities.
4. The parameter $\gamma\sigma_0 = \eta/\delta$) has dimensions of (length)$^{-1}$. It measures the probability per unit length that a second crack appears a long distance away from an existing fracture. As such it is a direct measure of joint density.

Equations 1.157 (or 1.160) can be viewed as the one and only joint spacing distribution both for low and high densities of joints. It is thus neither a "transition" between exponential, log normal to normal distributions, nor is it the CPP distribution of Fig. 1.37. The exact distribution, Eq. 1.157, is shown in Fig. 1.40 for several joint densities. The latter are represented by the parameter η. Note that the shapes of these distributions could easily mislead one to assume that the formerly mentioned "transitions"

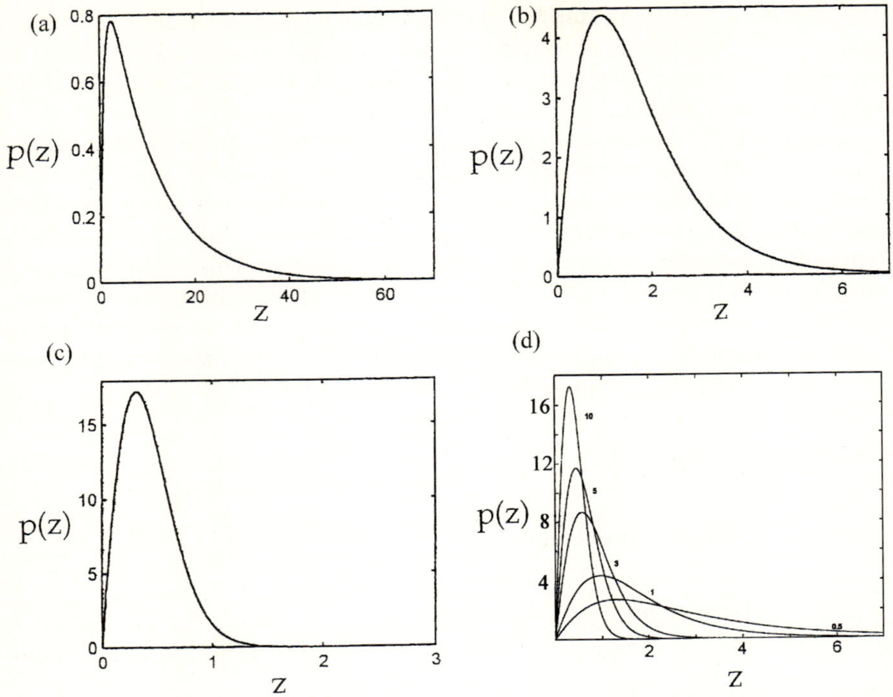

Fig. 1.40. Theoretical spacing distributions, $p(z)$ (arbitrary units) for the Cox-Hobbs case (Eq. 1.157). Lengths z are measured in dimensions of δ: $z = l / \delta$. **a** for $\eta = 0.1$, **b** for $\eta = 1.0$, **c** for $\eta = 10.0$. Note that η is proportional to the remote stress in the layer and to δ. **d** for $\eta = 0.5, 1, 3, 5, 10$; here distributions are shown on the same scale to emphasize the fast change of the peak with η between 0.5 and 3 and the slow change for η between 5 and 10

does occur. Now, before comparing this result to creep measurements, we want to pause a little at this point in order to derive some interesting features of this distribution and discuss the significance of the only two parameters that appear therein, namely δ and η. First, let us remark that since all lengths are scaled by δ, any statistic such as the average or the median should also be proportional to δ. We next discuss the behavior of the distribution median, its peak and the related interesting question of joint "saturation".

We start with the question of what "saturation" is and what is its origin. Saturation arises by the following two opposing influences. On the one hand, a larger stress results in smaller joint spacing. On the other hand, the "shadow" prohibits spacing from going to zero. To better understand this question, consider Eq. 1.157 and Fig. 1.40. It is seen that the distribution becomes more and more concentrated as η (or σ_0) increases. Yet the position of the peak and the mean of the distribution only slightly decrease with η (for high η values), and it seems that they would never become zero. Actually, as σ_0 increases, the *additional* amount of stress needed for the same incremental decrease in spacing greatly increases. To calculate the position of the peak of the distribution as a function of η, we equate to zero the derivative of $p(l)$ with respect to l. This yields $\exp(-z) - \eta(1 - \exp(-z))^2 = 0$ or, denoting $\nu = \exp(-z)$ and solving for the quadratic equation, we get

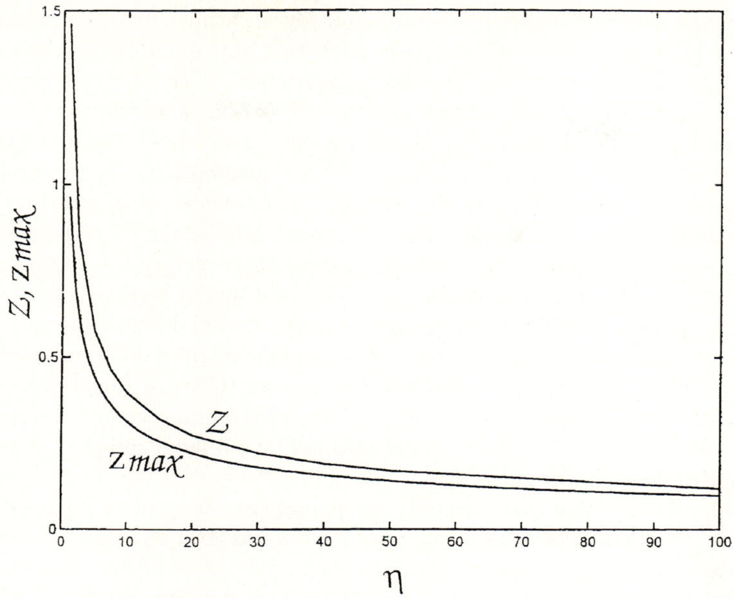

Fig. 1.41. Distribution peaks, z_{max}, and medians Z (both measured in dimensions of δ) as functions of η. Note the fast change of z_{max} and Z values for small η values, where new cracks are relatively easy to form for even a small change of stress; and the very slow (logarithmic) change of these values for larger η (effective saturation) where increasing the crack density becomes harder and harder

$$\nu = 1 + \frac{1}{2\eta} \pm \sqrt{\frac{1}{\eta} + \frac{1}{2\eta^2}} \tag{1.158}$$

hence, the distribution peak is

$$z_{max} = -\ln\left[1 + \frac{1}{2\eta} - \left(\frac{1}{\eta} + \frac{1}{4\eta^2}\right)^{1/2}\right] \tag{1.159}$$

where only the negative sign was chosen, in order to agree with the fact that $\exp(-z_{max})$ should be less than one. Figure 1.41 shows the change of z_{max} as a function of η. It is seen that although z_{max} tends to zero with η (spacings decrease with stress increase), this tendency is very slow, logarithmic in fact (but not hyperbolic). Therefore, although *no real saturation occurs*, z_{max} presents an "*effective saturation*" behavior, (which is in line with experimental results (quoted by Becker and Gross 1996, Fig. 2) on polystyrene plates and cross-ply glass where the mean fracture spacing was measured as a function of stress (strain) and deemed to be hyperbolic (Rives et al. 1992)). In these experiments, excessive additional stresses are required for the attainment of additional cracks. Note that the change of z_{max} and median values (see below) are gradual and the transition from the range of "easy" crack density increase to the range of "effective saturation" also occurs gradually, (say, in the range $5 < \eta < 15$) rather than through a threshold value of $\eta(\sigma_0)$ (as suggested by Bai and Pollard 2000a).

The median of the distribution $L = \delta Z$ can easily be calculated numerically. Results are shown in Fig. 1.41, where Z is shown as a function of η. It is seen that the behavior of the median is similar to that of the mean, as expected. Since $Z = L / \delta$, which is proportional to L / d, Z is actually proportional to the FSR parameter (Gross 1993) (called by Bai and Pollard the "fracture spacing to layer thickness ratio"). The parameter Z is (within the Cox-Hobbs's model) the "correct" normalized fracture spacing ratio. If Z would have been a constant, then the slope of L as a function of d (more accurately of δ) would have been constant, leading to a constant value of the FSI (Narr and Suppe 1991) which is the slope of the best fit line for points depicting L vs. d for layers of different thicknesses. According to Fig. 1.41, this is not so. The median is proportional to δ only as long as η does not change significantly between different layers. This implies that the straight line obtained for FSI measurements from different layers (Narr and Suppe 1991) is somewhat fortuitous. For an almost "exact" linear dependence, both σ_0 and δ would have to be the same for the different layers, and the spread of layer thickness for the different layers should not be too large and should lie in the larger η ("effective saturation") region.

This study shows that there are only two parameters that can be obtained by the analysis of spacing distributions, i.e., δ and η (or δ and $\gamma\sigma_0$). The value of δ can also be calculated by the layer thickness and the elastic properties of the rock. Therefore, η becomes the most important measurable parameter. Its value

1. Determines the position of the fracture population of a particular layer with respect to effective saturation.
2. On comparing different layers, can provide ratios of remote stresses causing fracture in them. It is seen (Fig. 1.40 and 1.41) that spacing distribution depends on the maximum stress that has acted in the rock layer for a significant period of time.

We now show that the distribution of Eq. 1.157 is indeed compatible with the histograms of joint spacing measured by Becker and Gross (1996, Fig. 9) of the Turonian Gerofit Formation, southern Israel. The comparison is outlined here for their Sect. I, and the results of all sections are given subsequently. Consider the histogram of Sect. I (Fig. 1.42a). The l-axis stretches between zero and 80 cm, in increments of 3 cm. The bars have heights (number of points) as follows (obtained from Becker and Gross 1996, Fig. 9): 1, 4, 9, 14 etc. (since we would like to treat bars with widths of 1 cm each (unit length), each number should actually appear three times with 1 / 3 height). The sum of all spacing is 110. Now, in order that the distribution be normalized (the sum of all bars be equal to one), each number should be divided by $3 \times 110 = 330$. These numbers (* in Fig. 1.42a) are compared to Eq. 1.157 written in the form

$$P(l) = \gamma\sigma_0\left[1 - \exp\left(-\frac{l}{\delta}\right)\right]\exp\left\{-\gamma\sigma_0\left[l - \delta\left(1 - \exp\left(-\frac{l}{\delta}\right)\right)\right]\right\} \tag{1.160}$$

in a "least square" fashion (such as "curve fit" in Matlab). Here, the two parameters sought are $\gamma\sigma_0$ and δ. Results for all four sections are shown in Fig. 1.42 and the values of $\gamma\sigma_0$, δ and $\eta = \gamma\sigma_0\delta$ are summarized in Table 1.2.

As seen, the agreement of the data to Eq. 1.160 is very good, especially the shapes of the distributions. However, the parameters $\gamma\sigma_0$, δ and η of sect. II are seen to be far

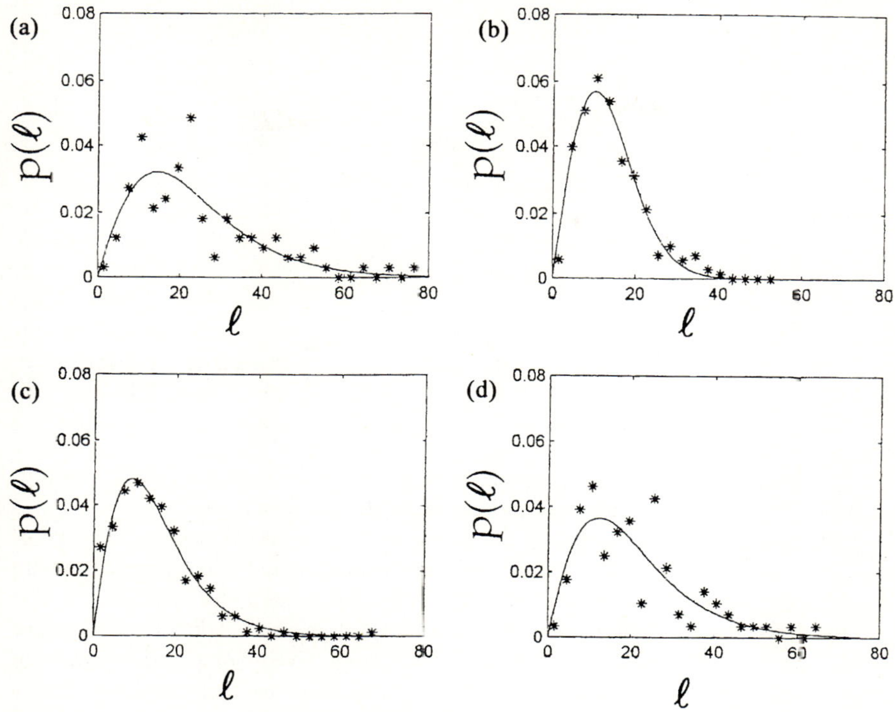

Fig. 1.42. Analysis of results of Becker and Gross (1996) spacing measurements. Sections I–IV are given in **a–d**, respectively. * are measurements and *full lines* are fits (Eq. 1.160). Note that spacing is not normalized to δ (measured in cm) and that $p(l)$ is normalized (see text)

Table 1.2. Physical parameters of joint spacing (based on data from Becker and Gross 1996)

Section	$\gamma\sigma_0$ (1 cm^{-1})	δ(cm)	η (dimensionless)
I	0.089	19.39	1.725
II	0.637	65.29	41.59
III	0.124	11.37	1.41
IV	0.091	14.15	1.288
IIa	0.276	27.76	6.312

off from all the rest. As discussed by Becker and Gross (1996), this section experienced extreme stress conditions and the data thereof is therefore exceptional. Moreover, these data are composed of several subsections, and such composition is not advisable (see below). Some remarks are in order.

1. The "normalized layer thickness", δ, of sections I, III and IV is ~15.3 ±4 cm and the stress field in sect. III is somewhat higher than these in sect. I and IV (which are rather similar). In order to better understand the results of sect. II, the histogram of sect. IIa was analyzed separately in Fig. 1.43 (the histogram of sect. IIb is dubious since *(a)* this is the section with the largest spread of σ, and *(b)* the number of events does not add up to the quoted 120). As seen from Table 1.2, even sect. IIa

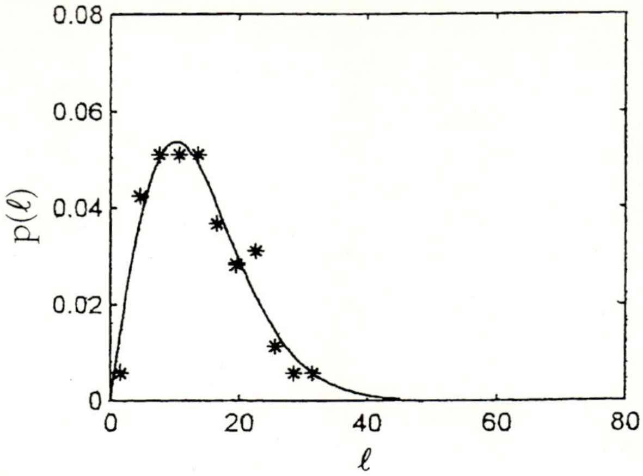

Fig. 1.43.
Section IIa of Becker and Gross (1996). Signs as for Fig. 1.42

shows already much higher parameters than the other three, suggesting unevenness of layer thickness, stresses and/or elastic parameters.

2. As mentioned, Table 1.2 shows that δ for sect. I, III and IV has an average of 15.3 ±4 cm. The latter value could be used as the range of error in the measurement. To check this result, we have calculated δ by its definition $\delta = (d/2)(E/G_N)^{1/2}$. The values of E and G_N for limestone and shale, respectively were taken from Engelder and Peacock (2001) to be 66 ±10 and 6.5 ±0.9, respectively. For a layer thickness of 19 cm (Becker and Gross 1996) the value of δ, calculated by its definition, becomes 29 ±4.2 cm. The discrepancy between $\delta = 15$ cm and $\delta = 29$ cm can be interpreted to represent the deviation of the elastic properties of the rocks in the Gerofit Formation from those borrowed from Engelder and Peacock (2001).
3. It is seen that the only genuine length-normalizing coefficient is δ, being the inherent physical variable of the Cox-Hobbs spacing. Nevertheless, regarding the question of the addition of distributions from different layer positions and different lithologies, it is clear (from Fig. 1.40) that such a procedure is not advisable. For example, an addition of two, even normalized, distributions of types appearing in Fig. 1.40a and 1.40c would create a "distribution" having a fictitious double peak.
4. The values of η obtained for sect. I, III and IV are all around 1.5, placing these sections deep inside the "unsaturated" region in Fig. 1.41. The high η value of sect. II indicates that this section has "suffered" a much higher stress (strain) during its history, bringing it (or at least the IIb part) into the saturated region.
5. A simple test of the self-consistency of the results can be carried out. Consider, e.g., sect. III. A direct measurement from the histogram yields the median to be $L \approx 11.5$ cm. Hence, $Z = L/\delta \approx 11.5/11.37 = 1.01$. By Fig. 1.41, this Z value leads to $\eta \approx 1.7$, which is close to the 1.41 obtained by the least square method.

1.5.3.5
Creep Experiments, a Close Analogy to Slow Geological Fracture

We turn now to creep experiments (see e.g., Poirier 1985). These experiments are usually conducted at high temperatures and low stresses. Due to the limited period of measurement, however, stresses are usually not as low as those prevailing in geological layers. In creep experiments, the measured quantity is the creep (strain) rate, $\dot{\varepsilon}$, measured by a scanning laser technique. Three stages of creep are usually discerned (primary, secondary and tertiary) by the magnitudes of the respective $\dot{\varepsilon}$. But since in geological processes, the stresses generally involved in single-layer jointing are assumed to be small (see below), we restrict ourselves here to the primary stage. Usually for creep in ceramics, the dependence of creep rate on stress and temperature is represented by the so-called Norton equation:

$$\frac{\partial \varepsilon}{\partial t} = \dot{\varepsilon} = B\sigma^n \exp\left(-\frac{Q}{RT}\right) \tag{1.161}$$

where B is an empirical constant, Q an activation energy and n the stress exponent, which normally ranges between 1 and 2. However, as was pointed out by Yoon et al. (2000) this empirical formula is inadequate to describe low stress creep both in tension and in compression. Although the experiments of Yoon et al. (2000) were carried out only for silicon nitride, we assume that they hold for other ceramic materials as well as for polycrystalline rocks. Their results show that although the creep processes within a material under compression were different from those operating under tension, in both cases the first part of $\dot{\varepsilon}$ (i.e., for small strains, in the "primary" stage) behaved in a similar way with respect to stress: $\dot{\varepsilon}$ was in fact proportional to the stress σ. Note that the obtained behavior for tension was actually

$$\dot{\varepsilon} = a\sigma \exp(\alpha\sigma) \exp\left(-\frac{Q}{RT}\right) \tag{1.162}$$

but for low stresses, $\exp(\alpha\sigma) \approx 1$ and $\dot{\varepsilon}$ is again proportional only to σ.

To recapitulate, according to Yoon et al. (2000) for a brittle material both under tension and under compression, the response in creep is that $\dot{\varepsilon}$, the rate of change of strain, is proportional to the stress σ, when σ is small enough. Since for many joints fracture processes have continued for very long times, it is conceivable that the stresses present were quite small; otherwise, all jointings would have been in the "extreme saturation" state (see above, Sect. 1.5.3.4), which is not the case. We therefore assume that the layer creep rate is given by

$$\dot{\varepsilon} = \sigma a \tag{1.163}$$

where a is a constant. Consider now a layer in which a single joint exists and is under a stress $\sigma(t)$, which is quite small but can change with time. Hence, after eons have passed, the present strain in the layer is given by

$$\varepsilon = \int_0^T \dot{\varepsilon} \mathrm{d}t = a \int_0^T \sigma(t) = b\overline{\sigma} \tag{1.164}$$

where $\overline{\sigma}$ is the average stress.

This relation should also be true for each segment Δl around the point x. The strain in the vicinity of x, $\varepsilon(x)$ can be approximated by

$$\varepsilon(x) = \Delta N(x)\frac{\Delta}{\Delta l} \tag{1.165}$$

where $\Delta N(x)$ is the number of fractures in the segment Δl, Δ is the opening of a crack, and it is assumed that the increase in length of the segment is due only to crack openings. Measuring x from an existing joint, $q(x)$ is evidently proportional to $\Delta N(x) / \Delta l$ and hence proportional to $\varepsilon(x)$, which is in turn proportional to $\sigma(x)$.

This argument is obviously somewhat crude, since we have neglected the changes in σ by the newly added joints. If, however, the density of cracks is not very high, this approximation is reasonable.

1.5.3.6
Summary

Several important results have been achieved regarding joint spacing.

1. The Rives et al. (1992) conjecture that the joint spacing distribution (JSD) is changing with joint density from negative exponential to log-normal and to normal was shown to be incorrect.
2. The 1D car parking process distribution, which is a derived distribution supposed to simulate the joint spacing one, is shown to be inadequate since it completely ignores the shadow effect.
3. The JSD is shown to be dependent only on $q(x)$, where $q(x)\mathrm{d}x$ is the conditional probability that if a crack exists at $x = 0$, another crack would appear between x and $x + \mathrm{d}x$ away from it.
4. The JSD is exactly calculated from $q(x)$ (Eq. 1.149).
5. An empirical method to analyze experimental joint spacing results is presented. Using the method for several experimental cases revealed that only two processes lead to jointing, the so-called Cox-Hobbs and Pollard-Segall. The method is shown to *possess the ability* to find out which process was operative in a specific layer.
6. A bold assumption is made, partly based on indication from a recent creep experiment under extremely low stress loads, that $q(x)$ is proportional exclusively to the stress level $\sigma(x)$. This assumption is verified by comparing the analytical JSD calculated under this assumption with field data (Becker and Gross 1996). The JSD analytical calculations were carried out only for the Cox-Hobbs case, while the treatment of the Pollard-Segall process is to be carried out in the future.
7. For the Cox-Hobbs case, a complete JSD is obtained, valid both for high joint densities as well as for low ones. Contrasting a recent work by Bai and Pollard (2000a,b), the obtained JSD does not discard Hobbs calculation. It just amends his infilling approach and leads to a new understanding of the saturation phenomenon.
8. Saturation is shown not to be a certain lower level threshold below which joint spacing would never occur. It is rather an infilling process in which joint spacing always decreases when strain (or stress) increases, but when joint spacing is high, the rate of change of spacing with stress is high, while, when it becomes lower, the rate of change decreases logarithmically, i.e., it "costs" higher, and higher stress increments for the same amount of joint spacing decrease. There is a clear distinc-

tion between unsaturated and saturated jointing, although the transition from the first to the second occurs gradually.

9. It is shown that the only two parameters governing the JSD are δ, which is a normalized bed thickness and η, which is proportional to the highest stress level prevailing in the layer. Experimental results can yield only values of these two parameters. The parameter η can be used to compare stress levels at different layers as well as to locate the position of the joint spacing in the measured layer with respect to saturation (that occurs for η values above ~10).

Elements of Fracture Geology

2.1
Introduction

It is often quite a challenge to distinguish between fractures induced in rocks by shear or by tension when examined in micro scales (features that are unseen or almost unseen with the naked eye) or in outcrop scales (between centimeters and tens of meters in size). We take this challenge and concentrate in this chapter on fractures that are predominantly induced by mode I. Thus, fracture geology is limited in the present context to various phenomena of fracture growth, primarily by tension (or extension). This is a constraint that we essentially impose on the other chapters in this book, i.e., Chap. 2 stems from a treatment of principles of fracture physics in Chap. 1 and provides a base for the treatment of fracture provinces on regional scales in Chap. 3 and 4, also concentrating on tensile fracture.

The issues of fracture nucleation and growth in rocks will be raised first, followed by descriptions of fourteen experimental investigations of fractography in technological materials that have bearing on fractography in rocks. The presentation of the data will emphasize the distinction of morphological features in the mirror plane from those in the fringe (Fig. 2.1a–d). Observations on fracture surface morphology from sedimentary rocks as well as from granites will come next. The treatment of two outstanding questions in structural geology and in rock mechanics, namely, conjugate hybrid joints and longitudinal splittings, will end Sect. 2.2. The classification of systematic joints into the burial, syntectonic, uplift and post-uplift four genetic groups (Bahat 1991a) provided reference for subsequent studies on fracture in rocks, for us and for other authors (e.g., Engelder et al. 1993; Ghosh 1993; Bankwitz et al. 1994, 1997). We expand on the nomenclature of uplift and post-uplift joints and introduce the class of "surface joints" in Sect. 2.3. The fascinating multi relationships between primary joints, faults and secondary fractures are far more understood today than one or two decades ago due to recent publications on this issue (Peacock 2001a and references therein). This topic is addressed in Sect. 2.4.

2.2
Fractography and Tectonofractography

2.2.1
Introduction

Fracture initiates at stress concentrators in rocks and in synthetic, brittle and semi-brittle materials (as well as in some non-brittle materials, Bahat 1991a, p. 67). The frac-

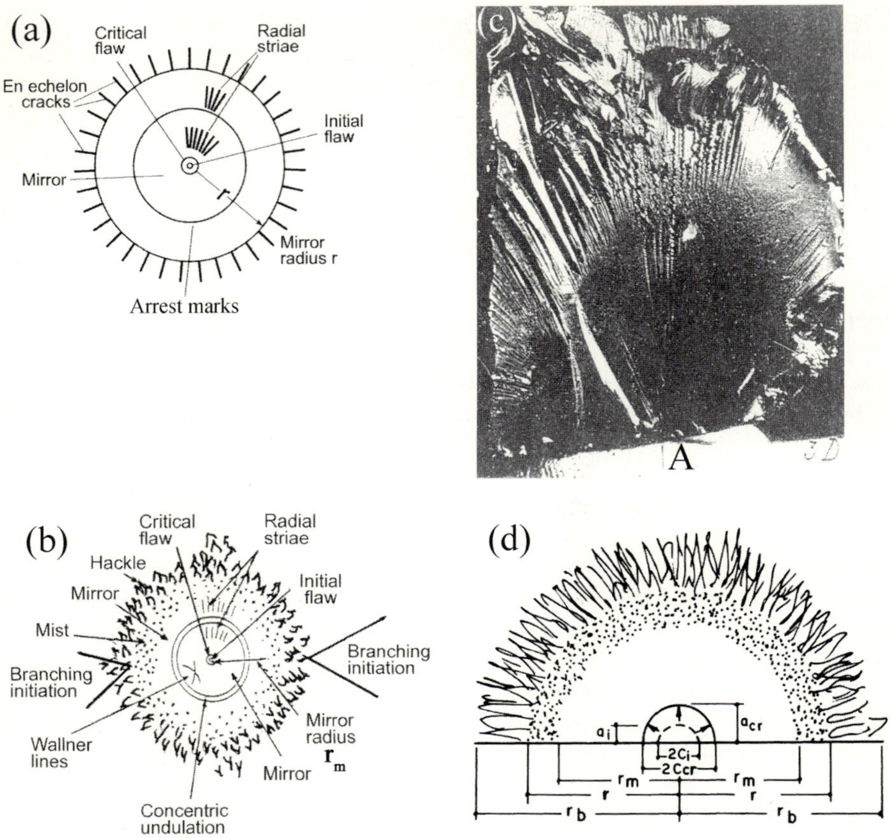

Fig. 2.1. Schematic representation of various fractographic elements on a fracture surface and distinguishing the mirror from the fringe. **a** The mirror radius, r (*arrow*) is measured from the critical flaw to the inner boundary of the *en echelon* fringe. **b** The mirror radius, r_m (*arrow*) is measured from the critical flaw to the inner boundary of the mist. Branching may initiate either from the outer rim of hackles at the right or from the inner rim of hackles, at the left (forming hackle and branching initiation together). **c** A fracture surface in bitumene (from De Freminville 1914). Note a few radial striae within the mirror plane and a multitude of hackles beyond it. Many hackles appear as echelon cracks, extensions of the striae. **d** Schematic fracture surface of brittle materials showing idealized initial flaw length $2c_{ci}$ and depth a_i, critical flaw length $2c_{cr}$ and depth a_{cr}. The three mirror radii, r_m (mirror-mist boundary) r (mist-hackle boundary) and r_b (initiation of microscopic crack branching) are shown as well (modified from Mecholsky and Freiman 1979)

ture then propagates through a great number of disturbances caused by material heterogeneities (e.g., flaws) and inhomogeneous stress fields, emitting sound waves. The sound waves interact with the advancing fracture front, and the results are recorded on the fracture surface by fracture surface morphologies, also termed fracture surface topography, fracture markings, or in short, fractography (e.g., Kerkhof 1973; Hull 1999). Every individual fracture is characterized by its specific fractography. Woodworth (1895, 1896) set the stage for the science of fracture surface morphology by astute observations of the morphology of joints in geologic exposures. He also noted the essential morphologic similarities between fracture markings in glass samples and rock outcrops. Fractography is that branch of science that analyses fracture surface morphology and related features and their causes and mechanisms in technological materials. Fractography also contributes significantly to the understanding of fracture processes that form joints. The fractographic analysis helps to define on the fracture surface the following (Fig. 2.1a–d):

1. Fracture origin.
2. Properties of the initial flaw and of the critical flaw.
3. Modes and directions of fracture propagation.
4. Stress configurations.
5. Failure types (including, fatigue, brittle, ductile, impact and thermal influences).
6. Fractographic symmetric elements.
7. The distinction between the mirror plane (the parent joint) and the fringe.
8. Estimating the local paleostress.

This analysis establishes whether the surface features reflect a single or multistage fracture process and provides tools for fracture mechanics analysis, fracture toughness properties and related parameters of joints. From all the excellent photographs of mirror planes found in the literature (e.g., Holloway 1986), we select to show the ingenious discovery by De Freminville (1914, Fig. 2.1c). Tectonofractography is a branch of tectonics and tectonophysics, which applies fractographical analysis to rock fractures and to regional fracture systems (joint sets). The objectives of tectonofractography are to determine the fracture paleostress directions and the mechanical conditions involved and identify the tectonic processes that produced the fractures. Both fractography and tectonofractography were discussed extensively in Bahat (1991a). This discussion is expanded in subsequent sections of this volume, while minimizing repetitions from the previous volume to mere essentials.

2.2.2
The Origin: Crack Nucleation via Fractographic Techniques

2.2.2.1
Different Applications of the Term Nucleation

The terms crack (or fracture) origin, crack initiation, nucleation and "Griffith flaws" are often used interchangeably in the literature, relating to vast scale ranges. Lawn (1993, p. 307) distinguishes two ranges of "Griffith flaws" dimensions from which failure of brittle materials initiates. In homogeneous materials like monocrystalline quartz, flaws range from 1 nm (in pristine fibers and whiskers) to 1 µm (in aged, as-handled solids) and usually occur at the surface. In heterogeneous polycrystalline materials like ceramics, flaws range from 1 µm (in high-density, fine grained, polished materials) to 1 mm and above (in refractories and concrete) and occur in both the surface and the bulk. Flaws vary in size, shape and orientation such that the coexistence of more than one flaw type often gives rise to a bimodal or even multimodal flaw population. Lawn (1993, Table 9.1) provides a rather liberal-size classification of "Griffith flaws", inversely correlative to strength. In very strong materials (10 GPa for whiskers) the flaw size is in the nm range, whereas in very weak materials (1 MPa for the Earth's crust) the flaw size is in the m range. An experimental study of fracture in a transparent glass ceramic reveals two oppositely propagating striae from an origin that cannot be precisely determined in a spot. An area of 1–2 × 1–2 mm is measured instead and termed "nucleation zone", where the bilateral fracture originated (Sect. 2.2.3.7). See a further elaboration on nucleation in micro scales in Sect. 1.2.

Joints. There are earth scientists who search fracture nucleation in the rock microcrack (Simmons et al. 1975) grain size scale (e.g., Segall and Pollard 1983). In the Devonian layered siltstones and shales of the Appalachian Plateau, central New York, and in the Eocene chalks from the Beer Sheva syncline, many joint initiation points are located along bedding boundaries (Bahat and Engelder 1984; Helgeson and Aydin 1991; Lacazette and Engelder 1992; Weinberger 2001a). These initiations occurred at stress concentrators such as free surfaces, fossil inclusions, pyrite concretions, voids, cusps, and burrows along the boundaries. Many other joint initiations in these two provinces occur from stress concentrators within the rock layer, remote from its boundaries (Weinberger 2001a). The exact location of joint nucleation within granite bodies can be fractographically identified on the Borsov joints that formed by slow propagation (Sect. 4.4.2) but not on the Mrákotin joints that formed rapidly (Sect. 4.4.3). On a larger scale, early fan-shaped cracks that were marked by plumes ranging in size from a few meters to tens of meters on surfaces of exfoliation joints interacted with each other and coalesced into composite fractures on vertical cliffs of granitic rocks at Yosemite National Park, California (Sect. 4.5.3.3). The orientation and arrangement of the fans indicate that they initiated in "nucleation zones" at certain heights, from where they propagated upwards and downwards parallel to the cliff of El Capitan monolith.

The treatment of joint initiation has been idealized from a Griffith internal crack model in glass (either elliptic or slit shapes). In moving from a model of a slit (Griffith 1924), the shape of the crack has been thought to be an oblate (Sneddon 1951, p. 486) or a prolate spheroid. The "empty" crack model (that does not have a mass) was modified to one in which fluid pressure is activated from the interior on the crack walls. Secor's theory (1965, 1969) suggests that the growth of an extension fracture at depth in the Earth's crust is a slow process, consisting in detail of numerous short episodes of crack propagation that alternate with longer periods of quiescence. Pore fluids from the surrounding rock percolate into the crack during quiescence and then wedge it open. Secor (1969) considers that the failure criteria for tension fracturing of rocks with internal pore pressure P in plane strain is

$$\sigma_{e3} = -S = -\left[\frac{\pi E \gamma}{2c(1-\nu^2)}\right]^{1/2} \tag{2.1}$$

where σ_{e3} is the least effective principal stress and $-S$ is the tensile strength of the rock (tension considered negative) for the general pore pressure condition $\sigma_{ej} = \sigma_j - P$, (Secor 1969), where c, σ_{ej}, σ_j, γ and ν are the penny-shaped crack radius, the effective principal stress, the principal stress, the fracture surface energy and the Poisson ratio, respectively. Fracture initiates when a fraction of P exceeds σ_3 and becomes effective in wedging open the crack. This will occur when fluid in the crack reaches a "critical volume" (Secor 1969):

$$v_{crit} = \frac{16}{3}\left[\frac{\pi\gamma(1-\nu^2)}{2E}\right]^{1/2} c^{5/2} \tag{2.2}$$

The cancellation of the traction parallel to the crack is done for convenience. The resemblance of Eq. 2.1 to the known Griffith equation (Eq. 1.21) implies fracture by extension that according to Secor can form joints at considerable depths. Field observations support Secor's model of a fracture process by alternating episodes of crack propagation and periods of quiescence (Bahat and Engelder 1984). A remaining outstanding question, however, is what the range attained by K_I in jointing is. Slow fracture processes in sedimentary rocks cannot be well characterized by K (Sect 1.5.3), while in granites, in some cases there are indications of jointing under $K_I \geq K_{IC}$ conditions (Chap. 4).

Faults. Segall and Pollard (1983) investigated nucleation and growth of strike slip faults cutting granite. They suggested that faults nucleated on pre-existing joints that acted as weak surfaces capable of accommodating later shear deformations and modeled fault nucleation from sub parallel arrays of joints. Triaxial tests have been used to study the nature of the "damage zone" that propagates ahead of the growing fracture (Hoagland et al. 1973). While the "damage zone" describes a zone at the tip of an existing fracture that consists of multiple cracks oriented with respect to the fracture, 'nucleation zone" (see above) relates to an area where fracture nucleation unrelated to an existing fracture occurs. These two terms are occasionally combined. Lockner et al. (1991) and Reches and Lockner (1994) applied triaxial tests to simulate nucleation of faults by the coalescence of axial microcracks into an inclined "process zone" under mixed mode conditions. The latter triaxial experiments were performed on intact granite cylinders 190.5 mm long and 76.2 mm in diameter, and planar quasi-circular "process zones" were suggested to be the nucleation sites. In both the outcrop scale and the triaxial test (Segall and Pollard 1983; Reches and Lockner 1994), the nucleation zones consisted of arrays of cracks, and in the two models, the fault growth continues along the orientation of the array.

Studies of nucleation of earthquakes (e.g., Rice 1979; Rudnicki and Olsson 1998; Gudmundsson and Homberg 1999) are carried out on seismograms (e.g., Umeda 1990; Ellsworth and Beroza 1995). Based on friction parameters, Scholz (1998) distinguishes two regimes, stable and unstable, and suggests that earthquakes can nucleate only in those regions of a fault that lie within the unstable regime. During the inter-seismic period, the fault is loaded by steady slip on the deep, stable portion of the fault. Just before the earthquake, a pre-seismic phase occurs. In this phase, known as nucleation, slip accelerates until instability results in coseismic motions. The nucleation phase is sometimes associated with precursory phenomena, such as foreshocks. A model of earthquake nucleation is considered for the weak crust of California by Bahat et al. (2001a). They suggest that the multi fault rupture pattern in California shows greater resemblance to multi fracture patterns of weak chalk cylinders broken in triaxial tests than to a single process zone in fractured strong granite. The rupture of the chalk cylinders does not start from a localized nucleation; it rather develops from many distinct sites by a coalescence process. Possibly, the nucleation of earthquakes in a weak terrain occurs from many centers rather than a single site.

2.2.2.2
Fractography

We restrict the treatment of crack initiation into a rigid framework, which is limited by visible and measurable fractographic parameters. A key feature is the "critical flaw" (Fig. 2.1a,b), which is measured as $2c_{cr}$ (Fig. 2.1d). When $2c_{cr}$ is attained, crack growth occurs spontaneously, whereas below this length nucleation is unstable and the crack is expected to heal (in glass in vacuum). Thus, the critical flaw is the upper boundary of the crack nucleation stage. The relationship in Fig. 1.12 is analogous to the maximum in free energy plotted against the radius of a nucleated new crystal. Below a "critical radius size" the nucleus tends to be unstable and dissolve, and beyond it crystal growth occurs spontaneously (Carmichael et al. 1974, Fig. 4–14). This analogy should be useful particularly for demonstrating the distinction between rates of nucleation and growth (Bahat 1970).

The "critical flaw" can be identified experimentally, as it grows from the "initial flaw" $2c_i$ (Fig. 2.1d and 2.2), showing that $2c_{cr}$ may reach sizes around 10 mm for an opening mode fracture in glass. It was found (Mecholsky and Freiman 1979) that the ratios of the mirror radii (r_m, r and r_b, Fig. 2.1d) to the critical flaw radius have corresponding constants. However, fracture results obtained by various investigators of glass and ceramics have yielded considerable discrepancies in these ratios, leading to the suggestion that a poor distinction between $2c_i$ and $2c_{cr}$ was possibly an important factor contributing to these discrepancies (Bahat 1991a, p. 109). Fortunately, occasional difficulties in defining the boundaries of $2c_{cr}$ in geological outcrops may be trigonometrically overcome (Fig. 6.18).

Fig. 2.2. Microphotograph of an initial flaw (*white arrow* pointing upward) growing to concentric ripple marks *A*, and radial scars in glass. These markings reveal that the growth of the initial flaw is controlled by a heterogeneous stress field. Curve *B* marks the transition to the mirror plane at the tip of the critical flaw (*black arrow* pointing downward). Lengths of diameters of initial flaw and critical flaw from ends to ends are 0.9 mm and about 1 cm, respectively (after Bahat et al. 1982)

Thus, in an attempt to provide a common denominator for comparison, where possible, we make the following distinction. The transition from sub-visible defects or singularities (undetected in the optical microscope) to the initial flaw $2c_i$ is at 2 μm (Lawn 1993, p. 13). The initial flaw may reach several mm (Segall and Pollard 1983). Its transition to the critical flaw $2c_{cr}$ varies in the mm range but can reach the cm scale on joints (Bahat and Rabinovitch 1988). These two transitions belong to the nucleation stage, while propagation from $2c_{cr}$ is taken to mark the initial growth (see also Doremus and Johnson 1978; Rabinovitch 1994). Occasionally, initial flaws may deviate from a single plane and can lead to different parts of the flaw acting essentially independently, creating more than one critical crack, such that each one would generate its own part of a mirror plane. The latter effect seems to increase with the decrease of the average angle of the initial flaw relative to the tensile axis (Bahat 1991a, Fig. 2.33d and 2.35c), resulting in the increase of shearing and the reduction of relative tension on the growing crack during the nucleation stage (Fig. 2.2).

2.2.2.3
Fracture Initiation in Stratified Muddy Sediments

In this and in the next section we cite studies by Weinberger (1999, 2001a,b) on fracture initiation in sedimentary rocks. These systematic studies improve our understanding of fracture origin in rocks and help to focus on the difficulties involved in differentiating joint nucleation from joint initial growth.

Persistent desiccation and contraction of muddy sediments give rise to mud cracks. The fracture characteristics of mud cracks were studied in dehydrating mud puddles in the Dead Sea region, Israel (here we cite extensively Weinberger 1999). The mud consists mainly of carbonate and clay particles typically displaying fining upward. There is a distinct surface discontinuity between the upper desiccated layers (that tend to contract and crack) and the uncracked lower ones. This surface is referred to as the bottom of the polygons. Mud cracks with delicate plumes formed several days after a rainstorm and ceased to propagate a few days to weeks after initiation. Several polygons of convenient size and weight (mostly containing a pair of desiccated layers) were then systematically lifted out of the dried puddles and their walls examined under optimal oblique illumination. In this way, the surface morphology of hundreds of mud cracks were studied in precise detail, based on fractographic techniques. The mud cracks form networks of interconnected tension fractures arranged in polygonal patterns. Because tensile stress due to drying declines downward through the sediment, mud cracks have generally been theorized to nucleate near the surface, propagate downwards, and terminate at depth.

Weinberger (1999) found that systematic nucleation at the bottom of the polygons and upward propagation of mud cracks are much more prevalent than previously postulated. The mud-crack initiation points are often located along or near the base of the desiccated sediment, where potential flaws (e.g., long grain boundaries) are more abundant because of the natural fining-upward sorting of grains (Fig. 2.3a). Consequently, these large flaws become critical before small flaws at the top (e.g., short grain boundaries), even though the tensile stress caused by drying declines downwards through the mud (see below). On growing, the cracks predominantly rupture the des-

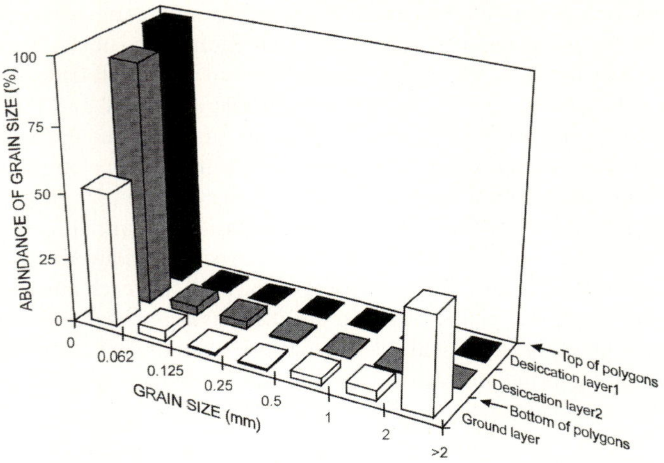

Fig. 2.3. a Grain-size distribution in the studied sediment displaying fining upward (sorting effect) and stratification. This distribution was obtained by gently disintegrating the grain aggregates of each layer and sieving the grains through screens of decreasing mesh size (from Weinberger 1999)

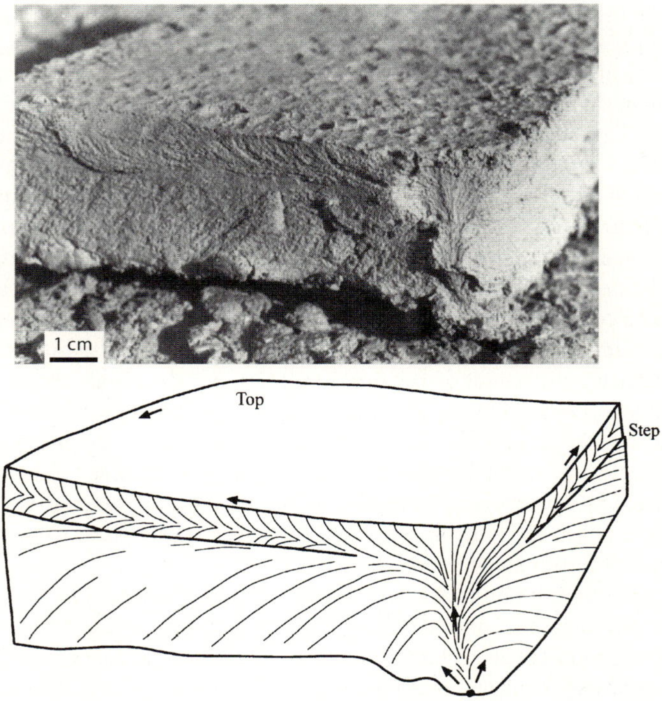

Fig. 2.3. b Oblique view of a polygon, a photograph above and a drawing below, explaining propagation directions by *arrows*. The plume shows that the crack nucleates at depth, ruptures the desiccated layers, and subsequently propagate bi-laterally away from the origin along curved paths. Minor damage caused to the polygon wall during recovery (*left corner* in photograph) is not presented in the surface-morphology drawing. Note in the photograph the dendrite-like plume that is divided into branches that are further sub-divided into minute barbs (from Weinberger 1999)

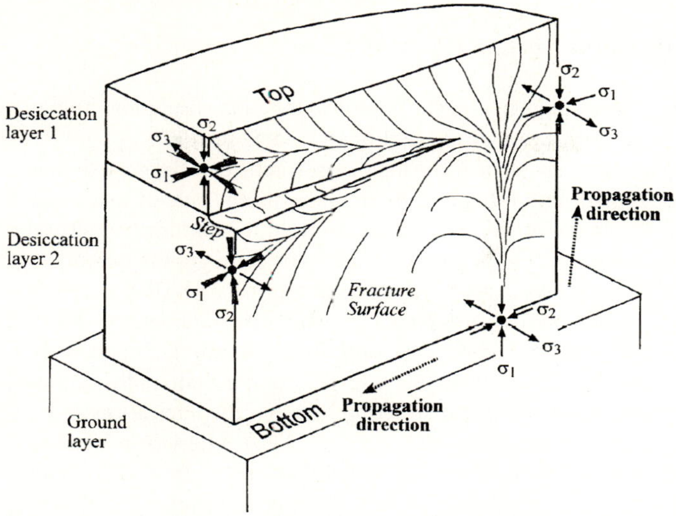

Fig. 2.3. c Schematic illustration of stress distribution throughout a polygon wall. The least compressive principal stress σ_3 is everywhere perpendicular to the crack plane, but the greatest compressive principal stress σ_1 and the intermediate compressive principal stress σ_2 vary in orientation from place to place on that plane. Next to the crack origin, where the crack extends predominantly vertically, σ_1 is vertical and perpendicular to bedding, on account of the weight of the overlying sediment. Away from the crack origin at the region where the crack extends predominantly laterally, the in-plane horizontal stress due to contraction seems to be more significant than the in-plane vertical stress due to the weight of the sediment, and σ_1 and σ_2 switch directions. Plume formation and segmentation along layer boundaries are associated with the indicated rotations of the local principal stresses (from Weinberger 1999)

iccated layers before they significantly extend laterally and sequentially form the polygonal patterns (Fig. 2.3b,c).

Thus, the consistent location of crack origins at the bottom of the polygons strongly suggests that stress concentration due to flaw discontinuities and layer boundaries play a fundamental role during mud-crack nucleation. Consequently, the stress gradient due to drying (possibly related to capillary forces) may be less important for mud-crack nucleation than has been previously postulated. Returning now to the Griffith model, Weinberger explains that mud-crack propagation consumes energy in the form of surface energy for the creation of a new crack surface. This energy comes from the release of elastic strain energy within the drying mud. In this mechanism, the only mechanical energy available to drive a crack is the change in elastic strain energy (Eq. 1.19), which must decrease while the surface energy increases (Eq. 1.20) during crack growth. Adhesive forces along the bottom of the polygons resist the horizontal contraction of the mud. This resistance gives rise to stresses along the bottom and causes the elastic strain energy to be stored. Since crack growth strongly depends on the stored energy, the boundary effect probably plays a key role not only for crack nucleation but also for crack propagation. Hence, Weinberger (1999) describes nucleation and growth processes where actual c_i and c_{cr} (Fig. 2.1) are not identified. It is, however, quite possible that nucleation in the mud took place at sites that ranged between c_i and c_{cr}. Note that the large grains along the bottom of the polygons exceed the size of 0.062 mm (Fig. 2.3a). Thus, his interpretation of both the nucleation and growth processes is quite convincing. See the experiment on profiles of "mud cracks" in Sect. 2.2.3.9.

2.2.2.4
Fracture Initial Growth in Dolomite Layers

Weinberger (2001a) investigates the well-developed surface morphology of joints in the Soreq Formation, Judea Group (Arkin and Hamaoui 1967; Upper Cretaceous, central Israel) that uniquely records the history of fracture initial growth in layered rocks. Weinberger (2001a) uses fractography to elucidate the role of macroscopic flaw distribution in the shape of cavities, on the jointing mechanism in natural conditions, and to examine the effect of bedding interfaces on joint growth and arrangement. The formation of the cavities is related to the dissolution of earlier anhydrite nodules in the dolomite by marine-derived water. The study by Weinberger (2001a) is very detailed and is cited extensively below. We differ from Weinberger in using the term nucleation. For us, joint initiation from $2c_{cr}$ is growth rather than nucleation (as mentioned above). There are cases where a distinction between the nucleation and initial growth stages can be made reasonably well (e.g., Fig. 2.2b), whereas in other instances this distinction is unclear. It could be that nucleation started from certain $2c_i$ sites within the cavities while the anhydrite nodules still resided in them (Weinberger 2001a). It appears more likely however that both nucleation and initial growth occurred from $2c_{cr}$ locations around the peripheries of the cavities after the removal of the anhydrite nodules.

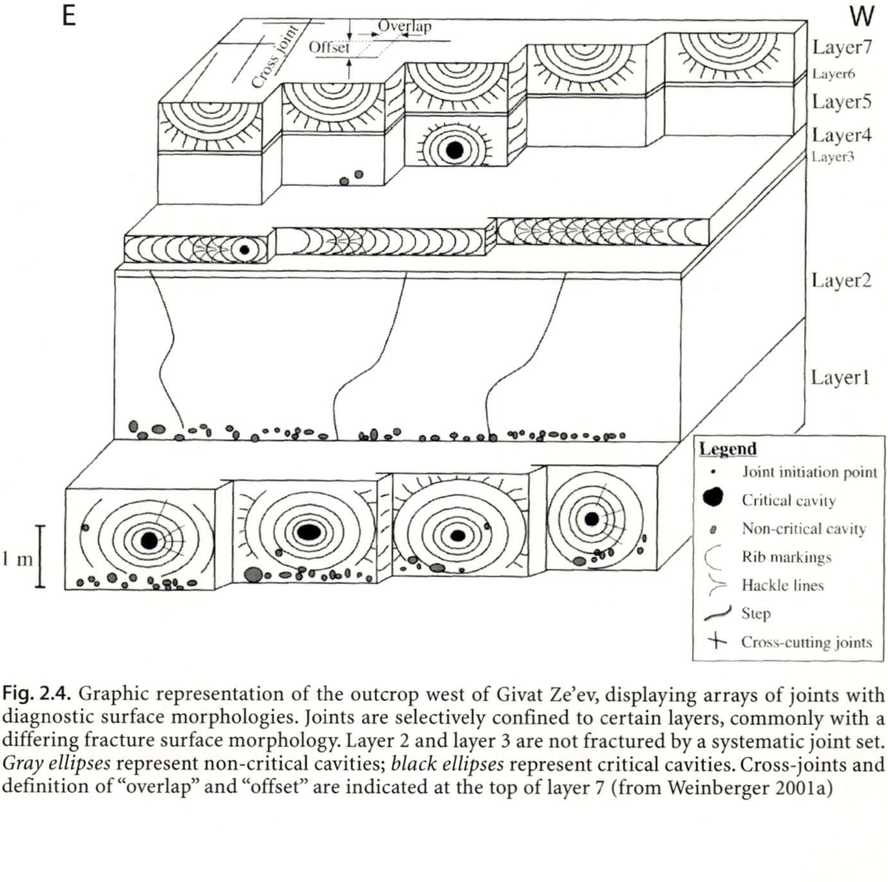

Fig. 2.4. Graphic representation of the outcrop west of Givat Ze'ev, displaying arrays of joints with diagnostic surface morphologies. Joints are selectively confined to certain layers, commonly with a differing fracture surface morphology. Layer 2 and layer 3 are not fractured by a systematic joint set. *Gray ellipses* represent non-critical cavities; *black ellipses* represent critical cavities. Cross-joints and definition of "overlap" and "offset" are indicated at the top of layer 7 (from Weinberger 2001a)

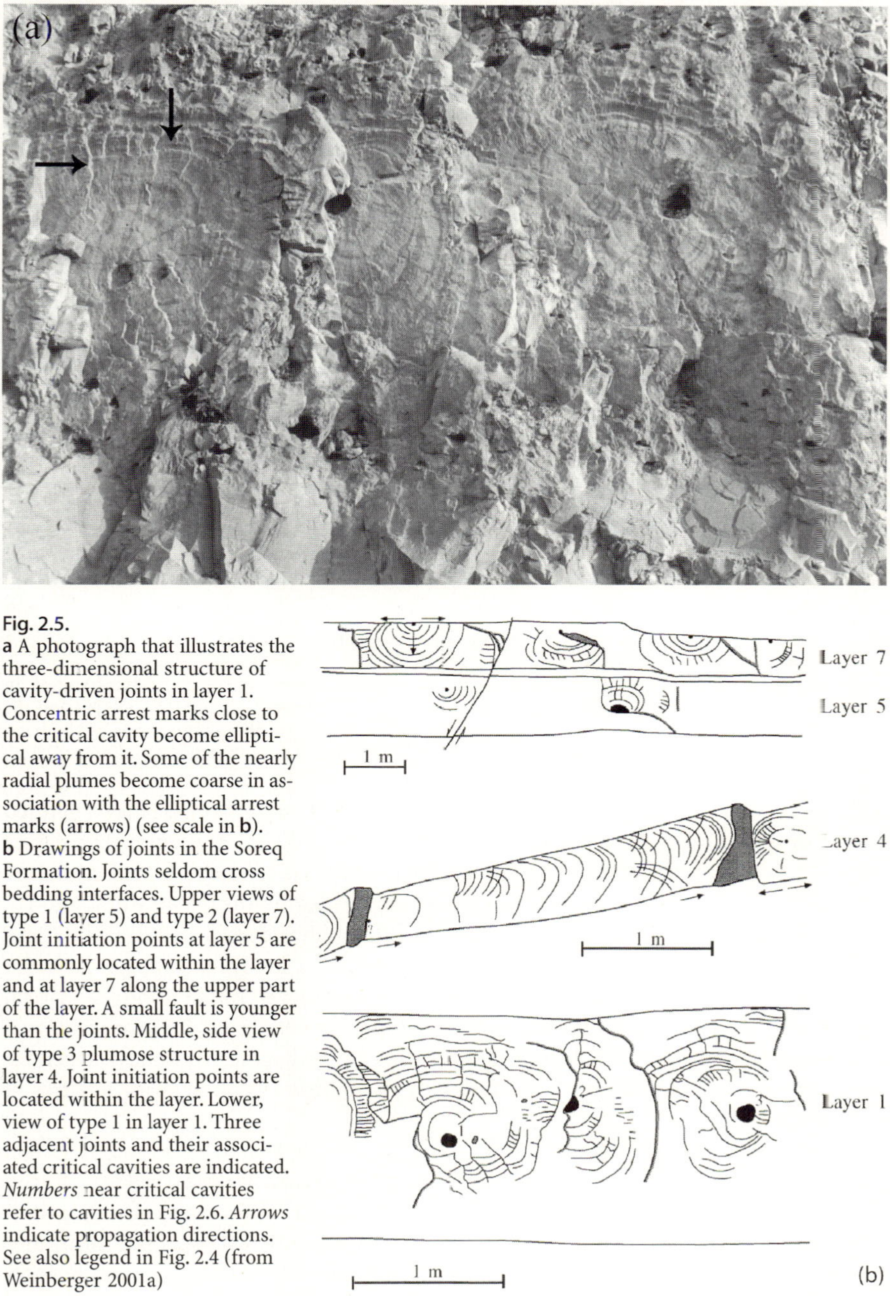

Fig. 2.5.
a A photograph that illustrates the three-dimensional structure of cavity-driven joints in layer 1. Concentric arrest marks close to the critical cavity become elliptical away from it. Some of the nearly radial plumes become coarse in association with the elliptical arrest marks (arrows) (see scale in **b**).
b Drawings of joints in the Soreq Formation. Joints seldom cross bedding interfaces. Upper views of type 1 (layer 5) and type 2 (layer 7). Joint initiation points at layer 5 are commonly located within the layer and at layer 7 along the upper part of the layer. A small fault is younger than the joints. Middle, side view of type 3 plumose structure in layer 4. Joint initiation points are located within the layer. Lower, view of type 1 in layer 1. Three adjacent joints and their associated critical cavities are indicated. *Numbers* near critical cavities refer to cavities in Fig. 2.6. *Arrows* indicate propagation directions. See also legend in Fig. 2.4 (from Weinberger 2001a)

Cavities associated with joint initial growth in the dolomite layers are termed hereafter "critical cavities" and their respective fractures are termed "cavity-driven joints". A cavity in the present context approximates the "critical flaw" (Fig. 2.1a,b,d): The cavity's horizontal semi-axis, C_h (corresponding to c_{cr}) and vertical semi-axis, C_v (cor-

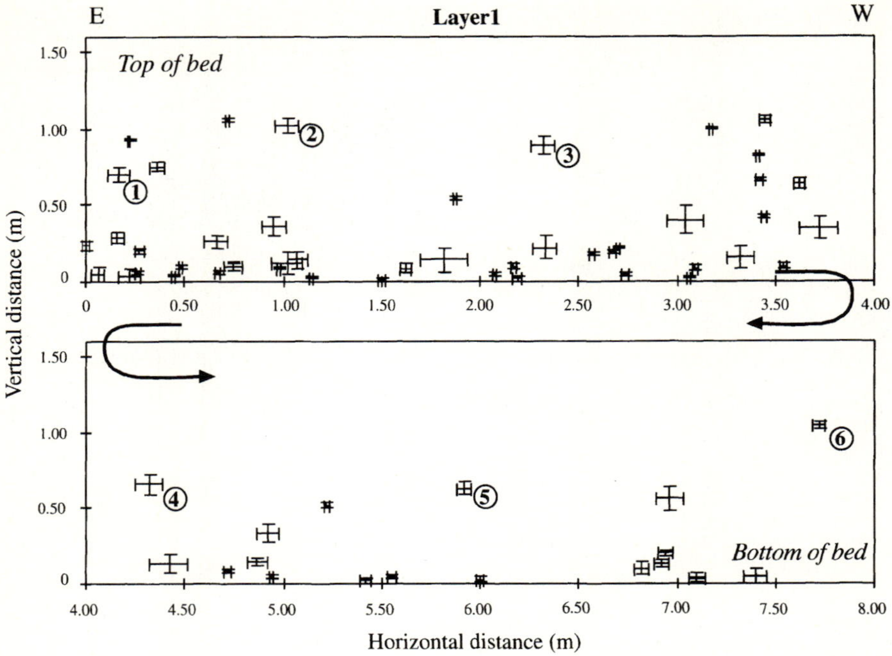

Fig. 2.6. Representative distribution and dimensions of cavities in layer 1 (Fig. 2.4 and 2.5b) along an 8 m E-W traverse. The small steps along the traverse (Fig. 2.4) are not indicated. *Bars* represent the in-plane axes of the ellipsoidal cavities. Cavity centers are located at the intersections of the bars. *Numbers* denote critical cavities that are associated with joint initial growth (from Weinberger 2001a)

responding to a_{cr}) are measured from the center of a cavity to its periphery, defined by the intersection of the respective fracture plane and the cavity rim. The cavity "depth", C_d, is measured from the center of the cavity to its rim normal to the fracture plane. The characteristic size of a cavity, Z, is defined as the mean of C_h and C_v. Fourteen adjacent critical cavities in layer 1, one in layer 5, and five in layer 7 and their associated joints were carefully measured for the analysis (Fig. 2.4–2.5). For comparison, all cavities along an 8 m long traverse in layer 1, including six critical cavities and 58 non-critical cavities, were also studied in detail (Fig. 2.6). An additional interdependent variable, the degree of isolation, can be derived from these measurements. These variables are analyzed below in an attempt to characterize features of flaw criticality.

Shape. Critical cavities are approximately spherical, with $C_h = 45 \pm 20$ mm, $C_v = 43 \pm 20$ mm, $C_d = 45 \pm 23$ mm, and in-plane $C_h / C_v \approx 1$ ratios. Non-critical cavities are also approximately spherical, with $C_h = 35 \pm 27$ mm, $C_v = 30 \pm 23$ mm, and $C_d = 32 \pm 23$ mm (Fig. 2.7a). Noticeably, several eccentric cavities ($C_h / C_v \approx 2$) appearing close to the base of layer 1 are non-critical, e.g., an ellipsoidal cavity at a distance of 4.40 m along the traverse (Fig. 2.6). If a cavity deviates from a perfect sphere, the major semi-axis commonly coincides with C_h. This is particularly the case for critical cavities, where only in one case the difference between C_v and C_h is larger than 3 mm.

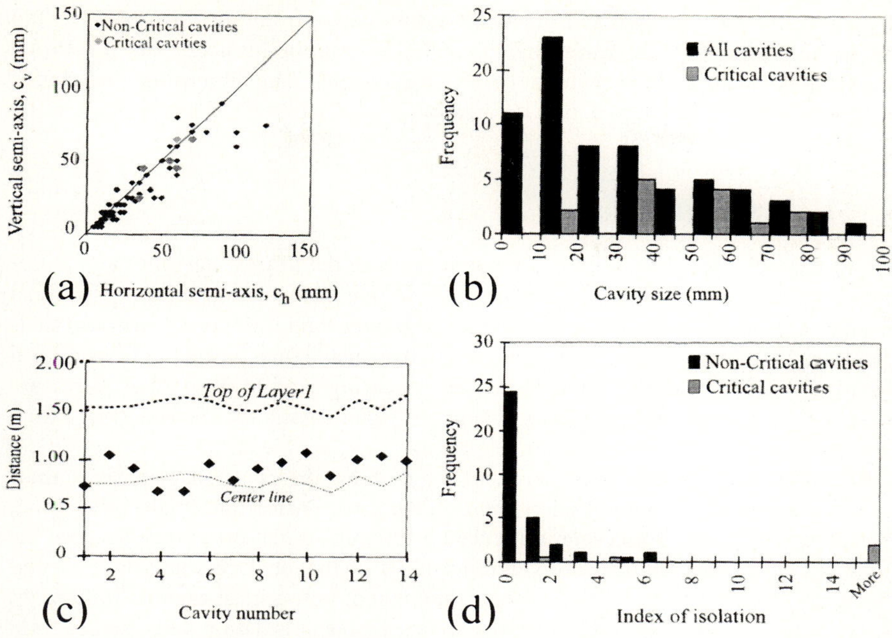

Fig. 2.7. Analysis of cavity shape, size, and distribution in layer 1. **a** In-plane horizontal vs. vertical semi-axes of critical (*gray*) and non-critical (*black*) cavities along an 8 m E-W traverse in layer 1 (Fig. 2.6). **b** Histograms of sizes of critical (*gray*) and non-critical (*black*) cavities in plane 1. **c** Origin location of fourteen cavity-driven joints with respect to the local layer thickness. **d** Histograms of isolation indices of critical (*gray*) and non-critical (*black*) cavities along an 8 m E-W traverse, layer 1 (Fig. 2.6) (from Weinberger 2001a)

Size. The mean characteristic size of all critical cavities ($Z = 44$ mm) is larger than that of non-critical cavities ($Z = 33$ mm). More details are gained from histograms of cavity sizes in layer 1 (Fig. 2.7b). The frequency of critical cavities is different from that of all cavities; relatively small and very large cavities are not associated with joint initiation. Indeed, no critical cavity is among the top five largest and the fifteen smallest cavities.

Location within layer. To test the null hypothesis that cavities (critical and non-critical) are randomly distributed with layer 1, the section presented in Fig. 2.6 was divided into 48 area units of 0.5×0.5 m each, and the number of cavities in each unit counted. This procedure provided the observed data for a standard Chi-square test on this cavity population and helped to calculate the expected data based on a Poisson distribution (Davis 1973), with λ equal to the mean size of all cavities. The test concluded that the null hypothesis should be rejected at 95% confidence. Thus, cavity distribution is non-random, reflecting the tendency of cavities to be more abundant toward the layer base. Additionally, of all cavities, critical cavities are generally located closest to the top of the layer. Only 5 out of 58 non-critical cavities are located more closely to the top of the layer than the critical cavities (Fig. 2.6 and 2.7c). In addition, all five are smaller in size than the smallest critical cavity ($C_b = 15$ mm, $C_v = 15$ mm and $C_d = 10$ mm).

Isolation. For each cavity, the distance, Dj, was measured from the center of cavity j to the center of the in-plane nearest neighbor, and the number of neighbors Sj within a radius of 0.5 m around the center of cavity j was counted. The dimensionless index of in-plane isolation for cavity j, Ij, is defined by

$$Ij = \frac{Dj}{Zj} \frac{1}{Sj+1} \tag{2.3}$$

A high isolation index implies a large distance to the nearest neighbor and a limited number of neighbors. The results indicate that the isolation indices of critical cavities are typically one order of magnitude greater than those of non-critical cavities (Fig. 2.7d), i.e., critical cavities are much more isolated than most of the non-critical cavities. Noticeably, several isolated cavities appearing close to the base of layer 1 are non-critical, e.g., a small cavity at a distance of 6.00 m along the traverse (Fig. 2.6).

Initial growth of joints. Weinberger (2001a) found that the cavity shape plays a minor role in governing the joint initiation points in the Soreq Formation. On the other hand, the cavity size has a noticeable influence on the joint initiation points, because the frequency of critical cavities differs significantly from that of all cavities (Fig. 2.7c) and the mean size of critical cavities is larger than that of non-critical cavities. Noticeably, for two closely located cavities of similar shape, degree of isolation and distance from the layer base, the largest cavity is the one associated with joint initiation (e.g., compare critical cavity 2 and its nearest neighbors to the east at a distance of 0.7 m along the traverse, Fig. 2.6). These observations suggest a cavity size effect, whereby larger cavities preferably fail before smaller ones. This size effect is complicated by post-fracturing dissolution processes that might obliterate the exact cavity size at initiation. Nonetheless, this later dissolution might enlarge all cavities, including the non-critical cavities. Although theoretically the stress concentration factor (SCF-q in Eq. 1.22) of circular hole is 3.0, independent of the size (Lawn 1993, p. 3), actual strength measurements in rock mechanics experiments indicate that sample strength decreases with increasing hole size (Anderson 1995, p. 339). Therefore, the SCF effects of the larger hole act over a wider distance, indicating that the volume over which the stress acts is important for hole failure and rock strength. Other studies have also shown that rock strength decreases with increasing flaw size and flaw density (Wong et al. 1996; Wong and Chau 1998; Bazant and Planas 1998). These size effects, however, are obscured by the fact that they depend on the deformation processes, which in turn depend on the loading conditions (Li et al. 1999). Indeed, in the Soreq Formation, only one critical cavity is among the top five largest cavities, indicating that other factors play a more significant role than the size during jointing.

The distribution of cavities in layer 1 is non-uniform, with cavities clustered toward the layer base. This, in turn, leaves relatively isolated cavities near the middle of the layer, many of which are critical (Fig. 2.6). For two cavities of similar shape, size and distance from the layer base, the more in-plane isolated cavity is the one associated with joint initiation (e.g., compare cavity 6 and non-critical cavity at a distance of 3.45 m along the traverse, Fig. 2.6). The analysis of the stress in the vicinity of a hole in an elastic material indicates that the stress distribution is markedly affected by the hole only within an area of about three radii from the center of the hole (Timoshenko and Goodier 1951). Weinberger (2001a) arrives at the interesting implication that closely located cavities may inhibit joint initiation because of the stress perturbation induced by adjacent cavities. On the other hand, in more plausible configurations, cavity-driven joints in less dense regions near the middle of the layer might communicate with other cavities, facilitating joint growth caused by stress enhancement.

Three types of ripple marks. Weinberger (2001a) distinguished three types of ripple marks (Fig. 2.5). Type 1 consists of circular-to-elliptical marks and is formed in layers with plentiful cavities. Type 2 consists of semi-circular-to semi-elliptical marks and is formed only in "cavity-free layers". Type 3 consists of marks with approximately similar curvature and is formed in relatively thin layers (<0.4 m).

Summary: Joint growth after initiation from c_i and c_{cr} sites. In summarizing the observations from the study by Weinberger (2001a), the analysis of cavity geometry and distribution reveals that joint initiation at c_{cr} locations is governed by isolated, relatively large cavities, preferably located close to the bedding top and by stress gradients during jointing. In the absence of macroscopic cavities, sites of joint initiation vary in size from c_i to c_{cr} along bedding interfaces (layer boundaries). Consequently, joints typically show several forms of growth, depending on the abundance and spatial distribution of cavities within the layers. In layers with plentiful cavities, joints preferably initiate at critical cavities in the layer interior and propagate vertically upward and downward and horizontally and form circular to elliptical fractures that abruptly end (Fig. 2.4) or more commonly arrest either at layer boundaries or at surfaces of previous joints (Fig. 2.5b). In layers free of cavities, joints commonly nucleate at layer boundaries and propagate downward toward the layer base and adjacent joints (Fig. 2.4, layer 7). Alternatively, they may nucleate at surfaces of previous joints or contacts between joints and layer boundaries (Bahat 1991a, Fig. 4.8).

The various forms of growth in cases where cavities are not apparent are summarized schematically in Fig. 2.8a,b A–H, first defining joint dimensions (Fig. 2.8a). Figure 2.8b A–D show fractographic manifestations of various initiation and growth styles without fringes in single-layer joints. These joints often form fringes along one or two layer boundaries (Fig. 2.8b E, F). Less common are *en echelon* fringes along the distal continuations of joints (Fig. 2.8b G), possibly because the fracture slanting mechanism (Sect. 2.2.5.2) does not promote segmentation there. Very rare are hackle fringes or cuspate hackle fringes (Fig. 2.42) along the distal continuations of single-layer joints in sedimentary rocks (Fig. 2.8b H). In a rare case of a hackle fringe in a sedimentary layer (Fig. 2.30a), the fringe does not occur in a distal position. This general observation, based on thousands of inspections of joint fractographies, provides a strong argument in suggesting that the joints under consideration mostly form under low or very low effective stresses (Sect. 1.5.3).

The division into two forms of fracture growth by Weinberger (2001a) corresponds quite well to an analogous division into two types of fracture growth in technological materials. Breakage in good quality glass bottles most often initiates at the outside wall and propagates inside (towards the center of the bottle), forming semi-circular fracture (Fig. 2.1d), because stress concentrators (e.g., cavities) inside the body are absent. On the other hand, in low-quality materials that contain discontinuities (e.g., cavities/voids), fracture may start from inside the glass body (e.g., Bahat 1991a, Fig. 2.34d) and propagate along directions complying with constraints imposed by the boundary conditions, much as they control fracture in sedimentary layers (Fig. 2.8).

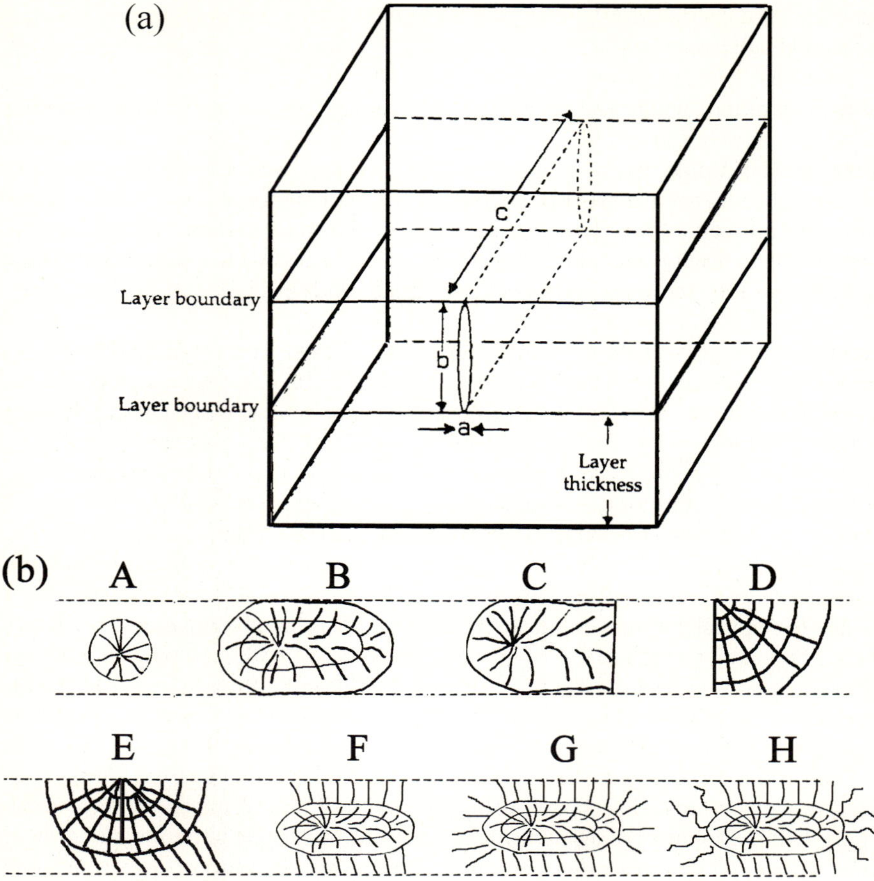

Fig. 2.8. Several styles of joint growth in a rock layer. **a** Definition of thickness, width and length dimensions (*a–c*, respectively) of a single-layer joint. **b** Several growth forms of joints, as described by plumes and ripple marks, from left to right: *A*: An initial circular joint; *B*: A full joint growth to an elliptical mirror, limited by the layer boundaries; *C*: An advanced quasi-elliptical joint that arrests along an earlier fracture; *D*: An advanced joint that initiates at a corner formed by a layer boundary and a previous joint; *E*: A semi-elliptical joint initiation at the upper layer boundary and propagation downward creating an *en echelon* fringe; *F*: A full joint growth to an elliptical mirror, forming *en echelon* fringes along the layer boundaries; *G*: A joint like *F* that forms *en echelon* segmentation at its distal ends; *H*: A hypothetical joint *F* that forms hackles at the distal ends (modified after Bahat 1998a and Weinberger 2001a)

2.2.3
Fractographic Experiments

This section brings together a series of fractographic experimental studies (mostly since 1991) that concern fracture areas beyond the critical flaw (Fig. 2.1). The first studies (Sect. 2.2.3.1–2.2.3.7) pertain to fractures that are contained within the boundaries of the mirror plane, while the other studies (Sect. 2.2.3.8–2.2.3.13) relate to fractures that reach the fringe beyond these boundaries. The fractographic results of these studies have implications regarding various tectonofractographic issues, which are discussed in subsequent sections.

2.2.3.1
Experimental Arrest Marks

According to Guin and Wiederhorn (2003), Michalske (1977) carried out on soda lime silicate glass microscope slides the following experiment (Fig. 2.9a–e). First, he propagated a crack in water at a stress intensity factor K_I of 0.375 MPa $m^{1/2}$, which was clearly above the static fatigue limit K_0 (Fig. 4.34). Then, he reduced the applied stress intensity factor to 0.225 MPa $m^{1/2}$, a value clearly below the static fatigue limit, which was inferred from crack growth studies on soda lime silicate glass (Wiederhorn and Bolz 1970). He held the stress intensity for a period of 16 h. Finally, he increased the stress

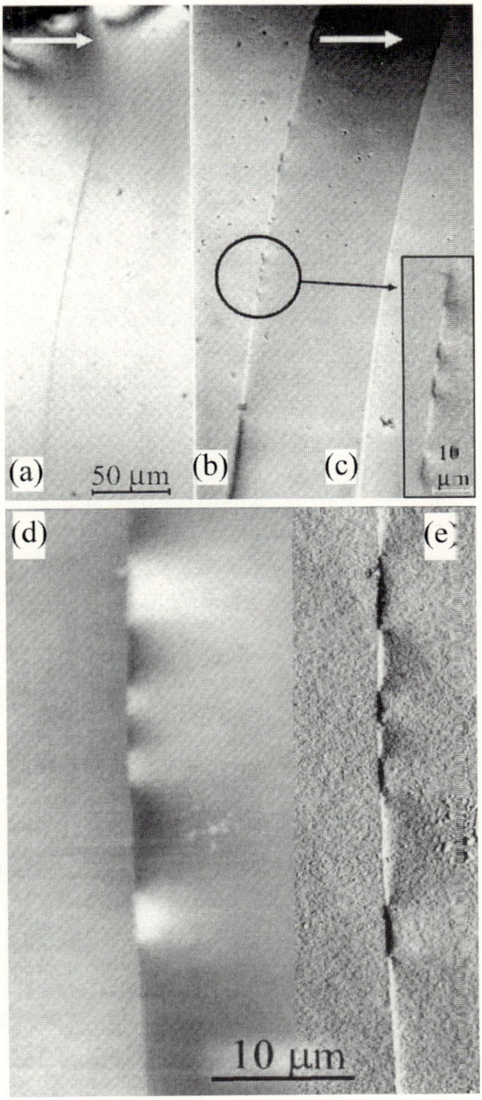

Fig. 2.9.
Optical micrographs obtained by using the height mode, to illustrate the crack deflection along the crack front of crack arrest marks left on fracture surfaces as a consequence of holding the crack for a period of time at a stress-intensity factors that lies below the "fatigue limit". **a** $K_{Ih} = 0.24$ MPa $m^{1/2}$, $t_h = 67$ h, $K_{Ir} = 0.37$ MPa $m^{1/2}$; **b** $K_{Ih} = 0.12$ MPa $m^{1/2}$, $t_h = 48$ h, $K_{Ir} = 0.37$ MPa $m^{1/2}$; **c** $K_{Ih} = 0.12$ MPa $m^{1/2}$, $t_h = 1$ h, $K_{Ir} = 0.35$ MPa $m^{1/2}$, where K_{Ih}, t_h and K_{Ir} are the hold K_I below the static fatigue limit, the hold time and the hold K_I above the static fatigue limit, respectively. The *white arrows* on the figure give the direction of crack propagation. **d** and **e** images by atomic force microscope from the *circled region* of the crack arrest mark shown in **b** using the deflection mode to show the change in slope along the crack front. The crack approaches the arrest mark on a common plane. On *repropagation*, the crack has discontinuously split, some areas along the crack front lying above (*white*) the incoming surface, some areas lying below (*dark gray*) the incoming surface. In **e** the *bright* or *dark hands* indicate a sharp alternate change in crack growth (from Guin and Wiederhorn 2003)

intensity factor to the value used for crack propagation, 0.375 MPa $m^{1/2}$. The time to repropagate the crack was about 2 000 s. Michalske found crack arrest marks on the fracture surface at the site of the arrested cracks that he attributed to restarting crack growth. No such marks were observed for a crack that had its stress intensity factor momentarily reduced and then increased again.

Following up on Michalske's (1977) study and duplicating his experiment, Wiederhorn et al. (2002) and Guin and Wiederhorn (2003) found that the arrest marks appeared not to be the result of blunting, but represented what we term a "split-discontinuity" in the crack surface so that after the hold period below the crack growth threshold, the crack developed waviness that was not there when the crack was arrested (Fig. 2.9b,d,e). By matching both halves of the fracture surface, these authors showed that the split upper and lower surfaces matched across the marks left at the crack front. Within the experimental accuracy of the technique, ±2 nm, crack blunting did not occur. The discovery by Wiederhorn et al. (2002) and Guin and Wiederhorn (2003) of the split-discontinuity shows a subtle but significant mixed mode I and III loading, additional to mode II along arrest marks that form when the crack *restarts* growing (Fig. 2.9e). The implication of this section is discussed in Sect. 2.2.4.2.

2.2.3.2
Experimental Undulations

Richter and Kerkhof (1994) used a method of superposing shock waves or continuous waves onto a static or quasi-static tensile field, causing a crack to propagate in soda lime silicate glass from a starter crack (edge-notched) induced into the glass by thermal or mechanical techniques. This procedure allowed them to create visible, concentric stress wave markings (undulations) on fracture surfaces. The undulations formed in successive positions of crack fronts at definite points in time by applying the stress waves at different modulated frequencies, and this enabled them to calculate the velocity of crack propagation from 10^{-6} m s^{-1} up to the maximum velocity of 1 540 m s^{-1} (Kerkhof 1975). This method has been successfully applied to other materials as well (Richter and Kerkhof 1994).

Figure 2.10 shows a fracture surface of a single edge-notched tensile specimen generated at a stress wave frequency of 5.02 MHz. The fracture surface displays a blunt, starter crack, SC and zones of mist, M, hackle, H, and crack branching, B. The undulations convex in the direction of propagation; they are superimposed by thinner and sharper Wallner lines at different orientations. Also visible are a few straight striae in the mist and hackle zones. In this method, an increase in distances between undulations indi-

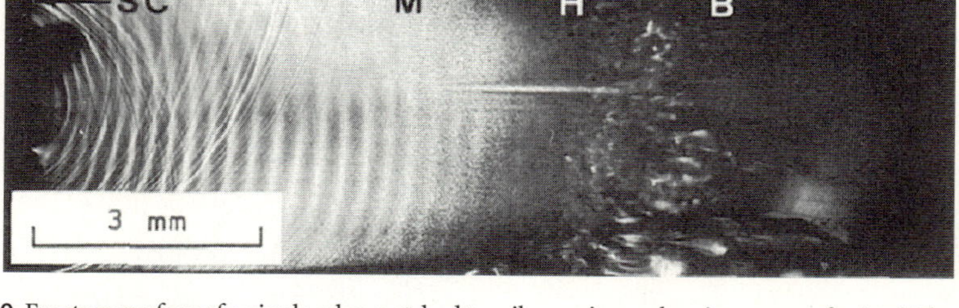

Fig. 2.10. Fracture surface of a single edge-notched tensile specimen showing zones of mist, *M*, hackle, *H*, and crack branching, *B*. Stress wave frequency: 5.02 MHz. *SC* = starter crack (after Kerkhof and Richter 1969)

cates increase in crack velocity. In this example, a relatively high load had to be applied to start propagating the blunt starter crack. Consequently, the crack acceleration was high, leading to wide intervals between the undulations almost throughout the fracture surface, and the maximum crack velocity of 1 540 m s^{-1} was attained. The crack branched beyond the hackle zone after having passing a certain distance at almost a constant crack velocity. After a phase of crack branching, a region of smoother fracture surface occurred

2.2.3.3
Oscillating Cracks

Yuse and Sano (1993, 1997) applied thermal stress along thin borosilicate cover glasses that varied in size around 100 × 20 × 0.1 mm by heating them to high temperature, T_h (maximum 400 °C), in a hot region. The cover glasses were then moved slowly down into a cold region, T_l (18 °C) at velocity v (maximum 50 mm s^{-1}), while the distance between the upper hot region and the lower cold region was fixed. If v was slow enough, the part of the sample in the hot region was at temperature T_h and the part in the cold region at temperature T_l, imposing a temperature difference $\Delta T \approx T_h - T_l$, and maximum thermal stress between the two regions (Fig. 2.11a). Thus, the control variables of the quasi-static crack propagation were ΔT, v and W, the sample width. Various types of crack patterns were formed by different control parameters. Generally, increasing the values of the control parameters changed the crack morphology from simple to complex, falling into three types with increasing complexity:

1. Straight cracks (Fig. 2.11b).
2. Oscillating cracks (Fig. 2.11c).
3. Branched cracks (Fig. 2.11d,e).

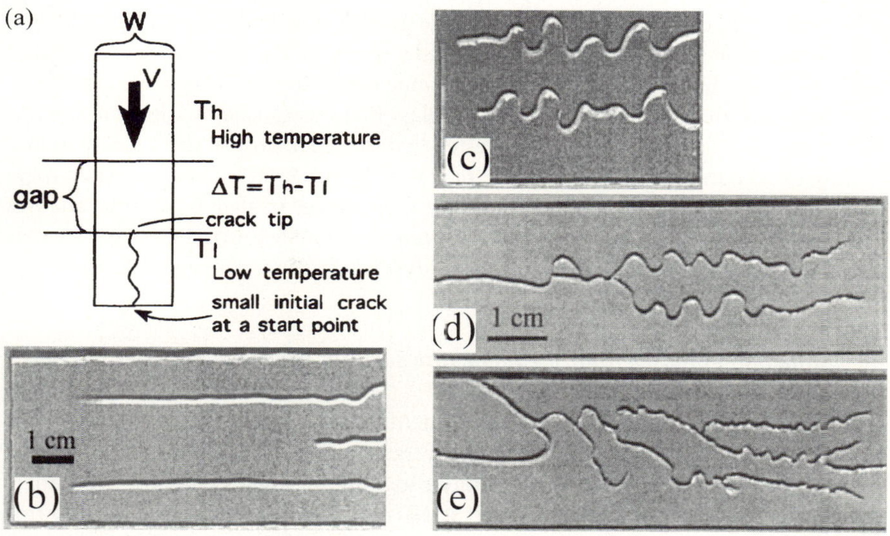

Fig. 2.11. a Configuration of the experiment. A thin sample plate moves slowly across the temperature gradient gap in the direction of the *arrow* with velocity v (after Yuse and Sano 1993). **b–e** Several crack patterns. **b** Straight cracks; **c** two oscillating cracks; **d, e** branched cracks (after Ferney et al. 1999)

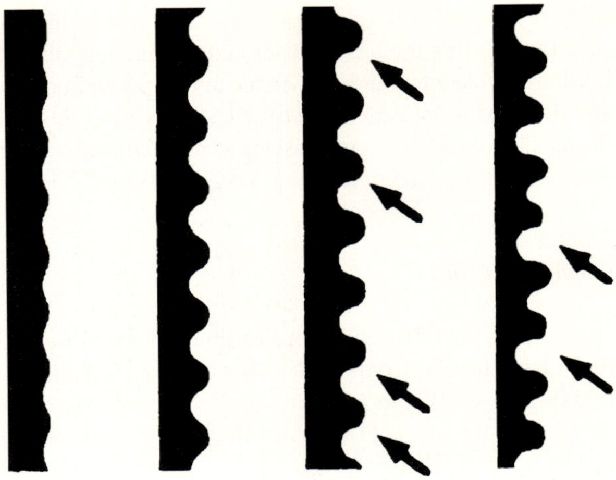

Fig. 2.12. The morphology of oscillating crack changes from simple to complex with increasing control parameters. *Arrows* show initial cusps. The width is scaled by two (after Yuse and Sano 1993)

We shall concentrate on the oscillating cracks. While both undulations (Sect. 2.2.3.2) and oscillating cracks bend along the crack propagation direction (Fig. 2.11 and 2.12, respectively), undulations appear to superpose as small, secondary bends on primary crack surfaces, while oscillating cracks are larger and exhibit the actual shape of the primary crack. For v and ΔT near the onset of oscillation, the oscillating cracks were almost sinusoidal (Fig. 2.12a). With increasing v or ΔT, the shapes resembled a series of semi-circles (Fig. 2.12b). After that, the crack became asymmetric in the propagation direction. For large v or large ΔT the crack propagated in an almost horizontal direction at a very high speed until it stopped at some distance from the side edge, then grew slowly upward, forming a cusp (Fig. 2.12c). For larger parameters, additional oscillations appear quasi-periodically or double periodically with cusps (Fig. 2.12d).

Yuse and Sano (1997) demonstrate that under severe thermal stresses, cusps replace oscillations within the range where they become asymmetric. Murgatroyd (1942) observed symmetric cusp marks in fractured glass that was caused by a low-energy impact (Bahat 1991a, Fig. 2–20a). He suggested that the crack propagated upward before coming to rest and resumed its course in a slightly downward direction when it recommenced. An asymmetric cusp-like profile was induced in glass by thermal stresses (Bahat 1977). These observations suggest a need to reconcile the occurrences of asymmetric ripple marks and cusps in both slow and fast fracture propagation.

2.2.3.4
The Distinction of Tensile from Shear Fracture Surfaces in Laboratory Tests of Chalk

A full understanding of developments in fracture geology is often hampered by our limited ability to distinguish between "pure" tensile loading (mode I) and processes controlled by shear stresses and mixed mode loading (of modes I, II and III in different combinations), particularly in small sizes (mm–cm). This section relates to this problem. A detailed fractographic technique that was applied on experimentally fractured chalk enabled the establishment of a clear distinction between tensile and shear surfaces (Fig. 2.13–2.17). In their study, Bahat et al. (2001a) tested chalk cylinders

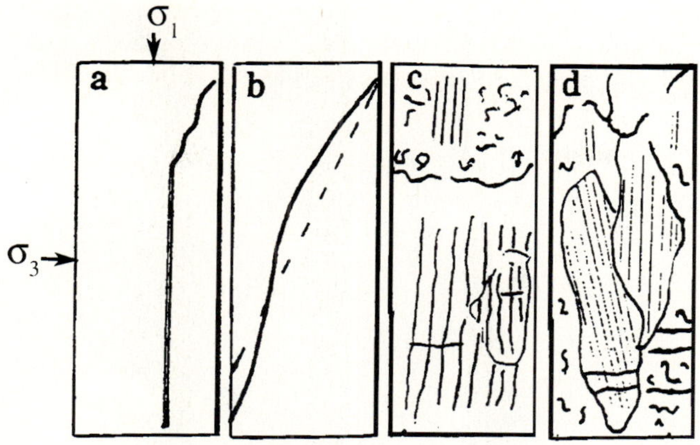

Fig. 2.13. Various types of brittle fracture developed experimentally in chalk. **a**, **b** Profile sections. **a** Longitudinal splitting by uniaxial stress in continuation of initial conic fracture induced by boundary effects at the end of the sample. **b** Shear fracture by a triaxial test, showing formal straight conjugate cut (*dashed line*) and actual curved fracture. **c**, **d** Parallel sections. **c** Derived by uniaxial test, showing a shear surface above, separated by a boundary from a tensile surface subdivided into domains below it. **d** Meso-fracture, cutting throughout the sample, formed by a triaxial test, showing adjacent domains marked by ridges and grooves in different orientations (from Bahat et al. 2001a)

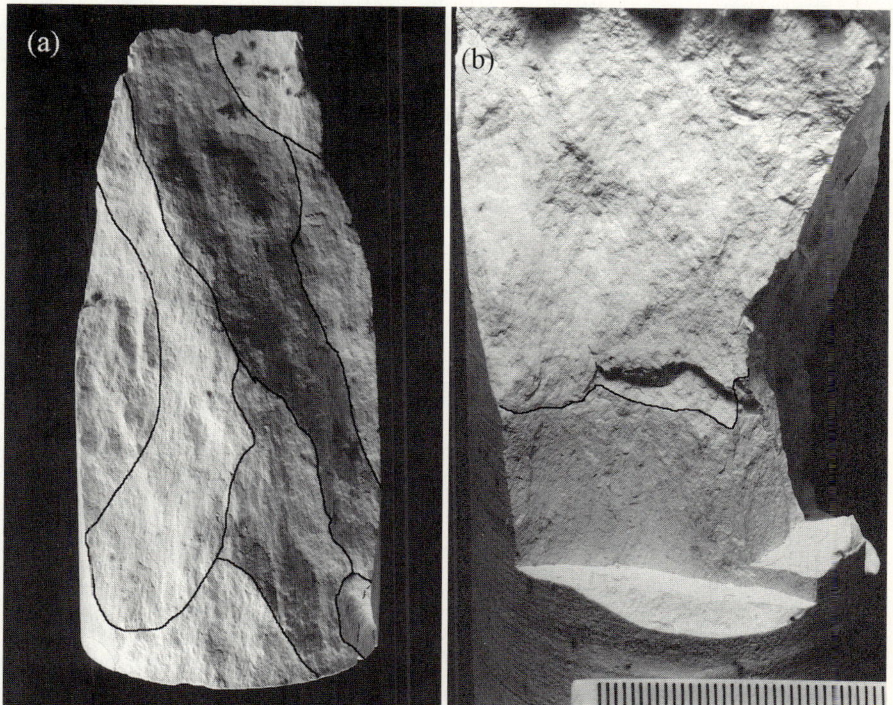

Fig. 2.14. Photographs of fractured chalks under triaxial stresses (cm scale). **a** A shear surface divided into distinct domains in which ridges and grooves slightly change their orientation. **b** A section shows three parts from bottom to top, the original cylinder, a tensile crack (*gray*) and a large shear crack (*white*), respectively. The shear part displays conjugate Lüders' bands (from Bahat et al. 2001a)

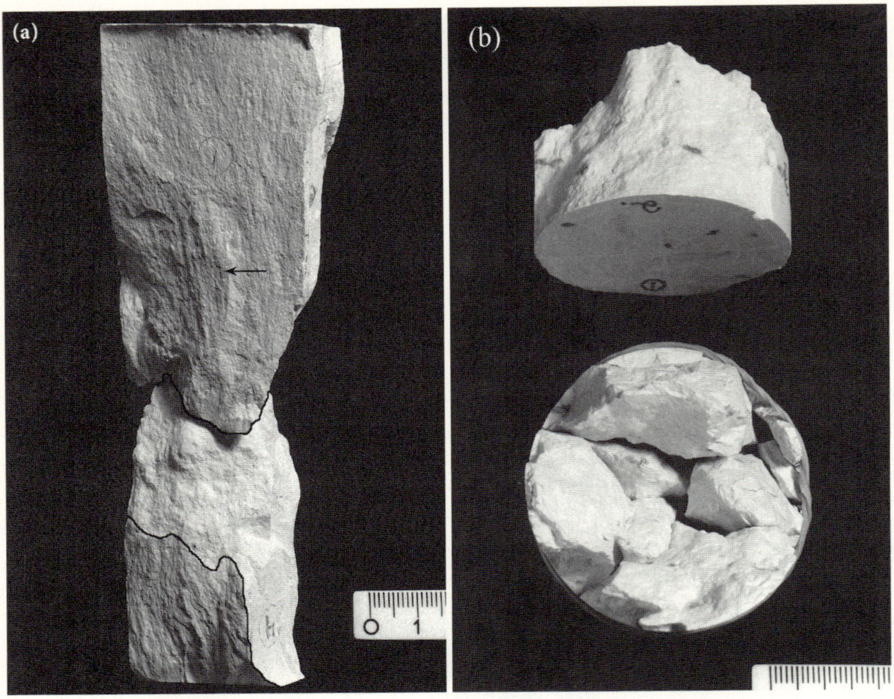

Fig. 2.15. Photographs of fractured chalks under uniaxial stresses (cm scale). **a** A longitudinal split dominated by an axial tensile fracture. The two tensile parts are displaced in order to show a late inclined shear fracture (between upper and lower boundaries) that crossed the split. Sub-axial plumes mark the split. **b** Reconstruction of an original cylinder (at upright position, below) and the quasi-cone (axis parallels to **a**) (from Bahat et al. 2001a)

100 mm long and 52 mm in diameter under both uniaxial and triaxial loading. At the end of this section, we present results of a recent study on the pore radii of this chalk (Rabinovitch et al. 2003a).

Summary of results. The results of the above study are condensed below as follows.

1. The drying and desiccation procedures of the chalk used in the study resulted in high compressive strength results ($\sigma_1 - \sigma_3$) ranging from 37.7 MPa to 52.6 MPa.
2. With the exception of a few cylinders that fractured either by a single shear fracture or by strongly dominated longitudinal splittings, most samples under both triaxial and uniaxial loading failed by a combination of shear and tensile fracture (Fig. 2.13–2.15).
3. The authors identified several "key surface morphologies" according to shear and tensile categories (Fig. 2.13). Shear features include a sugary texture, ridges and grooves (RG) and steps (Fig. 2.14). The RG appear on sheared white sugary, ragged surfaces that form curved topographies within boundaries of individual domains (of about 3 to 10 cm^2 in size). RG in neighboring domains may form various angular relationships with the cylindrical axis and are either parallel or sub-parallel to the fracture dip. An individual domain on a shear surface is often subdivided into sub-domains (of about

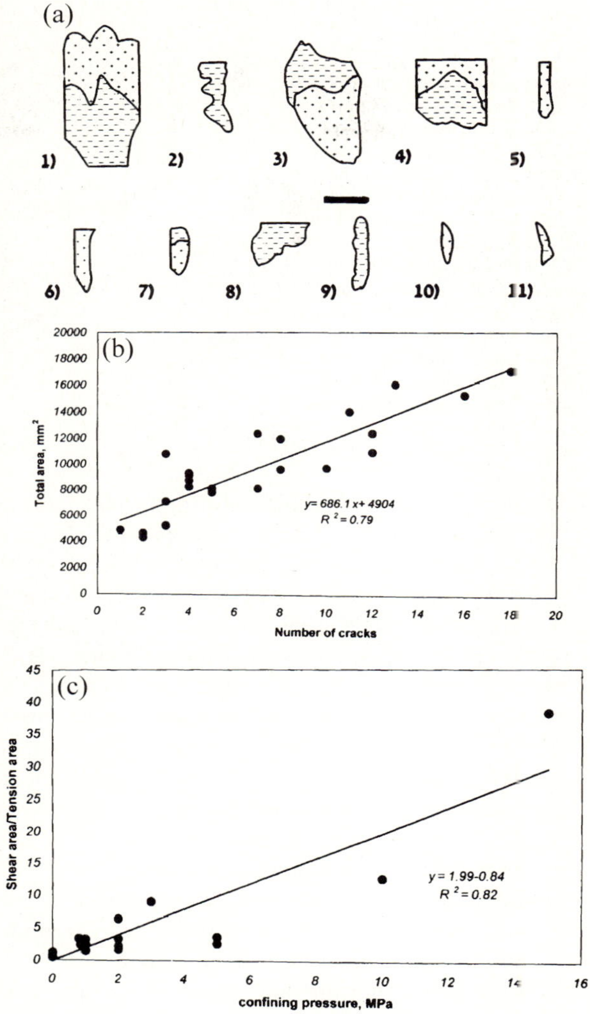

Fig. 2.16.
a A display of half of the total number of fracture surfaces (one of each matching couple) from a sample. Shear surfaces are marked by '–' and tensile surfaces by '+'. *Scale bar* is 4 cm. **b** Relationship between the number of fractures (one from each two matching fractures) and total area of all fractures. **c** Relationship between the ratio of (fractured area by shear) / (fractured area by extension), and lateral compression, σ_3 (from Bahat et al. 2001a)

0.5–1 cm^2 in size) that occur at different topographic elevations. RG maintain the same orientation in passing from sub-domain to sub-domain, indicating the continuity of fracture. A sub-domain consists of a set of microcracks (≤ 0.1–4 mm long) that are aligned along the RG that cross sub-domain boundaries within individual domains. No wing cracks (Sect. 2.4.2) were found on the fracture surfaces.

Tensile features include plumes (or striae) and stairs (Fig. 2.15). The plumes appear as braids on planar tensile surfaces of a matt white-grayish color and on the micro-scale the barbs are sub-parallel to the cylindrical axis. In tensile domains (that are of approximate sizes of shear domains), "stairs" result in a division of large surfaces into smaller planar sub-domains at different elevations, ranging in size from 0.5 cm^2 to 2 cm^2. Sub-domains maintain the general plume orientation within a domain. Domains of all tensile surfaces are planar and smooth, either parallel or sub-parallel to the cylindrical axis.

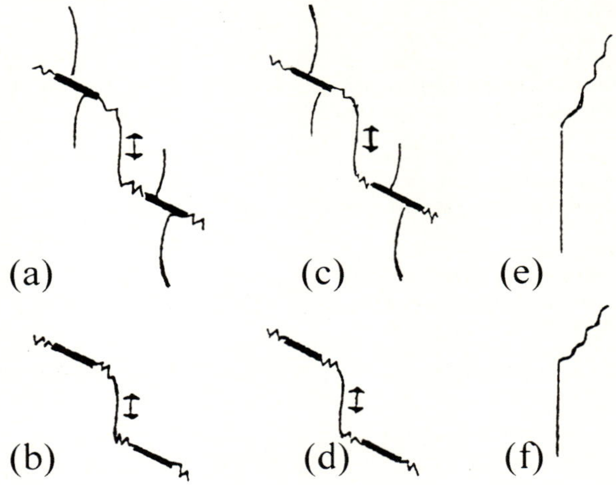

Fig. 2.17.
Schematic summaries of six modes of crack propagation from two experiments on weak rocks. **a, b** In uniaxial compression and **c, d** in biaxial compression. *Thick inclined lines* are pre-cuts, *zigzag inclined lines* are new shear fractures and *subvertical, slightly curved thin lines* are wing cracks (from Bobet and Einstein 1998, Tables 3 and 4). **e** In uniaxial compression, and **f** in triaxial compression from the present study (without pre-cuts): Shear fractures show ragged profiles and are inclined, and tensile fractures are straight and parallel to σ_1, not showing wing curvature (from Bahat et al. 2001a)

4. Whereas tensile longitudinal splitting (Sect. 2.2.13) can stretch continuously on a parallel tensile plane or on adjacent parallel planes, shear failure into meso-fractures (that divides the cylinder into two pieces) combines a series of distinct domains on separate surfaces that slightly differ in curvature and orientation. Thus, a key feature of single shear fractures (Fig. 2.14a,b) is their strong deviation from planar into concave and convex topographies (Fig. 2.13b). Therefore, surfaces of shear fractures have considerably larger areas than the "nominal" ones, whereas in tensile fractures "actual" and "nominal" surfaces are similar.
5. The reconstruction of the original cylindrical samples was accomplished by assembling the pieces (Fig. 2.15b), which enabled the determination of the fracture sequences that resulted in our experiments by the cross cutting relationships. Four fracture sequences were distinguished:
   a several stages of longitudinal splitting,
   b initial quasi-cone and subsequent longitudinal splitting,
   c longitudinal splitting that initiated from around the sample peripheries and subsequent inclined shear fracture, and
   d longitudinal splitting that initiated from around the sample center and subsequent inclined shear fracture. Generally, the sequence on shear surfaces is incipient shear fracture displayed by RG that occurs before total loss of cohesion, which is then followed by displacement shown by steps.
6. Conic and incomplete "quasi-cones" (Fig. 2.15b) may be obtained under both uniaxial and triaxial tests. The patterns of the tensile fractures often vary with the shape of the quasi-cones from two large matching longitudinal splits to repeated subdivisions into radial longitudinal splitting.
7. The tensile and shear areas that formed were quantified on each fractured sample (Fig. 2.16a). The total fracture area increases with the number of cracks (and fragments) (Fig. 2.16b). The ratio of "fractured area by shear" to "fractured area by tension" tends to increase with confining pressure σ_3 (Fig. 2.16c). Triaxial failures resulted in a few fractures (some 1–12 pieces) and in ratios of total shear area to tensile area >1, while uniaxial failure resulted in many fractures (some seven to eighteen pieces) and in ratios of total shear area to tensile area <1 (Fig. 2.16c).

8. There is a distinct trend of length increase of the longitudinal splits with the decrease of the confining pressure, as shown below by three samples: From 1 cm (when $\sigma_3 = 5$ MPa) through 2 cm (when $\sigma_3 = 2$ MPa) to 4 cm (when $\sigma_3 = 1$ MPa). This correlation shows that tensile longitudinal splitting occurs also under triaxial stresses, where the extent of the tensile fracture is inversely controlled by the confining pressure.
9. Most large fractures are composites of fracture domains. Fractures formed under both uniaxial and triaxial compression show that there is a strong correlation between the orientation of cracks (small fractures) and their mode of propagation: Shear cracks are inclined to σ_1 (Fig. 2.13b), while tensile cracks are parallel or subparallel to σ_1 (Fig. 2.13a and 2.17) The transitional angle between the two is $\alpha_c = 8 \pm 2°$. This correlation possibly supports previous suggestions that early microcracks are linked by inclined shear cracks (e.g., Hoek and Bieniawski 1966) rather than by tensile cracks (see implications of this observation in Sect. 2.2.12).
10. The authors correlated variations of the electromagnetic radiation (EMR) with changes of crack sizes in the fractured chalk throughout the investigated stress range (see elaboration in Chap. 5).
11. The multi-fracture complexes exhibited by the chalk samples bear a greater resemblance to surface ruptures that display wide belts of shear zones (associated with earthquakes that occurred in weak crusts) than to single slip surfaces simulated in strong rocks. The possible implication is that fracture in weak crusts is simulated better on weak rocks than on strong ones.

Microcrack populations and orientations. Stress induced microcracks that constitute RG on shear fractures generally are parallel or sub-parallel to the fracture dip, i.e., they may be tilted up to about 10° right or left with respect to the fracture dip (on parallel sections) (Fig. 2.13c–d). Likewise, barbs that constitute plumes on tensile surfaces are sub-parallel to the fracture dip. Whereas microcrack lengths can readily be measured between their tips on RG, barb lengths on tensile surfaces are not clearly defined, because they are continuously connected to form plumes. Microcracks may vary from sub mm to several cm in length, where the upper sizes are limited by the sub-domain and domain boundaries. The cm-long cracks and the microcracks commonly maintain the same sub-axial orientation.

All the meso-fractures that cut through the chalk cylinders (and have surfaces on the order of 50 cm^2 in size) display fracture-composites of domains and sub-domains consisting of microcrack coalescence. These assemblages are more common on the three-dimensional shear fractures (Fig. 2.14a) but also occur on planar tensile surfaces. Note, for example, that even the "uniform" large tensile fracture shown in Fig. 2.15a is divided into a "lower" surface marked by a delicate plume and an "upper" surface (partly shaded) marked by a coarser one (black arrow). Both plumes have the same orientations on the two sub-domains, suggesting that the planar fractures propagated in "sets", much like joint sets in the field.

The large EMR pulse populations at low stresses are obtained from many microcracks. Some of these microcracks possibly reflect dilatancy under both uniaxial and triaxial loadings. Dilatancy is generally understood to occur between one-third and two-thirds of the macroscopic fracture stress and is associated with an opening of mirocracks throughout the sample normal to σ_1 (Brace et al. 1966). Sprunt and Brace

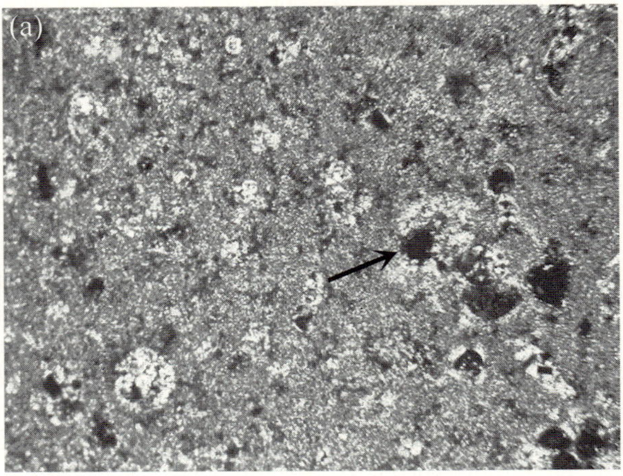

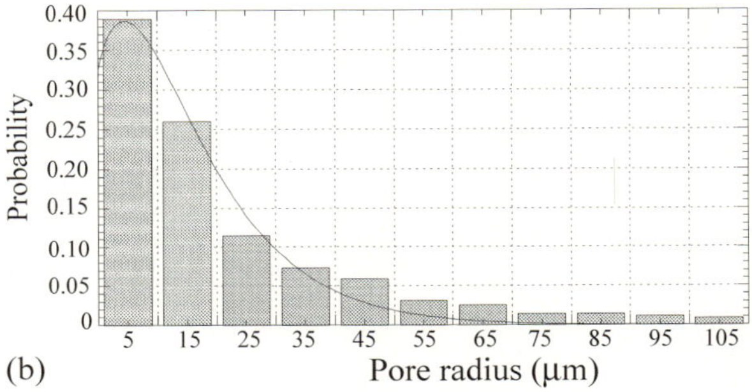

Fig. 2.18. **a** Photomicrograph of a Middle Eocene (Horsha Formation) chalk (width of picture is 1.2 mm). *Large white spots* are plankton skeletons, *small white spots* are microsparites, and *gray patches* are micrites (calcite). *Black spots* are pores (the one shown by *arrow* is 84 μm in diameter) and iron oxide grains. **b** Measured pore radii distribution in **a**, fitted to $P'(r) = a^2 r \exp(-ar)$; best fit obtained for $a = 0.105 \pm 0.004\ \mu m^{-1}$ ($R^2 = 0.98$) (**a** and **b** after Rabinovitch et al. 2003a)

(1974) observed that stress-induced microcracks are distinguishable from the initial defects in the rock in their shapes and orientations. The stress-induced microcracks are sharp-ended and elongated sub parallel to σ_1, whereas the initial defects are blunt and are oriented in all directions. Our results show that EMR pulses can be correlated with microcracking throughout most of the stress range, i.e., much earlier than one-third of the peak stress (Chap. 5). This includes pulses at the lower range of σ_1, probably in association with the deformation of "initial defects" (pores). Note that since EMR pulses are created solely by cracks and not by existing defects, this deformation must consist of some cracking. The measured pore radii distribution in the chalk is presented here for reference (Fig. 2.18).

Ridges and grooves and incipient shear before failure. Ridges and grooves (RG) are elongated lineations (slickenlines/slickensides) parallel to the direction of relative slip on fault surfaces. However, their origin is not clear (Davis and Reynolds 1996 p. 277). It is not understood why continuous lineations are longer than the known displacement along faults (Hobbs et al. 1976, p. 303) and how these lineations can be formed when the displacement is negligible (Means 1987). The results obtained by Bahat et al. (2001a) only exacerbate this dilemma. They obtained RG on fractured surfaces that had practically zero displacement. These surfaces did not cut through their respective cylinder peripheries; instead, they were arrested within the cylindrical samples (Fig. 2.14b). The close resemblance of the RG in the latter samples to RG that occur on surfaces that were relatively displaced (e.g., Bahat et al. 2001a, sample 16) is also intriguing. The only fractographic difference is that steps occur on the fracture surface of sample 16 but not on that of sample 15 in which the cut was arrested within the sample. The implication is that RG develop by the shear process during a decrease in sample cohesion due to microcracking prior to failure. The clear morphology of the plumes in these cases is a compelling indication of the lack of relative displacement on them because such displacements would have erased the delicate plumes.

It is possible therefore to distinguish an incipient shear when some but not total cohesion is lost. This is the time when the grooves and ridges are formed. The displacement that occurs subsequently is recorded by steps that can be assigned to a total loss of cohesion. These observations partly resemble results obtained by Wibberley et al. (2000), who identified an "incipient breakdown zone" at the immediate tip of the micro-fault. This zone consisted of early, long, pervasive tensile micro-cracks, oriented parallel to σ_1 that were later cross-linked by a set of *perpendicular short tensile* microcracks (unlike the results by Bahat et al. 2001a), leading to a subsequent failure.

Inclined shear and axial tensile fractures. The strong correlation between crack orientation and mode of propagation mentioned above relates to both cracks several mm to several cm in length and fractures up to about 10 cm in length. This correlation does not automatically apply to microcracks below 0.1 mm in size, for which Bahat et al. (2001a) did not have accurate fractographic observations. If, however, this correlation could be confirmed for micro-sized cracks, it might have important implications regarding the transition from microcracks to meso-fractures close to the triaxial peak stresses. This would imply that the initial axial tensile fractures (longitudinal splits) were connected by oblique shear microcracks. It would correspond to the observations by Shen et al. (1995) and Bobet and Einstein (1998) on fractures in gypsum, and by us on inclined shear fractures in chalks (Fig. 2.17). Note that Hoek and Bieniawski (1966) also suggest that early microcracks are linked by inclined shear cracks, with the difference that the latter authors assumed the early microcracks to be inclined to σ_1. The "connection" by oblique shear microcracks would differ from the models by Horii and Nemat-Nasser (1985) and by Reches and Lockner (1994), who advocate that such a connection is carried out in granite by tensile microcracks. Additional experiments backed up by close fractographic examination of induced tensile/shear fractures should help to ascertain whether the application of results on microcrack linkage obtained on weak materials (gypsum and chalk) to strong ones (e.g., granite) is justified.

2.2.3.5
Experimental Fracture on Five Sedimentary Rocks

Berea sandstone, Marianna limestone, Hasmark dolomite, Repetto siltstone and Muddy shale have been subjected to triaxial compression tests, measuring the confining pressures, the compressive strength (maximum ordinate of the stress-strain curve) and the dihedral angle 2α (measured relative to the maximum principal stress direction) by Handin et al. (1963). They show that both the compressive strength and dihedral angle results varied considerably in the five tested rocks. Generally, these rocks were weak (not stronger than 6 MPa), and the dihedral angles range was large (ranging between 0° and 80°) (see Table 2.2). No correlation between the dihedral angle and the confining pressures can be detected for these rocks. We are not aware of significant deviations from the results presented here that have been found for similar rocks by additional investigators (e.g., Friedman and Logan 1973).

2.2.3.6
Experimental Fracture on the Chelmsford Granite

The compressive strength of the Chelmsford granite ranges from 106 MPa to 200 MPa (Peng and Johnson 1972). Peng and Johnson (1972, Fig. 27) observed strong preferred orientation of the growing microcracks under stress along the splits. Characteristically, about 70–80% of the microcracks were oriented within 10° off the maximum principal stress direction σ_1. Some of the microcracks were straight, while others were partly curved. Holzhausen and Johnson (1979) also observed that a characteristic mode of failure for uniformly stressed specimens of Chelmsford granite was evidence of longitudinal splits that were inclined up to about 10° from σ_1. We interpret the straight splits to be essentially extensile (mode I) even if they form small angles 5 ±5° with the load direction (Peng and Johnson 1972, Fig. 18), whereas curved microcracks reflect local growth under shear. This interpretation follows the criteria derived from experimental observations by Bahat et al. (2001a) that shear cracks are inclined (also curved) to σ_1, while tensile cracks are (straight) parallel or sub-parallel to this direction. However, since these criteria concern the behavior of microcracks in a weak rock (chalk), the analogy to a strong rock (granite) should be presently qualified until more evidence becomes available to support it.

2.2.3.7
Experimental Fracture on Glass Ceramic

The fracture properties of glass ceramic induced by compression were investigated by combined electromagnetic radiation (Chap. 5) and fractographic methods (Bahat et al. 2002a). The study of a transparent sample enabled the fractographical elucidation of the sequence of crack nucleation, growth and interaction, and the ultimate longitudinal splitting under the incremental increase of uniaxial stress in five stages. Sample failure by longitudinal split occurred under uniaxial compression of 112 MPa in the fifth stage. A split into two slightly unequal parts (termed larger and smaller) took place along a straight, smooth plane parallel to the sample axis such that different fracture events were identified on the surfaces of the two parts (see more about the sample shape and experimental procedure in Bahat et al. 2002a). Experiments on transparent samples (Germanovich and Dyskin 2000 and references therein, and Sect. 2.4.2.2) were aimed at different objectives.

Striae (plumes) stretch continuously along the fracture surface of the large part almost from end to end of the sample (Fig. 2.19a). They start from an origin that is located about 18 mm from the F end (see below) and 86 mm from the B end, and occur at about mid distance (10 mm) from the two lateral edges of the main fracture. The striae radiate in two directions towards the two ends of the specimen. The bulk of the striae propagate towards the B end such that with the increase of distance from the origin

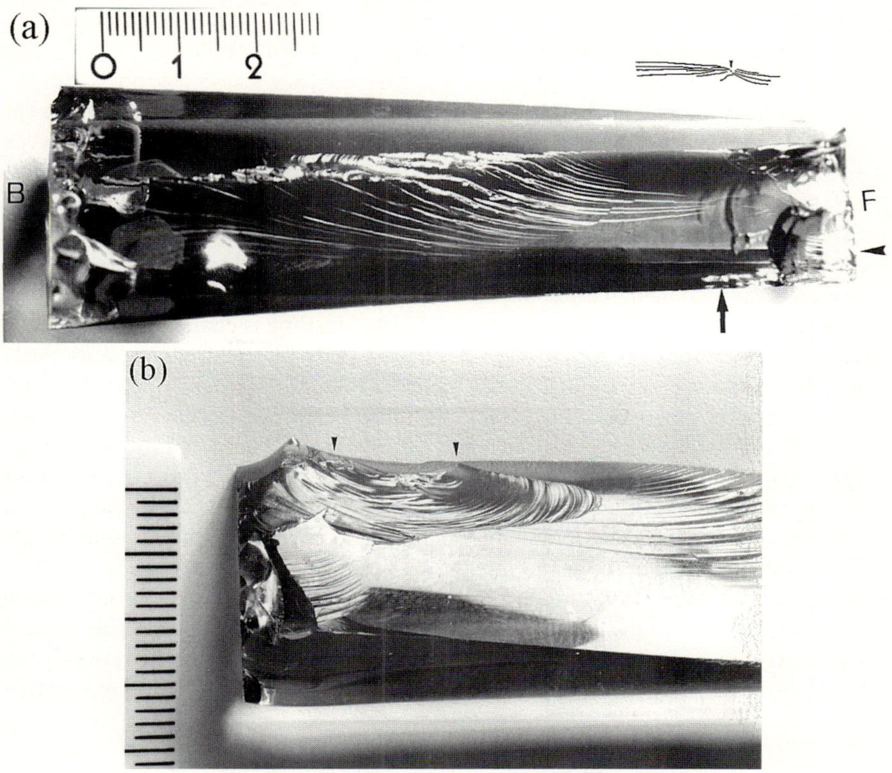

Fig. 2.19. Longitudinal split in a transparent glass ceramic. **a** Striae initiate above *vertical arrow* and propagate bilaterally (see *inset* above photograph) toward *B* and *F* ends of specimen (*horizontal arrow* shows striae at the *F* end) in one part of the split. Striae initiate within the specimen. They propagate along straight trajectories towards the *B* end at first. When gradually growing in size they approach the upper surface of specimen and curve towards that boundary. **b** Two successive cracks that formed after the specimen failure and split formation, when stress has relaxed. Two *arrows* mark the size of the first, successive crack (after Bahat et al. 2002a)

gradual coarsening of the striae morphology occurs. Only three-four short (6–7 mm long) striae propagated towards the F end (inset in Fig. 2.19a). Accordingly, the exact location of the origin, that is, the focus where the radial striae originated, is within an area of 1–2 × 1–2 mm, which we term the "nucleation zone". A microscopic examination indicates that the nucleation zone is not associated with any morphological discontinuities (wing cracks, glass defects etc.). Moreover, the transparency of the sample enables us to verify that no microcracks occur underneath the origin. Various shades in axial and transversal-concentric orientations around the origin are reflections from the outer surface of sample and do not belong to the main fracture surface. Hence, a longitudinal split along almost the entire length of the sample was partly straight and parallel to the compression axis before being deflected by the sample boundaries in the strong glass ceramic material. The second part of the split contains two successive cracks that formed after the specimen failure and split formation, when stress has relaxed, as indicated by the electromagnetic radiation results (Sect. 5.3.4) (Fig. 2.19b).

2.2.3.8
Inducement of En Echelon *Fractures by Mixed Mode I + III in PMMA*

Cooke and Pollard (1996) experimented on rectangular blocks (15 × 15 × 5 cm, ASTM E399-83) of polymethyl methacrylate (PMMA or Plexiglas), resembling the conceptual fracture mechanics models of mixed mode loading by Lawn (1993, Fig. 2.3) and Broek (1991, Fig. 1.3). They built a system that enabled them to load cracks in mixed mode I + III. Initially, a starter crack was grown from a chevron notch at a rate of about 1 cm in 10 s up to the peak load. Then, when the crack has reached the desired length, the sample was unloaded in order to begin the mixed mode portion of the experiment. They made the distinction that if mixed mode I + II loading is changed abruptly on a quasi-static fracture, the propagation path follows a sharp kink (Erdogan and Sih 1963) (Fig. 2.20a). On the other hand, if the K_{II}/K_I ratio is smoothly changed on a continuously propagating crack, laboratory fractures follow a continuously curving crack path (Cotterell and Rice 1980; Thomas and Pollard 1993) (Fig. 2.20b). A similar relationship between the loading history and the smoothness of the path is expected for mixed mode I + III loading (Fig. 2.20c,d). Cooke and Pollard (1996) considered that fracture propagation paths depend not only on the load ratio applied but also on sample geometry, loading configuration, and interaction among growing fractures.

Cooke and Pollard (1996) found that both the angle of twist ω of the *en echelon* cracks (Fig. 2.21a) and the number of segments formed increase with the increase of the ratio K_{III}/K_I. They however observed that the increase of ω with the ratio K_{III}/K_I falls below theoretical predictions (Cooke and Pollard 1996, Fig. 14). They propose that

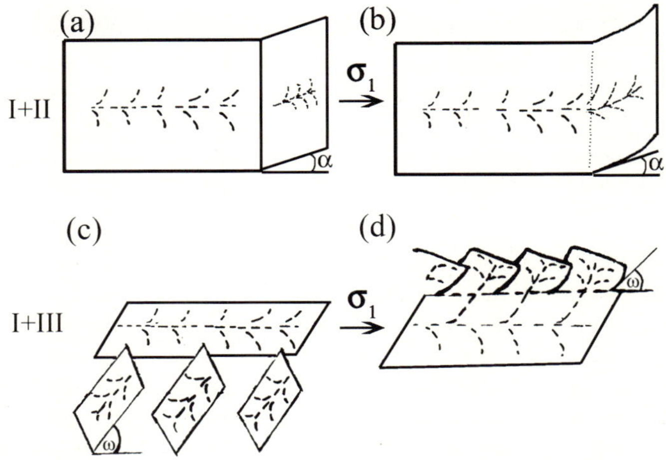

Fig. 2.20. Mixed mode fracture propagation paths for various loadings, when the maximum principal stress σ_1 is horizontal. **a** If a crack is loaded abruptly in mode I + II, a sharp kink is produced, forming the α angle at the distal part of the parent joint; the plume on the parent joint discontinues and a new plume forms on the kink. **b** If mixed mode I + II loading is increased gradually, as the fracture propagates, a path that follows a curve results and the plume on the parent joint continues to the curved kink that forms. **c** An abrupt I + III loading produces twisted echelon segmentation at ω on fringes that form along one or two layer boundaries, the plume on the parent joint discontinues and new plumes form on the fringe. **d** If mixed mode I + III loading is increased gradually, *en echelon* segments that gradually curve result, and the plume on the parent joint continues to the curved segments (modified after Cooke and Pollard 1996)

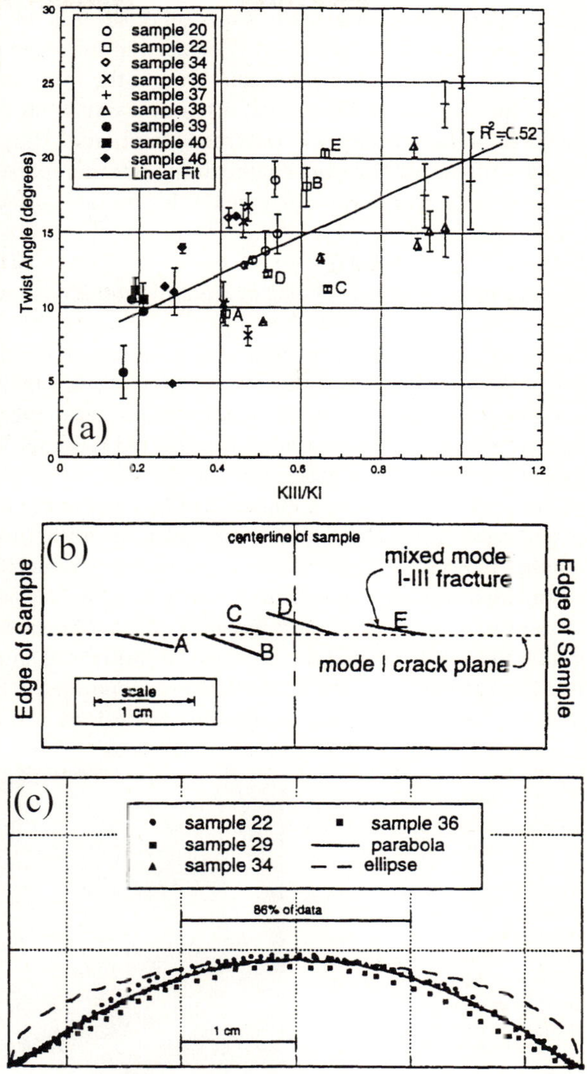

Fig. 2.21.
a Twist angle between parent crack and incipient fracture after break-down vs. stress intensity ratio for nine samples. A linear best fit through all measurements shows a trend of increasing twist angle with K_{III}/K_I. For each sample a similar trend can be observed as K_{III}/K_I varies with echelon fracture location along the breakdown front. Data from sample 22 are labeled with letters corresponding to digitized fractures in Fig. 2.21b and show a trend of increasing twist angle with K_{III}/K_I. **b** Digitized data from an experimental result showing breakdown geometry. Fracture *C* is sandwiched between fractures *B* and *D*. While the overlap with *B* is minor, there is considerable overlap between fractures *C* and *D*. The mechanical interaction between adjacent cracks may act to decrease the angle of twist in experimental and natural mixed mode I + III fractures. **c** Breakdown front shape for four samples compared with half ellipse with major axis of 5.0 cm and half mirror axis of 0.92 cm. The elliptical shape matches best at the center of the sample where most of the data are collected. The difference between the half ellipse approximating the crack front and a best fit through the crack front data is less than 0.1 cm (12%) in the region $-1\text{ cm} < z < 1\text{ cm}$ where 86% of the data are collected (from Cooke and Pollard 1996)

interaction among growing fractures (Fig. 2.21b) may contribute to the discrepancy between theoretically predicted twist angles and those observed in these and other mixed mode I + III experiments. Cooke and Pollard (1996) noted that the majority of the *en echelon* fractures form at the center of the sample where the ellipse best approximates the parabolic breakdown fronts (Fig. 2.21c). Such breakdowns often occur in nature as well (e.g., center of Fig. 2.35c), suggesting that their results are applicable in geological outcrops.

Cooke and Pollard found that the crack front advanced further than expected compared to the trailing edges at the sample sides (Fig. 2.21c). They suggested that when the sample sides are oppositely displaced in a tearing motion by mode III, a component of mode II stress intensity K_{II} is applied to the curved crack front near the sides of the sample. This is because the direction of shear is not perfectly parallel to the fracture front. Thus, a local mode II is induced in the system loaded by mode III. Similar load superpositions probably occur around the tips of all normal and reverse faults. This experimental finding is supported by a similar study that was carried out on the behavior of the crack front in fractured glass bottles. "Edge effects" caused the mist and hackle zones to advance more at the front center compared to the trailing edges of these zones at their intersection with the bottle's outside surface (Bahat et al. 1982, Fig. 4).

2.2.3.9
Fractography of Tensile Cracks in Starch-Water Mixtures

We briefly cite below a study on desiccation cracks in starch-water mixtures of 50–100 mm in diameter and 2–40 mm in thickness that were investigated by Müller and Dahm (2000) and Müller (2000). Fracture velocities were measured from videos and estimated from photos. Fractures covered the range from spontaneous nucleation, dynamic cracks with velocities of 100–200 mm s^{-1} to quasi-static cracks with velocities of 0.1 mm s^{-1} and less. Plumes that formed on the fracture surface gave the fracture direction (Fig. 2.22a), and their relation to fracture velocity was similar to the relation between seismic rays and seismic wave velocity. A ray-tracing method from seismology was used to calculate plume lines for depth-dependent rupture velocity. Moreover, an inverse method, based on finite difference travel times and conjugate gradients, was developed to invert a set of measured plume directions into a rupture-velocity distribution, which could also depend on the horizontal coordinate on the fracture surface. The main results and conclusions of this study are as follows.

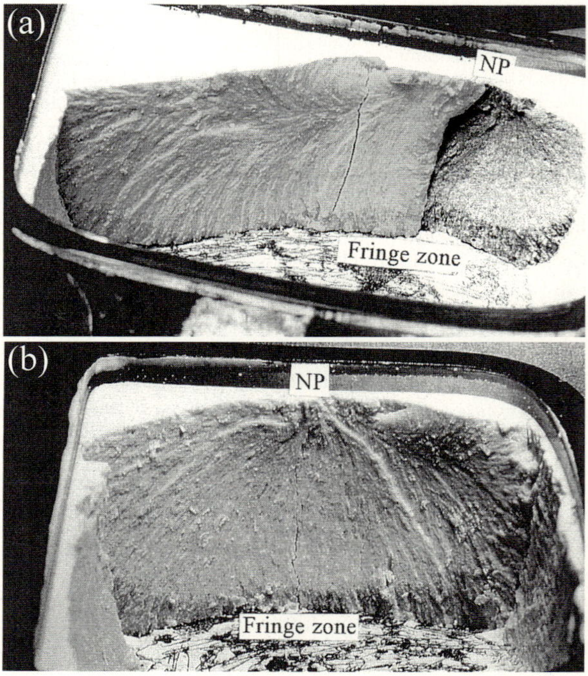

Fig. 2.22.
Fracture surface in starch. **a** A coarse plume propagates on the parent joint from the nucleation point *NP*. A fringe zone shows an angular tilt from the parent joint at the lower right. **b** A coarse fringe zone, where fracture velocity was lowest, formed beyond a mirror boundary. Length (diameter) of specimen is 65 mm (after Müller 2000)

1. Plume lines can successfully be inverted into relative fracture velocity.
2. In thin starch layers (thickness less than 0.2 times the diameter), fracture velocity decreases from top to bottom by a factor of 2–5, following a decrease of tensile stress due to the increase in water concentration.
3. Horizontal variation of fracture velocity reflects horizontal variation of stress, including stress relaxation due to the propagating crack, and ranges from dynamic to quasi-static velocities.
4. In thick starch layers (thickness about 0.5 times diameter), fracture is predominantly quasi-static.
5. Starch cracks sometimes have a fringe zone where topographic amplitudes of the *en echelon* segments are higher and rupture velocities lower than on the main part of the crack (Fig. 2.22b); this probably also applies to joints in rocks and their fringe zones.
6. Starch-water mixtures at rupture have a Poisson ratio close to 0.5. Cracks in starch are closest to subsidence or diagenesis joints in sedimentary racks.

There is some reservation regarding the interpretation of the fringe shown in Fig. 2.22b, because its morphology deviates from the common perfect shingling of *en echelon* segmentation, often seen in nature (e.g., Fig. 2.35). This deviation should possibly be assigned to the material properties and the difficulties involved in its maturing during the fracture process. There is no doubt, however, regarding the distinction between the mirror and the fringe in Fig. 2.22b.

2.2.3.10
Formation of Fringe Cracks in Silicon Single Crystal

Cramer et al., (1999, 2000) conducted dynamic fracture experiments on silicon single crystal plates that were loaded to force a {110} cleavage crack in a <110> direction. The fractographic features that resulted vary with the changes in the average crack velocity v_{av} and the energy release rate G. The fracture surface is smooth and mirror-like at the lowest fracture stress, corresponding to $G = 7\ \mathrm{J\ m^{-2}}$ (*circles* in Fig. 2.23a). At $G \le 14\ \mathrm{J\ m^{-2}}$ and above $v_{av} = 3\,000\ \mathrm{m\ s^{-1}} = (2/3)v_R$, the Rayleigh wave velocity (triangles in Fig. 2.23a). Atomic Force Microscopy (AFM) reveals that the crack partially deviated from the initial (110) plane and displays hills and valleys extending in the crack propagation direction. Facets appear on {111} planes next to the mirror with further increase in G. The size of the facets increases with increasing G. Instability, exhibited by an abrupt onset of rough hackle zones, starts at $G > 40\ \mathrm{J\ m^{-2}}$ and at velocities close to V (or v_{ter}), the terminal crack velocity (squares in Fig. 2.23a). The instabilities are manifested by the cracks on the {111} facets to "overgrow" each other, displayed by the simultaneous propagation of two crack fronts and in secondary shear fracture of the material placed between them (Fig. 2.23d).

Thus, there are three modes that show how the excess energy formed with the fracture growth is spent. Initially, the excess energy contributes to the increase in crack velocity. More excess energy results in faceting. Finally, great excess energies lead to path instabilities, resulting in the propagation of multiple cracks. The path instabilities in the silicon plates bear similarities to the bifurcation instability found in PMMA by Sharon and Fineberg (1999). The important difference, however, between the latter

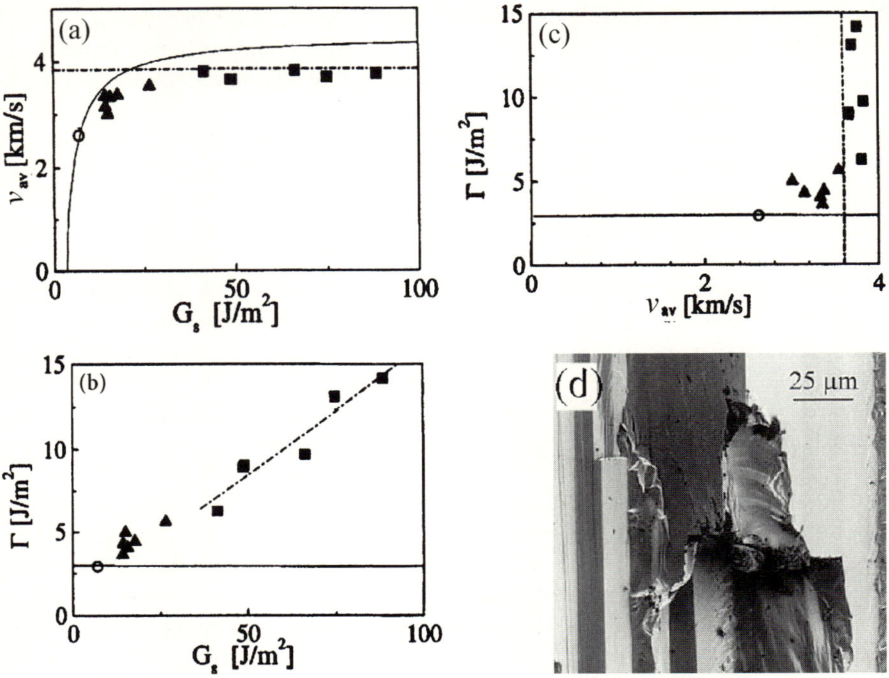

Fig. 2.23. a Dependence of the average fracture velocity v_{av} on the steady state energy release rate G_s. At the lowest G_s (*open circle*) the fracture surface is smooth. A faceted fracture surface is observed at higher G_s (*triangles*). The fracture surface is very rough at the highest G_s (*squares*). The *solid line* follows their Eq. 1, which is given by our Eqs. 1.134–1.136, using $\Gamma = 3$ J m^{-2}. **b** This equation is used to determine Γ as a function of G, and **c** as a function of v_{av}. **d** SEM micrograph, showing smooth *en echelon* segments cleaved from top to bottom at {111} planes (*light*) that are bridged by steps of shear cracks (*dark*) (from Cramer et al. 2000)

and the result by Cramer et al. (2000) (Fig. 2.23d), is that the bifurcation in the isotropic amorphous PMMA occurs along various alternative fracture paths, whereas in the silicon crystal instability develops by tensile cleavage on particular planes that are essentially *en echelon* cracks, which are connected by shear fractures, often termed steps or bridges (Fig. 2.39). Hence, under constraints of anisotropy, *en echelon* segmentation may occur at fracture terminal velocities.

2.2.3.11
Simulation of En Echelon*/Hackle Fringes and Longitudinal Splits in Glass*

A brief description of the investigation. We introduced new loading techniques (Fig. 2.24a–c) in investigating the formation of transitional *en echelon*/hackle fringes and longitudinal splits in soda-lime glass samples. Previous investigators combined sequential loading of mode III after mode I for obtaining *en echelon* fringes (e.g., Sect. 2.2.3.7). In the new experimental setting, a remote uniaxial compression was applied on cylinders (10 cm in length and 3 cm in diameter) that contained pre-cuts along the sample walls. Therefore, the compression was transformed into local mixed mode I + III

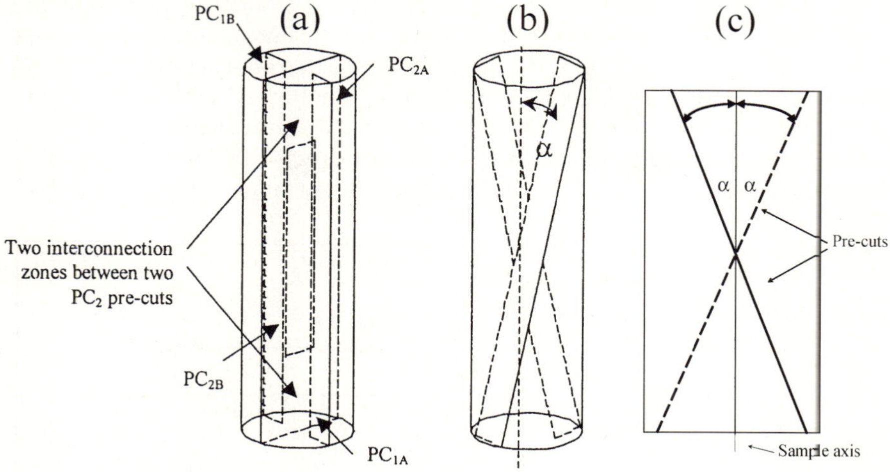

Fig. 2.24. Two pre-cut types. **a** The first type consists of two pairs of pre-cuts (PC_1 and PC_2) parallel to the sample vertical axis. The PC_1 pair is normally oriented to PC_2. PC_{1A} and PC_{1B} are the first and second pre-cuts, respectively, of the first pair PC_1, while PC_{2A} and PC_{2B} are the first and second pre-cuts, respectively, of the second pair PC_2. PC_{2A} and PC_{2B} are interconnected at the top and at the bottom ends of the samples. Depth of the interconnection zones at each end of the sample is 30 mm. **b** The second type consists of two pre-cuts inclined ("X"-wise) to the sample vertical axis. Both pre-cuts create approximately the same inclination angles α to the vertical axis. **c** Measurement of α is carried out between the pre-cut and the vertical axis of the sample

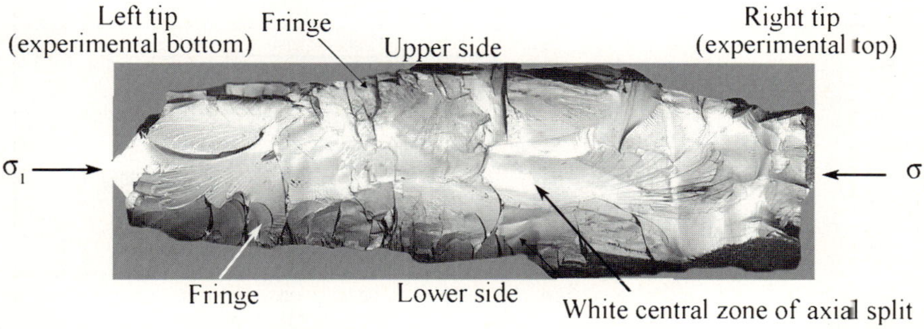

Fig. 2.25. A photograph of a fracture surface of glass that resulted by failure of the second type (135 MPa). Sample length is 100 mm. The sample is rotated 90° clockwise into a horizontal position and *thick arrows* show the direction of the uniaxial compression. Note the plumes that extend close to the two tips in continuation with the axial split at the center

loading, resulting in segmentation into two fringes along the pre-cuts. Subsequently, a longitudinal split of complex fractography formed along the fringes (Fig. 2.25).

Experiment. Two types of crack pre-cut geometries were introduced into the sample walls. The depth of all pre-cuts was 10 mm and aperture 2 mm. The first type consists of two pairs of pre-cuts parallel to the sample vertical axis (Fig. 2.24a, with one pair (PC_1) normally oriented to the other (PC_2)). Two pre-cuts, PC_{2A} and PC_{2B}, of the second pair were also interconnected at the top and bottom ends of the sample. The depth of the interconnection zones was 30 mm at both end parts of the sample. The second type consists of two pre-cuts inclined (X-wise) to the vertical sample axis (Fig. 2.24b), creating a series of pre-cut angles α, while maintaining about the same α (±2°) for the two pre-cuts (Fig. 2.24c). Note that for the first pre-cut type, $\alpha = 0°$. Glass cylinders in the size tested are very strong and generally resisted stresses beyond 200 MPa that we could achieve with our instrument. The introduction of pre-cuts to the samples weakened them. Another method of reducing the effective strength of the material is applied by using truncated elliptic cone shaped samples (Bahat et al. 2002a).

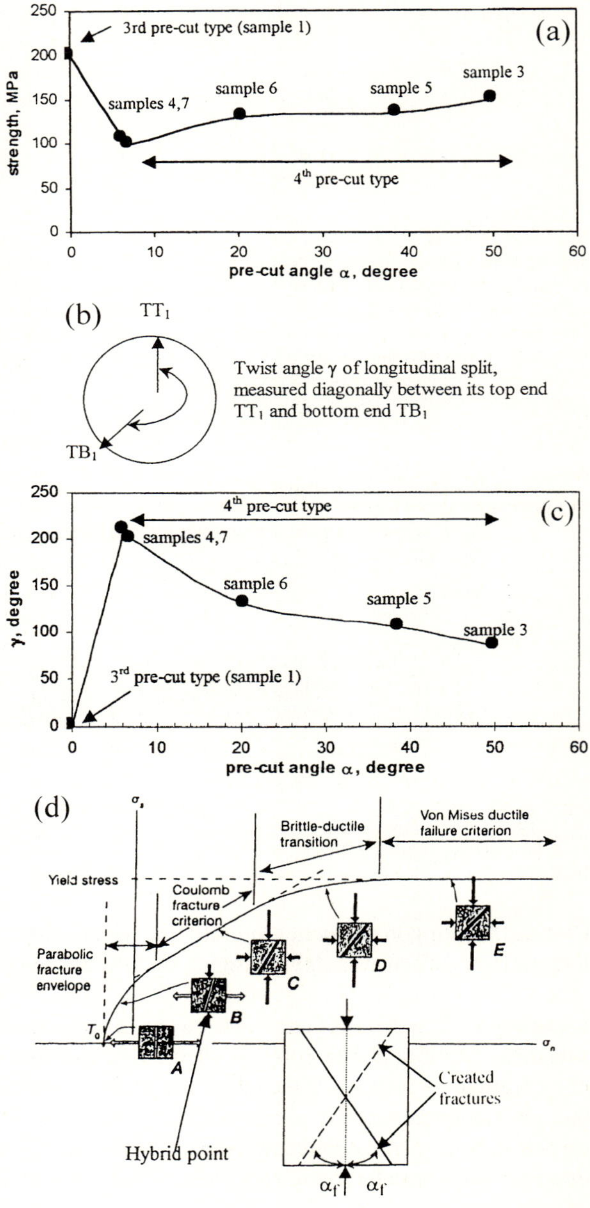

Fig. 2.26. Fracture parameters of the cylindrical glass. **a** The relation between the pre-cut angle α and sample strength. **b** A look at the sample along the axis. **c** The relation between α and the twist angle of longitudinal splits γ. **d** Schematic picture of a failure envelop, showing an increase stress, corresponding to the increase in the fracture angle α_f created by joints/faults (modified from Twiss and Moores 1992, Fig. 9.9)

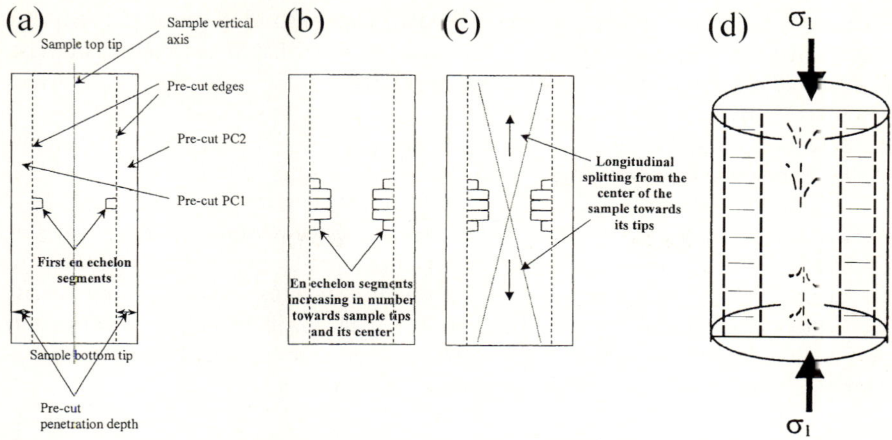

Fig. 2.27. A flat 2D projection of the 3D (twisted splits) surface created during failure of glass samples of the second type. The fracture process consists of three main stages. **a** Nucleation of *en echelon* segmentation at the edges of both pre-cuts at about the middle of sample height. **b** An increase of the *en echelon* dimensions and number, towards the sample center and along both edges of pre-cuts towards sample tips. **c** Propagation of axial splits from *en echelon* fringes (at about the mid sample height) to sample tips (see *arrows*) along the sample vertical axis. **d** Schematic arrangement of three zones, pre-cut, *en echelon* fringe and longitudinal split in the sample

Results. Generally, longitudinal splits do not start from the sample center, because this is a zone of great strength. They often initiate from the sample tip peripheries and propagate axially to the center. Weakening of the sample center can however initiate a longitudinal split there due to local tension induced by buckling (Holzhausen and Johnson 1979) or as in the present study by *en echelon* segmentation. The introduction of the second type of pre-cuts to the samples enables one to study the systematic dependence of strength on α (Fig. 2.26a), showing minimum glass strength at 6.5–7 $\pm 2°$, and then a rise in α requires an increase in load. The enlargement of α augments the angle produced between the *en echelon* fringes that form along the pre-cuts in the "X"-wise structure (Fig. 2.26b). The longitudinal splits that form along the fringes interact with each other creating a twist along the cylindrical axis. The twist measured by γ (Fig. 2.26b) increases with α, and the shape of the curve in Fig. 2.26c produces a mirror image of Fig. 2.26a. It is quite intriguing that the minimum and maximum in Fig. 2.26a and c, respectively, are close to the maximum deviation of splits from the sample axis in conventional tests (in granite, Sect. 2.2.3.6, and in chalk, Sect. 2.2.3.4). The fracture sequence is diagrammatically shown in Fig. 2.27a–c.

Discussion. Contrary to previous experiments and many field exposures in sediments that exhibit transitions from parent fractures to fringes, the present study simulates a reverse transition. We demonstrate a transition from fringes to parent fractures by changing local mixed mode I + III to a local single mode I. Monitoring of the electromagnetic radiation (FMR) that was induced by the fractured samples enabled us to determine in real time the sequence of events, the fracture size and the fracture velocities along the various failure stages (Sect. 5.3.4). Very often geological descriptions trail behind descriptions of experimental results. Presently, the fracture sequence described here may be reasoned to be an "unlikely geological process". However, new, "unexpected" fracture phenomena are repeatedly discovered in granites (Chap. 4), and who knows what else is still concealed from us there? The stress analysis of these unexpected results is currently under investigation.

2.2.3.12
Hackle Formation in Polymers

Arakawa and Takahashi (1991) compared fractographic parameters to several mechanical ones in polymers. Fractography was measured by two methods, pit density and surface roughness λ. They scanned the fracture surfaces with a scanning instrument at an interval of 0.5 mm using a needle (10 µm in tip radius) in the direction perpendicular to the direction of crack propagation and defined surface roughness using the term "root mean square roughness" (RMS):

$$\mathrm{RMS} = \left[\frac{1}{L} \int_0^L |f(x)|^2 \mathrm{d}x \right]^{1/2} \tag{2.4}$$

where L is the scanned length and $f(x)$ is the roughness height at the point x. Arakawa and Takahashi (1991) found a good correlation between changes in pit density and surface roughness. They employed the shadow optical method of caustics to evaluate K_d and K_c, the dynamic tensile stress intensity factor during crack propagation and the tensile stress intensity factor for the arresting crack, respectively, where a Cranz-Shardin type high speed camera was used. K_d was evaluated by

$$K_d = \frac{2\sqrt{2\pi}}{3Z_0 d c^* \eta^{3/2}} \left(\frac{\phi}{3.17} \right)^{5/2} \tag{2.5}$$

where ϕ is the caustic diameter, Z_0 is a distance between the specimen and the image plane, d is the specimen thickness, c^* is the stress-optical constant for dynamic conditions and η is a convergence factor for incident light rays.

Arakawa and Takahashi (1991) show that the curve linearly correlating crack velocity, v with K_d, may be divided into three regions of different slopes in Hommalite-100 (Fig. 2.28a). There is a gradual increase in K_d with v in the low velocity region A, where λ remained relatively smooth. In region B, there is a rapid increase in K_d along very short intervals of v and a corresponding increase in λ. The curve slope in region C exhibits an additional rapid increase in K_d and an extremely rough surface that reached its peak prior to branching. The latter slope clearly shows that branching did not occur at the peak crack velocity but at the maximum value of K_d, i.e., branching is K_d dependent. The dotted curve connects the peak velocity points, separating the accelerating and decelerating areas in the diagram (see below).

Arakawa and Takahashi (1991) show that maximum λ is best correlated with G_v. The physical meaning of G_v is energy per unit crack width per unit time, and it is considered to relate to the flow rate of energy into the crack tip region. A very good fit among the maxima of the curves of λ, G and G_v occurs for epoxy (araldite D), while it is shown that crack lengths giving peak values of λ and v differ from each other (Fig. 2.28b). A slight discrepancy between crack lengths also seems to exist, giving peak values of λ

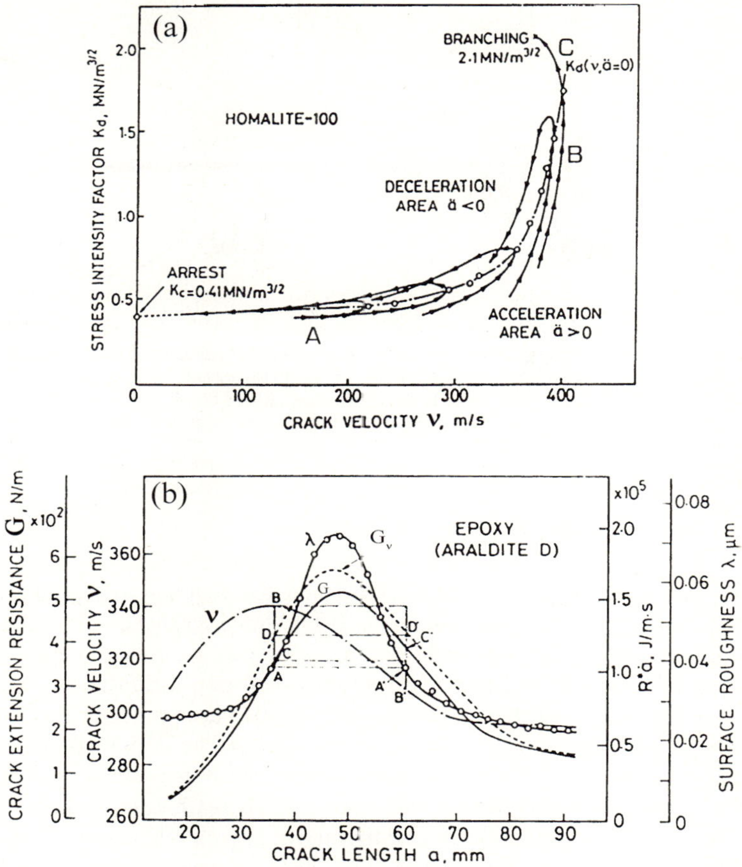

Fig. 2.28. **a** Dependence of stress intensity factor K_d, fracture velocity v and fracture acceleration $\ddot{a}$ in Homalite-100 on fracture length (after Arakawa and Takahashi 1991). **b** Surface roughness λ, specific crack extension resistance G and the product of G_v for epoxy as a function of crack length a (after Arakawa and Takahashi 1991)

and G: For the same λ values indicated by points A and A', the corresponding values of v (at points B and B') or G (at points C and C') slightly differed. On the other hand, a good agreement is shown between crack lengths giving the peak values of λ and G_v, and the values of G_v are almost equal at points D and D', which correspond to points A and A', respectively. These results support the suggestion that λ is dependent on G_v. Hence, the implication is that in the examined polymeric materials, λ behavior (presumably along the traverse through the mirror-mist-hackle-branching, Fig. 2.1) is correlated with G_v, K_d, G, and v in a decreasing order. The results by Arakawa and Takahashi (1991) were confirmed by Arakawa et al. (2000), using different experimental conditions on PMMA. These studies demonstrate that maxima in Fig. 2.28a and b (particularly the latter) show strong fluctuations in the various parameters along the crack length. The *association* of v with the formation of hackle in polymeric materials shows, however, the subordinate *dependence* of this morphology on v. On the other hand, these results indicate the important influence that G_v and K_d have on the formation of hackles.

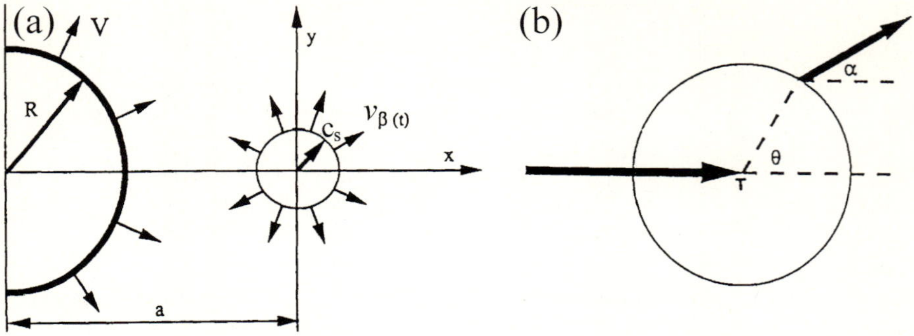

Fig. 2.29. Primary and secondary cracks. **a** A schematic picture of the geometrical setup. A secondary crack of length $2c_s$ at a distance a from the origin of the primary crack of a semicircular radius R starts to grow at velocity $v_{\beta(t)}$ when $t = 0$. **b** Side view of the primary crack, T its tip, while θ and α are the starting growth angle of an existing flaw and the direction of its growth into a secondary crack, respectively (after Rabinovitch et al. 2000a)

2.2.3.13 *Hackle Formation in Glass*

Previous views about the mist and hackle zones maintain that they are "identical in appearance but different in scale" (e.g., Mecholsky 1991). More recently, however, Rabinovitch et al. (2000a) suggested a fundamental difference between these zones, as explained below. The secondary cracks (SC) start to grow from an existing "Griffith flaw" (Sect. 1.1.2 and 2.2.2) when the critical stress intensity K_{IC} is attained in front of the tip of the primary crack (PC) cutting the glass (Fig. 2.29 and Sect. 1.4.4). The SC move under a changing stress field caused by the PC, whose distance from the SC is continuously changing. Hence the SC start from zero velocity, and asymptotically reach v_t. In the mist zone, the PC overtakes the SC almost instantly after they begin to grow. Therefore, the SC in this zone are short (for details see Rabinovitch et al. 2000b). However, towards the end of the mist zone, the growing SC become longer before the PC overtakes them, and they manage to attain larger sizes and rougher morphologies. Apparently, the transition to the hackle zone is related to the inability of the PC to catch up with the SC. In this case, the SC grow separately compared to those SC that are to a large extent "swallowed up" by the PC and form the mist (Rabinovitch et al. 2000a). Previous authors made the (erroneous) assumption that the SC move from their very incipience, with the same v_{ter} of the PC, resulting in parabolic or hyperbolic shapes throughout. Such shapes actually appear in the hackle zone but not before. Hence, hackle formation is associated with the distinct attainments of v_{ter} by both the primary and the secondary cracks. This *association* however does not imply a *dependence* of hackle formation on v_{ter}. It has been suggested that hackle formation is rather stress intensity dependent (e.g., Bahat et al. 1982). This dependence seems to be more intricate than the latter, as is shown in Sect. 2.2.3.12.

2.2.3.14 *The Sequence of Mirror, Mist, Hackle Fringe and Branching on Fracture Surfaces of Glass*

It has been accepted by many authors (e.g., Rice 1984, Fig. 1; Hull 1999, Fig. 5.3 and 5.31) that fracture growth from the origin to its termination is in the following "conventional order":

The mirror plane, mist, hackle and branching (branching at the right side of Fig. 2.1b and Fig. 2.10). Branching involves the creation of two secondary fractures, at least one of which is non-coplanar with the primary fracture. Accordingly, a distinction has been made between the radii from the origin to the mist boundary, the hackle boundary and to branching (Fig. 2.1d). This view is quite surprising, considering the existence of many observations that are unlike the conventional order. Rice (1984, Fig. 66b), Ravi-Chandar and Knauss (1984d IV, Fig. 11), Holloway (1986, Fig. 13), Hull (1999, Fig. 5.4) and Bahat et al. (2001b, Fig. 2b) obtained results of a different "fractographic sequence". They show that the mist zone is followed by a transition zone from mist to small hackles, beyond which a drastic change of superposing large hackles on new branches occurs. This indicates simultaneous growths of these two features, i.e., branching occurs before large hackles (branching at the left side of Fig. 2.1b). The same fractographic sequence occurs on joint surfaces in chalks and geological exposures of granites (Fig. 2.30a,b and 2.56). These observations should equate the radii from the origin to the hackle boundary and to branching. What could be the reason for the discrepancy between the "conventional order" and the "fractographic sequence"?

Possibly, the results of repeated four cycles from the mirror plane to branching shown below can be used for explanation. A soda lime silicate glass bottle was fractured by the internal pressure technique (Bahat 1991a, p. 121). The saddle-shaped fragments that characteristically result in such fractures (Fig. 2.30c), which show repeated branching in profile (Fig. 2.30d), were investigated. A saddle-shaped fragment shows the initiation of four cycles from mirror to successive branchings in plan view (Fig. 2.30e). A primary mirror M1 passes on to mist m and then to hackle h, ending in fracture-branching that initiates in the outer boundary of the hackle, at h. The second cycle starts in secondary mirrors that form a shingle-like *en echelon* segmentation M2 that transforms into secondary mist m and hackle h, where a new branching occurs. The third cycle starts by a series of tertiary mirrors M3 in a disturbed shingle-like segmentation. On one mirror, there are undulations that convex in the direction of fracture propagation. The fourth cycle (only part of it has been retrieved) is "disorganized" such that tertiary and quaternary mirrors M4 have no distinct boundaries.

Most intriguing is the change from cycle to cycle when each branching starts with new mirrors. The first branching occurs at the outer boundary of the first hackle (before M2), following the "conventional order". However, the second hackle is already divided into two sub-zones, such that the first one (before M3) belongs to the second cycle, while the second hackle sub-zone rests on the early parts of the third cycle mirrors, M3. The next hackle zone occurs beyond a mist zone at the ends of M3 but also appears to have initiated the fourth cycle, M4. Thus, one observes superposing hackles on the new branch, indicating simultaneous growths of these two features according to the above "fractographic sequence". Hence, both "conventional order" and "fractographic sequence" occur adjacent to each other, forming during the early (second cycle) and late (fourth cycle) stages of the same experiment, respectively, while the fractography of the third cycle, M3, shows a transition between the two.

The occurrence of the "fractographic sequence" in geological exposures (Fig. 2.30a and 2.56) encourages efforts to comprehend the meaning of cycling fractography that may help to improve the understanding of the paleostress conditions leading to dynamic jointing (Sect. 2.4.6.1 and Chap. 4). A possible future discovery of a "conventional order" in geological exposures may add important information for detailed interpretation of the fracture process.

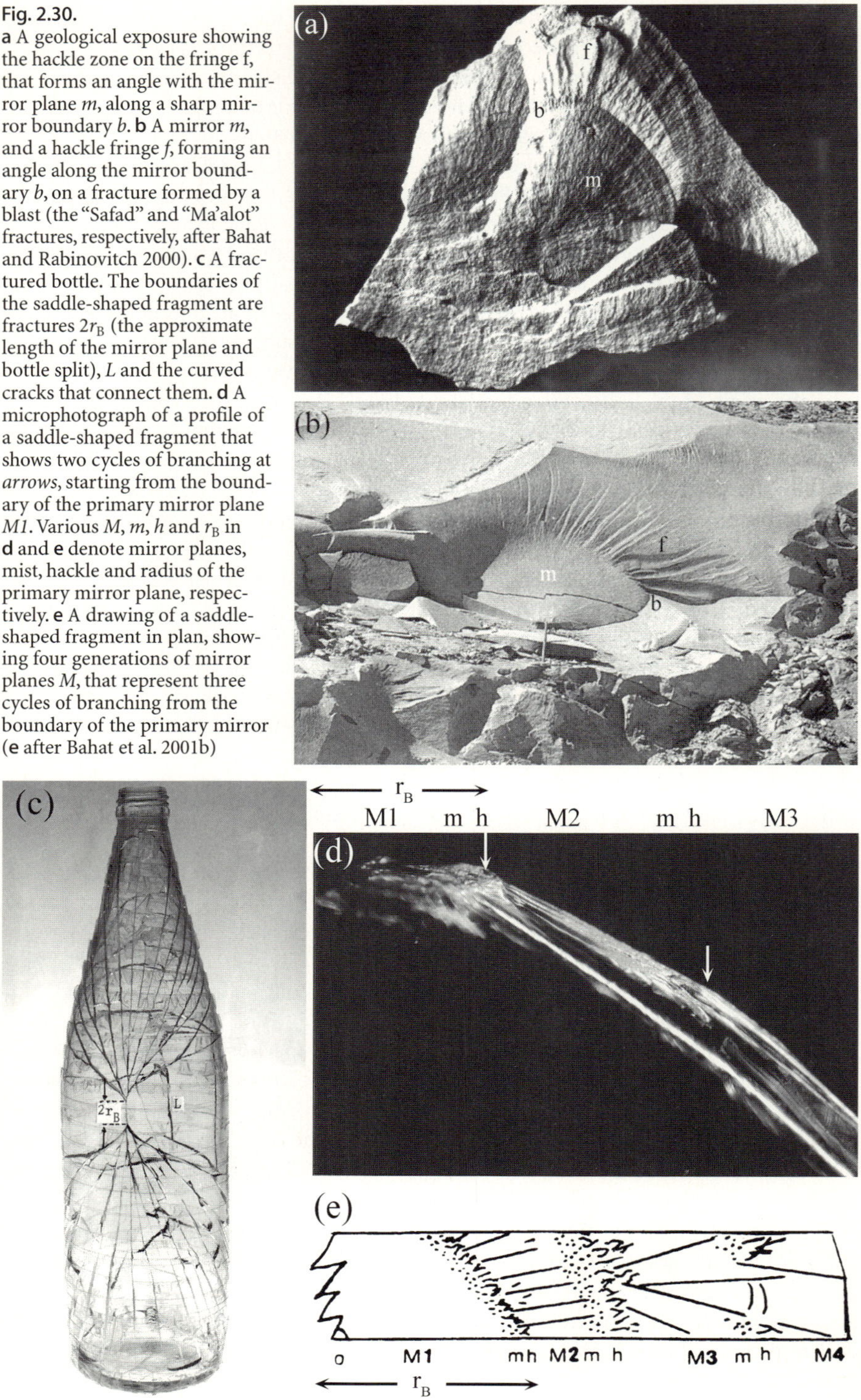

Fig. 2.30.
a A geological exposure showing the hackle zone on the fringe f, that forms an angle with the mirror plane *m*, along a sharp mirror boundary *b*. **b** A mirror *m*, and a hackle fringe *f*, forming an angle along the mirror boundary *b*, on a fracture formed by a blast (the "Safad" and "Ma'alot" fractures, respectively, after Bahat and Rabinovitch 2000). **c** A fractured bottle. The boundaries of the saddle-shaped fragment are fractures $2r_B$ (the approximate length of the mirror plane and bottle split), *L* and the curved cracks that connect them. **d** A microphotograph of a profile of a saddle-shaped fragment that shows two cycles of branching at *arrows*, starting from the boundary of the primary mirror plane *M1*. Various *M*, *m*, *h* and r_B in **d** and **e** denote mirror planes, mist, hackle and radius of the primary mirror plane, respectively. **e** A drawing of a saddle-shaped fragment in plan, showing four generations of mirror planes *M*, that represent three cycles of branching from the boundary of the primary mirror (**e** after Bahat et al. 2001b)

2.2.3.15
Fracture Interaction at Contacts along Layer Boundaries by a Numerical Experiment

The nature of interaction between adjacent fractures has been investigated for decades both in the laboratory and in geological outcrops (e.g., Yokobori et al. 1971; Olson and Pollard 1989, 1991; Bahat 1991a, p. 317). This section relates to fracture interaction with *layer boundaries*, which is fundamental to fracture geology. Furthermore, within layered sedimentary rocks, the termination of fractures at *layer boundaries* can limit vertical flow (particularly in the unsaturated zone, Sect. 6.9) and produce highly tortuous flow paths (e.g., Tsang 1984). In contrast, fractures that propagate straight-through *layer boundaries* (vertically and at various oblique angles) (e.g., Bahat 1991a, Fig. 4.5) provide well-connected pathways for vertical fluid flow (Cooke and Underwood 2001). A potential intermediate case for fluid flow is a fracture that jogs or steps over a few centimeters along successive *layer boundaries* (see the profile in Fig. 2.31). Composite joints, such as those sketched in Fig. 2.31 are believed to form by repeated step-over of a propagating fracture across the *layer boundaries* (Helgeson and Aydin 1991; Cooke and Underwood 2001).

A numerical experiment was carried out by Cooke and Underwood (2001) exploring the influence of *deformation at* contacts along layer boundaries, in the form of local sliding and/or "debonding" and subsequent opening, on the fracture intersection with

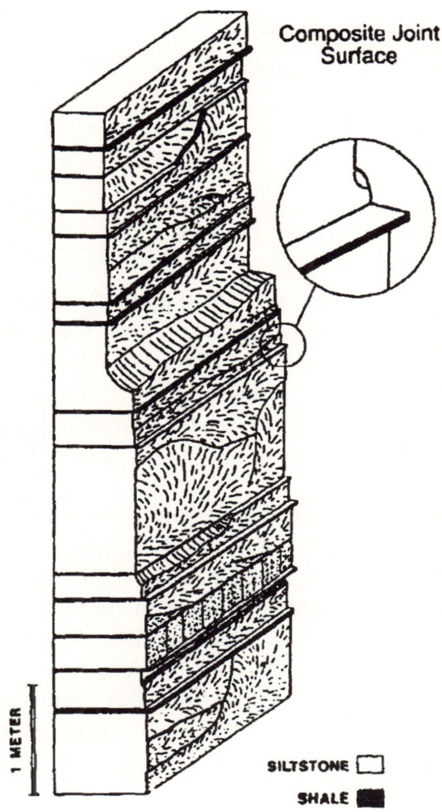

Fig. 2.31. Sketch of composite joint within interbedded shale and sandstone sequence in New York State. The overall vertical joint trace has a discontinuous nature (after Helgeson and Aydin 1991)

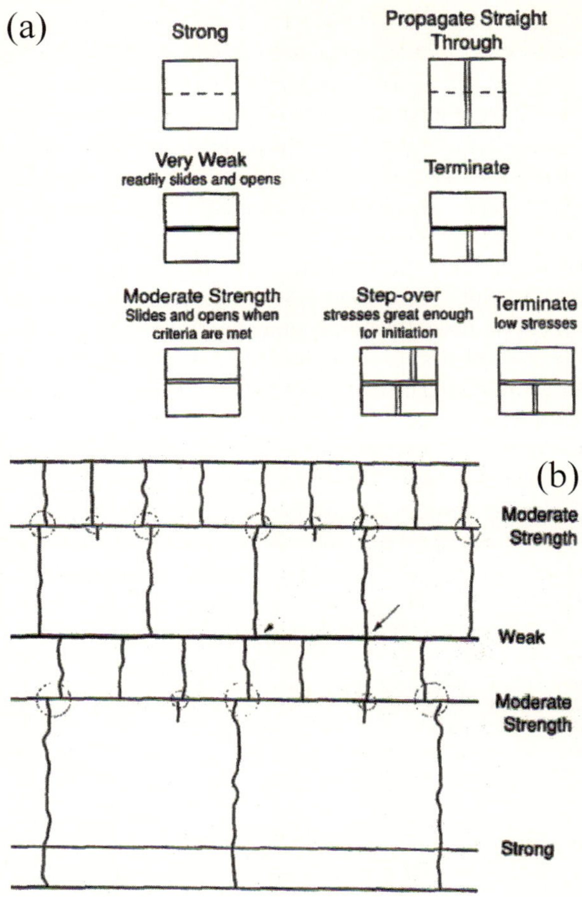

Fig. 2.32.
a Inferred variations in geometry of fracture-bed contact intersection for bedding contacts with different strengths. **b** Hypothetical pattern of fractures within a series of layers separated by different strength interfaces. *Circles* highlight occurrences of step-over fractures due to opening along the moderate-strength interfaces and *circle size* correlates with distance of fracture step-over. The spacing of fractures is approximated from the layer thickness. Apparent step-overs along the weak interface (indicated with *arrows*) result from the coincidental alignment of fractures among layers. The distinction between coincidental alignment and step-over is difficult to assess without characterization of fracture surface textures (i.e., signatures of propagation direction), relative interface strength (to guide probable intersection mechanisms), and consistencies in fracture pattern along bedding contacts (Fig. 2.32a,b from Cooke and Underwood 2001)

layer boundaries (see also Thomas and Pollard 1993). Their conclusions are cited below. Fractures propagating toward *layer boundaries* may either

1. Terminate at the boundary.
2. Propagate straight through the boundary.
3. Step-over to the left or right at the *layer boundaries* (Fig. 2.32a).

Based on the model results, Cooke and Underwood 2001 suggest that local interface opening, rather than sliding, is primarily responsible for the termination and step-over of fractures. Furthermore, the model results suggest that the strength of contacts controls the type of resulting fracture intersection. Fracture termination is favored at very weak contacts, whereas fractures propagate straight through strong contacts (Fig. 2.32b). Most sedimentary contacts will fall between these two end members. Such moderate-strength contacts may develop step-over fractures due to a local opening along the contact, or if the stresses are not great enough to produce new fractures, the parent fracture will terminate at the moderate-strength contact. Fracture termination is more likely under conditions of shallower burial depth, lower effective layer-parallel tension, and fluid-driven propagation, rather than equivalent remote layer-parallel tension. Thicker layers and greater effective layer-parallel tension may produce greater amounts of step-over than thinner layers and more compressive layers. Fractures aligned within several centimeters across a contact may be coincidentally aligned or result in fracture step-over. Careful characterization of the fractures and analysis of the pattern may distinguish whether the fractures are the result of coincidental alignment or fracture step-over.

2.2.4 Joint Mirror Planes

Two end members of mirror planes are characterized in this section, drawing on both laboratory and field observations.

2.2.4.1 *Two Types of Mirror Planes*

We distinguish between two types of fracture mirror planes. One type characterizes fractographies that form dynamically by rapid fracture when high stresses are exerted. This type has been studied for decades in the laboratory on glass, ceramics, metals and polymeric materials (e.g., most of Chap. 2 in Bahat 1991a), and has recently been characterized in outcrops from the South Bohemian Pluton (Chap. 4). The second mirror type characterizes fractographies that form quasi statically by slow fracture when low stresses are exerted. This type is far less known from laboratory investigations than the first type, but it can be identified on innumerable surfaces of systematic joints that cut sedimentary rocks.

The dynamic mirror plane. When fractured primarily by high tensile stress the mirror plane exhibits characteristic morphological features (Fig. 2.1b–d), including the fracture origin at an initial flaw that develops into a larger critical flaw, concentric undulations, striae, Wallner lines, and the mirror boundary. This mirror plane is smooth, and often distinguished as a polished-like surface. Undulations (note below a distinction from ripple marks) develop on this surface by local mixed modes I + II due to stress fields that progressively bend the crack front around the axis normal to the direction of propagation when the two larger principal axes are on the fracture surface. On the other hand, striae form by local mixed modes I + III in a stress field that rotates the fracture front about the axis of the direction of crack propagation (e.g., De Freminville 1914; Smekal 1953). The distance between the origin to the mirror boundary is the mirror radius. Three mirror radii are legitimate, r_m, r and r_b, at the inner boundaries of the mist zone, the hackle zone and branching initiation, respectively (Fig. 2.1a,b,d). In multi-crystal ceramics and rocks, the micro-features of the mist are "camouflaged" by the grain boundaries and only the hackle zone may be recognized. Practically, all the above-mentioned fractographic features are generally restricted to the mirror plane. There is a distinction between primary mirrors that are related to the parent fracture and secondary mirrors that form on subsequent fringes (Fig. 2.30d,e, 4.9c,d and 4.11a–c).

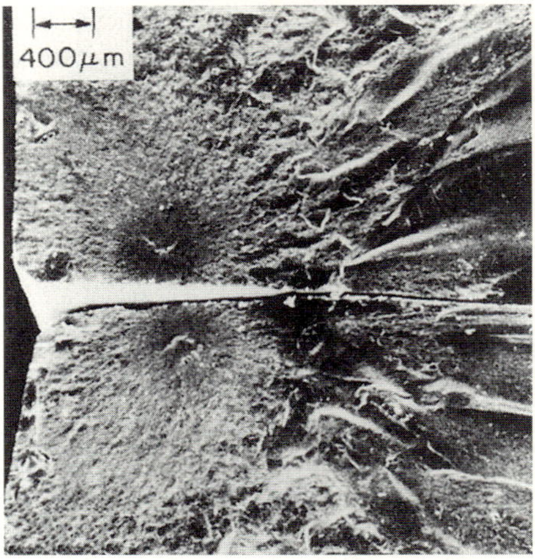

Fig. 2.33a. Two matching halves of a fracture in a ceramic body made of $BaTiO_3$, showing a smooth mirror plane around the fracture origin (*dark gray*), developing towards right into sharp hackle ridges which are termed cuspate-hackles (from Rice RW 1974)

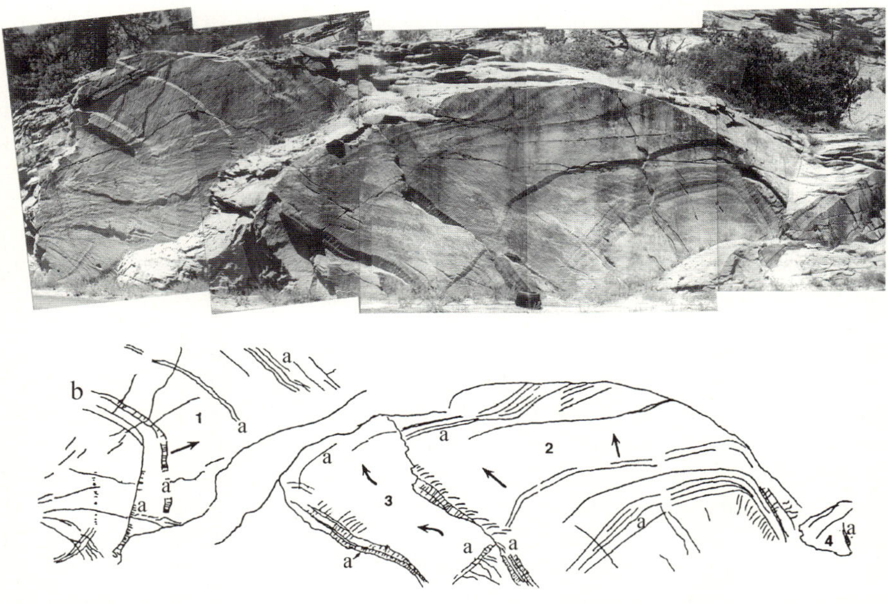

Fig. 2.33b,c. Photograph and drawing of a "wall" consisting of four joints (1–4), that propagate to various directions as shown by *arrows*, arrest marks *a*, occasionally show "box" profiles *b*, and striae *s*, in Zion National Park. A camera bag 38 cm in length is used for scale

The quasi-static mirror plane. When fractured under low stresses, propagation is slow and sub-critical (Sect. 1.5), and a different mirror plane forms. On this mirror, the location of fracture origin can be identified by the convergence of plumes at the center of concentric ripple marks, but the distinction between an initial flaw and the critical flaw is vague. Additional morphological features are arrest marks and the mirror boundary. Experimental sub-critical surfaces are briefly described in Sect. 2.2.3.1. Single-layer joints that cut sedimentary rocks often fall into this category. Quasi-static mirror planes form matte,

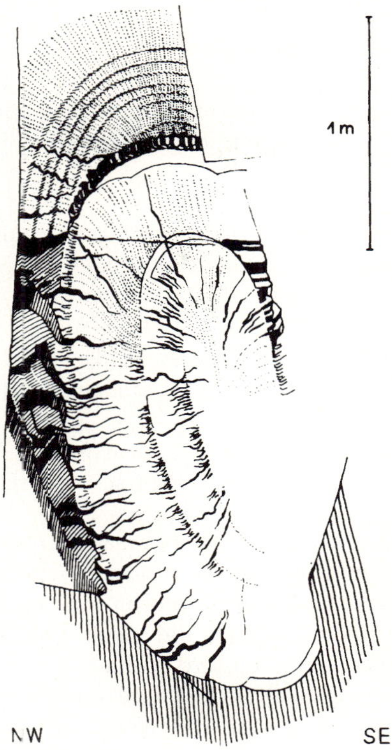

Fig. 2.33d.
An elliptical mirror plane decorated by concentric arrest marks and radial striae in Ordovician pencil slate, Thuringia, Germany. Many short striae superpose the arrest marks, and some striae merge into a single, large stria (from Bankwitz 1965). Recall that those striae are products of mixed modes I and III, and the arrest marks are products of mixed modes I, II and III

Fig. 2.33e.
Photograph of a mirror plane and fringe on an exfoliation joint from Zion National Park. The somewhat hackled *en echelon* segments display different lengths. Note the exact, curved mirror boundary. A page at left is used for scale (after Bahat et al. 1995)

rather than polished-like surfaces. While rapid fracture through grainy ceramic material or rock form transgranular, smooth mirror planes (Fig. 2.33a and 4.11a, respectively), slow, intergranular fractures (Sect. 4.5.2.2) maintain a grainy, matte surface morphology. On such surfaces, arrest marks often replace undulations, plumes replace striae, and minute features including Wallner lines on the mirror plane and mist along the mirror boundary would be unrecognizable. Distinctions between arrest marks and undulations are given below, and distinctions between striae and plumes follow.

2.2.4.2
Fractography on Mirror Planes

We extend below the treatment of fractography on mirror planes from Bahat (1991a) and make some changes along the following lines:

1. Ripple marks on mirror planes.
2. Arrest marks and striae on geological exposures.
3. Undulations on geological exposures.
4. Striae and plumes.

Ripple marks on mirror planes. Ripple marks are systematized on mirror planes according to two categories, type and shape. Two types of ripple marks are distinguished:

1. Arrest marks that signify crack arrest.
2. Undulations that form on the fracture surface while it propagates.

Undulations may form in different fracture velocities, often starting when propagation is slow, and continue at greater velocities (Fig. 2.10), or even maintain slow propagation. Ripple marks occur in various shapes, including curved cross sections (Fig. 2.10), cusps (Fig. 2.12), and angular cross sections that may attain "box" (defined by four kinks) and "semi-box" (defined by three kinks) structures (see below). These distinctions are important for the quantitative interpretation of the mechanism of fracture in geological exposures by using v and K parameters (Sect. 4.8.3).

Arrest marks and striae on geological exposures. The distinction between undulations and arrest marks has been obscured in the past, seriously hindering the application of fractography in engineering materials and structural geology for the interpretation of fracture processes. Particularly, it was impossible to distinguish between ripple marks that formed by rapid fracture (undulations) and ripple marks that reflected slow propagation (arrest marks). Here are a few examples and rules that can help in identifying arrest marks both in experimental specimens and in rock outcrops:

1. Arrest marks may occur in all shapes and sizes, including angular cross sections (Fig. 2.33b,c). Asymmetric cross sections are particularly common.
2. Following the results obtained by Guin and Wiederhorn (2003) (Sect. 2.2.3.1), an arrest mark is associated with an inducement of mixed mode I + III loading at its front (Fig. 2.9), such that striae would be induced on the mirror while *commencing* from this front. Thus, an arrest mark is frequently associated with intense striae morphologies, characterized by either multiple striae that superpose arrest marks or by the increase of the intensity of individual striae, or both (Fig. 2.5a and 2.33b–d).
3. Quite often striae on the mirror plane and particularly in association with arrest marks resemble fringe cracks and may erroneously be interpreted as *en echelon* segments or hackle cracks. The distinction between these two features may be primarily based on two striae types. First, many short striae start when superposing an arrest mark but they fade away after a small distance from it (Fig. 2.33b–d), compared with fringe cracks that maintain their lengths, or even increase it (Fig. 2.34,

2.35, 2.33e). Second, when striae start between arrest marks they merge into a single stria of intense morphology at the arrest mark, compared to *en echelon* segments that maintain their number or hackles that even multiply (Fig. 4.11a). Figure 2.33d shows quite well the two morphological types. Both morphologies signify slow fracture propagation, and an immediate transition from an arrest mark to a hackle fringe that signifies rapid fracture would be an unlikely occurrence (this criterion may be used for examining this theory in future observations).

4. Arrest marks may form in the shape of "box"-shaped structures that superpose mirror planes (Fig. 2.33b,c) or as "semi-box" structures that occur even on fringes (Fig. 2.30a). In the latter case, the right side of the semi-box on the fringe indicates reduction in fracture velocity, compared to its left side, which is marked by hackles that indicate rapid fracture. Note that the striae and the "semi-box" on the fringe are significantly coarser than the striae and undulations on the mirror (see more in Bahat and Rabinovitch 2000).

Finally, many, but not all arrest marks may be recognized. Some more arrest marks that can be identified on rock fractures are presented in Bahat (1991a, Fig. 3.42c, 4.2a,b, 4.4 and 4.11), they all exhibit striae according to the above rules. Occasionally, transitions between areas of multiple striae into mirror areas that are not imprinted by arrest marks and do not contain striae (Fig. 4.36) possibly suggest a reduction in fracture velocity but not quite an arrest.

Undulations on geological exposures. How do undulations differ from arrest marks in geological exposures? Two examples of undulation propagation are described and interpreted below. As mentioned above, when distances between undulations increase, fracture velocities increase (Fig. 2.10). The undulations in Fig. 3.24c display fracture "diffraction" along boundary A, such that they change orientation and trail behind along boundary B. At some stage, however, drag is reduced, resulting in velocity jump (at point C) to conform to a uniform velocity in area D. There is a difference between the shapes of the upper ripple mark and the lower ones at the right side of Fig. 3.24d. The upper one is convex towards the direction of propagation, which is faster midway between the two bounding sub-vertical surfaces, and trailing behind along the surfaces. This is a characteristic feature of undulation propagation on experimental fractures (e.g., Bahat 1991a, Fig. 2.1c). The lower ripple marks are straighter, implying an approach to equal velocity along the fracture front. The ripple marks on the left side of Fig. 3.24d are inclined as they come closer to their arrest at the older horizontal joint, suggesting that these are arrest marks. Note that the undulations are approximately symmetrical and the arrest marks are asymmetric.

Striae and plumes. Striae that form by rapid fracture in glass or glass ceramic penetrate into the material and propagate as sharp, continuous cuts along straight trajectories, unless influenced by boundaries that may deflect them. For instance, there is a change from approximately straight striae within the material in their early part (closer to the F end) and a gradual curving in their continuation towards the B end as they approach the sample upper surface (closer to the viewer, Fig. 2.19a). They may lose some of their sharpness while cutting granular rocks when moving fast (see delicate straight strie and hackle in Fig. 2.30a). They may completely lose their sharpness and sub-divide into dendrite-like plumes, such that every dendrite branch would be further sub-divided into minute barbs (Fig. 2.3b) when propagating slowly (Fig. 2.22). This may explain why the topography of plumes on mirrors that form dynamically either in the laboratory (Fig. 2.10), by explosives (Fig. 2.30b), or in geological outcrops (Fig. 2.33e, 2.56, 4.11a) is often delicate (or hardly visible), whereas on slowly forming joints, or when they curve (Fig. 4.22), plumes may cover significant areas and attain coarse morphologies (e.g., Fig. 4.7b).

Summary. Two types of mirror planes are distinguished. One type characterizes fractographies that form dynamically by rapid fracture when high stresses are exerted. Such mirror planes are smooth, often distinguished as polished-like surfaces that reflect light. The second mirror type characterizes fractographies that form quasi statically by slow fracture when low effective stresses are exerted. This mirror type characterizes surfaces of many systematic single-layer joints that cut sedimentary rocks; they maintain grainy, matte surface morphologies.

We divide ripple marks in geological exposures into two categories. First, "undulations" generally deviate slightly from symmetrical cross sections (Fig. 3.24c) that signify fracture propagation in various velocities. Second, "arrest marks" often are cut characteristically by striae and deviate considerably from symmetrical shapes, with the occasional tendency to form "box" or "semi-box" cross sections (Fig. 2.33b,c). These marks signify slow propagation down to arrest.

Striae that form by rapid fracture in glass or glass ceramic penetrate into the material and approximate sharp, continuous cuts (Fig. 2.19a). However, when striae propagate slowly in granular rocks, they may completely lose their sharpness and form dendrite-like plumes (Fig. 2.22a, 2.45).

2.2.5 Joint Fringes

2.2.5.1 *Introduction*

We start with a first order distinction between two end member fringe types, *en echelon* fringes (Fig. 2.1a) and hackle fringes (Fig. 2.1b). *En echelon* segmentation from parent fractures generally occurs in reduced velocities (Kulander et al. 1979; Müller and Dahm 2000), standing for the "quasi-static end member", at a velocity of $\leq 10^{-1}$ m s^{-1} (Wiederhorn et al. 1974) in the "fringe spectrum". Hackles form at high velocities (e.g., Kerkhof 1975), representing the "dynamic end member" at a velocity of $\geq 10^{-1}$ m s^{-1} (Wiederhorn et al. 1974). Between these two end members, there are two distinct fringes that bound mirror planes on joint surfaces. One of them consists of somewhat hackled *en echelon* segments (Fig. 2.33e) that display different lengths, resulting in no uniform width of the fringe (Fig. 2.34a). The other one consists of *en echelon* segments of equal lengths that are not hackled (Fig. 2.35a,b), often resulting in a uniform fringe width (Fig. 2.34b).

The need for this distinction has been previously raised, but it has been based on limited data derived from rock exposures in sedimentary rocks (Bahat 1991a, Table 3.3; Bahat 1997). Recent studies on fracture in granites enriched the information on the properties of geological hackle fringes (Chap. 4) so that a sound distinction between the two fringe types becomes possible. Understanding the full mechanical significance of the spectrum between hackle fringes and *en echelon* fringes around mirrors is quite a challenge. This understanding would improve the tectonofractographic understand-

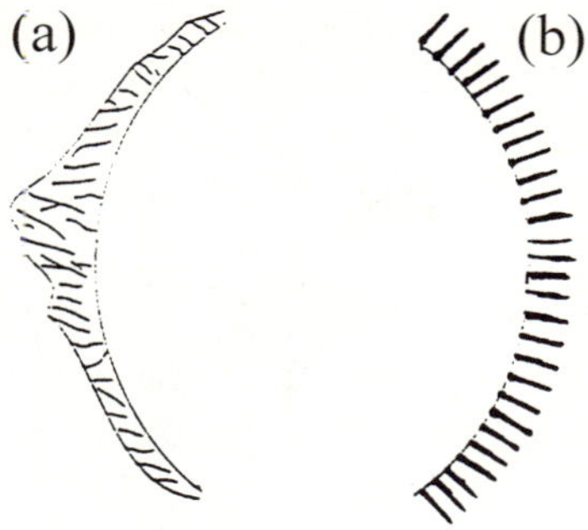

Fig. 2.34.
Diagrams of two types of fringes. **a** A no uniform width of the fringe of hackly *en echelon* segments that have different lengths. **b** A uniform width of the fringe of *en echelon* segments of approximately the same lengths (after Bahat 1998b)

ing of joint formation, including the quantitative estimation of paleostress conditions leading to fracture, as well as applications to other research fields, such as petrology (Chap. 4 and 6). It is appreciated, however, that the above distinction between the two end member fringes is not always simple, and the question whether a particular marking is to be classified as a hackle fringe or *en echelon* fringe may arise in some cases. Occasionally *en echelon* segments and hackle cracks populate the same fringe (Fig. 2.1c). Also, there is an experimental finding of fast propagation of *en echelon* cracks (Sect. 2.2.3.10). Thus, an awareness of occasional difficulties in the distinction between the two end member fringes is required. *En echelon* fringes are better known to geologists than hackle fringes. Also not well known are cuspate hackles, but these form a small subgroup that is not central in this chapter. Accordingly, the next two sections (Sect 2.2.5.2 and 2.2.5.3) present a recap on *en echelon* fringes and on the sub-group cuspate hackles, respectively (partly from Bahat 1991a, pp. 170–183). The issues of hackle fringes and transitions between the two end member fringes are expanded in Chap. 4.

2.2.5.2
The En Echelon *Fringe*

General. *En echelon* segmentation in non-coplanar planes is ubiquitous on various scales from micrometers to kilometers in geological exposures occurring in different terrains. This implies that breakdowns may be the consequences of various mechanisms (see a comprehensive description of *en echelon* distribution in Pollard et al. 1982). Interaction of non-coplanar cracks was observed by De Freminville (1914) in bitumene. Following that, Preston (1931) modeled the process leading to striae and *en echelon* segmentation in glass. He pioneered the explanation that when a change in the direction of the principal tension occurs, the crack must break down into a number of segments so that each segment readjusts its orientation to maintain normality with respect to the new direction of the principal tension. This mechanism has since been reaffirmed (e.g., Sommer 1969) and was formulated into a sophisticated fracture mechanics theory of mixed mode I + III loading (Lawn and Wilshaw 1975a; Ryan and Sammis 1978; Pollard et al. 1982; Cooke and Pollard 1996; Hull 1999), essentially following the rules leading to the formation of the plumes by the same mixed mode (Fig. 2.20c,d).

Starting with some of the earlier works (e.g., Woodworth 1895, 1896; De Freminville 1914; Murgatroyd 1942; Hodgson 1961a,b; Bankwitz 1965, 1966; Sommer 1969; Kulander et al. 1979; Bahat 1979a, 1986b; Pollard et al. 1982), the study of *en echelon* fringes involved various fracture aspects that have been outlined by Bahat (1991a).

These include fracture geometry, breakdown styles, *en echelon* segments and connecting bridges (or steps), fracture markings on segments, breakdown mechanisms, the different roles of the dilatant *en echelon* cracks vs. their bridges, and the distinction between the influences of remote and local stresses on the evolution of the fringe.

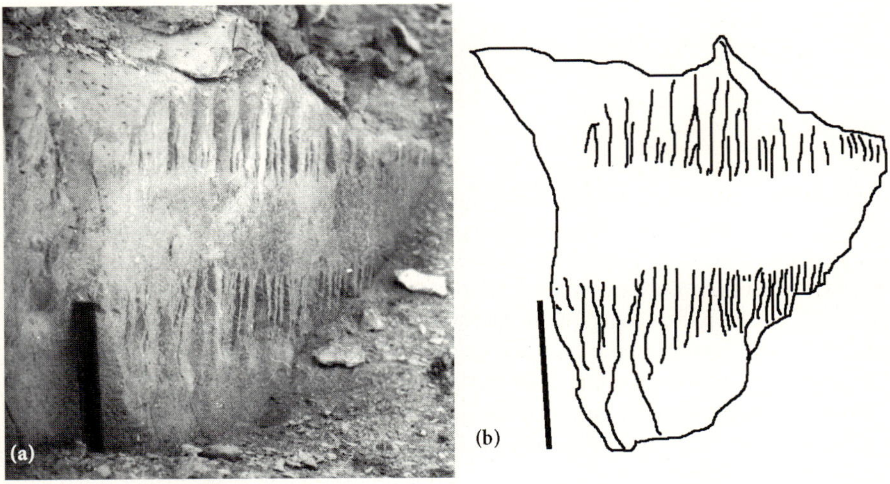

Fig. 2.35a,b. Styles of *en echelon* segmentation. **a** Photograph (from Bahat 1997) and **b** drawing of discontinuous breakdown of *en echelon* fringes both above and below the parent joint. The sense of rotation of the segments in the two fringes is counter clockwise. Note similarity in shape and size of segments in both fringes. *Scale* is 15 cm

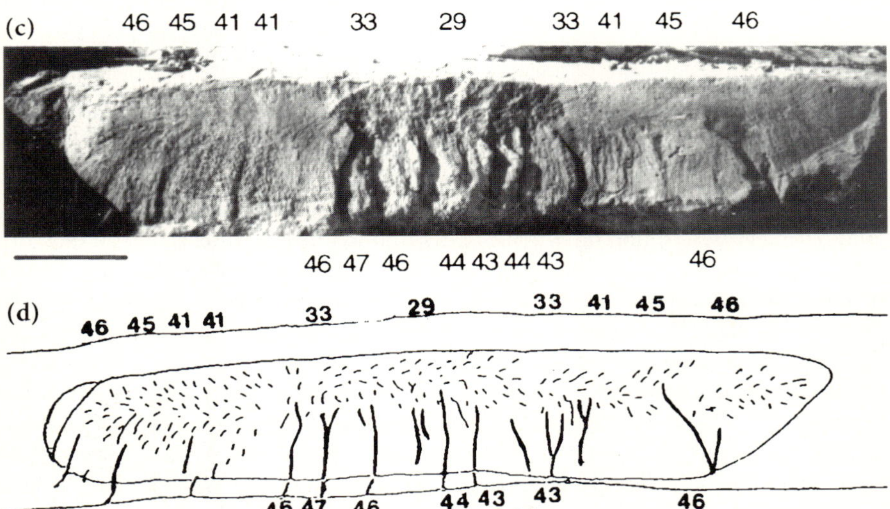

Fig. 2.35c,d. Styles of *en echelon* segmentation. Photograph (**c**) and drawing (**d**) of the curved-joint with an intense discontinuous breakdown into many segments, below the elliptical perimeter of the early "embryonic" joint at center, and a continuous breakdown at the distal sides. The joint undulates along the strike. The *upper numbers* are measured azimuth degrees (from 029° to 046°) on the parent joint. The *lower numbers* are azimuth degrees (from 043° to 047°) on *en echelon* segments. The delicate barbs on the joint surface represent a bilateral plume. *Scale bar* is 50 cm (after Bahat 1997)

Several aspects of *en echelon* segmentation are expanded below, making the distinction that fracture markings on the parent joint record events from the early history of the joint, while fringes reveal processes that were associated with advanced and final stages of fracture.

Fig. 2.35e.
Styles of *en echelon* segmentation. "Double fringes" beyond the boundary of the mirror plane on a joint cutting limestone. Note the reverse sense of stepping on the two fringes. The first fringe shows *en echelon* segmentation on which there is an "upstairs climbing" from right to left, while the second fringe shows climbing in the opposite direction, implying change in the stress field between the formation of two fringes (a *non-in situ* Cenomanian sample from Iben-Dib, south Jordan)

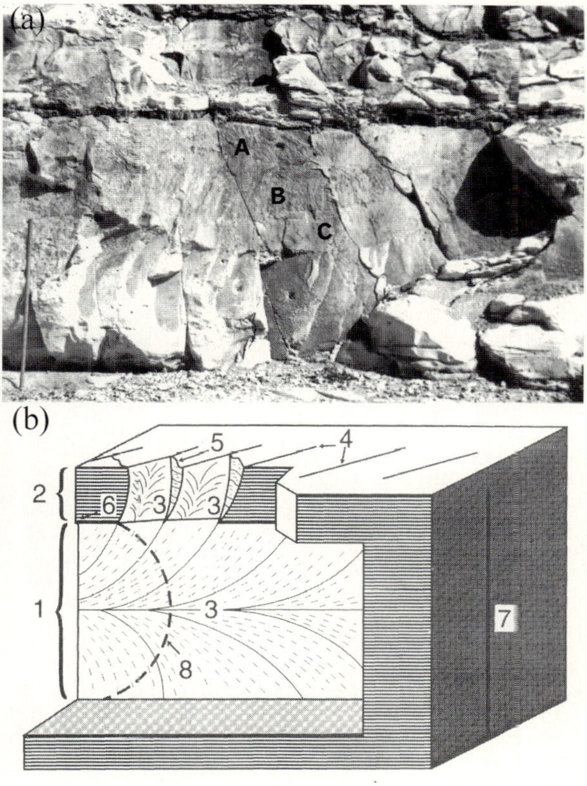

Fig. 2.36.
a A joint from outcrop R showing a division into two parts. An upper semi-ellipse parent joint concave downward is separated by a curved shoulder from a fringe of *en echelon* segments below. There is a (hardly visible) division of the parent joint into three concentric zones (*A–C*, see text for more information). Note the steps of different sizes at a discontinuous breakdown along a curved shoulder. *Scale* is 1 m (after Bahat 1997). **b** Schematic block diagram showing primary surface markings of a systematic joint: *1*: Main joint face; *2*: Fringe; *3*: Plumose structures (plumes); *4*: F-joints (b-planes); *5*: Cross-fractures (c-fractures); *6*: Shoulder; *7*: Trace of main joint face. Note that cracks 5 are steps that bridge adjacent *en echelon* cracks in the fringe. Note in **a** and **b** vertical and lateral fracture fronts, respectively (modified by Roberts (1995) from Hodgson (1961a))

Breakdown styles. Two breakdown styles are distinguishable: Discontinuous (Fig. 2.35a and b) and continuous (Fig. 2.35c and d). In layered rocks, the discontinuous breakdown starts along a shoulder at the edge of the parent joint from where the steps start and approach the layer boundary. When the shoulder is straight and parallels the adjacent layer boundary, steps of a given fringe would reach approximately the same maximum size (Fig. 2.35b). However, a discontinuous breakdown with curved shoulders may have steps of different sizes (Fig. 2.36a). Note that in Hodgson's diagram (Fig. 2.36b), *echelons* appear in a discontinuous style and the plumes initiate on them at the fringe of the parent joint (not on the parent joint itself). A discontinuous breakdown is often characterized by a "root zone" (Bankwitz 1966). This zone is a narrow strip of small *en echelon* cracks along the joint shoulder to which larger *en echelon* cracks are added as the fringe develops. The continuous breakdown starts on the joint plane at various distances from the layer boundaries and the steps have no shoulder as a starting line. Such breakdowns often display fringes with steps of quite different shapes and sizes (Fig. 2.35d). On one continuous breakdown, Bankwitz (1966) shows plumes that initiate on the parent joint and continue to propagate along the *en echelon* crack.

Abrupt and gradual loadings. It is inferred from the experimental results by Cooke and Pollard (1996) (present Fig. 2.20c,d) that generally, breakdowns that are manifested by discontinuous *en echelon* segments reflect processes of abrupt twisting. These are associated with abrupt mixed mode I + III loading that occurs between the development of the parent joint and the fringe and produces *en echelon* twisting (e.g., Fig. 2.35a,b). On the other hand, continuous breakdowns are associated with gradual twisting, responding to mixed mode loading as the joint propagates and segmenta-

Fig. 2.37.
Mineralization on *en echelon* cracks (*dark color*) but not on the steps that connect them (alternating white fractures) demonstrates two separate phases of failure. The sequence of events in this chalk sample (from the Middle Eocene) was fracture of the parent joint (note delicate plumes), *en echelon* cracking, mineralization of iron oxides that was followed by dendrites of manganese oxides on the parent joint and *en echelon* cracks, and finally, delayed step formation, after flow of solutions had seized (sample length from top to bottom, about 12 cm, after Bahat 1986b)

tion occurs (Fig. 2.35c,d). The gradual mixed mode loading probably does not involve any time-break during the process. It is, however, possible that time breaks do occur during discontinuous processes, but their extents are not known (Fig. 2.35e). Quite possibly the different fracture modes of the *en echelon* cracks and steps that cause propagation vertically downward and laterally, respectively, may be associated with a time interval between them (Fig. 2.37). Two fracture paths are manifested by part D of the composite fracture (Fig. 2.38, see also Fig. 3.40a) along Wadi Secher (Fig. 3.33c). An abrupt clockwise breakdown occurred in part D, striking into a path of twisted segments (azimuths 080–083°). However, as the loading increased while these segments propagated downward beyond the arrest mark, a curved path took over, and a gradual counter clockwise rotation of the segments occurred (into azimuths 072–076°).

In outcrop Y (Fig. 3.33c) continuous and discontinuous breakdowns occur in the same layer (Bahat 1986a). Furthermore, the "curved joint" was first segmented by a discontinuous process underneath the early elliptical fracture on the curved joint, and the second breakdown was continuous along the distal edges (Fig. 2.35c,d). The close proximity of these two breakdown styles suggests that their fracture conditions were not drastically different. The curved joint (Fig. 2.35c,d) and the composite fracture (Fig. 2.38 and 3.40a) demonstrate that the continuous breakdown may have two manifestations. First, a gradual growth of *en echelon* segments from distal locations on the parent joint (Fig. 2.35d) and second, a gradual curving of growing segments (D in Fig. 3.40a). These two fracture manifestations are associated with relatively small twist angles ($0° \leq \omega \leq 11°$). On the other hand, discontinuous breakdowns in the curved joint and in other joints from stations Y and R (Fig. 3.33c, 2.35d, 2.36a) are connected with larger angles ($12° \leq \omega \leq 25°$) (Bahat 1997).

Fig. 2.38. Photograph of a composite fracture from Wadi Secher (see explanation in the text and in Fig. 3.40a) (after Bahat 1997)

Breakdown mechanisms. We discuss below two fracture mechanisms that form *en echelon* fringes by *local* mixed mode loading. The "breakdown" model by Pollard et al. (1982) concentrates on the mixed mode I + III that develops by an abrupt process during the transition from the parent joint to the fringe (Fig. 2.39a–c and 2.40a,b). The "fracture slanting" model (Bahat 1991a, 1997) has some resemblance to the "breakdown" model and supports its basic principles. An alternative model by Ramsay and Lisle (2000) that applies the mixed mode I + III by *remote* extension and shear stresses is presented in Sect. 6.10. See also reservations regarding the "breakdown" model by Lazarus et al. (2001a,b).

Pollard et al. (1982) examined the mechanism of a fracture breakdown into echelon cracks, and this section basically is cited from their work. They use three idealized block diagrams (Fig. 2.39a and 2.40a,b) and show that

$$s = c \sin \omega \tag{2.6a}$$

and

$$m = \frac{W}{2} - k = \frac{W}{2} - c \cos \omega \tag{2.6b}$$

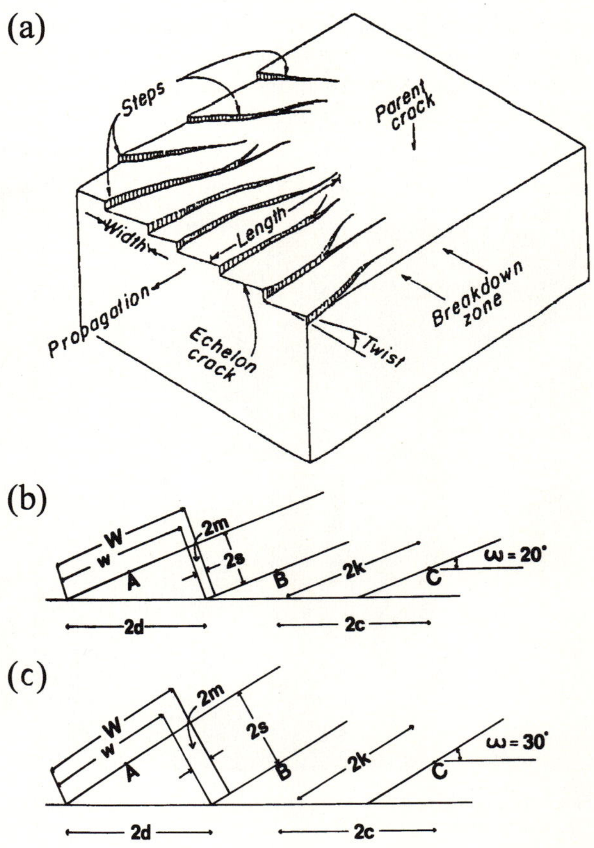

Fig. 2.39. **a** Block diagram illustrating the geometry and mutual relations between parent crack and *en echelon* cracks (after Pollard et al. 1982). **b**, **c** Schematic sections of a discontinuous fringe normal to *en echelon* crack lengths. W = crack width, w = part of the width which is not overlapped, 2 m = the overlapped portion of the width, when $W = w + 2$ m; $2s$ step, the normal distance between two adjacent cracks, $2k$ = the distance between centres of two adjacent cracks parallel to cracks, $2c$ = the distance between centers of two adjacent cracks along the shoulder, $2d$ = the distance between initiation points of two adjacent cracks along the shoulder, and the twist angle ω, between the shoulder and *en echelon* cracks. **b** $\omega = 20°$; **c** $\omega = 30°$. A comparison between **b** and **c** shows that overlapping increases with ω. At low ω, $W \cong w$ (after Bahat 1991a)

where ω is the twist angle, $2c$ and $2k$ are different distances between centers, and $2s$ is a step. Pollard et al. (1982) adapt Sommer's (1969) explanation on fracture breakdown, but add the internal pressure P to the normal stress σ_{yy}^{∞} acting on the fracture plane (tensile stress is positive) (Fig. 2.40a,b).

Pollard et al. consider an elastic body subject to remote principal stresses σ_1^{∞} and σ_2^{∞} acting in the y-z-plane, and a fracture of width $2b$ subject to pressure P lying in the x-z-plane with straight tips parallel to z (length and width of parent fracture by Pollard et al. are reversed in the present text, to be consistent with other descriptions). Breakdown into echelon segments is independent of the third remote principal stress, which acts in the propagation direction x. A uniform rotation Ω of the remote stresses about the x-axis introduces a normal stress σ_{yy}^{∞} and a shear stress σ_{yz}^{∞} acting in the fracture

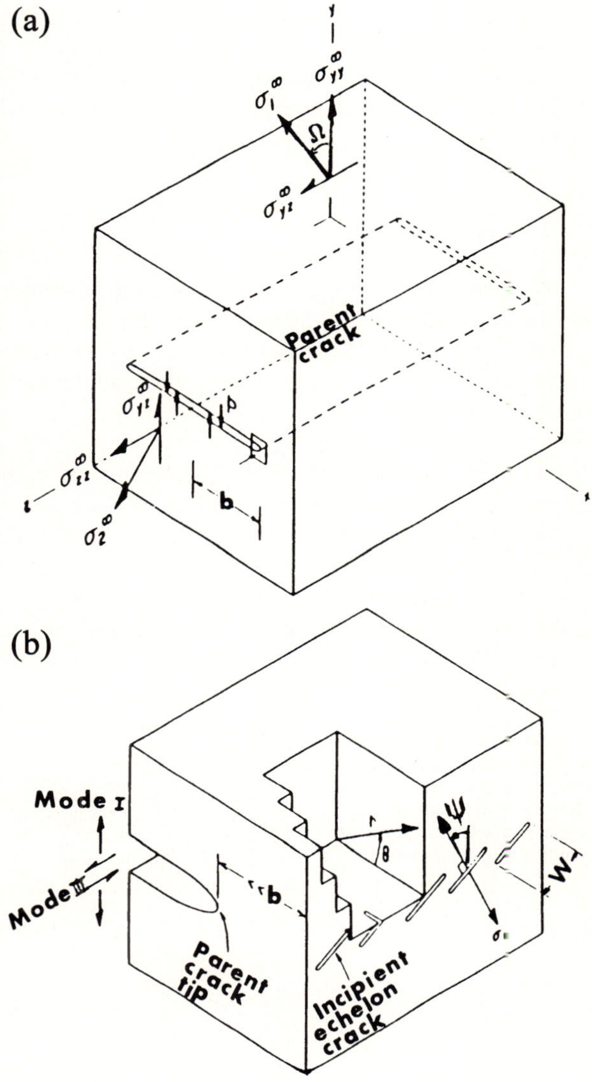

Fig. 2.40. Block diagrams illustrating relations among applied stresses and cracks. **a** Parent crack of width $2b$ and infinite length in z is loaded by pressure P and remote principal stresses σ_1, σ_2. Rotation Ω of principal stresses about x introduces stresses σ_{yy} and σ_{yz} acting on the crack plane. **b** Small element cut from near parent-crack tip (*small arrowed square* in **a** with incipient echelon cracks of width W whose normal, the maximum local tension σ_1 making angle ψ to the y-axis. Polar coordinates centered at parent-crack tip are r, θ. *Heavy arrows* indicate relative motion of parent crack walls when subject to mode I and mode III deformation (modified after Pollard et al. 1982)

plane. According to Pollard et al., breakdown of the parent fracture depends upon Ω and the applied stress ratio

$$R = \frac{2P + \sigma_1^\infty + \sigma_2^\infty}{\sigma_1^\infty - \sigma_2^\infty}, \qquad P > -\sigma_1^\infty \tag{2.7}$$

The inequality relating P and σ_1^∞ is required for the parent fracture to be open. The numerator in Eq. 2.7 is the difference between the internal pressure and the remote mean stress or pressure, $-(\sigma_1^\infty + \sigma_2^\infty)/2$. The denominator in Eq. 2.7 is the remote maximum shear, $(\sigma_1^\infty - \sigma_2^\infty)/2$.

Near the crack tip the stress field is proportional to the stress intensity factors. Pollard et al. resolve that the remote normal stress and internal pressure contribute to a mode I stress intensity, $K_\mathrm{I} = (\sigma_{yy}^\infty + P)(\Pi b)^{1/2}$, and the remote shear stress introduces a mode III stress intensity, $K_\mathrm{III} = \sigma_{yz}^\infty(\Pi b)^{1/2}$. The tendency for breakdown increases as the ratio of stress intensities $K_\mathrm{III}/K_\mathrm{I}$ differs from zero. Writing σ_{yy}^∞ and σ_{yz}^∞ in terms of Ω and the remote principal stresses by resolving stress in the x-z-plane, Pollard et al. obtain

$$\frac{K_\mathrm{III}}{K_\mathrm{I}} = \frac{\sin 2\Omega}{R + \cos 2\Omega}, \qquad R + \cos 2\Omega > 0 \tag{2.8}$$

Inequality of R and $\cos 2\Omega$ is required for the parent fracture to remain open. For a particular rotation angle, the stress intensity ratio increases as the stress ratio R decreases. For $R \cong 1$, which is reasonably close to the range of R in nature, $K_\mathrm{III}/K_\mathrm{I} \cong \tan\Omega$, and Pollard et al. find that relatively small rotations of the principal stresses could initiate breakdown.

The fracture slanting mechanism. Various fracture properties of both edge and central cracks were investigated in metallic plates as a function of their thickness (Bank-Sills and Schur 1989). It was found that for both geometries, as plate thickness increases the ratio of the critical stress for the slant crack divided by the critical stress for the flat (unslanted) crack increases at failure, and the crack propagation rate for the slant crack is slower than that for the flat crack (Fig. 2.41). Bank-Sills and Schur (1989) observed that for both edge and central slant cracks, the $K_\mathrm{III}/K_\mathrm{I}$ ratio increases as plate thickness de-

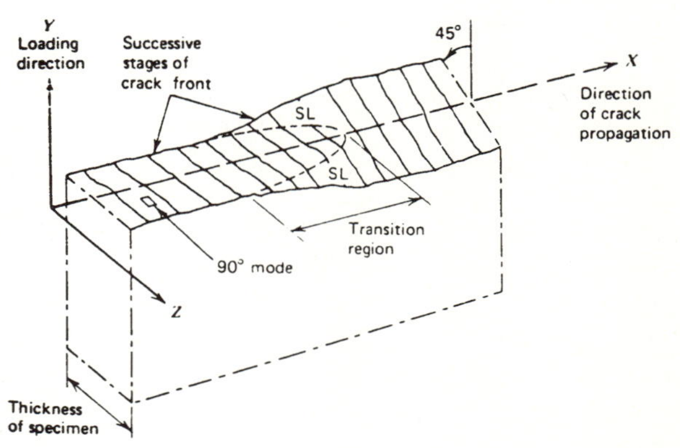

Fig. 2.41.
Diagram showing fracture mode transition from flat fracture under plane strain conditions (*left*) to slant fracture under plane stress conditions (*right*) (after Hertzberg 1976)

creases and that the greatest value of this ratio occurs at the plate surface. Furthermore, slant cracks in thinner plates will propagate more rapidly than those in thicker ones. Thus, as thickness increases the tendency for slanting decreases, and the tendency for slanting becomes greater in thinner plates as plane stress conditions are attained.

These experimental results reaffirm previous observations (Hertzberg 1976; Broek 1984) that cracks in tensile fields under fatigue loading propagate out of their plane in thin plates, and they also show that this property is thickness dependent. The tendency for slanting that is associated with the increase in the K_{III}/K_I ratio becomes greater with the decrease in thickness of the plate and from the center of the plate towards its edges. Cooke and Pollard (1996) investigated crack segmentation in PMMA under various K_{III}/K_I conditions. They found that for ω of the *en echelon* cracks and the number of segments formed, both increase with the increase of the ratio K_{III}/K_I (Sect. 2.2.3.8, see also Pollard et al. 1982). Hence, the main resemblance between the "breakdown" and "fracture slanting" models is in the dependence of *en echelon* segmentation on the K_{III}/K_I parameter (Sommer 1969). The fracture slanting model, however, links (and limits) the increase in K_{III}/K_I to plane stress conditions, i.e., to fringes along joints in layered rocks, and it does not apply to *en echelon* segmentation in massive rocks like granite, a limitation not articulated by Pollard et al. 1982. An extension to other quasi-plane stress settings of stratified structures, such as segmentation of large igneous dikes near the ground surface (Anderson 1951) or *en echelon* fracturing on the ground along branches of major strike-slip faults (e.g., Allen et al. 1972), should not be ruled out.

Non fracture slanting mechanisms. Fringes along curved boundaries often exhibit changes in the senses of segment stepping, implying slight deviations from orthogonality of the maximum extensile direction with respect to the parent joint. This mechanism does not operate along straight layer boundaries, but it may operate in connection with *en echelon* segmentation that often occurs on large joints in sandstone at remote locations from straight boundaries (e.g., Fig. 3.23a). Scarcity of hackles on such surfaces seems to suggest sub-critical conditions. Another type of *en echelon* segmentation occurs along fracture fronts when critical stress intensity $K_I \geq K_{Ic}$ and rapid crack propagation conditions are attained (Chap. 4). *En echelon* fringes around exfoliation joints may (Fig. 2.33e, 3.24b and 3.26c) or may not (Fig. 3.24c, 3.25a) attain critical fracture conditions. On fringes along curved surfaces see Sect. 4.5.3.2.

2.2.5.3
The Cuspate Hackle Fringe

Cuspate hackles are distinguishable from *en echelon* and hackle fringes. Cuspate hackles have sharp ridges of approximately constant width that radiate in a fringe around the main fracture. They are common on blasted fracture surfaces (Fig. 2.30b) and rare on joints (Fig. 2.42). The fracture marking from the Charing chalks in Kent, England displays a curious type of fringe. The echelon cracks occur above and below the plume, whereas cuspate hackles continue the trend of the plumes along the distal direction (left side of joint). This is interpreted as greater stress intensity along the lateral fracture propagation as compared to the upward and downward directions. The common occurrence of cuspate hackles in quarries (Bahat 1991a, Fig. 3.60a–c), imply fracture by high energy conditions (fast crack propagation).

Fig. 2.42. Cuspate hackles. A plume developed into right-stepping and left-stepping echelons at the upper and lower parts of the marking, respectively, but transformed laterally towards left into cuspate-hackles, suggesting a lateral cracking under conditions of greater stress intensity. From an Upper Cretaceous chalk in south England (after Bahat 1991a)

The sharp peaks of cuspate hackles are distinct from the flat surfaces of *en echelon* cracks. Crack overlapping, which is a typical feature of *en echelons*, does not occur in cuspate hackles. They have not been identified on fracture surfaces of technical glasses. When they occur on joint surfaces they seem to serve the role of hackles on fracture surfaces of glass. There is a strong resemblance between cuspate hackles on joints and fractures that form by blast to fracture surfaces of some rapidly fractured ceramics (Fig. 2.33a), even if size scales are an order of magnitude different. As mentioned above, hackle fringes are described in Chap. 4.

2.2.6 Mirror Boundaries

We characterized in the preceding sections the properties of the mirror plane and the fringe. Another important parameter is the boundary that separates these two parts of the joint. The importance of finding well-defined critical flaws when coupled with well-defined mirror boundaries provide the necessary conditions for estimating the local tensile paleostress on a joint surface (Bahat and Rabinovitch 1988) (Sect. 6.5). In plutonic rocks, there are occasionally favorite conditions for finding clear mirror boundaries (e.g., Bahat et al. 1999, 2003, see present Chap. 4), no such clarity has been so far obtained in sedimentary rocks. We now consider the criteria for elucidating mirror boundaries of single-layer joints. At least one of the following two criteria must be identified on a joint surface for determining a mirror boundary, while the third and fourth criteria can increase the credibility of the definition:

1. There is a stepwise morphological increase in the size of the cracks across the boundary from plumes on the mirror to fringe cracks (Fig. 2.22b and 2.33a,e). This has to be consistent along at least 50% of the suspected mirror boundary of the investigated joint.
2. The tilt angle ϕ is formed between the imaginary continuation of the mirror plane and the fringe (Bahat and Rabinovitch 2000). This angle is better seen in profile than in plan view (Fig. 4.11c).
3. The ratio of mirror radius (Fig. 2.1d) to the radius of the critical flaw should be 15 ±5 (Bahat and Rabinovitch 1988).
4. Apart from size difference between plumes and fringe cracks, there are occasionally changes in the sense of stepping in the transition from the plume to the *en echelon* segmentation on the fringe (Fig. 3.40b).

2.2.7 Recent Studies on *En Echelon* Fringes from the Beer Sheva Syncline

This section relates to the occurrence of fringes on joint surfaces from sedimentary rocks in the Beer Sheva syncline, based on the study by Bahat (1997) that we cite below. Here we concentrate on fractographic relationships, while the geological implications of this study are presented in Sect. 3.4. The Eocene chalk layers from the Beer Sheva syncline (Fig. 3.33) are only slightly deformed platform cover sediments. Chalk is a weak rock with considerable change in properties when exposed to aqueous solutions. Therefore, fracture may be readily induced in this material, even in very low differential stresses. The focus here is on bed restricted uplift joints (Bahat 1999a). Jointing in Eocene chalks around Beer Sheva is enriched with various phenomena of *en echelon* segmentation that bear relationships to diverse fracture styles and multiple joint systems (Chap. 3).

2.2.7.1 *The Relationship of* En Echelon *Length to Layer Thickness*

Lengths of *en echelon* segments cutting layers of various thicknesses were measured along stations Y, R and Q in Wadi Naim (Fig. 3.33c). In these locations, *en echelon* fringes occur only below the parent joint (Bahat 1986a). The length of the longest segment for each joint was measured from its initiation point to the lower layer boundary. The layer thickness at that location was measured as well (estimated error for both measurements is ±2 cm). Five continuous and fourteen discontinuous segmentations were studied, and the results are summarized in Fig. 2.43. There is a positive linear fit ($R^2 = 0.928$) of the readings for discontinuous segmentation, which is given by $l = 0.270T + 0.829$, where l is segment length and T is bed thickness. R^2 is reduced to 0.836 when one reading ($T = 210$ cm, $l = 55$ cm) is omitted. The readings for the continuous segmentations scatter above the curve. There is no evidence of any layer-parallel slip, along either chalk-chalk or chalk-chert boundaries linked to *en echelon* fringes in the stations investigated in this study. On the other hand, there are many inter-layer radial markings on horizontal layer boundaries that reflect tensile vertical strains perpendicular to layer boundaries in the chalk (Bahat 1991a, p. 298).

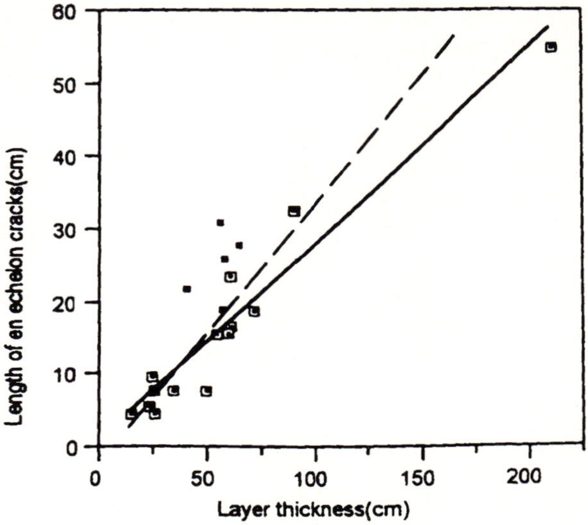

Fig. 2.43.
A plot of layer thickness vs. length of *en echelon* cracks in chalk outcrops along Wadi Naim. The large fourteen *open squares* and the five *solid squares* designate respectively discontinuous and continuous breakdowns. The *solid* and *dashed plots* are the products of 14 and 13 readings of discontinuous breakdowns, respectively. The last reading ($T = 210$ cm, $l = 55$ cm) is omitted from the latter. The former is considered to be a better representation of the results (after Bahat 1997)

2.2.7.2
Similar Fringes below and above the Parent Joint

En echelon segments maintain counter clockwise rotation in fringes close to both the lower and upper boundaries in outcrop X (Fig. 2.35a,b), and the two fringes, above and below the parent joint, have similar geometries and shapes. This coupling implies that *en echelon* segmentation occurred simultaneously in the two fringes. It took place after the end of the required diagenetic process for the equalization of the lithological properties along both the lower and upper layer boundaries. The fractographic features of the joints in outcrop X are quite distinct from the fractographies of all other joints in the investigated area.

2.2.7.3
Opposite Senses of Segment Rotation in a Single Fringe

The observation of the close proximity of both clockwise and counter clockwise rotations of *en echelon* cracks in a single fringe is quite intriguing. Based on experimental results (Cooke and Pollard 1996), it follows that the clockwise twisted segments in fringes reflect right-lateral shear, and the counter clockwise twists disclose left-lateral shear. There is a need to distinguish between two cases of opposite senses of segment rotation in a single fringe: First, in a curved fringe, and second, in a straight fringe.

Opposite senses of segment rotation in curved fringes are quite common in fractured engineering materials and in association with discoid joints. Segment rotation is counter clockwise on the left side of a rounded parent fracture in steel and is clockwise on the right side of the fracture (Fig. 2.44a, after Yukawa et al. 1969). Exactly the same pattern occurs around a discoid joint in the Frauenbach quarzite from the Thuringian Schiefergebirge (Fig. 2.44b, after Bankwitz and Bankwitz 1984). In fact, Bankwitz and Bankwitz (1984, Fig. 2) observed a symmetry that was created by a reverse segmentation. The same segmentation pattern is repeated on the surface of a fractured glass rod that failed under combined operation of modes I and III (Sommer 1969, Fig. 2).

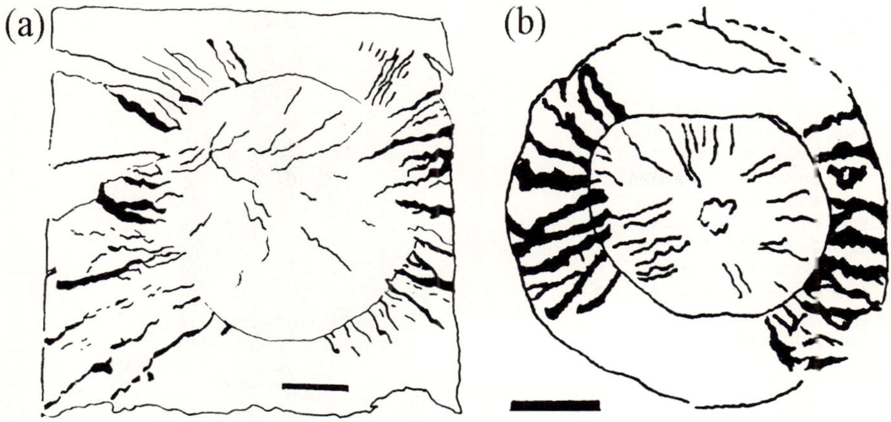

Fig. 2.44. Drawings of fringes around circular parent fractures. **a** Segment rotation is counter clockwise on the left side of a fracture in steel and is clockwise on the right side of the fracture (after Yukawa et al. 1969, Fig. 1). *Scale bar* is 12 mm. **b** A repeated pattern from **a**, around a discoid joint in the Frauenbach quarzite from the Thuringian Schiefergebirge (after Bankwitz and Bankwitz 1984, Fig. 3). *Scale bar* is 5 mm

This fracture pattern is a result of a slight deviation of the tensile axis (by not more than several degrees, Sommer 1969) from normality to the parent fracture. It caused two segmentations advancing in opposite directions, rather than a single propagation that surrounded the entire boundary of the circular parent fracture. The latter propagation would occur under "pure tension", which is probably difficult to achieve, and therefore would be less common. On the discoid from Mt. Carmel (Fig. 3.40c), there is no segment overlapping in certain parts of the concentric fringes. It is assigned to "pure tension", which was applied on the central part of the discoid but not on its right and left parts, suggesting that even locally the stress field was not homogeneous. The "concentric joint" displays opposite senses of segment rotation in a single curved fringe (Fig. 3.40b) induced by a local stress field. A remote deviation of the minimum principal direction would result in similar segment rotations in neighboring joints, typical of many joint fringes along Wadi Naim (Bahat 1986a), which is not the case shown in this outcrop.

2.2.7.4
Opposite Rotations of En Echelon *Segments in Fringes below and above the Parent Joint*

Significant dissimilarities in size and shape of different fringes of the same joint should reflect a deviation from simultaneous *en echelon* segmentation. The two fringes in Fig. 2.45 differ in the directions of segment rotation and in segment size and geometry, quite possibly suggesting that the two fringes responded to different loading. It appears that the fracture sequence of this joint was initially:

1. Propagation of the parent joint from left to right, as shown by the plume.
2. Segmentation of the lower fringe in a rational continuation with the plume.
3. Segmentation of the upper fringe follows a change in the loading direction.

Fig. 2.45.
A fracture cutting Lower Eocene chalk in the Shephela syncline. *En echelon* fringes occur on both the upper and the lower sides of the parent joint. There is a clockwise rotation of the *en echelon* segments with respect to the parent joint in the upper fringe and a counter clockwise rotation in the lower one. However, cracks in the two fringes differ considerably in shape. Length of plume is about 30 cm (after Bahat 1997)

2.2.7.5
Timing Relationships of Plumes, Arrest Marks and En Echelon *Segmentation*

The contact relationships of rib marks and barbs, combined with their interactions with crack segmentation, are useful in estimating changes in local stresses and enable the distinction of different stages in the fracture process. The plume on the "curved joint" (Fig. 2.35c) is bilateral, and the segmentation shows a uniform clockwise rotation, which corresponds to one direction of the plume but not to the other. The implication is that breakdown occurred after the completion of the growth of the bilateral joint. Hence, the growth of the curved joint was in several stages:

1. Fracture into an early elliptical joint.
2. Growth along the upper part of the layer into a bilateral joint that curves along the strike.
3. Breakdown of the joint into segments that propagated downward.

Some joint surfaces show divisions into concentric zones that are bounded by arrest marks (Sect. 2.2.4.1). These arrest marks do not only represent the fracture front at "rest" but they also mark the loci of stress redistribution (Murgatroyd 1942). This is revealed by changes in the sense of rotation of overlapping barbs with respect to the *en echelon* segments (Fig. 3.40b) or by changes in the segment azimuths (Fig. 2.38 and 3.40a, part D). Arrest marks (Sect. 2.2.4.1) can be useful in the interpretation of the interrelationships of individual joints associated in a common composite fracture. The series of sub-horizontal arrest marks that continue through parts B, D and E (Fig. 3.40a) suggest an early synchronization of the downward propagation of these parts.

What was the growth rhythm of the various segments along the "curved joint"? There seem to be conflicting observations. On the one hand, the breakdown into segments along the curved joint occurred under various K_{III}/K_I ratio conditions, which may imply that segmentation was not synchronous along the joint, but it occurred at the center before developing at the distal ends. On the other hand, it seems quite likely that the various segments reached simultaneously the lower part of the joint, as demonstrated by an arrest mark that decorates all the segments at the same height and practically appear to be a monotonous morphological feature. Thus, the possibility of a non synchronous segment initiation but a synchronous completion of segmentation is not ruled out.

2.2.7.6
The Spreads of Azimuths in Joint Sets versus En Echelon *Fringes*

The spread of azimuths in individual joints and in joint sets is often greater than the spread of azimuths of the *en echelon* cracks associated with them. One example is shown in Fig. 2.35d. Other joints in outcrop Y show similar patterns. A summary of fracture orientations in outcrop X shows that the mean azimuth of the parent joints is N 03° W with standard deviation ±12° (63 measurements). The corresponding result for the *en echelon* cracks along the fringes of these joints is N 23° W ±6° (91 measurements), i.e., considerably lower spread (Bahat 1986a).

Geologists often appreciate the difficulty in estimating paleostress directions because of the uncertainties involved in determining the spread of the joint population into sets and establishing the average azimuth of a given set. If the spread of azimuths of *en echelon* segments in a set may be considerably smaller than the spread of azimuths of the parent joints associated with that set, it should enable quite readily the distinction between neighboring joint sets and their associated *en echelon* segmentation.

This is well exemplified in Table 2.1. It would be difficult to distinguish between joint sets X and O-Y3 because of the overlapping of ranges of fracture azimuths (350–014° and 012–035°, respectively). It is, however, considerably easier to distinguish these sets on the basis of their segmentation distinct ranges (338–342° and 026–053°, respectively). Hence, the styles of *en echelon* segmentation and rotation can considerably improve the confidence in the division of joints into sets and upgrade the analysis of paleostress directions.

Table 2.1. Orientation of joints and associated *en echelon* segments in the Beer Sheva syncline and corresponding compression

Station	Name of joint or set	Azimuth of joint (°)	Azimuth of segment (°)	Segmentation		Rotation of segment		Corresponding compression
				above joint	below joint	CW	CCW	
1) S	Part D	073	073–083		+	+	+	?
2) Y	The curved joint	029–046	043–047		+	+		Local
3) Y	Set O-Y3	012–035	026–053	+	+		+	Local
4) X	Set X	350–014	338–342	+	+		+	N–S, regional[a]
5) T	Mixed stepping	062–064	047–049 073–080	+ +			+	Local ?

Stations correspond to those in Fig. 3.33c and Bahat (1996a). All results are from vertical outcrops. *Set S* is from unit III and *sets X, Y* and *T* from unit IV. Clockwise is *CW* and counterclockwise is *CCW*.
[a] Southern part of the Dead Sea rift.

2.2.7.7
An Ascending Order of Fracture Morphologies on Joint Surfaces

It is useful to set the various fracture morphologies on joint surfaces in an ascending order of compounded elements. These can be correlated with various stress scenarios, from 1 to 12 (Fig. 2.46). In a crude way, scenarios 1 to 7 represent various manifestations of fracture morphologies that respond to changes in local stresses, whereas scenarios 8 to 12 show fracture morphologies that may help to evaluate changes in directions of remote stresses.

Scenarios 1 to 3 represent unilateral, bilateral and radial modes of joint propagation, respectively. Common to these morphologies is the perpendicularity of the minimum principal stress with respect to the parent joint, with possible slight deviations from it (Bahat 1997). Local constraints dictate the choice among the three, and a prolonged fracture in a thick layer would favor the third one (scenarios 1 and 2 are well known, and scenario 3 is manifested by the parent joint shown in Fig. 2.36a, displaying three concentric plume series.

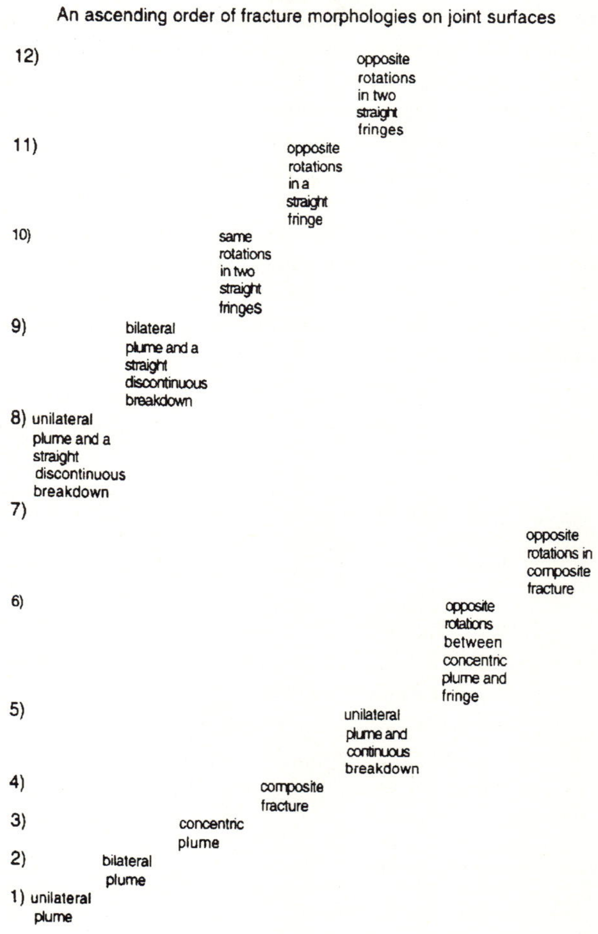

Fig. 2.46. Fracture morphologies on joint surfaces in an ascending order of compounded elements which correlate with twelve stress scenarios; see text for explanation (after Bahat 1997)

Scenario 4 represents a composite fracture that developed by the merging of a series of sub-parallel smaller joints, propagating in the same approximate stress field (Fig. 2.38). Scenario 5 describes the propagation of one part of the composite fracture (Fig. 3.40a, part D), which is the only joint that deflects in response to a slight deviation of the local minimum principal stress from perpendicularity. A deviation of the remote principal stress from perpendicularity to the composite fracture would have resulted in the deflection of additional parts of the composite fracture as well

Scenario 6 results when the local minimum principal stress tilts somewhat from perpendicularity to the parent joint, so that a left-lateral shear changes to a right-lateral shear (Fig. 2.44a,b). It appears that segmentation without overlapping (Fig. 3.40d) signifies a transitional stage between left-lateral and right-lateral shears. Scenario 7 concerns alternating shears among neighboring joints that constitute a composite fracture. This situation occurs in massive sandstones (Fig. 3.23b,c), but is not known in chalk.

Scenarios 8 and 9 are recognized by the abrupt transitions between the parent joint and the *en echelon* straight fringe. These structures are good candidates for the identification of remote stress directions. Scenario 10 (Fig. 2.35a) may be considered as an "upgrading" of scenarios 8 and 9, also leading to the identification of remote stress directions. Scenarios 11 and 12 are related to opposite segment rotations in straight fringes (Bahat 1997, Fig. 9 and Fig. 2.45, respectively). They also provide indications for the distinction between different remote stress directions.

2.2.7.8
Local Stresses versus Remote Stresses

We summarize below the distinctions between the influences of local and remote stresses on the formation of *en echelon* cracks (Pollard et al. 1982; Bahat 1986a, 1997).

Local stresses. Field observations indicate that much of the interrelationship of the parent joint with the fringe and the behavior of the *en echelon* segments are influenced by local conditions, as itemized below:

1. There is a correlation between the size of *en echelon* cracks formed by the discontinuous breakdown and layer thickness (Fig. 2.43).
2. The various measurements of *en echelon* lengths vs. the layer thickness in the thickness range of 55–65 cm (Fig. 2.43) are all from layer 5 (Bahat 1986a, Fig. 2) and they show both continuous and discontinuous breakdowns that cannot be explained by remote influences.
3. Occasionally, specific timing relationships exist among plumes, the vertical *en echelon* cracks and the horizontal segments of arrest marks. These relationships may occur on surfaces of individual joints or composite joints (Fig. 2.35d and 2.38) but not in association with any of their neighbors along the same layer.
4. The sense of rotation of the segments in the fringe is clockwise, compared to counter clockwise rotation of the barbs on the parent joint (Fig. 3.40b). It seems highly unlikely that these two rotations reflect remote stresses. Quite possibly the counter clockwise rotation of the barbs is in response to local stress.
5. The reverse segment rotation along the fringe of the concentric joint reflects local conditions, rather than two distinct regional loadings.

Remote stresses. In spite of the important influence of local conditions on the behavior of *en echelon* segments, there are certain parameters that can quite clearly be correlated with remote stresses:

1. The decoupling of the fringe from the bilateral plume described above (Fig. 2.35c,d) suggests that these two morphologies were formed in response to distinct stress conditions. Here, the preference of the plumes to propagate bilaterally rather than unilaterally demonstrates a response to local conditions. On the other hand, the azimuth uniformity of the *en echelon* segmentation in the fringe and other fringes of neighboring joints in a set conforms to the influence of a remote stress. The decoupling of a fringe from a unilateral plume is as much an indication that the segmentation in the fringe represents a remote stress (as in the previous case related to the bilateral plume). A good example is provided by Younes and Engelder (1999). The uniformity in the sense of rotation of *en echelon* cracks with respect to their parent joints over great distances is a manifestation of fracture induced by remote stresses.
2. The conformity of clockwise rotation of *en echelon* segments with the NNE joints in station Y, on the one hand, and the strict counter clockwise rotation associated with the N-S joints in outcrop X on the other hand suggest two sets responding to distinct remote stresses (Table 2.1).
3. Specifically, fractures in outcrop X differ from all other fractures in the investigated area by the uniformity of their *en echelon* cracks in terms of their sense of rotation, their size and shape, and in the fringes that occur both below and above the parent joint. This distinct assemblage of properties categorizes a distinct remote stress field.
4. When a set of joints demonstrates a gradual azimuth rotation, the *en echelon* cracks at the fringes of these joints rotate in covariance with the joints (Bahat 1986a).
5. Discontinuous breakdowns are distinct from continuous breakdowns by often having a series of small *en echelon* cracks along the shoulder distributed among the larger *en echelon* cracks. Such series of small cracks have been named the "root zone" (Bankwitz 1966). They generally maintain their azimuths in alignment with their larger neighbor *en echelon* cracks, responding to remote stresses.

Spatial and temporal misalignments. The reader should be aware of differences in terminology. There are investigators that prefer to use different terms in their distinction between the remote and local categories. Younes and Engelder (1999) assert that if an advancing joint enters a volume of rock subject to a stress field misaligned relative to that guiding the parent joint, transverse adjustments in crack propagation due to mixed-mode loading cause the joint to break down into fringe cracks. If the misaligned stress field was present prior to joint propagation, the misalignment is spatial. If the misalignment took place after propagation and arrest of the parent joint, the misalignment is temporal. Due to occasional unclear cases regarding the propagation and arrest of the parent joint, we are not in favor of using this distinction and prefer the terms remote and local according to the description presented above.

2.2.7.9
En Echelon *Segmentation under Constraints*

Dilatational *en echelon* segmentation occurs under favorable conditions for mode III operation, and it is limited when constraints are imposed on this mode. As was mentioned above, *en echelon* segmentation is very common in rocks. It is also frequent in fractured polycrystalline metals (Hertzberg 1976; Tschegg 1983; Broek 1984 and references therein) and in single-crystals (Gilman 1958). On the other hand, *en echelon* segmentation is rare in (non-polymeric) glass, and the widely cited experiment by Sommer (1969) is related to segmentation in glass under fluid pressure.

The explanation to these differences is that defects in the structure continuity, like screw dislocations in single-crystals (Gilman 1958) and inclusions and grain boundaries in steel (Tschegg 1983) and in rocks, promote the development of local, heterogeneous stress fields, resulting in mixed mode loading that induces segmentation. The much smaller population of such discontinuities in non-polymeric glass is possibly the reason for the limitation of *en echelon* cracks in this material. *En echelon* segmentation does occur in polymeric glasses like PMMA (e.g., Cooke and Pollard 1996). However, fracture propagation in this material must involve an early strain and a breakdown of craze fibrils (Maccagno and Knott 1989). The effect of "mode III crack closure" (Tschegg 1983) imposes an important constraint on segmentation. Namely, *en echelon* cracking along layer boundaries would be minimized under compression by the overburden, and therefore, even if found in rocks in considerable depths, an early failure at shallow depths should be suspected. On the other hand, high liquid pressures could overcome the constraint imposed by the lithostatic pressure and induce extensile *en echelon* segmentation at considerable depths, in a similar manner by which they form joints at great depths by mode I (Secor 1965). The latter hypothesis still needs to be verified.

2.2.7.10
The Influence of Ductile Conditions on En Echelon *Segmentation*

We have seen above that mode III loading is a major promoter of *en echelon* segmentation. The plastic zone of mode III loading is several times larger than that for mode I for similar materials and at the same stress intensity values (Tschegg 1983). Thus, segmentation is associated with the increase in the plastic zone. This observation coincides well with the plane strain/plain stress relationship in a plate, showing that the plastic zone gradually increases from the plane strain size in the interior to a plastic zone of a larger size under plane stress at the surface of the plate (Hertzberg 1976; Broek 1984). Analogously in geological terms, conditions along the layer boundaries are more conducive for segmentation than in the layer interior.

2.2.7.11
Limitations and Future Research

The study presented in this section has certain deficiencies that are addressed below. First, most of the information presented here has been gathered by one researcher (the first author), and some subjective interpretation is perhaps unavoidable. Therefore, the various observations and subsequent rules that were determined in this study should be considered tentative until they are verified in similar sedimentary fracture provinces. Second, the present section essentially relates to a single lithology, chalk. Therefore, some of the relationships that were established here may have to be modified when applied to multi-lithological provinces (such as the Appalachian Plateau). Third, there is a need to further investigate the nature of the continuous breakdown. What causes breakdown to start from the boundary of the parent joint (shoulder in Fig. 2.35a) and when does it to start from within the parent joint (e.g., Fig. 2.35c)? Fourth, alternative theories of fracture mechanisms that exist today regarding the formation of fringes (Sect. 6.10) need to be evaluated, in order to develop a strong framework into which additional field and experimental information would be added and interpreted.

2.2.8 Straight Kinks and Curved Kinks

A kink is generally perceived as a relatively short crack that occurs at the termination of a straight parent fracture that started as a joint. Due to mixed mode I + II loading, kinks deviate from the strike of the parent fracture, pointing towards the continuations of the extensional quadrants (Cotterell and Rice 1980) (Fig. 2.47a). Cruikshank et al. (1991) used a mixed-mode linear elastic fracture mechanics model and applied the maximum circumferential stress criterion (Erdogan and Sih 1963) in order to ac-

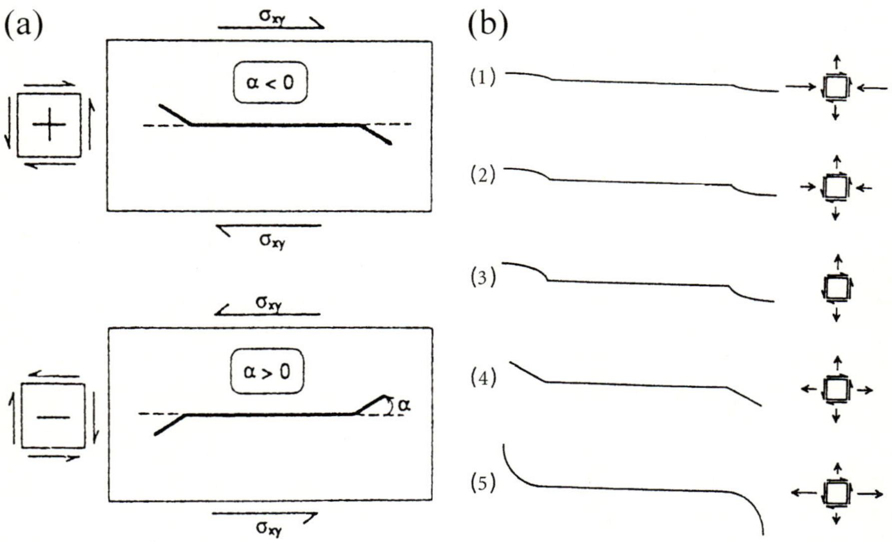

Fig. 2.47. a Relations between sense of shear and direction of kinking. For positive, right-lateral shear, the kink angle is negative. For negative, left-lateral shear the kink angle is positive. **b** Examples of idealized tail cracks for various far-field loading conditions. In all cases, pressure within the fracture is adjusted to cause propagation. In *1* and *2*, the stress parallel to the fracture is compressive and in *3* it is zero. Tail cracks kink and then converge. In *4* the stress parallel to the fracture is tensile and equal to that normal to the fracture. The fracture merely kinks. In *5*, the stress parallel to the fracture is a larger tensile stress than that normal to the fracture, and the tail rack diverges (from Cruikshank et al. 1991)

count for the orientation of cracks near the ends of small-displacement faults. They assumed that the faults were subject to uniform driving stresses and that the cracks opened perpendicular to the greatest circumferential stress ($\sigma_{\vartheta\vartheta}$) near the fault tip (see Lawn and Wilshaw 1975a, Fig. 3.4, for general conditions, and present Fig. 1.25 for $\vartheta = 0$). In their model, the sense of kinking indicates the sense of shear during fault-

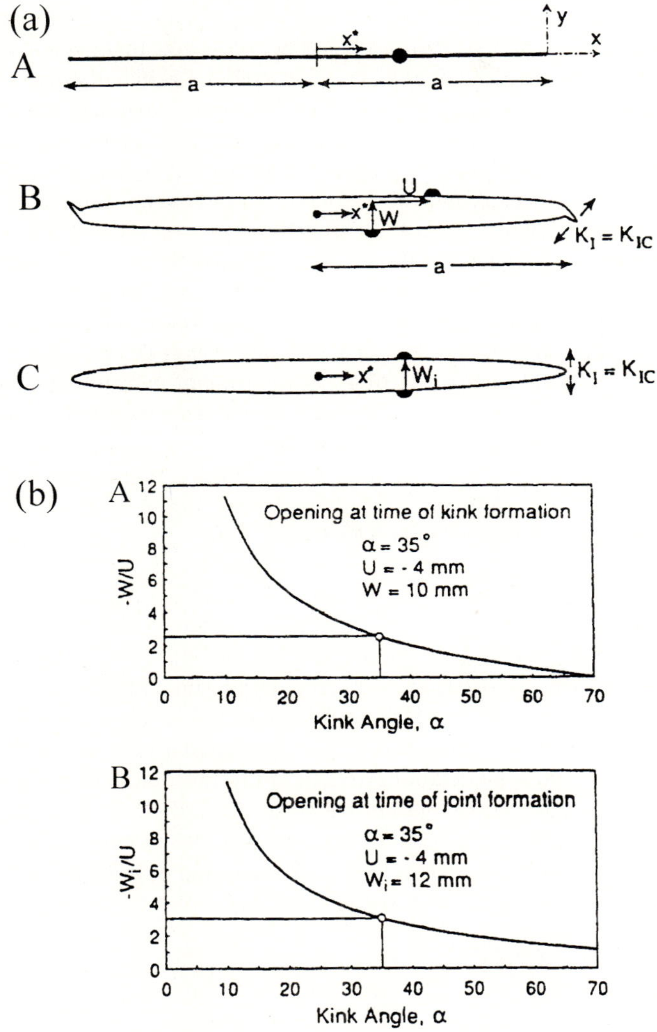

Fig. 2.48. **a** Definitions of kinematic properties of a crack. *A* Crack length is $2a$. y-coordinate normal to crack, x^*- and x-coordinates parallel. *Dot* is a material particle split by the crack. *B* Displacement components, U and W, of part of a material point along a crack wall relative to the other part of the material point along the opposite crack wall. Mode I stress intensity equal to critical stress intensity for mode I fracturing at the time a kink forms. *C* Displacement in mode I loading at the time the initial fracture formed. Material point split and parts displaced only normally relative to one another. Mode I stress intensity equal to critical stress intensity at the time the fracture stops lengthening. **b** Relations between kink angle α and, *A* ratio of opening, W, to slip, U along fracture, and *B* relation between kink angle and ratio of initial opening, W_i to subsequent slip, U, along fracture (from Cruikshank et al. 1991)

ing: A kink that turns clockwise with respect to the direction of the main joint is a result of right-lateral shear, and a kink that turns counter clockwise is a result of left-lateral shear. Furthermore, the kink angle α, the angle that the secondary fracture forms with the extension of the primary fracture, is related to the ratio of the shear stress responsible for the kinking to the normal stress responsible for the opening of the fault (Fig. 2.47a).

Cruikshank et al. (1991) consider both sharp/straight and curved kinks and show a span of orientations of a straight kink and curved kinks that can be digitally obtained under various loading conditions (Fig. 2.47b). In their model, which limits stresses to principal directions, pressure within the parent fracture is adjusted to cause propagation. Thus, while Cruikshank et al. (1991) concentrate on stress differences (Fig. 2.47b), Cooke and Pollard (1996) study the influence of loading rate on the shape of kinks (Fig. 2.20a,b). Cruikshank et al. (1991) suggest that the amount of opening of a joint at the time it faulted or even at the time the joint itself formed can be estimated by measuring the kink angle and the amount of strike-slip at some point along the faulted joint (Fig. 2.48a,b). They relate the ratio of fault aperture, W, to slip, U, at the time of faulting to α. For a pure mode I fracture ($W/U = \infty$), $\alpha = 0°$. For a pure mode II fracture ($W/U = 0$), $\alpha = \cos^{-1}(1/3) \approx 70.5°$. For angles between 35° and 55°, they calculated that the ratio of opening to offset at the time of faulting was between 3.02 and 1.70, respectively. These angles hold no matter how large the relative displacements of the fracture walls are, provided they are not zero.

2.2.9 Recent Studies from the Appalachian Plateau

2.2.9.1 *General*

This section relates to the occurrence of fringes and kinks associated with joints in sedimentary rocks from the Appalachian Plateau, based on the study by Younes and Engelder (1999) that we cite below. Here we concentrate on fractographic relationships, while the wide geological implications of their study are cited in Chap. 3. Younes and Engelder (1999) consider three types of fringe cracks in their study area: Gradual *en echelon* fringes (that they term gradual twist hackles), abrupt *en echelon* fringes (that they term abrupt twist hackles), and kinks. A description of these three types with a subsequent discussion is presented below.

2.2.9.2 *Gradual* En Echelon *Fringes*

Gradual *en echelon* fringes occur within the same bed and lithology that hosts the parent joint (Fig. 2.49a,b). The surfaces of the *en echelon* segments occasionally show plumes that indicate an overall vertical propagation direction. Younes and Engelder (1999) observe that this surface morphology is often continuous with that of the parent joint, giving no indication that the parent joint arrested before the gradual *en echelon* fringe propagated. In their model, however, the plumes on the fringe are separated from the plume on the parent joint (Fig. 2.49a), as compared with our model

(Fig. 2.20d) that shows plume continuity. Upon breakdown at the joint tip line (equivalent to the shoulder in Fig. 2.36b), the *en echelon* segments twist away from the parent. The segments realign normal to the local maximum tensile stress, propagating perpendicular (i.e., vertically) to the overall lateral propagation direction of the parent joint. This is clearly seen on the diagram in Fig. 2.49a (not as clear in the photograph of Fig. 2.49b).

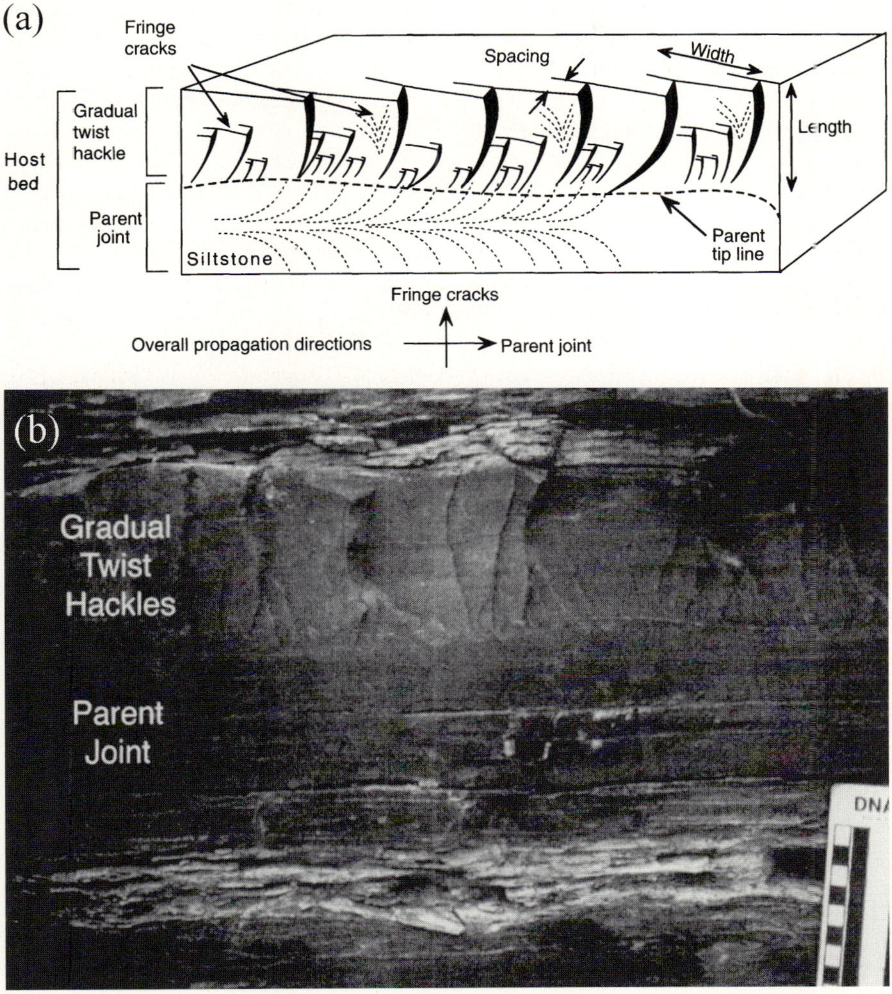

Fig. 2.49a,b. *En echelon* segmentation in the Appalachian Plateau. **a** Schematic of a gradual *en echelon* fringe. The parent joint and *en echelon* fringes are carried within a single host bed. The number of fringe cracks decreases away from the parent joint. The spacing of the cracks changes from small at the tip line of the parent joint to large at the end of the fringe. **b** A photograph of a gradual *en echelon* fringe. A parent joint and an *en echelon* fringe are carried within a single bed of siltstone of the Ithaca Formation at Taughannock Falls State Park, New York. Propagation direction for the *en echelon* fringe is upward. The sense of stress field rotation in this example is clockwise. The *scale* to the lower right is divided into centimeters (from Younes and Engelder 1999, no changes are made in the wording of original figures)

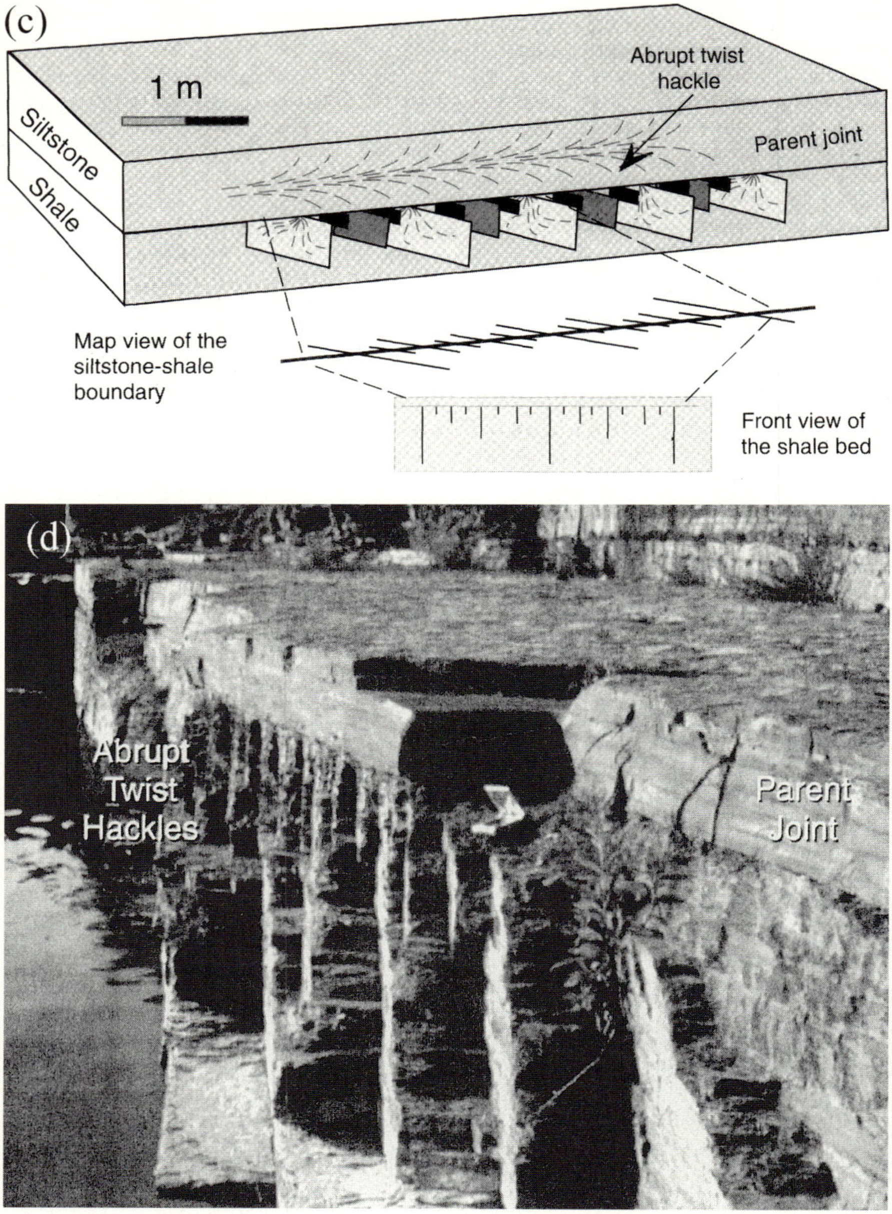

Fig. 2.49c,d. *En echelon* segmentation in the Appalachian Plateau. **c** Schematic of an abrupt *en echelon* fringe. These fringe cracks can propagate downward or upward from a siltstone bed into adjacent bed. They have a characteristic sequence of large and small cracks, indicative of suppressed growth where crack stress shadows limit the growth of some cracks. **d** A Photograph of an abrupt *en echelon* fringe. This set of fringe cracks propagated downward into a thick shale bed from a thinner siltstone bed hosting the parent joint at Taughannock Falls State Park, New York. These rocks are part of the Ithaca Formation. The sense of stress field rotation in this example is clockwise. The *scale* is a geologic compass with an 8-cm base (from Younes and Engelder 1999, no changes are made in the wording of original figures)

2.2.9.3
Abrupt En Echelon *Fringes*

According to Younes and Engelder (1999), abrupt *en echelon* fringes are commonly found in shale and initiate from the tip line of dip joints in siltstone layers that are typically less than 30 cm thick (Fig. 2.49c,d). Because abrupt *en echelon* segments abut the parent joint at a point, they appear in map view to grow across the edge of the parent joint. In contrast with gradual *en echelon* fringes that grow within the same bed as the parent joint, abrupt *en echelon* fringes are separated from parent joints by a layer boundary (Younes and Engelder 1999). These *en echelon* cracks are equally common, propagating upward, downward, or in both directions from siltstone beds.

2.2.9.4
Kinks

According to Younes and Engelder (1999), in the Appalachian detachment sheet kinks occur exclusively in siltstone beds (Fig. 2.50), and no evidence for slip was observed on any parent joint carrying kinks in the detachment sheet. They make the point that kinks occasionally show a surface morphology of plumes that are continuations of the plume morphology from the parent joint. Therefore, they suggest that kinks occur as mode I cracks and the tilt in a crack path is driven by a shear traction superimposed on the parent joint after *arrest*, following a finite amount of mode I propagation (Erdogan and Sih 1963; Cotterell and Rice 1980).

Fig. 2.50. An example of fringe cracks in a siltstone bed of the Ithaca Formation found near Whitney Point, New York. The shale above and below the siltstone bed carried an abrupt *en echelon* fringe. The sense of stress field rotation in this example is counter clockwise, both for the kink and the abrupt *en echelon* fringe. The parent joint is found in a bed roughly 20 cm thick (from Younes and Engelder 1999, no changes are made in the wording of original figures)

2.2.9.5
Tilt and Twist

Younes and Engelder (1999) observe that the angle of twist or tilt of a fringe crack set changes according to the host rock, the type of the fringe crack and the sense of rotation. For simplicity, they use the same angle sign (β) for the twist angle (equivalent to ω in this book), the tilt angle and the kink angle. Younes and Engelder found that the twist angles for clockwise *en echelon* cracks are generally smaller than for counter clockwise *en echelon* cracks. Gradual *en echelon* cracks have small twist angles, with a mode at 3° for clockwise *en echelon* cracks and 6° for counter clockwise *en echelon* cracks. Abrupt

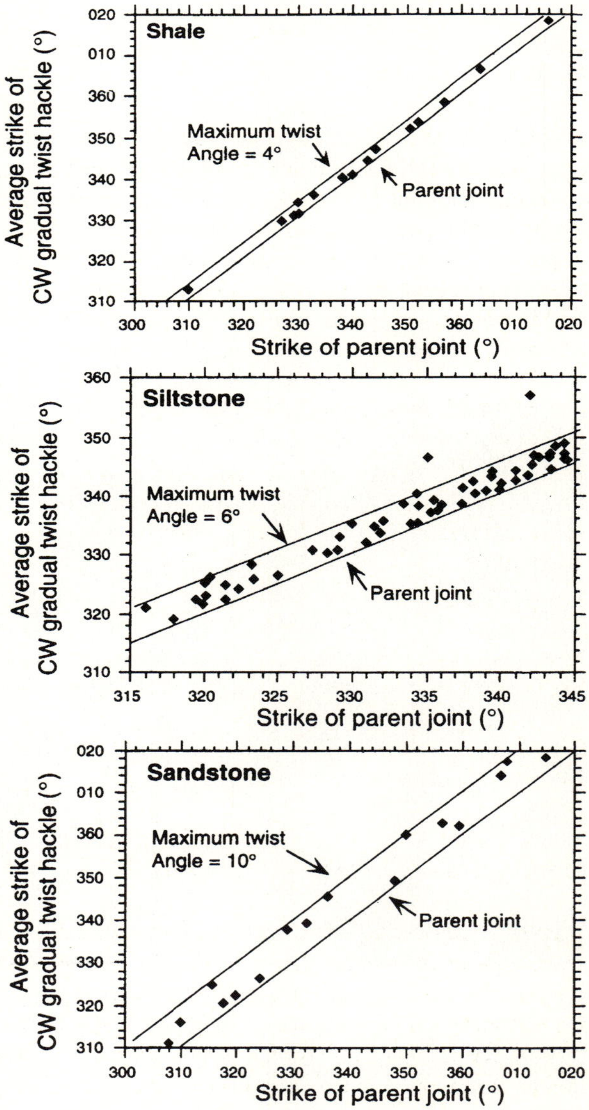

Fig. 2.51. Change in the angle between parent joint and the clockwise (*CW*) gradual *en echelon* fringes according to lithology. The maximum twist angle increases as the clastic grain size becomes larger (from Younes and Engelder 1999, no changes are made in the wording of original figures)

en echelon cracks have larger twist angles, with a mode at 13° for clockwise *en echelon* cracks, and 22° for counter clockwise *en echelon* cracks. Kinks have tilt angles with a mode of 16° for clockwise *en echelon* cracks and 14° for counter clockwise *en echelon* cracks. The maximum twist angle for gradual *en echelon* cracks varies according to the lithology hosting the parent joint. This is seen in a plot of the strike of fringe cracks vs. the strike of parent joint (Fig. 2.51). In this plot, data for counter clockwise fringe cracks would plot below a line with a slope of one, whereas data for clockwise fringe cracks are plotted above this line. By drawing delimiting envelopes parallel to this line, Younes and Engelder (1999) found the maximum twist angles in coarser beds. For example, *en echelon* cracks in shale layers have a maximum twist angle of 4°, whereas those in sandstone layers reach 10° for *en echelon* cracks with a clockwise twist angle.

2.2.9.6
Discussion on Kinking

While Cruikshank et al. (1991) restrict kinking to faulting on existing joints, Younes and Engelder (1999) consider kinking on joints without slip. The common gradual curving of joints as they approach earlier joints (e.g., Fig. 3.36b) and curved mud cracks that maintain continuous plumes along the strike (Weinberger 2001b, and pers. comm.) imply that curved (gradual) kinking (Fig. 2.20b) is possible. A more difficult issue is whether straight (abrupt) kinks (Cruikshank et al. 1991) exist on joints, i.e., whether kinking occurs in continuation with the propagation of the parent joint (in one event) or after parent joint arrest (in more than one event, Fig. 2.20a). If continuities of plumes are observed on parent joints and straight kinks, single event processes may be suspected. On the other hand, if discontinuities are observed between the plumes on the parent joints and kinks, more than one event should be considered. The latter could be the growth of a kink after a certain time interval following a fracture arrest. A careful inspection of both the continuous and discontinuous plumes should be exercised on suspected straight kinks, because a cutting relationship between two distinct joints of different sets can be mistaken for straight kinking (see Sect. 2.2.10.2 below). It appears to us that presently, no compelling evidence for straight kinking on joints exists, and until such evidence is provided, straight kinking by faulting is the only known mechanism. Finally, the important role of plumes in the interpretation of the genesis of kinks has a certain similarity to their role in the study of pinnate joints (Fig. 2.68d).

2.2.10
On the *En Echelon* Fringe Terminology

2.2.10.1
The Parent Joint is a Part of the Mirror Plane

We note that the gradual *en echelon* fringe type of Younes and Engelder (1999) (Fig. 2.49a,b) resembles the famous diagram of surface markings on a systematic joint by Hodgson (Fig. 2.36b), where an abrupt transition from the parent joint to the fringe occurs along the shoulder. Also, in both diagrams (Fig. 2.36b and 2.49a), new plumes start at the fringes. The gradual *en echelon* fringe of Younes and Engelder (1999) is also similar to the discontinuous fractography (Fig. 2.35a,b), which displays an abrupt transition of the parent joint to the fringe. The above examples illustrate only part of the confusion regarding joint terminology (see more in Bahat 1991a, Sect. 2.2).

The main deficiency of the existing descriptive schemes of *en echelon* fringe terminology is that actually, they have no compelling limitations. A classification with compelling limitations that would minimize confusions may be achieved by forming a terminology based on superposing crack descriptions on fractographic features that correspond to fracture mechanic rules (Sect. 4.8). We suggest that the key to a comprehensive terminology should be the notion that the parent joint is part of a mirror plane. This relationship will automatically define all the morphological features that reside on the parent joint and on the fringe (Fig. 2.1). This is fairly clear in granites (Chap. 4), but it is occasionally hindered when joints cutting sedimentary rocks are considered, because in the latter rocks the parent joint is often a portion of a mirror plane, which is constrained between layer boundaries and cannot achieve full growth (Bahat 1998a and present Fig. 2.8). This terminology is a part of a wider classification that distinguishes between *en echelon* fringes and hackle fringes (Fig. 2.1, 4.35 and 4.37–4.40).

2.2.10.2
A Fringe of a Parent Joint Can Only Form from "Continuous Material"

The study by Cooke and Underwood (2001) (present Fig. 2.32) seems to justify the re-evaluation of previous discussions on three cases of *en echelon* segmentation at different lithological interfaces (Bahat 1997; Younes and Engelder 1999). The first case concerns fracture in Devonian sediments from the Appalachian Plateau (Younes and Engelder 1999) that relates to two joints in siltstone (relatively smooth surfaces) alternating with shale beds (rough surfaces) (Fig. 2.50). While the joint surfaces in the siltstone layers change orientation along the strike (three sectors in the lower layer and two sectors in the upper one) the fracture surfaces in the lower shale bed maintain the same azimuth. We suggest that whereas the sectors of different orientation in the siltstone represent two distinct joint sets, the fractures cutting the shales represent a single, distinct joint set, i.e., the fractures in the shales do not form an "abrupt" *en echelon* fringe to the parent joint in the siltstone. Possibly the set in the shales is Ia and one of the sets in the siltstone is Ib (Engelder and Geiser 1980, the former set is clockwise with respect to the latter). This relationship is repeated in additional outcrops (e.g., Fig. 2.49d). Quite intriguing is the decrease in spacing of the joints in the shales as they approach the lower boundary of the siltstone bed. The reduction in spacing implies an increase in the tensile stress concentration along the boundary between the shales and siltstone. The multiplied tensile stress resulted in a repeated growth of (short) joints in the shales that were added between the earlier long joints of an existing set (Fig. 2.49d and 2.50). Hence, the short joints *are not en echelon segments* in a fringe and are unconnected to remote extension normal to the parent joint in the siltstone, but rather, they relate to the energy released at the contact between the two rocks that affected fracture only in the shales.

The second case is about fracture in chalk adjacent to a fracture in a chert bed (Fig. 2.52a). The interpretation here is the same as for the first case: The short fractures represent a joint set cutting the chalk; they were not formed in a fringe of the

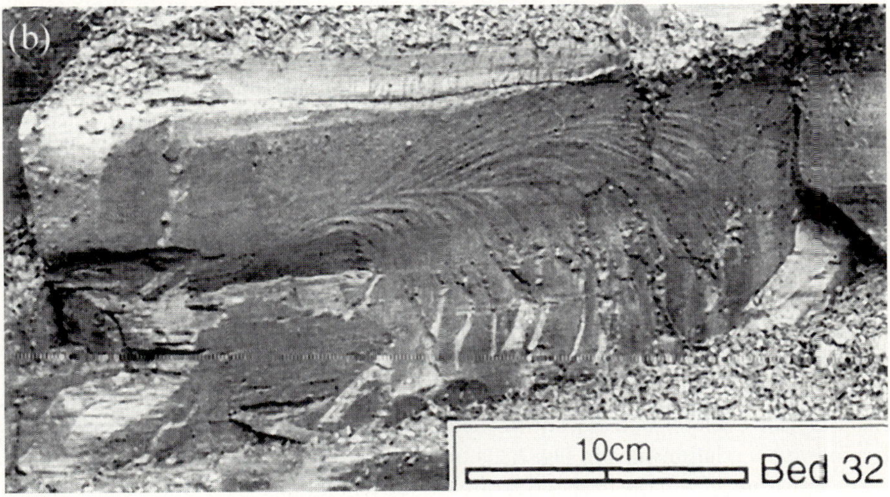

Fig. 2.52. **a** A sandwich of two chalk layers and a chert bed in the middle from the Mor Formation near Beer Sheva. *En echelon* segmentation occurs in the chalk only along contacts with joints cutting the 15 cm thick chert bed, but not along contacts with unfractured chert nodules (after Bahat 1997). **b** Plumes that developed within the limestone layers appear to have continued, forming a fringe of *en echelon* segments in the interbedded shales underneath the limestone (after Roberts 1995, Fig. 5d)

parent joint in the chert. Quite possibly the interface contacts in the first and second cases were not welded enough to act as a "continuous material", which is a required condition to form a fringe of a parent joint. In these two cases, separate fracture processes took place in the two lithologies.

The third case concerns a study on jointing in interbedded limestones and shales that appear to be welded together (Fig. 2.52b). While plumes developed within the limestone layers, a fringe of *en echelon* segments occurs in the interbedded shales just underneath the limestone in an apparent continuation from the plume in the limestone. However, it appears that both fracture surface morphologies occur on a vertical veneer of secondary mineralization that covers the two lithologies and acts as a "continuous material". Thus, we interpret the sequence under consideration to be: Early vertical fracture through both lithologies, secondary mineralization, and then, late fracture along the contact between the veneer and the early vertical fracture.

Our interpretation of the three cases needs to be verified in additional fracture provinces. Particularly, the first case needs to be carefully examined (we take it as a speculation that could be wrong). If this interpretation is proven correct, it would limit or question the validity of the category of abrupt *en echelon* fringes in the Appalachian Plateau.

2.2.11
Summary of Five Categories of Breakdown Styles

The various extension paths from parent joints by tilt and twist joints to kinks and fringes may be condensed into five categories that reflect different mixed mode loadings and propagation rates. Four categories are summarized in Fig. 2.20a–d. Note that the scheme given in Fig. 2.20c,d concentrates on discontinuous breakdowns that initiate close to the layer boundary (Fig. 2.35a,b), excluding the continuous one that initiates between the layer boundaries (Fig. 2.35c,d). The continuous-gradual style may be considered a variation linked to the gradual, the fourth category, and a full analysis of joint fringes should concern this style as well.

While studying rare hackle fringes in sedimentary rocks, Bahat and Rabinovitch (2000) observed angular relationships that differed from the above four categories. The tilt angle ϕ produced between the mirror plane and the fringe varied from 0° to 30 ±5° along the mirror boundary of a natural joint (Fig. 2.30a), ϕ, of an artificial fracture that formed by blast varied from 30 ±8° to 50 ±8° (Fig. 2.30b). Thus, the tilt angle ϕ appears to be larger on the artificial fracture that was dynamically blasted by high stresses than on the joint surface that was probably created under moderate stress conditions. The fractographic features on the mirrors and fringes of both fractures seem to have been formed discontinuously, suggesting abrupt transitions. Hence, the fractographic relationships in Fig. 2.30a,b suggest a fifth category. This category concerns transitions from the parent fractures to the fringes that are influenced by mixed I + II + III modes. The tilt angle ϕ is induced by mixed I + II modes, and the twist angle ω (although can hardly be measure in the hackle fringes of Fig. 2.30a,b) is formed by mixed I + III modes.

The fifth category is rare in sedimentary rocks but seems to be relatively common in granites. Note that in Fig. 2.30a,b and 2.56a,b, which display semi-circular mirror planes, there are no distinctions between fringes that tilt in the distal direction and perpendicular to it, i.e., normal to the larger or smaller radii of the semi-circular mirror planes. Thus, the fifth category that relates to a rim (consisting of quasi-fringes and quasi-kinks) of secondary cracks that surround the mirror plane is clearly a prod-

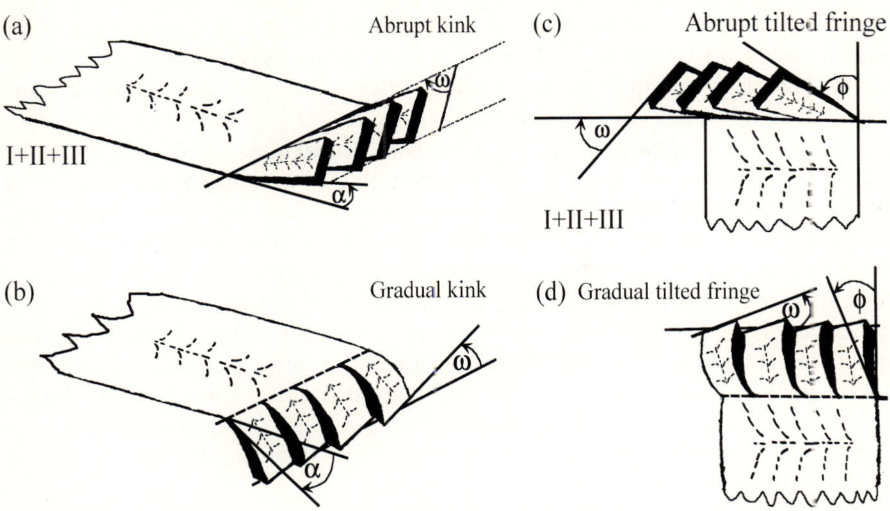

Fig. 2.53. Four styles of breakdown by mixed mode I + II + III. Breakdowns occur in the kink along the distal continuation of the parent joint and in the fringes parallel to layer boundaries in sedimentary rocks. **a** and **b** relate to abrupt and gradual tilted kinks, respectively. The segmented kinks are described by α and ω angular relationships. **c** and **d** relate to abrupt and gradual tilted fringes, described by ϕ and ω angular relationships. ω is formed by the edge of the mirror plane (normal to the main plume) and the segment width, α is formed by the edge of the mirror plane (parallel to the main plume) and the segment length (or its tangent), and ϕ is formed by the edge of the mirror plane (normal to the main plume) and the segment length (or its tangent)

uct manifestation of mixed I + II + III modes. Nevertheless, we generalize and divide this category in our scheme (Fig. 2.53a–d) into four corresponding groups that would also fit hypothetical cases in sedimentary rocks. Groups a and b relate to abrupt and gradual tilted fringes, respectively that form at the distal part of the parent joint; these groups concern segmented kinks that are described by α and ω angular relationships. On the other hand, groups c and d relate to abrupt and gradual tilted fringes, respectively that form along one or two layer boundaries; these groups display ϕ and ω angular relationships. These fractographies are treated in great detail by Bankwitz and Bankwitz (2004) (see more in Sect. 4.4.4).

2.2.12
Recent Studies from the South Bohemian Pluton

2.2.12.1
Introduction to Joints from Two Granitic Quarries

The bulk of the existing knowledge on jointing is related to sedimentary rocks. Most of these joints develop under anisotropic constraints imposed by rock discontinuities such as layer boundaries and fabrics of different origins. In undeformed quasi-isotropic plutonic rocks on the other hand, such discontinuities are rare, and even flow fabrics mostly do not create loss of cohesion in the rock. Therefore, fracture in the latter rocks would be expected to propagate in an unrestricted manner, following fracture mechanic criteria: Initial elliptical cracks would tend to grow in isotropic materials under tension into larger circular fractures, because the stress intensity factor is greater for the portion of the curve that has a larger radius, resulting in the decrease of ellipticity (Irwin 1962; Roy and Saha 1995).

To our knowledge the granites of the South Bohemian Massif in the Czech Republic exhibit one of the best spectra of joint fractography in the world by showing a great fracto-graphic diversification. Most of these granites had fractured during the late cooling stages of the pluton (Sect. 4.4.5) and were not deformed after solidification (apart from later fracture) before or after their uplifts, at least during the Upper Cenozoic. Therefore, they maintain the original shapes of early fractures from periods before the uplifts. The unfissured properties of these rocks were beneficial during more than a hundred years in the stone industry and construction purposes throughout Europe, leaving behind large quarries, many of which are still active. These quarries display large fractographic features on joints that cut the granites (Bankwitz and Bankwitz 1984). Therefore these granites are most suitable for studying joint fractography in plutonic rocks.

Section 2.2.12.2 presents a description of characteristic fractographic features of the mirror planes and fringes (Fig. 2.1) of ten selected joints from the Borsov granite quarry in the South Bohemian Massif. An additional fractographic treatment that includes geological interpretation of these joints is presented in Chap. 4. Accordingly, the fractographic descriptions are presented in two parts: Photographs are shown in Chap. 2 (Fig. 2.54–2.56) and their corresponding drawings (Fig. 4.6–4.8) and explanation, supplemented by Tables 4.1 and 4.2 are given in Chap. 4. A study on mirror planes and fringes from the Mrákotin granite quarry in this massif is presented in Sect. 4.4.3.1(the deferred treatment of this quarry is done in order to avoid its excessive partition among sections). The Borsov and Mrákotin quarries were chosen from many others in the massif since they best represent fractographic series between end members from "quasi-static" to "dynamic" joints (Sect. 4.9.3.3).

2.2.12.2
Fractography in the Borsov Quarry

Fractographic diversity. In their full exposure, the Borsov joints are divided into four groups:

1. Mirror planes without fringes (J1–J4).
2. Mirrors and simple *en echelon* fringes (J5–J7).
3. Mirrors and non-uniform/complex *en echelon* fringes (J8–J9).
4. A mirror and a hackle fringe (J10). They are briefly itemized below, citing Bahat et al. (2003).

The simplest joint surfaces (J1 and J2) show a superposed single plume on an array of arrest marks (Fig. 2.54a) and a single radial plume (Fig. 2.54b). More complex textures are the series of superposed concentric arrest marks and a radial plume that exhibits considerable changes in barb coarseness (width and height) on J3 (Fig. 2.54c), as well as the outcrop that reveals a set of joints (J4) that exhibit plumes and arrest marks of different styles and intensities (Fig. 2.54d). These four exposures display incomplete mirrors without fringes (therefore they do not exhibit mirror boundaries).

There are four additional examples of joints that display both mirrors and fringes: First, the heterogeneous combination of arrest marks and plumes with the small fringe on J5 (Fig. 2.55a); second, a series of elliptical arrest marks superposing a multi-texture plume and a triangular fringe on J6 (Fig. 2.55b), and third, a fully developed circular en-echelon fringe of a uniform width on J7 (Fig. 2.55c). The fourth example (J8) consists of a circular mirror boundary with a non-uniform *en echelon* fringe (Fig. 2.55d). Still more complex are the heterogeneous and noncircular mirror boundaries, which are surrounded by complex fringes on the "trefoil joint" (J9) (Fig. 2.56a). The tenth joint, J10, displays a large fringe (relative to the exposed mirror) with transitions from *en echelon* to hackly morphologies (Fig. 2.56b). Fringe morphologies may vary very considerably in comparison to the plumes on the mirrors. Whereas fringe morphologies on J5–J8 (Fig. 2.55a–2.55d, respectively) are generally only slightly stronger than the corresponding plume morphologies, on J9–J10 (Fig. 2.56a–2.56b, respectively) the fringe morphologies are much more pronounced than the corresponding plumes. Finally, note the coarse, cusp nature of some of the arrest marks that decorate some of the "slow joints" (Fig. 2.54c, 2.55a, 2.55b), and lack of coarse ripple marks on the "rapid joints" (Fig. 2.56a–2.56b). A full understanding of this difference requires further studies (Sect. 2.2.4).

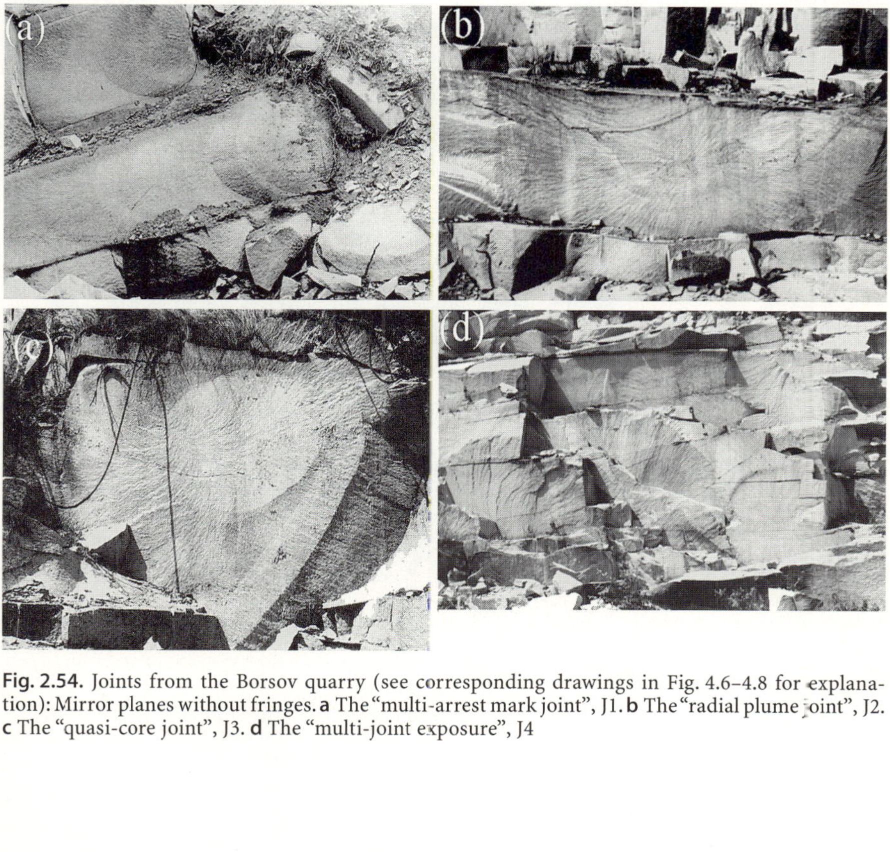

Fig. 2.54. Joints from the Borsov quarry (see corresponding drawings in Fig. 4.6–4.8 for explanation): Mirror planes without fringes. **a** The "multi-arrest mark joint", J1. **b** The "radial plume joint", J2. **c** The "quasi-core joint", J3. **d** The "multi-joint exposure", J4

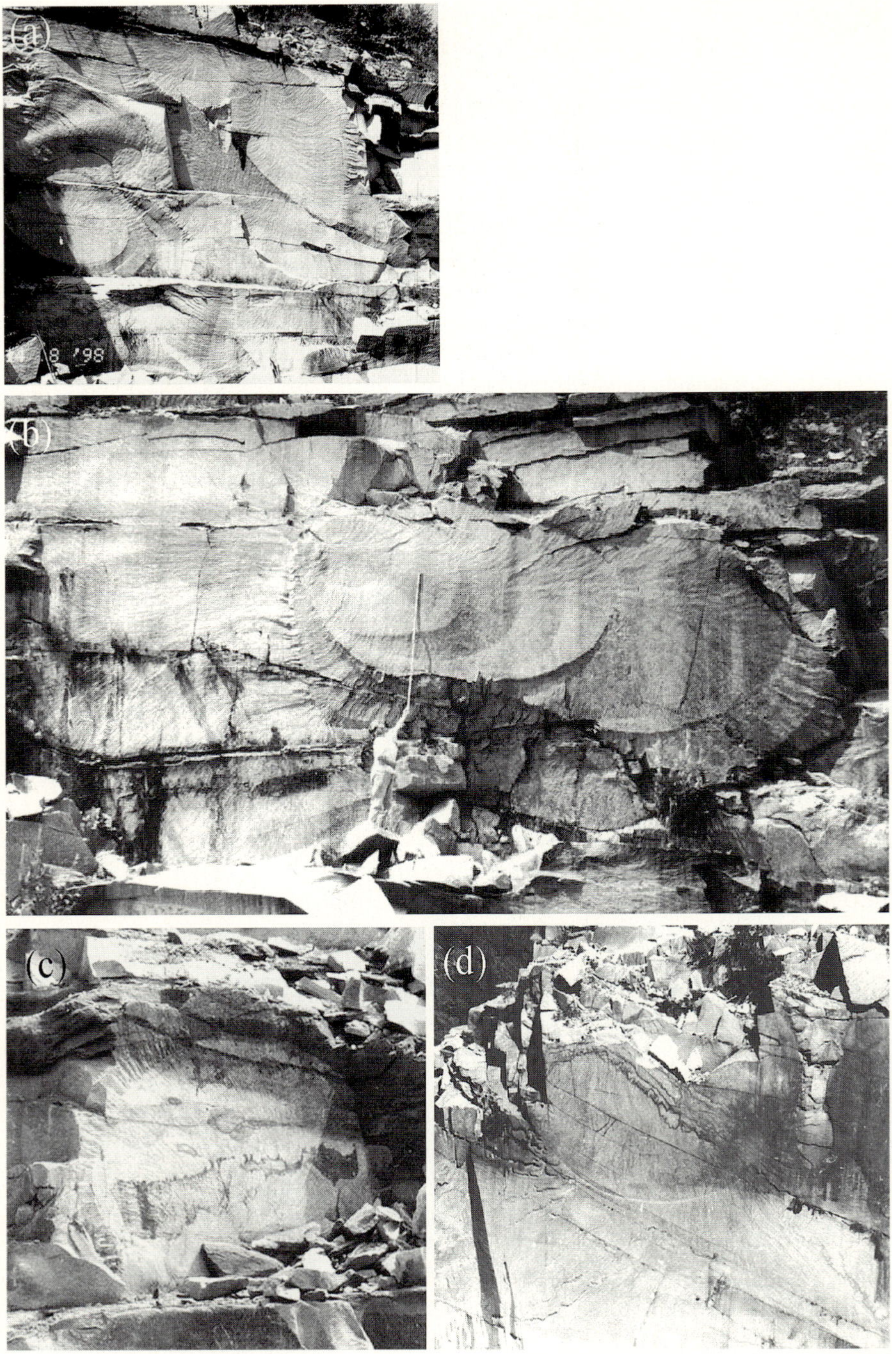

Fig. 2.55. Joints from the Borsov quarry (see corresponding drawings in Fig. 4.6–4.8 for explanation): Mirror planes that contain *en echelon* fringes. **a** "The divided arrest mark joint", J5. **b** The "asymmetric joint", J6. **c** The "uniform fringe joint", J7. **d** The "partly erased joint", J8

Fig. 2.56. Joints from the Borsov quarry (see corresponding drawings in Fig. 4.6–4.8 for explanation): Mirror planes with diversified fringes. **a** The "trefoil joint", J9. **b** The "hackled fringe joint", J10

Mirror ellipticities. Practically all the joints at Borsov with exposed boundaries show mirrors that deviate from circularity into various ellipses. Some of these ellipses slightly rotate, i.e., mostly maintaining the large axes of their ellipses at various acute angles with respect to the horizontal orientation. Furthermore, often fracture ellipticity increases (the ratio of long to short axis increases) with joint growth (Table 4.2).

Incremental crack growth displayed by arrest marks. Periodicity and incremental crack growths are common. The arrest marks exhibit two common periodicity styles:

1. Uniform radius of curvature on all arrest marks (J1 in Fig. 2.54a).
2. Concentric growth of arrest marks (J6 in Fig. 2.55b).

Joint J6 displays a new type of incremental fracture of a strict growth of the arrest marks radii along the inclined symmetric axis at a dip of 8°, without propagation along the sub-vertical directions.

Central vs. eccentric fracture origins. Whereas certain joints initiate at the focus of radiating plumes or approximately at the center of the smallest arrest mark (e.g., Fig. 2.54a,b and 2.55b), some other joints initiate close to one side of the elliptic arrest mark (e.g., Fig. 2.54c and 2.55b). It is noticeable that the latter joints that display eccentric origins also show small angle plunge of their symmetry axes.

Variations in plume intensity. The Borsov joints mostly reveal three plume styles (Table 4.1): First, plumes with a uniform intensity, that is, a monotonous barb coarseness (relief); second, plumes that display alternating increases and decreases in their intensities along their imaginary symmetry axis; third, plumes that display an increase in their intensities as they curve away from their symmetry axis.

The first style is represented by J2 (Fig. 2.54b) and J7 (Fig. 2.55a). Whereas the plume of J2 is relatively coarse, that of J7 is weak and is hardly visible. The multi-joint exposure (Fig. 2.54d) displays adjacent joints decorated by plumes of various intensities. However, plumes of individual joints generally maintain uniform barb coarseness. The second style is represented by J3 (Fig. 2.54c), J5 (Fig. 2.55a) and J6 (Fig. 2.55b). On these joints, plumes change their intensities along their lengths at various locations on the mirror. Periodic changes in plume intensity are also known from sedimentary rocks (Fig. 4.36). The large plume of J6 initiates at the edge of the *en echelon* "triangle" fringe (Fig. 2.55b) and propagates horizontally towards the left at much lower intensities than towards the right. A transition of a fringe into a plume on the parent joint is rare, i.e., in all other known descriptions the transition is from plume to fringe. The third style is observed at the lower part of J1 (Fig. 2.54a), on the first and some subsequent elliptical growth stages of J6 (Fig. 2.55b) and on J8 (Fig. 2.55d).

Deviation from orthogonality. Previous studies (e.g., Hodgson 1961a) reported that plumes and ripple marks superpose each other orthogonally at every intersection on a given joint, indicating a simultaneous development of the two features. We found a deviation from the 90° pattern to smaller angles on J6 (see insert to Fig. 4.7b). While plumes generally form by local mixed mode I / III, in the present case the local mixed mode was possibly I / II / III.

2.2.13 Oblique Extension Joints and Longitudinal Splits (or Splittings)

2.2.13.1 *Introduction*

Conjugate hybrid joint sets. Occurrences of "conjugate hybrid joint sets" that form at small angles (≤45°) in the field have been considered by many investigators (e.g., Muehlberger 1961; Price 1966; Stearns 1968; Ladeira and Price 1981; Etheridge 1983). The same observation was termed differently by various authors, e.g., "oblique extension fracture" (Dennis 1972), "conjugate hybrid joints" (Hancock and Al-Kadhi 1978), "hybrid extension/shear fractures" (Ladeira and Price 1981), "transitional tensile behavior" (Suppe 1985, p. 154), hybrid joints (Bahat 1989), and transitional-tensile fracture (Engelder 1999). According to Hancock (1985), systematic joints can form an angular continuum from a coaxial relationship with the maximum principal compression (0°) to a maximum dihedral angle of c. 45°. Price (1977) and Hancock (1985) interpreted the "hybrid-shear fractures" as belonging to a failure class transitional between extension fractures and Navier-Coulomb shear fractures showing a conjugate set with a dihedral angle greater than 45°. However, explaining the structure of "conjugate hybrid joints" has been an outstanding problem in structural geology. For instance, Engelder (1999) suggests that based on assessments from laboratory (e.g.,

Brace's 1964 tests on a dog-bone-shaped sample), field and petrofabrics (e.g., a fracture system from Price and Cosgrove 1990) and theoretical observations (e.g. Reches and Lockner 1994), the transitional-tensile fracture propagation is unlikely to occur in homogeneous, isotropic rock, and that it is not explained by a parabolic Coulomb-Mohr failure envelope. We examine this suggestion below.

The outstanding question is whether oblique extensile fractures (joints) exist, i.e., whether a parent fracture (a vein) showing shear offset initiated as such within an isotropic, homogeneous rock, and then propagated in its own plane while that plane was subjected to a shear couple as well as an effective crack-normal tension (Engelder 1999). Several observations suggest that under certain conditions such results are possible. The occurrence of straight dilatational *en echelon* cracks at angular relations to the remote compression (Pollard et al. 1982) implies that such a simulation is possible. Willemse and Pollard (1998) observed this in nature and provided an explanation why straight wing cracks (Sect. 2.4.2) are relatively common (Fig. 2.57a). Perhaps another relevant case is bifurcation into two tensile branches that form small angles, bisected by the imaginary continuation of the parent fracture. This is known to occur in various scales, from small size crack bifurcation in glass bottles (Fig. 2.30c) to large ones along faults in California (e.g., faults that branch from bifurcation 6 in Fig. 2.57b are about 500 m long). We are aware that the above three examples can be "excused" on different grounds (mixed mode and pore pressure, the straight wing crack is short, and dynamic fracture, respectively), but they cannot be ignored.

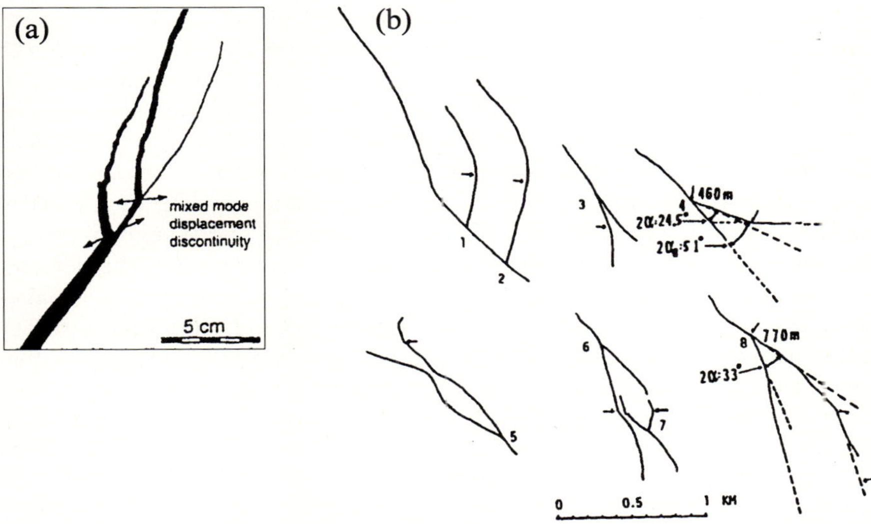

Fig. 2.57. a Wing cracks near tip of a fracture that experienced mixed mode loading i.e., both sliding and opening. Calcite cement (*black*) now fills the flaw and wing crack. Note the sharp angles at which the (*thick*) wing cracks deviate from the parent fracture, the curved shape of the wing cracks, as well as the (*thin*) straight one. Matching up of both surfaces of parent fracture allows determination of the displacement discontinuity as shown. Example from Lower Jurassic limestones between East Quantoxhead and Lilstock, Somerset, England (after Willemse and Pollard 1998). **b** Eight fault terminations by branching (1–8) along segments of the Coyote Creek fault in south California. Branchings 3, 5, 6, 7 and 8 show approximate symmetries between the two branches (forming the angle 2α, that are initially straight, reflecting local tensile stress. Branchings 1, 2 and 4 show deviations from symmetry, implying significant mode II/mode I (modified after Clark 1972 by Bahat 1982)

Let us examine another set of observations. These are related to *en echelon* cracks, which are equivalent to tension gashes and stepping joint segments at parent fracture terminations (Petit and Barquins 1987, 1988; Petit 1988; Price and Cosgrove 1990; Cruikshank et al. 1991) (Fig. 2.58a–d). Petit and Barquins (1987, 1988) induced a wing crack w (that they term branch fracture), and a shear zone of *en echelon* cracks e in PMMA and in sandstone under uniaxial compression, when applied on inclined pre-cuts (Fig. 2.58a,b). The wing crack that showed an opening resulted from local mode I loading below peak stress, and the *en echelon* cracks were induced by local mode II loading under maximal stress conditions. The sandstone revealed cataclastic deformation along the *en echelon* cracks in the shear zone. The wing crack strongly deviated from the direction of the pre-cut, whereas the shear zone was aligned approximately in continuation with it.

The various structures (Fig. 2.58a–d) show that the *en echelon* cracks do not maintain a uniform orientation; instead, they systematically rotate counter clockwise with their distance from the tip of the parent fracture. While in the field where two adjacent parent fractures occur, this rotation could be viewed as a consequence of interaction by the tips of the two fractures (Cruikshank et al. 1991) (Fig. 2.58d), i.e., caused by the interaction of two stress fields. However, this *en echelon* rotation is unrelated to such an interaction in an experiment where only one parent fracture is simulated (Fig. 2.58a,b) and also in the outcrop shown by Fig. 2.58c (assuming an isolated structure), questioning the validity of the interpretation of interaction by the two fractures.

There is a systematic rotation of the *en echelon* cracks towards parallelism to the maximum principal compression (Fig. 2.58a–c), and towards co-axiality with the parent fracture (Fig. 2.58d). Thus, in a given remote stress field, a local predominant mode II close to the tip of the parent fracture gradually changes to a local predominant mode I loading with the distance from the tip reaching the maximum rotation at a *particular distance* from the tip, in a fracture termination zone. This is measurable in Fig. 2.58a–d (see more in Sect. 6.6). Compared with the systems shown in Fig. 2.58a–d, no significant rotation is discerned in the *en echelon* sets that occur beyond the tips of joints along which no strain has occurred (Bahat 1991a, Fig. 4.22). Thus, w and e formed by the same remote stress field (Fig. 2.58a,b) and a reorientation of the remote stress field (Engelder 1999) was not necessary for their creation. They formed by a stable process under conditions of increasing stresses (e.g., Brace and Bombolakis 1963; Jayatilaka and Trustrum 1978). Some of the *en echelon* tensile cracks in the arrays formed oblique to the maximum principal compression in a transient rotation (Fig. 2.58a–c). However, the last *en echelon* crack in the Garden Area (horizontal arrow in Fig. 2.58d) was formed at the end of the transient rotation, possibly suggesting

▶

Fig. 2.58. Rotation of *en echelon* cracks along their arrays, in different systems. **a** Induced wing crack *w* (branch fracture), a shear zone of *en echelon* cracks *e* and *uf* in PMMA under uniaxial compression, when applied on an inclined pre-cut *p* (from Petit and Barquins 1987, 1988). **b** *En echelon* superficial tensile fracture formed from a pre-cut under biaxial loading in a plexiglas plate (after Petit 1988). **c** Rotation of *en echelon* tension gashes at the tip of a tension gash (from Price and Cosgrove 1990). **d** Stepping joint segments (*long*) and echelon cracks (*short*) in the Garden Area, Arches National Park, USA. Two ways that joint segments terminate in the vicinity of neighboring joints. Tips of a segment may simply veer toward a neighbor. Or, beyond the tips of a segment may be "array" of echelon cracks. The segments step right and are twisted a few degrees counter clockwise relative to the trend of the zoned joint, whereas the echelon cracks associated with the segments step left and are twisted a few degrees relative to the trend of the segment. Note that in the sheared stress field shown here, the *en echelon* cracks rotate counter clockwise along their arrays (modified after Cruikshank et al. 1991; see text for explanation)

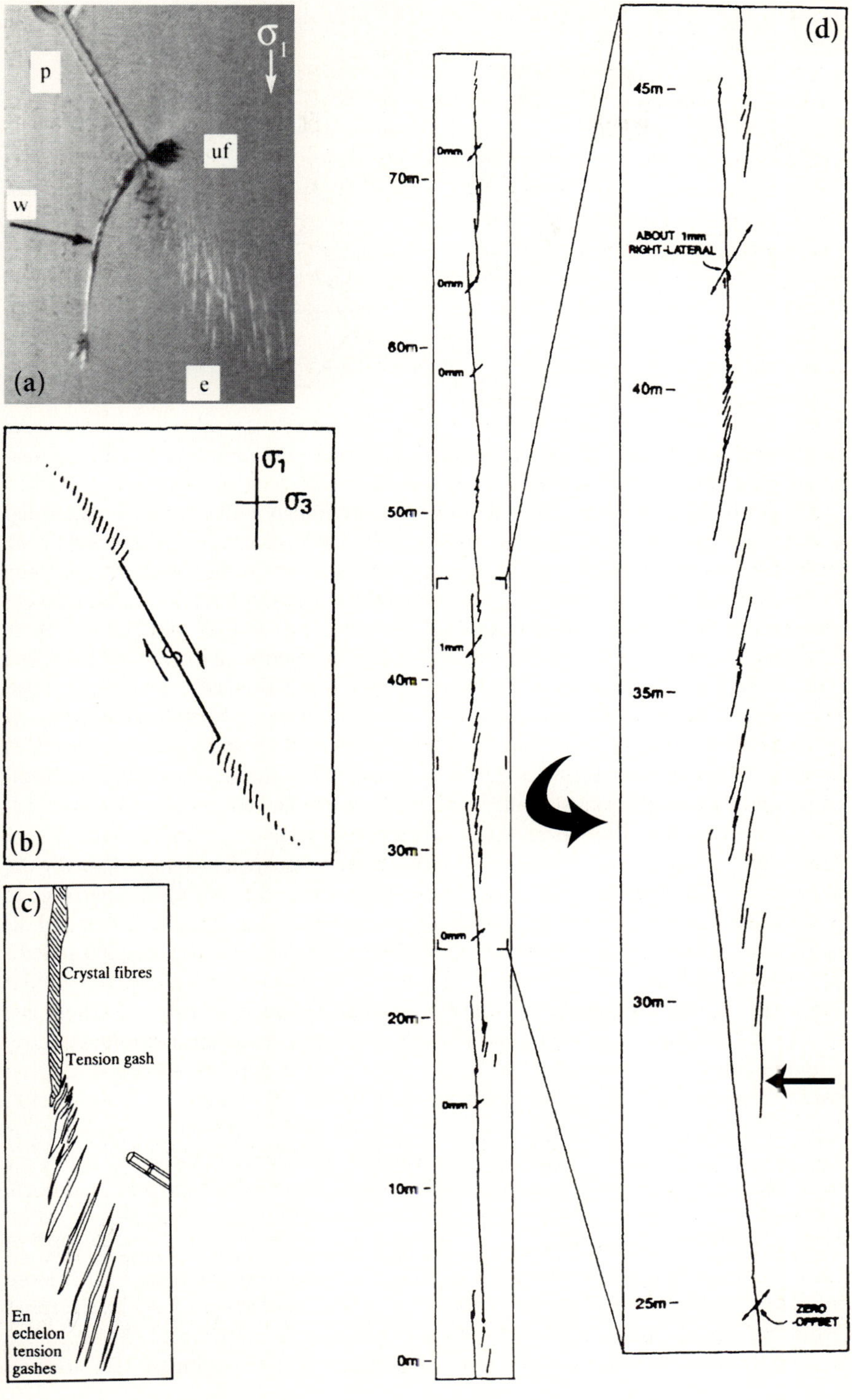
σ1
p
uf
w
e
(a)
σ1
σ3
(b)
(c)
Crystal fibres
Tension gash
En echelon tension gashes
(d)
0mm
1mm
70m
60m
50m
40m
30m
20m
10m
0m
45m
ABOUT 1mm RIGHT-LATERAL
40m
35m
30m
25m
ZERO OFFSET

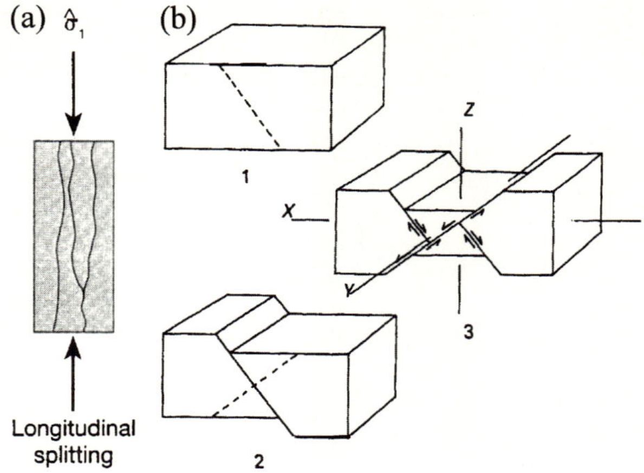

Fig. 2.59.
a Longitudinal splitting along wiggly curves (modified after Twiss and Moores 1992, Fig. 9.2B). **b** A model of a sequential fracture in a conjugate fault set (from Park 1983, Fig. 9.4)

a general condition for forming joints oblique to the maximum principal compression. Hence, oblique extensile fracture is possible.

The problem of oblique extensile joints may be linked with another outstanding enigma: The true nature of the longitudinal splitting. Some authors (e.g., Engelder et al. 1993, p. 46) contemplate the idea that longitudinal splitting is the mechanism leading the formation of regional jointing. At the base of this idea is the rationale that uniaxial compression is coaxial with longitudinal splitting (e.g., Griggs and Handin 1960; Holzhausen and Johnson 1979) and analogically, the maximum horizontal principal stress is coaxial with strikes of individual systematic joints (Engelder and Geiser 1980). It is important to take into account that experimental splitting has been observed by many investigators to be wiggly rather than sharp straight fractures (e.g., Jaeger and Cook 1979, Fig. 4.5.1; Twiss and Moores 1992, Fig. 9.2B; Cosgrove 1995) (Fig. 2.59a). What is the meaning of these deviations from straight fractures? Experimental results obtained from different materials under diverse fracture conditions as well as observations made in geological outcrops are presented in the next sections and compared, seeking to improve the current understanding how oblique extensile joint sets and longitudinal splitting are formed. These sections do not deal with situations of changes in magnitude of initial dihedral angles during fracture growth (e.g., Kalthoff 1973; Ramsay and Lisle 2000, p. 968).

Conjugate vs. contemporaneous, simultaneous and sequential fracture. There is a need to make a few term distinctions. A convenient term in structural geology is contemporaneous. This term is applied when a field relationship reveals that in a given research area, in some locations a joint of set A repeatedly arrests at a joint of set B, suggesting that set A is younger than set B, and in other locations the sequence is reversed. This is taken to indicate with reasonable confidence that the two sets were formed contemporaneously during the same geological period, which could be "short" or "long". Thus, reverse sequences are implied by the term contemporaneous. More challenging is the term simultaneous. It is quite possible to claim that a series of fractures associated with a certain earthquake was formed simultaneously. It is more difficult to prove that a series of joints of a particular (systematic) set was formed together. Probably joints from some sets do not form simultaneously (Bahat 1991a, Fig. 4.8). It is far more difficult to prove that joints of "two conjugate sets" formed simultaneously. Quite possibly, even two sets that exhibit the closest to ideal field relationships that would justify the term conjugate would be a product of a sequential process. In a way this perplexity resembles uncertainties that are related to conjugate fault sets: Do they really form simultaneously or possibly sequentially (Fig. 2.59b)? We apply here the issue of simultaneously vs. sequentially to two topics, oblique extensile joints and longitudinal splits, and to the problem of two contradicting interpretations regarding the mode of formation of sets 309° and 344° in the Beer Sheva syncline (Sect. 3.4.8 and 6.3.9.1).

2.2.13.2
Fracture Results from Different Sources

This section provides observations from geological outcrops, which are supplemented by experimental results that are presented in Sect. 2.2.3.

Regional joint patterns from chalks in East England. Goodwin (1995) investigated the natural fracture behavior of chalks in south England. He found two sets of regional distribution, striking NW and NE. Both joint sets comprise steep conjugate (ignoring the sequential issue) and vertical joints, everywhere consistent with bedding (Fig. 2.60). In the southern investigated area (Norfolk), traces of the NW set on the vertical NE-NW section typically show conjugate steeply inclined joints that form dihedral angles between 25° and 45°. Vertical joints occur less commonly. In the northern area (south Humberside/north Lincolnshire) by contrast, vertical NE-NW sections across the joint sets show mainly vertical extensional joints with few steeply inclined planes occurring, i.e., the dihedral angle of the Humberside joints <20°. The joints of Norfolk are interpreted by Goodwin (1995) as the conjugate hybrid joints of Hancock (1985, see above). This would include "hybrid joints" that allow for loading of both, shears along, and extension across the plane to take place.

Goodwin (1995) considers two alternative explanations for the differences in size of the dihedral angle in the southern and northern areas. First, the dominance of extensional joints in the north is correlated with "hard" chalk of lower porosity and greater rock strength (dry density 1.78–2.27 g cm^{-3}), compared with the conjugate characteristics in the south that are correlated with "soft" chalk of high porosity and lower rock strength (dry density 1.40–1.72 g cm^{-3}). Second, the above differences reflect a possible northward decrease in differential stress across eastern England. This dilemma was investigated in our laboratory, and the results are in favor of the first explanation (Sect. 6.2.2).

Regional joint patterns from chalks around Beer Sheva, South Israel. Three types of surface markings occur on the single-layer, cross-fold joint surfaces in the Lower Eocene chalks from the Beer Sheva and Shephela synclines in south Israel:

1. Horizontal axial plumes.
2. Circular rib markings.
3. Transitional variations between the two.

The three types may occur together in the same outcrop (Bahat 1991a, Fig. 3.28). About half (56%) of the 225 FSM examined on joints in four outcrops around Beer

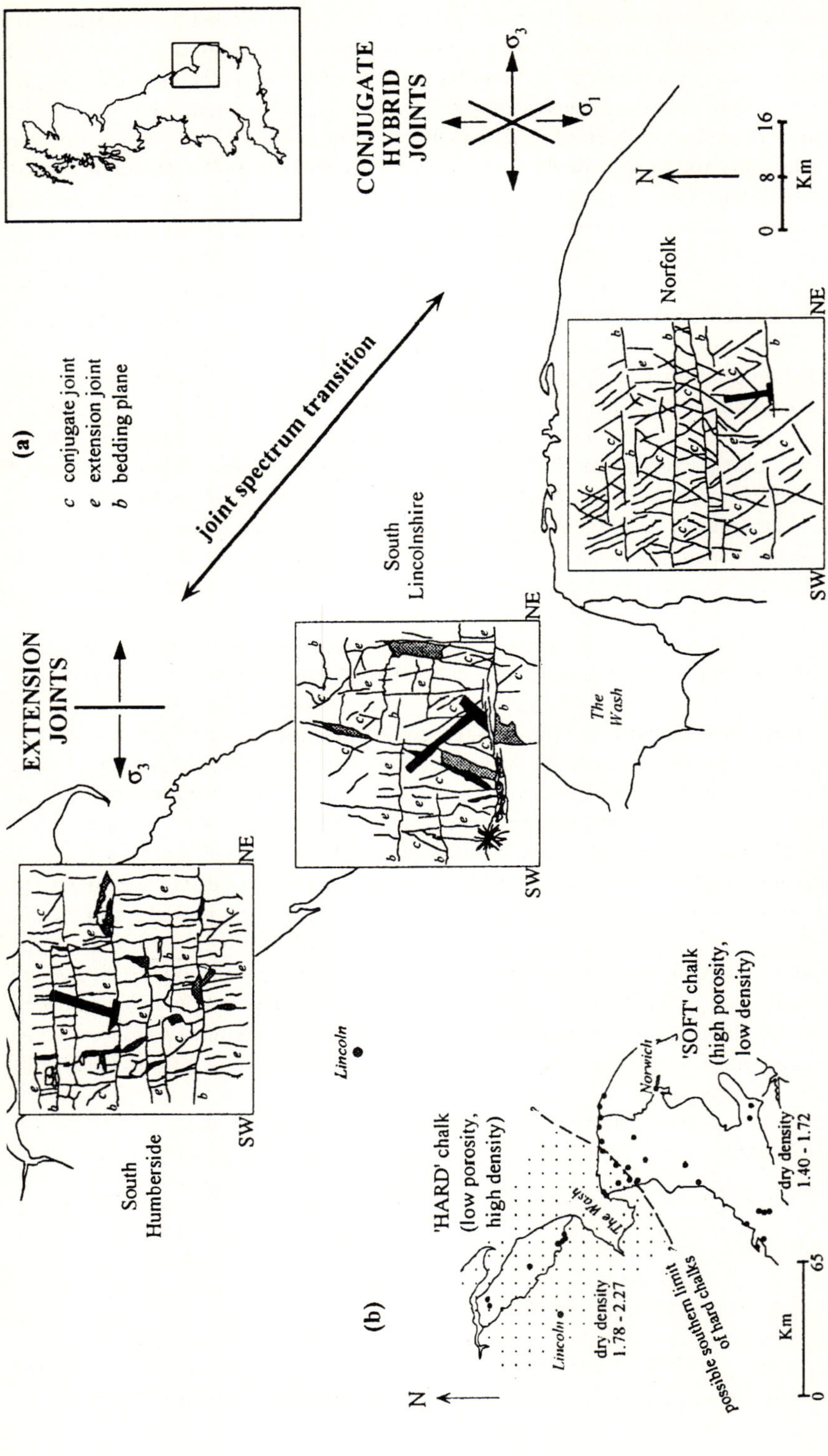

Fig. 2.60. Natural fracture behavior of chalks in south England, see text for explanation (after Goodwin 1995)

Sheva bear circular markings, and these occur on joints of a set of average strike 328° (N 32° W). Sixty five samples (29%) are axial plumes, and these occur on joints belonging to sets of average 309° and 344° strike. Transitional markings make up 12%, and about 3% are other combinations. Joint sets with circular markings and joint sets with plumes occur in separate rock layers, even in the same outcrop.

It was interpreted (Bahat 1987a, 1991a) that possibly the circular markings on the 328° joints have been formed along the direction of the horizontal principal stress, and the axial plumes on sets 309° and 344° were formed on joints at low angles to that principal stress in a conjugate pattern (Stearns 1968). The circular markings developed when the minimum principal stress σ_3 was horizontal and normal to the joint surface, and the maximum horizontal σ_1 equaled the vertical principal stress σ_2, i.e., exerting "pure extension" on the joint. The axial plumes represent the approximate conditions with respect to σ_3 like for the circular markings, when the horizontal principal stress different from the vertical, $\sigma_1 \neq \sigma_2$, exerting some differential stresses on the joints and resulting in superposing a slight shear stress on the joint. The three sets were formed in the Lower Eocene chalks during the burial stage while still immersed in water. Therefore, fracture occurred while the chalk was weak, possibly having the tensile strength of about 0.5 MPa (Bahat 1991a, p. 243). Hence, a dihedral angle of 35° was formed by the strikes of the 309° and 344° sets that resulted from the regional maximum compression in the 328° direction, and the plumes on the surfaces of the former two sets may reflect some mixed mode loading.

2.2.13.3
The Behavior of the Dihedral Angles from Different Sources

Plots of the dihedral angles. The dihedral angles obtained from the different sources presented above are plotted in the range $2\alpha = 0–65°$ vs. the compression strength and dry density of the material (Fig. 2.61a and b, respectively, see Fig. 2.24c for the angular geometry). The decrease in 2α with the increase in compressive strength is correlative for various materials. Longitudinal split, probably of an extensile nature, occurs in the range $2\alpha \sim 0–20°$ in "strong" glass ceramic, glass and rocks of medium strength. We speculate that in weak rocks, slow (geological rate) fracture may result in extensile fractures that attain 2α values up to 35° by an unclear mechanism. The value of the dihedral angle may stretch up to 65° in "weak" rocks, involving the magnification of the tendency of fracture by mixed modes I + II. There seems to be a general trend of decrease in the dihedral angle with the increase of the dry density of chalks in east England (Fig. 2.61b), resembling the quantitative curve (Fig. 2.61a). However, due to the limited number of points on the latter curve, its value is only qualitative.

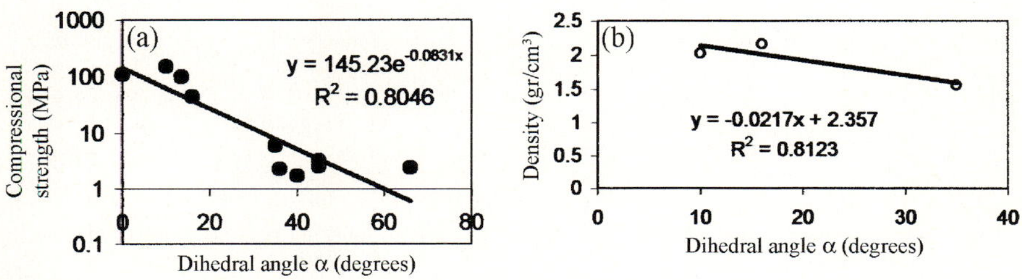

Fig. 2.61. Dihedral angle 2α plot vs. material properties. **a** Compression strength; **b** dry density

The reliability of the results. The comparison of the various dihedral angles obtained from different sources (Table 2.2, Fig. 2.61) needs to be qualified, because the comparison relies on different degrees of certainties regarding the various results. Three reliable results are the ones obtained experimentally on the chalk, glass ceramic and glass (Sect. 2.2.3.4, 2.2.3.7 and 2.2.3.11, respectively), because direct fractographic examinations of their splits clearly suggest that these fractures are essentially extensile. These results suggest that extensile longitudinal split maintains 0° deviation from the compression direction in a strong material (glass ceramic) but may deviate up to $\alpha = 8 \pm 2°$ from this direction in materials of reduced-strength. Included in the latter category are the intensely dried chalk from the Beer Sheva syncline and the glass samples to which pre-cuts were introduced. The results on longitudinal split obtained in fractured Chelmsford granite (Sect. 2.2.3.6) support this suggestion.

Table 2.2. Dependence of the fracture dihedral angle on material/rock properties

No.	Dihedral angle 2α (°)	Compressional strength (MPa)	Dry density for chalk (g cm^{-3})	Material/rock	Source
1	52 – 80 (66)	0.5–4.3 (2.4)		Sandstone Berea weak	Handin et al. 1963
2	20 – 70 (45)	0.4–4.6 (2.5)		Limestone Marianna weak	Handin et al. 1963
3	10 – 80 (45)	0.7–5.9 (3.3)		Dolomite Hasmark weak	Handin et al. 1963
4	0 – 80 (40)	0.3–3.1 (1.7)		Siltstone Repetto weak	Handin et al. 1963
5	0 – 72 (36)	0.4–4.17 (2.3)		Shale Muddy weak	Handin et al. 1963
6	35	6[a]		Chalk Beer Sheva weak	Bahat 1987a, 1989
7	16 ±4	45	2.16	Chalk Beer Sheva medium strength	Bahat et al. 2001a
8	10 ±10[b]	106–200[b]		Granite Chelmsford strong	Peng and Johnson 1972
9	13 – 14 ±2	105–98		Soda lime Glass strong	Section 2.2.3.11
10	0	112		Glass ceramic strong	Bahat et al. 2002a
11	25 – 45	1.40–1.72		Chalk East England	Goodwin 1995
12	0 – 20	1.78–2.27		Chalk East England	Goodwin 1995

[a] This compressional strength is obtained by multiplying the tensile effective fracture stress 0.5 MPa by 12 (Jaeger and Cook 1979).
[b] Interpretation is not clear.
Parenthesis gives mean.

It is not clear how reliable the other results are that are not supported by direct fractographic observations. One is convinced by Goodwin's (1995) results that the stronger chalks in east England have a greater tendency to fracture in extension than the weak ones. The weak chalks (from Norfolk) on the other hand, provide a large range of dihedral angles (25–45°). Fracture at Norfolk probably went far beyond extension, but how dominant the superposed shear mode was in the failure of these chalks is unknown. Muraoka and Kamata (1983) observed that the shear angle ranges between conjugate faults in the Kusu Group were clearly lower in the generally more brittle siltstone (17–45°) than in the sandstone of lower competence (45–69°). Thus, the general trend of the increase of the size of the dihedral angle with the material weakness (incompetence) is also shown for faults.

2.2.13.4
Conjugate Extension Joints and Inclined Longitudinal Splits that Form by Mode I

The results that are summarized in Table 2.2 and ranked in Table 2.3 suggest a new classification that distinguishes four groups. The extension by mode I at $2\alpha = 0°$ for a strong material (item 10 in Table 2.2) is obvious. The interpretation of the results obtained for the materials of reduced strength (items 7 and 9 in Table 2.2, second group in Table 2.3) is that joints may form exclusively by mode I loading, even at small dihedral angles $2\alpha \leq 20°$. This interpretation is supported by the observation made on the tendency of tensile *en echelon* cracks to rotate, so that some of the cracks form oblique to the maximum principal compression under mixed mode I + II loading (Fig. 2.58a–d). With certain qualifications, this interpretation is possibly sustained by fracture in granite (item 8 in Table 2.2). The suggestion given here lacks a statistical support and stays tentative until additional results will confirm it. In the meantime $2\alpha \sim 20°$ will mark the boundary between groups 2 and 3 (Table 2.3). The lack of fractographic criteria to distinguish between these two groups remains a challenge for future research. Finally, the composite failure envelope formally distinguishes between groups 3 and 4 (Fig. 2.26d, B and C, respectively). Additional studies may possibly identify fractographic criteria to distinguish between the latter two groups. The angular boundaries among the four groups are not expected to be "critical angles"; they are likely to be transitional ranges that depend on rock properties (competence/strength) and the rate of applied strain. But the existence of four groups possibly suggests some "spectrum of discontinuities" of dihedral angles, rather than a "continuous spectrum" (Hancock 1986).

Table 2.3. Four fracture groups with varying dihedral angles

Group	Fracture type	Dihedral angle (°)	Loading modes	Offset
1[a]	Axial extension joints	0	Exclusively mode I	None
2[a]	Inclined extension joints	Up to about 20	Exclusively mode I	None
3	Hybrid joints	Up to about 35	Mixed modes I and II[b]	None
4	Conjugate shear fractures	Various dihedral angles, generally >20 (Coulomb fracture criterion)	Mixed modes I and II	Yes

[a] Sets 343° and 350° (Sect. 6.2.2) belong to groups 1 and 2, respectively.
[b] The presence of plumes on the surfaces of hybrid joints testify to lack of offset.

2.2.13.5
The Mechanisms of Plume and Striae Formation on Various Fractured Surfaces

We can now extend the discussion on plume straightness and coarseness (Sect. 2.2.4.2). Striae occur on fractured surfaces from groups 1, 2 and 3 that form under unlike conditions and mechanisms (Tables 2.2 and 2.3). On the extension joints from group 1 (fracture in glass ceramic) the striae are essentially straight, consisting of single fractures, and close to their ends they curve towards one of the sample boundaries and slightly divide into secondary small microcracks (Fig. 2.19a). On the individual surfaces of the fractures from group 2, the micro-plumes are straight (Fig. 2.15a). Straight plumes indicate straight fracture propagation predominantly by mode I that occurred on the relatively small fracture surfaces in the chalk. Almost no plume curving and division into smaller microcracks occurred, because fracture propagation took place on surfaces that were isolated from influences of adjacent sample boundaries. On the other hand, plumes cutting joints from group 3 were strongly affected by layer boundaries, and their division into secondary branches and barbs is quite pronounced (Fig. 2.22a); curving of plumes is more noticeable in thick rock layers.

The plumes in the Lower Eocene chalks (Fig. 2.45) are considerably coarser than the plumes in the Middle Eocene chalks (Fig. 2.35a,c, 2.36a and 2.38). Burial single-layer jointing occurred in the Lower Eocene chalks under increasing overburden stresses, whereas uplift single-layer jointing took place under increasing tensile stresses in the Middle Eocene chalks (Bahat 1999a). These distinct conditions quite possibly resulted in different fracture velocities, which were greater in the chalks of the Middle Eocene than in the chalks of the Lower Eocene for several reasons. First, it is quite likely that fracture velocity will be greater close to the ground surface under tensile conditions, compared to conditions of increasing compression with depth. Second, experiments by Müller and Dahm (2000) (Fig. 2.22) show the correlation of increase in plume coarseness with the reduction in fracture velocity. Third, we use a tentative assumption that still requires confirmation (Sect. 6.3.9.1) according to which the plumes on the conjugate sets 309° and 344° from group 3 were formed under mixed modes I + II, which were associated with the remote compression from 328° compared with the joints cutting the Middle Eocene chalks that do not show a conjugate relationship and appear to have responded to rotating co-axial compression. The possible implication is that while shear modes were minimal on the latter joints, more significant shears were exerted on the conjugate sets 309° and 344°. Possibly, the overburden constraints imposed on the Lower Eocene chalks caused some of mode II to locally transform into mode III (Sect. 2.2.3.8), leading to plume coarsening.

Even more intriguing is the fact that no fringes of *en echelon* segmentation occur on joints cutting Lower Eocene chalks, which are marked by coarse plumes, while fringes of the type shown in Fig. 2.35a are abundant on joints that cut the Middle Eocene chalks, which are marked by delicate plumes. This shows that coarse plumes are not always more likely to transform to *en echelon* segments than delicate ones. This "paradox" supports the suggestion that a time interval may have occurred between the fracture of the parent joint and its fringe of the type shown in Fig. 2.20c, and this could involve significant changes in fracture conditions.

2.2.13.6
Conclusion

1. Our interpretation of the systematic rotation of the *en echelon* cracks towards parallelism to σ_1 (Fig. 2.58a–d) is that in a given remote shear stress field, a local predominance of mode II gradually changes to a local predominant mode I loading. In this setting, a reorientation of the remote stress field is not necessary for creating extensile cracks (mode I) oblique to σ_1 in certain locations. Hence, extensile cracks can form in inclined orientation with respect to σ_1.
2. Based on experimental evidence (Table 2.2), extensile longitudinal splitting can be extended from a dihedral angle of $2\alpha = 0°$ in "strong" materials to $2\alpha \sim 20°$ in materials of "medium strength", supporting the interpretation in conclusion 1.
3. Field evidence possibly suggests that in "weak" materials the dihedral angle of conjugate hybrid joints (mixed modes I + II) may be extended to about 35°.
4. The wiggly longitudinal splits that have been observed by various investigators represent composite fractures consisting of combined extensile and shear growths that interacted with each other.

2.3
On Uplift, Post-Uplift, Neotectonic and Surface Joints

Joints in sedimentary rocks are genetically divided into four groups: Burial, syntectonic, uplift and post-uplift (Bahat 1991a; Engelder et al. 1993; Ghosh 1993; Bankwitz and Bankwitz 1994; Bankwitz and Bankwitz 1997). In this section we refer again to the differences between uplift and post-uplift joint groups and further elaborate on the latter. We briefly discuss the diverse characteristics of "neotectonic joints", "exfoliation joints" (that are distinct from "sheet joints"), "cross joints" and introduce the "surface joints" group.

2.3.1
Neotectonic Joints

2.3.1.1
General

An inventory of neotectonic joint attributes was compiled by Hancock and Engelder (1989) in various terrains containing flat-lying sedimentary sequences in the platform covers of cratons from the Appalachian Plateau, southeast England-northeast France, the Arabian platform, and the Ebro basin in Spain. Systematic neotectonic joints are the youngest tectonic structures in all of these study areas. They are, however, older than superficial non-systematic cross-joints of non-tectonic origin that abut and link them. The next section is a citation from Hancock and Engelder (1989).

2.3.1.2
Fracture Styles of Neotectonic Joint Systems

The geometry of the systematic neotectonic joint systems is characterized by simplicity, irrespective whether they cut previously intact or already fractured rocks

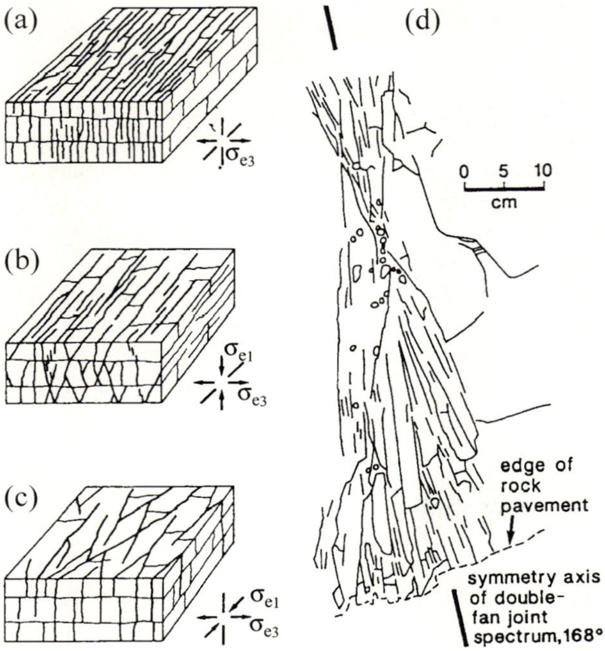

Fig. 2.62. Characteristic neotectonic joint systems. **a** Single set of systematic vertical extension joints linked by non-systematic cross-joints. **b** A spectrum of systematic joints comprising vertical extension fractures and steep conjugate fractures enclosing a range of dihedral angles of less than 45°. Two steep fracture directions are expressed by arrays of *en echelon* vertical joints. Non-systematic joints link systematic joint. **c** A spectrum of systematic joints comprising vertical extension fractures and vertical conjugate fractures (in plan) enclosing a range of dihedral angles of less than 45°. Non-systematic joints link systematic joints; σ_1, maximum effective principal stress; σ_3, minimum effective principal stress. **d** Neotectonic joints cutting horizontal limestones. Line drawing from a photograph, looking down on the double-fan joint spectrum. The symmetry axis of the spectrum trends sub-parallel to the mean strike of all neotectonic joints in the investigated area (after Hancock and Engelder1989)

(Fig. 2.62a). The commonest assemblage comprises vertical joints that are either parallel to each other or dispersed less than 10° about the mean orientation of the set (Fig. 2.62b). Where the vertical set occurs in conjunction with other neotectonic joints, it bisects the acute angle between conjugate sets (Fig. 2.62c). In all the above mentioned four case-study terrains, a less abundant, but nevertheless typical neotectonic joint pattern comprises steep conjugate joints striking parallel to a neighboring single vertical set (Fig. 2.62b). Steep conjugate joints generally enclose dihedral angles (2α) in the range of 10–45°. The latter joint sets appear in some upper but muddier parts of the Devonian succession of western New York State, whereas in the central part of the state, lower silt and shale units are cut by a single set of joints.

Some conjugate joints strike parallel to locally developed late Cenozoic topography. They occur in western New York and postdate systematic cross-fold joint sets (Engelder 1989). Those joints with a dip direction pointing downslope become more prominent in the near surface. This correlation of dip with Cenozoic topography is the strongest evidence supporting a neotectonic age for the east-northeast joints of the Appalachian Plateau.

At a few stations in the Arabian platform and Ebro basin, well-ordered conjugate steep or vertical sets are replaced by a continuum of joint orientations with an angular dispersion of as much as 45° about a symmetry axis that is either vertical or trends parallel to the mean strike of a nearby single joint set (Fig. 2.62d). Hence, Hancock and Engelder (1989) distinguish several fracture styles in the systematic neotectonic joint systems: Vertical single and conjugate sets, steep conjugate sets, conjugate joints which strike parallel to locally developed late Cenozoic topography and joints that their dip orientation is characterized by angular dispersion. In addition to the systematic neotectonic joint systems there are non-systematic fractures, termed "cross joints" (distinct from systematic cross-fold joints) that extend across intervals between earlier systematic joints (Fig. 3.36d) and may be categorized as products of the neotectonic stress field (Engelder and Gross 1993).

2.3.1.3
Neotectonic, Uplift and Post Uplift Joint Systems

According to Hancock and Engelder (1989), burial followed by unloading as a result of denudation and lateral relief consequent of uplift are prerequisites for the formation of shallow neotectonic joints. A major conclusion in their paper is that shallow (at depths of less than 1 km) neotectonic joint systems are the most recent joint system. Therefore, they are of potential value for tracking the orientation of the contemporary stress field even in structurally complex terrains, although there may be a slight misalignment between joint strike and the direction of the maximum horizontal stress. Some regional joints, interpreted as neotectonic, strike parallel to or approximately parallel to the direction of contemporary horizontal maximum compression known from in situ stress measurements or fault plane solutions of earthquakes (Hancock and Engelder 1989; Gross and Engelder 1991). This provides an important motivation for studying neotectonic joints. We argue below that the application of these useful relationships needs to take into consideration a sub-grouping within the broader group of neotectonic joints.

It appears that the "neotectonic joints" (Hancock and Engelder 1989) embrace two distinct categories by Bahat (1988a, 1991a) of shallow fractures (depth down to about 0.5 km) without differentiating between them: The "uplift joints" and the "post uplift joints". The attributes that are common to both, uplift and neotectonic joint systems (in addition to shallow depths) include multilayer and large size characteristics (i.e., larger spacing and apertures than in neighbor joint sets of older generations), significant misalignments between joint strikes, and their occurrence in repeated failure events.

While many attributes are common to both neotectonic joints and uplift joints, post uplift joints seem to be analogous to the conjugate joints that strike parallel to locally developed late Cenozoic topography described by Hancock and Engelder (1989). The latter two are the most recent joint systems. On the other hand, "uplift joints" need not be the most recent joint system. Uplift joints may populate the same outcrop that exhibits "post uplift joints" of a younger generation, and they entirely differ in their strikes (e.g., Bahat 1991a, outcrop 3 in Fig. 5.29). Strike scattering of uplift joints (Bahat 1991a, Fig. 5.28) suggests that while some uplift joints may reflect the contemporary (remote) stress field, others may not. The post uplift joints are geo-morphological products of local stress fields (Bahat et al. 1999) and some of them do not reflect the contemporary stress field. Hence, when neotectonic joints are characterized, their affiliation to either the uplift or post-uplift groups should be considered.

2.3.2
Exfoliation Joints and Sheet Joints

Exfoliation fractures (e.g., Sect. 3.3 and 4.5.3) and sheet fractures (e.g., Twidale et al. 1996) reflect fracture propagation in principal planes normal to the minimum principal stress and form approximately by the same mechanisms. However, they seem to genetically and morphologically represent different categories as will be explained below. Gilbert (1904) believed that sheeting is *older than the topography*. Various authors (e.g., Dale 1923 and Holzhausen 1989) made comments on "generations" of sheet fractures (e.g., preglacial and postglacial). Similarly, Holman (1976) observed hairline cracks, which are evident only where weathering has etched them into relief that commonly subdivide sheets into two to four secondary slabs. Hence, the term sheeting should be applied to horizontal and sub-horizontal fractures that maintain approximately uniform spacing with their neighbor (lower and upper) sheets, but generally increase their spacing with depth. Sheeting does not have to develop parallel to and close to a previous fracture, although sheet fractures often occur in sets. Commensurate with the above, Holzhausen (1989) made the observation that sheet fractures must be products of states of stress that are approximately uniform over distances of at least tens to hundreds of meters in order to explain the regularity and size of the fractures.

Exfoliation joints, on the other hand, are generated by fracture parallel to existing surfaces in all attitudes. They generally develop at short distances from the exposed surface (tens of centimeters), in small spacing, and produce rock slices of both uniform thickness and lenticular shape. In a way, sheeting responds to unloading (before the topography sets in), whereas exfoliation results from loading that takes place following the establishment of the topography. Hence, sheets and exfoliation fractures correspond respectively to uplift and post-uplift fractures. We refrain here from further addressing sheet joints and refer the reader to the literature on this subject (e.g., Johnson 1970). Concentric exfoliation cracks in basalts belong to a different fracture category, which is not treated in this book.

2.3.3
Surface Joints

We introduce a new category of fractures that we term "surface joints". These joints mostly occur in the upper parts of rock outcrops. They may vary in length from several centimeters (Fig. 2.63a) to several meters and can appear in various styles, from straight shapes to twisted forms that transform into non-systematic anastomotic networks (Fig. 2.63b). Occasionally, they occur in chalks of Eocene age in the Beer Sheva region, Israel and are filled with coarse sparitic gypsum. These veins are connected with the loess-soil cover of the Upper Pleistocene, which overlay the chalks, possibly concentrating by roots of sulfate tolerant trees and depositing from groundwater that became over-saturated with $CaSO_4$ (Issar et al. 1988). Particularly characteristic are the sub-horizontal surface joints that occur in "transitional layers" that underlay the upper, eroded soil layer and overlay an uneroded jointed chalk layer (Fig. 2.63a). These surface joints possibly formed perpendicular to tensile stress that was induced by overburden reduction due to the density decrease of the overlying soil, which formed from the eroded chalk (see more in Sect. 6.8).

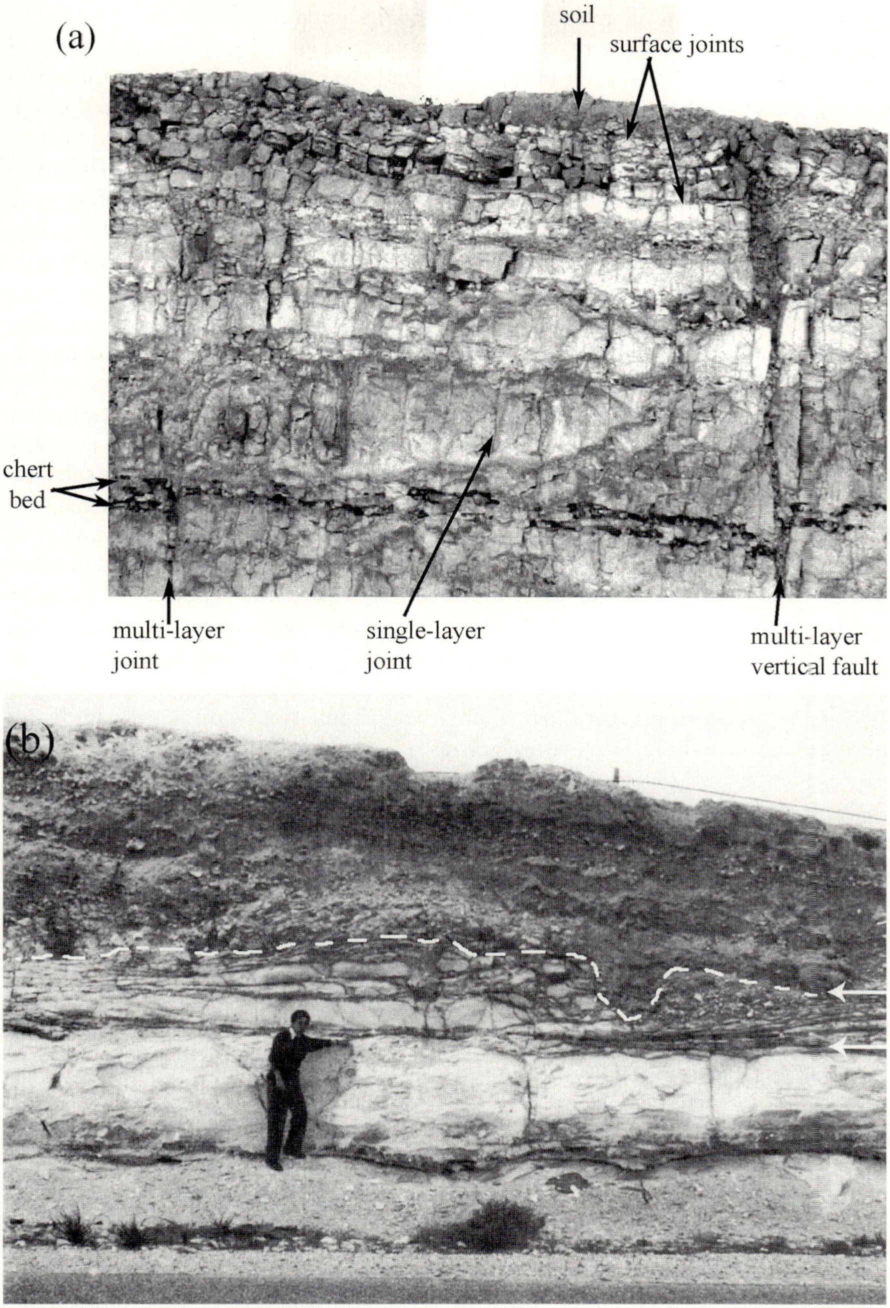

Fig. 2.63. Surface joints. **a** A section that consists of characteristic fracture elements that occur in association with a "transitional layer" of surface joints (bounded by the two *upper arrows*) in Lower Eocene chalks around Beer Sheva, Israel. The thickness of chert bed is 10 cm. **b** The "transitional layer" bounded by the two *arrows*, that underlies the upper, eroded soil layer and overlay an uneroded jointed chalk layer (after Issar et al. 1988)

Surface joints often fill spaces between previous joints and mimic the older joints. Their identification is done by comparing the new, smaller joints that lose some of their straightness and spacing to the larger and straighter, earlier joints that form greater spacing in lower layers where the surface joints do not occur. When they cut rectangular blocks that were formed by orthogonal joints, they further subdivide them to into smaller blocks. Thus, surface joints can be distinguished by their relative irregularities, reduced spacing, non-uniformity in spacing, exaggerated apertures and by their containment of secondary mineralization, erosive rock and soil. These fractures may concentrate along trenches and canals. Although they are essentially non-systematic, they may occasionally appear systematic.

Surface joints have been mistaken for burial joints by some environmental specialists, and this justifies their distinct characterization. The recognition of the importance of surface joints may be critical in planning and evaluation engineering and environmental projects. They are not only a subject of the past; these fractures may develop in the future as well. Joint intensity (fracture area per rock volume) is a basic parameter in geological engineering. The value of this parameter may change substantially between two locations that are only a few meters laterally apart in the same rock layer and differing in depths from the ground surface, along an eroded slope. The engineer may like to consider the minimum depth that a survey trench should reach below the depth of the "transitional layer" for obtaining "clean" information. The hydrologist may be misled by the attitudes and spacing of joints in this zone in predicting the flow properties of water. Finally, surface joints differ from the sub-surface, non-systematic joints that form in association with rockburst in mines (e.g., Jaeger and Cook 1979, p. 505, see also present Chap. 5) and there should not be difficulties in distinguishing between these two fracture groups.

2.4 Primary and Secondary Fractures

2.4.1 Introduction

This section is about secondary fractures that form adjacent to primary fractures when the latter are activated or reactivated by tensile or shear stresses. The geometric relationship of tensile secondary fractures in relation to a primary fracture on a principal plane is relatively simple (the primary fracture is given for simplicity as $2r_B$ in Fig. 2.30c). More complicated is the system of a primary fracture in the shape of a parent joint or a fault in the field, or a pre-cut in the laboratory (Sect. 2.2.3.11) under remote compression at a small angle α to the primary, causing local shear and tensile stresses that result in shear and tensile fracture (Fig. 2.64). In the latter diagram, we show various tensile fracture types, omitting a certain type of joints from a fault termination zone (Fig. 6.19) and shear fractures known from Riedel experiments (e.g., Fig. 2.68). In most cases, these pre-cuts (occasionally also termed slits, out-of-plane fracture, pre-existing crack, plane of weakness, initial crack or initial flaw) are introduced to the material by design (Germanovich and Dyskin 2000). We therefore use the term "pre-cut" in our descriptions. Pre-cuts are distinguished from initial flaws (Sect. 2.2.2.2), which occur as unintentional defects in the material, commonly on sub-

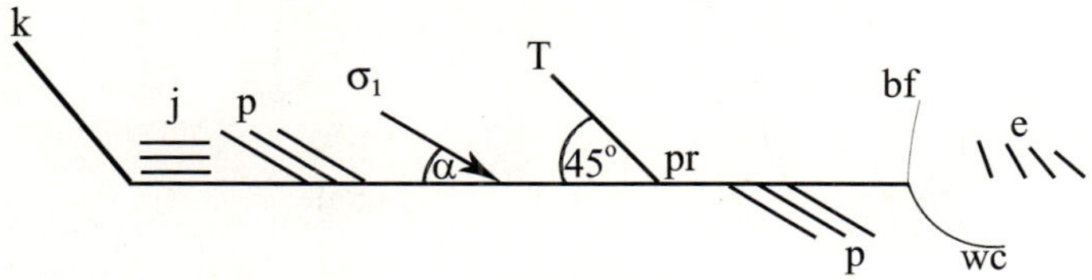

Fig. 2.64. A schematic, idealistic fracture pattern which forms by compression. Compression is applied on the primary fracture or a pre-cut *pr*, at a small angle α. Several types of secondary fractures may form, including, wing cracks *wc*, sets of horsetails (or pinnate joints) *p*, kinks *k* and coplanar joints *j*, at the two extensional quadrants, *and* a tensile fracture *T* that forms at 45° along the primary. In addition, *en echelon* sets *e*, as well as curved branching faults *bf*, may form at the two compression quadrants

millimeter scales. The geometric distinctions and the positioning of the secondary fractures along the primary fracture and next to their tip are described below in relation to the various fracture mechanisms that are involved.

The literature is profuse with terms assigned to secondary fractures by various authors, including wing cracks, curved branches, branch fracture, kinks, curved kinks, pinnate joints, splay cracks, horsetail fractures, bifurcation fractures, overstepping joints, *en echelon* cracks, and more. We concentrate below on several types of secondary fractures, including the wing crack, pinnate joints and horsetail fractures, fracture polarization, termination-zone joints and fracture branching. Additional descriptions of secondary fractures, such as kinks and stepping joints are presented elsewhere in this chapter (Sect. 2.2.8 and 2.2.13).

2.4.2 The Wing Crack Model

2.4.2.1 *Summary of the Conventional Concept*

Different models were proposed to explain the phenomenon of crack growth in compression. We cite below the study by (Germanovich and Dyskin 2000) on this subject. The majority of works relate the crack growth to the action of a stress concentrator situated in the tested sample. The role of such a stress concentrator is frequently played by pre-cuts (Fig. 2.65). The pre-cuts with contacted faces can be considered as the strongest source of the secondary crack growth, and the wing crack geometry has been a convenient modeling tool in describing fracture in compression (Fig. 2.66).

We adhere to the convention of many experimentalists that the pre-cut of length $2c_0$ is inclined with respect to the compression direction at an angle α (Fig. 2.66) (Fairhurst and Cook 1966). One notes, however, that this is the geometry often used for the angle of hade θ, complementary to the dip angle α (e.g., Park 1983, p. 4), a potential source of confusion. A rectilinear pre-cut is opened by a pair of concentrated forces simulating the action of the inclined contact area (Fig. 2.66). The value of these forces (per unit length of the inclined contact area) is assumed to be equal to the horizontal projection of the shear force, F, tending to displace the opposite faces of the pre-cut:

$$F = 2c_0(\sigma - q)\beta(\alpha)\,, \qquad \beta(\alpha) = C\sin^2\alpha\cos\alpha \qquad (2.9)$$

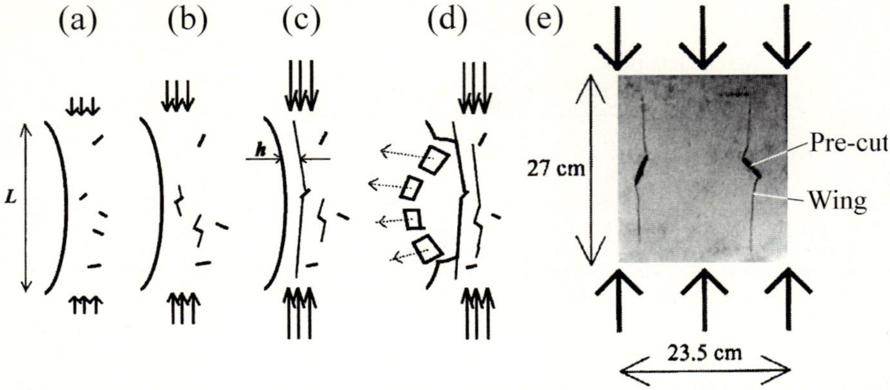

Fig. 2.65. Mechanism of rock fracture near a free surface: **a** Loading begins, **b** crack growth initiates and propagates stably while load is increased, **c** crack starts growing unstably when reaching a critical length ($2l$), while the influence of the free surface is increased, **d** repeated buckling and creating new free surfaces when $L > h$, and **e** growth of initial cracks in the direction of compression in Tennessee Sandstone (after Germanovich and Dyskin 2000)

Fig. 2.66. Wing crack growth under compression (modified from Fairhurst and Cook 1966 by Germanovitch and Dyskin 2000)

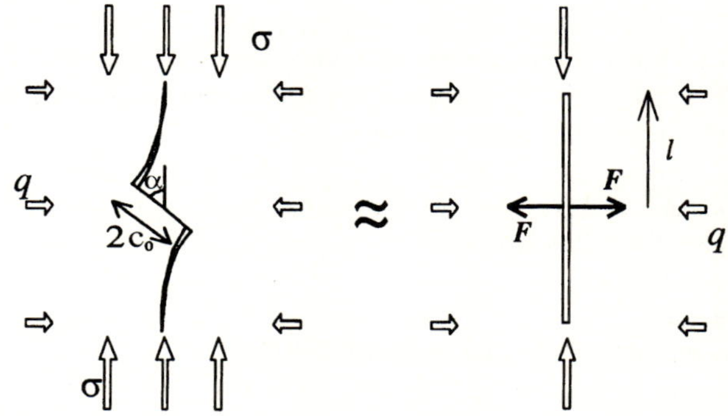

Here σ is the applied compressive stress acting along the free surface, q is the confining pressure and C is a factor accounting for the friction and influence of curvature of the secondary cracks. Using finite clement modeling, Germanovich et al. (1995) calculated C that happened to be of the order of unity. The propagating crack $2l$ is governed by the intensity of the stress concentrations at the crack tip, which is expressed through the stress intensity factor K. As the crack model is symmetrical, only K_{I} is accounted for (under compressional stress, Fig 2.65a,b).

$$K_{\mathrm{I}} = \frac{F}{(\Pi\, l)^{1/2}} - q(\Pi\, l)^{1/2} \tag{2.10}$$

The influence of mode II loading in the case of actual wing crack (Fig. 2.66) is assumed to be suppressed by the compressive stress acting along the crack. Hence, this mode is neglected in the crack growth consideration. The direct numerical modeling reported by Germanovich et al. (1995) confirmed this assumption to a sufficient degree of accuracy.

The conventional criterion of crack growth is

$$K_I = K_{IC} \tag{2.11}$$

where K_{IC} is the rock fracture toughness. It is seen that the crack growth is stable: As the crack elongates, K decreases; hence, greater loads are required to maintain the propagation.

The presence of q significantly reduces K and hence, drastically hampers the crack growth. The influence of the lateral pressure manifests itself in:

1. Reducing the crack opening.
2. Reducing the force, F, by decreasing the shear stress acting on the inclined crack.

The first type of influence is (l / c_0) times stronger than the second one. Since the discussed model accounts for only the main asymptotic term with respect to $l / c_0 \ll 1$, the influence of the lateral pressure on F should be neglected. Therefore, it will be assumed that

$$F = 2\sigma\beta(\alpha)c_0 \tag{2.12}$$

Note application of Fig. 2.65 in Sect. 4.5.3.3.

2.4.2.2
The Three-Dimensional Wing Crack Model

According to Germanovich and Dyskin (2000), the above wing crack model represents the experiments and considerations that are essentially 2-D. The real fracture mechanism is, however, 3-D. Numerous experiments on the growth of internal 3-D wing cracks in various transparent and non-transparent brittle materials under uniaxial compression revealed that there is a strong limitation on the extent a single wing crack can grow: Its wings can reach about the size of the pre-cut (Fig. 2.67a), which is insufficient to cause failure. This limitation seems to be universal and independent of the shape and orientation of the pre-cut. Therefore, the actual 3-D crack growth in uniaxial compression differs radically from simple 2-D mechanism crack growth. The main reason for that limitation is in the shape of the wings as they grow, wrapping around the pre-cut.

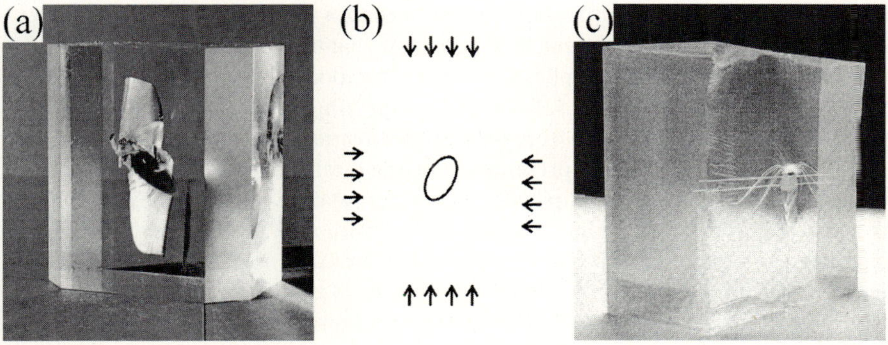

Fig. 2.67. Three-dimensional wing crack growth. **a** A wing crack in PMMA. **b** Schematic diagram of initial flaw in biaxial compression. **c** Extensive wing crack growth in biaxial compression (after Germanovich and Dyskin 2000)

It is this "surrounding" that prevents the wings from extensive growth into plane vertical fractures, such that the ability of the material itself to withstand load gets exhausted for most brittle materials. Accordingly, once the induced stress exceeds the compressive strength of the loaded material, the latter fails in an "inherent mode" regardless of presence or absence of the pre-existing defect. However, under biaxial compression (when one direction is still free of load) an extensive wing crack growth is formed (Fig. 2.67b,c) at a load much below the material inherent strength limit. The reason is that the action of compression in the intermediate principal stress direction prevents wings from wrapping around the pre-cut, enabling them to grow extensively in a plane perpendicular to the unloaded direction.

2.4.3
Pinnate and Horsetail Fractures

There is some uncertainty whether pinnate fractures that form adjacent to faults are shear or tensile structures. Hancock (1972) describes several types of secondary fractures that are associated with shear zones, including veins, arrays of tension gashes in *en echelon* arrangements, Riedel shear structures and pinnate, antithetic conjugate shear fractures at 60°to a fault (Fig. 2.68a,b). There are, on the other hand, descriptions of tensile pinnate fractures (Fig. 2.68c). These tensile sets that form in small angles (<45°) are different from the straight *T* tensile fracture that forms at 45° to the general direction of Riedel shear structure (Fig. 2.68a,b). Hancock et al. (1982) describe an intriguing outcrop (Fig. 2.68d), showing an *en echelon* array of tensile pinnate fractures that are related to a small dextral strike slip fault, where many of the pinnate joints bear plumes with nearly vertical axes aligned perpendicular to the bedding boundary.

The interpretation of the mechanism that created the *en echelon* array of tensile pinnate fractures (Fig. 2.68d) follows from the model depicted in Fig. 2.68a. A primary, remote compression created the stress field of the primary shear couple, which induced the strike slip fault in Fig. 2.68d, i.e., the dextral wrench fault in the latter figure is equivalent to Sc' in the former one. But the plume orientation cannot be explained as a direct consequence of lateral shear by Sc' (that would have formed horizontal plumes); it rather suggests two alternative explanations. First, σ_1'' was vertical (parallel to the plume axes) rather than horizontal. Second, a vertical σ_2'' created the *en echelon* set and plumes. In any event, it is quite clear that the vertical propagation direction of the *en echelon* cracks implies a local stress rotation. While according to the "fracture slanting" model (Sect. 2.2.5.2) the *en echelon* fringe forms by a local rotation of the maximum principal stress from horizontal direction to a vertical one, it appears that in the pinnate joints case presented here, the local stress rotation was driven by mode II exerted by the strike slip fault. The former rotation concerns the creation of a fringe, whereas the latter seems to relate to the entire rock layer, i.e., to what should have been a parent joint if no breakdown had occurred. A mechanical model depicting the difference between the above two systems is a challenge for future studies. Figure 2.68d possibly suggests that vertical plumes on pinnate joints are more common than previously thought: Perhaps they characterize many other conjugate shear zones (e.g., Shainin 1950). If this would be shown, it should support the concept of "tensile pinnate fractures", suggesting that shear pinnate fractures (Fig. 2.68b) are created from joints by a secondary process.

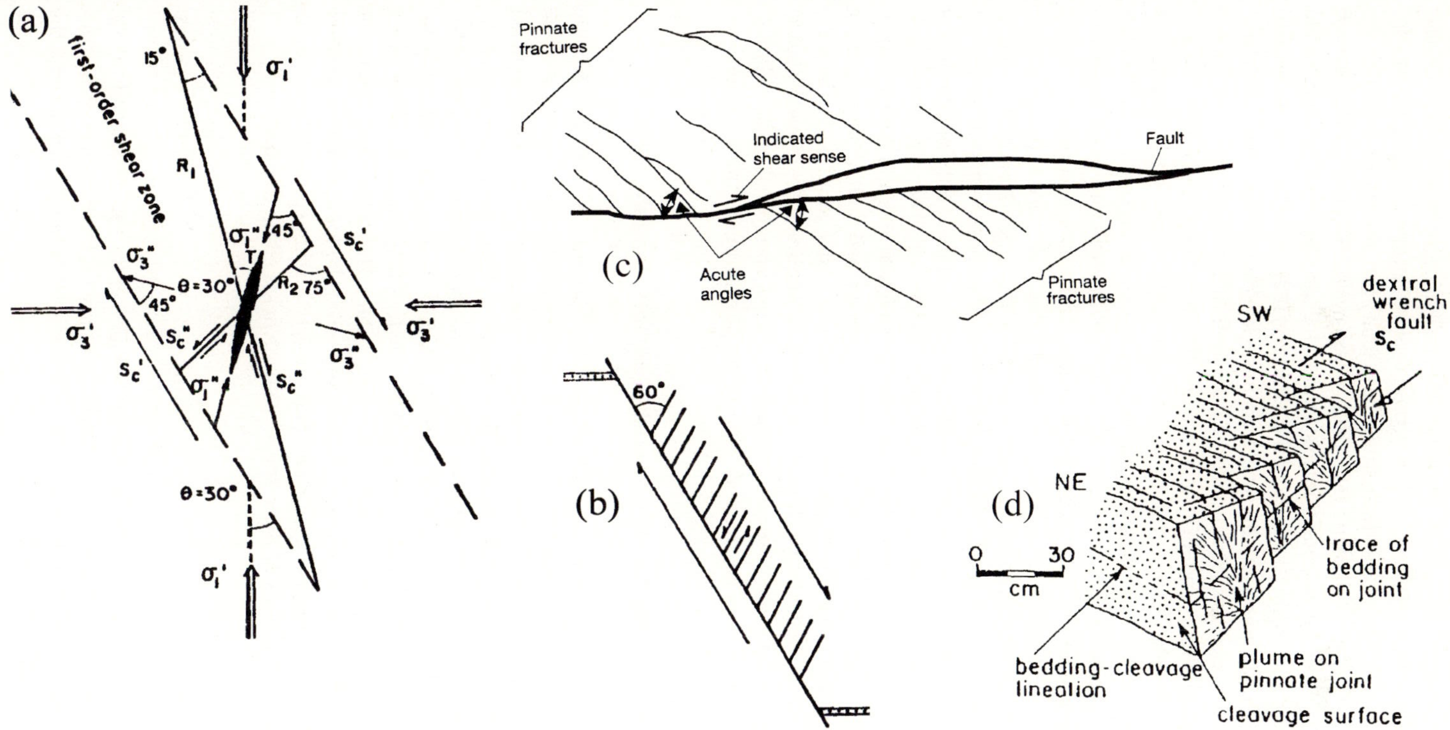

Fig. 2.68 a Conventional interpretation of *en echelon* and pinnate fractures with respect to primary and secondary stresses. *T*: Tension fracture at 45° to an array; R_1 and R_2: Riedel shears at 15° and 75° to an array, *Sc*': Primary shear couple, *Sc*'': Secondary shear couple which may give rise to third order fractures; σ_1' and σ_3': Primary maximum and minimum principal stresses; σ_1'' and σ_3'': Secondary maximum principal stresses; θ: Angle between shear fracture and σ_1'. **b** Pinnate, antithetic conjugate *shear* fractures at 60° to a fault (modified after Hancock 1972). **c** Pinnate fractures (or feather fractures) along a brittle fault. The acute angle between the fractures and the fault points in the direction of relative motion of the fault block containing the fractures (from Twiss and Moores 1992). **d** Plumes on pinnate joints at St. Ann's Head, Dale, Dyfed southwestern Wales (see text for explanation) (after Hancock et al. 1982)

Horsetail secondary fractures are approximately straight, sub-parallel fractures near the ends of primary fractures that form in response to shearing along primary joints, apparently initiating at roughness elements along the primary fracture, where tension is developed as a result of slip (Cruikshank et al. 1991). Renshaw and Schulson (2001) propose that bending in compression of slender microcolumns that form *between* horsetails could result in instability and localized failure of the structure. It appears from these descriptions that tensile pinnate joints and horsetail fractures relate essentially to the same structure. Thus, we focus here on tensile fracture sets that are confined (but unpolarized, Sect. 2.4.4) to the extensional quadrants and form acute angles with the primary fault (Fig. 2.68c).

Martel (1997) points out that the uniform shear stress along faults applied by Cruikshank et al. (1991) in simulating secondary fracture should result in a single fracture to open precisely at the end of a fault trace. This opening will take place owing to a singular, infinitely strong stress concentration there, and a new, secondary fracture should extend from the fault trace at an angle of about 70° (e.g., Pollard and Segall 1987). According to Martel (1997), this theory encounters two difficulties. First, real materials cannot sustain infinite stress levels. Second, this theory is inconsistent with numerous field observations. Often, multiple secondary horsetails open near, but not precisely at the tip (end) of fault traces (Brace and Bombolakis 1963, Fig. 2). These fractures commonly form behind fault tips. They often have nearly straight traces and intersect at an angle of 50° or less with the fault. Martel (1997) models a small fault along which the shear stress is *non-uniform* when it slips. He postulates a "cohesive zone" where the shear stress is elevated near the fault tip. This zone will help to eliminate the stress singularity and would account for the observed orientation, shape, and distribution of horsetails. Martel's concept is challenging and requires support through future studies.

2.4.4 Fracture Polarization in the Two Extensional Quadrants along the Fault

It has been observed that the occurrence of secondary fracture sets at various acute angles with respect to some primary faults may apparently be polarized in the two extensional quadrants (Sect. 6.7). One set of fractures assumes the form of a joint set that dips approximately in the direction of the primary fault at the upper part of the hanging-wall (Fig. 6.24c). They differ from pinnate fractures and from T fractures that create small angles and 45°, respectively, with the fault (Fig. 2.68a,b). The other set consists of twisted and tilted faults that resemble Riedel shear fractures in the lower part of the foot-wall of a normal fault (1–3 in Fig. 6.24c). Hence, the appearance or absence, the differences in the angular relationships, and the locations of secondary fractures *in* the two extensional quadrants may vary, reflecting subtle differences in the histories and dispositions of the respective stress fields along the fault.

2.4.5 Termination-Zone Joints

Termination-zone joints is the term assigned to a joint set that starts around the tip of a strike slip fault and is associated with the secondary faults that form in its continuation along a termination zone (Sect. 6.6). This set curves in conformity with the primary fault and secondary faults.

2.4.6
Fracture Branching

2.4.6.1
Dynamic Joints

Dynamic jointing can be identified by systematic fracture branching (bifurcation). There is a need to distinguish between systematic and non-systematic branching. Systematic fracture branching implies successive branches that form in a single, *continuous* (rapid) process in which the branches propagate in particular directions, away from the fracture origin (Fig. 2.30c) (see more in Bahat 1991a, p. 112). Systematic dynamic branching joints are very rare in sedimentary rocks, but they are known to have been formed in great depths, probably by extreme fluid pressures in granites (Fig. 2.69a,b, see Chap. 4). Their branching faces one direction, and commonly branching fractures are fairly straight.

Anastomose fracture systems are occasionally wrongly interpreted as bifurcations. Anastomose fractures often form by long, *discontinuous* processes that contain intermittent pauses (e.g., Fig. 3.36f). Numerous repeated "branching" that may be suspected to represent dynamic fracture in such systems is fairly erratic and inconsistent in its directions. This "rebranching" is often the result of adjacent existing fractures that interact due to later stress release processes, rather than a single fracture process. Anastomose fracture systems often display (on both horizontal and vertical outcrops) erosional and colored surfaces recording repeated precipitations from liquid flows, particularly in carbonate rocks.

Fig. 2.69a. Joint bifurcation by sudden stress release. Lausitz, Cadomian biotite granodiorite, Klosterberg quarry near Bautzen vertical section; *scale* is 1 m (Fig. 4.3 and Lobst 2001, Fig. A-1)

Fig. 2.69b. Joint bifurcation by sudden stress release. South Bohemian Pluton. The large block at right bifurcates towards left (note optical distortion)

2.4.6.2
Branching Faults

Branching faults are secondary fractures, often rectilinear in shape, which propagate unstably from the tip of the primary fracture due to predominant mode I loading, mostly under dynamic conditions. These secondary fractures may be symmetrically displayed at both sides of primary faults in their two termination zones (Fig. 2.57b). It has been suggested that such structures reflect predominant tensile mode even in overridingly shear regimes (Bahat 1982, 1991a). However, deviations from rectilinear to asymmetric "almost straight" or "curved" branching faults may occur when the primary fracture is influenced by significant mode II / mode I ratios. Rectilinear branching faults generally differ from curved splay faults (e.g., Cruikshank et al. 1991, Fig. 12; Twiss and Moores 1992, Fig. 7.10) such that the former type develops unstably whereas the latter may form stably.

Chapter 3

Four Fracture Provinces in Sedimentary Rocks

While Chap. 2 is devoted to principles of fracture geology on all scales from laboratory to outcrop (up to 1 km long), the present chapter deals with fracture provinces on regional scales (tens of km in length). An ideal study of a fracture province should present all the province's fracture elements in comprehensive detail: Meso-structures and regional structures complete with their spatial relationships, whether they are interdependent or not. This would throw light on the strain and stress histories of the province, besides providing useful data for applied projects. To date, such ideal studies are practically unavailable; no large fracture province is sufficiently exposed or probed to provide what may be regarded as a full structural picture. Every well-investigated fracture province is "specialized" in certain structural features but nevertheless keeps some of its secrets. The objective of this chapter is, rather than providing a comprehensive review of the literature, to present a brief account on four fracture provinces in sedimentary rocks that display distinct fracture elements that would complement each other. By mentally compounding the characteristics of these separate provinces, a realistic picture may emerge on the nature of large-scale fracture patterns and relationships.

First to be described is a Devonian fracture province of alternating siltstone and shales in the Appalachian Plateau of eastern USA, where the classic work by Sheldon (1912) has demonstrated the concept of regional jointing. This will be followed by a study of the Somerset Buttress anticline in the Jurassic limestone along the Bristol Channel in England, with references to fracturing in some other well-known anticlines. Next we shall bring a description of exfoliation joints that cut Jurassic sandstone at Zion National Park, Utah, in western USA. The fourth region of interest relates to the Eocene chalks around Beer Sheva, southern Israel. A specific discussion is devoted to each province, as well as, a comparative review at the end of this chapter.

3.1 The Devonian Fracture Province of the Appalachian Plateau, New York

3.1.1 General Geology

Engelder and Geiser (1980) mapped 20 000 km^2 of systematic joint sets in the Upper Devonian rocks of the Appalachian Plateau, New York (Fig. 3.1a). In this area, Wedel (1932) mapped very gentle anticlinal folds, with limbs dipping less than one degree and axes whose strikes vary over 30°, from azimuth 090°, in the east to 060° in the west of

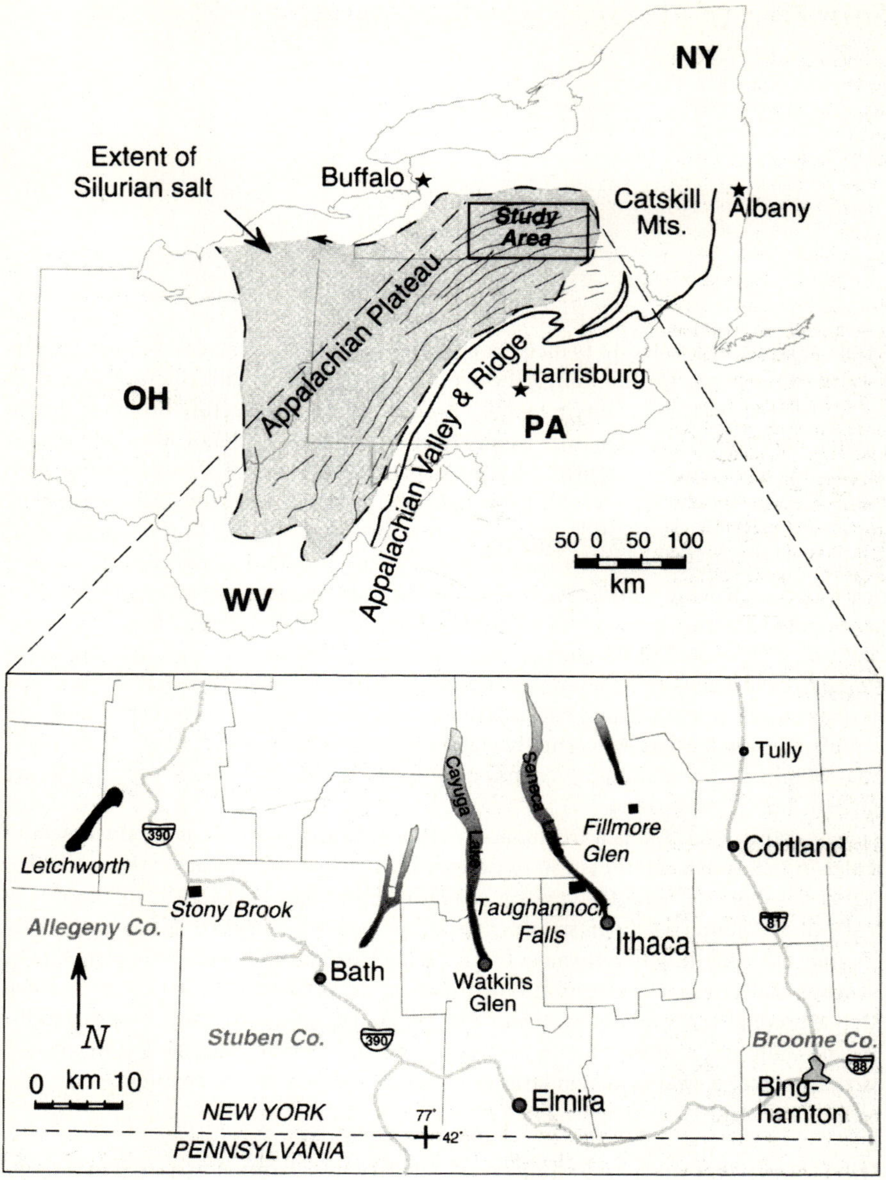

Fig. 3.1a. Geological maps from the Appalachians. Location map of the Finger Lakes district in New York State. The extent of the Silurian salt is *shaded* and *outlined* by *dashes* in the states of New York (*NY*), Pennsylvania (*PA*), Ohio (*OH*), and West Virginia (*WV*) (from Younes and Engelder 1999). *Dark lines* depict the stress trajectories of S_{Hmax} according to the distribution of cross fold joints in the Appalachian Plateau

the region. Engelder and Geiser (1980) discerned two different cross-fold joint sets (or cross-strike sets, approximately perpendicular to the fold axis). These sets preserve their approximate cross-strike position from east to west, with angles between the sets ap-

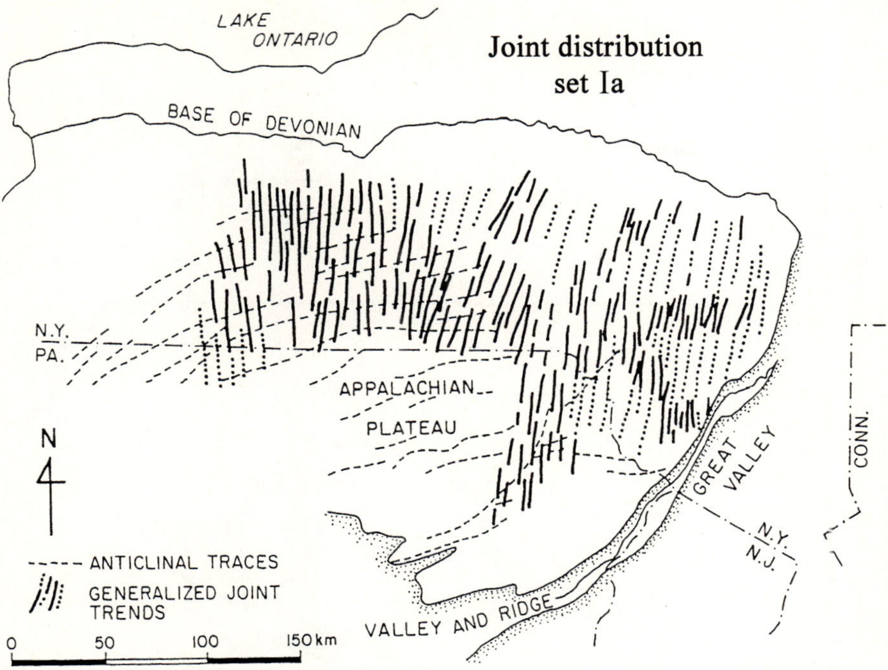

Fig. 3.1b. Geological maps from the Appalachians. Set Ia (from Geiser and Engelder 1983)

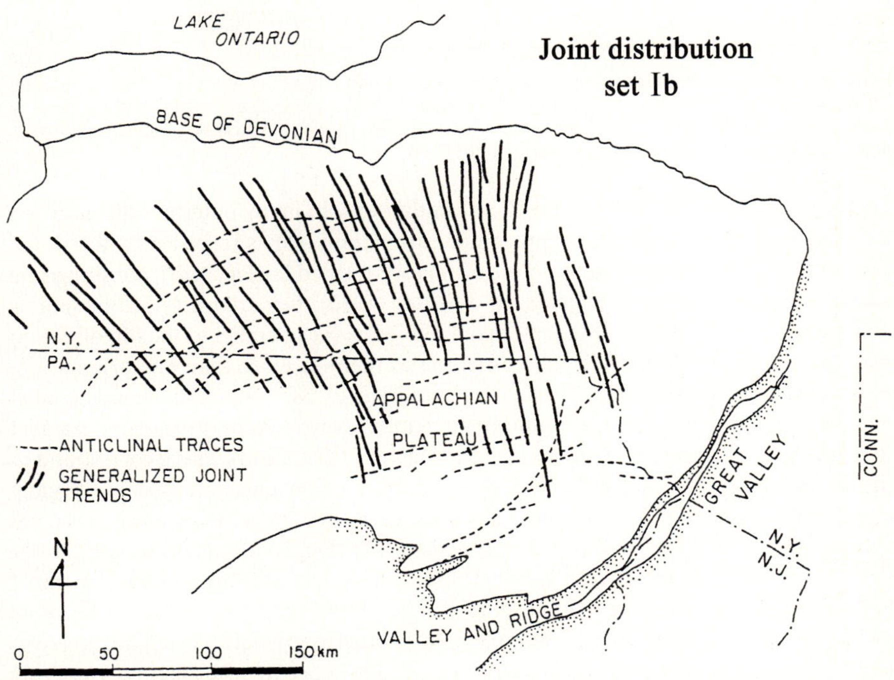

Fig. 3.1c. Geological maps from the Appalachians. Set Ib (from Geiser and Engelder 1983)

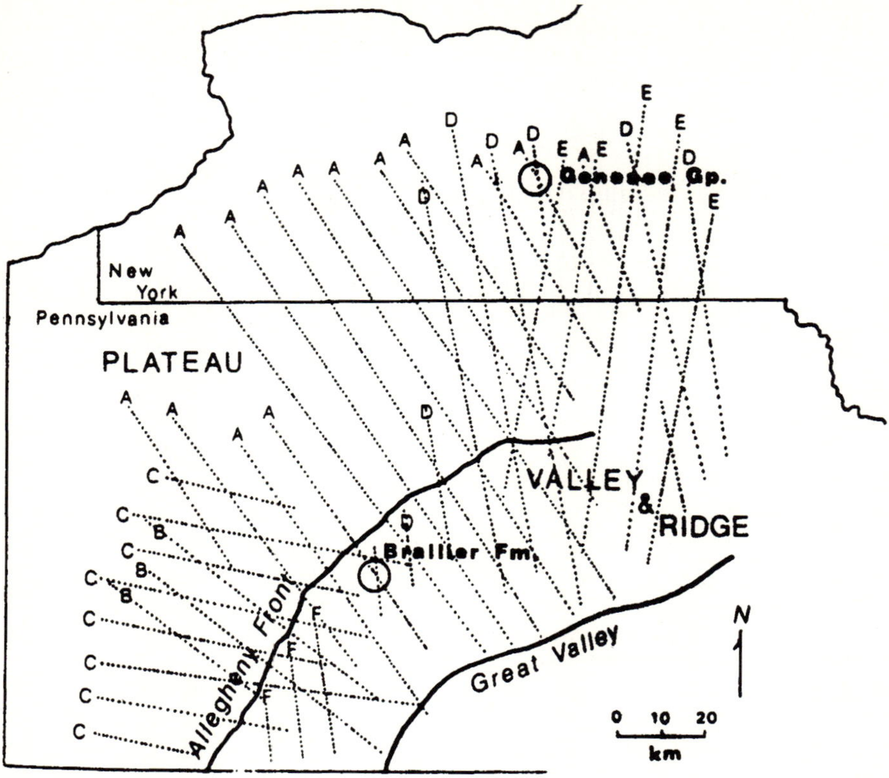

Fig. 3.1d. Geological maps from the Appalachians. Orkan and Voight's (1985) map of the trend lines for regional joint sets within the central Appalachian fold-thrust belt. Sets *A* through *E* are those of Nickelsen and Hough (1967). Regional joint sets are based on the data of Nickelsen and Hough (1967), and Engelder and Geiser (1980) (from Engelder 1993)

proximating 18 ±2° in the east and 30 ±4° in the west. In many outcrops, one joint set parallels the direction of maximum compressive strain (ef) as recorded by fossil distortion (parallel to the bedding). Joints of this set, termed Ia (Fig. 3.1b), are sometime calcite-filled. According to Engelder and Geiser (1980) they are part of joint set I as defined by Parker (1942), of the type called dip joints (i.e., parallel to the dip direction of bedding) by Sheldon (1912) who often found them in more than one orientation. Joints of the other set, termed Ib (Fig. 3.1c), strike at 16–34° counter clockwise from ef and are parallel to eoc, the maximum in situ compressive strain as measured parallel to bedding by overcoring techniques (in the Machias Formation). These correspond to the compressive direction indicated by a penetrative fabric in outcrops in Allegany County (Fig. 3.1b). Sets Ia and Ib are equivalent to the systematic joint sets E and A respectively, described by Nickelsen and Hough (1967) (Fig. 3.1d). In addition, Engelder and Geiser (1980) characterized joint sets II and III. Joints of set II strike sub-parallel to the fold axes. They belong to joint set II of Parker (1942) and were called strike joints by Sheldon (1912). They have the same orientation toward the fold axes as the non-systematic joints described by Nickelsen and Hough (1967). Set III, which belongs to set III of Parker (1942), is not obviously related to either ef, eoc, or to the fold axes.

The above discrimination of joint sets illustrates how complicated the "regional joint pattern" of a given area can be, and how complex the challenge is to try and form a sensible tectonic picture from it. The above references and subsequent studies have made the Appalachian Plateau into a classic arena of fracture studies, where many basic concepts of jointing were first recognized.

A comprehensive updated account of the Devonian Fracture Province of the Appalachian Plateau is presented by Younes and Engelder (1999) and their results and discussion are quoted extensively here. Fringe cracks on joint surfaces and kinks are considered in the context of individual outcrops, in Chap. 2. The application and conclusions obtained from parent joints, fringe cracks and kinks and their relation to the regional tectonics in the eastern United States are given in the present chapter.

Younes and Engelder (1999) document the spatial distribution and nature of joints within the northeastern portion of the Appalachian Plateau detachment sheet (Fig. 3.1a). The area, about 3 000 km^2, includes the Finger Lakes district of New York State, which extends from Broome County to Allegany County (Fig. 3.1a). The detachment sheet in this area consists of clastic sedimentary rocks of the Devonian Catskill Delta, transported from the Acadian highlands to the east (Ettensohn 1985). The delta prograded from east to west, attaining its maximum thickness east of the study area. The delta complex consists of packages grading from black and gray shales through siltstone to sandstone. The clastic Hamilton, Genesee, Sonyea, and West Falls Groups (Fig. 3.2 and 3.3) are separated by three black shale formations: Genesee, Middlesex, and Rhinestreet shales, which reflect anoxic conditions at the time of deposition. The Tully limestone is an important marker bed at the base of the Genesee Group.

The detachment sheet was deformed into a series of low-amplitude folds (mapped in outcrop by Wedel (1932), and in the subsurface by Bradley and Pepper (1938) and Murphy (1981)). Folds of the detachment sheet are broad, persistent, and can be traced farther to the southwest in Pennsylvania and into West Virginia. In the study area, the fold axes trend east-west in the eastern part and northeast-southwest in the western part. The detachment sheet contains layer-parallel shortening structures of Alleghanian age above a decollement within the Silurian Salina salt (Rodgers 1970; Engelder 1979; Murphy 1981). Much of the layer-parallel shortening was accommodated by pressure solution and the formation of solution cleavage (Engelder and Geiser 1979; Geiser and Engelder 1983). The amount of slip on the decollement is as much as 22 km to the north-northwest as estimated at the Allegheny front (Engelder and Engelder 1977). Mapping of the detachment uses strain markers such as deformed fossils and subsurface data (Engelder and Engelder 1977; Engelder 1979; Engelder and Geiser 1980; Beinkafner 1983; Geiser 1988; Hudak 1992). The foreland limit of the detachment sheet and the region of folds in the post-Silurian rocks coincides with the limits of the Silurian salt (Frey 1973).

Joints of Alleghanian age strike approximately normal to the axes of the folds within the detachment sheet (Fig. 3.1b,c) (Parker 1942; Ver Steeg 1942; Nickelsen and Hough 1967; Engelder and Geiser 1980; Geiser and Engelder 1983; Lacazette and Engelder 1992). Note that clockwise rotation from west to east is observed both among individual cross fold and within each set (Fig. 3.1b,c). Multiple sets of cross fold joints are particularly well developed in interlayered siltstone-shale beds, where the earlier dip joints favored siltstone beds. Clear fractographic distinctions can be made among the various cross fold sets (Bahat and Engelder 1984; Bahat 1991a, p. 248), which moreover are occasionally decorated with fringe cracks (Younes and Engelder 1999).

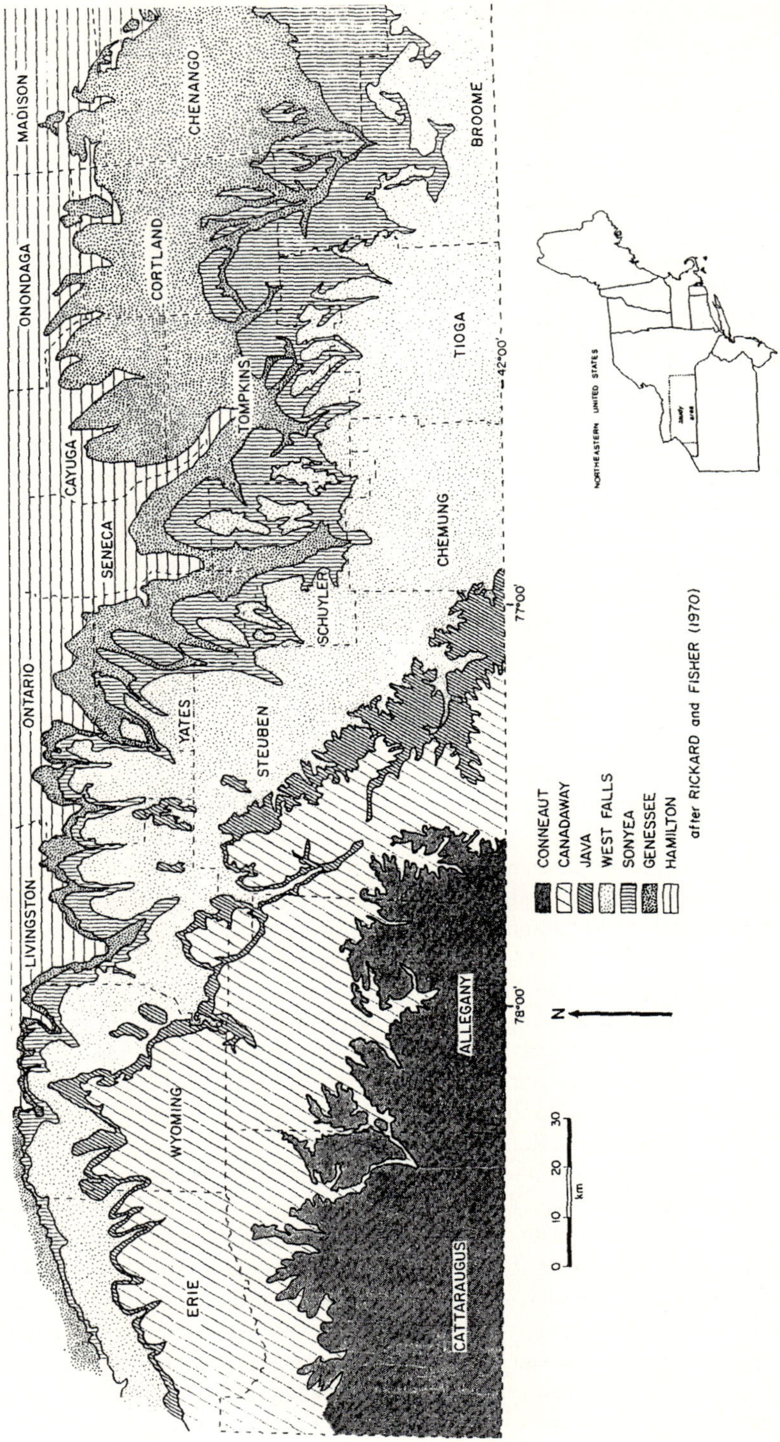

Fig. 3.2. The stratigraphy and geography of the southern tier of New York State (from Engelder and Geiser 1980)

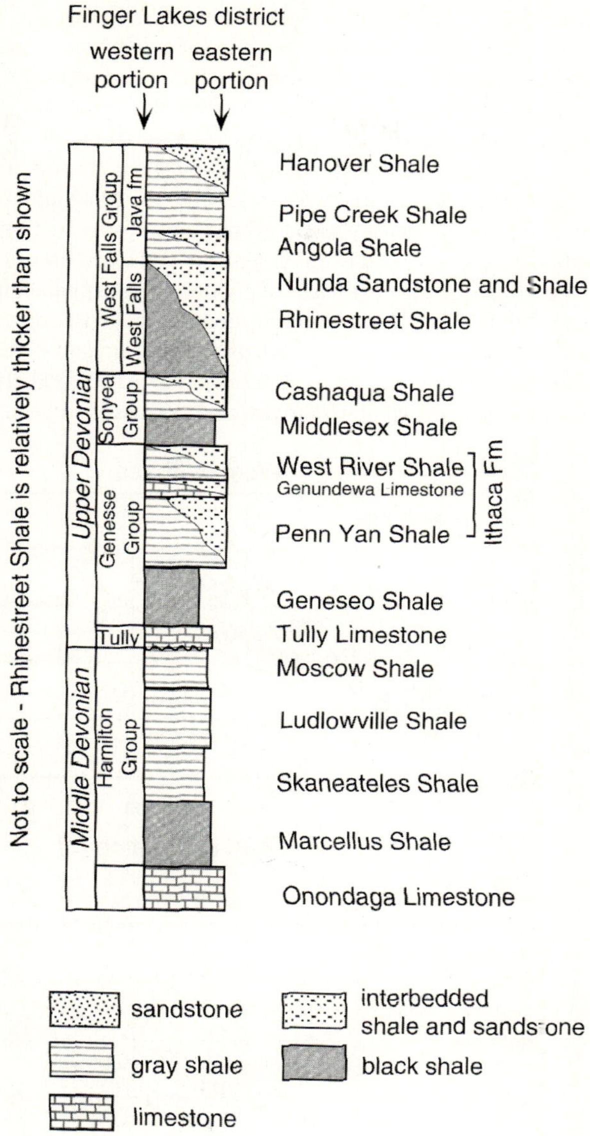

Fig. 3.3.
A simplified stratigraphic column of major rock units in the eastern and western portions of the study area in the Finger Lakes district of New York State (modified from Van Tyne 1983). The facies become progressively coarser to the east as the proximal parts of the Catskil Delta are approached. The unconformity at the base of the Tully limestone extends up to the base of the Cashaqua shale in western New York (from Younes and Engelder 1999)

3.1.2
The Tectonic Problem

Nickelsen (1979), in an analysis of the Bear Valley strip mine in the Appalachian Valley and Ridge recognized a group of structures that testify to a clockwise rotation of maximum horizontal stress (S_{Hmax}) during the Alleghanian orogeny (Younes and Engelder 1999). Engelder and Geiser (1980) also found joint evidence on the Appalachian Plateau of a clockwise rotation of the Alleghanian stress field. Rocks with two distinct cleavages or commonly displaying two dip joint sets led Geiser and Engelder

(1983) to conclude that the Appalachian Plateau was affected by two different tectonic phases: The Lackawanna phase and the Main phase. However, outcrops in the anthracite coal district of the Appalachian Valley and Ridge of Pennsylvania, including the Bear Valley strip mine, indicate that the rotation of the Alleghanian stress field produced structures with a broad range of orientations. From these, Gray and Mitra (1993) concluded that the Alleghanian orogeny was a continuous series of structural events reflecting *a gradual* clockwise rotation in the Alleghanian stress field, not punctuated by two tectonic phases as suggested by Geiser and Engelder (1983). Bahat (1991a, Tables 5.1 and 5.2) and Evans (1994) showed structures in more than two orientations, suggesting that the northeastern portion of the Appalachian Plateau detachment sheet was also deformed by a more continuous clockwise rotation of the Alleghanian stress field. If so, the tectonic evolution of the Appalachian Plateau detachment sheet is consistent with that of the Valley and Ridge. These two conflicting concepts (discrete phases vs. gradual stress rotation) are further discussed in Sect. 6.3.

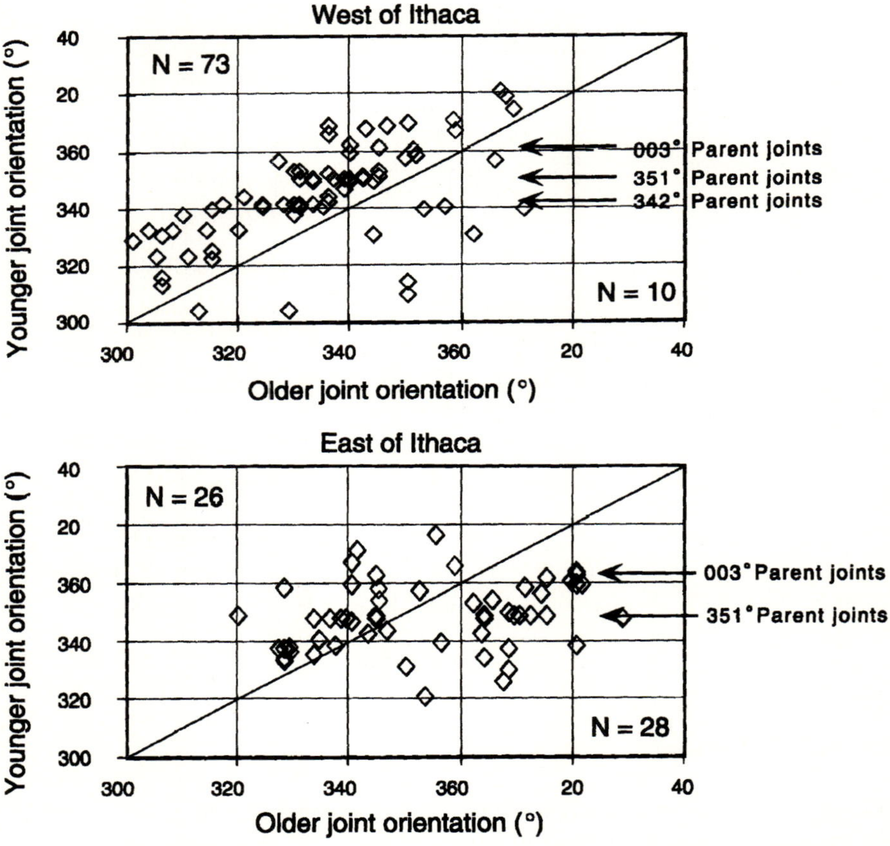

Fig. 3.4. Abutting relationships west and east of Ithaca, New York. The *diamond symbols* above the *straight line* indicate a clockwise abutting relationship, whereas those under the line are counter clockwise rotations. The abutting of joints west of Ithaca is dominated by clockwise rotations, but mixed rotations are found east of Ithaca

3.1.3
Sequence of the Dip Joint Development

Evidence for the timing of joint development is visible throughout the detachment sheet (Younes and Engelder 1999). It consists of curving or kinking of younger joints as they approach the pre-existing joints, or abrupt termination of younger joints against older joints. Unambiguous cases of abutting are less common than mutual cross cutting. Nevertheless, abutments between dip (i.e., parent) joints in the detachment sheet show west-to-east constancy. The later dip joints, particularly in the western part of the study area, strike a few degrees clockwise from earlier dip joints; younger dip joint sets are defined by the horizontal clustering of data in Fig. 3.4. Where present, fringe cracks are always counter clockwise from these parent joints (Fig. 3.5a–c). Clustering is observed in abutting dip joints at strikes 342°, 351°, and 003°, suggesting that younger dip joints have a consistent orientation throughout the region, even though their predecessors show widely varying orientations. The earliest dip joint set, in the range of 320–330°, does not cluster. The same appears to be true for another dip joint set, striking 006° to 021° range in the easternmost part of the study area. These data indicate that joints striking at 342° generally abut joints striking at 320–330°, and 351° joints generally abut 342° joints, and so forth. Nevertheless, an exception to this general rule of clockwise sequence of younger joints is found at the eastern edge of the study area, where 003° joints abut parent joints striking roughly 020°, indicating in this area a counter clockwise rotation of the stress field with time. Thus, the sequence of dip joint

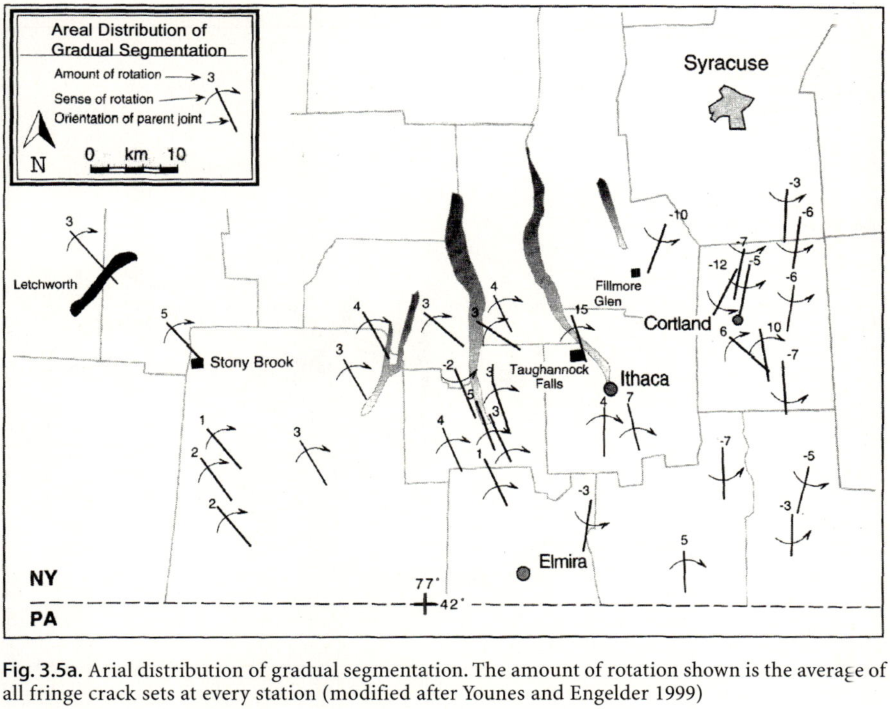

Fig. 3.5a. Arial distribution of gradual segmentation. The amount of rotation shown is the average of all fringe crack sets at every station (modified after Younes and Engelder 1999)

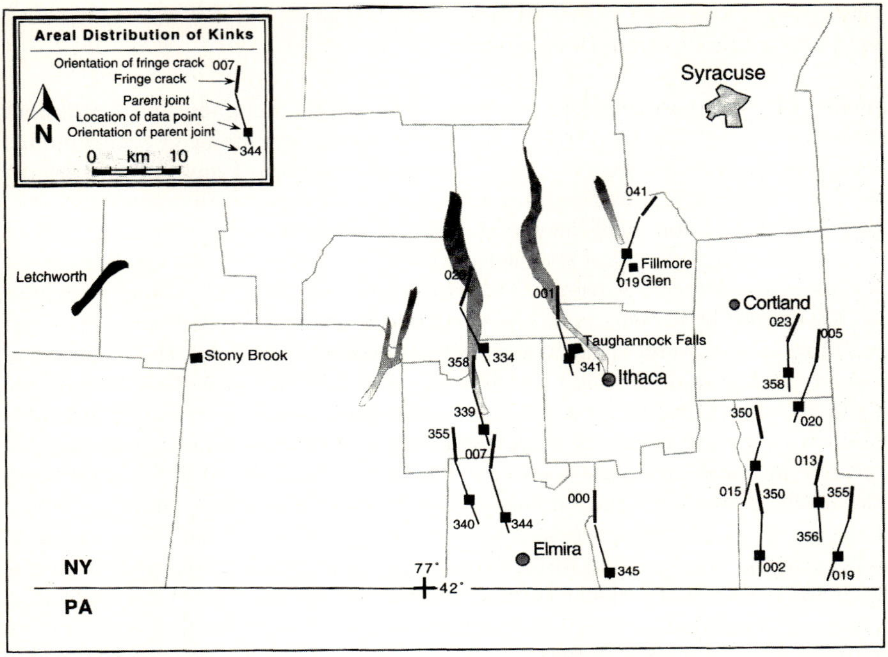

Fig. 3.5b. Arial distribution of kinks (after Younes and Engelder 1999)

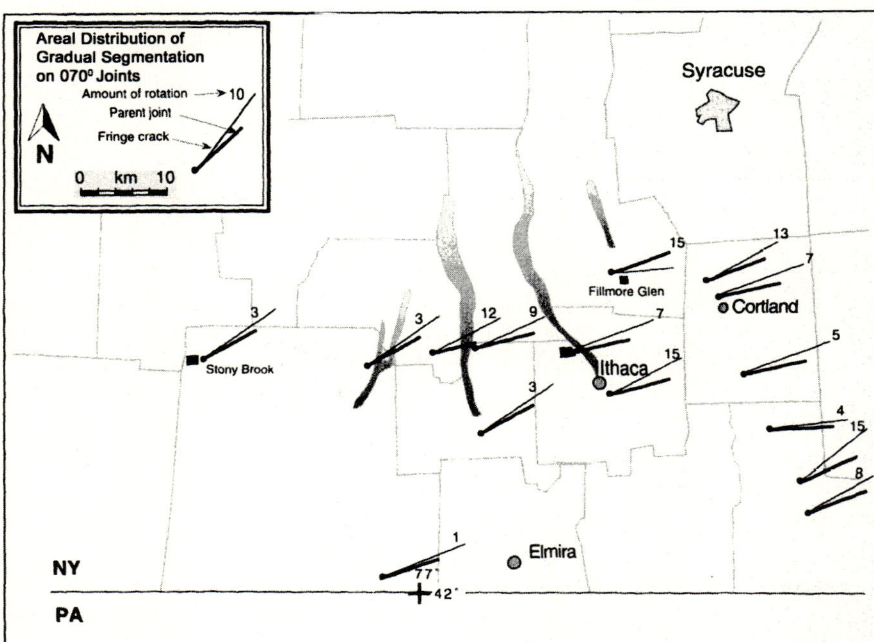

Fig. 3.5c. Arial distribution of of gradual segmentation on 070° joints. The amount of rotation shown is the average of all fringe crack sets at every station (modified after Younes and Engelder 1999)

development in the detachment sheet appears to be as follows (parentheses indicating sets of joints clustered about that orientation):

$$(320^\circ - 330^\circ) \rightarrow \begin{bmatrix} \text{East Ithaca} \\ (006^\circ - 021^\circ) \\ \\ (342^\circ) \\ \text{West Ithaca} \end{bmatrix} \rightarrow (352^\circ) \rightarrow (003^\circ)$$

Other post-Alleghanian joint sets are also found throughout the detachment sheet. One post-Alleghanian 070° set is best developed in black shale below the Rhinestreet Formation (Loewy 1995). The 070° joints are cut by Middle Jurassic kimberlite dikes (Kay et al. 1983) in the vicinity of Ithaca, and are therefore older than Middle Jurassic. Another post-Alleghanian 000° set consists of short, curviplanar-to-planar joints that occasionally abut earlier 070° joints. The 000° set striking parallel to the Mesozoic kimberlite dikes (Martens 1924; Parker 1942), is the youngest joint set found by Younes and Engelder (1999).

3.1.4
Progressive Development of the Regional Stress Field

3.1.4.1
Patterns of Parent Joint Sets

Data on the abutting of parent joints of the Appalachian Plateau detachment sheet show that there are at least two joint sets, which according to Engelder and Geiser (1980) extend several hundred kilometers along strike in a fan-shaped pattern. Four development stages are discerned in the pattern of dip joints, as is indicated by both the abutting of parent joints and the rotation angle of fringe cracks. The regional pattern of parent joints is due to the overlapping of joint sets with different strikes and developed from west to east as progresively younger joint sets propagated within a clockwise rotating stress field. Throughout this development, the Alleghanian stress field sweeps through roughly 40 degrees from 320° to 360°. To the southeast in the anthracite region of the Valley and Ridge, the Alleghanian stress field rotated 30 degrees clockwise, from 336° to 006° (Gray and Mitra 1993). Although the stress fields of the Plateau and of the Valley and Ridge differ slightly in magnitude, the correlation is close enough to suggest that both areas underwent similar tectonic development, marked by a clockwise rotation of the horizontal stress field.

3.1.4.2
The Orientation of Fringe Cracks Relative to Their Parent Joints

Fringe cracks occur in most lithologies that are found in the Finger Lakes district of New York State. More than 225 sets of fringe cracks were mapped by Younes and Engelder (1999) in the Finger Lakes district of New York State on the northeastern portion of the Appalachian Plateau detachment sheet. Although there are millions of joints throughout the detachment sheet, joints decorated with fringe cracks are rare, and within individual outcrops their number varies from one to several dozen. Maps in Fig. 3.5a–c show the orientation and aerial distribution of fringe cracks and their parent joints as expressed by their average strikes. Younes and Engelder (1999) observe that the general orientation of all fringe crack sets throughout the detachment sheet follows the systematic change in fold trend and strike of the parent joints from east to west. In general, and depending on the type of fringe crack, the angle between parent joint and fringe cracks remains about the same across the area. However, there is a pattern to the sense of fringe crack rotation across the region. East of Ithaca, fringe cracks strike counter clockwise from the parent joints, while to the west they strike clockwise from their parents, and both types overlap in the Ithaca area. Because loading conditions and the timing of propagation are different for each fringe crack type, there are characteristic differences in twist and tilt angles (Fig. 2.20). Kinks are concentrated in the eastern half of the study area and tend to strike within a few degrees of 002°, seemingly independent of the parent's orientation.

3.1.4.3
Counter Clockwise Propagation of Fringe Cracks near the Eastern Boundary

According to Younes and Engelder (1999) the abutting geometries of parent joints, coupled with fringe cracks, provide a record of the Alleghanian stress field as it evolved across the Appalachian Plateau detachment sheet. The fidelity of the record of stress field rotation as derived from the abutting geometries and fringe cracks can be evaluated by comparing it with the rotation of the Alleghanian stress field and strain within

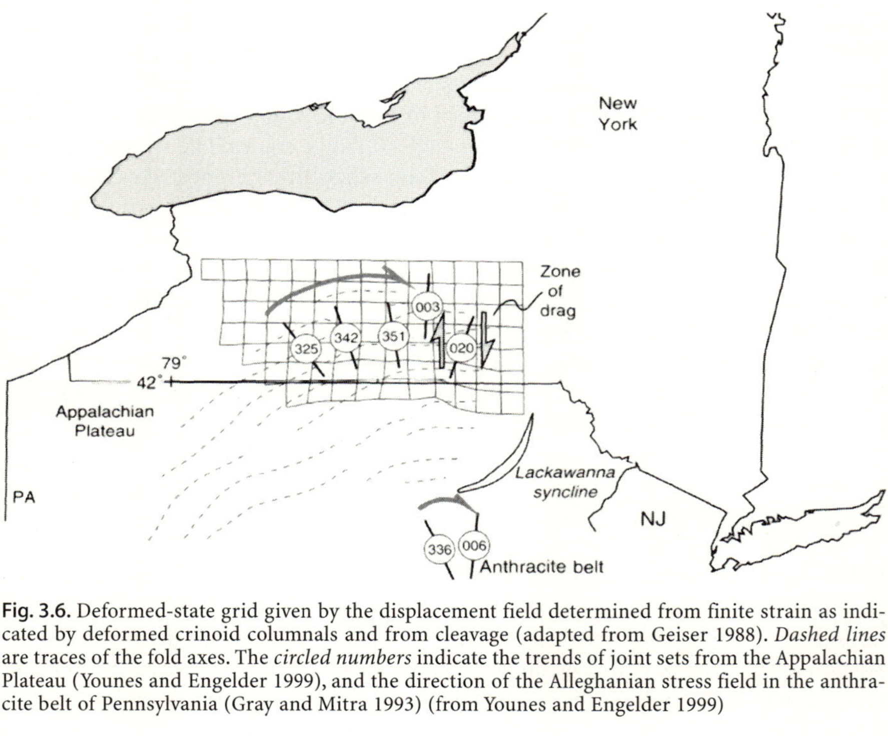

Fig. 3.6. Deformed-state grid given by the displacement field determined from finite strain as indicated by deformed crinoid columnals and from cleavage (adapted from Geiser 1988). *Dashed lines* are traces of the fold axes. The *circled numbers* indicate the trends of joint sets from the Appalachian Plateau (Younes and Engelder 1999), and the direction of the Alleghanian stress field in the anthracite belt of Pennsylvania (Gray and Mitra 1993) (from Younes and Engelder 1999)

the detachment sheet. An isostrain map for layer-parallel shortening of the detachment sheet was constructed by Geiser (1988) from orientation data of cleavage and fossil deformation. Layer-parallel shortening shows a radial fan pattern from west to east (Fig. 3.6). The east-west lines display a uniform spacing (i.e., strain gradient), except east of Ithaca to the eastern part of the study area, where the rectangular grid is distorted and the east-west grid lines are more closely spaced. Greatest grid distortion coincides with the region containing parent joints decorated with fringe cracks, reflecting a counter clockwise rotation on the Alleghanian stress field (Fig. 3.5a). This conformance of fringe cracks to strain distribution provides evidence for constructing the tectonic history of the detachment sheet.

In the eastern part of the study area, evidence is strong for counter clockwise rotation of the stress field after the formation of an initial set of dip joints striking between 006° and 021°. This is indicated by both abutting geometries and fringe cracks. Relative to its orientation farther west (i.e., 342°), the 20° to 30° misalignment of the early stress field to the east is significant. Younes and Engelder (1999) note that this misalignment occurs in the pinchout area of the Silurian salt, which served as a convenient detachment surface. They attribute the misalignment of the stress field to rock drag, produced by the lack of salt detachment east of Ithaca. Drag at the edge of the detachment sheet sets up a local remote stress field, controlling the orientation of early parent joints striking between 006° and 021° (see Geiser 1988). A later stress field producing both the 352° and 003° sets reversed the sense of shear on the parent joints along the eastern boundary of the detachment sheet. This causes the later fringe cracks to form counter clockwise from parent joints.

3.1.5
Tectonic History of the Appalachian Plateau Detachment Sheet

Younes and Engelder (1999) maintain that any tectonic history of the Appalachian Plateau detachment sheet must take into account the following observations:

1. Whenever observed in the same outcrops parent joints in the shale are oriented clockwise relative to parent joints in the siltstone layers.
2. The sequence of formation of parent joints, as indicated by the abutting relationships, shows a clockwise rotation of regional stress from west to east, except in the eastern part of the study area.
3. Fringe cracks on dip joints show a clockwise rotation of the stress field in the west, a counter clockwise rotation in the east, and both kinds of rotation near Ithaca (Fig. 3.5a).
4. The rotation angle for counter clockwise fringe cracks is generally larger than for clockwise fringe cracks.

3.1.5.1
The Proposed Sequence of Stress Directions in the Northeastern Part of the Detachment Sheet

The 320–330° parent joint set. The earliest parent joints in the study area are limited to the western region, the area (Fig. 3.6) that is closest to the extension of the Juniata culmination into the plateau. Joints belonging to this set range over about 10° (320–330°)

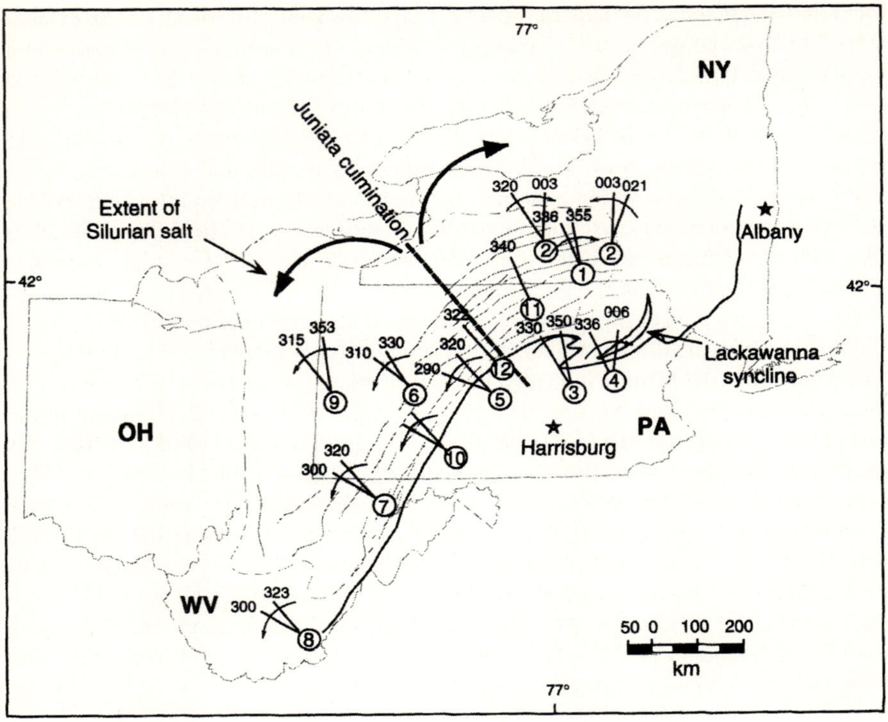

Fig. 3.7. A model for the rotations of the Alleghanian stress field in the Central Appalachian Mountains. The *heavy black line* is the boundary between the Valley and Ridge and Appalachian Plateau. The *short straight segments* indicate the first and last stages of deformation as determined by the following authors indicated by the corresponding number: *1:* Engelder and Geiser (1980); *2:* Younes and Engelder (1999); *3:* Nickelsen (1979); *4:* Gray and Mitra (1993); *5:* Nickelsen (1988); *6:* Nickelsen and Hough (1967); *7:* Dean et al. (1984); *8:* Dean et al. (1988); *9:* Evans (1994); *10:* Nickelsen (1996); *11:* Spiker and Gray (1997); *12:* Nickelsen and Engelder (1989). The *thin arrows* indicate the local sense of rotation, and the *thick arrows* indicate the regional sense of rotation proposed by Younes and Engelder (1999). The *heavy dashed line* marks the location of the Juniata culmination (from Younes and Engelder 1999)

(see explanation below). This same joint set is found west of the Juniata culmination and in the detachment sheet as far south as West Virginia (Fig. 3.7). However, to the west and southwest of the Juniata culmination, the earliest parent joints are characterized by a counter clockwise sequence of development. Zhao and Jacobi (1997) reported that west of the study area, joints striking 322° to 340° predate 312–320° joints and a second group of 280–305° joints. Evans (1994) reported that west of the Juniata culmination, the earliest joint set strikes 350° and is followed by the 320° to 330° joints. In central New York State, the earliest structures include folding (042° fold axes) of the Tully limestone (Younes and Engelder 1995). Because the shortening direction indicated by the deformation of fossils is at a significant clockwise angle with the earliest joints (Engelder and Geiser 1980), these earliest joints apparently predate layer-parallel shortening. The earliest joints in the Valley and Ridge also predate appreciable layer-parallel shortening (Nickelsen 1979). Early joints occur throughout the stratigraphic section near the Juniata culmination, yet farther east they are found only below the Tully limestone. The orthogonal relationship between the fold axes of the Tully limestone and the early joints suggests synchronous formation of the two features. At this stage, detachment within the salt decollement may have begun but was not well developed. Farther southeast, early fold axes of the Lackawanna syncline (Fig. 3.6) are parallel to those found in the Tully limestone. Thus, we suggest that set 320–330° appears to be of the burial type which forms prior to the onset of intense tectonics (Bahat 1991a, p. 240). The early age of this set fits that of the early fold axes of the Lackawana syncline before the actual detachment. Possibly, the wide range of strike in this set reflects small differential stresses, compared to relatively large differential stresses in the later, syntectonic sets, resulting in narrower strike ranges.

The 342° parent joint set. The propagation of the 342° set marks the initiation of significant detachment along the Salina salt, as indicated by deformed fossils. Clockwise rotation of the stress observed in the early stages of layer-parallel shortening was recorded as cleavage in the Tully limestone. As the amount of layer-parallel shortening increased, the eastern edge of the detachment sheet began to drag where the eastern salt pinch-out prevented easy detachment. As a consequence of this drag, a shear couple developed within the eastern region of the detachment sheet so that the local stress field was rotated considerably clockwise relative to that found in the rest the detachment sheet (more about the association of shear couple and local stress field rotation in Sect. 3.2). Joints in the eastern region propagate in directions ranging from 006° to 021°. During this tectonic stage, folding continued in the section above the Tully limestone. A clockwise stress field rotation east of the Juniata culmination is reflected in the initial development of clockwise fringe cracks in the western region of the study area. The overall uniform strike of set 342° suggests that it is of the syntectonic type.

The 351° parent joint set. A third stage is marked by the propagation of 351° joints. By this stage, parent joints had propagated throughout most of the northern detachment sheet, and drag along the eastern edge of the detachment sheet was at a maximum. At this stage, counter clockwise fringe cracks began to be formed along the eastern edge of the detachment sheet. Kinks equivalent to the 351° parent joints are also found throughout the western parts of the sheet.

The 003° parent joint set. The continuing clockwise rotation of the Alleghanian stress field is next indicated by 003° joints. It is possible that the kinks showing a counter clockwise angle developed during this stage as a result of north-south compression. The general north-south orientation of the kinks (Fig. 3.5b) and their constant angles to the parent joint for both clockwise and counter clockwise sets (Fig. 3.8) indicate that they are related to the same event. This is also supported by the mechanical requirement that kinks will only form after arrest of their parent joints (more in Sect. 2.2.9.6).

The 070° parent joint set. This stage of jointing is best developed in black shales of the Catskill Delta (Loewy 1995) and is the most difficult to date. The 070° joints are a result of tectonic relaxation before abnormal pressure within the delta shales could leak off (McConaughy and Engelder 1999). The 070° set carries fringe cracks that are always rotated counter clockwise (Fig. 3.5c).

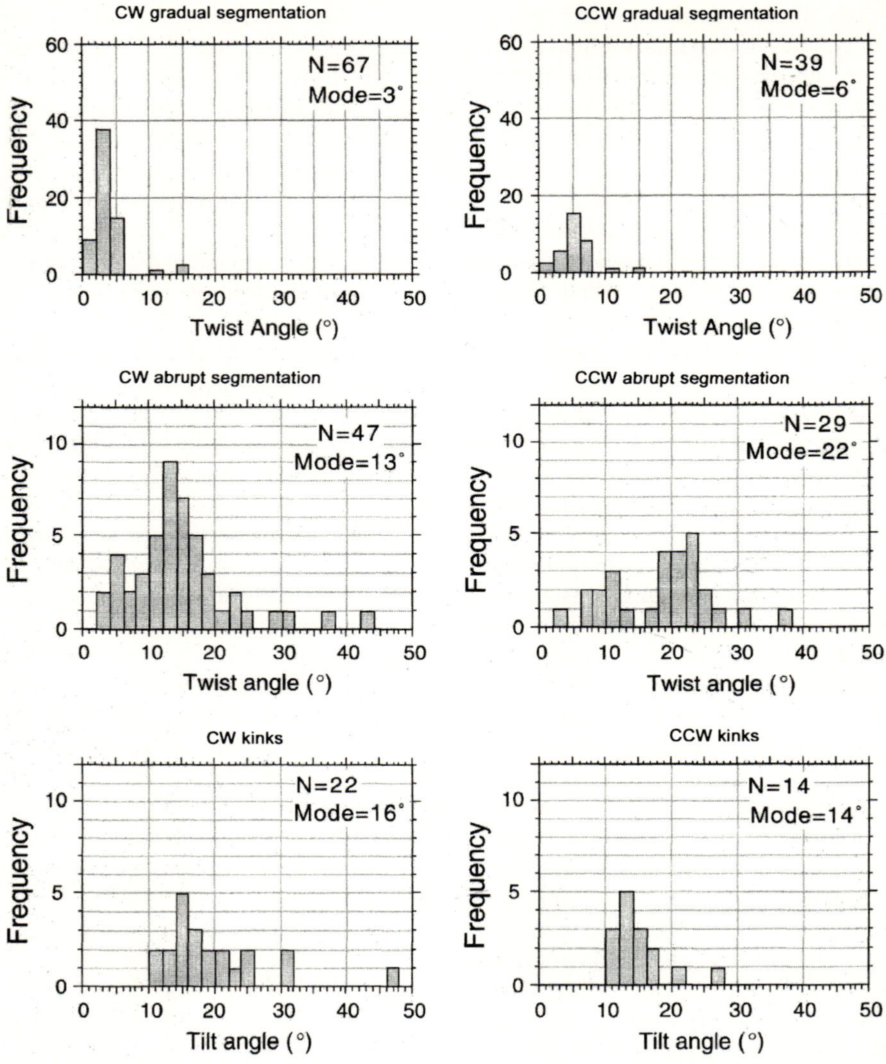

Fig. 3.8. Histograms of the magnitude of clockwise (*CW*) and counter clockwise (*CCW*) rotations for gradual and abrupt fringe cracks and kinks. The number of fringe sets and the statistical mode of orientation are shown inside each histogram. The amount of CCW rotation for fringe cracks is larger than the CW rotation. Kinks have a more consistent orientation (mode is almost the same for CW and CCW rotations) (from Younes and Engelder 1999)

The 000° parent joint set. The propagation of a second north-south joint set is post-Alleghanian. These late joints are parallel to the kimberlite dikes cutting the detachment sheet. These dikes were dated as Early Jurassic, and therefore this set may also be early Mesozoic in age. This later 000° set may be associated with continuing slip on faults of the Clarendon-Linden zone (R. Jacobi 1998, pers. comm. to Younes and Engelder). Joints oriented 003° and 000° are also distinguishable by differences in length and planarity.

3.1.5.2
Tectonics on Either Side of the Juniata Culmination

According to Younes and Engelder (1999), the Juniata culmination correlates with a regional lineament known in central Pennsylvania from surface geology and magnetic and gravity anomalies (Gold and Parizek 1976). The joint propagation sequence around the Juniata culmination indicates that paleostress trajectories rotated away from parallelism with the culmination in opposite directions, leading to a clockwise sequence of structures to the east and north of it and a counter clockwise sequence to the west and south of it (Fig. 3.7 and 3.9) (Younes 1996). As mentioned, the trend of the Juniata culmination is parallel to the oldest joints in the detachment sheet (320–330°).

Nickelsen and Hough (1967) mapped five sets of dip joints in the Pennsylvania portion of the detachment sheet (Fig. 3.1d, sets A–E). Set A, oriented 330°, is parallel to the Juniata culmination and perpendicular to the Lackawanna syncline. It is flanked by joint sets D and E to the northeast, and joint sets B and C to the southwest. Based on the similarities between the joint orientations in central New York State and those in northeastern Pennsylvania, and also on the similarities between those in southeastern Pennsylvania to those in West Virginia (Dean et al. 1984, 1988), Younes and Engelder (1999) relate set D to clockwise stress field rotation in the northeast, and sets B and C to counter clockwise rotation in the southwest. This interpretation agrees with Evans' (1994) analysis of joints west of the culmination. Zhao and Jacobi (1997) reported a clockwise superposition of joints to the east of the Juniata culmination and a counter clockwise superposition of joints west of it.

Exposures of clockwise and counter clockwise fold rotations in the Antes shale of the Sinking Valley fault zone, Pennsylvania, show a transport direction that is compatible with transport parallel to the Juniata culmination (Nickelsen and Engelder 1989). The direction of transport was estimated as 322°, similar to the oldest trends in

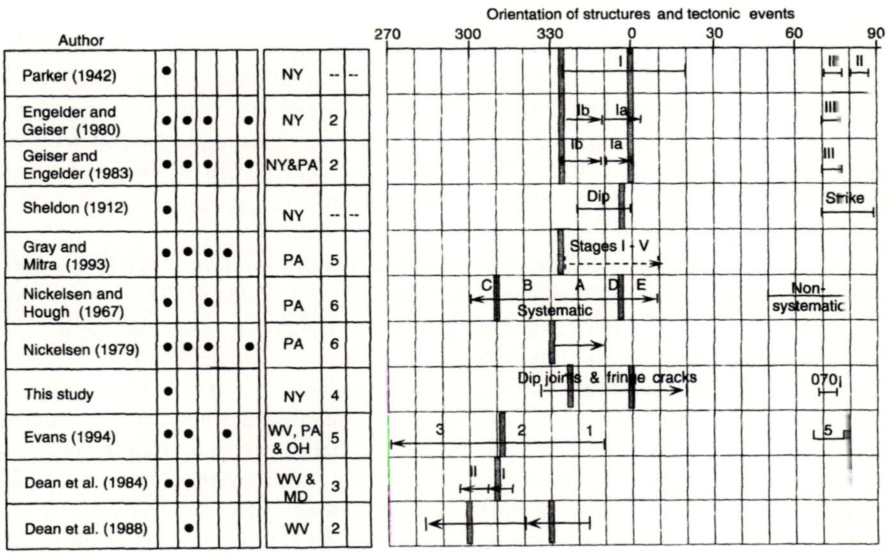

Fig. 3.9. Selected data on the sense of rotation of the regional stress field in the central Appalachian Mountains during the Alleghanian orogeny (from Younes and Engelder 1999)

the study area. Deformation at the Allegheny front near Williamsport, Pennsylvania, shows a single transport direction at 340° (Spiker and Gray 1997). Nickelsen (1988) noted that the Jacks Mountain fault, farther south, is intersected by a series of small wrench faults from which he was able to determine the directions of compression. These directions fall into orientation about 322°, then 335°, and then 312°, which indicates both senses of stress field rotation that might be expected near the culmination. Viewed on a larger scale, the Juniata culmination divides both the Appalachian Plateau and Valley and Ridge into an eastern area of clockwise paleostress rotation and a western area of counter clockwise paleostress rotation (Fig. 3.7).

3.1.6
Three Main Concepts of Regional Cross-Fold Joint Patterns

Based on the previous studies of fracture in the Appalachian Plateau, we distinguish three main concepts relating to regional patterns of cross-fold joints. These concepts apply to the different geometries of joint sets that form during *various time spans*. We distinguish:

1. The "strict orientation pattern" of joint sets (adopted from Nickelsen and Hough 1967) (Fig. 3.10a).
2. The "change in orientation within a joint set" (adopted from Engelder and Geiser 1980) (Fig. 3.10b).
3. The "simultaneous fan-shaped pattern" (adopted from Younes and Engelder 1999) (Fig. 3.10c).

Engelder (1993) notes that Nickelsen and Hough's (1967) outcrop-to-outcrop correlation of joints depends largely on the assumption that the Earth's stress field is "nearly homogeneous", and therefore joints of one set have similar orientations over large regions. If a set of joints at an outcrop differs by, say, 15° from a joint set at other outcrops, then this local suite of joints belongs to a second joint set, regardless of its orientation with respect to local structures. The implication is that joints of one set do not change orientation even as fold axes swing and curve through a mountain range such as the central Appalachian Plateau (superposition of Fig. 3.1d on anticlinal traces of Fig. 3.1c). On a regional scale, the change in strike of fold axes is accommodated by the overlap of joint sets of different orientations.

Engelder (1993) maintains that Nickelsen and Hough (1967) developed their correlation strategy based on the observation that first-order folds in the Pennsylvania Valley and Ridge are kink folds (e.g., Faill 1973) with straight axes. Nickelsen proposed (personal communication to Engelder) that the Valley and Ridge developed as overlapping thrust sheets cored with duplexes moving toward the craton, with straight-axis kink folds delimiting the thrust duplexes. The curvature of the central Appalachian Valley and Ridge is accommodated by abrupt changes in the orientation of first-order folds. Strictly parallel joint sets are a kinematic feature of thrust sheets associated with straight-axis kink folds. Presumably, the motion of various sheets is independent, thereby setting up homogeneous stress fields that produce parallel joint sets, unrelated to others in time and space. If this is indeed the case, then stress trajectories cannot be correlated throughout the whole mountain belt (Engelder 1993).

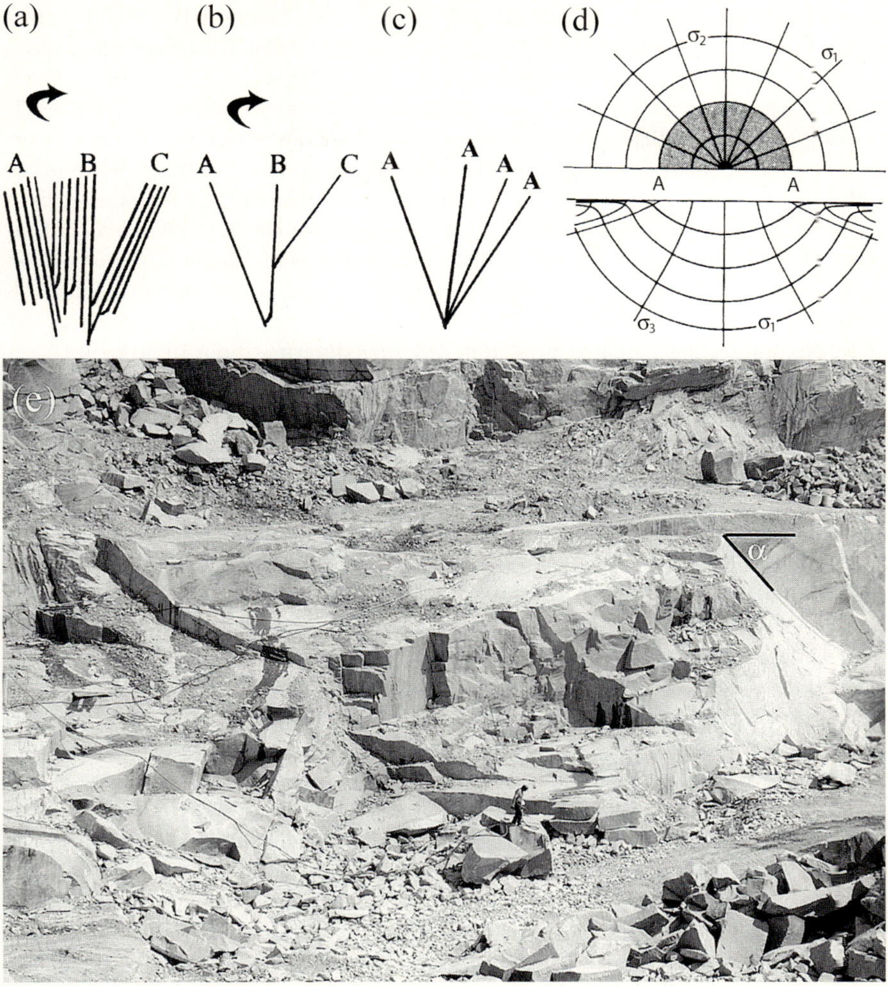

Fig. 3.10. Three concepts of regional cross fold joint formation. **a** Three joint sets (*A–C*) form sequentially in a rotational manner (similar to Fig. 3.1d). **b** Joints of a given set rotate sequentially from *A* to *C* (Fig. 3.1b,c). **c** Joints form concurrently in a radial pattern (Fig. 3.12a). **d** Contours of principal stresses strongly deviate from an orthogonal pattern in a Hertzian field induced by indentation. Half-surface view (*top*) and side view (*bottom*) of stress trajectories, plotted for Poisson ratio $u = 0.33$. *AA* denotes diameter of indenter contact (after Lawn and Wilshaw 1975b). **e** Hertzian fracture in granite from the South Bohemian Pluton, showing the cone opening downward (see the α angle), note a person for scale

While remapping cross-fold joints on the Appalachian Plateau, Engelder and Geiser (1980) developed the second concept, using the assumption that joint sets change orientation gradually so as to remain roughly perpendicular to local fold axes. Their analysis implicitly assumes that inhomogeneous stress fields accommodate stress trajectories that curve on a regional scale during an orogenic pulse.

Tracing joint sets in one lithology allows for the possibility that a single joint set can change orientation along with local fold axes (Engelder 1985) (Fig. 3.1b,c). We suggest that in the Appalachian Plateau, the distinction between the first and second concept is based primarily on criteria of fracture surface morphology: Fractography changes from set to set (Bahat and Engelder 1984), but it remains the same in a given set, even if joints within the set display some change in strike.

While evaluating the concepts that were introduced by Nickelsen and Hough (1967) and by Engelder and Geiser (1980), Younes and Engelder (1999) coined the third concept, the simultaneous fan-shaped pattern. They suggested that if the early joints in the Tully limestone at Taughannock Falls correlate with the 320–330° set found more than 100 km to the west, then the earliest 320–330° set does not fan, but rather is an indication that the same joint set appears in different lithologies in different locations. If, on the other hand, the 342° set in the western part of the study area is simultaneous with the 006° to 021° joint set in the east, then it reveals a fan-shaped pattern. Superimposed on this fanning pattern are two more joint sets, each differing about 10° in strike. Measurements show that individual sets are most recognizable in the abutment of parent dip joints (Fig. 3.4) where later joints cluster at 342°, 351°, and 003°. This joint pattern is consistent with data presented by Nickelsen and Hough (1967), which show how joints of one set cluster about one orientation, not taking part in the fanning pattern.

The first and second concepts require changing stress fields, by rotation mechanisms of S_{Hmax}. The two concepts complement each other in the joint pattern of New York: The various joint sets were formed by the mechanism described in the first concept, showing the rotation from set to set (Fig. 3.1d and 3.10a). The rotation within each set (Fig. 3.1b,c) was effected by the mechanism outlined in the second concept (Fig. 3.10b). Both rotations took place at intermittent time intervals. The third concept, on the other hand, postulates the simultaneous formation of a fan pattern of radial joints over long distances (Fig. 3.10c). We agree with Younes and Engelder (1999) that this concept cannot be rejected, though it would be better supported if fan patterns were found to exist in other fracture provinces. The third concept can be explained by mechanisms that relate to "moderate" heterogeneous stress fields. The radiating joints in Fig. 3.12a seem to be a good example. The formation of conjugate joint patterns may also be connected with simultaneous growth of joints in different directions. For comparison, a "strong" heterogeneous stress field may be induced by Hertzian indentation (Fig. 3.10d). Such extremely heterogeneous stress fields can be postulated for Hertzian fractures (e.g., Fig. 3.10e) found in granites and other magmatic rocks (Bahat 1979c) but rarely for fracturing in sedimentary rocks.

3.2 Fracturing in The Bristol Channel Basin

3.2.1 General Geology and Previous Investigation

The Bristol Channel Basin (BCB) extends on both sides of the Bristol Channel, from west England to South Wales. Unlike jointing in the other three provinces described in this chapter, which affects horizontal or slightly folded layers, joints of the Bristol Channel Basin occur in steeply inclined layers, some dipping more than 30°, due to asymmetric anticlinal folding. Fracturing in this province was described by Rawnsley et al. (1998) who worked on four edge locations of the basin (Fig. 3.11a), and by Engelder and Peacock (2001) who concentrated on the eastern side of the channel (Fig. 3.11b).

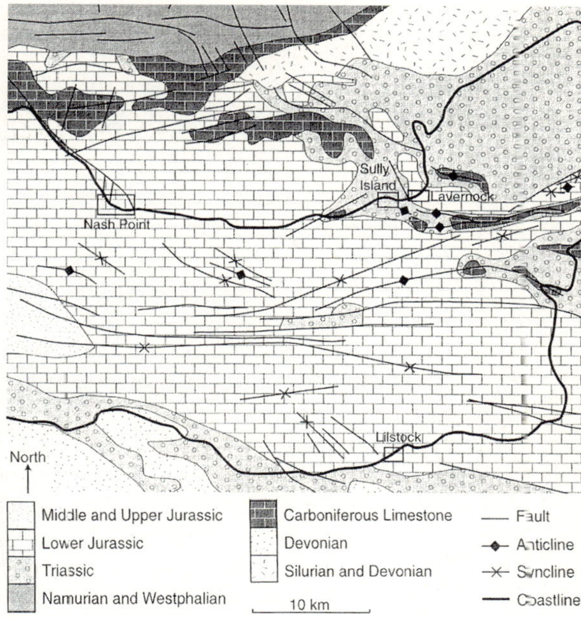

Fig. 3.11a. Map of the Bristol Channel. Map showing the main geological features of the eastern Bristol Channel, with positions of locations studied (modified from Rawnsley et al. 1998, based on the Bristol Geological Survey sheet 51° N–04° W, Bristol Channel)

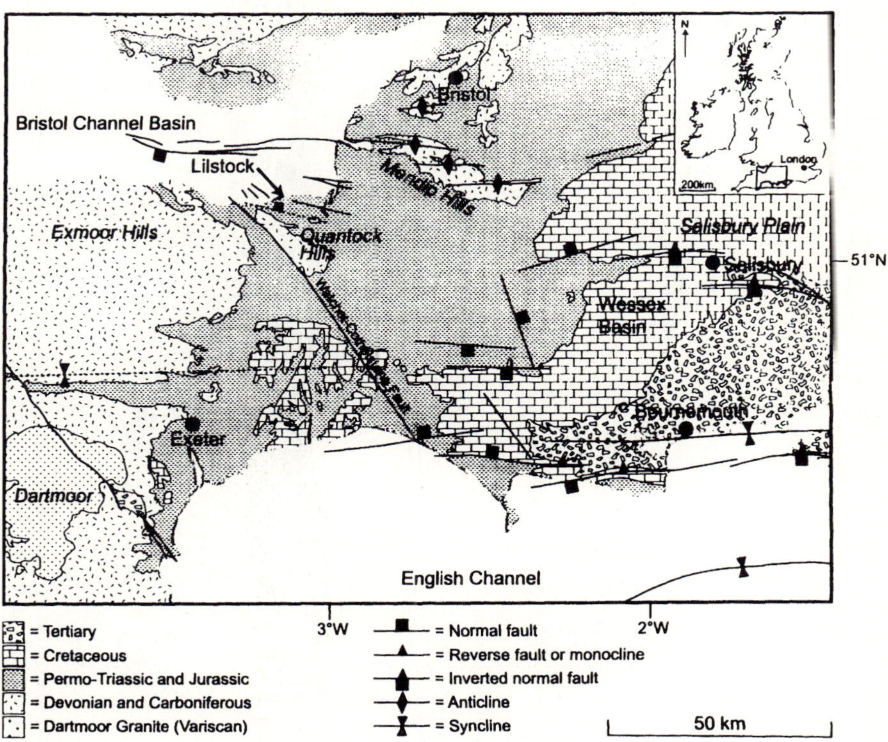

Fig. 3.11b. Map of the Bristol Channel. Geological map of south-western Britain, showing the location of the Bristol Channel basin and Lilstock (from Engelder and Peacock 2001)

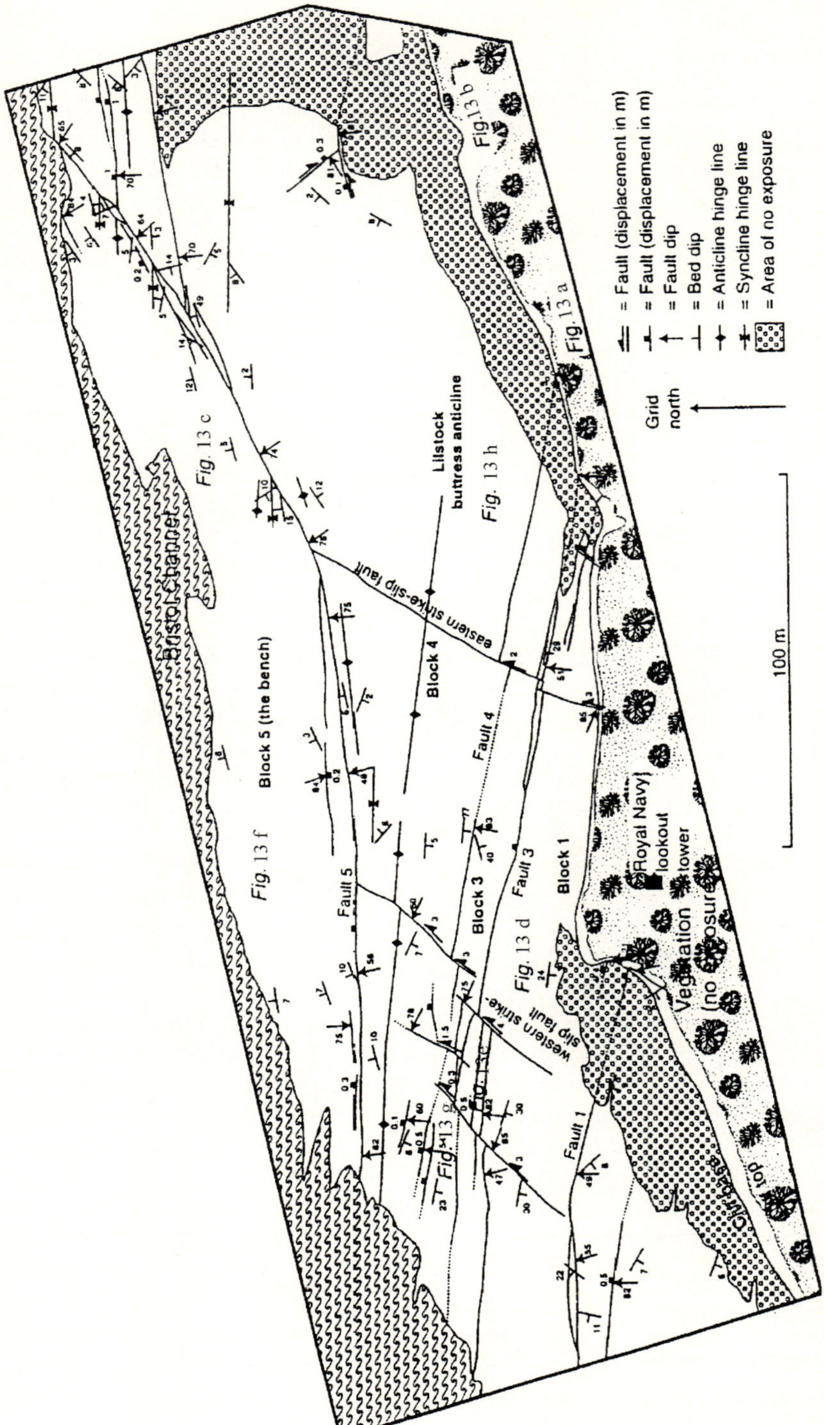

Fig. 3.11c. Map of the Bristol Channel. Geological map of Lilstock Beach (see location in Fig. 3.11a, adaptation from Rawnsley et al. 1998). The locations of photos in Fig. 3.13 are indicated (modified from Engelder and Peacock 2001)

The latter study focuses on the south-dipping beds at Lilstock (Fig. 3.11c), which these authors regard as particularly informative. These two papers, particularly the one by Engelder and Peacock (2001), are quoted extensively in the present section.

The Blue Lias Formation of Lilstock Beach consists of more than 50 limestone beds, 5–40 cm thick (Whittaker and Green 1983), interlayered with shales that range from a few centimeters in thickness to over 300 cm. The order of layers is consistent and can be followed even in separate fault blocks (Engelder and Peacock 2001). According to Rawnsley et al. (1998) (Fig. 3.11a), the BCB is an exhumed E-W-trending Mesozoic basin. Basin formation started as early as the Permian, with N-S extension causing reactivation of south-dipping Variscan thrusts (Van Hoorn 1987; Brooks et al. 1988; Roberts 1989). Extension and subsidence continued through the Triassic and Jurassic. To this stage belong E-W-striking normal faults and calcite veins, which are well exposed on the Somerset coast (Peacock and Sanderson 1991, 1992, 1994). Evidence for Alpine N-S contraction on the Somerset coast includes E-W-striking thrusts and reactivated normal faults (Peacock and Sanderson 1992; Dart et al. 1995) and strike-slip faults which conjugate about N-S (Peacock and Sanderson 1992, 1995). E-W-trending folds are related to wall-rock deformation around the E-W-striking normal faults and were probably tightened during the N-S Alpine compression. NNW-striking faults, e.g., the Sticklepath fault zone, appear to belong to the Variscan phase and may have been active at various times (Holloway and Chadwick 1986; Lake and Karner 1987; Van Hoorn 1987; Roberts 1989). Sinistral displacement occurred on these faults during the early Tertiary, so σ_1 was approximately NW-SE across the BCB at that time (Van Hoorn 1987). Dextral displacement occurred during the late Oligocene to Miocene (Arthur 1989), so σ_1 was approximately N-S across the BCB (Van Hoorn 1987).

Rawnsley et al. (1998) found that the joint patterns at the four edge locations in the BCB all post-date the high-density veins (HDVs) and the faults. The joints appear to have been formed in five main phases:

Phase 1: These are sub-parallel to a regional Alpine compression direction (about 340–350°) and are almost unperturbed by faults. At Nash Point, faulting started to perturb the stress during the development of phase 1 joints, i.e., there is a transition from phase 1 to phase 2 joints. Phase 1 joints were reactivated after phase 3 at Lavernock.

Phase 2: These are joints that were perturbed by faults and by phase 1 joints. The phase 2 joints follow the perturbed stress trajectories and curve from points of stress concentration along fault surfaces, e.g., at fault bends. The transition from non-perturbed (phase 1) to perturbed (phase 2) joints reflects a decrease in the regional compression and an increase in residual tensile stress. Joints of this phase are absent where no faults are present, as at Lavernock.

Phase 3: At Lavernock and Lilstock, phases 1 and 2 seem to have left no traces, and the only joint set found there belongs to phase 3, which is parallel to the 290° striking fold axes. These joints cut the veins that formed during folding and result from the relaxation of the Alpine compressional stresses within the folds. Phase 3 joints typically rotated counter clockwise through time to become parallel to the fold axes, perhaps as a result of diminishing the regional influence. This phase is absent where there are no folds, as at Nash Point.

Phase 4: This phase consists of cross-joints.

Phase 5: This phase consists of polygonal joint patterns. Phases 4 and 5 occur at all the locations studied. They are related to relaxation or contraction of the rock at a late stage. Rawnsley et al. (1998) show that their model applies well to the Liassic exposures around the basin's edge, although different patterns may be found in the unexposed center of the BCB. Such may be the case for a different orientation of joints linked to the regional Alpine compression, which e.g., at Lavernock are 010–020° (Rawnsley et al. (1998, Fig. 7a).

Engelder and Peacock (2001) assume that the present-day lithospheric stress field in the NW European platform has the same orientation as the Alpine stress field of the Miocene (Brereton and Evans 1987; Zoback 1992). Furthermore, the youngest joint set recorded throughout southern England correlates with the contemporary tectonic stress field (Bevan and Hancock 1986; Hancock and Engelder 1989). Hancock and Engelder (1989) argue that exhumation in a late-stage "Alpine" stress field is responsible for these youngest NW-striking joints in northwestern Europe. Thus, the post-Miocene tectonic event caused a clockwise rotation of the European stress field, only to have it return to its "Alpine" orientation, as is testified by the latest jointing in northwestern Europe.

Engelder and Peacock (2001) emphasize that the Alpine deformation, or "Alpine inversion", was accommodated by reverse-reactivation of normal faults, new thrust faults, new strike-slip faults, and by hanging wall buttress anticlines. Beck et al. (1993) define the "buttress effect" as resistance to displacement arising from a space problem; a body of rock can move only if it has somewhere to go. Buttress folds of Eocene to Oligocene age show a parallel geometry that is produced by flexural flow with bed-parallel slip (Nemcok et al. 1995). It should be noted that reverse faulting was more prevalent along the southern margin of the Bristol Channel Basin than on the north margin (Nemcok et al. 1995). Only one case of minor thrust faults associated with folding was found along the north margin in the southern limb of the Trwyn-y-Witch anticline in south Wales (Roberts 1974).

3.2.2 Fracture Description

Following are some fracture descriptions according to Rawnsley et al. (1998), specializing on phenomena that are less common in other provinces, including high-density veins, perturbations of joints near faults, "ladder" patterns and polygonal joints.

High density veins (HDV). Thin, closely-spaced calcite veins appear to be present everywhere in the limestone beds at Lilstock. In the east they mostly strike 270–290°, their common width is about 0.08 mm, and average spacing is less than 10 mm. They are often only visible where differential erosion has enhanced the contrast between the calcite HDVs and the limestone country rock, especially around the edges of joints. In thin sections they can be seen to persist throughout the rock, often forming *en echelon* patterns. HDVs either pass through the fault-related calcite veins or are slightly deviated by them. The HDVs are not perturbed by joints, and some joints may follow HDVs. In the west of the area, the HDVs strike 082–088° and have an average spacing of about 20 cm. Most of the calcite has been removed, with oxidization fronts about 30 mm wide. In the cliff, the HDVs strike about 110° and dip about 80° more steeply to the north than the bedding. HDVs strike 350–010° at Nash Point, which is the approximate Alpine compression direction. These are followed by 350–010° striking faults. The 010° faults appear to be dextral and the 350° faults appear to be sinistral, so they are conjugate about N-S. Thus, the HDVs represent a pre-joint phase that possibly was associated with faulting that accompanied the Alpine compression.

Converging joints perturbed by the faults. At least three main zones of converging joints occur along the 065° striking section of fault F5 at Lilstock (see also Swaby and Rawnsley 1997, Fig. 6). They are elongated in the 320–360° range. Joint spacing decreases slightly towards the center of the convergence as the joints curve towards the fault (Fig. 3.12a). These converging joints are the longest at Lilstock. For a detailed description, see Rawnsley et al. (1992). According to Engelder and Peacock (2001) the most plausible interpretation of the perturbed joints is that limestone-limestone contact across the fault generated a local stress field during or after the period of high-angle reverse faulting on reactivated normal faults, i.e., joint propagation followed fault initiation but not necessarily total slip. Curved trajectories of phase 2 joints in proximity to phase 1 joints are interpreted by Engelder and Peacock to mean that phase 1 joints were open and also perturbed the stress field during stage 2 joint propagation (Dyer 1988; Rawnsley et al. 1992, 1998).

Ladder joint patterns. "Ladder" patterns occur at Nash Point and at Lilstock. The joint pattern at Nash Point varies across the area. In the east, the main set strikes 155–160°. There is a two-level hierarchy, with 160° joints connected by orthogonal joints to form a "ladder" pattern that crosses layer boundaries. Within these ladders, shorter 160° joints and non-cross-cutting orthogonal joints form a smaller-scale system of ladders (Fig. 3.12b), which mostly do not cross bedding boundaries. Oblique ladders occur at Lilstock over a 095–135° range in all limestone beds within about 30 m of fault F3 (Loosveld and Franssen 1992). They usually rotate counter clockwise with time (i.e., 135° joints are the oldest and 095° joints are the youngest), forming oblique ladders. Left-stepping arrays spaced about 15 m trend 160–170° between two zones of convergent joints. Multidirectional joints also occur between zones of convergent joints and may be formed over extensive areas. Limestones appear to be more jointed than the shales. Joints often pass from the limestones into the shales, suggesting that they initiated in the limestones.

Polygonal joints. These enigmatic joints are found throughout the BCB, typically surrounding blocks. They are believed to indicate horizontally isotropic effective tension. On the other hand, they may represent thermal contraction following uplift, which may also be responsible for non-cross-cutting perpendicular joints between closely-spaced joints. The latest, isotropic joints, are related to biaxial effective tension in the plane of bedding, especially in "shadow zones" between convergence points. Between the convergent zones, an almost isotropic joint pattern was produced by a low horizontal differential stress. Impedance of later joints on earlier joints indicates that the latter were always open. Deciphering the origin of these joints remains a challenge.

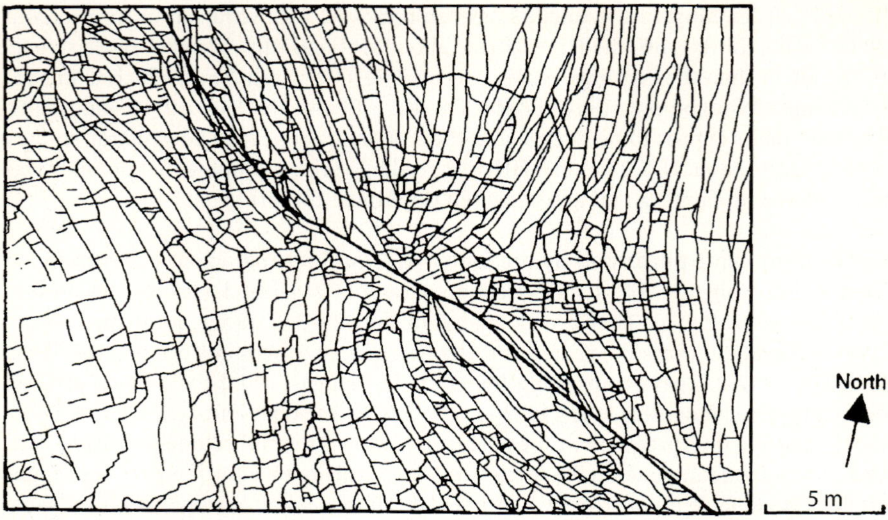

Fig. 3.12a. Line drawing of fracture patterns at Nash Point (see Fig. 3.11a). Perturbations in joints (modified from Rawnsley et al. 1998)

Fig. 3.12b. Line drawing of fracture patterns at Nash Point (see Fig. 3.11a). Joints displaying a ladder pattern of joints (modified from Rawnsley et al. 1998), note a certain resemblance to cross-joints (Gross 1993)

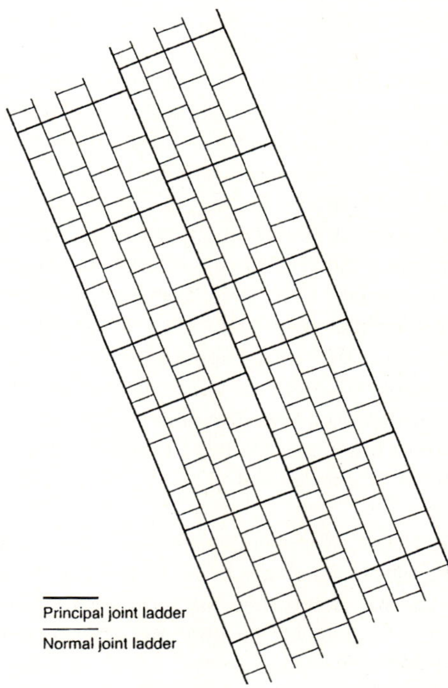

Engelder and Peacock (2001) note that at Lilstock Beach joints within limestone beds are generally contained within these beds and rarely pass into adjacent shales. They also note that patterns of systematic jointing differ from one limestone bed to the next, which explains the different descriptions of the joint sets at Lilstock Beach in earlier studies. All that can be said about the sequence of jointing at Lilstock Beach is that the joints postdate early normal faulting and vein development. Commonly, joints propagate after normal fault development, as is indicated by abutting and cutting relationships (e.g., Fig. 3.13d,e). Some curving joints radiate from points of stress concentration on faults (e.g., Fig. 3.13c), which are a natural consequence of slip on an irregular fault. These joints could have propagated during a later stage of slip (Engelder and Peacock 2001). By way of comparison, perturbed curving joints that are associated with the Naim fault in the Beer Sheva syncline were formed under a secondary stress field, after cessation of movement along the primary strike-slip fault (Sect. 6.6).

Faults and folds. The structures on Lilstock Beach include a set of four north-dipping, high-angle faults that are offset by later strike-slip faults (Fig. 3.11c). Engelder and Peacock (2001) divided Lilstock Beach using three of these high-angle faults (faults 3, 4, and 5 of Rawnsley et al. 1998) and two strike-slip faults. The area north of fault 5 is called block 5 (or "the bench", in recognition of Loosveld and Franssen's (1992) mapping there). The other three blocks are hanging walls to faults 4, 3 and 1. Engelder and Peacock (2001) refer to the eastern, central, and western portion of each "normal" fault block, depending on outcrop position relative to two prominent strike-slip faults (Fig. 3.11c).

Engelder and Peacock found that separation at any point on the faults can be measured with confidence by juxtaposition of bedding (Fig. 3.11c). Throw on at least one of the high angle faults (fault 4) was reversed during Alpine inversion. Reverse slip may have occurred on the other faults, but the total throw was not reversed. The steep dip of fault 4 and its parallelism with normal faults in the area indicate that it originated as a normal fault, which inversed together with the other reverse faults in the region (e.g., Whittaker and Green 1983; Dart et al. 1995). Net throw (the combination of pre-Alpine normal slip and Alpine reverse slip) on the four faults with throw >2 m is: Fault 1 ~ 18.7 m (normal), fault 2 ~ 8 m (normal), fault 4 ~ 8 m (reverse), and fault 5 ~ 14 m (normal). The outcrop is also cut by numerous high-angle faults with <2 m throw, which mostly exhibit normal slip and are parallel to the strike of bedding. Strike-slip faults displace the high-angle faults and the limbs of the Lilstock buttress anticline. The strike-slip faults are cut off at fault 4 (Fig. 3.11c), indicating that strike-slip faulting and some high-angle faulting may have been contemporaneous (Kelly et al. 1998).

Block 4 consists of a large buttress fold, the Lilstock buttress anticline, with a gently dipping north limb and relatively steep south-dipping beds (Fig. 3.13a,b). The Lilstock buttress anticline originated as a reverse-drag fold (Hamblin 1965) during the Mesozoic normal faulting and was tightened during reverse slip on fault 4. Engelder and Peacock focused their study on joints in the south-dipping limb of this anticline. Beds in the northern half of block 4 dip gently to the north (<5°), whereas beds in the southern half of the outcrop dip to the south, and in places the dip is more than 30°. Along the south side of the Bristol Channel, dips of more than 30° are restricted to the flanks of buttress anticlines. Diagrams of the tight folding typical of buttress folds in the Mesozoic strata of the Somerset coast are presented in Peacock and Sanderson (1992, Fig. 7a) and Dart et al. (1995).

Fig. 3.13a–d.
Outcrops from Lilstock. **a** View looking NNW from the eastern cliff along faults 3 and 4 with the hinge of the buttress anticline shown in the background. **b** View looking NNW from the eastern cliff along the hinge of the Lilstock buttress anticline. Bed 0497 is visible in the foreground. A control sample from bed 0497 was taken across the hinge of the anticline and Fig. 3.13h was taken at the point where bed 0497 rolls over into the limb of the anticline. **c** Curving joints in bed 1848 of fault block five. Photographer is standing on fault 5 (location shown in Fig. 3.11c). **d** J2 joints cutting two normal faults within bed 2429 of fault block 1 (from Engelder and Peacock 2001)

Fig. 3.13e–h.
Outcrops from Lilstock.
e J3 joints abutting a small (throw <1 m) normal fault in bed 1921 in the western portion of fault block three. A sinistral strike-slip fault offsets bed 1921 (in the foreground) by 2 m. That fault is internal to the western portion of fault block 3. **f** Joint set development within bed 1848 on the bench north of fault 5. Joint set J2 is parallel to the direction of view. Sets J4 and J5 are oriented anticlockwise from set J2, with J6, anticlockwise from J4. The view is looking south-east toward the cliff from where the photographs for Fig. 3.13a,b were taken. View across the bottom of the photograph is about 3 m. **g** J3 joints cutting veins and transecting beds dipping in a relay ramp developed during early extension in the western portion of fault block 3 (coin diameter = 27 mm). **h** The counter clockwise relationship between J2 and J3 in bed 0497 where it rolls over into the south limb of the anticline (from Engelder and Peacock 2001)

3.2.3
Data Collection and Joint Identification

Engelder and Peacock (2001) studied joint development in 25 out of 46 carbonate beds in a 40 m thick Liassic section (British Geological Survey stratigraphic column, Whittaker and Green 1983) (Fig. 3.14). To indicate position in the vertical section, they labeled beds by height (in cm) above an arbitrary datum. For example, bed 1848, 18.48 m above the datum features prominently in the studies of Loosveld and Franssen (1992) and Rawnsley et al. (1998). Engelder and Peacock sampled systematic joints at Lilstock Beach in a volume of Blue Lias measuring 40 m thick by 200 m long (E-W). The repetition of the section by high angle faults permitted sampling up to 50 m along the N-S direction of some beds. Focusing on the effect of Alpine inversion on joint development, they assumed that joints of this stage are likely to be found in the south-dipping limb of the Lilstock buttress anticline in block 4, since the relatively steep dip of these beds indicate they are the most strained rocks at Lilstock Beach. To confirm

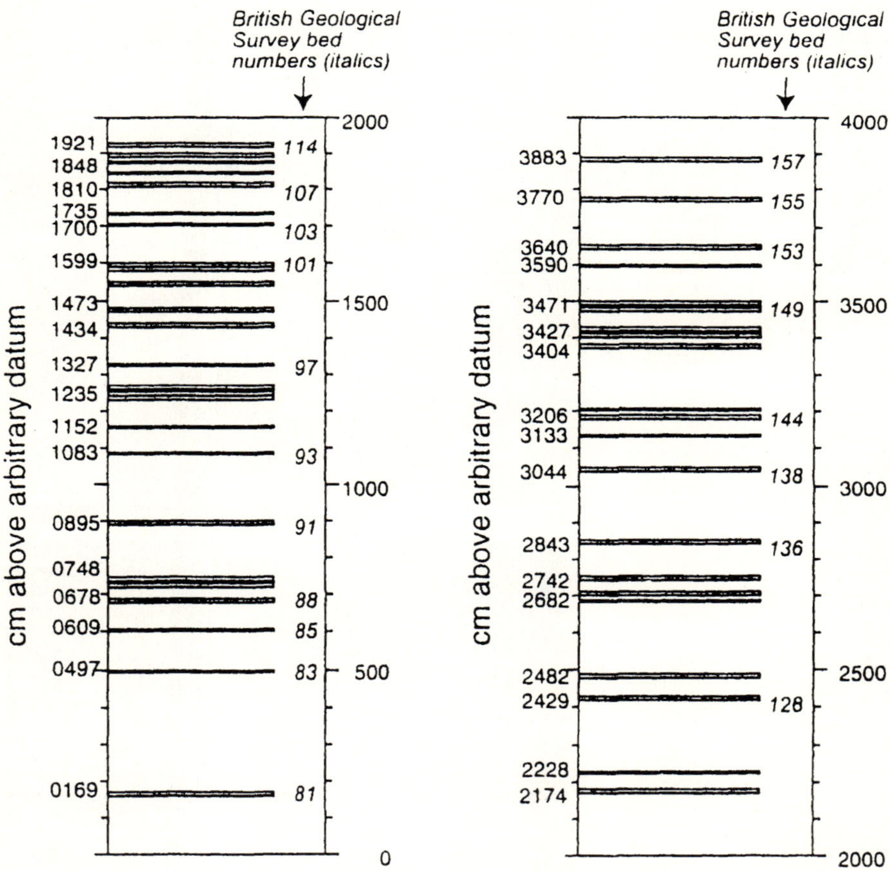

Fig. 3.14. Stratigraphic section at Lilstock Beach showing the position of 46 carbonate beds above an arbitrary datum. The beds are labeled according to their position relative to the datum. The British Geological Survey (Whittaker and Green 1983) designation for each bed is also given (from Engelder and Peacock 2001)

this, they measured the joint patterns on sub-horizontal surfaces in block 5 as well as in the hinge area of the Lilstock buttress anticline of block 4, to serve as control samples. The joint patterns in dipping and sub-horizontal limestone beds were compared. Standard scan line techniques (LaPointe and Hudson 1985) were then used to record the location and orientation of each systematic joint. The joint sets in each bed, based on abutting and orientation characteristics, were then correlated from bed to bed by means of subsequent data reduction. The most subjective assignments concerned curving joints that on the base of orientation might be included in more than one set.

A collection of parallel systematic joints counts as a set (Hodgson 1961b). According to Engelder and Peacock (2001), however, the poles to joints of a systematic set may cluster through several degrees, because individual joints are never perfectly planar and measurements cannot be precise (Fig. 3.15). Therefore, joint sets that differ by less than 10° may have overlapping clusters in a stereographic plot, and not each individual joint can be assigned to a particular set even by post-fieldwork analysis. Hence, Engelder and Peacock rely on field notes, where joints were presorted into sets

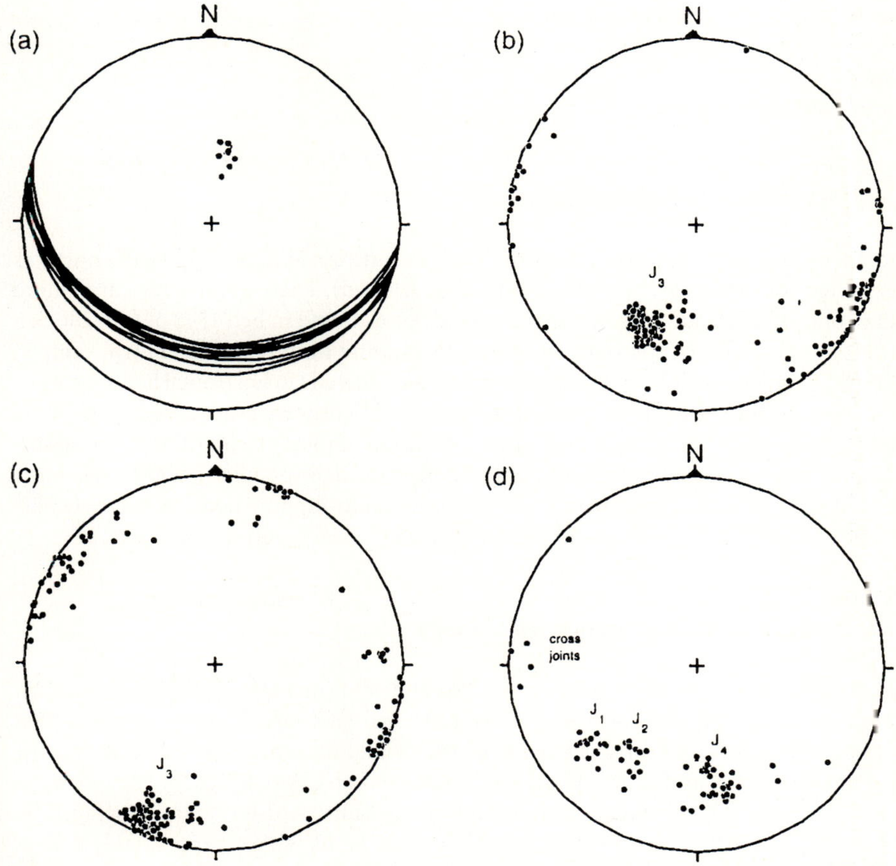

Fig. 3.15. Lower hemisphere stereographic projections for beds and joints in fault block 1. **a** Bedding traces and poles for bed 2429. **b** Present-day poles to joints in bed 2429. **c** Poles to joints in bed 2429 with bed dip removed. **d** Present-day poles to joints in bed 3883 (from Engelder and Peacock 2001)

within each particular bed. The grouping of joints into distinct sets was most difficult where several beds are involved and where joint sets tend to curve, as they do in the south-dipping limbs near the faults of Lilstock Beach. In these cases the control samples (see above) served as a guide. Using this procedure, six joint sets were distinguished, falling in the following ranges: J1, (330–310°), J2 (310–295°), J3 (295–285°), J4 (285–275°), J5 (275–265°), and J6 (<265°)(where 265° is equivalent to 085° in the upper hemisphere). It should be noted that joint distribution in the western part of the BCB differs somewhat from the findings in England. Between Mumbles Head and Rhossilli Bay in south Wales, Roberts (1979) found six sets striking 360°, 340°, 290°, 270°, 240°, and 200°, as perhaps can be expected over such distances and changes in the regional structure. The challenge to the investigator is how to sieve out homogeneous domains from a heterogeneous province for deciphering the tectonic history of the region.

Engelder and Peacock (2001) used a simplified AVTD plot (see Wise and McCrory 1982) for obtaining a qualitative sense of relative joint density by normalizing the data on joint occurrence and comparing joint development in sub horizontal beds with beds that have been tilted by Alpine inversion. This procedure is not given here, and the interested reader is referred to the original paper.

3.2.4 Joint Sets in the Blue Lias

3.2.4.1 *Joint Sets in Steeply-Dipping Beds*

Joints in the south-dipping limb of the Lilstock buttress anticline dip to the north as expected for beds that dip to the south (Fig. 3.15a–d). These commonly carry more than one joint set, but no beds carry as many sets as found in bed 1848 of fault block 5 (Fig. 3.16a). To correlate joints in the south-dipping limb with the control sample, Engelder and Peacock (2001) rotated joints to their attitude in horizontal bedding. After rotation, the joint-frequency data for steeply-tilted beds were normalized in the same manner as for sub-horizontal beds. The population of joint orientations in the south-dipping beds of Lilstock Beach is indeed different from the population in the sub-horizontal beds. The most prominent joint set in the south-dipping beds, J3, is nearly parallel to the strike of bedding and the hinge of the Lilstock buttress anticline.

3.2.4.2 *The Orientations of Joints as Related to Faults*

Joints of set J3 (strike 290°) transect the fold and veins in a relay ramp in the western portion of fault block 3 (in bed 1735) (Fig. 3.13g). This relay ramp is evidence that J3 joints post date the early extension of the Bristol Channel. Various relationships between J3 joints and normal faults are observed at Lilstock Beach. For example, some J3 joints cut through incipient high-angle faults without being displaced by them (Fig. 3.13d). Other J3 joints abut high-angle faults without appreciable deflection (Fig. 3.13e). Both types of fault are early calcite-filled normal faults with no evidence of reverse reactivation during Alpine contraction. This is like the relay ramp in bed 1735.

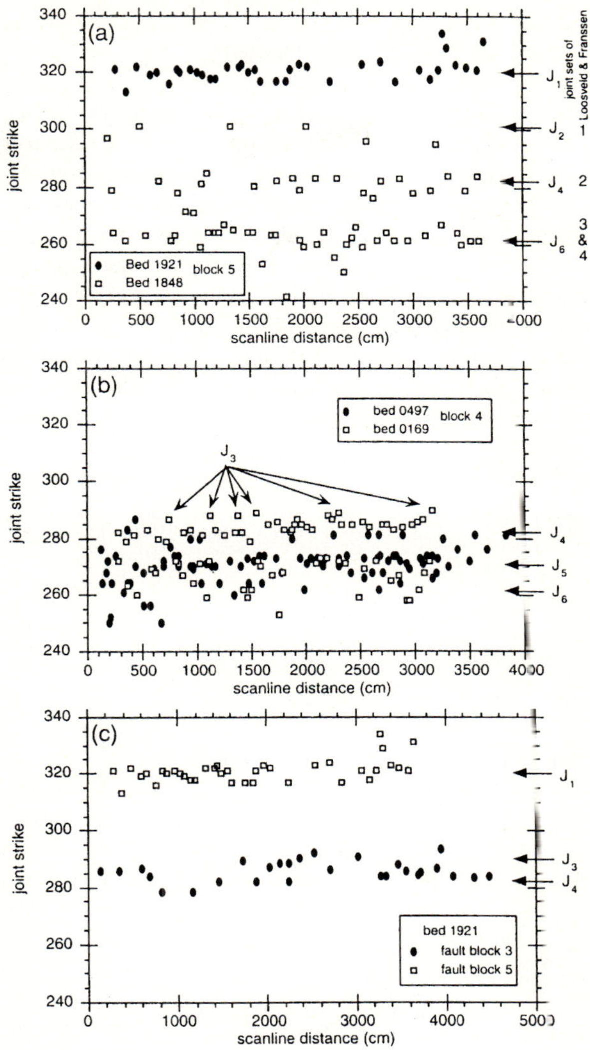

Fig. 3.16.
a Joint strike vs. distance for a scan line corrected to a scan line azimuth of 040° for beds 1848 and 1921 on the bench north of fault 5. **b** Joint strike vs. distance for a scan line azimuth of 044° for beds 0169 and 0497 in eastern fault block 4. **c** Joint strike vs. distance for bed 1921 for a scan line azimuth of 040° for the bench north of fault 5 and 110° in fault block 3 (from Engelder and Peacock 2001)

The orientation of the joint population varies from fault block to fault block. For example, bed 1921 in western block 3 dips up to 43° to the south compared with its slight northward dip in the bench of block 5. Within western block 3, J3 and J4 appear without J1, which is the only set in bed 1921 on the bench (Fig. 3.16c). This example is typical of the difference in joint patterns within the same bed from block to block. Engelder and Peacock (2001) conclude from these frequent differences that jointing took place after beds became isolated within individual fault blocks following at least some fault displacement. The widespread occurrence of systematic joint sets within fault blocks suggests that the internal portions of fault blocks were not affected by stress reorientation during fault slip. They also suggest that, alternatively, these differences mean that different mechanisms operated in the south-dipping limb of the Lilstock buttress anticline and in the sub-horizontal beds.

3.2.4.3
The Sequence of Joint Development

The abutting sequence within sub-horizontal bed 1848 indicates that J2 propagated before J4, which propagated before J6 (Fig. 3.13f). Assuming that joints are created parallel to regional S_{Hmax} directions, this sequence shows the same counter clockwise reorientation of the remote stress field as seen throughout the eastern part of the Bristol Channel (Rawnsley et al. 1998). The same counter clockwise sequence may be traced from the sub horizontal bed 0497, from block 4 into the south-dipping limb of the Lilstock buttress anticline (Fig. 3.13h). This is a rare case of J3 joints abutting joints of J2. Where bed 0497 steepens, J2 is absent and J4 abuts J3. Another case of this counter clockwise joint formation is found in bed 3883, which dips up to 30° in the central and western parts of fault block 1 (Fig. 3.15d). Other beds also show a clear counter clockwise age sequence, although this is found in other joint sets (Table 3.1).

Table 3.1. Joint development: All stations in south limb of buttress anticline

Bed	BGS Beds	Fault block	J1	J2	J3	J4	J5	Sequence[a]
4074	161	1	×					
4005	159	1	×	×				CCW
3883	157	1	×	×			×	CCW
3770	155	1				×		
3590	151	1	×					
3492	149	1	×				×	Separate
3427	147	1	×	×				CCW
3206	144	1				×		
3185	142	3				×		
3133BC	140	1			×		×	Separate
3133A	140	3		×				
3044	138	3			×			
2742D	134	3			×			
2742ABC	134	1			×	×		CCW
2429B	126	4			×		×	CCW
2429AC	126	1		× or	×			Separate
2228	121–123	4		×				
1921A	114	3			×	×		Separate
1848E	109	3		×				
1848CD	109	3			×		×	Separate
1848A	109	4			×		×	CCW
1735C	105	3			×			
1735AB	105	4		×			×	Separate
1700	103	3				×		
1599AB	101	4		×				
1152	95	4					×	
0678	88	4			×		×	Separate
0609	85	4					×	
0497D	83	4				×	×	Separate
0497B	83	4			×		×	CCW

[a] CCW–joints that abut with an counter clockwise sense of development.
Separate–joints that have developed in separate regions of a bed (from Engelder and Peacock 2001).

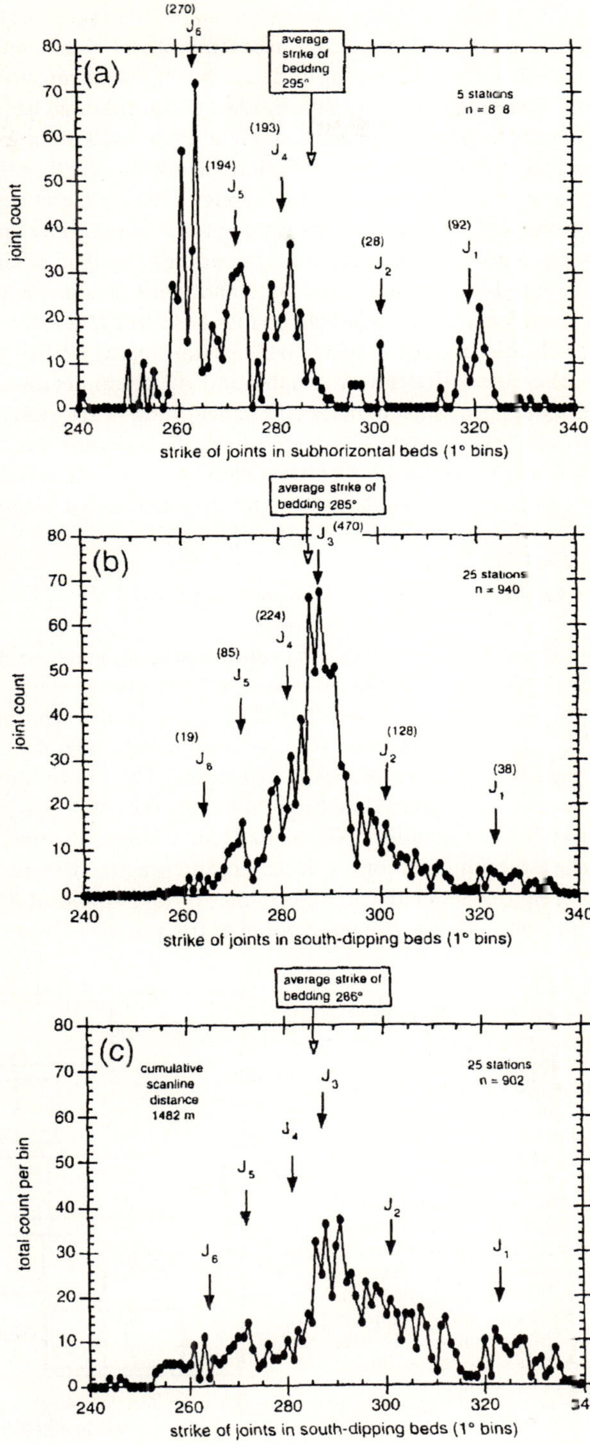

Fig. 3.17.
Joint frequency vs. joint strike. Joint count normalized to a 250 m scan line with the data binned into 1° intervals. **a** Subhorizontal beds in blocks 4 and 5. **b** Dipping beds of the Lilstock buttress anticline, in blocks 1, 3 and 4. **c** Binned data without normalizing joint data for either scan line length or scan line orientation, in blocks 1, 3 and 4 (from Engelder and Peacock 2001)

The counter clockwise age sequence found in south-dipping beds is more clearly observed on sub-horizontal surfaces (Table 3.1). Determination of sequence requires two or more joint sets abutting at a station, but most exposures are too small to display the required joint contacts (e.g., Fig. 3.13d,e,h), although beds may contain two groups of joints with different orientations on different parts of the exposed bedding surface along strike. In such cases, it is difficult to determine whether the groups belong to different sets or whether they belong to a single curving set, as is seen near faults. More than half the samples from the south-dipping beds display only one set (Fig. 3.17), showing no time sequence. As a rule, beds containing J1 or J2 joints lack J3 joints. Where J3 joints are found, they are the oldest joints and strike closest to bedding strike (Table 3.1). Thus, the evidence for the age of J3 relative to J1 and J2 is rather too meager to be definitive.

The distribution of joints in the south-dipping limb of the Lilstock buttress anticline is also correlative to stratigraphic and structural position. J1 and J2 joints (found only in block 1) are high in the section. Sets J3 and J4 are most common in the middle of the section (blocks 3 and 4), with the greatest development of J3 joints near the reverse fault that created the Lilstock buttress anticline. J5 and J6 joints are commonest in the lower parts of the section and nearest the anticline-hinge, where dips are shallow.

3.2.4.4
The Attitude of Joints in Relation to Bedding

Joints in competent beds are typically perpendicular to the bedding (Hancock 1985). However, this is not the case for sets J1 to J5 in the south-dipping beds of the Lilstock buttress anticline. When bedding is restored to the horizontal (e.g., J3 joints in Fig. 3.15c), these joints consistently dip to the north, with their poles plunging south, towards the European Alpine hinterland. The joint-pole plunge angle is used to define the "tilt towards the hinterland" for the joint sets (Fig. 3.18). The tilt increases from 3° for the older J1 joints to 7° for younger J5 joints. J3 joints occur only in the south-dipping limb of the anticline, and are therefore of particular interest for interpreting the geometry and mechanics of fold and joint development; their geometric characteristics are used as a guide in the search for a cause of joint tilt.

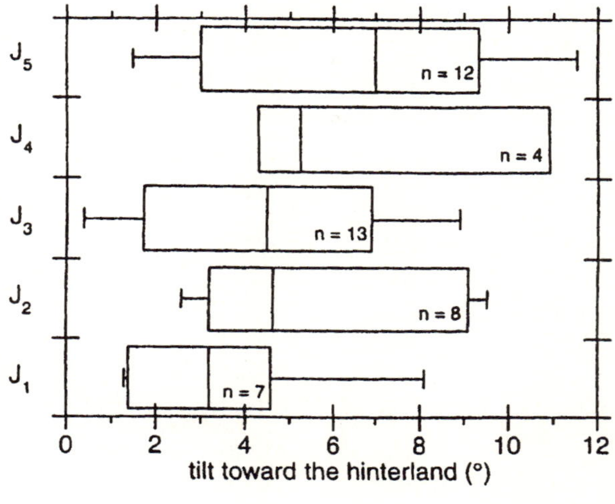

Fig. 3.18. Box plot for the tilt of the vector mean pole to joints from 44 joint populations in south-dipping beds of the Lilstock buttress anticline. The tilt is measured downward in the direction of the hinterland (i.e., southern Europe) relative to bedding when it is restored to horizontal. The *center line* of each box represents the 50th percentile of the data while the left and right edges of the boxes represent the 25th and 75th percentile of the data, respectively (from Engelder and Peacock 2001)

3.2.5
Stresses in the South-Dipping Limestone Beds during Alpine Inversion

3.2.5.1
J3 Joints in the Lilstock Buttress Anticline

According to Engelder and Peacock (2001), the J3 joints in the south-dipping beds at Lilstock Beach were formed during the development of the Lilstock buttress anticline, reflecting Alpine inversion. They base this conclusion on six observations:

1. J3 joints are rare in the sub-horizontal beds of Lilstock Beach but are the most common set in the steep limb of the Lilstock buttress anticline. The mechanism driving J3 jointing must involve higher strain on the limbs of folds and little strain in the hinge area.
2. J3 joints are most common in the vicinity of fault 4, which was reversed to form the Lilstock buttress anticline. This spatial association suggests that the formation of set J3 with a high-angle reverse fault was synchronous with Alpine inversion.
3. Poles to joints in the southern limb of the Lilstock buttress anticline consistently plunge toward the hinterland, their azimuth parallel to the inferred direction of compression (019°) during Alpine inversion (Kelly et al. 1998). The consistent relation of tilt to bedding suggests that joints developed by shear couple on bedding.
4. The strike of J3 joints is nearly parallel to the strike of beds in the south-dipping limb of the anticline. Strike-parallel joints are commonly associated with outer-arc extension by bending of beds during folding (e.g., Van Hise 1896; Hancock 1985) (Fig. 3.19a), but outer-arc extension by tangential longitudinal strain folding is greater in the hinge area and absent at inflection points in the limbs (Price and Cosgrove 1990).
5. J3 joints cut completely through beds and thus differ from joints that form by outer-arc extension above a neutral fiber (Van Hise 1896).
6. J3 joints are not present in beds with J1 or J2 joints, and vice versa (Table 3.1). Whatever mechanism was responsible for J3, it is clear that they propagated neither in sub-horizontal beds, nor in previously jointed beds.

Alpine inversion was driven by a remote S_{Hmax} directed about 019° (Kelly et al. 1998). This presents a mechanical paradox if J3 joints propagated during parallel folding, because they opened in a direction that nearly coincides with that of S_{Hmax}. If controlled by the remote stress, J3 joints would open in a plane that contains S_{Hmax}. Engelder and Peacock (2001) therefore suggested that the opening must be controlled by a local stress. Accordingly, they sought a set of boundary conditions that generate a local tensile stress on the limestone-shale package when a regional compressive stress is superimposed on the Lilstock buttress anticline.

To understand the tilt of joints in the south-dipping beds of the Lilstock buttress anticline and their ability to open sub-perpendicular to S_{Hmax}, Engelder and Peacock apply a simple finite element model, which involves the development of a shear couple on limestone beds during parallel folding by flexural flow of the shale interlayers. If there indeed is a genetic relationship between the orientation of J3 joints and the shear couple on limestone beds, the model proposes a possible orientation for S_{Hmax} that fits these boundary conditions. To anticipate their result, remote S_{Hmax} controlling J3 joints should be the same direction as the remote S_{Hmax} that caused the conjugate strike-slip faults during Alpine inversion.

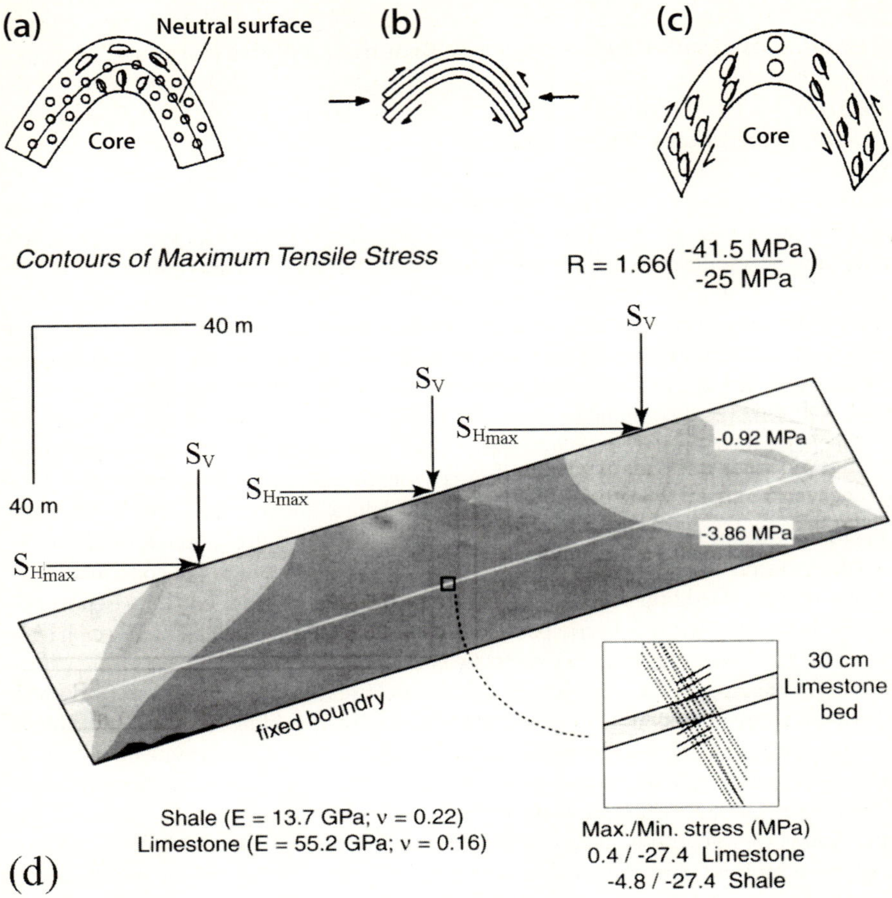

Fig. 3.19. Fold mechanisms. **a** Outer-arc (buckling or tangential longitudinal strain), showing outer-arc stretch, inner-arc contraction and no distortion on limbs. **b** Flexural slip, successive layers are displaced upwards towards the anticline crest with respect to the layer below. Individual layers are unstrained. **c** Flexural flow or flexural shear, showing no shear on hinge and shear within beds on the limbs (Fig. 3.19a–c from Park 1983, p. 72). **d** A three-bed model of the limb of the Lilstock buttress anticline contoured for σ_3 (compressive stress is negative). The top surface is loaded as indicated with the bottom boundary fixed and the left and right boundaries free to shear. The thin middle bed is a 30 cm limestone with elastic properties as indicated. In close-up inset the orientation of the local principal stresses are indicated with $\sigma_1 = -27.4$ MPa and $\sigma_3 = 0.4$ MPa (limestone), and $\sigma_3 = -4.8$ MPa (shale). Tensile stress is indicated by a *solid line* and compressive stress is indicated by a *dashed line* (after Engelder and Peacock 2001)

3.2.5.2
Stress Configuration in Limestone Beds of the Lilstock Buttress Anticline during its Development as a Flexural-Flow Fold

According to Engelder and Peacock (2001), the Lilstock buttress anticline, a sequence of interbedded limestones and shales, should display a significant contrast between the elastic properties of the limestone and shale (e.g., Hatheway and Kiersch 1982).

This lithological pair is commonly associated with flexural slip folding (Fig. 3.19b) in which the more brittle limestone beds slip on ductile shale interlayers (e.g., Ramsay and Huber 1987). Displacement also occurs by layer-parallel shearing of the shale during flexural-flow folding (Fig. 3.19c) (flexural flow is also termed flexural shear, e.g., Twiss and Moores 1992, Fig. 12.7). Often, the shale interlayers are thin as compared to the carbonate beds, but at Lilstock the opposite is the case. Engelder and Peacock suspect that thick shale layers obscure all evidence for slip by simple shear strain within the shale, if present, largely because the shear strain could be relatively small and because such strain markers as stretched ammonites are rare. Nevertheless, the regional S_{Hmax} that tilted the beds of the anticline induced a shear couple in them (assuming for simplicity that S_{Hmax} and S_v are principal stresses). This assumption of a remote horizontal principal stress is supported by horizontal slip on the conjugate strike-slip faults active during Alpine inversion and by slickenside lineation orientations on the inverted normal faults (Kelly et al. 1998).

Engelder and Peacock (2001) mapped the stress fields induced in an elastic medium under loading configurations consistent with a shear couple developed during layer-parallel shear, applying 2D finite element analyses, using the interactive fracture analysis program FRANC (Wawrzynek and Ingraffea 1987). They used a model mesh of quadratic, isoparametric elements (Fig. 3.19d) applying boundary conditions to simulate a three-layer model of a limestone bed of the Blue Lias between two shale layers. Figure 3.19d shows a contour map of the maximum tensile stress (i.e., σ_3 with tensile stress positive) generated in beds dipping at 17° and subject to a stress ratio (i.e., $R = S_{Hmax}/S_v$) of 1.66. Their model has a length to height ratio of 5, which is sufficient to carry end conditions away from the central part of the model, which portrays a 3 m^2, centrally situated square element. The initial versions show that the elastic contrast between the limestone and shale beds leads to a tensile σ_3 within the limestone layers for values of R between one and four and $\nu_{sh} > \nu_{ls}$ (Poisson ratios of shale and limestone, respectively). Tensile stress is generated via the flexural flow as permitted under the tractions specified in Fig. 3.19d. Fixing the ends of the model to prevent layer-parallel shear strain would of course be inconsistent with folding by flexural flow (Price and Cosgrove 1990) and it would not permit the generation of tensile stresses in the stiff beds.

Not only is the local σ_3 tensile in the stiff beds under conditions of flexural-flow folding, but σ_3 cants downward relative to bedding in the direction of the horizontal stress (Fig. 3.19d). When local σ_3 is tensile, joints propagate parallel to local σ_1, which is always compressive in the models (the dashed lines in Fig. 3.19d). Local σ_1 refracts slightly at the boundary between the limestone and the shale beds, becoming steeper in the stiffer limestone bed.

Engelder and Peacock (2001) ran several versions of the model to test the effect of R on the tilt of the local σ_1, and hence the predicted attitude of J3 joints with respect to bedding. As R is increased from 1 to 4, the local σ_1 tilts at an incrementally larger angle toward the foreland in both the shale and the limestone beds. Tensile σ_3 in limestone beds becomes larger (0.5 MPa to 5.2 MPa) as R increases from 1 to 3. At $R \approx 4$, σ_3 in the shale near the limestone bed also becomes tensile. When running the same series of tests for beds dipping at 3°, 5° and 9°, the tilt of the joint plane toward the hinterland (i.e., local σ_1) increases for increasing R, with higher rates of increase in the more steeply dipping beds. Engelder and Peacock also ran tests (not cited here) to determine the effect of the Poisson ratio on the model. They concluded that ν was smaller in limestones than in shales because more joints would otherwise occur in the shales than in the limestones and because $\nu_{ls} > \nu_{sh}$ is inconsistent with in situ geophysical logging data.

3.2.5.3
Joint Propagation at Depth Not by Means of High Fluid Pressure

Joint development at depth is often attributed to high fluid pressures (Secor 1965; Bahat and Engelder 1984) except in cases of composite layers with different elastic properties (Hobbs 1967; Gross et al. 1995). The above described model for the generation of tensile stresses during flexural flow is a variation on the composite layer model first proposed by Hobbs (1967). An important consequence of the model is that compressive stresses may generate local tensile stresses in the stiff layers of a composite stack where bounding surfaces are subject to shear traction, as would be required for flexural flow folding. This mechanism, acting at depth, can produce jointing also in the absence of high fluid pressure. Engelder and Peacock favor the flexural flow mechanism over hydraulic fracturing because: (1) joints with tilt occur only in dipping rocks of the fold limbs and (2) as predicted by their model, joints are not orthogonal to bedding. Engelder and Peacock's (2001) model for the Lilstock buttress anticline determines that joint (and vein) growth in the limbs of flexural-flow folds should have several distinct characteristics. The joints should extend through the entire bed, should be non-orthogonal to the bedding, and dip toward the anticlinal axial plane, while the poles tilt away from it.

3.2.5.4
Stress Ratio and Frictional Slip

Laboratory values of friction (Zoback and Healy 1984) predict that fault slip occurs when $R = 3$, where R is the ratio of the remote principal compression S_{Hmax} divided by the global vertical stress S_v. Engelder and Peacock (2001) use this stress ratio and the hinterland tilt of joints to predict the dip of bedding at the time of the onset of jointing. Their model predicts that J3 joints with a tilt of about 4° should have propagated by the time beds had been upfolded to have a dip of 3°. If in the course of Alpine inversion faults slipped under low friction, R could have been <3. For example, if $R = 1.33$ due to low-friction faults, a hinterland tilt of 4° could cause jointing in beds dipping as much as 17°. They suggest that the flexural flow model may be applied outside the immediate area of the Lilstock buttress anticline. During the extensional phase of the Bristol Channel tectonics, large blocks throughout the basin were slightly tilted (Dart et al. 1995), entering the Alpine inversion phase when already possessing a modest dip. The model allows tensile stress to be generated at dip angles smaller than 2°, particularly since the Poisson ratio of limestone is lower than that of shale. Many of the jointed strata cropping out on the north and south shores of the Bristol Channel have a gentle dip that would favor joint formation if submitted to inversion tectonics. Joints not orthogonal to the bedding have been recorded at Lavernock Point, South Wales, by Rawnsley et al. (1998).

3.2.6 Conclusions by Engelder and Peacock (2001)

Engelder and Peacock (2001) established the presence of six joint sets in the limestone beds of the Jurassic Blue Lias, numbering them J1 to J6. They all cluster about the axial strike of the 290° striking Lilstock buttress anticline, including J1 (330–310°), J2 (310–295°), J3 (295–285°), J4 (285–275°), J5 (275–265°), and J6 (<265°). In sub-horizontal beds, including those near the hinge surface of the Lilstock buttress anticline, J3 joints are rare. Abutment of joints belonging to J2, J4, J5 and J6 indicates a counter clockwise sequence of development. In the south-dipping beds of the anticline, J3 joints are most frequent in the vicinity of the reverse fault responsible for the anticline. J3 joints are absent where J1 and J2 joints are present and diminish with distance from the anticlinal hinge and higher in the stratigraphic section. J3 joints in the south-dipping beds are not perpendicular to the bedding but have poles that plunge several degrees towards the south when bedding is restored to horizontal.

According to Engelder and Peacock, the mechanism that formed the J3 joints must explain:

1. Their tilt relative to bedding on the steep south flank of the anticline.
2. Their greater frequency in the south-dipping limb as compared to the axial region.

These two characteristics strongly suggest that the J3 joints were formed during folding, but before beds reach their present tilt. These conditions are met in their shear model for interlayered limestone-shale beds during flexural flow folding (Fig. 3.19c), where local tensile stresses are generated within a regional stress field capable of driving inversion tectonics. Such tensile stresses will create through-cutting joints rather than arresting at a neutral surface associated with outer-arc extension (Fig. 3.19a). In this model, joints can open against the direction of the maximum compressive stress under conditions where external tractions are compressive. This indicates that at least some of the joints at Lilstock formed during the main Alpine inversion (cf. Rawnsley et al. 1998). It also shows that joints that are parallel to a fold hinge may form on the limbs during flexural-flow folding and not in the hinge region as a consequence of outer-arc extension during tangential longitudinal strain folding.

3.2.7 Discussion by Engelder and Peacock (2001)

3.2.7.1 *The Problem of Delayed Fracture*

Both Rawnsley et al. (1998) and Engelder and Peacock (2001) accept that jointing in the Jurassic (Blue Lias) rocks at Lilstock is associated with Late Oligocene to Miocene Alpine compression, i.e., that there was a long time interval between deposition of the limestones and their eventual fracturing. Very often single-layer joints form in sedimentary rocks during their burial stage. There must be a reason why jointing at Lilstock skipped the burial and did not develop during this long time interval. A similar question applies to the "delayed fracturing" in the Middle Eocene chalks of the Beer Sheva syncline, Israel (Bahat 1999a), where the time interval between deposition and fracturing, all within the Tertiary, was much shorter than at Lilstock Beach. At Lilstock Beach, fracturing began in the syntectonic stage (or during relaxation of tectonic loading). With few notable exceptions (e.g., Becker and Gross 1996; Weinberger 2001a) systematic joints are rare in limestone and dolomite strata in anticlines in Israel (unpublished data by Bahat). What is special about these exceptions, for instance, the Tully limestone (e.g., Younes and Engelder 1999) that contains systematic joints? Why are systematic joints uncommon in the Upper Cretaceous chalks of South England (Bahat 1991a, Sect. 5.3.3), while they are very abundant in Santonian and Eocene chalks in Israel? Research combining limestone petrology and structural geology may shed more light on these questions.

3.2.7.2
Opening of Joints Counter to Remote Compression, and Deviation from Perpendicularity to Layer Boundaries

Engelder and Peacock (2001) found that J3 joints open against the remote compressive S_{Hmax} instead of the expected opening in a plane that contains S_{Hmax}. They consider this to be a "mechanical paradox" and constructed a flexural flow model to explain it. We propose an alternative mechanism of local stress field rotation that would produce fracture patterns similar to those described by Engelder and Peacock.

Previous investigators have also reported on opening of joints against the remote compression. Dieterich and Carter (1969) investigated a numerical model of the stress field in two-dimensional, large-amplitude folding of a viscous layer in a less viscous matrix, using a modification of the finite element method. They found that large magnitudes of σ_1 act parallel to the layer throughout the early stages of folding. With increasing amplitude,

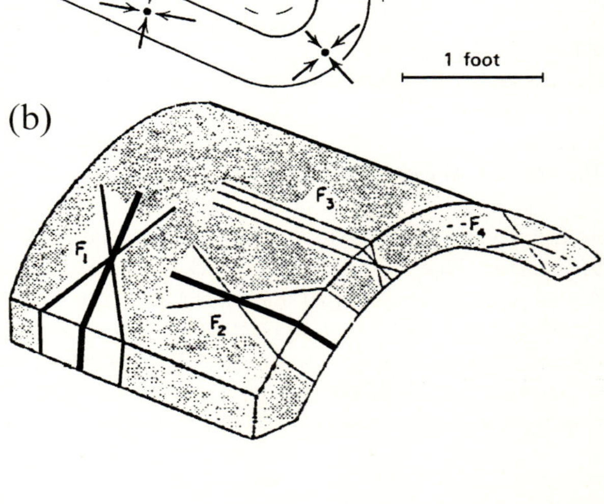

Fig. 3.20.
a A sketch showing geometry of folded quartzite layer. *Dots* indicate locations of specimens used for dynamic analysis of quartz lamellae. *Long arrows* show orientation of σ_1, and *short arrows* show orientation of σ_3 (all principal stresses are compressive and positive) (after Dieterich and Carter 1969). **b** A schematic symmetric fold, showing basic four joint systems, F1–F4. Each system ideally consists of three sets. Fractures from the systems F1 and F2 are more common than others, and in the latter two, the sets co-axial with the principal stresses are more common than the inclined sets

σ_1 decreases in magnitude along the limbs and rotates into increasingly higher inclinations to the layer. In the upper hinge area, σ_1 decays rapidly with increasing amplitude, and the deviatoric stresses parallel to the layer rapidly become low-magnitude tensile stresses. Their analyses generally show that in the limbs of open folds, σ_1 is inclined at low angles to the layer, σ_2 is parallel to the fold axis, and σ_3 is nearly normal to the layer (Fig. 3.20a). Figure 3.20a also implies that the stress axial cross may deviate from orthogonality with respect to the bedding. Thus, the fractures that may develop during the local rotation of the principal stresses need not be normal to the layer boundaries.

According to Dieterich and Carter (1969), similar results were obtained by other studies (Hansen and Borg 1962; Carter and Friedman 1965; Scott et al. 1965), which focused on intragranular flow of calcite and quartz during folding. Burger and Hamill (1976, Fig. 14A) performed petrofabric stress analysis on the Dry Creek Ridge anticline in Montana and also found rotation of principal stress during folding. Initially, the entire multilayer package was subjected to compression with σ_1 parallel to layering and orthogonal to the fold axis. As shortening continued, a neutral surface formed, dividing the hinge and upper limb areas into domains. The compression domain implies σ_1 parallel to layering and orthogonal to the fold axis, and the extension domain implies σ_3 parallel to layering and σ_1 orthogonal to layering.

Reches (1976) identified five joint sets (J1–J5) that cut two asymmetric anticlines in southern Israel, the Hathira and Hazera monoclines. None of the sets display symmetry with the fold axes, and as such they differ from the idealized joint geometry suggested by Stearns (1968) for symmetric folds (Fig. 3.20b). Four sets (J1–J4) are normal to the bedding and are thought to have developed prior to or at the beginning of the folding, and their present orientation reflects their rotation by folding. On the other hand, joint set J5, considered to be related to advanced folding, is inclined to the bedding. Its average strike is 030°, which is sub-parallel to the trend of the monoclines, parallel to the bedding and normal to the axes of the monoclines. The deviation of the five joint sets (J1–J5) of the Lilstock Beach anticline from normality to the bedding resembles that of the J5 joint set from Hathira and Hazera. Tilted joints were observed in the Santonian chalks from the "Arad syncline, Israel. These are apparently related to an intense deformation phase of the chalks (Bahat 1991a, p. 259). It appears that all the above-mentioned tilted sets in both anticlines and synclines were formed during advanced folding, while rotation of the local principal stresses underwent rotation. Hence, the local rotation of the principal stresses during advanced folding in Dieterich and Carter's model and in field observations by others point to an alternative mechanism to the flexure flow model (Fig. 3.19c) proposed by Engelder and Peacock (2001). This mechanism is a more general one and may operate in formations that do not necessarily consist of alternating lithologies (e.g., limestone and shale). It may also explain why there is no actual evidence of slip by simple shear strain, as expected in the flexure flow model within the shales at Lilstock. Finally, Engelder and Peacock (2001) argue that in their model, tensile stresses will drive joints completely through beds, and will not be arrested at a neutral surface in the outer-arc extension model (Fig. 3.19a). Chinnery (1966), on the other hand, maintains that a tensile fracture, once started, could continue to propagate even where entering a region of compression. Thus, the flexure flow model is not the only mechanism to explain the formation of set J3 at Lilstock. For a discussion on fracture in symmetric folds and in slightly folded structures, see Sect. 3.4.

3.3 Post-Uplift Exfoliation Joints and Arch Formation in Sandstone at Zion National Park, Utah, USA

3.3.1 Introduction

3.3.1.1 *General*

Exfoliation in the massive sandstones of the Colorado Plateau, and at Zion National Park in particular, has drawn the attention of many investigators (e.g., Gregory 1950; Holman 1976). Exfoliation in rocks has been debated since Gilbert (1904) proposed a connection between sheeting and the expansion of the rock body in response to unloading by erosion. Several natural mechanisms have been suggested, including contraction of magma during cooling, regional tectonic stresses and residual stresses, fluctuations of ground and surface temperature and differential weathering (Johnson 1970; Holzhausen 1989).

Various stress conditions have been proposed to explain the formation of exfoliation joints. The stress conditions close to the ground surface, particularly the high compressive stresses that parallel the exfoliation, are of great importance (Dale 1923, p. 34; Holzhausen 1989). Fracturing by residual compressive stresses (Bradley 1963; Holman 1976) develops when critical stress difference occurs between the principal stresses parallel to the surface (e.g., White 1946) and if these are sufficiently high so that additional stress will fracture the rock (Cadman 1969) (for further discussion on exfoliation, see in Holzhausen (1989). Although it is clear that exfoliation in Zion National Park generally parallels earlier fractures, it has been debated whether exfoliation joints curve to remain faithful to the new valley sides, independent of the earlier regional jointing (Bradley 1963), or they rather follow the trends of prominent earlier vertical prominent joints (Gregory 1950).

Sandstone arches in Zion National Park were considered by Gregory (1950) to be primarily the result of erosion. Cruikshank and Aydin (1994) explained the arches of Arches National Park in southeastern Utah as products of localized erosion that occurred at zones of intense fracturing due to shear on existing discontinuities. Robinson (1970) explained the formation of large arches in Zion Canyon by modeling the arch as a plate with a semicircular hole, which is subjected to uniaxial vertical compression. When the rock above the center of the arch is under tension, it would tend to fail along curvilinear stress trajectories.

Apart from scientific interest, exfoliation must be reckoned with in engineering and architecture. Exfoliation results from a rock's tendency to fracture at the appropriate stress, parallel to a previous surface, like cliffs or the topography. It is therefore an important consideration in the construction and maintenance of roads, canals, dams etc. Quarrying and pit mining deal with it daily. Carefully planned excavation, taking account of exfoliation, can overcome some acute safety problems and prevent economic losses (Niles 1871; Cadman 1969; Holzhausen 1989). Knowing a terrain's exfoliation features is essential to hydrological projects of waste disposal and remediation (e.g., Cohen 1993). A distinction is made between the terms "exfoliation" and "sheeting" (see end of chapter). Section 4.5.3 deals with post-uplift exfoliation joints in granites.

3.3.1.2
Geologic Setting

Zion National Park, a fracture province marked by a multitude of exfoliation joints and arches, is located in southwestern Utah, between the Hurricane and the Sevier fault zones (Fig. 3.21a). This area occurs within the transitional region between the Colorado Plateau in the east and the Basin and Range province in the west (Grater 1945; Kurie 1966; Stewart and Taylor 1996). Mesozoic rocks of the Colorado Plateau in southern Utah bear structural evidence of three major regional tectonic events (Davis 1999). The first de-

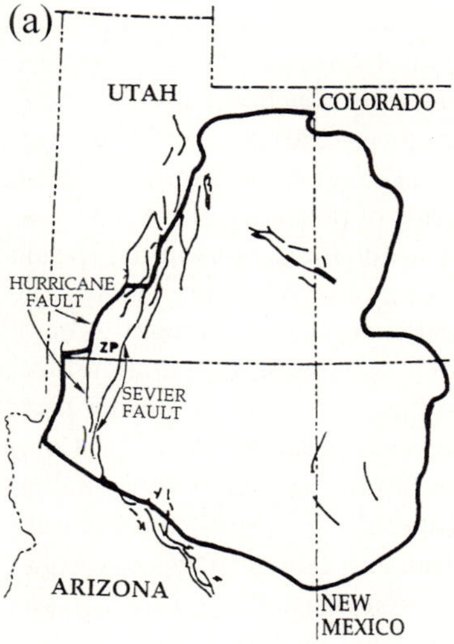

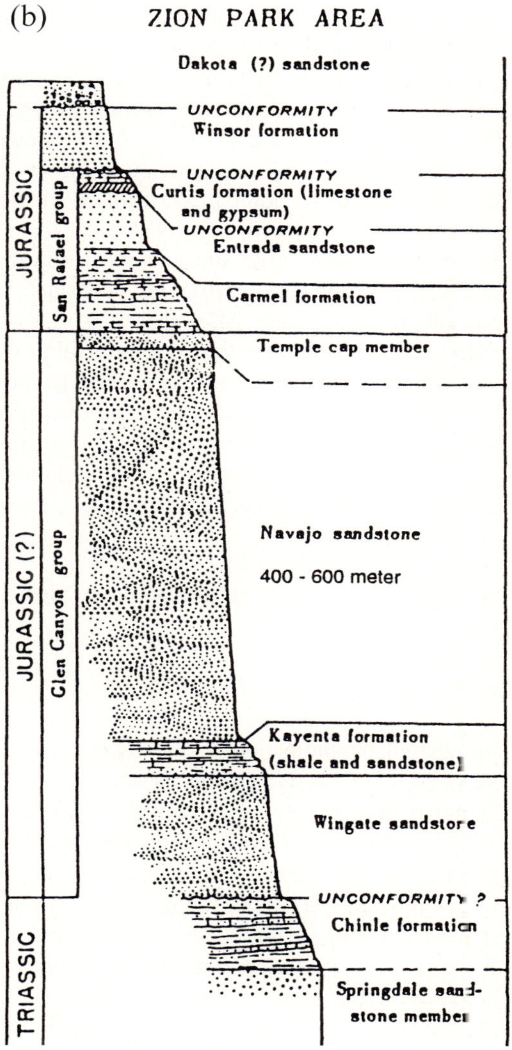

Fig. 3.21. **a** Simplified structural map of the Colorado Plateau (*thick lines*). *Thin lines* represent major faults; *ZP* designates Zion National Park (modified after Hunt 1956). **b** Stratigraphy of the Navajo sandstone in the Zion National Park area (modified from Gregory 1950)

formation was during the Laramide orogeny, from ca. 90–50 m.y., when an ENE-WSW convergence between the North America plate to the east and the Farallon plate to the west generated horizontal compression that contracted the Colorado Plateau region. The second deformation took place during the Miocene, from ca. 25 to 19 m.y., when the rate of plate-tectonic convergence had slackened from ~100 mm yr^{-1} during the Laramide orogeny to ~50 mm yr^{-1}, and the angle of subduction was becoming steeper. The third deformation was due to the Basin and Range extensional faulting, which began at about 15 m.y. and is still active today. This extension produced three major high-angle normal faults along the southwestern margin of the Colorado Plateau. They are (from west to east): The Hurricane, Sevier and the Paunsagunt faults.

Two regional joint sets striking about NW-SE and NE-SW are known from the Colorado Plateau in northeast Arizona and in southeast Utah (Hodgson 1961b). Three Neogene sets striking 335° (N 25° W), 355° and 030° were identified by Bergerat et al. (1992) in the massive sandstones of the Colorado Plateau. Gregory (1950) observed two series of regional joints striking NNW and ENE in the Zion Park region. No clear correlation can be established between the various joint sets and any of the above major deformations. The present section concentrates on a series of exfoliation joints in the Navajo sandstone associated with the NNW and ENE sets observed by Gregory (1950). The Navajo sandstone is about 400–600 m thick in the park area, and is thought to be of Middle Jurassic age (Fig. 3.21b). The rock is quartz-rich, massive, friable, fine-grained, and equigranular (Grater 1945; Gregory 1950). Various investigators (e.g., Gregory 1950; Bradley 1963) have studied the exfoliation joints in the massive sandstones of the Colorado Plateau and at Zion National Park in particular. Bradley considered the exfoliation to be of Pleistocene age. Based on studies by Bahat et al. (1995, 1999) and Grossenbacher et al. (1996), we characterize below the initiation and growth stages of exfoliation joints in relation to previous regional joints. The role of exfoliation fractures in arch formation is explored, including a mechanical model of arch formation.

3.3.2
Fracture Characteristics at Zion National Park

3.3.2.1
Exfoliation Joints at Zion National Park

Two types of exfoliation joints were discerned at Zion National Park. The first type consists of large fractures that form vertical cliffs. These joints occur mainly along straight tributaries, mostly striking NNW (Fig. 3.22). The tributaries trending NNW are subparallel to each other. Most of them run for 2 000 m or more, and many of them are deep and narrow canyons only tens of meters across with straight walls hundreds of meters high (USGS 1 : 24 000 topographic maps of the Zion Park area, 1980), displaying extensive exfoliation (Fig. 3.23a). Many narrow canyons are elevated from the Virgin River and form large stairs down to the main canyon. They belong to the more generalized regional pattern of prominent fractures that mostly strike NNW or NE at Zion National Park (e.g., Gregory 1950). The second type of exfoliation joints occur in various sizes and orientations that conform to present-day surface morphology (e.g., Fig. 2.33e). Joints of this type occur in various parts of the main canyon as well as along the tributaries.

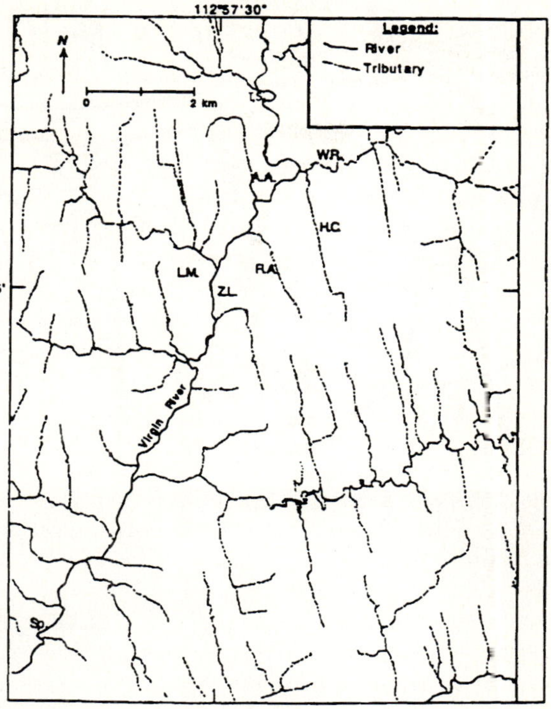

Fig. 3.22a. Tributary map in Zion National Park, Angle Arch, Red Arch, Lady Mountain, Weeping Rock, Hidden Canyon, Springdale and Zion Lodge are designated respectively, by *A.A.*, *R.A.*, *L.M.*, *W.R.*, *H.C.*, *Sp.* and *Z.L.* *Dash-lines* represent tributaries (after Bahat et al. 1995)

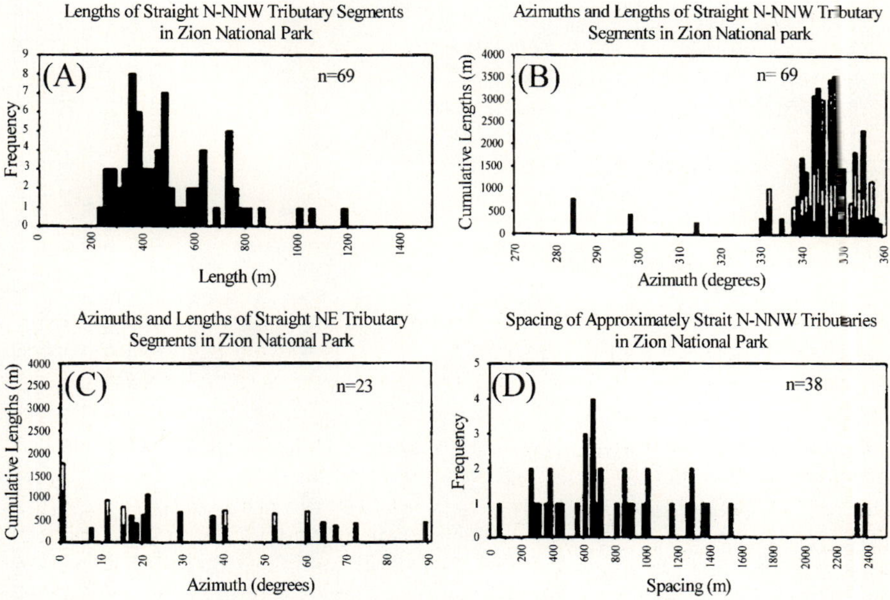

Fig. 3.22b. Histograms of parameters on exfoliation joints along approximately straight tributaries in Zion National Park. *A* Lengths along N-NNW tributaries. *B* Azimuths and cumulative lengths of N-NNW tributaries. *C* Azimuths and cumulative lengths of NE tributaries. *D* Spacing of N-NNW tributaries (after Bahat et al. 1995)

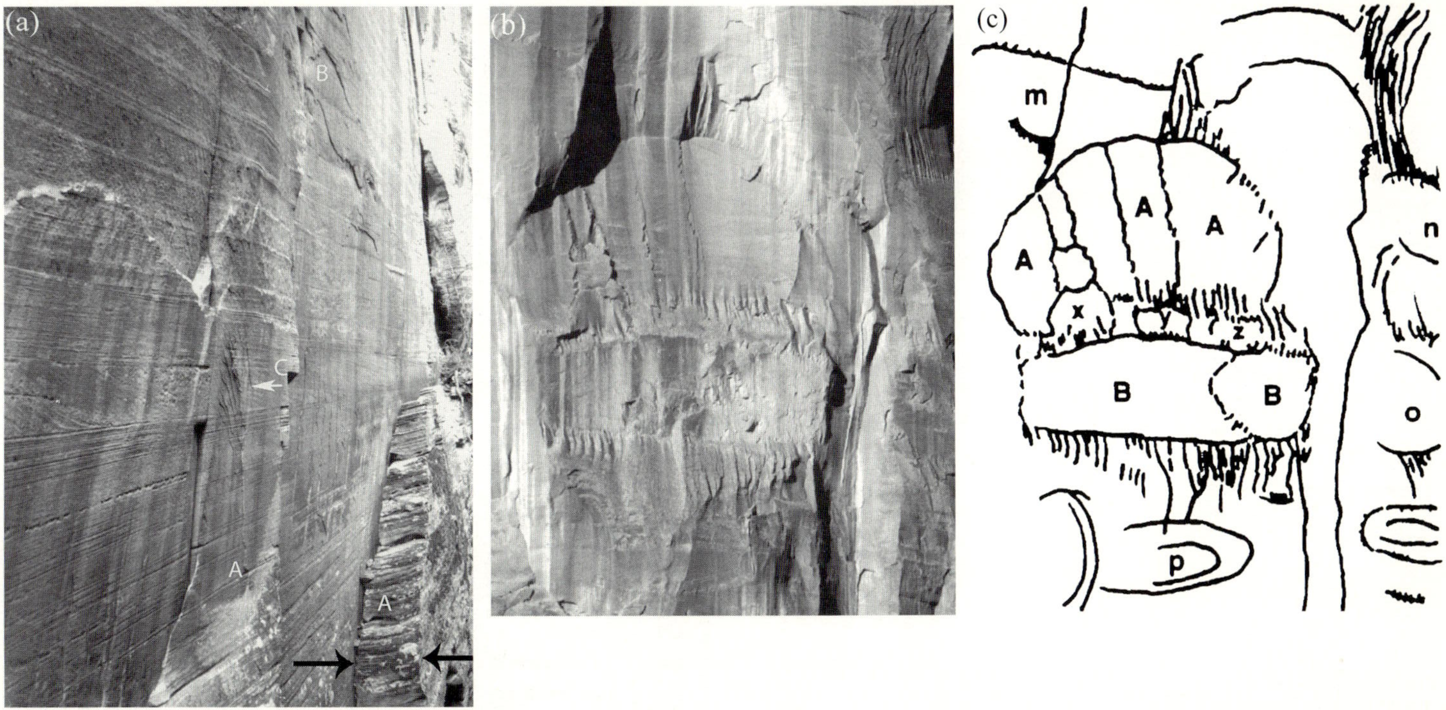

Fig. 3.23. a The lower part of a vertical NNW exfoliation joint in Hidden Canyon, showing two vertical remnants of exfoliated slices (marked *A*), several meters high and up to 1 m thick. Such slices occasionally show *en echelon* relationship. Right-stepping *en echelon* segments cut the exfoliation at upper center (*B*). At the center is another third *en echelon* array (*C*) with inclined stepping, tens of cm long and tens of mm thick, spreading upward from a single center. Note sub-horizontal bedding of the sandstone, which does not interfere with *en echelon* cracks and exfoliation. **b** Interaction of adjacent quasi circular/elliptical cracks of different size that grew independently and impinged on each other on an exfoliated joint. **c** Diagram of **b**. Fracture *A* is ornamented by several long vertical striae that arrest at the *en echelon* fringe below *A*. Fracture *B* is a combination of two smaller fractures and the *en echelon* fringe below *B*. Three smaller fractures, *x*, *y*, *z*, lie along the contact between *A* and *B*. Additional fractures around *A* and *B* are marked by letters *m*–*p*. The height of *ellipse B* is on the order of 4 m (after Bahat et al. 1995)

3.3.2.2
Circular/Elliptical Cracks and Composite Fractures

Joints of the first exfoliation type are often marked by quasi-elliptical or quasi-circular cracks (QCC), which mostly range from one meter to several tens of meters in diameter, occasionally interacting with each other by partial penetration (Fig. 3.23b,c). They grow approximately coplanar to each other and often merge in a variety of styles, producing composite fractures (Bankwitz and Bankwitz 1984; Helgeson and Aydin 1991; Bahat 1991a, p. 268). Some composite fractures form either partial or full ellipses or circles, occasionally rimmed by *en echelon* fringes (Fig. 3.24a,b).

Ripple marks. Ripple marks are of two types: Undulations and arrest marks (Sect. 2.2.4.2). Occasionally, undulations occur on large exfoliation joints unrelated to QCC. A spectacular series of more than forty undulations (Fig. 3.24c) occurs on the western wall of the "Hidden Canyon" (Fig. 3.22a), where they cover a cliff surface tens of meters wide, and about as high. This feature marks the propagation of a large, concentric, single-fracture front that is convex to the north. The undulations show that the joint grew horizontally from south to north. They also show how rock anisotropy influences fracture propagation: The fracture cuts across bedding, but one boundary was resistant, and refracted the fracture front into another direction. Arrest marks propagate in all directions but seem to advance particularly downward (Fig. 3.24d). Arrest marks also occur on fractured side-walls of arches (Fig. 3.24b).

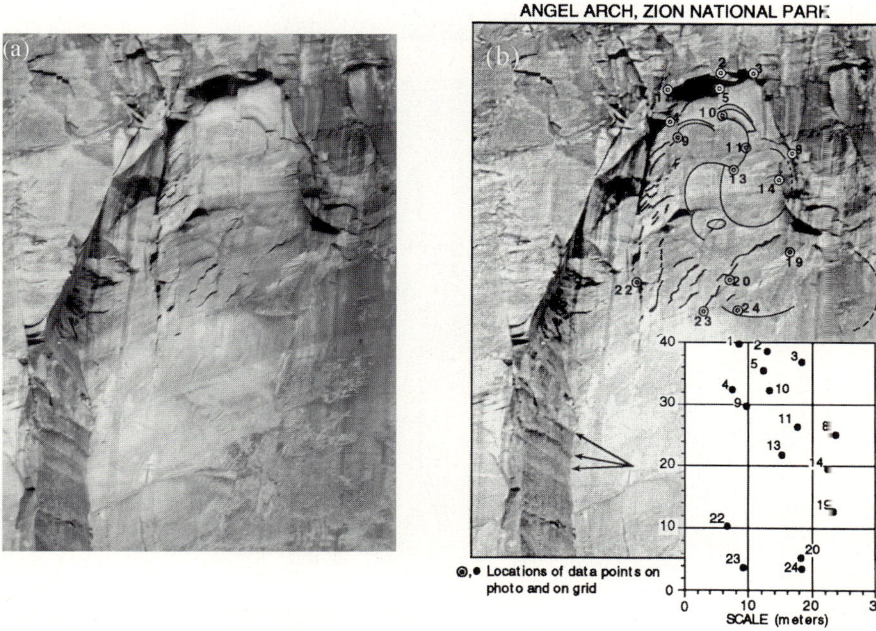

Fig. 3.24a,b. **a** The Angel Landing Arch. **b** Scaled photograph of the same arch, showing detailed measurements. Many circular/elliptical cracks are surrounded at left by a fringe of *en echelon* radial cracks. Note vertical discontinuity (at left of *points 1* and *4*), and arrest marks on left side-wall (*three arrows*). (after Bahat et al. 1995)

Fig. 3.24c,d. c A central part of a nearly planar, vertical NNW-striking exfoliation joint. Undulations several tens of meter long convex towards right (north). The fracture front is interpreted to have changed its propagation direction at a resistant bedding boundary (*inclined arrow*). The occurrence of two superposing exfoliation joints with similar ripple marks is shown by two sub-horizontal arrows. Letters *A–D* relate to locations of changes in fracture velocity. **d** Different characteristics of ripple marks, undulations – *vertical arrow*, and arrest marks – *inclined arrow* (Sect. 2.2.4.2) (after Bahat et al. 1995)

Radial plumes (striae). Quasi-circular cracks may be decorated by radial striae that indicate the sites of fracture initiation and illustrate the fracture growth mode. Occasionally, radial striae decorate a niche (see below) and constitute the only part remaining of the QCC. Striae are not common. They occasionally occur either as straight individuals, often in radial setting, or as straight strands bent in places (Fig. 3.25a,b).

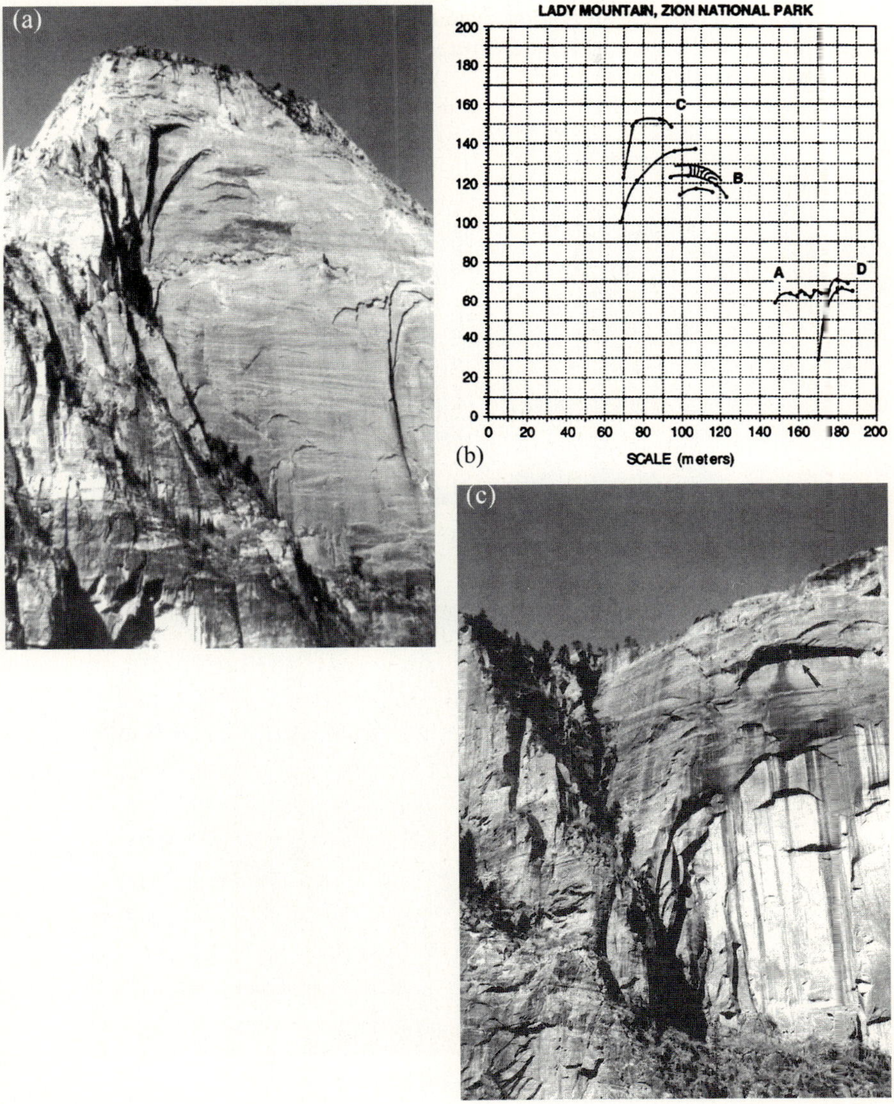

Fig. 3.25. a A series of large *en echelon* cracks on a NE striking cliff at Lady Mountain. Note previous exfoliation slices at left. **b** Scale diagram of **a**, showing crack dimensions. *A* and *B* are series of upward convex niches. Radial striae are marked at *B*. *C* and *D* are two sub-horizontal, straight niches formed by a series of *en echelon* fractures, propagating leftward and producing partial arches. **c** Exfoliations on a cliff, exhibiting a series of small niches (at center), with a large niche or a small arch about 20 m long above (*arrow*). A series of partial arches is seen at lower-center of picture (after Bahat et al. 1995)

3.3.2.3
En Echelon *Cracks, Niches and Arches*

***En echelon* cracks.** Many fractures deviate a few degrees from the cliff plane, and appear as *en echelon* segments, ranging in size from tens of cm to tens of meters. The *en echelon* segments occur in various shapes and orientations. Occasionally three or more *en echelon* sets of different sizes and orientations occur in close proximity (Fig. 3.23a). Other cliffs exhibit large *en echelon* fractures next to niches (Fig. 3.25a,b) or along the mirror boundary (Fig. 3.24a,b).

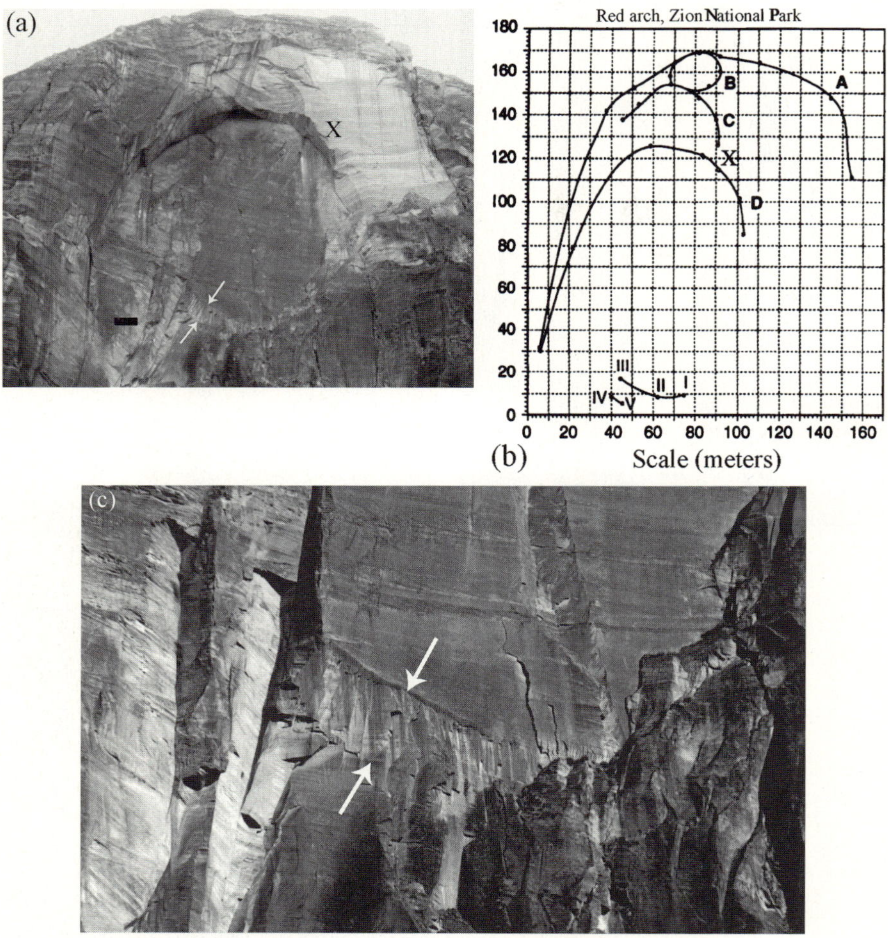

Fig. 3.26. a The Red Arch, showing sharp fracture contours in the upper parts (*x*). Fringe of *en echelon* segments (two *arrows*) marks the lower boundary of the quasi-circular arch and at lower-left side some remnants of exfoliated slices. *Scale bar* is about 15 m. **b** Scale diagram, showing measured dimensions of **a**. *A*, *B* and *C* contours describe earlier large and small "attempts" of arches, while *D* follows the contours of the main, Red Arch quasi-circular contours, surrounding the large, flat back-wall at the center. This back-wall is a mirror plane. Numbers *I*, *II* and *III* mark the upper boundary of the lower fringe of the mirror, and *IV* and *V* occur on the lower boundary of the fringe. *x* in **a** and **b** depict the same spots. **c** The fringe in close-up (after Bahat et al. 1995)

Niches. Occasionally, individual QCCs or composite fractures cut into the rock and deviate several degrees from the cliff face. At these localities, receding bays formed in the cliff are termed niches (Fig. 3.25a–c). The niches commonly appear in crescent shapes that concave downward (distinguished in the outcrop by their shades). Occasionally, they tend to be straight. Sizes vary from several meters to several tens of meters. Two types of niches, A and B, are distinguished. Niches of type A occur close together at about the same topographic elevation (Fig. 3.25b). Type B niches are grouped one above the other. Adjacent niches in an outcrop may show shape and size transitions to "small arches" (Fig. 3.25c). The majority of QCCs, composite fractures, *en echelon* segments and niches occur within the middle third of the cliff face (Fig. 3.24a,b, 3.25a,b, 3.26a,b and 3.27).

Fig. 3.27. Side-wall of a partial arch with convex ripple marks indicating downward fracture propagation. The upper mark coincides with a bedding boundary, which influenced its location (below *arrow*). See a rappelling person (*arrow*) for scale (after Bahat et al. 1995)

Arches. Arches are found by the hundred in Zion National Park in a variety of sizes (Table 3.2) and shapes (Fig. 3.26–3.28). Arch sizes range from several meters to many tens of meters. Some are full arches (Fig. 3.26a,b and 3.28a), either symmetric or asymmetric (Fig. 3.28a); others are incomplete arches (Fig. 3.27 and 3.28b). A full arch exhibits a ceiling and two side walls. Height is measured from base to the highest point on the ceiling, and the width is measured between opposite side walls at the base. In some arches, width is larger than the height (Fig. 3.28a), in others this relationship is reversed (Fig. 3.24a,b). Sub-circular arches (Fig. 3.26a) are rare. A partial arch consists of a side wall and a fraction of ceiling (Fig. 3.27 and 3.28). Many ceilings are not fully concave, but partly coincide with straight, sub-horizontal niches or other bedding oriented boundaries (right side of Fig. 3.24a and 3.32b). Occasionally vertical fractures cut ceilings of full arches or form boundaries of partial arches (left side of Fig. 3.24b).

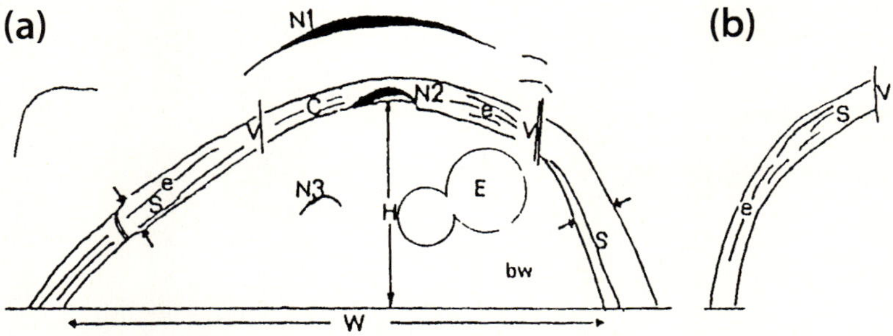

Fig. 3.28. **a** Diagram showing elements of an asymmetric arch as follows. Side-walls *S*, ceiling *C* of the arch (shown between two *arrows* that indicate thickness on both sides of the arch), the vertical fracture *V* cuts the arch at the ceiling, and exfoliations e divide the rock into slices. The height of the arch *H*, and the width, *W*. *N1*, *N2* and *N3* are niches associated respectively, with external exfoliation, at the ceiling, and cutting the back-wall of the arch. Two QCC, marked *E*, occur on back-wall, *bw*. For simplicity, niches are mostly crescent and not straight. **b** Diagram of a side-wall of a partial arch

Table 3.2. Measured parameters of arches in Zion National Park

Parameter	Angel Landing Arch (Fig. 3.24a,b)	Lady Mountain (Fig. 3.25a,b)	Red Arch (Fig. 3.26a,b)
Strike of exfoliation joint (°)	334	352	042
Dip of exfoliation joint (°)	87 NE	82 NE	67 NW
Arch			
Height (m)	37		56
Width (m)	28		86
Angular partial arch			
Height (m)		50	
Width (m)		20	
Niche			
Height (m)		6	
Width (m)		15	
Elevation of base station (m)[a]	1467	1480	1440
Elevation above sea level (m)[b]	1504	1542	1492

[a] Approximately at the foot of measured cliff (m).
[b] At zero line of grids (from Bahat et al. 1995).

We interpret the external exfoliation above the arch (N1 in Fig. 3.28a) to be a relict of an earlier cliff face. N3 marks the back wall of the arch, which is the youngest, internal exfoliation joint (Bahat et al. 1999) (Fig. 3.26a and 3.28a). Side walls and ceilings of arches penetrate into the cliff, perpendicular to the external and internal exfoliations and connect them. Side walls and ceilings of arches are generally angular toward the external exfoliation, indicating that the side-walls and ceilings were formed by brittle fracture rather than by erosion. Also, concentric arrest marks on side-walls (Fig. 3.24b, 3.27) are interpreted to indicate that the arches were formed as mode I fractures (Bahat 1979a). The lower parts of arches have various shapes. In some arches, the side walls reach close to the ground surface, whereas in others (Fig. 3.26a) they merge into the cliff. The Red Arch is a sub-circular arch, and the lower boundary of the mirror plane (that coincides with the back-wall) is marked by a concave-up *en echelon* fringe (Fig. 3.26a–c). The rock-slice that separates between the external and internal exfoliation joints has remained partly in place, closing the lower left side of the arch.

3.3.3 Fracture Mechanisms at Zion National Park

3.3.3.1 *Circularity of QCCs on Principal Planes*

When tensile stress operates normal to an elliptical crack, it tends to grow into a circular fracture because the stress intensity K_I is greater at the end of the minor axis of the ellipse. This trend toward greater circularity is apparently the cause of the common occurrence of QCCs and some composite circular fractures (Fig. 3.23c and 3.24). The planarity of both isolated and linked QCCs is consistent with mode I fracturing, with the fracture planes essentially normal to the least compressive stress.

3.3.3.2 *Fractography of Exfoliation Joints*

The QCCs at Zion National Park resemble the fan-shaped early cracks of the El Capitan granite at Yosemite National Park, which merge into exfoliation joints (Fig. 4.27). The merging fracture styles at the two localities in different rock types is characteristic of post-uplift joints and was first described as such from Eocene rock strata near Beer Sheva (Bahat 1991a, p. 301). The merging style is unknown on surfaces of earlier burial, syntectonic or on uplift regional joints (Bahat et al. 1999). Hence, one good criterion for distinguishing exfoliation joints from other joint types cutting sandstone is their specific fractography of QCC merging (Fig. 3.23b,c and 3.24b).

3.3.3.3 *The Growth of Quasi Elliptic Cracks to Composite Exfoliation Fractures*

The growth of QCCs into composite exfoliation fractures proceeds through four steps (Fig. 3.29). They begin as QCCs (as shown in Fig. 3.23b,c and upper part of Fig. 3.29a), or as early fans in granodiorite of El Capitan at Yosemite National Park (Fig. 4.27 and

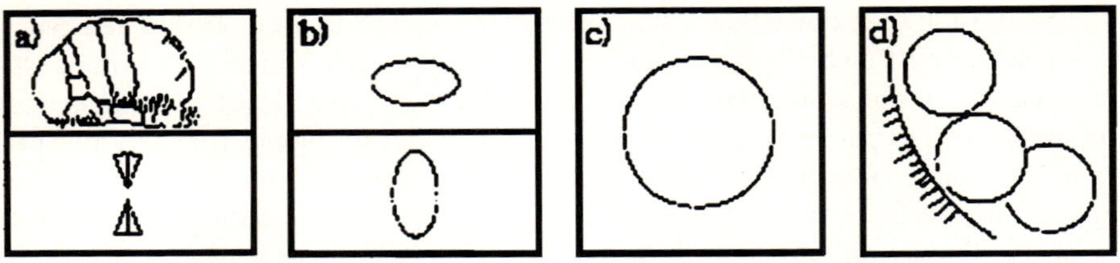

Fig. 3.29a–d. The development of composite fractures from early cracks (see text for explanation)

lower part of Fig. 3.29a). These cracks grow into elliptic fractures, with their long diameter either horizontal or vertical (Fig. 3.29b). The elliptic fracture grows into a circular shape (Fig. 3.29c) by the mechanism described above (Sect. 3.3.3.1). Finally, the fractures grow into a composite exfoliation joint (e.g., Fig. 3.24b and Fig. 3.29d), where they may attain critical conditions (around K_{IC}, Fig. 4.34 and 4.35), leading to the development of an *en echelon* fringe (see lower left side of composite in Fig. 3.26). The latter is also seen at Half Dome at Yosemite National Park (Bahat et al. 1999) (Fig. 6.18).

3.3.3.4
The Evolution of Arches through Niche-Forming Exfoliation

We favor Robinson's (1970) explanation of arch formation by a mechanical process. Figure 3.30 shows the stress trajectories calculated around a circular hole in a plate of elastic rock subjected to uniaxial compression (Means 1979, p. 118). The σ_1 stress trajectories surrounding the hole resemble magnetic force lines around a "circular" magnet or hydraulic streams around a circular obstacle. When a vertical plate with a circular hole is subjected to a biaxial stress with vertical maximum compression σ_1 exerted by the weight of the overlying rock, the resulting strain may lead to an arch. In an isotropic rock, tensile stress will concentrate on point A at the apex of the circular hole, resulting in a vertical fracture normal to the stress trajectories of σ_3 and to the cliff. However, on many full arches (Fig. 3.24) and partial arches (Fig. 3.27) at Zion National Park, vertical fractures (Fig. 3.28a,b) occur at locations like B and C, which are not at the exact apex of a circular hole (Fig. 3.30). Due to the imperfect isotropy of natural rock, arch ceilings are often not rounded at their tops, but are partly shaped by straight niches that follow essentially horizontal sedimentary bedding. Such influence of bedding on fracture growth is seen in many outcrops: Undulations deviate from smooth curvature in Fig. 3.24c, the location of at least one arrest mark coincides with bedding in Fig. 3.27, and niches that follow bedding are shown in Fig. 3.25a. That the modification of arch curvature is caused by unequal strength of the rock is supported by experiment. Robinson (1970) found four values for tensile strength of Navajo sandstone cores determined by uniaxial tension tests. The values found for two cores parallel to the bedding were 3.0 MPa and 1.2 MPa, while values for two cores normal to the bedding were significantly lower, 0.5 MPa and 1.0 MPa. This shows that bedded sandstone is not isotropic. Resistance to fracturing is greater across the bedding than parallel to it.

Arch formation proceeds through several stages of tensile fracture (Fig. 3.31a–d). Vertical fractures (Fig. 3.28a,b) form in zones of tensile stress, which is usually concen-

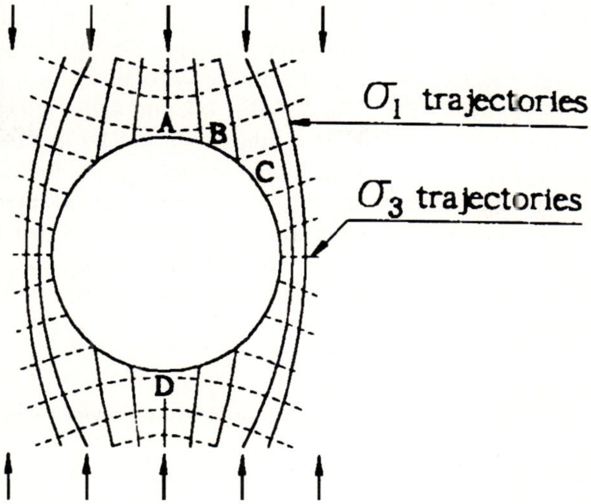

Fig. 3.30. Stress trajectories near a circular hole in a plate subjected to biaxial compression. *A*, *B* and *C* are possible locations of tensile stress concentration (modified after Means 1987) (see text for explanation)

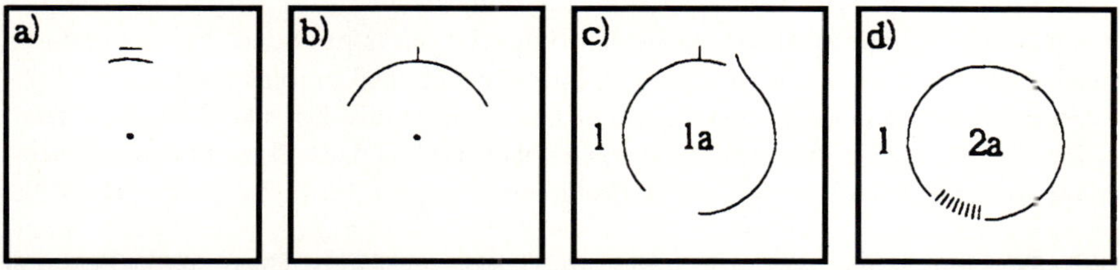

Fig. 3.31. The evolution of an arch from initial niches. **a** Two incipient niches. **b** A vertical fracture forms in the apical part of a growing niche. **c** Vertical fractures grow at the apex and off apex, in analogy to *A* and *B* from Fig. 3.30. **d** The two curves meet and create an arch circle. The area *2a* can form a mirror, rimmed by a fringe, like the back-wall of the Red Arch

trated at apices of niches or along their sides at their lateral continuations (Fig. 3.31a,b). The vertical fractures propagate downward along the curvilinear stress trajectories σ_1 (Fig. 3.30) almost completely separating area 1a from area 1 on the cliff (Fig. 3.31c). This curvilinearity varies in shape between the quasi-circular geometry of the Red Arch (Fig. 3.26) and the more elongated shape of the Angle Arch (Fig. 3.24), depending on local conditions. Such a fracture would also propagate into the interior of the rock, perpendicular to the cliff surface, even where entering a region of compression (Chinnery 1966). The deeper it propagates into the rock and the more exfoliation joints it traverses, the thicker will become the arch. The thickness of the arch is therefore independent of the depth of the initial niche. Arch thickness is also independent of niche shape, whether semi-circular or straight, because the fracture follows the σ_1 stress trajectories. The ceiling, the side walls and the back wall are fractures, which together with the cliff face separate the arched block from the surrounding sandstone, leading to its collapse by gravity. An arch like the Red Arch (Fig. 3.26a,b) is formed when a rock slice, 2a, is detached from the inner exfoliation and collapses (Fig. 3.31d). This description means that arches may be formed entirely by fracturing, and undercutting by erosion is not required in these cases.

The common occurrence of arches in anisotropic sandstone formations in Zion Canyon and in other provinces (e.g., Fig. 3.32a,b) and their relative rarity in isotropic granitic rocks (Sect. 4.5.3.5) favor the role of rock anisotropy in arch formation by tensile fracturing (Fig. 3.31). Cruikshank and Aydin (1994) explained arch formation in Arches National Park in southeastern Utah by localized erosion along zones of intense shear fracturing. Such fracture zones are not found near arches at Zion Canyon, so that this mechanism probably does not apply to these arches. We did not identify the Zion exfoliation joints and other arch characteristics on the two arches found at Timna in south Israel. These arches are poorly shaped compared with many arches at Zion Canyon, probably because the Timna arches were formed by erosion and not by fracturing.

3.3.3.5
Parallelism between Exfoliation Joints and Previous Regional Joints

The position of the QCCs at Zion Canyon strongly resembles that of the fan-shaped early cracks which appear about halfway up the cliff of El Capitan in Yosemite National Parks (Bahat et al. 1999 and Fig. 4.27). This suggests also similarity in the fracture mechanism. At both localities, earlier cliff-parallel joints provided the free surfaces needed for buckling and longitudinal splitting, which lead to joint exfoliation (Holzhausen and Johnson 1979). Exfoliation at Zion Canyon displays two directions, NW-NNW and NE, occasionally producing square blocks (Fig. 3.32d). These directions correspond to trends of earlier, regional, vertical joints (Hodgson 1961b; Bergerat et al. 1992).

Fig. 3.32a. Arch from thick Paleozoic sandstone, near Petra, south Jordan

Fig. 3.32b,c.
Arches from: **b** An Arch and niches from thick Jurassic sandstone, in Chapada dos Guimaraes, about 50 km NEE of Cuiaba, central Brazil. Note approximately straight ceilings that conform to bedding; **c** Sub-orthogonal jointing at Zion National Park. The surfaces marked *A* and *B* strike NE-ENE and NNW, respectively. Note arrest marks that convex downward (*arrows*)

3.3.4
Discrimination between Sheeting and Exfoliation

The terms exfoliation (Cadman 1969) and sheeting (Holman 1976; Twidale et al. 1996) have been used interchangeably by many authors. Both sheets and exfoliation fractures reflect fracture propagation in principal planes, normal to the minimum principal stress. However, they belong to genetically and morphologically different categories. Citing Gilbert (1904), who believed that sheeting was older than the topography, we can add that exfoliation follows topography. Therefore, sheeting and exfoliation correspond, respectively, to uplift and post-uplift fractures.

The term sheeting should be applied to series of horizontal and sub-horizontal fractures that maintain approximately uniform spacing but generally with a trend to increase their spacing with depth (Johnson 1970, p. 356). Various authors (including Dale 1923 and Holzhausen 1989) distinguished "generations" of sheet fractures (e.g., preglacial and postglacial). Holman (1976) further distinguished two to four secondary sheets or slabs. Holzhausen (1989) notes that the sheet fractures must be products of states of stress that are approximately uniform over distances of at least tens to hundreds of meters in order to explain the regularity and size of the fractures.

Exfoliation joints, on the other hand, are generated by fracturing parallel to existing surfaces at various orientations, particularly vertical. These joints generally start to develop at a short distance (tens of centimeters), from the exposed surface and form closely spaced sets in small spacing (e.g., Fig. 4.29), producing rock slices that vary from prismatic to lenticular. Thus, whereas sheeting responds to unloading of the overburden, exfoliation results from loading of the overburden.

3.3.5
Summary

The joints at Zion National Park, USA initiated as circular cracks and grew to dimensions as great as several tens of meters. Interaction between these cracks produced composite exfoliation joints tens to hundreds of meters across. Large concentric undulations on some of the large exfoliation fractures indicate a single rather than a composite origin. Generally, exfoliation joints develop parallel and adjacent to joints formed by earlier tectonics. Where the exfoliation joints bend into the cliff, niches are formed, often having crescent shapes that concave downward. Such niches are sites of stress concentration that leads to arch formation. However, the essentially stratified structure of many sandstones causes horizontal extension of the niches, displacing arch formation laterally from the point of stress concentration at the original apex of the niche (Fig. 3.30). Thus, an arch-bounding joint often starts tangent or vertical to a niche and propagates downward, creating a curved surface that approximates the trajectory of σ_1 that is induced by the weight of the overlying rock. The propagating curviplanar fracture also cuts into the rock body, traversing parallel joints that divide the rock into slices parallel to the cliff. Fracture-bounded blocks that detach and fall from the rock body leave an arch-shaped hollow.

3.4 Jointing and Faulting in Eocene Chalk Formations in the Beer Sheva Syncline, Israel

3.4.1 The Beer Sheva Syncline

The Beer Sheva syncline is an asymmetrical fault-fold basin within the Syrian Arc (Krenkel 1924), a sigmoid fold system that stretches from Syria in the north, through Israel to Egypt in the south (Fig. 3.33a). The Beer Sheva syncline, approximately 100 km long, is oriented about 050° (Fig. 3.33b,c). The width of the syncline is not well defined on its western side due to a Cenozoic cover, but is estimated to be about 35 km or less. The structural geology of the Beer Sheva syncline, particularly its northern part, which is rich in Eocene chalk outcrops, was studied by Bahat (1985–2001). Research was also carried out in the Shephela syncline north of Beer Sheva (Fig. 3.33b,c). Fractures, i.e., joints and faults associated with joints, were investigated in two formations: The Mor Formation and the Horsha Formation, lower and lower-middle Eocene, respectively, each of them about 100 m thick. The present account uses data from about one hundred stations (Fig. 3.33d), quoting extensively publications by Bahat.

The Lower (or Early) Eocene Mor Formation, also termed Adulam (Fig. 3.34), is exposed along the peripheries of the syncline. It consists of thin (40 to 90 cm) chalk layers alternating with beds of chert nodules, up to 10 cm thick (Bahat and Grossmann 1988). Two types of joints are common: Single-layer (s.l.) joints and multilayer (m.l.) joints, i.e., that cut through many layers (Bahat 1988a). The cross-fold s.l. joints of set 328° (normal and sub-normal to the synclinal axis) arrest at the boundaries of the chalk layers with chert beds. At one locality, a normal fault that displaces these joints contains in its fracture zone fragments of chert that were broken during early offsetting of the fault, and unfractured nodules that cumulated after fault movement. The solidification of chert (Knauth and Lowe 1978) is associated with the burial phase that includes sedimentation, downwarping and diagenesis, all pre-dating syntectonic deformation and uplift. Chert does not occur in the overlying Middle Eocene strata (Horsha Formation, also termed Maresha). Therefore, both the cross-fold joints and the fault must have evolved before the Middle Eocene, under a tectonic regime that prevailed in the Lower Eocene. The strike-parallel s.l. joints, oriented about 059°, were formed partly intermittently with the cross-fold joints (Bahat 1991a, Fig. 4.8) and partly at a later stage (Gross et al. 1997).

The Horsha Formation, which is exposed at the center of the syncline (Fig. 3.34), differs from the Mor Formation by the absence of chert beds. Single-layer joints in both formations show a general similarity, with several exceptions. Joints of the Horsha Formation vary considerably in orientation, plumes on the joint surface are generally more delicate (Bahat 1987a) and *en echelon* fringes are common (Bahat 1985a). In addition, joints of the Horsha Formation show indications of "delayed fracture" (Bahat 1991a, p. 245, 1997).

The patterns of fold-joint relationships of the Beer Sheva syncline are considerably more diversified than those described in the literature, although they doubtlessly are to be found in other locations also. The detailed planning of environmental engi-

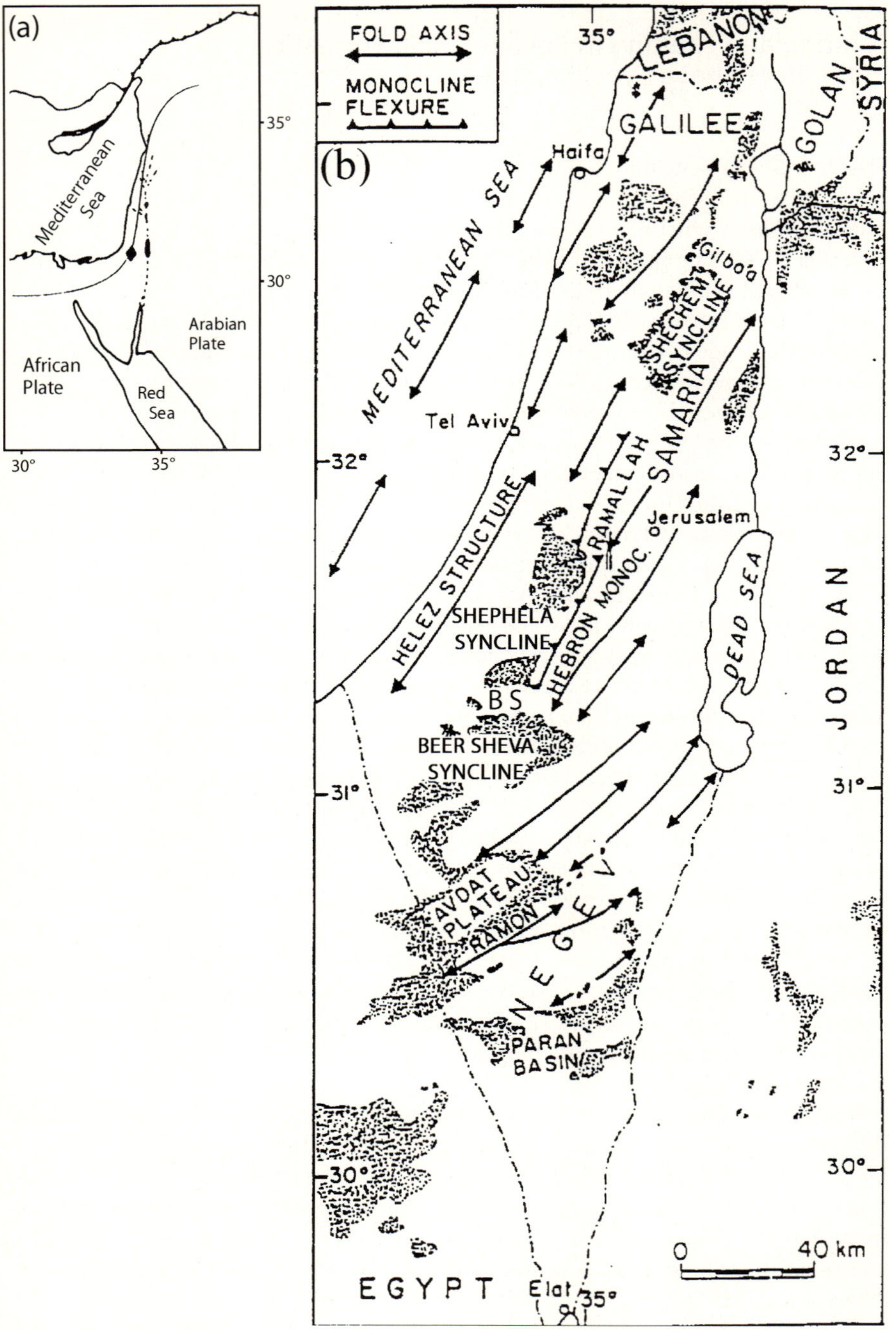

Fig. 3.33a,b. a Location map. The N-S sigmoid trace delineates the Syrian Arc, and the rhomb marks the investigated area in Fig. 3.33b–d. **b** Regional map, showing Eocene outcrops (gray) and major structures of the Syrian Arc system in Israel and adjacent countries. The city of Beer Sheva is marked BS (modified after Buchbinder et al. 1988)

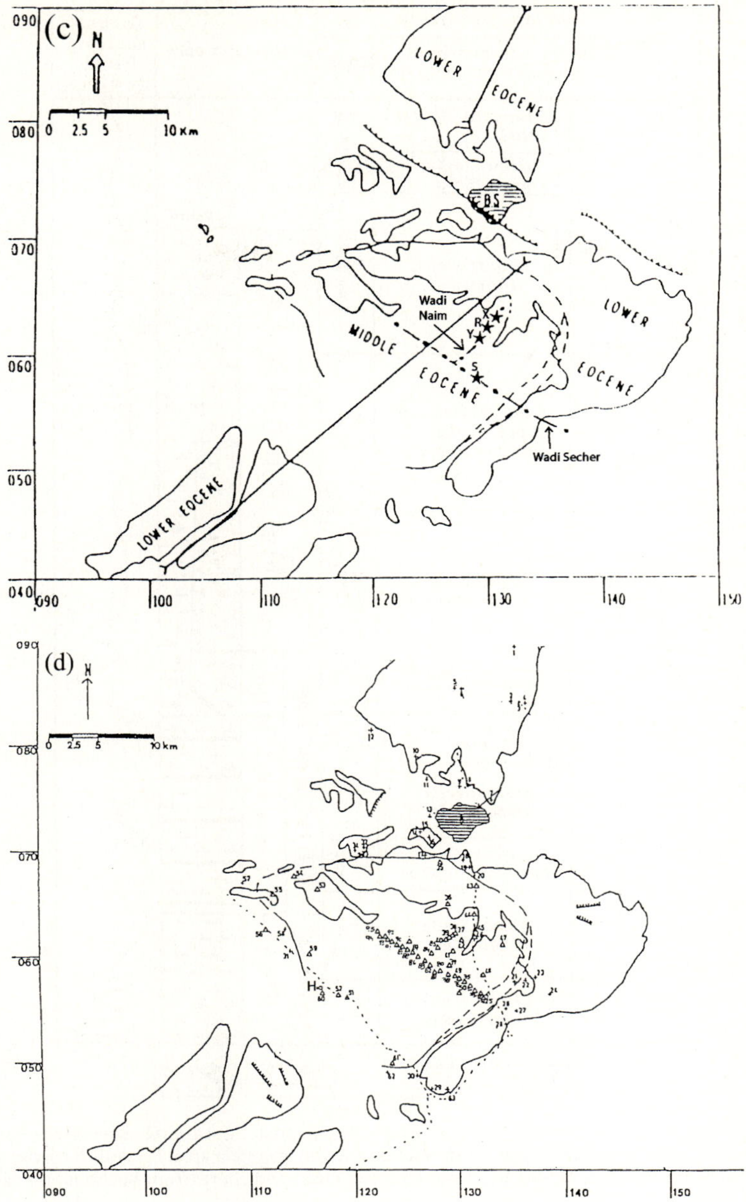

Fig. 3.33c,d. c The city of Beer Sheva (*BS*) is located at the center, between the southern tip of the Shephela syncline in the north and the northern part of the Beer Sheva syncline in the south, axes symbolized by straight lines. *Curved solid lines* are boundaries of exposed Lower Eocene rocks (Mor Formation) after Bentor et al. (1970) and *dashed lines* are approximate boundaries between the Lower and Middle Eocene (Horsha Formation). *Barbed lines* are inferred faults south of Beer Sheva, after Gvirtzman (1969). See references to stations *X*, *Y*, *R* and *S* in the text. **d** Station-map. Stations designated "+" are in the Mor Formation and those marked by "Δ" are in the Horsha Formation. The Δ stations from 70 to 95 are located along Wadi Secher (Fig. 3.37). Stations 37–40 are on outcrops along Wadi Naim. Maps are marked by the Israeli grid (modified after Bahat and Grossmann 1988) (modified after Bahat and Grossmann 1988)

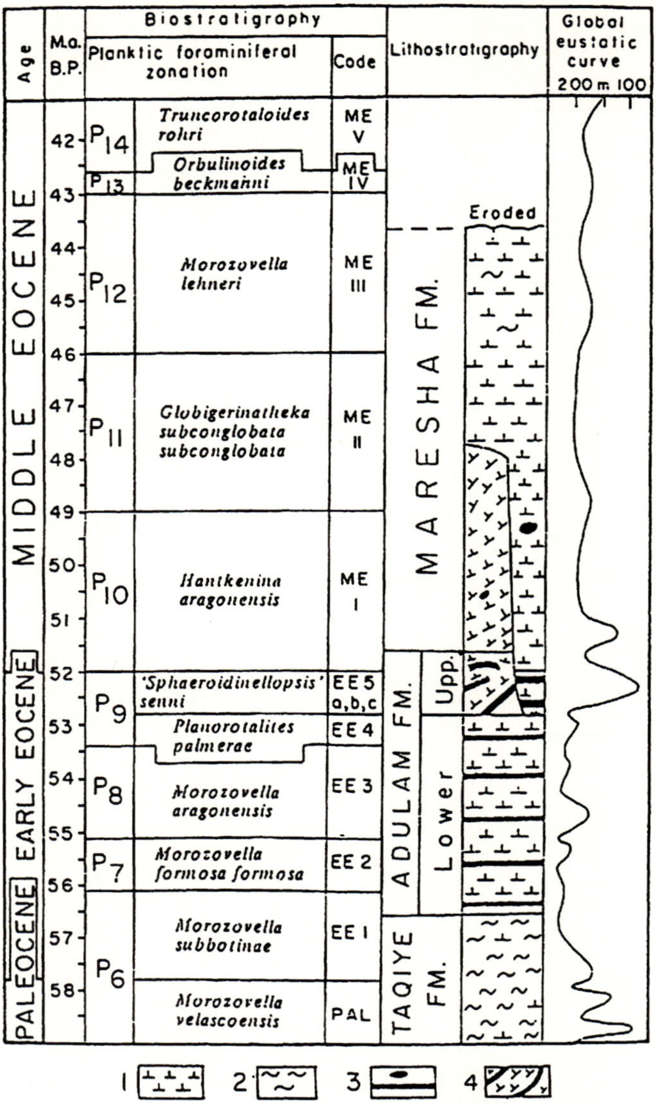

Fig. 3.34. Lithostratigraphy, biostratigraphy, time scale and eustatic sea-level curve of the Early (approximately Lower) to Middle Eocene with reference to the Shephela area. Legend: *1:* Chalk; *2:* Marl; *3:* Chert beds or nodules; *4:* Mass-transported units of chalk and chert (from Buchbinder et al. 1988)

neering projects requires basic data on the whole spectrum of structural relationships. The objective of this section is to synthesize the new results with previous findings and to report on the diversity of fracture in the Beer Sheva syncline. Also, the problems of joint set classification will be considered in some cases, and the various factors that could have contributed to the fracture complexity in this fold will be discussed. The methods of investigation that were applied in the investigation of the syncline are presented in Sect. 6.2, in connection with additional aspects of this topic.

3.4.2
Fractures in the Beer Sheva Syncline

3.4.2.1
General

Folding of the syncline is slight, and joints are mostly vertical or near-vertical. Twenty three fracture sets were distinguished in the Beer Sheva syncline, which are arranged in sixteen groups (Table 3.3). Each group commonly represents a set, except groups 1, 4, 7 and 11 that consist of two or more sets, which have specific genetic affinities. The groups are assembled in four rock units (I–IV in Table 3.3) on the basis of stratigraphy, lithology (Bentor and Vroman 1960; Braun et al. 1977; Buchbinder et al. 1988) and fracture appearance (Fig. 3.35a). The joints in the Mor Formation are represented by unit I, while units II, III and IV represent the joints of the Horsha Formation. Unit II is composed of thick layers (≥5 m), while units I, III and IV are composed of thin layers (≤1 m).

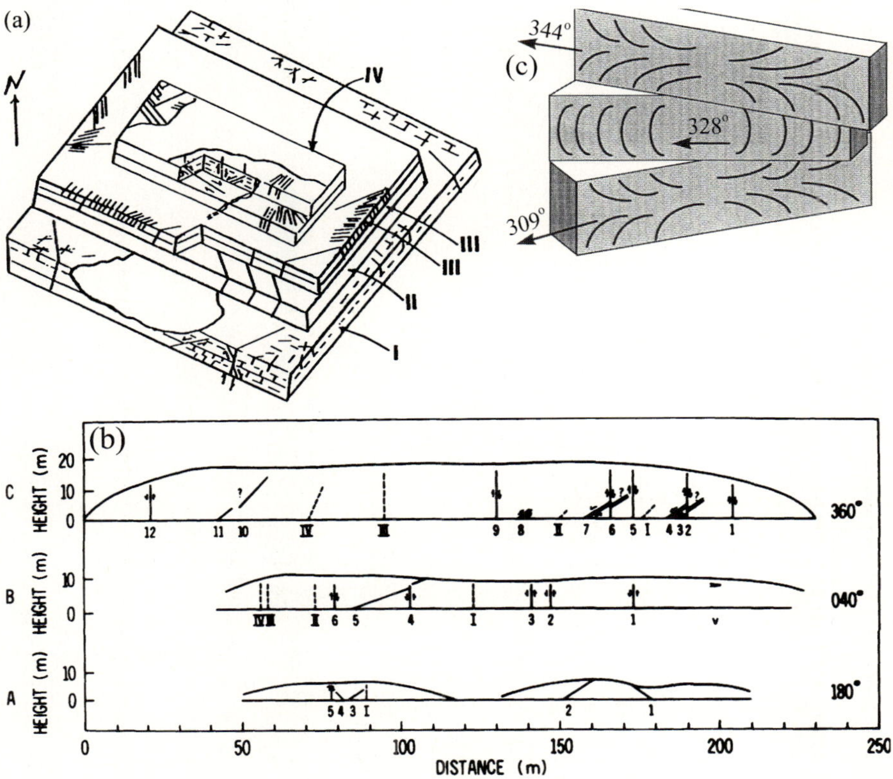

Fig. 3.35. a Diagrammatic (not to scale) illustration of the various fracture elements which are distinguishable in the Mor and Horsha Formations in the Beer Sheva syncline. The diagram consists of four stacked rock units (I–IV), see text for explanation. **b** Three sections from outcrops in unit I showing fracture relationships: Multilayer joints (*dashed lines*), vertical faults (*solid vertical lines*) and normal faults (*inclined lines*); the latter are emphasized where displaced by vertical faults. Numbers at the right indicate strike of outcrops. Stations 20, 63 and 58 (Bahat and Grossmann 1988) are marked by *A*, *B* and *C*, respectively. **c** Diagram of the conjugate three sets from the Mor Formation (see text for explanation)

Table 3.3. Distinct fracture phenomena in Lower and Middle Eocene chalks from the Beer Sheva syncline

Group and set	Eocene period of chalk	Formation and unit	Fracture strike azimuth (°)	Fracture style[a]	Fracture category (location in syncline)	Stress	Source
1a	Lower	Mor (I)	328	s.l.	Burial (distal end)	Regional	a
1b	Lower	Mor (I)	059	s.l.	Burial (distal end)	Regional	a, b
1c	Lower	Mor (I)	309	s.l. hybrid	Burial (distal end)	Regional	a, c
1d	Lower	Mor (I)	344	s.l. hybrid	Burial (distal end)	Regional	a, c
2	Lower	Mor (I)	055	s.l.	Burial (SE flank)	Synclinal	a
3	Lower	Mor (I)	294	Faults	Normal dips < 45° (throughout the fold)	Synclinal	d
4a	Lower	Mor (I)	335	m.l.	Uplift (distal, central)	Regional	a, e, f
4b	Lower	Mor (I)	335 305	Faults	Uplift vertical (distal, central)	Regional	f
5	Lower	Mor[b] (I)		m.l. composite	Post-uplift dips 55°–75° (distal)	Local	f
6	Middle	Early Horsha (II)	336	m.l.	Uplift (distal)	Regional	g
7a?	Middle	Horsha (III)	NE–N	s.l. rotational	Burial?	Synclinal	n.o.
7b?	Middle	Horsha (III)	N–NW	s.l. rotational	Uplift	Synclinal	n.o.
8	Middle	Horsha (III)	055	m.l.	Uplift (transversal)	Regional/synclinal	n.o.

Table 3.3. *Continued*

Group and set	Eocene period of chalk	Formation and unit	Fracture strike azimuth (°)	Fracture style[a]	Fracture category (location in syncline)	Stress	Source
9	Middle	Horsha (III)	040, 130	Faults	Normal dips > 45° < (transversal)	Synclinal/regional	n.o.
10	Middle	Horsha (III)	N–NW	s.l.	Neotectonic[c] (cross joints) (transversal)	Regional/synclinal	i
11a	Middle	Horsha (IV)	318	Fault primary	Strike-slip (axial)	Synclinal	f
11b	Middle	Horsha (IV)		Faults secondary	Strike-slip (axial)	Local	f
11c	Middle	Horsha (IV)	344	s.l.	Confined to faults (axial)	Local	f
12	Middle	Horsha (IV)	012–035	s.l. rotational en echelon single fringe	Uplift (axial)	Synclinal	j, k
13	Middle	Horsha (IV)	062	m.l. rupturing	Syntectyonic-rebound (axial)	Synclinal	f
14	Middle	Horsha (IV)	040–060	m.l. planar	Uplift (axial)	Synclinal/regional	f
15	Middle	Horsha (IV)	358[d]	Faults		Regional	f
16	Middle	Horsha (IV)	350–014	s.l. rotational en echelon two fringes	Uplift (axial)	Regional	j, k

Group 1 consists of *sets 1a–1d*. *Groups 8* and *14* consist of *sets 8* and *14*, respectively. Single-layer joints: *s.l.*; multi-layer joints: *m.l.* Locations in the syncline include: *distal* (northeastern part), *lateral edge* (southeastern part), *central* (close to the centre of the fold), *transversal* (along Wadi Secher that approximately crosses the fold) and *axial* (along Wadi Naim that approximately parallels the synclinal axis). In the *series 1a–d*, *4a–b* and *11a–c* the various individual structures are genetically linked. *Sources: a*: Bahat and Grossmann (1988); *b–g*: Bahat (1989, 1987a, 1985, 1988, 1991a, 1991b); *i*: Engelder and Gross (1993); *j–k*: Bahat (1986, 1997). *n.o.*: (from unpublished report by Bahat and Shavit 1997).

[a] Joints are generally vertical.
[b] Supplemented by data from the Shephela syncline.
[c] Tentative definition based on preliminary studies.
[d] 358°±10° mean of three conjugate faults.

Fractures in the syncline include s.l. joints, m.l. joints and normal faults. Each of these features occurs in several variations with different stratigraphic-structural associations. Some of the s.l. joints are associated with the burial stage, while the m.l. joints are considered to be uplift joints (Bahat 1991a). Other s.l. joints were clearly formed during uplift (Bahat 1999a). Vertical faults occasionally occur in association with m.l. joints, and there are some small strike-slip faults. Wherever two types of fracture strike in different directions, they are assigned to different sets, since their propagation (which is normal to the least principal stress S_{Hmin}) could not be the same for both structures at the same time.

3.4.2.2
Fractures in Unit I (Mor Formation)

Nine joint sets are identified in the Mor Formation (Table 3.3). Those that are associated with the same stress system are grouped together and marked by letters within the group (sets 1a–1d and 4a–4b). Three typical outcrops are shown in Fig. 3.35b.

Group 1. These include s.l. joint sets that are associated with burial. Four sets are identified according to characteristic orientation and appearance (Bahat and Grossmann 1988). Sets 1a and 1b consist of a cross-fold set striking 328°, and approximately orthogonal to it a set striking 059°. Set 328° generally predates set 059°, but there are also indications of reverse order (Bahat 1988a, Fig. 5). These sets are observed particularly in the northeastern distal part of the syncline. In some outcrops, both orthogonal sets are exposed abutting each other, in others only one of the sets is exposed. The orthogonal sets are also concentrated in the southwestern distal part of the Shephela syncline and along the southeast flank of this fold (Bahat and Grossmann 1988). As far as can be seen from exposures, cross-fold joints are rare in the northwestern flank of the Beer Sheva syncline.

Sets 1c and 1d are hybrid s.l. joint sets (Hancock 1985; Bahat 1987a): 1c strikes 309°, and 1d strikes 344°. Both sets occasionally occur in association with set 328° (Fig. 3.35c and 3.36a), but they appear mostly in separate outcrops. The age relation between sets 1a–1b vs. 1c–1d is not clear, although in one outcrop there is evidence that set 328° preceded set 344° (Bahat 1991a, p. 242).

At rare localities, the time sequence of adjacent joints of a given set can be determined based on the curving of joint terminations, where late joints curve toward early joints and arrest at approximately right angles (Bahat 1983; Dyer 1988). In the exposure shown in Fig. 3.36b, joint C postdates joint B, which is younger than joint A. In other exposures, the sequence is more difficult to determine (e.g., Bahat 1988b, Fig. 5; Bahat 1991a, Fig. 3.34).

Group 2. These include joints that are concentrated on the southeastern flank of the Beer Sheva syncline, trending predominantly 055° (Bahat and Grossmann 1988, stations 26, 27, 48, 49 and 50). Cross-fold joints are very rare in these stations. Morphologically, they do not resemble joints from set 1b, but look more like joints of group 7 (Sect. 3.4.2.4). There is a slight, but perhaps significant age difference between the layers that host fracture groups 1 and 2. Both groups belong to the Mor Formation, but in the Beer Sheva syncline the layers that host group 2 occur in an area of a gradual transition to the Horsha Formation along Wadi Secher (Fig. 3.33c) (Bahat and Grossmann 1988, Fig. 3).

Fig. 3.36a–d. Photographs of outcrops. **a** Joints of sets 328° (marked by *horizontal arrow*) and 309° (marked by *vertical arrow*), cutting the Mor Formation (unit I). *Scale bar* is 25 cm. **b** Joints of set 328° cutting the Mor Formation (unit I), showing the sequence of formation: *A* postdates joint *B*, and *C* is younger than *B*. **c** Multilayer joints cut thick layers in the Horsha Formation (unit II). *Scale* is 90 cm. **d** Cross joint cutting the Horsha Formation (unit III). *Scale bar* is 5 cm

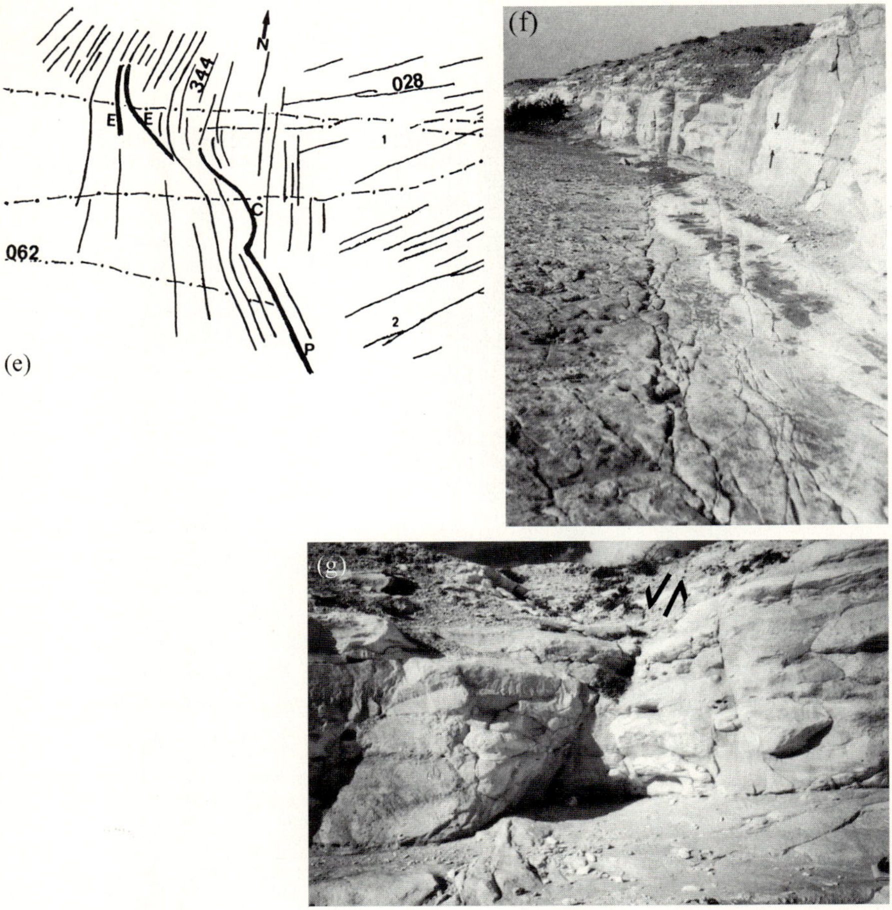

Fig. 3.36e–g. Photographs of outcrops. **e** A syntectonic joint set curves sympathetically with a primary A and secondary faults *B* and *C* in the Horsha Formation (unit IV). **f** Joints from set 062° dissect unit IV along Wadi Naim. They cut the wadi floor anastamotically and become planar m.l. joints as they cut the cliff (unit IV), layer marked by *arrows* is 90 cm thick. **g** Joints of group 12 are displaced by a normal fault from group 15 (unit IV)

Group 3. These include low-angle (dip ≤ 45°) normal faults that strike about 294° and are closely associated with s.l. joint sets 328° and 059° (Bahat 1985; Gross et al. 1997). These faults are thought to be a result of intra-synclinal seismic tremors (Bahat 1985; Buchbinder et al. 1988) (Fig. 3.35bA).

Group 4. This group, trending about 335°, includes vertical m.l. fractures that cut many layers (Bahat 1988a; Bahat 1991a, p. 306). The m.l. fractures are divided into two sets (Fig. 3.35bC). Set 4a consists of non-displacing m.l. joints, occasionally adjacent to vertical faults of set 4b, which developed from the former as a result of changing stress. Set 4b consists of vertical faults, one of which displaces a low-angle normal fault of group 3 (Bahat 1991a, p. 283), suggesting a sequence of (1) early, low-angle normal faults from group 3, (2) set 4a, and (3) set 4b.

Group 5. These include post-uplift joints that occur in the Shephela syncline (Bahat 1991a, Fig. 5.24). They cross the local Lower Eocene chalks at 55° to 75° dip angles and show a relationship to the local topography. The surface morphology of these joints is quite different from burial, syntectonic and uplift joints. They are composite radial fractures, interacting with each other, resembling other post-uplift composite fractures in sandstone and in granite (Bahat et al. 1995, 1999) (Fig. 3.23c and 4.27).

3.4.2.3
Fractures in Unit II (Horsha Formation)

Rocks of this unit are exposed in the northeastern part of the syncline, particularly along the Beer Sheva-Nitsana road (Bahat and Grossmann 1988).

Group 6. Included here are joints oriented 336°, cutting thick chalk layers of about 6–8 m. These joints are flat and smooth, not marked by plumes or *en echelon* fringes, or any other fracture markings. Joint spacing is irregular, varying from 50 cm to 20 m (Fig. 3.36c) (Bahat 1991b). The relation between fracture spacing and bed thickness (e.g., Ladeira and Price 1981) suggest that this set formed after, rather than in the course of sedimentation (Hodgson 1961b). By way of comparison, periodic arrest marks on syntectonic joints in strongly deformed, thick Senonian chalk layers (Bahat 1991a, Fig. 3.22) imply rhythmic fracture propagation, most likely induced by pore pressure (Secor 1969) under intense tectonic strain of the kind that never affected the almost horizontal sediments of the Eocene. These layers never attained over-pressure of fluids during the burial stage, and the m.l. joints of group 6 could develop only after the chalk was sufficiently dewatered and lithified, possibly not before uplift.

3.4.2.4
Fractures in Unit III (Horsha Formation)

Rocks of this unit are exposed along Wadi Secher, which runs down the southeastern flank towards the synclinal center.

Group 7. This group consists of s.l. joints that display strike rotation along Wadi Secher (Fig. 3.33d and 3.37), from about 090° to 020° along its southeastern part, through the 300–360° range along the sector of the wadi that crosses the central part of the syncline further west. Fracture patterns were categorized along Wadi Secher (Fig. 3.37) (with the exception of station 83), according to the "selection method" (Sect. 6.2.1.2). The following features stand out:

1. In the eastern area (stations 71 to 80), most outcrops are dominated by ENE striking joints (050–083°). Further west, the joints strike gradually more to NNE, and there is an increase in the number of N-S and NNW trending joints.
2. Adjacent stations usually have similar joint sets, with subtle transitions. For instance, the major set in the eastern part of station 86B is 060°, and the minor set is 330°; a little further west set 040° becomes dominant, and more to the west set 330° becomes the more important set. In station 86C, set 025° is more important on the eastern side and set 320° is dominant on the western part of the outcrop. However,

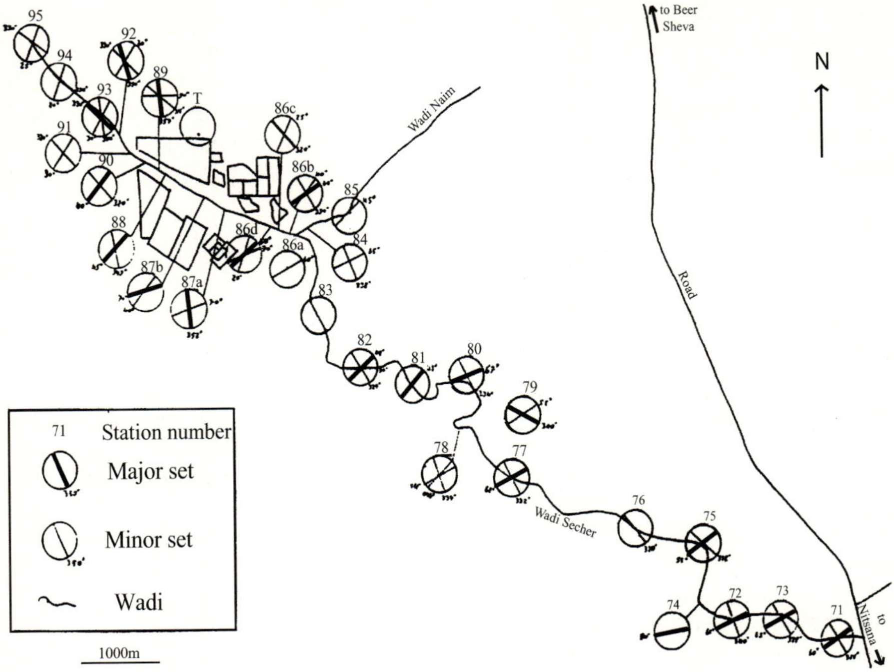

Fig. 3.37. Map-diagram summarizing joint orientations at stations 71–95 along Wadi Secher. *Circles* show the azimuths of sets at each station (with the exception of station 83) (after Bahat and Shavit 1997)

set 060° unexpectedly returns as the dominant set in station 86D, where 020° and 330° are the two minor sets.

3. In spite of the many "unexpected" returns of ENE trending joints, the overall emerging pattern is the increase in importance of NNE and NNW trending joints towards the west.
4. A summary of joint orientation in stations 71–95 (not including station 83) shows the general increase in importance of NNE and N-NNW trending joints relative to ENE trending joints, when moving from east to west along Wadi Secher (Fig. 3.38a–c). Hence, there is a subtle counter clockwise joint rotation in Wadi Secher.

Station 83 (in Fig. 3.37) was studied by the "comprehensive method" (Sect. 6.2.1.2) in greater detail, along a 700 m exposure in Wadi Secher. Joint strikes were measured separately for five "layer groups" (Fig. 3.39a). Average joint orientations in the various layer-groups show distinctive traits (Table 3.4). The combined strikes of s.l. joints from all layer groups of the Horsha Formation at station 83 shows a single maximum at 48 ±16° (F in Table 3.4). Average trends rotate gradually counter clockwise from the older group-layer B through C and D to the younger group-layer E (from 63° through 55° and 51° to 42°, respectively). An exception is the average 42° in layer-group A, which contradicts the trend that was through layer-groups B–E. Joints fringed with *en echelon* segmentation occasionally occur in upper parts of the western cliff of the wadi resemble the *en echelon* fringes that occur in Wadi Naim (see group 12 below).

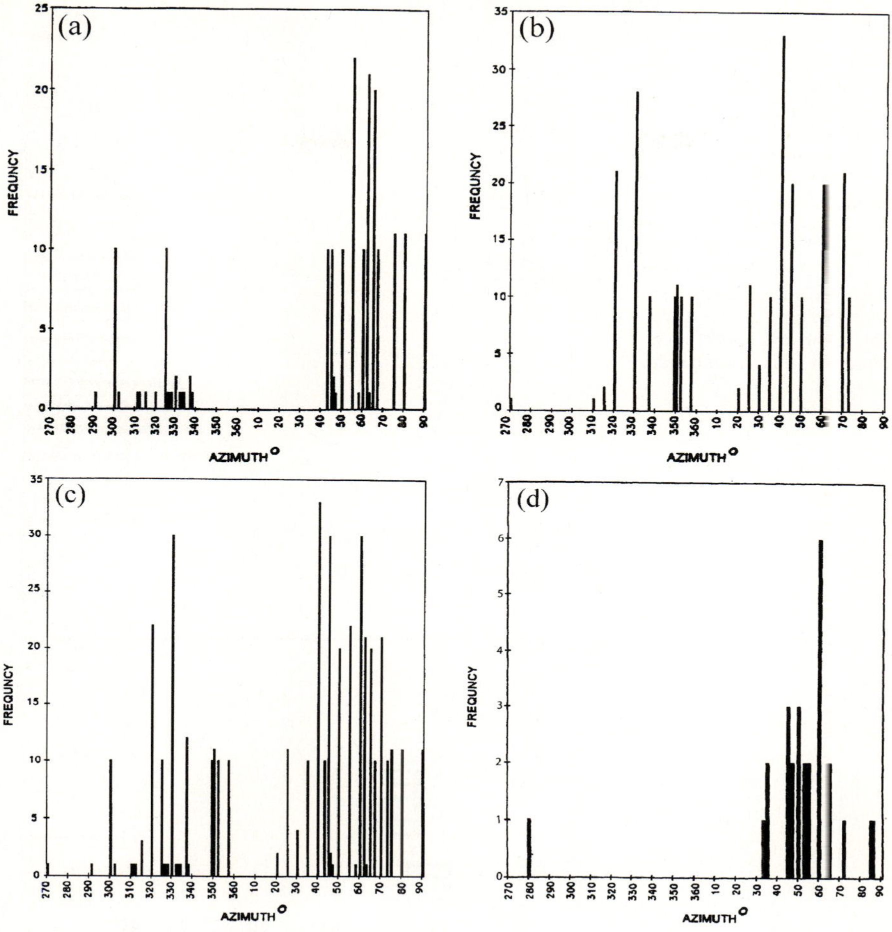

Fig. 3.38. Histograms summarizing data on s.l. joints from all stations in Fig. 3.37, not including station 83. **a** Stations 71 to 86A. **b** Stations 86B to 95. **c** Stations 71 to 95. **d** Multilayer joints in station 83 (after Bahat and Shavit 1997)

Table 3.4.
Strikes and properties of joints trending NE and NW in station 83, Wadi Secher

Layer group	Average of NE joints			Average of NW joints		
A	41.7°	(16.9)	[57]	–	–	–
B	62.7°	(15.0)	[39]	281.3°	(4.8)	[4]
C	55.0°	(7.7)	[244]	–	–	–
D	51.1°	(7.4)	[51]	329.5°	(3.7)	[19]
E	41.9°	(16.1)	[122]	321.1°	(16.5)	[62]
F	47.6°	(15.7)	[670]	318.1°	(17.2)	[85]
G	55.0°	(13.1)	[25]	280°		[1]

( ) and [ ] designate respectively, standard deviation and number of measurements, *A–E* are layer-groups in Fig. 3.39a, *F* is a summary of *A–E*, and *G* is the summary of m.l. joints in station 83 in Fig. 3.37 (after Bahat and Shavit 1997).

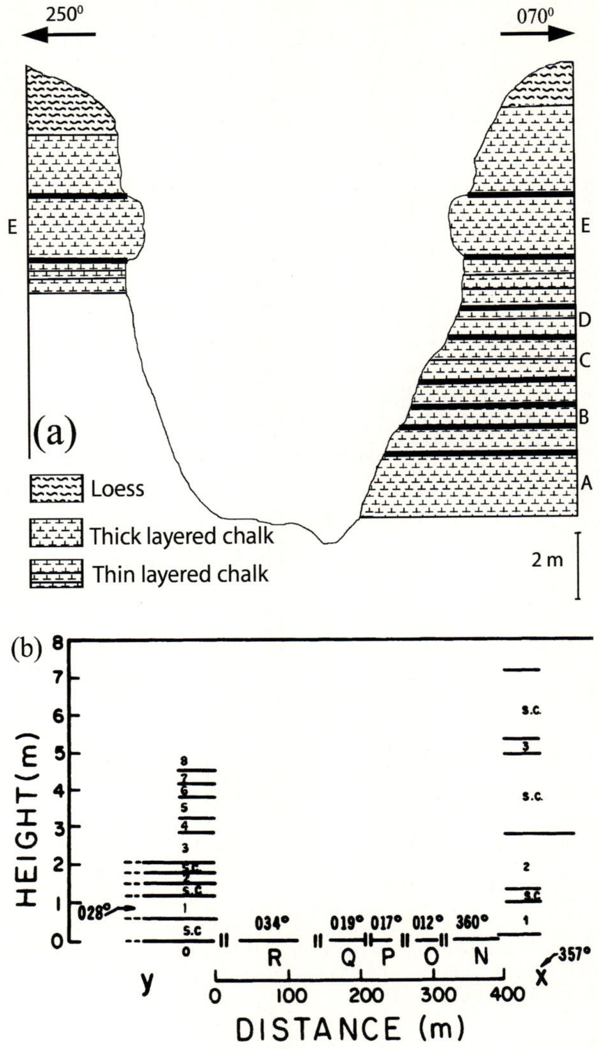

Fig. 3.39.
a Section across Wadi Secher at station 83, showing layer-groups *A* to *E* on the cliffs (note vertical scale) (after Bahat and Shavit 1997). **b** Diagram of outcrops *X* (set 16, Table 3.3) and *Y* (set 12), including floors *N–R* between them, showing rotation in mean azimuths of joints from 357° to 034°. Beds of soft chalk (*s.c.*) alternate with hard chalk beds designated by numbers. The two small vertical lines between floors *N* and *Y* represent talus (from Bahat 1986a)

Group 8. Twenty five m.l. joints constitute a set along a 600 m exposure at station 83, with an average trend of 055.0 ±13.1° (Fig. 3.38d). Their average differs by 7.5° from the s.l. average (F in Table 3.4). This relative orientation matching is noticeable, compared with the more pronounced deviations of m.l. azimuths from those of s.l. in the Mor Formation (Bahat 1991a, p. 306). Like m.l. joints in units I and II (Bahat 1991a, Fig. 5.16), group 8 often occurs in fracture zones with closely spaced joints.

The discrimination between s.l. burial joints and m.l. uplift joint in unit I is quite simple due to considerable morphological differences between the two fracture types. On the other hand, the distinction between s.l. and m.l. uplift joints in unit III is not always clear, because in some outcrops there are "transitional" fractures between the s.l. joints of small widths (heights, Fig. 2.8a) and m.l. joints of great widths. Single-layer joints and "transitional" joints are offset by normal faults, but no contacts between m.l. joints and normal faults were found in unit III. Therefore, it appears that these faults are younger than the s.l. and the "transitional" joints. It is however not clear that the normal faults are younger than the m.l. joints.

Group 9. This group consists of four normal faults, dipping above and below 45°, and striking 052° and 130°, south of station 83. A study of the largest and best exposed fault of this series is discussed in Sect. 6.7).

Group 10. This is a distinct s.l. group which consists of "cross-joints" (as distinguished from "cross-fold joints") that are straight along their central parts, oriented in the N-NW direction, but their tips gradually curve orthogonally towards joints of an earlier set (Fig. 3.36d). Their shape resembles the geometry of curving cross joints whose central straight part parallel to the S_{Hmax} of the neotectonic stress field or its Tertiary predecessor in eastern North America (Engelder and Gross 1993). These cross-joints presumably represent the latest s.l. joints in unit III, in association with uplift and erosion (Hancock and Engelder 1989). There is no indication, however, that these joints are younger than joints of unit IV.

3.4.2.5
Fractures in Unit IV (Horsha Formation)

This unit is exposed in Wadi Naim, along the synclinal axis (Fig. 3.33c,d).

Group 11. Fractures of this group, which cut unit IV, exhibit a unique fault termination zone that consists of three major elements:

1. A primary fault.
2. Secondary faults.
3. A set of fault-associated joints (Fig. 3.36e). A description of group 11 is presented in Sect. 6.6.

Group 12. This group consists of s.l. joints oriented in the 012–035° range, rotating counter clockwise from younger to older layers upstream Wadi Naim (Bahat 1986a, 1999a) (Fig. 3.39b). *En echelon* segmentation occurs intensively along the lower fringe of these joints, and rotates covariantly with the parent joints. Note that *en echelon* fringes are almost exclusive to groups 12 and 16 (see below), and occasionally in group 7 (see above).

Group 13. Joints of set 062° occur in various parts of Wadi Naim, along tens and even hundreds of meters. Unlike the straight m.l. joints of groups 5, 8 and 14, these joints undulate and "branch" from each other anastomotically (viewed in horizontal sections on a platform) (Fig. 3.36f), but on the average strike 062°. They display irregular spacing, and their openings vary from about 1 mm to several centimeters. Intensive irregular cracking occurs at and adjacent to these fractures (as revealed by opening with a chisel, and on vertical exposures). This set was apparently formed by breakage during an extended buckling-rebound when the compression normal to the syncline axis relaxed (Bahat 1987b), implying an important difference between the genesis of set 062° as compared with the other joint sets (e.g., group 12). Unlike those other sets, which were formed by remote compression co-axial to their strike, set 062° was formed by remote tension. The multilayer joints in general and set 062° in particular are the most important vertical conduits of ground water in the syncline (Sect. 6.9).

Group 14. These m.l. joints strike in the 040–060° range and form remarkably planar cliffs in Wadi Naim, and they are straighter than m.l. joints that cut other units in the syncline. These m.l. joints overprint s.l. joints of group 12 or cut them at small angles (less than 30°). The wadi exposures provide a perfect view of fracture, both at the cliff and in the adjacent floor, revealing all details of the transition from group 13 to group 14. Joints of set 062° significantly change their appearance from an anastomose style when dissecting layers 1–2 in the wadi floor (Fig. 3.39b) to planar joints of group 14 that cut the upper layers of the cliff (Fig. 3.36f) (Bahat 1987b). Hence, following the buckling-rebound that created set 062°, the m.l. joints of group 14 were formed during an advanced uplift process, preserving the same approximate strike for both joint groups. Compared with the profusion of fracture markings on s.l. burial joints cutting Lower Eocene chalks and syntectonic joints cutting Santonian chalks, fracture markings are rare on all m.l. joints. Perhaps it reflects fracture by fluid pressure in the former two cases and tension unrelated to fluid pressure in the latter.

Group 15. These are three neighboring conjugate normal faults (dipping both above and below 45°), striking 358 ±10°. These faults offset s.l. joints from group 12 along Wadi Naim (Fig. 3.36g). There are no contacts between normal faults and m.l. joints from group 14, leaving no clue as to their age relationship (see also Sect. 6.7).

Group 16. This is a 350–014° striking joint set, with *en echelon* fringes both at the top and bottom of the parent joint (Fig. 2.35a) (see group 12 for comparison, and elaboration in Bahat 1986a, 1997).

3.4.3 Fracture Diversity in the Beer Sheva Syncline

The high fracture diversity, allowing the distinction of no less than sixteen groups, may appear bewildering to those not familiar with the Beer Sheva syncline, making one wonder if all the above described twenty three proposed fracture sets in Table 3.3 are really distinct sets, particularly since there are certain proposed sets from different fracture groups that have similar orientations. This issue is addressed below.

3.4.3.1 *Strike Superposition of Different Joint Sets*

Seven strike superpositions of suspected different joint sets (Table 3.3) are examined for the possibility of genetic linkages, which if found, would reduce the actual number of sets.

Superposition 1. The distinction between s.l. sets 1b and 2 (striking 059° and 055°, respectively) is mostly based on considerable morphological differences. Set 1b characteristically occurs in orthogonal relationship with a cross-fold set 1a, has very small apertures (<0.1 mm) and is confined to the distal end of the syncline. Set 2, on the other hand, does not have an orthogonal counterpart set, has variable openings (apertures up to 10 mm and more) and occurs in the southeastern flank of the syncline. These differences imply different genetic origins for the two sets despite their almost identical orientations: Set 1b is probably an early burial set, whereas set 2 was possibly formed later (Sect. 3.4.2.2).

Superposition 2. Sets 8 and 14 strike about 055°. These m.l. joints are similar in some respects, and they may be genetically related. It is possible that fracturing in these sets took place in different times, or perhaps, fracture occurred in units III and IV at about the same time. There is no compelling evidence either way.

Superposition 3. There is a genetic linkage between joints of sets 4a and 4b trending 335° (Fig. 3.35bC). They are geographically adjacent and possibly the 4b vertical faults developed from the 4a vertical multilayer joints when local stress conditions changed. Thus, they formed in two distinct stress fields (S_{Hmax} parallel and inclined to sets 4a and 4b, respectively) into two fracture sets. Early joints that developed into faults are well known from other fracture provinces (e.g., Segall and Pollard 1983).

Superposition 4. There is no relative age indication for (group) set 6 m.l. joints. Their orientation (336°) suggests genetic association with set 4 m.l. joints that trend in the same direction. This possibility is however challenged by the presence of Upper Pleistocene gypsum mineralization along joints of set 6 (Issar et al. 1988). Some of these gypsum crystals are orthogonal to the jointing, which is not characteristic of set 4. This orthogonality suggests that opening of set 6 joints might be related to the age of mineralization (Ramsay 1980). Thus, in spite of uncertainties of relative age, it appears that set 6 is associated with processes younger than those associated with set 4.

Superposition 5. The strike slip fault (set 11a) preceded the adjacent set 028° that developed during an early stage of the uplift (Bahat 1987b). This fault was formed by horizontal compression along 344°, as evidenced by the orientation of joint set 11c (Fig. 3.36e). The direction of compression is at an approximate 30° angle to the strike of the strike-slip fault. Group 11 is found in a small area only about 16 m long and 7 m wide in Wadi Naim. Thus, a synclinal stress is more likely to be the cause of group 11 formation rather than a regional one.

However, compression from 344° coincides with the strike of set 1d, which occurs in various areas of the syncline and is considered to be a regional direction (Eyal and Reches 1983; Bahat 1999b). Although there is probably no genetic connection between groups 1d and 11, such a connection may still be proven by new findings. A regional compression from 344° could be tectonically effective in the syncline, forming group 11, a short time before the synclinal rebound and stress rotation (Bahat 1999a).

Superposition 6. Joints of set 10 are oriented N-NW and may be suspected to have been produced by the stress field that formed some of the joints of set 7 of a similar trend. However, their distinctive morphologic features, the rotational style of set 7 and the characteristic curvature of set 10 (Fig. 3.36d) all point to different genetic groups.

Superposition 7. Both groups 15 and 16 strike approximately N-S, yet they represent two different mechanisms: One that produced normal faults, and one that that produced joints with two *en echelon* fringes. Nevertheless, their genetic linkage cannot be excluded (Sect. 3.4.6.2).

In summary, we tried to show in the above review how fracturing episodes may be fitted to the documented sets. There is no compelling evidence to rule out genetic linkage between sets 4 and 6, 8 and 14, 1d and 11c, as well as 15 and 16. Combining representative sets from these four couples would reduce the total number of sets in Table 3.3 from 23 to about 19. Yet it is equally likely that superpositions 1, 3 and 6 combine sets of different genetic origin, and mere similarities in strike directions are not a sufficient reason for grouping them together. The great structural diversity in the Beer Sheva syncline may therefore be real.

3.4.3.2
Single-Layer Burial Joints versus Single-Layer Uplift Joints

Burial s.l. joints form early in the history of the rock and typically arrest at layer boundaries. Uplift m.l. joints, on the other hand, develop late in the history of the rock and cross layer boundaries. The discrimination between these two groups is relatively simple, because the m.l. joints, cutting many layers (or thick strata), have wider open-

Table 3.5. Fracture properties of single-layer burial and single-layer uplift joints

	Single-layer burial joints (Mor Formation)	Single-layer uplift joints (Horsha Formation)
1	Orthogonal cross-fold and strike sets	There is no orthogonal jointing
2	There are conjugate joint sets	There are no conjugate joint sets
3	Joints generally are pre-early normal faulting	Joints are post early srike-slip faulting
4	Not contemporaneous with any m.l. set	Contemporaneous with the m.l. 062° set
5	There is no joint rotation	Joint rotation is extensive
6	Older layers fracture first	Younger layers fracture first
7	No fracture interaction among joints	Considerable fracture interaction among joints of set 028°
8	No en echelon segmentation	Fringes of en echelon segmentation associated with NE trending joints
9	Closed joints (aperture <0.1 mm)	Joint opening varies considerably (from 0 to >10 mm)

Sources for data and explanation are given in text.

ings and their spacing is irregular and considerably larger than of the s.l. joints (Fig. 3.36c) (Bahat 1991a; Ghosh 1993, p. 492). Much more difficult is the distinction between burial s.l. joints and uplift s.l. joints, because very often they display similar outcrop features. The distinction is important because it can be useful in deciphering geological processes. Table 3.5 lists nine diagnostic properties of s.l. burial joints and uplift joints in the Mor and Horsha formations (see also Bahat 1999a). One of these properties, the rotation of joint strikes under regional stresses, is treated in Sect. 6.3.

3.4.4 The Structural Interpretation of *En Echelon* Fringes in the Beer Sheva Syncline

En echelon fringes in the Beer Sheva syncline are characterized in terms of style and in terms of paleostress directions (see Sects. 2.2.5 and 2.2.7).

3.4.4.1 *Three Styles of* En Echelon *Segmentation*

Three styles of *en echelon* segmentation can be distinguished in the Beer Sheva syncline. They are represented by set X (unit IV), set O-Y3 (unit IV) and joint S (unit III) (Table 2.1). Set X has *en echelon* fringes both below and above the parent joint (Fig. 2.35a), and in both, segment rotation is counter clockwise, ranging in strike between 338° and 342°. Quite distinct from set X is set O-Y3 in station Y. In this set, a fringe occurs only below the parent joint and segment rotation is clockwise, ranging in strike between 026° and 053° (corresponding to the clockwise rotation of the parent joints). The curved joint in station Y (Fig. 2.35c,d) undulates considerably along the strike but the segmentation is confined to a small azimuth range (043–047°), which is only a limited spread within the range 026° and 053° mentioned above. The concentric joint (Fig. 3.40b) is a subvariation in set O-Y3. Joint S is best represented by part D that strikes 073°, whose general clockwise segment rotation is in the 073–083° range (Fig. 2.38 and 3.40a). However, some counter clockwise segment rotation also is observed on this joint. These three styles differ from the Mt. Carmel style (Fig. 3.40c) (Bahat 1997).

3.4.4.2 *Three Paleostress Directions Set*

Set X consists of joints that strike between 350° and 014° in outcrop X, and seems to reflect a regional N-S post Middle Eocene paleostress. This stress corresponds to the "last phase" N-S regional compression reported by Letouzey and Tremolieres (1980) from southern Israel. Mart and Horowitz (1981, Table I) also described a regional N-S trending joint system along the western shoulder of the Dead Sea rift. Joints from set O-Y3 on the other hand strike over a wide range, from 012° to 035° and are apparently associated with synclinal stresses (Bahat 1997). The joint direction (073°) and *en echelon* characteristics found in part D from station W.S. (Fig. 3.40a) are quite different from X and O-Y3 sets, suggesting three different stress fields.

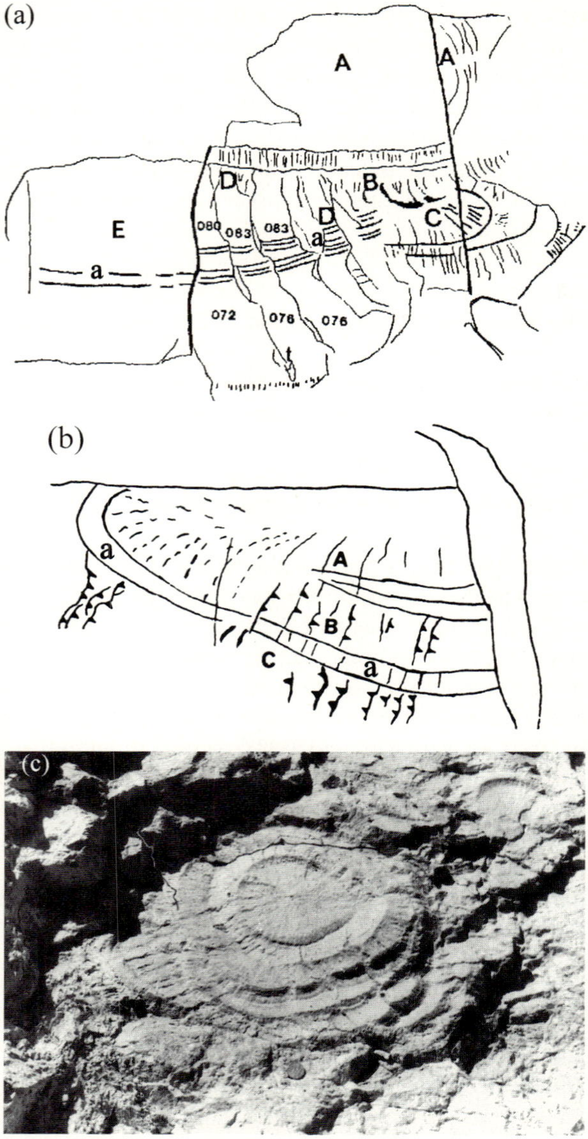

Fig. 3.40.
a Morphological sketch of a composite joint in Wadi Secher (station S in Fig. 3.33c, see Fig. 2.38). The fracture surface is divided into five parts, *A* to *E*. Lower edges of tear surfaces and arrest marks are marked *t* and *a*, respectively. *Numbers* on segments are azimuths (°). Two late fractures are marked by *thick, sub-vertical lines*. *Scale bar* is 15 cm. **b** The "concentric joint". A discontinuous breakdown occurs below the outer arrest mark *a*. The sense of rotation of overlapping barbs is counter clockwise in zones *A* and *B*. Segments at the right side of fringe rotate clockwise with respect to the parent joint (*triangles* along *heavy lines* point to direction of overlapping segments). This rotation conforms to the other rotations in outcrop *Y*. However, segments at the left side of fringe show counter clockwise rotation, which is in conformity with the rotation of overlapping barbs on the joint surface. *Scale bar* is 20 cm. **c** A discoid joint on Cenomanian chalk Mount Carmel, northern Israel cutting. The discoid displays several concentric fringes with a non-uniform breakdown. Segment rotation is counter clockwise on the left side of the discoid and is clockwise on the right side of the joint. There is no overlapping in the upper and lower parts of the discoid (see coin for scale)

3.4.5
Fold-Crossing Joints

3.4.5.1
General

The geometric relationships between joints and asymmetric folds are discussed in Sect. 3.2. Quite different are the geometries of joints that develop across symmetric folds (Fig. 3.20b). These have been studied by many investigators (e.g., Price 1966, Fig. 43;

Stearns 1968). De Sitter (1964) demonstrated a 90° rotation of the S_{Hmax} between symmetrical syncline and anticline folds. The rotation occurred due to local tension at the crest of the anticlines and compression at the trough of the synclines, like in a bent beam (outer-arc, Fig. 3.19a) (e.g., Hobbs et al. 1976, Fig. 4.33a; Burger and Hamill (1976); Ghosh 1993, p. 254). All these studies show relatively simple fold-fracture structures, with only a few fracture sets (Fig. 3.41). For instance, Stearns and Friedman (1972) found only five major fracture patterns associated with folds. Ramsay and Huber (1987, p. 655) consider the combination of four sets, two extensile sets (in cross-fold and in strike orientations) and two conjugate sets to be "classical" (which approximates F_1 and F_2 in Fig. 3.20b). When this classical view is "maximized" by additional sets, one obtains all the sets in Fig. 3.20b and 3.41 that still maintain a "simple" fold-joint geometry.

This simplicity often reflects a compromise between deviations in fracture direction that are dictated by local stress field changes around singularities (e.g., Rawnsley et al. 1992) and by the ongoing process of folding (Dieterich and Carter 1969; Burger and Hamill 1976), acting against the tendency of the remote stress to avoid changes by "self-correction". Self-correction, also known as "local symmetry", is a term adapted here from small-scale cracking mechanisms. It is a property that distinguishes between fracture behavior in isotropic and anisotropic materials (Ravi-Chandar and Knauss 1999). While arbitrary loading on a structure could result in a mixed mode loading, a growing crack quickly turns so as to reorient itself into mode I symmetry in isotropic materials. On the other hand, anisotropy forces the crack to grow under mixed mode loading along specific planes in crystals and along faults in earthquakes. Occasionally, however, self-correction into mode I may manifest itself along specific planes in crystals (Cramer et al. 2000 and Fig. 2.23d) as well as along layer boundaries in stratified formations (e.g., Bahat 1991a, Fig. 5.22) and along faults in earthquakes (Bahat 1982). Thus, fold-crossing joints tend toward simplified patterns, due to control by some self-correction mechanism.

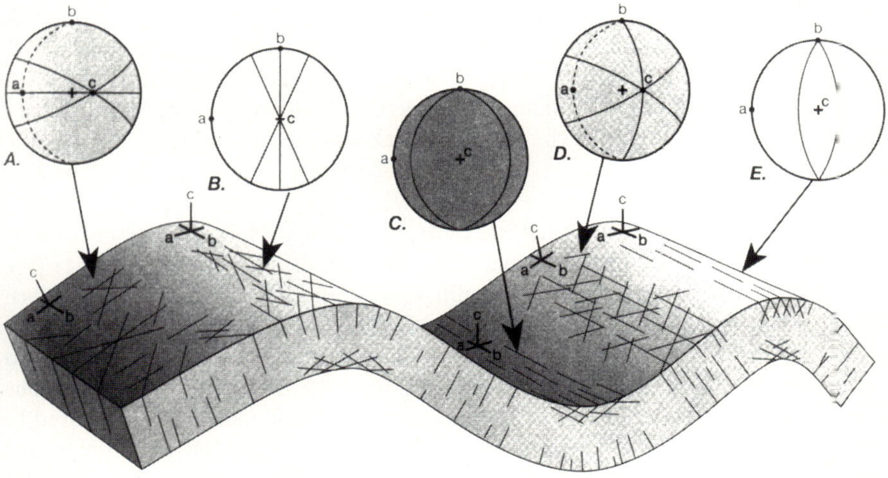

Fig. 3.41. Simple fold-fracture relationship. *a*, *b* and *c* are three tectonic axes. The orientation of *b* is constant, parallel to bedding, while *a* and *c* change their attitudes in different parts of the fold. The stereographic projections show orientations of the coordinate system, of the bedding where it is not horizontal (*dotted great circles*) and of the fractures (*solid great circles*) (from Twiss and Moores 1992)

3.4.5.2
Fold-Joint Relationships in the Beer Sheva Syncline versus the Simple Fold-Fracture Concept

Structural features of the Mor and Horsha Formations. We compare the "simple" fold-fracture pattern in Fig. 3.41 to the fracture pattern in the Beer Sheva syncline (Table 3.3, Fig. 3.35a).

1. Outstanding in unit I are two orthogonal sets of s.l. burial joints, trending NW and NE. These are only found in the distal parts of the Beer Sheva and the Shephela synclines. Less common are non-orthogonal s.l. joints, striking NE-ENE. In some outcrops, NNW or WNW joints are added to or occur instead of the more common NW joints. Uplift m.l. joints are often oriented NW-WNW. In one station, a normal fault is displaced along a sheared, vertical m.l. fault (set 4b).
2. NW trending uplift joints cut massive chalks of unit II in the Horsha Formation, above the Mor Formation.
3. In the eastern flank of the Beer Sheva syncline, the most common fractures in unit III are NNE-NE-ENE s.l. joints and m.l. uplift joints trending NE. Further west towards the synclinal center, NW trending s.l. and m.l. joints occur in increasing abundance along Wadi Secher. A series of conjugate, normal faults also occurs along the wadi.
4. An early feature in unit IV is a right-lateral strike slip fault system, striking NNW. A syntectonic, anastomosing joint set sub-parallel to the synclinal axis occurs along Wadi Naim. Almost exclusive to this unit is the abundance of *en echelon* segmentation in NNE-N trending s.l. uplift joints. Single-layer joints reveal a strike rotation. Multilayer joints overprint some of the s.l. joints. A normal fault from a NS striking conjugate set initiates on top of the earlier right-lateral strike-slip fault.

Similarities between the above two patterns. Single-layer joints in unit I and m.l. joints in units II and III generally correspond to the pattern in Fig. 3.41, in the sense that they follow the orientations of elements from groups A and B. Sets 328°, 309° and 344° correspond to the pattern of conjugate hybrid joint structure (Fig. 3.35a), and m.l. joints in unit I partly follow elements from both groups A and B. Similarity also applies to the syntectonic 062° and other m.l. sets in unit IV that fit the extensile direction of group B.

Dissimilarities between the above two patterns. Multilayer joints in unit I occasionally develop in other orientations that differ from the "simple" view (Fig. 3.41) (see Bahat 1991a, p. 306). Neither the post-uplift joints in unit I nor the rotational pattern of s.l. joints in units III and IV have parallels in Fig. 3.41. Groups C and E from Fig. 3.41 are not found in the Beer Sheva syncline. A remarkable difference between the two systems is the considerable vertical continuity of joints as is visible in photographs and diagrams (Price 1966; Stearns 1968; Stearns and Friedman 1972), compared to the short vertical continuity of fractures in the Beer Sheva syncline. With the exception of set 062°, the vertical extent of most fracture groups, including m.l. sets from Table 3.3, is limited to one thick layer (Fig. 3.36c) or to a few tens of thin layers (Bahat 1991a, Fig. 5.16). No m.l. joints have been observed to cross the boundaries between units I–IV, or to reach rocks below the Eocene chalks.

3.4.6
Factors Influencing Fracture Diversity in the Beer Sheva Syncline

The great diversity of fracture structures in the chalks of the Beer Sheva syncline (Table 3.3) points to a considerable diversity of fracture mechanisms. The role of several factors that influenced this diversity is discussed below.

3.4.6.1
The Influence of Paleostress on Fracture Diversity

General. Paleostress is a prime factor in the fracture evolution of the Beer Sheva syncline. Several scales of stress are distinguished. First are the regional stresses of the Mediterranean continental collision area (e.g., Letouzey and Tremolieres 1980) which prevailed from the Upper Cretaceous to the Quaternary. Second are synclinal stresses that originate within the fold and cause structural changes in the syncline (Bahat 1986a). Third, also important, are local stresses that are confined to a particular outcrop (Bahat 1991a, p. 301) or to a particular lithology at a given location (Gross et al. 1997) or to layers of particular thickness (Bahat 1997). In the following discussion, these scales are referred to as regional, synclinal and local.

Regional influences in unit I. The orthogonal sets 328° and 059° (1a and 1b in Table 3.3) cut Lower Eocene chalks and maintain these exact strikes and fracture characteristics in outcrops that are 17 km apart in two distinct folds, the Beer Sheva and the Shephela synclines (Fig. 3.33b,c). The two synclines differ considerably in their orientation (striking about 050° and 028°, respectively). The implication is that sets 328° and 059° were formed in response to large-scale (inter-continental) S_{Hmax} stresses and stress relaxation (Bahat 1989), which were "indifferent" to the already existing shape of the Syrian Arc (e.g., Letouzey and Tremolieres 1980; Zoback 1992; Buchbinder and Zilberman 1997).

In places, the extensile set 328° is associated with conjugate hybrid joint sets (1c and 1d) (Fig. 3.35c and 3.36a), reflecting mixed modes I + II (Sect. 2.2.13). Therefore, sets 1a–d are genetically related to a particular regional stress. Bahat (1999b) suggests an alternative possibility that the three sets 1a, 1c and 1d are genetically unrelated, and that they were formed at different times under different regional stress conditions (Sect. 6.3.9.1, see also Goodwin 1995). According to both models, however, sets 1a–d are products of regional stresses.

Synclinal influences in unit IV. This unit is exposed along the Beer Sheva synclinal axis and consists of Middle Eocene rocks. Sets 12–16 that succeeded group 11 in time, display rock responses to synclinal rebound and uplift. The s.l. joints of set 12 partly interact with joints of set 062° from group 13 that were associated with the synclinal rebound, exhibiting a concurrent relationship (Bahat 1987b). Those s.l. joints of set 12 that show no interaction with set 13 possibly suggest that some of them had developed before set 062°, i.e., before synclinal rebound, and that they continued a prolonged process (Sect. 3.4.6.2 and 3.4.6.3).

Local influences in unit I. The normal faults cutting the Mor Formation (set 3) show evidence of repeated slip, mostly during the Lower Eocene. However, some fracturing can be assigned to local stresses responding to specific lithological conditions (Gross et al. 1997). Unlike post-uplift joints in sandstone, which tend to follow the orientation of earlier joints (Sect. 3.3), the post-uplift joints of set 5 (Table 3.3) developed in response to local stresses, which show affinity to the erosional morphology of the rock surface (Bahat 1991a, p. 301).

3.4.6.2
Rotations of Synclinal Stresses in Units III and IV

The systematic azimuth rotation of the s.l. joints in Wadi Secher (Fig. 3.37–3.39a and Table 3.4) and in Wadi Naim (Fig. 3.39b) describes fracture sequences, which strongly suggest that these joints were formed by linked mechanism(s), rather than erratically. A mechanism of clockwise stress field rotation was suggested for the formation of the s.l. uplift joints in Wadi Naim (Bahat 1986a, 1999a, present Fig. 3.42a–c). The joints oriented 360° and 012° occur in upper layers, upstream along Wadi Naim, compared to the joints oriented 019° and 028° (wadi slope is exaggerated in Fig. 3.42d). Since the layers are approximately horizontal, the joints oriented 012° fractured earlier than the joints oriented 028°. This sequence is consistent with the propagation of uplift joints. During uplift the stress gradient is such that the least principal stress decreases downward, causing the greatest strain at the surface. Figure 3.42d shows that jointing along Wadi Naim that straddles sub-parallel to the fold large axis occurred in older layers, closer to

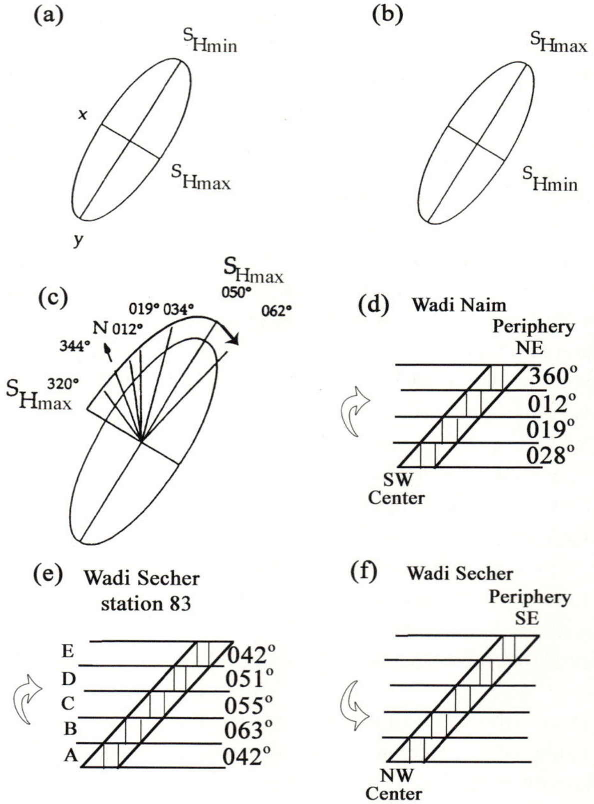

Fig. 3.42. Proposed evolution of fracture in the Beer Sheva syncline under changing synclinal stresses. **a** During buckling, maximum horizontal stress S_{Hmax} parallels the short axis *x* of the syncline, and the minimum horizontal stress S_{Hmin} parallels the long axis *y*. **b** Buckling rebound occurred when the compression along *x* relaxed and became S_{Hmin}, and the compression along *y* became S_{Hmax}. **c** Clockwise paleostress rotation is shown sequentially in Wadi Naim (from Bahat 1999a). **d–f** Diagrammatic wadi section (angle of slope is exaggerated) showing different joint rotation directions (*opened arrows*); see text for explanation

the synclinal periphery, before taking place in younger layers closer to the synclinal center. The detailed results from station 83 in Wadi Secher (Fig. 3.39a and Table 3.4) show that the joints oriented 042° fractured in a younger group-layer (E) before the joints oriented 063° in the older group-layer (B) (Fig. 3.42e), revealing a clockwise joint rotation. Thus, a clockwise joint rotation occurred both in Wadi Secher and in Wadi Naim, implying that jointing along the two wadies probably was associated with the same process of stress field rotation at about the same time. This interpretation does not fit, however, the general fracture pattern from the twenty-eight stations along Wadi Secher (Fig. 3.42f). Here, joints oriented ENE formed earlier in older layers, closer to the synclinal periphery, and a shift towards NNW joint orientation took place later, in younger layers closer to the synclinal center (Fig. 3.37). This shows a counter clockwise rotation (Fig. 3.38a,b and 3.42f), which differs from the clockwise rotation displayed in Fig. 3.42d,e. How can this difference be explained? One possible interpretation is that the s.l. joints in Wadi Secher did not form in a single continuous process. Perhaps, in the southeastern part of the wadi, which is dominated by Lower Eocene chalks (see description of groups 2 and 7 above, Table 3.3), jointing occurred earlier, by a mechanism that differed from the one causing jointing in a later stage, in the Middle Eocene chalks, towards the synclinal center, as shown in station 83. Hence, a unique model of joint rotation in Wadi Secher is uncertain, but this scenario seems to fit the data best.

There are deviations from simple stress rotations, particularly, the "unexpected" departures from the rotation pattern in two scales along Wadi Secher. First, there are repeated azimuth fluctuations in joint rotation along the wadi (Fig. 3.37). Second, the average azimuth of one particular joint set in group-layer A at station 83 is reverse to that of the general rotation in station 83 (Table 3.4). There is also some fluctuation in joint rotation along Wadi Naim: Joints cutting floor R diverge from the general rotation pattern (Fig. 3.39b). These deviations, however, do not invalidate the joint rotation model suggested here; they merely demonstrate the versatility of these processes.

The exclusion of groups 16 and 15, both striking about N-S (Table 3.3) seems inevitable, not least because group 16 is linked with regional stresses (Sect. 3.4 4.1 and 6.3) and group 15 is associated with probably the youngest fracturing along Wadi Naim (Sect. 6.3). The sum of the evidence suggests that sets 15 and 16 were formed after the synclinal stresses had completed their rotation.

3.4.6.3
Additional Factors of Fracture Diversity in the Syncline

The observed fracture diversity in the Beer Sheva syncline is linked to several additional factors:

1. The Beer Sheva fold is a synform.
2. The low strength of chalk.
3. Developments during various stages of burial, unfolding and uplifting.
4. Different lateral position in the fold.

It is often difficult to separate the effect of single factors from the influences of other factors. Accordingly, combined factors are discussed below.

Synforms vs. antiforms and rock strength. Interestingly, systematic joints are rare in Cenomanian chalks, anticlinally folded and uplifted in northern Israel – Mount Carmel and the cliffs of Rosh Hanikra. Particularly intriguing is the absence of m.l. uplift joints comparable to groups 6 or 8 (Table 3.3) from the cliffs of Rosh Hanikra, which extend along hundreds of meters. On the other hand, systematic joints are quite abundant in the synclinal early Senonian (Santonian) and Eocene chalks in Israel. This comparison suggests that joints are abundant in chalks when they are part of synforms but not when they are part of antiforms.

The influence of lithology on jointing is well demonstrated by the profusion of joints in Eocene chalk formations in synclines of south and central Israel, whereas in synclines in northern Israel where the Eocene is dominated by limestone (the Gilboa' and Rosh Pinna synclines, Flexer et al. 1984), jointing is much less developed. It seems that the combination of synform structures with weak rock is particularly conducive to intense and diversified fracturing. The dramatic reduction of strength of wet chalk (Hayati 1975; Bahat et al. 2001a) probably also contributed to differential fracturing, because various fracture episodes are expected to have developed under different degrees of wetness.

The influences of burial, folding, unfolding and uplifting on fracturing in the Beer Sheva syncline. Buchbinder and Zilberman (1997) present a detailed account of the uplifts that affected the eastern Mediterranean margin during the Miocene-Pleistocene. Since Oligocene times, the eastern Mediterranean paleo-shelf (today coinciding with the Judean foothills and extending north-south through the Beer Sheva syncline) has been serially uplifted. Early Miocene, Middle Miocene and Pleistocene uplifts are recognized by sedimentary hiatuses during the periods 25–17 m.y., 14–9 m.y., and 1.5 m.y. to Recent.

The overall character of jointing makes it fairly clear that joints of groups 1 and 3 (Table 3.3) were formed during burial stages (Bahat 1985) and groups 4 and 5 were created during the uplift and post-uplift stages, respectively (Bahat 1991a). Joint rotation in units III and IV is best correlated with unfolding and uplifting that were more influential in the central parts of the syncline, and in various cases specific fracture episodes can be related to particular tectonic phases. Some fracture groups, however, such as 2, 8 and 14, cannot be fixed with certainty in the frame of events (Sect. 3.4.3.1).

Limited sedimentary cover. The young age of the Eocene sediments and the fact that they never were covered by a thick overburden was possibly significant in controlling fracturing processes in the syncline.

3.4.7 Summary

The Beer Sheva syncline is an asymmetric basin in which the sediments are only slightly folded. Nevertheless, the chalky Eocene strata of this syncline display a great diversity of fractures, not visibly related to the basin's asymmetry. The Lower and Middle Eocene chalk formations of the Beer Sheva syncline are divided into four rock units (I–IV), which display considerable differences in fracturing mechanisms. Table 3.3 presents a summary of twenty-three sets in the syncline.

No diversity of joints as rich as in the Beer Sheva syncline has been reported from other folds in rocks of comparable strength.

Different fracture episodes relating to joint sets of similar strikes are presented in "strike-superposition", a method that helps to verify the genetics of sets.

Three scales of paleostresses regimes can be distinguished: Regional (extra-synclinal), synclinal (intra-synclinal) and local (confined to a particular site). The different paleostress scales are most clearly identified in different parts of unit I, particularly in the distal part of the fold, and to a lesser extent in the other units.

Orthogonal sets of single-layer burial joints (328° and 059°) occur in unit I at the northeastern distal part of the syncline which trends about 050°, while the predominant set in the southeastern flank is oriented NE.

Nine criteria of distinction were established between s.l. burial joints in unit I and s.l. uplift joints in unit IV.

Single-layer joints cut units III and IV. Two cases of s.l. joint strike rotation are measured in unit III and one in unit IV. Two measurements, in Wadi Naim and Wadi Secher, show clockwise rotation, and one in Wadi Secher shows a counter clockwise one. These, are best explained by intra-synclinal paleostress rotations: A simultaneous clockwise rotation along Wadi Naim and the younger layers of the Horsha Formation along Wadi Secher. On the other hand, a different, counter clockwise paleostress rotation took place in the older layers of the Horsha Formation and younger layers of the Mor Formation.

Single-layer joints of unit IV, which are the youngest in the syncline, exhibit en echelon segmentation along layer boundaries. Some en echelon segmentation occurs sporadically along boundaries of s.l. joints in unit III, and in one outcrop, in connection with composite fracturing. Styles of en echelon segmentation and particularly their limited spreads of azimuth helped to identify three sets of joints in unit IV. These include, a set produced by a regional N-S post Middle Eocene paleostress, a set with wider azimuth range (012° to 035°) which is connected to stress rotation, and a third set oriented 073°, whose origin is not clear.

Multi-layer uplift joints overprint syntectonic joints (the 062°) in unit IV.

All multi-layer joints cutting units I–IV appear to be related to uplift processes. Their "fine tuning" however is not clear. Strike patterns vary in the four units: A considerable range of strikes in unit I, NNW jointing in unit II, NE jointing in unit III, and joint spreading around NE in unit IV. Joint morphologies vary as well, particularly in unit IV.

The multi-layer joints are the most important vertical conduits of ground water in the syncline (Sect. 6.9).

Compared with the profusion of fracture markings on s.l. burial joints cutting Lower Eocene chalks, and syntectonic joints cutting Santonian chalks, fracture markings are rare on all m.l. joints. Perhaps it reflects fracture by fluid pressure in the former two cases, and tension unrelated to fluid pressure in the latter.

A vertical fault, genetically linked to multi-layer uplift joints, offsets a normal fault in unit I.

In unit I, low angle (dip ≤ 45°) normal faults interacted with s.l. burial joints. Generally, the interaction took place during the sedimentation stage. High angle (dip ≥ 45°) normal faults interacted with uplift single-layer joints in unit IV during advanced uplift stages of the syncline.

Unit IV exhibits a unique fault termination zone of a right-lateral strike slip fault. This zone displays primary and secondary faults, and a joint set exclusive to the zone.

Additional factors that contribute to fracture diversity in the Beer Sheva syncline, besides paleostress regimes, include: The synform structure, the low strength of the chalk (especially when wet), and differential developments during stages of burial, unfolding and uplifting.

3.4.8 Open Questions and Future Studies

In spite of detailed studies, many problems relating to fracturing in the Beer Sheva syncline are still unresolved. Following is a brief review of these open questions.

Sets 1a–d are possibly conjugates that formed in one stress field (Bahat 1987a). Alternatively, the three sets 1a, 1c and 1d (Table 3.3) evolved in a process of clockwise stress rotation (Bahat 1999b). Both processes, however, would be caused and sustained by regional stresses (Sect. 6.3.9.1).

If both sets 1d and 11c were formed by regional compression from 344°, future fieldwork may yet discover a genetic relationship between them.

An intriguing issue is fracture by stress relaxation. Currently, it is considered that strike-parallel joints (set 1b) are more widely spaced because they were formed in response to smaller stresses relieved by recurring phases of relaxation, as compared to the cross-fold joints (set 1a) (Bahat 1991a, Table 5.4). The latter were formed by greater, periodically recurring increases in fluid pressures associated with regional stresses.

According to Rawnsley et al. (1998), the joints in the Bristol Channel basin were formed in five main phases, at least partly during relaxation of the Alpine stresses (Sect. 3.2.1). Engelder and Peacock (2001) do not argue against relaxation but assert that the J3 joints in the south-dipping beds of the anticline at Lilstock Beach propagated during folding (before relaxation). Although jointing during stress relaxation has been proposed by various authors (e.g., Price 1974, Bahat 1989) the available arguments are circumstantial and require more compelling geological evidence, though there is experimental evidence of fracturing under this condition (e.g., Sect. 2.2.3.7).

To date, no study has dealt with the connection between "transitional" joints from unit III (group 8) with layer boundaries (Underwood and Cooke 2001 and Sect. 2.2.3.15). Such investigations may help to understand what prevents such joints from becoming m.l. joints.

The unclear genetic relationship between groups 2 and 7 needs to be investigated.

There is currently no proof that s.l. joints from group 10 in unit III are younger than s.l. joints from unit IV.

The structural and genetic relationship of the four adjacent normal faults in Wadi Secher (group 9) needs to be established.

Three adjacent normal faults strike 358 ±10° (group 15). These faults offset s.l. joints from set 12 along Wadi Naim. There are however no contacts between the normal faults and m.l. joints from group 14, leaving their age relationship open.

Field data enabling the timing of groups 8 and 14 are still lacking: Were groups 8 and 14 formed simultaneously, or separately? Were they formed earlier in unit III and later in unit IV, or *vise versa*?

The concept of regional stress meets a difficulty: If the stresses that induced joints were not controlled by the early folding pattern of the Syrian Arc, as suggested above, why are their products (group 1, sets 1a–1d) clearly associated exclusively with the distal parts of both the Beer Sheva and the Shephela synclines? Should not these sets be uniformly dispersed throughout the two synclines during the Lower Eocene?

More studies are required to explain how lithological differences between chalk and limestone, other than strength (and possibly related to their history of lithification) may influence their modes of fracturing.

The young age of the Eocene sediments and the fact that they never were covered by a thick overburden is most probably significant for the development of fracturing in the syncline, but this topic needs be further investigated.

Finally, a micropaleontological study along Wadi Secher might confirm that unit II is lacking there and establish the rate of joint rotation in station 83.

3.5 Comparison of Unrelated Fracture Provinces

The four different fracture provinces that were selected for presentation in this chapter, the Appalachian Plateau, Bristol Channel, Zion National Park in the Colorado Plateau and the Beer Sheva syncline, do not represent a comprehensive review of the subject. Certainly there are additional such provinces that are worth concentrating on. Nevertheless, even in the present scope it may be instructive to bring up the similarities and differences between these provinces. Soon it will be seen that in spite of being universally subject to the physical laws of rock mechanics and fracture mechanics, the fracture products differ considerably in the four provinces due to the diversity of geological conditions in which they evolved and the unique properties and material individuality of any rock package.

The following topics might be a good basis for inter-regional comparison of jointing.

1. Rock age, lithology and age of fracturing.
2. Genetic classification.
3. Changes in stress fields, structural complexity and overburden.

3.5.1 Rock Age, Lithology and Age of Fracturing

Fracturing took place on the Appalachian Plateau, northeastern United States, in Devonian multi-rock formations of siltstone, shales and limestone. Most foreland fracturing occurred in the Upper Paleozoic during four phases of the Alleghanian orogeny. Later, post-Alleghanian joint sets developed. In the Bristol Channel province, southwestern Britain, Lower Jurassic limestone underwent fracturing during several tectonic stages. The important joint set J3 was formed during the compressional Alpine orogeny that started in the Eocene to Oligocene, and continued through the Late Oligocene to Miocene. In the Beer Sheva province, south Israel, fracturing started during sedimentary burial in the Lower Eocene and continued throughout the Oligocene/Miocene, or later uplift periods, cutting Lower Eocene and Middle Eocene chalks. In the Jurassic sandstone of the Zion province, northwestern U.S., fracturing probably started in the Mesozoic, but main geo-morphological developments occurred at younger Pliocene-Pleistocene ages. In all these events, in unconnected provinces, different time intervals elapsed between sedimentation and rock fracture, and in each province fracturing took place in different multi-stages, lithologies and stratigraphies. Apparently, the numbers of fracturing styles and packages of fracture-sediment type are limited, but their combinations are numerous.

3.5.2 Genetic Classification

The attempts at genetic classification of joints in the various provinces (i.e., distinguishing between burial, syntectonic, uplift and post-uplift joints) are not simple. Notwithstanding, efforts for obtaining such a classification are justified because they enable one to discriminate between tectonic phases and to establish the actual circumstances of fracturing, such as depths, and sequence of paleostresses.

Fracturing in the Appalachian Plateau probably started in the burial joints of set 320–330° (Younes and Engelder 1999), and these can be linked quite likely to the Ib (325–337°) burial set (Bahat 1991a, p. 251). Fracturing continued with three to four syntectonic stages during the Alleghanian orogeny. Young, post-Alleghanian joints are possibly related to post-uplift events (Hancock and Engelder 1989). In the Bristol Channel province, all jointing seems to be syntectonic (Engelder and Peacock 2001). The Beer Sheva province presents joints of all the four genetic groups, while the magnificent fractures at Zion National Park are mostly post-uplift joints.

3.5.3 Changes in Stress Fields, Structural Complexity and Overburden

Fracture-producing stresses were notably dissimilar in the various provinces. Post-uplift jointing at Zion National Park takes the form of parallel, simple longitudinal splitting (Bahat et al. 1995, 1999), mostly preserving the orientation of older fractures. In this case, maximum compression was vertical, induced by overburden weighing down on present-day plateau rocks consisting of strata that were not folded.

Horizontal compression of the Alleghanian orogeny induced two sequences of rotation (Sect. 3.1) in the Appalachian Plateau, and these were characterized by small differential stresses (6 MPa, Younes and Engelder 1999) and only slight folding. In the Bristol Channel province, joints propagated in response to Cenozoic Alpine compression. This compression also generated the growth of buttress anticlines through the reactivation and reversed pre-Alpine normal faults. The Alpine inversion, which was accompanied by intense deformation, reflects large regional, differential stresses (27.8 MPa in limestone and 32.2 MPa in shale, Engelder and Peacock 2001, Fig. 9). On these buttress anticlines, local stresses were superimposed, forming joints along faults, and other singularities.

In the Beer Sheva syncline, two stress types are distinguished: A regional stress that activated various synclines in the Syrian Arc and affected mainly burial joints, and an intra-synclinal stress, confined to the Beer Sheva syncline that was connected with the formation of syntectonic and uplift joints in the syncline. Young uplift joints superimposed on burial or syntectonic joints are common. Single layer fractures in the slightly folded Eocene chalks were created by very low differential stresses (–0.5 MPa extensile effective stress, Bahat 1989).

When comparing the four provinces, the degree of structural complexity increases from Zion through the Appalachian Plateau, to the Bristol Channel buttress anticline, and finally, the most complex pattern is found in the Beer Sheva syncline. Overburden was thin and fracturing took place close to the ground surface in the canyon at Zion National Park. In the Beer Sheva syncline, overburden was also thin, particularly when fractures developed during and after uplift (possibly tens of meters, probably not more than several hundred meters). The question at what depths fracturing of the various sets occurred, remains partly enigmatic and justifies further research.

Chapter 4

Jointing in Granites

Whereas Chap. 3 concerns fracture provinces in sedimentary rocks, Chap. 4 relates to fracture provinces and domains in granites. While a fracture province may concern many plutons that maintain geographic and genetic affinities, a fracture domain often relates to a particular pluton or a pluton-complex in a province. There are many publications on granite that relate to different fracture aspects of the rock (e.g., Johnson 1970; Segall and Pollard 1983). This chapter does not provide a review of the subject; rather it focuses on certain topics. We divide Chap. 4 into two parts. Part 1 (Sect. 4.1–4.6) relates to geological and fractographic aspects of joints; it presents a characterization of joints cutting granites in central Europe and in California, showing the diversity of joint formation in different fracture provinces and domains. The descriptions relate to joints throughout various stages in the granite histories, from fracture during the cooling history of the pluton to fractures that formed after the pluton had arrived at the ground surface. Part 2 (Sect. 4.7–4.10) concerns fracture mechanic aspects of joints, particularly problems of dynamic fracturing by drawing on fractographic properties of joints as follows:

1. Plotting joints on the velocity v vs. stress intensity K curve.
2. Presenting a general discussion of various manifestations of v and K in granites.
3. Calculating the new fracture areas on *en echelon* vs. hackle fringes.
4. Estimating the "index of hackle raggedness" of joints.

We present at the end of this chapter (Sect. 4.11) a description of a method how to drill in situ parts of fractures from the rock.

Part 1

4.1 Background

Bankwitz and Bankwitz (1984, Fig. 8 and 13) describe joints about 40 m long in granite from the South Bohemian Pluton near Mrákotin, west of Telc in the Czech Republic. These joints were considered to be representatives of the uplift stage (Bahat 1991a) because they resembled uplift joints cutting granite in Sinai (Bahat and Rabinovitch 1988). According to the latter authors, the presence of markings on the fractures in those granites was promoted by the rapidity of "late fracture" propagation during the uplift process, despite the coarse textures of the rocks. Recent studies of fracture in the South

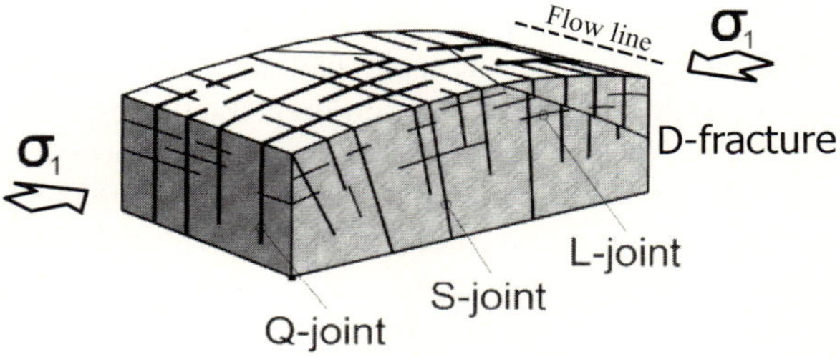

Fig. 4.1. The Cloos fracture model. Orientation of primary, near vertical Q- and S-joints and oblique faults D, as well as sub-horizontal L-joints with respect to the regional maximum horizontal principal stress σ_1, which is perpendicular to flow lines in the country rock (modified after Müller et al. 2001)

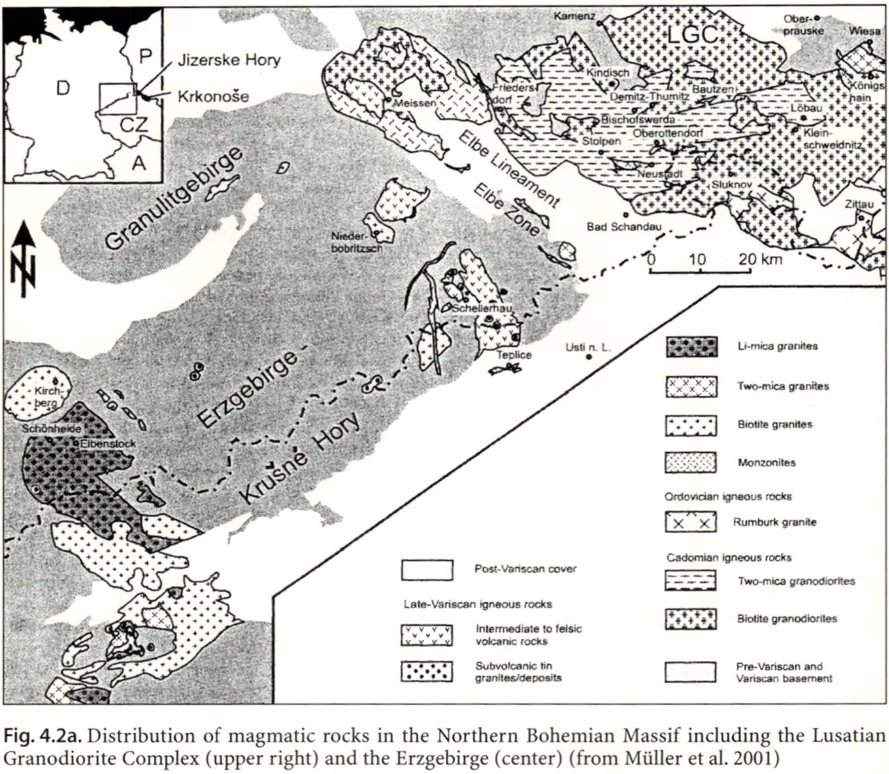

Fig. 4.2a. Distribution of magmatic rocks in the Northern Bohemian Massif including the Lusatian Granodiorite Complex (upper right) and the Erzgebirge (center) (from Müller et al. 2001)

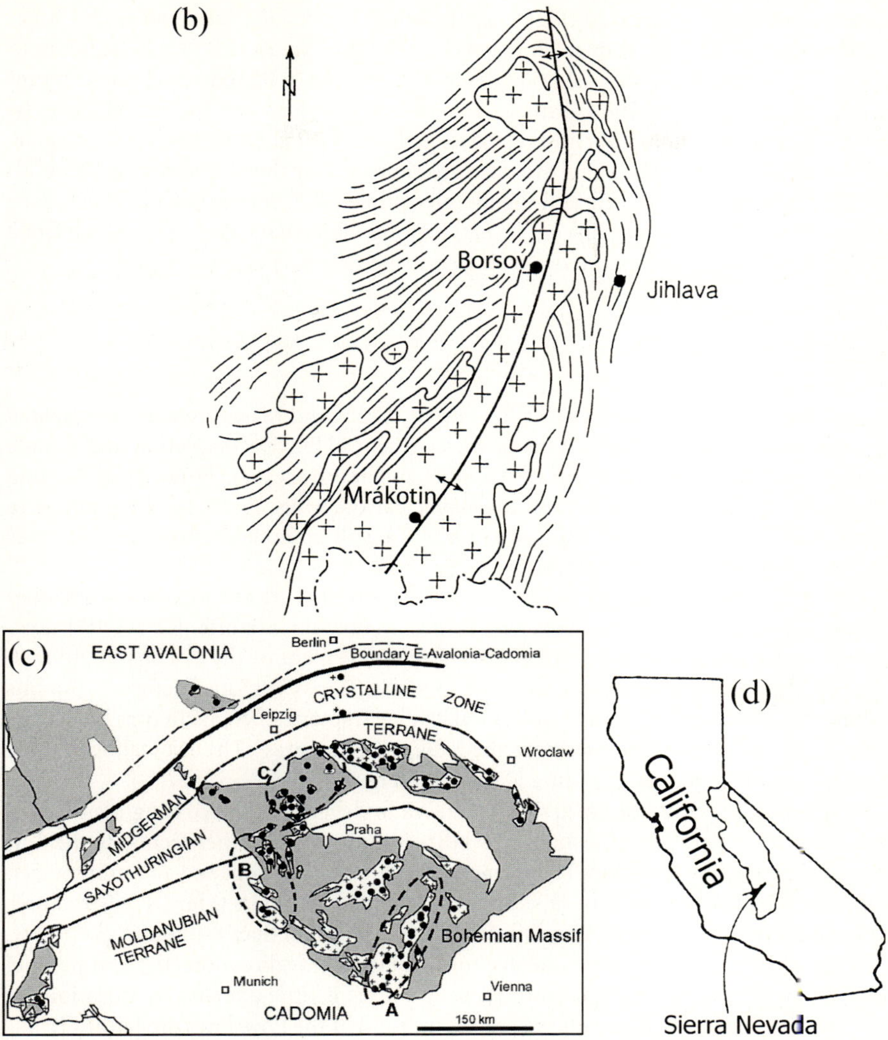

Fig. 4.2b–d. b A simplified geological map of the South Bohemian Pluton (marked +), the axis of major fold the orientation of foliation of gneiss county rock on various sides of the pluton (modified after Benes 1971). **c** Late to post-Variscan granites of the Bohemian Massif: *A:* South Bohemian Pluton; *B:* Oberpfalz plutons; *C:* Erzgebirge plutons; *D:* Lusatian Granite Massif. *Dots* mark localities of investigation by Bankwitz et al. 2004 (from Bankwitz et al. 2004). **d** The Sierra Nevada Batholith from central California (modified after Segall and Pollard 1983)

Bohemian Pluton reveal however that many vertical joints are in fact pre-uplift as will be demonstrated in the following sections. These studies suggest that rapid fracture can occur by "early fracture" in granite while this body still exists at great depths

We begin by presenting the Cloos (1921, 1922) pioneering model of fracture in granites (Fig. 4.1). The Cloos model provides a sound basis for the classification of fracture in granites and for distinguishing between "early fracture" and "late fracture". A series of recent investigations on fracture in granites from three different provinces, the North

Bohemian Pluton in Germany (Fig. 4.2a), the South Bohemian Pluton in the Czech Republic (Fig. 4.2b), both in central Europe (Fig. 4.2c), and the Sierra Nevada Batholith in central California, USA (Fig. 4.2d) are described. The North Bohemian Pluton is represented by two domains, the Lusatian Granodiorite Complex and the Erzgebirge. The South Bohemian Pluton is portrayed by data gathered particularly in two distinct quarries (Mrákotin and Borsov). The two domains representing the Sierra Nevada Batholith are the Knowles Granodiorite and the pluton complex in Yosemite National Park. Synthesized into these descriptions are several investigations that have not been published before. A section on comparative jointing ends the first part of this chapter.

4.2 The Cloos Model

According to Müller et al. (2001), Cloos (1921, 1922) and his associates investigated the development of fracture in igneous rocks in central Europe, in relation to the mode of emplacement of an intrusive mass. Cooling and hence crystallization, shrinking and degassing of the igneous melt first takes place at the walls and roof of the intrusive mass. Continued movement and intrusion of the still liquid core give rise to stresses in the outer shell of the intrusion.

A system of three main types of "primary fractures" that are orthogonal to each other is recognized and defined by Cloos with respect to flow-lines of oriented crystals, flow-planes and *schlieren*, which develop during the movement of the viscous liquid melt through the process of intrusion. These are Q-joints, ("cross", or "dip fractures"), S-joints ("longitudinal fractures"), and L-joints ("flat-lying fractures"). A fourth type that forms later than the "primary fractures" is termed "diagonal fractures". The orientation of these fractures with respect to flow-lines is shown in Fig. 4.1.

Q-joints are among the earliest of fractures to develop in the cooling mass. Typically, they form parallel to the main stress direction (σ_1) and perpendicular to the flow lines, i.e., stretching lineation. These fractures are sometimes opened and mineralized (aplite and hydrothermal minerals), and correspond to ac-joints (Fig. 4.1). Q-fractures are less jointed (in terms of divisibility, Müller et al. 2001) than S-joints.

S-joints are oriented perpendicular to the main stress direction (σ_1) and parallel to the flow-lines. The orientation of these fractures is little affected by variations in pitch of the flow-lines. However, variations in trend of the flow-lines are faithfully followed by changes in the strike of the S-joints. S-joints are closed, rarely filled with aplite or "dike" material, and the minerals are usually different from those found in the other kinds of primary fractures.

L-joints tend to develop sub-horizontal near the apex, or dome, of an intrusion or in flat sheets and laccoliths. It is difficult to see how these structures can he interpreted on dynamic grounds. It has, however, been suggested that they form when the center of an intrusion shrinks due to cooling. These structures are also filled with hydrothermal minerals, and have been referred to as primary flat-lying fracture so that they may be distinguished from exfoliation joints that may be much younger.

Diagonal fractures form at 45° or more, to the trend of the flow-lines, resulting from compression normal to, and extension in the azimuthal direction of flow-lines. These fractures are also commonly filled with aplite or hydrothermal minerals. Displacements along the diagonal fractures are clear, indicating that they are shear phenomena.

4.3
Fracture in Granites from the North Bohemian Massif in the Lusatian Granodiorite Complex and the Erzgebirge

Bankwitz and Bankwitz (2004) describe various European granites (mostly Variscan), with a special reference to their age and some fracture properties. This section is mostly cited from Müller et al. (2001). The study areas are the Lusatian Granodiorite Complex (LGC) and the Erzgebirge in SE Germany (Fig. 4.2a), with the main emphasis of the study on the LGC. Kossmat (1927) assigned both the LGC and the neighboring Erzgebirge to the Saxothuringian Zone of the Variscan orogen (Fig. 4.2c).

4.3.1
The Lusatian Granodiorite Complex

4.3.1.1
Geological Background

Based on geological considerations, Watznauer (1954) wrote of a "granite flood", which was assumed to extend from the LGC across the Erzgebirge into Thuringia. In contrast, Suess (1926) proposed to separate the LGC from the Saxothuringian Zone due to the different structure and development of the basement in both domains. Since then, a number of differences and similarities of the crustal structures and their development have been found. The principal difference between the provinces is the main age of intrusion, which is Cadomian (540 to 530 Ma) in the LGC, and late-Variscan (325–318 Ma) in the Erzgebirge (Bankwitz and Bankwitz 2004).

The LGC is separated from the Erzgebirge by the Elbe lineament and by the rock collage of the Elbe Zone (Fig. 4.2a). The LGC is bordered by major faults and over thrusts. The virtually undeformed Cadomian granitoids cover an area of about 2 500 km^2 and constitute a genetically related, mostly peraluminous magmatic suite (Krauss et al. 1992). Most of the contacts of the granodiorites and wall rocks are defined by faults resulting from uplift since the Late Proterozoic (Eidam et al. 1992). The late-Variscan granitoid activity in the LGC is represented by the Stolpen and Konigshain monzogranites, which intruded as stock-like bodies along large deep-seated regional faults (e.g., Hirschmann 1966).

4.3.1.2
The Gravity High of Lusatia

Müller et al. (2001) investigated the gravity field of Lusatia and of the Erzgebirge, using the tomographic method of Linsser-filtering (Linsser 1967; Conrad et al. 1996; Trzebski 1997). The Linsser-filtering provides an alternative technique to the conventional selective gravity field methods since it allows depth selective gravity gradient analyses down to 20 km depth, based on the Bouguer gravity field. Thus revealing information about the three-dimensional extent of steep-dipping (<45°) rock density contrasts in selective depth levels. The LGC and the Erzgebirge are distinguished by the most significant gravity anomalies of the Bohemian Massif: The gravity high of Lusatia and the gravity low of the Erzgebirge. The widespread occurrence of high density rock beneath the LGC is interpreted in terms of the presence of a large-scale layered gabbroic intrusion ("gabbroic layer", Lindner 1972; Behr et al. 1994) of unknown age associated with a mantle upwelling. This poses the question of whether a gabbro intrusion beneath the LGC could generate around 25 000 km^3 (2 500 $km^2 \times 10$ km thickness) of magma from metagraywacke-metapelitic turbidites (anatexis of late Proterozoic, e.g., Eidam et al. 1992), which are exposed at the NW margin of the LGC.

4.3.1.3
Results and Discussion

According to Müller et al. (2001) the strike of the interpolated Q-jointing is harmonic over the whole area of the LGC and forms a sigmoidal pattern (Fig. 4.3a). The general parallelism of the strike of the Q-jointing and the axis of the gravity high is notable. Subsequent to the formation of the LGC, gabbroic magmas mostly used these Q-joints as ascent paths in the western and central part of the LGC (Fig. 4.3b). In the east of the LGC the mafic magmas intruded diagonal joints. The sigmoidal striking of the Q-joints is parallel to the indicators of 5, 7.5 and 10 km depths, which mark the southern boundary of the Lusatian gravity high. The second vertical joint direction, the S-jointing, forms angles between 90° and 60° with the Q-jointing (Fig. 4.3c). Angles of more than 80° between the S- and Q-joints are in accord with Cloos's joint model. The primary orthogonal joint system is overprinted by sets of diagonal joints. This occurs in association with large-scale fault zones at the margin of the LGC. Generally, the primary joint sets Q and S are best preserved in the core zone of the southern part of the gravity high in the area of Bischofswerda and Bautzen (Fig. 4.2a). Thus, it is interpreted by Müller et al. (2001) that since the high density rock beneath the LGC is considered to be a large-scale layered gabbroic intrusion and the pattern of Q- and S-joints fit the high gravity pattern quite well, the updoming of the postulated gabbroic intrusion could have induced these fractures during the solidification of the LGC between 520 and 590 Ma. Lobst (2001) examined field relationships of fractures in the Klosterberg quarry, some 10 km SW of Bautzen, and found that steep E-W oriented joints were older than N-S joints. Apparently, the Klosterberg quarry is located geographically, such that the E-W and N-S joints fit the directions of the Q-joints and S-joints, respectively, in the maps drawn by Müller et al. (2001) (Fig. 4.3). This suggests that the Q-joints were formed before S-joints.

Müller et al. (2001) could not confirm the existence of late-magmatic, flat-lying L-jointing (Fig. 4.1) in the Lusatian granite. On the other hand, they identified a joint set subparallel to the surface topography that was related to late processes of surface exposure of the granite body, which they consider to be an exfoliation set. They observe that the exfoliation is best developed in granites and granodiorites with large joint spacing (>5 m). Such conditions are given in the top area of the Lusatian gravity high between Bautzen, Bischofswerda and Kamenz (Fig. 4.2a). The exfoliation is poor and irregular in regions where joint spacing is <1 m. The distance between the surface parallel joints increases with increasing depth, as observed in other granite quarries (e.g., Johnson 1970, p. 358).

In summary, Müller et al. (2001) show that the primary joint system of the Cadomian Lusatian granodiorite fits the model of updoming by an underneath intrusion: In this case, the presumed gabbroic intrusion at depth. The fractures under investigation formed during the solidification of the LGC between 520 and 590 Ma, such that the Q-joints were formed before S-joints. However, in contrast to the Cloos model it has been found that the formation of sub-horizontal joints in the granite is related to the external process of exfoliation and has no relationship to the primary joint set.

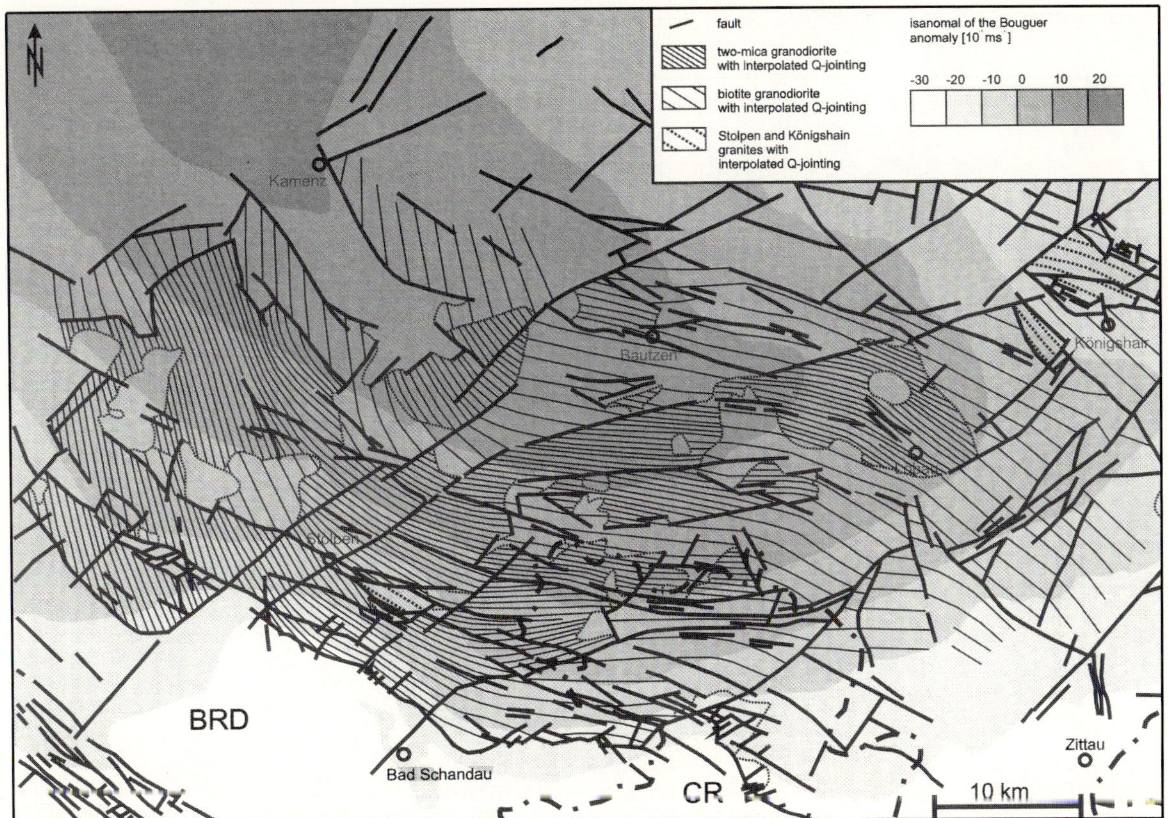

Fig. 4.3a–c. Structural maps of the Lusatian Granodiorite Complex. The maps are underlain by isoanomals of the gravity high of Lusatia with the local maximum NW Kamenz. **a** Main faults (*thick lines*) and interpolated striking Q-joints (*thin lines*) (from Müller et al. 2001)

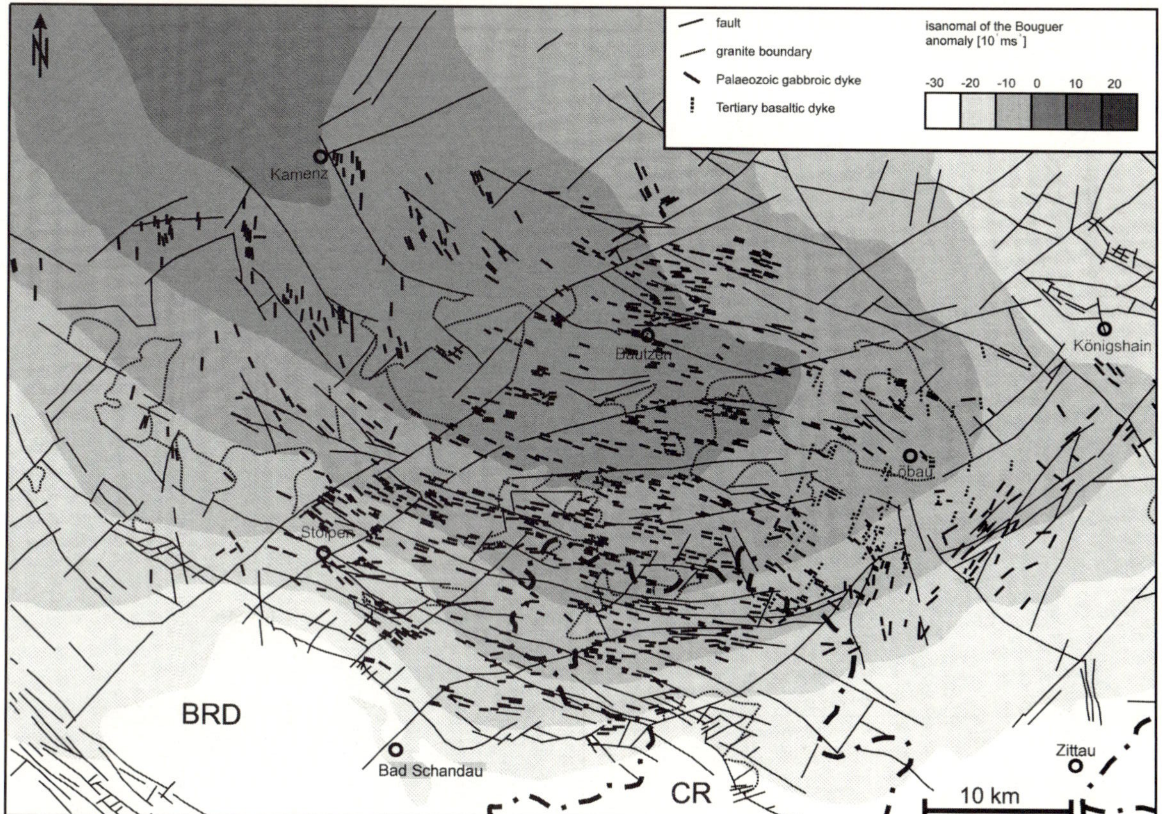

Fig. 4.3b. Structural map of the Lusatian Granodiorite Complex. The map is underlain by isoanomals of the gravity high of Lusatia with the local maximum NW Kamenz. Devonian gabroic dikes used predominantly Q-joints as ascent paths, whereas Tertiary basaltic magmas intruded along N-S to NE-SW trending faults (from Müller et al. 2001)

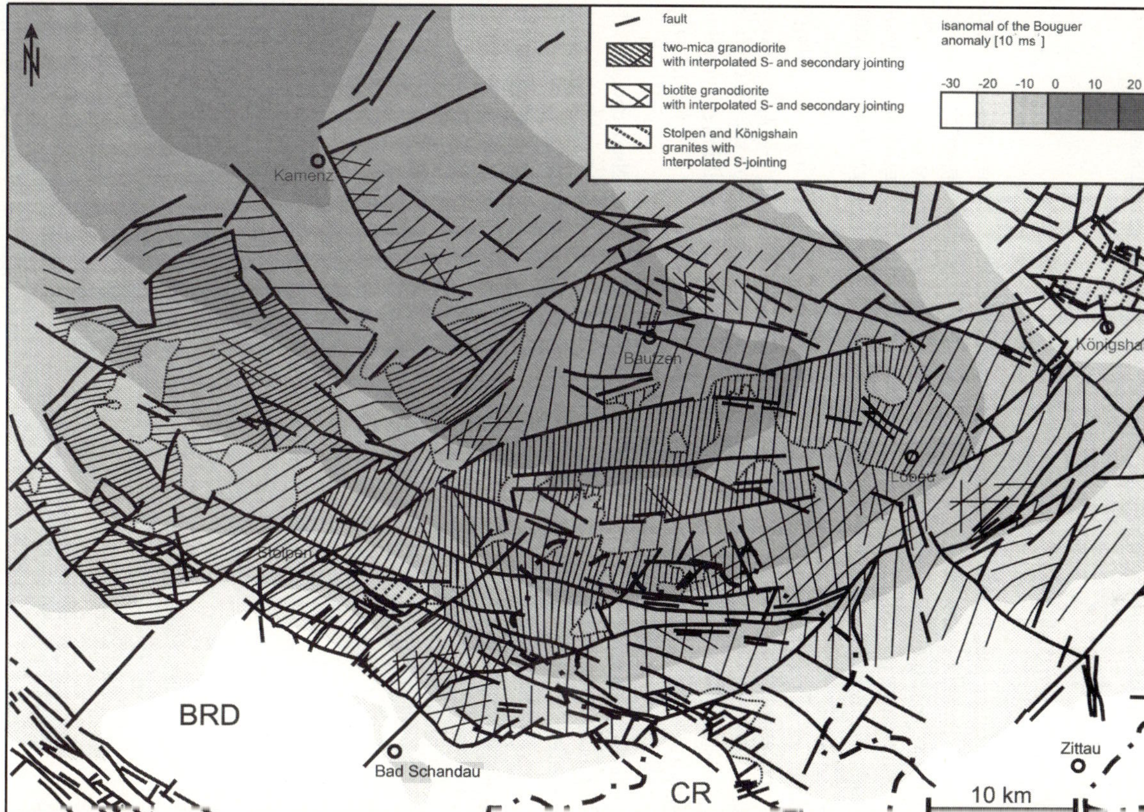

Fig. 4.3c. Structural map of the Lusatian Granodiorite Complex. The map is underlain by isoanomals of the gravity high of Lusatia with the local maximum NW Kamenz. Main faults (*thick lines*) and interpolated striking S-joints (*thin lines*) and additional steep-dipping joints (from Müller et al. 2001)

4.3.2
The Erzgebirge

We consider previous investigations on two plutons in the Erzgebirge, the Niederbobritzsch granite (northeastern Erzgebirge, Fig. 4.2a), studied by Müller et al. (2001) and the Eibenstock granite (southwest Erzgebirge), studied by Bankwitz and Bankwitz 2004).

4.3.2.1
The Niederbobritzsch Granite

According to Müller et al. (2001), the Erzgebirge covers an area of about 6 000 km^2. This area is characterized by Late Carboniferous plutons that intruded Upper Proterozoic to Lower Palaeozoic metamorrphic sequences. The plutons are composed of spatially separated multiphase intrusions with complex textural and field relationships. The granites are traditionally divided into the older (330–320 Ma), and the younger (300–290 Ma) intrusive complexes (Laube 1876; Breiter and Seltmann 1995).

The Bouguer gravity low of the Erzgebirge is the most significant negative gravity anomaly in Central Europe with an extreme of –76 mgal near Karlovy Vary. It extends along the northern margin of the Bohemian terrain and is limited by the Oberpfalz block in the west. The 250-km-long and 40-60-km-wide NNE-SSW striking gravity low continues south of the gravity high of Lusatia to the Jizerské Hory and Krokonose Mountains, where the minimum turns to an E-W direction.

Müller et al. (2001) point out that the NNE-SSW trending joints of the Niederbobritzsch pluton (Fig. 4.4a) are concordant to the ac-plane of the Proterozoic gneisses. These joints are interpreted as Q-joints, because they are opened joints partly filled with quartz. The S-joints strike in the direction of the main stretching axis of the wall rock. Thus, the development of the Q- and S-jointing of the Erzgebirge granites is adapted to the syn-intrusive stress field of the intrusion frame (wall rock), which corresponds to Cloos's joint model. On the other hand, the primary joint sets of the Cadomian granodiorites in the Lusatian Granodiorite Complex reflect the updoming of a postulated gabbroic intrusion, which underlies the granodiorites (as mentioned above).

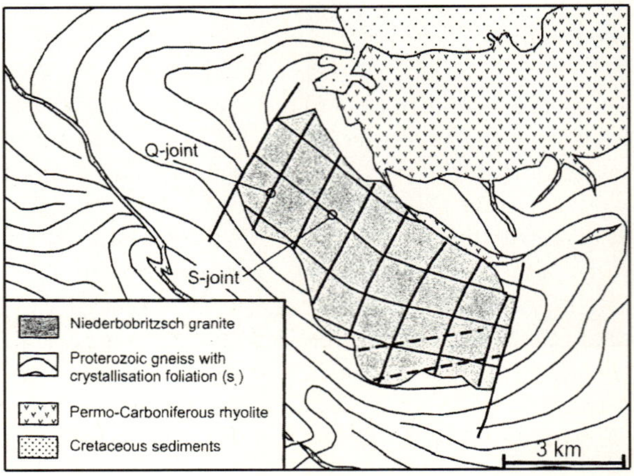

Fig. 4.4a. The orthogonal Q- and S-jointing of the late-Variscan Niederbobritzsch granite in the eastern Erzgebirge (from Müller et al. 2001)

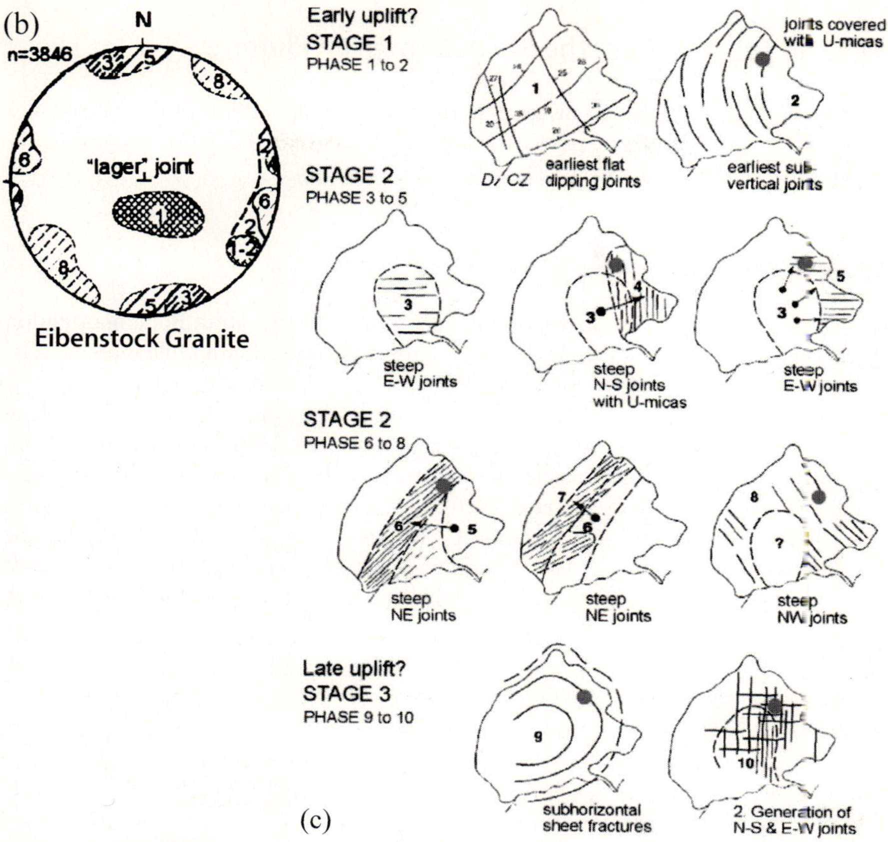

Fig. 4.4b,c. b Synoptic diagram of lower hemisphere stereographic projection of joints in the Eibenstock granite. Numbers *1–8* of maxima indicate the relative sequence of joint formation (from Bankwitz and Bankwitz 1995, 2004). **c** Schematic illustration of the mapped joint sequence in the Eibenstock pluton, see text for explanation (from Bankwitz and Bankwitz 1995, 2004)

4.3.2.2
The Eibenstock Granite

The Eibenstock granite is a large laccolithic (exposure of 20 × 40 km) with complex fracturing. Bankwitz and Bankwitz (1995, 2004) distinguish a series of joint sets in this granite, which they describe in great detail. Jointing started with flat to oblique dipping (18–35°) "lager" joints (Cloos 1921) (Fig. 4.4b,c). The other sets were steeply dipping with the exception of set 9. The first subvertical joints (phase 2) followed the shape of the exposed pluton. The relative timing of the various other sets suggests fracture migration with time, e.g., from phase 3 to phases 4 and 5, and from phase 6 to phases 7 and 8. Sheet fractures formed in stage 3, and finally, a second generation of smaller, N-S and E-W-trending joints was created, terminating at the sheet fractures. A dense deposition of greenish Uranium-micas on joints of sets 2 and 4 helps to distinguish the two generations of orthogonal sets.

4.4
Joints in Granites from the South Bohemian Pluton

This section concentrates on investigations that were carried out in the Borsov and the Mrákotin quarries, citing primarily Bahat et al. (2001a, 2003), Bankwitz and Bankwitz (2004) and Bankwitz et al. (2004).

4.4.1
Geological Setting of the South Bohemian Pluton

The study areas under consideration here are located in the South Bohemian Pluton (SBP) (Fig. 4.2b), which is the northeastern part of the large South Moldanubian Massif that belongs to the internal zone of the Variscan belt of Europe (Fig. 4.2c). The country rock consists predominantly of kyanite-sillimanite bearing gneisses and schists of Late Proterozoic to Early Paleozoic age (Suk 1984; Matte et al. 1990). The metamorphism occurred at 370–350 Ma. The stretching lineation trends NW-SE with indications of a south-eastward shearing (Matte et al. 1990). The SBP intruded the core of an antiform composed of gneisses and schists (Benes 1971). The SBP trends NNE-SSW through the Czech part of the Bohemian Massif. According to Bankwitz et al. (2004),the SBP extends over a distance of 150 km between Havlíckuv Brod in the north and the Danube River,

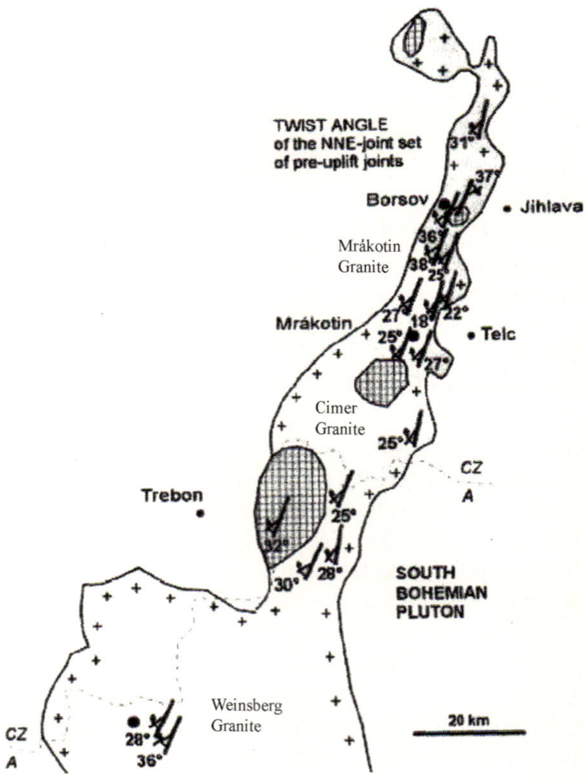

Fig. 4.5.
Geological sketch map of the South Bohemian Pluton, showing the dominant first vertical fracture orientation, set NNE. *Dots:* Locations of investigated fluid inclusions. *Cross hatched areas:* Granite bodies in the axial part of the SBP (modified Breiter 2001 by Bankwitz and Bankwitz 2004)

and further south, recognized from drillings below the Molasse Zone of the Alps (Scharbert 1998). The root zone of the SBP occurs in its southern part between the Danube and the Pfahl faults. The granitoids were emplaced at mid- to upper-crustal levels into hot country rocks (~15–18 to 7–10 km, ~450–650 °C) shortly after the thermal peak of regional metamorphism (Büttner 1997). *P*-*T*-paths in the country rocks suggest that granite formation, low-*P*/high-*T* metamorphism and extensional thinning were preceded by a phase of intense crustal thickening that occurred during late Palaeozoic continent-continent collision (Gerdes et al. 1998; Matte et al. 1990)

The SBP is a large late-orogenic batholith, which consists of several phases of intrusion with isolated sub-types (e.g., Cimer, Mrákotin, Eisgarn, Weinsberg, Freistadt, Mauthausen types) (Breiter and Koller 1999; Breiter and Sokol 1997; Bankwitz et al. 2004), one of which is the Borsov granite that has a close affinity to granodiorite. Its geochemistry is strongly influenced by the metamorphic country rock. The granitic rocks have developed from a high-*K* peraluminous melt (Breiter and Koller 1999; Matejka and Janousek 1998) with the presence of biotite and biotite-muscovite petrographic manifestations in the rock. According to Matte et al. (1990), the intrusion age is about to 330 Ma. Field contacts in the SBP at the Borsov quarry show that the main fracture systems occur in the following sequence:

1. Early sub-horizontal tension fractures.
2. Vertical or sub-vertical E-W and NNE trending joints.
3. Late, sub-horizontal joints that are represented by the uplift sheeting (e.g., Dennis 1987), cutting the other joints without influencing them, i.e., the joints of the second group had ended their formation before the late joints started.

Both early horizontal and vertical tensile fractures are expected to form in association with the increasing magma pressure in the pluton (Knapp and Norton 1981). This section focuses on the NNE vertical set (Fig. 4.5) due to its particular regional significance, minimizing the treatment of other sets in the pluton.

The South Bohemian Pluton was chosen for investigation for several reasons (Bahat et al. 2003). First, these granites are cut by joints that exhibit a great variety of fracture surface morphologies (displaying most of the features in Fig. 2.1a,b). Second, the fracture morphologies of the joints are usually well preserved, because most of these granites were not deformed in post-Variscan (Late Paleozoic) times. Therefore, the fractures maintained their original shapes from the time before the uplift. Third, the bulk of the existing knowledge on jointing is related to sedimentary rocks, and field observations are normally made along road cuts or natural exposures, which are shallow and seldom allow observations of large joints to their natural ends. These granite rocks have been quarried for more than a hundred years for construction purposes throughout Europe. The large quarries expose large walls required for obtaining good perspective views of large joint fractographic features that granites often exhibit. Fourth, and perhaps most important, the unique exposure of the Borsov granite allows the study of paleo-jointing at the intrusion depth before uplift and erosion. This provides an insight into fracture processes at great depths in the crust. Finally, the SBP is an important part of the Variscan orogen in central Europe and our study may add important information for use in future investigations of the orogen.

It is convenient to distinguish two classes of rock fracture: Those formed by shear stress and those by tensile stress. The vast majority of experimental work in fracture mechanics has concentrated on mode I, tensile fracture propagation (Atkinson 1987). The results of these studies, quite apart from their importance in understanding joint formation, may be applied to a wide spectrum of other geological problems. These include hydrofracturing, microcracks associated with larger structures, tensile fractures of macro and mega-scales (or extensile at great depth), dike formation, magmatic intrusion, boudinage and crevasse penetration (Atkinson 1987). Earthquake studies relate more to mode II shear operations (Scholz 1990), but may have mode I manifestations as well (e.g., Bahat 1982).

4.4.2 Fractographic Versatility on Pre-Uplift Joints in the Borsov Quarry

Existing fractographic studies on fracture in granites generally concern late histories of the granites, following their cooling and uplift (e.g., Holzhausen and Johnson 1979; Bahat et al. 1999), whereas investigations of fractures that formed during the cooling stages (e.g., Knapp and Norton 1981; Lee and Diehl 1989) commonly have not used fractographic techniques. The following sections concern fractures in two outcrops, the Borsov and the Mrákotin quarries from the South Bohemian Pluton in the Czech Republic (Fig. 4.2b), which reveal joints that formed during the early history of the granite.

4.4.2.1 *Fractographic Characterization of Ten Joints from the Borsov Quarry*

The ten joints with highly diversified fracture surface morphologies (Bahat et al. 2003) are briefly described photographically (Fig. 2.54–2.56), and in their corresponding drawings (Fig. 4.6–4.8). Their properties are summarized in Tables 4.1 and 4.2. The geological interpretation of the fractographies of these joints is presented in Sect. 4.4.5.1, and the fracture mechanic analysis is given in Sect. 4.8.

▶

Fig. 4.6. Drawings of joints from the Borsov quarry (see photographs in Fig. 2.54–2.56, modified from Bahat et al. 2003). Mirror planes that do not contain fringes. **a** The "multi-arrest marks joint", J1. Some fifteen arrest marks superposed by a faint plume. The fracture markings are confined in a "wedge" between two sub-horizontal boundaries, above and below the fracture pattern. Thickness of wedge at the left side is about 90 cm. **b** The "radial plume joint", J2, showing the fracture origin at the center and a radial plume representing the propagation mode. The joint is bounded by two horizontal partings. **c** The "quasi-core joint", joint J3. A mirror plane is decorated by elliptical arrest mark which is partly surrounded by additional partial arrest marks. The origin is situated in an eccentric position with respect to the ellipse, and a continuous plume that starts at the origin propagates in a radial manner to all directions, with various intensities. A symmetric axis passes through regions of greatest intensity plumes. Sub-horizontal fracture *p*, cuts the mirror. **d** The "multi-joint exposure", J4. The plume m1, of the "semi-circular joint" reveals a three-stage concentric growth into a semi-ellipse.The long axis of this semi-ellipse approximately coincides with an horizontal plane p1, that forms its upper boundary. The "circular joint" is characterized by a circular single arrest mark and a faint, single continuous plume, m2, that starts at an eccentric position with respect to the arrest mark and spreads approximately radially. Two upward propagating plumes occur on distinct parallel planes, m3 and m4 at the left side of the "circular joint". Note also the plume m5. Sub horizontal fractures p2 and p3 cut m3 and m2, respectively. However, m4 and m5 seem to truncate against the continuation of p2, suggesting that p2 predates m4 and m5

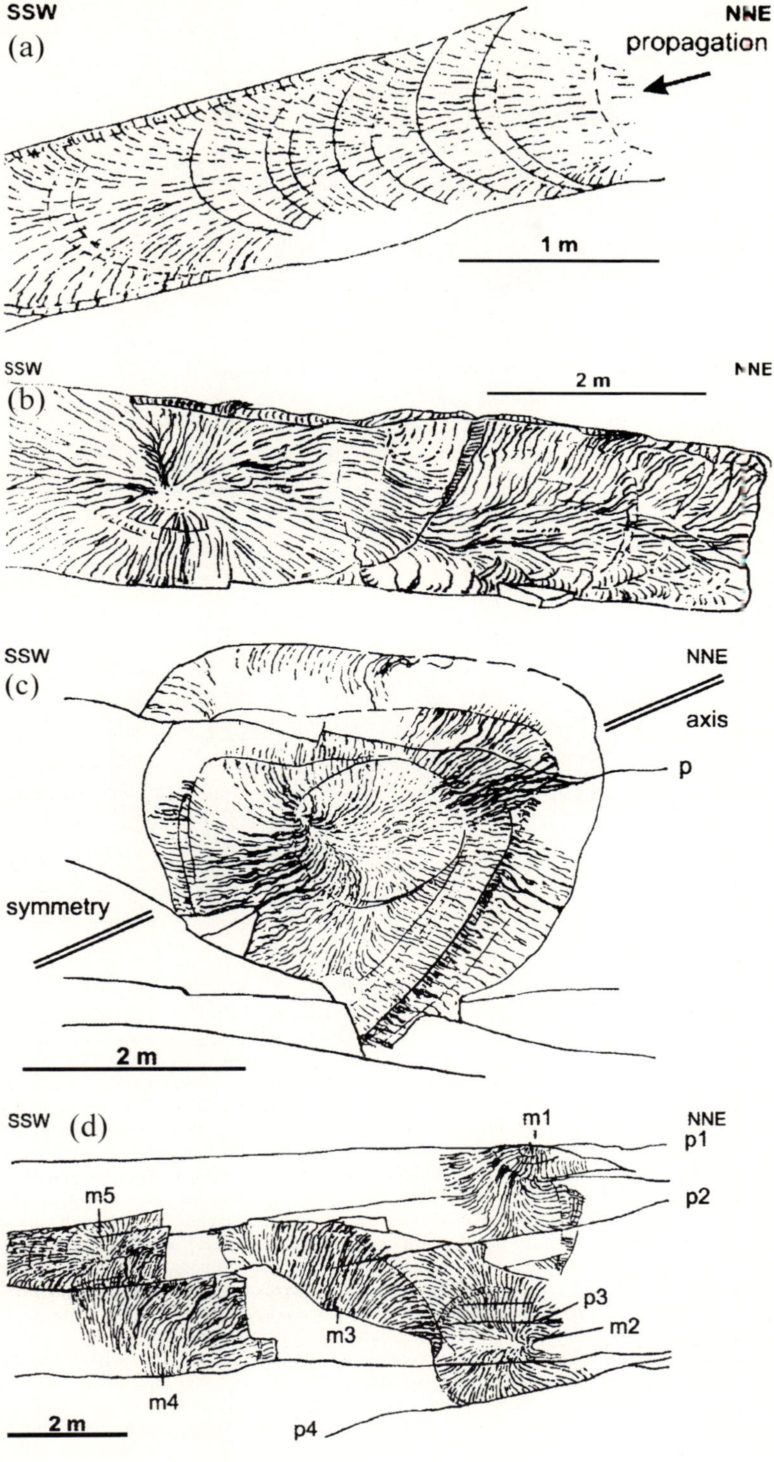
SSW
NNE
(a)
propagation
1 m
SSW
NNE
2 m
(b)
SSW
NNE
(c)
axis
p
symmetry
2 m
SSW
(d)
m1
NNE
p1
m5
p2
p3
m2
m3
m4
p4
2 m

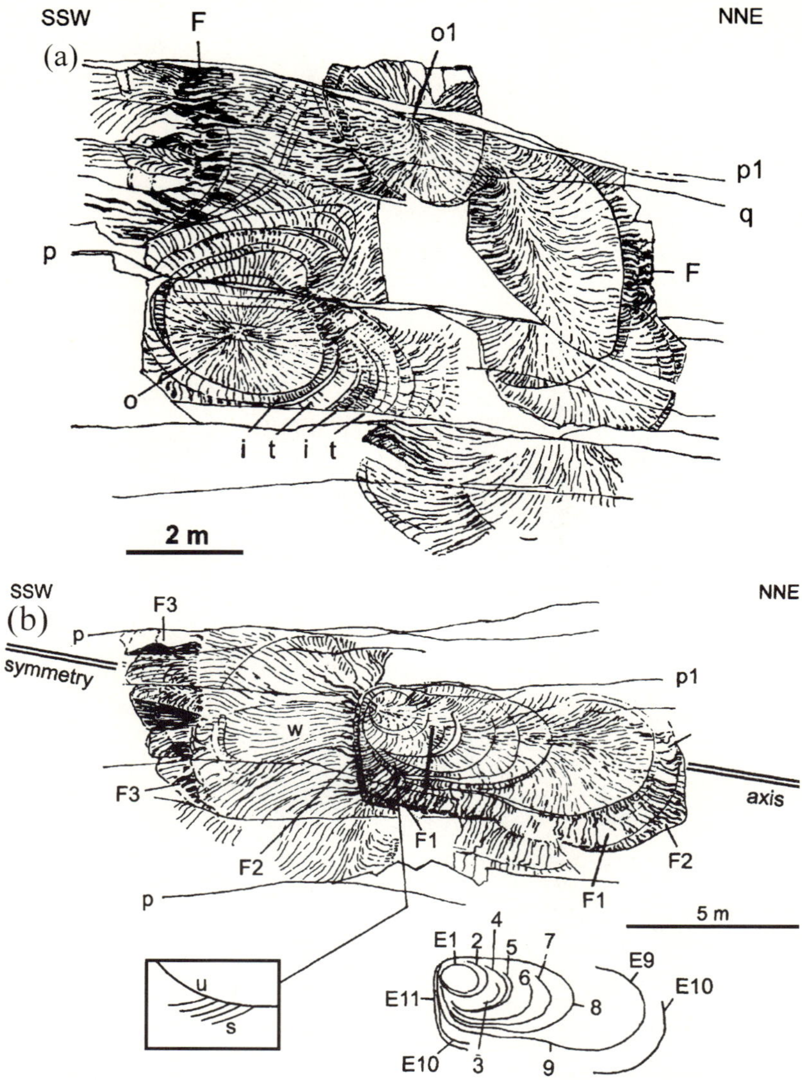

Fig. 4.7a,b. Drawings of joints from the Borsov quarry (see photographs in Fig. 2.54–2.56, modified from Bahat et al. 2003). Mirror planes that contain small *en echelon* fringes. **a** The "divided arrest mark joint", J5 starting at o. A major horizontal parting, p divides the lower semi-circular series of arrest marks from the upper ones of the "divided arrest mark joint". Note a rhythmic increase and decrease in barb-coarseness in certain spacings between the arrest marks, including an inner spacing, *i* which has a greater barb-coarseness than its adjacent outer one, *t*. Another plume started at o1 on parting p1 and was cut by later q. Two plumes terminated in small *en echelon* fringes F. **b** The "asymmetric bilateral joint", J6 displays some ten concentric arrest marks (E1–E10), that propagated from the origin at an eccentric position in ellipse E1 (*inset* at right). The arrest marks are superposed by a continuous plume divided by a symmetry axis. At the left side of the origin there is a large wavy plume, w almost without any arrest marks. A "triangle" fringe of *en echelon* cracks, F1 is separated from w by a "girdle" F2 of short *en echelon* segments. Inset at left shows *en echelon* segmentation, s that initiates from arrest mark u that corresponds to E9. F1 and F2 at the right side are interpreted to be the continuations of F1 and F2 at left. E11 and F3 are additional hypothetical arrest marks and fringe, respectively. J6 is limited by partings p and is cut by a series of sub-horizontal fractures (like p1). Vertical scale at the center is 2 m

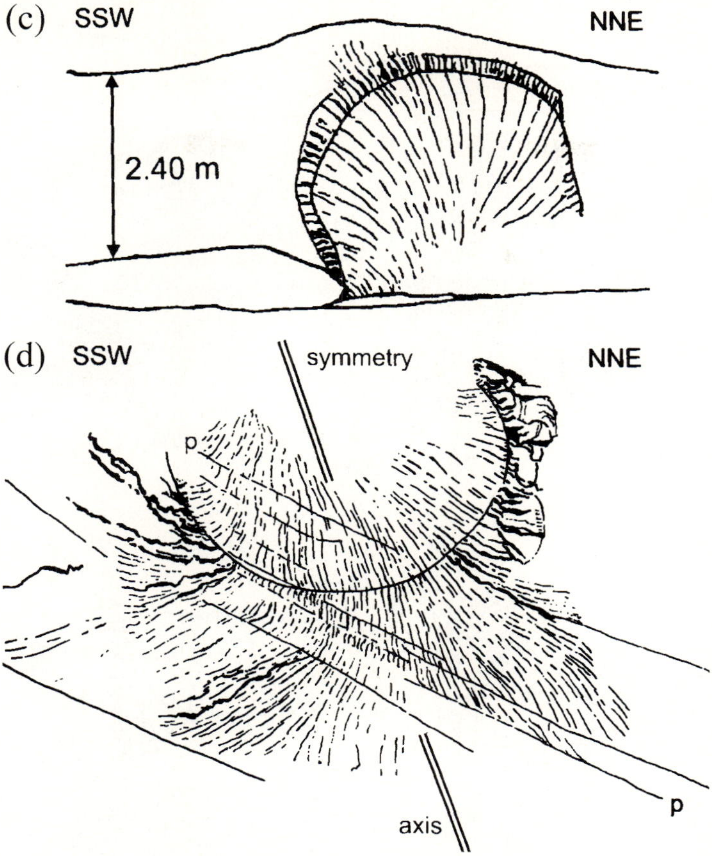

Fig. 4.7c,d. Drawings of joints from the Borsov quarry (see photographs in Fig. 2.54–2.56, modified from Bahat et al. 2003). Mirror planes that contain *en echelon* fringes. **c** The "uniform fringe joint", J7. The boundary between the mirror that is marked by a faint radial plume and the fringe of *en echelon* segments of approximately equal lengths is well defined. **d** The "partly erased joint" J8 shows a mirror that is surrounded by a convex downward circular boundary and a fringe of *en echelon* segments of non uniform sizes. Diameter of the mirror is approximately 10 m

▶

Fig. 4.8. (*see next page*) Drawings of joints from the Borsov quarry (see photographs in Fig. 2.54–2.56, modified from Bahat et al. 2003). Mirror planes with diversified fringes. **a** The "trefoil joint" J9, that forms the shape of a three leaflets flower, showing two types of ripple marks, delicate concentric undulations u (hardly visible), at the mirror center, and two lateral arrest marks a (showing pronounced morphologies), at the left side of mirror. Plumes with a general clockwise, cw stepping are superposed on the latter. The two sides of the mirror contain intensely fractured fringes. The fringe at the left side of the mirror is divided in correspondence with the zigzag of the mirror boundary, into five sectors by the designations *i*, *n*, *t*, *r*, and *m*, where segmentation is counter clockwise, ccw (for details see Bahat et al. 2001). Frame shows location of **b**. **b** The fringe at the right side of the mirror has the shape of a fan which starts at a low side next to the mirror, and opens upward away from the mirror, divided to five distinct sub-fringes. The sense of stepping is not uniform along the various sub-fringes. **c** The "hackled fringe joint" J10, showing a joint on which the fringe is particularly large compared to the mirror size (partly removed), and is heterogeneous with respect to having adjacent hackles and *en echelon* segments. **d** The large diameter in the simplified diagram of (**c**) is about 2 m

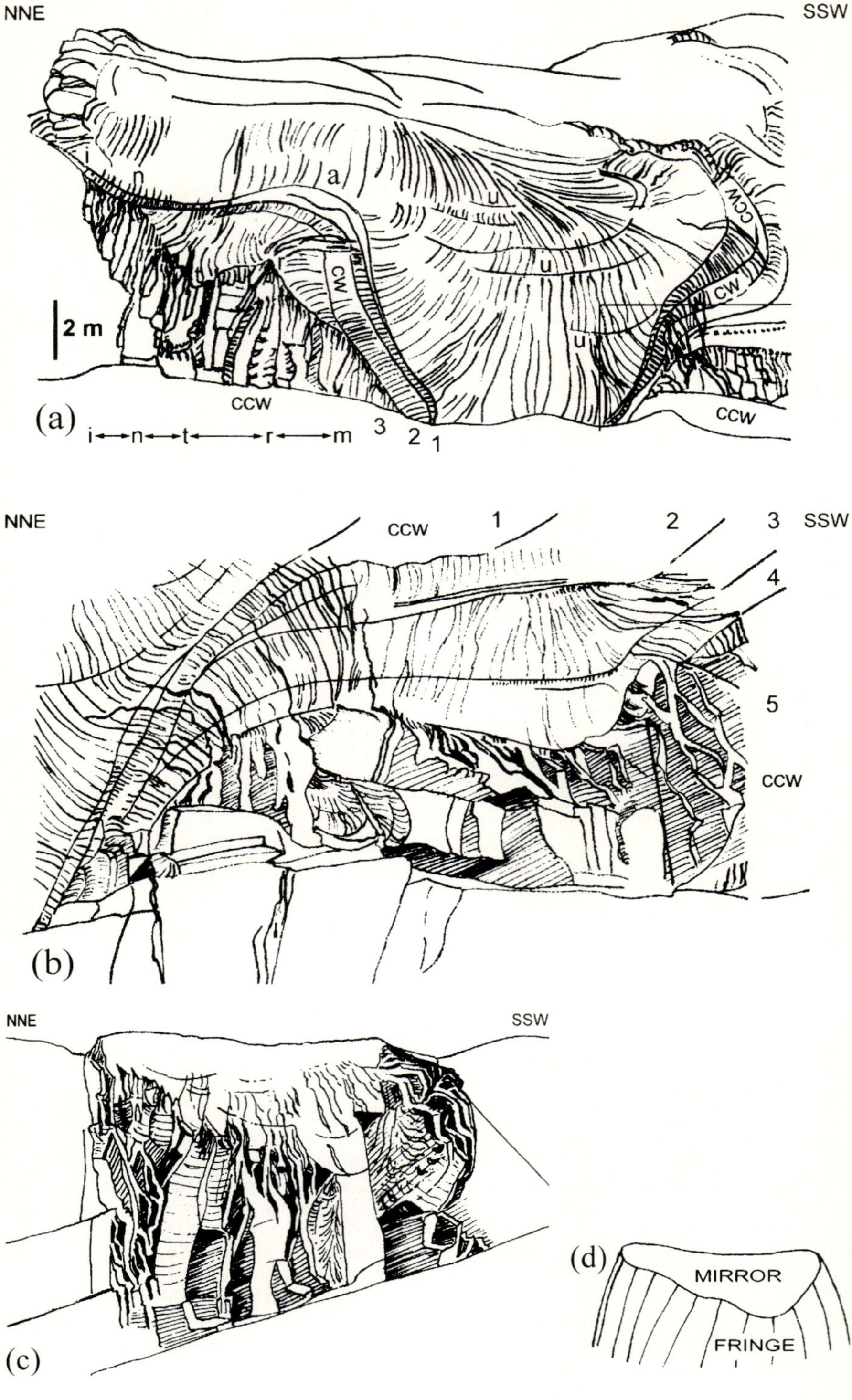

NNE
SSW
a
u
u
u
CCW
CW
CW
2 m
CCW
CCW
(a)
i n t r m 3 2 1
NNE
CCW
1
2
3
SSW
4
5
CCW
(b)
NNE
SSW
(c)
(d)
MIRROR
FRINGE

Table 4.1. Characteristics of the north-northeast-trending set of joints in the Borsov quarry (example: Ten joints, J1–J10)

Origin point	Mirror shape	No.	Ripple marks		Propagation direction	Superposed features	Fringes	Additional remarks
			Properties	Curvature				
Group 1: Mirrors without fringes								
J1: Multiarrest marks								
At the center of a block between two earlier joints	(Center erased)	15	Increasing width	About half a circle	Laterally, northward	Faint plume on the ripple marks	–	The joint is confined between two flat-lying earlier joints, narrowing northward and connected with a slight decrease of the ripple mark curvature (Fig. 4.6a).
J2: Radial plume joint								
At the center of the mirror	Elliptical	4	Increasing width		Laterally, northward and southward	Coarse plume	–	The joint is confined between two flat-lying earlier systematic joints separated by 186 cm (Fig. 4.6b).
J3: Quasi-core joint								
Excentric, close to the first undulation (left)	Deviation from circularity into an ellipse	4	Some are only partly observable; increasing width	First circle, then ellipse	First oblique to the center (right down) of the ellipse, then radially	Continuous plume; coarse barbs propagating perpendicular to the initial direction	–	One of the postuplift sheeting joints cuts joint J3 horizontally without influencing it. The coarse plume is restricted on the oblique direction (right upward and left downward) in zone of low-ripple mark relief (Fig. 4.6c).
J4: Multi-joint								
a. Semicircular joint								
At the center of the plume	m1: circular	3	Increasing width	Concentric half a circle	Radially	Plume with three-stage growth	–	An earlier horizontal joint forms the upper boundary (p1)
b. Circular joint								
Close to the right side, within the single ripple mark	m2: circular	–	–	–	First laterally southward, then radially	Faint continuous plume	–	Three subhorizontal postuplift joints p2–p4 (sheeting fractures) cut the joint J4 without influencing its pattern.

Table 4.1. *Continued*

Origin point	Mirror shape	No.	Ripple marks		Propagation direction	Superposed features	Fringes	Additional remarks
			Properties	Curvature				
c. Three adjacent joints								
Not completely observable	m3: curved	1			Upward radially	Uniform barbs and coarseness	–	The m3 plane interacts with m2. Great variations in plume morphology: m4 coarser than m2, and both differ from m1. m1 to m5 are not more than 2–5 m apart on three distinct planes that belong to a single set with remarkable variations (Fig. 4.6d).
Not observable	m4	2	Only partly observed		Upward	Plume		
Center	m5	5	Increasing width		Radially	Faint plume		
Group 2: Mirrors with diversity of fringes								
J5: Divided arrest mark joint								
At the center, remote from the adjacent partings	Quasi-circular	11	Increasing; width above p:	Below p: circular; lower right elliptical	Radially; continuations to the upper left and in barb coarseness	Faint radial plume; rhythmic increase (I) and decrease (t)	A small fringe	A horizontal parting (p: emplacement shear band with gneissic fabric) separates the lower quasi-circular arrest mark from the upper asymmetric curved one (Fig. 4.7a).
Neighboring additional joints: Radial plume at the top								
At the center	Circular	3	Increasing width	Quasi-circular	Radially; upper continuation eroded	Faint radial plume; increasing coarseness outside the first arrest mark	Two fringes (F) with en echelon segments at both sides	The radial plume joint interacts with others, e.g., the "divided arrest mark joint" J5, to form a "composite fracture". The subhorizontal parting seems to be younger, without restriction for the upward propagation of the mirror.
J6: Asymmetric bilateral joint								
Close to the left side of the first undulation	Elliptical	10	Increasing width; concentric on a symmetry axis	Circular to elliptical	Laterally, first northward, then southward	Continuous plume northward; wavy plume southward	Three fringes (F1–F3) with sinistral en echelon segments	See explanations in the text (Fig. 4.7b).

Table 4.1. *Continued*

Origin point	Mirror shape	No.	Ripple marks		Propagation direction	Superposed features	Fringes	Additional remarks
			Properties	Curvature				
J7: Uniform frinqe joint								
Near the bottom	Circular	–	–	–	Radially upward	Faint radial plume	One fringe of uniform width, with en echelon segments	All en echelon segments have nearly the same length and twist with respect to the parent joint (Fig. 4.7c).
J8: Partly erased joint								
(Erased)	Circular	–	–	–	Radially downward; upper part unknown (erased)	Radial plume; continues beyond the boundary as plume in the middle part or as coarse relief	One fringe and segments (mainly at both sides) in diversified morphologies	Mirror and fringe are divided by a symmetry axis: the twist of the en echelon segments changes from dextral at the left side to sinistral at the right side of the symmetry axis (Fig. 4.7d).
Group 3: Mirror with complex fringes								
J9: Trefoil joint								
About the center	Elliptical with a large "tongue" downward	5	Increasing width, mainly to the right (southward)	Trefoil-like curved; in the inner part concentric	Radially laterally downward	Faint plume on the mirror and undulations	Three fringes to the left and five subfringes to the right of the tongue are intensely fractured (twist angles). The fringes at both sides differ considerably. Several en echelons and their steps in between are covered with secondary mirrors and fringes, indicating dominant mode I development. For further explanation, see text (Fig. 4.8a,b).	
Group 4: Mirror with fringe of en echelon and hackly morphologies								
J10: Hackled fringe joint								
Not observable	Elliptical	–	–	–	Downward	Secondary mirrors with plumes and fringes on a few en echelons	One large fringe compared to the mirror size, with heterogeneous hackles and en echelons. A few are covered by secondary fractographic patterns, comparable to J9. For further explanation, see text (Fig. 4.8c,d).	

Note: The letter *p* is used in figure captions and figures and in this table to designate subhorizontal partings (that are often fractures). For explanations of other letter designations, see figure captions.

Table 4.2. Ellipticities of the north-northeast-trending Borsov joints

Joint		Strike azimuth (°)	Mirror size, L × S (m)	Ellipticities (L / S)			Joint size (L × S): mirror and fringe (m)
No.	Name			Initial mirror	Final mirror	Mirror and fringe	
J1	Multiarrest marks joint	23					
J2	Radial-plume joint	25	1.63 × 1.26				
J3	Quasi-core joint	23	2.60 × 1.90				
J4	Circular joint	~30	2.70 × 2.37	1.14			
J5	Divided arrest mark joint	25	2.61 × 1.90	1.37	1.27	1.67	
J6	Asymmetric bilateral joint	~20	1.40 × 1.10	1.31	2.56	2.29	15.30 × 6.67
J7	Uniform fringe joint	~2	2.90 × 2.60	1.29	1.43	4.29 × 3.00	
J8	Partly erased joint	~25	~6.00 × 5.00	1.20			
J9	Trefoil joint	25					
J10	Hackled fringe joint	~25	2.00 × 0.93	~2.14			

Note: *L* = long axis; *S* = short axis. Initial and final mirrors represent mirrors at their initial and final growth states.

4.4.2.2
The Unique Features of the Mirror Plane, Undulations, Arrest Marks and Fringes of the Trefoil Joint

General. The Trefoil joint is one of the ten Borsov joints described in Tables 4.1 and 4.2 and it displays a unique fracture surface morphology. The mirror plane of this joint deviates considerably from the common circular shape. Part of the mirror boundary extends downward as a "big tongue" beyond the common circular perimeter, and it forms the shape of a three-leaflet flower (Fig. 2.56a and 4.8a). The mirror is marked by two types of ripple marks (Sect. 2.2.4 2). The concentric set of delicate undulations (hardly noticeable on Fig. 2.56a) that imply relatively rapid fracture propagation, differs considerably from the set of lateral arrest marks, which reflect slow fracture propagation (u and a, respectively, in Fig. 4.8a). The arrest lines occur on the left side of the mirror but not at its right side. A faint set of radial striae (a plume) superposes the undulations on the mirror, indicating downward and radial joint propagation. A set of intense striae, emphasize the "semi-box" superpose orthogonally the lateral arrest marks (Sect. 2.2.4.2).

The two sides of the "tongue" contain large fringes that are intensely fractured, and their fracture styles differ considerably from each other. There is a resemblance in shape between the curvatures of the arrest lines and the series of sub-fringes at the right side of the mirror. These sub-fringes show contrasting senses of *en echelon* stepping (see Fig. 2.35e), implying a change of fracture mode under the influences of modified stress fields. The fringe on the left side of the mirror shows a complex fracture (Bahat et al. 2001b) on which we elaborate below. Finally, the mirror is jointed by sub-horizontal parting fractures that generally slope gently from left to right; several of them cut the mirror and the two fringes.

The left fringe. Here we focus on a joint with a particular relationship between a "primary mirror" (the mirror) and a "secondary mirror" that occurs on the fringe. The boundary between the mirror and the fringe is not a continuous curve but a zigzag one (Fig. 4.8a). The fringe is divided into four sectors that generally correspond to the zigzag boundary. These are from left to right, i–n, n–t, t–r, and r–m. We concentrate on the structure in the n–r sector. Two growth styles of *en echelon* segments and steps can be distinguished: First, vertical downward growth of alternating segments and steps from the boundary of the mirror (Fig. 4.9a,b) can be discerned; second, some six segments merge into one step that displays a tensile, secondary mirror plane with semi-circular boundary and a fringe that convex downward in continuation with the lengths of the segments that form the merger (Fig. 4.9c,d). This morphology of a tensile-step differs known fractographies in sedimentary rocks, where mirrors and other tensile fracture markings occur on *en echelon* segments but are absent on steps. Hence, the mirror of the trefoil joint was fractured into a segmented fringe that then continued to fracture into a secondary mirror, which also was fractured into a fringe. We discuss the conditions under which this joint formed in Sect. 4.9.3.1.

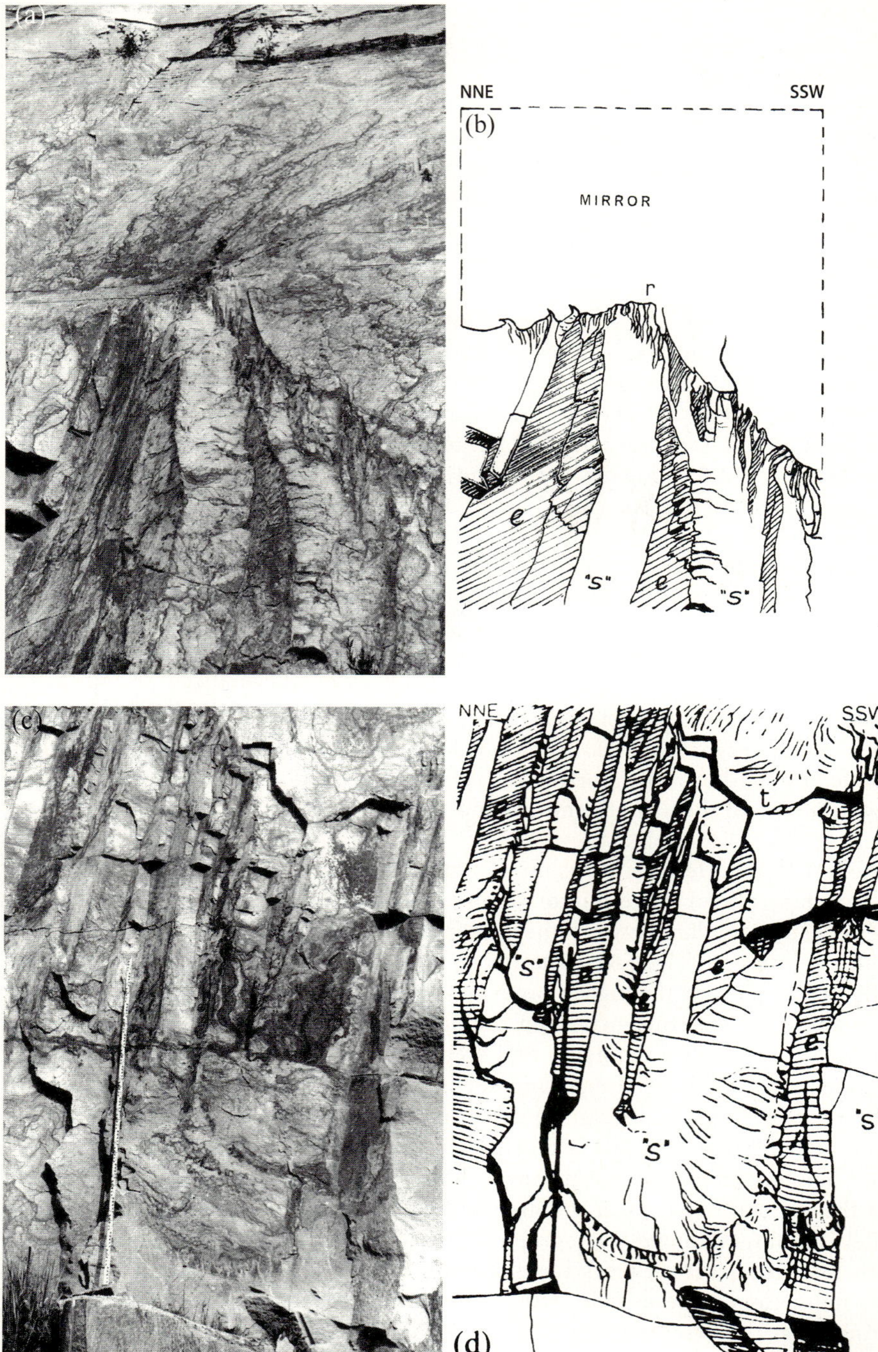

Fig. 4.9. a–d The left fringe of the "trefoil joint", where *e* and *"s"* are *en echelon* segments and steps, respectively (see text for explanation) (after Bahat et al. 2001b)

4.4.3
"Dynamic Joints" in a Fracture Zone from the Mrákotin Quarry

4.4.3.1
Characterization of Mirror Planes and Fringes in the Mrákotin Quarry

According to Bahat et al. (2001c; Bankwitz and Bankwitz 2004), two joint types are distinguished in the Mrákotin quarry (Fig. 4.2b), both of them trend NNE-SSW. First, vertical individual joints that are spaced several meters apart, and second, fracture zones that contain several parallel joints spaced about 5–10 cm (Fig. 4.10a). We concentrate on a set, which is exposed in a fracture zone about 30 m long and about 18 m high. These joints parallel each other on a single wall. The "Mrákotin set" consists of nine distinct neighbor joints, all striking about 025° (Fig. 4.10b,c). Joints A and B occur on one plane, joints C–H form a composite fracture on another plane that is further inside the rock body, and joint I forms a third plane still deeper in the rock. Spacing between the three planes varies from 5 to 20 cm, while joint fringes partly touch neighbor planes. Joints A–H with the exception of C contain mirrors that have one or two hackle-fringes above and below them (Fig. 2.1b). Joint C on the other hand exposes only a fringe at its lower part, which consists of *en echelon* segments (Fig. 2.1a). The mirrors of joints A–H deviate from the conventional penny shape into various ellipticities. The hackles of all fringes have the same sense of stepping in that they show clockwise rotation both above and below the mirrors; that is, they all are arranged such that walking across the hackles is downstairs from left to right (from south to north). Joint I exposes a large faint plume, and its mirror boundaries are not revealed. The nine individual joints are described below from south to north (Tables 4.3 and 4.4).

Joint A. Two large hackle fringes occur above and below the mirror of joint A. Whereas the upper joint boundary is somewhat curved, the lower boundary is fairly straight (Fig. 4.11a–c). Quarrying has removed the southern (left) mirror half and a larger part of the upper fringe so that its length parallel to the joint length is not fully revealed. The length of the lower fringe is 9 m. It becomes gradually wider (the lengths of individual hackles increase) from south (left side of Fig. 4.11a,b) to north (right side of figure) towards the center of the fringe. Hence, maximum width of the fringe occurs at about the middle of its length. There is correspondingly an increase in hackle coarseness in this direction.

In profile, the hackles at the lower fringe show strong deviations from the common uniform "shingle-structure" of *en echelon* fringes. Hackles vary considerably in lengths and show extreme variations in thickness among individuals and along the length of an individual hackle, appearing as wedges (Fig. 4.11c). They are not tabular, but some of them are rather distorted as if responding to a twist mechanism. They have similar strikes, but they differ in their dips. Note that whereas most of the lower hackles dip away from the mirror (towards the left, Fig. 4.11c), some hackles close to the mirror boundary of A dip towards the parent joint (to the right in this figure). Generally, the fringe forms an angle (ϕ) of about 10° with the parent joint. An intriguing feature in this fringe is one hackle, A* that strikes 290° and dips 65° towards NE, i.e., this joint rotates 85° with respect to the strike of A. This hackle has its own mirror and is fringed with a secondary hackle-fringe (Fig. 4.11a–c, Tables 4.3 and 4.4).

Fig. 4.10. Joints in the Mrákotin quarry. **a** A large joint (18 m high), roughly trending E–W, with vertical traces of approximately NNE trending individual joints and fracture zones (*arrowed*). The latter corresponds to the discussed Mrákotin set in the next figures. (from Bahat et al. 2001c). Photograph (**b**) and drawing (**c**) of the Mrákotin set of nine joints (*A–H* and the large plume at the upper right). The sub-vertical fracture *V1* separates joints *A* and *B* and *V2* cuts joint *D*. The fringe of jont *C* is somewhat erroded. Vertical *scale bar* at **c** is 2 m (from Bahat et al. 2002b)

Table 4.3. Characteristics of the 025°-set of joints in Mrákotin (example: Nine joints A–H)

Joint	Max. length (m)	Mirror propagation	Fringe	Boundary	Fringe cracks	Fringe propagation	Hackles and en echelons				Related figures
							Attitude	Strike	Dip	Type	
A	17 17	Laterally Laterally	Upper Lower	Curved Partly straight	Hackles Hackles (twist 25°)	Vertically (tilt 10°)	Hackles vary considerably in length and extremly in thickness (partly appearing as wedges) increasing towards the center of the fringe	Ca. 50°NE	Differing	Not tabular (deviation from common "shingle structure"	4.11
A*	2.7	Oblique			Hackles	Vertically Vertically	Differs from other hackles	290°WNW	65°NE	More tabular	4.11
B		Laterally	Upper		Hackles	Vertically	Similar to joint A fringe, only upper fringe is preserved			More tabular	4.11a; 4.12a
C	13	Laterally	Upper Lower	Covered by B Straight	En echelon segments (twist 20°)	Vertically		Ca. 45°NE		More tabular, but curved	4.12a
D	2.6	Laterally	Only upper Lower mirror boundary interacts with mirror C and F (which are much larger)		Hackles	Vertically	Interacted with the lower fringe of mirror E. This interaction promoted a local enlargement of the upper fringe			Not tabular	4.12a
E	3.4	Laterally	Upper Lower		Hackles	Vertically	Occurs above D				4.12a
F	10	Laterally	Upper Lower	Straight Straight	Hackles Hackles	Vertically Vertically					4.12b,c
G	1.8	Laterally	Upper		Hackles	Vertically	Occurs above joint D and at the right side of H				4.12b,c
H	1.5	Laterally	Upper		Hackles	Vertically	Occurs above joint E				4.10c

A* is one large fringe-hackle of joint A with its own mirror and its own secondary hackle-fringe.

Table 4.4. Fracture parameters of the Mrákotin 025°-joint set

I Joint	II Mirror (m) $L \times W$	III Mirror ratio L/W	IV Lower fringe (m) $L \times W$	V Upper fringe (m) $L \times W$	VI Joint ratio L/W	VII Mean ω echelons/hackles	VIII Range ω
A	17.0 × 4.0	4.4	16.0 × 4.0	16.0 × 4.0	1.3	26	0–28 ($n = 10$)
A*	2.7 × 0.6	4.5	2.3 × 1.6	2.7 × 1.2	1.3		
B							
C	13.0 × 1.6	8.1	13.0 × 8.0		1.4	20	0–26 ($n = 12$)
D	2.6 × 0.6	4.3		2.6 × 1.0	1.3		
E	3.4 × 1.0	3.4		3.4 × 1.5	1.4		
F	10.0 × 0.5	4.3	10.0 × 4.0	10.4 × 1.0	1.4	25	0–27 ($n = 12$)
G	1.8 × 0.5	3.6		1.8 × 1.8	1.3		
H	1.5 × 0.5	3.3		1.5 × 1.6	1.4		

Explanation: $L \times W$ designates (length × width), i.e. lengths of long and short axes, respectively. L/W designates ratio of long / short axes; ω is the twist angle; A* is a secondary mirror at the fringe of joint A.

Fig. 4.11a. Joints in the Mrákotin quarry. Photograph of the 025° trending joint *A*. *Vertical ruler* is 2 m (from Bahat et al. 2001c)

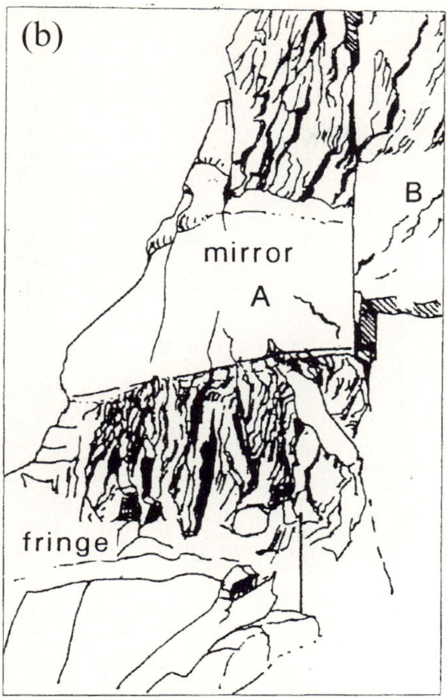

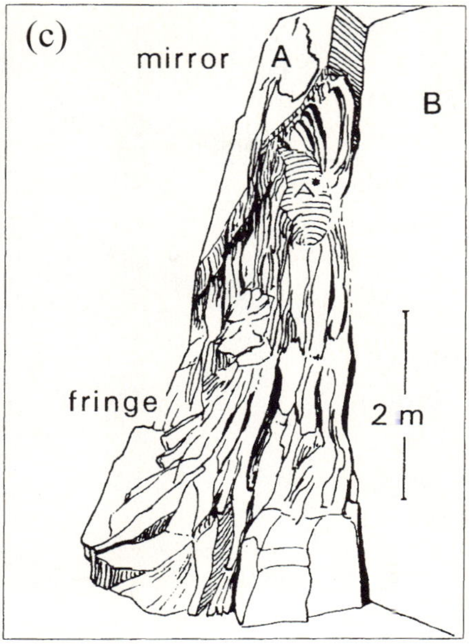

Fig. 4.11b,c. Joints in the Mrákotin quarry. **b** Drawing of the 025° trending joint *A*. *Vertical ruler* is 2 m. **c** Profile drawing of joint *A*, where *A** is possibly an equivalent of *M2* in Fig. 2.30e (after Bahat et al. 2001c)

Joints B–H. Joint B reveals only the upper part of the mirror and the upper fringe. Mirrors A and B are separated by a vertical fracture (V1 in Fig. 4.10c), which had preceded the mirrors because the mirrors and fringes of these joints look as if their shapes were modified along fracture V1 (Bankwitz 1966). This fracture seems to belong to the set that includes the large joint (Fig. 4.10a). The upper part of the parent joint C is partially covered by joint B, but the lower boundary of the mirror of joint C is exposed, and it is approximately straight (Fig. 4.12a). The visible part of joint C has only one fringe, below the mirror, consisting of *en echelon* segmentation. The small mirror D has a fringe only at its upper side. In its lower part, mirror D interacts with mirrors C and F (which are much larger) (Fig. 4.12a–c). Joint E occurs above joint D and has fringes both above and below the mirror. The upper fringe of joint D is relatively large, and it appears that it interacted with the lower fringe of mirror E, penetrating somewhat into the E mirror.

By far the longest mirror in the composite is F (Fig. 4.12b,c). The F-joint seems to have propagated from north to south. There are large fringes above and below the parent joint and both boundaries are fairly straight. The lower fringe is partly erased by the quarrying process. The upper fringe, on the other hand, is well preserved and it has its maximum width at about its mid-length. Joint G is the smallest one; it occurs north of joint E and above the upper fringe of joint F. Joint H is located above joint G and is a little smaller than that joint.

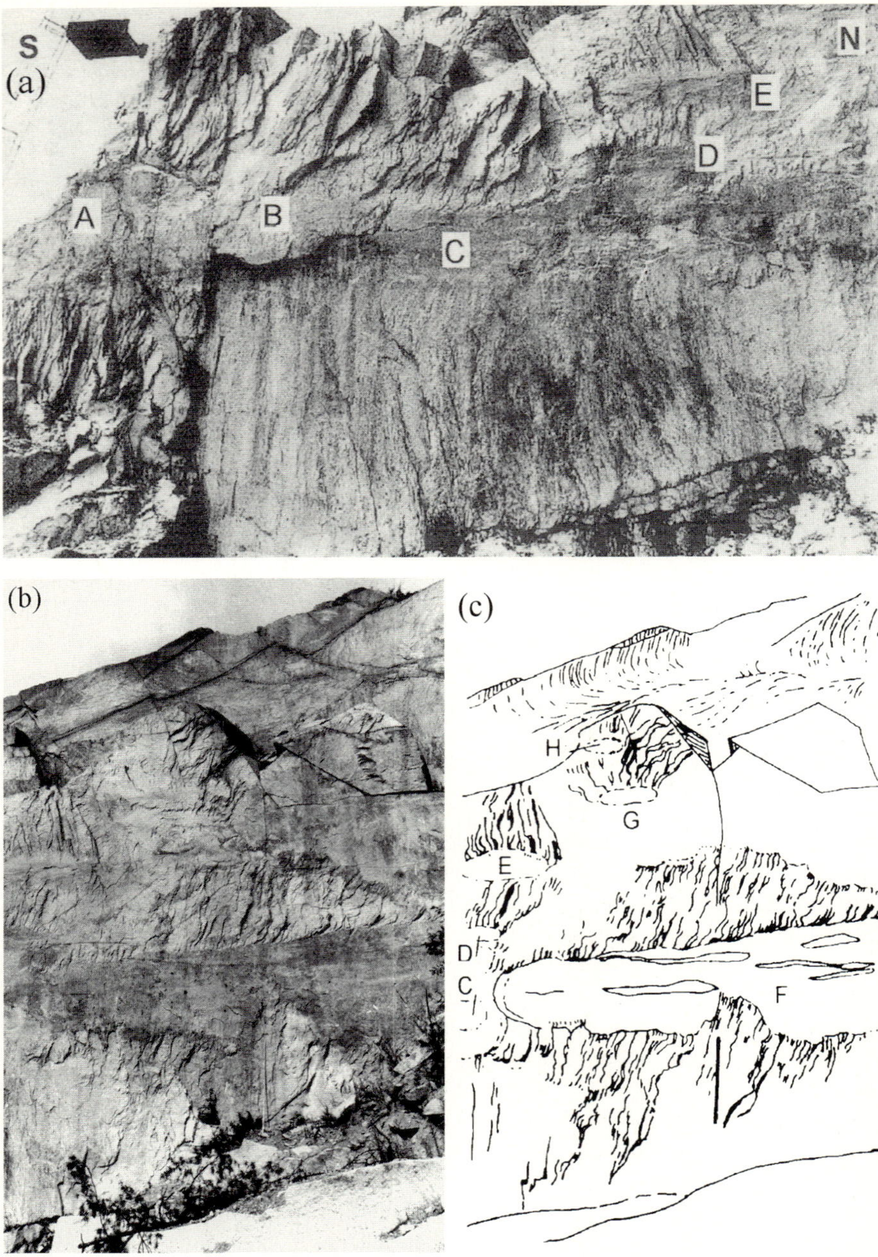

Fig. 4.12. Joints in the Mrákotin quarry. **a** Joints *A–E*. **b** Photograph and **c** drawing of a part of the "composite fracture" wall with joints *C–H*. *Ruler* is 2 m (from Bahat et al. 2001c)

Ellipticity of mirrors and joints. The mirrors of joints D and E are surrounded by curved boundaries. On the other hand, the boundaries of joint A and the lower boundary of joint F approach straightness, and the upper boundary of joint F is straight.

Our measurements of the Mrákotin set (Table 4.4) reveal two features. With the exception of C, for all other joints the ratio of long to short axes is similar for the various elliptical mirrors, ranging from 3.3 to 4.5, independent of elliptical size. However, this ratio is reduced to the range from 1.3 to 1.4 for the full joints (combined mirrors and maximum widths of the fringes for every joint).

Shape of fringes. Joints C and F have only lower fringes, whereas joints D, E, G and H have only upper fringes. The increase in fringe width to a maximum (towards a shape of quasi-triangle) seems to repeat itself in the upper fringes of joints D, E and F, and possibly also in the lower fringes of joints A and F, and the upper fringe of joint B. This characteristic is different from the conventional fringes of uniform width in synthetic materials (e.g., Mecholsky 1991, Fig. 1) and many rocks (e.g., Fig. 2.34b and 2.35a,b).

4.4.4 Twist and Tilt Angles on Joints Cutting Granites in Europe

Two angular relations that fringes form with mirror planes of joints are of particular interest: The twist angle ω and the tilt angle ϕ (Fig. 2.53). The characterization of ω and ϕ may be used in the interpretation of fracture conditions (Cooke and Pollard 1996 and Bahat and Rabinovitch 2000, respectively). Bankwitz and Bankwitz (2004) measured ω and ϕ on joint surfaces that cut various granites in Europe. Their results are summarized in Fig. 4.5, 4.13a,b and 4.14a, providing the first comprehensive account of the tilt angle ϕ in a series of rocks. They illustrate the steady NNE orientation of seventeen pre-uplift joints throughout the SBP and the consistent sense of *en echelon* segmentation with respect to the parent joints, and a large, 16° spread of ω values (between 22° and 38° (Fig. 4.5). Larger joint populations show even greater spreads of ω value, i.e., 40° in the Borsov quarry, and some 35° in the Mrákotin quarry (Fig. 4.13b). Values of ω from the Erzgebirge vary in the 9–22° range (Fig 4.14a). The spreads of the ϕ values vary considerably from pluton to pluton and the spreads are particularly large in the Borsov quarry (Fig. 4.13a) and in the Erzgebirge (Fig. 4.14a).

Bankwitz and Bankwitz (2004) plotted the logarithmic tilt/twist ratio vs. palaeo-depth of pluton emplacement (Fig. 4.14b). They suggest a correlation in which the shallowest intruded pluton (Eibenstock, Germany) represents the largest ratio range and the deepest pluton (now exposed) is characterized by the smallest ratio range. Apparently, the depth of joints from Borsov is 7.4 km, and the ratio range is very large, while the depth of joints from Weinsberg is 15 km and the ratio range is much smaller, raising the suspicion that in shallow depths fracture conditions are variable, while great depths unify them, and this occurs between 7.4 and 15 km depths. This however does not coincide with the great spread of ϕ values observed in the shallow Erzgebirge (Fig. 4.14a). Quite intriguing is also the low spread of ω on joint surfaces in the Erzgebirge. Perhaps it suggests that fracture took place under relatively uniform stress conditions, which maintained only small variations in the local K_{III}/K_I ratio (Fig. 2.21a). Hence, it appears that generally the spreads of ϕ and ω are quite large (Fig. 4.13a,b) and for no clear reasons. Also noticeable are the ϕ values of some joints that vary very considerably in a given fringe (see gray markings in Fig. 4.13a). This suggests that an interpretation of the petrological conditions (e.g., depth) of these granites during jointing should be carried out with great caution. The great angular ranges found at the Borsov quarry is not surprising in view of the extreme diversified fractography found in this quarry (Fig. 4.6–4.9).

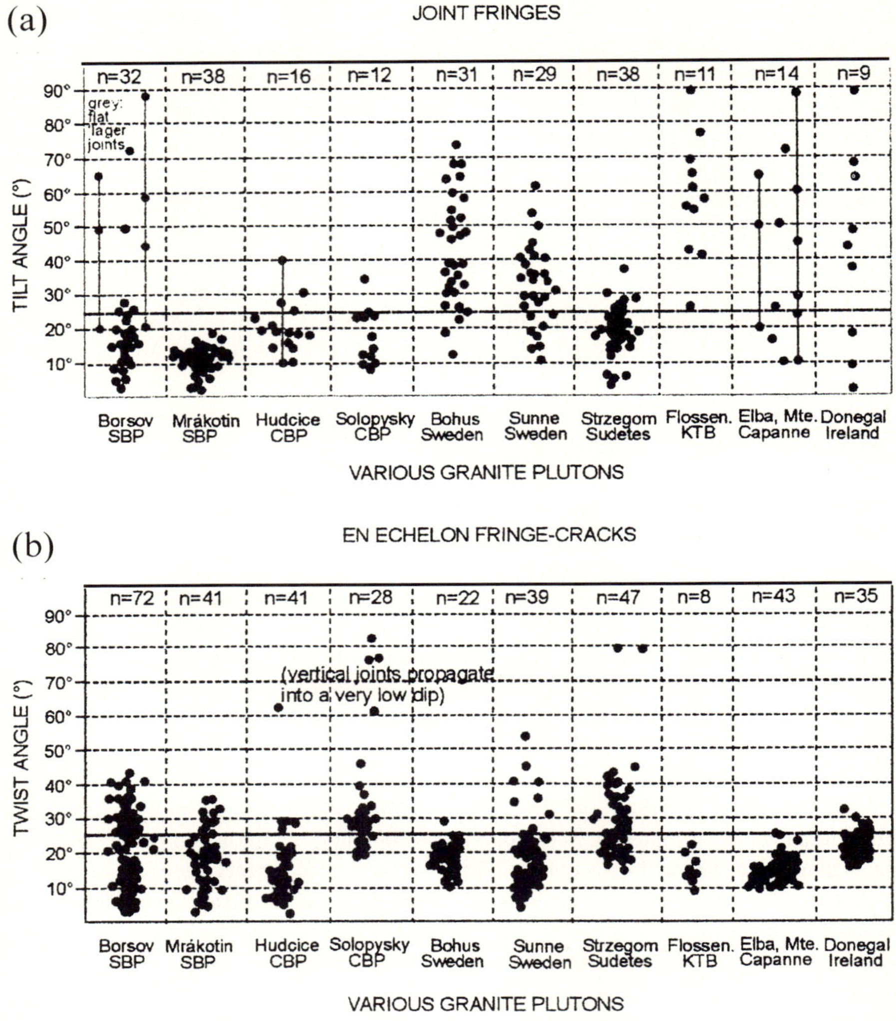

Fig. 4.13. **a** Tilt, ϕ and **b** twist, ω angles on joint surfaces from various plutons in Europe: *SBP:* South Bohemian Pluton; *CBP:* Central Bohemian Pluton. *KTB:* Near the KTB (Continental Super-Deep Borehole); Flossen: Flossenbürg Granite, Bavaria (from Bankwitz and Bankwitz 2004)

4.4.5 Geological Interpretation of the Joint Fractographies in the Borsov and Mrákotin Quarries

While the set described here from Borsov consists of joints that occur in various outcrops within the large quarry, the investigated set from the Mrákotin quarry consists of joints that are displayed on a single wall. The sets from the two quarries display different fractographies, and their respective interpretations focus on different issues complementing each other.

Fig. 4.14.
a Relationship between twist, ω tilt, ϕ angles of early granite joints in different plutons in Europe. **b** The logarithmic ratio of ϕ/ω vs. palaeo-depth of emplacement of the plutons plotted in Fig. 4.14a. The most shallowly intruded pluton (Eibenstock) shows the widest range of ratios; the deepest pluton (Weinsberg, now exposed) is characterized by the smallest range of ratios, between ~0.5 and 1.0. Symbols as in Fig. 4.14a (from Bankwitz and Bankwitz 2004)

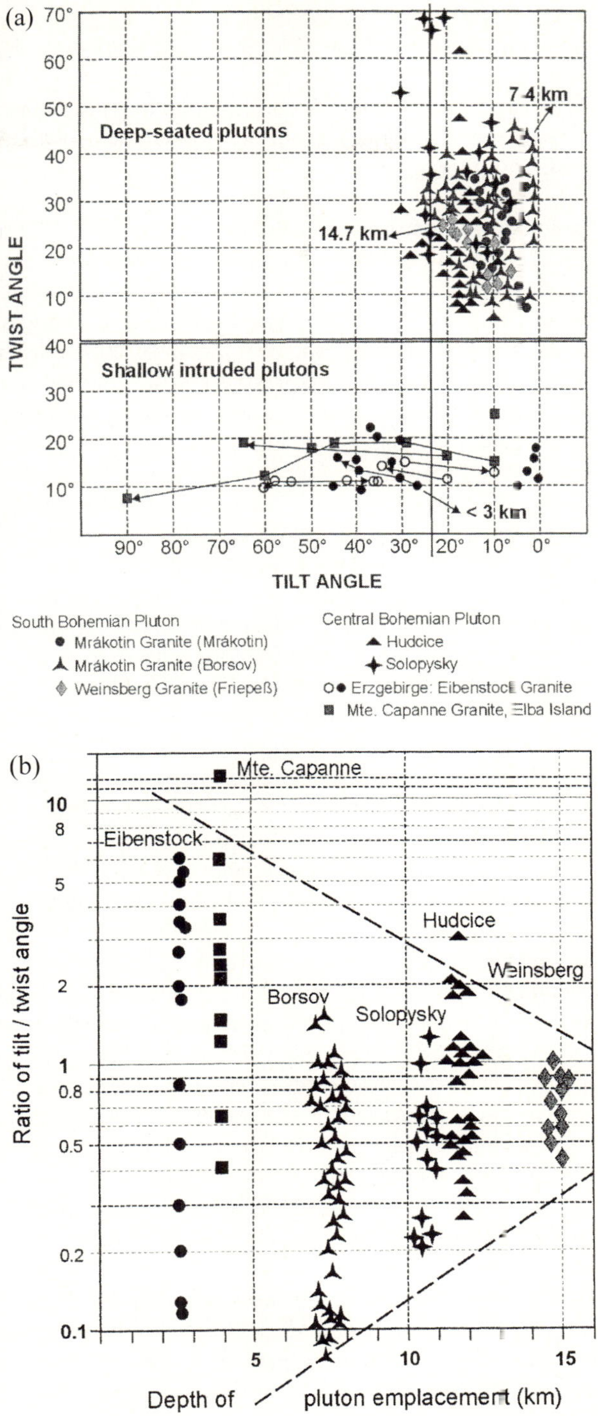

4.4.5.1
Geological Interpretation of Joint Fractography in the Borsov Quarry

In this section, we limit our discussion to three aspects of jointing: Fluid-driven fracturing, mirror ellipticities and their unilateral growth, and plutonic vs. sedimentary rock's fractographies.

Fluid-driven fracturing. The Borsov joints cover very wide v and K_I ranges (Fig. 4.35). Particularly intriguing is the question of what could cause these fractures to grow under such extreme different conditions, from sub-critical to post-critical. These conditions could be changes in local stresses due to changes in the magnitude of remote stresses that would alter the component of local normal stress. It appears however that drastic changes in fluid pressures in the granite history could have been the main mechanism leading to joint fracture surface diversification at Borsov. The rhythmic arrest marks (particularly in Fig. 4.6a–c and 4.7a) are linked to the fundamental concept in the Secor theory (1969) of fracturing by pore pressure (Bahat and Engelder 1984). This process involves "fluid-driven fracturing" (Engelder et al. 1993, p. 44) and "rhythms of fluid-pressure build-up and fracture-induced release" Davis and Reynolds 1996, p. 248). Thus, the frequent appearance of numerous arrest marks (Sect. 2.2.4.2) probably indicates fluid-driven fracture processes, at least during the early stages of joint growth at Borsov. This interpretation fits well the presence of biotite and biotite-muscovite petrographic manifestations in the rock.

The elongated axis of the South Bohemian Pluton coincides with the dominant strike direction of the flat dipping main foliation in the neighboring metamorphic country, and structurally coincides with the crest line of the pluton (Benes 1971). This suggests that the shape of the pluton could have controlled significant thermal and fluid-pressure gradients, which in turn influenced the orientation of the joints. It could also influenced one of the regional, horizontal principal stresses in the NNE direction (see below). Such relationships most likely have existed during the late stage of the granite solidification before its uplift. Field contacts at the Borsov quarry support this interpretation: Aplite thin dikes that are associated with two joints seem to register melt fillings along the existing joints, i.e., the NNE joints were available when this late fluid of the granite injected. If aplite represents the fluid that created the joints, this rock should have been associated with all (or most) joints, which is not the case. See an alternative interpretation in Sect. 4.4.6. These joints are less likely to be the result of mechanical stresses induced by thermal gradients during early stages of the cooling process. Such gradients are likely to be influenced by local conditions, causing considerable strike deviations from regional uniform joint patterns. This could not be recognized. It is also unlikely that the NNE fractures are late (Tertiary) uplift joints, mainly because uplift horizontal sheets cut the earlier vertical joints, and the latter do not show fracture morphologies indicating downwards propagation, which characterize uplift joints.

Mirror ellipticities and their unilateral growth. Practically all the joints at Borsov that reveal their boundaries show that their mirrors deviate from circularity into various ellipticities. The incremental crack growth exhibited by J6 shows a pronounced unilateral growth of the radii of arrest marks (Fig. 4.7b), suggesting a general slow fracture growth but a relative faster fracture propagation along the 8° dip of the "mirror symmetry axis" than normal to this direction. Note that the dip of the symmetry axis of J3 is even greater (24°) (Fig. 4.6c); both deviate from either horizontal or vertical directions. Apparently, the same joints that display eccentric origins also show small angle dips of their symmetry axes on the mirror planes. Hence, some earlier fractures that constrain the Borsov joints are not quite horizontal, leading to occasional inclined mirror symmetry axes. The constraints that caused this unilateral growth could be sub-horizontal unisotropies imposed by either petrographic boundaries (Bankwitz and Bankwitz, unpublished), differences between horizontal and vertical compression, or earlier fractures. It appears that constraints by earlier fracture influenced the growth of most Borsov joints as seen on joints J1, J2, J4, J5, J6 and J7.

Plutonic vs. sedimentary rocks. The wide spectrum of fracture surface morphologies displayed by the Borsov joints that reflect different fracture processes in a single set seem to be exclusively associated with jointing in granite during its late solidification stage and do not occur in sedimentary rocks. Fracture markings from a given joint set in sedimentary rocks are commonly limited to simple combinations of one to three features shown in Fig. 2.1a, because they form by a restricted number of fracture mechanisms. Rare complex combinations often reflect special processes. A case in point is displayed by "mark overprinting", where a superposition of a late uplift joint on an early burial (diagenetic) joint shows different fracture surface markings, but these different markings occur along two different sets of joints (Bahat 1991a, p. 288). Unfortunately, only partial mirror boundaries are known from sedimentary rocks, and these are not sufficient for comparison with the present results. The eccentric origins and small dips of the symmetry axes on the mirror planes are rare in sedimentary rocks. In horizontal sedimentary layers, plumes generally are parallel to layer boundaries.

4.4.5.2
Geological Interpretation of Jointing in the Mrákotin Quarry

The joints of the Mrákotin set display composite fractures on three adjacent closely spaced parallel planes in a fracture zone. The joints on each plane interacted intensively with each other such that fringes between mirrors are either missing or increase to large sizes. This interaction resembles the interacting early joints cutting the El Capitan granite at Yosemite National Park (Bahat et al. 1999) (Sect. 4.5.3.3). There are however several important differences between the two interactions.

Dynamic growth. Compared to the El Capitan set that did not develop full mirrors and fringes, the Mrákotin set formed full mirrors and hackle fringes. The hackle fringes imply dynamic growth, i.e., rapid growth of most joints under high stress intensity conditions (Sect. 4.8.4), with the consequence of fracture interaction. Apparently, joints of strongly reduced spacing within fracture zones can be distinctly parametrized from joints in sets that do not occur in fracture zones (Rabinovitch and Bahat 1999). The series of distinct, adjacent joints implies independent fracture nucleation from many sites. The considerable size difference of the mirrors (Table 4.4) suggests that the growth of these joints was strongly affected by local fluid pressure conditions. While most of the Borsov joints (J1–J7) propagated sub-critically during most of their fracture histories (Sect. 4.8.5), most of the Mrákotin joints essentially formed dynamically.

Ratios of long to short axes of joints. The reduction in the ratio of long to short axes of the mirrors from the 3.3 to 4.5 range to the ratio of almost 1 (Table 4.4) for the full joints including fringes is intriguing. The growth towards the ratio 1 is expected from the theory that an ultimate growth of a fracture should follow the trend of decrease in ellipticity (e.g., Irwin 1962; Roy and Saha 1995). The long axes of the elliptical mirrors are horizontal. What stopped the mirrors from growing to a circular (penny) shape? We did not identify early sub-horizontal fractures comparable to those found to have influenced the growth of some of the Borsov joints. The depth of emplacement of the Mrákotin granite was determined by Büttner and Kruhl (1997) to be ~15–18 to 7–10 km (compare to Sect. 4.4.6). Great overburden pressure at these depths may explain what prevented the mirrors from propagating vertically, upward and downward and induced lateral propagation of the mirrors. The question still remains of what has changed when growth occurred beyond the mirror boundaries, and if it was gradual or abrupt. It appears that the change was violent and abrupt, as suggested by the hackles and the quasi-triangle fringe of joint F (Fig. 4.7b). Possibly, the change was caused by a sudden rise in fluid pressure that increased the extensile stress on the joint mirrors, such that the overburden pressure was overcome.

Joints A and C. Joint C differs from all the other fractures in the Mrákotin set in having a fringe consisting of *en echelon* segments, compared to the hackle fringes that characterize the others. Furthermore, joint C is surrounded upward and sidewise by the other joints (Fig. 4.10c and 4.12a). This again implies that the joints did not form by a uniform stress field. It is suggested that fluid pressures were smaller or less effective at the center on joint C and greater around it.

4.4.5.3 *Summary*

This section focuses on the Mrákotin quarry, where a single "vertical wall" exhibits a set of nine joints in a fracture zone. These joints nucleated independently from many sites.

The considerable size difference of the mirrors probably suggests that the growth of these joints was strongly affected by local fluid pressure conditions.

The hackle fringes that characterize these joints imply dynamic growth, i.e., a rapid growth of most joints under high stress intensity conditions, with the consequence of fracture interaction.

For most joints, the ratio of long, sub-horizontal to short axes of the elliptical mirrors ranges from 3.3 to 4.5. However, this ratio is reduced to the range from 1.3 to 1.4 for the full joints (combined mirrors and maximum widths of the fringes for every joint).

The sub-horizontal mirror boundaries of the joints, particularly joint F, imply growth constraints above and below the mirror possibly due to great overburden pressure, which prevented the propagation of the mirror to a circular shape. At some stage the growth was resumed, as demonstrated by the fringes, possibly by a rise in fluid pressure.

It is expected that the fluid pressures that induced fracture in the Mrákotin quarry had to surpass the overburden pressure, and they reached maxima that were greater than the pressures calculated by the fluid inclusion analysis (see below). Possibly, the fluid pressure was smaller or less effective on joint C at the center and greater on the joints around it.

There is a genetic linkage between the joints investigated in the Mrákotin and Borsov quarries. The strike of all mirrors at the Mrákotin quarry is roughly uniformly directed at NNE, characteristic to the SBP, with the implication that this direction coincides with one of the regional horizontal principal paleostresses (Sect. 4.7, below).

4.4.6 Timing, Depth and Temperature of Joint Formation

This section is cited from Bankwitz et al. 2004, who combined several techniques in studying the timing, temperature and depth of NNE trending joints in the SBP (Fig. 4.5).

Methods. Samples of the Mrákotin granite were taken from fringes of the "Trefoil joint" and the "Bilateral asymmetric joint" that outcrop in the Borsov quarry (Fig. 4.8a and 4.7b, respectively) and from the the Weinsberg granite (Fig. 4.5) at the Friepeß quarry in Austria. The K-Ar method was used on these samples for age determination and for fluid inclusion analysis (applied on muscovites) of the temperature and depth of intrusion of the granites (Scharbert 1998). The latter analysis was also applied on Erzgebirge plutons (Fig. 4.2a).

Temperature and depth of intrusion of the Mrákotin granite and a late dike. According to Bankwitz et al. 2004, the minimum intrusion depth was estimated from the *P-T* conditions of the trapping data of fluid inclusions (Table 4.5), using a lithostatic model with a rock density of 2 600 kg m^{-3}. In the Mrákotin granite, the average temperature of homogenization in the fluid phase of seventeen fluid inclusions is $T = 620$ °C, i.e., the trapping temperature of the inclusions that are under pressure conditions of $P = 1.89$ kbar. Under the assumption of a lithostatic pressure at this crustal level, the depth at the time of inclusion formation of these samples is estimated to be 7.4 km. For the inclusions in the Weinsberg granite, the level of generation was much deeper, at 14.3 km with associated higher temperature and trapping pressure of 3.65 kbar.

Temperature and depth of intrusion of plutons in the Erzgebirge. The heating-quenching microthermometric method was applied by Bankwitz et al. 2004 in the analysis of 1 386 melt inclusions in quartz, apatite and feldspar of several plutons (e.g., Eibenstock, Kirchberg, Bergen and Ehrenfriedersdorf granites) of the Erzgebirge (Fig. 4.2a). According to the deep seismic reflection profile, most of these plutons belong to a shallow granite "sheet" within the upper crust without deep roots (Behr et al. 1994). The estimated trapping pressures of melt inclusions in the various plutons range between 1.1 ±0.1 kbar and 1.0 ±0.1 kbar, and trapping temperatures of 562 ±2 °C to 577 ±7 °C correspond to intrusion levels of about 3 km. Geological evidence for shallow intrusion levels comes from vitreous glass inclusions that occur together with

Table 4.5. Trapping data of fluid inclusions in quartz of a late-granite dike, intersecting the Mrákotin granite (*FK-1*), and in quartz of the Weinsberg granite (*FK-11*). Both granites are granite-types of the South Bohemian Pluton; *n:* Number of determinations

Sample	Salinity (NaCl-equiv. %)	Homogenization temperature (°C)	*n*	Trapping temperature (°C)	Trapping pressure (kbar)	Depth (km)
FK-1	5.8 ±0.6	375 ±6	17	620	1.89	7.4
FK-11	13.8 ±0.2	332 ±10	20	662	3.65	14.3

normal crystallized melt inclusions in granitic quartz of several Erzgebirge plutons.. Glass inclusions demonstrate rapid cooling or quenching that is not compatible with a deep intrusion level (Thomas and Klemm 1997). Therefore, the first fractures could be initiated during granite cooling below 560 °C at a depth of about 3 km, where the trapping of the investigated inclusions of the Erzgebirge samples occurred.

Interpretation of timing, depth and temperature of joint formation. Bankwitz et al. 2004 arrived at the following sequence of events. The intrusion age of the Mrákotin granite is 327 ±4 Ma, and cooling took place during the 327–325 Ma interval. According to the cooling age of muscovite extracted from a late-granite dike associated with jointing, fracturing started at 324.9 Ma. After joint initiation cooling of the SBP continued to 320 Ma though, 312 Ma was measured in the southern part of the Cimer granite (Mrákotin granite is the local name of the Cimer granite in the more northern SBP intrusion area). The younger granite Homolka intruded at 320–314 Ma, and then the deep-rooting granite bodies intruded in the axial part of the SBP (the Zvule granite, Breiter 2001). The Mrákotin granite intruded in about 800 °C at a deeper level during 327–328 Ma and exhumed to 7.4 km at 325 Ma. The NNE oriented joints in the Borsov quarry formed at 7.4 km depth in 324.9 Ma by the late-granite dikes that induced natural hydraulic fracturing. The fluid inclusions of the dikes were trapped in 620 °C. The dike then cooled to about 350 °C, given by the cool-

Fig. 4.15. Model of early deep-seated fracturing in solidifying granite magma of the South Bohemian Pluton, summarized from fractographic investigations on fluid and melt inclusions (from Bankwitz et al. 2004)

SUCCESSION OF EVENTS WITHIN SOLIDIFYING AND FRACTURING FELSIC MAGMAS:

Cooling of the magma to the solidus P-T conditions at the emplacement level
(evidence: melt & fluid inclusions)

▽

Increase of the H_2O pressure near the solidus by crystallization of anhydrous phases
(evidence: P-T conditions of melt & fluid inclusions)

▼

Fracture initiation
(age & depth evidence)

▼

Cyclic propagation by input of a mobile low-viscosity "residual melt" into the natural hydraulic formed fracture slow subcritical crack growth
(evidence: spot-like penny-shaped joints & late-granite dyke)

▼▼

Associated generation of small portions of new granitic melt (evidence: overpressure near the solidus)

▽

Continuing slow sucritical crack growth
(evidence: fractographic features & late-granite dyke)

▼

Reaching a critical point of fracture propagation, further autonomous crack propagation occur
(fractographic evidence: fringe with straight en echelon cracks)

▼

Crystallization of the new melt or residual melt as late-granite dykes along the fracture
(evidence: inclusions & age determination)

ing temperature of the muscovite in the dike in 325 Ma (Scharbert 1998). Finally, most intriguing is the occurrence of NNE oriented joints, presumably of the same set, in plutons that originated in extremely different depths: Joints from Borsov at 7.4 km and joints from Weinsberg (Fig. 4.5) at 15 km.

According to Bankwitz et al. (2004), the sequence of intrusions in the SBP (Table 4.6) combined with the above set of depths and temperature conditions, provide evidence only for weak exhumation during the cooling stage between 327 Ma and 325 Ma. A recognizable exhumation that started at only 310 Ma is known from the Bohemian Massif (west of the SBP where analogous intrusions exist), based on results derived from apatite-fission tracks. This process lasted during the 310–280 Ma period in association with the late-Variscan exhumation (Wagner and Van Den Haute 1992; Menzel and Schröder 1994). The exhumation was not a continuous process. Exhumation of 3.3 km is estimated to have taken place during the first stage (Coyle and Wagner 1995). The second period of exhumation occurred in Lower Triassic and the third in Upper Cretaceous/Tertiary times. Finally, Bankwitz et al. (2004) conclude that the sequence of major events was:

1. Cooling of the Mrákotin granite melt down to the solidus.
2. Injection of residual melt into the solid phase of the Mrákotin granite, thus forming late-granite dikes and creating the NNE-trending hydraulic fractures.

They hypothesize a multi-stage process of repeated cycles of melt generation and fracture (Fig. 4.15), a model that may guide a future stimulating research.

Table 4.6. Succession of early main joints in granites of the SBP

	Early joints		Later joints
Quarry	**First sets (Ø)**	**Second sets (Ø)**	**Third sets**
Mrakotin granite (SBP-N)			
Boršov	Subhorizontal NNE: 025°/090° + ESE:110° to 120°/090° fringe cracks: 060°/090°	NNE + ESE 042°/040°	ENE ~120°/090° Subhorizontal
Mrákotín	Subhorizontal to W-dip NNE: 025°/085° + ESE:120°/090°		
Rásná	Subhorizontal to W-dip NNE: 010°/090° + ENE:080°/090°	E–W	N–S
Mislotice	NNE	NNW to N–S	
Sumrakov	NNE: 020°/090° + ESE:~115°/090		
Olši	NNE		
Weinsberg granite (SBP-S)			
Friepeß	NNE: 025°/085° fringe cracks: 060°/090°	ESE: ~120°/090°	
Hartelberg	ENE: 080°/080° ESE: 120°/085° NNE: 018°/090° + ESE: 115°/090° Fringe cracks: 060°/090°		
Hartberg	NNE: 010°/090° + ESE: 105°/090°		
Windhaag	NNE: 030°/088° + ESE: 100°/085°		

4.5 Joints in Granites from the Sierra Nevada Batholith in California

4.5.1 Introduction

Bateman and Wahrhaftig (1966) discriminate between several types of joints in the Sierra Nevada (Fig. 4.16a). These include early "cooling", later "steep", presumably near vertical (including vertical), and younger "gently dipping sheeting" as well as "exfoliation joints" that are oriented like the steep joints. They found that generally it is possible to identify two principal groups of steep joints almost at right angles, one group striking NE and the other NW. Bateman and Wahrhaftig (1966) did not distinguish different sets within the groups of steep joints. However, in some places nearly parallel joints cross one another, and where a significant change takes place in the strike of one joint set, the joints of one strike generally interfinger with the joints of another strike; a description that raises suspicion about more than one set (Younes and Engelder 1999 and present Sect. 3.1.2 and 3.1.6). Furthermore, Bateman and Wahrhaftig (1966) note that spacing between joints ranges from less than an inch to hundreds or even thousands of feet. Such changes in spacing often provide guidelines in separating joints into distinct categories and possibly distinct sets (Bahat 1991a, p. 308). Thus, it seems quite possible that the near-vertical joints belong to more than two orthogonal sets.

Bateman and Wahrhaftig (1966) consider that the continuity of joints across boundaries between individual plutons indicates that the joints formed after the consolidation of the entire Sierra Nevada batholith, before the Eocene. The abundance of regional near-vertical joint sets striking N-NW and E-NE that are approximately perpendicular to each other in the Sierra Nevada Batholith was noted by additional others (e.g., Becker 1891; Huber 1987, Fig. 33). Lockwood and Moore (1979) observe that strike-slip microfaults (original joints on which some displacement is measurable) are pervasive throughout the granitic rocks of the eastern Sierra Nevada. Offsets typically range from less than a millimeter to several tens of centimeters, but exceed 100 m in some places. The microfaults are oriented in two nearly vertical conjugate sets: A N-NE striking set showing right lateral offset and an E-NE striking set showing left lateral offset, the latter is more common. They also plotted and measured lineaments (Fig. 4.16b,c) and found that most lineaments visible on aerial photographs are microfaults. The age of this microfaulting is not precisely known. It developed after consolidation of the youngest granitic plutons in the Sierra (79 Ma) and is known to cut a late Miocene volcanic dike in one area. Slickensides along the microfaults are sub-horizontal but show an average plunge of about 3° westward, suggesting that much of the deformation occurred prior to the westward tilting of the Sierran block in late Tertiary time. The direction of maximum horizontal extension strain (determined as the bisector of average microfault trends) changes (rotates) systematically from north to south (WNW at 38.5° N; NW at 36.5° N). A detailed study of microfaults at 37°20' N indicates a maximum extension of 2.3% in a N 61° W direction. These extension directions are remarkably parallel to late Mesozoic to present-day tectonic extension directions in the Basin and Range province. Thus, Lockwood and Moore (1979) differ in the description of fracture from Bateman and Wahrhaftig (1966); they found two major conjugate sets of microfaults, (N-NE and E-NE) rather than two major groups almost at right angles, one set striking NE

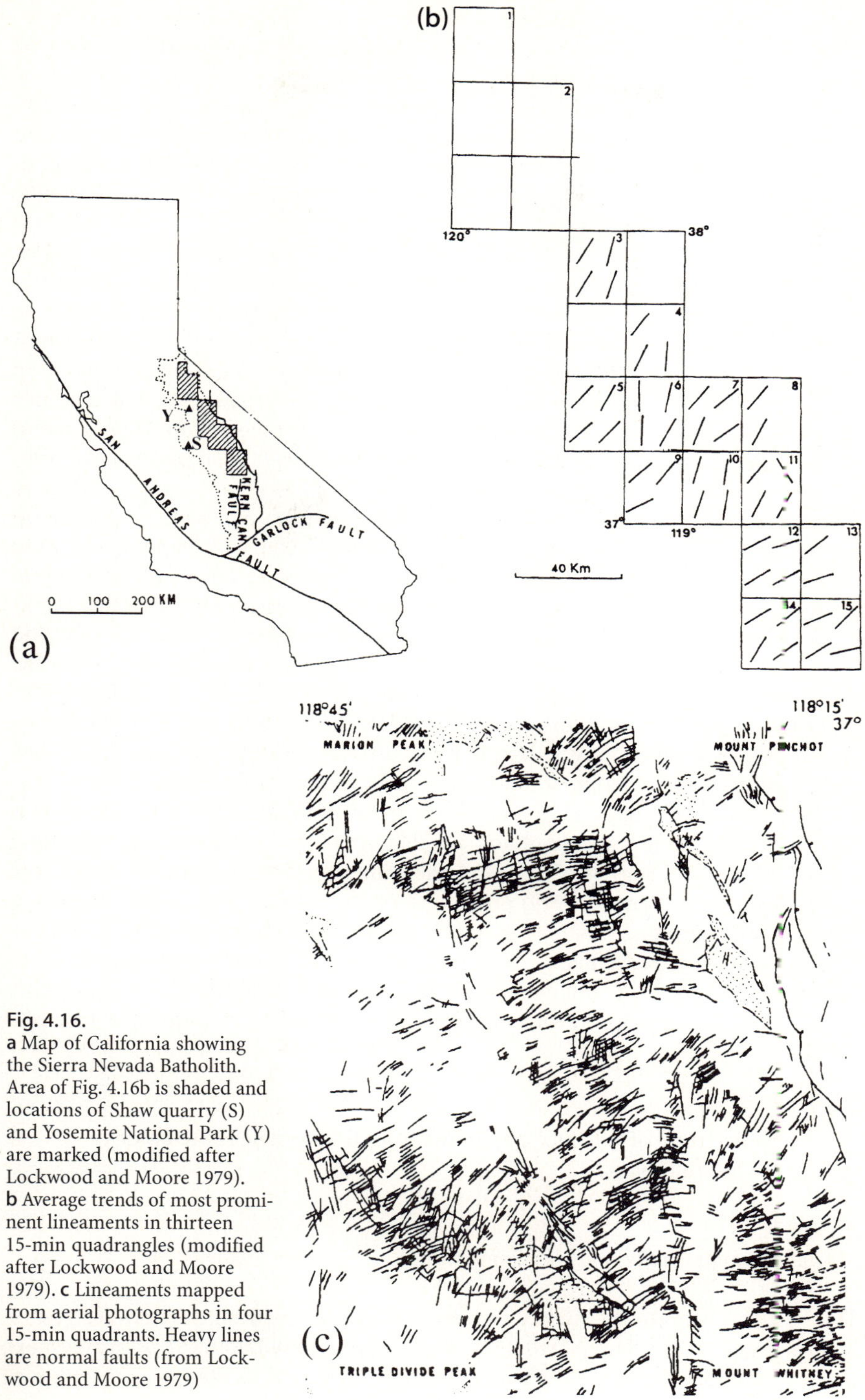

Fig. 4.16.
a Map of California showing the Sierra Nevada Batholith. Area of Fig. 4.16b is shaded and locations of Shaw quarry (S) and Yosemite National Park (Y) are marked (modified after Lockwood and Moore 1979). **b** Average trends of most prominent lineaments in thirteen 15-min quadrangles (modified after Lockwood and Moore 1979). **c** Lineaments mapped from aerial photographs in four 15-min quadrants. Heavy lines are normal faults (from Lockwood and Moore 1979)

and the other NW. However, Lockwood and Moore (1979) agree with the interpretation of Bateman and Wahrhaftig (1966) regarding the "post consolidation" timing of fracture. They conclude that the pattern of microfaulting demonstrates that the (supposedly monolithic) Sierran terrane was affected by the late Cenozoic and possibly earlier regional extension of western North America. Huber (1987) considers that jointing could have occurred during late tectonic events, such as tilting of the Sierra region, 25 to 15 million years ago. Accordingly, these fractures reflect *syntectonic* (possibly *uplift*) processes rather than *cooling* ones.

In contrast to Bateman and Wahrhaftig (1966) and Lockwood and Moore (1979), Balk (1937) and Mayo (1941) consider that the steep orthogonal fracture systems in the Sierra Nevada developed during the emplacement of the plutons, and Segall et al. (1990) suggest that they formed soon after the host pluton was emplaced, between 85 and 79 Ma. Zawislanski (1994) suggests that in the Knowles granodiorite that has been dated as 111 Ma (see below), pegmatites, representing late granite differentiates, may have filled early tension-fractures related to the cooling of the pluton. Thus, it seems justified that the fractures described by Balk (1937), Mayo (1941), Segall et al. (1990) and Zawislanski (1994) be termed *cooling joints*.

Restating the problem, it is possible that different authors may have assigned steep joints to distinct times in relation to the petrological history of the pluton (during emplacement vs. post consolidation of the pluton), and that they consider different fracture systems to represent three or four regional sets rather than two that developed during different geological periods. Alternatively, as inferred from Lockwood and Moore (1979), there was a systematic change in stress direction that resulted in the formation of a series of orthogonal sets of joints and faults (Segall and Pollard 1983), which underwent rotation with time. This problem that confronts the concepts of prolonged paleostress directions vs. rotating paleostresses is analogous to a similar one that relates to the distinction of joint sets in sedimentary rocks (Sect. 6.3).

There are in addition, very young joints that seem to fall into the two categories, termed, "gently dipping sheeting" and "exfoliation joints" (Bateman and Wahrhaftig 1966). Becker (1891) suggests that (presumably the same fractures) are Pliocene in age and Lockwood and Moore (1979) propose that they are post-Pliocene. In summary, there are still many uncertainties about the multi-fracture behavior in the Sierra Nevada Batholith that await further investigation. The present section narrows its interest into only two fracture domains in the Sierra Nevada Batholith, the oblique joints in the Knowles Granodiorite and the exfoliation joints that cut plutons in Yosemite National Park.

4.5.2 Petrographic Study of Two Oblique Joints from the Knowles Granodiorite in Western Sierra Nevada

4.5.2.1 *Previous Studies on the Knowles Granodiorite*

This section relates to the surface properties of two joints from the Knowles granodiorite (Bateman and Sawka 1981). These joints are exposed in the Shaw quarry, next to the Raymond field site (Karasaki et al. 2000), which is located near the edge of

Rte 411, 5 km east of the town of Raymond and 70 km north of Fresno, at the western foothills of the Sierra Nevada Mountains (Fig. 4.16a). The Knowles granodiorite pluton is 15 km by 5 km, elongating in the NS direction and is fine to medium grained, generally not deformed, and has an isotopic age of 111 Ma (Stern et al. 1981; Bateman and Sawka 1981). It intruded the tonalite of Blue Canyon that has been dated at 114 Ma.

Zawislanski (1994) measured 438 fractures in the area around the Shaw quarry and grouped them into four: Two orthogonal, near vertical sets striking at N 30° W and N 60° E and dipping between 75° and 90°, a sub-horizontal set that dips to the west at about 5°, and a less prominent set, termed "oblique", striking at about N 45° E and dipping between 30° to 60° to the NW. He observed that pegmatite dikes are pervasive in the granodiorite. The dikes are oriented in two sets roughly 90° apart, one striking N 35° E, the other striking N 70° W. Dikes in both sets dip between 20° and 60° to the NW and NE. Thus, the orientations of the pegmatite dikes do not coincide with either the near vertical or sub-horizontal fractures, but are similar in orientation to some of the oblique joints. Based on the lack of well-developed chilled margins in the pegmatite dikes, Zawislanski (1994) considers that the dikes were emplaced shortly after the granodiorite emplacement. The dikes are routinely cut by the two near vertical sets, generally with very little to no displacement. Also, the rough coincidence of the NW dipping dikes with oblique joints suggests that the pegmatites may be fillings of early cooling joints. Hence, the sequence that followed the rock emplacement seems to have been oblique joints, pegmatite dikes and near vertical fractures. Zawislanski (1994) also found that sub-horizontal fractures cut the sub-vertical ones. Since the dip of the sub-horizontal fracture roughly coincides with that of the regional slope, he concludes that they are young, unloading fractures.

Zawislanski (1994) also observed that most fractures are very tight and have no significant secondary mineralization, although apertures are often ambiguous in surface exposures due to the tendency of weathering processes to erode and "open" fractures. Displacements on the order of 1 to 2 cm were observed on some sub-horizontal and oblique fractures. However, no measurable displacement was observed on near vertical fractures. Spacing is irregular, with some fractures as closely spaced as a few centimeters (in fracture zones), and some as widely spaced as tens of meters. The average spacing of the near vertical fractures is 0.95 m (standard deviation = 0.85; $n = 85$).

4.5.2.2
Joints C and D in the Knowles Granodiorite from the Shaw Quarry

The fracture classification in granites by Cloos (Fig. 4.1) that divides fractures into near vertical, sub-horizontal and diagonal (oblique) categories provides a good basis for comparison of fractures from different domains and provinces. Most previous efforts of fracture investigation concentrated on the first two categories, almost ignoring the oblique fractures. However, some intriguing observations that were made on diagonal fractures in plutons (e.g., Cloos 1921, 1922; Zawislanski 1994) (Sect. 4.5.2.1) seem to justify their detailed study. Furthermore, near vertical fractures seem to be rare in the Shaw quarry (Fig. 4.17). Accordingly, the present section focuses on two diagonal joints that cut the Knowles granodiorite in the Shaw quarry. Joints C and D that are described in this section differ in their orientations. Joint C strikes 078° (N 78° E) and dips at 42° to the SE and joint D strikes 045° and dips at 36° to the NW

Fig. 4.17. Photographs of a part of the Shaw quarry, showing joint D in three scales. **a** The joint dips from left to right at the lower center of picture and appears to be sub-parallel to several joints above it, that belong to the same set (*D* is about 3 m long). The joints partly show concentric arrangement. An oblique plume (*horizontal arrow*) is seen on a joint from set C. **b** Three drills along joint D from which rock cylinders have been removed. **c** Part of the joint in a close up (note two syringe needles). **d** A large plume (*delicate lines*) is examined by students on a "oblique-concentric joint" in a granite outcrop from the Bohemian Massif

(Fig. 4.17a). The detailed petrography of the fracture apertures of joints C and D is the main subject matter here. In studying the joints, the "epoxy procedure" was applied in order to obtain in-situ samples for preparing thin sections of the fractures (Sect. 4.12).

Microphotography. A series of photo micrographs each covering a square area of 4.5 × 4.5 cm were taken by both plane light and plane polarized transmitted light from the examined thin sections. These micrographs, partly overlapping each other in order to minimize photograph distortion, were assembled into mosaics covering sections along the joint apertures. A comparison between images obtained by the two lights enables one to distinguish the boundaries of the primary fractures C and D (Fig. 4.18–4.20) and branching of a secondary fracture from joint C (Fig. 4.18a and 4.19c), as well as secondary precipitates, additional very thin cracks, and other common petrographic features.

Distinction between transgranular and intergranular fracture. Both transgranular and intergranular fractures were observed on joints C and D. Several criteria for transgranularity were defined. First, a mineral that shows a uniform extinction at both margins of the joint is more likely to represent a transgranular fracture and is less likely to represent late crystallization of two independent minerals in precisely the same orientation. A further confirmation of transgranularity is obtained when corresponding locations of a certain boundary and/or polysynthetic twins of the mineral

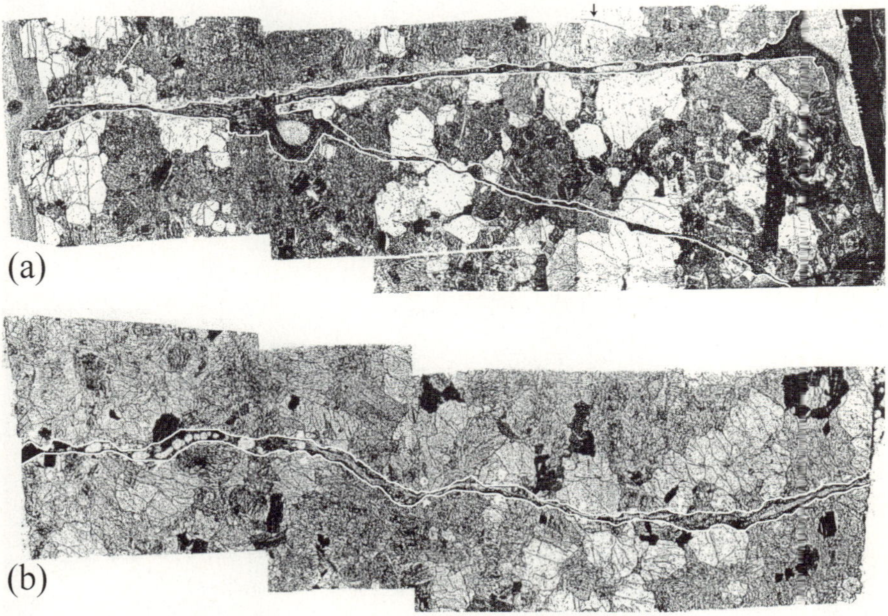

Fig. 4.18. Thin sections of joints C and D constructed by mosaics of micro-graphs. Photographed sections are 4.5 cm long. **a** Joint C, showing straightness of the primary fracture (sub-horizontal) that branches at a large embayment (*dark*) to the secondary fracture, which inclines downward, along a transverse section (parallel to the face of the original rock). **b** Joint D, showing curvatures along a longitudinal section (parallel to the hole formed by drilling). Quartz appears in white and gray, feldspars and muscovite in gray and biotite in black colors

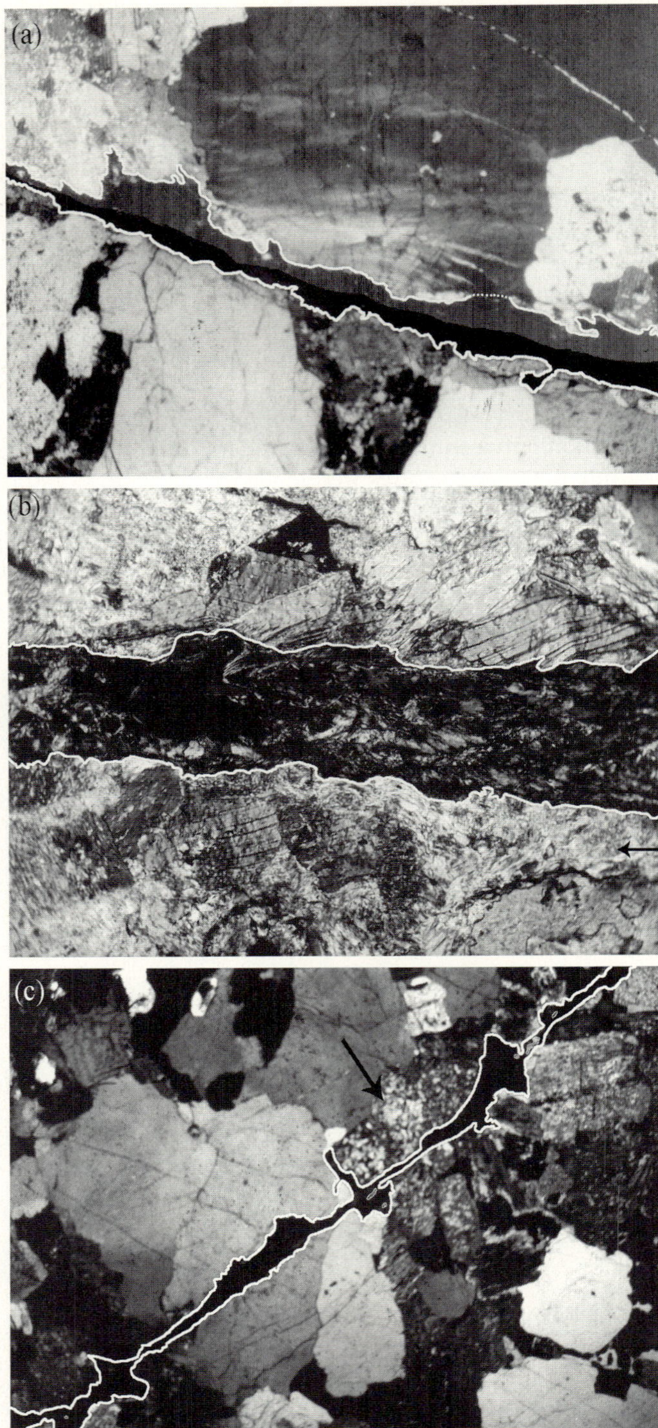

Fig. 4.19.
Profiles of fracture C. **a** Waviness in double refraction occurs on a quartz crystal that is partly cut radially by microcracks. Aperture at center is 0.5 mm (longitudinal section). **b** Inclined crystals of micas along the upper margin of the primary fracture. An angular embayment at lower right of fracture (*arrow*) is filled with unoriented, altered micro muscovite grains. A possible third generation of micro micas, quartz and sericite are scattered along the fracture. Aperture at right is 0.5 mm (transverse section). **c** Transgranular fracture by mode I (without shear displacement) along the branching secondary fracture. Matching is very good in quartz two grains and biotites (scale of several mm) but is poor in the submillimeter scale. Aperture at center, next to the biotites (*arrow*) is 0.1 mm (transverse section)

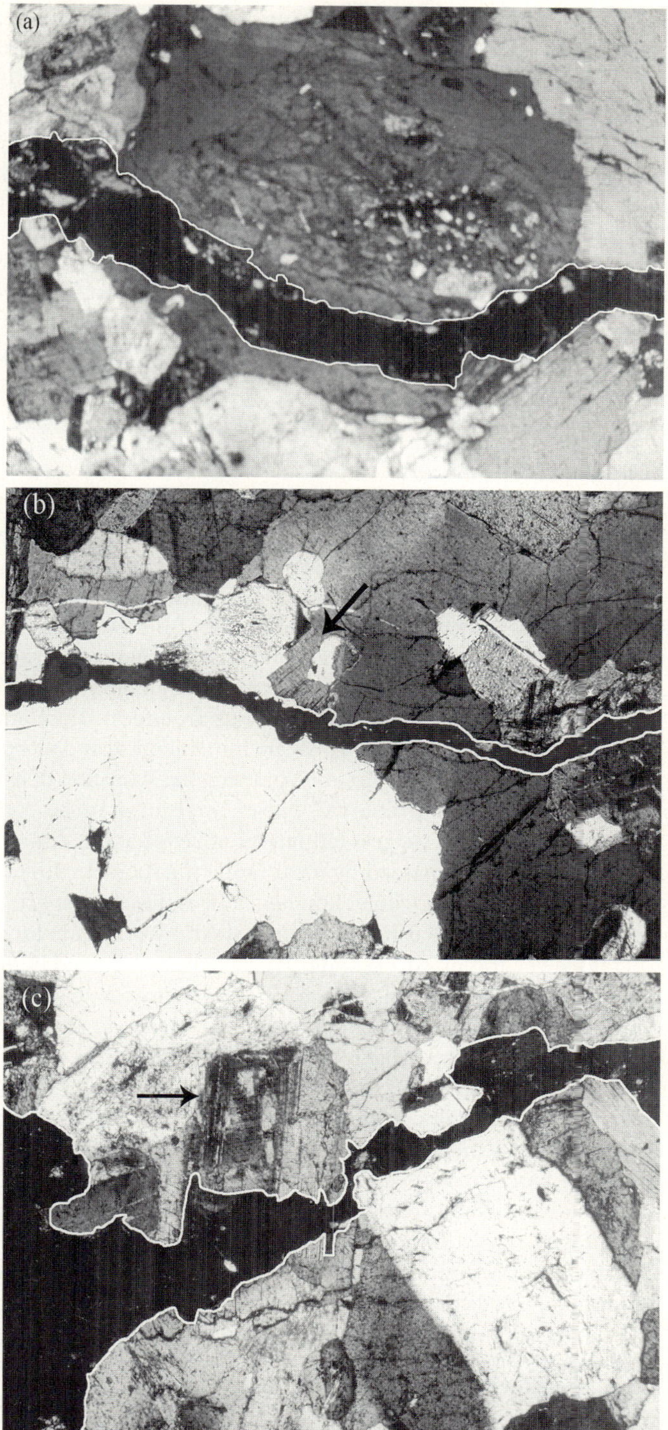

Fig. 4.20.
Profiles of fracture D. **a** A curved transgranular fracture that cuts a K feldspar 5.5 mm long, but propagates around the boundary of a biotite, at lower right (longitudinal section). **b** Two adjacent large quartz crystals at different orientations, appearing as dark and light patches. The dark one is cut by a transgranular fracture. An intergranular fracture propagated along the grain boundaries of medium size plagioclase and biotite at left of center. The length of biotite at center (*arrow*) is 0.8 mm (transverse section). **c** A transgranular fracture through a twinned plagioclase (*arrow*) and K feldspars (light and dark colors), showing considerable variation in aperture and size of mineral fragments along the fracture. The length of twinned plagioclase is 1.6 mm (traverse section). Matching is good in the feldspar scale (several mm) but is poor in the sub-millimeter scale where fragments required to reconstruct the original mineral are missing

can be traced on both sides of fracture (Fig. 4.20b,c). A morphological matching on the sub mm scale, additional to above criteria, provides the highest degree of certainty for transgranularity. Such matchings, however, were rare (see below). Transgranular fracture is a useful tool in determining the extents of opening by mode I vs. shear offsets along the joint walls. When above transgranular criteria are established on a series of different adjacent minerals, their reconstruction to "zero displacement" reveals the extent of opening. A reconstruction that shows deviations from "mode I closing" reveals the extent of shear offset. In contrast to transgranular fracture that cuts through the grain along curved (e.g., Fig. 4.18b and 4.20a) or more often, along straight (e.g., Fig. 4.18a and 4.19c) trajectories, intergranular fracture curves along grain boundaries (Fig. 4.20a). Accordingly, different extinction angles of two grains of the same species on both sides of fracture imply intergranular fracture, particularly when the grain boundaries are not straight.

4.5.2.3
Results

Fracture C. The primary fracture C branches out into a secondary one so that it is represented by two fractures (Fig. 4.18a). Both primary and secondary fractures are straight and maintain approximately uniform apertures, ranging from about 0.1 mm to about 0.5 mm, but the joint walls recede considerably from each other into a large embayment at the branching site and in several other locations along the fractures. Fracture is characteristically transgranular, mostly cutting through the larger feldspar and quarts crystals. Occasional intergranular fracture cuts around medium and small size minerals (feldspars, quartz and biotite). There are mineralogical and morphological mismatches along the section shown in Fig. 4.19a, raising the suspicion of a lateral offset along the longitudinal section of primary C. One or two transgranular contacts of quartz grains (arrows in Fig. 4.18a) suggest left lateral offset of about 4 mm. Wave extinction in a quartz crystal, accompanied by local radial microcracks reveal strain in the quartz, possibly caused by an inclined impingement on the primary fracture wall.

The space along the primary fracture is empty in certain locations, but two or possibly three generations of secondary muscovite are aligned in others (Fig. 4.19b). The first generation consists of 0.7 mm long fresh muscovite grains that display high order double refraction colors. Some of them form small angles (0–20°) with respect to the fracture walls. These micas are kinked, reflecting shear strain during their history. The second generation occurs as small altered patches 1–2 mm in size that often fill small embayments along the fracture walls (making it look straighter than it originally was). These muscovites are often altered to sericite, and their double refraction colors are hardly recognizable. The joint space is filled in some locations by late, micro crystals ($\ll 1$ mm) of mica/sericite and xenomorphic quartz.

Uniform extinction and grain-boundary fit of several minerals along parts of transgranular fracture surfaces of the branching joint testify to an exclusive mode I fracture without any discernable lateral offset (Fig. 4.18a and 4.19c). Matching of the fracture walls along a transgranular fracture in quartz occurs in a scale of several millimeters. However, characteristically, some quartz is missing, thereby negating an unequivocal match at the 0.5 mm scale (left center of figure). A considerably altered grain of biotite 3 mm long is cut by the fracture without lateral displacement (at right side of fracture) (Fig. 4.19c).

Fracture D. Fracture D occurs along a single structure (Fig. 4.18b). The transverse and longitudinal sections reveal curvy fractures to various extents. The longitudinal section exhibits curvatures at two wave lengths. The larger one is about 5 mm (Fig. 4.18b) and the smaller one is about 1 mm (Fig. 4.20a), although there are some straight parts as well. The aperture size deviates considerably from uniformity, between 1.2 mm (Fig. 4.20a,c) down to 0.2 mm (Fig. 4.20b). Fracture is transgranular through the larger crystals of K feldspar (Fig. 4.20a), quartz (Fig. 4.20b) and plagioclase (Fig. 4.20c), along both straight and curved traces. But it is occasionally intergranular around medium- and small-sized minerals (feldspars and biotite) (Fig. 4.20a,b). The transverse section shows a very slight left lateral offset (Fig. 4.20c). Secondary micas are rare along this section, but fragments from broken quartz and feldspars are common along trans-granular parts of the fracture. There is a mismatch on the scale comparable to the joint aperture. No shear offset can be identified in the longitudinal section, and matching improves considerably at 5 mm and above (see good matching in the large K feldspar, in Fig. 4.20a). Thus, the longitudinal and transverse sections both show evidence for just mode I operation, or nearly so.

4.5.2.4
Fracture Behavior in the Two Joints from the Knowles Granodiorite

Comparison with a previous study. Moore and Lockner (1995) investigated the microcrack density (crack length per unit area) in a 19.5 cm-long cylinder of Westerly granite, after inducing a shear fracture by an axial load and 50 MPa confining pressure, at rates less than 10 mm s^{-1}. Their results show that three-quarters of the examined fault length follows intragranular cracks, with only one fourth along grain boundaries. Also, microcrack densities are consistently higher on the dilational side of the shear than on the compressional side. The uneven distribution of microcracks along the "fault" tends to pull the propagating fracture tip towards the dilational side, even though the trend is away from the overall fault strike. As a result, the propagating shear follows the microcrack trend for some distance and then changes direction in order to maintain an overall in-plane propagation path, resulting in a zigzag or saw tooth fracture pattern. Although no systematic investigation of microcrack density was carried out on fractures C and D, an overall impression is that also in the latter fractures, intragranular cracking is more common than intergranular (Fig. 4.18–4.20). However, no distinctions can be made between zones of higher and lower microcrack densities, i.e., no separation between "dilational sides" and "compressional sides" is possible. The waviness along fracture D (Fig. 4.18b) does not resemble the angularity of the zigzag fracture pattern reported by Moore and Lockner (1995, Fig. 14), and fracture C is straight. This comparison possibly suggests that the fracture mechanisms in these two studies were different, shear in the Westerly granite and essentially tensile in the Knowles granodiorite. Nevertheless, in both cases intragranular cracking was dominant.

Fracture sequence. While there seems to have been some shear offset along the primary fracture, no offset is identified in the secondary fracture of joint C. The differences in the modes of displacement between the primary and secondary parts of the joint and lack of muscovite along the secondary fracture suggest that fracture branching occurred after displacement(s) along the primary had taken place. The small angle between the two fractures suggests that a reverse sequence would have caused offsets along the secondary fracture as well. This implies that the geometric splay (Fig. 4.18a) does not reflect a dynamic, single branching of the secondary from the primary in the strict mechanical sense of the term, i.e., the two fractures did not propagate simultaneously under conditions of $K_I > K_{IC}$ (Sect. 4.8.1 and Fig. 2.30c). Thus, a slight opening of the joint enabled the introduction of K-silica rich solutions that produced muscovite, sericite and quartz along the primary joint. Next, a minor shear offset took place along the primary, and then, slightly tilted tensile stress resulted in the formation of the branching fracture.

Straightness characterizes both the primary fracture and the secondary branch of joint C, regardless of the fracture mode, i.e., the offset (s) along the primary that amounted to about 4 mm did not influence the fracture straightness. Quite possibly during the early history of the rock while at depth, lithostatic stresses could have closed the fracture, inducing an impingement of the two fracture walls (Fig. 4.19a), which produced the fragments along the joint, even without lateral offset. This could be an important mechanism contributing to the formation of embayments along the fractures.

The curved part of fracture D reveals a typical transgranular mode I loading (Fig. 4.18b and 4.20a). This precludes the possibility that fracture occurred by curving due to local shears, e.g., due to two approaching non axial fractures from opposite directions (e.g., Yokobori et al. 1971; Acocella et al. 2000). The lack of muscovite flakes along joint D implies that joint C is younger or older than joint D, i.e., joint D formed before the initiation of the hydrothermal activities or after their termination, respectively. However, if analogy is made to the lack of muscovite in the secondary joint C compared to the older, primary C, it appears that joint D is younger than joint C.

It is possible that the relative multitude of joints deviating considerably from verticality in the Shaw quarry may have been the reason for the rarity of vertical or near-vertical fractures in the quarry. A rational for this would be that early oblique fractures released a major amount of stored energy in the rock, minimizing the driving force of subsequent, vertical fracture. This would imply that joints C and D are older than the vertical joints in the area. Such a sequence is supported by the observations made by Zawislanski (1994) in the nearby Raymond field site (Karasaki et al. 2000) that the formation of oblique (diagonal) joints and intrusion of pegmatite dikes preceded the formation of vertical dikes. Thus, further studies are needed to establish whether systematic diagonal fractures have preceded vertical fracture sets in the Sierra Nevada. Finally, some of the oblique straight joints in the Shaw quarry appear to be oblique-concentric joints (Fig. 4.17a), and there is no clear categorization of these fractures. Similar dilemmas occur in other granite provinces as well (Fig. 4.17d), possibly suggesting a distinct class of oblique-concentric joints.

4.5.3 Joint Exfoliation and Arch Formation at Yosemite National Park, California

4.5.3.1 *Introduction*

The granitic rocks at Yosemite National Park constitute part of the Sierra Nevada batholith (Fig. 4.16a). The "monoliths" El Capitan and Royal Arches are part of the "intrusive suite of Yosemite Valley" (Huber 1987). The El Capitan granite intruded older plutons about 108 million years ago and now makes up the bulk of the western half of the valley area (Fig. 4.21). Huntington (1966) and Huber (1987) found that fracture in

RS	Rockslides
RF	Ribbon Fall
EC	El Capitan
TB	Three Brothers
EP	Eagle Peak
YF	Top of Yosemite Falls
YV	Yosmite Village
IC	Indian Canyon Creek
R	Royal Arches
W	Washington Column
TC	Tenaya Canyon
ML	Mirror Lake
ND	North Dome
BD	Basket Dome
MW	Mount Watkins
E	Echo Peaks
C	Clouds Rest
SM	Sunrise Mountain
Q	Quarter Domes
HD	Half Dome
M	Mount Maclure
L	Mount Lyell
F	Mount Florence
CC	Cascade Cliffs
LY	Little Yosemite
LC	Liberty Cap
B	Mount Broderick
SD	Sentinel Dome
G	Glacier Point
SR	Sentinel Rock
FS	Fissures
T	Taft Point
CS	Cathedral Spires
CR	Cathedral Rocks
BV	Bridalveil Fall
LT	Leaning Tower
DP	Dewey Point
MR	Merced River

Fig. 4.21. Bird's-eye view of Yosemite Valley, with selected landforms. Note particularly locations of El Capitan, Royal Arches and Half Dome (from Huber 1987, Fig. 18)

the Yosemite Valley varies with the rock. Both composition and texture influence the spacing of joints, such that the finer-grained diorite is the most closely jointed rock and the El Capitan granite is the least fractured (Huber 1987). Huber (1987) suggests that the vertical cliffs of Yosemite may have hidden vertical fractures behind and parallel to the cliff face, i.e., vertical exfoliation possibly takes advantage of free surfaces provided by previous jointing. Important studies on exfoliation in granitic rocks from Yosemite National Park have also been carried out by Cadman (1969) and Holman (1976).

Arches also occur in granitic rocks. However, arches in granites are far less abundant than in sandstone at Zion National Park (Sect. 3.3). Moreover, their occurrence is selective. We divide arches for convenience in this section into "small" and "large" ones. Small arches exclusively develop at the foot of the cliff of El Capitan. Large arches "decorate" selectively one monolith at Yosemite National Park with the majestic "Royal Arches" (Fig. 4.21) but almost no others. Hence, there is some intriguing variability in exfoliation and arch formation in granitic rocks.

Qualitative and quantitative fractographic and electronic surveying methods have been applied in studying exfoliation joints in middle elevations of the cliff of El Capitan and on the cliff of Royal Arches (Bahat et al. 1995, 1999; Grossenbacher et al. 1996), and these are cited here. Characterization of arches at the foot of El Capitan and a comparison of their fracture mechanisms to those that formed arches at Zion National Park are included as well.

4.5.3.2
Joint Exfoliation and Arch Formation at the Foot of El Capitan Cliff

We start in characterizing the exfoliation joints at the foot of El Capitan cliff, by fractographically describing two types of plumes, the "curved-concentric plume" that exhibits some unique features and the "sub vertical" one, which displays rather common features. We then describe the "full arch" and the "series of partial-arches".

a. The curved-concentric plume. This plume occurs on a curved section of the cliff along which there is a gradual strike change of approximately 2–4° (a precise measurement in this location is technically very difficult). The strike change is greater in the upper part of the plume than in the lower one (Fig. 4.22a,b). The curved-concentric plume is divided into three areas (Fig. 4.22b):

1. The plume at the center, p.
2. A series of sub-concentric fractures, cf, at its southwestern side (at left upper part of figure).
3. *En echelon* cracks in a fringe, e, at its northeastern side (at the right side of figure), as itemized below.

▶

Fig. 4.22. a A photograph consisting of a mosaic of five photos and **b** drawing, showing the curved concentric plume on a curved surface (resulting in some optical distortion), at the foot of El Capitan cliff. In the drawing, *p*, *cf*, *e* and *n*, designate the curved-concentric plume, sub-concentric fractures, *en echelon* cracks and niches, respectively. Also, *small B* designates fracture boundaries of the fringe, f, and *z* marks merge zones along the boundary fracture, *BF*. Numbers *1–3* designate the three distinct areas. *Arrows* show directions of fracture propagation. The line *AB* marks a section, which is rotated at 90° in **c**, to show (exaggerated) third dimension of the joint surface

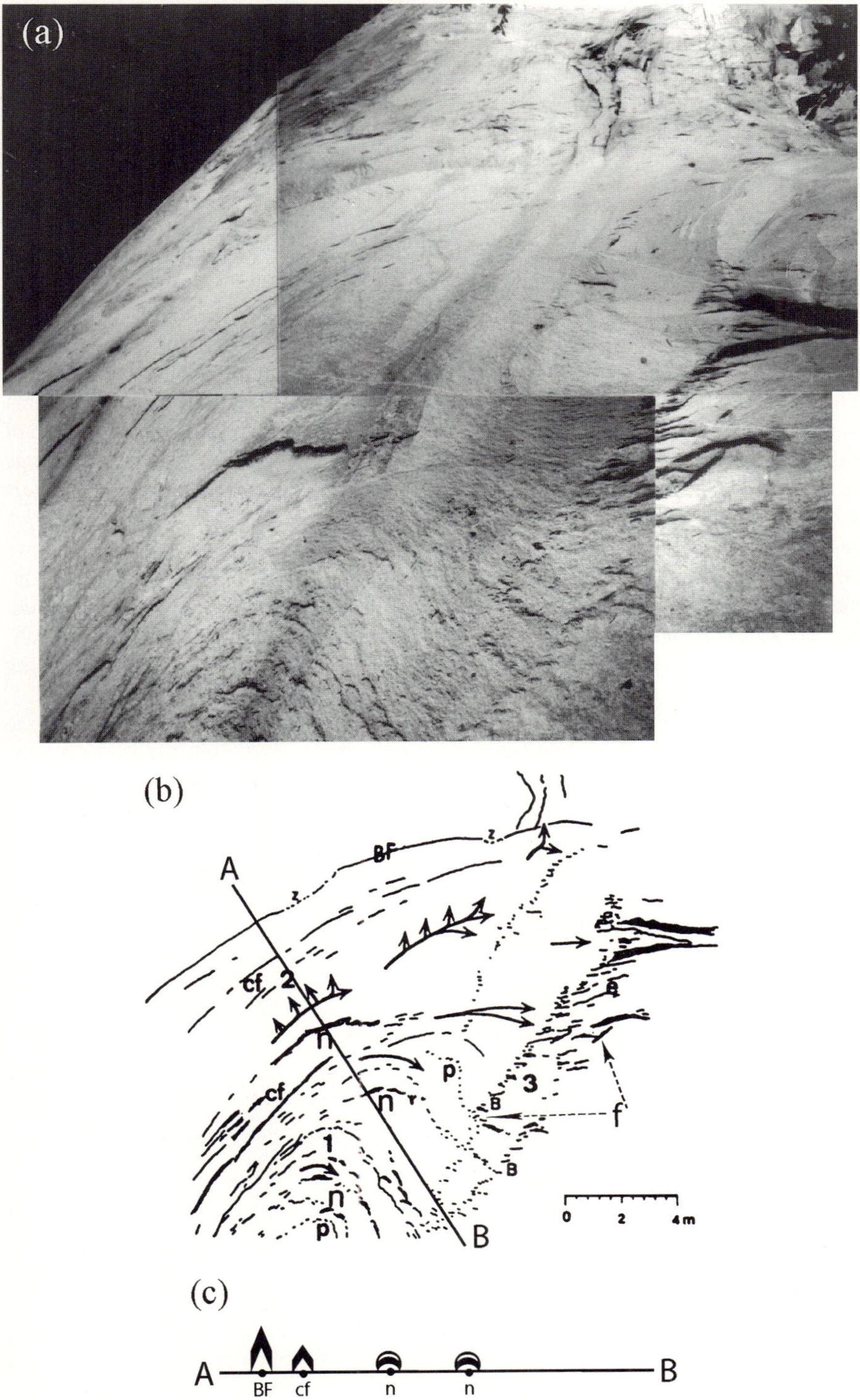
(a)
(b)
A
BF
cf
2
n
cf
n
1
n
p
p
3
B
B
f
B
0 2 4 m
(c)
A
B
BF
cf
n
n

Area 1. The most striking feature in this fracture morphology is the curvature of the plume that propagates upward, and after reaching a peak it bends and the barbs (individual segments of the plume) turn sidewise and then downward, revealing a rotation of the fracture front. The barbs are not uniform in shape along their lengths. As the plume curves, the barbs bend slightly away from the cliff, forming small niches in certain intervals (*n* in Fig. 4.22b,c), as arches of a centimeters or tens of centimeters in size. The bends become particularly prominent at locations of strong surface curvatures, where barbs at higher elevations partly superpose those below them, such that they form a "staircase" climbing up the cliff. There is a gradual increase of the radius of curvature of the plume in higher elevations.

Area 2. The curved-concentric plume becomes coarser, transforming into a series of large niches and concentric fractures at the meter scale (n and cf at the left sides of Fig. 4.22b,c). These fractures cut laterally (asymptotically) into the rock, partially assuming the shape of partial-arches at meter or tens of meter scales. The upper end of this structure consists of three lateral boundary fractures, BF that meet at two "merge zones" (z in Fig. 4.22b) that appear as triangle-like discontinuities along the otherwise continuous curved boundary (Fig. 4.22a).

Area 3. The termination of the concentric plume at its right side is not abrupt; it occurs along two, parallel, straight boundaries (B in Fig. 4.22b) that separate the parent joint from the fringe (Hodgson 1961a). The plumes approach these boundaries suborthogonally, and beyond, *en echelon* cracks, e, maintain the same sense of stepping

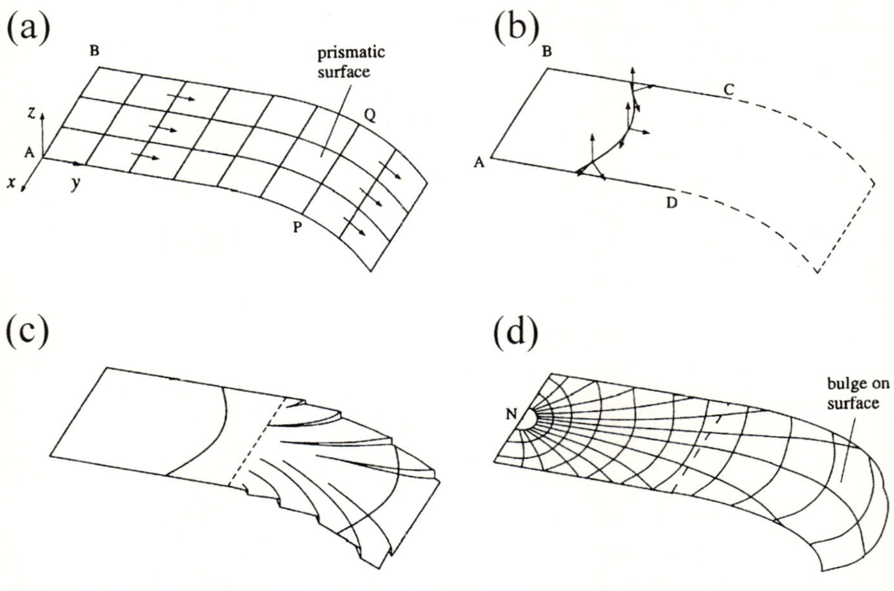

Fig. 4.23. Growth of cracks round by point after bends: **a** Straight-fronted crack moving on a prismatic surface with crack tip parallel to axis of prism. **b** Principal frame fields on a curved crack front. **c** Continued growth of a crack, with a curved front, by twisting into numerous small *en echelon* cracks. **d** Continued growth of a crack with a curved front, by bulging (from Hull 1999)

shown in the plume (the same "staircase" as in area 1). Many *en echelon* cracks are straight and are slightly open. The propagation directions (arrows) are marked in the various transitional parts from area 1 to areas 2 and 3 (Fig. 4.22b).

The fractography displayed by the curved-concentric plume reflects the mechanism of exfoliation around the curved foot of El Capitan (Fig. 4.22a). The mechanism of crack growth into round bends has been investigated by Hull (1999, p. 88) and is cited below. According to Hull (1999, Fig. 3.25), a curved surface is formed by crack tilting (mode II), provided the crack front remains straight and normal to the plane of the specimen. For a variety of reasons, real cracks do not have straight fronts, even in flat plate specimens tested in uniaxial tension. When changes in the local externally applied stress fields cause a crack with a curved front to bend, more complicated crack movements are required. Figure 4.23 is a 3-D representation of the simple case for a straight-fronted crack. The crack front AB is always parallel to the x-axis (Fig. 4.23a). A curved surface is formed by tilting about this axis. At any position PQ, the crack tip is always straight and parallel to the x-axis. For the realistic case (Fig. 4.23b), the crack front is curved (see ripple marks in Fig. 3.24). Suppose that initially, the crack is in a planar surface and that as it grows the stress fields caused it to bend on a path described by the broken lines. The principal frame fields marked at the tip of the crack show that the crack can continue to propagate in the planar surface without any geometrical constraint. Any stresses that cause the crack to bend, as illustrated, result in local stress fields at the crack tip with a mode III component. This applies to all elements of the crack tip not parallel to the x-axis. The crack tip experiences twist forces. However, to maintain a smoothly curving surface, the crack front cannot move by twisting.

There are four possible ways for a crack that is forced to bend to follow the general path marked by the broken lines in Fig. 4.23b. First, the curved crack front straightens and becomes parallel to the x-axis. The crack then follows a prismatic surface for continuous tilting at the straight crack tip (Fig. 4.23a). Second, the crack front remains curved and the crack is driven by the stress fields to bend on a path equivalent to the prismatic surface. This inevitably means that the surface cannot remain smooth and continuous. The local mixed I / III stress fields at the crack tip nucleate an array of *en echelon* segments, separated by steps on the fracture surface, parallel to the direction of crack growth (Fig. 4.23c). Third, the crack front remains curved and follows a smooth 3-D double curvature path. This is achieved by the crack front bulging out of the prismatic surface Fig. 4.23d). This diagram shows striae and undulations on the surface with the undulations mapping out the successive positions of the crack tip for a non-twisting crack. Fourth, the crack front moves by a combination of the first three mechanisms, such that the relative contributions depend on the ease of initiating arrays of secondary cracks at the tip of the parent crack.

The propagation of the curved-concentric plume (Fig. 4.22) approximates Hull's (1999) 3-D model (Fig. 4.23). The curved front of the exfoliation joint propagates around the curved surface of the cliff (see the curvilinear shapes of the long arrows in Fig. 4.22b). Breakdowns due to local mixed I / III stress fields occur at the joint tip, initially into delicate plumes, then to coarse plumes and finally, to *en echelon* segmentation. The breakdown of the joint front (Fig. 4.23c) and the bulge (Fig. 4.23d) reflect 3-D dimensions of Hull's model, while the niches of plumes and steps of the *en echelon* segmentation represent the actual 3-D on the cliff (Fig. 4.22a,b).

b. The sub-vertical plume. This plume occurs on a planar section of the cliff on a back wall of a full arch (Fig. 4.24). The convex side of an undulation faces downward, indicating that the plume propagated downward. The plume produces an asymptotic fracture at an angle ≥ 1° with the cliff. This undercutting reveals its tensile (Bahat 1979) role in the formation of joint exfoliation.

c. The full arch. A few tens of meters from the curved-concentric plume southwestwards along the cliff, there is a fully developed arch that consists of a ceiling and of two sidewalls (Fig. 3.28 and 4.25). The height of the arch from base to ceiling is about 7 m. At the top of the upper part of the arch is a merge zone of two boundary fractures, which assume the shape of partial-arches. Each partial-arch consists of parts of the ceiling and of a sidewall, such that the concave side of the partial-arch on the right faces towards the left, and on the left the concave side of the partial-arch faces towards the right. Hence, the partial-arches behave independently of each other when forming a full arch.

Fig. 4.24.
A coarse, vertical plume undercuts thin slices from the back wall of an arch (at both sides of the plume) at the foot of El Capitan cliff; the sidewall of the arch is shaded at right. The undulation (*u*) maintains orthogonal relation with the plume. Note hammer for scale at lower left

Fig. 4.25.
A full-shape arch at the foot of El Capitan cliff. A little off the top of the arch is a merge zone (*arrowed*) of two partial-arches. Each partial-arch consists of part of the ceiling and a sidewall. Picture is taken looking upward the cliff (note an optical distortion). Size of arch approximates those of the sidewalls in Fig. 4.24 and 4.26

d. The series of partial-arches. There are many partial-arches that do not match with other partial-arches to form full arches. A series of some five individual left partial-arches that are convex in the same direction is shown in Fig. 4.26. Some partial-arches are subdivided into parallel thin slices, about 1 to 10 cm in thickness that form a small angle ≥ 1° with the cliff and laterally undercut into the rock. The sidewalls of the various partial-arches resemble the sub-concentric fracture cf (Fig. 4.22a–c).

Discussion and conclusion. The gradual increase in the topographic coarseness from delicate barbs, through small *en echelon* segments into large ones, can be correlated with the local increase of the K_{III}/K_I ratio on the curved section of the cliff. Particularly, large segments occur at the upper part of the fringe (around e in Fig. 4.22a,b). These are also the most twisted and opened cracks, implying the greatest ratio of K_{III}/K_I (Cooke and Pollard 1996). Sommer (1969) observed segmentation in AR glass at as little as 3.3° rotation of the parent fracture from normality to the tensile stress, and this is approximately the change in strike occurring across the areas 1 and 3 (Fig. 4.22b). The joint growth upward and downward (as demonstrated by the curved-concentric plume) along a curved front resulted in plume coarsening and *en echelon* segmentation, following Hull's model (Fig. 4.23c).

Fig. 4.26.
A series of partial-arches that concave towards right at the foot of El Capitan. Thickness of sidewall (between *arrows*) is about 80 cm

The plume trajectories mark the circular path of the joint front (arrows in Fig. 4.22b). This path differs radically from the unilateral propagation of the vertical plume on the back wall of an arch (Fig. 4.24); both are at approximately the same height on the cliff, only tens of meters away, implying different fracture mechanisms, possibly in different times. It demonstrates the small changes in local stress fields (across tens of meters) within larger stress fields (across hundreds of meters) dominated by remote stresses. There seems to be a gradual structural transition from curved-concentric plumes through niches and sub-concentric fractures into partial-arches. They form sets of various sizes that undercut laterally into the rock, forming small angles of several degrees with the cliff. When matching (left and right) partial-arches merge, full arches develop. A merge often occurs at one (Fig. 4.25) or two zones along the ceiling. Thus, the various fracture features shown in Fig. 4.22, 4.25 and 4.26 provide a complete evolutionary sequence of exfoliation and arch formation at the foot of El Capitan. Note that the mechanisms leading to the formation of the latter arches differ from those that created arches in sandstones (Fig. 3.31).

4.5.3.3
Joint Exfoliation at the Middle Height of El Capitan Cliff

Methods of investigation. Exfoliation surfaces were examined in the middle height of the approximately 1 000 m high granitic cliff of El Capitan (Fig. 4.21), within the frame shown in Fig. 4.27a (this section is cited from Bahat et al. 1999). An electronic total station, which incorporates a laser range-finder, provided the means to survey points and measure angular and size parameters on the cliffs of El Capitan, the Royal Arches and the southwestern side of Half Dome (Sect. 6.5) at Yosemite National Park (see also Grossenbacher et al. 1996). A principal assumption in this technique is that these cliffs are planar, which applies to the investigated exfoliation joints. For the geometries used, a deviation of 3 m from a true planar surface yields an error of at most 1 m in a calculated position on a cliff. With distances measured of about 50–100 m and a distance from base station to cliff of over 300 m, this does not appear to involve significant error, as the relief on the cliff is less than 5 m (as verified by examination of topographic base maps). Taken together, these yield an error of less than 3% which was deemed acceptable, considering the large distances in the park.

Fractography. Vertical fracture surfaces on the N 32° E oriented exfoliations are marked by tens of early cracks, which are shaped as fans of radial plumes. Each fan is comprised of an origin and radial plumes. In some cases, a single or several concentric ripple marks (Fig. 4.27a,b) form orthogonal relationships with the plumes at their intersections on the fans. Distorted fans manifested by plumes deviating from radial structures are common as well. The size of a fan varies from several meters to several tens of meters. Occasionally, only partially developed radial plumes appear.

These fans cut country-rock and dikes of the Taft Granite (Fig. 4.27b) without apparent change in direction. Early circular cracks resembling in shape and size similar fractures from Zion National Park (Bahat et al. 1995 and present Fig. 3.24b) also occur sporadically on the cliff of El Capitan. The fans are more common than the early circular cracks. Many neighboring fans interacting with each other merge into vertical composite joints, which are several hundreds of meters in size.

The fan population produces a specific fracture pattern at the middle height of the cliff. This pattern is not so clear for fractures A and B (Fig. 4.27c). However, a particular preferred orientation appears for fractures C and D, where fans indicate upward propagation above the approximate height of D (where ripple marks in the fans convex upward), and fans below this height show downward propagation (where ripple marks in the fans convex downward). Hence, fans cutting the granite at El Capitan propagated upward and downward from initiation zones on individual exfoliations. These zones occupy similar heights on the cliff (Fig. 4.27d).

A few exfoliation joints occur parallel to each other on the cliff, such that each two adjacent joints contain a rock "slice" between them. Some of the slices do not maintain a constant thickness, causing deviations from parallelism of the exfoliation joints. Letters A–D represent surviving remnants of distinct exfoliation surfaces, such that D reveals that the previous more external slices (A–C) have been removed at this location (Fig. 4.27c). Exfoliated slices apparently detach selectively from the cliff in a gravity driven process, as revealed by certain undetached parts from the cliff (dash lines in Fig. 4.27c).

Fig. 4.27a,b.
Exfoliating joints in the El Capitan granite at Yosemite National Park (after Bahat et al. 1999). **a** Photograph of a cliff on El Capitan striking N 32° E, showing the frame which is detailed in **b. b** Photograph of the exfoliated cliff within the frame, showing multiple groups of radial striae. Some groups occur in association with concentric ripple mark, together forming fan-shapes. The cliff is divided by sub-vertical fractures into distinct exfoliations. Fractures cut sub-horizontal dikes of the Taft granite (which have a light color)

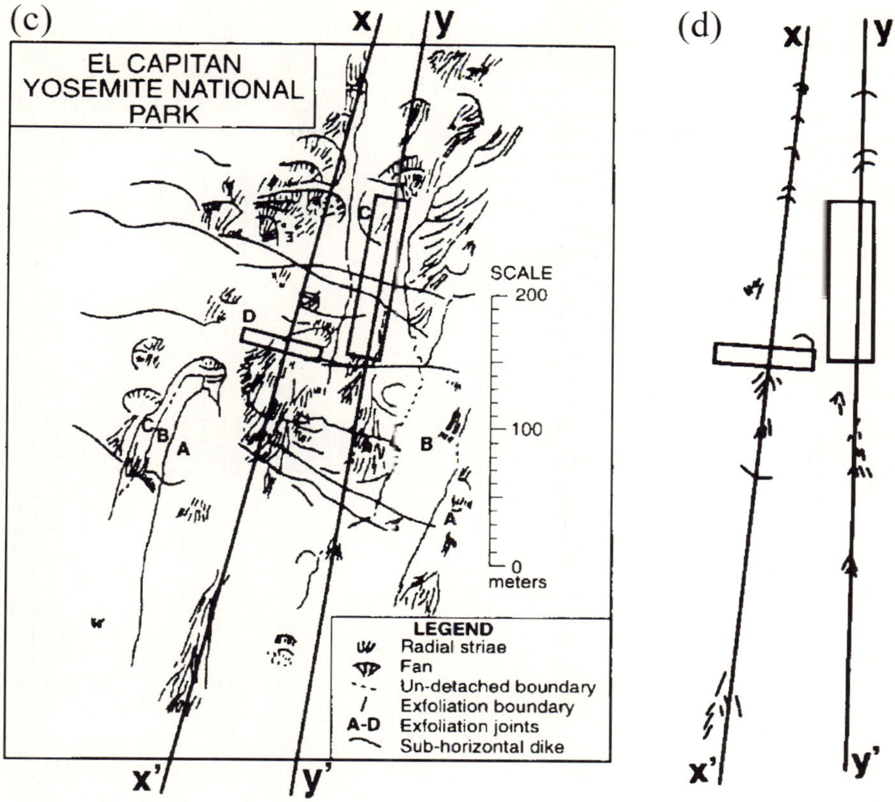

Fig. 4.27c,d. Exfoliating joints in the El Capitan granite at Yosemite National Park (after Bahat et al. 1999). **c** Scaled diagram of **b**, showing exfoliation boundaries between slices belonging to distinct joints *A–D*. *Rectangles* designate initiation zones at which the propagations depart to opposite directions, as shown by traces of convex ripple mark or radial plumes. **d** Traverses *x–x'* and *y–y'* on joints *C* and *D* from **c**, depicting fans. The fans are either radiating barbs or convex ripple marks pointing to the direction of fracture propagation upwards in the upper part of diagram and downwards in the lower one. Rectangles are as in **c**

Mechanism of exfoliation. According to Holzhausen and Johnson (1979) longitudinal splitting occurs parallel or sub-parallel to the direction of loading in uniaxially compressed rock cylinders. Although longitudinal fracture often originates at the ends of the specimens (due to sample end-effects) and propagates toward the center, uniform loading also results in fracture origins within the specimens (Peng 1970). Holzhausen and Johnson (1979) postulate that an "internal fracture" is formed by the coalescence of earlier axial cracks (analogous to the term composite fracture used in this section), which have created stably under high loads (local compression in Fig. 4.28a,b). An extension of the internal fracture to longitudinal splitting may be caused by buckling of the rock away from this surface. This would happen when the two separated parts of the cylinder bend outward (local tension in Fig. 4.28c). Tension at the tips of the longitudinal split may cause unstable growth to the ends of the specimen. Holzhausen and Johnson (1979) consider several tensile and shear criteria but refrain from advocating one mechanism leading to splitting (Sect. 2.2.3.11 and 2.2.13, see also Fig. 2.65).

Fig. 4.28.
Schematic representation of how buckling causes longitudinal splitting of cylindrical specimens. **a** Axial cracks coalesce under applied axial stress σ_a to an "internal fracture" of length 2*c* at a distance d from the free surface. **b** The internal fracture shown at 90° position to *a*, exhibiting early cracks in shapes of fans pointing upward and downward, before coalescence. **c** Axial propagation of the internal fracture (in the *a* position) occurs when it buckles outward (from Bahat et al. 1999, modified after Holzhausen and Johnson 1979)

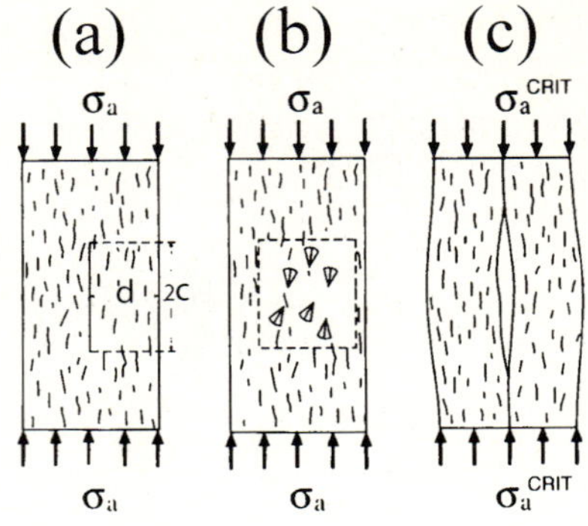

Fig. 4.29.
a A section model showing the profile of joint exfoliation by longitudinal splitting and buckling of the "internal fracture" from Fig. 4.28a. *Vertical arrows* indicate overburden load. Fracture (longitudinal splitting) occurs in association with buckling (an exaggerated buckling is shown by the *dash line*), which develops sub-normal to σ_3 (*horizontal arrow*). **b** An early slice collapsed (at *0*) and washed away, and new exfoliations develop (earlier ones closer to the free surface at *arrow* numbered *1* to *arrow* numbered *10*). **c** A profile of an arch at the lower part of the cliff, which forms when part of the slice collapses (*white arrow*)

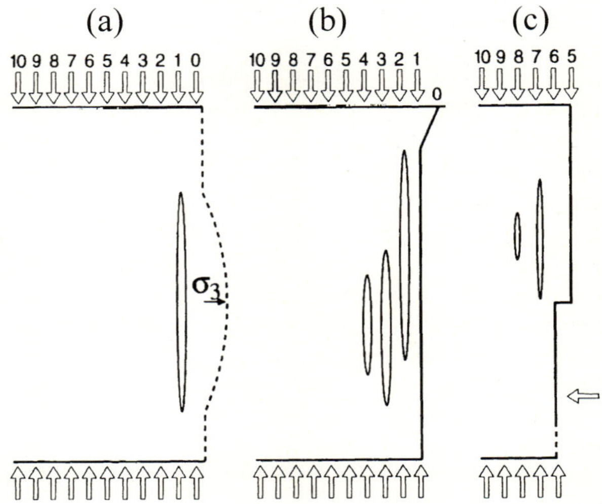

The findings in the field are in harmony with various experimental observations made by Holzhausen and Johnson (1979). Coalescence of coplanar fans (Fig. 4.28b–d) that are close to and parallel to free surfaces (the outer surface of the cliff) is the main fracture process in the middle height of El Capitan cliff that leads to exfoliation. Exfoliation developed along principal planes populated by multiple early cracks by a mechanism of longitudinal splitting, which was caused by the overburden load. Splitting occurred upward and downward from an initiation zone at an elevation where the normal resistance to buckling was minimal (Fig. 4.28) next to a free surface in analogy to Fig. 4.29. Fracture propagation on El Capitan cliff was polarized upwards and downwards from this zone. It followed the creation of multi-nucleation sites of independent cracks that had developed into fans and then merged with each other into composite joints (Fig. 4.27c,d).

Considering tension vs. shear mechanisms causing splitting (Holzhausen and Johnson (1979), the fractographic observations suggest the following. First, the ripple marks and plumes on the surfaces of the early joints reflect mode I opening as the predominant mode (Bahat 1979), i.e., on principal planes normal to the minimum principal stress. Second, minor local shear was superposing the tensile mode (Bahat 1987b). Third, significant shear strain can be ruled out because such a process would have erased the fracture markings, and fourth, the lack of hackle or exfoliation-branching (Fig. 2.1) raises questions about an unstable fracture analogous to that observed experimentally (Holzhausen and Johnson 1979). These observations imply stable joint propagation at the middle height of El Capitan cliff that may be correlated with the compressive state in the uppermost lithosphere. If however buckling is promoted close to the ground surface of the granite, unstable exfoliation may occur (Fig. 2.65).

A formula used by Holzhausen and Johnson (1979) is adapted for the analysis of the critical conditions for buckling:

$$\frac{\sigma_{\mathrm{a,critical}}}{\pi^2 \frac{E}{3}} = \left(\frac{d}{2c}\right)^2 \tag{4.1}$$

where $\sigma_{\mathrm{a,critical}}$ is the critical axial stress for buckling of a beam with fixed ends, E is Young's modulus, d is the distance between the internal surface (nucleation of the longitudinal split) and the free surface, and $2c$ the internal fracture length (Fig. 4.28). This formula shows that $\sigma_{\mathrm{a,critical}}$ increases with d^2. It explains the observations both in granite and in sandstone (Fig. 4.30 and Fig. 3.25c) that vertical dark streaks produced by dripping solutions mark external exfoliations and not internal ones. That is, the earlier longitudinal splits develop close to the free surface at small d values (Fig. 4.29), and then new ones gradually form further inside the rock. Also, Eq. 4.1 implies that rock slices produced between successive exfoliations should be much thinner than the lengths of the associated "internal fractures". This may explain the great ranges of exfoliation spacing depicted in various publications (e.g., Holman 1976, p. 2; Twiss and Moores 1992, Fig. 3.5).

Joints are classified into four fracture categories: Burial, syntectonic, uplift and post-uplift groups. These four groups differ in their various properties, including their fractographies. It turns out that the merger of early circular cracks (Fig. 3.23b,c) and early fans (Fig. 4.27b–d) is common to exfoliation joints in both sandstone and granite. These are characteristic fractographic features exclusive to post-uplift joints, which develop close to a "free boundary" above the ground surface. A similar merger of fans into a larger post-uplift joint has been observed in chalk (Bahat 1991a, p. 301). Mergers of early circular cracks or early fans that occur on post-uplift joints are not known from surfaces of burial and uplift joints, and they are rare on syntectonic joints. As mentioned above, Huber (1987) suggested that the vertical exfoliations possibly took advantage of free surfaces provided by previous jointing. These observations support his suggestion.

4.5.3.4
The Royal Arches

The series of Royal Arches is located between El Capitan and Half Dome in Yosemite Valley (Fig. 4.21) and provides an exceptional occurrence of very large arches (Fig. 4.42a,b). They are considerably larger than arches in Zion National Park (e.g., Fig. 3.24a,b) and are by far larger than arches at the foot of El Capitan. These arches are confined to a single cliff, cutting the entire monolith, in contrast to the division into distinct structures at different heights on El Capitan and the limitation of arches to its lower elevations.

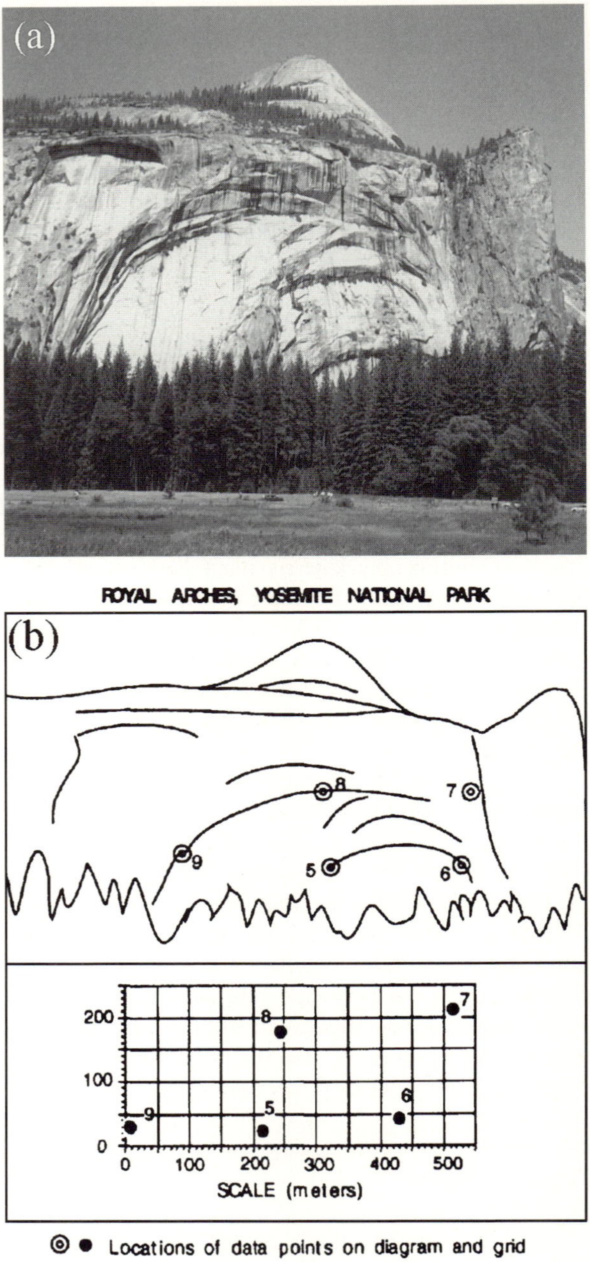

Fig. 4.30.
a Photograph of the Royal Arches. **b** Scale diagram of the Royal Arches showing measured dimensions. There are angular merges of ceilings with the sidewalls at points 5 and 9 and in the upper left arch. Note the division of parallel thick slabs along side-walls and ceilings into series of thin slabs (particularly along arch 7–8–9), often combined with sharp contours that outer surfaces make with sidewalls and ceilings. Note vertical, dark streaks of dripping solutions between 8 and 9

The Royal Arches occur in a concentric structure and show both symmetric and asymmetric features (Fig. 3.28a). The lower and smaller arches tend to be symmetric, whereas the upper and larger arches are asymmetric. The latter arches display angular merges of the ceiling with the sidewalls at one side of the arches (their western side) and an arcuate continuity of the ceiling with the sidewalls of the arches on the other side (the eastern one). The close association of exfoliation with arch formation observed at the foot of El Capitan is repeated in the Royal Arches. However, the ceilings and sidewalls in the Royal Arches are subdivided by more undercut exfoliation joints (e.g., between points 8 and 9 in Fig. 4.30b).

4.5.3.5 *Arch Formation in Granitic Rocks and Sandstones*

Structural features common to the various arches. *(a)* Intimate associations of arches with exfoliation joints are almost invariable, suggesting that joint exfoliation is a prerequisite for arch formation. *(b)* Many contacts between the outer surface of cliffs and the ceilings and sidewalls are angular (in some cases close to an angle of 90°), implying a development by fracture, rather than by erosion. *(c)* There are both symmetric and asymmetric arches. *(d)* On many arches, particularly asymmetric ones, there are merge zones of sidewalls and ceilings or matching partial arches, implying that these arches generally form by the connection of distinct structural components. *(e)* Niches occur in multiple sizes at bending barbs on centimeter scale and at sub-concentric fractures on meter scales at the foot of El Capitan (Fig. 4.22). They also appear on large scales of tens of meters on partial-arches and arches in Zion National Park. Niches appear to be initiation sites of many symmetric and asymmetric arches. The shape and symmetry of an arch depends on the number of initial niches and their location along the arch perimeter. A single niche at the highest position of the arch is likely to result in a symmetric arch. On the other hand, niches at positions below the arch summit are likely to produce asymmetric arches.

Structural features that differ in the various arches. *(a)* There are many more arches in sandstone at Zion National Park (Sect. 3.3) than in granitic rocks at Yosemite National Park. *(b)* Arches at the foot of El Capitan differ significantly from the Royal Arches. *(c)* The mechanisms of partial-arch formation at Zion National Park and at the foot of El Capitan are different, as follows: In the sandstone, the ceilings and sidewalls often start tangent or vertical to a niche and propagate downward to yield a curved surface normal to the trajectories of least compressive stress induced by the gravitational load of the overlying rock. This propagating curviplanar fracture often cuts *normally* into the rock body across a series of rock "slices" sandwiched between exfoliation joints that develop parallel to the cliff and to each other. On the other hand, partial arches at the foot of El Capitan are formed by curved-concentric plumes (Fig. 4.22) that undercut laterally and form small angles (1–4°) with the cliff. Correspondingly, arches at Zion National Park are commonly impressively thick, since they consist of many slices of rock, whereas arches at the foot of El Capitan are considerably thinner.

Open questions. The differences in fracture mechanisms suggest that additional to the driving exfoliation, a certain set of conditions needs to be provided for arch formation. The available information does not warrant a speculation on what these combined conditions at each location should be. This leaves several open questions as summarized below.

1. Huntington (1966) observed that for unknown reasons, sheeting developed most abundantly in quartz monzonite and granite, "spottily" in granodiorite, and never in quartz diorite or gabro. That is, arch formation is dependent on the petrography of the rock, which in turn should significantly influence the strength and resistance properties of the rock to fracture. Thus, it is pertinent to enquire how such differences resulted in distinct exfoliation and arch formation on the cliff of El Capitan vs. the Royal Arches.
2. Moreover, there are significant differences in fracture behavior at different heights of El Capitan (Fig. 4.22, 4.24–4.26 vs. Fig. 4.27). These differences cannot merely be assigned to differences in rock properties. We need to determine what the distinct fracture stress conditions were in such cases.
3. Dry conditions probably prevailed during fracture at the middle heights of El Capitan, whereas pore pressure activities quite likely, influenced exfoliation in the sandstone at Zion National Park (Bahat et al. 1995). In view of the multitudes of arches both at Zion National Park and at the Royal Arches, may one assume that fracture conditions (dry vs. wet) have been similar at the two sites?
4. The exfoliations investigated at El Capitan and at Half Dome generally follow the NNE directions of earlier regional fractures in Yosemite National Park (Bahat et al. 1999). The implication is that previous fractures provided free surfaces parallel to which exfoliation occurred in these two sites. This fracture overprinting pattern is repeated at Zion National Park (Bahat et al. 1995). The Royal Arches are approximately E-W and deviate from the regional NNE direction in Yosemite National Park. This could be related to an exfoliation that relates to a different regional-structural setup.

4.6 Fractographies that Pertain to both Quasi-Static and Dynamic Fractures

We start by raising the question of how *en echelon* segments (generally reflecting quasi-static propagation) and hackles (generally reflecting dynamic propagation (Sect. 2.2.5) can occur adjacently on a single fringe (e.g., Fig. 2.1c). The answer is based on the assumption that only "slight deviations" of the direction of the minimum principal compression (or maximum principal tension) from orthogonality to the fracture surface would induce *en echelon* segmentation (Sommer 1969). Accordingly, fringes that form by both quasi-static and dynamic propagations may be influenced by such "slight deviations". Fractures formed quasi-statically when the ratio K_{III}/K_I is high will be dominated by *en echelon* fringes (Fig. 2.1a and 2.35), and these fringes will have the tendency to assume simple, uniform widths (Fig. 2.34b). On the other hand, fractures formed dynamically when K_{III}/K_I is low will be dominated by hackle fringes, on which, however, *en echelon* segmentation in a shingle-like arrangement may be superimposed due to this ratio. Such superposition can explain the frequent occurrence of fringes that show hackles in a shingle-like arrangement (Fig. 4.12b). The latter fringes often exhibit non-uniform widths (Fig. 2.34a and the center of Fig. 4.12b). Hence, it is quite possible that on the fracture surface obtained by De Freminville (1914) (Fig. 2.1c), the two fringe types reflect local slight changes in the extent of normality of the maximum principal tension: The fringe area that displays hackles reflects the "pure" mode I, whereas the fringe area that displays hackles in a shingle-like arrangement reflects a mixed mode I + III loading.

In summary, most fringes from sedimentary rocks suggest quasi-static growth, while rare fringes display cuspate hackles (Bahat 1991a, Fig. 3.41a) or hackles in shingle-like arrangements (e.g., Fig. 2.33e). Note the resemblance between the latter fracture morphology and the fracture surface of a turbine that is interpreted to have initiated in fatigue growth and possibly approached catastrophic fracture when $K_I \sim K_{Ic}$ conditions were attained (Fig. 2.44a) (Bahat 1991a, p. 132). In many outcrops, the distinction between these fringe types would be relatively easy, whereas in others that exhibit transitional fractographies, this distinction is expected to be more problematic. Hackle fringes are far more abundant in granitic rocks than in sedimentary ones, but even in granites hackles often show shingling. Note that in Fig. 4.11 in the lower fringe shingling is almost absent, whereas in the upper fringe shingling is more pronounced.

4.7 Comparative Jointing

4.7.1 Cooling Joints in Granites

4.7.1.1 *Regional Axial Fracture in Two Provinces*

Maps of the two fracture provinces, the Sierra Nevada Batholith (Fig. 4.31) and the South Bohemian Pluton (Fig. 4.32), both show that these are regional, elongated intrusion-complexes consisting of series of large and small discrete plutons that are aligned along the complex axes (Bateman and Wahrhaftig 1966 and Breiter 2001, respectively). According to Breiter (2001), the evolution of the South Bohemian Pluton started with the intrusion of the peraluminous Lasenice granite. Next was the intrusion of high-K peraluminous melt that crystallized as porphyritic Cimer granite in the south and non-porphyritic Mrákotin granite in the north. Further fractionation of the "Cimer melt" produced the Eisgarn granite in the central part of the intrusion-complex. Additional small bodies were then produced by intense fractionation of the "Eisgarn melt" and a still younger magmatic input formed the Zvule, Melechov and Cerinek deep-seated stocks.

A detailed geochemical investigation of the Eisgarn, Zvule and Melechov granites enabled Breiter (2001) to construct the inner structures of these plutons. Breiter (2001) shows that the Eisgarn (Fig. 4.33a), and Melechov (Fig. 4.33b) granites are elongated bodies and their long axes are oriented NNE, whereas the Zvule granite (Fig. 4.33c) has a slight elliptical shape with the longer axis slightly oriented E-W. It has been found that NNE is an important direction of a set of cooling joints along the South Bohemian Pluton (Fig. 4.5). The possible implication of the corresponding direction of the inner structures of the Eisgarn and Melechov granites to the regional direction of the NNE joint set is that there is a genetic connection between the two. Possibly, the cooling joints are aligned in the South Bohemian Pluton along certain discrete plutons as long as they also maintain the regional orientation, i.e., the regional influence is more pronounced than the one exerted by the individual plutons on joint orientation. This interpretation would be challenged if future studies will prove the non-existence of NNE oriented joints in the Zvule granite, which is approximately oriented E-W (Fig. 4.33c).

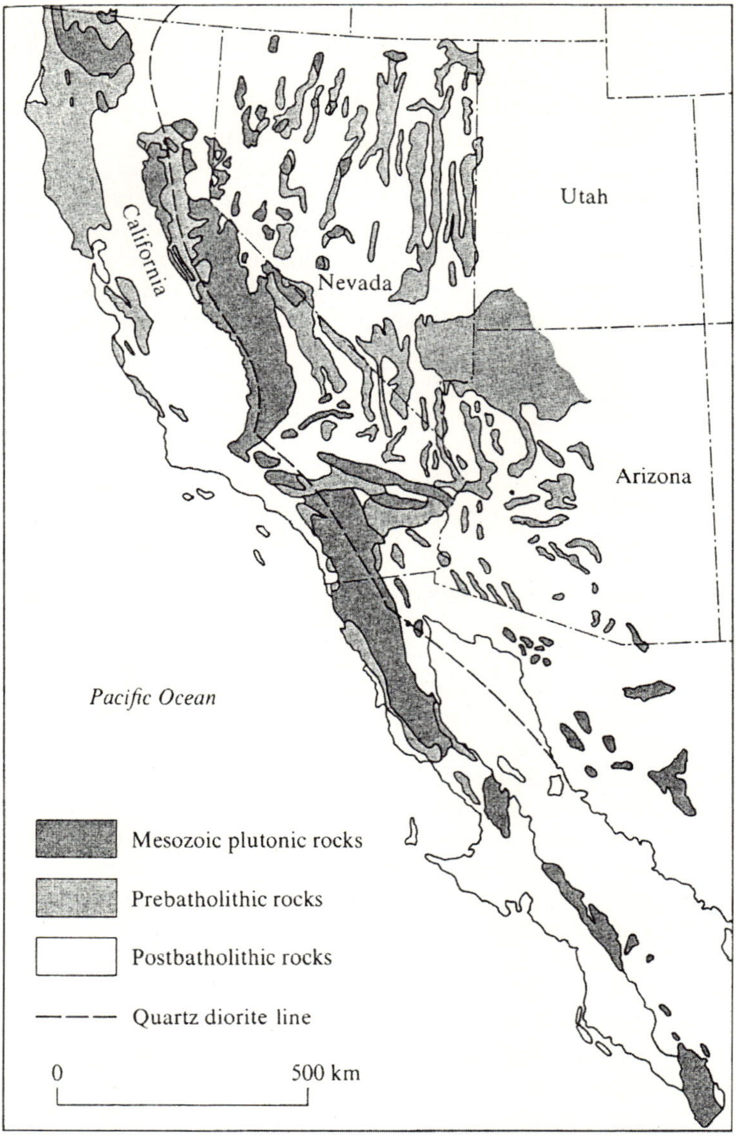

Fig 4.31. The general distribution of Mesozoic plutonic rocks in California and Lower California. The Peninsula Ranges are located in Southern and Lower California. The Sierra Nevada Mountains are in east-central California (modified from Ehlers and Blatt 1982)

To the best of our knowledge, no special treatment of cooling joints exists for the Sierra Nevada Batholith. Furthermore, the classification into distinct genetic groups/stages of joints in granites is challenged by existing investigations. For instance, Davis and Reynolds (1996, p. 265) cite works by Rehrig and Heidrick (1972, 1976) regarding jointing in Late Cretaceous to Eocene plutons in southern Arizona. The latter authors concluded that these plutons were intruded into the upper levels of the crust and as they cooled and congealed,

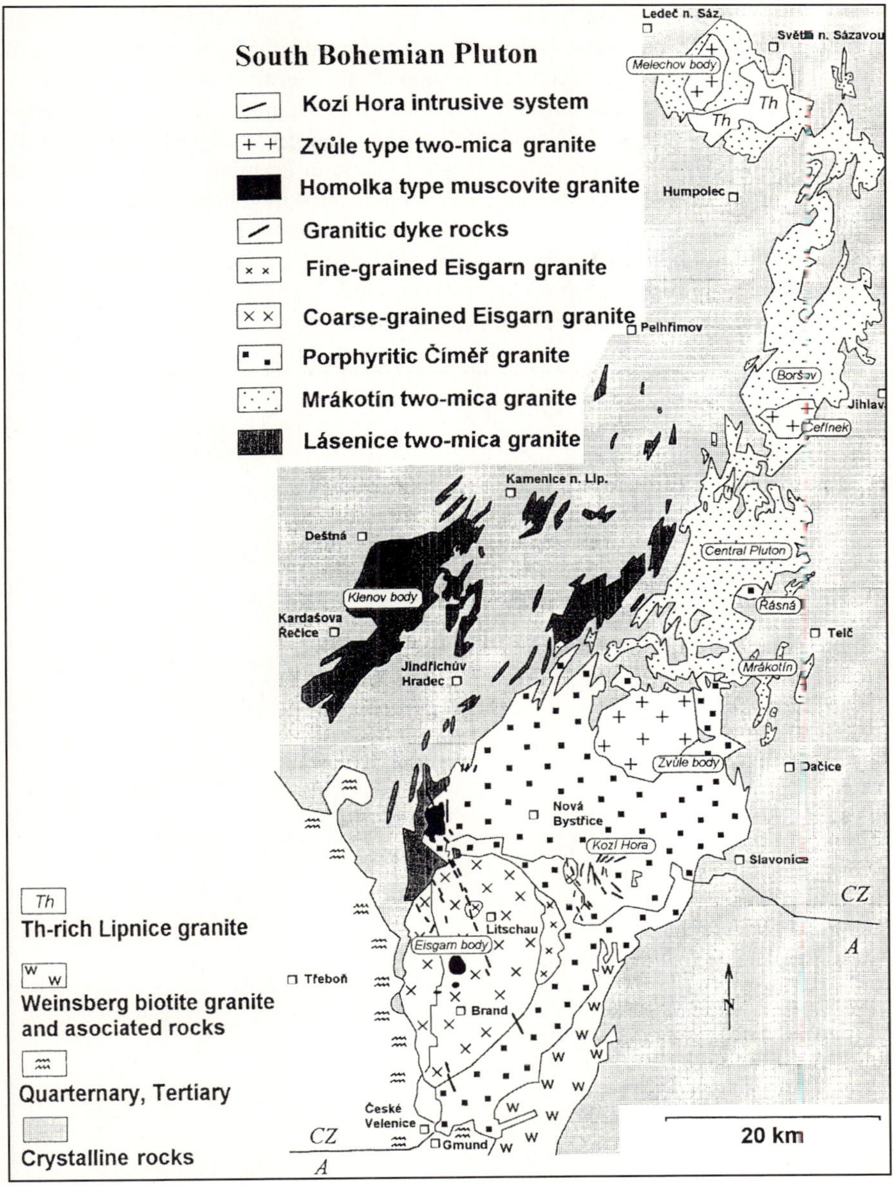

Fig. 4.32. Simplified geological map of the South Bohemian Pluton (after Breiter 2001)

and cracked in directions systematically related to the regional tectonic stress field that existed at the time. Thus, the interpretation seems to be that these joints formed on cooling (like cooling joints) by regional tectonic stress (syntectonic joints) at the upper levels of the crust (as uplift joints?), i.e., this explanation combines three distinct stages and mechanisms in one process. This interpretation may well be legitimate for south Arizona, but it may or may not be applicable to other fracture provinces, like the Sierra Nevada Batholith.

We now return to the debate about the age of the joints in the Sierra Nevada Batholith (Sect. 4.5.1). Suppose some of the joints formed during the emplacement of the plutons (e.g., Balk 1937; Mayo 1941) or soon after it (Segall et al. 1990), so that they might be considered to be cooling joints. According to (Bateman and Wahrhaftig 1966) the granitic rocks of the Sierra Nevada Batholith occupy the axial part of the N 40° W trending, faulted synclinorium, which began to form in Permian or Triassic time in central Sierra Nevada. Intense deformation is manifested by fold axes, steep or vertical beds, cleavage, and lineations that took place, both before and during the magma emplacements. Folds, at least, in the western part of the Ritter Range pendant may be related to the axial region of the synclinorium. Thus, it appears that an early NNW-NW trending structure may have significantly influenced the emplacement of the Ju-

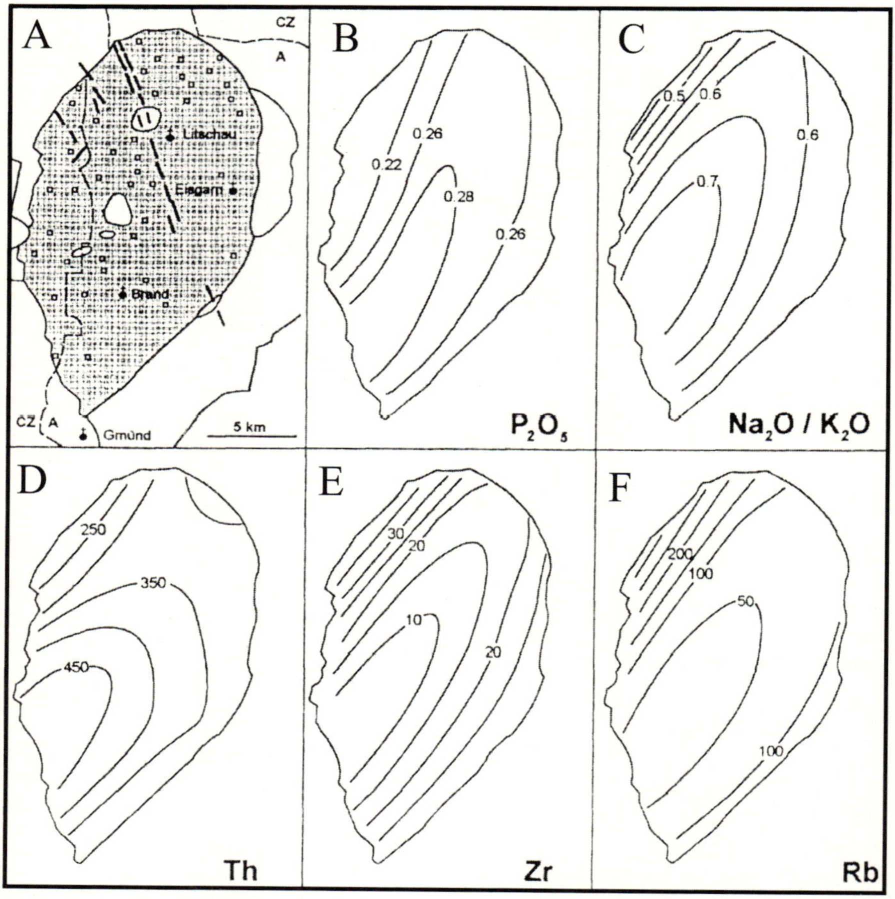

Fig. 4.33a. A detailed geochemical investigation of the Eisgarn, Zvule and Melechov granites by Breiter (2001). An idealized arial distribution of selected elements within the Eisgarn body. The isolines of element contents or ratios were constructed by methods of data interpolation by polynomial regression of third order. *A:* Simplified geological sketch of the interpreted body. The interpreted area is dotted; localization of used samples is expressed by small squares. *B:* Idealized distribution of P_20_5 in wt.%; *C:* Idealized distribution of Na_2O / K_2O-ratio in wt.%; *D:* Idealized distribution of Th in ppm; *E:* Idealized distribution of Zr in ppm; *F:* Idealized distribution of Rb in ppm

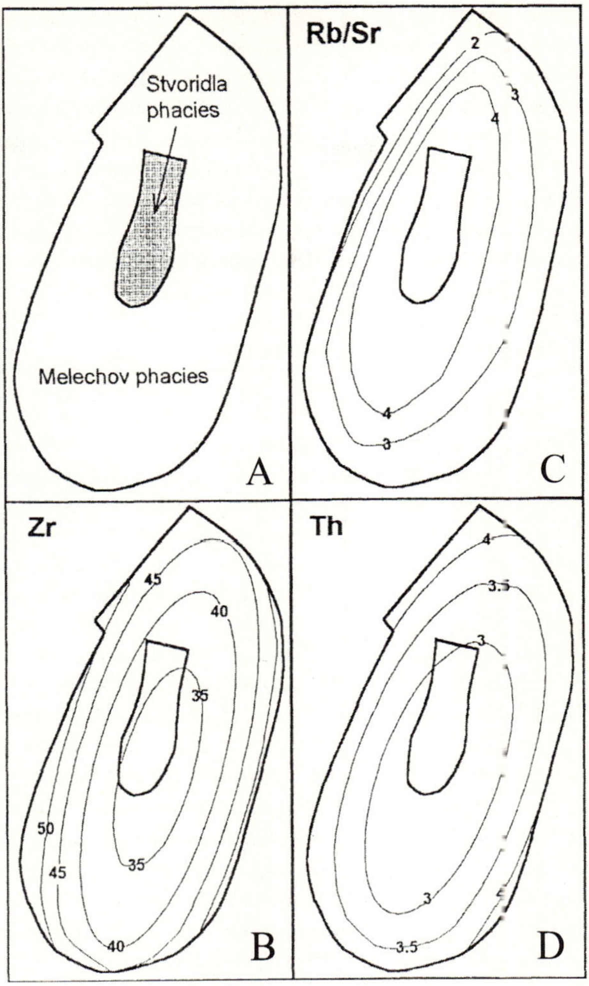

Fig. 4.33b.
Idealized arial distribution of selected elements within the Melechov body. *A:* Simplified geological sketch of the interpreted area; *B:* Idealized distribution of Zr in ppm; *C:* Idealized distribution of Rb/Sr in ppm; *D:* Idealized distribution of Th in ppm (from Breiter 2001)

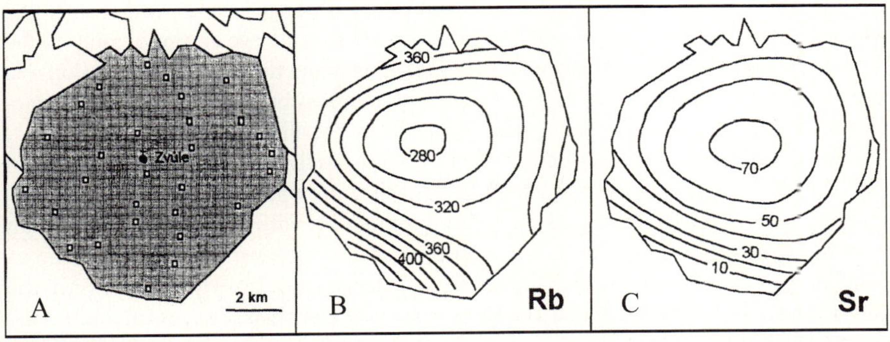

Fig. 4.33c. Idealized distribution of selected elements within the Zvule body. *A:* Simplified geological sketch of the interpreted area. The interpreted area is *dotted*; localization of used samples is expressed by *small squares*. *B:* Idealized distribution of Rb in ppm; *C:* Idealized distribution of Sr in ppm (from Breiter 2001)

rassic granites in central Sierra Nevada. This structure may also have controlled fracturing during cooling of the plutons and possibly also subsequent fractures along this direction. In support of this suggestion, Tobisch et al. (1989, Fig. 7) show a definite NNW-NW trend of a continuous, planar cleavage, predominant throughout the foothills terrain in central Sierra Nevada, where it is heterogeneously developed, occasionally in shear and mylonitic zones in the Upper Jurassic granitoids. Hence, there are similarities in the European and Californian fracture provinces, where in both there are indications that emplacement of granites was influenced by earlier, regional elongated structures, that also controlled later regional jointing.

4.7.2 A Comparison to the Cloos Model

In view of the general NNW-NW orientation of the elongated Sierra Nevada Batholith (Fig. 4.31), the NNW-NW cooling joints in the batholith (Sect. 4.5.1 and 4.5.2.1) seem to have significant similarities to the NNE cooling joints that are oriented along the elongated South Bohemian Pluton which is oriented NNE (Fig. 4.5 and 4.32). It appears that common to these two fracture groups is their fit to the S-joint category in the Cloos model (Fig. 4.1). The NNW-NW fractures developed perpendicular to the main horizontal compression that created the synclinorium and axial structures that subsequently controlled the formation of these fractures in the Sierra Nevada plutons. Furthermore, Lockwood and Moore (1979) found that the direction of maximum horizontal extension strain in the Sierra Nevada Batholith varies between WNW and NW, precisely fitting Cloos's criterion for S-joints, if applied to the fractures in this direction.

Sub-parallelism between fractures and isoanomals of the gravity high of Lusania expressing structural contours of the country rock has convinced Müller et al. (2001) that these fractures are S-joints (Fig. 4.3). The elongated South Bohemian Pluton sub-parallels the earlier structural contours of the country rock (Fig. 4.2b). The similarities in these two sub-parallelisms support the suggestion that the NNE set represents S-joints. It has been suggested that the shape of the South Bohemian Pluton could have controlled significant thermal and fluid-pressure gradients that in turn influenced the orientation of the NNE cooling joints (Bahat et al. 2003). This interpretation would coincide with the Cloos model, if proven in the future that the fluid-pressure gradients were not merely local on the surfaces of the individual joints (Sect. 4.4.5.2), but rather regional, i.e., perpendicular to the whole elongated pluton. The present interpretation relating the two groups of joints to the S-joint category would be challenged if proven that the main horizontal compression was axial during the cooling of the respective plutons, i.e., possibly implying that these fracture groups are Q-joints, which does not seem a likely possibility.

The descriptions of joints C and D (Sect. 4.5.2.2) partly resemble Cloos's "diagonal fractures" (Fig. 4.1). They dip at about 40° and they are partly filled with hydrothermal minerals. However, their strikes are more compatible with that of the Q-joints than with the strike of S-joints (Sect. 4.5.2.2 and Fig. 4.1). Moreover, the present observations suggest that they *predate* the formation of steep joints. If so, they differ from the Cloos model, in which the diagonal fractures are younger than the vertical joints. Furthermore, Cloos's diagonal fractures are clearly faults, whereas fractures C and D (D in particular) seem to fit better the term joints. Perhaps C and D represent a new fracture type that does not appear in the Cloos model?

4.7.3 Joint Genetic Grouping in Sedimentary Rocks and in Granites

The genetic division of systematic joints that occur in sedimentary rocks is into four groups: The burial, syntectonic uplift and post-uplift (Bahat 1991a, pp. 239–322). There are certain indications that the above classification of joints in sedimentary rocks may be partly applicable to joints cutting granite. Bankwitz and Bankwitz (1994, 1997) characterized syntectonic joints in two granite massifs from Europe. Bahat and Rabinovitch (1988) described uplift joints in granite from Sinai, Egypt. Holzhausen (1989) observed fractographic features in sheet joints cutting granite in Massachusetts, and Bahat et al. (1999) described post-uplift joints in granites from California. Thus, there seem to be resemblances in tectonic setting for the last three joint groups (syntectonic, uplift and post-uplift) between granites and sedimentary rocks. On the other hand, there are no corresponding similarities between early joints. The earliest systematic joint sets in sedimentary rocks are burial ones that maintain parallel orientation and approximate uniform dimensions and spacing. Probably the earliest joints in granites form during the cooling process as soon as some critical range of viscosity develops. The cooling of a pluton may be a long, multistage process, consisting of different thermo-mechanical modes. Hence, a correlation between "cooling joints" in granites to burial joints in sedimentary rocks is not possible, and a special genetic classification of cooling joints in plutonic rocks is required.

Part 2

Part 2 concerns fracture mechanic aspects of joints by drawing on fractographic properties of joints as follows:

1. Plotting joints on the velocity v vs. stress intensity K curve (Sect. 1.3–1.5).
2. Presenting a general discussion of various manifestations of v and K in granites.
3. Calculating the new fracture areas on *en echelon* and hackle fringes.
4. Estimating the "index of hackle raggedness" of joints.

4.8 Analysis of Fracture Velocity versus Stress Intensity Factor in the Borsov Joints Based on Fractographic Criteria

This section presents six aspects of the v vs. K_I diagram:

1. Physical meaning.
2. Justification for application to geological fracture.
3. Construction for geological application.
4. General criteria and method of placing the joints on the diagram.
5. Plotting of the Borsov joints on the v vs. K_I diagram.
6. Present limitations of the method.

The basic ideas of this scheme were outlined in Bahat (1991a, p. 234), and these were elaborated in Bahat et al. (2003).

4.8.1 The Velocity versus Stress Intensity Factor Curve

Wiederhorn (1967) and Wiederhorn and Bolz (1970) experimented on glass and showed the dependence of crack velocity v on the applied force and the tensile stress intensity

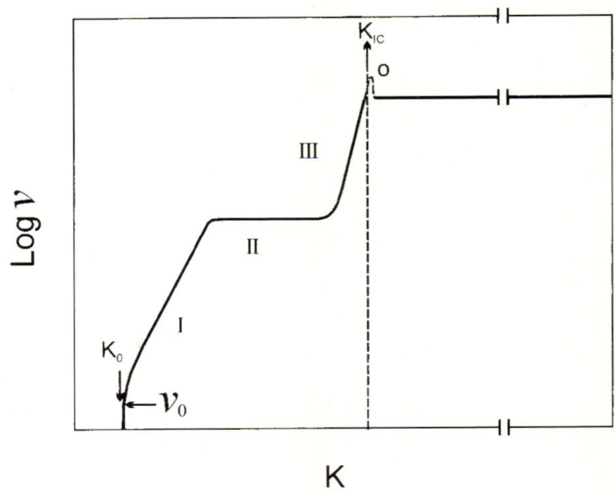

Fig. 4.34. Schematic variation of stress intensity factor K with crack velocity v (modified from Evans 1974). The subcritical curve left of K_{Ic} is divided into three regions I, II and III. K_0 and v_0 are the stress corrosion limit and the crack velocity at this limit (Wiederhorn and Bolz 1970). Supercritical growth occurs mostly along a terminal velocity plateau following a rapid increase in crack velocity at K_{Ic}. A velocity overshoot o, immediately after K_{Ic} and before the plateau has been predicted by Rabinovitch and Bahat (1979)

factor K_I in the sub-critical range (below K_{Ic}, the critical stress intensity factor). Wiederhorn and Bolz (1970) obtained a curve that was characterized by three distinct regions (Fig. 4.34). The slow crack growth in region I is manifested by a linear logarithmic plot. The behavior of region I is attributed to stress corrosion. The rate of crack growth is controlled by the chemical reaction between the reactant (solution) and glass at the crack tip when the system is essentially at equilibrium. Region II is controlled by the rate of reactant transport to the crack tip. In this region V almost does not change and is independent of K_I. In region III crack velocity is quite high; it is insensitive to the reactant and to stress-corrosion. It was also established that regions I–III were sensitive to water environments (vs. vacuum conditions) and were influenced by chemical conditions such as pH. The K_{Ic} line shows that at a critical stage in region III, the crack undergoes a catastrophic growth associated with a sudden jump in V, i.e., at $K_I \geq K_{Ic}$ post-critical crack propagation occurs.

4.8.2 Application of Fractography and Fracture Mechanics to the Study of Joints

Fractographic investigation of joints rests on analogies between the fracture surface morphologies of joints in geological exposures to fractures in glass, ceramics, metals and polymeric materials (e.g., Woodworth 1896; Bankwitz 1965, 1966; Kulander et al. 1979; Bahat 1979a, 1991; Cooke and Pollard 1996). It has been shown that K_{Ic} estimates derived from geological measurements are compatible with existing data from laboratory fracture experiments (Segall and Pollard 1983). Quantitative tectonofractography is a method that was developed to estimate paleostresses from strain measurements on joint surfaces, and this method stems from principles of fracture mechanics (e.g., Bahat 1979a, 1991; Bahat and Rabinovitch 1988; Weinberger 2001a). A strong correlation between the behavior of *en echelon* segmentation in polymethyl methacrylate (PMMA) (by mixed modes I + III) and *en echelon* segmentation in field outcrops was demonstrated by Cooke and Pollard (1996) and Bahat (1997). Sophisticated experimentation by Müller and Dahm (2000) further developed this concept and showed that knowledge of rupture velocities in sedimentary rocks can be gained from experimentation on starch-water mixtures. Particularly important were a series of demonstrations by various investigators (e.g., Atkinson and Meredith 1987a,b) that plots of minerals and rocks on V vs. K_I diagrams greatly resembled analogous plots of glasses. All these (as well as other sources) justify the suggestion by Pollard (2000) that the mode is eminently suitable for the taxonomic classification and description of fractures in the field.

4.8.3 The Velocity versus Stress Intensity Curve for Granite

The curve in Fig. 4.35a is a synthesis of nine elements:

1. The basic concept stems from the sub-critical V vs. K_I curve determined on glasses (Fig. 4.34).
2. Section A–B on the curve corresponds to the sub-critical V vs. K_I curve determined for Westerly granite, under 15 kPa p(H_2O), at 300 °C by Atkinson and Meredith (1987b).
3. From the two hypothetical curves, B–C and B–D, the former assumes the existence of regions I–III and the latter favors the existence of only region III (Fig. 4.35a).

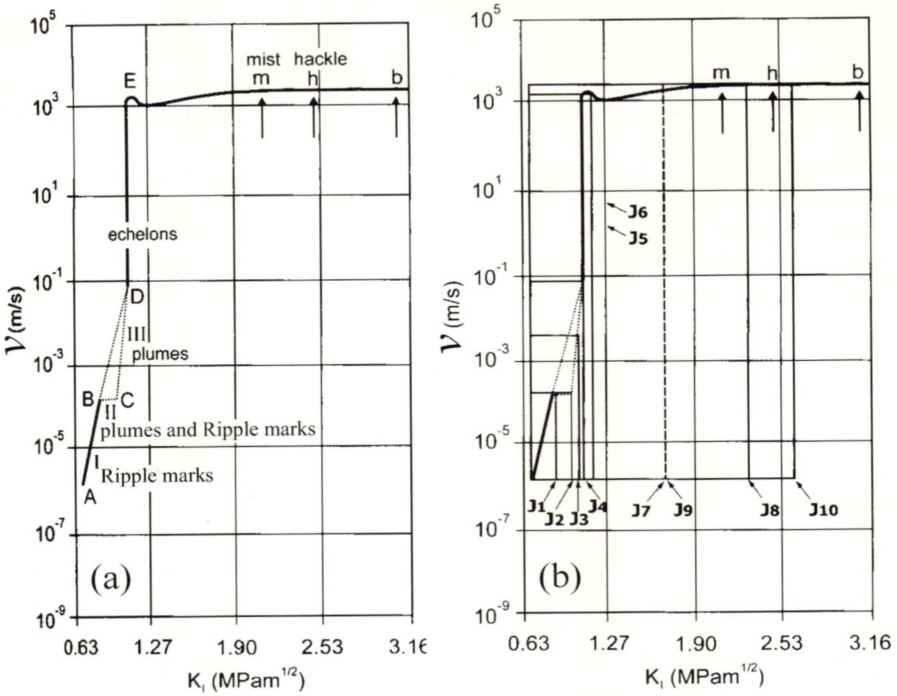

Fig. 4.35. **a** The V vs. K_I curve for joints in granite from the South Bohemian Pluton, synthesized from laboratory experiments. **b** The estimate of a range of crack velocities for ten joints in granite from the South Bohemian Pluton. For each joint the range of fracture velocity is given in a *rectangular frame* constructed on the curve from Fig. 4.35a, where the vertical right side of each frame gives the maximum fracture velocity that was attained by the joint

4. The position of D approximates $K_{Ic} = 1.09$ MPa $m^{1/2}$ (the K_{Ic} value of Westerly granite at 300 °C, Atkinson and Meredith 1987a).
5. The position of D independently fits well the definition of K_{Ic} at crack velocity of $\geq 10^{-1}$ m s^{-1} (Wiederhorn et al. 1974).
6. A velocity overshoot immediately beyond K_{Ic} at E was mathematically predicted by Rabinovitch and Bahat (1979) and was recently verified experimentally (Cramer et al. 2000, Fig. 1A).
7. Kerkhof (1975) proposed a unified curve for both sub-critical and post-critical conditions (at both sides of K_{Ic}) and suggested that the positions of mist and bifurcation initiations correspond to m and b, respectively. These positions are approximated in the post-critical part of the present curve.
8. We add h (Beauchamp 1996) representing an approximate location for hackle formation between m and b.
9. Finally, the two ends of the curve are at K_0, signifying zero slow crack growth below point A, and at b where the fracture bifurcates along the terminal velocity plateau (Evans and Wiederhorn 1974).

Hence, Fig. 4.35a is essentially based on experimental data from glass but it is constrained by results from granite and the definition of K_{Ic} (elements 2–5).

In further justification of these elements for the construction of Fig. 4.35a, we need to add that the Westerly granite was chosen because of its textural similarity to that of the granite under investigation (fine to slightly medium grain size). Curve A–B was selected because we consider joint fracture under considerable pore pressures at depths of about 10 km, so that assuming 300 °C due to the thermal-gradient seems reasonable. There is no compelling data that indicates which curve, B–C or B–D is more realistic, so we use both. A velocity overshoot immediately beyond K_{Ic} is anticipated to be an inherent property of fracturing in elastic materials following great velocity "jumps" such as the one from D to E. Although a realistic velocity jump is expected to occur along a slope (through a certain range in K_I), a conventional vertical C–D is adapted (Atkinson and Meredith 1987b, Fig. 4,15).

4.8.4
Methodology for Plotting Joints onto Velocity versus Stress Intensity Factor Curve

Various sub critical and post critical growth conditions of joints may be inferred on the v vs. K diagram (Fig. 4.35a) by two fractographic distinctions: First, plumes and ripple marks (see Sect. 2.2.4.2 for definitions) vs. hackle cracks (or hackles) and second, *en echelon* cracks vs. hackles. The first distinction is between characteristic fractographic features below and above K_{Ic} (Fig. 4.34 and 4.35a). Both plumes (striae) on the mirror within the sub critical regime and hackles on the fringe in the post critical regime often occur radially, such that the latter often appear to be in continuation with the former (Fig. 2.30a). However, they can be distinguished quite readily. Whereas striae are delicate and approximate unidimensional textures by their barb lines (Fig. 2.55b and 4.8a), hackles are three-dimensional, coarse complexes consisting of superimposed flakes of material that are separated by microcracks in all orientations. In fringes formed in glass by extremely intense fracture (high K_I values), each flake may display a miniature assemblage of fractographic features shown in Fig. 2.1a. On surfaces cutting granite that represent moderately intense fracture conditions. the differences between plumes and hackles are less discernable than in glass, but they are quite pronounced.

Kerkhof (1975) observed that arrest marks (ripple marks) were induced in plate glass at propagation rates below $v = 4 \times 10^{-5}$ m s^{-1} and $K_I < 0.73$ MPa m$^{1/2}$ that correspond to the range of regions I and II in Fig. 4.35a. Accordingly, we propose that fracture markings in rocks that display arrest marks reflect the range of regions I and II. The frequent occurrence of ripple marks superposed by radial plumes on joint surfaces suggest that the plumes are formed under fracture conditions similar to those that produce ripple marks but under slightly more intense ones, i.e., closer to region II. Fracture experiments on PMMA have shown that circular ripple marks develop before radial plumes on the fracture surface (Bahat 1991a, Fig. 2.50c), i.e., these ripple marks developed at lower K_I values than the plumes. Striae developed on a smooth fracture surface of soda-lime silica glass in water when the velocity of fracture propagation reached about 10^{-2} m s^{-1} at around $K_I = 0.7$–0.8 MPa m$^{1/2}$ (Michalske 1984, Fig. 1). This was above the mid range between the stress corrosion limit, $K_I = 0.3$ MPa m$^{1/2}$, and the fracture toughness $K_{Ic} = 0.9 \pm 0.1$, i.e., striae formed in the range of regions II to III (Fig. 4.35a). We propose that markings like the short rhythmic plumes from the Appalachian Plateau (Fig. 4.36) that show periodic increase of barb intensities reflect the range of regions II to III, whereas long uniform plumes free of arrest lines relate more

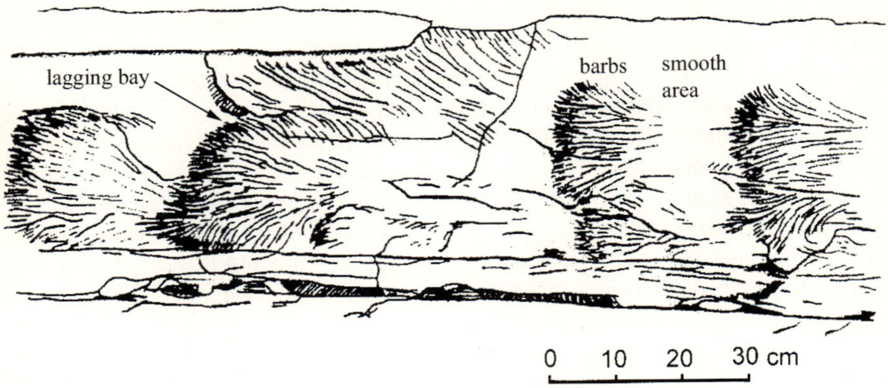

Fig. 4.36. Rhythmic plume growth. Fan perimeters designate loci of arrest marks where barbs reach their maximum intensity and vanish into smooth areas. Barbs then resume their growth as fine lines within the plume that mark the local direction of fracture propagation. The perimeters convex toward the direction of fracture propagation (from right to left) and the barbs may be partly "branched" due to lagging bays (modified after Bahat and Engelder 1984)

to region III (Engelder et al. 1987 and Bahat 1994a, Fig. 3.18). On the other hand, the periodic smooth areas where a new cycle of fracture begins (Fig. 4.36) represent *V* and K_I conditions within region I and possibly below it.

The second distinction relates to the range of K_I conditions under which *en echelon* cracks develop. This range is unclear because it varies considerably. *En echelon* cracks have been observed in borosilicate glass at very low K_I (Wiederhorn and Johnson 1973 Fig. 5). There are some indications that *en echelon* cracks formed in rocks and rock-like material under low K_I values (Kulander et al. 1979; Bahat 1997; Müller and Dahm 2000). On the other hand, Hertzberg (1976, Fig. 14.9) analyzed the fracture conditions of a turbine rotor made of steel and found that the *en echelon* cracks developed at close to K_{Ic} conditions. This problem is further complicated by the occasional presence of hackle cracks (Fig. 2.1a) and *en echelon* cracks (Fig. 2.1b) at close proximities, suggesting some transitional relationship at the fringe (Bahat 1991a, p. 68). All these support early observations that *en echelon* segmentation depends on the K_I / K_{III} ratio (Sommer 1969). Thus, plotting *en echelon* segmentation (Fig. 2.1b) on the *V* vs. *K* curve is still problematic, and a transitional range of some uncertainties may exist up to and around K_{Ic} conditions for *en echelon* cracks (Hertzberg 1976). On the other hand, the situation that concerns hackle cracks (Fig. 2.1a), which are expected to occur above K_{Ic} up to branching, is clearer (e.g., Kerkhof 1975; Rabinovitch et al. 2000a).

4.8.5
Plotting Joints of the Borsov Pluton

It is unrealistic to determine the points of joints J1–J10 (Fig. 4.6–4.8, respectively) along the curve of Fig. 4.35a. A more reasonable approach is to estimate the range of crack velocities for each joint (Bahat et al. 1999), given in rectangular frames, where the vertical right side of each frame gives the maximum fracture velocity that was attained by the respective joint (Fig. 4.35b). It is assumed that joints generally start in low velocities before reaching fast propagation (e.g., Bahat 1991a, p. 58).

We first consider the four joints of group 1 (Fig. 4.6). Based on the criteria in the previous section, the arrest marks superposed by radial plumes on joints J1 and J2 (Fig. 4.6a,b) signify maximum velocities close to region II and region III. There is a considerable variability in the plume intensity of the quasi-core joint, J3 (Fig. 4.6c), implying that fracture stresses, as well as V and K_I were not uniform on the joint surface. The long plumes not associated with arrest marks signify monotonous propagation (approximately constant velocity) that could reach region III. However, the plumes of J3 that are associated with arrest mark seem to signify vacillation between region I and part of region III. These plumes did not develop into fringes. The joints from the multi-joint exposure, J4 (Fig. 4.6d), mostly exhibit fractographies that occur on joints J1–J4, placing them between regions I and III.

The fringes in group 2 display a series of fringes between the two end members shown in Fig. 2.1a,b). The divided arrest mark joint, J5 (Fig. 4.7a) exhibits a small *en echelon* fringe, which may indicate that K_I temporarily reached K_{Ic}, because these fringes resemble the features of the fringe studied by Hertzberg (1976) that formed close to critical conditions (at K_{Ic}). The extended fringe F1 of J6 (Fig. 4.7b) relates to the entire structure consisting of the ten (or eleven?) concentric arrest marks. Accordingly, it appears that both the right side of the joint and the triangle were formed before the propagation of the large wavy plume w from the triangle towards the left. Therefore, the large wavy plume represents an intriguing example of transition from *en echelon* fringe in F1 and F2 to a low intensity plume. We assign to J6 a frame similar to that of J5 (Fig. 4.35a,b).

The *en echelon* segmentation is clearly defined on the fringe of J7 (Fig. 4.7c) and plotting this joint is somewhat problematic, as mentioned above. Prolonged growths under sub-critical conditions, i.e., through regions I–III, may have transformed to the mode of post-critical growth, i.e., possibly beyond K_{Ic} (dashed line in Fig. 4.35b). The extent of this frame is still a topic for future research.

The partly erased joint, J8, shows a well-defined circular mirror boundary, but the fringe is very irregular (Fig. 4.7d). A delicate radial plume reaches the boundary from an inside location of the joint and continues radially downward and sidewise beyond the boundary as irregular *en echelon* segments in diverse morphologies. The *en echelon* segments are delicate in certain areas, while in other locations they assume coarse relieves of hackles. This diversity fits between the two end members shown in Fig. 2.1a,b, suggesting a range in K_I between m and h (Fig. 4.35a,b).

The trefoil joint, J9, has the two most complex fringes on the right and on the left sides of the mirror (Fig. 4.8 and 4.9). However, these fringes fit the pattern of Fig. 2a and resemble the fringes of J7. Therefore their position along the V vs. K_I curve is further left from that of J8 (dash line in Fig. 4.35b). The joint with the hackled-fringe, J10 (Fig. 4.8c), shows features resembling both Fig. 2a and Fig. 2b, but closer to Fig. 2b than to Fig. 2a. We therefore place the right side of the frame further right from h (Fig. 4.35a,b).

4.8.6 Limitations of this Methodology

The curve shape in Fig. 4.34 depends on various parameters, including material properties as well as environmental conditions such as temperature, fluids and pH (e.g., Wiederhorn and Bolz 1970; Guin and Wiederhorn 2003); it is particularly sensitive to stress and stress rate. Therefore, Fig. 4.35a,b may be modified in future research along the various segments of the curve. Changes may occur along the sub-critical range due to different water/vapor contents, such that various rocks would fit different region II–III relationship. This will come with a deeper understanding of the properties of undulations vs. arrest marks and plumes (striae), i.e., when criteria of distinguishing them on mirrors that form sub-critically from mirrors that form post critically are established (Sect. 2.2.4). Also, plotting of joints in the sub-critical range would be influenced by the dependence of K_I on v that may change along different v ranges (Guin and Wiederhorn 2003). Additionally, the K_{Ic} for granite may change, as K_{Ic} values are refined.

The shape of the post-critical "terminal velocity" segment may vary to fit different dynamic fracture data. Results generally show a strong increase in fracture velocities at the transition from the mirror to the hackle and branching zones (in glass) (Cramer et al. 1999). But velocity fluctuations can also be expected (Sharon and Fineberg 1999). The plot of hackles within a limited range close to h (Fig. 4.35b) would be quite reliable. However, the plot of *en echelon* fringes remains problematic until further information on the dependence of these fringes on K_I (rather than K_I / K_{III}) will become available. Meanwhile, a useful criterion would be (Sect. 2.2.4.2): Faint plumes and lack of arrest marks imply relatively high K_I (Fig. 2.56a), whereas pronounced plumes and arrest marks suggest relatively low K_I (Fig. 2.54a,b). All these may somewhat shift v and K_I values for particular joints, but they would not challenge the principles of the method. Finally, note that while in Fig. 4.34 the curve of terminal velocity is horizontal, it has a slight slope in Fig. 4.35. The actual shape of this curve is still unknown, and it may vary in various materials. Moreover, the exact conditions of tension (or extension) are important. While under pure tension the terminal velocity may be attained rapidly (perhaps on the mirror plane), deviations to bending by slight lateral compression may delay the attainment of terminal velocity (Shand 1954, also Bahat 1991a, Fig. 2.19).

4.9 Velocity and Stress Intensity Manifestations of Fracture Propagation in Granites

The plot in Fig. 4.35b offers a generalized quantitative frame within which various other natural and experimental fractures can be examined; it also demonstrates some of the limitations of this methodology. When using V and K in analyzing the fracture history of joints, one needs to bear in mind several points:

1. Fracture velocity changes all the time, generally by growing up to the terminal velocity, but also up and down.
2. *En echelon* segmentation is a complicated indicator of crack velocity.
3. When quantitative estimate of fracture behavior is impossible (Fig. 4.35b), it is convenient to describe qualitatively the relative increase or reduction in either v or K of specific fracture exposures.
4. There are significant differences in the fitting of curves that correlate v and K with fractographic features (Arakawa and Takahashi 1991; Sect. 2.2.3.12).
5. There are theoretical disagreements regarding the issue of which parameter actually controls dynamic fracture (aspects of fracture velocity in Sect. 1.4, and aspects of stress intensity (Sects. 1.4.3.1 and 1.4.3.2).
6. Both remote and local forces that act on the fracture, influence v and K.

We present below two examples of fracture in granites that touch on these six points. The first relates to fractographic features on the mirror plane of a post-uplift joint from California (after Bahat et al. 1999), while the second concerns the fringes of cooling joints from the Czech Republic (after Bahat et al. 2001a,b, 2003).

4.9.1 Fracture Velocities of Exfoliation Joints Cutting the El Capitan Granite and Comparable Fractures

Fan-like markings on exfoliation surfaces from the middle height of El Capitan are made of radial plumes and concentric ripple marks orthogonal to each other at their intersections, which imply simultaneity (Fig. 4.27). This fracture morphology suggests a sub-critical propagation rate below 4×10^{-5} m s^{-1}, assuming dry conditions, and below 10^{-2} m s^{-1}, if the fracture were immersed in water (Bahat 1991a, p. 234 and present Fig. 4.35). Accordingly, it would take longer than fourteen days to fracture a fan having radial plumes 50 m long in dry conditions, and a fan immersed in water, whose plume is 36 m long would fracture in more than an hour. Due to the uplift of the El Capitan block, the middle height of the cliff is expected to be above the water table. In addition, due to the low porosity of the granite, the above dry conditions are preferred. If however buckling and splitting (Fig. 4.28) would enable water flow into the fracture zone, a more rapid fracture would occur, requiring a much stronger preference of downward fracture propagation compared with an upward propagation, due to the flow of more water towards lower elevations. This is not supported by the fan pattern on the cliff (Fig. 4.27) so that propagation under dry conditions is favored. Olson's (1993) calculations for joints from Sierra Nevada granites (generated by a mechanism other than exfoliation) suggest that fractures of some 10 m long were formed in about three minutes or faster, possibly under conditions approaching critical.

4.9.2 Fracture Velocities Relative to Estimated Paleostress Magnitudes

The calculated tensile paleostresses that initiated joints in granitic rocks under three different geological conditions are presented in Table 4.7. The results compared here were obtained by different procedures. The method applied to the analysis of a fracture on Half Dome (Fig. 6.18) and granite from East Sinai (Bahat and Rabinovitch 1988) is based on measuring fracture markers on surfaces of individual joints. The method used by Segall and Pollard (1983) relies on measurements of extension strain accommodated by the dilation of many joints (and was used for the analysis of fracture in granodiorites from Mt. Givens). Accordingly, these two methods pertain to local and remote stresses, respectively.

Segall and Pollard (1983) assumed quasi-static crack propagation conditions in their analysis of the initial tensile stresses that induced regional joints in the Mt. Givens Granodiorite from the Sierra Nevada. The relatively high fracture stress values resulting in jointing at Mt. Givens (ranging from 1.2 ±1.0 MPa to 26 ±11 MPa) seem to suggest syntectonic fracture (Bahat 1991a). Fracture was enhanced by pore pressure, as testified by mineral fillings (Segall and Pollard 1983), but under considerable depths,

Table 4.7. Tensile paleostress conditions for jointing in granodiorite and in granite

Example			Relative fracture velocities[a]
Post-uplift exfoliation joint in granodiorite from the SW cliff of Half Dome in Yosemite Valley, California			
The stress intensity method Range of fracture stress σ_f		0.01 to 1.48 MPa	Long growth under sub-critical conditions, and short post-critical growth
The surface energy method Range of fracture stress σ_f		0.06 to 0.37 MPa	
Syntectonic joints in granodiorites from Mt Givens (Segall and Pollard 1983)			
	Ward Lake	Florence Lake	Long growth under close to critical conditions, but never reaching them
Initial stress σ_i	≤26 ±11 MPa	10 ±6 MPa	
Initial stress σ_i	≥1.7 ±0.9 MPa	1.2 ±1.0 MPa	
Uplift joint in granite from East Sinai (Bahat and Rabinovitch 1988)			
The stress intensity method Range of fracture stress σ_f		2.9 to 15.8 MPa	Relatively short growth under sub-critical conditions before assuming post-critical growth
The surface energy method Range of fracture stress σ_f		2.4 to 6.0 MPa	

[a] See text for discussion.

that prevented fracture branching (Segall and Pollard 1983). Such conditions would fit sub-critical fracture velocities (Table 4.7). Greater fracture stress results than those mentioned above for the granodiorites were obtained for an uplift joint in granite from Sinai. The latter paleostress results are less, but comparable to those obtained by tensile experiments on granite (from 7 MPa to 18 MPa, Segall and Pollard 1983). Quite possibly, the Sinai joint started to propagate from certain depths below the water table while being exposed to excessive pore pressures that resulted in high fracture stress values.

In summary, fracture velocities may change considerably during the fracture process. A two stage process is assumed for the three granitic rocks:

1. Sub-critical fracture growth, which had occurred before critical conditions were attained.
2. Post-critical fracture velocities that could have initiated under reduced overburden pressures. The difference is in the amount of time needed for fracture growth during the first stage.

The results suggest that the first stage was relatively long for the exfoliation in the cliff of Half Dome, whereas for the joint in East Sinai this stage was relatively short. However, many joints, possibly including the joints from Mt. Givens (Table 4.7) never reach post-critical fracture velocities along their history, mostly because of overburden constraints.

4.9.3 Cycling Fracture

The previous sections relate to the general property of fracture to increase its fracture velocities up to the terminal velocity (Fig. 4.34 or 4.35). Quite common however are alternating changes in fracture velocity (and stress intensity) that are reflected from fractographies, which are displayed on mirror planes (e.g., Fig. 4.7b, 4.36). Rarer and more intriguing are cycles that are exhibited by both mirrors and fringes (e.g., Fig. 4.8a,b). We now interpret the observations from the mirror-left fringe relationship of the trefoil joint (Fig. 4.9c,d) in light of some experimental results.

4.9.3.1 *Fracture Velocities During Growth of the Trefoil Joint*

Müller and Dahm (2000) formed plumes and fringe zones on surfaces of desiccation tensile cracks that were created by drying starch-water mixtures (Sect. 2.2.3.9). Particularly relevant to the present section were their observations that at transitions from the parent joints to fringes, topographic amplitudes occurred, i.e., enlargements in the crack area, were associated with a reduction in fracture velocities. The latter findings resemble the experimental results obtained by Sharon et al. (1995). They applied tensile stresses perpendicular to a thin plate (0.8 or 3 mm in thickness) made of cast PMMA. These authors found that cracks discontinue their straight propagation and become "unstable" beyond a critical velocity, v_c, of $0.36v_R$, where v_R is the Rayleigh wave speed in the material (the velocity at which a wave moves along a free surface). This transition from quasi-static velocities to dynamic ones involves two coupled phenomena, velocity oscillations and secondary micro-branching from the primary parent crack (Fig. 1.33). Sharon et al. (1995) observed that the increase in new crack area by the formation of the micro-branches was associated with reduction in fracture velocities.

The fractography of the trefoil joint (Fig. 4.9c,d) seems to fit the fracture principles derived by Sharon et al. (1995). The trefoil primary mirror plane and *en echelon* segments (first fringe) are analogous to the primary and secondary (branch) cracks (Sharon et al. 1995), respectively. The trefoil secondary mirror plane and secondary *en echelon* segments reflect the stages of resumption of rapid propagation of the primary crack and subsequent appearance of secondary cracks (Sharon et al. 1995), respectively. The implication is that the remote stress operating on the trefoil joint did not subside (or perhaps was resumed after a quiescent period due to local conditions). Therefore, exerted tension resulted in the creation of a new, secondary mirror that propagated in the same downward direction, as shown by its downward facing convex boundary. The increase in length of the secondary mirror was associated with increasing its fracture velocity in analogy to the above mentioned fracture principles (Sharon et al. 1995). This fracture velocity was reduced again, resulting in the secondary fringe, i.e., the energy dissipation that occurred by creating the new segment surfaces (tensile and shear) was associated with a reduction in fracture velocity, as also observed by Müller and Dahm (2000). Thus, the primary mirror of the trefoil joint was segmented into a fringe that then continued to fracture into a secondary mirror, which also was segmented into a secondary fringe.

4.9.3.2
The Formation of the Trefoil Joint under Mixed Modes Loading

Couplings of velocity decrease with the increase in new fracture surface area were observed both under quasi-static conditions (Müller and Dahm 2000) and through dynamic ranges (Sharon et al. 1995). What are the differences between the two? Sommer (1967, 1969) simulated *en echelon* segmentation on the fracture surfaces of circular glass rods loaded in tension by the superposition of a small amount of torsion, creating conditions of increasing mode III / mode I ratios. On the other hand, high K_I / K_{III} ratio increased crack velocity, leading to crack branching. Bank-Sills and Schur (1989) showed the inherent tendency of increasing mode III / mode I ratios from the center of metallic plates towards their boundaries, and Bahat (1997) reported that the latter tendency occurred also in sub-horizontal chalk beds.

All these experimental and field observations match into the following scenario of the trefoil joint growth. The primary joint propagated down under conditions of increasing K_I magnitudes and very low K_{III} / K_I ratios to the mirror boundary at which an abrupt increase in the mode III / mode I ratio occurred, leading to the breakdown into *en echelon* cracks (Cooke and Pollard 1996). This was linked with energy dissipation on many new crack surfaces and a reduction in crack velocity. The crack speed reduction was probably associated with the "mode III crack closure" effect, which was linked with friction and interlocking interferences along the fracture surfaces (Tschegg 1983). Crack speed reduction can be very substantial under these conditions: *En echelon* cracks have been observed in borosilicate glass after application of stress with an intensity factor less than the fatigue limit at very low crack velocities (Wiederhorn and Johnson 1973, Fig. 5). As mentioned above, recent experiments on rupture velocities in starch show similar results (Müller and Dahm 2000). The second mirror reflects resumed conditions of increasing K_I magnitudes. Hence, the micro-branching processes described by Sharon et al. (1995) relate to crack velocity oscillations under high K_I / K_{III} ratios, whereas the experiments conducted by Müller and Dahm (2000) took place under increasing K_{III} / K_I ratios.

4.9.3.3
Fractographies that Pertain to both Quasi-Static and Dynamic Fractures

How can *en echelon* segments (generally reflecting quasi-static propagation Sect. 4.7.3.1) and hackles (generally reflecting dynamic propagation) occur adjacently on a single fringe? (e.g., Fig. 2.1c). The answer is based on the assumption that even "slight deviations" of the direction of the minimum principal compression (or maximum principal tension) from orthogonality to the fracture surface would induce *en echelon* segmentation (Sommer 1969). Accordingly, fringes formed by both quasi-static and dynamic propagation may be influenced by such "slight deviations". Fractures formed quasi-statically when K_{III} / K_I are high will be dominated by *en echelon* fringes (Fig. 2.1a), and these fringes will have the tendency to assume simple, uniform widths (Fig. 2.34b). On the other hand, fractures formed dynamically when K_{III} / K_I are low will be dominated by hackle fringes, on which however *en echelon* segmentation in a shingle-like arrangement may be superimposed in accordance with this ratio. Such superposition can explain the frequent occurrence of fringes that show hackles in a shingle-like arrangement (center of Fig. 4.12b). The latter fringes often exhibit non-uniform widths (Fig. 2.34a). Hence, it is quite possible that on the fracture surface obtained by De Freminville (1914) (Fig. 2.1c) the two fringe types reflect local slight changes in the extent of orthogonality of the maximum principal tension: The fringe area that displays hackles reflects "pure" mode I, whereas the fringe area that displays hackles in a shingle-like arrangement reflects mixed mode I + III loading.

In summary, most fringes from sedimentary rocks suggest quasi-static growth, while rare fringes display cuspate hackles (Fig. 2.30a and 2.42) or hackles in shingle-like arrangements (e.g., Fig. 2.33e). Note the resemblance between the latter fracture morphology and the fracture surface of a turbine that is interpreted to have initiated in fatigue growth and possibly approached catastrophic conditions, i.e., under $K_I \sim K_{Ic}$ conditions (Fig. 2.44a) (Bahat 1991a, p. 133). In many outcrops the distinction between these fringe types would be relatively easy, whereas in others that exhibit transitional fractographies, this distinction would be more problematic. Hackle fringes are far more abundant in granites than in sedimentary ones, but even in these rocks hackles often show shingling. Note that in Fig. 4.11a in the lower fringe shingling is almost absent, whereas in the upper fringe shingling is more pronounced.

4.10 New Fracture Area in the *En Echelon* and Hackle Fringes on Joint Surfaces from the Mrákotin Quarry

4.10.1 Fracture Area Increase in the Transition from *En echelon* to Hackle Fringe

A basic physical tendency of a cracked body is to minimize its stored strain energy (Eq. 1.19), which is done by the creation of new crack area (Eq. 1.20). Therefore, measuring changes in crack area is important for the characterization of the fracture process. A comparison of the relative flat areas of the "segments" on the fringe of joint C (Fig. 4.12a and 4.37a) to the extremely wrinkled "flake" surfaces of the hackles on the fringe of joint A (Fig. 4.11a–c and 4.37a,b) intuitively suggests that the overall fracture area of the latter fringe is significantly larger than the former fringe. This intuitive observation regarding the difference in new area in the two fringe types needs to be proven, and meeting this requirement is the main objective of this section, citing Bahat et al. (2002b).

The extent of superposition of two adjacent segments (or flakes) in a fringe is often an enigma. The best way of resolving this question is by removing the upper segment and examining the area underneath the upper segment (Bahat 1986a, Fig. 7). This provides two measurable parts: The revealed area of the upper segment before its removal and the hidden area that belongs to the lower segments. The ratio between the two can then be determined. This determination is however difficult to accomplish because in many outcrops the segments are inaccessible, requiring the pursuit of estimation methods that rely on certain fracture mechanic considerations.

The *en echelon* segments are approximately flat on the fringe of joint C and are not divided into multiple smaller cracks. Therefore, the actual fracture area is given by

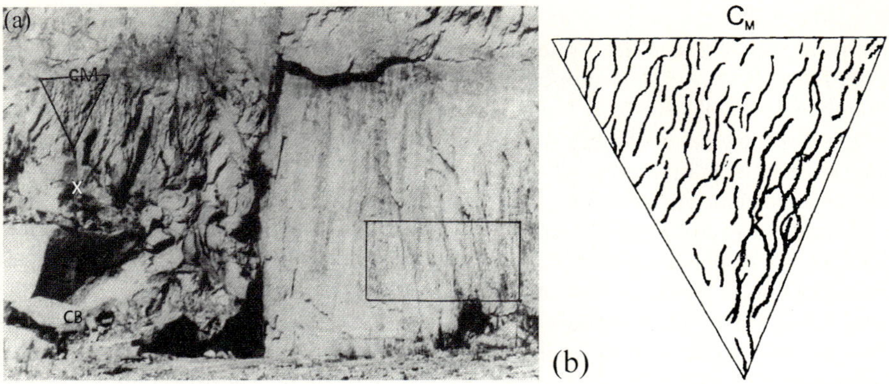

Fig. 4.37. Joints *A* and *C* from the Mrákotin quarry (Fig. 4.10). **a** The "triangle" within the hackle fringe of joint *A* (left), where c_M is on the mirror boundary, and *X* is the most distant point on the lower fringe, at a continuation of the triangle, and *CB* is remote from *x*. The "rectangle" within the *en echelon* fringe of joint *C* (right). **b** An expansion of the triangle, showing the drawn traces of flake boundaries

the sum of the measured, revealed segment areas plus the sum of the estimated underlying, hidden surfaces due to shingling. The number of the hidden parts equals that of the segments, and they are approximately the same for all the observed segments of similar size, which maintain the same twist angle.

The fracture surface is however considerably more complex in the hackle-fringe of joint A. A superposition in an *en echelon* fringe involves the shingling of two adjacent segments, whereas in a hackle-fringe, several (from two to five) small hackle-like flakes often superpose a single "large hackle" (Fig. 4.11a). Furthermore, in this flaky texture, the multi-surfaces that often deviate from planarity overlap each other on a given nominal area such that the actual crack area in the hackle-fringe is a greater sum than the shingling surfaces.

The estimation of the difference between the ratios of the hidden surface area (HA) divided by the revealed surface area (RA) in the two fringes is presented below. We recognize that the difference in the nature of the steps between adjacent segments is different in the two fringes. The step areas are similar for all the adjacent segments in the *en echelon* fringe, whereas in the hackle-fringe, some steps are much larger than others. However, for the sake of simplicity we concentrate on the fracture areas of segments and hackles and do not enter the contribution of the steps and secondary segmentation into our calculation, which are far smaller than the segments. This would reduce the accuracy of our measurements by about 10%.

4.10.2 Estimation of the Fracture Areas in the Fringes of Joints A and C

The *en echelon* fringe on joint C. The exposure of the fringe of joint C is not of a uniform quality throughout its entirety. Therefore we selected a representative area of a rectangle shape for the detailed study. The "rectangle" is 8 m^2 on which 19 ±1 m fracture length of *en echelon* segments were measured (Fig. 4.37a). The rectangle was selected as a representative of the fringe, because:

1. This is a nominal fringe area on which the distribution of fracture length per area is relatively large compared to other fringe areas of this joint.
2. The appearance of fractures is clear, rendering them to be measurable with a reasonable confidence.
3. The rectangle is located at a far end of the fringe, at a relatively long distance from the mirror boundary, such that it fits our "assumption" (see below).

The total fracture area in the rectangle consists of the revealed (nominal) area plus the hidden area. The hidden area between two neighboring segments is not known. In two locations, the maximum superposition at the bottom of the fringe is measured to be (0.2 ±0.1 m) (looking upward towards the mirror). However, this superposition does not necessarily represent the hidden area throughout the various segments, because there is no certainty that the same relative superposition is maintained along the segments. *En echelon* segmentation forms by mixed mode I + III (Sect. 1.1.3). Mode III contributes to tearing the rock in a scissors-like operation. This requires that shingling is zero at the segmentation initiation and the hidden area is increased gradually, stepwise or otherwise. Our "assumption" is that there is a linear increase in the hidden area with distance from the location of superposition initiation (O in Fig. 4.38a).

At the end of the *en echelon* fringe, the average overlap of two adjacent segments (AB) is 0.2 ±0.1 m. Therefore, the angle $\varphi = 1.2 \pm 0.6°$ (Fig. 4.38a, see also difference between l and li). The hidden area (HA_E) is produced by two adjacent segments along the width of the rectangle, and is given by

$$HA_E = S_{ABCD} = S_{ABO} - S_{CDO} = \frac{1}{2}(l^2 - l_i^2)\sin 2\varphi \quad (4.2)$$

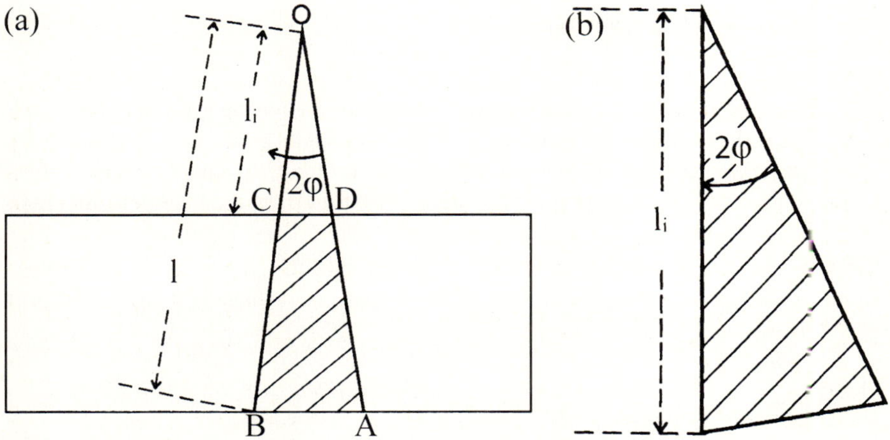

Fig. 4.38. Diagrams for calculating crack areas in fringes. **a** A schematic picture of the overlap development from the location of superposition initiation *O* on the rectangle in Fig. 4.37a. *AB:* The amount of overlap at the end of the rectangle. 2φ is the opening angle of overlap (exaggerated); $l = OB$; $l_1 = OC$; *dashed area*: Overlap inside rectangle; $\varphi = 1.2°$; therefore, $\sin \varphi/2 = 1/2 \times AB/l = 1/2 \times 0.2/4.8 = 0.0208$. **b** Schematic picture of overlap of a single flake. l_i: Length of flake (sub orthogonal to mirror boundary) inside the hackle triangle. *Dashed area* designates overlap = *HPHi*; 2φ is the opening angle (exaggerated). $HPHi = 1/2 l_i \sin 2\varphi$ [area of triangle = 1/2base × height = $1/2(2l_i \sin\varphi(l_i \cos\varphi))$]

Accordingly, $HA_E = 1/2(4.8^2 - 2.8^2)\sin 2.4° = 0.32 \pm 0.16\ m^2$. The hidden area per 1 meter segment boundary length is therefore $0.16 \pm 0.08\ m^2 m^{-1}$. The revealed area ($RA_E$) of the rectangle is 8 m^2, and the total segment boundary length in the rectangle is 19 ±1 m. Hence, the total hidden area is $2 \times 0.16 \times 19 = 6.08\ m^2 \pm 3.00\ m^2$ within the nominal area (the factor of 2 comes from the mirror image to the fracture on its other side that was lost by quarrying) and the ratios of the hidden surface area (HA_E) divided by the revealed surface area (RA_E) is: $u_E = 0.76 \pm 0.4$.

The Hackle fringe on joint A. The "triangle" was selected as a representative of the fringe, because it contains a fairly uniform and clear fracture distribution, rendering them to be measurable with a reasonable confidence. Note that the base of the triangle is on the mirror boundary (Fig. 4.37a). In analyzing the "triangle" we use a similar approach to that applied for the "rectangle" in that we assume that each flake boundary (equivalent to segment boundary) overlaps its neighbor by the same angle as that of the *en echelon* segments, i.e., $\varphi = 1.2 \pm 0.6°$ (Fig. 4.38b). Table 4.8 summarizes the measurements and calculations of the flake boundaries across the triangle (Fig. 4.37b). Fractures (flake boundaries) l_i are measured along their trends sub orthogonal to the mirror boundary. These measurements do not include flake boundaries that curve into orientations that form angles greater than 45° with these trends. Accordingly, the total hidden area is

$$HA_H = \frac{1}{2}\sin 2\varphi \sum_i S_i l_i^2 \tag{4.3}$$

Accordingly, (see Table 4.7 for the calculation of $\sum_i S_i l_i^2$)

$$\frac{1}{2}\sin 2.4° \times 16.92 = 0.35\ m^2 \pm 0.17\ m^2$$

The revealed area (RA_H) of the triangle is ~1.86 m^2; hence the ratio of hidden area to the revealed area in the triangle, $u_H = 2HA_H / RA_H = 0.38 \pm 0.2$.

For a fracture growing in tension under a constant load, the energy release rate G is defined as the elastic energy (per unit width) released from the field per unit crack

Table 4.8. Flake boundaries and hidden area in the triangle of joint *A*

Number of cracks n_i	Max. length l_i in Fig. 4.8b (mm)	Flake length l_i real (m)	$n_i l_i$ (m)	$n l_i^2$ (m^2)
18	10	0.142	2.556	0.363
27	20	0.284	7.668	2.178
4	30	0.426	1.704	0.726
3	40	0.568	1.704	0.968
2	50	0.710	1.420	1.008
2	60	0.852	1.704	1.452
3	130	1.846	5.538	10.223
Sum			22.290	16.918

length (Sect. 1.3.3.5). G is usually proportional to $\sigma^2 c$ where σ is the stress field value and c is the crack length: $G = n\sigma^2 c$. On the other hand, by the Griffith criterion, when the crack becomes critical, i.e., for $c = c_{cr}$ (at the critical flaw, Fig. 2.1a), the energy release rate is equal to 2Γ, where Γ is the energy per unit area, $2\Gamma = n\sigma^2 c_{cr}$. Beyond c_{cr}, the excess energy released per unit crack length is therefore $n\sigma^2(c - c_{cr})$. This energy is manifested either as new created area (as in the hackle and bifurcation zones, and to a lesser extent in the mist zone), or as kinetic energy, heat, deformation, radiation etc.

We assume that the linear relationship between G and c continues throughout the fracture length. Therefore, if the fracture had bifurcated, at say, c_B (an imaginary point that should be further away, beneath point X (Fig. 4.37a), G there would have to be close to twice that of G at c_M (Beauchamp 1995) (to create twice the fracture area, provided that the kinetic energy per unit crack length changes very little at the bifurcation); then we can estimate the maximal additional (hidden) area ratio as a function of crack length, c, to be $\Delta = (c - c_M) / (c_B - c_M)$, where c_M is on the mirror boundary (Fig. 4.37a). In the fringe of joint A, the most distant point, which can be observed on the triangle (point X) is at a distance of 5.20 m from the origin of the crack, and it is clear that no bifurcation has occurred down to this point. Therefore, $\Delta \leq (c - c_M) / (X - c_M)$. At the hackle region at the center of the triangle $c \sim 3.06$ m, which limits the hidden area ratio u_H (see below) at this point to be less than $(3.06 - 1.65) / (5.20 - 1.65) = 0.4$.

4.10.3 A Comparison between the Fringes of Joint C and Joint A

The part of the energy release rate, which is used to create the overlapping area is $G - 2\Gamma - qc - p$, where qc is the kinetic energy per unit crack length (q is proportional to v^2, where v is the crack speed). For simplicity, we neglect the plastic energy, p. At c_M, $G(c_M) = 2\gamma + qc_M$. Therefore, for $c_M < c < c_B$ we get that the ratio of the additional area to nominal area is proportional to $c / c_M - 1 = (\eta)$.

In order to compare hidden area ratios between the *en echelon* and the hackle fringes, they should be both at the same scaled distance from the respective crack origins. Since they are situated at different distances, the following procedure is conducted. For the *en echelon* segments in the rectangle, its average distance from the origin is 5.88 m and the distance from the origin to the mirror boundary is $c_M = 0.94$ m; therefore, $\eta_E = 5.88 / 0.94 - 1 = 5.26$. For the hackles in the triangle, a similar calculation yields $\eta_H = 3.06 / 1.65 - 1 = 0.85$. If it is assumed that the hidden area is linear with crack length, the hidden area should be normalized by division by η for an appropriate comparison. We therefore compare the ratio of hidden area to revealed area at the "same distance" for both fringes. For the hackle fringe we have $r_H = u_H / \eta_H = 0.38 / 0.85 = 0.45$, while for the *en echelon* fringe it is $r_E = u_E / \eta_E = 0.76 / 5.26 = 0.14$, which is 3.14 times smaller. In summary, the normalized (real) ratios of hidden area divided by revealed area on the fringes of joints A and C are respectively, $r_H / r_E = 0.44 / 0.14 = 3.14$.

4.10.4 How was Energy Released Differently in the Two Fringes?

It turned out that energy release in the actual formation of new fracture area (per a given nominal fringe area) is substantially greater in the hackle-fringe than in an *en echelon* one.

What is the reason for this? We follow the principle that $G = G_I + G_{II} + G_{III}$ (Lawn 1993, Eq. 2.18) and we neglect G_{II} because of its minor importance in relation to fracture conditions in the two fringes under consideration. The G relationship then becomes

$$G_{IH} = G_{IE} + G_{IIIE} \tag{4.4}$$

That is, the energy release rate by mode I loading in the hackle fringe equals the energy release rate by modes I + III loading in the *en echelon* fringe. How do we explain the reduction of the "effective" G to produce new fracture area in the transition from the hackle fringe to the *en echelon* one?

The state of stress influences the size of the plastic zone, and shear stresses promote the enlargement of this zone (Hahn and Rosenfield 1965). More specifically, the plastic zone associated with the mode III loading is several times larger than that for mode I for similar materials and at the same stress intensity values (Tschegg 1983). Plastic deformation is associated with the loss of considerable energy into heat. The latter energy transition may reach extreme values at rapid fracture velocities (up to 60% or even 90% of G in PMMA and some metals) (Döll 1976), but even at low crack velocities heat losses may be significant under induced plastic conditions. Hence, the transformation of some G_I into G_{III} in the *en echelon* fringe involves an increase in plastic deformation and subsequent heat loss, which is not used in the creation of new fracture surfaces.

4.10.5 Summary

The distinction between fringes of *en echelon* segmentation and hackles on joint surfaces is important because they reflect different fracture conditions. It was found that the normalized (real) ratios of hidden area divided by revealed area on the hackle and *en echelon* fringes of joints A and C are respectively, $r_H = 0.44$ and $r_E = 0.14$, hence, $r_H / r_E = 0.314$. This shows that considerably more new area is formed in hackle zones than in *en echelon* zones per any given nominal area of the fringe. Possibly, the transformation of some G_I to G_{III} in the transition from hackle to *en echelon* fringe involves an increase in heat loss, which is the portion of energy not used in the creation of new fracture surfaces.

4.11 The Index of Hackle Raggedness (*IHR*) on Joint-Fringes of the Mrákotin Joint Set

This section concentrates quantitatively on the fractographic differences between the fringes of joints A, C and F (Sect. 4.4.3), citing Bahat et al. (2004). It has been shown that considerably more new area is formed in hackle sections than in *en echelon* sections per given nominal area of fringe (Sect. 4.9). Since the crack area is a fundamental parameter in fracture physics, we seek to devise a method of correlating changes in crack area by tectonofractographic techniques, i.e., by the characterization of the secondary fracture on joint fringes. The two distinct morphologies manifested by the *en echelon* and hackle fringes (Fig. 2.1) represent contrasted fracture mechanisms, reflecting different local stress conditions (Sect. 4.8.4). We start dealing with this relationship by formulating a series of criteria for estimating the value of the Index of Hackle Raggedness (*IHR*) on a given fringe, based on geometric distinction of crack features between the hackle and *en echelon* end members in the fringe. We then apply it to joints A, C and F from the Mrákotin set, summarizing the results in Table 4.9.

4.11.1 Criteria of *IHR* and Application to the Mrákotin Joint Set

Before getting into the actual definition and procedure, one has to consider three limitations that qualify the rules. First, there is a need to establish which of the seven key criteria presented below is applicable to a given outcrop. Nature is quite selective in revealing good fracture markings and very often only small parts of the ideal fractography (Fig. 2.1a,b) is exposed. Occasionally, only one fringe is well displayed (the upper or lower). Therefore, in many outcrops only a certain number of these criteria would be useful, according to the quality of the crack exposure. Second, it is required to apply independent estimations of the *IHR*, ranging from 1 to 10 for each of the above criteria. Third, since the *IHR* is new (even for the engineering literature), lack of previous experience justifies at this stage the assumption of equal weight for the various criteria. Therefore, the *IHR* for a given joint will be the mean of *IHR* values from all measurable parameters (criterion 3 is an exception, where the angle of "triangularity" $\sin\alpha$ is converted to *IHR* by multiplication by 10). The total *IHR* error per joint, consisting mainly of measurement errors is thought to be ±0.7.

4.11.1.1 *Length Variability*

In many fringes *en echelon* segments maintain uniform lengths, if microcracks in root zones (Bankwitz 1966) and elsewhere on the segments are excluded. Hackles on the other hand show great length variability even among neighbor cracks. Hence, increasing deviation of length ratios (of maximum to minimum length in a given fringe) from length uniformity (Fig. 2.34) produces an enlargement of *IHR*. The length ratios of fringe-cracks for joints A, C and F are 6, 1 and 3, corresponding to the respective *IHR*.

4.11.1.2 *Rectangularity*

En echelon segments approximate rectangular shapes whose aspect ratios can be readily measured. Hackles on the other hand deviate considerably from such shapes. Measurement of the deviations from straight lines (*DSL*) along the rectangle length and width can provide information regarding the departure of the fringe crack from its rectangular shape. We use a set of profiles that resembles the scale of fracture-surface roughness constructed by Barton and Choubey (1977), which shows a gradual *DSL* ranging from 0 to 10 (Fig. 4.39). The *DSL* value is the mean of combined deviations along the length and width of three fringe-cracks, assessed manually. Thus, the *DSL* values of A, C and F are 6.5, 2.5 and 3.5, respectively.

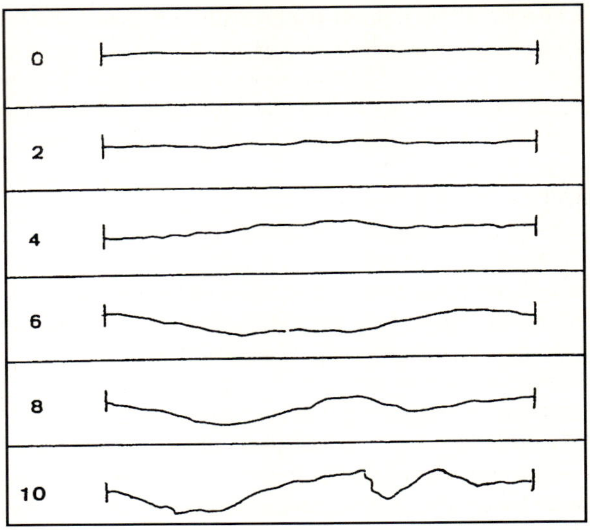

Fig. 4.39. A set of profiles that show gradual deviations from straight lines on a scale ranging from 0 to 10 (modified from Barton and Choubey 1977)

4.11.1.3 *Fringe Width*

The tendency of *en echelon* fringes (to be distinguished from individual segments) to maintain uniform widths, compared to non-uniform widths of hackle fringes (Fig. 2.34) has been repeatedly observed (Bahat 1998b). This is also demonstrated by the "quasi-triangular" shapes of the two fringes of joint F (Fig. 4.12c) compared to the approximate parallelism of the mirror and fringe boundaries at the lower side of joint C (Fig. 4.10c). A measure of deviation from a uniform fringe width is given by the angle α that increases with the "triangularity" of the fringe. Unfortunately, this parameter cannot be measured with high confidence for joint A because only certain parts of the fringe lengths of joint A are revealed (sufficient for the study of criteria 1–5 but not for this parameter) so that an appropriate index cannot be established for the latter joint. Accordingly, the sin α values for joints C and F were estimated to be 0.17 and 0.34 ($\alpha = 10 \pm 2°$ and $20 \pm 2°$), respectively, and the corresponding *IHR* for these joints are 1.7 and 3.4. When clear triangularity is recognized this method would be more reliable than the one given in Sect. 4.11.1.1.

4.11.1.4 *Fringe Thickness*

En echelon segments have approximately (quasi two-dimensional) planar surfaces and display uniform thickness along their lengths. On the other hand, hackles consist of multi-fracture surfaces superposing each other not uniformly along the crack lengths, often leading to the increase in thickness of individual cracks and in the fringe thickness (*FT*) with distance from the mirror-fringe boundary. Therefore, *FT* is normally maximum at the far end of the fringe and minimum close to the mirror boundary. *FT* is abruptly reduced when fracture branching occurs, but there is not clear evidence that such a process actually took place in this outcrop. The change in *FT* often renders the hackle-fringe to look quasi three-dimensional (Fig. 4.11c), such that *IHR* increases with the ratio of maximum / minimum *FT*. Changes in thickness were measured on an enlarged profile of a photograph of joint A. These ratios for A and C were found to be 6.5 and 1 ±1, respectively (no measurements for joint F are available).

4.11.1.5
Fringe Angularity

There are indications that the angle ϕ that the fringe (as a set of cracks, rather than an individual one) forms with the mirror, increases with the "fracture-dynamics" of the process, i.e., roughly corresponding to the increase in fracture velocity. Bahat and Rabinovitch (2000) found that in the same rock (chalk) the ϕ varies from 0° to 30 ±5° on a natural joint and from 30 ±8° to 50 ±8° on an artificial fracture along a road cut (that was formed by explosion). It is assumed that this relationship may change significantly from rock to rock (see more in Bankwitz and Bankwitz 2004 and Sect. 4.4.4). It was found that $\phi = 5 \pm 2°$ for joint A, while $\phi = 0–1°$ for joint C (it could not be measured for joint F). We therefore assign *IHR* values of 5 and 1 ±1 to joints A and C, respectively.

4.11.1.6
Crack Dipping

Deviation of individual fringe-cracks from a uniform dip also increases the *IHR*. Individual *en echelon* segments in a fringe commonly dip in the same direction of the mirror, i.e., the "normal direction", forming zero to small $+\phi$ angles. However occasionally, different hackle flakes may also dip into the mirror plane, i.e., in the "opposite direction", forming $-\phi$. For determining the *IHR* value we combine the maxima of $+\phi$ with those of $-\phi$ for different cracks on the same fringe to a single angle ϕ_f representing a fringe. Our results show that the ϕ_f values for joints A, C and F are 10 ±5° ($+\phi = 7 \pm 3°$ plus $-\phi = 3 \pm 2°$), ~0.5°and 2–3°, respectively. Accordingly, the corresponding *IHR* are 10, 1 and 2–3.

Table 4.9. Summary of IHR indices for joints A, C and F

Criterion number	Criterion	A	C	F
1	Length variability	6	1	3
2	Rectangularity	6.5	2.5	3.5
3	Fringe width	n.m.	1.7	3.4
4	Fringe thickness	6.5	0.5	n.m
5	Fringe angularity	5	0.5	n.m
6	Crack dipping	10	0.5	2.5
7	Secondary mirror	n.m.	n.m.	n.m.
Sum		34	6.7	12.4
IHR(mean)		6.8	1.1	3.1

n.m.: not measured.

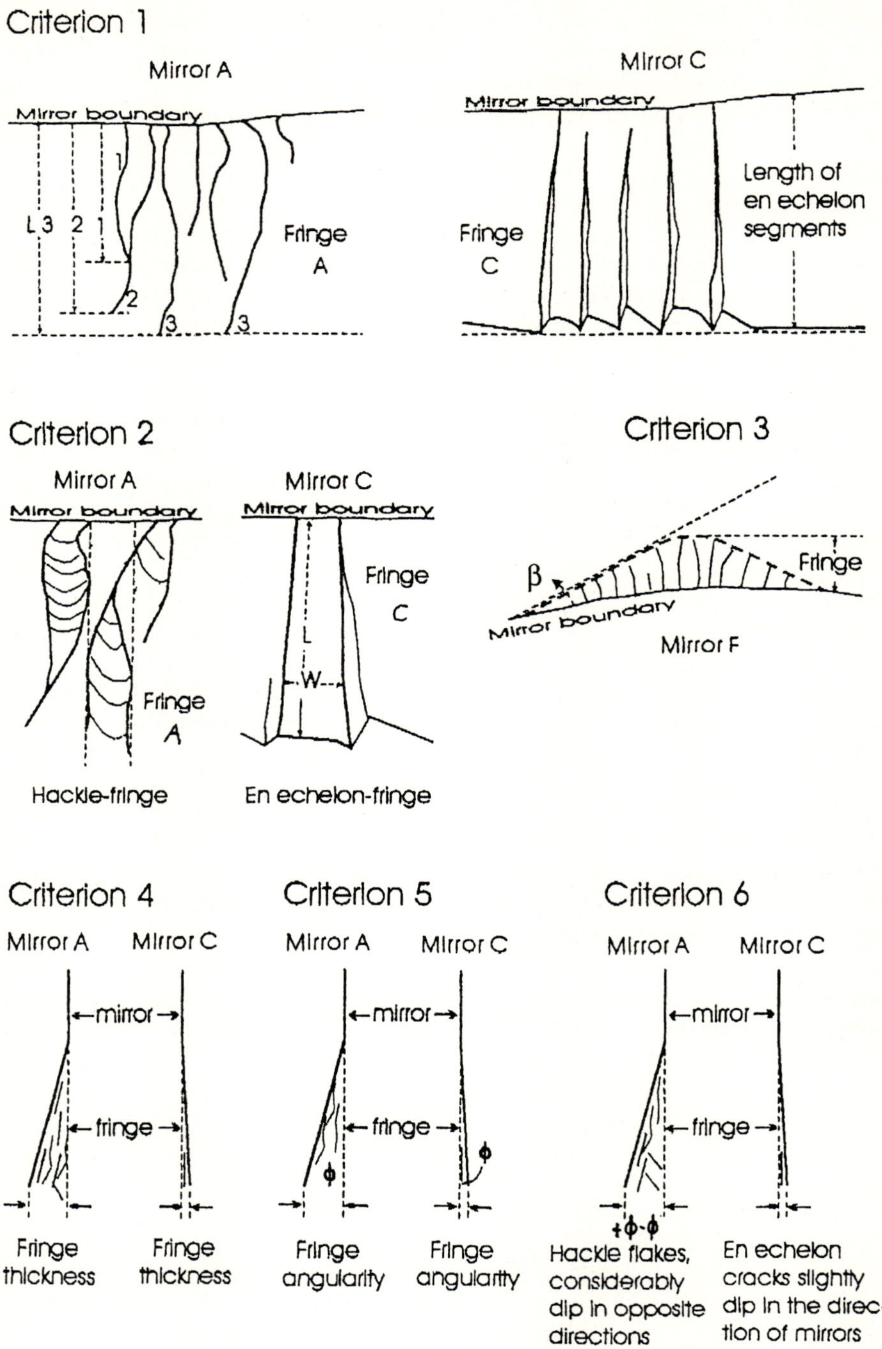

Fig. 4.40. A diagrammatic summary of six criteria for estimating the value of the index of hackle raggedness on joint fringes (see text for explanation)

4.11.1.7
Secondary Mirror

Gross deviations from "normal fringe morphologies" require special attention. They are very rare on *en echelon* fringes and are rare on hackle fringes. Quite intriguing is the rarity of "secondary mirrors" in natural exposures (Fig. 4.11a–c), compared to their frequent appearance in dynamically fractured glass (e.g., Fig. 2.30c). It is presently impossible to construct a scale for secondary mirrors. Quantification of this parameter remains a challenge for future study. Qualitatively, however, this distinction supports the relatively high *IHR* value assigned to joint A that exhibits a secondary mirror.

4.11.2
Discussion

The seven criteria. Criteria 1–6 are summarized diagrammatically in Fig. 4.40. Note that criteria (1–3) relate to changes observed parallel to the mirror, while criteria (4–6) concern parameters perpendicular to the mirror. Criterion 7 requires measurements both parallel and perpendicular to the mirror. We did not use deviations of fringe cracks from a uniform strike as an *IHR* criterion, because we unexpectedly found it not to be diagnostic. The gap between the top value of the scale 10 and the value 6.8 found for joint A probably suggests that joint A was not formed by the most dynamic conditions, and the 10 *IHR* value should be obtained under more extreme dynamic conditions, such as intense fracture of glass would occur in the laboratory (e.g., Fig. 2.30, see also Bahat et al. 1982, Fig. 2).

4.11.3
Summary

A method of quantifying the "spectrum" of secondary fractures on joint fringes ranging from *en echelon* segments to hackle cracks is presented. The method is based on a set of seven criteria that enable one to place every fringe on an "index of hackle raggedness" (*IHR*) along a scale from 0 to 10. This method is applied to the "Mrákotin joint set" from a granite quarry in the Czech Republic part of the South Bohemian Pluton. The results show an *IHR* range between 1.2 ±0.7 and 6.8 ±0.7. The gap that exists between 10, the top value of the scale and 6.8 the maximum obtained value, suggests that the joint for which *IHR* = 6.8 was not formed by the most dynamic conditions, and the 10 *IHR* value should be obtained under more extreme dynamic conditions.

4.12
A Method of Drilling in situ Fracture from the Rock

An important structural problem is the determination of shear offsets along joints, which often are concealed in an outcrop scale, particularly when offsets are limited to short distances (millimeter scale). Thus, transitions from joints to faults are often obscured. This enigmatic situation may be fully or partly resolved by resorting to petrographic methods. A detailed study of millimeter and sub-millimeter scales of past natural offsets along the fracture requires the assurance that in situ conditions are being maintained in the laboratory.

The objectives in sample preparation are to keep intact both margins of the examined joints following their removal from the outcrop so that the original aperture (or thickness, Fig. 2.8a) would be examined. The sample preparation process consists of several stages:

1. Coat the trace of the fracture with gray epoxy.
2. Inject red epoxy into the aperture to fill the space and stick to the two margins of the fracture.
3. Core drill samples.
4. Vacuum imbed selected samples with blue epoxy.
5. Prepare thin sections.

The "epoxy procedure" consisting of different epoxy colors helps to follow the timing of shear offset in the thin sections.

A 25 gauge 5/8" syringe needle is inserted into the selected fracture (Fig. 4.17c). Then, a Hardman brand double-bubble epoxy is smeared on the front of the fracture. This secures the needle in place, and prevents the forthcoming red epoxy from flowing out of the front of the fracture upon injection. Also, it affixes the two margins of the fracture at the rock surface together to maintain it in the original juxtaposition after sample removal. This epoxy sets in about 5 min.

The next step is to prepare the red epoxy mixture for injection into the fracture. Twenty drops of toluene are mixed into 4 pouches (totaling 8 g) of Hardman epoxy, to which fifteen drops of Morefast brand red epoxy die are added. All of this mixture is stirred for 1.5 min. The epoxy mixture is poured into a 5 cc Luerloc syringe or a 1 cc slip-tip syringe with their plunger removed. The plunger is put back in, and the syringe is then attached to the needle that had previously been imbedded in the rock. The plunger is then depressed, causing the red epoxy to flow into the fracture. Because of the tightness of the aperture, it is impossible in some fractures to obtain penetration of large quantities of injected epoxy. However, even little penetration may occasionally be enough to secure original juxtaposition in a few cylinders. Handling time for the red epoxy mixture is 10–15 minutes, and the setting time, 1 hour.

Chapter 5

Electromagnetic Radiation Induced in Fractured Materials

The key to a successful application of electromagnetic radiation (EMR) to rock fracture investigation is a good understanding of the basic physics of EMR emission. Such an understanding is expected to provide useful universal tools or "fingerprints" of the process involved. In this chapter we describe our efforts in the last decade investigating EMR in the radio frequency range (1 kHz to 50 MHz, Fig. 5.1) emitted by fracture. These investigations were conducted in three consecutive principal stages:

1. "Small-scale" laboratory investigations, which consisted of two sub-stages, i.e., quasi-static and dynamic. The aim of this stage was to derive the basic physical principles operating behind EMR emission from microcracking excited by stiff compression and tension and to find the similarities and dissimilarities between EMR emissions from different dynamical and quasi-static cracking.
2. "Medium-scale" pilot investigations carried out both in open quarries during blasting and in underground mines prior to rock bursts and rock and gas outburst. The goal of this stage was to understand the differences in EMR during "low scale" and

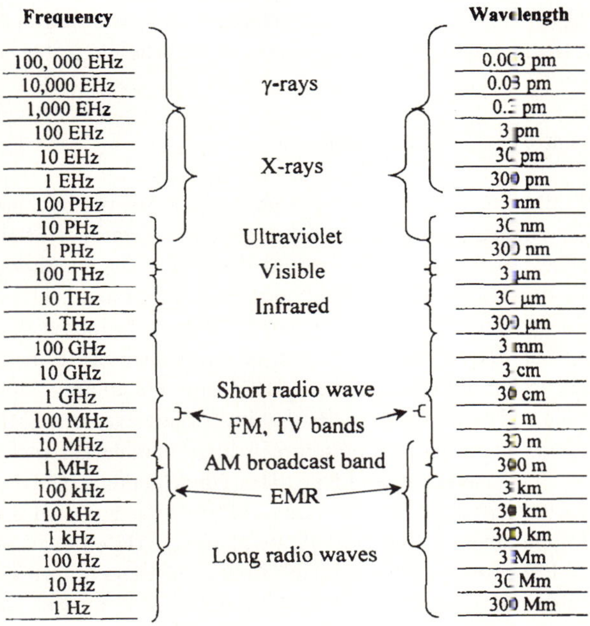

Fig. 5.1. Electromagnetic wave spectrum (E (Exa): 10^{18}, P (Peta): 10^{15}, T (Tera): 10^{12}, G (Giga): 10^{9}, M (Mega): 10^{6}, k (kilo): 10^{3}, c (centi): 10^{-2}, m (milli): 10^{-3}, µ (micro): 10^{-6}, n (nano): 10^{-9}, p (pico): 10^{-12} (Rabinovitch et al. 2000c)

"medium scale" cracking and thereby to pave the way for planning "large-scale" investigations.

3. The "large-scale" stage is meant to study EMR behavior during earthquake nucleation. Since earthquake monitoring and prediction by seismological techniques is "still in its infancy" (Wyss 2001), the possible addition of more sophisticated EMR methods should be welcome. In particular, the EMR method measures micro-cracking, i.e., the initial process in an earthquake, a process occurring much *earlier* than the actual earth movements; the study of which might thus be beneficial for earthquake forecasting.

Accordingly, this chapter *(a)* summarizes our investigations in the small and medium scales, *(b)* reports some improvements suggested thereby for monitoring and prediction in the mining industry and *(c)* points out guidelines for large-scale experiments (earthquake forecasts) based on our acquired theoretical understanding and the "fingerprints" accumulated from tensile and shear experiments.

We hope that this chapter, by describing the current state of the art, could be interesting not only to EMR specialists, but also to rock physicists, geoscientists, and mining engineers who always seek new methods for multi-scale failure investigation.

We begin the chapter with a review of previous EMR investigations (Sect. 5.1), then present our new EMR model (Sect. 5.2) and describe in detail its experimental verification (Sect. 5.3). A comparison of this model with previous ones is presented in Sect. 5.4. Individual EMR pulses and their classifications are given in Sect. 5.5 followed by Sect. 5.6, where influence of the elastic properties of rocks on EMR activity is analyzed. A novel EMR phenomenon, polarization induced in materials during percussion drilling, is examined in Sect. 5.7. EMR effects induced by blasting in open quarries and appearing prior to volcanic eruptions and earthquakes are considered in Sect. 5.8, while universal EMR features are discussed in Sect. 5.9. In Sect. 5.10, which we believe would be interesting mostly for mining engineers, we focus on underground EMR observations prior to rock burst and gas outburst deriving criteria of their forecasting. Finally, a brief summary of the whole chapter is presented.

5.1 History of EMR Research

Emitted EMR from fractures was firstly observed by Stepanov (Urusovskaja 1969). He loaded samples of KCl and detected an appearance of electromagnetic pulses during specimen failure. At that period, the EMR did not gather any implementations, and investigations of this phenomenon were continued basically on samples of alkaline halide crystals (Nabaro 1967; Kornfel'd 1971, 1977; Gol'd et al. 1975; Finkel et al. 1975; Golovin and Shibkov 1986a,b). It was observed that the creation of new microcracks stimulated excitation of EMR (Gol'd et al. 1975; Golovin and Shibkov 1986a,b). These studies were followed by numerous ones that investigated various EMR aspects. The EMR phenomenon was measured for fracture in different materials, including metals and alloys (Misra 1975; Jagasivamani and Iyer 1988), single crystals (Gol'd et al. 1975; Khatiashvili 1984), rocks (Nitsan 1977; Warwick et al. 1982; Ogawa et al. 1985; O'Keefe and Thiel 1995; Rabinovitch et al. 1995, 1996, 1998) and ice (Fifolt et al. 1993; Petrenko 1993).

EMR investigations during rock specimen compression, which in the majority of cases were combined by acoustic emission (AE) measurements (Nitsan 1977; Goncharov et al. 1980; Warwick et al. 1982; Sobolev et al. 1982; Ogawa et al. 1985; Cress et al. 1987; Yamada et al. 1989; Frid 1990), showed that EMR was indeed detected at the same moment that microcracks were formed. EMR features on different scale levels (earthquake (EQ), rock burst (RB) and rock sample fracture) were studied by Khatiashvili (1984). In all of these cases, EMR anomalies were observed preceding the final rock failure.

According to the theory of material destruction (Zurkov et al. 1969; Regel' et al. 1972; Kuksenko and Mansurov (1986); Kuksenko et al. 1985, 1987; Petrov and Gorobets 1987; Gueguen and Palciauskas 1994), the process of rock fracturing consists of two stages: The first one consists of an accumulation of microcracks, and the second consists of microcrack clustering and final failure (Sect. 1.2.3.3). Rock destruction processes on all scale levels (from rock samples up to crust fracture) adhere to this "scheme" (Kuksenko et al. 1987; Mansurov 1994). As hypothesized by Gershenzon et al. (1987), EMR is initiated in the first stage, appears before final failure and could be an EQ forecast indicator.

Large-scale EMR studies prior to rock burst (RB) in mines are also registered. Nesbitt and Austin (1988) registered EMR in a gold mine (2.5 km depth). Frid (1997a,b, 2000, 2001) observed EMR anomalies before RBs and gas outbursts. Vozoff and Frid (2001) measured EMR anomalies prior to a roof fall in a coal mine. Sakai et al. (1992); Tomizawa et al. (1994); Tomizawa and Yamada (1995); Rabinovitch et al. (2002a) investigated EMR induced by explosions.

During the seventies and eighties, the interest in EMR emanating from failure of rocks was renewed in connection with the problem of EQ prediction. Numerous investigations (King 1983; Sadovskii et al. 1979; Gokhberg et al. 1979, 1982, 1985; Morgunov 1985; Gokhberg et al. 1986; Gershenzon et al. 1987; Yoshino et al. 1993; Yoshino and Tomizawa 1989), a significant part of which was carried out in Japan and in the former USSR, showed that the magnitude of EMR sharply increased several hours or even days before EQ and quickly decreased at the moment of EQ or directly before it. The magnitude of the detected EMR anomalies was of up to 10–15 dB over the usual noise level. Thus, EMR is associated with fractures on multi-scale lengths and is connected with various applications.

Several atomic-scale models have hitherto been suggested to explain the EMR phenomenon. All of these models were qualitative, and efforts to use EMR quantitatively were unsuccessful, due to the lack of a detailed understanding of the EMR phenomenon (King 1983; Rabinovitch et al. 1998). Until quite recently, the physical mechanism of EMR was unknown (Freund 2000, 2002). An essential requirement for this type of knowledge is a careful laboratory investigation of rock failure. Such an investigation has been carried out by the group of Rabinovitch et al. (1995, 1996, 1998, 1999b, 2000a,b, 2002a,b, 2003a,b).

5.2 Foundations of EMR Model

A general shape of an EMR pulse induced during fracture experiments is shown in Fig. 5.2.

It can be characterized by the following general relationship (Rabinovitch et al. 1998):

$$A = \begin{cases} A_0 \sin(\omega(t-t_0))\left(1-\exp\left(-\dfrac{t-t_0}{\tau}\right)\right), & t < T \\ A_0 \sin(\omega(t-t_0))\exp\left(-\dfrac{t-T}{\tau}\right)\left(1-\exp\left(-\dfrac{T-t_0}{\tau}\right)\right), & t \geq T \end{cases} \tag{5.1}$$

where t is the time, t_0 is the time from the origin up to the pulse beginning and T is the time from the origin up to the EMR pulse envelope maximum. Thus, $T' = T - t_0$ is the time interval to reach pulse maximum; τ is the rise time and the fall time (RFT), which turn out to be the same; A_0 is the pulse amplitude and ω is the frequency. The signal's general shape including its envelope, frequency and duration can be obtained by our basic (see below) theory, while the parameters themselves can be derived by least squares fitting of the experimental results (Rabinovitch et al. 1998, see Fig. 5.16).

Four basic elements of a physical model, which would result in such simple relationships, are (Frid et al. 2003):

1. Time of growth (Rabinovitch et al. 1998): The pulse amplitude, A, increases as long as the crack continues to grow, since new atomic bonds are severed and their contribution is added to the EMR. When the crack halts, as it does if created under a compressive load, the pulse amplitude starts to decay. The time from the start of the pulse up to its maximum, T', is proportional to the number of severed atomic bonds and thus to the crack length ℓ (the crack velocity v_{cr} is almost constant).

$$T' \sim \frac{\ell}{v_{cr}} \tag{5.2}$$

2. Dynamics of growth and decay of the amplitude: In the time increment, dt, 'A' decreases by dissipation (to phonons, see Sect. 5.3.7) to

$$A(t)\left(1 - \frac{dt}{\tau}\right)$$

but it is "replenished" by a term proportional to the number of severed bonds, αdx, where dx is the length increment of the crack in the interval dt ($dx = v_{cr}dt$) and α is a

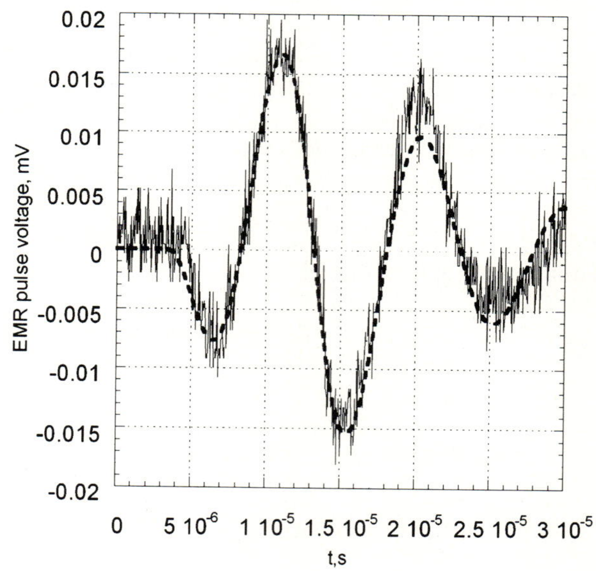

Fig. 5.2.
An experimental pulse shape (output voltage against time) and a numerical fit according to Eq. 5.1 (*dashed line*) (Rabinovitch et al. 1998)

coefficient relating the antenna output to dx. α is assumed proportional to the width of the crack, b, (the following calculations assume a constant crack width), to the number of bonds per unit area, to the emittence of a bond, to the solid angle of the crack seen by the antenna and to the latter's conversion factor. It follows that:

$$\frac{\partial A}{\partial t} = -\frac{A}{\tau} + \alpha v_{cr} \tag{5.3}$$

Hence,

$$A = \alpha v_{cr} \tau (1 - e^{-(t-t_0)/\tau}) , \qquad t_0 \leq t < T \ (= t_0 + T') \tag{5.4}$$

For $t > T$ the crack stops and the amplitude thereafter diminishes resulting in

$$A = A(T) e^{-(t-T)/\tau} , \qquad t \geq T \tag{5.5}$$

3. Regarding the oscillatory parts of Eq. 5.1, we assume that these EMR waves are created by charge oscillations at both sides of the propagating crack (Fig. 5.3).
   Consider the line of bonds located at the front (tip) of the propagating crack (Fig. 5.3a). These bonds break when the front moves to the next line. Following this

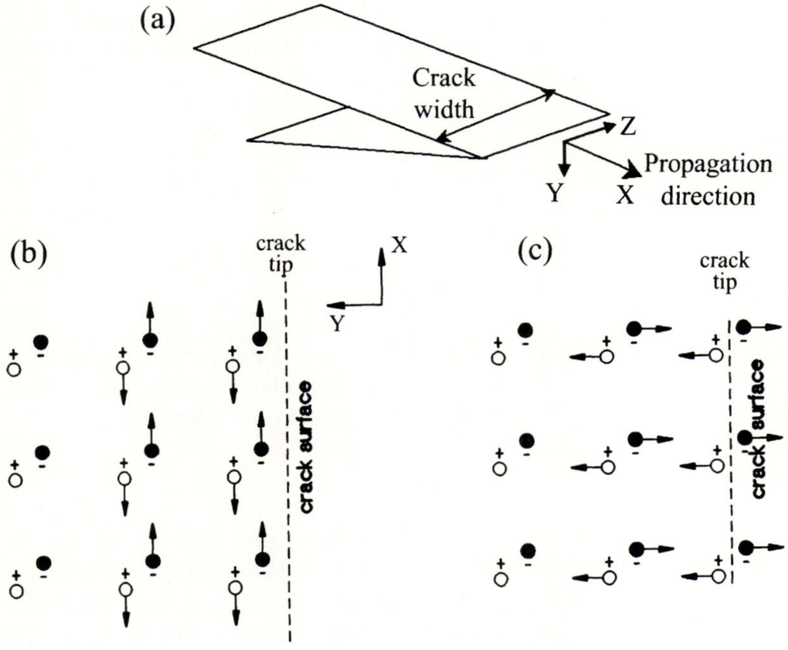

Fig. 5.3. **a** A schematic picture of crack propagation; **b**, **c** schematic "optical surface wave" at crack surface (a similar wave propagates on the other surface) at a specific time. Crack surface is in the x-z-plane and the crack moves in the x-direction. Note that charge separation can either be longitudinal (**b**) or transverse (**c**) with respect to the surface, with appropriate EMR polarizations. Note also the exponential decay of the wave into the material. Charge separation is oscillatory so that at a later time the dipole directions are reversed (Frid et al. 2003)

break, the atoms on both sides of the bonds are moved to "non equilibrium" positions in relation to their (new) steady state ones and will perform oscillations around them. If each atom (ion) vibrated individually, the situation would resemble the Einstein model of lattice vibrations and the frequency of these oscillations would approach 10^{15} Hz. Since, however, the lines of vibrating atoms move together and are also connected to atoms around them (in the forward direction and also to atoms on their side of the two surfaces newly created by the fracture), the ensuing vibrations are similar to those obtained for the bulk by the Debye model, having much lower frequencies. It is the oscillations stimulated by this process at the new surfaces, which give rise to the EMR, as follows. The positive charges on these surfaces move together in one direction away from the equilibrium plane (one crack side), while the negative charges move in unison in the other direction from the equilibrium plane (the same crack side), and vice versa, retaining an overall charge neutrality throughout. These surface oscillations, similar to "optical" Rayleigh waves, decay exponentially into the bulk (Fig. 5.3b,c) – "surface vibrational optical waves (SVOW)" (e.g., Srivastava 1990), much like "bulk optical (phonon) waves" (e.g., Kittel 1987) observed in vibrating crystals.

4. It has been shown elsewhere (e.g., Srivastava 1990) that surface waves decay in time as a result of an interaction with bulk phonons. We therefore (Rabinovitch et al. 2003a) consider τ to be the relaxation time of such a (Rayleigh-like) surface wave, which interacts with a bulk phonon, leading to the creation of another bulk phonon (a three-phonon process), and use Eq. 13 of King and Sheard 1970) to characterize the process. The rate of occurrence of the process per unit time is given by the golden rule formula (Schiff 1986):

$$P_i^f = \frac{2\pi}{\hbar}\left|\langle f|H_3|i\rangle\right|^2 \delta(E_f - E_i)$$

where the initial and the final states are

$$|i\rangle = |n_R, n_{b1}, n_{b2}\rangle, \; |f\rangle = |n_R - 1, n_{b1} - 1, n_{b2} + 1\rangle$$

H_3 is the time dependent anharmonic part of the crystal Hamiltonian, n_R, n_{b1} and n_{b2} are the numbers of surface phonons, initial and final bulk phonons, respectively. E_i and E_f are the initial and the final energies of the three-phonon system, so that

$$E_f - E_i = \hbar(\omega_{b2} - \omega_R - \omega_{b1})$$

where ω denote the related frequencies. The relaxation time τ, which we identify with the rise and fall time of the EMR pulse, is obtained from the golden rule formula ($1/\tau$ being proportional to the transition probability) using the explicit expression for H_3 (Eq. 6.47 in Srivastava 1990) and the displacement field due to the surface modes written in second quantized notation. Integrating over the states of the initial and final bulk phonons b1 and b2, one obtains (Eq. 8.36, Srivastava 1990)

$$\frac{1}{\tau_R} \propto \frac{\omega_R T^4}{\rho^3 v_R^2} \tag{5.6}$$

where T is the temperature, ρ is the material density, and v_R is the Rayleigh wave velocity. The proportionality coefficient contains data of bulk phonons and of the crystallographic orientation and is considered to be constant for the same material. Note that T is the local temperature at the crack tip. It is much higher than room temperature (Fuller et al. 1975).

In a developing crack, a surface waves propagate along the crack surfaces. Their emitted EMR frequency is the same as that of the oscillating ions of the crack sides (Rabinovitch et al. 1998), ω. Therefore we equate $\omega = \omega_R$.

Several important results can be obtained even from these basic ideas.

1. Equations 5.4 and 5.5 together with oscillations part $\sin \omega(t - t_0)$ immediately lead to the correct empirical shape (5.1).
2. If the half wavelength, $\lambda / 2$, of the surface waves creating the EMR is limited by the crack width b (since at both sides of the crack, atomic movements are restricted), then (Frid et al. 2000; Rabinovitch et al. 2000c)

$$b \approx \frac{\lambda}{2} = \frac{\pi v_R}{\omega} \tag{5.7}$$

3. From Eq. 5.2 and Eq. 5.7,

$$\frac{T'}{\omega} \sim \frac{1}{\pi v_{cr} v_R} S \tag{5.8}$$

   where $S = \ell b$ is the crack area. Since T' and ω are measurable quantities (from the pulse shape, using Eq. 5.1), the ratios of the crack surface areas, if not their exact values, can be calculated using Eq. 5.8.
4. Since EMR pulse amplitude A is proportional to α (Eq. 5.4), which was assumed to be proportional to crack width b, it is expected that EMR amplitude should be inversely proportional to the EMR signal frequency $\omega = 2\pi f$.
5. The rise and fall time, τ, is expected to be inversely proportional to the pulse frequency ω (Eq. 5.6).

5.3 EMR Model Verification

5.3.1 Experimental Arrangement

In order to investigate EMR properties through the range of the whole stress strain curve, a triaxial load frame (TerraTek stiff press model FX-S-33090; axial pressure up to 450 MPa; confining pressure up to 70 MPa; stiffness 5×10^9 N m^{-1}) was used (Fig. 5.4).

It is combined with a closed-loop servocontrol (linearity 0.05%), which is used to maintain a constant axial piston rate of displacement. The load was measured with a sensitive load cell (LC-222M, maximum capacity 220 kN, linearity 0.5% full scale). The confining pressure was continually controlled by a clock-type sensor to preserve its

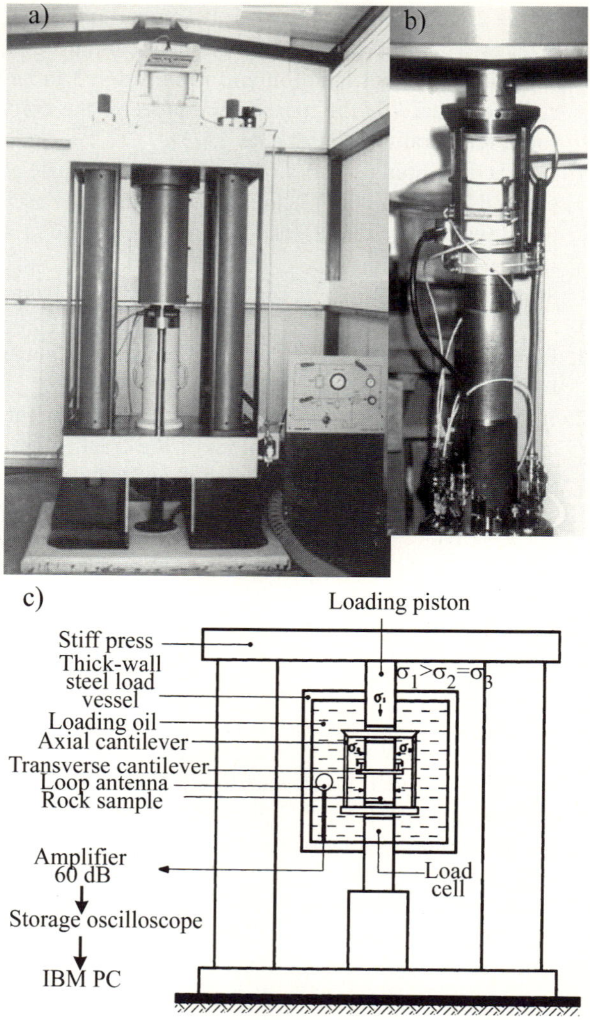

Fig. 5.4.
Compression press: **a** the triaxial load frame, **b** the sample stack, **c** schematic diagram of the EMR experimental arrangement

preset value through volumetric changes of the sample during the loading process. The cantilever set (consisting of axial and lateral detectors; strain range about 10%; linearity 1% full scale) enables us to measure sample strains in three orthogonal directions parallel to the three principal stresses.

A magnetic one-loop antenna (EHFP-30 Near Field Probe set, Electro-Metrics Penril Corporation) 3 cm in diameter was used for the detection of the EMR. It is wound within a balanced Faraday shield, so that its response to external electric fields is negligibly small. A low-noise micro-signal amplifier (Mitek Corporation Ltd, frequency range 10 kHz to 500 MHz, gain 60 ±0.5 dB, noise level 1.4 ±0.1 dB across the entire frequency band) and a Tecktronix TDS 420 digital storage oscilloscope connected by way of a General Purpose Interface bus to an IBM PC completes the detection equipment.

5.3.2
Materials

For our uniaxial and triaxial compression investigation we used the following materials:

1. *Granite.* A large Eilat granite "block" from the Nahal Shelomo area of Sourthen Israel, nearly 3 km from Eilat was used. This granite is gray and consists of K-feldspars (about 40%), quartz (about 35%) albite-oligoclase (about 20%), and biotite (about 3–4%). It is medium to coarse-grained (feldspars and quartz, 2–4 mm; biotite 0.5–1 mm) and non-porphyritic (Bogosh et al. 1997).
2. *Chalk* from the Horsha Formation (Middle Eocene), sampled in the Beer Sheva syncline (Ramat Hovav, about 9 km from Beer Sheva, Israel, Bahat 1991a). Chalk is a pure-white, soft and friable rock consisting of calcitic (nannoplankton) microfossils. The color, purity, texture, cementation and hardness of the rock may vary due to additives such as water, oxides, marl, clay and flint, and by diagenetic and tectonic processes (Mortimore 1990). The strength of chalk under compression depends on the above properties, particularly on the dryness of the sample, and varies from values of ~1 MPa when wet, to some 50 MPa when extremely dry. A strict drying process was therefore applied to the samples, which involved a cycle of heating to 110 °C for 24 hours, and then an immediate transfer to a desiccator to avoid water absorption by the samples. All samples were cut from one chalk layer, with unified co-orientation within the rock, into standard cylinders of 100 mm in length and 53 mm in diameter. The density of all investigated samples was $2.16 \pm 0.01 \times 10^3$ kg m^{-3}. The ends of the samples were scrupulously polished to minimize heterogeneity of the stress field under compression (Jaeger and Cook 1979).

   The endcups had a diameter identical to the samples. To prevent the hydrostatic loading oil from penetrating into the sample pores, all samples were enclosed in plastic jackets (ALPHA FIT-221-3), and the contacts of their ends with the endcups were carefully closed. Each chalk sample was tested axially with a strain rate of 1×10^{-5} s^{-1} and laterally by a different hydrostatic oil pressure. The axial pressure varied from 37 MPa up to 65 MPa and the confining pressure from 0 to 15 MPa.
3. *β-quartz solid solution glass ceramic.* The composition of this material (VISION by Corning) in weight percent is $SiO_2 = 68.8$, $Al_2O_3 = 19.2$, $Li_2O = 2.7$, $MgO = 1.8$, $ZnO = 1.0$, $TiO_2 = 2.7$, $ZnO_2 = 1.8$, plus additional traces (Beall 1989). The glass-ceramic is close to monophase crystals of β-quartz solid solution about 600 Å (50 nm) in diameter and a small amount of zirconium titanate crystals (<100 Å in diameter) which allow efficient light transmission (Beall 1989). The resistance to failure under uniaxial compression of samples, which comply with the standard specification dimensions (106 mm length and 52 mm in diameter), is very high (greater than 450 MPa) and was beyond the load capacities of our press machine. Since our purpose was to overcome the resistance to fracture and to induce controlled cracking (rather than to obtain accurate strength results), we selected a sample geometry that deviates from standard specifications. The best results were obtained for a sample with the following specifications: Length 104 mm and having a frustum shape (the part of a cone left after cutting off a top portion by a plane parallel to the base) and an elliptical cross section, such that the large and small diameters at the base end

are 33 mm and 21 mm, respectively, and the two diameters at the top end of the frustum are respectively 33 mm and 10 mm. The abraded flexural strength of the β-quartz solid solution glass-ceramic is 69.3 MPa (Beall et al. 1967). The axial pressures varied from 0 up to 112 MPa (there was no confining pressure in this case).

4. *Glass.* Cylindrically shaped soda-lime glass samples 10 cm in length and 3 cm in diameter, which were uniaxially loaded. Soda-lime glass is homogeneous and transparent. Its structure consists of glassy matrix consisting of silicon, oxygen and calcium atoms, which are strongly bound, with weakly bound sodium ions. The density of all investigated glass samples was $2.6 \pm 0.01 \times 10^3$ kg m^{-3}.

5.3.3 Amplitude Investigation

The amplitude of the envelope of the EMR pulse (Eq. 5.4) is determined by the following relation:

$$\overline{A} = A_0\left(1 - \exp\left(-\frac{T'}{\tau}\right)\right) \tag{5.9}$$

As we noted above (Eq. 5.7), the frequency of the EMR pulses is inversely proportional to the crack width. EMR pulse amplitude increases as long as new atomic bonds are severed (Rabinovitch et al. 1998, 1999b); hence, this amplitude is related to the crack area, and it is expected that $\overline{A} \propto 1/\omega$ (Rabinovitch et al. 1998, 1999b).

The voltages of the EMR pulses depend on the antenna reaction (antenna efficiency), which changes with frequency. Compensating for this factor, $E = h(\overline{A})$ (E being the field amplitude reaching the antenna), by the appropriate antenna efficiency chart (h-EHFP-

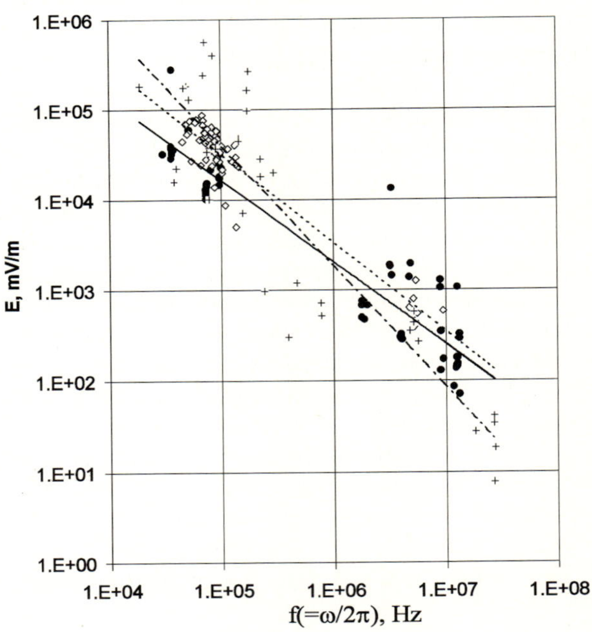

Fig. 5.5. Amplitude-frequency relation of EMR pulses associated with chalk, granite and glass-ceramic compression: Chalk (*filled circles*); granite (*diamonds*); glass-ceramic (*plus signs*); chalk trend line (*unbroken line*); granite trend line (*dashed line*); glass-ceramic trend line (*dashed dotted line*) (Rabinovitch et al. 1999b)

30 Near Field Probe set, Electro-Metrics Penril corporation), we were able to compare the heights of the various EMR signals that have different frequencies.

Figure 5.5 (Rabinovitch et al. 1999b) shows the compensated amplitudes (E) of the electromagnetic field signals induced by the fracture of chalk, granite and glass ceramic samples. Analysis of about 160 pulses shows that the amplitude-frequency ratio of each of the three materials can be fitted by a power-law type relation. Thus, for chalk, the relation is $E \cong 6 \times 10^8 f^{-0.91 \pm 0.04}$ (squared regression coefficient $R^2 = 0.87$), for granite $E \cong 3 \times 10^9 f^{-0.99 \pm 0.04}$ ($R^2 = 0.89$), and for the glass ceramic $E \cong 2 \times 10^{11} f^{-1.32 \pm 0.11}$ ($R^2 = 0.82$).

The exponent of these three relationships is very close to −1, although for the glass ceramic the value is somewhat scattered. EMR pulses induced by chalk and granite fracture could generally be fitted by a single frequency. In contrast, some EMR pulses associated with glass ceramic fracture were "multi-frequent". Therefore the amplitude-frequency ratios of these pulses were corrected in the following way. The contribution of a specific frequency peak of the "fast Fourier transformation" (FFT) to the compensated amplitude was calculated by

$$E(f_i) = \frac{EA_i}{\sum_i A_i}$$

where A_i is the area under the ith peak and E is the measured pulse amplitude. Such a procedure evidently adds errors; it increases the spread of glass ceramic results and could also be the reason for the deviant exponent. If we collect all EMR data on one figure (Fig. 5.6, Rabinovitch et al. 1999b), a single combined relation can be obtained: $E \cong 5 \times 10^9 f^{-1.06 \pm 0.04}$ ($R^2 = 0.84$).

These results imply that the amplitude of the EMR field is indeed inversely proportional to the signal frequency and is hence proportional to the crack width (Rabino-

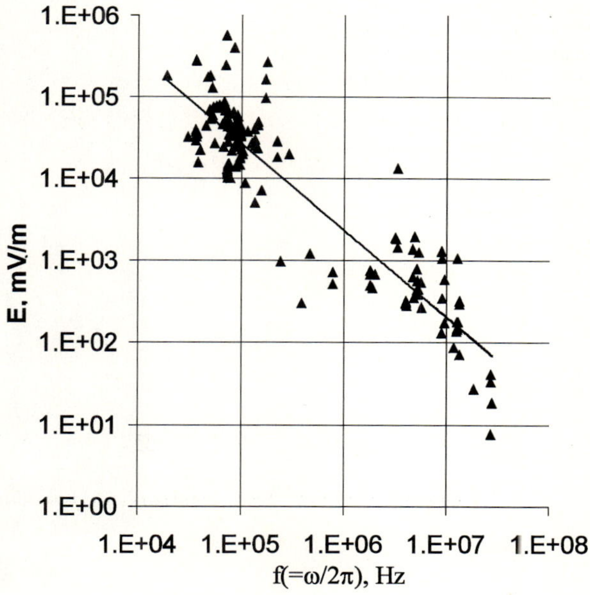

Fig. 5.6. General amplitude-frequency relation of chalk, granite and glass-ceramic samples (Rabinovitch et al. 1999b)

vitch et al. 1998). The final measured amplitude obviously depends also on other parameters such as fracture length, geometry of antenna-fracture orientation, attenuation along the path, etc. These factors constitute the source of the (rather large) vertical spread in the results of Fig. 5.5 and 5.6.

A controversy exists in the literature as to the magnitude of EMR from tensile and shear fracture. While Yamada et al. (1989) maintain that tensile cracking excites more intensive EMR than shear cracking, Petrenko (1993) states that the EMR amplitude is rather much greater for the shear fracture. Here we consider the influence of tensile vs. shear fracture modes on EMR. Since in our model (Sect. 5.2) no distinction is drawn between shear and tensile fracture modes, it is expected that magnitudes of EMR from both would be comparable.

We selected chalk for our study due to its micro-texture and low strength leading to a relatively small number of fragments (between 1 to 18) at failure, which can be analyzed fractographically (Bahat 1991a; Bahat et al. 2001a).

As expected from our previous studies (Bahat 1988c), the fractured surfaces exhibited a useful fractography. Tensile fractures (Fig. 5.7a) were planar and intensely marked by "striae" (plumes) (Bahat 1979a). These surfaces reflected a light grayish color. The shear fractures, on the other hand, had rough morphologies that were marked by white coarse "sugary" textures (Fig. 5.7b). The discrimination between these very distinct surfaces provided us with a fractographic criterion to distinguish between the two modes of

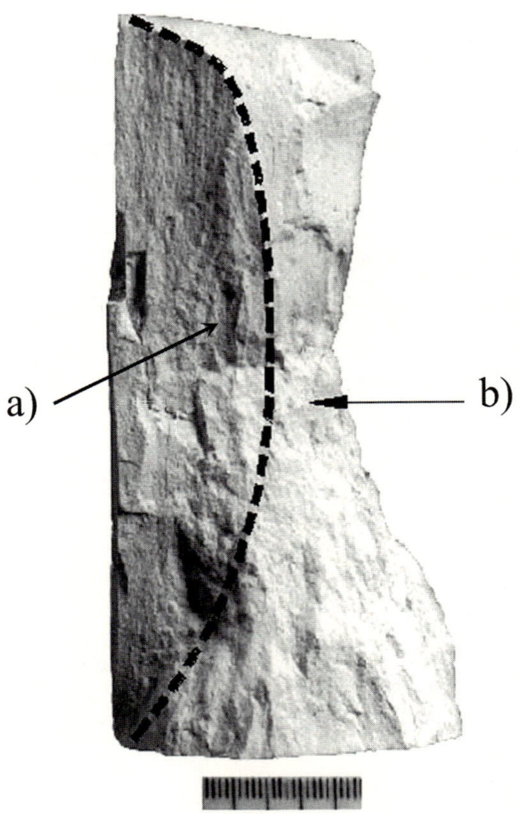

Fig. 5.7. A photograph of a fractured surface on which the *dashed line* marks the boundary between the tensile part (**a**, showing planarity, gray tint and strong striae) and the shear part (**b**, showing coarse topography, white color and sugary texture). The tensile part (*'a' arrow*) is still covered by a piece of the shear part at the center of the sample. *Scale* in cm (Frid et al. 2000)

fracture and helped us to develop a method for the determination of tensile vs. shear fracture zones.

The area of each fracture was manually measured as the sum of all crack surfaces obtained for each sample for each type (Fig. 5.7, Frid et al. 2000). Measurement errors varied from 3 to 8%. Twenty-three chalk samples were investigated under different loads. The first eleven samples were loaded uniaxially, while the twelve following ones were loaded triaxially. During the triaxial test, the lateral load was changed from 1 to 15 MPa, which caused a monotonous rise of sample strength from 37.7 MPa up to 80.1 MPa. Young moduli for all tested samples ranged from 6.2 up to 11.3 GPa, while the Poisson ratio changed from 0.07 up to 0.37. Since brittle-ductile transformation after the Coloumb-Mohr region diminishes EMR activity (Frid et al. 1999), this region was not treated. Tensile fracture was the preferable mode of failure during uniaxial compression, while the shear mode became significant only during triaxial loading: The "(shear area) / (tension area) ratio" changed from 0.5–0.8 during uniaxial compression to 2–35 during triaxial loading (Bahat et al. 2001a).

We compared the total amplitude (sum of compensated amplitudes) of the EMR pulses from each sample (measured above the sample's elastic limit) with the total respective areas (sums of crack areas of a given sample) of tensile and of shear cracks (Rabinovitch et al. 1999b). Figure 5.8 (Frid et al. 2000) shows a 3D graph of total compensated EMR pulse amplitudes vs. total areas of tensile and shear cracks of all investigated samples. The total EMR amplitude E was fitted by the linear equation: $E = -33.39 + 0.00655\,S_t + 0.00596\,S_s$ (with a squared regression coefficient $R^2 = 0.86$), where S_t and S_s are the total areas of tensile and of shear cracks of a given sample, respectively. The difference between the two coefficients multiplying S_t and S_s is of the order of ±5%, which is rather small and might be attributed to the error in area measurements (3–8%). This result shows that it is only the size of the *entire* area of the crack, irrespective of its mode, which governs the amplitude of the EMR.

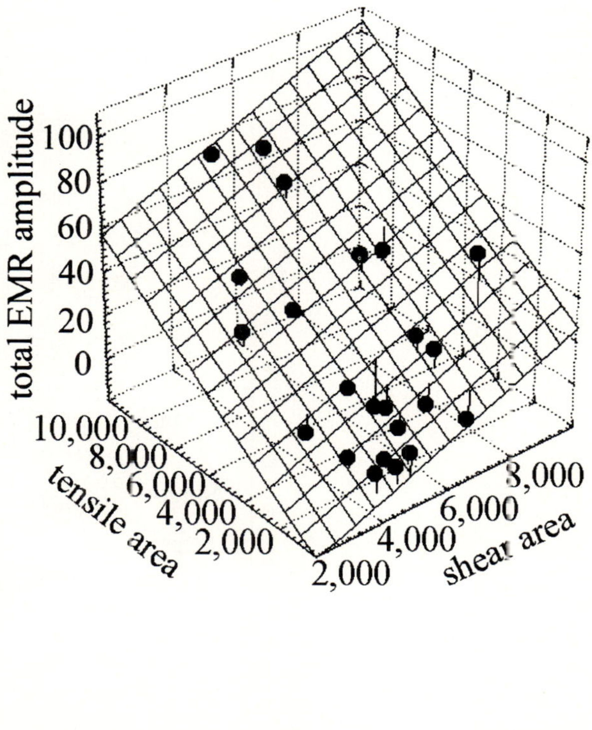

Fig. 5.8. A 3D experimental graph of total compensated EMR pulse amplitudes (V m^{-1}) vs. total areas (mm^2) of tensile and shear crack areas of all investigated samples and its fit (Frid et al. 2000)

5.3.4
Investigation of the Time up to Envelope Maximum and its Relation with Frequency

To check the validity of Eq. 5.2, two experimental tests were carried out:

1. Mutual measurement of T' and crack lengths (Bahat et al. 2002a) in transparent glass ceramics under compression. A glass ceramic sample was gradually loaded uniaxially. Loading was stopped as soon as EMR pulses were recorded. The sample was removed for a check and then the procedure started afresh. The procedure was repeated five times (five stages-Fig. 5.9). Characteristic pulses of different modes observed during the entire process are presented in Fig. 5.10 and Table 5.1 (Bahat et al. 2002a).

   Figure 5.11 shows an example of an EMR pulse observed during the collapse of a glass ceramics sample with its least square fit to Eq. 5.1. As can be seen, both the accuracy of measurement and that of the parametrization procedure are adequate ($R^2 = 0.96$).

   The results of parametrization of this EMR pulse are marked by arrows in Fig. 5.12, where we summarize our results that pertain to the correlations between T' and crack length. Note that T' values varied between 0.8 and 1.5 μs under low stresses (0.36–1.7 MPa) in association with microcracks of several millimeters in length. T' values of 5–15 μs were associated with stresses of up to 65 MPa and were correlated with cracks of up to several centimeters in length. A value of T' larger than 20 μs occurred under greater stresses (112 MPa) and was correlated with the longitudinal splitting at failure (~10 cm). The correlation between T' and crack length shows a positive trend (Fig. 5.12). A straight line can be drawn through the points representing the stages from 3 to 5 (Fig. 5.9). The dashed line indicates our assertion,

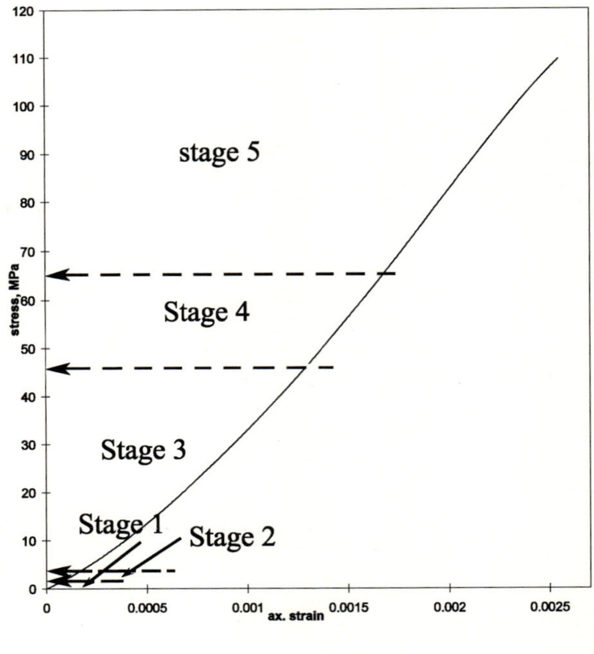

Fig. 5.9.
Stress vs. strain for the fifth stage of the incremental stress increase. *Dashed arrows* indicate stress levels achieved during stages 1–4 (Bahat et al. 2002a)

$T' = 8.15 \times 10^{-7} L$ ($v_{cr} \approx 1\,200$ m s^{-1}), where L is the crack length in mm and T' is in seconds. The R^2 coefficient obtained was 0.73. Considering the small number (five) of points, this coefficient is considered reasonable.

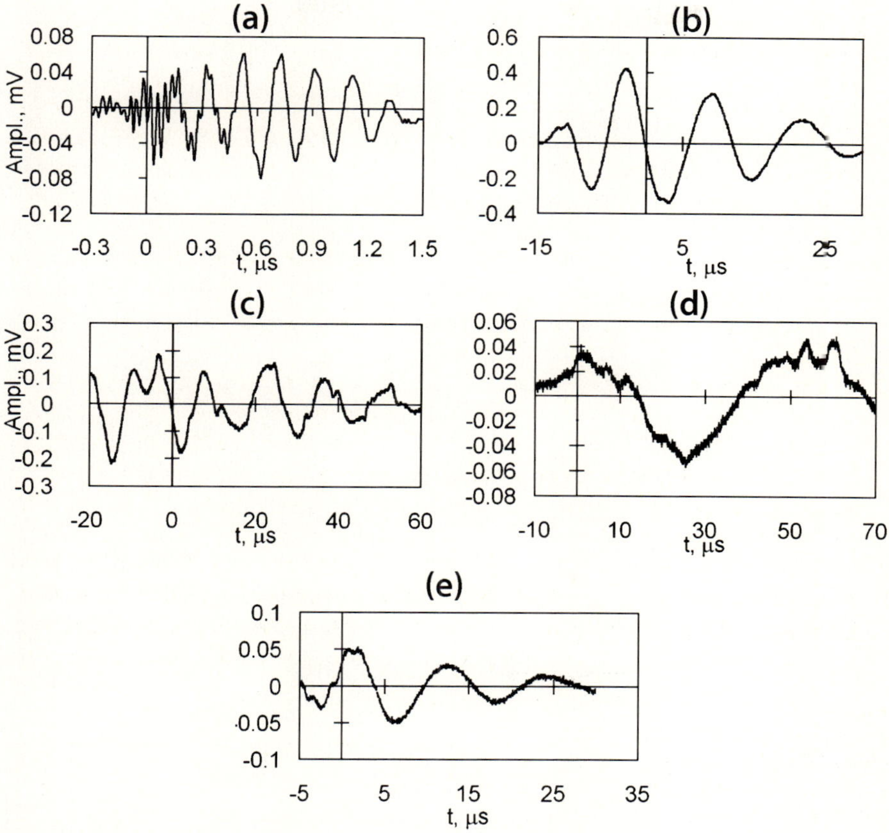

Fig. 5.10 Characteristic pulses of the different modes (Table 5.1). **a** A short pulse (stage 1, pulse 1). **b** A medium, individual pulse (stage 4, pulse 7). **c** A medium, uneven pulse (stage 4, pulse 10). **d** A lengthy pulse (stage 5, pulse 1). **e** A medium, individual pulse (stage 5, pulse 2) (Bahat et al. 2002a)

Fig. 5.11. An example of an EMR pulse of the fifth stage with its least square fit to Eq. 5.1 (*dashed line*). Results of fit: $T' = 42 \pm 0.3$ μs, $f = 18.7 \pm 0.012$ kHz, $R^2 = 0.96$ (Bahat et al. 2002a)

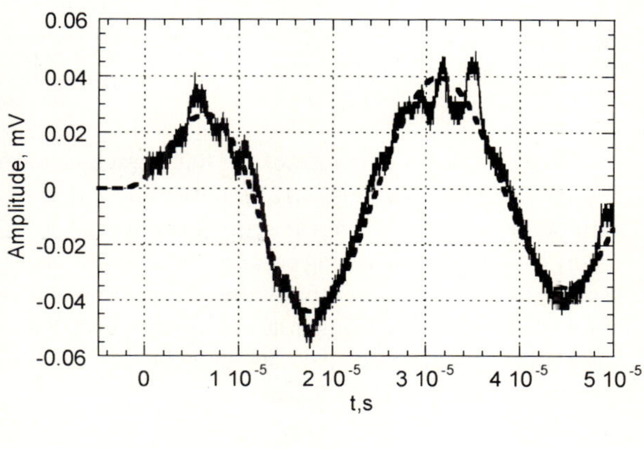

Table 5.1. EMR pulse parameters vs. stress and strain in fractured glass-ceramic

Stage	Stress (MPa)	Strain	Pulse number	Frequencies		T' (µs)	Pulse length	Crack size (mm)
1	0.36	0.00042	1	5.2, 27	MHz	0.95	Short	3.5 × 7.5
			2	5.7, 27	MHz	1.5	Short	
2	1.7	0.0011	1	1.2, 5.1, 18.4, 27.5	MHz	0.8	Short	2.0 × 7.0
			2	4.8, 19, 27	MHz	1.5	Short	3.5 × 10
3	47	0.0011	1	49.3, 156	kHz	25	Medium	2 × 5 × 19
4	0.9		1	46.7	kHz	15	Medium	10 × 10
			2	51.8	kHz	15	Medium	
	1.9		3	53, 85	kHz	25	Medium	
	3		4	41, 70, 122	kHz	17	Medium	
	5		5	72	kHz	20	Medium	
	12.5		6	100, 177	kHz	20	Medium	
			7	172	kHz	20	Medium	
	65	0.01850	8	172, 292, 468, 776	kHz	15	Medium	
			9	144, 225	kHz	15	Medium	
			10	75, 144, 225, 780	kHz	15	Medium	
5	112	0.025	1	18.7, 37.9	kHz	42	Lengthy	20 × 100
	<112		2	90	kHz	17	Medium	2.5 × 12
			3	40, 78, 242, 391	kHz	20	Medium	7 × 29

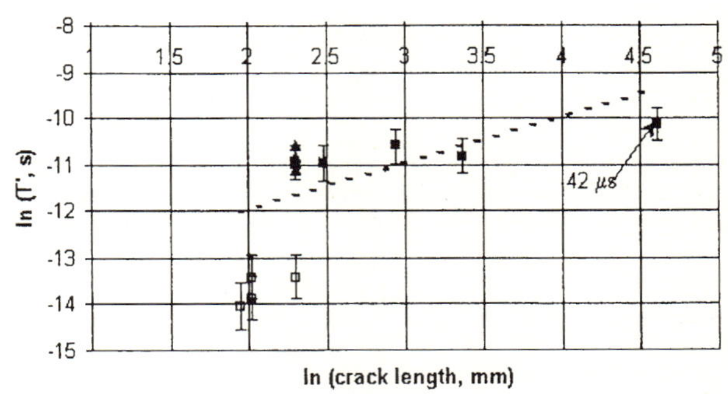

Fig. 5.12.
Correlation between T' and crack length. *Open squares* show EMR-fracture data during the first two stages; *closed squares*: Stages 3–5; *closed triangles* show spread of EMR data during the forth stage. *Error bars* here were estimated by the spread of the EMR data of the forth stage of loading to be ~3.5% for $\ln T'$ (Bahat et al. 2002a)

Note that the dimensions of the flaky cracks (stages 1 and 2, Fig. 5.9) calculated by EMR parameters (Fig. 5.12) are much less than the measured ones (Table 5.1) and hence deviated from the regression lines in Fig. 5.12. We assume that the initially created cracks during compression were in fact of small dimensions (as measured by the EMR). Eventually these fractures caused flaking off the external surface of the sample of much larger dimensions. The reason that the initial cracks were optically undetected stems from the fact that their apertures, were apparently below the wavelength of light.

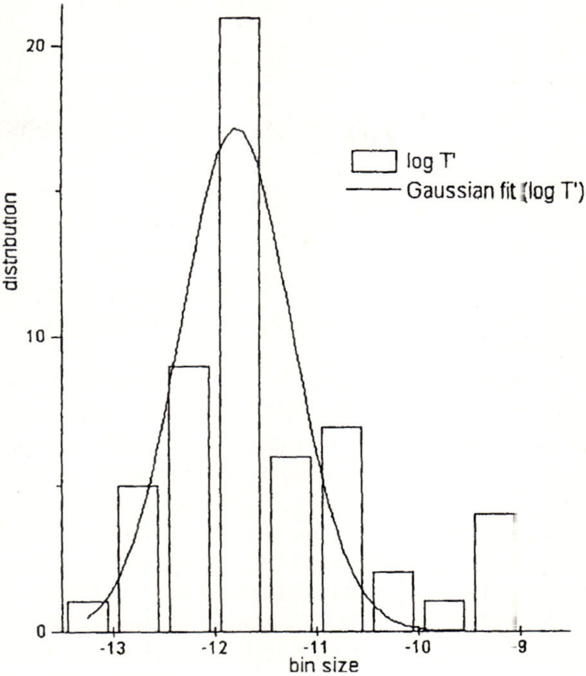

Fig. 5.13.
Histogram of log T' (where T' is the time interval to reach the pulse maximum) and its Gaussian fit Rabinovitch et al. 1998)

2. T' values distribution (Rabinovitch et al. 1998). If we assume that the fracture process develops incrementally and that each new increment is proportional to the existing crack length (as previously observed by Gillespie et al. (1992) and by Cowie and Scholtz (1992a,b), then a log-normal distribution should be expected (Aitchison and Brown 1976). Results show that T' indeed obeys a log-normal distribution (Rabinovitch et al. 1998) (Fig. 5.13).

   Another problem is the possibility of a scaling relation between crack length and width. Walmann et al. (1996) showed that the aperture, u, and the length ℓ of the crack were related by a fractal relation, $u \sim \ell^{\beta}$, where β ranges between 0.5 and 1.

   Although the relation was observed for u and not for crack width, we tried to correlate between T' and ω: $\omega \sim (T')^{-\nu}$, which would have implied a scaling between crack length and width. Results, however, show that the obtained ν is of the order of zero, i.e., crack length is independent of its width, implying that the width is constrained by an auxiliary mechanism, such as grain boundary, intergranular spacing, etc. A similar constraint was already noted (Bahat 1988c) in geological fractures in chalks, where in a given layer of a constant width, there is a wide range of fracture lengths.

5.3.5
Frequency Investigation

Equation 5.7 reveals that for the same crack width, the EMR frequency for stiffer materials would be higher than that for the weaker materials. Indeed, since the Rayleigh wave velocity in a material with a Young modulus E, Poisson ratio μ and density ρ, is given by

$$v_R = \frac{0.87 + 1.12\mu}{1+\mu}\sqrt{\frac{E}{2\rho(1+\mu)}}$$

the ratio of EMR frequencies for two materials is given by

$$\frac{\omega_1}{\omega_2} = \frac{(0.87+1.12\mu_1)(1+\mu_2)}{(0.87+1.12\mu_2)(1+\mu_1)}\sqrt{\frac{E_1}{E_2}\frac{\rho_2(1+\mu_2)}{\rho_1(1+\mu_1)}}\frac{b_2}{b_1} \tag{5.7'}$$

To check the validity of Eq. 5.7, we compared EMR frequencies and fracture widths for four different materials, the elastic properties of which are shown in Table 5.2 (Frid et al. 2003).

Only during glass uniaxial compression experiments was an exact one-to-one correspondence between the EMR signals and the longitudinal splits causing them obtained. Even during the transparent glass-ceramic experiments not all pairs of EMR signal-fracture (S-F) were perfectly related. In our analysis, we included only those pairs of S-F that could be accurately "linked". During chalk and granite (that are non-transparent materials) experiments, a one-to-one link between EMR signals and fractures was not possible. Moreover, fracture widths and lengths were uncertain since the fracture directions were usually undefined. Hence we included in our analysis only "maximal" fractures, which are, of course, easily discerned, and of those only the ones that had a "stretched shape", where one dimension (considered as length) was significantly larger than the other (considered to be the fracture width). Figure 5.14 (Frid et al. 2003) shows the relation of $\pi v_R / \omega$ (where v_R is calculated by the relation leading to Eq. 5.7') vs. crack widths, b. The slope of this relation is 0.93 ($R^2 = 0.82$), being very close to 1, in agreement with the theoretical prediction (Eq. 5.7 and 5.7').

Table 5.2. Experimental results of material properties

	Young modulus E (GPa)	Poisson ratio μ	Density ρ (kg m^{-3})	Rayleigh wave velocity v_R (m s^{-1})
Glass	45	0.2	2600	2448
Glass ceramics	51.6	0.25	2370	2715
Chalk	8.5	0.16	2604	2377
Granite	44.8	0.32	2160	1178

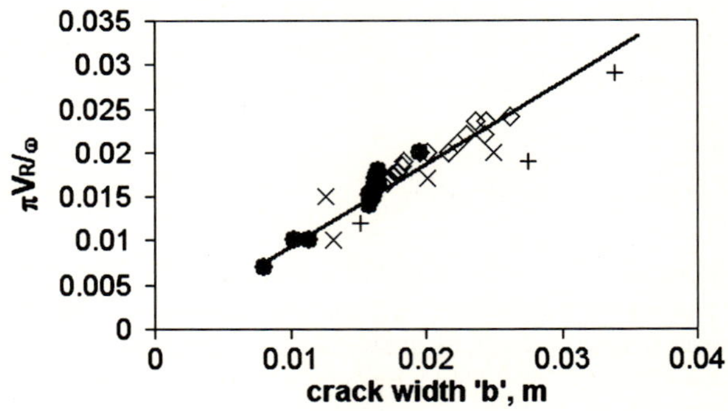

Fig. 5.14. The relation between reciprocal EMR frequency and fracture width for chalk, granite, glass and glass ceramics under compression (• chalk, ◊ granite, + glass ceramic, × glass) (Frid et al. 2003)

5.3.6
T'/ω Investigation (Eq. 5.8)

The T'/ω parameter is directly calculable from the parametrized data, and hence is independent of different auxiliary factors, which can be sources of large fluctuations (errors) if we would like to relate fracture areas to A_0 (Eqs. 5.1 and 5.9).

5.3.6.1
Experimental Procedure and Results

Twenty-four chalk samples were investigated in the conventional compression test (Rabinovitch et al. 2000c). Nine samples were loaded uniaxially and fifteen samples were loaded triaxially. Twenty-one of the twenty-four samples showed a similar EMR behavior, while three (not treated here) behaved differently. Table 5.3 presents rock-mechanics properties of all investigated samples. Figure 5.15 shows a representative normalized stress-strain curve (For the sake of comparison, stress values were normalized by the peak stress value, while the axial strain values were normalized by the axial strain at peak stress) of a chalk sample together with its T'/ω data. The latter was calculated from the EMR data by a least square method as shown in the caption of Fig. 5.16. T'/ω data are shown in Fig. 5.15a in a semi-logarithmic scale and in Fig. 5.15b in a normal scale. As is seen from Fig. 5.15, an increase of stress (strain) is generally followed by an increase of the T'/ω factor, indicating an increase of the area of newly formed cracks. The range of

Table 5.3. Rock-mechanics parameters of investigated chalk samples

Sample number	Confining pressure (MPa)	Strength (MPa)	Differential stress (MPa)	Poisson ratio	Young modulus (GPa)	Shear modulus (GPa)
1	0.00	47.60	47.60	0.21	8.50	3.51
2	0.00	45.20	45.20	0.16	9.00	3.88
3	0.00	50.40	50.40	0.07	9.30	4.35
4	0.00	51.50	51.50	0.15	9.20	4.00
5	0.00	46.20	46.20	0.19	9.70	4.08
6	0.00	41.90	41.90	0.23	7.30	2.97
7	0.00	37.70	37.70	0.14	6.10	2.68
8	1.00	47.60	46.60	0.20	9.30	3.88
9	1.00	39.70	38.70	0.19	7.60	3.19
10	5.00	56.20	51.20	0.15	9.30	4.04
11	10.00	54.20	44.20	0.08	7.70	3.56
12	15.00	65.10	50.10	0.14	9.40	4.12
13	3.00	48.40	45.40	0.20	8.90	3.71
14	0.80	46.00	45.20	0.16	8.70	3.75
15	0.85	47.50	46.65	0.14	9.10	3.99
16	2.00	48.80	46.80	0.13	8.80	3.89
17	2.00	45.80	43.80	0.20	7.70	3.21
18	2.00	47.90	45.90	0.11	8.60	3.87
19	2.00	48.70	46.70	0.19	9.10	3.82
20	5.00	43.50	38.50	0.18	7.80	3.31
21	5.00	44.00	39.00	0.18	7.80	3.31

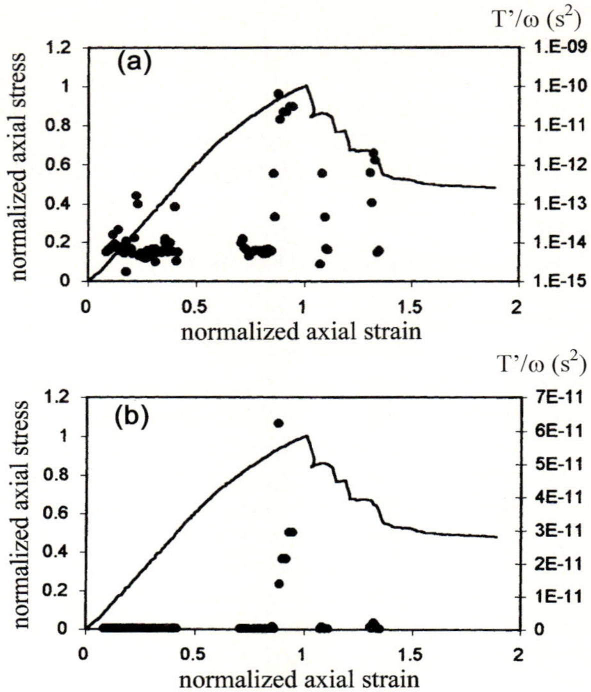

Fig. 5.15. A representative normalized stress-strain curve of a chalk sample together with its T'/ω data (T'/ω data are shown in a semi-logarithmic scale (**a**), and in a normal scale (**b**)) (Rabinovitch et al. 2000c)

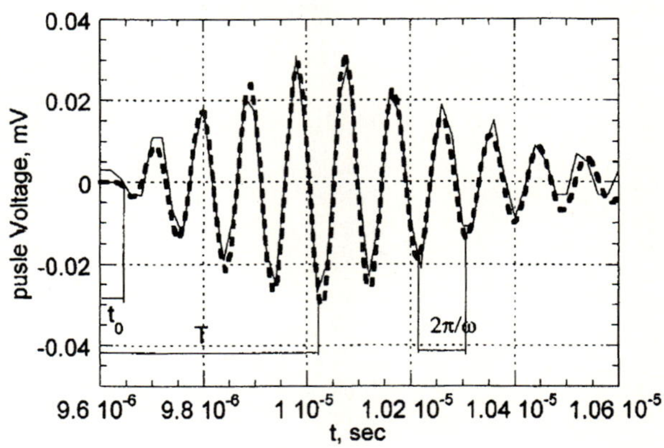

Fig. 5.16. An experimental EMR pulse and its numerical fit. For this pulse, t_0 is 9.63 µs, T is 10.07 µs. Hence, $T' = T - t_0$ is 0.44 µs. The frequency ω is 6.76×10^7 s^{-1}. Hence, T'/ω is 6.51×10^{-15} (Rabinovitch et al. 2000c)

the value of T'/ω stretches from 10^{-15} to 10^{-10} (s^2) (5 orders of magnitude!). The minimum values of T'/ω were observed at the beginning of the stress-strain curve. Later, the elastic and elastic-plastic zones before the peak of the stress-strain curve, T'/ω values of the order of 10^{-14}–10^{-11} (s^2) were measured, and the T'/ω value jumped to a maximum (of the order of 10^{-10}) in the zone immediately preceding the peak stress. This jump is indicative of the collapse failure. After the peak stress, T'/ω amplitudes (and the corresponding crack areas) decrease back to values of 10^{-14}–10^{-11} (s^2) ("relaxation" region). A similar behavior of T'/ω vs. stress-strain (namely, a gradual increase of T'/ω with stress increase followed by an abrupt jump) was observed during uniaxial compression of glass ceramics (Bahat et al. 2002a), suggesting that such behavior could be of general validity (see below). These

Fig. 5.17.
Fractography: Various fracture types and sizes are shown on specimen No. 1 (Table 5.3). Two large fractures (~4 × 4.5 cm and ~5 × 1.5 cm) are separated by a ridge at the lower part of the picture, whereas the upper part of the picture reveals length variations from 0.1 mm (magnified), through 2–4 mm up to 1–4 cm (cm *scale* at top of picture) (Rabinovitch et al. 2000c, Bahat et al. 2001a)

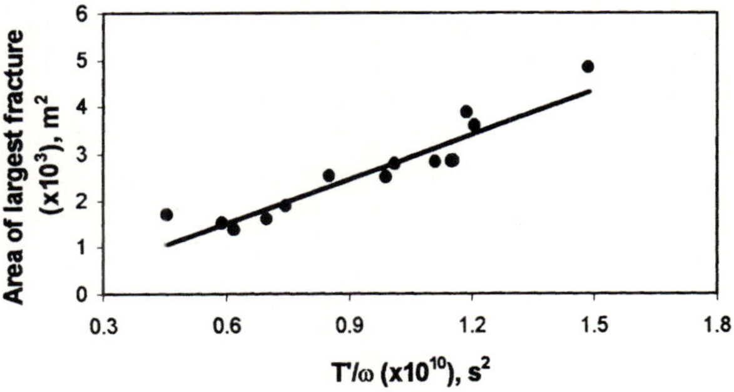

Fig. 5.18.
Experimental largest T'/ω values vs. maximal fracture areas and a linear fit (squared regression coefficient $R^2 = 0.9$) (Bahat et al. 2001a)

results imply that following the early pore-closure zone of the stress-strain curve, microcracking and possibly coalescence occurred, while collapse happened just before peak stress and was followed by smaller relaxation cracking in the post-peak zone.

Fractographic examination of the failed chalk samples revealed fractures ranging in size from 0.1 × 0.1 mm up to about 40 × 50 mm. The failed surface (Fig. 5.17) consists of composites of cracks that vary in size from several mm to several cm. This range of crack areas (10^{-2} mm² to ~2 000 mm²) is in accordance with the variation by 5 orders of magnitude of the T'/ω values mentioned above.

Since a one-to-one correspondence between the EMR pulses and the causative fractures is not possible in the chalk measurement, only the cracks possessing the largest areas were measured and correlated with the largest T'/ω values (Fig. 5.18, and see also Bahat et al. 2001a). As can be seen, very good agreement exists between Eq. 5.8 and the experimental results both in the range of values and in the correlation of the largest values ($R^2 = 0.9$). In a recent experiment on transparent glass ceramics (Bahat et al. 2002a) and glass (Frid et al. 2003), even a one-to-one correlation was shown to exist between T'/ω and crack area.

The order of magnitude of the velocities in Eq. 5.8 can also be estimated from these measurements. Thus $v \approx [S/(\pi T'/\omega)]^{1/2}$ and, e.g., for the 0.1×0.1 mm case, we have $T'/\omega = 10^{-15}$ s^2 whereupon $v \approx 1.4 \times 10^3$ ms^{-1}. This is of the correct order of magnitude (since the average Rayleigh velocity calculated for chalk (Bahat et al. 2002a), at the known values of its dry density, Young modulus, and Poisson ratio is 1.2×10^3 ms^{-1}).

5.3.7 Decay Time Investigation (Rabinovitch et al. 2003a)

5.3.7.1 *Experimental Results and Analysis*

We have analyzed EMR pulses of chalk, granite, glass and glass ceramics by fitting them to Eq. 5.1 and deriving their parameters (in particular τ and ω). Figure 5.19a shows the dependence of τ on ω for glass, glass ceramics and granite. The slopes on a logarithmic scale are within -1 ± 0.1, with $R^2 = 0.96$, 0.85 and 0.91, respectively, agreeing with the theoretical prediction (Eq. 5.6). For chalk, however, an "effective" slope of -0.7 is obtained (Fig. 5.19b). It cannot be attributed to the low accuracy of the fitting parameters: The error for τ ranges between 2% and (seldom) 30%, which is quite small in a logarithmic scale, while the error for ω does not exceed a few percent, and is usually only a fraction of a percent. We would like to attribute this "change of slope" to the spread in temperatures of the small cracks there (see below).

The distribution of $(\tau\omega)^{-1/4}$, which according to Eq. 5.6 should be proportional to the absolute temperature for small and large crack widths in granite is shown in Fig. 5.20a and b, respectively. We denote $(\tau\omega)^{-1/4}$ values henceforth as "temperatures". The mean values of the crack "temperatures" in relative units are respectively 0.67 and 0.65 for small and large cracks (in granite), which are considered to be the same within the accuracy for "temperature" values (10%). Similar distributions for chalk are shown in Fig. 5.21a and b. Here, however, the mean values are different, being 0.47 for small crack widths and 0.67 for large ones. As mentioned, the error for ω is not more than a few percent, mostly a fraction of a percent, while for τ it ranges between 2% and 30%. Thus the error of $(\tau\omega)^{-1/4}$ may be estimated as ~10%, yielding "temperatures" of 0.47 ± 0.05 for small cracks and 0.67 ± 0.07 for large cracks.

Figure 5.22 shows the distribution (histogram) of $\log(1/\omega)$ (where $1/\omega$ should be proportional to crack width, Eq. 5.7, for both small and large cracks). We denote by small (large) cracks those cracks that have ω values larger (smaller) than 2×10^7 s^{-1} (e.g., Fig. 5.19). If one assumes the Rayleigh velocity in chalk to be (see above) 1 200 m s^{-1}, then the calculated small crack widths in chalk (by Eq. 5.7) range between 52.5 and 130 μm. For large cracks, the corresponding range is from 130 μm to 1.8 cm.

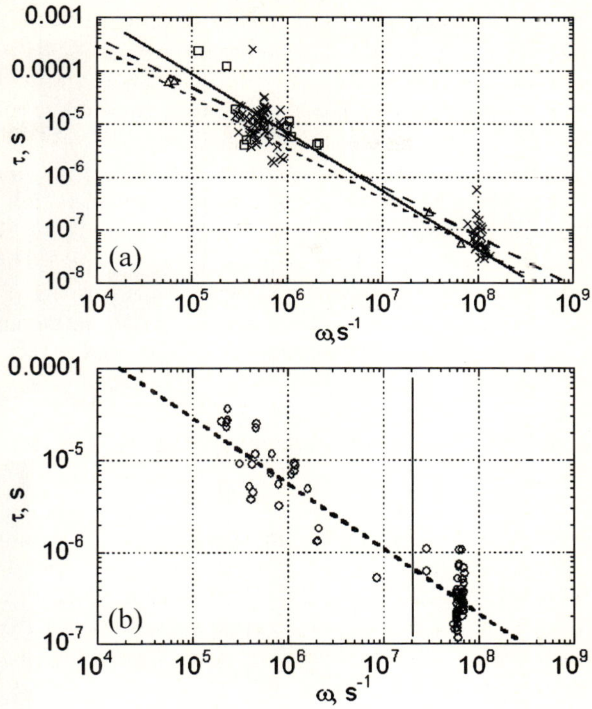

Fig. 5.19. Rise and fall time τ as a function of frequency ω: The *points* are the experimental results of individual pulses, and the *line* is the power law fitting result: **a** for glass (*triangles*), fitting $\tau = 1.87\omega^{-0.95}$ (*short-dashed line*); for glass ceramics (*squares*), fitting $\tau = 28.1\omega^{-1.1}$ (*solid line*); for granite (×), fitting $\tau = 2.23\omega^{-0.93}$ (*long-dashed line*); **b** for chalk (*circles*), fitting $\tau = 1.87\omega^{-0.7}$ (Rabinovitch et al. 2003a)

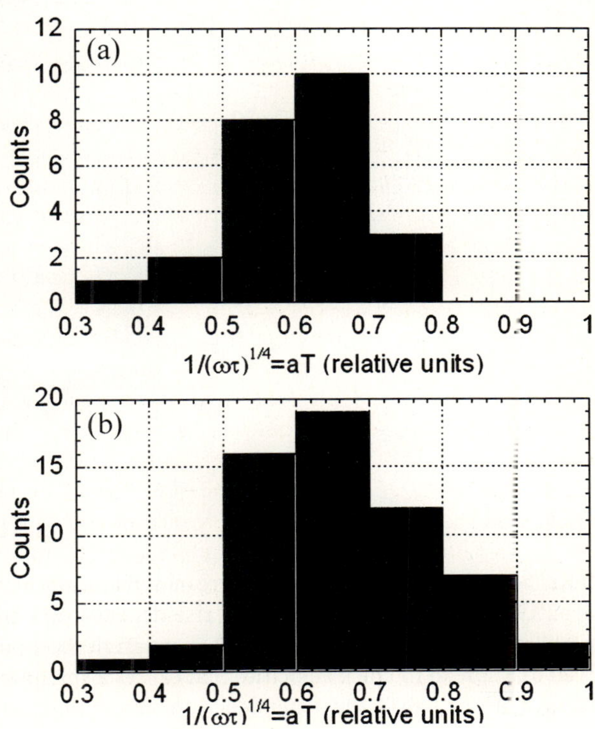

Fig. 5.20. Frequency counts of "temperatures" $(\omega\tau)^{-1/4}$ for cracks in granite: **a** short cracks; **b** long cracks (Rabinovitch et al. 2003a)

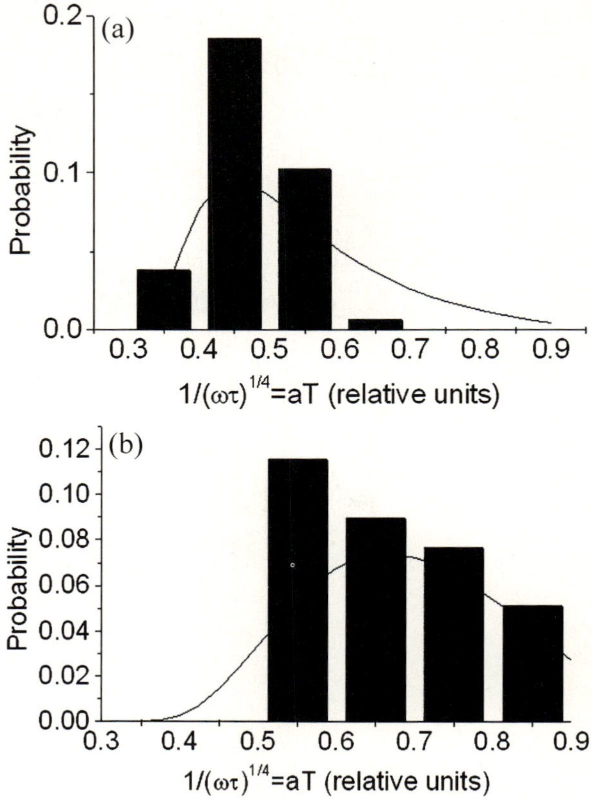

Fig. 5.21. Frequency counts of "temperatures" $(\omega\tau)^{-1/4}$ for cracks in chalk (*columns*) and calculated (*solid line*) (Rabinovitch et al. 2003a): **a** for short cracks; **b** for long cracks; crack aperture is assumed to behave as $\varepsilon \sim kx^{\alpha}$, where x is the crack half-width (Rabinovitch et al. 2003a)

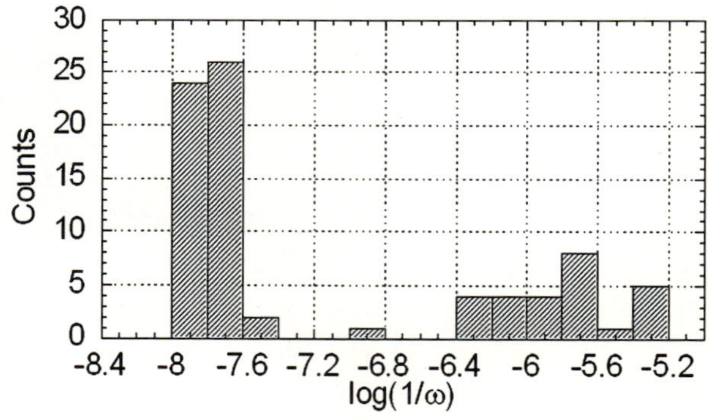

Fig. 5.22. Frequency counts for $\log(\omega^{-1})$ (ω^{-1} is proportional to crack widths, $b = \pi v_R / \omega$) in chalk fit (Rabinovitch et al. 2003a)

5.3.7.2
Discussion

It has been shown that the temperature of dynamically propagating cracks rises (Fuller et al. 1975). The actual temperature rise depends on crack velocity and on material properties (Yatomi 1981). The spread of experimental points in Fig. 5.19 can be attributed to a spread in crack velocities that causes a spread in temperatures. Since we assume that $1/\tau \sim \omega T^4$, the dependence on temperature is very strong. However, cracks in glass, in glass ceramics (which are not porous) and in granite (the porosity of which is about 5%) lead to almost uniform distribution of crack "temperatures" around their mean (Fig. 5.20, resulting in the almost exact $\omega\tau = \text{const.}$ relation of Fig. 5.19a). In chalk, however, $\omega\tau \neq \text{const.}$ The "temperature distribution" (histograms) for small and large cracks are given in Fig. 5.21a and b. It can be seen that "temperatures" of small cracks are significantly lower than those of large cracks.

We assume that this temperature difference in chalk is due to the interaction between the propagating cracks and the existing material pores. Chalk is the only material of the four analyzed that has large porosity (of about 40%). The pore radii in chalk range (Gueguen and Palciauskas 1994) between 0.05 and 100 μm. The results of our measurements (see Rabinovitch et al. 2003a), together with their exponential fitting, are given in Fig. 2.28b. Note that although the range of observed pore sizes was between 2 μm and 210 μm, their 3D fitting, being exponential, includes all pores, also those of lower sizes. The cracks in chalk encounter one or more pores during their development. Thus crack heat can be partially spent on raising the temperature of the air inside the pores; in addition, air from the pore can enter the crack and expand adiabatically; this process could also lower the latter's temperature. Although all cracks in chalk are cooled down in this way, the temperatures of small-sized cracks are more strongly influenced by these processes than those of large cracks, since the final temperature depends on the ratio between the pore's volume and the volume of the crack. The greater this ratio, the stronger is the temperature decrease. To show the self-consistency of our assumptions, we have theoretically derived the probability distribution of final temperatures (Rabinovitch et al. 2003a). The experimental and the calculated distributions of final "temperatures" for small and large cracks are shown in Fig. 5.21a and b. The agreement is adequate.

5.3.8 Summary of the EMR Model

It is seen that all experiments undertaken to check the validity of the model turned out to yield positive results. Thus,

1. The amplitude $\bar{A}$ was shown to be indeed proportional to $1/\omega$.
2. EMR amplitudes from shear fractures are similar (for similar areas) to those from tensile fractures.
3. T' is proportional to crack length ℓ.
4. The distribution of T' values follows a lognormal curve.
5. No correlation between length and width of cracks is obtained.
6. ω is proportional to $1/b$ and scales with Young modulus.
7. T'/ω is indeed proportional to crack area with the right proportionality constant.
8. τ is proportional $1/\omega$ and reasonable "temperature" distribution can be deduced for porous and non-porous materials.

All of these results support the contention that the model of EMR, although laid out only in a simplistic manner, is the right one. In the next section, we analyze the hitherto existing models of the origin of EMR and show that they all fail to stand experimental or self consistent tests.

5.4
Comparison of the New EMR Model with the Previous Ones

5.4.1
Dislocations and Charged Electrons

Misra (1977) and Misra and Ghosh (1980) suggested that "during non-uniform distribution of dislocations, which occur at the transition stages of elastic-plastic deformations under tension, namely, yield point, end of Luders' strain, and crack propagation, mobile dislocations arrange themselves into some configuration, that is mechanically stable". If, at the halting position of a dislocation, its energy is reduced, the dislocation can become self-trapped. "Conduction electrons (CEs) associated with such a dislocation would be stopped and trapped relative to the positive ions". This CE braking process would be similar to a "Bremsstrahlung" mechanism and would result in the emission of EMR. Misra (1977) further assumed that "during each transition stage (like a yield point and/or fracturing) there must be a re-adjustment of the CE distribution within the microscopic system, and the latter creates an oscillating Hertzian dipole. Thus, the EMR must appear only at transition deformation stages".

In a further analysis Molotskii (1980) pointed out that in the adjustment of the CEs to the slowly moving dislocations, their acquired energy would be of the order of 10^{-11} eV per mean free path, so that the maximum frequency of emitted EMR would be of the order of 10^3 Hz, i.e., significantly lower than the value predicted and measured by Misra. The acceleration of CEs by moving dislocations could not therefore be the cause of the EMR from metals, and Molotskii suggested therefore that the EMR was due to the increase of the total dislocation length and velocity, which occurred at transient deformation stages. Since these dislocations acted as electric dipoles, this mechanism would lead to an accelerated rise in the dipole moment of the material with an accompanying emission of EMR.

Both explanations however relating EMR to dislocation phenomena seem questionable. As is well known the motion of dislocations can be totally neglected in the cracking of brittle materials, and this mechanism therefore cannot explain EMR from glass and other brittle materials (e.g., most geological materials). A typical EMR pulse from the failure of a glass cylinder under uniaxial compression is shown in Fig. 5.23. The shape of EMR signals is found to be invariant for different brittle materials (glass, glass ceramics, granite, rhyolite, limestone, and chalk; Rabinovitch et al. 1995, 1996, 1998, 1999b, 2000a,b, 2002a; Frid et al. 1999, 2000; Bahat et al. 2002a), and even to be inde-

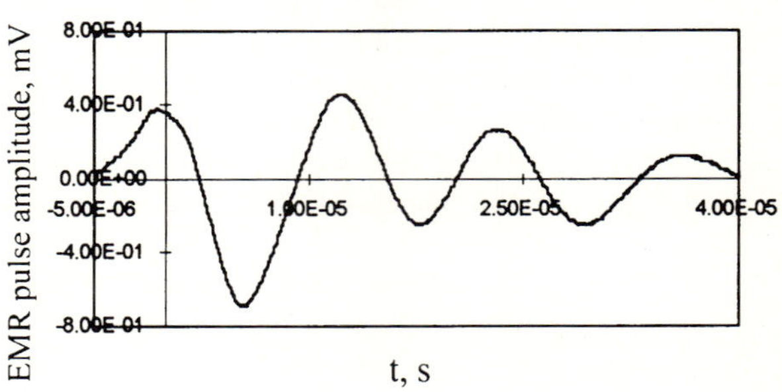

Fig. 5.23. A typical EMR pulse observed during glass cylinder failure under uniaxial compression (Frid et al. 2003)

pendent of the type of loading (compression, drilling and blasting (Goldbaum et al. 2001; Rabinovitch et al. 2002b) and mining (Vozoff and Frid 2001)).

The basic weakness of the "dislocation movement-hypothesis" was also pointed out by Jagasivamani and Iyer (1988), who showed experimentally that the EMR amplitude even increased with the brittleness of the investigated metals. Indeed, this latter result is in line with our measurements (Frid et al. 1999), which also showed that EMR activity increased with the brittleness of materials and decreased in the case of a transition from a brittle to a ductile behavior.

5.4.2
Discharge

Finkel et al. (1975) demonstrated that the splitting of alkali-halide crystals creates a mosaic of positive and negative charges, which appears on both sides of the fracture as it is formed. Since such charge separation can create an electrostatic field of the order of 10^7 V cm^{-1}, an electric discharge may occur, which was suggested as the origin of EMR (Finkel et al. 1975). However, Miroshnichencko and Kuksenko (1980) and Khatiashvili (1984) have already noted that the spectrum of discharge radiation is known to be of a "white noise" type and to be independent of the mechanical properties of the materials; the observed EMR behaves in an entirely different manner (Miroshnichencko and Kuksenko 1980; Rabinovitch et al. 1998, 1999b; Frid et al. 2000). Our results show that the EMR appears as individual pulses or as clusters of pulses caused by the different (individual or group) fractures (Rabinovitch et al. 1999b, 2000c; Frid et al. 2000). The properties of the pulse are influenced by the dimensions of the fracture (Rabinovitch et al. 1998, 1999b, 2000c; Frid et al. 2000) and by the elastic properties of the materials (Khatiashvili 1984; Frid et al. 1999). Moreover, the high resolution of our experimental system (Rabinovitch et al. 1998) allows us to obtain the exact shapes of the EMR pulses (Figs. 5.2, 5.10, 5.23, 5.24a). These definitely do not behave as "white noise" but rather exhibit a very distinct character (Eq. 5.1). These results confirm those observed by Miroshnichencko and Kuksenko (1980) and Khatiashvili (1984). Note in particular that the EMR spectrum (Fig. 5.24b) is highly localized around a single frequency.

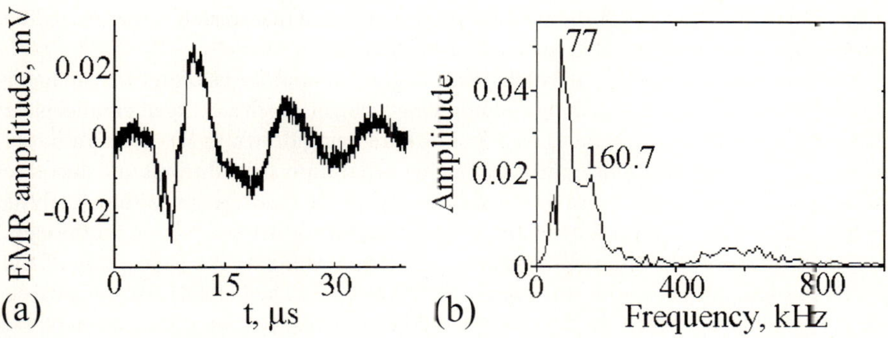

Fig. 5.24. An example of EMR signal **a** excited by fracture of granite in compression and its spectrum **b** (Frid et al. 2003)

5.4.3
Movement of Fracture Tips

EMR is observed (Gershenzon et al. 1985) during the propagation of cracks introduced by cleavage of LiF crystals with a knife. The developing crack (tip) propagates in the direction of indentation. According to their assumption, negative electrical charge moves with the crack tip while positive charge was assumed to accumulate at the region of the indenter-material contact. Dipole radiation should occur from these opposite charges, separated by the length of the crack, by the acceleration of the latter. Based on these assumptions, the power of the emanating EMR was calculated (Gershenzon et al. 1985), further assuming that the charging mechanism arises from moving charged dislocations.

As already pointed out, however, charge transfer by dislocations cannot constitute a general explanation of the origin of EMR. The "movement of fracture tips" explanation is questionable even for a dielectric like LiF, since the velocity of dislocations in alkali halide crystals (Tetelman and McEvly 1967) is *smaller* than the velocity of the crack, implying that charged dislocations cannot be the reason for crack tip charging.

An even stronger adverse argument to this model is that there is no clear reason for any "symmetry breaking", i.e., there is no known mechanism that would select a certain tip (side) to become negative while the other tip (side) becomes positive and not vice versa. In particular for the case discussed by Gershenzon et al. (1985), it is not clear why the crack tip should accumulate negative and not positive charge.

5.4.4
Movements of Fracture Sides (the "Capacitor" Model)

In order to determine whether movement of the crack sides during cracking could cause EMR with the appropriate characteristics, Miroshnichencko and Kuksenko (1980) used the acoustic wave emitted by a fracturing material to drive the plates of a specially built auxiliary-charged capacitor. The EMR from the capacitor was monitored by an antenna. Since the measured signal "shapes" were comparable to those obtained by the "directly" observed EMR, they concluded that the latter was caused by similar movements of the charged sides of the crack.

A decade later, O'Keefe and Thiel (1995) suggested another version of a capacitor model (for EMR from the cracking of compressed ice), in which a charged parallel plate capacitor is created whose plates (crack sides) are being drawn apart. After an initial charge is formed on the crack surfaces, further separation should result in a decrease of capacitance and an increase of the voltage across the crack. A space/time analysis of the emanating radiation can be carried out using the diffusion equation for the electric field in the crack. Although it proved possible to simulate the time decay of the EMR, no oscillatory behavior was predicted. O'Keefe and Thiel (1995) also considered the many routes by which the net charge could be created, i.e., pre-polarization of the material or applied physical gradients due to piezoelectricity or pseudo-piezoelectricity, temperature, deformation, impurity concentration gradient effects, etc. Petrenko

(1993) claimed that the electrification of crack sides could be caused by the surrounding non-homogeneous elastic strains, and Ogawa et al. (1985) assumed that crack side electrification was due to piezoelectrification and to contact (or separating) electrification.

However there are problems with this scenario.

1. In this model the EMR is assumed to be caused by an accelerated dipole created by charged crack sides. This assumption implies that the EMR can arise only from tensile cracks and not from shear ones. In contrast, our measurements of chalk under uniaxial and triaxial compression (Frid et al. 2000) showed that the EMR amplitude is independent of crack mode, tensile or shear and is related only to the entire area of the crack (see Sect. 5.3.3).
2. The capacitor model implies a correspondence between EMR and acoustic emission (AE) signal appearance. However, our own measurements and those of Yamada et al. (1989) show that although there are failure events for which AE and EMR signals are measured together (e.g., Rabinovitch et al. 1995), there do exist events for which AE is detected while EMR is absent and events for which EMR is observed while AE is completely missing. These results therefore disagree with the capacitor model. Note that the last discord between AE and EMR signal appearance is probably due to the difference of mechanisms leading to the two types of radiation and is not yet completely understood.

   In addition, the general no "symmetry breaking" argument mentioned above evidently applies here as well.
3. Even under the assumptions that the two crack sides could be charged in a charge mosaic manner (Finkel et al. 1975), thus retaining an overall charge neutrality and that the EMR was induced by the dipoles consisting of pairs of oppositely charged mosaic "elements" on the two crack sides, the EMR induced by the moving crack would be expected to be very weak due to the cancellation of the radiation originating from mutually oscillating oppositely charged dipoles (random distribution of the mosaic elements).

5.4.5 Summary of the New Model and its Properties

The suggested model of EMR origin can be described as follows: Following the breaking of bonds, the atoms on both sides of the severed bonds are moved to "non equilibrium" positions relative to their steady state ones, and oscillate around them. Lines of oscillating atoms move together, and by being connected to atoms around them (in the forward direction and also atoms on their side of the two surfaces newly created by the fracture), the latter also participate in the movement. The larger the number of cut bonds, the larger is the number of excited atoms, and hence the greater becomes the EMR amplitude. These oscillations behave like surface vibrational optical waves, where positive charges move together in a diametrically opposite phase to the negative ones and decay exponentially into the material like Rayleigh waves. The resulting oscillating electric dipole is the source of the EMR. The pulse amplitude decays by an interaction with bulk phonons.

Our model does not suffer from the defects of the former ones since

1. No dislocations are included in the model, and therefore it can be applied also to brittle and amorphic materials. EMR frequencies obtained experimentally do agree with those of the surface vibrational optical wave (SVOW) model.
2. The spectrum of the SVOW is definitely not of a white noise type. Moreover, it is in line with given in Eq. 5.1 and the measured ones (see, e.g., Fig. 5.24b). In particular the peak frequency depends on crack width according to Eq. 5.7.
3. Note that no symmetry breaking mechanism appears in the SVOW model.
4. Our model is applicable to shear, tensile and even mixed modes of fracture.
5. The exact correlation between AE and EMR predicted by the "capacitor" model is not always found experimentally. No such correlation arises in the SVOW model.
6. Charge neutrality is assured here via a different mechanism ("optical" oscillations). Therefore no cancellation is expected.

5.5 EMR Pulses Induced by Rock Fracture

5.5.1 Previous EMR Pulse Investigations

All previous investigations considered EMR activity (number of EMR pulses) or EMR amplitude (intensity). Several investigators tried to measure EMR signals in different materials: Metals, glasses, ionic crystals and different rocks.

Misra (1975) investigated EMR induced in samples of carbon steel, aluminum, brass, zinc, and lead under tensile loads. The scheme of Misra's experimental arrangement is shown in Fig. 5.25.

Samples were 9 mm in diameter and 160 mm in length. An EMR sensor (a co-axial semi-cylindrically curved foil 27 mm in internal diameter, 50 mm in length and 0.3 mm in thick-

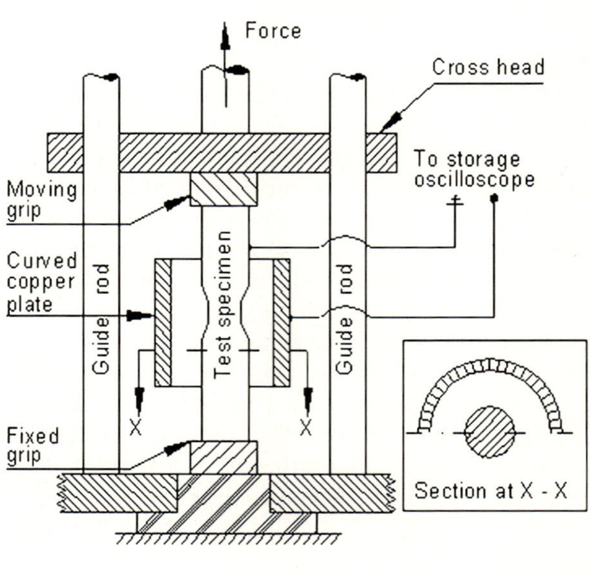

Fig. 5.25. Schematic diagram of the Misra (1975) experimental arrangement

ness) was placed near the fracture zone. EMR signals were passed through an amplifier operating at a frequency of 30 MHz (with a frequency band of 0.7 MHz) and were picked up by a storage oscilloscope. Figure 5.26 shows four examples of EMR signals observed during fracture of samples of carbon steel *(a)*, aluminium *(b)*, brass *(c)*, and zinc *(d)*. EMR signals from different materials were similar in shape but varied in their amplitude.

Jagasivamani and Iyer (1988) also noted that heat-treated high carbon spring steels subjected to tensile loadings emitted EMR during fracture. Figure 5.27 shows an example of EMR bursts measured by Jagasivamani and Iyer (1988).

It was found that the bulk of the EMR energy was in the range of 200 Hz to 10 kHz.

Nitsan (1977) investigated EMR signals induced by the fracture of quatrz-bearing rocks (several different granites, granodiorites, sandstones and quartzites) and single crystals of quartz and tourmaline. Materials were fractured under uniaxial loads between two parallel flat plates or by a small metal ball against a flat surface. The EMR signals were measured by a radio-frequency coil, the effective bandwidth of which was 1–10 MHz. Figure 5.28 shows an example of such EMR pulses observed by Nitsan

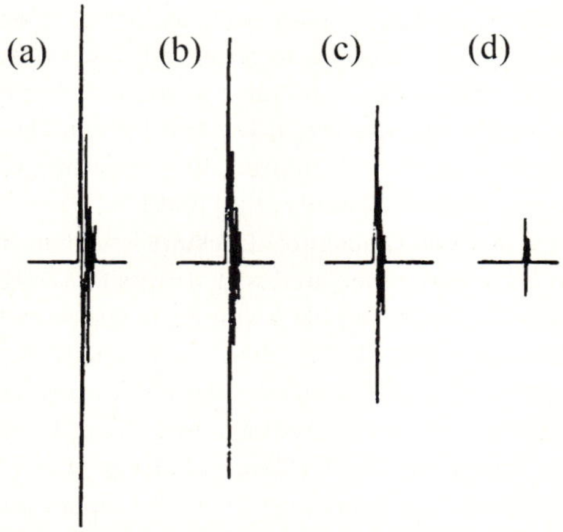

Fig. 5.26. Four examples of EMR signals observed during fracture of samples of **a** carbon steel, **b** aluminium, **c** brass and **d** zinc (Misra 1975)

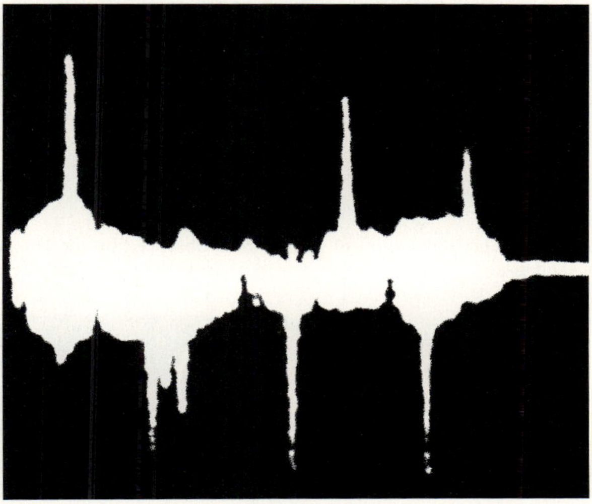

Fig. 5.27. An example of EMR signals measured by Jagasivamani and Iyer (1988)

(1977). He noted that small cracks induced in quartz and tourmaline crystals were followed by short radio frequency signals with an approximately exponential decay with a time constants of about 10^{-5} s for quartz (Fig. 5.28a) and about half of this time for tourmaline (Fig. 5.28b). Figure 5.28c shows a radio-frequency signal with a more complex shape. This signal was excited by the fracture of a sandstone sample.

Miroshnichencko and Kuksenko (1980) investigated EMR pulses irradiated by ionic crystals, glasses and different rocks during fracture. They detected EMR by an inductance coil of 100 kHz resonance frequency and low Q factor ($Q = 2$). An example of EMR signals induced in diabase is shown in Fig. 5.29.

This signal was induced by a crack of 1 mm in length. Assuming that the creation of the crack took time that was equal to the duration of the first "pulse" namely was 1 μs, Miroshnichencko and Kuksenko (1980) found that the average crack growth velocity was about 1 000 m s^{-1}, which was about one-third of the velocity of sound in this material.

Yamada et al. (1989) observed EMR and acoustic emission during the entire deformation of a rock using a standard uniaxial deformation tests at a constant strain rate of 10^{-6} s^{-1}on cylindrical cores of Indian granites 25 mm in diameter and 62.5 mm in length. In the first experiment, they used five acoustic sensors to detect acoustic emission and one coil antenna to observe EMR. In the second experiment, one acoustic transducer and five coil antennas were used. The EMR antennas consisted of a coil of 20–40 turns of enameled copper wire 0.2 mm in diameter. The coil antennas were tuned in rather narrow bands. Figure 5.30 shows the overall frequency response of the experimental system (Yamada et al. 1989).

An example of Yamada et al. results is shown in Fig. 5.31. They claimed that their EMR signals were associated with cracks of 1–100 μm length. Note that the shape of Yamada et al. (1989) signals is similar to the ones observed by Nitsan (1977) for polycrystaline rocks (Fig. 5.28c).

Ivanov et al. (1990) investigated EMR pulses induced in rock samples from the Konstatinov, Saralin and Tashtagol ore deposits. All samples were investigated under uniaxial compression with different strain rates (10^{-4}–10^{-7} m s^{-1}). The initial signals were detected by a capacitance antenna (copper foil placed at the sample faces), trans-

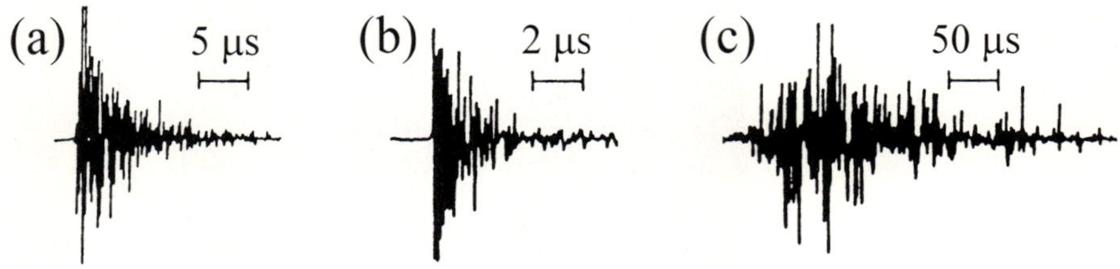

Fig. 5.28. Three examples of EMR pulses observed by the Nitsan (1977): **a** quartz, **b** tourmaline, **c** sandstone

Fig. 5.29.
An example of EMR signals induced in diabase (Miroshnichencko and Kuksenko (1980)

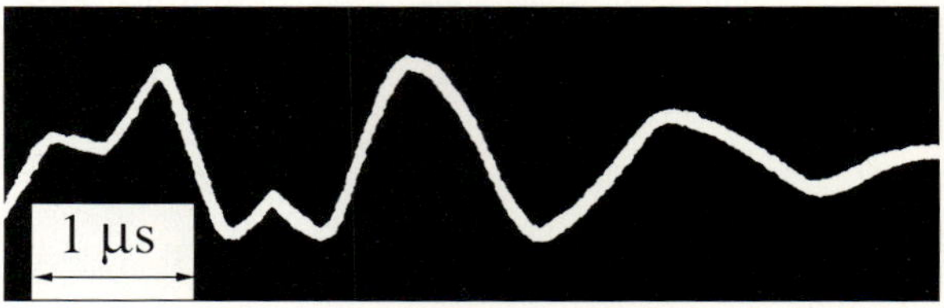

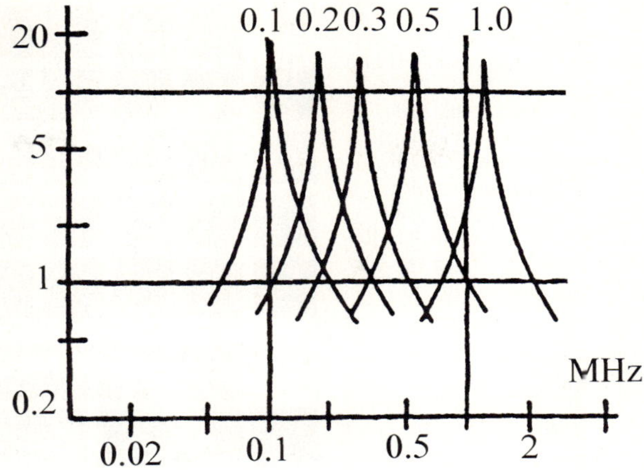

Fig. 5.30.
The overall frequency response of the experimental system (Yamada et al. 1989)

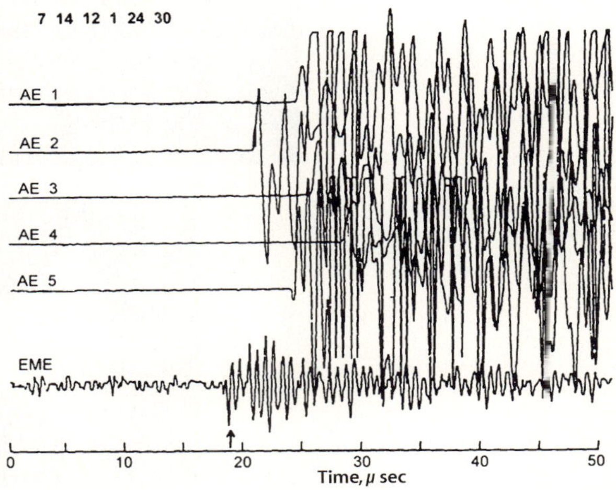

Fig. 5.31.
An example of EMR and AE signals observed by Yamada et al. (1989)

formed by an amplifier and a filter block and stored by a storage oscilloscope, from which screen they were photographed. The filter block consisted of a high frequency filter, which had ten fixed cut-off frequencies: 300 Hz, 1, 3, 10, 30, 100 and 300 kHz, and 1 and 10 MHz. The low-pass filter had seven fixed frequencies: 300 Hz, 1, 3, 10, 30, 100 and 300 kHz. An optimal working frequency band was selected by this filter and henceforth enabled the improvement of the signal / noise ratio. A total of 140 rock samples were studied. Figure 5.32 shows two examples of EMR signals observed by Ivanov et al. (1990).

These authors linked the rise-time of the pulse front with the crack propagation time and the decay time with a charge relaxation process.

Frid (1990) investigated EMR signals during stiff uniaxial compression. He examined rocks from Khibin Deposit (Kola Peninsula), Ural bauxite mines, Shpitzbergen and North Kuzbass coal deposits. EMR was detected by a narrow band resonance antenna (resonance frequency of 100 kHz and a *Q* factor of 2), passed through a 30 dB narrow band amplifier and stored by a storage oscilloscope, from the screen of which signals were photographed. Figure 5.33 shows an example of an EMR pulse obtained in this investigation.

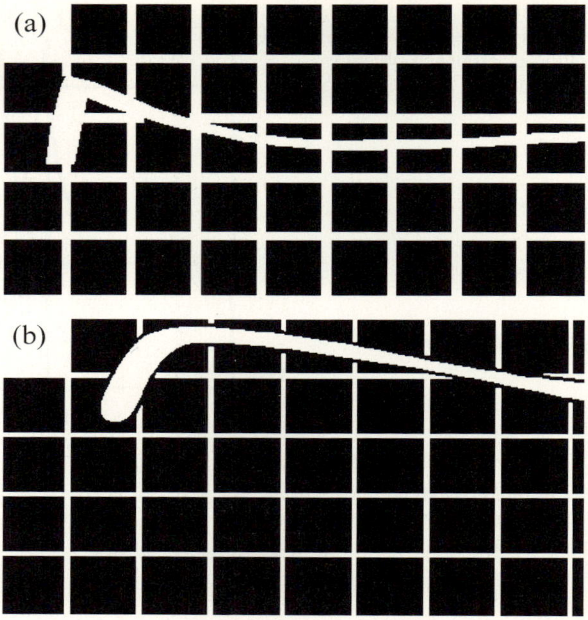

Fig. 5.32.
Two examples of EMR signals observed by Ivanov et al. (1990) induced by gabbrodiorite fracture: **a** horizontal scale 20 μs / division, vertical scale 1 V / division, crack length 2.5 cm; **b** horizontal scale 2 μs / division, vertical scale 20 mV / division, crack length 3.5 mm

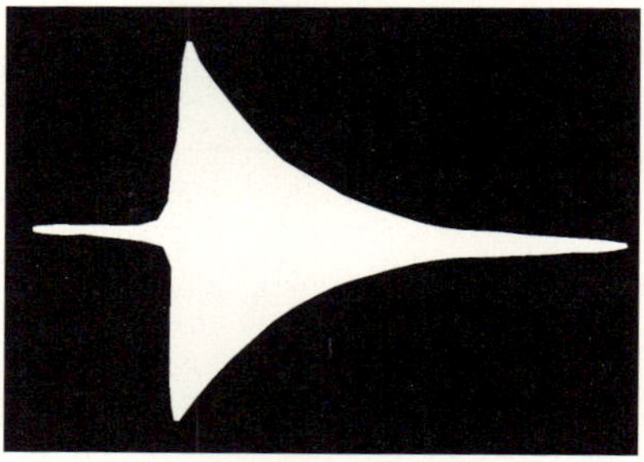

Fig. 5.33.
Typical EMR pulse measured by Frid (1990) during rock fracture

The shapes of signals from all investigated rocks were shown to be similar, whereas amplitudes and decay times were different. Note that the shapes of the EMR signals here are similar to those observed by Nitsan (1977) for a mono-crystalline material (Fig. 5.28a–c).

O'Keefe and Thiel (1995) performed EMR measurements on samples of dimensions $100 \times 100 \times 100$ mm. A schematic illustration of their experimental arrangement is shown in Fig. 5.34. The samples were stressed by uniaxial compression. The signals were detected by a 100 mm dipole antenna placed 100 mm from the sample. The signal was passed through an amplifier (54 dB) and an anti-alias filter (cut-off 160 kHz) and was connected to a PC, where signals were digitized and saved.

The digitization rate was 500 ks s^{-1}. An example of the electric signals observed by O'Keefe and Thiel (1995) is shown in Fig. 5.35.

It is seen that this signal is very similar to those obtained by Ivanov et al. (1990) (Fig. 5.32).

An analysis of all of these results shows that the EMR pulses consist of a rising part and a falling part. The majority of cases also show oscillations under the whole rising and fall-

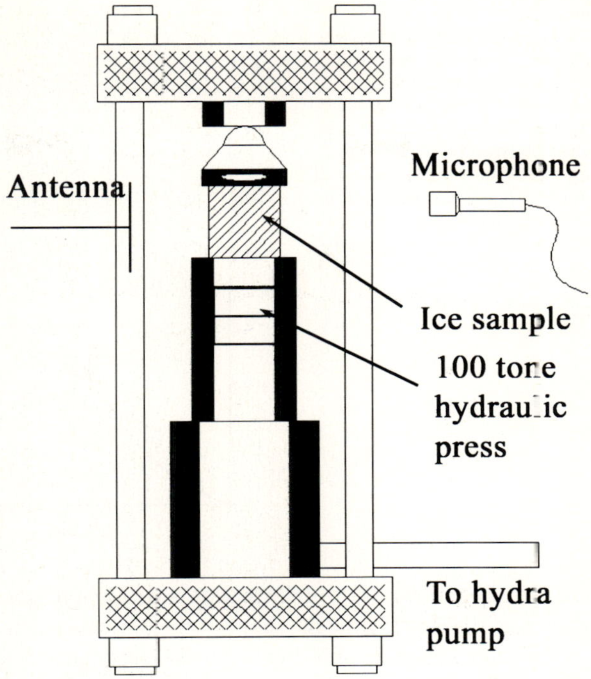

Fig. 5.34.
A schematic experimental arrangement of O'Keefe and Thiel (1995)

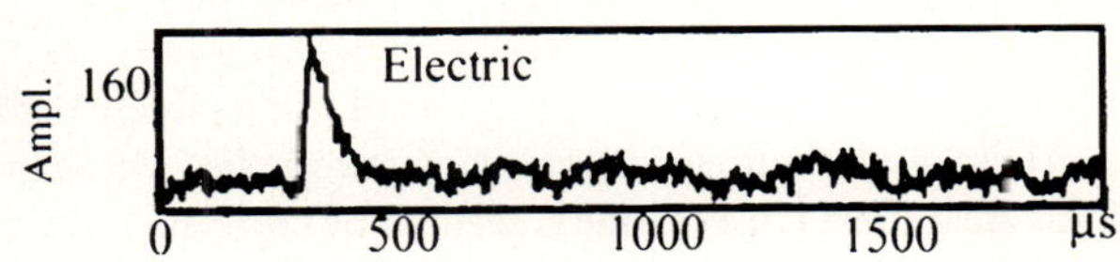

Fig. 5.35.
An example of electric signals observed by O'Keefe and Thiel (1995)

ing envelope. However, these investigations were conducted under different conditions and most of them with the help of a narrow band antenna, which makes a quantitative comparison between them very hard. Moreover, the shape of the signals depends to a large extent on the digitization rate, and the latter must be significantly higher than the 500 ks s^{-1} used, e.g., by O'Keefe and Thiel (1995) in order to obtain an accurate picture.

5.5.2 Our EMR Pulse Classifications

Our investigations (Rabinovitch et al. 1996) showed that

1. EMR pulses appear in two main different forms: "Lengthy" (total duration 30–110 µs and up to 400 µs, e.g., Fig. 5.11) and "short" (0.5–6 µs, e.g., Fig. 5.10a).
2. An analysis of the EMR pulses emanating during uniaxial failure showed that the deformation process can be divided into three zones according to the EMR features (Fig. 5.36, Rabinovitch et al. 1996):
   Zone I: Of individual EMR pulses;
   Zone II: Of EMR pulses with an uneven shape and/or groups of pulses;
   Zone III: Of lengthy EMR pulses.

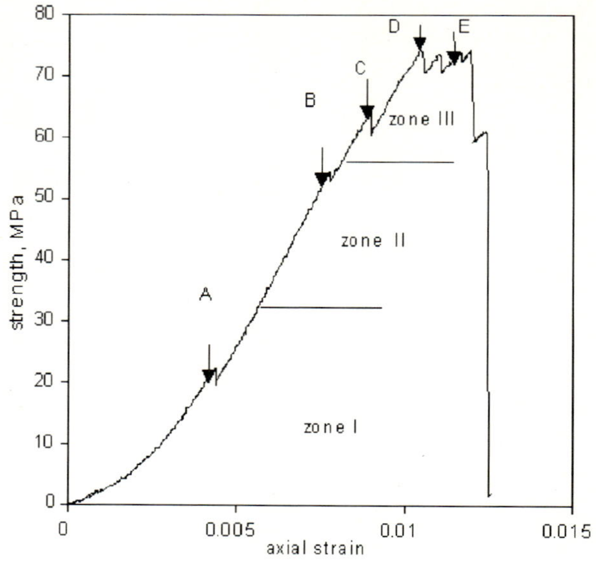

Fig. 5.36. Stress-strain curve of rhyolite sample. *Arrows* on the graph show moment of EMR pulse excitations. Zone I is defined by individual pulses; zone II is characterized by pulses with uneven shape and/or groups of pulses; zone III shows particularly lengthy pulses (see elaboration of *arrows* (*a–e*) significance in Rabinovitch et al. 1996)

3. In the beginning of the deformation individual pulses occurred, while at the upper part of the elastic zone the shape of the EMR events became more complex indicating a superposition of individual pulses (sequences), with internal interpulse spacing of less than 3 μs. The interval between sequences was 40–80 μs.

Lengthy EMR pulses were registered immediately before and at the peak zone.

These results indicate a correlation between the EMR shapes and the failure stages. Thus, individual short pulses (zone I) seem to be correlated with the stage of individual microcrack formation, multi-pulse sequences (zone II) can be correlated with the crack coalescence stage, and the lengthy pulses (zone III) can be correlated with the sample faulting.

5.6 EMR and Material Elasticity

Gol'd et al. (1975) found differences in activity of EMR emanating from different materials in a row of materials investigated under uniaxial compression. Materials were ordered (from "high" to "low") according to the number of EMR signals registered during cracking as follows: Quartz, orthoclase, plagioclase, granites, sandstones, and metamorphosed aleurolites. They noted that a decrease of the number of EMR signals could be due to a decrease of rock/mineral hardness (i.e., elasticity) from quartz via granites to metamorphosed aleurolites. Khatiashvili (1984) investigated EMR induced by splitting of LiF, NaCl, KCL and $CaCO_3$ crystals. He showed that EMR frequency spectra are specific for each crystal (Fig. 5.37a). The spectra differ in the positions and widths of the radiation bands. He also observed that the intensity (amplitude) of the EMR induced by alkali halide crystal failure was proportional to the squares of the elastic moduli and the crack length (Fig. 5.37b). These investigations (Gol'd et al. 1975; Khatiashvili 1984) were carried out only under uniaxial loading.

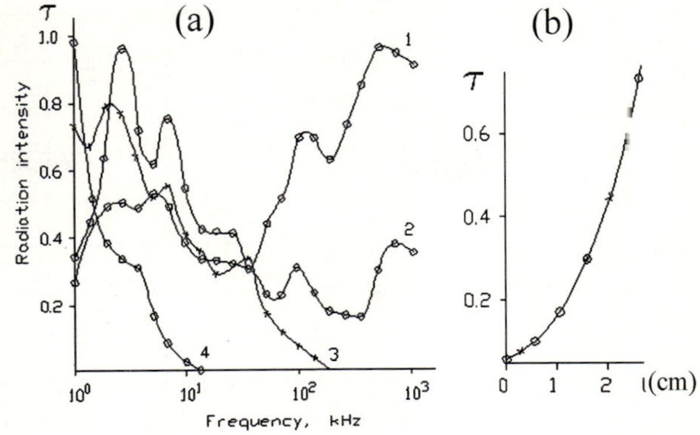

Fig. 5.37.
a EMR spectra from splitting 1-LiF; $0.1I = 1$ mV, 2-NaCl; $0.1I = 80$ µV, 3-KCl; $0.1I = 15$ V, 4-CaCO; $0.1I = 0.1$ mV; **b** EMR intensity as function of fracture length (Khatiashvili 1984)

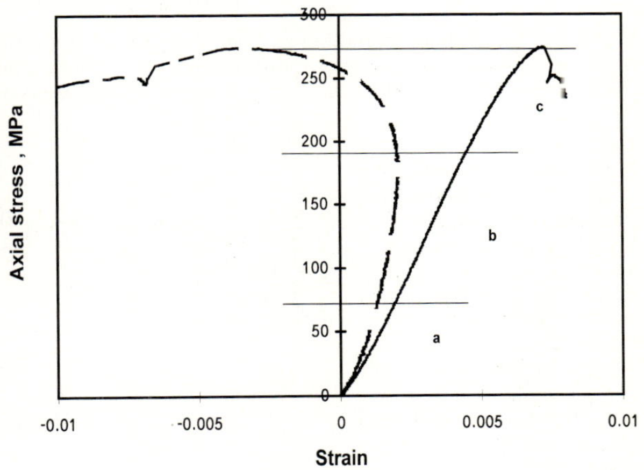

Fig. 5.38.
An example of granite triaxial deformation (confining pressure: 10 MPa); *a* the pore closure region (10 EMR pulses were measured); *b* the elastic region (6 EMR pulses were measured); *c* the nonlinear region before the peak stress (8 EMR pulses were measured); *unbroken line:* Axial strain, *dashed line:* Volumetric strain (Frid et al. 1999)

For our investigations, we used granite samples (see above). Each sample was tested by an axial strain rate of 1×10^{-5} s^{-1} and laterally by different hydrostatic oil pressures. The axial pressure varied from 110 MPa up to 284 MPa and the confining pressure from 0 to 14 MPa.

Figure 5.38 shows an example of a stress-strain curve of a representative granite sample (confining pressure 10 MPa). All rock samples went through the "normal" three deformation stages (Jaeger and Cook 1979) during compression up to the peak stress: *a* the nonlinear region of pore closure (axial strain curve here is slightly bent down, *b* the elastic region (the stress-strain curves are quasi linear), and *c* the nonlinear region before the peak stress.

As seen from the Mohr-Coulomb diagram (Fig. 5.39), the increase of confining pressure from 10 MPa (third circle) to 14 MPa (fifth circle) is not accompanied by an appreciable increase of shear strength (the Mohr-Coulomb envelope becomes much less steep than for smaller confining pressures).

Since, the confining pressure does not cover a wide range of pressures, the ductility stage has not yet been completely reached. However, the failure plane inclination to the axial load axis was measured to be 41 ±1°. Thus, our results are reasonably close to obeying the von Mises ductile failure criteria.

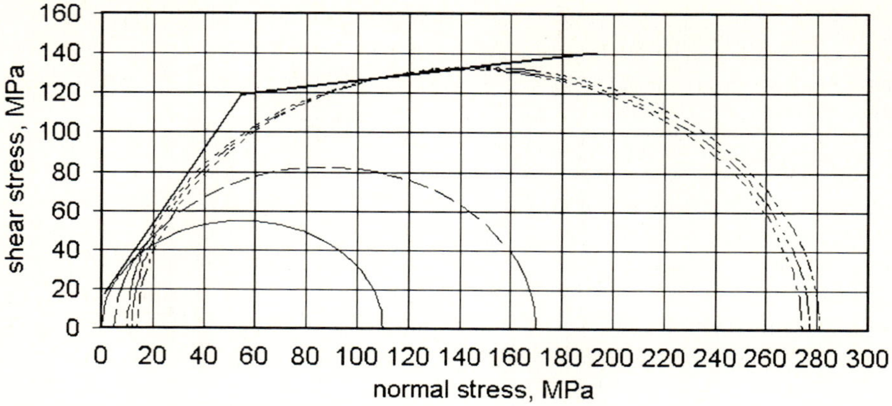

Fig. 5.39. Two stages of the failure envelope (*thick line*) around Mohr circles which represent the different confining pressures, —— uniaxial test, – – – 5 MPa confining pressure, - - - 10 MPa confining pressure, – – – 12 MPa confining pressure, – – – – 14 MPa confining pressure (Frid et al. 1999)

Fig. 5.40.
Relationship of the following characteristics with confining pressure: **a** Poisson ratio; **b** Young's modulus; **c** EMR activity. *Error bars* on these figures are determined by the accuracy limits of our rock mechanics equipment; that is ±2% for Poisson ratio and Young's modulus and by the dead time of the storage oscilloscope-IBM PC system (the period it takes for registration, digitizing and memorizing an EMR sequence, during which time the system is shut down. The error in EMR activity is of the order of +2% (Frid et al. 1999)

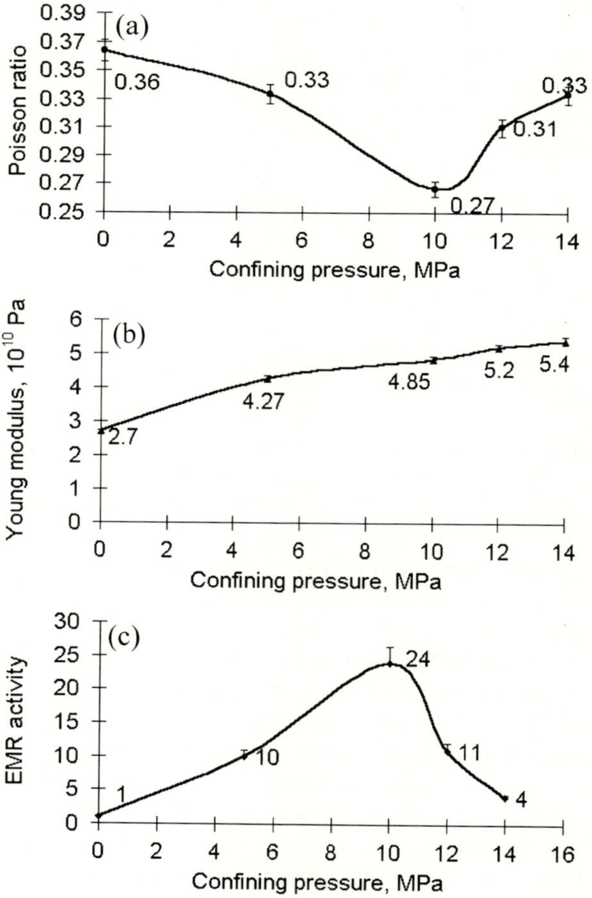

We denote here the EMR activity as the number of individual EMR pulses emitted during tracing of the entire stress-strain curve. Its value is seen to change with the loading type (Fig. 5.40c), increasing first with the confining pressure and then decreasing again.

We checked the following stress-strain characteristics to investigate possible correlations with EMR activity: Axial strain and its increment; confining pressure; deformation modulus (ratio of axial stress increment to rise of axial strain – the secant modulus) and its absolute value); differential stress; differential volumetric compression work (product of differential stress increment and volumetric strain changes) and its absolute value; elastic strain work (product of average differential stress and elastic volumetric strain changes); elastic volumetric strain (part of volumetric strain that returns to its original value after applied stresses are removed); lateral deformation modulus (ratio of axial deformation increment to rise of lateral deformation – the secant lateral modulus) and its absolute value; maximal stress; Poisson ratio (ratio of lateral elastic deformation to an axial elastic one, measured in the elastic zone of deformation); residual axial strain (remaining axial deformation after removal of external stress); residual strain work (strain work minus elastic work); residual volumetric strain (remaining value of the sum of the three principal strains after removal of external stress) and its absolute value; strain work (product of average differential stress and volumetric strain changes) and its absolute value; volumetric strain (sum of the three principal strains) and its absolute value; Young's modulus (ratio of axial stress increment to rise of elastic axial strain, measured in the elastic zone of deformation).

Note that these stress-strain characteristics can be classified as either "elastic" or "general". For example, the Poisson ratio and the Young modulus are defined here as elastic characteristics and measured in the elastic zone only. Lateral deformation modulus and deformation modulus, on the other hand, are similar in definition to the Poisson ratio and Young modulus, respectively, but they are "general" characteristics: They can change during a loading process.

Our analysis shows that out of all these stress-strain characteristics the Poisson ratio is the *only* one that shows good correlation with EMR activity. The Poisson ratio (Fig. 5.40a) changed from 0.36 (for uniaxial test) via 0.27 for 10 MPa confining pressure to 0.33 for 14 MPa confining pressure. Its squared regression coefficient was –0.93. An increase of the Poisson ratio (Fig. 5.40a) is accompanied by a decrease of EMR activity (Fig. 5.40c) and vice versa. Note that since all samples here are made of the same material (granite), the Poisson ratio varies only with loading conditions, and does not vary with material properties, on the other hand e.g. the Young's modulus (Fig. 5.40b) increases monotonically from 27 GPa to 48.5 GPa with the confining pressure, and therefore it is obvious that its correlation with EMR activity should be poor (the obtained squared regression coefficient was actually 0.18).

The result of a poor correlation between Young's modulus and EMR activity is different from that found previously (Gol'd et al. 1975; Khatiashvili 1984) in which an increase of Young's modulus of a material was associated with an increase in EMR activity. In the previous experiments, however, the correlation between Young's modulus and EMR activity was investigated only under uniaxial compression. Our loading conditions were different. We studied changes of EMR activity during the transition from tensional fracture (uniaxial test) via brittle shear fracture according to the Coulomb criterion, to shear fracture in the brittle-ductile region under an increase of confining pressure (see, for example, Twiss and Moores 1992, Fig. 9.9). As noted by Jaeger and Cook (1979), Young's modulus increases with the confining pressure, while the Poisson ratio could both decrease and increase. We suggest that our sample deformed with the confining pressure as follows: An increase of the confining pressure from 0 to 10 MPa closed granite microcracks and hence the Poisson ratio was decreased to its intrinsic value of a material devoid of openings. Under a greater confining pressure (12–14 MPa), however, the deformation changed from a brittle to a ductile one causing an increase of Poison ratio. The brittle-ductile transition is confirmed by the large angle (41 ±1°) produced between the axial load and the failure plane and in the insignificant increase of shear strength (Fig. 5.39). Correspondingly during the brittle deformation, EMR activity increased with confining pressure while in the brittle-ductile region EMR activity decreased.

The correlation (or rather anti-correlation) of EMR activity with the Poisson ratio seems plausible. The Poisson ratio measures the compliance of the material in the transverse direction when stressed axially. The lower the Poisson ratio, the harder it is for the material to strain transversally, and hence the higher is the probability of new fractures (especially parallel to the axis) and of the ensuing EMR. On the other hand, the higher the Poisson ratio the easier it is for the material to strain transversally, and accordingly, fewer fractures and lower EMR activity should be expected. The key elastic parameter for EMR characterization during triaxial compression is therefore the Poisson ratio and not the Young's modulus. Note that our results are in line with those of Gol'd et al. (1975) and Khatiashvili (1984) for uniaxial load. Our results, however, being triaxial, do show the correct global parameter to choose. They also indicate that the EMR activity correlates well with the lateral resistance to axial fracture, which is compensated by an increase in the number of new cracks.

5.7 EMR during Percussion Drilling

Recently, Rossmanith et al. (1996, 1997) carried out experimental and numerical investigations in order to understand the mechanism of material damage under percussion drilling. They found that the stress applied during drilling, besides inducing elastic waves emission, also resulted in defect creation: Every time a percussion occurs, existing defects grow, microcracks are formed and propagate, and a "damage zone" around the drilling tip is thus created (Rossmanith et al. 1996, 1997), a zone consisting of a network of many small and a few larger cracks. Another observed defect consists of circumferential cracks around the hole and appears when weak regions exist in the material (Ravishankar and Murthy 2000).

Since fracturing also emits EMR, we discuss here an investigation of the EMR emitted from fractures that develop during percussion drilling.

We used cylindrical samples 100 mm in length and 30 mm in diameter (Fig. 5.41) that were percussion drilled. Drilling was carried out at about the mid length of the cylinder, perpendicular to the cylinder axis. EMR was measured in the frequency range 1 kHz to 50 MHz with 1 μV sensitivity throughout. Radiation was detected by a one-

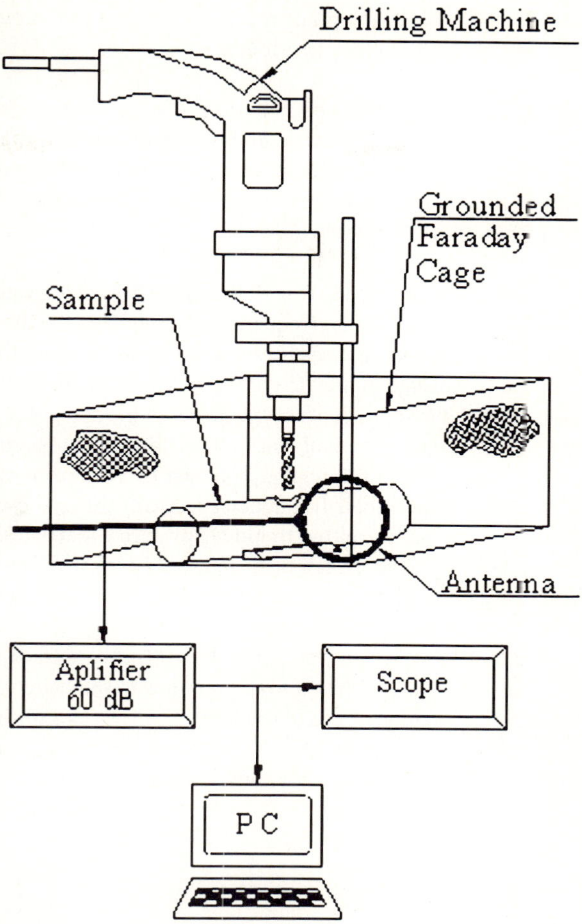

Fig. 5.41. Schematic diagram of experimental arrangement during drilling (Goldbaum et al. 2001)

loop magnetic antenna 3 cm in diameter, which is electrically "small" and exhibits negligible response to foreign electric fields. The antenna was placed at a distance of 3 cm from the drilling hole. External EMR disturbances were further decreased by the two following methods:

1. The sample together with the antenna were placed in a grounded Faraday cage.
2. The plane of the antenna was aligned perpendicular to the drilling direction, so that the influence of electromagnetic disturbances emitted by the motor of the drilling machine was minimized.

All EMR signals were electrically amplified by 60 dB, digitized and collected at a PC hard disk. The data was analyzed after test completion.

The materials used in the study were chalk from Middle Eocene layers in the Beer Sheva syncline, Eilat granite from the Nahal Shelomo area of southern Israel (Rabinovitch et al. 1999b), Solenhofen limestone from Germany, PMMA and soda-lime glass.

The 700-odd EMR signals measured during percussion drilling of all five materials were classified into four groups:

1. The first group: "Short single pulses" of 0.3–1.5 μs duration.
2. The second group: "Short chain of single pulses" of 2–15 μs duration.
3. The third group: "Extended chain of pulses" of 15–60 μs duration.
4. The fourth group: "Pulses on top of base-line voltage changes" whose duration varies between 10 and 800 μs.

Pulses of the first group (Fig. 5.42a) probably correspond to single cracks and are characterized by a single main frequency ranging between 10 and 25 MHz. It means (Eq. 5.7) that the corresponding cracks are very small, and indeed their T'/ω values (Eq. 5.8) indicate that they range in area from 0.005 mm^2 up to about 0.05 mm^2. Pulses of the second group (Fig. 5.42b) consist of several pulses of the first group type and probably correspond to several overlapping small cracks. Their frequency spectrum is thus a bit more complicated. Pulses of the third group consist of numerous pulses of the first two groups (Fig. 5.42c). Hence, their frequency spectrum shows a broad band (11–23 MHz). These pulses are probably emitted by tiny "powdered" particles fragmented from the walls of the drilled hole, where each particle has its own size and emits its own frequency. Note that like that of the first group, the area of the "powder" particles here ranges from 0.005 mm^2 up to about 0.05 mm^2.

Signals of the fourth group (Fig. 5.42d) (never seen in the compression experiments) show baseline voltage changes with very low frequencies of several kilohertz and in addition contain lengthy groups of the third type. Cracks emitting such radiation should range in area from 1 cm^2 up to 16 cm^2. The additional presence of the third group pulses indicates that "powder" is simultaneously created.

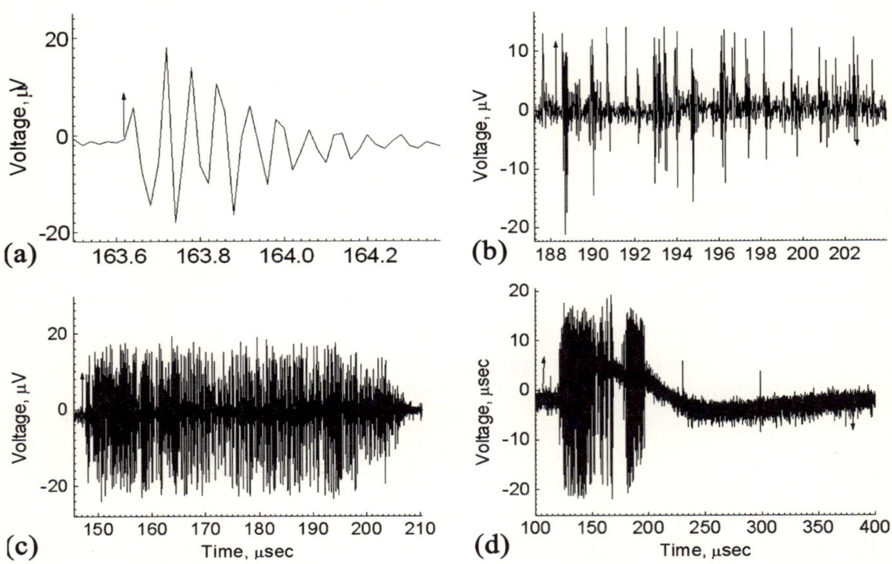

Fig. 5.42. Experimental EMR pulses induced by percussion drilling of the first group (**a**), second group (**b**), third group (**c**) and forth group (**d**) (Goldbaum et al. 2001)

The first three group types have been observed in our compression experiments, where however single pulse frequencies never exceed 10 MHz (Rabinovitch et al. 2000c; Frid et al. 2000).

They provide a new method for a general identification by EMR profile of the wide population of "powdered" and "chipped" particles of various sizes that form by percussion drilling and a one to one identification of all the defects involved in the process. The fourth group is discussed next.

5.7.1
The 4th Group and Polarization

The fourth group includes two different subgroups (4a): "Lengthy forms" (e.g., Fig. 5.42d), whose duration varies between 10 and 800 μs and which show low frequencies (several kilohertz) baseline voltage changes and also contain high frequency third type pulses and (4b): A different type of "Lengthy forms" again decorated by high frequency strings (Fig. 5.42, 5.43).

Here we present the results of investigation of the 4a and 4b signal types (Goldbaum et al. 2001), where the "4" is dropped in following. The samples tested were cylinders of soda-lime glass, 10 cm in length and 3 cm in diameter that were percussion drilled. The experimental arrangement was described in Sect. 5.7.

Signals of groups a and b occurred in 5 glass samples (2, 3, 4, 5, 6). Group 'a' signals are similar to individual EMR pulses (Rabinovitch et al. 1998), but their frequencies are smaller (several kHz). Group 'b' signals (Fig. 5.43) consist of a sharp voltage "jump" (of a duration of ~3% of the whole pulse duration) and range in size between 6 and 12 μV; the high voltage remains more or less constant for a period that varies between 10 and 1 000 μs and is terminated by a rapid falloff (of about the same duration as that of the rise time). Group 'a' signals are frequently followed by group 'b' signals after a time interval of several hundreds of microseconds. This occurrence is denoted by "(a/b) pair". If a 'b' signal has not been detected within this time interval, then an 'a' or a 'b' signal appears after a time period ranging between 0.2 s and 2 s. Sometimes a string of several 'a' signals appears and is followed by a 'b' signal.

Fig. 5.43.
Typical pulse of group 'b' (Rabinovitch et al. 2003b)

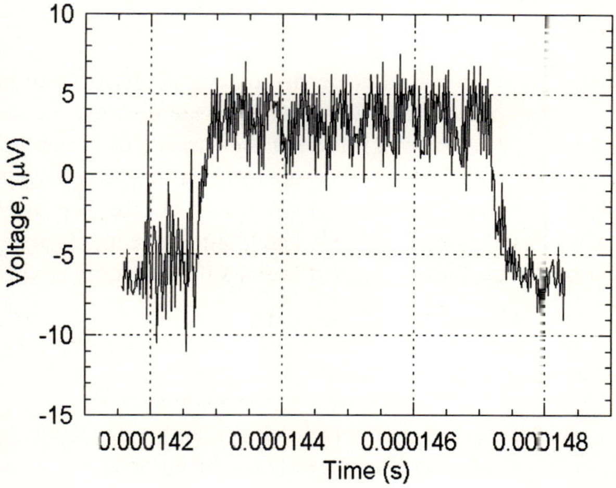

5.7.2
Fractographic Examination

Fractographic examination showed that in addition to glass powder and a network of small cracks (Rossmanith et al. 1997), some large "chips" were formed. These are large cracks, which originated at some depth and propagated until reaching the surface of the sample. Their observed fracture areas vary between 1 and 30 cm^2.

As shown in Goldbaum et al. (2001), the 'a' type pulses are excited by the creation of these flakes and the 'b' type pulses are derived from a polarization process, as shown presently.

5.7.3
Brief Theoretical Consideration

The measured quantity in our experiments is the electromotive force in the loop antenna, which is given by

$$V = -\frac{d\Phi}{dt}$$

the time derivative of the magnetic flux

$$\Phi = \int B dS = BS$$

Here B is the magnetic field through the antenna and S is its (constant) area. Since $r/c = 10^{-10}$ s it can be neglected with regard to the apparatus sensitivity (2×10^{-8} s) in comparison and we have (Bleaney and Bleaney 1965):

$$\begin{aligned} B &= \frac{\mu_0 lI \sin\theta}{4\pi r^2} - \frac{\mu_0 l\dot{I} \sin\theta}{4\pi c r} \\ \frac{dB}{dt} &= \frac{\mu_0 l\dot{I} \sin\theta}{4\pi r^2} - \frac{\mu_0 l\ddot{I} \sin\theta}{4\pi c r} \end{aligned} \tag{5.10}$$

where μ_0 is the magnetic permeability, c is the speed of light, l is the length of the element in which current is present, r is the distance between the antenna and the current, $I = I(t)$ is the magnitude of the current, the angle θ is that created by the current direction and the radius-vector to the antenna relative to the current, and a dot denotes time differentiation. For EMR, the current is produced by the oscillating surface waves discussed in Sect. 5.2. Since our experiments were carried out in the so-called "short" zone, we can neglect the second terms in (Eqs. 5.10) and write

$$V = -S\frac{dB}{dt} = -\frac{S\mu_0 l \sin\theta}{4\pi}\frac{\dot{I}}{r^2} \tag{5.11}$$

Consider now a changing dipole $p(t)$ with a constant length l and changing charge $q(t)$, $p(t) = lq(t)$. The current (Bleaney and Bleaney 1965) is $I = dq/dt$, $lI = dp/dt$, and

$$l\dot{i} = \frac{d^2 p}{dt^2} = \ddot{p} \tag{5.12}$$

Thus, the electromotive force V in the antenna is proportional to the second derivative of the dipole moment:

$$V = -\frac{S\mu_0 \sin\theta}{4\pi r^2}\ddot{p} \tag{5.13}$$

If instead of a dipole moment one considers the sample's polarization P, which is defined as its dipole moment per unit volume, the electromotive force would be proportional to the second derivative of the polarization:

$$V = \frac{-\ddot{P}}{C} \tag{5.14}$$

Here $1/C > 0$ is the proportionality constant. The polarization current is given by the integral of (5.14):

$$\dot{P}(t) = \dot{P}_0 - \int_0^t CV(t')\mathrm{d}t' \tag{5.15}$$

where $\dot{P}_0$ is a constant, and the polarization P is given by the integral of (5.32):

$$P(t) = P_0 + \int_0^t \dot{P}(t')\mathrm{d}t' \tag{5.16}$$

5.7.4 Examination of 'b' Type Signals

We assume that the 'b' type signals (Fig. 5.43) originate by a change of material polarization (Eq. 5.14). By Eq. 5.15, the first integral of the signal should be proportional to $\dot{P}$, while the second integral should be proportional to the polarization (P) itself. This sequence is schematically shown in Fig. 5.44. A smoothed 'b'-like signal is shown in Fig. 5.44c, where units of amplitude and the time are evidently different from those of Fig. 5.43, but signal shape is preserved. The first and second integrals appear in Fig. 5.44b and 5.44a, respectively.

Ogawa et al. (1985) measured EMR from impacted materials using an electric sensor. Hence the signals obtained by them (see Fig. 2 and 3 of Ogawa et al. (1985), redrawn here as Fig. 5.45a and b) should be proportional to the depolarization current or to the first derivative of the polarization, and the time derivative of their signal (Fig. 5.45c) should be proportional to $\ddot{P}$.

Comparing the pulse shown in Ogawa et al. (Fig. 3 of Ogawa et al. 1985, redrawn here as Fig. 5.45b) with the integral of a 'b' pulse (Fig. 5.44b), it is clear that they are similar, indicating that both stem from a similar process, namely polarization changes. The shape of the polarization itself can be obtained by the appropriate integrals.

Fig. 5.44.
Schematic polarization accumulation and depolarization process: **a** polarization: *P*. Note: Only the final part of the polarization increase is shown. **b** First derivative of *P*, and **c** second derivative of *P*, as functions of time. Note the negative tail appearing in **c** (Rabinovitch et al. 2003b)

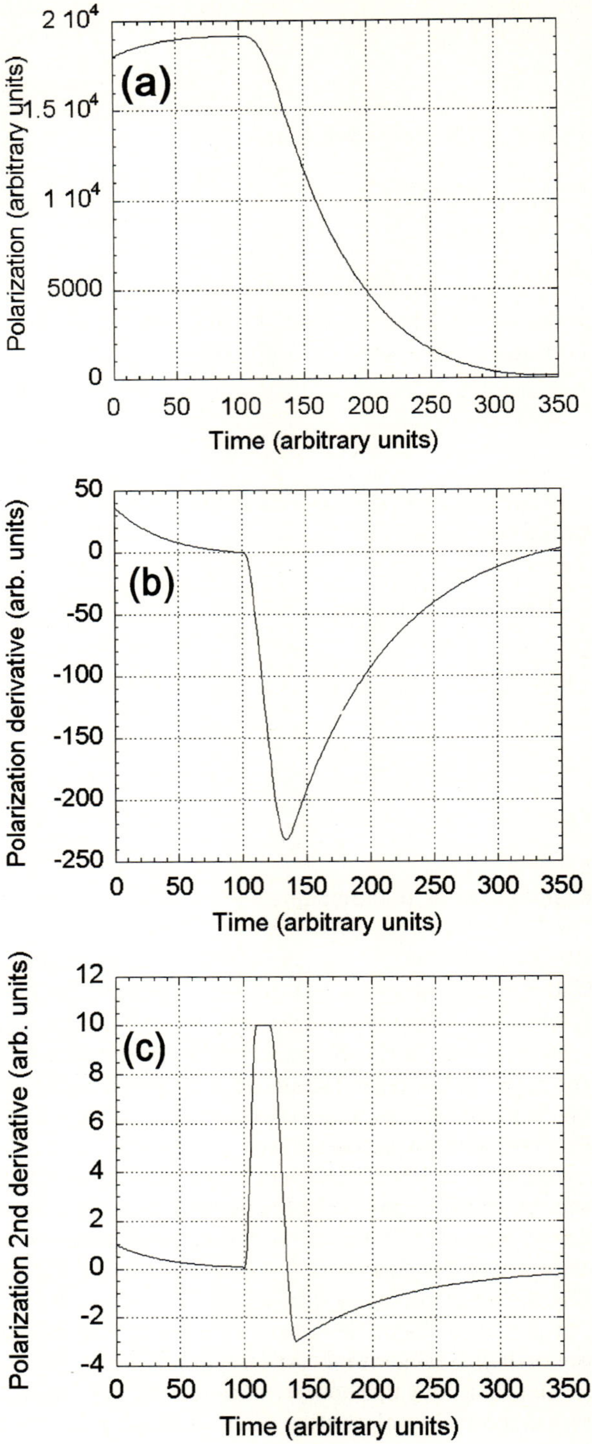

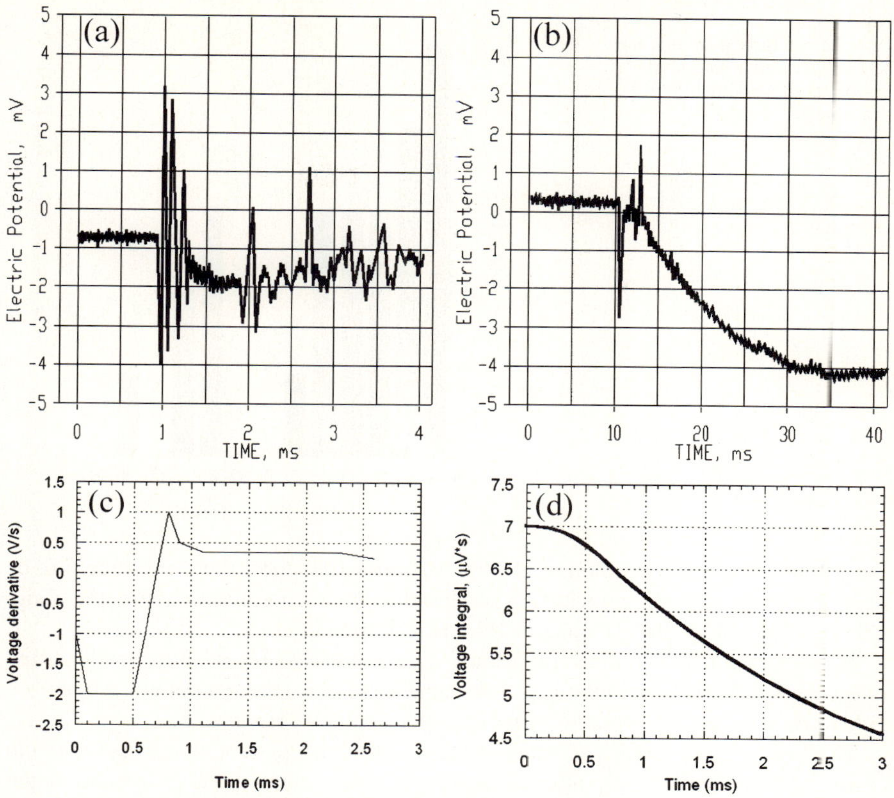

Fig. 5.45. Pulses obtained by Ogawa et al. (1985): **a** their Fig. 2, **b** their Fig. 3, **c** smoothed time derivative of 5.45a, and **d** the integral of 5.45a (Rabinovitch et al. 2003b)

5.7.5 The Polarization Model

The emerging picture of the process is as follows: During drilling a stress-strain field is slowly accumulated in the sample by numerous accumulated impacts (percussions), followed by an increase in polarization (Varotsos et al. 1999) (see Fig. 5.44a) for the final stage of this process). This accumulation of stress and polarization can take place only sufficiently ahead of the drill bit, since the region close to the latter is obviously reached by it, releasing the stress, before the latter grows sufficiently. During polarization increase, $\dot{p}$ is too small to be detected (Fig. 5.44c, the region up to $t = 100$). Note that electric polarization (i.e., the appearance of a dipole moment in a material) can be caused (Khesin et al. 1996, 1997; Rzhevskii and Novik 1978) by four basic processes: Electronic, ionic, dipolar and space charge polarization. Since relaxation times of the first three processes are less than 10^{-6} s and since measured relaxation times for glass polarization are of the order of 10 s (Kuksenko et al. 1990), the process here can only be due to space charge polarization.

Crack development begins when the accumulated stress becomes sufficiently large. Crack propagation (indicated in our measurement by an 'a' signal) is accompanied by stress relaxation, and the latter causes depolarization and an appearance of a 'b' signal.

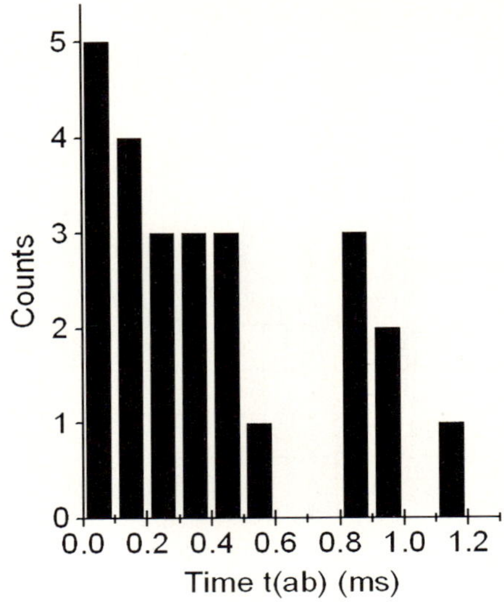

Fig. 5.46.
Distribution histogram of time lags between an 'a' and a following 'b' signal (Rabinovitch et al. 2003b)

Natural polarization relaxation in glass is very slow (see above) and can therefore be neglected. Here however polarization is relaxed by fracturing, and therefore the 'a' signals are followed by the 'b' signals, within a small time interval.

However, polarization does not fall off at once, because not all ions participating in the depolarization current begin their movement at the same moment: An ion begins to move when the acoustic waves excited by fracture reach it.

The time it takes a crack to travel through the entire sample whose size is 10 cm (with a velocity of 300 m s^{-1}, which is low but possible in glass) is 300 μs. We have registered thirty-three couples in which 'a' signals were followed by 'b' ones, and twelve of them have a time lag of less than 300 μs (Fig. 5.46). There are seven couples with a time difference between the 'a' and 'b' pulses of up to 600 μs. These longer times may be due to the fact that not all ions are situated along the way of the propagating crack. Some of them are placed at rather large distances from the crack, so stress relaxation must first reach a definite point on the crack route, and then be transferred by acoustic waves to the distant ions. For larger measured differences, it is assumed that there was an additional signal in between those observed that has not been detected by our equipment possibly because of a very small solid angle at the antenna.

Note that the spread in the histogram (Fig. 5.46) can also be due to another reason. To understand it, let us consider Fig. 5.47a and *b* that schematically show two examples of possible depolarization events. The calculated 'b' signals (second derivatives) (Fig. 5.47c,d) show that the second one (Fig. 5.47d) is shifted by ten time units with respect to the first one (Fig. 5.47c). This means that a negligibly small change in the depolarization shape may cause a significant shift in the 'b' signal origin, and hence to a spread of delay times between the 'a' and the 'b' signals.

Flake cracking induced by the drill bit thus leads to the measurement of a group 'a' and group 'b' signal. When cracking stops, stress relaxation and depolarization current increase also halt. Thus "immediate" depolarization can be incomplete, and some polarization could still remain in the material. The latter continues to decrease rather slowly to zero (Fig. 5.44a). Depolarization current (Fig. 5.44b) generally changes too slowly to

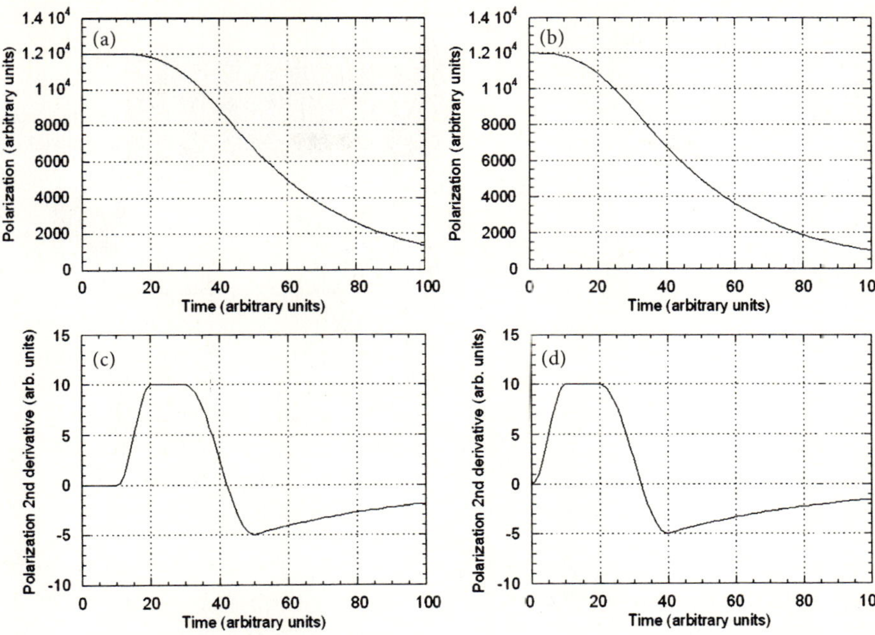

Fig. 5.47. (Schematic) Slight differences in depolarization shapes (**a**, **b**) can cause large shifts in the time lags between the observed signal pair (**c**, **d**) (Rabinovitch et al. 2003b)

give a measurable second derivative (Fig. 5.44c), but in some pulses a "tail" appears (Fig. 5.48a, arrow). Since measurements were carried out only up to 1.3×10^{-3} s we artificially continued this "tail" (for the specific pulse of Fig. 5.48a) and evaluated its first and second integrals (Fig. 5.48b,c). Note that the derivative of the pulse obtained by Ogawa et al. (1985) (Fig. 5.45c) shows a similar "tail".

The estimated decay (relaxation) time for such a "tail" is several milliseconds, which is much shorter than the ~10 s natural relaxation times observed in glass (Kuksenko et al. 1990). This decay time decrease can be due to material heating. It was shown (Kuksenko et al. 1990) that the EMR pulse decay time decreases significantly with a rise of temperature. Measurements of temperature rise during fracture have been carried out for different materials yielding very high values. For instance, for PMMA the temperature increase was about 500 °C (Swallowe et al. 1986), while for glass it was ~840 °C for a 200 m s^{-1} crack velocity (Li et al. 1988).

Comparison of the general shape and value of the rise-fall times of group 'b' pulses enables us to assume that these pulses are excited by space-charge polarization (Rzhevskii and Novik 1978; Gueguen and Palciauskas 1994). The charge carriers causing polarization are probably the weakly pinned Na^+ ions (Kuksenko et al. 1990) of the glass. Hovestadt (1902) described a change in the position of temperature markers on a thermometer over time. A structural change in a silicate mineral (feldspar) at room temperature was ascribed to a Na^+ displacement under strain (Bahat 1968). These relatively heavy, weakly bound ions (Babcock 1977) can leave their equilibrium positions and move under stress, leading to polarization. When stress is removed or relaxed, these ions return to their equilibrium positions.

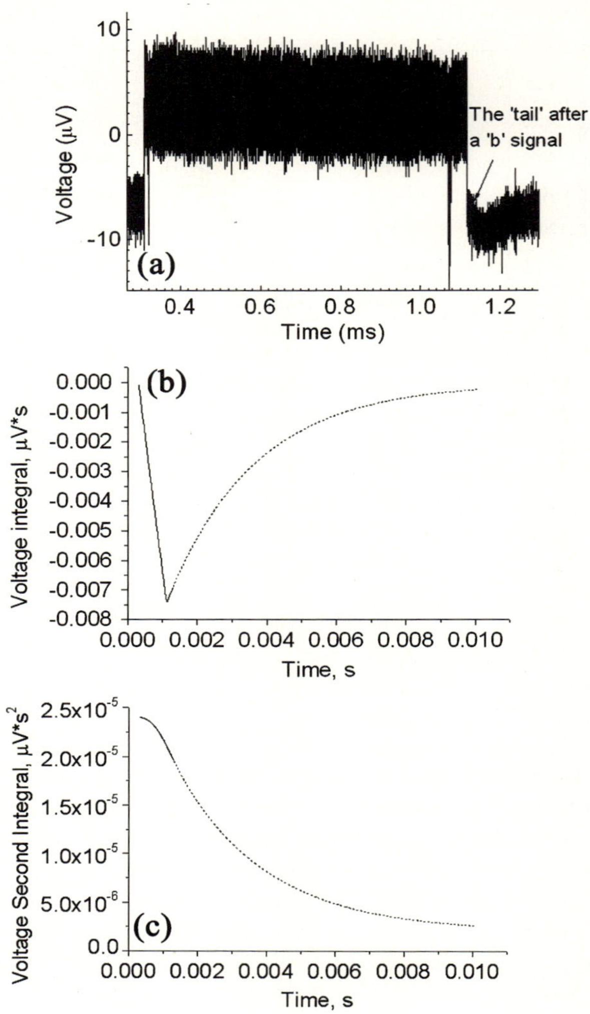

Fig. 5.48. A 'b' type pulse with a clearly visible "tail" (*arrow*) (**a**), the first integral (**b**) and the second integral (**c**) of the extended pulse (see text) (Rabinovitch et al. 2003b)

One remark is in order. The high frequency "decorations" of both the 'a' and the 'b' signals are evidently due to the powder and small cracks generated by the drill simultaneously with the flaking and depolarization process discussed here.

The flake creation process itself is accompanied by an 'a' pulse, while the depolarization process is accompanied by a 'b' signal. The latter, being fast enough, is registered by our equipment. The flake/depolarization pairs should occur in a cyclic manner, since following depolarization it takes some time for a repolarization process ahead of the drill bit to reoccur. Hence also the emission of 'a'/'b' signals should be cyclic and not continuous. However, since following the first 'a'/'b' pair creation the stress does not fall off to zero, it should take less time for the stress to increase to the critical value than the time to build up the first 'b' pulse. This phenomenon was actually observed as a gradual decrease in the time periods between 'a'/'b' pairs.

5.8
Intermediate- and Large-Scale EMR Detection

5.8.1
Blasting Experiments

5.8.1.1
Experimental Arrangement and Results

Blasting experiments were carried out in Turonian age limestone in an open quarry near the village Omarim, located about 25 km north-east of Beer Sheva (Israel). Two blasts of 5 and 1.4 t of explosive material respectively were performed. In each blast, one slope of the quarry was exploded. The slopes are of a height of about 11 m and of length of about 100 and 20 m, respectively. A magnetic one-loop antenna (Electro-Metrics Penril Corporation) 0.5 m in diameter was used for the detection of the EMR. A low-noise micro-signal amplifier (Mitek Corporation Ltd, frequency range 10 kHz to 500 MHz, gain 60 ±0.5 dB, noise level 1.4 ±0.1 dB throughout) and an analog-to-digital converter connected to a triggered PC completed the detection equipment. We placed our magnetic antenna in front of the blasted slope at a distance of about 100 m from it (Fig. 5.49).

Figure 5.50 shows ~5% (in time) of the EMR-record that was observed during blasting. A detailed analysis of the EMR record shows that the main frequency of the observed pulses lies in the range 2–8 MHz, and that the intervals between individual pulses are of about 1–5 μs (Fig. 5.51). The shapes of individual pulses (Fig. 5.51) that were formed by a digital expansion of Fig. 5.50 are very similar to those observed during compression and drilling (Fig. 5.52a,b). Our analysis shows that it is only the entire durations of the EMR-records that were different. The durations of the EMR records from blasts were similar to those of the explosions themselves ~0.5 s for the more powerful blast (5 t) and ~0.3 s for the second one (1.4 t).

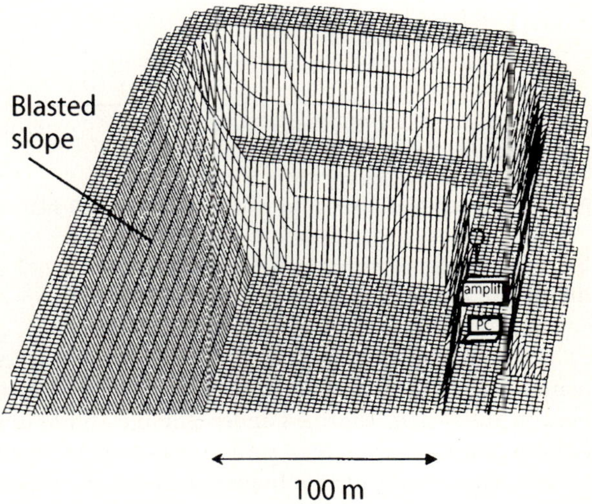

Fig. 5.49. Schematic diagram of experimental arrangement during blasting

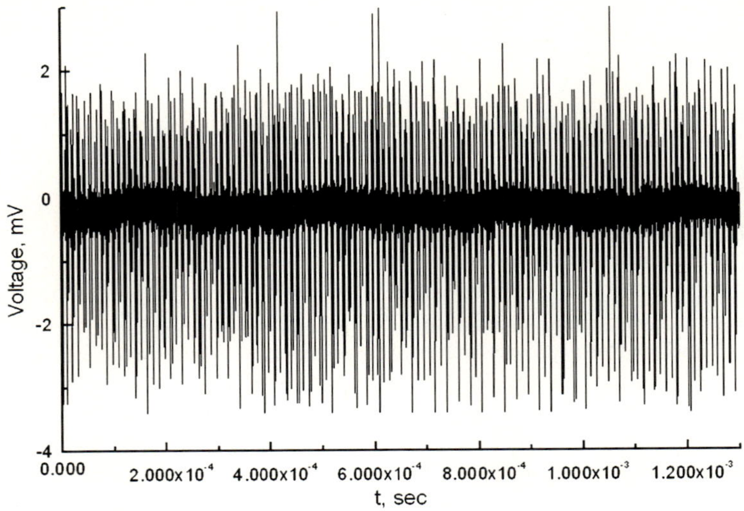

Fig. 5.50. The EMR string induced by blasting (Rabinovitch et al. 2002a)

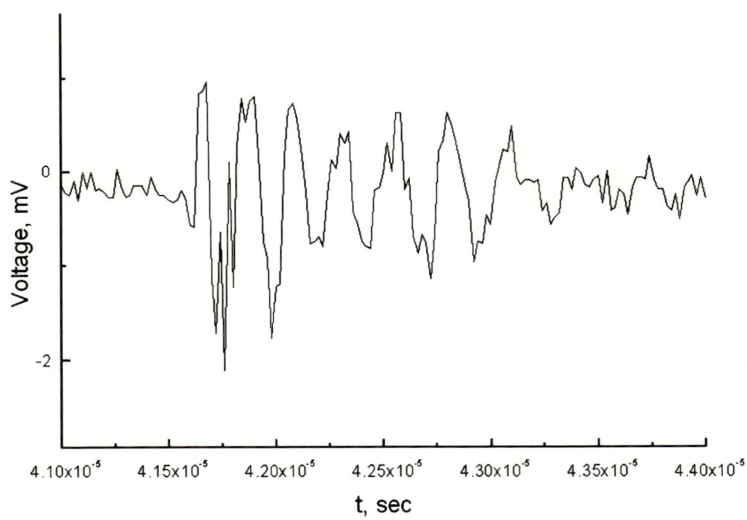

Fig. 5.51. An individual EMR pulse induced by blasting (Rabinovitch et al. 2002a)

5.8.1.2
Analysis of Results

As is known (Rossmanith et al. 1997), when stress waves propagate through a brittle material, a population of micro-defects are generated and extended. In drilling and blasting the rock is deformed under dynamic conditions, which create high intensity stress waves. A stress wave of sufficient amplitude (like those created during blasting or drilling) can initiate rock fragmentation at the expense of wave energy (Gueguen and Palciauskas 1994). As is also known (Rabinovitch et al. 1998, 1999b, 2000a, 2000b),

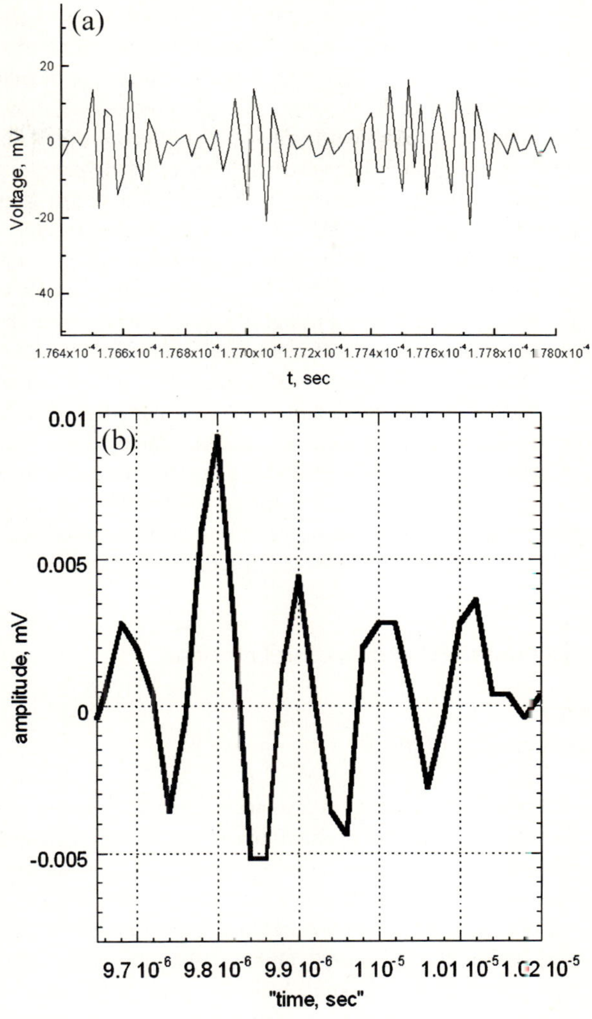

Fig. 5.52.
a Three individual EMR signals from chalk drilling. **b** An experimental EMR pulse induced by chalk compression (Rabinovitch et al. 2002a)

each crack is a source of an individual EMR pulse. Hence, a fragmentation process, such as drilling or blasting, where numerous cracks are created, is an origin of strings of EMR pulses. A larger stress wave energy leads to a larger fragmentation volume in the rock and hence to a larger duration of these strings (and to a larger number of pulses per strings). Therefore EMR pulse strings, which represent a small volume of rock in drilling, are much shorter (2–60 μs) than those induced by blasting (~0.3 and 0.5 s). In comparison, during compression, where these strings stem merely from crack coalescence, the duration of each EMR string never exceeded 30 μs. A zooming in on the EMR strings induced by blasting shows that both the amplitude and the frequency of individual pulses are different from those obtained in drilling and in compression. Nevertheless an overall comparison of EMR signals excited by compression, drilling and blasting shows their basic shape to be invariant to the dynamic/quasi-static loading mode. Note that this basic shape was shown previously to be invariant to the fracture mode

(Rabinovitch et al. 2000c, Sect. 5.2.3). The similarity of shape of single EMR pulses induced by these three processes therefore enables us to analyze the EMR signals induced by drilling and blasting by the same technique we developed earlier for compression (Rabinovitch et al. 2000c, Sect. 5.2). This technique allows us to calculate the dimensions of the crack, which was the source of the measured EMR signal. Our analysis shows that the time from EMR pulse origin up to its envelope maximum T' induced by blasting ranges between 0.01–0.5 µs. The ω value of the signals ranges between 16 and 50×10^6 s^{-1}. Hence their ratio T' / ω will be of about $0.06–1 \times 10^{-14}$ s^2. According to Bahat et al. (2001a), the proportionality coefficient between T' / ω and crack area S is $\sim 3 \times 10^7$. Hence, crack areas induced by blasting include a major fraction whose individual areas are estimated to be of about of 0.02–0.3 mm^2. Note that cracks of similar areas are created at the beginning of compression loading (Rabinovitch et al. 2000c, Sect. 5.3.4).

As noted by the Rossmanith et al. (1997), the set of fractures induced by blasting can be divided into three zones: A crush zone (near the borehole) of short uniformly distributed radial cracks; an intermediate zone, where short cracks from the crush zone extend; and an outer zone with long radial cracks. Comparison of crack areas observed during our blasting with this model shows that our EMR signals are induced by the cracks from the two first zones (crush and intermediate) and can indicate the beginning of a dynamic large-scale failure.

5.8.2
EMR Prior to EQ and Volcanic Eruptions

5.8.2.1
Current State of the Art

As noted above, during the nineteen seventies and eighties, interest in EMR increased in connection with the problem of earthquake (EQ) prognosis. Several investigators observed the appearance of a disruption of radio communications, atmospheric luminescence and perturbations before EQs (Davis and Baker 1965; Finkelstein and Powell 1970; Drobdev et al. 1978; Galperin et al. 1985; Chmyrev et al. 1986; Kustov and Liperovsky 1988; Bilichenko et al. 1990; Ouellet 1990; Johnston 1991; Derr 1986; Molchanov and Hayakawa 1998a,b; Liu et al. 2000; Lee et al. 2000). Sadovskii et al. (1979) observed an irregularity of EMR before some Karpatian EQs. Gokhberg et al. (1979) investigated EMR prior to an Iranian EQ. This study consisted of EMR registration at frequencies of 25, 385 and 1 630 kHz, for a period starting 55 min before the first EQ shock. EMR perturbations were observed to gradually increase up to the first shock moment. Following this shock, the levels at frequencies of 27 and 1 630 kHz sharply decreased, while those at 385 kHz remained high up to the last "aftershock". These authors proposed that if the EMR sources were deposited in the Earth's crust at the EQ epicenter, the original amplitude of the EMR signal would have to be of the order of hundreds of Volts per meter. Similar works were also conducted in Japan (Gokhberg et al. 1982). EMR was registered at a distance of 250 km from the epicenter of an EQ of magnitude $M = 7$. Since an analysis of "industrial" noise showed its absence at a frequency of 81 kHz, that frequency was chosen for EMR studies. It was shown that 30 min before the EQ, the EMR amplitude rose by up to 15 dB and then sharply decreased at the EQ moment. EMR intensity (voltage) changes in a tunnel at a depth of about 60 m were also checked. Analysis of diurnal cycles of EMR intensity variations showed that they were significantly disturbed a few days before an EQ. Gokhberg et al. (1986) carried out measurements of EMR features prior to an EQ in Gasli. They found that 2 h 40 min before the EQ ($M = 5$), perturbations above the original noise level were increased by up to 6 dB. Yoshino and Tomizawa (1989) measured EMR (at a frequency of 82 kHz) before a volcanic eruption (VE). The first EMR anomaly was registered two weeks before the VE, and another powerful EMR emission (20 dB higher than the usual noise level) was observed one day prior to the VE. Rikitake (1997), analyzing 60 EQ events measured in Japan, decided that EMR was a "short-term" precursor, with an estimated mean time prior to an EQ of ~6 hours. Nikiforova and Yadakhin (1989) investigated EMR in the Issik-Kul seismic region at a frequency of 15 kHz. They obtained a significant increase of EMR intensity in the immediate zone of active geological faults, and claimed that the EMR sources were of ionosphere origin. On the other hand, Morgunov and Matveyev (1991) conducted an EMR investigation during the Spitak EQ and claimed that EMR could not be explained as an ionosphere disturbance; its origin could only be ascribed to crust destruction. Gershenzon et al. (1987, 1989) and Gokhberg et al. (1982, 1986) proposed that the anomalies of EMR prior to EQ were originated by a mechanic-electric transformation. They noted that before an EQ, a deformation of the Earth's surface took place. Changes in the stress state of surfaces resulted in local destruction, in the formation of micro-fractures and in friction of the nearby rock blocks. Each of these processes could lead to EMR generation. Bella et al. (1992), Morgunov (1985), Morgunov et al. (1988) and Morgunov and Matveyev (1990) also favored the idea that EMR was excited by surface deformation in the vast zone of the EQ vicinity. Gokhberg et al. (1985) noted the following principal features of EMR occurrences:

1. Abnormally high EMR levels, which occur hours or even days before an EQ.
2. The EQ itself takes place during a decrease of this EMR anomaly or immediately after it.
3. The EMR spectrum has a wide frequency range (from several Hz to hundreds kHz).

The third conclusion was also endorsed by Khatiashvili (1984) (who registered EMR in the range of 1–1 000 kHz) and Bella et al. (1992) (who studied EMR in the range of 300 Hz to 300 kHz). It is clear from the above sections that there is no uniquely accepted mechanism regarding the mechanism of EMR prior to an EQ. Parrot et al. (1993), after a detailed consideration of a large number of presently known EMR-EQ investigations, remarked that although the existence of EMR in relation to seismic and/or volcanic activities was clear, EMR selection out of a host of artificial signals remained a significant problem. Nevertheless, investigations of EMR as a precursor to EQ are still in progress (Drakopoulos et al. 1993; Hayakawa et al. 1993; Yepez et al. 1995; Fujinawa and Takahashi 1998) and presently there is an agreement in the literature that EMR might be a prospective forecaster for EQ (King 1983; Mogi 1985; Yoshino et al. 1993). However, as we noted above, all efforts to use EMR for earthquake prediction met with very meager success, due to the lack of a detailed quantitative understanding of the EMR mechanism (King 1983; Rabinovitch et al. 1996).

5.9 Global Features of EMR

5.9.1 EMR versus Gutenberg-Richter Relation

5.9.1.1 *Fractal Nature of Earthquake*

A fundamental observation in seismology is the Gutenberg-Richter law (Gutenberg and Richter 1954):

$$\log(N) = a - bM \tag{5.17}$$

where M is the earthquake magnitude (which is defined as the logarithm of the integral of slip along the fault during an earthquake), N is the number of earthquakes having magnitudes greater than M, and a and b are constants. This power law or fractal distribution is valid both for main events and for aftershocks (Nanjo et al. 1998). It is presently thought that such a distribution can be a fundamental result of "multiple fracturing" (Turcotte 1992), when spontaneously occurring microcracks tend to coalesce, leading, by numerous upscalings, to a catastrophic failure. A similar relation is also observed in laboratory studies of acoustic emission (Lockner et al. 1991; Lockner 1993). It is even true for energy distribution of neutron star-quakes (Jagasivamani and Iyer 1988) (the source of a star-quake is a fracture in its neutron crust, which may release strain energies of up to 10^{46} erg! (Link and Epstein 1996)), which is many orders of magnitude larger than that of an earthquake.

Recently, enough data has been collected to extract statistics on individual systems of earthquake faults (Dahmen et al. 1998), and it was found that the distribution of earthquake magnitudes may vary substantially from one fault system to another and for different Earth regions.

Regarding the exponent '*b*' of the power-law distribution, it was previously claimed to be universal and close to 1 (Gutenberg and Richter 1954). In Molchan et al. (1997), the related *b*-values (for many earthquakes in Italy calculated for the period 1900–1993) are in the range 0.7–1.35. Elgazzar (1998) estimated the *b*-value to be 0.85 ±0.2. Nanjo et al. (1998) give b to be between 0.5 and 1.5. It has recently been claimed that '*b*' fluctuates in time and depends on the earthquake magnitude (Olami et al. 1992; Carlson et al. 1991).

In recent years, numerous investigators tried to explain b theoretically. For example, "self-organized criticality" models were proposed (Bak and Tang 1989; Ito and Matsuzaki 1990; Matsuzaki and Takayasu 1991; Olami et al. 1992) to explain this fractal law as being a result of the extremal nature of the dynamic rules governing the system. There are also other deterministic models (Carlson et al. 1991; Scott 1996) describing earthquake dynamics by the friction and elastic forces acting in the fault zone. For example, the one-dimensional Burridge-Knopoff model (Carlson and Langer 1989) leads to the Gutenberg-Richter law with $b \approx 1$.

Anton (1999) proposed an analytic model based on a relation between stress release rate and kinetic energy loss by seismic waves and noted that the relation $b < 1$ implied that the stress release rate was greater than the loss of kinetic energy by seismic waves.

Gabrielov et al. (2000a,b) developed a "colliding cascades model" consisting of a hierarchical structure. They proposed that an external load applied to the largest block is transferred hierarchically to the smallest components. Fracture processes expand in an inverse manner. The two processes "collide" and interact. On the basis of this model, they argued that the *b*-value should be 0.53 for main earthquakes and 0.69 for aftershocks. Nanjo et al. (1998), Kossobokov et al. (2000), Elgazzar (1998), Anton (1999), and Gabrielov et al. (2000a,b) for the distribution of EQ magnitudes in that the *b* value is of the order of 2/3.

5.9.1.2
Comparison of Fractal Behavior of EMR and EQ

As we noted above, amplitude changes of acoustic emission from fractures during laboratory compression tests of brittle materials were considered by Lockner (1993) and Lockner et al. (1991) to be similar to those occurring in EQ. This kind of similarity can also be gleaned from a comparison between energy release during starquakes (Fig. 5.53a, Kossobokov et al. 2000) and our measurements of amplitude of EMR induced by chalk compression failure (Fig. 5.53b). In this section this similarity is investigated, focusing on the possibility of a power law for EMR induced during conventional uniaxial and triaxial fracture.

Figure 5.54 shows the cumulative number (the number of signals having amplitudes larger than *E*) of EMR signals (measured additively from twenty-four chalk samples during uniaxial and triaxial conventional tests) vs. EMR pulse amplitude.

Since the voltage output of the EMR pulses '$\overline{A}$' depends on the antenna reaction (antenna efficiency), which changes with frequency, it was compensated (Sect. 5.3.3). We were thus able to compare heights of EMR signals of different frequencies.

EMR pulse amplitudes *E* changed by 5 orders of magnitude (Fig. 5.54): From 0.001 V m^{-1} to 100 V m^{-1}. The figure consists of four parts. Its main part is consistent with a Gutenberg-Richter type law with a *b*-value of 0.62 ($R^2 = 0.95$).

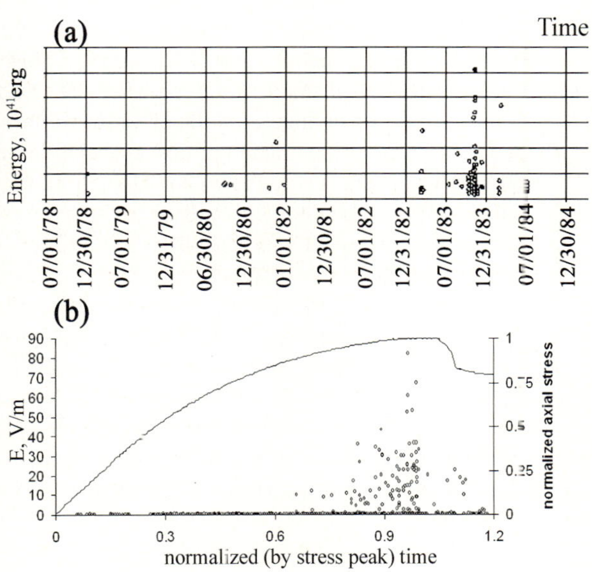

Fig. 5.53. A qualitative comparison between energy changes during starquakes (**a** after Kossobokov et al. 2000) and **b** amplitude changes of electromagnetic radiation induced by rock compression (Rabinovitch et al. 2002b)

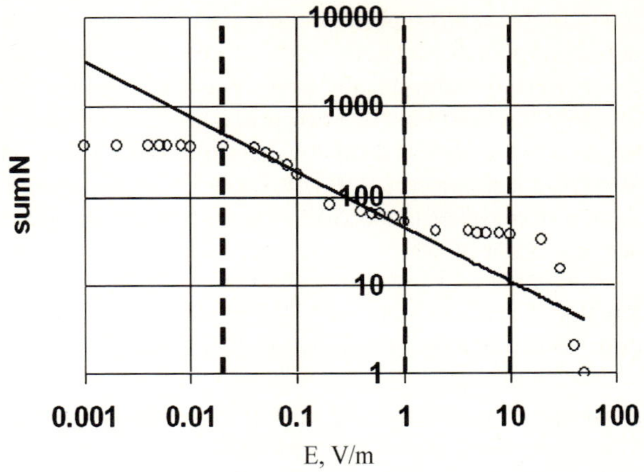

Fig. 5.54.
The number of signals having amplitudes larger than *E* of EMR signals (measured additively from twenty-four chalk samples during uniaxial and triaxial conventional tests) vs. EMR pulse amplitude (Rabinovitch et al. 2002b)

$$\text{sum}N = 51.44A^{-0.62} \tag{5.18}$$

Note that this *b*-value is close to 2 / 3 (see above). On both ends of the graph there is a deviation from the Gutenberg-Richter type law. The *b*-value for the small amplitude range is very low (0.02, $R^2 = 0.76$). This small value might be related to either an incomplete sampling of small events, to noise or to a new physical effect governing the process. In the range of large amplitudes $1 < A < 10$ V m^{-1}, the *b*-value is also significantly lower (0.08, $R^2 = 0.76$) than in the main part, which might possibly be due to the finite size of the sample. Note that the EMR amplitude in the range of $1 < A < 10$ V m^{-1} is measured when the external stress is higher than the sample's elastic limit (Fig. 5.53b). Figure 5.54 has also a fourth part (A > 10 V m^{-1}) with $b = 2.3$ ($R^2 = 0.77$). This is the range immediately before sample collapse (Fig. 5.53b).

Several investigators have noted this slope change effect for seismic data as values in the smallest and largest ranges in their investigations also deviated from the Gutenberg-Richter type law. For example, Lockner (1993) and Lockner et al. (1991) showed a roll-off in the distribution of AE signals at low amplitudes and explained it as incomplete sampling of small events (Lockner 1993; Lockner et al. 1991). Kossobokov et al. (2000) noted this change of slope effect in the high range of earthquakes and explained it as being due to the maximum energy release being limited by the size of the crust and by the energy density. They also noted that for the largest earthquakes the downward slope may altogether disappear or even turn into an upward slope.

A change of slope effect was also observed by Molchan et al. (1997) for induced seismicity both in the range of small and of large earthquakes.

The decrease of the *b*-value with stress, as happens here between 1 and 10 V m^{-1} (Fig. 5.54), is also a known effect (e.g., Lockner et al. 1991).

Figure 5.53 exhibits the escalation of the fracture process before collapse. Figure 5.53a (Kossobokov et al. 2000) shows the energy of starquakes vs. time, which occurred at a distance of about 40 000 light years from Earth. Figure 5.53b gives an EMR amplitude-stress-time graph of all chalk samples. Stress values (Fig. 5.53b) were normalized by the peak stress value. Visual comparison indicates that the two graphs are very similar.

5.9.2
Benioff Strain Release

Fracturing processes can be measured either by intensity changes (see, e.g., Fig. 5.54) or by their energy release. The latter is usually represented by the so-called "cumulative Benioff strain release"(Newman et al. 1995; Kossobokov et al. 2000; Gabrielov et al. 2000a,b), which relates the total sum of the square root of the energy released for sequential fracture events to the time prior to the collapse failure. Hence, in addition to the Gutenberg-Richter purely statistical law, the "cumulative Benioff strain release" relation enables us to monitor the continuous development of the upscaling fracture process through time.

The measured EMR amplitude is proportional to the magnetic field intensity (H) reaching the antenna. Since the energy of the electromagnetic field is proportional (Pointing vector) to $|H|^2$, the EMR amplitude is proportional to the square root of the energy. Figure 5.55 shows a bilogarithmic graph of a cumulative amplitude of all EMR signals registered during all samples' compression vs. time before their collapse failure. Since maximal values of EMR amplitude are excited by the samples' collapse failure, their occurrence time was taken as the zero time for each sample.

The graph (Fig. 5.55) shows an almost constant slope in its middle part (about a whole decade). It was fit to a power law ($R^2 = 0.987$ and slope = 1.42) and is seen to be accompanied by some logarithmic-periodic variations. This graph is very similar to the usual "Benioff strain release" ones (e.g., Kossobokov et al. 2000) and the slope of its main part is close to the usual Benioff slope of 1.35 (Newman et al. 1995).

Results obtained here show that both the b value of the Gutenberg-Richter type law and the slope of the Benioff strain release relationship of EMR signals obtained during chalk fracture in the laboratory are similar to those measured in earthquakes and starquakes. The qualitative similarities and even more so the almost exact power laws are striking. The fractal nature of the processes controlling earthquakes and starquakes can therefore be extended to a microscale regime. This "global" nature of multiple fracture effects evidently implies that a basic general process is "acting behind" all these phenomena.

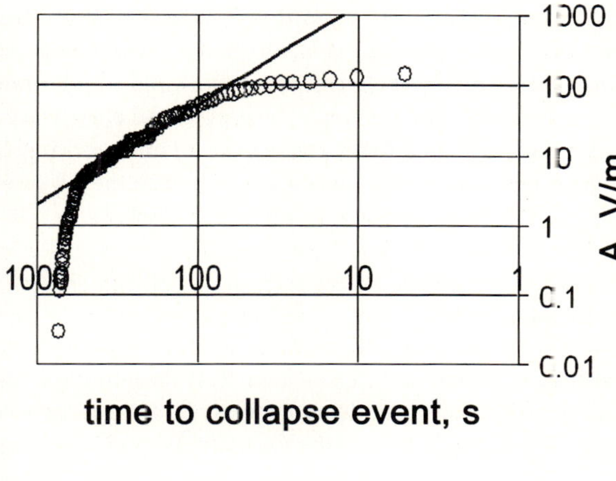

Fig. 5.55. The "Benioff strain release" relationship of a cumulative amplitude of all EMR signals registered during all samples' compression vs. time before their collapse failure (time goes backwards from the moment of collapse) (Rabinovitch et al. 2002b)

5.10 EMR Induced by Underground Rocks under Stress

5.10.1 General Background

Nesbitt and Austin (1988) registered EMR in a gold mine (2.5 km depth). An EMR signal (1.2 mA m^{-1} amplitude) was generated some seconds prior to the micro seismic event (magnitude of −0.4). Registration of EMR activity in Ural bauxite mines showed (Scitovich and Lazarevich 1985) that its value sharply increased with rock burst (RB) hazard increase. Analogous works in Noril'sk polymetal deposit (Krasnoyarsk region) revealed an increase of EMR amplitude (up to 150–200 mV m^{-1}) and activity in the RB hazardous zones (Red'kin et al. 1985). Markov and Ipatov (1986) investigated EMR activity changes in an apatite underground mine (Khibin deposit, Kola Peninsula) and found out that EMR amplitude in RB hazardous zones was in the range of 8–25 mV m^{-1} and EMR activity there was significantly higher than the regular noise level.

A survey of this general background shows that the EMR investigations in mines are rare and non-systematic. We conducted careful investigations in underground mines in different mining and geological situations associated with an increase of RB and rock-gas outburst hazard. The main aim of these investigations was to correlate between specific EMR patterns and an increase of RB and outburst hazard in underground mines and hence to develop EMR criteria for rock bursts and rock-gas outburst. We hope that a good understanding of these relations would create new avenues for successful "large-scale" EMR investigations of EMR pulses prior to ensuing earthquakes-enabling earthquake monitoring.

5.10.2 EMR in Mines

5.10.2.1 *Criterion of Rock-Burst Hazard in Coal Mine and Short Description of the "Gum" Method*

RB hazard increases if the load on a given part of a coal seam exceeds some critical level, while the distance from this part to the stress maximum (in the zone of influence of a mine-working) is lower than a critical value (Petukhov and Lin'kov 1983). Note that we use by the term "mine-working" for all underground openings irrespective of their size and purpose. If the potential energy accumulated in the zone of high stress is enough to destroy rocks near a mine-working, RB takes place.

As shown by Petukhov and Lin'kov (1983), the RB safety condition is the following:

$$\begin{cases} P_\mathrm{n} < P_\mathrm{m} \\ \dfrac{M_\mathrm{c}}{E_\mathrm{r}}\dfrac{a_\mathrm{n}}{h} < 1 \end{cases} \tag{5.19}$$

where P_n is the external applied load, P_m is the limiting load (strength), M_c is the coal drop modulus (descending modulus of the stress-strain curve beyond the peak point under rigid compression, Fig. 5.56), E_r is the Young modulus of adjoining rocks, a_n is the distance from

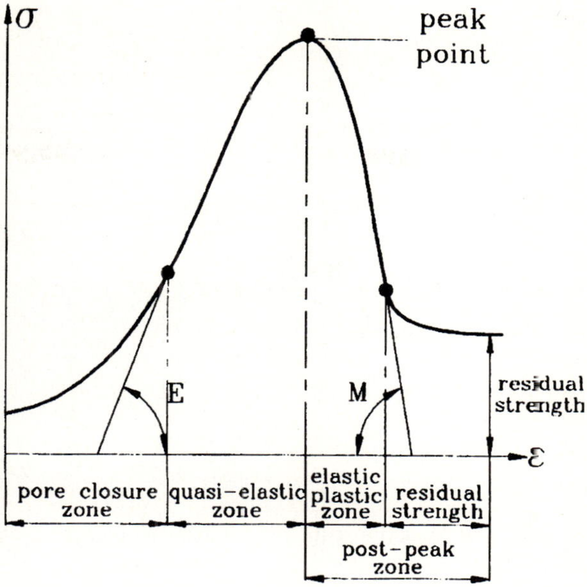

Fig. 5.56. Typical stress-strain curve during stiff sample compression (Frid 2000)

the mine-working face up to the position where there exists a peak stress, and h is the half thickness of the coal stratum. If coal is under residual strength (non-hazardous condition), both inequalities comply with the following: Load on the given zone is lower than the limiting value (first inequality) and the left side of the second inequality is smaller than 1 (the drop modulus in the residual strength zone is equal to 0).

The RB hazard is usually determined by some standard geomechanical method, e.g., volume measurement of gum (dry coal "powder"), measurement of the diameter of the hole created or the number of disks that are created due to core fracturing as a result of drilling in a highly stressed zone, etc. (Petukhov 1972).

Drilling of a highly stressed coal seam leads to an intensive fracturing process in the zone influenced by the drill hole. The volume of this highly cracked zone depends on the hole diameter, the drilling rate and especially, the stress level. Hence, if the first two parameters remain invariant for a given coal seam, the stress value in the coal seam is responsible for the volume of gum that is recovered from the hole (i.e., from the highly stressed zone drilled by the hole). This method is widely used for RB prediction, although it is time consuming and in some conditions even dangerous.

The normalized ratio, β, of the highly stressed zone (ratio of the non-elastic deformation zone, diameter D, to the hole diameter $d = 0.043$ m, Fig. 5.57, Frid 2001) can be calculated from the following formula (Petukhov 1972) is given by:

$$\beta = -0.5 + \sqrt{0.25 + \frac{3\frac{M_0 + M_s}{M_0} n_r - n_r + 2}{n_r - 1}} \tag{5.20}$$

where n_r is the coefficient of coal loosening on the borehole wall that is usually equal to 1.3–1.4, M_0 is the gum volume in the borehole ($M_0 = \Lambda \pi d^2 / 4$, Λ is the borehole

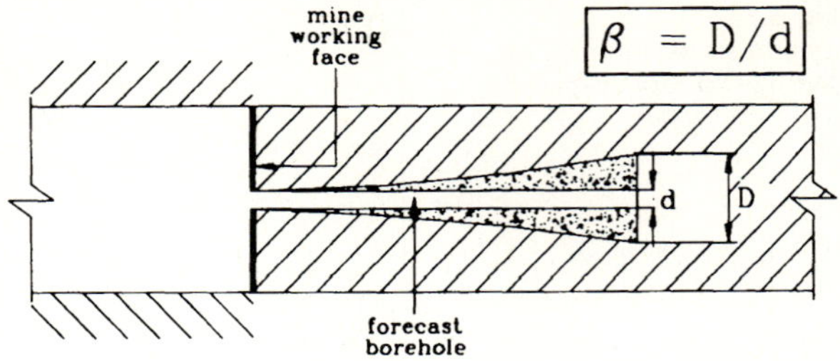

Fig. 5.57.
The zone of non-elastic deformation excited by drilling in the high stressed zone (β is the non-dimensioned diameter of this zone: The ratio of the non-elastic deformation zone diameter D to the hole diameter $d = 0.043$ m) (Frid 2001)

Table 5.4.
Calculation of rock-burst hazardous zone parameters

a' (m)	Gum volume ($l\,m^{-1}$)	β	$\sigma_{k\gamma H}$ (MPa)	X_m (m)	ξ (m)
4.00	5.00	5.03	54.98	6.35	2.35
5.00	6.00	5.71	58.28	6.98	1.98
6.00	6.50	6.02	59.67	7.43	1.43
7.00	7.50	6.60	62.08	7.96	0.96

length), and M_s is the gum volume induced by drilling in a stressed zone) (Petukhov et al. 1976).

The vertical stress in the coal seam can be determined as follows (Petukhov et al. 1976):

$$\sigma_{k\gamma H} = k_*(1 + 2\ln\beta) \tag{5.21}$$

where k_* is the coal shear strength. Forecast boreholes are usually drilled in successive depths (the length of each beeing 1 m). Hence, we can estimate the vertical stress distribution in the coal seam near the mine-working face.

After the drilling of each interval, the gum volume is measured and if it exceeds a definite limiting value (the experience at the mining works in North Kuzbass shows that the limiting values are 5 to 8 liters per meter at the fourth and the seventh drilling meter from the drift face, respectively, (Table 5.4, Frid 2001)), drilling is stopped, and that part of the mine-working is considered RB hazardous.

As noted by Lockner (1993, 1996), there is a strong parallelism between the well-known Gutenberg-Richter relation for seismic events (from macro (earthquake) to micro (RB)) and power law frequency-magnitude relationship for acoustic emission events. This analogy suggests that micro shocks (high frequency and small magnitude) are precursors of macro failure (large magnitude and small frequency). This is the theoretical basis for RB forecasting by the acoustic emission method (Kuksenko et al. 1982; Mansurov 1994). The EMR frequency range is close enough to the acoustic emission band. Moreover, Gutenberg-Richter type relationships similar to the seismic ones were also observed for EMR (Rabinovitch et al. 2002b, Sect. 5.9.12) Hence, it would be correct to assume that EMR could be useful for RB hazard forecasting along with acoustic emission. Moreover, being non-contact, the EMR method has some advantages over acoustic emission. For example, when a rapid and comprehensive prognosis of a short-term mine-working region (for example, in a drift face) is needed, the roughness of the mine walls becomes a marked problem for the acoustic emission method for rapid data acquisition due to inferior contact between the acoustic emission transducer and the mine wall.

5.10.2.2
EMR Measurements in Coal Mines

EMR pulses excited by rock fracture show up as a "bell-shaped" oscillation characterized by amplitude, a frequency and an individual rise and fall time (Rabinovitch et al. 1995, 1996, Sect. 5.2). As a rule, EMR is measured in mines within the narrow band in the 30–150 kHz range. The middle part of this range (100 kHz) is the most convenient one in coal underground mines (Frid et al. 1992). This range is conditioned both by the industrial electromagnetic noise level in underground mines and the electromagnetic pulses' amplitude. The Electromagnetic pulses' field amplitudes excited by coal failure near the face of a mine-working range are between 5 and 50 μV m^{-1} (Frid et al. 1992) that is at least one order of magnitude larger than mine industrial noise levels. An industrial noise level at the Earth's surface is usually not lower than 5 mV m^{-1}. Hence, an industrial electromagnetic noise in coal mines springs from underground mining machines, but not from the Earth's surface. The daily work cycle of a mine consists of four six-hour shifts: three extraction (mining) ones and one repair shift. The probability to pick up an electromagnetic transient process pulse due to the starting/stopping of electrical mining machines during extraction shifts is large enough, while the electrical noise level during the repair shift is as noted above. Thus, we carried out our measurement during the repair shift. As shown by Frid et al. (1992), EMR registration near a face of coal drift enables us to obtain EMR signals excited by a coal failure up to a depth of 10 m from the drift face. The daily drift moving in North Kuzbass conditions was usually of about 6 m. Hence, a measurement of EMR during the repair shift can characterize at least the daily activity at a coal drift face.

The EMR was measured by a resonance 100 ±1 kHz antenna that was located 1 m from the mine-working face and was connected to a counter. We defined the EMR activity as the number of crossings of the EMR voltage signals (per unit time) through a given voltage level of a special counter (Fig. 5.58, Frid 2001). At all studied points 30 EMR activity readings were taken, each with a duration of 10 s. For simplicity, we named units of EMR activity "pulse per 10 seconds", and they were symbolized by

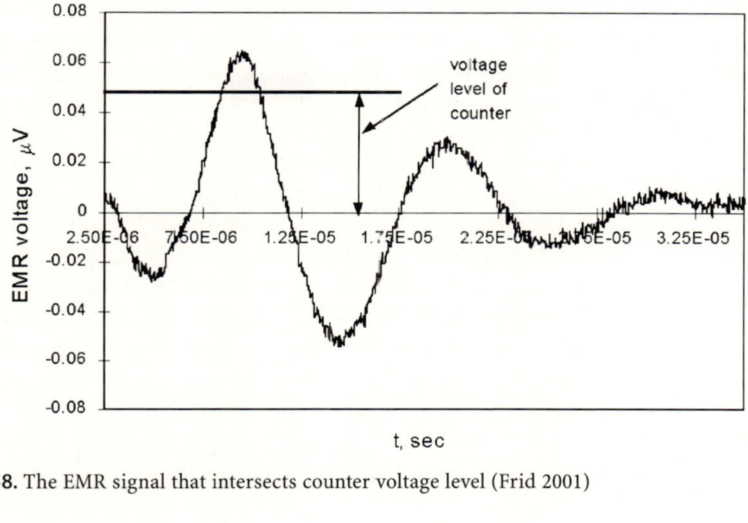

Fig. 5.58. The EMR signal that intersects counter voltage level (Frid 2001)

"pulse/10 s". A complete cycle of EMR measurements at a single drift face is of 5 min duration. This method allowed us to determine a stable mean that characterized EMR activity at a given point.

Our preliminary mine estimation of EMR compatibility conditions showed that choosing the given resonance frequency would allow us to accurately extract the useful signal from the industrial background noise. As is known, the use of the shortest wavelength band of the measuring system is a necessary condition to increase the system's sensitivity. The *Q* factor in our case is 50, resulting in a system sensitivity not lower than 0.1 mV m^{-1}.

5.10.2.3
Mine Investigations prior to Rock Bursts

An increase of RB hazard is usually associated with the existence of pillars, geological faults, sharp changes of inclination angle or thickness of coal seams (Petukhov 1972). We conducted our investigations in coal mines of the North Kuzbass, which exhibited all of these conditions.

An example of EMR registration at the zone of pillar influence is shown in Fig. 5.59 (Frid 1997a) The study was conducted in a mine-working where the influence of pillars was found at a distance of 70 m from the conveyer drift 13. Profiling of the mine-working (in spacing of 10 m) showed an anomalous EMR activity at stressed regions under pillars.

A second example of EMR registration at a zone of pillar influence is shown in Fig. 5.60 (Frid 2001). The EMR sequential measurements (profiling with 3 m spacing) were carried out in an extraction gallery that partly drifted into the zone of pillar influence. As can be seen, EMR activity is anomalous under the pillar (120–140 pulse/10 s). Gum value measurements showed the existence of a RB hazard in this region (gum value was about 7–12 l m^{-1}).

A third example of EMR registration is shown in Fig. 5.61 (Frid 2001). In this case, drift was in a zone of an intensively variable thickness of coal seam and a sharp change-

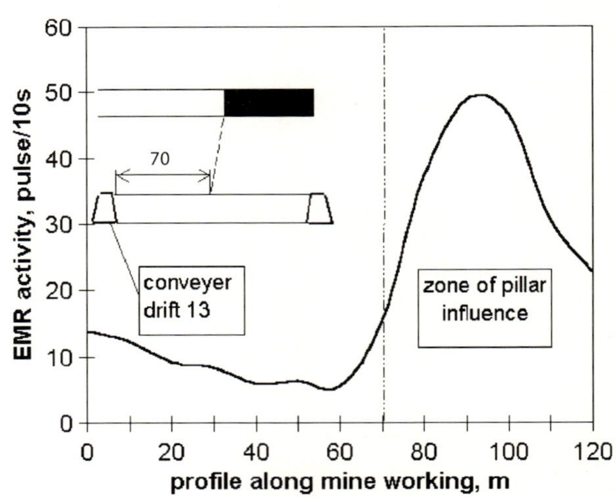

Fig. 5.59. EMR profiling under pillar (Frid 1997a)

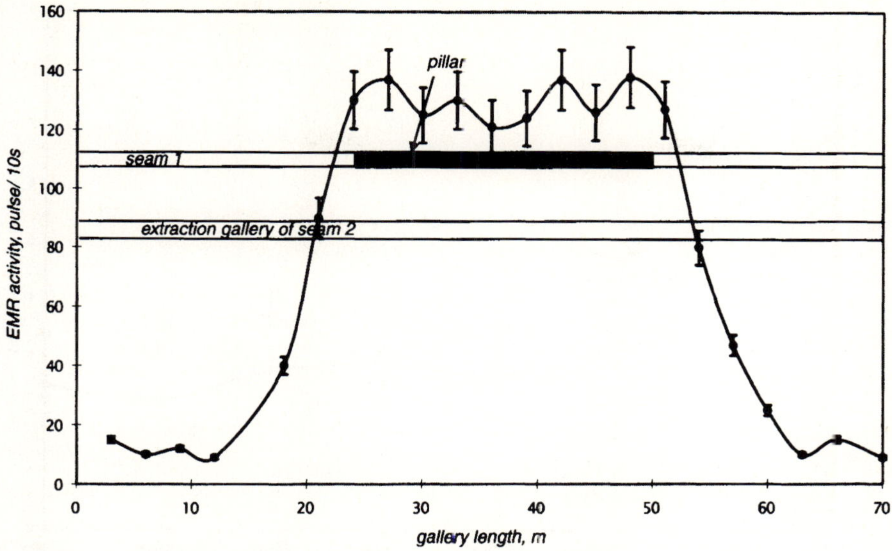

Fig. 5.60. EMR profiling at the zone of pillar influence. The error in EMR activity is of the order of ±7.5% (Frid 2001)

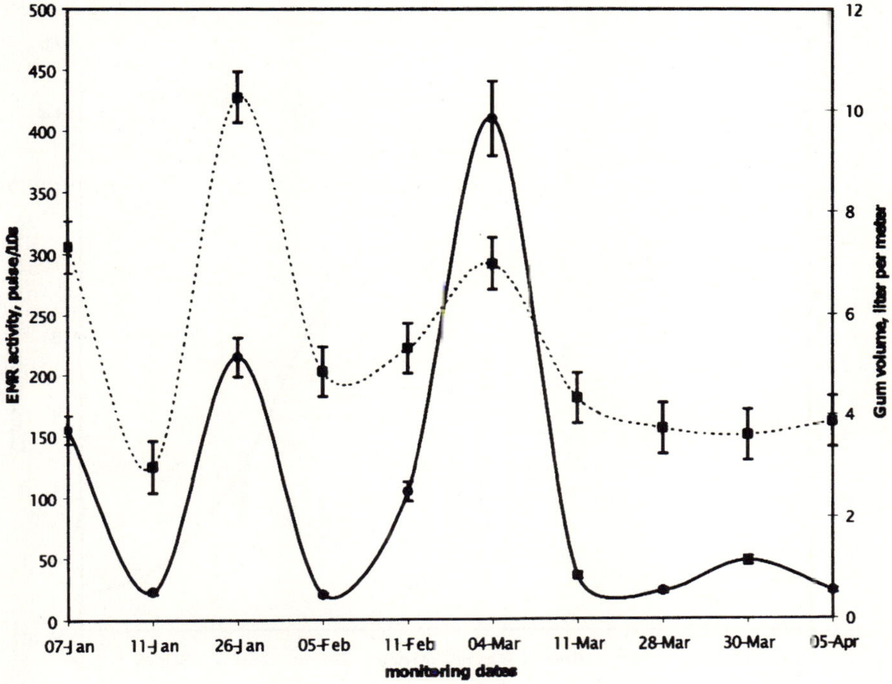

Fig. 5.61. EMR monitoring during drift excavation in an intensively changeable thickness of coal seam. The error in EMR activity is of the order of ±7.5%, while in gum value is of the order of ±0.5 l m^{-1} (Frid 2001)

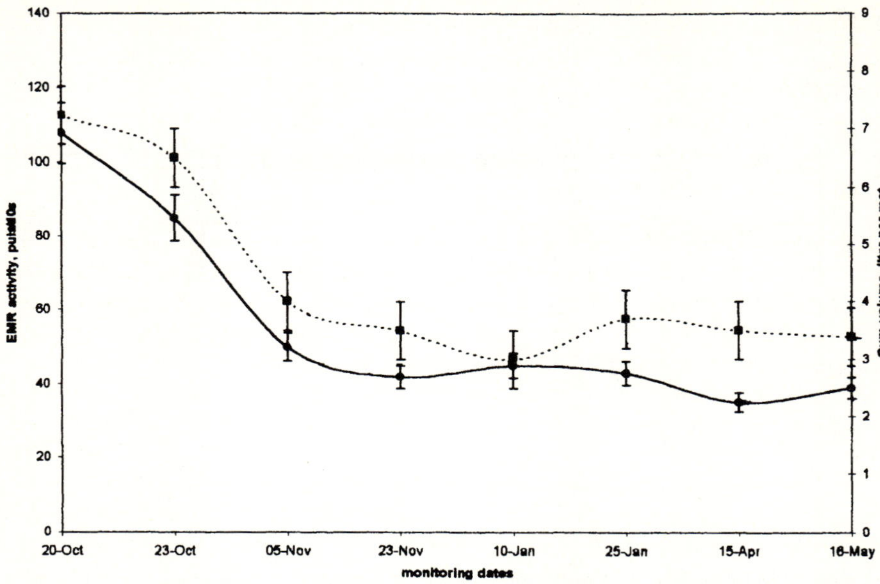

Fig. 5.62. Anomalous EMR activity and gum value in the mine excavation near the uplift. The error in EMR activity is of the order of ±7.5%, while in gum value is of the order of ±0.5 l m^{-1} (Frid 2001)

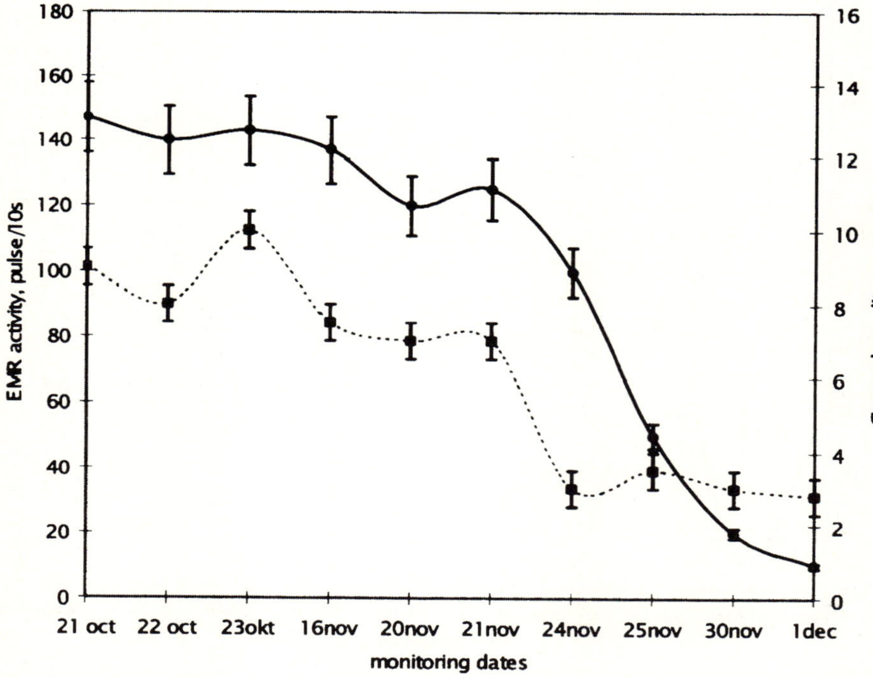

Fig. 5.63. EMR monitoring during pillar unloading. The error in EMR activity is of the order of ±7.5%, while in gum value is of the order of ±0.5 l m^{-1} (Frid 2001)

able inclination angle. The regular thickness of the coal stratum was 1.5–1.7 m. A sharp decrease of the thickness of the coal stratum was associated with a RB hazard increase (a decrease of coal stratum thickness induces a stress maximum nearer to the drift face and hence a RB hazard increase). For example, on 7 and 26 January and on 4 March, the coal stratum thickness decreased to 0.5–0.7 m, with a corresponding gum volume and EMR activity increase of up to 7–10 l m^{-1} and 150–400 pulse/10 s respectively.

Figure 5.62 (Frid 2001) shows an example of an anomalous EMR activity and gum value in a mine excavation that was opened at a distance of 8 m from an uplift with a 2 m amplitude. As the excavation moved away from the uplift, the EMR and the gum monitoring showed gradual decrease. Unloading of RB dangerous rocks near mine-workings is usually performed by the following methods: Hydroprocessing, drilling of boreholes of large diameter and blasting. We carried out EMR activity investigations both before and after these unloading operations. If these operations were successful, i.e., mine-working became safe, EMR activity should have decreased to its usual level.

Figure 5.63 (Frid 2001) shows an example of an EMR registration during pillar unloading. The average level of EMR activity during the period of 21 October to 16 November was 130–150 pulse/10 s. Measurement of the gum volume showed that this pillar was RB hazardous (gum volume was 8–10 l m^{-1}). Unloading was performed by drilling 100 mm diameter holes. The holes enabled us to decrease the pillar brittle response. Drilling began at 17 November. This led to pillar unloading, and both gum volume and EMR activity gradually decreased.

Figure 5.64 (Frid 1997a) shows an example of EMR profiling in a drift that was initially RB dangerous along all of its length after it was unloaded. Unloading was conducted by hydroprocessing that includes the drilling of a borehole of 6 m length at intervals of 30–50 m followed by water pumping into the rock. Hydroprocessing leads to controlled rock cracking and removes the zone of increased stress from the mine-working wall deeper into the rock-mass. Figure 5.64 (Frid 1997a) shows an example of EMR measurement when one part of drift (from 0 to 70 m) was previously hydroprocessed and became safe (gum value was lower than 5.5 l m^{-1}). The average level of EMR activity at this region was not more than 40 pulse/10 s. However, the unprocessed drift zone was very hazardous (gum value of up to 8–9 l m^{-1}), a fact that was reflected by a high level of EMR activity (of up to 250 pulse/10 s).

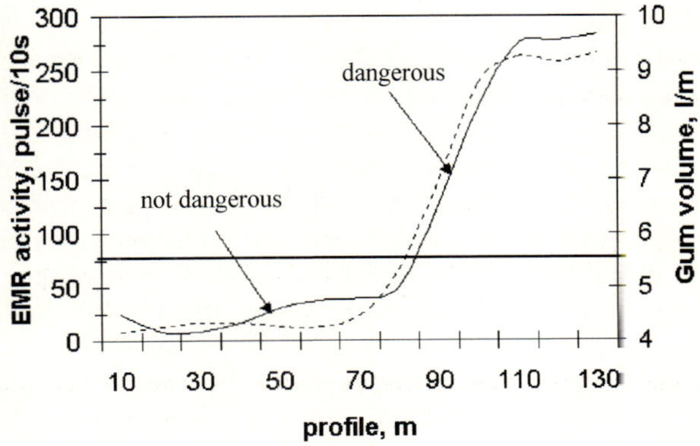

Fig. 5.64. EMR registration by drift profiling (Frid 1997a)

We carried out EMR investigations in more than 400 different RB situations in underground coal mines (North Kuzbass deposit) (Frid 1990; Frid et al. 1992; Frid 1997a,b). Our investigations showed that an increase of RB hazard, accompanied by an increase of EMR activity, occurred when pillars were found above mine-working, where geological faults existed in the area, and/or where seam thickness sharply diminished (Frid 1997b). These investigations also showed that the highly stressed zone near mine-workings is the source of the EMR (Frid 1990, 1999). An analysis of all the data enabled us to determine the following empirical criterion of RB hazard:

$$N > N_{\mathrm{emp}} \tag{5.22}$$

where N is the value of EMR activity registered at a mine-working, and N_{emp} is the limiting value of EMR activity (115–120 pulse/10 s, Frid 1997b). If the value of the EMR activity, N, exceeds N_{emp}, a RB hazard existed.

The amount of time needed to determine the degree of a RB hazard by the EMR and the gum methods at a given point is about 5 and 30 min, respectively. The reduced time is a significant advantage of EMR forecasting. However, the amount of time for data collection to establish an empirical N_{emp} in a specific mine is much longer, even up to several years. The only way to reduce this time is to develop a theoretical model of this criterion. The following section establishes an empirical EMR criterion of RB hazard and compares it with the theoretical criterion.

5.10.2.4
Theoretical EMR Criterion for Rock-Burst Hazard in a Coal Mine

Zone of EMR formation under external stress. Goncharov et al. (1980) carried out an EMR study with a stiff press. They loaded concrete samples of $0.55 \times 0.55 \times 0.65$ m dimensions under compression. The EMR activity was measured with a magnetic antenna that had a resonance frequency of 800 kHz (±1 kHz). The first significant increase of EMR activity occurred at the post-peak zone under a load of 0.95 σ_c, where σ_c is the limit of strength, and a second rise of EMR activity was noted for a load of 0.9 σ_c (in the post-peak zone). Single pulses of EMR accompanied the subsequent decrease of the load until 0.75 σ_c. In the zone of residual strength, EMR activity ceased.

EMR studies with a stiff press were also performed on ores and rocks from the Khibin apatite deposit (Kola Peninsula, Ipatov 1989), on magnetic ores, granites, white marble and coal from Shpitsbergen (Frid 1990). EMR activity was registered with a 100 kHz (±1 kHz) resonance magnetic antenna. The maximum EMR activity was measured in the range of (1.0–0.8) σ_c in the post-peak zone, while in the zone of residual strength EMR activity stopped. Note that wide frequency band EMR measurements (Rabinovitch et al. 2000c) confirmed these results. An analysis showed that the following EMR features were common to all rocks:

1. A sharp increase of EMR activity near the peak stress.
2. An absence of EMR in the zone of residual strength.

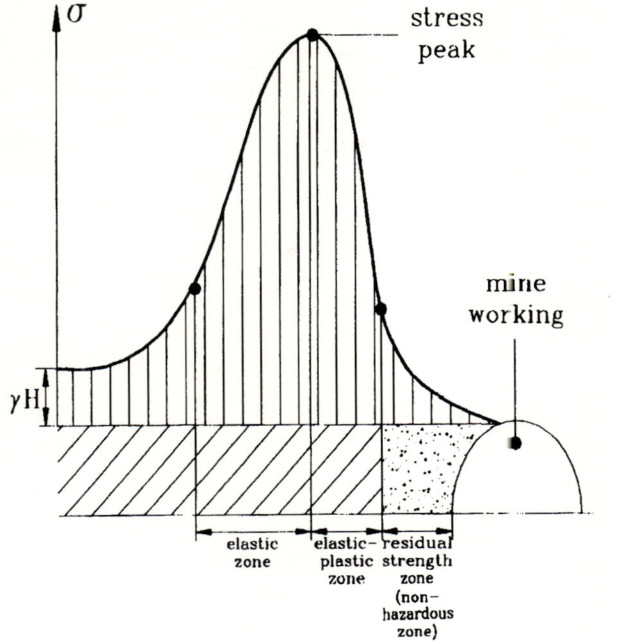

Fig. 5.65. Typical stress distribution curve in the zone of mine-working influence (Frid 2000)

To check these conclusions in situ we investigated the EMR activity during borehole drilling (hole diameter was 42 mm). Note that the residual strength zone (the last zone during sample deformation on a stiff press, Fig. 5.56) is the nearest zone to the mine-working face (Fig. 5.65).

All holes were drilled at 1 m separation intervals. During all interval drilling, a cumulative value of EMR activity (summarized value of EMR activity per drilling interval) was measured, and after drilling each interval the gum value was measured. Forecast drilling in the mine does not generally achieve stress peak due to instrument "gripping" in a borehole, and stops if gum volume (stress value) becomes higher than a predetermined value. Forecast drilling (in North Kuzbass mines) is generally conducted up to 6 m depth. Figure 5.66 shows two examples of EMR observation during borehole drilling:

1. In the RB hazardous zone (gum value is higher than the limiting value, see below).
2. In the non-hazardous zone (gum value is lower than the limiting value, see below).

As is seen from Fig. 5.66a, gum value increases up to 12 l m^{-1} in the 5th drilling meter. This value is significantly higher than the limiting value (5.5–6 l m^{-1}), and hence the forecast drilling was stopped. It evidently indicates the existence of a RB hazardous zone at the 5th meter depth from the mine-working face. Figure 5.66a shows that when a hole face approaches a stress peak (gum increase above the limiting value), EMR activity sharply increases, and vice versa: If the gum value doesn't change drastically (gum value lower than 5.5–6 l m^{-1} at the 5–6 drilling meter non-hazardous zone, Fig. 5.66b), EMR activity does not essentially increase.

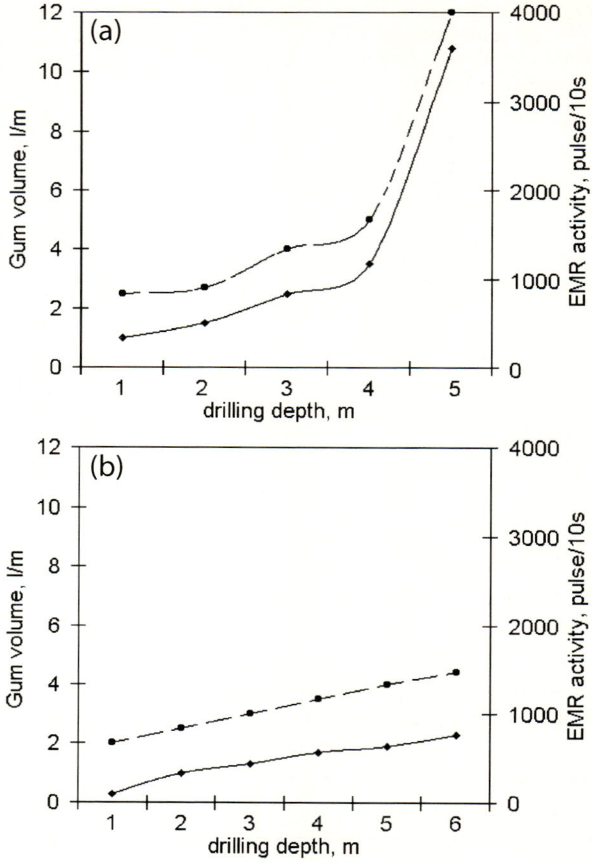

Fig. 5.66. Two examples of EMR study during forecast drilling: **a** in the rock-burst hazardous zone, and **b** in the non-hazardous zone. *Full-drawn curve* is the Gum value, while the *hatched curve* is the EMR activity (Frid 2000)

The analysis of more than 150 drillings conducted on three different coal stratums (Desyatiy, Koksoviy and Andreevskiy) enables us to summarize some general features of EMR activity associated with borehole drilling:

1. The EMR activity sharply increases if the hole face approaches a zone of highly stressed coal (that is near the stress peak, e.g., Fig. 5.66a).
2. Drilling in a non-hazardous zone does not tangibly affect EMR activity (e.g., Fig. 5.66b).
3. The EMR activity value induced by drilling is lowest nearest to the mine-working face, i.e., in the zone of crushed coal or residual strength (see, e.g., first drilling meter from the mine-working face, in Fig. 5.66).

Calculation of rock-burst hazard zone volume. Drilling for gum volume measurements in a mine is usually conducted up to a depth of increased stress $\sigma_{k\gamma H}$, but stress maximum is usually not measured due to instrument gripping (as a result of the hole squeezing around the drill rods) in the borehole. However, for the theoretical EMR analysis it is important to know the volume of the highly stressed zone, and hence the distance from the mine-working face up to the stress maximum.

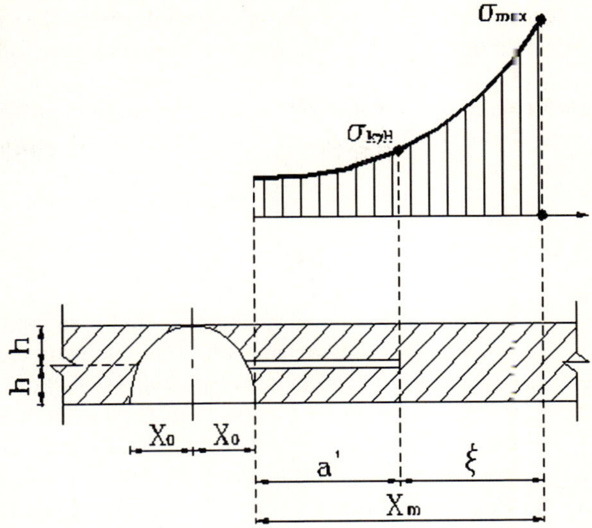

Fig. 5.67. The vertical stress distribution in the zone of influence of the mine-working (all parameters are discussed in the text) (Frid 2001)

The distance to the stress maximum (Fig. 5.67, Frid 2001) can be estimated as follows (Petukhov et al. 1976):

$$X_{\mathrm{m}} = \frac{1.23X_0}{\left(\dfrac{\sigma_{\mathrm{c}}}{\sigma_{k\gamma H}}\dfrac{X_0 + a'}{h}\right)^{2/3}} \qquad (5.23)$$

where X_0 is the mine-working half width, a' is the distance from the mine-working face to a boundary of the zone of increased stress ($\sigma_{k\gamma H}$, Fig. 5.67), h is the coal stratum half thickness, and σ_c is the coal uniaxial compressive strength (11 MPa; Frid 1990). If we assume that the drift width, $2X_0 = 3$ m (typical for vent and tramming drifts in North Kuzbass mines) and the coal stratum thickness $2h = 2$ m (also a representative value), we can calculate the distance to the stress maximum by Eq. 523. Table 5.4 shows that for all RB hazardous conditions (4–7 m from the mine-working face, Frid 1997a) the stress maximum is about $\xi = 1$ m from the commencement of the stressed zone ($\sigma_{k\gamma H}$ zone, Fig. 5.67).

Let us assume that the RB hazard zone before coal drifting has a parallelepiped form with a width equal to the drift width ($2X_0 = 3$ m), a height equal to the coal stratum thickness ($2h = 2$ m), and a length equal to the distance from the origin of the $\sigma_{k\gamma H}$ stress zone to the stress maximum ($\xi = 1$ m). Hence, the volume of the RB hazard zone would be 6 m^3.

A formation of a 6 m^3 stressed zone cannot be developed instantly. As noted by Kuksenko et al. (1982, 1987), the time to form a highly stressed zone is related to its dimensions. This relation is valid for laboratory fracture experiments, mine pillar loading, RBs and even for earthquakes (Kuksenko et al. 1982; Mansurov 1994) and can be approximated by the following formula:

$$T = 10^{3.75}L \qquad (5.24)$$

where T is the time (s) to form the stressed zone, and L is an average dimension of the stressed zone (m).

If the volume of the stressed zone is 6 m^3, its average dimension, L, would be about 2.5 m, and according to Eq. 5.24, the time to develop this zone will be approximately 4 hours.

EMR source size estimation. Assuming that the cracking rate n (number of cracks per unit time) is nearly constant during the whole period of the highly stressed zone formation, then

$$n = \frac{N'}{T} \tag{5.25}$$

where N' is the total number of cracks created in the volume V.

The crack length is associated with the time from the EMR pulse origin up to its envelope maximum (Rabinovitch et al. 1998, 1999b, Eq. 5.2). For 100 kHz pulses (Frid 1997a,b), this time is about $T' = 10\text{–}15\ \mu s$ (Frid 1990; Rabinovitch et al. 1998).

As noted by Marder (1996a), crack velocity does not exceed the Rayleigh wave velocity v_R. For coal, v_R is of the order of 1 100 m s^{-1} (Frid 1990). Hence, the length of the crack, ℓ, that radiates a 100 kHz EMR pulse would be

$$\ell = v_R T' \tag{5.26}$$

Estimation of EMR activity in the hazardous zone. As follows from the kinetic theory of failure (Zurkov et al. 1969; Regel' et al. 1972; Petrov and Gorobetz 1987), the maximum number of cracks of length ℓ in a volume V is limited by the following relation:

$$N' = \frac{V}{(k\ell)^3} \tag{5.27}$$

where k is about 3, the concentration factor. N' ranges between 45 000 and 170 000 and its average value is 107 500. Hence, the limiting cracking rate (according to Eq. 5.25) will be $n_{lim} = 7.7$ events per second. Hence, a RB hazard situation is predicted when the cracking rate, n, is larger than this limit value:

$$n > n_{lim} \tag{5.28}$$

The final step is to associate this criterion of cracking rate with a specific parameter of the EMR. There are numerous EMR parameters that could be measured in a mine, for example, EMR activity (see Sect. 5.10.2.2), the number of EMR events (often registered as the number of EMR peaks) per unit time, or the energy of EMR pulses per unit time determined, for instance, by the summation of pulse amplitude squared multiplied by pulse duration, per unit time, etc. All of these parameters characterize only the EMR rate (number of EMR events per unit time), and can specify only the cracking rate.

As noted above, the EMR activity is the number of crossings (per unit time) of the EMR signal amplitude a specific voltage level of the counter. The number of crossings formally depends on the EMR amplitude, i.e., the higher the EMR pulse amplitude, the larger is the number of intersections with a defined voltage threshold. However, our experience shows that the EMR amplitude is relatively stable (5–50 mV m^{-1}) and does not deviate by more than one order of magnitude from its average level. This phenomenon may be explained in the following manner: EMR amplitude is a function of crack area (Rabinovitch et al. 1998, 1999b) and this implies that this amplitude is inversely proportional to the EMR frequency (Rabinovitch et al. 1999b). Hence, if we select a specific frequency range to measure EMR activity (i.e., a definite crack width), we limit the range of EMR amplitudes that will be registered (the EMR amplitude span observed in laboratory conditions by Rabinovitch et al. (1999b) is also not more than one order of magnitude for a given specific frequency).

The amplitude of EMR pulses quickly decreases after 2–3 oscillations (Fig. 5.58, Frid 2001). To differentiate the EMR data from mine industrial noise, we preset the voltage level of our counter to register only the maximal half-wave of the signal (Fig. 5.58). The maximal half-wave therefore intersects a set voltage level twice (Fig. 5.58), and the value of the EMR activity (especially the limiting value) is twice as high as the cracking rate:

$$N_{th} = 2n_{lim} = 15.7 \text{ EMR pulses/s (157 EMR pulses/10 s)} \tag{5.29}$$

Hence, the theoretical criterion of RB hazard is the following: A RB hazard situation is predicted if the EMR activity, N, is larger than $N_{th} = 157$ pulse/10 s:

$$N > N_{th} \tag{5.30}$$

As seen, the calculated N_{th}=157 pulse/10 s and the empirical N_{emp}=115–120 pulse/10 s criteria (see above following Eq. 5.22) are close enough.

Generalizing formulas (5.22, 5.28–5.30), we can formulate a theoretical EMR criterion via the theoretical value of cracking rate for a given mine situation:

$$N > N_{th} = kn_{lim} \tag{5.31}$$

where N is the measured EMR parameter in a mine-working, N_{th} is the theoretical limiting value of this EMR parameter, k is a coefficient that depends on the type of parameter (e.g., k for activity equals 2, but if we measure the number of EMR peaks, $k = 1$), and n_{lim} is the limiting cracking rate.

Our theoretical approach based on a minor number of parameters (limiting value of gum volume, typical mine-working width, coal seam thickness, and coal uniaxial compressive strength) enables us to estimate a value for EMR criterion that must be verified by mine investigations. Its adoption would significantly reduce the time to use the EMR criterion.

5.10.2.5
Precursors to Rock-Gas Outburst

Another dangerous phenomenon in mines is the "rock and gas outburst" in which a burst is assisted by liberated gas (CH_4 in coal mines, N_2 in rock salt mines, etc.). The gas is liberated by the rock fracturing and acts (like pore pressure) to decrease the critical stress level needed for destruction of the mine-working walls (Petukhov and Lin'kov 1983).

At present the standard geo-mechanical method of rock-gas outburst prognosis in coal mines consists of measurement of volumes of gum and gas liberation rate (Petukhov and Lin'kov 1983). Forecast boreholes are usually being drilled at intervals of 1 m. After drilling of each interval, the volume of gum and the quantity of gas liberation are measured. If both of these parameters exceed a definite limiting values, drilling stops, and the given part of the mine-working is considered to be hazardous to rock-gas outburst. An experience of mining works in North Kuzbass shows that a formation of rock-gas outburst hazard zones is reliably fixed at the quantity of gum of more than 5–6 $l\,m^{-1}$ (liter per meter) and a gas liberation rate value of more than 5 $l\,min^{-1}$ (liter per minute) at the fourth to sixth intervals (depths) of the forecast borehole (from the fourth to the sixth meter from a mine-working face). This method is widely used, although sometimes bore-holing can itself initiate rock-gas outburst.

EMR observations in coal mines. After registrations of natural EMR activity, the degree of rock-gas outburst hazard at that point of mine-working was detected by the standard method of measurement of the volumes of gum and gas liberation rate.

The first example of the EMR registration is shown in Fig. 5.68 (Frid 1997b). In this case, drift was conducted in a zone of an intensively variable thickness of coal seam and a sharp changeable inclination angle. Figure 5.68 shows that a decrease of seam thickness from 2.6 m (18 April) to 1.1 m (from 17 May up to 30 May) corresponded to rock-gas outburst hazard increase (an increase of gum value of up to 6 $l\,m^{-1}$ and gas liberation rate quantity of up to 7–12 $l\,min^{-1}$) with a simultaneous increase of the natural EMR activity level. A restoration of coal seam thickness to its original value (15 June–29 June) caused a decrease of gum and gas liberation rate parameters to their regular levels and a decrease of EMR activity. An increase of EMR activity was also obtained during a drift conducted through an inclination angle raise (of up to 60°) of the coal seam (5 April, Fig. 5.68).

An example of EMR registration in a zone of geological fault influence is shown in Fig. 5.69 (Frid 1997b). At a distance of 8 m from a geological fault, an anomalous EMR activity was registered (20 October). Controlled boreholing showed that the value of gum was of the order of up to 6 $l\,m^{-1}$ and the gas liberation rate was of the order of up to 16 $l\,m^{-1}$. A further conduction of this drift left the fault influence zone, and EMR activity returned to normal (from 16 May up to 28 October).

EMR monitoring was also conducted in a mine-working where the influence of pillars stretched along 150 m of a drift. Profiling of the mine-working showed an EMR activity at the stressed region under the pillar where the values of gum and gas liberation rate were also irregular: More than 7 $l\,m^{-1}$ and 8 $l\,min^{-1}$ respectively.

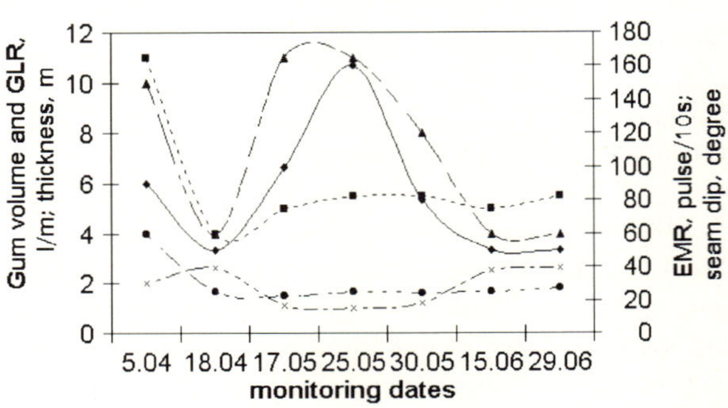

Fig. 5.68. EMR monitoring in mine-working. Note that *GLR* on this figure is gas liberation rate (■ Gum volume, ▲ GLR, × seam thickness, ● seam dip, ◆ EMR) (Frid 1997b)

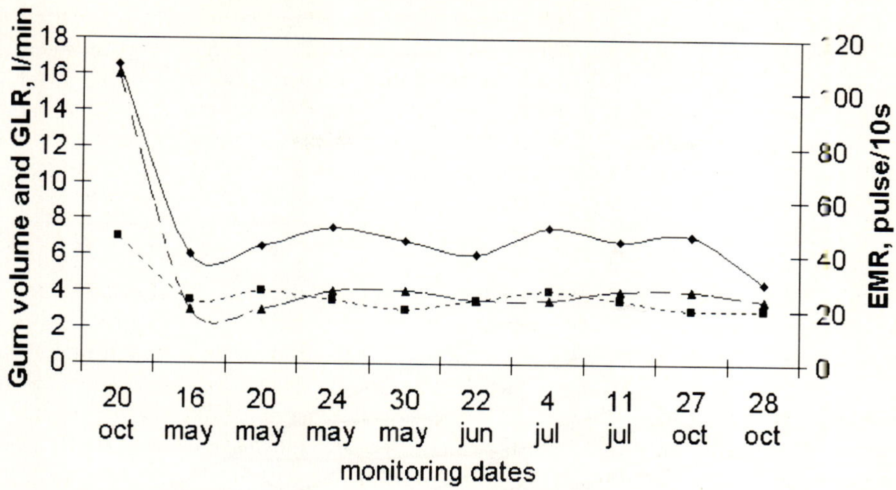

Fig. 5.69. EMR monitoring in drift (■ Gum volume, ▲ GLR, ◆ EMR) (Frid 1997b)

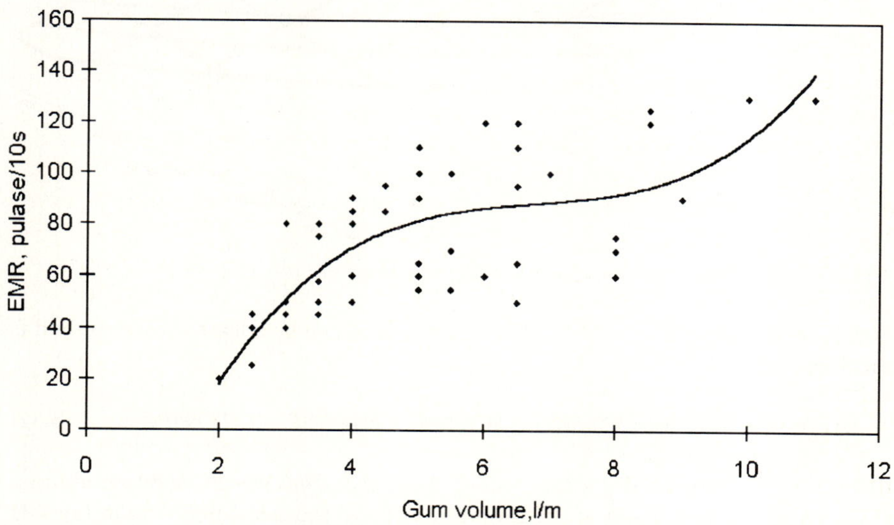

Fig. 5.70. EMR activity vs. Gum volume. Experimental EMR data are marked by ◆ (Frid 1997b)

Electromagnetic radiation forecast criterion. We have conducted more than fifty investigations in rock-gas outburst dangerous and safe mine situations (Frid 1997b). Figure 5.70 depicts a scatter of EMR activity data vs. gum values, while Fig. 5.71 shows EMR activity vs. gas liberation rate values. Significant data dispersion is seen in both cases, yielding regression coefficients in both cases of not more than 0.7. However, both show a monotonic increase of EMR activity with a rise in both gum and gas liberation rate. We hypothesized that the low correlation coefficients imply that the value of EMR activity depends on both parameters together rather than on a single one. To examine this hypothesis, we drew all experimental data (fifty-eight experimental points) in the form of EMR activity vs. both gum volume and gas liberation

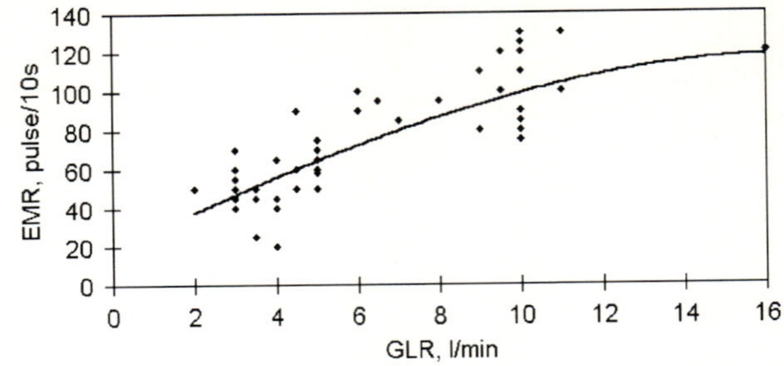

Fig. 5.71. EMR activity vs. gas liberation rate. Experimental EMR data are marked by ◆ (Frid 1997b)

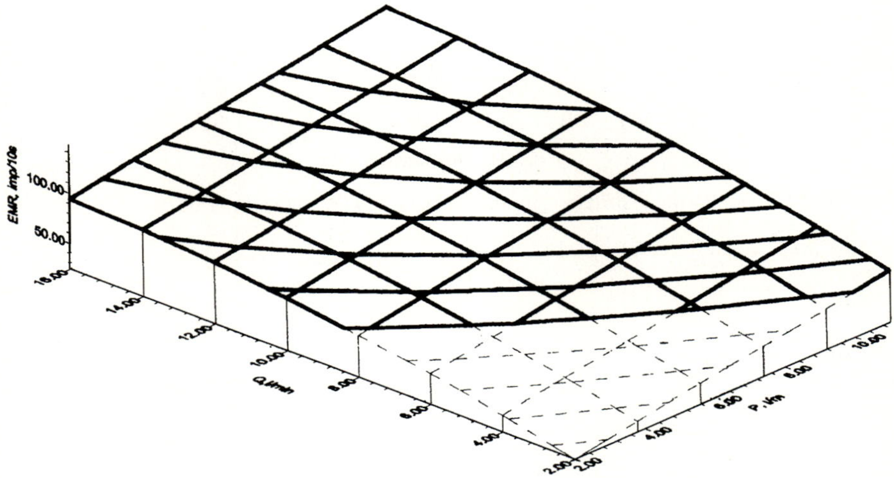

Fig. 5.72. EMR activity surface vs. Gum (P) and gas liberation rate (Q) (Frid 1997b)

rate quantities (Fig. 5.72). A fit of this surface calculated by a least square method is the following:

$$N(P,Q) = -16.1651 + 13.2875Q - 0.361033Q^2 + 6.44858P - 0.227596PQ \tag{5.32}$$

where N is the EMR activity, while P and Q are gum and gas liberation rate quantities, respectively. A comparison of the experimental and the calculated data in Eq. 5.35 shows a very good correlation (Fig. 5.73). The correlation coefficient is 0.908. Thus, Eq. 5.32 reflects the existence of a relationship between the EMR activity of a coal seam and the gum and gas liberation rate quantities. Hence, an increase of EMR activity can indicate rock-gas outburst hazard development.

We used this equation for an EMR criterion forecast. As was noted above, rock-gas outburst hazard zones are correlated with a quantity of gum of more than 5–6 $\mathrm{l\,m^{-1}}$ and a gas liberation rate of more than 5 $\mathrm{l\,min^{-1}}$. The related EMR quantity by Eq. 5.32 is 67–73 pulse/10 s. Statistical parameters of gum, gas liberation rate and EMR activity data are shown in Table 5.5. Analysis of this table shows that the EMR criterion yields mean and median values that are very close to the experimental data. We also carried out a correlation analysis of rock-gas outburst hazard estimation by the gum-gas liberation rate method and by the proposed EMR activity criterion. The correlation co-

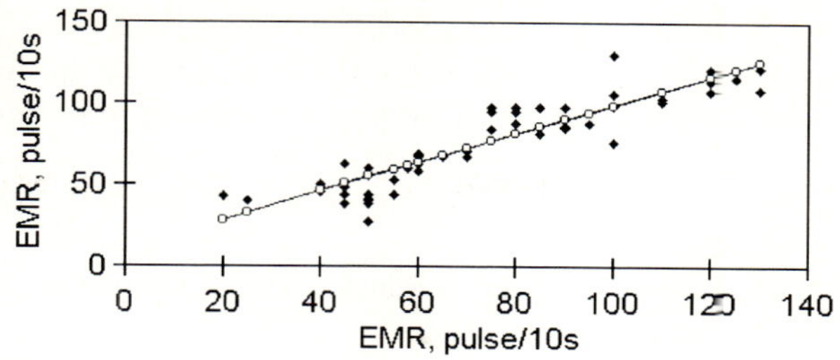

Fig. 5.73. Regression between EMR activity experimental (◆) and calculated data (Frid 1997b)

Table 5.5. Statistical analysis experimental and calculated EMR data

Statistical parameters	Gum ($l\,m^{-1}$)	GLR ($l\,m^{-1}$)	EMR (pulse / 10 s)	
			Experimental	Calculated
Mean	4.91	6.43	72.72	74.77
Median	4.00	5.00	70.00	69.66
Minimum	2.00	2.00	20.00	26.95
Maximum	11.00	16.00	130.00	129.71
Count	58.00	58.00	58.00	58.00

efficient is 0.966, indicating a very high forecast probability. Thus, if the EMR activity is higher than 70 pulse/10 s in a zone of a mine-working, this zone should be declared to be rock and gas outburst dangerous.

Equation 5.32 allows us to explain a relationship between the build-up of stress and gas liberation in a mine-working face (and hence, an increase of rock and gas outburst hazard) and the increase of EMR activity. As is seen, EMR activity level is affected by both stress level and gas liberation rate. An increase of stress level near a mine-working causes an intensive fracturing of the extremely stressed coal. This fracturing process is the source of EMR activity increase. Fracturing of the coal seam in a zone of higher stress excites an intensive process of gas liberation and in turn leads to an additional internal (pore pressure like) pressure. This pressure is akin to a tensional stress in the coal mass that enhances a fracturing catastrophe simultaneously leading to a magnified EMR activity.

EMR control of rock-burst and gas-outburst hazard elimination. Both RB and gas outburst hazards in a mine-working face are frequently eliminated by water infusion, the performance of which is usually verified after its completion by the gum method. During infusion, water "impregnates" coal near the mine-working face. If a control hole intersects a water saturated zone, it excites water spouting out. Then gum turns into slime, and it becomes impossible to measure its volume. This situation makes it necessary to drill additional control holes. However, the existence of several control holes (intended to estimate water infusion performance) diminishes the influence of the water infusion itself because the increase of water pressure sometimes causes breaching to a neighboring hole. Hence, we have a vicious cycle: To estimate performance we need a hole, and the holes negatively influence performance. The only way out is to estimate water infusion performance during its execution without drilling holes.

The analyses of safety conditions (see Sect. 5.10.2.1.), EMR features induced by sample compression and associated with borehole drilling (see above), enables us to state two principles of EMR used for water infusion control:

1. RB is impossible if the mine-working face is transferred back to the residual strength.
2. If a material is under residual strength, its loading is not accompanied by EMR excitation.

EMR-water infusion model. As is generally known, rock stress condition may be described by the Coulomb-Mohr envelope (Twiss and Moores 1992). Let us assume that the coal stress condition in the face of a mine-working is described by the Mohr circle (Fig. 5.74, 1, Frid 2000) and by the Coulomb-Mohr envelope below which the material is stable (Fig. 5.74, 1'). A water pressure increase forms a pore pressure, *P*, that is equivalent to an additive tensional stress. This tensional stress shifts the Mohr circle to the left by *P*. Hence, a part of the circle will now be above the Mohr-Coulomb envelope (Fig. 5.74, 2), i.e., coal becomes unstable, and will fracture. This fracturing results in an increase of EMR activity. As a result of coal fracturing, its strength decreases. However, if the stress level in a coal stratum is still high, a sharp decrease of water pressure (for example, due to pump stoppage) will excite crack closure and EMR excitation. Hence, a water pressure cycle (both increase and decrease) will affect EMR activity.

Thus, water pressure changes are accompanied by EMR activity up to the transition of coal to its residual strength (Fig. 5.74, 3), and vice versa: An absence of EMR activity variation during water pressure changes during the "loading-unloading" cycle could be a criterion of the water infusion performance or in other words, a check of hazard.

EMR during water infusion in coal mines. EMR measurements during water infusion were carried out on Koksoviy, Andreevsky, and Desyatiy coal stratums in North Kuzbass (Russia) (Frid 2000). Water infusion was conducted in the following manner:

The length of the holes was 6 m and the sealed length was 5–5.5 m (a mining experience in North Kuzbass shows that the existence of a 5–6 m safeguard zone (residual strength zone) in a mine-working face indicates a non-dangerous condition);

Water pressure was increased in steps of 0.2–0.3 γH to provide coal "impregnation" with water, where γ is the rock density (kg m^{-3}) and H is the stratum depth (m).

A new step of water pressure increase took place only after EMR activity stabilization. After reaching 0.8 γH, water pressure was sharply decreased to 0.3 γH; as our experience showed that if the water pressure exceeds a value of 0.8 γH, the probability of the water breaching to the mine-working face becomes very high.

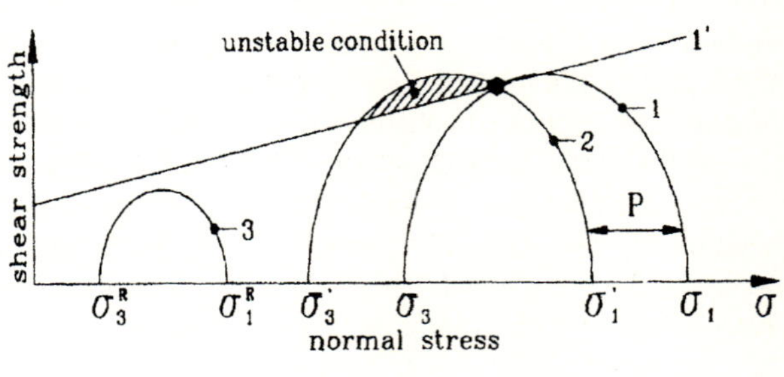

Fig. 5.74. Mohr-Coulomb failure envelope around Mohr circles which represents the different coal conditions during water infusion (Frid 2000)

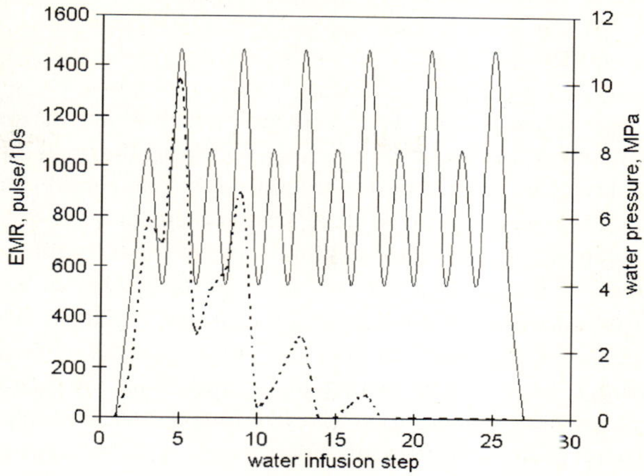

Fig. 5.75.
An example of EMR registration during water infusion. *Full-drawn curve* is the water pressure, while the *hatched curve* is the EMR activity (Frid 2000)

Cycles of pressure increase–decreases were repeated as long as the water pressure changes initiated an EMR activity. When EMR activity induced by water pressure changes ceased, water infusion was terminated.

An example of EMR registration during water infusion is shown in Fig. 5.75. Water infusion was carried out in the drift face of Andreevsky stratum (at 700 m depth, average rock density is about 2×10^3 kg m^{-3}, the step size of water increase is 4 MPa (0.3 γH), and the maximal water pressure value is 11 MPa (maximal water pressure must be less than 0.8 γH, see above). Gum value measured at the drift face was 6 l m^{-1}; hence, a RB hazardous situation existed.

Figure 5.75 shows that an increase of water pressure of up to 4 MPa during the first cycle is associated with EMR activity excitation (of 500 pulse/10 s); an increase of water pressure to 8 MPa initiated an additional increase of EMR activity (to about 800 pulse/10 s). The next decrease of water pressure to 4 MPa caused an insignificant decrease of EMR activity, and a following water pressure increase to 11 MPa resulted in an additional EMR activity jump (to 1 350 pulse/10 s). The next decrease of water pressure to 4 MPa led to a significant decrease of EMR activity, but its level remained high. Therefore, the water pressure rise during the second cycle also induced an increase of EMR activity. After the second cycle of water infusion, EMR activity decreased to 50 pulse/10 s. Drilling of a forecast borehole showed that gum value was about 4.5 l m^{-1}. This value is a little lower than the critical level, although it also shows that the zone of drift influence is still quite stressed. The EMR activity was large up to the third cycle of water infusion, while the fifth and sixth cycles of water infusion were not associated with any EMR appearance. Drilling of a forecast borehole after the sixth cycle of water infusion showed that it was impossible to determine gum value, due to very high coal humidity. However, measurement of natural EMR activity in this zone showed that the average value of natural EMR activity was about 50 pulse/10 s; that is significantly lower than the limiting value of general EMR activity (120 pulse/10 s, Frid 1997a, 2001) and implied that the zone near the mine-working face was safe.

Note that the absence of EMR excitation due to water pressure changes is a qualitative criterion. It does not determine, for example, the number of water pressure cycles still needed for water infusion performance. Our research (more than 70 EMR investigations during water infusion) has shown that the EMR criterion really characterizes water infusion performance.

5.11 Summary

A comprehensive model of EMR emission mechanism from cracks was developed. This model can be described as follows: Following the breaking of bonds, the atoms on both sides of the severed bonds are moved to "non equilibrium" positions relative to their steady state ones and oscillate around them. Lines of oscillating atoms move together and by being connected to atoms around them (in the forward direction and also atoms on their side of the two surfaces newly created by the fracture) the latter also participate in the movement. The ensuing vibrations are similar to those calculated in the Debye model for bulk oscillations. The larger is the number of cut bonds, the larger is the area of excited atoms, and hence the greater becomes the EMR amplitude. These oscillations behave like surface optical waves, where positive charges move together in a diametrically opposite phase to the negative ones and decay exponentially into the material like Rayleigh waves. The resulting oscillating electric dipole is the source of the EMR. The pulse amplitude decays by an interaction with bulk phonons.

The model leads to an invariant EMR 'basic pulse shape' that repeats itself in all examined materials when failed by different testing techniques (including triaxial compression, percussion drilling and large-scale rock explosion). The model enables us to use EMR results for calculating the length, width and area of cracks and to compare them to actual crack sizes measured by fractographic techniques. Thus, a "finger-print" of initial micro-cracking was established. The 'basic shape' appears also in pilot experiments in open quarries during rock explosion.

An overall comparison of EMR signals that were excited by compression, drilling and blasting shows that their basic shape is invariant under both dynamic and quasi-static loadings. The similarity in shape of single EMR pulses, induced by the three processes, therefore indicates that a unified EMR emitting mechanism is operating behind them all and therefore enables us to analyze EMR signals induced by all three processes by the same model.

Moreover, our investigations showed that global relationships (Gutenberg-Richter and Benioff strain release) of EMR signals obtained during rock fracture in the laboratory are similar to those measured in earthquakes.

EMR measurements were conducted in mines. Criteria for rock-bursts and rock-gas outbursts were developed based on results from these measurements and on theoretical models. These criteria are easier and faster to implement than the existing ones, which necessitate drillings and gum and GLR measurements.

Based on these achievements, possible experiments were proposed as guidelines that can lead to earthquake forecasting via EMR detection and monitoring.

Chapter 6

Assorted Problems in Fracture Geology

This chapter consists of a variety of basic issues in fracture geology. Section 6.1 explains the importance of studying regional jointing. Section 6.2 outlines some necessary procedures in studying joints, from data collection in field outcrops to joint set discrimination and determination of paleostress directions. Section 6.3 relates to studies of paleostresses in the Israeli Negev and considers regional jointing in the broad context of plate tectonics. Section 6.4 presents a new, perhaps controversial probe into certain properties of joint length. Section 6.5 briefly mentions the method of local paleostress estimation by fractographic techniques. Sections 6.6–6.9 concern four investigations from the Beer Sheva and Shephela synclines. Section 6.6 contains information about a multi-fracture assemblage in a fault termination zone, where a joint set exhibits an intricate pattern of the paleostress trajectories. Section 6.7 describes a variety of fault-joint geometric and genetic relationships across the local stratigraphic column. Section 6.8 explores the formation of "surface joints". Section 6.9 relates to studies on several aspects of fracture hydro-geology. Finally, Section 6.10 presents a new approach to *en echelon* segmentation.

6.1 The Importance of Studying Regional Jointing

Regional vs. local joint distribution. Sections 6.1–6.3 relate to the discrimination of systematic joint populations (Hodgson 1961b) into distinct sets, which is often termed "the study of the regional joint pattern". These joints fall into four groups (burial, syntectonic, uplift and post-uplift) that generally have large, regional spreads and reflect slow, quasi-static fracture (Sect. 1.5), particularly in the early burial stage. They are distinguished from dynamic joints that form rapidly in granites (Sect. 2.4.6.1) and occassionally in connection with earthquakes, and have limited, local distribution. This distinction is an approximation; it is made for convenience and is not always accurate: Some joints may start slowly and reach high velocities (Sect. 4.9) and the identification of dynamic joints is not always simple (Sect. 2.4.6.1).

Definitions and discrimination of joint sets and motivation for their study. There are different definitions for sets. Whereas Twiss and Moores (1992, p. 38) consider that a set consists of joints that have a "similar geometry", for Ghosh (1993, p. 478) this term applies to "an array of parallel joints". It was shown that the task of distinguishing a joint set from others may be simple in some areas (e.g., Gross 1993), but occasionally, can be quite difficult (e.g., Nickelsen and Hough 1967; Bahat 1991a, p. 251–255; Younes and Engelder 1999). However, a reliable division of joint sets in a fracture province provides the key to identifying directions of paleo-principal directions and supplies a major tool in the investigation of local as well as regional aspects of structural geology. If it is fractographically clear that the joint surface forms normal to the direction of the maximum principal tension or extension (Bahat 1979a), it becomes also clear that a systematic joint set is coaxial with that of the direction of maximum paleo-compression (Engelder and Geiser 1980).

The investigation of paleostress directions. It is quite remarkable that similar dilemmas regarding paleostress directions have intrigued investigators of regional fracture in two remote fracture provinces: The Appalachian Plateau in the eastern USA and the Beer Sheva syncline in the Israeli Negev (Chap. 3). This issue is discussed below (Sect. 6.3) by elaborating on results from Israel, with reference to results from the Appalachian Plateau. We note that in the context of identifying directions of paleo principal directions, "cross fold joints" that strike approximately orthogonal to fold axes (Engelder and Geiser 1980) are particularly important in these two provinces and in other investigated areas that exhibit "regional joint patterns".

At the heart of this investigation lies a formidable difficulty: Which methods should one use in discriminating joint sets and what relative weight should be assigned to the results obtained by the various methods? We deal below with this issue in two sections:

1. Qualitative, measurements, classification, plotting and division into sets.
2. Mapping joint distribution.

Experience has shown that the best results derive from a sensible utilization of all qualitative and quantitative methods applicable to a specific research area.

6.2 The Procedures of Studying Systematic Joints

6.2.1 Qualitative, Measurements, Classification, Plotting and Division into Sets

Carrying out the various steps in the procedure of studying joints involves a somewhat complex procedure. The first efforts should be spent on acquiring a general familiarity with the scope of outcrops in the investigated area (e.g., different lithologies and major structures). There seems to be no alternative to a good vehicle and a lot of travelling in the search for appropriate outcrops. Air photographs of various magnifications can help but do not provide reliable substitutes in this pursuit. One starts with a qualitative fracture study. Measurements would be the next step, as a prelude to classification. Classification, plotting and division into sets need to be done in alternating steps when progress of each step is helping to examine the validity of estimations of the preceding ones. A useful procedure of joint investigation is given by Goldstein and Marshak (1988); we try to supplement their manual.

6.2.1.1
Qualitative Description

The qualitative study consists of two parts. The first one involves a general description of the physical features of fracture. These include fracture (three) dimensions, characteristic spacing, straightness vs. curvature, general deviation from verticality, possible orthogonal relationships, anastomotic appearance as well as joint contacts and fracture interaction styles. The second part relates to fractography. Numerous investigators report about the lack of fractographic features in their research areas. However, in many other areas rocks maintain a record of their fracture history by their fracture surface morphology. While the mirror plane (Fig. 2.1) may disclose details of early events, the fringe often tells about late developments of joints. Familiarity with the fractographic descriptions in Chap. 2–4 may help in obtaining this information. First, take photographs of the selected joints (close-ups at various extents), paying particular attention to having the object lighted properly. Photographs taken under different light conditions will show different fractographies of a given fracture surface. It is occasionally advisable to wet the joint surface or increase the light on it. Do not forget to place a scale that does not obscure the features you want to show. Second, make accurate drawings of your selected photographs. Remember that drawing is far from being a trivial undertaking; it is already an important part of the fractographic interpretation. The radial features of the plume will lead to the fracture origin, provide information regarding changes in propagation directions and tell about the sequence of fracture generations when cut by other fractures. The ripple marks will inform something about the growth rhythm of the joint. Perhaps the most difficult task is marking the boundary between the mirror plane and fringe; it can often be quite controversial (Sect. 2.2.6).

6.2.1.2
Measurements

Strike measurements. The number of strike measurements will often be a compromise that takes into account field conditions, joint dips, the objectives of the study, economy of time and method (Goldstein and Marshak 1988). Generally, no dip needs be measured when joints deviate only several degrees from verticality. However, when bedding dip exceeds five degrees, joint strikes may be evaluated only after rotation of bedding to the horizontal position. We describe below several techniques that have been used in studying joints cutting chalk horizontal layers in the Beer Sheva syncline (Chap. 3) that are thought to be adequate.

At least a hundred strike measurements were taken in large outcrops in which thousands of joints were exposed (e.g., stations 2, 3, 4 in Bahat and Grossmann 1988). In such outcrops the "discrimination method" was used, whereby measurements were carried out only on "reliable", original joint surfaces that could be clearly identified by fractographic ornamentation, yielding the accuracy of ±2–3°. This method provided the most trustworthy results. The "comprehensive method" was used in many medium size outcrops in which about a hundred or two hundred joints were exposed, whereby all "reliable", systematic joints were measured (e.g., stations 20 as above, and 83 from present Fig. 3.37). This method provided quite dependable results. Sector intervals of 1°

were used in all the histogram plots, assuming that inaccuracies would balance each other. The "selection method" (Goldstein and Marshak 1988) was used in twenty-nine stations along Wadi Secher (Bahat and Shavit 1997). In this method, we visually scanned each outcrop and selected representatives of the prominent joint sets. We then measured a few tens of joints in each station. This method provided a good "general picture" but its reliability for statistical analysis would be limited. An obvious question that follows this description is: What about taking measurements in rocks where joints are not fractographically decorated, fractures are not quite straight, strike and dip spreadings are significant in suspected individual sets and joint surfaces are weathered and/or splintered (like in many limestones and dolomites throughout Israel and in some chalks in England, Bahat 1991a, Table 5.5)? The answer is yes, take measurements, but be aware of the reduced reliability of the results and stay conservative in data processing and interpretation of the results.

Length measurements. Scaneline (traverse) techniques (e.g., Goldstein and Marshak 1988) are often practiced in studying joint length distribution. We recommend however that a fractographic method should be used where possible. This can be accomplished on vertical exposures along road cuts and wadies. An example is given from Wadi Naim, where the lengths of forty four joints were accurately measured from tip to tip of unilateral and bilateral plumes that mark the joint surfaces along a horizontal 60 cm thick layer (Bahat 1988b). In this method, fracture length is actually measured where the plume tips show the exact ends of the joint (Fig. 6.1). Scaneline techniques are less accurate because they rely on measurements of many fractures even where a large proportion of their traces terminates beyond the boundaries of the typical exposure. This difficulty is overcome by making measurements only at the intersections of the fractures with grids of parallel scanlines without attempting to identify the

Fig. 6.1. An exact joint tip is shown by the sharp distal boundary (*arrow*) of the plume, for measuring joint lengths. Width of joint is 20 cm

ends of the fracture traces. Fracture trace length is then estimated from the censored data by statistical methods (e.g., Panek 1985).

Additional aspects of joint-set discrimination. There are situations where additional methods are worth considering. "Strike superposition" results are also useful for set discrimination, whereby results obtained in a given area by different investigators (Bahat 1991a, p. 251) or by a given investigator (Sect. 3.4.3.1) are compared. Comparing strikes of *en echelon* segments to strikes of the parent joints (Sect. 2.2.7.6) can also improve the reliability of joint set differentiation. Furthermore, certain joint sets can be identified by particular fracture markings (Fig. 3.36a). Thus, some initial results on the regional joint pattern can be obtained by combined qualitative methods. Often, new research areas require new approaches to the problem. In the Bristol Channel (Sect. 3.2), the regional joint pattern has some unique complications that required the application of sophisticated techniques specially designed for that area (Engelder and Peacock 2001). The study of the gradual development of "surface joints" was carried out by combined photographic and statistical techniques (Sect. 6.8).

A Genetic classification into joint groups. The genetic classification of joints that occur in sedimentary and plutonic rocks is important in fracture analysis, in order to avoid mistakes in the resolution of the regional joint pattern and in the tectonophysics interpretation of the investigated areas (e.g. Sects. 6.9.3 and 4.7.1.1). The genetic classification of joints must be carried out before plotting the results so that joints of different groups should not be plotted unintentionally together. The genetic characterization of various fault-joint types (Sect. 6.7) can also help in understanding the fracture history of investigated areas.

Plotting diagrams. Plotting the sum of measurements following separation into genetic groups is the first step toward interpreting the results. The best way of plotting results for vertical joints (either originally measured as such or after rotation) is by histograms. Bahat (1997) showed that results displayed on a rose diagram may give a different impression from the same results when plotted on a histogram. In extreme cases, while a histogram can show "extra" peaks that possibly suggest a large number of sets, a rose diagram for the same joint population may "hide" an essential peak or two. Thus, there are various degrees of certainty regarding the division into joint sets when all measurements are summarized in histograms or stereographic nets or in both (e.g., Grossmann 1983; Bahat and Grossmann 1988), frequently due to unclear maxima and minima on the plots. Additional statistical techniques are often used in order to increase the credibility of joint set discrimination.

6.2.2 Statistical Analysis in Support of the Division of Joints into Sets

We refer below to two analytical tracks in support of the division of joints into sets; first, methods of separation between two adjacent populations of joint azimuth, and second, by a special reference to joint spacing. Often one selects the technique according to the access to a computer distribution program, and these become more available with time (e.g., Exell, jmp, Rock Works). A few choices are briefly mentioned below.

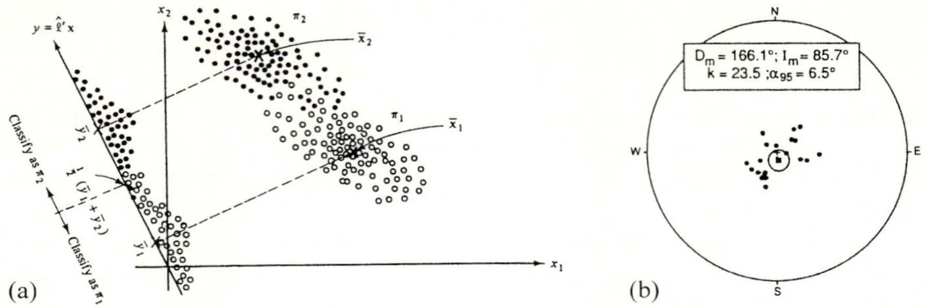

Fig. 6.2. a Fisher's (1938) solution to separate joint populations of two sets when their measurements partly overlap each other is by projecting the sample raw data to improve the definition of the boundary between the two. After rotation of the coordinate system of the original π_1 and π_2 populations, their distinction is significantly improved on the projection (from Johnson and Wichern 1988). **b** Equal-area projection of synthetic directional data that were randomly selected from a Fisherian population with true mean direction $I = +90°$ and precision parameter $k = 20$; individual directions are shown by *black dots*, mean direction is shown by a *solid square* with surrounding α_{95} confidence limit (circle), where D_m, I_m and k are, the calculated mean direction, the calculated mean inclination and the best estimate of k, respectively (from Butler 1992)

Joint set discrimination. A recurrent problem is how to separate two joint populations when measurements of these populations partly overlap each other. Fisher's (1938) solution is by projecting the sample raw data, as illustrated schematically in Fig. 6.2a. This projecting is based on the "Fisher sample linear discrimination function", and the method is explained in textbooks of statistics (e.g., Johnson and Wichern 1988). Smoothing techniques (Wise and McCrory 1982; Goldstein and Marshak 1988) can also help to make peaks stand out more clearly.

Additional contributions which are applicable in the earth sciences stemmed from the need for statistical analysis of paleomagnetism (Fisher 1953; Fisher et al. 1987). Fisher and coworkers developed a method by which each measured direction is given unit weight and is represented by a point on a sphere of unit radius. This method is known as the "Fisher distribution" and occasionally also termed "α_{95}". The confidence limit α_{95} is a measure of the precision at which the true mean direction has been estimated. One is 95% certain that the unknown true mean direction lies within α_{95} of the calculated mean. The inference is that there is a 5% chance that the true mean lies more than α_{95} from the calculated mean (Fig. 6.2b). Thus, the "Fisher distribution" helps to estimate the mean direction of given population. However, this method does not discriminate between populations, i.e., the directions in Fig. 6.2b may represent more than one population, which is disregarded by the small circle of α_{95} confidence limit.

Good computer distribution programs (e.g., jmp) help to calculate many statistical parameters like the arithmetic mean, median, mode, standard deviation, skewness (sk) and kurtosis (ku) (see Nomenclature). In addition, they can divide given collected data of azimuths into sub-populations and offer alternative combinations of sub-populations for comparison. It is up to the researcher to correlate the various combinations with additional qualitative (fracture physical properties and fractography) and quantitative (statistical parameters like kurtosis, spacing and FSR, see below) results in attempting to achieve high resolution in joint set discrimination. Such careful procedures can lead to satisfactory results and improve the credibility of the division into sets, based on the interpretation of histograms.

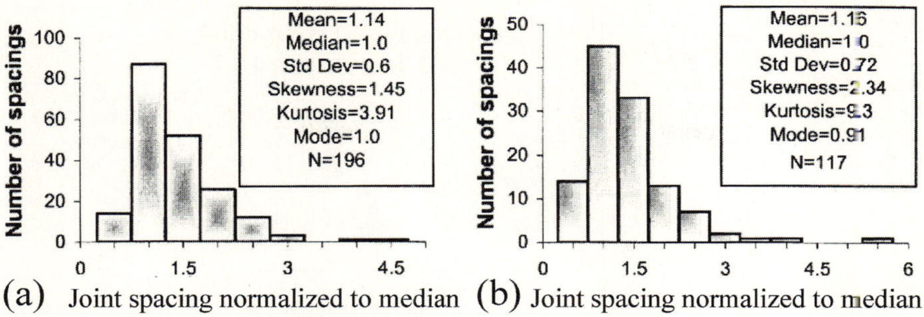

Fig. 6.3. Joint spacing distribution from the Mor Formation. **a** Set 343° (normalized joint spacing histogram of joint set B). **b** Set 350° (normalized joint spacing histogram of joint set C) (from Levi 2003)

Division into joint sets by a special reference to joint spacing. The importance of joint spacing for various geological and other applications is well known (e.g., Price and Cosgrove 1990, p. 54). Spacing is commonly measured between neighboring fractures of the same set along a line orthogonal to the average orientation in the set. We show below how a statistical analysis of joint spacing can improve both, the characterization of joint sets in relation to rock lithologies and the credibility of grouping into sets, citing joint investigation in two quasi-synclines from the south Negev (Levi 2003).

Sets 343° (B) and 350° (C) from the Mor Formation (Fig. 6.3a,b) raise the suspicion of a genetic linkage between the two (Levi 2003 and present Sect. 6.3.6). The high kurtosis value of set 350° suggests the possibility of a connection between the change in joint direction (from 343° to 350°) and differences in mechanical properties of the rocks cut by the two sets, a possibility raised by Goodwin (1995) (see also present Fig. 2.60). Like in the Beer Sheva and the Shephela synclines (Bahat 1987a), the two sets do not mix in the same chalk layer, rather they appear in separate layers (resembling Fig. 3.36a). This enabled the following experiment to be run with a high degree of credibility. Two sets of cylinders were drilled from the corresponding layers and were examined by an ultrasonic method (ASTM: D2845-90) for their wave velocity v_p, Young modulus E, shear modulus G_m and dry density d_d. The results (Fig. 6.4) clearly show a difference between the mechanical properties of the two joint sets. It has therefore been suggested (Levi 2003) that both 343° and 350° sets were formed in one stress field. However, the joints of set 343° formed in co-axial orientation with S_{Hmax} (the maximum horizontal compressive stresses) in the rock of higher density, whereas the joints of set 350° that formed in the rock of lower density deviated from parallelism to S_{Hmax}. This finding corresponds quite well with additional sets of results (Fig. 2.61 and Tables 2.2 and 2.3).

Levi (2003) compares the FSR values (Gross 1993, see Nomenclature) of all the investigated sets discriminated in his study area. The FSR value of set 326° is distinguished among all the other sets in having low values and a very small spread (lower left corner of Fig. 6.5). The other sets have high FSR values that scatter considerably. Levi (2003) considers that these results possibly reflect fracture under different tectonic conditions. Set 326° formed during the burial stage under increasing overburden conditions, resulting in relatively low strains under approximately uniform conditions. On the other hand, the characteristics of the other sets generally fit the criteria of fracture during uplift (Tables 3.3 and 3.5). Single-layer uplift joints form consequent to denudation under conditions of reduced overburden (Bahat 1999a) and are highly

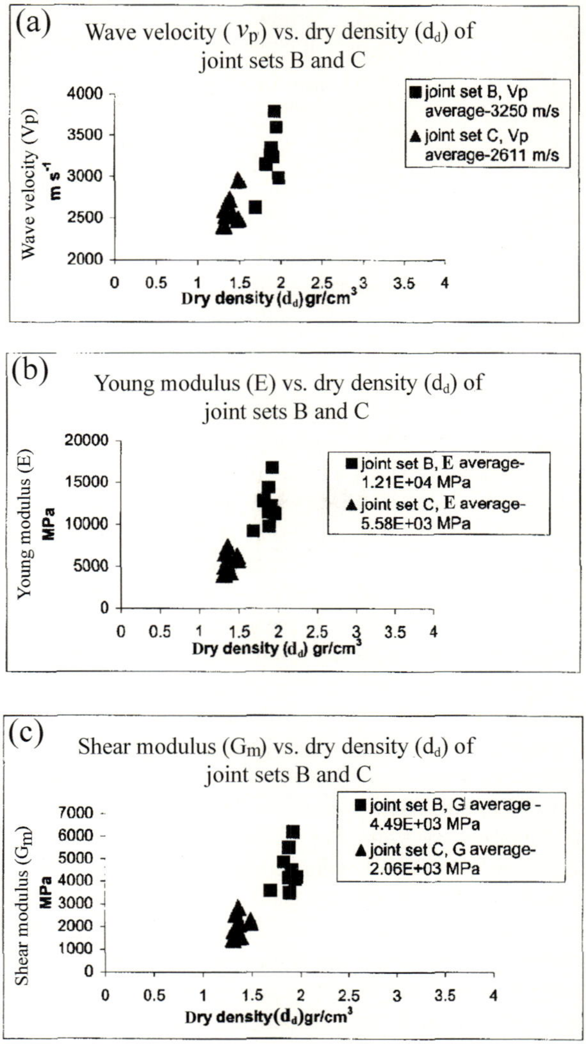

Fig. 6.4. Plots of elastic properties vs. dry density for sets 343° (B) and 350° (C) from the Mor Formation, when wave velocity is v_p, Young modulus is E, shear modulus is G_m and dry density is d_d (from Levi 2003)

strained, probably in tensile rather than extensile conditions. Such conditions may result in high FSR values (Bai and Pollard 2000a). Fracture of uplift joints occurs in shallow depths, in stress fields that often compromise between regional and local stresses, due to influences from surface conditions (Bahat 1991a, p. 297). Therefore, spread in FSR values (Fig. 6.5) may be expected to characterize uplift joints. Hence, the determination of FSR values and their spread can help in the process of differentiating joint genetic groups.

It was found that the ratio of joint length to joint spacing (L/S) (Bahat 1991a, p. 309) characteristically varies according to joint grouping: Between 2 to 10 in burial joints and in the 20 ≥100 range in syntectonic joints. In uplift joints, L/S values generally range between those of the burial and syntectonic joints. Wu and Pollard (1995) distinguished two kinds of fracture patterns in sedimentary layers and in laboratory speci-

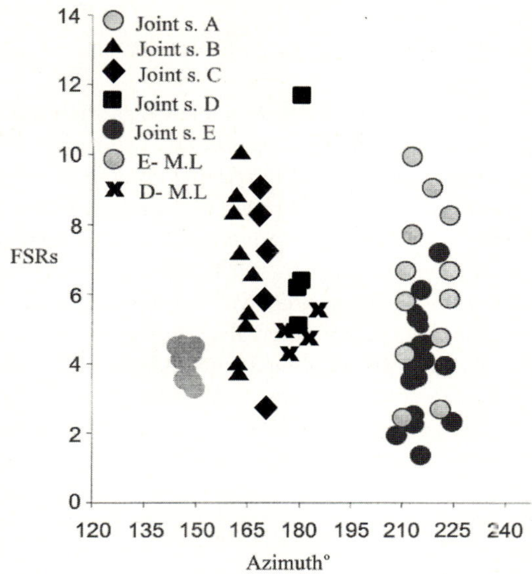

Fig. 6.5. Spread of FSR values in different joint sets (see text for explanation) (from Levi 2003)

mens. One, termed *poorly-developed*, is characteristic of the early stages in the development of a fracture set when typical fracture lengths are roughly equal to or less than typical spacings, i.e., $L/S \leq 1$. The other, termed *well-developed*, is characteristic of the later stages when fracture lengths are much greater than spacings, i.e., $L/S \gg 1$. Wu and Pollard (1995) suggest that characteristic fracture growths occur under various degrees of strain. Fracture opening occurs during an advanced stage of jointing without significant fracture propagation. As the applied strain increases, the spacing decreases, because existing fractures increase in length and new fractures begin to propagate from flaws. However, when the applied strain reaches a certain limited value, the spacing stops evolving and remains nearly constant as the strain continues to increase. Levi (2003) found that the L/S values for sets 326° and 343° in the Mor Formation are 4.4 and 15.4, respectively, and that these values suggest that set 343° represents a more advanced strain than set 326°. This interpretation corresponds to the distinction made above, based on the FSR criterion. Thus, FSR and L/S are complementary parameters that correlate joint spacing characteristics with strain magnitudes.

6.2.3 Mapping Joint Distribution

Once the classification into joint sets is generally established, maps of joint set distribution can be produced. "Trajectory maps" were successfully produced by Engelder and Geiser (1980) and Geiser and Engelder (1983), showing the regional trajectories of S_{Hmax} inferred by the extrapolation of joint directions of particular sets from station to station (Fig. 3.1b,c). If in addition, one is interested in showing the relative concentration of the various joint sets in particular areas on a structural map, "maps of set distribution" may be used. Such maps may help in the interpretation of the effects of local and regional stress fields (e.g., due to neighboring faults) on the formation of the various sets (Levi 2003, p. 25).

6.3 Paleostresses in the Israeli Negev

We present below two different models regarding mesostructures and paleostresses in the Israeli Negev. One model considers two major stress field regimes that fluctuated through time in Israel (Eyal and Reches 1983; Eyal 1996; Eyal et al. 2001). The other model maintains that the sum of systematic joint sets in the Israeli Negev reflects a system of multi paleostresses (more than two major ones) that generally shows a clockwise rotation with time (Bahat 1997, 1999b). The difference between the two concepts is a reminder of the difficulties involved in interpreting the regional joint distribution in the Appalachian Plateau and the paleostress evolution in that region (Bahat 1991a, p. 251) (Sect. 3.1). A similar dilemma is connected with the interpretation of joint mechanisms in granites (Sect. 4.5.1). Hence, the two concepts represent a basic debate in fracture geology. We present below brief accounts from various studies that relate to this debate and then suggest our interpretation.

6.3.1 Jordan

6.3.1.1 *Clockwise Rotation of Compression in Jordan*

Burdon (1959) suggests three compression phases in Jordan. The first phase 'a' was minor, started in the Maestrichtian and came from southeast, approximately in the 285° direction. These forces were light, and the stresses they induced were relieved by the minor folds seen in the chert beds of the Maestrichtian of the Belqa Series. The second phase 'b' was major, probably in the Lower Miocene, or even a little earlier, approximately in the 322° direction. In north Jordan the Ajlun dome was formed and in the crystalline basement multi fracture systems occurred, including tension fractures in the 322° direction. The third phase 'c' was major. Compression came from the SSE, approximately in the 349° direction and took place in two sub-stages (C1 and C2), which were separated by a comparative quiet stage in the Pleistocene. These two phases were associated with a 107 km displacement along the Dead Sea rift, C1 of 62 km and C2 of 45 km. Salameh and Zacher (1982) measured horizontal stylolites in limestones of the Upper Cretaceous in Jordan. They found two dominant paleostress directions, 310–320° and 350°, which seem to be close to the above phases 'b' and 'c'.

6.3.1.2 *Southwestern Jordan*

Eldeen et al. (2000) present results obtained from kinematic slip data in the eastern margins of the Dead Sea rift in southwest Jordan, and their results are cited below. Stress inversion of fault slip data was performed by using an improved right-dieder method, followed by rotational optimization (Delvaux, TENSOR program). Fault slip data (totaling 2 773 data) include fault planes, slickensides and sense of movements. These results were obtained from rocks ranging in age from Precambrian crystalline basement to Oligocene/Pleistocene sediments and were inverted to determine eighty-eight different paleostress tensors. Fault slip data allowed several paleostress tensor groups (stages) from the Precambrian to Post-Pleistocene period to be pointed out and were correlated with the tectonic evolution of the Dead Sea rift. The first two tensor groups were obtained from the Late Proterozoic (between 570 and 560 Ma) dike swarm, and the fault slip data related to the dike emplacement. Both show a transtensional stress regime, the first one (T1) with NE-SW S_{Hmax} corresponding to the major NE-trending dike system, and the second one (T2) with N-S S_{Hmax} corresponding to a minor and later N-S trending dike system.

The other tensors are related to Cretaceous or younger deformation stages, roughly coinciding with the three compression directions observed by Burdon (1959). The oldest stage (T3) has a strike-slip stress regime with E-W S_{Hmax} that affected all Pre-Tertiary series, but not the younger rocks present. The next stage (T4) is also characterized by a strike-slip stress regime, but with NW-SE S_{Hmax}, affecting the Palaeocene-Eocene rocks but not Pleistocene rocks. Eldeen et al. (2000) relate NW-SE trending dykes to the Miocene, including them in stage T4. A transpressional stage (T5), with NNW-SSE to N-S S_{Hmax} affected rocks with an age up to the Pleistocene. In addition, Late Pleistocene extensional stage (T6) with E-W S_{Hmin} (the minimum horizontal compressive stresses) stress regime was evidenced from the fault slip data that affected the Dana Conglomerate. The youngest tensor group obtained from the Dana Conglomerate points to an extensional regime with N-S S_{Hmin} (T7). Eldeen et al. (2000) suggest that the Late Pleistocene-Holocene sinistral strike-slip deformation is still active.

6.3.2
The East Mediterranean

Letouzey and Tremolieres (1980) conducted an extensive investigation of paleostress directions by measuring a large number of tectonic stylolites, joints, shear planes and faults in sedimentary rocks around the Mediterranean Sea. They plotted their results on a series of maps, displaying a complex pattern of paleostress directions that were correlated with different stages of the structural evolution in the region. These tectonic changes were linked to the Europe-African- Arabian indentation in the collision zone since the Upper Cretaceous and closure of the Tethys Ocean (Fig. 6.6a–e). Particularly relevant to the present discussion are the two maps from Turkey (Fig. 6.7a–d) and from Israel (Fig. 6.8). Most intriguing is the great variety of paleostress directions found in south Turkey that were categorized into three phases (we use below only upper semicircle azimuth numbers in describing directions of phases).

Letouzey and Tremolieres (1980) assign to the Late Cretaceous shortening (the "first phase") an early 075° compression direction associated with the Alpine zone that changed to 280° west of the belt of Levantin faults in north Syria (along the continuation of the Dead Sea rift up to south Turkey). The 315° shortening, taken to be the north Arabian average direction, was prominent east of this belt (Fig. 6.7a). Letouzey and Tremolieres (1980) consider that the first phase could correspond to a change in orientation of the stress due to the indentation process or to sinistral rotation of the Anatolian microplate during collision. The 280° corresponds reasonably well to the 290° compression direction found in Israel by them (Fig. 6.8).

The Late Eocene-pre-Middle Oligocene "second phase" is related to a tectonic event well known from the Pontids, Taurus and north Cyprus Mountains. This phase involved

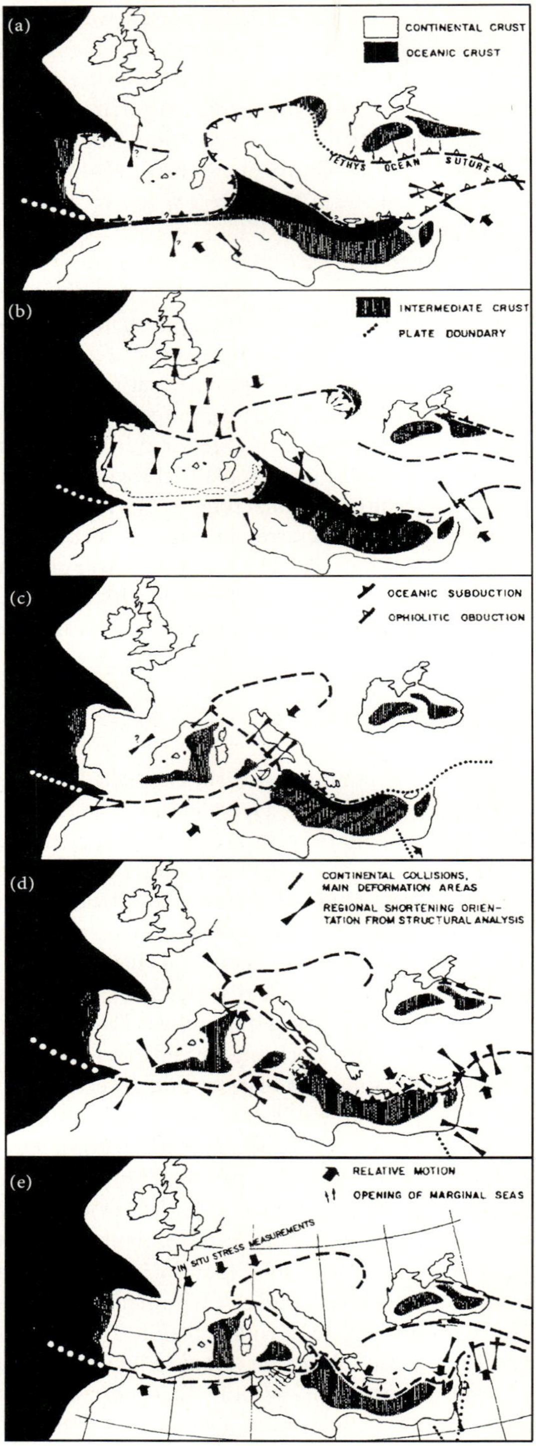

Fig. 6.6.
Sketch of evolution of the Mediterranean basins and paleo-stress field orientations. On all figures Europe was arbitrarily fixed. Regional paleo-stress orientations are in fossil positions. **a** Late Cretaceous; **b** Eocene; **c** Late Oligocene-Early Miocene; **d** Late Miocene; **e** Plio-Quaternary (from Letouzey and Trémolières 1980, see text for explanation)

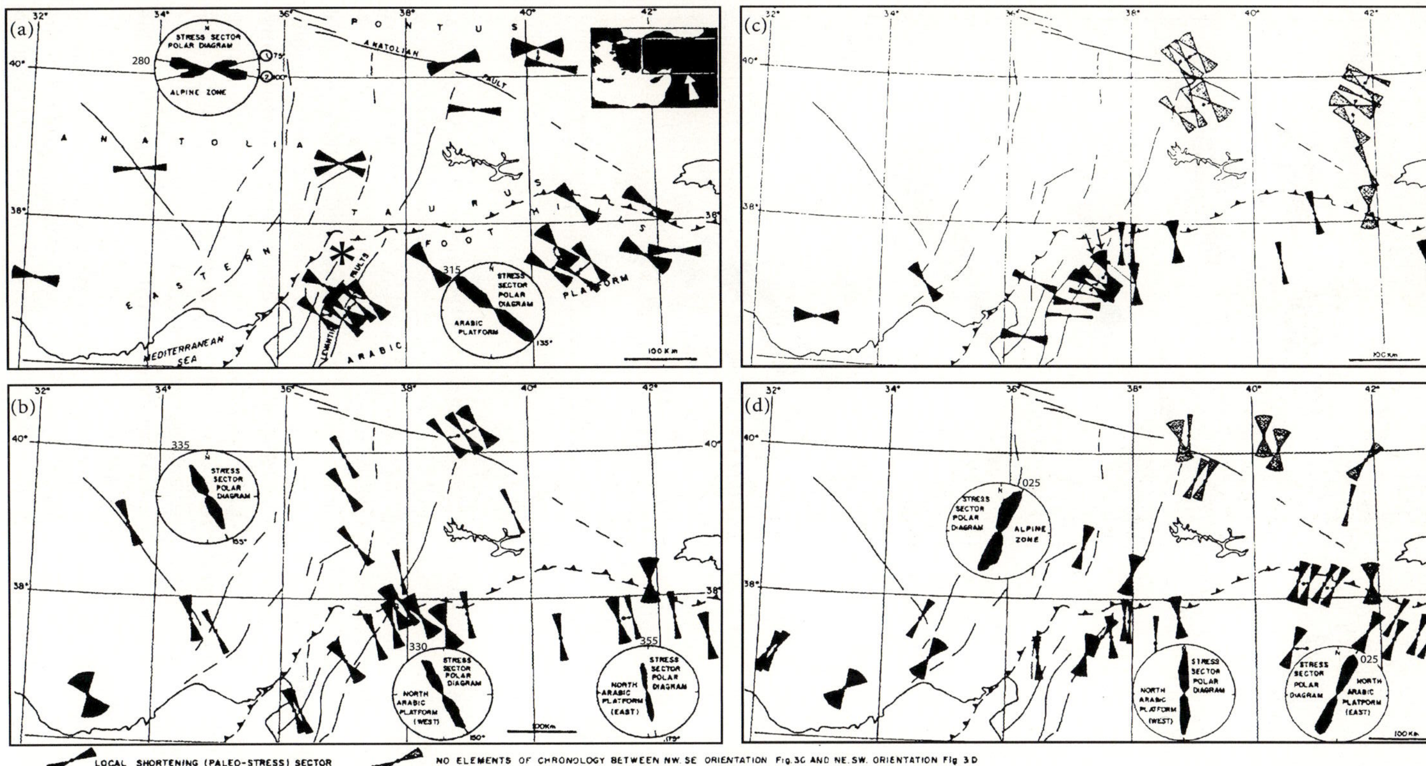

Fig. 6.7. Shortening phases in Eastern Turkey, where circle diagrams give the paleostress directions (PD) discussed in the text. **a** Late Cretaceous phase. At left PD from the Alpine zone, at right PD from the Arabic platform. Note the belt of Levantin faults south of the star (*). **b** Late Eocene-Early Oligocene phase. At lower left PD from the north Arabic platform (west), at lower right PD from the north Arabic platform (east). **c** Late Miocene phase (two small *arrows* show the 343° and 350° shortening directions). **d** Plio-Quaternary phase. At left PD from the Alpine zone. At right PD from the Arabic platform (modified from Letouzey and Tremolieres 1980, see text for explanation)

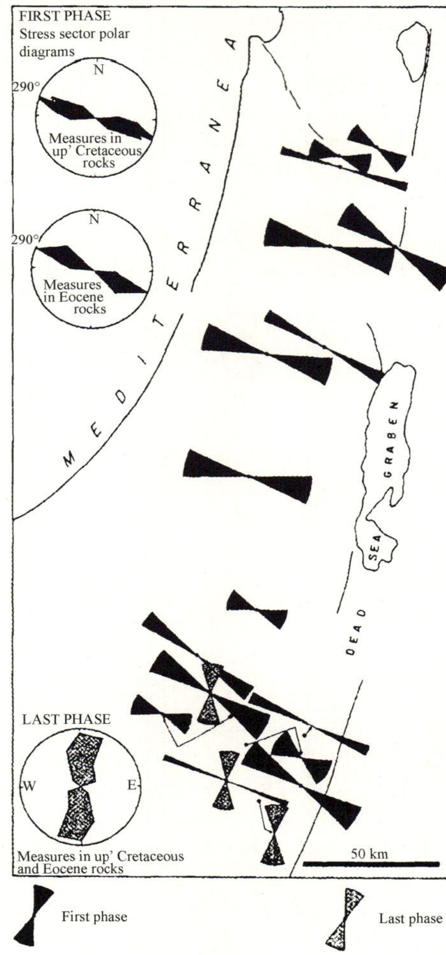

Fig. 6.8. Measurements of shortening phases in Upper Cretaceous rocks and Eocene rocks from Israel. The first phase (290°) is interpreted to have preceded the last phase (approximately N-S) (from Letouzey and Tremolieres 1980, see text for explanation)

a 25° spatial change in the stress direction observed in the northern part of the Arabian plate, along its collision contact with the Anatolian microplate. The difference is between the 330° direction in the western side and the 355° direction in the eastern side of the northern Arabian plate (Fig. 6.7b).

The tectonics of the Late Miocene (the "third phase") affected all the geological formations that Letouzey and Tremolieres (1980) studied. It was associated with folding, thrusting and breaking of continental blocks and exhibited large changes in orientation of the stress field in different structural areas, showing changes in compression (or partition of) directions between E-W in the western side and 355° in the eastern side of the northern part of the Arabian plate (Fig. 6.7c).

The Plio-Quaternary phase is the continuation of the Late Miocene tectonics in the north of the Arabian platform, including the large strike-slip Levantine and North Anatolian fault systems (Fig. 6.7d). The recorded paleostress compression directions from this tectonics are 025° in the Alpine zone, N-S in the western boundary of the northern Arabian platform, and 025°, east of the boundary of the northern Arabian platform. The sequence of the latter two directions is not clear, and it is not certain that there was a continuation of the clockwise rotation during this phase. However, Letouzey and Tremolieres (1980) implicitly integrate the influence of the Dead Sea transform into the general pattern of the Europe-African-Arabian collision, not earlier than the Plio-Quaternary (Fig. 6.6e), exhibiting the new plate boundary in the Plio-Quaternary phase.

6.3.3 The Sinai-Israel Sub-Plate

Eyal (1996, Fig. 1) finds that almost all middle Miocene to Recent S_{Hmax} and S_{Hmin} data compiled from the Sinai-Israel sub-plate cluster in two "distinct" orientations. This observation is based on an investigation of both mesostructures (not including joints) and macrostructures (including dikes and focal mechanism data). One S_{Hmax} trend striking 344 ±3° (ninety-eight sites) is consistent with the Dead Sea stress field (DSS), and the second, S_{Hmax} trend striking 282 ±5° (twenty-six sites), is consistent with the Syrian Arc stress field (SAS). These trends correspond to the SAS and DSS by Eyal and Reches (1983) and according to these authors, reflect an "outstanding homogeneity" of the strain in the region. The SAS direction resembles the "first phase" (290°) by Letouzey and Tremolieres (1980) (Fig. 6.8).

6.3.4 Two Synclines in the Northern Negev

Results from the Beer Sheva and Shephela synclines around Beer Sheva enabled Bahat (1997) to show that the regional joint pattern in the Lower Eocene chalks differs significantly from that of the Middle Eocene, indicating heterogeneous strain in Eocene rocks in the area. In addition, the outstanding maxima 334° and 328° shown by Flexer et al. (1984) and by Bahat and Grossmann (1988), respectively (which are based on thousands of measurements in three synclines along the Syrian Arc), strongly point to a NW S_{Hmax}, which is in variance with the 282° SAS direction discussed by Eyal (1996). The maxima 328° and 334° populate the angular zone between the distinct SAS and DSS. In addition, Bahat (1991a, Fig. 5.8) related to an intensely jointed outcrop of Santonian chalk on the rift shoulder east of Arad (next to the Dead Sea) and found that by far the dominant set was oriented about 320°. This direction corresponds far closer to the NW S_{Hmax} than to the DSS. Thus, the combined observations made by Eyal (1996) Flexer et al. (1984) and Bahat and Grossmann (1988) show more than two major paleostress directions (Eyal 1996) since the Upper Cretaceous in Israel.

Combining the above observations with additional results by various authors, Bahat (1999b) suggested an alternative model to the one pertaining to two distinct SAS and DSS stress fields, as follows. A common feature of mesostructures that show WNW-WSW directions of S_{Hmax} is that they generally occur in limestones and dolomites in Cenomanian-Turonian monoclines, whereas those mesostructures that point to NW-NNW S_{Hmax} directions are mainly joints that cut Eocene chalks in synclines. These observations were interpreted according to a "stress rotation" model, which considered an early remote S_{Hmax} that was oriented WSW to WNW during the Cenomanian-Turonian period and was later rotated clockwise to NW-NNW in the Senonian-Eocene.

6.3.5 The Neqarot Syncline in the Central Negev

Eyal et al. (2001) found four joint sets in carbonate beds from upper Turonian rocks within the Neqarot syncline of south-central Israel:

1. A joint set with a mean strike of 160–340° and $\alpha_{95} < 3°$ found in forty beds at thirty-two stations.
2. A joint set with a mean strike of 108–288° and $\alpha_{95} < 5°$ found in twenty-one beds at nineteen stations.
3. A joint set with a mean strike of 036–216° and $\alpha_{95} < 9°$ found in nine beds at eight stations.
4. A joint set with a mean strike of 136–316° and $\alpha_{95} < 6°$ found in six beds at five stations, where α_{95} is the radius of confidence cone of Fisher (1953) vector distribution.

Eyal et al. (2001) also found that the two most prominent joint sets observed in carbonate beds of the Gerofit Formation in Nahal Neqarot (N in Fig. 6.12a) are aligned NNW (340°) and WNW (288°) and are compatible with the DSS and SAS, respectively (Eyal 1996 and Sect. 6.3.3). They found that cross-cutting relations of these joint sets and the other less prominent joint sets imply different timing relationships. In some beds, the WNW set consistently predated the NNW set, whereas in other beds (altogether six stations out of some forty), the NNW set predates the WNW set. Based on these findings Eyal et al. (2001) came up with the following paleostress model. First, these results support the Eyal (1996) model of two major stress field regimes in Israel that fluctuated through time since the middle Miocene, and second, the joint data suggest abrupt changes rather than gradual rotations in stress field orientations. In seeking to establish the validity of the stress model by Eyal et al. (2001), we revisited the Nahal Neqarot outcrop with the objective of examining joint abutting contacts and their sequence as well as their strike spreading and fracture properties. We concentrated on stations that show the succession of DSS compatibles that formed before SAS compatibles as suggested by Eyal et al. (2001, Fig. 2 and Table 1, stations 12, 13, 15, M-4, 26 and 28). New measurements were carried out at specific locations where joints of different sets showed clear contacts (Fig. 6.9a–f), itemizing the results below.

Joints striking 032° arrest at joints that strike 292° in station 28; in these sets spacing is limited to several cm (Fig. 6.9a). In a neighbor outcrop, some 40 m upstream the wadi, a 090° striking joint arresting at a 330° joint and a 297° arresting at a 340° striking joint, forming a 43° angle (Fig. 6.9b). In station 13 we found a 084° joint arresting at a 356° striking joint, creating an 88° angle (Fig. 6.9c). In station M-2 we found in bed 1 a joint striking 286° arresting in joints striking approximately 333° at 47° angle, but the reverse abutting relationship seems to occur as well. While spacing in the former set was some 80 cm, spacing in the latter set was very low, occasionally about 1 cm (Fig. 6.9d). In station M-2, joint strikes in the NNW set range between 325° and 338°. Contacts suggesting the two opposite sequences similar to those shown in station M-2 are repeated in station 15 (Fig. 6.9e). A joint striking 270° abuts two joints that strike 330° and 340° in station 14, forming 60° and 70° angles, respectively (Fig. 6.9f).

Fig. 6.9. Joint traces in Nahal Neqarot (see text for descriptions). **a** The dominant joint set 292° appears from left to right; **b** A joint oriented 297° parallels to *pen*; **c** A 084° joint arresting at a 356° striking joint; **d** A 286° joint (parallel to *pen*) arrests at joints striking approximately 333°; **e** A contact similar to the one in **d**; **f** A joint striking 090° (parallel to *pen*) abuts two joints that strike 330° and 340°

There are indeed indications in Fig. 6.9a–f that sets grouped in the range 264(084)–297° are younger than sets grouped in the range 330–356° as suggested by Eyal et al. (2001). What remains unclear however is whether this sequence sufficiently justifies the con-

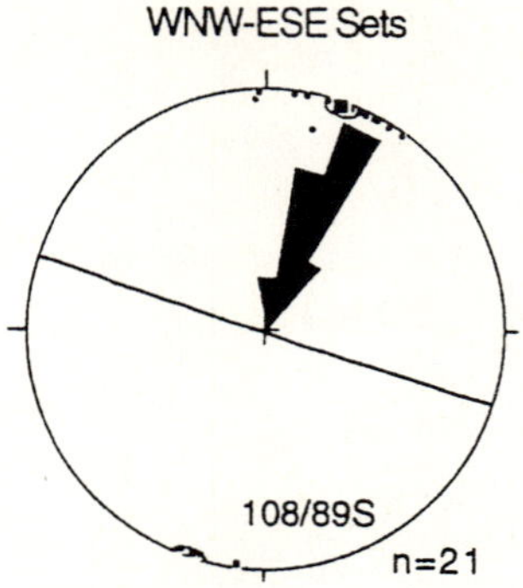

Fig. 6.10.
Summary equal area stereographic projection of the WNW-ESE sets at Nahal Neqarot. Each data point (i.e., pole) is the mean orientation for a single set of systematic joints within an individual bed. The mean orientation for each combined plot is a great circle whose pole is represented by a square. Rose petals indicate the relative frequency of the poles (after Eyal et al. 2001)

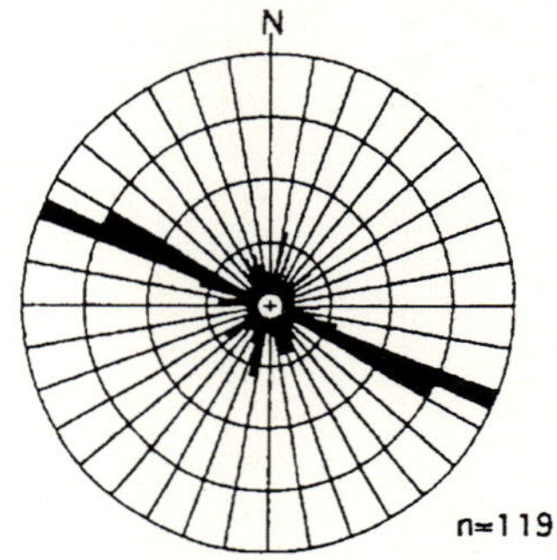

Fig. 6.11.
Rose diagram of joint orientations in the vicinity of a scanline. Sector length is proportional to number of joints (from Becker and Gross 1996)

clusion that these two groups are SAS- and DSS-compatible. The two groups (264–297° and 330–356°) show considerable spreading in Fig. 6.9a–f. Correspondingly, these joints produce a wide range of angles at their contacts (from 43° to 88°). Angular spreading is even greater in Fig. 2 and Table 1 from Eyal et al. (2001), e.g., in station 28, the average of set DSS compatible strikes 328°, which is far from the DSS average. Several joints are oriented reasonably close the specified SAS and DSS directions (282 ±5° and 344 ±3°, respectively, Eyal 1996). These include 286° (Fig. 6.9d), 279° (Fig. 6.9e), 340° (Fig. 6.9b) and 340° (Fig. 6.9f). However, the orientations of most other joints deviate from the specified directions by more than 10°. The strike spread of the WNW set (Fig. 6.9 and 6.10, also Eyal et al. 2001, Table 1) is significantly greater than the strike spread of the SAS compatible set (293 ±3°) obtained by Becker and Gross (1996) (Fig. 6.11). Both data were collected in carbonates of the Turonian Gerofit Formation in the same general area (note the difference in number of measurements in Fig. 6.10 vs. 6.11). This raises the possibility that the WNW oriented joints consist of more than one set. Furthermore, there is a considerable heterogeneity in the physical properties of the joints that represent each of the above two groups (as reflected in Fig. 6.9a–f and other contacts). Particularly, joints of similar trends have considerable spacing differences in different locations in a given outcrop. For instance, in station M-2, joints trending 286° are spaced approximately 80 cm in bed 1 (Fig. 6.9d), while joints trending about 285° have spacing of some 15–20 cm in bed 5, albeit their occurrence in rock layers of a similar thickness (about 40 ±10 cm) and lithology. In addition, the implication that each group is a set, representing a unique paleostress field (Table 1 in Eyal et al. 2001), does not take into account other genetic possibilities: Some of the joints could have been created by influences exerted by the adjacent large Makhtesh Ramon fault (see the neighboring

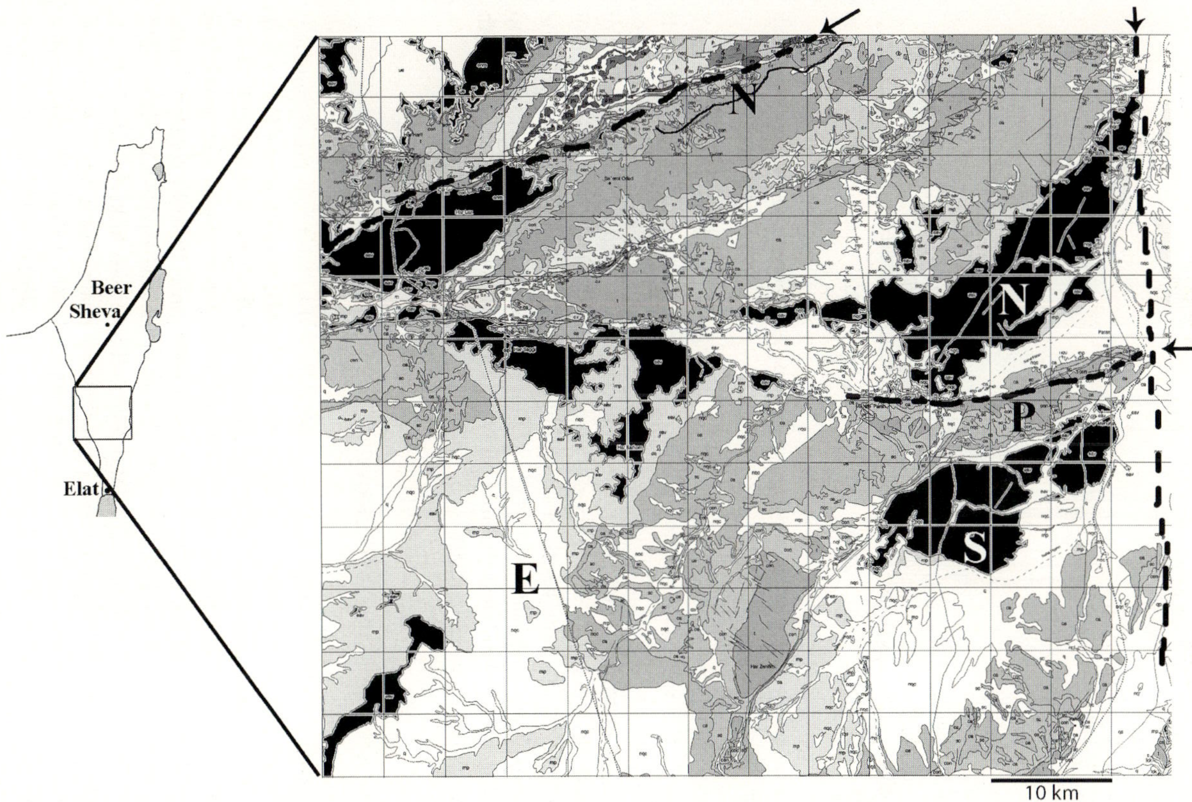

Fig. 6.12a. Geological map of part of southern Israel. The *black areas* marked by *N* and *S* are northern and southern quasi-synclines, separated by the Paaran fault, *P* (*horizontal arrow*). *Vertical* and *inclined arrows* relate to the approximate western boundary of the Dead Sea rift, and the Ramon fault respectively. *E* and *N* neighbor the boundary with Egypt and the Neqarrot outcrop, respectively (from Levi 2003)

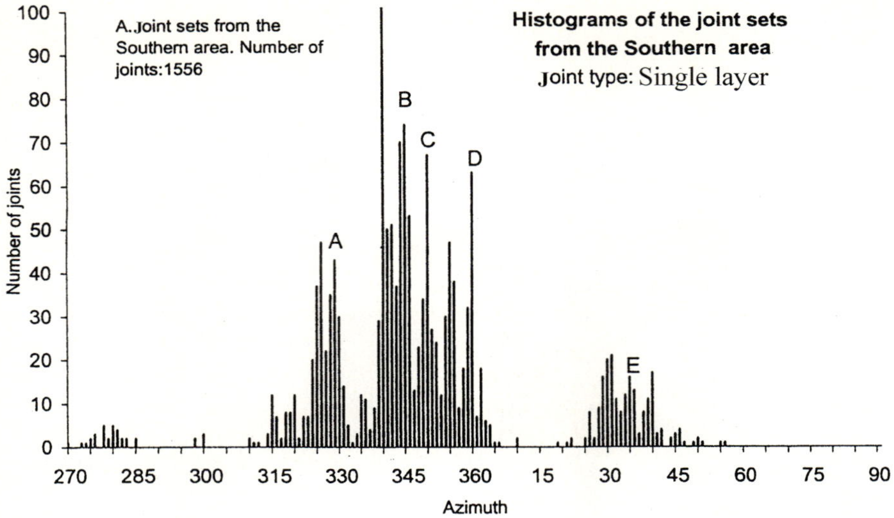

Fig. 6.12b. Histograms of five joint sets (*A–E*) from the quasi-synclinal area south of the Paran fault in Fig. 6.12a (modified from Levi 2003)

Neqarot outcrop N and the Ramon fault, marked by an inclined arrow in Fig. 6.12a), or some of the DSS compatible joints could have been formed by additional stress fields in the area, such as N-S S_{Hmin} (T7, Sect. 6.3.1.2). Finally, Eyal et al. (2001, Table 1) relied on the α_{95} method in their set division, not taking into account the limitations of this method (Sect. 6.2.2). In summary, the repeated observations that sets grouped in the range 264–297° are younger than sets grouped in the range 330–356° support the intriguing sequence suggested by Eyal et al. (2001). However, the strike spreading and other fracture properties stress the possibility that these sets could be associated with more than two unique stress fields.

6.3.6
Two Quasi-Synclines in the South Negev

Five joint sets. Levi (2003) investigated jointing in and around two quasi-synclinal areas (folds that are structurally not well defined), north and south of the Paran fault in south Israel, along the west side of the Dead Sea rift, some 30 km south of the Ramon fault (Fig. 6.12a). His research concentrated on fracture in Santonian and Eocene chalk formations and younger sediments. Based on 2646 strike measurements, he distinguished five single-layer joint sets (A–E): A) 326 ±3°, B) 343 ±2°, C) 350 ±1°, D) 360 ±2°, and E) 033 ±4° (Fig. 6.12b), and two multilayer sets linked to sets D and E. He found that set 326° fractured Eocene Lower Eocene chalks in the burial stage, during rock consolidation. Joint set 343° fractured after set 326°, and Levi (2003) linked it to a clockwise stress rotation during the Miocene. The difference in strike between sets B and C was assigned to lithological differences (Fig. 6.4). Both single-layer and multi layer D and E sets were interpreted to be associated with Pliocene and younger tectonics. However, set D was found to be younger than set E representing a local counter clockwise stress rotation with time.

6.3.7
Summary of Results that Evidence S_{Hmax} Clockwise Rotation

The observations by Eldeen et al. (2000) show evidence for a general clockwise rotation with time of the S_{Hmax} axis from E-W trend in the Cretaceous (their T3), through NW-SE S_{Hmax} that affected the Palaeocene-Eocene rocks (T4) and NNW-SSE to N-S S_{Hmax} affecting rocks with an age up to the Pleistocene (T5). This three-stage (T3–T5) rotation from southwest Jordan fits previous observations quite well. They approximately correspond to the clockwise rotation of the 285°, 322° and 349° compression directions suggested by Burdon (1959) (Sect. 6.3.1.1). Stages T3 and T4 fit the paleo-stress pattern suggested by Bahat (1999b) in the northern and central Negev parts of Israel (WNW-WSW S_{Hmax} in Cenomanian-Turonian monoclines and NW-NNW S_{Hmax} in Eocene chalks from synclines) (Sect. 6.3.4). Clockwise rotation of paleostress S_{Hmax}, was also demonstrated by Levi (2003) in southern Israel (Sect. 6.3.6), with some similarities and a difference compared to the data given by Eldeen et al. (2000). Levi's set A, implying S_{Hmax} 326° in the Lower Eocene and set D, implying S_{Hmax} 360° in Pliocene and younger tectonics correspond well to T4 and T5, respectively. The difference relates to results that concern the Miocene. While set B striking 343 ±2° and set C striking 350 ±1° are connected to the Miocene by Levi (2003), Eldeen et al. (2000) assigned the more westerly direction NW-SE to that period.

The NW-NNW S_{Hmax} (approximately the 326° direction) that correlates with the tectonics that took place during the Eocene (Bahat 1999b; Levi 2003 and possibly Eldeen et al. 2000 as well) matches reasonably well the 330° direction linked with Late Eocene-pre-Middle Oligocene by (Letouzey and Tremolieres 1980) (Sect. 6.3.2, Fig. 6.7b). The compression direction 355° in the eastern side of the northern part of the Arabian plate (Letouzey and Tremolieres 1980) correlates fairly well with Levi's (2003) set C (350 ±1°). Sets B and C are oriented quite close to certain shortening directions measured in the area along the Levantin faults (arrows in Fig. 6.7c). Quite significant are also the similarities of the youngest stress directions. The recorded compression directions by Letouzey and Tremolieres 1980) from the Plio-Quaternary tectonics are N-S in the western side of the northern Arabian plate and 025° from the eastern side of the plate (Fig. 6.7d). Both are almost precisely the same paleostress directions found by Levi (2003) that were implied by sets 360° and 033°, respectively. The N-S compression matches T5 by Eldeen et al. (2000).

In conclusion, the results obtained by Burdon (1959), Letouzey and Tremolieres (1980), Bahat (1999b), Levi (2003) and Eldeen et al. (2000) that were conducted in different areas, i.e., south Turkey as well as south Israel and Jordan, all point to a clockwise rotation of the paleostress field from the Upper Cretaceous to Pleistocene. This rotation possibly started and continued before and beyond these times. Also, with a few exceptions, the linkages of compression directions to specific periods observed by the various authors are fairly close. These correlations support the ideas formulated by Letouzey and Tremolieres (1980), pertaining to the connection between the tectonics along a wide belt east and west of the Dead Sea rift and the Europe-African-Arabian plate collision. These studies particularly demonstrate a linkage between processes that took place in areas from South Turkey and jointing in Israel.

6.3.8 Beyond the Mediterranean Area

6.3.8.1 *The Appalachian Plateau*

Three cross fold joint sets were distinguished in two outcrops near Watkins Glen in upstate New York (Bahat 1991a, Tables 5.1 and 5.2). This suggested deviation from the previous concept of two major cross fold sets and demonstrated the considerable difficulties that were involved in joint set differentiation. Clearly, any changes in the differentiation of cross fold joint sets would alter the interpretation of both changes in paleostress mechanisms and fracture history in the region. Younes and Engelder (1999) resolved this problem (Sect. 3.1). A review of this subject distinguishes three main concepts of regional cross-fold joint patterns in the Appalachian Plateau (Sect. 3.1.6 and Fig. 3.10a–c). Two of these concepts relate to rotational mechanisms of the horizontal remote compression, and the third one concerns a fan pattern of radial joints that form contemporaneously by a given stress field. The observations from the Appalachian Plateau are combined below with the results from the Israeli Negev in formulating a model of regional jointing.

6.3.8.2 *Systematic Jointing and Plate Collision*

The lateral movements that took place along the Dead Sea transform ended at the collision front along the Taurus belt in south Turkey. It appears from the above comparison of regional jointing that it is south of this collision front (Fig. 6.6) where strain records from various times tell about the paleostress histories of different regions along the transform, including Israel. Therefore, the sum of systematic joint sets reflecting paleostresses in Israel and its vicinity is best explained via the history of the Europe-African-Arabian collision and closure of the Tethys Ocean zone since the Upper Cretaceous (Letouzey and Tremolieres 1980). A similar explanation of jointing in the Appalachian Plateau can be linked to the closure of the old Atlantic Ocean between the continents of Africa and North America (e.g., Twiss and Moores 1992, Fig. 6.11 and 6.12).

Quite possibly, such correlations may be made in other regions on the glob, where analogous regional closures (e.g., the Canadian Cordillera) can be compared to recorded regional joint patterns in their respective forelands, in accordance with the plate tectonic theory. Ultimately, set directions derived from regional joint measurements in "large forelands" (that formed by regional closures) and in "small forelands" (e.g., Arlegui and Simón 2001) will be linked to a global map of paleostresses analogous to other global stress patterns (Zoback 1989). Such linkages would be particularly relevant to burial and syntectonic joints cutting sedimentary rocks and to lesser extents to uplift and post uplift joints. Repeatedly reactivated burial joints are recognizable even in thrust-and-fold belts (possibly parts of present Fig. 3.1d and Tokarski et al. 1999). Comparing the characteristics of these joints to those associated with forelands would be quite an intriguing project.

A fascinating question is what are the mechanisms of paleostress changes and how fast changes occur by shortening processes. Probably the modes of changes reflect the variable boundary conditions in collision zones that may differ from time to time along each zone. The interaction of two plates (or microplates) involves some rotation at the contact, but this would be irregular, somewhere between rigid and plastic compliance (Molnar and Tapponier 1975) and occasionally involve back and forth motions. Changes in some strains recording these processes might be gradual, and others will be abrupt. An abrupt change will occur when a previous process ends, enabling the new process to take place. Deformations are observed far away from the collision zone and orientations of paleostresses deduced from structural analyses are surprisingly constant over wide areas on continents (e.g., Sbar and Sykes 1973; Letouzey and Tremolieres 1980). Accordingly, there is justification to look for possible connections between paleostress records in Israel and in the northern part of the Arabian plate in south Turkey.

Is it appropriate to term the sequence described above, from the 326–330° range in the Upper Cretaceous-Eocene through 343° in the Oligocene-Miocene to 360° and 033° in the Plio-Quaternary, to be a clockwise stress rotation? Most probably yes. Was this rotation a gradual progressive process? It probably was not, because it contained deviations from a simple process. Deviations may record local rotations in an opposite direction from the general direction in a region. For instance, while a general clockwise rotation applies to Levi's (2003) joint sets A–E, a local counter clockwise rotation took place between sets D and E (set D was younger than set E) (Sect. 6.3.6). Quite interestingly, gradual paleostress rotations are also recorded on much smaller scales in the Beer Sheva syncline. These rotations were superimposed by deviations from the general trends with back and forth motions (Sect. 3.4, Fig. 3.39b), suggesting diversions from simple, gradual processes.

6.3.9 Open Questions

We consider below several open questions that relate to jointing and paleostresses, including the rotation vs. conjugate concepts, the NNW (approximately 344°) compression, the Zoback (1992) model of global stress in the lithosphere, the timing of strike parallel joints, and the rarity of fracture surface morphology on joints cutting limestone.

6.3.9.1 *The Rotation versus Conjugate Concepts*

The 309° and 344° joint sets were interpreted by Bahat (1987a) to be a conjugate couple associated with the extensile set 328°, which indicated a regional compression, approximately along the bisector of the angle opened between 309° and 344° (sets 1c and 1d in Table 3.3). On the other hand, the idea that the 309° set had been an expression of a "transient stage" of a stress clockwise rotation in the region was entertained as a possible alternative interpretation (Bahat 1999b). This disagreement has been perplexing for some time.

An interpretation that might resolve this disagreement is given below. This explanation stems from results that pertain to the possible existence of extensile fracture at deviating angles from $\alpha = 0$ up to some 10° (Sect. 2.2.13.4). In the process of a clockwise regional rotation of the S_{Hmax} in Israel, from WSW-WNW, say 293° (Becker and Gross 1996) through 309° to 328° and 344° (Bahat 1987a, 1999b), joints in the following orientations were formed sequentially (Fig. 6.13). Set 293° was formed parallel to S_{Hmax}. Set 309° formed during the period of rotation from 293° to 328° and was

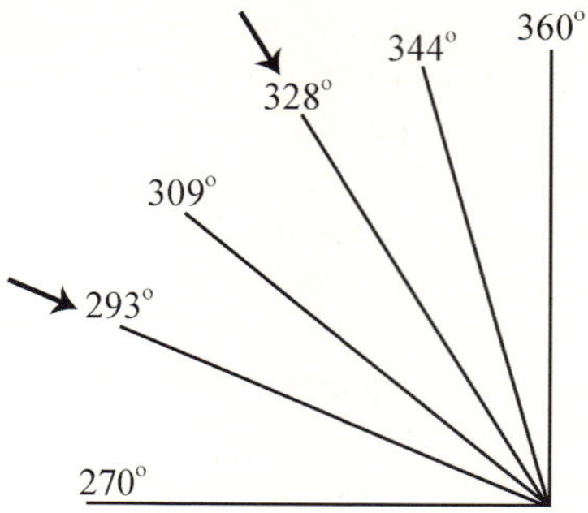

Fig. 6.13.
Formation of a series of joint sets by a hypothetic clockwise stress rotation model. Two *arrows* mark paleostress directions that were coaxial with the joint sets 293° and 328°, while the other sets formed at small angles to rotating horizontal compression

mostly decorated by horizontal plumes). Set 328° was created parallel to S_{Hmax} (and was mostly decorated by concentric rib marks). Set 344° formed during the rotation from 328° to 344° (and was mostly decorated by horizontal plumes). Perhaps fracture of the latter set took place when S_{Hmax} approximated the 330° direction (e.g., the second phase by Letouzey and Tremolieres 1980) or 334° (Flexer et al. 1984). The rate of this rotation is not known, but it is estimated that sets 309° and 328° were created in the Lower Eocene and sets 334° and 344° formed somewhat later. Thus, the similar angular relations between sets 309°and 344° with respect to S_{Hmax} at 328° do not necessarily imply that these three sets formed simultaneously (Fig. 3.10). The speculation depicted in Fig. 6.13 needs to be further investigated in additional fracture provinces before becoming a sound model. The investigation is particularly necessary because this scheme considers creation of joints at small angles with respect to principal stress directions, which is a deviation from the conventional concept of co-axiality of joint formation with respect to the direction of S_{Hmax}.

6.3.9.2
The Significance of FSR Values

It is not really clear what is the genetic significance of the high FSR values and the high scatters of FSR values obtained for set 344° by Levi (2003) (Sect. 6.2.2). One explanation relates them to accumulated strains after repeated stress events (earthquakes) along an extended period of time (Eyal et al. 2001). Levi (2003) shows high FSR values for both B (set 344°) and D (set N–S) (Fig. 6.5). He considers that the latter two sets originated in the early Miocene and Plicetocene, respectively. This raises the question of how the high FSR value could be formed in set D during such a short time, since the Plicetocene, if the mechanism is repeated stresses along an extended period of time (Eyal et al. 2001). In addition, both Eyal et al. (2001) and Levi (2003) found high scatters of FSR values for set 344° (Fig. 6.5), implying that non uniform processes exerted on the rocks and perhaps suggesting more than a single mechanism.

Perhaps an alternative explanation is more plausible (Levi 2003). Set A (326°) formed in the burial stage under increasing overburden conditions, while the rock was not fully solid, and therefore, displayed a fairly uniform FSR distribution. On the other hand, the other sets (B–E) developed during uplift processes (Levi 2003). The uplifts took place under reduced overburden loads and increased tension (Bahat 1999a), so that the rock encountered lower resistance to jointing, with consequent increase of both the FSR values and their scattering.

6.3.9.3
The Global Stress Model

Zoback (1992) presented a model of global stress in the lithosphere consisting of first- and second-order patterns. The first-order midplate stress fields are believed to be largely the result of compressional loads applied at plate boundaries, primarily ridge push and continental collision. These regional uniform stress orientations often extend 20–200 times the approximately 20–25 km thickness of the upper brittle lithosphere. The second-order stresses are local perturbations that are associated with specific geologic or tectonic features, including lithospheric flexure, lateral strength contrasts, as well as lateral density contrasts, which give rise to buoyancy forces. These second-order stress patterns typically have wavelengths ranging from 5 to 10 times the thickness of the brittle upper lithosphere. Often the second-order stress field results in a rotation of the horizontal stresses. The distinction between first- and second-order scales and mechanisms is appealing. Such ideas have been entertained by various investigators in resolving "misalignments" in the field (Currie and Reik 1977).

It appears that on account of great scale differences, there is no clear correlation between the rotation model linked with the continental collisions in the Mediterranean region (Sect. 6.3.2) and the global stress model. The scale of the first-order, which may fit that of the Mediterranean region, does not involve a rotation of the horizontal stresses, and the Mediterranean region is far greater than the scale of the second order. However, a "compromise model", pertaining to a "local perturbation" the size of the Mediterranean region may, with some terminological adaptations, fit the global concept by Zoback (1992)?

6.3.9.4
The Timing of the Strike Parallel Joints

Much emphasis has been placed in previous sections on various aspects of cross fold joint formation. Also important is the relative timing of cross fold and strike parallel joints in a given fold (F_1 and F_2 joints, respectively in Fig. 3.20b) or a slightly folded area, like the Appalachian Plateau or the Beer Sheva syncline. It is generally accepted that F_1 joints form in response to σ_1 compression parallel to layering and normal to the fold axis, during early stages of folding in both (approximately) symmetric (Burger and Hamill 1976) and asymmetric anticlines (Reches 1976). What about the timing of F_2 joints? Burger and Hamill (1976) argue that calcite data that reveal a state of stress associated with the time of F_1 joints formation on the Teton anticline (Friedman and Stearns 1971) contain no evidence for stresses related to F_2 fractures that were also developed. Therefore, it is possible that either the stresses associated with the development of F_2 joints were maintained for too short a time for calcite to undergo appreciable twin gliding or that F_2 fractures formed at very low mean stress. There are, on the other hand, indications of time alternation of F_1 and F_2 joints in the two sets, in slightly folded synclines (Bahat 1991a, Fig. 4.8), and late reoccurrence of F_2 joints when local stresses were resumed (Gross et al. 1997). Many F_2 joints probably formed during relaxation of the remote stress (Price 1974; Bahat 1989, see also Dunne and North 1990). It seems that generally, local principal stress rotation during advanced folding of asymmetric anticlines (Fig. 3.20a) may be sufficient to open joints against the remote compression (i.e., strike parallel joints). Hence, the timing and sequence of F_2 joints can be quite variable in a fold, and decisions regarding their timing should be taken with great caution until more deterministic criteria become available.

6.3.9.5
Jointing in Limestone

It is our impression that the frequent inferior qualities of joints in limestone, in terms of joint straightness and flatness, azimuth spreading in a given set and rarity of fracture surface morphology, seems to be significant compared with those in chalk. If this impression is sound, it would remain a challenge for combined efforts of experts in structural geology and limestone petrology to resolve this issue.

6.4
Analysis of the "Fat Tail" of Joint Length Distributions

6.4.1
Joint Length Distributions

A "fat tail" is a characteristic of statistical distribution, where the decay of the distribution for "large values" shows a power law behavior and does not conform to either negative exponential or Gaussian forms. Although the statistical "fat tail" data treatment characteristically consists of small populations, these populations are exceptionally important because they are represented by the largest individuals in the system that for certain applications are the dominant ones. For instance, Bahat and Adar (1993) showed that from a series of joint sets around Wadi Naim in the Beer Sheva syncline, Israel, one particular set (that was oriented at azimuth 062°) provided by far the best water flow conditions. This set consisted of the longest fractures in the area, through which most of the water drainage took place (Sect. 6.9). What is the connection between the Naim set 028° (Table 3.5 and Fig. 3.36a,b) and the fat tail (FT) distribution?

We study the fracture length histograms of two joint sets in detail: First, a set cutting Middle Eocene chalks along Wadi Naim in the Beer Sheva syncline, Israel (Fig. 6.14a), and second, a set from Ward lake, California (Fig. 6.14b). We focus on certain characteristics of the longest joints at the "end" of the distribution curve. The joints of the first set that were formed by extension during Tertiary uplifts (Bahat 1999a) are fully revealed along Wadi Naim, showing a bimodal distribution. Bahat (1988b) demonstrated bimodal distributions in additional joint sets, compared to previous suggestions that joint length in rocks had Weibull, normal, or power-law distribution (e.g., Segall and Pollard 1983; Kelner et al. 1999; Zhao et al. 2000 and references therein), and this issue remained an open question until now.

The second distribution (Fig. 6.14b) was considered by Segall and Pollard (1983) to conform to a power-law distribution with powers of ~1.8. The fits of their curve to the actual distribution is not clear. Trying to fit the whole distribution by a power law has two shortcomings:

1. The decrease of frequency for small values of joint lengths is ignored.
2. The bimodal character of the distributions is overlooked.

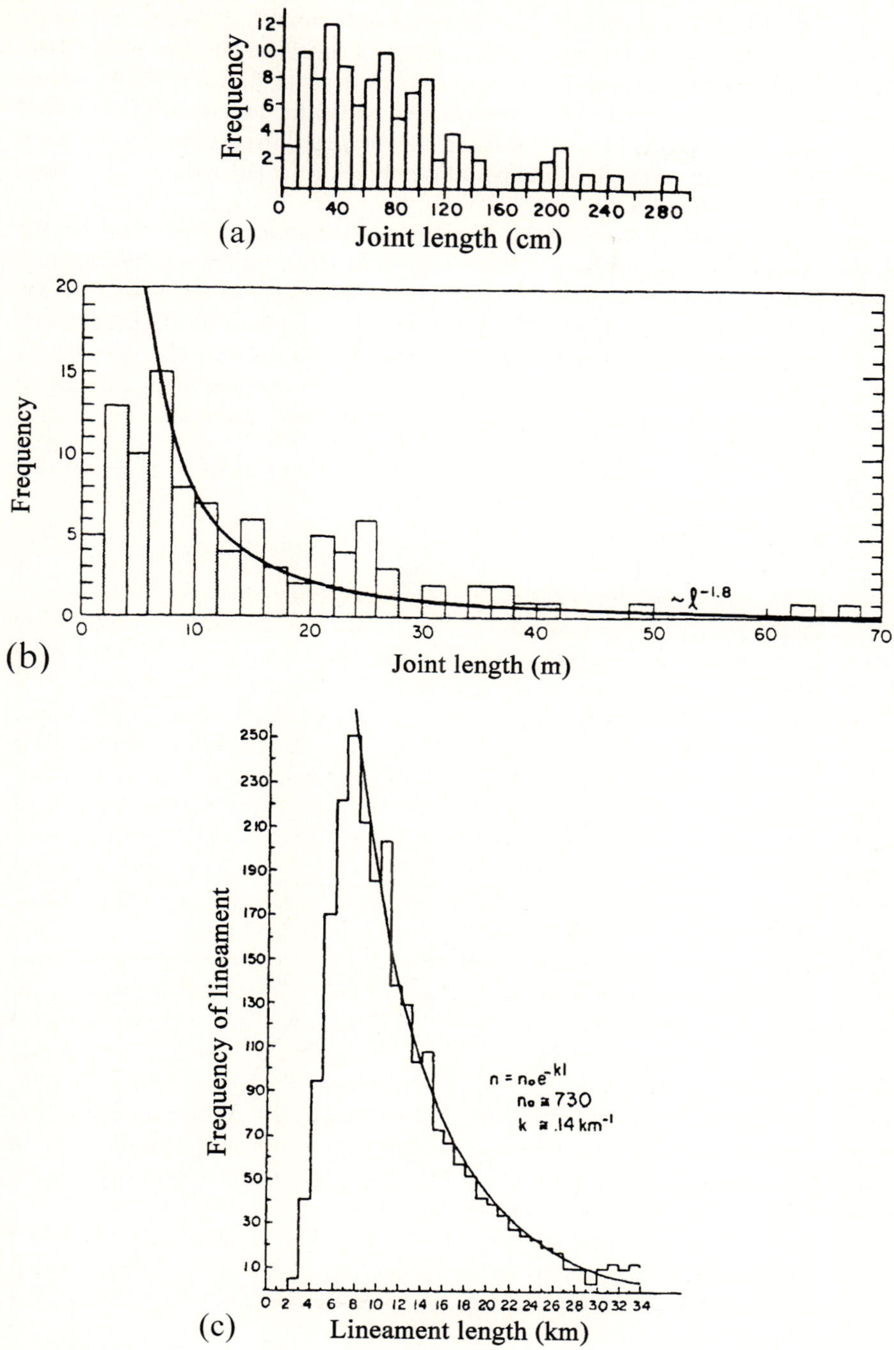

Fig. 6.14. Three histograms of joint lengths (note lithology and scale differences). **a** Set 023° from Middle Eocene chalks (Bahat 1988b); **b** From granodiorite Ward Lake (Segall and Pollard 1983, Fig. 8b). **c** Lineaments (modified from Nur (1982) after Kowalik and Gold (1976))

It is the second problem that is considered here. Apparently, a similar bimodality of fracture length population characterizes findings in much larger scales (Fig. 6.14c). This distribution was considered to be exponentially dependent (Nur 1982). As is presently shown, the second peak of the distribution marks the fat tail, which is indeed given by a power law distribution. The latter, however, has different power values than those used by Segall and Pollard and pertains only to the fat tail and not to the complete joint length distribution.

Drawing for the Wadi Naim set, the cumulative distribution of length, $P(x)$, i.e., the number of all crack lengths larger than or equal to x as a function of the fracture length x, on a log-log plot, one obtains (Fig. 6.15a) a reasonable FT (albeit only for one decade) with a slope of around 2.8. A similar graph (Fig. 6.15b) of the Segall and Pollard (1983) results shows a similar FT behavior with slopes of ~3.3 (for their Fig. 8a results) and ~2.8 (for Fig. 6.14b). Although FT results only continue along a decade or so, they definitely do not conform to exponential or to faster decreases.

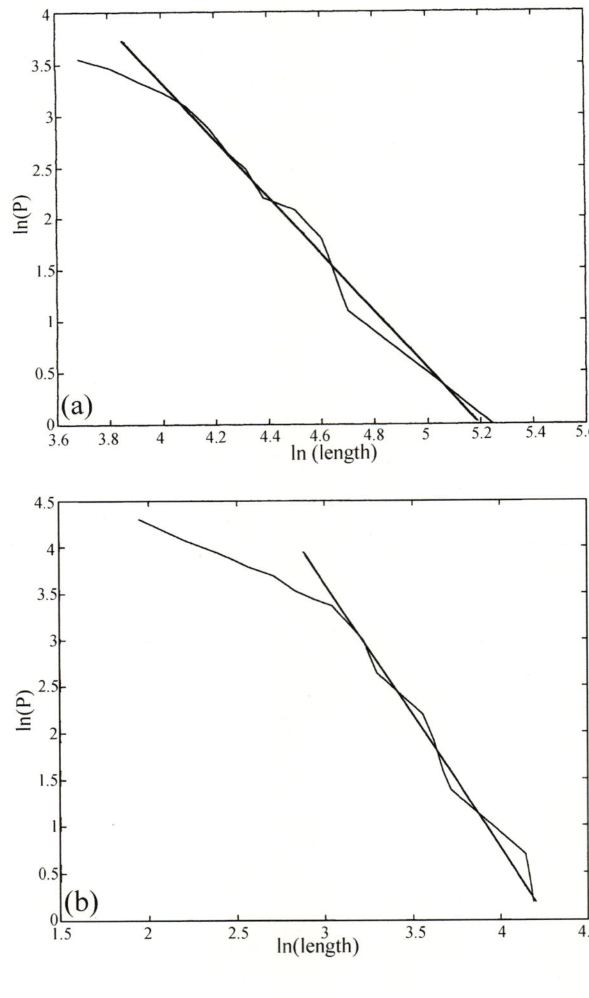

Fig. 6.15. The cumulative distribution of length, $P(x)$ = number of all crack lengths larger than or equal to x, as a function of the crack length x, on a log-log plot. Straight lines: Linear regression. **a** Results from Wadi Naim (length in cm); **b** results from Ward Lake (length in m)

6.4.2
The "Fat Tail" Phenomenon

There exist many situations that show distributions having FT. These include phenomena in economics (Mandelbrot 1960; Mirowski 1995), physics (Krapivski and Majumdar 2000), earth sciences (Sornette et al. 1996; Caers et al. 1999a,b; Kelly et al. 1994) and are distributed in books and review articles (Arnold 1983; Adler et al. 1998). The connection of FT with fractals has also been extensively discussed (Takayasu 1989). However, the exact origin of the FT behavior is hitherto unknown. Sand dune avalanche distribution was explained by the theory of self organized criticality (Jensen 1998), while some distributions with FT encountered a purely economic model for their origin. None of these models however can be used for the distribution of fracture lengths, which we consider here.

The term FT relates to the following phenomenon. Consider the frequency distribution of lengths of fractures found in a specific rock. The expected behavior of this (or any) distribution is that it should decrease (for large lengths) at least exponentially (say for a Poissonian distribution) or even normally (large numbers rule, Feller 1971, p. 219). There are, however, quite a few distributions for which the frequency, especially for large values (tails) of the measured quantity (large fracture lengths for the example), decreases as some power of this quantity – a FT. It is the purpose of this section to address such a behavior of the distribution of fracture lengths measured in rock layers and to discuss the possible origin of this FT.

6.4.3
The Model

According to Wu and Pollard (1992) (see also Sect. 1.2, second scenario), fractures in a layer develop in two stages. In the first stage, fractures appear at random sites until a certain distribution of spacing saturation (Sect. 1.5) is obtained. In the second stage, almost no new fractures are created but the already existing ones grow in length. We adapt this two-stage model (with some modifications) and treat the second stage here. Consider a fracture of length x. Suppose its closest neighbor is another fracture of length x'. According to Segall and Pollard (1983, Fig. 11) and (Nur 1982), the fracture of length x can (and will) grow if $x > x'$, while the x' crack will not grow. This interaction between x and x' is valid, provided both cracks belong to a set and the spacing between them is less than ~5 times the average inter-crack spacing in the material. Since we treat parallel cracks after they have reached saturation, we adhere to the above assumption: Namely, x grows if larger than x', its nearest neighbor (nn).

The probability that x will grow is therefore

$$\int_0^\infty p(y)\mathrm{d}y \int_0^x \bar{g}(x,y,x')\mathrm{d}x' \tag{6.1}$$

where $p(y)\mathrm{d}y$ is the probability that the nn be at a distance (spacing) of between y and $y + \mathrm{d}y$ from the x fracture, and $\bar{g}$ is the conditional probability that if the fracture is of length x and if its nn is at a distance y from it, than this nn is of length x'. We now assume that $\bar{g}(x,y,x')$ is independent of either x or y. This is a plausible assumption but is not always valid. Under this assumption, $\bar{g}(x,y,x') = g(x')$, the probability that a fracture is of length x'. Note that the integrals on y and on x' of Eq. 6.1 become disconnected and the first integral becomes equal to 1. The probability $g(x)$ evidently depends on time during the process, a dependence suppressed here but to be discussed later on.

Denote by

$$S(x) = \int_0^x g(x')\mathrm{d}x' = 1 - Q(x) \tag{6.2}$$

which under the previous assumption gives the probability that a fracture of length x grows. Note that

$$Q(x) = \int_0^\infty g(x')\mathrm{d}x'$$

Next, we calculate the amount of growth of an x-long fracture. The exact details of this calculation are not of major importance. It is only the scheme that is relevant. Assume that there is a minimal value of extension stress σ_{00} that only when $\sigma > \sigma_{00}$ the fracture can grow. Since the change in crack length Δx is small with respect to the fracture length itself, the criterion $\sigma > \sigma_{00}$ is equivalent to $K_\mathrm{I} > K_\mathrm{I0}$ where K_I0 is a minimal stress intensity factor below which fractures do not grow at all (K_0 in Fig. 4.34, see also Sect. 1.5). Note that we speak about extreme subcritical conditions in the sense that K_I (and, a fortuori, K_I0) are very much smaller than Griffith's K_Ic, the value of K_I above which the crack starts to run at great speeds. During a time interval Δt (which can be very long, in a geological process), the stress field has changed. In this interval, there are assumed to have occurred N subintervals, Δt_i, during which σ was larger than σ_{00} and cracks grew.

6.4.4 The Fractal Nature of the Tail

Suppose that the crack growth velocity (v) (Charles 1958, Sect. 1.5) is given by

$$v = \frac{\mathrm{d}x}{\mathrm{d}t} = AK_\mathrm{I}^\mathrm{n} \tag{6.3}$$

where A and n are constants, and x is the crack length at the time t.

This is the accepted form for the velocity changes during creep for a large series of materials. Note that the most important parameter in Eq. 6.3 is n, which characterizes crack growth for a specific material and can be obtained experimentally from the slope of a $\log v$ vs. $\log K_\mathrm{I}$ plot of actual data. Assuming that the crack growth, Δx_i, during the subinterval Δt_i, is small, namely $\Delta x_\mathrm{i} \ll x$, (Eq. 6.3) can easily be integrated. Since K_I is proportional to $\sigma\sqrt{x}$, we have

$$\Delta x_i \cong bx^{n/2} \int_{\Delta t_i} \sigma^n(t)\mathrm{d}t \tag{6.4}$$

and the total increase in length of a crack of length x during the complete time interval Δt is

$$\Delta x = \sum_{i=1}^{N} \Delta x_i = \mathrm{d}x^{n/2}\Delta t \tag{6.5}$$

where

$$d = b\sum_i \int_{\Delta t_i} \sigma^n(t)\mathrm{d}t$$

The balance equation for the probability (per unit time and per unit length) $g(x, t)$ to find at time t (measured along the process) a crack of length x, is as follows:

$$g(x,t+\Delta t) = g(x,t)[1-\beta S(x,t)] + g(x-\Delta x,t)\beta S(x-\Delta x,t) \tag{6.6}$$

Here, the first term on the right denotes the probability that the crack had already the length x at time t and stayed at this length, while the second term describes the probability that the crack had a length $x - \Delta x$ at time t, where Δx is given by Eq. 6.5 and was able to grow. This probability is proportional to S at that time. From Eq. 6.6 one gets,

$$\frac{\partial g}{\partial t} = -\beta\frac{\mathrm{d}x}{\mathrm{d}t}\frac{\partial}{\partial x}(gS) \tag{6.7}$$

Since by Eq. 6.2,

$$g = \frac{\partial S}{\partial x}$$

and using Eq. 6.5 the following nonlinear partial differential equation is obtained for S:

$$\frac{\partial^2}{\partial x \partial t}S = -\gamma x^{n/2}\left[\left(\frac{\partial S}{\partial x}\right)^2 + S\frac{\partial^2 S}{\partial x^2}\right] \tag{6.8}$$

A separation of variables $S = X(x)T(t)$ then yields

$$-\frac{T'}{T^2} = \frac{\gamma x^{n/2}(X'^2+XX'')}{X'} = u = \text{const.} \tag{6.9}$$

where γ is a constant and prime denotes a derivative with respect to the corresponding variable. The time-dependent part is therefore

$$T(t) = [\mathrm{a} - u(t-t_0)]^{-1} \tag{6.10}$$

where t_0 is the initial time and $a = 1/T(t=0)$. Equation 6.10 is evidently valid as long as $t < t_1$, where $t_1 - t_0 = a/u$. For "short" times $(t - t_0) \ll a/u$, T increases linearly with time: $T \cong 1/a + (t - t_0)\,u/a$. This linear increase is in agreement with Wu and Pollard (1992, Fig. 6c). The fractal nature of the tail of the distribution is obtained from the x-dependent part $X(x)$. Assuming $X(x) = \eta x^{\mathrm{m}}$, where η is a constant, Eq. 6.9 can be fulfilled only if

$$n = 2(-m+1) \tag{6.11}$$

A solution of Eq. 6.8 is therefore

$$S(x,t) = \int_0^x g(x',t)\mathrm{d}x' = \frac{Kx^m}{a - u(t-t_0)} \tag{6.12}$$

or, for "short" times, choosing $t_0 = 0$ as the starting time,

$$S(x,t) = (c_1 + c_2 t)x^m \tag{6.13}$$

hence

$$g(x,t) = (c_1' + c_2' t)x^{m-1} \tag{6.14}$$

Equations 6.11–6.14 establish for this model the "power" character of the distribution $g(x)$. Note that this character stems from the nonlinearity of Eq. 6.7, whose origin is the interdependence of growth of one fracture on its neighbors. Another contributing factor is the Charles "subcritical" fracture velocity-K_I dependence, Eq. 6.3.

6.4.5 Discussion and Conclusion

So far the final bimodal distribution (Fig. 6.14a–c) was considered. Here we explain the process of its creation, starting from a "one peak" distribution. We consider a simple example as follows. Assume that the normalized joint length distribution function *at the time of saturation* is of the form shown in Fig. 6.16a and expressed as

$$g(x) = \left(\frac{2x}{l_0^2}\right)\exp\left\{\frac{-x^2}{l_0^2}\right\} \tag{6.15}$$

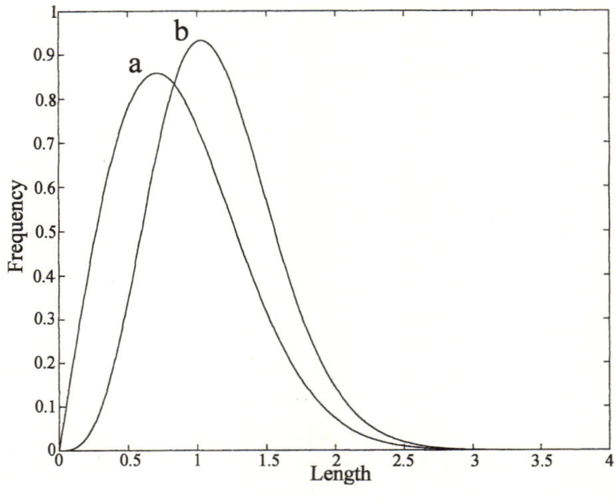

Fig. 6.16. **a** The crack length distribution example of Eq. 6.15 with $l_0 = 1$. **b** The probability per unit length that a crack of length x and with the distribution of Eq. 6.15 would grow (Eq. 6.16) with $\Delta y = l_0 = 1$

Here l_0 is a scaling length. Note that the maximum of this distribution lies at

$$\frac{l_0}{\sqrt{2}}$$

The probability p that a fracture of length between y and $y + \Delta y$ grows is the product of the probability that the fracture be of such a length, i.e., $g(y)\Delta y$ and the probability that its neigbours are shorter than y, namely,

$$S(y) = \int_0^y g(x)\mathrm{d}x$$

For the distribution of Eq. 6.15,

$$S(y) = 1 - \exp\left\{-\frac{y^2}{l_0^2}\right\}$$

and therefore,

$$p = \left(\frac{2y\Delta y}{l_0^2}\right)\exp\left\{-\frac{y^2}{l_0^2}\right\}\left[1 - \exp\left\{-\frac{y^2}{l_0^2}\right\}\right] \tag{6.16}$$

This relation is depicted in Fig. 6.16b. It is easy to calculate (numerically) that the maximum of p occurs at $y \sim 1.03\,l_0$. Fractures of this length have the highest probability to grow. One therefore expects that with time, a *decrease* of population (dip) would occur in the vicinity of this length and an *increase* of population would appear beyond this point leading to a bimodal distribution (Fig. 6.17), which eventually assumes the slope of Eq. 6.14. Note that for the distribution of Eq. 6.15 the ratio be-

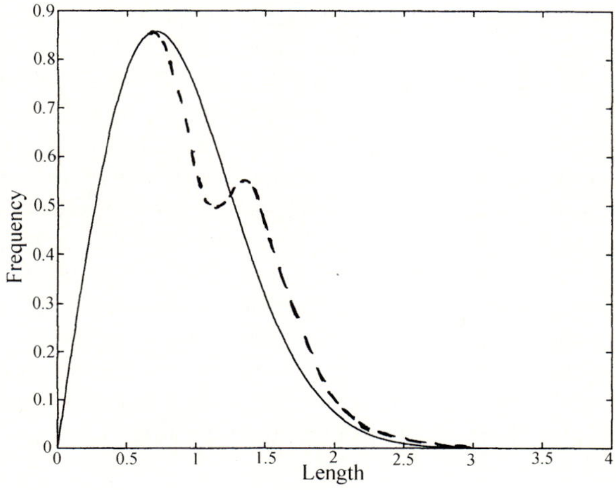

Fig. 6.17. *Full line:* The crack length distribution example of Eq. 6.15 with $l_0 = 1$. *Dashed line:* Schematic bimodal distribution obtained after length increase beyond saturation, according to the present model (see Fig. 6.16 and text)

tween the position of population dip ($1.03\,l_0$) and the position of the first maximum ($0.707\,l_0$) is 1.46. This ratio is characteristic of the specific joint length distribution and is not general.

Returning now to the field results, firstly it is clear that a power of −2.8 for $P(x)$ implies a power of $m = -3.8$ for $g(x)$ (Eq. 6.14). Hence, for the results of Segall and Pollard (1983), as well as for the results of Bahat (1988b), n of Eq. 6.3 is ~9.6, by Eq. 6.11. It is definitely within the lower range of n values obtained for rocks (Atkinson 1987, Table 11.6). Let us stress that this is an indirect method to obtain the characteristic creep value of n for these rocks. Since geological creep conditions cannot be reproduced in the laboratory, this is apparently the only method hitherto known to obtain these values.

Note also that the values of n obtained in this method for chalk and for granite are very similar in spite of rock differences. It might be argued that for such prolonged creep processes, n values are smaller than those obtained in the laboratory (Atkinson 1987), and probably these values are within the same range for different rocks. This general result should be tested for other bimodal crack length distributions.

In summary, the present section provides a new interpretation of the profile of joint length distributions. Here, two sets of fracture length measurements are discussed. It is shown that both show a FT behavior with a similar power. Next, a model based on creep growth of cracks is developed to account for the FT behavior and simultaneously explains the appearance of the bimodal behavior of the distribution. We hypothesize that this model should be applicable to additional joint length distributions.

6.5 Estimation of Paleo Fracture-Stress

The methods of paleo-fracture stress estimation on joint surfaces were summarized (Bahat and Rabinovitch 1988; Bahat 1991a, p. 229 and reference therein), and generalized for outcrops that do not display the fracture origin (Bahat et al. 1999). Weinberger (2001a) applied the method for estimating the paleostresses on joints from certain dolomite layers of the Judea Group in central Israel. When the fracture origin is unknown, the mirror radius (Fig. 2.1a,b) cannot be measured directly and needs be calculated. The location of the fracture origin of the joint that is represented by the relict fringe of *en echelon* cracks is uncertain (Fig. 6.18a) so that the mirror radius, r is determined trigonometrically (Fig. 6.18c). The measured distances are given by $2y$ between the two tips of the visible fringe and by x (Fig. 6.18c). Other relationships are $z / \sin\alpha = 2r$, where α is the angle produced by $2y$ and z; $z^2 = x^2 + y^2$ and $r = x^2 + y^2 / 2x$. For the derivation of r, two distances were determined at random locations along the fringe boundary (Fig. 6.18a), resulting in two r values differing 8% from each other, suggesting the error range in measurement. The error in r determination related to distance measurement and to a slight ellipticity deviation of the mirror from circularity is estimated as 15% (see detailed calculations in the above mentioned references). In these calculations, the local paleo-fracture stress is obtained, i.e., the stress applied on the entire joint surface. It is assumed that the stress is unchanged throughout the fracture process. This stress differs from both the (regional) remote stress and the magnified stress acting at the tip of the joint.

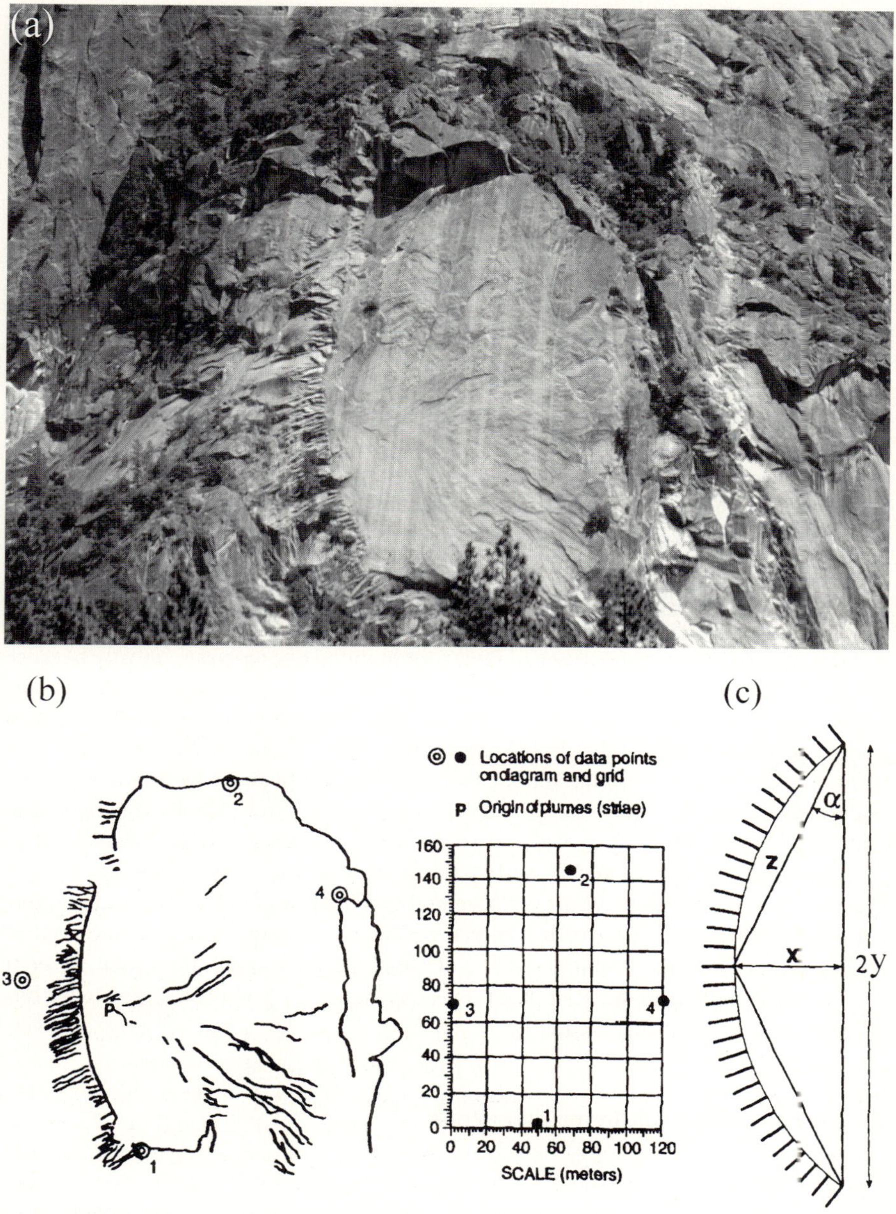

Fig. 6.18. The calculation of mirror radius when it is unmeasurable on the outcrop, applied on exfoliation joints striking N 27° E and dipping 42° NW on the southwestern side of Half Dome at Yosemite National Park. **a** Photograph of a curved fringe of *en echelon* cracks at left representing a remnant of an exfoliation joint (whose origin has been at the right side of the outcrop) on a slice that has been removed. **b** Scaled diagram and grid of **a**. The removal of the external slice exposed the surface of a second exfoliation joint whose origin is identified at *P* by the meeting area of radial barbs of a large plume. **c** Diagram showing the curved inner boundary of the fringe of *en echelon* cracks and the angle α between $2y$ and z (from Bahat et al. 1999)

6.6 Fracture Variability in a Fault Termination Zone

6.6.1 Introduction

A fault termination zone is a segment boundary, often related to earthquake dynamics (Schwartz and Sibson 1989), and is a potential major component in seismic activities (Aki 1979; King 1986). The maximum displacement in the middle of a strike-slip fault transforms into other manifestations of strain around the tip of the fault where the slip approaches zero (e.g. Muraoka and Kamata 1983). Chinnery (1966) shows that following this transformation, new regions develop around the tip of the fault in which the maximum shear stress, instead of being reduced, it may increase considerably. Changes lead to a complex geometry in the fault termination zone, including jogs, secondary faults, folds and rotation of fracture components (e.g. Trudgill and Cartwright 1994). A left-stepping bend with right-lateral shear produce contractional (anti-dilational) jogs, whereas stepping and lateral shear of the same sense produce extensional jogs. Contractional jogs may control rupture initiation and may impede fault growth (Sibson 1985; Scholz 1990, p. 155). Hence, the special conditions that develop in the fault termination zone may create both stabilities and instabilities, and understanding fault termination zones provides a key insight into rupture reoccurrence, fault growth, and growth impedance.

Detailed field characterizations of fault termination zones are rare, mostly because full exposures of such terminations are uncommon (e.g. Cruikshank et al. 1991; Bayasgalan et al. 1999 and references therein). This section concerns a detailed study of a termination zone of a contractional jog, characterized in chalk, which is a weak rock and its failure has resulted in a unique combination of a primary fault and eight secondary structures. The relative ages and interactions of joints and faults are often enigmatic. Previous analyses of curved joints near faults were presented by Rawnsley et al. (1992) in a scale of tens of meters, and by Castaing et al. (1996, Fig. 6) in scales of hundreds of meters. Castaing et al. (1996) observed that joints are both earlier and possibly contemporaneous with the faults. The present section contains a detailed account of relative ages of various components of a strike-slip fault system and a joint set on a small scale (length of 10–20 m in plan). This exposure may provide a template for terminations of larger strike-slip faults.

6.6.2 Description of the Fault Termination Zone

The fault termination zone is confined to a single layer (layer 1 in Fig. 3.39b and 6.19). The map of this zone (Fig. 6.20) is a product of a mosaic of forty overlapping photographs. The pictures were taken at 1 m height at an angle of 90° to the outcrop face. The resolution of the map is ≤10 cm. Details of fracture relationships are supplemented by photographs (Fig. 6.21a–f). The fault termination zone consists of eight fracture elements:

1. A straight fault and a gouge zone (Pr in Fig. 6.19, 6.20 and 6.21a).
2. A curved secondary fault S1.
3. Two left-stepping, secondary faults, S2 and S3.

4. A bulge caused by a local uplift in the northeastern area that surrounds the tip of Pr.
5. A set of curvilinear joints that vary in orientation sympathetically with the faults.
6. Cataclastic fault rock along certain parts of the faults.
7. Transverse cracks perpendicular to Pr in the bulged area.
8. *En echelon* cracks cutting the curved fault.

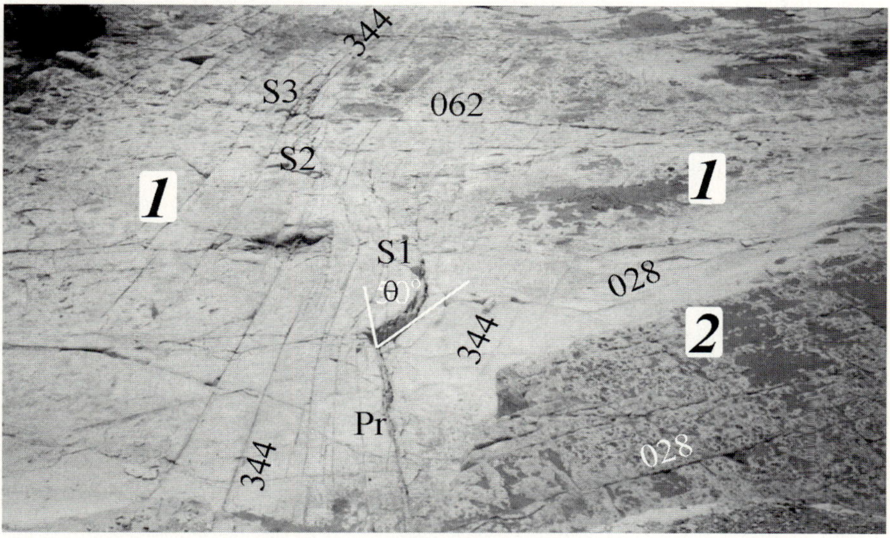

Fig. 6.19. A general view of the fault termination zone. The primary fault *Pr* (is 3.75 m long), the three secondary faults *S1*, *S2* and *S3*, set 344°, layers 1 and 2, joint set 028° in layers 1 and 2, and set set 062°: θ is the branching angle of *S1* from *Pr*

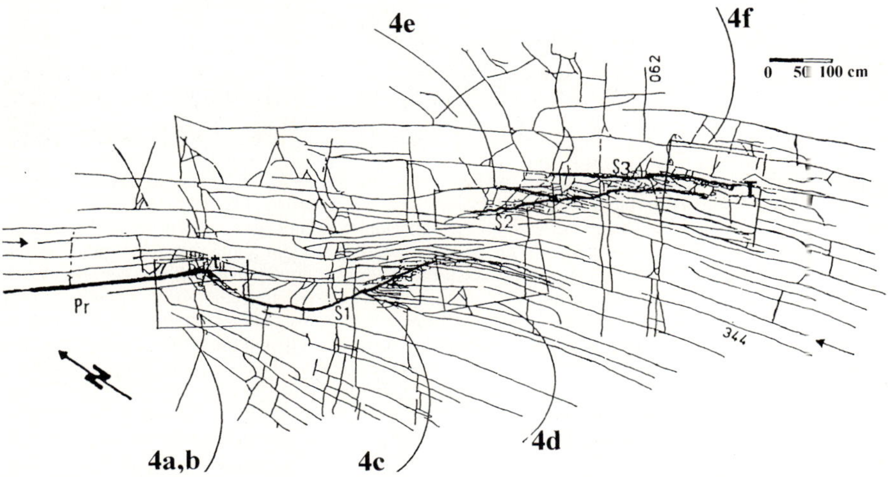

Fig. 6.20. A map of various components of the fault termination zone in layer 1: Curvilinear joint set 344°, the primary fault Pr, traverse cracks t, three segments (*S1–S3*) of the secondary faults and the termination at *T*. A single joint that crosses the termination zone is highlighted by two *lateral arrows*. The fault traces are slightly emphasized by rock splintering. The dominant joint set *344°* crossed by younger joint set *062°*. Note *frames* (4a–f) of Fig. 6.21a–f on map. Frames for Fig. 6.21a and b almost coincide

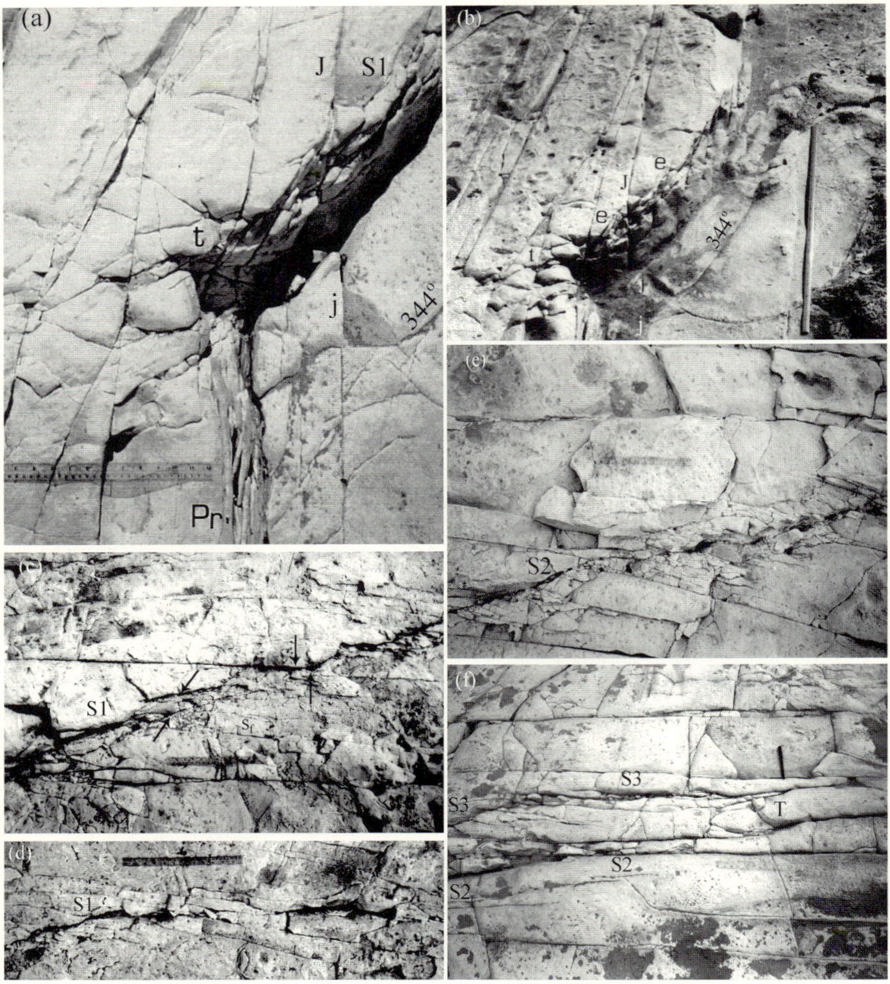

Fig. 6.21. Close-up photographs of parts of the fault termination zone. **a, b** The end of primary fault *Pr*, the curved fault S1 and a gap (*black shadow*) at the contact area between the two. The shear zone along the primary fault is populated by shear fractures (on the right side of 30 cm scale). Joints of set 344° are parallel to the primary fault. At the right side of the primary fault is a straight joint j that sub parallels to the joint J. At right-center of picture is a medium-sized joint trending approximately 344° and curves towards joint j. Transverse cracks *t* sub orthogonal to set 344° around the contact area of *Pr* with *S1*. Photographs **a** and **b** were taken from different angles. *S1* starts at the end of *Pr* where the maximum uplift of 18 cm (in shadow) occurs. *S1* is cut by two *en echelon* crack sets e (some of them sub parallel to the meter scale). Joint *j* next to *Pr* differs from Joint *J* next to *S1*. Transverse cracks *t* occur at left side of end of primary fault. **c** Two manifestations of cataclastic fault rock along the second part of *S1* (diagonal in picture). Rock fragmentation occurs between walls of a fault zone (between *inclined short arrows*) and many short joints (S_f) occur at right of fault but do not cross it. Right-lateral 2.5 cm displacement of an early straight, long joint along the fault (*vertical arrows*). An earlier fault would have prevented the long joint from crossing the fault zone. 30 cm *scale* parallels the 344° joints. **d** The curved fault *S1* terminates at an early joint where slight rock fragmentation ends (below 30 cm *scale*). **e** The initial part of the first of two left-stepping faults *S2*. Note intense rock mosaics of short joints and rock fragmentation along the fault. *Scale* is 30 cm. **f** A *pen* (13.5 cm long) points to the fault termination location *T* (see Fig. 6.20). *S2* and *S3* approach this location from left to right along joints. They bound a stretch of gradually reduced width with loss of cataclastic fault rock and increased joint spacing. At the right of *pen* jointing of set 344° initiates a "normal" distribution

The primary fault (Pr) and the gouge zone. This is a straight, vertical fault, which strikes 318° along 3.75 m. The extent of the fault continuation into the cliff is unknown (Fig. 6.19); it curves along the strike before entering into the cliff. There is a gouge zone along Pr that reaches the width of 7 cm at its tip (Fig. 6.21a). The gouge zone is sheared, resulting in slightly curved left-stepping R-shears resembling the right-lateral shear zones obtained experimentally by Bartlett et al. (1981, Fig. 6).

The fault S1: Its gap with Pr and the bulge. The curved fault S1 is vertical. It forms a sickle-like fracture in combination with the primary fault, such that S1 appears to be right stepping with respect to Pr, in a left-stepping system (see S2, S3 and Fig. 6.23 below). The northeastern fault wall (Fig. 6.20) is folded upward, bulging along the end of Pr and the beginning of S1 such that there is a gradual increase of vertical offset in layer 1 from 0 to 18 cm along the two faults, producing a crest at the contact area between them (at the Pr-S1 contact in Fig. 6.22). This crest is restricted to the "uplifted side" (note that layer 1 is higher on the left side of Pr than on its right side (Fig. 6.19)), as distinguished from the "lowered side" of the fault (in the southwestern side of the sickle-like fracture). Thus, bulging on one side of the fault introduced a distinction between uplifted and lowered sides along Pr and S1 and also along S2 and S3 (Fig. 6.22). The angle at branching initiation of S1 from Pr, θ, is 50 ±5° (Fig. 6.19).

There is a gap between the fault walls that continues laterally along S1 until it gradually dies out (Fig. 6.21a,b). There are some remnants of secondary material in this area (gouge?), suggesting that much more of it had filled the gap before being washed away. The gap is irregular and difficult to reconstruct to its original condition. This reconstruction may be achieved in two alternative ways. A good fit is obtained when the down thrown side is brought in contact with Pr by laterally overlapping the shear zone. This would close S1 and would leave about 0.5 cm right lateral displacement along Pr. On the other hand, a left lateral slip of 2.5 cm of the down thrown side along the gouge would also close S1 quite well. Both reconstructions would indicate an original right lateral displacement along Pr and S1.

S1 is divided into three parts that expose different styles of interaction with the joints of set 344° (see below). The first part, the one closer to the tip of Pr is 2.9 m long along its curved contour and has a smooth surface. Joints of set 344° seem to arrest at the first part of S1. Joint j (at the lower part of Fig. 6.21a) is straight while joint J at the upper part of the figure wiggles somewhat, and they do not appear to have been a single joint before separation by the fault. The second part of S1 is 2.4 m long and partially appears as a fault zone containing cataclastic fault rock (Fig. 6.21c) that becomes less abundant towards its end (Fig. 6.21d). A joint of set 344° appears to be right-laterally displaced 2.5 cm along the second part of S1. This joint is straight on both sides of the fault and does not show any interaction with the fault that would indicate any "aware-

Fig. 6.22. A simplified perspective of fault termination zone. The vertical displacements along *Pr*, *S1*, *S2* and *S3* are shown by *black areas*. Note bending along *Pr* and *S1* and rhythmic maxima for the various faults

ness" of the fault presence (Fig. 6.21c). Thus, the fracture contacts reveal that the first part of S1 preceded the jointing next to it, while jointing occurred before the second part of S1, implying intermittent growth of either faulting or jointing there. There are in addition short joints (see below) that arrest at the second part of S1 (Fig. 6.21c). Along its third part, which is about 35 cm long, S1 makes a slight right turn towards its end, cutting a few medium-sized joints (see below) before coinciding with the direction of one of the joints (Fig. 6.21d). Cataclastic fault rock is minimal along the third part of the curved fault.

Two left-stepping faults. Two left-stepping faults (S2 and S3 in Fig. 6.19 and 6.20) occur near the curved fault. S2 starts about 4 m and 0.5 m from the closest points of Pr and S1, respectively. S2 is slightly concave toward the southwest. It overlaps about 0.7 m with the curved fault. At 2.15 m distance from its northwestern end, the fault splays with a branch about 1.30 long at a 20° angle. S2 crosses joints of set 344° near its early part (Fig. 6.21e), and then it coincides with a joint. The splay of this fault also coincides with a joint. S3 is almost straight, overlaps S2 and runs entirely along a joint. The distance between S2 and S3 varies from about 40 cm at their northwestern ends to about 30 cm at their southeastern termination.

The two left-stepping faults terminate in close proximity to each other (Fig. 6.21f, at T). The intensity of cataclastic fault rock declines to almost zero beyond T. The number of joints diminishes from some ten before T to three beyond this termination. Also, the opening along the fractures reduced at T from 1–2 cm to 1 mm or less.

The folding to bulges along S2 and S3 have maxima (11 cm and 8.5 cm, respectively) at the northeastern up-thrown part of each fault, and they vanish along the vertical fault planes towards the southwestern ends (Fig. 6.22). S2 and S3 differ significantly in orientation from S1. Whereas S1 strongly deviates from the direction of the primary fault Pr, S2 and S3 start to propagate along strikes, approximately in continuation with Pr (Fig. 6.20 and 6.21). While fault S1 terminates with only little cataclastic deformation along a joint of set 344°, fault S2 initiates with abundant cataclastic fault rock (Fig. 6.21e).

The joint set 344°. This set displays a strong curvilinearity along which growth occurred in several stages (generations). The joint set is spatially restricted to layer 1 in a 7 m wide area around the faults. The area is bounded by talus along the northeastern side and by a joint set oriented 028° along its southwestern side (at the right of Fig. 6.19 and 6.20). The joints of set 344° change strike in "four zones" of the termination zone as follows. The joints are approximately straight and parallel to Pr along the fault. Beyond the tip of Pr the joints curve sympathetically with respect to the secondary faults, S1, S2 and S3, forming the "curvilinear zone". On the southeastern side, beyond the final fault termination (T in Fig. 6.21f) and throughout the sourthwestern "lowered side" of the fault all joints are straight with uniform 344° orientation.

A joint oriented 344° in the "lowered side" curves and "hooks" into a straight joint that parallels to the primary fault (Fig. 6.21a), suggesting a fracture sequence between the joints, because the former joint "knew" about the existence of the straight joint. At least one joint approximately follows the shape of S1: First, the concavity of S1 towards northeast and then, its concavity towards southwest, towards the termination of S1.

(Fig. 6.19 and 6.20). Some joints however are crossed by the secondary faults at acute angles ~40° or at smaller angles ~5–15°. Individual joints of set 344° may be up to 10 m long with only very little interaction with each other. Hence, although these joints curve and vary in orientation, they are termed set 344°.

We divide the joints of set 344° into three groups according to their lengths. The "long", "medium-sized" and "short" fractures are up to 10 m, 4 m and 1 m long, respectively. The long and medium-sized joints are parallel to each other. Near S1 and along S2 and S3, medium-sized joints fill gaps between long joints (Fig. 6.20), reducing spacing to about 2 cm. However, further away from the faults S2 and S3, towards the southeast end of the outcrop where long joints are oriented 344°, their spacing becomes larger and ranges from 10 to 25 cm, not having the medium-sized joints among them. Thus, the medium-sized joints appear to postdate the long joints.

The long and medium-sized joints cut by S1 are distinguished from the short joints not cut by S1 (Fig. 6.21c). The medium-sized joints appear to be quasi-systematic (straight and parallel to each other), whereas the short joints are non-systematic and somewhat distorted. Joints of set 344° are generally quite closed (an aperture of about 0.1 mm or less). However, some of them widen their apertures up to about 1 cm along the second part of S1, and particularly, along the stretch between S2 and S3.

Cataclastic fault rocks. In certain locations, the country rock is intensely ruptured into fragments that range in size from several millimeters to several centimeters. The "mosaic style" occurs along the second part of S1 (Fig. 6.21c), where rock fragmentation occurs between two walls of a fault zone. A mosaic is also spatially associated with short joints that either start or arrest at the southwestern side of the fault zone, on which perpendicular younger cracks may arrest, creating a rock mosaic of square and angular splinters. This style is also in association with S2, but the mosaic next to S2 (Fig. 6.21e) is more intensely populated. Mosaics extend up to 30 cm from a fault, mostly occurring along the concave side of the faults (Fig. 6.21e), where the secondary faults cross the 344° joints, particularly where joint spacing is small. Fragmentation also occurs as rock splintering not associated with rock mosaics. They are particularly found between S2 and S3. This splintering also extends to distances of up to about 12 cm across the two sides of the faults.

Transverse cracks perpendicular to Pr in the bulged area. The bulged area close to the tip of Pr is intensely fractured by transverse cracks. The cracks are short (about 12 cm), trending approximately orthogonal to the primary fault (t in Fig. 6 20 and 6.21a), and generally arrest at joints of set 344°. The transverse cracks cut part of the gouge along the primary fault (they are not visible in Fig. 6.21a).

***En echelon* cracks.** Two sets of *en echelon* cracks cut the bulged area of the first part of S1 at its continuation from the tip of Pr (e in Fig. 6.21b). Their azimuths are 330° and 350°, and their crack length averages are about 22 cm and 14 cm, respectively. Cracks of set 350° generally arrest at cracks of set 330°, implying that set 330° generally was older. Rock splintering associated with these cracks is not as intense and irregular as the cataclastic fault rock along the second part of S1 and the left-stepping faults.

Table 6.1. Ratios of fracture parameters in the fault termination zone

Fracture parameter	Ratio
1. [Length (max) S1] / [length (straight) S1]	455 / 435 = 1.05
2. [Length (straight) S1] / [length (straight) S2]	435 / 470 = 0.93
3. [Length (straight) S1] / [distance between edge of Pr to start of S2]	435 / 350 = 1.24
4. [Length (max) S1] / [vertical offset (max) of S1]	455 / 18 = 25.3

S1, and *S2* are the curved fault and the first of the two left-stepping faults, respectively. The terms *max* and *straight* designate the full curved length of a fault, and the length of a fault ignoring the curvature, respectively. Ratios are obtained by division of parameters given in centimeters.

6.6.3 Fracture Parametrization in the Fault Termination Zone

Ratios of fracture parameters. Assuming similarity of strike-slip faults in various scales and therefore similarity of their termination zones, ratios of fracture parameters are possibly more important than absolute values. Some ratios of fracture parameters are summarized in Table 6.1. It turns out that the "straight" lengths of S1 and S2 (ignoring curvature along the fault) and the distance from the tip of Pr to the initiation of S2 (Fig. 6.20) are about the same.

Fracture sequence in the fault termination zone. The great versatility of fracture styles and their clear exposure enables one to decipher the sequence of most fracture processes in the fault termination zone (Table 6.2). The right-lateral offset along Pr resulted in creating a wing crack (the first part of S1) and a gap at the contact area between Pr and S1. More than a single offset along Pr (repeated S2 and S3 faults and e cracks) resulted in the deformation in the termination zone. Bulging in the area close to the tip of Pr has resulted in transverse cracks oriented 025° (characteristic to crests of outer arcs, Fig. 3.19a). These cracks arrest at the 344° joints that parallel to Pr in the "uplifted side" of the fault, suggesting that maximum bulging occurred after the formation of 344° joints. On the other hand, joints of set 344° seem to arrest at the first part of S1, suggesting that the first part of S1 preceded the joints. Also, in the "lowered side" a joint oriented 344° curves and "hooks" into a straight joint that parallels Pr, suggesting a time interval between these joints. Assuming that the straight parallel joints on both sides of Pr were formed approximately simultaneously, it should follow that the lateral component of deformation creating S1 preceded the vertical offset associated with the bulging. That is, when the lateral offset(s) was exhausted, further stress release resulted in vertical stresses producing the bulging.

The onset of cataclastic deformation along the second part of S1 suggests a change in mode of fracture at the transition from the first part of S1. This change is not surprising, considering the change in orientation along S1 with respect to the 344° azimuth, which is taken to be the σ_1 trajectory of the stress field that created Pr (Fig. 6.20). In analogy to previous observations (e.g., Brace and Bombolakis 1963; Jayatilaka and Trustrum 1978), the second part of S1 was formed by a stable process under conditions of increasing stresses. This part of S1 postdated a 344° joint (Fig. 6.21c). The implication is that the second part of S1 developed discontinuously (incrementally) with

Table 6.2. Sequence of fracture episodes of joints and faults in the fault termination zone

Episode No.	Joints/cracks	Faults	Uncer-tainties
1	–	Offset along Pr	
2	–	First part of S1	
3	Parallel to Pr	–	
4	–	Vertical displacement along Pr and S1	
5	Long (up to about 10 m long) curvilinear and 344 oriented	–	
6	–	Transverse cracks near tip of Pr	
7	Medium (up to about 4 m long)	–	?
8	–	Second part of S1 and rock splintering along S1	
9	Short (up to about 1 m)	–	?
10	Cross (one cm) in mosaics	–	
11	–	Third part of S1 and rock splintering along S1	
12	Short (up to about 1 m)	S2 and rock splintering along S2	?
13	Cross (one cm) in mosaics	–	
14	–	S3 and rock splintering along S3	
15	Short (up to about 1 m)	–	?
16	–	Rock splintering between S2 and S3	

Pr, *S1*, *S2* and *S3* are, the primary fault, the curved fault, first of the two left-stepping faults and second of the left-stepping faults, respectively. Uncertainties in time relationship of joints and faults are marked by *?* and are discussed in text.

respect to the first part of S1. The curved fault S1 terminates with only little cataclastic deformation along a joint of set 344° (Fig. 6.21d), whereas the initiation of S2 is associated with intense cataclastic deformation (Fig. 6.21e). This suggests that an early episode of secondary faulting ended with the termination of S1, but the creation of S2 manifests a new fracture episode that occurred when high stresses were attained.

Many short joints arrest at S2. Since S2 postdated S1, the joints that interact with S2 quite likely postdated those that are associated with S1. Spacing of set 344° in various areas adjacent to the secondary faults is reduced down to about 2 cm. The reduced spacing reflects multi-stage jointing of medium-sized and short joints: Late joints fractured between and parallel to previously long joints. It has been observed in many outcrops of syntectonic joints (joints associated with intense deformation) cutting chalks that these fractures generally form by more than a single failure event (Bahat 1991a, p. 282).

Generally, cataclastic fault rock is thought to form concurrently with faulting. However, it should not be ruled out that the mosaics that occur close to one boundary of a second fault (center of Fig. 6.21c) could also reflect post fault energy release. The question marks in Table 6.2 indicate that certain parts of the fracture sequence are not certain. Episode 7 is marked by '?', because only one medium-sized joint is cut by the second part of S1. The reason for marking '?' in relation to episodes 9, 12 and 15 is that short joints may be contemporaneous with the adjacent faults but could also be post faulting.

Two gap types. The gap continuation along the first part of S1 (Fig. 6.21a,b) is a result of a mode I fracture. Closing the gap into its original condition before faulting by a left lateral slip of the down thrown side along the gouge would create an appearance of a continuation of j- and J-joints across S1 and may suggest that the joint predated S1. On the other hand, closing the gap by contacting sidewise the down thrown side with Pr (overlapping across the shear zone) may result in an offset of joints j and J across S1. This would fit the model of joint j arresting at the earlier S1, becoming the favored explanation, implying that the shear zone developed only after the separation of the two walls of Pr.

It may be speculated that the growth of some primary faults and their termination zones to large (continental) strike-slip faults can be explained by a stepwise process. This takes place by reoccurring bridging of the gaps between the tip of the primary fault and the initiation of the first or second *en echelon* faults (Pr, S2 and S3, respectively, in Fig. 6.20). Such a process might shift the location of maximum stress intensity from the tip of Pr to the termination of S2 and/or S3 (point T) before the next earthquake would take place.

Models of termination zones and experimental results. There is an important difference between two existing models of termination zones. In one model of a right lateral offset, all secondary faults maintain left-stepping configuration, splaying towards one side of the continuation of the primary fault (Fig. 6.23a) (Hobbs et al. 1976, Fig. 7.18d and Twiss and Moores 1992, Fig. 7.7). Chinnery (1966) suggests two possible theoretical combinations of straight primary and curved secondary faults. One of them consists of a sickle-like structure that resembles the present combination

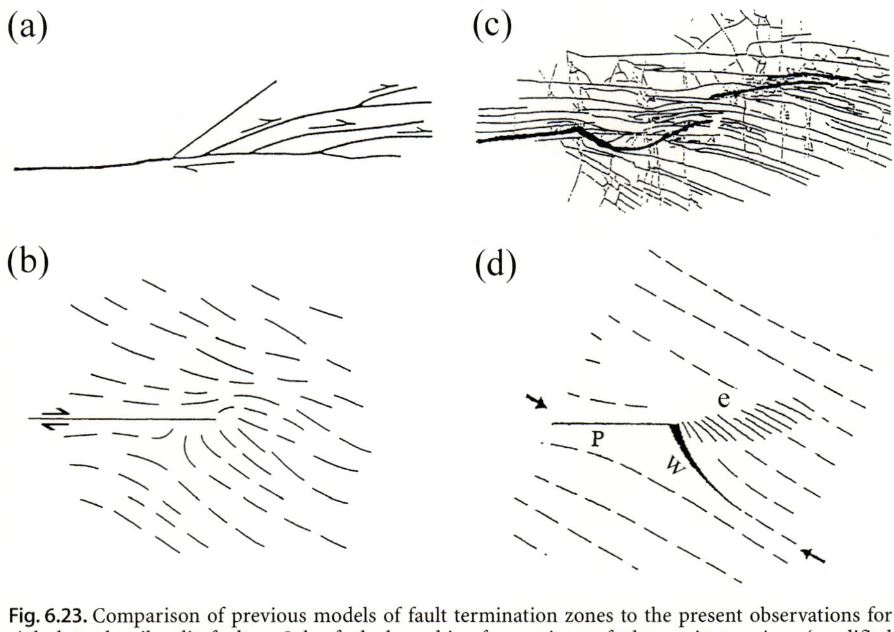

Fig. 6.23. Comparison of previous models of fault termination zones to the present observations for right lateral strike-slip faults. **a** Splay faults branching from primary fault near its terminus (modified after Hobbs et al. 1976). **b** Trajectories of maximum principal stress direction shown by *dashed lines* at the termination zone of a strike-slip fault (from Means 1979, after Chinnery 1966). **c** Miniature of fault pattern from present map in Fig. 6.20. **d** Compression applied at a small angle to primary fault *P*, resulting in secondary fractures *w* and *e* (see text for explanation) (modified after Petit and Barquins (1987, 1988)

of Pr and S1: The trajectories of the maximum principal stress at the termination zone of a strike-slip fault (Means 1979) resemble the contours of the present joint set 344° (Fig. 6.23b and c, respectively). In the latter, the curved fault S1 appears to be the result of right stepping, apparently in conflict with the requirement of an exclusive left-stepping arrangement of the first model (Fig. 6.23a).

Petit and Barquins (1987, 1988) demonstrate experimental results obtained from PMMA and sandstone under uniaxial compression, which strongly resemble the fracture geometry in the termination zone (Fig. 6.23d, see also Fig. 2.58a). In their experiments, the "sickle" formed by Pr and S1 and the secondary fault S2 are represented by P, w and e, respectively. The location and the angular relationship with respect to the primary fault Pr suggest that S2 and S3 are analogous to the "shear zone" e that Petit and Barquins (1987, 1988) produced under maximal stress conditions. They also found that the wing crack resulted from mode I operation, which shows opening, and the shear zone resulted from mode II loading (see Sect. 2.2.13). Also relevant is the experiment by Petit and Barquins (1987, 1988) on sandstone that revealed cataclastic deformation along the *en echelon* cracks at e, resembling the cataclastics found along S2 and S3. There is an additional resemblance between these experimental results and the fault termination zone, in that w deviates from the direction of P, and e strikes approximately in continuation with P (Fig. 6.23d).

Bulging and vertical offset. Spatial constraints cause shear stresses around the tip of the primary fault that are partly released by lateral displacement and partly by vertical offset. The area in the termination zone may be divided into compressional and tensional regimes. In a contractional jog, a vertical upthrow along one side of the fault would occur in the compressional regime and a downthrow would occur in the tensional one (Chinnery and Petrak 1968). Accordingly, bulging along the primary and secondary faults is exhibited in the upthrown block (Fig. 6.19) as expected. The rhythmic maxima in the vertical displacements along S1, S2 and S3 imply that these bulges occurred concurrently with each individual fault. A single vertical displacement at a later time would be expected to have been monotonous with a gradual change from maximum to minimum throughout the termination zone rather than a rhythmic one. This rhythmic mechanism supports the suggestion of an incremental growth of the secondary faults (Table 6.2).

6.6.4
Open Questions

The student may like to ask several questions that need to be further explored:

1. How do existing left-stepping configurations (Fig. 6.23a) reconcile with the field and experimental data shown in Fig. 6.23c,d?
2. How similar or different to this fault termination zone are termination zones of other contractional jogs?
3. To what extent fault termination zones change with size?
4. Set 344° was induced by a perturbation at the tip of the fault (rather than along it, Fig. 3.12a). There is a need to characterize the secondary stress field that formed around the tip of the primary fault and induced the three secondary fractures (far away from the Pr tip).

6.7 Fault-Joint Relationships in the Beer Sheva Syncline, Israel

6.7.1 Introduction

This section relates to eight distinct fault-joint systems in the Beer Sheva syncline from south Israel that differ from each other in their genetic affiliation and their geometric and age relationships, citing Bahat (2000, 2004).

6.7.1.1 *Joints and Fault Classifications*

Joints are the most ubiquitous structures in sedimentary rocks. This is because joints may be formed under small differential stresses, and therefore most environments are conducive to their development, both systematically and non systematically (Hodgson 1961b). Nelson (1979), Engelder (1985) and Bahat (1991a) proposed genetic classifications of joints. According to the latter classification, joints may be divided into four groups: Burial, syntectonic, uplift and post uplift. These can be distinguished in the field by both geometric and fractographic criteria. The four groups reflect four characteristic stages in basin evolution.

The last fifty years of fault investigation enjoyed the benefit of the pioneering classification of faults into three classes by Anderson (1951). The great power of Anderson's grouping stemmed from its simplicity. It helped to demonstrate the dependence of the three main fault types on systematic changes in the stress field that also reflected on the most likely tectonic conditions. However, fault classification cannot be applied to timing of basin histories. On the other hand, extended research on the temporal relationship between fault-joint relationships (FJR) (e.g., Sterns 1968; Pohn 1981; Cruikshank et al. 1991; Gross 1995; Martel 1997; Kattenhorn et al. 2000; Wilkins et al. 2001; Eyal et al. 2001) has accumulated a vast amount of knowledge that can be used in refining the fracture histories of basins. The work by Peacock (2001a) is a recent summary of some key observations related to this topic.

6.7.1.2 *The Importance of Temporal Fault-Joint Relationships*

The temporal fault-joint relationships are important for a variety of reasons that are specified below.

1. Sequences between faults and joints may help to decipher the tectonic history of the region (including earthquakes), and temporal relationships.
2. They also provide information on the stress history of the region.
3. Faults may affect later joint development and joints may transform to faults (Segall and Pollard 1983), and these processes are mechanically intriguing.
4. Distinct joint/fault genetics may be linked with different fracture parameters (aperture, length, spacing, interaction, etc.).

5. Fluid flow (water, hydrocarbon liquids and gas) depends on fault-joint relationships and their fracture parameters. For instance (Bahat 1991a, p. 239).
   a Burial pre-fault joints may be closed due to their development in a process of increasing lithostatic pressure;
   b Some syntectonic, syn-fault joints may be sealed into veins, while others strongly enhance water flow, like the strike parallel joints (062°) in the Beer Sheva syncline;
   c Uplift joints that form after burial-normal faults are opened fractures.

Unlike many other folds that exhibit relatively uniform structures, the Beer Sheva Eocene syncline is divided into four distinct stratigraphic units that display different fracture characteristics and store a record of many (more than twenty) different fracture episodes since the Lower Eocene (Sect. 3.4). One manifestation of this multi-fracture complex is the occurrence of at least eight distinct FJRs. Various aspects of FJRs in this syncline were introduced in previous publications, including pre and post-fault burial joints, syn-fault syntectonic joints and post-fault uplift joints that partly transform into faults (e.g., Bahat 1991a, pp. 241, 275 and 282; 1998a, 1999a; Gross et al. 1997). A synthesis of the previous observations constitutes the major part of this section, with the objective of demonstrating the extent to which the estimation of timing of fracture development, relative to the structural history of a syncline can be achieved. Price's (1974) model of fracture in basins is adapted here in analyzing fracture in the Beer Sheva syncline, as in previous publications (e.g., Bahat 1989). A distinction is made in the following description of the various FJR types between the "primary faults" and the various joints (or faults in one case) that are associated with them, which are collectively termed "secondary fractures". This distinction is not relevant to type 8 (see below).

6.7.2
Eight Types of Fault-Joint Relationships

Eight FJR types of fault-joint relationships from the Beer Sheva syncline (Fig. 3.33b) are characterized below. Table 6.3 summarizes FJR that take into account different parameters:

1. Temporal relationships.
2. Geometric contacts.
3. Tectonic setting.
4. Stresses.
5. Fracture system.

Six types relate to normal faults, one type concerns the terminal zone of a strike-slip fault and the eighth type is about vertical faults.

Type 1 joints occur in the Mor Formation of the Lower Eocene and is associated with a normal fault from the burial stage of the syncline (Bahat 1985). The fault strikes ~292° and dips ~45° N, cutting alternating chalk layers (about 90 cm thick) with chert beds (about 10 cm thick) (Fig. 6.24a,b), while the chalk is dissected by two orthogonal, vertical, single-layer joint sets oriented 328° and 059°. Joints of the cross-fold set 328° arrest at the boundary of the chalk layers with chert beds. The fault that dis-

Table 6.3. Eight types of fault-joint relationships (FJR) in the Beer Sheva syncline, Israel

FJR type[a]	Temporal relationship of joints	Joint attitude in relation to the fault one	Tectonic setting	Joint formation by stresses remote/local	Fracture system
1 (n)	Pre-fault	Different, independent	Burial, early	Remote	Closed, non-interacting joints, regular spacing (Fig. 6.24b)
2 (n)	Post-fault	Different, independent	Burial, early	Remote/local	Closed, non-interacting joints, decreased spacing (Gross et al. 1997)
3 (n)	Syn-fault	Correlative, dependent	Syntectonic early uplift	Local	Local fracture multiplication (in sector I, Fig. 6.24c and 6.25)
4 (n)	Syn-fault	Correlative, dependent	Syntectonic early uplift	Local/remote	Local twisted secondary faults (in sector III, Fig. 6.24c and 6.25)
5 (s)	Post-fault	Secondary Stress field	Syntectonic early uplift	Local	Non-interacting joints, variable spacing along primary and secondary faults (Fig. 6.20)
6 (n)	Pre-fault	Different, independent	Late uplift	Remote	Regular joint spacing (Fig. 3.36g)
7 (n)	Syn-fault	Correlative, dependent	Late uplift	Remote	Decreased joint spacing (Fig. 6.24d)
8 (v)	Transitional	Parallel	Uplift	Remote/local	Deep penetrating, creating vertical joints and faults (Fig. 3.35b)

[a] Normal fault (*n*), strike slip fault (*s*), vertical fault (*v*).

places these joints contains in its fracture zone fragments of chert that had been broken during early offsets of the fault and unfractured nodules that precipitated after the cessation of the fault activities. The precipitation of chert is associated with the diagenetic (burial) stage of the chalk (Knauth and Lowe 1978). Chert does not occur in the overlying Middle Eocene (Horsha Formation). Therefore, both the cross-fold joints and the normal fault occurred before the sedimentation of the Middle Eocene, i.e., they were formed during the burial stage by a process controlled by the tectonic regime that prevailed in the Lower Eocene (Gross et al. 1997). The joint orientations are unrelated to the fault attitude, requiring that the remote stresses had to change from the conditions under which the joints were formed to new conditions that induced the fault. Hence, during the burial stage of the syncline, differential stresses gradually increased, such that the formation of the cross-fold joints and some strike-parallel joints preceded the creation of the fault (Bahat 1985, 1991b; Gross et al. 1997). Generally, joint orientation and spacing is almost the same around the fault.

Type 2 relates to the same fault that is shown in Fig. 6.24a,b but it concerns a local spacing reduction in the hanging wall compared to the uniform spacing elsewhere around the fault. Gross et al. (1997) found that due to dragging of one of the chert beds, an unrelieved strain remained stored in the hanging-wall from the early stages of the fault slip. This unrelieved strain was subsequently released in the form of ~055° oriented joints. Hence, the fracture sequence during the burial stage was early cross-fold and strike-parallel orthogonal joint sets, normal faulting and subsequently some additional fold strike-parallel joints in the hanging wall.

Fig. 6.24.
Structural relationships of faults and associated joints from the Beer Sheva syncline. **a** Two conjugate faults. **b** An enlargement of the fault dipping towards left, to which types 1 and 2 relate. The faults cross vertical single-layer joints in Lower Eocene chalks near Beer Sheva. The two orthogonal joint sets and the faults strike in three distinct directions. Thus, joints and faults were formed in different remote stress fields (after Bahat 1985). **c** Types 3 and 4. Type 3 is represented by inclined joints that parallel the dip of a normal fault in Wadi Secher at the upper hanging wall (marked j). Type 4 is shown as inclined secondary faults (f1–f3, marked 1–3, respectively) at various acute angles to the primary fault in the lower footwall. **d** Type 7. Spacing of the vertical joint set decreases considerably as the joints come closer to the normal fault, which is a characteristic feature of syntectonic joints. Therefore, the primary fault and joints developed in the same remote stress field. The *black rectangle* is 10 cm long

Two more joint spacing peculiarities were observed in the same outcrop (Bahat 1991a, p. 255). First, joint spacing is wider in thinner chalk beds, i.e., the layer thickness-joint spacing plot shows a negative slope for both cross-fold joints and strike joints, which is in contrast to most observations by other investigators that show positive plots (e.g., Price and Cosgrove 1990, p. 54). An exceptional finding of a negative slope was also made by Eyal et al. (2001) in the chalky limestone beds Nahal Neqarot in southern Israel. Second, for a given layer thickness, joint spacing is a little wider in the strike-parallel joints (oriented 055°) than in the cross-fold joints (oriented 326°). An explanation for these two observations requires further studies. It is possible that spacing is wider in the strike-parallel joints because their formation was associated with stress relaxation, compared to the cross-fold joints that reflected regional stresses (Bahat 1989).

Types 3 and 4 joints outcrop along Wadi Secher (Fig. 3.33c) and occur in the Middle Eocene chalks of the syncline. The joints of Type 3 are inclined and are approximately parallel to the Secher fault dip (Fig. 6.24c). Fracture distribution around the fault is explained by Fig. 6.25a–e. The remote stresses (σ_1^∞, σ_2^∞, σ_3^∞) have induced the normal fault that dips towards left (Fig. 6.25a). A new, local stress field (σ_1, σ_2, σ_3 in the shape of a stress ellipsoid) was created by the slip (or slips) along the fault (Chinnery 1966). A damage volume was formed along the fault by its initial formation and by its subsequent slips. The damage volume is the rock volume that contains secondary structures (mostly fractures) that are confined to the fault, and it approximates the shape of an ellipsoid (Fig. 6.25b). Two orthogonal planes can schematically describe the distribu-

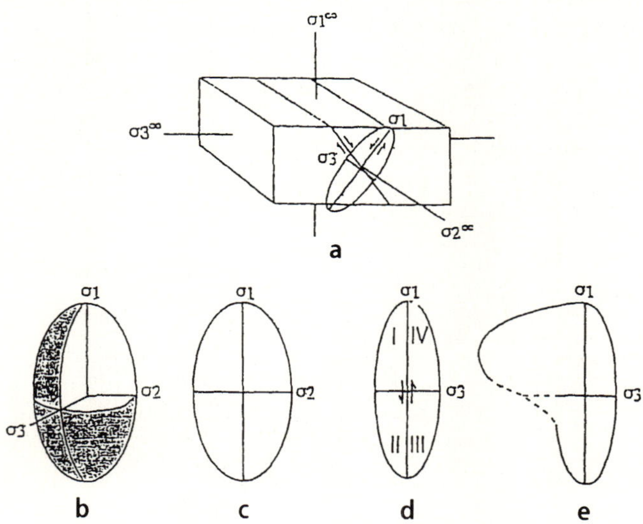

Fig. 6.25. Diagram of primary normal fault and secondary fractures in three dimensions. **a** The normal plane (*ellipse*) that describes the arrangement of local principal stresses is superposed on one of two conjugate faults in a rectangular block that describes the remote principal stresses. Orientations of remote and local principal stresses differ significantly. These figures contain geometric elements from Means (1979) and Park (1983) (see explanation to types 3 and 4). **b** Stress ellipsoid in perspective, showing its two principal planes, the fault plane σ_1–σ_2, which is termed the "parallel plane" with its elliptical tip line boundary, and the "normal plane" σ_1–σ_3. **c** The "parallel plane" that contains the fault plane with its elliptical tip line boundary. **d** The "normal plane" is divided into four sectors *I–IV*. **e** The approximate shape of sector *I* in the normal plane of the Secher fault A

tion of secondary fractures in a damage volume. The tip line forms an elliptical line (Walsh and Watterson 1988; Knott et al. 1996) that marks the loci on the fault plane (approximately the "parallel plane") along which displacement has decreased to zero (Muraoka and Kamata 1983; Walsh and Watterson 1988) (Fig. 6.25c). The "normal plane" is orthogonal to the parallel plane and is divided into four sectors (I–IV) by the trace of the fault and the tip line (Fig. 6.25d). For normal faults the extensional sectors I and III are widest (where rock is weak) and narrowest in the contractional sectors II and IV (where rock is strong) (e.g., Knott et al. 1996).

The direction of maximum shear stress in the initial field of remote stress ($\sigma_1^\infty, \sigma_2^\infty, \sigma_3^\infty$) is transformed to the direction of σ_1 (Fig. 6.23b) in a simple 45° rotation exercise (Means 1979, problem 12.4) during the formation of the local stress field. This explains how the local minimum principal stress (σ_3) was normal to the fault plane, creating type 3 joints (Fig. 6.24c). Some twelve joints of type 3 are spatially confined to sector I in the damage volume of the Secher fault (one or two joints occur in sector IV as well). Unexpectedly, the width of sector I of the Secher fault was found to be about 60% of the length of the fault, implying a strong deviation of the normal plane from an elliptical shape (Fig. 6.25e). Accordingly, the schematic shape of the damage volume (shown two dimensionally in Fig. 6.25d) is modified to a more realistic "quasi-ellipsoid damage volume" (Fig. 6.25e). Type 3 joints resemble the (unspecified) fractures that are parallel to the dip of the normal fault from the known results described by Stearns (1968) (the fractures dipping towards left in Fig. 6.26). They also resemble the FJR of large strike-slip faults in which joints cut parallel to the dip and strike of the primary fault close to the fault termination zone (Chinnery 1966, Fig. 5).

Type 4 joints consist of a series of three fractures (f_1–f_3) that formed in sector III of the normal Secher fault (Fig. 6.24c and 6.25d). Fractures f_1 and f_3 exhibit changes in orientation relative to that of the fault. The fault strikes 040° and dips 51° to NW, f_1 strikes 035° and dips 68° to NW, f_2 strikes 038° and dips 69° to NW and f_3 strikes 088° and dips 85° to NW. The changes reflect combined strike twist and dip tilting from an attitude relatively close to that of the fault to an almost vertical fracture that totally divorced from the trend of the fault. This reflects a response to local perturbed stresses

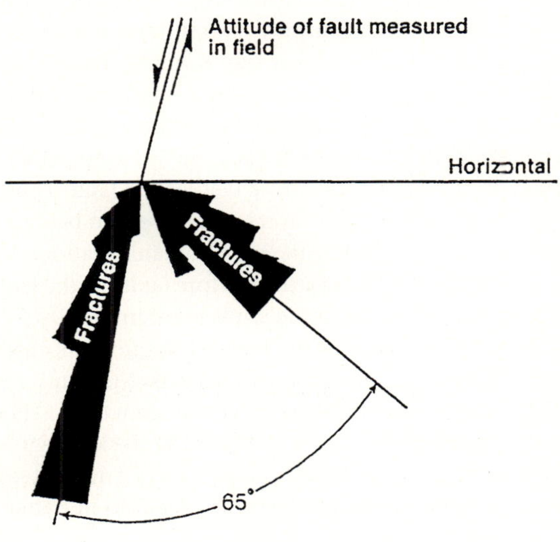

Fig. 6.26.
Rose diagram showing the distribution of dips of two sets of fractures (unspecified) associated with a normal fault (after Stearns 1968, and Twiss and Moores 1992)

along the fault (Rawnsley et al. 1992) that were effective in initiating fractures f_1–f_3, combined with a gradual increase of the influence of the remote stresses with distance of the fractures from the fault (up to about 3.5 m). There is some shear displacement along these fractures, which distinguishes them from the joints of the other types. Type 4 resembles the fracture set that occurs at a large angle (>45° but <90°) relative to the fault in the Stearns model (dipping to the right in Fig. 6.26). However, in the latter the angles created between the fault and joints are larger than in type 4 (see more about the model depicted in Fig. 6.25 in the discussion section). Type 4, like type 3 is thought to be genetically related to displacements along the fault, in association with an early uplift stage in the Beer Sheva syncline (see below).

Type 5 joints occur at a fault termination zone (Sect. 6.6), cutting a Middle Eocene chalk layer during the uplift stage. The termination zone consists of a primary fault, secondary faults, and a joint set in Wadi Naim (Fig. 6.19 and 6.20). The primary fault is a vertical right-lateral strike-slip fault, striking 318°. A set of three partly curved secondary faults initiate at the tip of the primary fault, and a joint set striking 344° curves sympathetically with the primary and secondary faults in a close proximity with the faults. Therefore, the 344° striking joints are considered to be genetically associated with the faults, and they are termed syntectonic joints. The 028° striking joint set is ubiquitous in Wadi Naim. However, where these joints approach the termination zone they arrest or interact (hooking style) with the 344° striking joint set, but they never penetrate to the termination zone. Therefore, the 028° striking joint set postdated the above three structures in the same chalk layer, i.e., they are younger than type 5 FJR joints. Thus, sets 344° and 028° belong to two separate fracture events: An early and an advanced one during the uplift process (Bahat 1999a).

Type 6 joints are cut by the Naim normal fault A (Fig. 3.36g). There are no visible differences in the behavior (spacing) of the joints from both sides of the fault. The strike of this set forms an angle larger than 45° with the strike of the fault. Unlike the system described by Kattenhorn et al. (2000) there are no genetic relations between the fault and joints. These are single-layer uplift joints that were associated with the advanced uplift in the syncline, but they had been preceded by the earlier 344° uplift set (type 5). Therefore, the fault of type 6 was associated with an advanced uplift stage of the syncline.

Type 7 joints are adjacent to Naim normal fault B that occurs some 125 m west of Naim fault A. The joints show significant decrease in spacing close to the lower part of fault B (Fig. 6.24d), and as such they comply with a basic criterion of syntectonic joints (Bahat 1991a, p. 279). Compared to types 1 and 6, which reflect joints and faults that formed by different stress fields, type 7 represents an assemblage of joints that were formed in the same local stress field as the fault. Thus, even if the faults from types 6 and 7 cut the same chalk layers and appear to belong to the same fault system, these two types seem to show different FJR manifestations. Whereas the fault of type 6 is associated with joints that formed before faulting, the fault of type 7 created a local stress field that formed syntectonic joints adjacent to it.

Type 8 joints occur in the Lower Eocene chalks around Beer Sheva in the form of vertical multilayer joints that often traverse the entire exposed outcrops; they may run alongside or across burial normal faults from type 1 (Fig. 3.35b). Often multilayer joints are associated with vertical multilayer faults in irregular 5–20 m spacing. These vertical fractures are considered to have been formed after the burial stage during a single or multi uplift episodes of the area (Fig. 5.15 in Bahat 1991a).

6.7.3 Discussion

6.7.3.1 *Genetic Relationships and Relative Ages of Fractures*

Types 1, 2 and 8 occur in Lower Eocene chalks. While types 1 and 2 are products of the burial stage, type 8 formed much later by uplift (Bahat 1985, 1991a). The primary faults of types 3, 4, 5 and 7 were created by remote stresses, but their secondary fractures resulted from local stresses that were associated with slip (or slips) along the respective faults, i.e., the primary fault and secondary fractures of these types are genetically related. On the other hand, type 6 concerns a fault and neighbor joints that were genetically unrelated, because they were formed by two distinct remote stresses. The genetically unrelated style of type 6 resembles that of type 1 from the Lower Eocene, whereas the genetically related style of types 3, 4, 5 and 7 resembles that of type 2. Hence, the joints of types 1 and 6 are pre-fault secondary fractures, whereas the joints of types 2, 3, 4, 5 and 7 are both post and syn-fault fractures.

These genetic relationships can help in working out the relative ages of types 3–7. The secondary fractures associated with types 3 and 4 formed in pristine chalks and predated the regional joints along Wadi Secher. On the other hand, types 6 and 7 occurred in chalk layers that had already been jointed in previous fracture episodes. The five FJR types 3 to 7 occur in Middle Eocene chalks and originated during uplift processes that took place in the syncline (Bahat 1999a). Also, types 3 and 4 are associated with the Secher fault as they cut older chalk layers at the flank of the fold (Wadi Secher, Fig. 3.35c), while types 5–7 are exposed in younger layers closer to the synclinal center (along Wadi Naim). Therefore, types 3 and 4 preceded types 6 and 7. Types 3 and 4 belong to an early uplift stage, whereas types 6 and 7 are correlated with an advanced uplift. Since type 7 joints are younger than type 6 joints, they may be categorized as late uplift products.

There are limitations in estimating the relative ages of certain FJRs. Type 5 preceded types 6 and 7, but no correlation can be made regarding the relative ages of types 3–4 and type 5 because they occur in remote locations and lack cutting relationships. Also, no comparison is possible regarding the timing of types 5–7 and type 8. It may however be suggested that the multilayer fracture of type 8 were products of stronger tensile stresses than types 5–7 (Bahat 1991a).

6.7.3.2 *Limitations of the Model Depicted in Figure 6.25*

The set of joints of type 3 formed in a principal plane sub-parallel to the plane of the primary fault (Fig. 6.24c) and the model in Fig. 6.25 explains how this was possible. This model is simple for normal faults that dip 45° or close to it, because for such faults the direction of maximum shear stress in the initial field of remote stress is easily transformed to the direction of σ_1 in the local stress field (Fig. 6.25a). The validity of this model would be verified if type 3 joints were exactly parallel primary faults in other fracture provinces. For normal faults whose dips deviate from 45°, additional studies will show the required changes and expected spreads in orientation of the secondary fractures of type 3. Perhaps the relatively large dip spread of the set that resembles type 3 joints in Fig. 6.26 is a good realistic example.

6.7.3.3
Speculation on Future Implications

Fractures in the Beer Sheva syncline display two FJR systems that are linked to burial, whereas six others relate to various syntectonic/uplift categories. Joint sets within these systems are categorized into pre-, syn- and post-fault associations. Early and late events that relate to pre-faulting and post-faulting, respectively can be distinguished in the burial system. In the uplift category, syn-fault early uplift and post-fault early uplift events can be distinguished from pre-fault and syn-fault late uplift events. Future studies may possibly show fuller ranges of FJR than the one described in the Beer Sheva syncline. These would include pre, syn and post fault joints in the three categories, burial, syntectonic and uplift and to lesser extents in the post-uplift category. Complex FJR structures in association with combined structures like burial-syntectonic or syntectonic-uplift can also be expected.

6.8
The Path from Geological Joints to Soil through a Transitional Layer of Surface Joints

This section concentrates on a "transitional layer" that lies above the uneroded chalk layers of the Mor Formation and below the soil cover that lies at the top of many outcrops around Beer Sheva, in south Israel (Sect. 3.4). The transitional layer reveals various extents of erosion in different locations and is characterized by an abundance of "surface joints" that were added to the earlier single-layer joints, which formed in the chalk of the Mor Formation (Fig. 2.63a) during the Lower Eocene (Bahat 1985). The surface joints may interest soil specialists, hydrologists and environmentalists, because they represent an important stage in the transition of rock to soil. In addition, these joints add significantly to the fracture population in the transitional layer and therefore increase the penetration of running water to the ground. An outcrop that shows the succession of several chalk layers, the transitional layer and the soil cover (Fig. 6.27a,b) was selected in the Shephela syncline (Fig. 3.33b) for detailed investigation, as described below.

6.8.1
Methods

For fracture analysis in Fig. 6.27b, photographs using Kodak Gold 200ASA film were taken, digitalized using an Epson Perfection 1600 scanner and panoramas built up using the photomerge utility of Adobe Photoshop Elements. These were entered into ImageJ (Rasband 2003). The scale was calibrated using a stick of known length (130 cm) that had been included in the photographs for this purpose. A grid 200 × 200 cm was then overlain on the pictures and then returned to Photoshop. Individual layers were divided into quadrants of 200 cm long. The photographs were again saved and re-entered into ImageJ. Using a Wacom graphics pad and pen, the vertical and horizontal fractures of each quadrant were outlined and measured separately using the software measuring utility. The area of each quadrant and the distance of the middle of the quadrant from the surface were similarly obtained. The results were recorded in Microsoft Excel and the total length of fractures was divided by the area to give fracture intensity (fracture length, given in cm, per unit area, given in m^2).

Fig. 6.27a. An outcrop that consists of six rock layers which are overlaid by a soil bed. **a** The examined outcrop, showing the frame which is detailed in **b**, where all fractures are traced

Fig. 6.27b. An outcrop that consists of six rock layers which are overlaid by a soil bed. Frame with traced fractures. *Numbers* are marked on each quadrant. The first digit marks the layer number and the other two relate to sequential quadrants to which the outcrop is divided

6.8.2 Results

Figure 6.27a shows a panorama of the examined outcrop, which is marked by a frame. All the fractures that are confined to individual chalk layers and the transitional layer are traced within the frame (Fig. 6.27b), revealing two distinct major fracture groups: Vertical and sub-vertical (VSV) joints and horizontal and sub-horizontal (HSH) joints. Figure 6.28a–c show plots of fracture intensity (cm m^{-2}) per quadrant vs. distance from the surface, i.e., every point on the graph represents a quadrant average. The curve that connects the averages of fracture intensities of the vertical fractures (VFI) undulates throughout the six layers (Fig. 6.28a), suggesting that vertical fractures were not significantly affected by the process leading to the development of the "transitional layer". On the other hand, a gradual, strong increase in fracture intensity of the horizontal fractures (HFI) is seen in layers 3–6 (Fig. 6.28b): From the average 122 cm m^{-2} (60.5 standard deviation) to

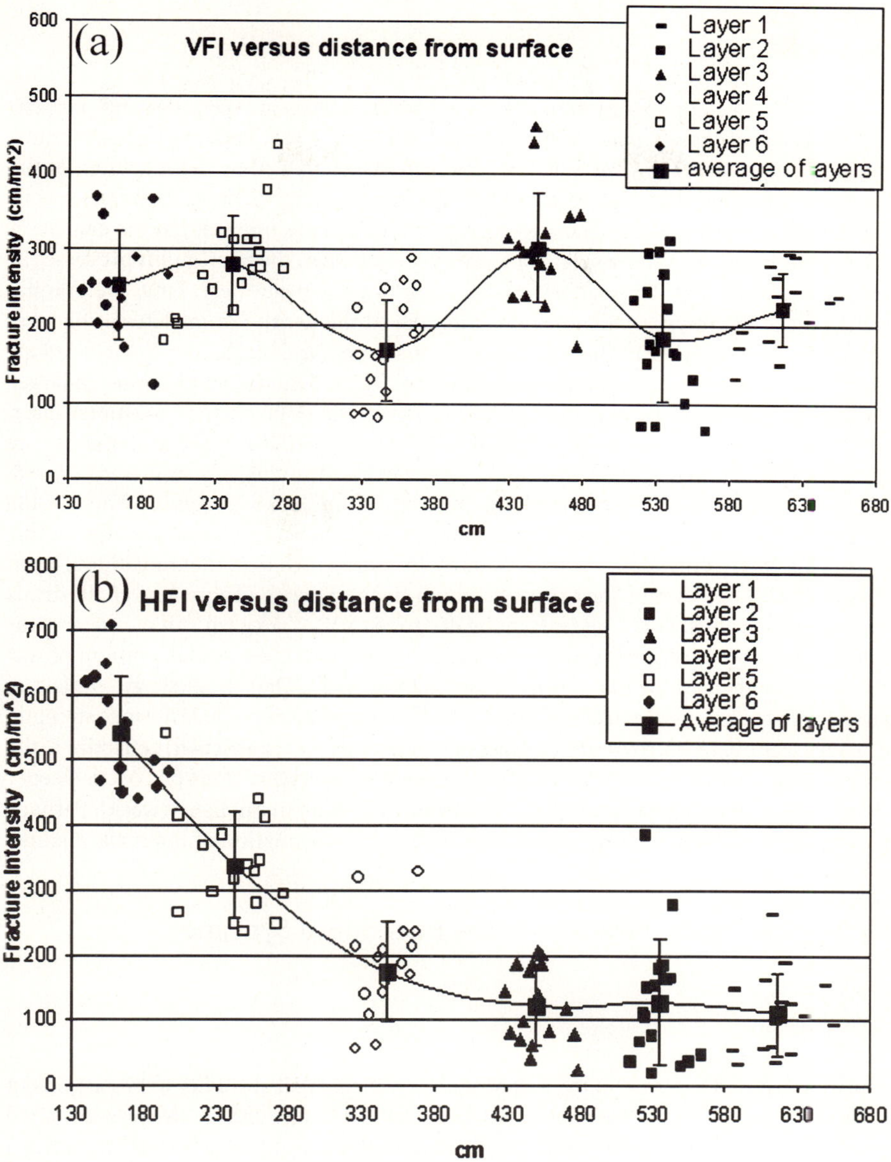

Fig. 6.28. Fracture parameters of single-layer fractures vs. vertical distance from the surface above the soil bed. **a** Fracture intensity of vertical joints. **b** Fracture intensity of horizontal joints

541 cm m^{-2} (85.5 standard deviation). The latter curve suggests that only the two upper chalk layers (4–5) were somewhat affected by the process leading to the development of the "transitional layer" in layer 6. Hence, the curve in Fig. 6.28b provides a "fracture profile" that characterizes the transition from rock (chalk) to soil in this outcrop (Fig. 6.27a).

6.8.3 Discussion

The increase of the population of the horizontal fractures in layers 4–6 and particularly in layer 6 suggests that fracture was induced by a stress field in which the minimum principal stress was vertical. This stress field was created by the reduction of the compression exerted by the weight of the overburden. The distinct, very strong concentration of HSH in layer 6, i.e., closest to the soil surface, implies that the change in stress field is much younger (Pleistocene or younger) than the latest uplift processes in the region (Sect. 3.4). Three mechanisms seem to be relevant here. First, the gradual transition of the upper rock layer into soil reduced the density of the latter, which resulted in some decrease in the vertical compression on layer 6, leading to the development of a vertical tension, much as it follows uplift and denudation processes in larger dimensions. The decrease in the vertical compression was strong in the transitional layer, moderate in layers 4 and 5, and practically absent in deeper layers. Second, what is particularly striking is that the transitional layer is almost exclusively limited to layer 6. This is such a common feature in all the outcrops of the region that it is inconceivable that the "right" amount of overburden was removed to produce cracking only in this layer. Thus, there must be some additional factor acting locally in the transitional layer. Haiaty (1975) shows that the tensile strength of the Eocene (Yoqneam) chalk drops dramatically from 4.3–27.5 kg cm^{-2} in the dry zone to 2.3–8.5 kg cm^{-2} in the water saturated zone. More recently Bahat et al. (2001a) showed that even a small amount of water markedly reduces the (compression) strength of chalk. Thus the exclusive contact of layer 6 with the water saturated loess soil was sufficient to critically reduce the rock strength, causing fracture. Third, rhythmic changes in temperature were particularly effective in the transitional layer: Diurnal/nocturnal variation of temperature does not occur deeper than 30 cm in these rocks (commun. from Howard C. B. on unpublished work). Perhaps a combination of these three mechanisms explains the formation of the surface joints.

6.9 Hydro-Geological Research in the Beer Sheva Syncline

6.9.1 Introduction

The well-exposed outcrops in the Negev desert enabled detailed studies to be carried out on fracture-geology in the Beer Sheva syncline (Sect. 3.4). This syncline also provides good grounds for hydro-geological investigation. Most of this investigation is concentrated in a relatively small area of some 25 km^2 in the center of the syncline around the large Ramat Hovav complex of chemical industry and the National Site for Treatment and Isolation of Hazardous Waste (Fig. 6.29). According to Nativ et al. (1995), the aridity of the area (180 mm yr^{-1} rainfall) and the low-permeability chalk matrix (~2 millidarcy; Dagan 1977a,b) were considered a natural barrier to potential groundwater contamination resulting from the industrial activities. However, the high concentrations of heavy metals and organic pollutants found in the local groundwater (Nativ and Nissim 1992) invalidated the concept of a natural barrier to contaminant migration across the vadose zone. These observations as well as others, such as seasonal fluctuations in groundwater lev-

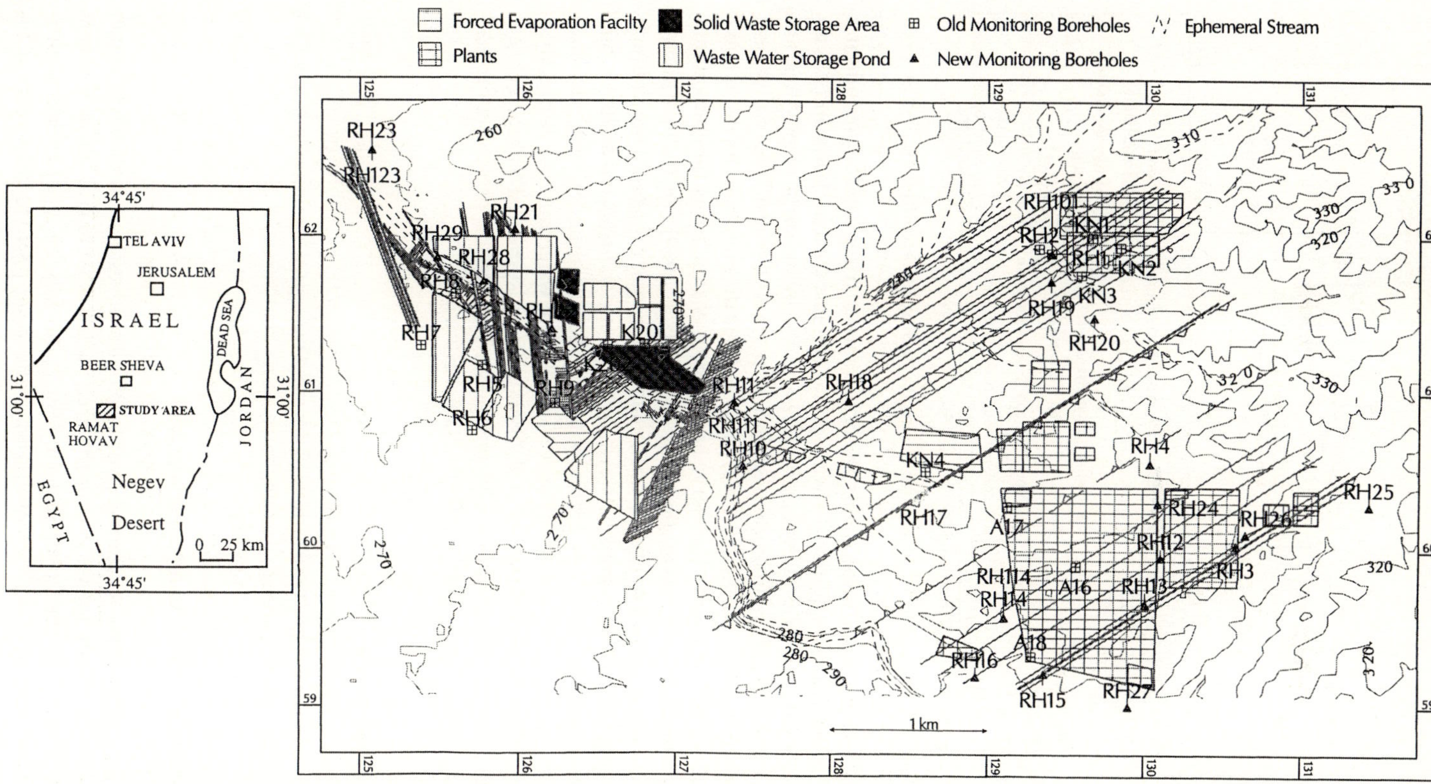

Fig. 6.29. Location map for the chemical industrial complex, facilities, ephemeral streams and observation holes. Old wells are represented by *squares* and new wells are represented by *triangles*. *Contours* describe topography and their interval is in meters. *Straight lines* are inferred fractures interpolated between measurement points along the ephemeral streams and man-made cuts. The trace of Wadi Secher is marked by *double dashed line*. *A*, *B* and *C* are locations of water level mounds (from Nativ et al. 1999)

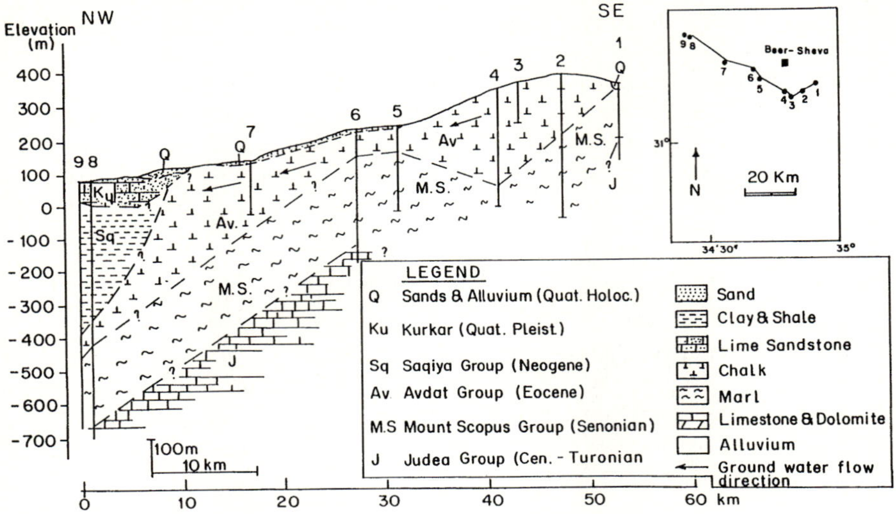

Fig. 6.30. Schematic geological cross section for the northern Negev Desert (from Nativ et al. 1999)

els, water table responses to flood events, the occurrence of tritium in the groundwater, as well as the behavior of additional chemical and isotopic tracers (Nativ and Nissim 1992), indicate that the groundwater is rapidly affected by on-surface activities, and infiltrating water and contaminants bypass the low-permeability chalk matrix via preferential flow paths. Nativ and Nissim (1992) suggested that the numerous fractures crossing the chalk matrix (Bahat 1988a) serve as preferential flow paths for contaminant migration from land surface to groundwater (Dahan et al. 1999) (Fig. 6.29). Although the Eocene (also termed Avdat) chalk is not a major water source (groundwater is brackish), the hydrological relationships between the chalk and the adjacent aquifers could result in the contamination of fresh groundwater (Fig. 6.30). The underlying Judea aquifer in the Cenomanian rocks provides over 30×10^6 m^3 of water annually to the city of Beer Sheva and its vicinity. Kronfeld et al. (1993) have documented leakage between adjacent aquifers(as a result of excessive pumping). This section presents a brief account of recent hydro-geological studies in Ramat Hovav.

6.9.2 Summary of the Results

Selected summaries are grouped into fluid flow and transport sections, ending with a section on open questions that stresses the awaiting challenges.

6.9.2.1 *Fluid Flow through Fractures Intersecting Chalk*

Preferred flow directions. A regional hydrogeological investigation of the flow pattern in the Eocene chalk formations in the area indicates a general subsurface flow in the 270° direction in the upper eastern basin following the topographic inclination (Nativ and Nissim 1992). However, toward the lower part (further west) of the

Beer Sheva syncline the groundwater flow is directed to the northwest (310°). Rophe et al. (1992) measured water levels in three observed wells in Ramat Hovav. The best fit with the obtained data indicates that the maximum hydraulic conductivity lies at an azimuth of 238° in which direction the IG (imaginary gradients) is minimal. Rophe et al. (1992) hypothesized the existence of a major set of fractures at this azimuth. Apparently, the 238° azimuth corresponds quite well to the direction of set 062° (242°), which consists of long and deep penetrating fractures in the syncline (Bahat and Adar 1993, present Fig. 6.29 and group 13 in Sect. 3.4.2.5). This finding has recently been confirmed by many hydraulic conductivity measurements (Nativ et al. 2003).

Water recharge and percolation through fractures. Four dry-drilling holes were bored by Nativ et al. (1995) through the vadose zone. Core and auger samples, collected at 30- to 50-cm intervals, were used for chemical and isotopic analyses, enabling the construction of the following:

1. A tritium profile to estimate the rate of water flow through the unsaturated zone.
2. Oxygen-18 and deuterium profiles to assess the evaporation of water at land surface before percolation and in the upper part of the vadose zone after infiltration.
3. Chloride and bromide profiles as tracers for inert solutes and pollutants.

The tritium and bromide profiles showed the rate of infiltration through the unsaturated matrix to be very slow (1.6–11 cm yr^{-1}). A conceptual model was proposed in which a small portion of the rainwater percolates downward through the matrix, while a larger percentage of the percolating water moves through preferential pathways in fractures. The water flowing through the fractures penetrates the matrix across the fracture walls where it increases the tritium concentrations, depletes the stable isotopic composition, and dilutes the salt concentrations. The observed rapid downward migration of tritium and heavy metals through the profuse fractures makes the chalk inefficient as a hydrologic barrier.

Hydraulic properties of fractures in chalk. One of the challenges of monitoring network design in a fractured rock setting is the heterogeneity of the rocks. Nativ et al. (1999) summarize the activities and problems associated with the monitoring of contaminated groundwater in porous, low-permeable fractured chalk in Ramat Hovav (Fig. 6.29). The fracture systems and their hydraulic properties were characterized by Nativ et al. (2003) to estimate contaminant release from Ramat Hovav. These included identification of the prevailing directions of fracture systems in outcrops, in cores retrieved from inclined coreholes, in coreholes using video logs, and in trenches. The orientation and inclination of these fracture systems were determined, and evidence of groundwater flow on the fracture surfaces was described and ranked. Their hydraulic conductivity was determined through slug and pumping tests performed at discrete intervals. Temperature, electrical conductivity, caliper, gamma and heat-pulse logs were conducted in the same coreholes. The results from the logs, tests, and core descriptions were compared to identify reliable and cost-effective tools for investigating the hydraulic characteristics of fracture systems. Nativ et al. (2003) concluded that in the study area:

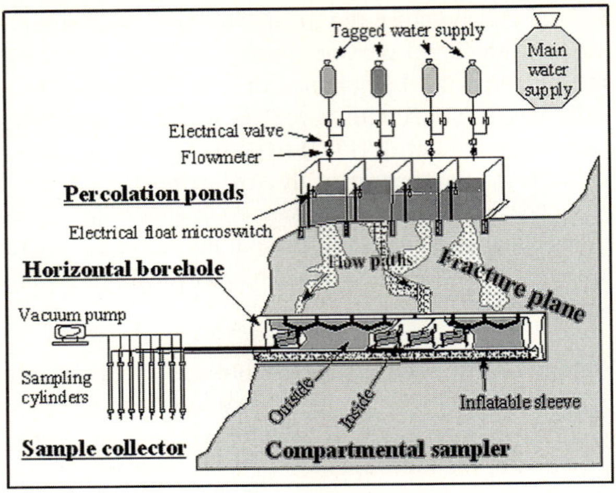

Fig. 6.31a. Schematic cross section of the field setup used to determine solute and particle flux through a single natural fracture (after Dahan et al. 1998 and Weisbrod et al. 2000a)

1. Fracture mapping in outcrops and coreholes (including downhole video and caliper logs) must be supplemented by hydraulic testing of the mapped fracture sets in the coreholes.
2. Inclined coreholes provide information regarding the orientation of the hydraulically active fracture systems that cannot be obtained from vertical boreholes.
3. Hydraulic testing of unpacked holes provides a reasonable estimate of the maxtmum hydraulic conductivity.
4. The hydraulic conductivity distribution with depth is log normal and all significant ground water flow takes place within the upper 25 m.

Percolation experiments in natural discrete fractures. The mechanisms controlling fluid flow through fractures intersecting chalk in the vadose zone were studied by Dahan et al. (2000) through water percolation experiments (Fig. 6.31a) in natural discrete fractures and by close examination of the inner structure of fracture voids. The percolation experiments showed that the flow is focused in dissolution channels along the fracture plane and that fluxes and flow trajectories within that net vary in both time and space. The locations of the dissolution channels, the main potential flow paths within the fracture plane, were generally associated with fracture intersections. The flow through these channels was governed primarily by the mineralogical composition of the filling material and the inner structure of the fracture voids. Salt dissolution, solid-particle migration and clay swelling were found to be the predominant processes controlling flow through the dissolution channels. These physical changes in the structure of the filling material in the dissolution channels accounted for the observed unstable flow behavior. The results suggest that models aimed at simulating water percolation through fractures in unsaturated chalk should consider the mapping of fracture intersections in addition to the commonly used mapping of fracture lineaments. Moreover, the detailed characterization of fracture apertures may not be the key parameter determining fracture flow, because in such formations the flow is controlled primarily by filling material. These materials undergo significant physical variations during wetting and drying cycles (Weisbrod et al. 1999).

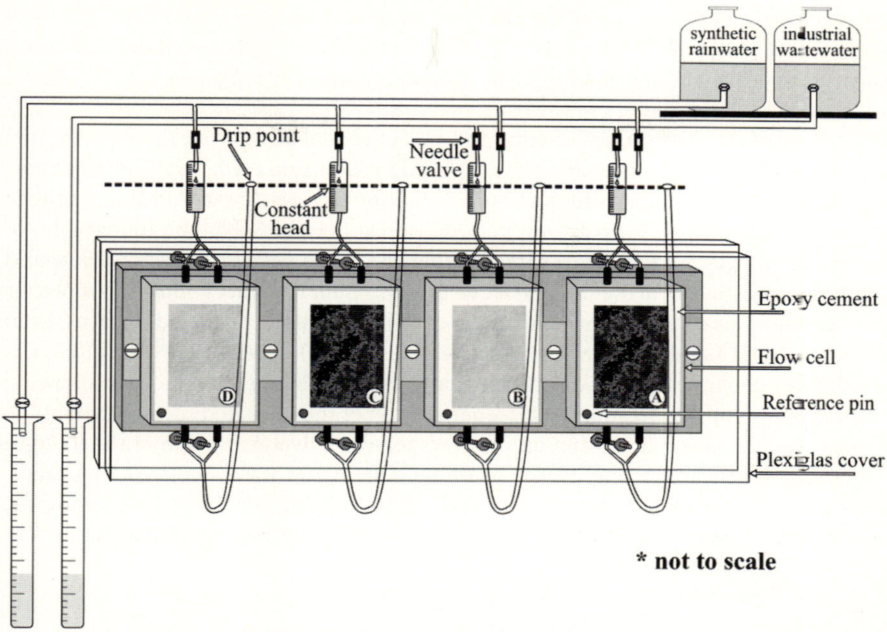

Fig. 6.31b. Schematic description of experimental setup. The solutions were drained from the dripping point into the gradual cylinders and then to the four flow cells (from Weisbrod et al. 2000b)

6.9.2.2
Transport of Solutes and Particles

Groundwater salinity. The processes affecting salinization of precipitation surface water, vadose water and groundwater in the chalk aquitard were studied by Nativ et al. (1997a,b). Observations spanning eighteen years included the collection of rainfall at three rain sampling stations, flood water at six flood stations, vadose water from four coreholes penetrating chalk formations, and groundwater from sixteen monitoring wells tapping the chalk aquitard. Dissolved carbonate dust and evaporation of the falling raindrops result in $Ca(HCO_3)_2$ facies and increased ion concentration of the rainwater with respect to inland, more humid regions. Based on isotopic compositions, the exposure of flood water to evaporation during flood events is minimal. The observed $Ca(HCO_3)_2$ facies and salt enrichment by a factor of three to five in the flood water with respect to precipitation, results primarily from interactions of the flood water with the chalk and limestone bedrock, including ion exchange on Na- and K-bearing minerals and the dissolution of calcite, gypsum and halite. The presence of these salts at and near land surface results from the complete evaporation of rainwater in land surface depression storage areas following most rain events. Water percolation and salt dissolution account for the NaCl facies of vadose water and the variable rates of isotopic depletion and salt dilution observed in the underlying heterogeneous groundwater in the saturated zone. Although the variable mixing with low-saline, isotopically depleted water percolating from the fractures accounts for the depleted isotopic composition of the groundwater, its relatively low solute content cannot modify the groundwater NaCl facies. Consequently, only groundwater salinity in the chalk is reduced by the preferentially flowing water, but the $Ca(HCO_3)_2$ facies prevailing in the rainwater and flood water disappears, and the NaCl imprint from the vadose zone prevails.

Fracture surface erosion. Two coated and two uncoated slices from the fracture surface of an unsaturated chalk core were exposed by Weisbrod et al. (1999) to short flow events (twenty-four, eight, and nine hours) of industrial wastewater and/or synthetic rainwater, followed by long drying periods (weeks) (Fig. 6.31b). The topography of the fracture surface was shown to be unstable due to the detachment of colloidal and large-sized particles during the first three to seven hours of flow. Following rainwater flow, erosion was more pronounced on the coated than on the uncoated surface (mean erosion of 0.313 and 0.134 mm, respectively) (Fig. 6.32). Interaction with industrial wastewater generated a skin of organic matter and gypsum that collapsed following contact with rainwater, leading to a deeper erosion of the uncoated surface than of the coated one (1.238 and 0.549 mm on the average, respectively). Erosion of the fracture surface was measured using a laser-scanning system and was calculated from high-resolution topographical maps (elevation $z \leq \pm 0.01$ mm) generated by Geographic Information System (GIS, ARCInfo) prior to and following the flow experiments. The

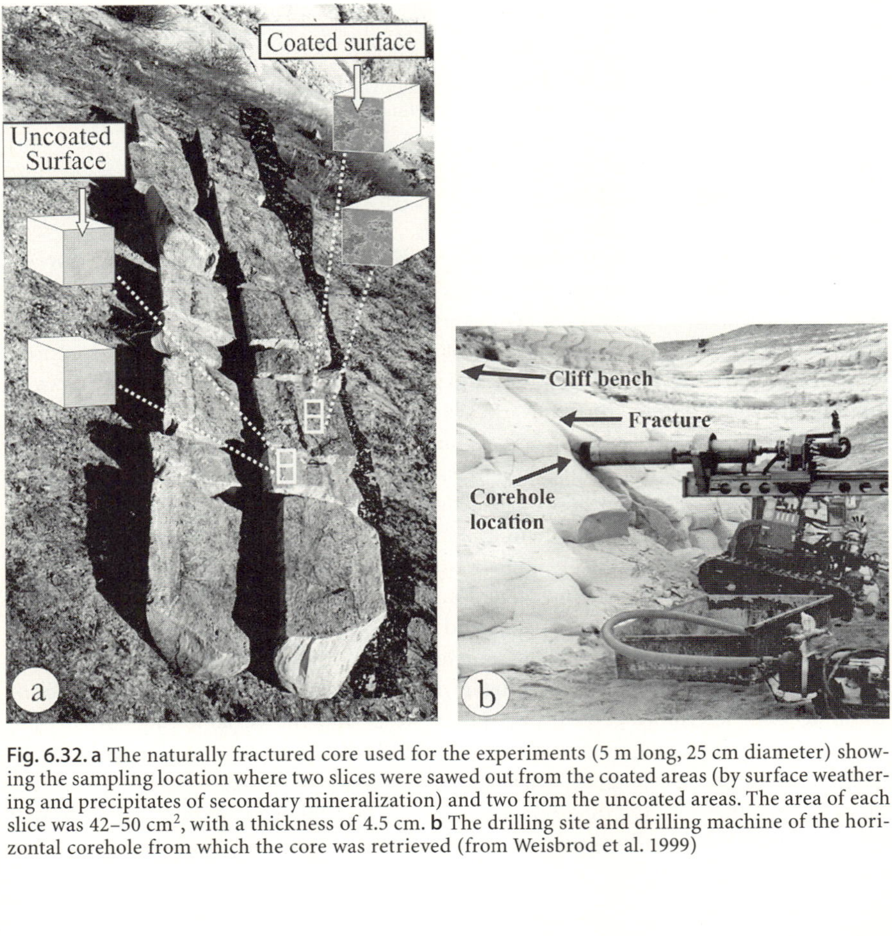

Fig. 6.32. a The naturally fractured core used for the experiments (5 m long, 25 cm diameter) showing the sampling location where two slices were sawed out from the coated areas (by surface weathering and precipitates of secondary mineralization) and two from the uncoated areas. The area of each slice was 42–50 cm^2, with a thickness of 4.5 cm. **b** The drilling site and drilling machine of the horizontal corehole from which the core was retrieved (from Weisbrod et al. 1999)

mean thickness of the eroded surface was found to be strongly correlated with the calculated total accumulated mass of particles and soluble salts released from the fracture surface. This relationship can be used to evaluate fracture surface erosion in large field and laboratory experiments.

Flow experiments on fracture surface properties. The surface topography of a natural fracture intersecting a chalk formation was mapped by Weisbrod et al. (2000b) (see also Weisbrod et al. 1998), using the aforementioned high-resolution laser-scanning device, and its roughness was evaluated using five different criteria. Two coated and two uncoated slices (Fig. 6.32) of the natural fracture surface were encased in flow cells and exposed to short (twenty-four and eight hours) interactions with percolating synthetic rainwater and industrial wastewater, followed by long drying periods (weeks) (Fig. 6.31b). These flow experiments simulated the intermittent infiltration of surface runoff and industrial effluents typical of the study area and resulted in erosion of the fracture surfaces and modification of their roughness. The temporal variations in surface roughness were evaluated using commonly used criteria. The surfaces coated by secondary mineralization (Avigur and Bahat 1990; Weisbrod et al. 1999) were significantly rougher, had greater relief, more heterogeneous topography, and a larger surface area than the uncoated surfaces (matrix mineralogy). The observations suggest that coating and fast weathering (caused by intermittent flow events) significantly change roughness, surface area, relief, and the heterogeneity of a fracture surface. These in turn may affect the spatial distribution of flow paths across that fracture surface.

Particle detachment and transport from fractures. A series of field and laboratory experiments were conducted by Dahan et al. 1999 and Weisbrod et al. 1999), respectively, to study the mechanisms of particle detachment and transport from fractures in vadose chalk. Experiments of intermittent flow events along fracture surfaces were carried out in the laboratory. In the field (Fig. 6.31a), water was percolated from land surface via a discrete fracture into a compartmental sampler installed inside a horizontal corehole located below the surface. The mass, size distribution and composition of the particles drained with the water from the fracture voids were examined along with flow rates and salt dissolution. Two boreholes penetrating the underlying saturated zone were sampled and analyzed for colloidal concentration and composition. The results of these experiments summarized by Weisbrod et al. (2002) suggest that:

1. Particle detachment causes flow-rate variability in the unsaturated fracture.
2. The mechanisms of particle detachment and salt dissolution within the fracture are linked.
3. Although most of the detached particles are large and likely to accumulate inside fractures, some colloidal particles also eroded from the fracture void and are likely to be transported to the groundwater.

Most of the detached particles had a mean diameter of >2 μm, while the mobile colloidal phase in the groundwater had a mean diameter of ~1 μm.

6.9.3
Open Questions Related to Water Flow in Heterogeneous Fracture Systems

The aforementioned summaries of recent studies from Ramat Hovav reveal a commendable progress in understanding water conduct (i.e., water flow and solute and particle transport) in the chalk aquitard. We return now to the work by Nativ et al. (2003) and take the risk of extrapolating it into six major challenges that are still awaiting the investigators in this area. The first challenge is differentiating the contributions of local geological structures to water conduct. The second is "getting the general picture" on water conduct by synthesizing these contributions. The third is minimizing the gaps along the interface between this "general picture" and adapted or newly devised conceptual and network fracture models on liquid flow (e.g., Nelson 1987; Berkowitz et al. 1988; Odling et al. 1999; Bonnet et al. 2001; Connoly and Cosgrove 1999). Contribution in this direction has been made by Berkowitz et al. (2001). The fourth is characterizing the differences between quasi-static and dynamic systems (such as intermittent changes in the apertures of joints) and applying them to new conceptual models. The fifth is devising engineering methods for eliminating water pollution, locally and regionally, and monitoring their execution. Finally, the sixth is comparing the Ramat Hovav system to other systems and models (e.g. Long et al. 1996).

We limit our comments to addressing the first challenge. This would involve differentiating the water conduct when flowing through different structures, as demonstrated by the two examples below. First, it is generally accepted that the major conduits in the area are the NE-ENE uplift multilayer joints (Fig. 6.29). It should be noticed, however, that the NE-ENE uplift fracture category consists of a series of fracture sets that formed in different geological associations when cutting Lower and Middle Eocene chalks (Table 3.3), and correspondingly, would have different drainage properties (including different affinities to coating and uncoating properties that were demonstrated above). Second, while many joints and faults provide poor drainage systems to water flow due to secondary minerals that often form along them (e.g., Avigur and Bahat 1990; Dahan et al. 1999), many other faults enhance rock permeability and water flow (e.g., Witherspoon and Newman 1967; Birkeland and Larson 1978). Quite important is the observation that liquid drainage may be considerably improved along fracture intersections (e.g., Hsieh et al. 1985; Dahan et al. 2000). The Beer Sheva syncline is profused with faults (Nativ et al. 2003) and joints (Sect. 3.4). Furthermore, there are numerous intersections and many styles of contact between faults and joints in the syncline (Sect. 6.7), and these should contribute significantly to the "general picture" of water conduct (see also Bahat 2004).

Finally, the investigation in Ramat Hovav involved recently an international participation of teams from Ben Gurion University of the Negev, The Hebrew University of Jerusalem (Israel), Bonn University, The University Of Karlsruhe (Germany), Penn State University, The University of California (Berkeley, USA), The Geological Survey of Denmark and Greenland (GEUS), The British Geological Survey (BGS) and The Volcani Research Center (Bet Dagan, Israel). These teams issued a series of annual reports (2000–2003), which were partly covered under the name FRACFLOW, a project promoted by the European Commission (see also Bloomfield 1999). These reports concerned various aspects of water flow through fractured rocks and contamination (note particularly Adar and Nativ 2003).

6.10 A New Fringe Characterization and Analysis

The mechanism of *en echelon* segmentation by local stress rotation (Fig. 2.4C) via the fracture slanting model (Sect. 2.2.5.2) is workable (see also Preston's (1931) model in Bahat 1991a, Fig. 2.5). However, some people still debate the exact gear that drives the local stress rotation (e.g., Lazarus 2001a,b). Ramsay and Lisle (2000) suggest a new mechanism, which differs from the model by Pollard et al. (1982) by exposing the rock to superposing *remote* shear stress on normal stress rather than *local* shear stress superposing on normal stress. The model by Ramsay and Lisle (2000, p. 948) is cited below with minor modifications, presenting one of their solutions (Fig. 6.33a–d).

When the *en echelon* segments have the same sense on either side of the parent joint (Fig. 6.33b), the situation is such that the layer boundaries are subjected to a shear strain (shear plane parallel to the bed boundaries, shear direction oblique to the main central fracture, and similar shear sense on both boundaries). The fracture tip propagates from its point of origin in such a way that the incremental fracture surface is oriented perpendicular to the maximum deviatoric tensile stress direction (and maxi-

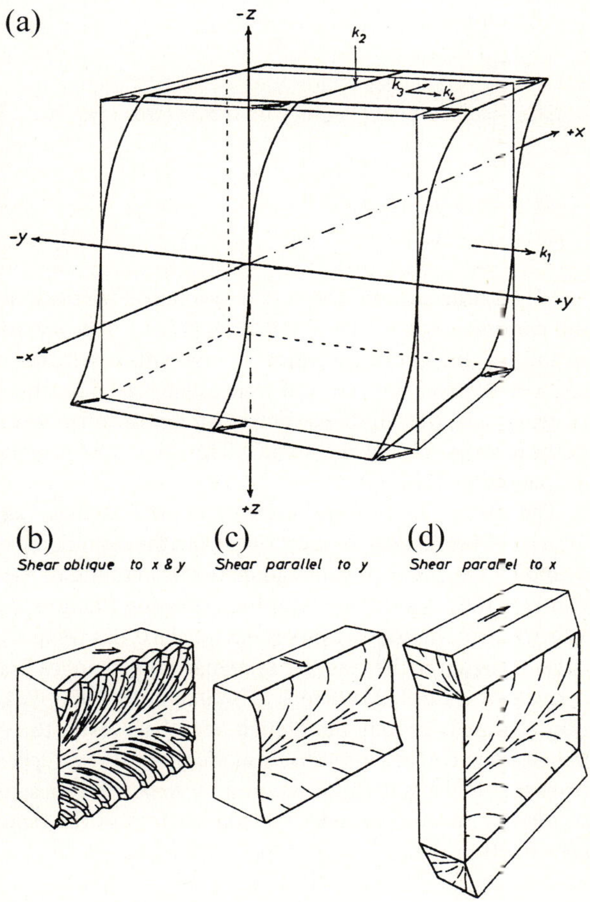

Fig. 6.33. Three-dimensional block model of *en echelon* segmentation, combining layer parallel extension, layer normal shortening with layer boundary shear. **a** The various *k*-parameters, see text for their significance. **b**, **c** and **d** show the predicted joint surface patterns arising with differing types of boundary shear (from Ramsay and Lisle 2000)

mum extensive elastic strain) that are normal to the parent joint. The layer is oriented parallel to the x-y-plane, with boundaries at $z = \pm 1.0$ (Fig. 6.33a). It has been subjected throughout to a uniform stretch parallel to the y-direction of k_1 and a uniform shortening in z of k_2, with no change in the x-direction (plane strain for this deformation component). The layer has then been displaced by a shear displacement with shear plane xy, with vector components at each level of k_3 parallel to x and k_4 parallel to y and which vary in shear intensity (γ) with distance from the central plane (i.e., with z) according to $\gamma = az^3$.

The odd power of z accords with the observation of shear sense symmetry based on similar *en echelon* symmetry on either side of the parent fracture. In this model, k_1, k_2, k_3, k_4 and a are all constant terms chosen to suit individual problems: k_1 and k_2 are paired parameters which describe the background bed-parallel, bed-perpendicular strains (and which include volume change $\Delta_v = 1 - k_1 k_2$): k_3, k_4 and a define the value of the shear at $z = \pm 1$ ($\gamma = a(k_1^2 + k_2^2)^{1/2}$), and orientation ($\alpha$) of the shear direction from the x-axis ($\tan\alpha = k_4 / k_3$). The equations which define the displacement of any point with initial coordinates (x, y, z) and final coordinates (x', y', z') are:

$$
\begin{aligned}
x' &= x + ak_3 k_2^3 z^3 \\
y' &= k_1 y + ak_4 k_2^3 z^3 \\
z' &= k_2 z
\end{aligned}
\tag{6.17}
$$

The displacement gradient matrix is (from Eq. 36.62 by Ramsay and Lisle 2000):

$$
M = \begin{vmatrix} 1 & 0 & 3ak_3 k_2^3 z^2 \\ 0 & k_1 & 3ak_4 k_2^3 z^2 \\ 0 & 0 & k_2 \end{vmatrix}
\tag{6.18}
$$

We cite Ramsay and Lisle's (2000) solution for shear at 45° to the x-(and y-)axis, for the parameters $k_1 = 1.10$, $k_2 = 0.90$, $k_3 = k_4 = -0.10$, $a = 1.00$, as applied to the fracture in Fig. 6.33b. These parameters give small strains with an increase in volume ($\Delta_v = +0.01$) and Ramsay and Lisle (2000) consider that they are compatible with an uplift type of overall displacement and with small shear strains along the boundary of the layer ($\gamma = \pm 0.14$ at $z = \pm 1.0$). The values and orientations of the principal strains are set out in Table 6.4.

The special interest in these data is the systematic change in the plane normal to the greatest extension as one moves from the central line of the layer, because this plane would be the one that would guide the orientation of the frontal fracture propagation plane. In the center of the layer the extension fracture would propagate parallel to the xz-surface of the coordinate system (strike 0, dip vertical), but as it propagates towards edge of the layer the fracture systematically changes both its strike and dip through the values, 0/1 to E, 179/5 to E, 175/10 to E, 166/18 to E, and 156/25 to E. This orientation change is exactly that which one observes in the *en echelon* segments in joint fringes (Fig. 6.33b). The stresses at the tips of the *en echelon* segments will be changed from that of the general situation, and there will be a strong tendency for interlinking of the tips of one segment with the surface of the next segment by a step (bridge) (Fig. 2.39).

Table 6.4.
The values and orientations of the principal strains

z	$1+e_1$	$1+e_2$	$1+e_3$
0.0	1.10 0/270	1.00 0/000	0.90 90/090
±0.2	1.10 1/270	1.00 2/180	0.90 87/025
±0.4	1.10 5/269	1.00 9/178	0.90 80/026
±0.6	1.11 10/265	1.01 16/172	0.88 71/027
±0.8	1.13 18/256	1.03 17/161	0.85 64/029
±1.0	1.17 25/246	1.04 14/149	0.81 61/032

See text for explanation (from Ramsay and Lisle 2000).

Although the values of the strains used in this example are perhaps a little large for elastic deformations, the orientation changes depend almost entirely upon the proportional values of the pairs of constants governing the relative values of the bulk distortion and the shear components, so the conclusions as to the main features of the geometry are valid for a large range of parameters. Finally, the program of Appendix 36.B TWISTHACKLES.BAS (given in Ramsay and Lisle 2000) enables the constant parameters to be varied as one wishes to explore the properties of different models.

Nomenclature

Physical Terms

Creep. A slow incremental crack increase under a constant or slowly varying load that is below that of the the Griffith criterion.

Critical dynamic stress intensity factor K_{ID}. The value that if the stress intensity factor is larger than it, dynamic (fast) fracturing would continue. K_{ID} is usually larger than K_{IC} (see Eq. 1.137).

Critical or post critical conditions. Are associated with crack growth at $K_I = K_{IC}$.

Critical stress intensity factor. A material-dependent property K_{IC}. When $K_I = K_{IC}$, the Griffith criterion is fulfilled and fracturing can commence.

Effective saturation. An assumption that beyond a certain value, crack density increases very slowly (logarithmically) with load increase.

Fatigue. Material failure by cracking under periodic or repeated stresses.

Flaws. General name for small heterogeneities in rocks (fossils, concretions, pores) that may initiate crack growth.

Griffith criterion. A necessary condition for fracturing of materials: The stress intensity factor K_I, becomes equal to the critical stress intensity factor K_{IC}.

Griffith flaws. Flaws in brittle material (glass) that have an effective length of ≈2 µm (Lawn 1993, p. 13), while in rocks they may range up to lengths of grain boundaries.

Linear elastic fracture mechanics (LEFM). The framework theory for linearly calculating stress fields in different configurations and deducing K_I values. Very valuable for static fracture problems, but fails for dynamic or extremely slow fracturing.

Saturation of crack spacing. An assumption that the crack spacing (No. of cracks per unit length) is limited. Once the spacing reaches this value, no increase is possible under any load.

Stable crack growth. Occurs under loads in which K_I decreases with the increase in crack length, and the crack arrests when the load is removed (also termed 'controlled').

Stress concentration. A situation where near to an abrupt geometrical change in the material, such as a flaw, crack, pore, vacancy, etc., a local increase of stress occurs in the stress field.

Sress intensity factor K_I. A measure of the ratio of the maximal stress concentration near a flaw to the nominal stress. In an infinite plate under remote tension, σ_0, having a sharp flaw of length c, expressed as $K_I = \sigma_0(\pi c)^{1/2}$.

Sub-critical conditions. Are associated with crack growth below the critical stress intensity factor K_{IC}.

Time to failure. The time interval from the minute of application of a load, smaller than that needed for instantaneous fracture, until the sample breaks.

Unstable crack growth. Occurs 'catastrophically' at $K_I = K_{IC}$ even when unloaded (also termed 'uncontrolled').

v vs. K_I diagrams. Correlate the changes in the instantaneous crack velocity v with values of the instantaneous tensile stress intensity factor K_I.

Nomenclature of Main Fractographic Features

Arrest marks. Ripple marks that form on the fracture surface and signify slow crack propagation down to arrest, or crack restarting growing.

Barbs. See plumes.

Dynamic mirror plane. Forms when fractured primarily by high tensile stresses; it propagates rapidly and exhibits characteristic morphological features.

***En echelon*-fringe.** A fringe populated by *en echelon* cracks.

***En echelon* segmentation.** Cracks that systematically relate to each other in a 'shingle-like' arrangement and are confined to the fringe.

Fractography. The branch of science that analyses fracture surface morphology and related features and their causes, as well as mechanisms, in technological materials.

Fringe. A zone that surrounds the mirror plane and contains cracks that display rough morphologies.

Hackle cracks. A series of cracks that relate to each other three dimensionally in various directions and are confined to the fringe.

Hackle-fringe. A fringe populated by hackle cracks.

Mirror boundary. A narrow morphological rim separating the mirror from the fringe.

Mirror plane. The crack surface that contains the initial flaw, the critical flaw, the striae (plume) and the ripple marks.

Oscillating cracks. Bends of the mirror plane around the axis normal to the direction of crack propagation.

Plumes. Form by slow fracture on the mirror plane due to stress fields which rotate the fracture front about the axis of the direction of crack propagation, and often are identified by their dendrite-like morphologies consisting of many splayed barbs.

Quasi-static mirror plane. Forms when fractured under low stresses; it propagates subcritically and exhibits characteristic morphological features.

Ripple marks. Form on the mirror plane due to stress fields that intermittently bend the crack front around the axis normal to the direction of propagation, when both axes are on the fracture surface. Two types of ripple marks occur, undulations and arrest marks.

Striae. Form by rapid fracture on the mirror plane due to stress fields which rotate the fracture front about the axis of the direction of crack propagation, and often are identified by sharp, continuous cuts in the material (glass or rock).

Undulations. Ripple marks that form on the fracture surface and signify crack propagation in various velocities.

Nomenclature of Joints

Burial joints, or subsidence joints. Form during the sedimentation and diagenetic processes, and often are single-layer joints. They may form in great depths.

Cooling joints. Form during various cooling stages of the pluton (or volcanic rock).

Crack. In the present text is almost synonymous with fracture. It is generally applied to small fractures. These are general terms, regardless of loading mode.

Cross-fold joints. Form sub-orthogonal to fold axes.

Cross-joints. A set of younger, nonsystematic joints that arrest at earlier, systematic, single-layer joints, seen in plan view (Gross 1993).

Dihedral angle. Produced between two genetically linked fractures or fracture sets (joints or faults) that maintain a conjugate relationship.

Fault. Of a primary nature, that originated by slip(s), which resulted in lost cohesion in the rock. Of a secondary nature, that formed by slip(s) on an existing joint.

Fractography. Fracture markings that describe the surface morphology that forms essentially by an opening mode. Every such surface (joint) has a unique fractography.

Group. This term is occasionally applied (rather than set or system) when fracture affinities are debatable.

Joint. A surface on which the rock has lost cohesion and does not show a discernible lateral slip.

Joint set. Consists of a series of joints of similar strikes that exhibit similar fracture surface morphologies (if present). This similarity often reflects on the same fracture conditions, even if in different fracture episodes.

Joint system. Consists of various joint sets that probably are genetically linked.

Multi-layer joints. Cut many layers, often dissecting the entire outcrop. They are widely and irregularly spaced.

Orthogonal joint sets. Form perpendicular to each other (in a map), but may deviate up to 10° from this relationship.

Post uplift joints. Post uplift fractures that form in association with erosion.

Single-layer joints. Cut vertically individual layers (or less commonly, two layers) but arrest at the layer boundaries. They maintain small, regular spacing.

Strike joints, or strike-parallel joints. Form sub-parallel to fold axes.

Surface joints. Recent fractures that are confined to the upper layer, below the soil cover.

Syntectonic joints. Form during intense deformation, occasionally associated with faulting. They may form in great depths.

Tectonofractography. The application of joint fractography to the interpretation of tectonophysical processes.

Uplift joints. Form during uplift. They can be both single-layer and multi-layer fractures and form in shallow depths.

Selected Supplementary Parameters

Skewness. The degree of asymmetry, or departure from symmetry, of a distribution. If the frequency curve of a distribution has a longer "tail" to the right of the central maximum than to the left, the distribution is said to be skewed to the right or to have positive skewness. If the reverse is true, it is said to be skewed to the left or to have negative skewness (Spiegel 1961).

Kurtosis. The degree of peakedness (between high and flat-topped peaks) of a distribution, usually taken relative to a normal distribution (Spiegel 1961).

Statistics of extremes. The statistics of the largest (or smallest) members of a population, e.g., the severest earthquakes, the highest waves, etc.

FSI. The fracture spacing index, defined as the slope of the best-fit line when joint spacing is plotted versus layer thickness (Narr and Suppe 1991).

FSR. The fracture spacing ratio defined as the layer thickness divided by the median joint spacing for an individual bed (Gross 1993).

***L* / *S*.** The ratio of joint length to joint spacing (Bahat 1991a, p. 309; Wu and Pollard 1995).

References

Abraham FF, Brodbeck D, Ridge WE, Xu X (1997) A molecular dynamics investigation of rapid fracture mechanics. J Mech Phys Solids 45:1595–1619

Abraham FF, Walkup R, Gao HJ, Duchaineau M, De la Rubia TD, Seager M (2002) Simulating materials failure by using up to one billion atoms and the world's fastest computer: Brittle fracture. Proceedings of the National Academy of Sciences of the United States of America 99:5777–5782

Acocella V, Gudmundsson A, Funiciello R (2000) Interaction and linkage of extension fractures and normal faults: Examples from the rift zone of Iceland. J Struct Geol 22:1233–1246

Adar E, Nativ R (2003) Contaminant transport, monitoring and remediation strategies in fractured chalk in the northern Negev. Annual Report for the year 2002, submitted to the Ramat Hovav Industrial Council

Adler RJ, Feldman RE, Taqqu MS (1998) A practical guide to heavy tails, statistical techniques and applications. Birkhauser, Boston

Ahlfors LV (1979) Complex analysis, 3rd edn. McGraw Hill Book Co, Aucland

Aitchison J, Brown JAC (1976) The lognormal distribution. Cambridge University Press, Cambridge

Aki K (1979) Geometric features of a fault zone related to the nucleation and termination of an earthquake rupture. U.S. Geological Survey Open-File Report 89-315, pp 1–9

Allen CR, Castle RO, Clark MM, Grantz AA, Sharp RV, Theodore TG, Wolfe EW, Young TL (1972) Map showing surface ruptures created at the time of and after the Borrego Mountain Earthquake of April 9, 1968 (GMT). U.S. Geological Survey, Professional Paper 78, Plate 1

Anderson EM (1951) The dynamics of faulting, 2nd edn. Oliver & Boyd, Edinburgh

Anderson TL (1995) Fracture mechanics: Fundamentals and applications. CRC Press, London

Anton L (1999) Simple equation for earthquake distribution. Phys Rev E 59(6):7213–7215

Arakawa K, Takahashi K (1987) Dependence of crack acceleration of the dynamic stress intensity factor in polymers. Exp Mech 27:195–200

Arakawa K, Takahashi K (1991) Relationships between fracture parameters and fracture surface roughness of brittle polymers. Int J Fracture 48:103–114

Arakawa K, Mada T, Takahashi K (2000) Correlations among dynamic stress intensity factor, crack velocity and acceleration in brittle fracture. Int J Fracture 105:311–320

Arkin Y, Hamaoui M (1967) The Judea Group (Upper Cretaceous) in central and southrn Israel. Isr Geol Surv Bull 42

Arlegui L, Simón JL (2001) Geometry and distribution of regional joint sets in a non-homogeneous stress field: Case study in the Ebro basin (Spain). J Struct Geol 23:297–313

Arnold BC (1983) Pareto distributions. International Co-operative publishing house, Fairland, Maryland, USA

Arthur MJ (1989) The Cenozoic evolution of the Lundy pull-apart basin into the Lundy rhomb horst. Geol Mag 126:187–198

Ashcroft NW, Mermin ND (1981) Solid state physics. CBS Publishing, Tokio, pp 398

Atkinson BK (1984) Subcritical crack growth in geological materials. J Geophys Res 89:4077–4114

Atkinson BK (1987) Introduction to fracture mechanics and its geophysical applications. In: Atkinson BK (ed) Fracture mechanics of rocks. Academic Press, pp 1–26

Atkinson C, Eshelby JD (1968) The flow of energy into the tip of a moving crack. Int J Fracture Mech 4:3–8

Atkinson BK, Meredith PG (1987a) The theory of subcritical crack growth with applications to the minerals and rocks. In: Atkinson BK (ed) Fracture mechanics of rock. Academic Press, London, pp 111–166

Atkinson BK, Meredith PG (1987b) Experimental fracture mechanics data for rocks and minerals. In: Atkinson BK (ed) Fracture mechanics of rock. Academic Press, London, pp 477–525

Avigour A, Bahat D (1990) Chemical weathering of fractured Eocene chalks in the Negev, Israel. Chem Geol 89:149–156

Babcock CL (1977) Silicate glass technology methods. John Wiley & Sons Ltd., New York

Bahat D (1968) Structural changes at room temperature of heated anorthoclases. Can Mineral 9:531–538

Bahat D (1970) Kinetic study on the hexacelsian-celsian phase transformation. J Mater Sci 5:805–810

Bahat D (1977) Thermally-induced wavy Hertzian fracture. J Am Ceram Soc 60:118–120

Bahat D (1979a) Theoretical considerations on mechanical parameters of joint surfaces based on studies on ceramics. Geol Mag 11:81–92

Bahat D (1979b) On the African rift system, theoretical and experimental study. Earth Planet Sc Lett 45:445–452

Bahat D (1979c) Interpretation on the basis of Hertzian theory of a spiral carbonatite structure at Homa Mountain, Kenya. Tectonophysics 60:235–246

Bahat D (1982) Extensional aspects of earthquake induced ruptures determined by an analysis of fracture bifurcation. Tectonophysics 83:163–183

Bahat D (1983) New aspects of rhomb structures. J Struct Geol 5:591–601

Bahat D (1985) Low angle normal faults in Lower Eocene chalks near Beer-Sheva, Israel. J Struct Geol 7:613–620

Bahat D (1986a) Joints and en echelon cracks in Middle Eocene chalks near Beer-Sheva, Israel. J Struct Geol 8:181–190

Bahat D (1986b) Criteria for the differentiation of en echelons and hackles in fractured rocks. Tectonophysics 121:197–206

Bahat D (1987a) Correlation between styles of fracture markings and orientation of cross-fold joints. Tectonophysics 136:323–333

Bahat D (1987b) Jointing and fracture interactions in Middle Eocene chalks near Beer-Sheva, Israel. Tectonophysics 136:299–321

Bahat D (1988a) Early single-layer and late multi-layer joints in the Lower Eocene chalks near Beer-Sheva, Israel. Annal-Tecton 2:3–11

Bahat D (1988b) Fractographic determination of joint length distribution in chalk. Rock Mech Rock Eng 21:79–94

Bahat D (1988c) Fractographic determination of joint length distribution in chalk. Rock Mech Rock Engin 21:79–94

Bahat D (1989) Fracture stresses at shallow depths during burial. Tectonophysics 169:59–65

Bahat D (1991a) Tectonofractography. Springer-Verlag, Heidelberg

Bahat D (1991b) Plane stress and plane strain fracture in Eocene chalks around Beer-Sheva. Tectonophysics 196:61–67

Bahat D (1997) Mechanisms of dilatant en echelon crack formation in jointed layered chalks. J Struct Geol 19:1375–1392

Bahat D (1998a) Four joint genetic groups and their distinct characteristics. In: Rossmanith HP (ed) Mechanics of jointed and faulted rock, MJFR-3, Vienna, Austria. Balkema, Rotterdam, pp 211–216

Bahat D (1998b) Quantitative tectonofractography – an appraisal. In: Rossmanith HP (ed) Mechanics of jointed and faulted rock, MJFR-3, Vienna, Austria. Balkema, Rotterdam, pp 59–67

Bahat D (1998c) On joints and paleostress associated with folds along the Syrian Arc in the Negev. Israel J Earth Sci 48:29–36

Bahat D (1999a) Single-layer burial joints vs. single-layer uplift joints in Eocene chalk from the Beer-Sheva syncline in Israel. J Struct Geol 21:293–303

Bahat D (1999b) On paleostresses associated with the Syrian Arc in Israel. Isr J Earth Sci 48:29–36

Bahat D (2000) The quasi-ellipsoid damage zone of faults and five types of fault-joint relationship in the Beer Sheva syncline, Israel. IX Congresso Geologico Chileno, pp 749–753

Bahat D (2004) Eight distinct fault-joint geometric/genetic relationships in the Beer Sheva syncline, Israel. In: Cosgrov JW, Engelder T (eds) The initiation, propagation and arrest of joints and other fractures. Spec Publ Geol Soc London (in press)

Bahat D, Adar EM (1993) Comment on "Analysis of subsurface flow and formation anisotropy in a fractured aquitard using transient water level data" by Rophe B, Berkowitz B, Magaritz M, Ronen D. Water Resour Res 29:4171–4173

Bahat D, Engelder T (1984) Surface morphology on cross fold joints of the Appalachian Plateau, New York and Pennsylvania. Tectonophysics 104:299–313

Bahat D, Grossmann HNF (1988) Regional jointing and paleostresses in Eocene chalks around Beer-Sheva. Israel J Earth Sci 37:1–11

Bahat D, Rabinovitch A (1988) Paleostress determination in a rock by a fractographic method. J Struct Geol 10:193–199

Bahat D, Rabinovitch A (2000) New fractographic aspets of natural and artificial fractures in chalks, from the Upper Galilee, Israel, and experimental fracture in Perspex. J Struct Geol 22:1427–1435

Bahat D, Shavit R (1997) Fracture in the Beer-Sheva Syncline (Ramat Hovav). Unpublished report submitted to the Ramat Hovav Industrial Municipality
Bahat D, Leonard G, Rabinovitch A (1982) Analysis of symmetric fracture mirrors in glass bottles. Int J Fracture 18:29–38
Bahat D, Grossenbacher K, Karasaki K (1995) Investigation of exfoliation joints in Navajo Sandstone at the Zion National Park and in granite at the Yosemite National Park by tectonofractographic techniques. LBL-36971; UC-400, 67 pp
Bahat D, Grossenbacher K, Karasaki K (1999) Mechanism of exfoliation joint formation in granitic rocks at Yosemite National Park. J Struct Geol 21:85–96
Bahat D, Rabinovitch A, Frid V (2001a) Fracture characterization of chalk in uniaxial and triaxial tests by rock mechanics, fractographic and electromagnetic radiation methods. J Struct Geol 23:1531–1547
Bahat D, Bankwitz P, Bankwitz E (2001b) Joint formation in granite plutons: En echelon-hackle series on mirror fringes (Example: South Bohemian Pluton, Czech Republic). Zeitschrift der deutschen geologischen Gesellschaft 152:593–609
Bahat D, Bankwitz P, Bankwitz E (2001c) Changes of crack velocities at the transition from the parent joint through the en echelon fringe to a secondary mirror plane. J Struct Geol 23:1215–1221
Bahat D, Frid V, Rabinovitch A, Palchik V (2002a) Exploration via electromagnetic radiation and fractographic methods of fracture properties induced by compression in glass-ceramic. Int J Fracture 116:179–194
Bahat D, Bankwitz P, Bankwitz E, Rabinovitch E (2002b) Comparison of the new fracture areas created by the formation of en echelon and hackle fringes on joint surfaces. Z Geol Wissenschaft 30:1–12
Bahat D, Bankwitz P, Bankwitz E (2003) Preuplift joints in granites: Evidence for subcritical and postcritical fracture growth. Geol Soc Am Bull 115:148–165
Bahat D, Bankwitz P, Bankwitz E (2004) Title?. In: Cosgrov JW, Engelder T (eds) The initiation propagation and arrest of joints and other fractures. Spec Publ Geol Soc London (in press)
Bai TX, Pollard DD (2000a) Fracture spacing in layered rocks: A new explanation based on the stress transition. J Struct Geol 22:43–57
Bai TX, Pollard DD (2000b) Closely spaced fractures in layered rocks: Initiation mechanism and propagation kinematics. J Struct Geol 22:1409–1425
Bak P, Tang C (1989) Earthquakes as a self-organized criticality. J Geophys Res 94:15635–15637
Bak P, Tang C, Wiesenfield K (1987) Self-organized criticality – An explanation of 1/f noise. Phys Rev Lett 59:381–384
Bak P, Tang C, Wiesenfield K (1988) Self-organized criticality. Phys Rev A 38:364–374
Balk R (1937) Structural behavior of igneous rocks. Geological Society of America Memoir
Bank-Sills L, Schur D (1989) On the influence of crack plane orientation in fatigue crack propagation and catastrophic failure. American Society for Testing and Materials, Standard Technical Publication 1020:497–513
Bankwitz P (1965) Über Klüfte I. Beobachtungen im Thüringischen Schiefergebirge. Geologie 14:241–253
Bankwitz P (1966) Über Klüfte II. Die Bildung der Kluftoberfläche und eine Systematik ihrer Strukturen. Geologie 15:896–941
Bankwitz P, Bankwitz E (1984) Die Symmetrie von Kluftoberflächen und ihre Nutzung für eine Paläospannungsanalyse. Z Geol Wissenschaft 12:305–334
Bankwitz P, Bankwitz E (1994) Event related jointing in rocks on Bornholm Island (Denmark) Z Geol Wissenschaft 22:97–114
Bankwitz P, Bankwitz E (1995) Aspekte der Entwicklung von Klüften in postkinematischen Graniten des Erzgebirges (speziell Eibenstocker Massiv). Z Geol Wissenschaft 23:777–793
Bankwitz P, Bankwitz E (1997) Fractographic features on joints in KTB drill cores as indicators of the contemporary stress orientation. Geol Rundsch 86:S34–S44
Bankwitz P, Bankwitz E (2004) The relationship of tilt and twist of fringe-cracks in granite plutons. In: Cosgrov JW, Engelder T (eds) The initiation, propagation and arrest of joints and other fractures. Spec Publ Geol Soc London (*in press*)
Bankwitz P, Bankwitz E (1994) Event related jointing in rocks on Bornholm usland (Denmark) Z Geol Wiss 22:97–114
Bankwitz P, Bankwitz E (1997) Fractographic features on joints in KTB drill cores as indicators of the contemporary stress orientation. Geol Rundsch 86 Suppl:S34-S44
Bankwitz P, Bankwitz E, Thomas R, Wemmer K, Kämpf H (2004) Age and depth evidence for pre-exhumation joints in granite plutons: Fracturing during the early cooling stage of felsic rock. In: Cosgrov JW, Engelder T (eds) The initiation, propagation and arrest of joints and other fractures. Spec Publ Geol Soc London (*in press*)
Bansal GK (1977) On fracture mirror formation in glass and polycrystalline ceramics. Phil Mag 35: 935–944

Bartlett WL, Friedman M, Logan MJ (1981) Experimental folding and faulting of rocks under confining pressure, Part IX. Wrench faults in limestone layers. Tectonophysics 79:255–277

Barton NR, Choubey V (1977) The shear strength of rock joints in theory and practice. Rock Mech 10:1–54

Bateman PC, Sawka WN (1981) Raymond Quadrangle, Madera and Mariposa counties, California; analytic data. U.S. Geological Survey Professional Paper 16:107–172

Bateman PC, Wahrhaftig C (1966) Geology of northern California. Bulletin – California, Division of Mines and Geology, pp 107–172

Bayasgalan A, Jackson J, Ritz JF, Carretier S (1999) Field examples of strike-slip fault terminations in Mongolia and their tectonic significance. Tectonics 18:394–411

Bazant ZP, Planas J (1998) Fracture and size effect in concrete and other quasi-brittle materials. CRC Press, Boca Raton

Beall GH (1989) Design of glass-ceramics. Rev Solid State Sci 3:333–354

Beall GH, Karstetter BR, Rittler HL (1967) Crystallization and chemical strengthening of stuffed quartz glass-ceramics. J Am Ceram Soc 50:99–108

Beauchamp EK (1995) Crack front stability and hackle formation in high velocity glass fracture. J Am Ceram Soc 78:689–697

Beauchamp EK (1996) Mechanics for hackle formation and crack branching. In: Varner JR, Frechette VD, Quinn GD (eds) Fractography of glasses and ceramics III. The American Ceramic Society, Ceramic Transactions 64:409–446

Beck ME, Rojas C, Cembrano J (1993) On the nature of buttressing in margin-parallel strike-slip fault system. Geology 126:755–758

Becker GF (1891) The structure of a portion of the Sierra Nevada of California. Geol Soc Am Bull 2:49–74

Becker A, Gross MR (1996) Mechanism for joint saturation in mechanically layered rocks: An example from southern Israel. Tectonophysics 257:223–237

Behr H-J, Dürbaum H-J, Bankwitz P, DEKORP Research Group (B) (1994) Crustal structure of the Saxothuringian Zone: Results of the deep seismic profile MVE-90 (East). Z Geol Wissenschaft 22:647–769

Beinkafner KJ (1983) Terminal expression of decollement in Chautauqua County, Now York. Northeastern Geology 5:160–171

Bella F, Bjadji PF, Della Monica J, Zilpimiani DO, Manguladze PV (1992) Observations of natural electromagnertic radiation during earthquakes of central Italy. Izv-Phys Solid Earth 28:88–94

Benes K (1971) Structure of plutonic bodies in the Bohemian Massif. Upper Mantle Project (UMP), Final Report Geology, Praha, pp 111–119

Bentor YK, Vroman A (1960) The geological map of Israel, scale 1:250000, sheet 16, Mt. Sodom, 2nd edn, with explanatory notes. Government Printer, Jerusalem

Bentor YK, Vroman A, Zak A (1970) Geological map of Israel, 1:250000. Survey of Israel

Bergerat F, Bouroz-Weil C, Angelier J (1992) Palaeostresses inferred from macrofractures, Colorado Plateau, Western USA. Tectonophysics 206:219–243

Berkowitz B, Bear J, Braester C (1988) Continuum models for contaminant transport in fractured porous formations. Water Resour Res 24:1225–1236

Berkowitz B, Nativ R, Adar E (2001) Evaluation of conceptual and quantitative models of fluid flow and chemical transport in fractured media. In: Panel on Conceptual Models of Flow and Transport in the Fractured Vadose Zone, U.S. National Committee for Rock Mechanics, Board on Earth Sciences and Resources, National Research Council (eds) Conceptual models of flow and transport in the fractured vadose zone. National Academy Press, Washington, pp 115–147

Bernstein N, Ness DW (2003) Lattice trapping barriers to brittle fracture. Phys Rev Lett 91:025501-1-4

Berry JP (1960) Some kinetic considerations of the Griffith criterion for fracture I. Equations of motion at constant force. J Mech Phys Solids 8:194–206

Bevan TG, Hancock PL (1986) A late Cenozoic regional mesofracture system in southern England and northern France. Journal of the Geological Society London 143:355–362

Bilichenko SV, Iljin FS, Kim EF (1990) ULF response of the ionosphere for earthquake preparation processes. Dokl Akad Nauk SSSR+ 311:1077–1080

Birkeland PW, Larson EE (1978) Putnam's Geology, 3rd ed. Oxford Univ Press New York

Bleaney BI, Bleaney B (1965) Electricity and magnetism. Oxford University Press, London

Bloomfield JP (1999) Fracflow-geological state-of-the art review. British Geol. Surv., Natural Envir. Res. Coun., Technical Report WD/99/13, 45 pp

Bobet A, Einstein HH (1998) Fracture coalescence in rock-type materials under uniaxial and biaxial compression. Int J Rock Mech Min 35:863–888

Bogosh R, Bourne J, Shirav M, Harnois L (1997) Petrochemistry of a Late Precambrian granitiferous granite, pegmatite and aplite, southern Israel. Mineral Mag 61:111–122

Bonn D, Kellay H, Prochnow M, Ben-Djemiaa K, Mounier J (1998) Delayed fracture of an inhomogeneous soft solid. Science 280:265–267
Bonnet E, Bour O, Odling NE, Davy P, Main I, Cowie P, Berkowitz B (2001) Scaling of fracture system in geological media. Rev Geophys 39:347–383
Brace WF (1964) Brittle fracture of rocks. In: Judd WR (ed) State of stress in the earth's crust. Elsevier, New York, pp 111–180
Brace WF, Bombolakis EG (1963) A note on brittle crack growth in compression. J Geophys Res 68: 3709–3713
Brace WF, Paulding JBW, Scholz C (1966) Dilatancy of the fracture of crystalline rocks. J Geophys Res 71:939–953
Bradley WC (1963) Large-scale exfoliation in massive sandstones of the Colorado Plateau. Geol Soc Am Bull 74:519–528
Bradley WH, Pepper JF (1938) Structure and gas possibilities of the Oriskany Sandstone in Steuben, Yates, and parts of the adjacent counties. New York Geological Survey Bulletin 899-A
Braun M, Flexer A, Lasman N (1977) Geological report – Ramat-Hovav underground oil storage project. Geoprospect LTD
Breiter K (2001) South Bohemian pluton. Overview. Exkursionsführer und Veröffentlichungen der GGW, Berlin, 212:130–140
Breiter K, Koller F (1999) Two-mica granites in the central part of the South Bohemian Pluton. Abhandlungen der Geologischen Bundesanstalt Wien 56:201–212
Breiter K, Seltmann R (1995) Ore mineralizations of the Krušné Hory Mts. (Erzgebirge). Third Biennial SGA Meeting, Excursion Guide, pp 1–200, Czech Geological Survey, Prague
Breiter K, Sokol A (1997) Chemistry of the Bohemian granitoids: Geotectonic and metallogenetic implications. Sbornik geologickych ved, 31, Praha, pp 75–96
Brereton NR, Evans CJ (1987) Rock stress orientations in the United Kingdom from borehole breakouts. Report of the Regional Geophysics research Group, British Geological Survey, no. R6 87/14
Broek D (1984) Elementary engineering fracture mechanics, 3rd edition. Martinus Nijhoff Publishers, Boston, p 469
Broek D (1991) Elementary engineering fracture mechanics. Kluwer Academic Publisher, Dordrecht
Brooks M, Trayner PM, Trimble TJ (1988) Mesozoic reactivation of Variscan thrusting in the Bristol Channel area, UK. J Geol Soc London 145:439–444
Buchbinder B, Zilberman E (1997) Sequence stratigraphy of Miocene-Pliocene carbonate-siliciclastic shelf deposits in the eastern Mediterranean margin (Israel): Effects of eustasy and tectonics. Sedimentary Geology 112:7–32
Buchbinder B, Benjamini C, Mimran Y, Gvirtzman G (1988) Mass transport in Eocene pelagic chalk on the northwestern edge of the Arabian platform, Shefela area, Israel. Sedimentology 35:257–274
Burdon D (1959) Handbook of the geology of Jordan. Benham & Co, Colchester
Burger HR, Hamill MN (1976) Petrofabric stress analysis of the Dry Creek Ridge anticline, Montana. Geol Soc Am Bull 87:555–566
Butler RF (1992) Paleomagnetism: Magnetic domains to geologic terranes. Blackwell Scientific Publications, Oxford, 319 p
Büttner S (1997) Die spätvariszische Krustenentwicklung in der südlichen Böhmischen Masse: Metamorphose, Krustenkinematik und Plutonismus. Frankfurter Geowissenschaftliche Arbeiten, A 16
Büttner S, Kruhl JH (1997) The evolution of the late-Variscan high-T/low-P region: The south-eastern margin of the Bohemian Massif. Geol Rundsch 86:21–38
Cadman J (1969) The origin of exfoliation joints in granitic rocks. Unpublished Ph.D. thesis. University of California, Berkeley, USA
Caers J, Beirlant J, Maes MA (1999a) Statistics for modeling heavy tailed distributions in geology. Part I. Methology. Geology 31:391–410
Caers J, Beirlant J, Maes MA (1999b) Statistics for modeling heavy tailed distributions in geology. Part II. Application. Geology 31:411–434
Carlson J, Langer J (1989) Mechanical model of an earthquake fault. Phys Rev A 40:6470–6484
Carlson J, Langer J, Shaw B, Tang C (1991) Intrinsic-properties of a Burridge-Knopoff model of an earthquake fault. Phys Rev A 44(2):884–897
Carmichael ISE, Turner FJ, Verhoogen J (1974) Igneous petrology. Mc-Graw Hill Book Company, New York, p 739
Carter NL, Friedman M (1965) Dynamic analysis of deformed quartz and calcite from dry Creek Ridge Anticline Montana. Am J Sci 263:747–785
Castaing C, Halawani MA, Gervais F, Chilès JP, Genter A, Bourgine B, Ouillon G, Brosse JM, Martin P, Genna A, Janjou D (1996) Scaling relationships in intraplate fracture systems related to Red Sea rifting. Tectonophysics 261:291–314
Castillo E (1988) Extreme value theory in engineering. Academic Press, Boston

Charles RJ (1958) Dynamic fatigue of glass. J Appl Phys 29:1657–1662
Chinnery MA (1966) Secondary faulting I. Theoretical aspects. Can J Earth Sci 3:163–174
Chinnery MA, Petrak JA (1968) Dislocation fault model with a variable discontinuity. Tectonophysics 5:513
Chmyrev VM, Isaev VN, Bilichenko SV (1986) Electric fields and hydromagnetic waves in the ionosphere above an earthquake center. Geomagn Aeronomy+ 26:1020–1022
Ciliberto S, Guarino A, Scorretti R (2001) The effect of disorder on the fracture nucleation process. Physica D 158:83–104
Clark MM (1972) Surface rupture along the Coyote Creek fault. US Geol Surv Prof Pap 787:190–207
Cloos H (1921) Der Mechanismus tiefvulkanischer Vorgänge. Friedrich Vieweg, Braunschweig
Cloos H (1922) Über Ausbau und Anwendung der granit-tektonischen Methode. Abh Preuss Geol Landesanstalt 89:1–18
Cohen AJB (1993) Hydrogeologic characterization of a fractured granitic rock aquifer, Raymond, California. Lawrence Berkeley Laboratory - 34838, UC-400
Coleman BD (1957) Time dependence of mechanical breakdown in bundles of fibers. I. constant total load. J Appl Phys 28:1058–1067
Connoly P, Cosgrove J (1999) Prediction of fracture-induced permeability and fluid flow in the crust using experimental stress data. AAPG Bull 83:757–777
Conrad W, Behr HJ, Trzebski R (1996) Die LINSSER-Filterung des Schwerefeldes der Süddeutschen Grossscholle und ihre Interpretation. Z Geol Wissenschaft 21:21–40
Cooke ML, Pollard DD (1996) Fracture propagation paths under mixed mode loading within rectangular blocks of polymethyl metha-crylate. J Geophys Res 101:3387–3400
Cooke ML, Underwood CA (2001) Fracture termination and step-over at bedding interfaces due to frictional slip and interface opening. J Struct Geol 23:223–238
Cosgrove JW (1995) The expression of hydraulic fracturing in rocks and sediments. In: Ameen MS (ed) Fractography: Fracture topography as a tool in fracture mechanics and stress analysis. Geological Society Special Publication 92:187–196
Cotterell B, Rice JR (1980) Slightly curved or kinked cracks. Int J Fracture 16:155–169
Cowie PA, Scholtz CH (1992a) Displacement-length scaling relationship for faults: Data synthesis and discussion. J Struct Geol 14:1149–1156
Cowie PA, Scholtz CH (1992b) Growth of faults by the accumulation of seismic slip. J Geophys Res 97:11085–11095
Cox HL (1952) The elasticity and strength of paper and other fibrous materials. Brit J Appl Phys 3:72–79
Coyle DA, Wagner GA (1995) Spaltspuren – Ein Beitrag zur postvariszischen Tektonik und Abtragung des KTB-Umfelds. Geowissenschaften 13:142–146
Cramer T, Wanner A, Gumbsch P (1999) Dynamic failure of glass and single-crystalline silicon. Z Metallkd 90:675–686
Cramer T, Wanner A, Gumbsch P (2000) Energy dissipation and path instabilities in dynamic fracture of silicon single crystals. Phys Rev Lett 85:788–791
Cress GO, Brady BT, Rowell GA (1987) Sources of electromagnetic radiation from fracture of rock samples in the laboratory. Geophys Res Lett 14(4):331–334
Cruikshank KM, Aydin A (1994) Role of fracture localization in arch formation, Arches National Park, Utah. Geol Soc Am Bull 106:876–891
Cruikshank KM, Zhao G, Johnson AM (1991) Analysis of minor fractures associated with joints and faulted joints. J Struct Geol 13:865–886
Currie JB, Reik GA (1977) A method of distinguishing regional directions of jointing and of identifying joint sets associated with individual geologic structures. Can J Earth Sci 14:1211–1228
Dagan G (1977a) Hydrological analysis of BS3, BS4, TH4 and TH5 exploratory wells data. Petrol Ser Ltd, Tel Aviv
Dagan G (1977b) Hydrological analysis of BS5 and BS10 exploratory wells data and summary of hydrological results for Ramat Hovav Area. Petrol Ser Ltd, Tel Aviv
Dahan O, Nativ R, Adar EM, Berkowitz B (1998) A measurement system to determine water flux and solute transport trough fractures in the unsaturated zone. Ground Water 36:444–448
Dahan O, Nativ R, Adar EM, Berkowitz B, Ronen Z (1999) Field observation of flow in a fracture intersecting unsaturated chalk. Water Resour Res 35:3315–3326
Dahan O, Nativ R, Adar EM, Berkowitz B, Weisbrod N (2000) On fracture structure and preferential flow in unsaturated chalk. Ground Water 38:444–451
Dahmen K, Ertas D, Ben-Zion Y (1998) Gutenberg-Richter and characteristic eathquake behaviour is simple mean-field modes of heterrogeneous faults. Phys Rev E 58(2):1494
Dale TN (1923) The commercial granites of New England. U.S. Geological Survey Bulletin 738
Dally JW, Fourney WL, Irwin GR (1985) On the uniqueness of the stress intensity factor-crack velocity relationship. Int J Fracture 27:159–168

Dart CJ, McClay K, Hollings (1995) 3D analysis of inverted extensional fault systems, southern Bristol Channel Basin, UK. In: Buchanan JG, Buchanan PG (eds) Basin inversion. Geological Society Special Publication 88:393–413
Davis JC (1973) Statistical and data analysis in geology. John Wiley & Sons Ltd., New York
Davis GH (1999) Structural geology of the Colorado Plateau region of southern Utah, with special emphasis on deformation bands. Geol Soc Am, special paper 342
Davis K, Baker DM (1965) Ionospheric effects observed around the time of the Alaskan earthquake of 28 March 1964. J Geophys Res 70:2251–2253
Davis GH, Reynolds SJ (1996) Structural geology of rocks and regions. John Wiley & Sons Ltd., New York
De Freminville MCh (1914) Recherches sur la fragilité –l'éclatement. Rev Metall-Paris 11:97_–1056
De Sitter LU (1964) Structural geology. McGraw Hill, New York
Dean S, Baranoski M, Bertoli L, Kribbs G, Stephan T, Kulander B, Sochman D, Mumpower D (1984) Regional fracture analysis in western Valley and Ridge and adjoining plateau' West Virginia and Maryland. AAPG Bull 67:448
Dean SL, Kulander BR, Skinner JM (1988) Structural evolution of the Alleghanian orogeny in southeastern West Virginia. Geol Soc Am Bull 100:299–310
Dennis JG (1972) Structural geology. The Ronald Press Company, New York, p 532
Dennis JG (1987) Structural geology: An introduction. Wm C Brown Publishers, Dubuque
Derr JS (1986) Luminous phenomena and their relationship to rock fracture. Nature 321:470–471
Dieterich JH, Carter N (1969) Stress history of folding. Am J Sci 267:129–154
Döll W (1976) Application of an energy-balance and an energy method to dynamic crack-propagation. Int J Fracture 12:595–605
Doremus RH, Johnson WA (1978) Depths of fracture-initiating flows and initial stages of crack propagation in glass. J Mater Sci 13:855–858
Drakopoulos J, Stavrakakis GN, Latoussakis J (1993) Evaluation and interpretation of thirteen official-telegrams for the period September 10, 1986 to April 28, 1988. Tectonophysics 224(1–3):223–236
Drobdev VI, Krasnov VM, Salihov RM (1978) About ionospheric perturbation associated with earthquake and explosions. Izvestia Vysshih Uchebnih Zavedeniy 21:1862–1978 (*in Russian*)
Dulaney EN, Brace WF (1960) Velocity behavior of a growing crack. J Appl Phys 31:2233–2235
Dunne WM, North CP (1990) Orthogonal fracture systems at the limits of thrusting: an example from southwestern Wales. J Struct Geol 12:207–215
Dyer R (1988) Using joint interactions to estimate paleostress ratios. J Struct Geol 10:685–699
Ehlers EG, Blatt H (1982) Petrology: igneous, sedimentary and metamorphic. WH Freeman and Company, San Francisco, 732 pp
Eidam J, Hammer J, Korich D, Krauss M (1992) Zur Abgrenzung altersverschiedener Granitoide innerhalb des Lausitzer Granodiorit-Massivs. Z Geol Wissenschaft 20:289–293
Eldeen ZU, Delvaux D, Jacobs P (2000) Tectonic and paleostress evolution in the Wadi Araba segment of the Dead Sea rift, SW Jordan. The first Stephan Müller Conference of European Geological Society (EGS). From Continental Breakup to Collision. Israel, Dead Sea. Abstract
Elgazzar AS (1998) An inhomogeneous self-organized critical model for earthquakes. Physica A 251:303
Ellsworth WL, Beroza GC (1995) Seismic evidence for a seismic nucleation phase. Science 268:581–858
Engelder T (1979a) Mechanisms for strain within the Upper Devonian clastic sequence of the Appalachian Plateau, western New York. Am J Sci 279:527–542
Engelder T (1979b) Nature of deformation within the outer limits of the Central Appalachian foreland fold and thrust belt in New York State. Tectonophysics 55:289–310
Engelder T (1985) Loading paths to joint propagation during cycle: An example of the Appalachian Plateau, USA. J Struct Geol 7:459–476
Engelder T (1989) The analysis of pinnate joints in the Mount Desert Island granite: Implications for post-intrusion kinematics in the coastal volcanic belt, Maine. Geology 17:564–567
Engelder T (1993) Stress regimes in the lithosphere. Princeton University Press, Princeton, N.J., 457 pp
Engelder T (1999) Transitional-tensile fracture propagation: A status report. J Struct Geol 21:1049–1055
Engelder T, Engelder R (1977) Fossil distortion and decollement tectonics of Appalachian Plateau. Geology 5:547–460
Engelder T, Geiser P (1979) The relationship between pencil cleavage and lateral shortening within the Devonian section of the Appalachian Plateau, New York. Geology 7:460–464
Engelder T, Geiser P (1980) On the use of regional joint sets as trajectories of paleostress fields during the development of the Appalachian Plateau, New York. J Geophys Res 85:6319–6341
Engelder T, Gross MR (1993) Curving cross joints and the lithospheric stress-field in eastern North-America. Geology 21:817–820
Engelder T, Peacock DCP (2001) Joint development normal to regional compression during flexural-flow folding: The Lilstock buttress anticline, Somerset, England. J Struct Geol 23:259–277

Engelder T, Geiser P, Bahat D (1987) Alleghanian deformation within shales and siltstones of the Upper Devonian Appalachian Basin Finger Lakes district, New York. Geol Soc Am Continental Filed Guide, Northeastern Section, pp 113–118

Engelder T, Fischer MP, Gross MR (1993) Geological aspects of fracture mechanics: A short course manual notes. Geological Society of America Annual Meeting, Boston

Erdogan F (1983) Stress intensity factors. J Appl Mech-T ASME 50:992–1002

Erdogan F, Sih GC (1963) On the crack extension plates under plane loading and transverse shear. J Basic Eng-T ASME 85:519–527

Etheridge MA (1983) Differential stress magnitudes during regional deformation and metamorphism: Upper bound imposed by tensile fracturing. Geology 11:231–234

Ettensohn FR (1985) The Catskill Delta complex and the Acadian orogeny: A model. In: Woodrow DL, Sevon WD (eds) The Catskill Delta. Geological Society of America Special Paper 201:39–49

Evans AG (1974) Role of inclusions in fracture of ceramic materials. J Mater Sci 9:1145–1152

Evans AG (1978) Microfracture from thermal expansion anisotropy – I. Single phase system. Acta Metall 26:1845–1853

Evans MA (1994) Joints and decollement zones in Middle Devonian shales: Evidence for multiple deformation events in the central Appalachian Plateau. Geol Soc Am Bull 106:447–460

Evans AG, Johnson H (1975) The fracture stress and its dependence on slow crack growth. J Mater Sci 10:214–222

Evans AG, Wiederhorn SM (1974) Proof testing of ceramic materials – An analytical basis for failure prediction. Int J Fracture 10:379–392

Eyal Y (1996) Stress field fluctuations along the Dead Sea rift since the Middle Miocene. Tectonophysics 15:157–170

Eyal Y, Reches Z (1983) Tectonic analysis of the Dead Sea rift region since the Late Cretaceous based on mesostructures. Tectonophysics 2:167–185

Eyal Y, Gross MR, Engelder T, Becker A (2001) Joint development during fluctuation of the regional stress field in southern Israel. J Struct Geol 23:279–298

Fabrocini A, Fantoni S, Eckhard K (eds) (2002) Introduction to modern methods of quantum many body theory and their applications. World Scientific

Faill RT (1973) Kink-band folding, Valley and Ridge-Province, Pennsylvania. Geol Soc Am Bull 84:1289–1213

Fairhurst C, Cook NGW (1966) The phenomenon of rock splitting parallel to the direction of maximum compression in the neighborhood of a surface. In: Proceedings of the 1st Congress International Society for Rock Mechanics, Lisbon, pp 687–692

Feller W (1971) An introduction to probability theory and its applications. John Wiley, New York

Ferney BD, DeVary MR, Hsia KJ (1999) Oscillatory crack growth in glass. Scripta Materialia 41:275–281

Fifolt DA, Petrenko VF, Schulson EM (1993) Preliminary study of electromagnetic radiation from cracks in ice. Phil Mag B 67(3):289–299

Fineberg J, Marder M (1999) Instability in dynamic fracture. Phys Rep 313:1–108

Fineberg J, Gross SP, Marder M, Swinney HL (1991) Instability in dynamic fracture. Phys Rev Lett 67:457–460

Fineberg J, Gross SP, Marder M, Swinney HL (1992) Instability in the propagation of fast cracks. Phys Rev B 45:5146–5154

Finkel VM, Golovin YI, Sereda VE (1975) Electrical effects accompaying fracture of LiF crystals and problem of rock control. Sov Phys Sol-State 17(3):492–495

Finkelstein D, Powell J (1970) Earthquake lighting. Nature 228:759–760

Fisher RA (1938) The statistical utilization of multiple measurements. Ann Eugenic 8:376–386

Fisher RA (1953) Dispersion on a sphere. P Roy Soc Lond A Mat 217:295–305

Fisher NI, Lewis T, Embleton BJJ (1987) Statistical analysis of spherical data. Cambridge, London

Flexer A, Diamant E, Polishook B, Livnat A (1984) Relation of joints in the 'Avedat group (Eocene) to the tectonic pattern of Israel. Israel J Earth Sci 33:12–25

Freund LB (1972a) Crack propagation in an elastic solid subjected to general loading – I. Constant rate of extension. J Mech Phys Solids 20:129–140

Freund LB (1972b) Crack propagation in an elastic solid subjected to general loading – II. Non uniform rate of extension. J Mech Phys Solids 20:141–152

Freund LB (1973) Crack propagation in an elastic solid subjected to general loading – III. Stress wave loading J Mech Phys Solids 21:47–61

Freund LB (1974) Crack propagation in an elastic solid subjected to general loading – IV. Obliquely incident stress pulse J Mech Phys Solids 22:137–146

Freund LB (1990) Dynamic fracture mechanics. Cambridge Univ Press

Freund F (2000) Time resolve study of charge generation and propagation in igneous rocks. Geophys Res Lett 105(5):11001–11019

Freund F (2002) Charge generation and propagation in igneous rocks. J Geodyn 33(4–5):543–570

Frey MG (1973) Influence of Salina salt on structure in the New York – Pennsylvania part of the Appalachian Plateau. AAPG Bull 57:1027–1037
Frid V (1990) Rockburst hazard forecast of coal seams by theirs electromagnetic radiation. Ph.D. thesis, The state research institute of mining geomechanics and mine surveying (VNIMI), St. Petersburg (*in Russian*)
Frid V (1997a) Rock-burst hazard forecast by electromagnetic radiation excited by rock fracture. Rock Mech Rock Eng 30(4):229–236
Frid V (1997b) Electromagnetic radiation method for rock and gas outburst forecast. J Appl Geophys 38:97–104
Frid V (2000) Electromagnetic radiation method water-infusion control in rockburst-prone strata. J Appl Geophys 43:5–13
Frid V (2001) Calculation of electromagnetic radiation criterion for rockburst hazard forecast in coal mines. Pure Appl Geophys 158:931–944
Frid V, Shabarov A, Proskurjakov V (1992) Formation of electromagnetic radiation in coal stratum. J Min Sci+ 28(2):139–145
Frid V, Rabinovitch A, Bahat D (1999) Electromagnetic radiation associated with induced triaxial fracture in granite. Phil Mag Lett 79:79–84
Frid V, Bahat D, Goldbaum J, Rabinovitch A (2000) Experimental and theoretical investigation of electromagnetic radiation induced by rock fracture. Israel J Earth Sci 49:9–19
Frid V, Rabinovitch A, Bahat D (2003) Fracture induced electromagnetic radiation. J Phys D Appl Phys 36:1620–1628
Friedman M, Logan JM (1973) Lüders bands in experimentally deformed sandstone and limestone. Geol Soc Am Bull 84:1465–1476
Friedman M, Stearns DW (1971) Relations between stresses inferred from calcite twin lamellae and macrofractures, Teton Anticline, Montana. Geol Soc Am Bull 82:3151–3162
Frost HJ (2001) Mechanisms of crack nucleation in ice. Eng Fract Mech 68:1823–1837
Frost HJ, Gupta V (1993) Crack nucleation mechanisms and fracture toughness measurements in freshwater ice. In: Proceedings of the AMD Symposium on Ice Mechanics held at the Joint SES-ASC-ASME Conference in Virginia (1993), June 6–9, AMD-Vol. 163, ASME, New York, p 255
Fujinawa Y, Takahashi K (1998) Electromagnetic radiations associated with major earthquakes. Phys Earth Planet In 105(3–4):249–259
Fuller KNG, Fox PG, Field JE (1975) Temperature rise at tip of fast-moving cracks in glassy polymers. P Roy Soc Lond A Mat 341:537–550
Gabrielov A, Keilis-Borok V, Zaliapin I, Newman WI (2000a) Critical transitions in colliding cascades. Phys Rev E 62(1):237–249
Gabrielov A, Zaliapin I, Newman WI, Keilis-Borok VI (2000b) Colliding cascades model for earthquake prediction. Geophys J Int 143:427–437
Galperin YI, Jorjro NV, Kavrazhkin RA, et al. (1985) VLF and ELF effects in the upper ionospheric caused by large-scaled acoustic waves in the lower ionosphere observed from Aurol-3 satellite. In: Results of the ARCARD-3 Project, Cepadues, Toulouse, pp 661–664
Gardiner CW (1990) Handbook of stochastic methods for physics, chemistry and natural sciences, 2nd edn. Springer-Verlag, Berlin
Geiser PA (1988) The role of kinematics in the construction and analysis of geological cross section in deformed terranes. In: Mitra G, Wojtal G (eds) Geometries and mechanisms of thrusting. Geological Society of America Special Paper 222:47–76
Geiser P, Engelder T (1983) The distribution of layer parallel shortening fabrics in the Appalachian foreland of New York and Pennsylvania – Evidence for 2 non-coaxial phases of the Alleghanian Orogeny. Geol Soc Am Mem 158:161–175
Gerdes A, Wörner G, Finger F (1998) Late-orogenic magmatism in the southern Bohemian Massif – Geochemical and isotopic constraints on possible sources and magma evolution. Acta Universitatis Carolinae Geologica Praha 42:41–45
Germanovich LN, Dyskin AV (2000) Fracture mechanisms and instability of openings in compression. Int J Rock Mech Min 37:263–284
Germanovich LN, Ring LM, Carter BJ, Ingraffea AR, Dyskin AV, Ustinov KB (1995) Simulation of crack growth and interaction in compression. Proc Inter Cong Rock Mech Balkema, Rotterdam, Brookfield, pp 1:219–226
Gershenzon N, Zilpimiani D, Maguladze P (1985) Electromagnetic radiation from crack tip during ionic crystals fracture. Dokl Akad Nauk SSSR+ 248:1077–1081
Gershenzon N, Gokhberg M, Morgunov V (1987) Sources of electromagnetic emissions preceding seismic events. Izv-Phys Solid Earth 23(2):96–101
Gershenzon N, Gokhberg M, Karakin A (1989) Modeling the connection between earthquake preparation process and crystal electromagnetic emission. Phys Earth Planet In 57(1–2):129–138

Ghahremani F, Hutchinson JW, Tvergaard V (1990) Three-dimensional effects in microcrack nucleation in brittle polycrystals. J Am Ceram Soc 73:1548–1554
Ghosh SK (1993) Structural geology: Fundamentals and modern developments. Pergamon Press, Oxford
Gilbert GK (1904) Domes and dome structures of the High Sierra. Geol Soc Am Bull 15:29–36
Gillespie P, Walsh JJ, Watterson J (1992) Limitations of displacement and dimension data for single faults and the consequences for data analysis and interpretation. J Struct Geol 14:1157–1172
Gilman JJ (1958) Creation cleavage steps by dislocation. Trans Metall Soc AIME 212:310–315
Gokhberg M, Morgunov V, Aronov E (1979) High frequency electromagnetic radiation during seismic activity. Dokl Akad Nauk SSSR+ 248:1077–1081
Gokhberg M, Yoshino T, Morgunov V (1982) Results of recording operative electromagnetic Earthquake precursor in Japan. Izv-Phys Solid Eart+ 18(2):144–146
Gokhberg M, Gufel'd I, Gershenzon N (1985) Electromagnetic effects during rupture of the Earth's crust. Izv-Phys Solid Eart+ 21:52–62
Gokhberg M, Morgunov V, Matveyev V (1986) On the observation of anomalous electromagnetic emission in the epicentral zone of an Earthquake. Izv-Phys Solid Eart+ 22(8):676–678
Gol'd RM, Markov G, Mogila PG (1975) Pulsed electromagnetic radiation of minerals and rocks subjected to mechanical loading. Izv-Phys Solid Eart+ 7:109–111
Gold DP, Parizek RR (1976) Field guide to lineamer and fractures in central Pennsylvania. International Conference on New Basement Tectonics, 2nd edn. Newark University of Delaware
Goldbaum J, Frid V, Rabinovitch A, Bahat D (2001) Electromagnetic radiation induced by percussion drilling. Int J Fracture 111:L15-L20
Goldstein A, Marshak S (1988) Analysis of fracture array geometry. In: Marshak S, Mitra G (eds) Basic methods of structural geology. Prentice Hall, Englewood Cliffs, NJ, pp 249–267
Golovin Y, Shibkov A (1986a) Dynamics of dislocation pile-ups and pulsed polarization of LiF single crystals in the case of single slip. Sov Phys Sol-State 28(9):1625–1626
Golovin Y, Shibkov A (1986b) Fast electrical processes and dynamics of dislocations in plastically deformed alkali halide crystals. Sov Phys Sol-State 28(11):1965–1968
Golubovic L, Feng S (1991) Rate of microcrack nucleation. Phys Rev A 43:5223–5227
Goncharov A, Korjakov VP, Kuznetzov VM (1980) Acoustic emission and electromagnetic radiation during uniaxial compression. Dokl Akad Nauk SSSR+ 255(4):821–824
Goodwin AM (1995) Spatial change in joint geometry in the Chalk of eastern England. In: Ameen MS (ed) Fractography: Fracture topography as a tool in fracture mechanics and stress analysis. Geological Society Special Publication 92:197–213
Gradshteyn IS, Ryzhik IM (1980) Table of integrals, series and products, fourth edn. Academic Press, New York, pp 80ff
Grater RK (1945) Landslide in Zion Canyon, Zion National Park, Utah. J Geol 53:116–124
Gray M, Mitra G (1993) Migration of deformation fronts during progressive deformation – Evidence from detailed structural studies in the Pennsylvania anthracite region, USA. J Struct Geol 15(3–5): 435–459
Gregory HE (1950) Geology and geography of the Zion National Park region Utah and Arizona. Geol Surv Prof Paper 220
Griffith AA (1920) The phenomena of rupture and flow in solids. Philos T Roy Soc A 221:163–198
Griffith AA (1924) The theory of rupture. Proceedings of the First International Conference of Applied Mechanics, Delft, pp 55–63
Griggs DT, Handin J (1960) Observations on fracture and hypothesis of earthquakes. In: Griggs DT, Handin J (eds) Rock deformation. Geol Soc Am Mem 79:347–364
Gross MR (1993) The origin and spacing of cross joints: Examples from the Monterey Formation, Santa Barbara coastline, California. J Struct Geol 15:737–751
Gross MR (1995) Fracture partitioning: Failure mode as a function of lithology in the Monterey Formation of coastal California. GSA Bull 107: 779–792
Gross MR, Engelder T (1991) A case for neotectonic joints along the Niagara Escarpment. Tectonophysics 10:631–641
Gross MR, Fischer MP, Engelder T, Greenfield RJ (1995) Factors controlling joints spacing in interbedded sedimentary rocks: Integrating numerical models with field observations from the Monterey Formation, USA. In: Ameen MS (ed) Fractography: Fracture topography as a tool in fracture mechanics and stress analysis. Geological Society (London) Special Publication 92, pp 215–233
Gross MR, Bahat D, Becker A (1997) Relations between jointing and faulting based on fracture-spacing ratios and fault-slip profiles: A new method to estimate strain in layered rocks. Geology 25:887–890
Grossenbacher K, Bahat D, Karasaki K (1996) Triangulator: Excel spreadsheets for converting relative bearings to XYZ coordinates, with applications to scaling photographs and orienting surfaces. Comput Geosci 22:1053–1059

Grossmann NF (1983) A numerical method for the definition of discontinuity sets. Proc. 5th Congress International Society for Rock Mechanics, Melbourne, Balkema, 1:B17–B21
Guarino A, Garcimartin A, Ciliberto S (1998) An experimental test of the critical behavior of fracture precursor. Eur Phys J B 6:13–24
Guarino A, Ciliberto S, Garcimartin A (1999) Failure time and microcrack nucleation. Europhys Lett 47:456–461
Guarino A, Ciliberto S, Garcimartin A, Zei M, Scorretti R (2002) Failure time and critical behaviour of fracture precursors in heterogeneous materials. Eur Phys J B26:141–151
Gucer DE, Gurland J (1962) Comparison of statistics of two fracture models. J Mech Phys Solids 10:365–373
Gudmundsson A, Homberg C (1999) Evolution of stress fields and faulting in seismic zones. Pure Appl Geophys 154:257–280
Gueguen Y, Palciauskas V (1994) Introduction to the physics of rocks. Princeton University Press, Princeton
Guin JP, Wiederhorn SM (2003) Crack growth threshold in soda lime silicate glass: Role of hold-time. J Non-Cryst Solids 316:12–20
Gutenberg B, Richter CF (1954) Seismicity of the Earth and associated phenomena. Princeton University Press, Princeton
Gvirtzman G (1969) The Saqiye Group (late Eocene to early Pleistocene) in the coastal plain and Hashephela regions. Geol Surv Isr Bull 51(2)
Hahn GT, Rosenfield AR (1965) Local yielding and extension of a crack under plane stress. Acta Metall 13:293–306
Hamblin WK (1965) Origin of Òreverse dragÓ on the downthrow side of normal faults. Geol Soc Am Bull 76:1145–1164
Hancock PL (1972) The analysis of en-echelon veins. Geol Mag 3:269–276
Hancock PL (1985) Brittle microtectonics: Principles and practice. J Struct Geol 7:437–457
Hancock PL (1986) Joint spectra. In: Nichol I, Nesbitt RW (eds) Geology in the real world – the Kingsley Dunham volume. Institution of Mining and Metallurgy, London, pp 155–164
Hancock PL, Al-Kadhi A (1978) Analysis of mesoscopic fractures in the Dhurma-Nisah segment of the central Arabian Graben system. J Geol Soc 135:339–347
Hancock PL, Engelder T (1989) Neotectonic joints. Geol Soc Am Bull 101:1197–1208
Hancock PL, Dunne WM, Tringham ME (1982) Variscan structures in south-west Dyfed. In: Bassett MG (ed) Excursion-guide on geological excursions in Dyfed, South-west Wales 12:215–248
Handin J, Hager RV Jr, Friedman M, Feather JN (1963) Experimental deformation of sedimentary rocks under confining pressure: Pore pressure tests. AAPG Bull 47:717–755
Hansen EC, Borg IY (1962) The dynamic significance of deformation lamellae in quartz of a calcite-cemented sandstone. Am J Sci 260:321–336
Harlow DG (1985) The pure flaw model for chopped fiber composites. P Roy Soc Lond A Mat397:211–232
Hatheway AW, Kiersch GA (1982) Engineering properties of rock. In: Carmichael RC (ed) Handbook of physical properties or rock. CRC Press Inc, Boca Raton, Florida, pp 289–311
Hauch JA, Marder MP (1998) Energy balance in dynamic fracture, investigated by a potential drop technique. Int J Fracture 90:133–151
Hayakawa M, Tomizawa I, Ohta K (1993) Direction finding of precursory radio emissions associated with earthquakes: a proposal. Phys Earth Planet In 77:127–135
Hayati G (1975) The engineering geology of some chalks in Israel. Ph.D. Thesis, Technion, Israel Inst of Technol, (*in Hebrew*)
Helgeson DE, Aydin A (1991) Characteristics of joint propagation across layer interfaces in sedimentary rocks. J Struct Geol 13:897–911
Hermann HJ, Roux S (eds) (1990) Statistical models for the fracture of disordered media. North Holland, Amsterdam
Hertzberg RW (1976) Deformation and fracture mechanics of engineering materials. John Wiley & Sons Ltd., New York, p 605
Hidalgo RC, Moreno Y, Kun F, Herrmann HJ (2002) Fracture model with variable range of interaction. Phys Rev E 65:046148-1–8
Hirata T, Satoh T, Ito K (1987) Fractal structure of spatial distribution of microfracturing in rock. Geophys J Roy Astr S 90:369–374
Hirschmann G (1966) Assyntische und variszische Baueinheiten im Grundgebirge der Oberlausitz. Freiberger Forschungsh C 212:1–146
Hoagland RG, Hahn GT, Rosenfield AR (1973) Influence of microstructure on fracture propagation in rock. Rock Mech 5:77–106
Hobbs DW (1967) The formation of tension joints in sedimentary rocks: An explanation. Geol Mag 104:550–556

Hobbs BE, Means WD, Williams PF (1976) An outline of structured geology. John Wiley & Sons Ltd., New York, p 571
Hodgson RA (1961a) Classification of structures on joint surfaces. Am J Sci 259:493–502
Hodgson RA (1961b) Regional study of jointing in comb ridge-Navajo Mountain area, Arizona and Utah. AAPG Bull 45:1–38
Hoek E, Bieniawski ZT (1966) Fracture propagation mechanism in hard rock. C R Congr Soc Int Mec Roches 3:243–249
Holloway DG (1986) The fracture behaviour of glass. Glass Technol 27:120–133
Holloway S, Chadwick RA (1986) The Sticklepath-Lustleigh fault zone: Tertiary sinistral reactivation of a Variscan dextral strike slip fault. J Geol Soc London 143:452–477
Holman WR (1976) The origin of sheeting joints. A hypothesis. Unpublished Ph.D. thesis. University of California, Los Angeles, USA
Holzhausen GR (1989) Origin of sheet fracture, 1. Morphology and boundary conditions. Eng Geol 27:225–278
Holzhausen GR, Johnson AM (1979) Analyses of longitudinal splitting of uniaxially compressed rock cilinders. Int J Rock Mech Min 16:163–177
Horii H, Nemat-Nasser S (1985) Compression-induced microcrack growth in brittle solids: Axial splitting and shear failure. J Geophys Res 90:3105–3125
Hovestadt M (1902) Jena glass and its scientific and industrial applications. Macmillan & Co, London
Hsieh PA, Newman SP, Stiles GK, Simpson ES (1985) Field determination of the three-dimensional hydraulic conductivity tensor of anisotropic media. 2. Methodology and application to fractured rocks. Water Resour Res 21:1667–1676
Huber NK (1987) The geologic story of Yosemite National Park. U.S. Geological Survey Bulletin 1595
Hudak P (1992) Terminal decollement in the Appalachian plateau of northwestern PA. Northeastern Geology 1:108–112
Hull D (1999) Fractography: Observing, measuring, and interpreting fracture surface topology. Cambridge University Press, Cambridge
Hunt CB (1956) Cenozoic geology of the Colorado Plateau. U.S. Geological Survey Professional Papers 279, 99 pp
Huntington HE (1966) The Yosemite story. Doubleday & Company, New York
Ipatov Y (1989) Fundamentals of electromagnetic radiation method for rockburst forecast on Khibin apatite mines, Ph.D. thesis, The State Research Institute of Mining Geomechanics and Mine Surveying (VNIMI), St. Petersburg (*in Russian*)
Irwin GR (1957) Analysis of stresses and strains near the end of a crack traversing a plate. J Appl Mech-T ASME 24:361–364
Irwin GR (1962) Crack-extension force for a part-through crack in a plate. J Appl Mech-T ASME 29: 651–654
Issar A, Bahat D, Wakshal E (1988) Occurrence of secondary gypsum veins in joints in chalks in the Negev, Israel. Catena 15:241–247
Ito K, Matsuzaki M (1990) Earthqaukes as self-organized critical phenomena. J Geophys Res 95: 6853–6860
Ivanov VV, Pimonov AG, Yegorov PV, Kolpakova LA (1990) Determination of the constants of termofluctuational equations for strength and crack parameters from electromagnetic pulse emissions of rocks. Izv-Phys Solid Earth 26(7):597–601
Jaeger JC, Cook NGW (1979) Fundamentals of rock mechanics. Chapman and Hall, London, pp 593
Jagasivamani V, Iyer KJL (1988) Electromagnetic emission during the fracture of heat-treated spring steel. Mater Lett 6(11–12):418–422
Jayatilaka A de S, Trustrum K (1978) Fracture of brittle materials in uniaxial compression. J Mater Sci 13:455–457
Jensen HJ (1998) Self-organized criticality. Cambridge Univ Press, Cambridge
Johnson RM (1970) Physical processes in geology. Freeman and Company, San Francisco
Johnson E (1992) Process region changes from rapidly propagating cracks. Int J Fracture 55:47–63
Johnson JW, Holloway DG (1966) On the shape and size of the fracture zones on glass fracture surfaces. Phil Mag 14:731–743
Johnson RA, Wichern DW (1988) Applied multivariate statistical analysis. Prentice Hall, Engelwood Cliffs, New York
Johnston AC (1991) Light from seismic waves. Nature 354:361
Kalthoff JF (1973) On the propagation direction of bifurcated cracks. In: Sih GC (ed) Dynamic crack propagation. Noordhoff, Leyden, pp 449–458
Kanninen MF, Popelar CH (1985) Advanced fracture mechanics. Oxford University Press, New York
Karasaki K, Freifeld B, Cohen A, Grossenbacher K, Cook P, Vasco D (2000) A multidisciplinary fractured rock characterization study at Raymond field site, Raymond, CA. J Hydrol 236:17–34

Kattenhorn SA, Aydin A, Pollard DD (2000) Joints at high angles to normal fault strike: An explanation using 3-D numerical models of fault-perturbed stress fields. J Struct Geol 22:1–23
Kay SM, Snedden WT, Foster BP, Kay RW (1983) Upper mantleand crustal fragments in the Ithaca Kimberlites. J Geol 91:277–290
Kelly PM, Leach CA, Pawlings RD (1994) Acoustic, emission studies of fatigue loaded SiC platelet-reinforced Y-TZP. J Mater Sci Lett 13:211–212
Kelly PG, Sanderson DJ, Peacock DCP (1998) Linkage and evolution of conjugate strike-slip fault zones in limestones of Somerset and Northumbria. J Struct Geol 20:1477–1493
Kelner S, Bouchon M, Coutant O (1999) Characterization of fractures in shallow granite from the modelling of the anisotropy and attenuation of seismic waves. B Seismol Soc Am 89:706–717
Kerkhof F (1973) Wave fractographic investigation of brittle fracture dynamics. In: Sih GC (ed) Dynamic crack propagation. Noordhoff, Int Pub Leyden, pp 3–35
Kerkhof F (1975) Bruchmechanische Analyse von Schadensfällen an Gläsern. Glastech Ber-Glass 48:112–124
Kerkhof F, Richter H (1969) Investigation of the influence of water vapour on crack velocities in glass by ultrasonic fractography. Proc. 2nd Intern. Conference on Fracture, Brighton 40:463–47?
Khatiashvili N (1984) The electromagnetic effect accompanying the fracturing of alkaline halide crystals and rocks. Izv-Phys Solid Eart+ 20:656–661
Khesin B, Alexeyev V, Eppelbaum L (1996) Interpretation of geophysical fields in complicated environments. Kluwer Academic Publishers, London
Khesin B, Alexeyev V, Eppelbaum L (1997) Rapid methods for interpretation of induced polarization anomalies. J Appl Geophys 37:117–130
King CY (1983) Electromagnetic emissions before earthquakes. Nature 301:377
King GCP (1986) Speculation on the geometry of the initiation and termination process of earthquake rupture and its relation to morphology and geological structure. Pure Appl Geophys 124: 567–585
King PJ, Sheard FW (1970) Unharmonic attenuation of surface waves on crystals at low temperatures. P Roy Soc Lond A Mat 320:175–179
Kittel C (1987) Quantum theory of solids. John Wiley & Sons Ltd., New York
Kloster M, Hansen A, Hemmer PC (1997) Burst avalanches in solvable models of fibrous materials. Phys Rev E 56:2615–2625
Knapp RB, Norton D (1981) Preliminary numerical analysis of crystallization and stress evolution in cooling pluton environments. Am J Sci 281:35–68
Knauss WG (2000) Prospectives in experimental solid mechanics. Int J Solids Struct 37:251–266
Knauth LP, Lowe DR (1978) Oxygen isotope geochemistry of cherts from Onverwacht group (3.4 billion years), Transvaal, South-Africa, with implications for secular variations in isotopic composition of cherts. Earth Planet Sc Lett 41:209–222
Knott SD, Beach A, Brockbank PJ, Brown JL, McCallum JE, Welbon AI (1996) Spatial and mechanical controls on normal fault populations. J Struct Geol 18:356–372
Kobayashi AS, Mall S (1978) Dynamic fracture toughness of Homalite-100. Exp Mech 18:11–18
Kobayashi A, Ohtani N, Sato T (1974) Phenomenological aspects of viscoelastic crack propagation. J Appl Polym Sci 18:1625–1638
Kornfel'd MI (1971) Electrical charges on the surface of alkaline-halide crystals. Sov Phys Sol-State 13(2):381–385
Kornfel'd MI (1977) Mechanism of electrification of cleaved crystals. Sov Phys Sol-State 19(4):648–649
Kronfeld J, Rosenthal R, Weinberger G, Flexer A, Berkowitz B 1993. The interaction of two major old water bodies and its implication for the exploitation of groundwater in the multiple aquifer system in the central and northern Negev. Journal of Hydrology 143: 169-190
Kossmat F (1927) Gliederung des varistischen Gebirgsbaues. Abh Sächs Geol Landesamt 1:1–39
Kossobokov VG, Keilis-Borok VI, Cheng B (2000) Similarities of multiple fracturing on a neutron star and on the Earth. Phys Rev E 61(4):3529–3533
Kotz S, Nadarajah S (2000) Extreme value distributions theory and applications. Imperial College Press, London
Kowalik WS, Gold DP (1976) The use of Landsat-1 imagery in mapping lineaments in Pennsylvania. Proc. 1st Int. Conf. on the New Basement Tectonics. Utah Geol Soc, pp 236–249
Krapivsky PL, Majumdar SN (2000) Traveling waves, front selection and exact nontrivial exponents in a random fragmentation problem. Phys Rev Lett 85:5492–5495
Krauss M, Eidam J, Hammer J, Korich D (1992) Die cadomisch-variszische Entwicklung des Lausitzer Granodiorit-Komplexes. Zbl Geol Paläontol Teil I 1/2:71–85
Krenkel E (1924) Der Syrische Bogen. Centrail Mineral 9:274–281, 10301–10313
Kuksenko VS, Mansurov VA (1986) Fracture localization in rocks at different scale levels. Physical-technical Problems of Economical Resources Mining 3:49–55

Kuksenko VS, Lyashkov AI, Mirzoev KM (1982) Connection between the sizes of cracks generated under loading and duration of elastic energy emission. Dokl Akad Nauk SSSR 246(4):846–848
Kuksenko VS, Manzikov V, Mansurov VA (1985) Regularities in the development of microfocal rupture. Izv-Phys Solid Earth 21(7):553–556
Kuksenko VS, Ingevatkin IE, Mahgikov BC (1987) Physical and methodical foundations of rockburst forecast. Physical-technical Problems of Economical Material Mining 1:9–21
Kuksenko VS, Kilkeev RS, Lyashkov AI (1990) Electrification of quartz glass by mechanical loading. Sov Phys Sol-State 32(8):1320–1322
Kulander BR, Barton CC, Dean SC (1979) The application of fractography to core and outcrop fracture investigations. Report to U.S. D.O.E. Morgantown Energy Technology Center, METC SP-79/3
Kurie AE (1966) Recurrent structural disturbance of Colorado Plateau margin near Zion National Park Utah. Geol Soc Am Bull 77:867
Kustov VA, Liperovsky T (1988) About perturbation in ionospheric F-region begore strong earthquake. Izv-Phys Solid Eart+ 4:12–20
Lacazette A, Engelder T (1992) Fluid-driven cyclic propagation of a joint in the Ithace siltstone, Appalachian Basin, New York. In: Evans B, Wong TF (eds) Fault mechanics and transport properties of rocks. Academic Press Ltd, London, pp 297–324
Ladeira FL, Price NJ (1981) Relationship between fracture spacing and bed thickness. J Struct Geol 3:179–183
Lake SD, Karner GD (1987) The structure and evolution of the Wessex Basin, southern England – an example of inversion tectonics. Tectonophysics 137:347
Landau LD, Lifshitz EM (1963) Theory of elasticity. Pergamon Press, London
LaPointe PR, Hudson JA (1985) Characterization and interpretation of rock mass joint patterns. Geological Society of America Special Paper 199
Laube GC (1876) Geologie des böhmischen Erzgebirges I. Archiv f Naturwiss Landesdurchforsch, pp 1–208
Lawn BR (1993) Fracture of brittle solids, 2nd edn. Cambridge Solid State Science Series, Cambridge University Press, Cambridge
Lawn BR, Wilshaw TR (1975a) Fracture of brittle solids. Cambridge University Press, London, pp 1–204
Lawn BR, Wilshaw TR (1975b) Review indentation fracture: Principles and applications. J Mater Sci 10:1045–1081
Lazarus V, Leblond JB, Mouchrif SE (2001a) Crack front rotation and segmentation in mixed mode I + III or I + II + III. Part I: Calculation of stress intensity factors. J Mech Phys Solids 49:1399–1420
Lazarus V, Leblond JB, Mouchrif SE (2001b) Crack front rotation and segmentation in mixed mode I + III or I + II + III. Part II: Comparison with experiments. J Mech Phys Solids 49:1421–1443
Lee FT, Diehl SF (1989) Geologic and geomechanical properties of the Mount Valdo granite, Maine. Maine Geological Survey, vol 2, pp 187–200
Lee CC, Liu JY, Pan CJ (2000) The height of sporadic-E layer simultaneously observed by the VHF radar and ionosondes in Chung-Li. Geophys Res Lett 27:641–644
Letouzey J, Tremolieres P (1980) Paleo-stress fields around the Mediterranean derived from microtectonics: Comparison with plate tectonic data. Rock Mech 9:173–192
Levi T (2003) Joint sets as a tool for analyses the tectonic deformation in the central Arava western margins. Ms Sc Thesis, Ben Gurion University, Israel (*in Hebrew*)
Li Z, Yang J, Lee H (1988) Temperature-fields near a running crack tip. Engng Fracture Mech 30:791–799
Li L, Aubertin M, Simon R (1999) Multiaxial failure criterion with time and size effects for intact rocks. In: Amadei B, Kranz RL, Scott GA, Smeallie PH (eds) Rock mechanics for industry. Balkema, Rotterdam, pp 653–659
Lindner H (1972) Ergebnisse der Gravimetermessungen im Bereich des Lausitzer Massivs und seiner Randgebiete. Geologie 21:927
Link B, Epstein RI (1996) Thermally driven neutron star glitches. Astrophys J 457:844–854
Linsser H (1967) Investigation of tectonics by gravity detailing. Geophys Prosp XV/3:480–515
Liu C, Knauss WG, Rosakis AJ (1998) Loading rates and the dynamic initiation toughness in brittle solids. Int J Fracture 90:103–118
Liu JY, Chu YH, Chen MQ (2000) Modeling and ground observations of the ionosphere related to the COSMIC project. Terr Atmos Ocean Sci 11:349–364
Lobst R (2001) Der Intrusionsverband cadomischer Granodiorite am Klosterberg bei Demitz-Thumitz. Exkursionsführer und Veröffentlichungen der GGW 212:67–69
Lockner DA (1993) The role of acoustic emission in the study of rock fracture. Int J Rock Mech Min 7:883–889
Lockner DA (1996) Brittle fracture as an analog to earthquakes: Can acoustic emission be used to develop a viable prediction strategy. J Acoustic Emission 14(3–4):S88–S101

Lockner DA, Byerlee JD, Kuksenko V, Ponomarev A, Sidorin A (1991) Quasi-static fault growth and shear fracture energy in granite. Nature 350:39–42

Lockwood JP, Moore JG (1979) Regional deformation of the Sierra Nevada, California, on conjugate microfault sets. J Geophys Res 84:6041–6049

Loewy S (1995) The post-Alleghanian tectonic history of the Appalachian Basin based on joint patterns in Devonian black shales. University Park, Pennsylvania, Pennsylvania State University, (*Master's thesis*)

Long JCS, et al. (1996) Rock fractures and fluid flow: Contemporary understanding and applications. Commission on Geosciences, Environment and Resources (CGER), The National Academies Press, 551 pp

Loosveld RJH, Franssen RCMW (1992) Extensional vs. shear fractures: Implications for reservoir characterization. Society of Petroleum Engineers 25017:23–30

Maccagno TM, Knott JF (1989) The fracture behaviour of PMMA in mixed modes I and II. Eng Fract Mech 34:65–86

Mandelbrot B (1960) The Pareto-Levy law and the distribution of income. Int Econ Rev 1:79–106

Mansurov VA (1994) Acoustic emission from failing rock behavior. Rock Mech Rock Eng 27(3):173–182

Marder M (1993) Simple models of rapid fracture. Physica D 66:125–134

Marder M (1996a) Energetic developments in fracture. Nature 381:275–276

Marder M (1996b) Statistical mechanics of cracks. Phys Rev E 54:3442–3454

Marder M, Gross SP (1995) Origin of crack tip instabilities. J Mech Phys Solids 43:1–48

Marder M, Liu XM (1993) Instability in lattice fracture. Phys Rev Lett 71:2417–2420

Markov GA, Ipatov Y (1986) Method of electromagnetic radiation for rockburst forecast on apatite mines. Eng Geol 3:54–57 (*in Russian*)

Mart Y, Horowitz A (1981) The tectonics of the Timna region in Southern Israel and the evolution of the Dead-Sea Rift. Tectonophysics 79:165–199

Martel SJ (1997) Effects of cohesive zones on small faults and implications for secondary fracturing and fault trace geometry. J Struct Geol 19:835–847

Martens JH (1924) Igneous rocks of Ithaca, New York and vicinity. Geol Soc Am Bull 3:305–320

Matejka D, Janousek K (1998) Whole-rock geochemistry and petrogenesis of granites from the northern part of the Moldanubian batholith (Czeck Republic). Acta Universitatis Carolinae Geologica 42:73–79

Matsuzaki M, Takayasu H (1991) Fractal features of the earthquake phenomenon and a simple mechanical model. J Geophys Res 96:19925–19931

Matte PH, Maluski H, Rajlich P, Franke W (1990) Terrane boundaries in the Bohemian Massif: Result of large-scale Variscan shearing. Tectonophysics 177:151–170

Mayo EB (1941) Deformation in the interval Mt. Lyell–Mt. Whitney, California. Geol Soc Am Bull 94:563–575

McConaughy DT, Engelder T (1999) Joint interaction with embedded concretions: Joint loading configurations inferred from propagation paths. J Struct Geol 21:1637–1652

Means WD (1987) A newly recognized type of slickensise striation. J Struct Geol 9:585–590

Means WD (1979) Stress and strain. Springer-Verlag, Heidelberg

Means WD (1987) A newly recognized type of slickenside striation. J Struct Geol 9:585–590

Mecholsky JJ (1991) Quantitative fractography: An assessment. In: Frechette VD, Varner JR (eds) Fractography of glasses and ceramics II. Ceramic Transactions 17:413–451

Mecholsky JJ, Freiman SW (1979) Determination of fracture mechanics parameters through fractographic analysis of ceramics. In: Freiman SW (ed) Fracture mechanics applied to brittle materials. ASTM Spec Tech Pub 678:136–150

Menzel D, Schröder B (1994) Geologische Kriterien zur Unterbau-Exhumierung im Naab-Gebirge. KTB Report 94-3:179–184

Michalske TA (1977) The stress corrosion limit: Its measurement and implications. In: Bradt RC, Evans AG, Hasselman DPH, Lange FF (eds) Fracture mechanics of ceramics: Surface flaws, statistics, and microcracking, vol 5. Plenum Press, New York, p 277

Michalske TA (1984) Fractography of slow fracture in glass. In: Mecholsky JJ Jr, Powell SR Jr (eds) Fractography of ceramic and metal failures. Philadelphia, American Society for Testing and Materials Special Technical Publication 827:121–136

Miroshnichencko M, Kuksenko V (1980) Emission of electromagnetic pulses during nucleation of cracks in solid insulators. Sov Phys-Solid State 22(5):895–896

Mirowski P (1995) Mandelbrot's economics after a quarter century. Fractals 3:581–600

Misra A (1975) Electromagnetic effects at metallic fracture. Nature 254:133–134

Misra A (1977) Physical model for the stress-induced electromagnetic effect in metals. Appl Phys 16:195–199

Misra A, Ghosh S (1980) Electromagnetic radiation characteristics during fatigue propagation and failure. Appl Phys 23:387–390

Mogi K (1962) Study of elastic shocks caused by the fracture of heterogeneous materials and its relation to earthquake phenomena. B Earthq Res I Tokyo 40:125–173

Mogi K (1985) Earthquake prediction. Academic Press, Tokyo

Molchan G, Kronrod T, Panza G (1997) Multiscale seismicity medel for seismic risk. Bull Seismol Soc Am 87:1220–1229

Molchanov OA, Hayakawa M (1998a) Subionospheric VLF perturbations possibly related to earthquakes. J Geophys Res 103(A8):17489–17504

Molchanov OA, Hayakawa M (1998b) On the generation mechanism of ULF seismogenic electromagnetic emissions. Phys Earth Planet In 105:201–210

Molnar P, Tapponier P (1975) Cenozoic tectonics of Asia: Effects of a continental collision. Science 189:419–426

Molotskii MI (1980) Dislocation mechanism for Misra effect. Sov Tekh Phys Lett 6(1):22–23

Moore DE, Lockner DA (1995) The role of microcracking in shear-fracture propagation in granite. J Struct Geol 17:95–114

Morgunov V (1985) Electromagnetic emission during seismic activity. Izv-Phys Solid Earth 21(3): 220–226

Morgunov V, Matveyev V (1990) Electromagnetic emission in aftershocks of Spitak earthquake. Izv-Phys Solid Earth 26:457–461

Morgunov V, Matveyev V (1991) Electric and electromagnetic effects in the epicentral zone of Spitak aftershocks. Izv-Phys Solid Earth 27:1002–1005

Morgunov V, Gerasimovich E, Matveyev V (1988) Deformation of the Earth surface and anomalous electromagnetic emission. Izv-Phys Solid Earth 24:924–929

Mortimore RN (1990) Chalk or chalks. Proc Intern Chalk Symp, Brighton Politechnic, Thomas Telford, pp 15–45

Mott NF (1948) Fracture of metals: Theoretical considerations. Engineering-London 165:16–18

Muehlberger WR (1961) Conjugate joint sets of small dihedral angle. J Geol 69:211–219

Müller A (2000) Cathodolumuniscence and characterization of defect structures in quartz with applications to the study of granitic rocks. Thesis, University Göttingen, pp 1–229, *http://webdoc.sub.gwdg.de/diss/2001/mueller/index.html*

Müller G, Dahm T (2000) Fracture morphology of tensile cracks and rupture velocity. J Geophys Res 105:723–738

Müller A, MŸller B, Behr HJ (2001) Structural constants in granitic rocks of the Lusatian Granodiorite Complex and the Erzgebirge, Germany – in commemoration of Hans Cloos. Z Geol Wissenschaft 29:521–544

Murakami Y (ed) (1987) Stress intensity factors handbook, vol 1, 2. The Society of Materials Sci, Japan, Pergamon Press, Oxford

Murakami Y (ed) (1992) Stress intensity factors handbook, vol 3. The Society of Materials Sci, Japan, Pergamon Press, Oxford

Murakami Y (ed) (2001) Stress intensity factors handbook, vol 4, 5. The Society of Materials Sci, Japan, Pergamon Press, Oxford

Muraoka H, Kamata H (1983) Displacement distribution along minor fault traces. J Struct Geol 5:483–495

Murgatroyd JB (1942) The significance of surface marks on fractured glass. J Soc Glass Technol 26: 155–171

Murphy PJ (1981) Detachment structures in south central NY. Northeastern Geology 3:105–116

Nabaro B (1967) Theory of crystal dislocations. Oxford University Press, Oxford

Nanjo K, Nagahawa H, Satomura M (1998) Rates of aftershocks decay and the fractal structure of active fault systems. Tectonophysics 287:173–186

Narr W, Suppe J (1991) Joint spacing in sedimentary rocks. J Struct Geol 13:1037–1047

Nativ R, Nissim I (1992) Characterization of a desert aquitard – hydrologic and hydrochemical considerations. Ground Water 30:598–606

Nativ R, Adar E, Dahan O, Geyh M (1995) Water recharge and solute transport through the vadose zone of fractured chalk under desert conditions. Water Resour Res 31:253–261

Nativ R, Adar E, Dahan O, Nissim I (1997a) Water salinization in arid regions – Observations from the Negev desert, Israel. J Hydrol 196:271–296

Nativ R, Halleran A, Hunley A (1997b) Evidence for ground-water circulation in the brine-filled aquitard, Oak Ridge, Tennessee. Ground Water 35:647–659

Nativ R, Adar EM, Becker A (1999) Designing a monitoring network for contaminated ground water in fractured chalk. Ground Water 37:38–47

Nativ R, Adar E, Assaf L, Nygaard E (2003) Characterization of the hydraulic properties of fractures in chalk. Ground Water 41:532–543

Nelson RA (1979) Natural fracture systems: Description and classification. Bull Amer Assoc Petrol Geolog 63:2214–2221

Nelson RA (1987) Fracture reservoirs: Turning knowledge into practice. J Petrol Technol 16470:407–414

Nemcok M, Bayer R, Miliorizos M (1995) Structural analysis of the inverted Bristol Channel Basin: Implications for the geometry and timing of fracture porosity. In: Buchanan JG, Buchanan PG (eds) Basin inversion. Geological Society Special Publication 88:355–392

Nesbitt AC, Austin BA (1988) The emission and propagation of electromagnetic energy from stressed quartzite rock underground. The Trans of the SA Inst of Electr Eng 79:53–57

Newman WI, Phoenix SL (2001) Time dependent fiber bundles with local load-sharing. Phys Rev E 63:0215071–19

Newman WI, Turtcotte DL, Gabrielov AM (1995) Log-periodic behavior of a hierarchical failure model with applications to precursory seismic activation. Phys Rev E 52(5):4827–4835

Nickelsen RP (1979) Sequence of structural stages of the Allegheny orogeny at the Bear Valley strip mine, Shamokin, Pennsylvania. Am J Sci 279:225–271

Nickelsen RP (1988) Structural evolution of folded thrusts and duplexes on a first-order anticlinorium in the Valley and Ridge Province of Pennsylvania. In: Mitra G, Wojtal S (eds) Geometries and mechanics of thrusting with special reference to the Appalachians. Geological Society of America Special Paper 222:89–106

Nickelsen RP (1996) Alleghanian sequential deformationon on the SW limb of the Pennsylvania Salient in Fulton and Franklin Counties, south-central Pennsylvania: Pennsylvania Geology Survey. Annual field conference of Pennsylvania geologists, 61st, guidebook, pp 1–9

Nickelsen RP, Engelder T (1989) Fold-thrust geometries of the Juniata culmination, Central Appalachians of Pennsylvania. In: Engelder T (ed) Structures of the Appalachian foreland fold-thrust belt. International Geological Congress, 28th, Fild Trip Guidebook T166

Nickelsen RP, Hough ND (1967) Jointing in the Appalachian Plateau of Pennsylvania. Geol Soc Am Bull 78:609–630

Nikiforova NN, Yadakhin TN (1989) Studies of electromagnetic emission of tectonic origin in the Kirghiz SSR. Phys Earth Planet In 57:68–74

Niles WH (1871) Peculiar phenomena observed in quarrying. Proc Boston Soc Nat Hist vol 14

Nitsan V (1977) Electromagnetic emission accompanying fracture of quartz-bearing rocks. Geophys Res Lett 4(8):33–336

Nur A (1982) The origin of tensile fracture lineaments. J Struct Geol 4:31–40

O'Keefe SG, Thiel DV (1995) A mechanism for the production of electromagnetic radiation during fracture of brittle materials. Phys Earth Planet In 89:127–135

Odling NE, Gillespie P, Bourgine B, Castaing C, Chils J-P, Christensen NP, Fillion E, Genter A, Olsen C, Thrane L, Trice R, Aarseth E, Walsh JJ, Watterson J (1999) Variations in fracture system geometry and their implications for fluid flow in fractured hydrocarbon reservoirs. Petrol Geosci 5:373–384

Ogawa T, Oike K, Miura T (1985) Electromagnetic radiation from rocks. J Geophys Res 90(d4): 6245–6249

Olami Z, Feder HJS, Christensen A (1992) Self-organized criticality in a continuous, non-conservative cellurar automaton modeling earthquakes. Phys Rev Lett 68:1244–1247

Olson JE (1993) Joint pattern development: Efects of subcritical crack growth and mechanical crack interaction. J Geophys Res 98:12251–12265

Olson JE, Pollard DD (1989) Inferring paleostress from natural fracture patterns: A new method. Geology 17:345–348

Olson JE, Pollard DD (1991) The initiation and growth of en-echelon veins. J Struct Geol 13:595–608

Orkan N, Voight B (1985) Regional joint evolution in the Valley and Ridge Province of Pennsylvania in relation to the Alleghany Orogeny: Guidebook for the 50th annual field conference of Pennsylvania geologists. Bureau of Topography and Geological Survey, Harrisburg, Pa, pp 144-164

Ouellet M (1990) Earthquake lights and seismicity. Nature 348:492

Panek LA (1985) Estimating fracture trace length from censored measurements on multiple scanlines. Proceedings of the International Symposium on Fundamentals of Rock Joints, Björkliden, pp 13–23

Paris P, Erdogan F (1963) A critical analysis of crack propagation laws. J Basic Eng-T ASME 85:528–534

Park RG (1983) Foundations of structural geology. Chapman and Hall, New York, p 135

Parker JM (1942) Regional systematic jointing in slightly deformed sedimentary rock. Geol Soc Am Bull 53:381–408

Parrot M, Achache J, Berthelier J, Blanc E (1993) High frequency electromagnetic effects. Phys Earth Planet In 77:65–81

Pauchard L, Meunier J (1993) Instantaneous and time-lag breaking of a two-dimensional solid rod under a bending stress. Phys Rev Lett 70:3565–3568

Pauchard L, Meunier J (1998) Experimental study of the breakage of a two-dimensional crystal. Phil Mag B78:221–224

Peacock DCP (2001a) The temporal relationship between joints and faults. J Struct Geol 23:329–341

Peacock DCP (2001b) Are the lower planes of double seismic zones caused by serpentine dehydration in subducting oceanic mantle? Geology 29:299–302
Peacock DCP, Sanderson DJ (1991) Displacements, segment linkage and relay ramps in normal-fault zones. J Struct Geol 13:721
Peacock DCP, Sanderson DJ (1992) Effects of layering and anisotropy on fault geometry. J Geol Soc London 149:793–802
Peacock DCP, Sanderson DJ (1994) Geometry and development of relay ramps in normal-fault systems. AAPG Bull 78:147–165
Peacock DCP, Sanderson DJ (1995) Strike-slip relay ramps. J Struct Geol 17:1351–1360
Peng S (1970) Fracture and failure of Chelmford Granite. Ph.D. thesis, Stanford University, Stanford, USA
Peng S, Johnson AM (1972) Crack growth and faulting in cylindrical specimens of Chelmsford Granite. Int J Rock Mech Min 9:37–86
Peterson RE (1974) Stress concentration factors. John Wiley & Sons Ltd., New York
Petit J-P (1988) Normal stress dependent rupture morphology in direct shear tests on sandstone with applications to some natural fault surface features. Int J Rock Mech Min 25:411–419
Petit J-P, Barquins M (1987) Formation de fissures en milier confine dans le PMMA (polymetachrylate de methyle): modele analogique de formation de structures tectoniques cassantes. C R Acad Sci Paris 305:55–60
Petit J-P, Barquins M (1988) Can natural faults propagate under mode II conditions? Tectonophysics 7:1243–1256
Petrenko VF (1993) On the nature of electrical polarization of materilas caused by cracks. Application to ice electromagnetic radiation. Philos Mag B 67(3):301–315
Petrov VA, Gorobetz LZ (1987) Size effect of the concentration threshold of destruction. Izv-Phys Solid Eart+ 23(1):75–77
Petukhov IM (1972) Rockbursts on coal mines. Nedra, Moscow
Petukhov IM, Lin'kov LM (1983) Mechanics of rock bursts and outbursts. Nedra, Moscow
Petukhov IM, Lin'kov LM, Sidorov VS (1976) Theory of protective stratums. Nedra, Moscow
Picu RC (1996) Singularities of an interface crack impinging on a triple grain junction. Int J Solids Struct 33:1563–1573
Picu RC (1997) Three-dimensional stress concentration at grain triple junctions in columnar ice. Phil Mag Lett 76:159–166
Picu RC, Gupta V (1996) Singularities at grain triple junctions in two-dimensional polycrystals with cubic and orthorhombic grains. J Appl Mech-T ASME 63:295–300
Pohn HA (1981) Joint spacing as a method of locating faults. GLGYB 9:258–261
Poirier JP (1985) Creep of crystals. Cambridge University Press, Cambridge, UK
Politi A, Ciliberto S, Scorretti R (2002) Failure time in the fiber-bundle model with thermal noise and disorder. Phys Rev E 66:026171–026176
Pollard DD (2000) Strain and stress: Discussion. J Struct Geol 22:1359–1367
Pollard DD, Segall P (1987) Theoretical displacements and stresses near fractures in rock: With applications to faults, joints, veins, dikes, and solution surfaces. In: Atkinson BK (ed) Fracture mechanics of rock. Academic Press, London, pp 277–349
Pollard DD, Segall P, Delaney PT (1982) Formation and interpretation of dilatant echelon cracks. Geol Soc Am Bull 93:1291–1303
Pomeau Y (1992) Brisure spontanée de cristaux bidimensionnels courbés. C R Acad Sci Paris, t314, Série II, pp 553–556
Pook LP (2000) Linear elastic fracture mechanics for engineering: Theory and applications. WIT Press, Southampton, Boston
Preston FW (1931) The propagation of fissures in glass and other bodies with special reference to the split-wave front. J Am Ceram Soc 14:419–427
Price NJ (1966) Fault and joint development in brittle and semi-brittle rock. Pergamon Press, New York, pp 110–161
Price NJ (1974) The development of stress systems and fracture patterns in undeformed sediments. Int Soc Rock Mech Cngr Proc pp 487–496
Price NJ (1977) Aspects of gravity tectonics and the development of listric faults. J Geol Soc London 133:311–327
Price NJ, Cosgrove JW (1990) Analysis of geological structures. Cambridge University Press, Cambridge, England
Rabinovitch A (1994) On the Mott derivation of crack velocity. Phil Mag Lett 70:231–233
Rabinovitch A, Bahat D (1979) Catastrophe theory: A technique for crack propagation analysis. J Appl Phys 50:321–334
Rabinovitch A, Bahat D (1999) Model of joint spacing distribution based on shadow compliance. J Geophys Res 104:4877–4886

Rabinovitch A, Belizovsky G, Bahat D (1993) On the direction of secondary cracks in brittle fracture. Int J Fracture 63:R25–R30
Rabinovitch A, Belizovsky G, Bahat D (1994) Length distribution of secondary cracks in brittle materials. Phil Mag Lett 70:277–288
Rabinovitch A, Bahat D, Frid V (1995) Comparison of electromagnetic radiation and acoustic emission in granite fracturing. Int J Fracture 71(2):R33–R41
Rabinovitch A, Bahat D, Frid V (1996) Emission of electromagnetic radiation by rock fracturing. Z Geol Wissenschaft 24(3–4):361–368
Rabinovitch A, Frid V, Bahat D (1998) Parametrization of electromagnetic radiation pulses obtained by triaxial fracture in granite samples. Phil Mag Lett 77(5):289–293
Rabinovitch A, Bahat D, Melamed Z (1999a) A note on joint spacing. Rock Mech Rock Eng 32:71–75
Rabinovitch A, Frid V, Bahat D (1999b) A note on the amplitude – frequency relation of electromagnetic radiation pulses induced by material failure. Phil Mag Lett 79:195–200
Rabinovitch A, Belizovsky G, Bahat D (2000a) Origin of mist and hackle patterns in brittle fracture, Phys Rev B 61:14968–14974
Rabinovitch A, Zlotnikov R, Bahat D (2000b) Flaw length distribution measurement in brittle materials. J Appl Phys 87:7720–7725
Rabinovitch A, Frid V, Bahat D, Goldbaum J (2000c) Fracture area calculation from electromagnetic radiation and its use in chalk failure analysis. Int J Rock Mech Min 37:1149–1154
Rabinovitch A, Bahat D, Frid V (2002a) Similarity and dissimilarity of electromagnetic radiation from carbonate rocks under compression, drilling and blasting. Int J Rock Mech Min 39(1):125–129
Rabinovitch A, Frid V, Bahat D (2002b) Gutenberg-Richter type relation for laboratory fracture induced electromagnetic radiation. Physical Review E 65:011401–011404
Rabinovitch A, Frid V, Bahat D, Goldbaum J (2003a) Decay mechanism of the fracture induced electromagnetic pulses. J Appl Phys 93(9):5085–5090
Rabinovitch A, Frid V, Goldbaum J, Bahat D (2003b) Polarization-depolarization process in glass during percussion drilling. Phil Mag 83(25):2929–2940
Ramsay JG (1980) Crack-seal mechanism of rock deformation. Nature 284(5752):135–139
Ramsay JG, Huber MI (1987) The techniques of modern structural geology, vol 2: Folds and fractures. Academic Press, London
Ramsay JG, Lisle R (2000) The techniques of modern structural geology, vol 3: Applications of continuum mechanics in structural geology. Academic Press, New York
Rasband W (2003) Image J 1.30V. National Institutes of Health, USA
Ravi-Chandar K (1998) Dynamic fracture of nominally brittle materials. Int J Fracture 90:83–102
Ravi-Chandar K, Knauss WG (1984a) An experimental investigation into dynamic fracture – I. Crack initiation and crack arrest. Int J Fracture 25:247–262
Ravi-Chandar K, Knauss WG (1984b) An experimental investigation into dynamic fracture – II. Microstructural aspects. Int J Fracture 26:65–80
Ravi-Chandar K, Knauss WG (1984c) An experimental investigation into dynamic fracture – III. Steady state crack propagation and crack branching. Int J Fracture 26:141–154
Ravi-Chandar K, Knauss WG (1984d) An experimental investigation into dynamic fracture – IV. On the interaction of stress waves with propagating cracks. Int J Fracture 26:189–200
Ravi-Chandar K, Knauss WG (1987) On the acceleration of the transient stress field near the tip of a crack. J Appl Mech-T ASME 54:72–78
Ravi-Chandar K, Knauss W (1999) Processes controlling fast fracture of brittle solids. Comput Sci Eng 1999 Sept/Oct:24–31
Ravishankar SR, Murthy CRL (2000) Characteristics of AE signals obtained during drilling composite laminates. NDT&E International 33:341–348
Rawnsley KD, Rives T, Petit JP, Hencher SR, Lumsden AC (1992) Joint development in perturbed stress fields near faults. J Struct Geol 14:939–951
Rawnsley KD, Peacock DCP, Rives T, Petit JP (1998) Joints in the Mesozoic sediments around the Bristol Channel Basin. J Struct Geol 20:1641–1661
Reches Z (1976) Analysis of joints in 2 monoclines in Israel. Geol Soc Am Bull 87:1654–1662
Reches Z, Lockner DA (1994) Nucleation and growth of faults in brittle rocks. J Geophys Res 99: 18159–18173
Red'kin V, Kuprijanov AS, Bufalov VV (1985) Geophysical devices for distance rock burst control. In: Smirnov V (ed) Geophysical methods of stress and deformation control. Novosibirsk, pp 81–82 (*in Russian*)
Regel' VR, Slutsker AI, Tomashevskii EE (1972) The kinetic nature of the strength solids Sov Phys Uspekhi 1:45–65
Rehrig WA, Heidrick TL (1972) Regional fracturing in Laramide stocks of Arizona and its relationship to porphyry copper mineralization. Econ Geol 67:198–213

Rehrig WA, Heidrick TL (1976) Regional tectonic stress during the Laramide and late Tertiary intrusive periods, Basin and Range Province, Arizona. Arizona Geological Digest 10:205–228

Renshaw CE, Schulson EM (2001) Universal behavior in compressive failure of brittle materials. Nature 412:897–900

Rice RW (1974) Fracture topography of ceramics. In: Frechette VD, La Course WC, Burdick VL (eds) Surfaces and interfaces of glass and ceramics. Plenum Publishing Corpration, New York, pp 439–472

Rice JR (1979) Theory of precursory processes in the inception of earthquake rupture. Gerl Beitr Geophys 88:91–127

Rice RW (1984) Ceramic fracture features, observations, mechanisms and uses. In: Mecholsky JJ, Powell SR (eds) Fractography of ceramic and metal failures. ASTM, Philadelphia, pp 1–51

Richter HG, Kerkhof F (1994) Stress wave fractography. In: Bradt RC, Tressler RE (eds) Fractography of glass. Olenum Press, New York pp 75–109

Rikitake TJ (1997) Nature of electromagnetic radiation precursory to an earthquake. J Geomagn Geoelectr 49:1153–1163

Rives T, Razack M, Petit JP, Rawnsley KD (1992) Joint spacing: Analogue and numerical simulations. J Struct Geol 14:925–937

Roberts JC (1974) Jointing and minor tectonics of the vale of Glamorgan between Ogmore-by-Sea and Lavernock Point, South Wales. Geol J 9:97–114

Roberts JC (1979) Jointing and minor tectonics of the South Gower Peninsula between Mumbles Head and Rhossilli Bay, South Wales. Geol J 14:1–14

Roberts DG (1989) Basin inversion in and around the British Isles. In: Cooper MA, Williams GD (eds) Inversion tectonics. Spec Publ Geol Soc London 44:131–150

Roberts JC (1995) Fracture surface markings in Liassic limestone at Lavernock Point, South Wales In: Ameen MS (ed) Fractography: Fracture topography as a tool in fracture mechanics and stress analysis. Geological Society (London) Special Publication 92:175–186

Robinson ES (1970) Mechanical disintegration of the Navajo Sandstone in Zion Canyon, Utah. Geol Soc Am Bull 81:2799–2806

Rodgers J (1970) The tectonics of the Appalachians. Wiley Interscience, New York

Rophe B, Berkowitz B, Magaritz M, Ronen D (1992) Analysis of subsurface flow and anisotropy in a fractured aquitard using transient water level data. Water Resour Res 28:199–207

Rossmanith HP, Knasmillner RE, Daehnke A, Mishnaevsky L (1996) Wave propagation, damage evolution and dynamic fracture extension. 1. Percussion drilling. Mater Sci+ 32:350–358

Rossmanith HP, Daehnke A, Knasmillner RE, Kouzniak N, Ohtsu M, Uenishi K (1997) Fracture mechanics applications to drilling and blasting. Fatigue Fract Eng M 20:1617–1636

Roy A, Saha TK (1995) Interaction of a penny-shaped crack with an elliptic crack. Int J Fracture 73:51–65

Rudnicki JW, Olsson WA (1998) Reexamination of fault angles predicted by shear localization theory. Int J Rock Mech Min 35: Paper No 88

Ryan MP, Sammis CG (1978) Cyclic fracture mechanisms in cooling basalt. Geol Soc Am Bull 89: 1295–1308

Rzhevskii VV, Novik GY (1978) Foundations of rock physics. Nedra, Moscow

Sadovskii MA, Sobolev GA, Migunov NI (1979) Changes of natural radiowaves radiation during Karpatian earthquake. Dokl Akad Nauk SSSR+ 244(2):316–319

Sakai H, Oda H, Nakayama T, Doi H (1992) Electromagnetic changes at explosion seismic experiment. J Phys Earth 40:447–458

Salameh E, Zacher W (1982) Horizontal stylolites and paleostress in Jordan. Neues Jahrb Geol P M 8:509–512

Sbar ML, Sykes LR (1973) Contemporary compressive stress and seismicity in eastern North America: An example of intra-plate tectonics. Geol Soc Am Bull 84:1861–1882

Scharbert S (1998) Some geochronological data from the South Bohemian Pluton in Austria: A critical review. Acta Universitatis Carolinae Geologica Praha 42:114–118

Schiff LI (1986) Quantum mechanics. McGraw Hills, London

Schoeck G, Pichl W (1990) Bond trapping of cracks. Phys Stat Solidi 118:109–115

Scholz CH (1990) The mechanics of earthquakes and faulting. Cambridge, Cambridge University Press

Scholz CH (1998) Earthquakes and friction laws. Nature 391:37–42

Schwartz DP, Sibson RH (1989) Fault segmentation and controls of rupture initiation and termination. U.S. Geological Survey Open File Report 89-13

Scitovich VP, Lazarevich LM (1985) Estimation of stress condition of rock massive by EMR. In: Smirnov V (ed) Geophysical methods of stress and deformation control. Novosibirsk, pp 65–66 (*in Russian*)

Scorretti R, Ciliberto S, Guarino A (2001) Disorder enhances the effects of thermal noise in the fiber bundle model. Europhys Lett 55:626–632

Scott DR (1996) Seismicity and stress rotation in a granular model of the brittle crust. Nature 381:592–595

Scott WH, Hansen EC, Twiss RJ (1965) Stress analysis of quartz deformation lamellae in a minor fold. Am J Sci 263:729–746
Secor DT Jr (1965) Role of fluid pressure in jointing. Am J Sci 263:633–646
Secor DT Jr (1969) Mechanics of natural extension fracturing at depth in the Earth's crust. In: Research in tectonics. Geol Surv Canada Paper 3–47
Segall P, Pollard DD (1983) Joint formation in granitic rock of the Sierra Nevada. Geol Soc Am Bull 94:563–575
Segall P, McKee EH, Martel SJ, Turrin BD (1990) Late Cretaceous age of fractures in the Sierra Nevada Batholith, California. Geology 18:1248–1251
Shainin VE (1950) Conjugate sets of en echelon tension fractures in the Athens limestone at Riverton, Virginia. Geol Soc Am Bull 61:509–517
Shand Shand EB (1954) Experimental study of fracture of glass: 2 Experimental data. J Am Ceram Soc 37:559–572
Shand EB (1959) Breaking stress of glass determined from dimensions of fracture mirrors. J Am Ceram Soc 42:474–477
Sharon E, Finberg J (1996) Microbranching instability and the dynamic fracture of brittle materials. Phys Rev B 54:7128–7139
Sharon E, Fineberg J (1998) Universal features of the micro-branching instability in dynamic fracture. Phil Mag B78:243–251
Sharon E, Fineberg J (1999) Confirming the continuum theory of dynamic brittle fracture for fast cracks. Nature 397:333–335
Sharon E, Gross SP, Fineberg J (1995) Local crack branching as a mechanism for instability in dynamic fracture. Phys Rev Lett 74:5096–5099
Sheldon P (1912) Some observations and experiments on joint planes. J Geol 20:164–183
Shen B, Stephensson O, Einstein HH, Ghahreman B (1995) Coalescence of fractures under shear stresses in experiments. J Geophys Res 100:5975–5990
Sibson RH (1985) Stopping of earthquake ruptures at dilatational jogs. Nature 316:248–251
Simmons G, Todd T, Baldridge WS (1975) Toward a quantitative relationship between elastic properties and cracks in low porosity rocks. Am J Sci 275:318
Smekal A (1953) Zum Bruchvorgang bei sprödem Stoffverhalten unter ein- und mehrachsigen Beanspruchungen. Oesterr Ing-Arch 7:49–70
Smith RL (1982) The asymptotic-distribution of the strength of a series-parallel system with equal load-sharing. J Appl Probab 10:137–171
Smith RL, Phoenix SL (1981) Asymptotic distributions for the failure of fibrous materials under series-parallel structure and equal load-sharing. J Appl Mech-T ASME 48:75–82
Sneddon IN (1951) Fourier transforms. McGraw Hill Company Inc., New York, p 542
Sobolev GA, Semerchan AA, Salov BG (1982) Precursors of the destruction of large rock sample. Izv-Phys Solid Eart+ 18(8):572–580
Sommer E (1967) Das Bruchverhalten von Rundstäben aus Glas im Manteldruckversuch mit Überlagerter Zugspannung. Glastech Ber-Glass 40:304–307
Sommer E (1969) Formation of fracture "lances" in glass. Eng Fract Mech 1:539–546
Sornette D, Knopoff L, Kagan YY, Vanneste C (1996) Rank-ordering statistics of extreme events: Application to the distribution of large earthquakes. J Geophys Res 101:13883–13893
Spiegel MR (1961) Statistics. McGraw-Hill Book Company, N.Y., 359 pp
Spiker EC, Gray M (1997) A study of progressive deformation at the Allegheny Front, Lycoming County, Pennsylvania. Geological Society of America Abstracts with Programs, vol 29, no 1
Sprunt ES, Brace WF (1974) Direct observations of microcavities in crystalline rocks. Int J Rock Mech Min 11:139–150
Spyropoulos C, Scholz CH, Shaw BE (2002) Transition regimes for growing crack populations. Phys Rev E 65:0561051–05610510
Srivastava GP (1990) The physics of phonons. Adam Hilger, Bristol
Stearns DW (1968) Certain aspects of fracture in naturally deformed rocks. In: Riecker RE (ed) Rock mechanics seminar, vol 1. Special rept., Terrestrial Sciences Lab Air Force, Cambridge Research Labs, Bedford, Mass, pp 27–116
Stearns DW, Friedman M (1972) Reservoirs in fractured rock. In: King RE (ed) Stratigraphic oil and gas fields. AAPG M Em 16:82–106
Stearns DW (1968) Certain aspects of fracture in naturally deformed rocks. In: Riecker RE (ed) Rock mechanics seminar, vol. 1, special rept. Terrestrial Sciences Lab Air Force, Cambridge Research Labs, Bedford, Mass, pp 27–116
Stern TW, Bateman PC, Morgan BA, Newell MF, Peck DL (1981) Isotopic U-Pb ages of zircon from the granitoids of the central Sierra Nevada, California. U.S. Geological Survey Professional Paper 17 pp

Stewart ME, Taylor WJ (1996) Structural analysis and fault segment boundary identification along the Hurricane fault in southwestern Utah. J Struct Geol 18:1017–1029

Suess FE (1926) Intrusionstektonik und Wandertektonik im variszischen Grundgebirge. Gebrüder Borntraeger, Leipzig, 268 pp

Suk M (1984) Geological history of the territory of Czech Socialist Republic. Geological Survey, Prague

Suppe J (1985) Principles of structural geology. Prentice-Hall, Englewood Cliffs, NJ

Suresh S (1998) Fatigue of materials, 2nd edn. Cambridge University Press, Cambridge

Swaby PA, Rawnsley KD (1997) An interative #D fracture modeling environment. Soc Petroleum Eng 9:88–92

Swallowe GM, Field JE, Horn LA (1986) Measurements of transient high-temperatures during the deformation of polymers. J Mater Sci 21:4089–4096

Takayasu H (1989) Fractals in the physical sciences. John Wiley & Sons Ltd., New York

Tetelman AS, McEvly AJ (1967) Fracture of structural materials. John Willey & Sons Ltd., New York

Thomas R, Klemm W (1997) Microthermometric study of silicate melt inclusions in Variscan granites from SE Germany. J Petrology 38:1753–1765

Thomas AL, Pollard DD (1993) The geometry of echelon fractures in rock: Implications from laboratory and numerical experiments. J Struct Geol 15:323–334

Timoshenko SP, Goodier JN (1951) Theory of elasticity. McGraw-Hill, New York

Timoshenko SP, Goodier JN (1984) Theory of elasticity, 3rd edn. McGraw-Hill Int Book Co, Auckland

Tobisch OT, Paterson SR, Saleeby JB, Geary EE (1989) Nature and timing of deformation in the Foothills terrane, central Sierra Nevada, California: Its bearing on orogenesis. Geol Soc Am Bull 101:401–413

Tokarski AK, Zuchiewicz W, Swierczewska A (1999) The influence of early joints on structural development of thrust-and-fold belts: A case study from the Outer Carpathians (Poland). Geol Carpath 50:178–180

Tomizawa I, Yamada I (1995) Generation mechanism of electric impulses observed in explosion seismic experiments. J Geomagn Geoelectr 47:313–324

Tomizawa I, Hakayawa M, Yoshino T (1994) Observation of ELF/VLF electromagnetic variations associated with seismic experimental explosion. In: Hayakawa M, Fujinawa Y (eds) Electromagnetic phenomena related to earthquake prediction. Terra Scientific Publishing Company, Tokyo, pp 337–347

Trudgill B, Cartwright J (1994) Relay-ramp forms and normal-fault linkages, Canyonlands National Park. Utah Bull Geol Soc Am 106:1143–1157

Trzebski R (1997) Morphogenesis, tectonic setting and intrusion dynamics of the late-Variscan granites at the northwest-margin of the Bohemian Massif. Göttinger Arb Geol Paläont 69:1–66

Tsang YW (1984) The effect of tortuosity on fluid-flow through a single fracture. Water Resour Res 20:1209–1215

Tschegg EK (1983) Mode 3 and mode 1 fatigue crack propagation behaviour under tortional loading. J Mater Sci 18:1604–1614

Turcotte DL (1992) Fractals and chaos in geology and geophysics. Cambridge University Press, Cambridge

Turcotte DL, Newman WI, Shcherbakov R (2003) Micro and microscopic models of rock fracture. Geophys J Int 152:718–728

Tvergaard V, Hutchinson JW (1988) Microcracking in ceramics induced by thermal expansion or elastic anisotropy. J Am Ceram Soc 71:157–166

Twidale CR, Vidal Romani JR, Campbell EM, Centeno JD (1996) Sheet fractures: Response to erosional offloading or to tectonic stress? Z Geomorph 106:1–24

Twiss RJ, Moores EM (1992) Structural geology. Freeman & Company, New York

Umeda Y (1990) High-amplitude seismic-waves radiated from the bright spot of an earthquake. Tectonophysics 175:81–92

Urusovskaja AA (1969) Electric effects associated with plastic deformation of ionic crystals. Sov Phys Uspekhi 11:631–643

Van Hise CR (1896) Principles of North-American Pre-Cambrian geology: U.S. geology. U.S. Geological Survey 16th Annual Report, pp 581–874

Van Tyne A (1983) Natural gas potential of the Devonian black shales of New York. NoE Geol 5:209–216

Varotsos P, Sarlis, N, Lazaridou M (1999) Interconnection of defect parameters and stress-induced electric signals in ionic crystals. Phys Rev B 59:24–27

Ver Steeg K (1942) Jointing in the coal beds of Ohio. Econ Geol 37:530–509

Vozoff K, Frid V (2001) Electromagnetic emissions monitoring to warn of wind blasts and gas outs. Results of ACARP Project at Moonee Colliery: C9005. In: Symposium on Geological Hazards, Rafferety's Resort Newcastle, 15–16 November 2001

Wagner GA, Van Den Haute P (1992) Fission track dating. Enke-Verlag, Stuttgart

Walmann T, Malthe-Sorenssen A, Feder J, Jossang T, Meakin P, Hardy HH (1996) Scaling relations for the lengths and widths of fractures. Phys Rev Lett 77:5393–5396

Walsh JJ, Watterson J (1988) Analysis of the relationship between displacements and dimensions of faults. J Struct Geol 10:239–247

Warwick JW, Stoker C, Meyer TR (1982) Radio emission associated with rock failure: Possible application to the great Chilean earthquake of May 22, 1960. J Geophys Res 87(b4):2851–2859

Watznauer A (1954) Die erzgebirgischen Granitintrusionen. Geologie 3:688–706

Wawrzynek PA, Ingraffea AR (1987) Interactive finite element analysis of fracture processes: An integrated approach. Theor Appl Fract Mec 8:137–150

Wedel AA (1932) Geological structures of the Devonian strata of southcentral New York, New York State. Mus Bull 294:74

Weinberger R (1999) Initiation and growth of cracks during desiccation of stratified muddy sediments. J Struct Geol 21:379–386

Weinberger R (2001a) Joint nucleations in layered rocks with non-uniform distribution of cavities. J Struct Geol 23:1241–1254

Weinberger R (2001b) Evolution of polygonal patterns in stratified mud during desiccation: The role of flaw distribution and layer boundaries. Geol Soc Am Bull 113:20–31

Weisbrod N, Nativ R, Ronen D, Adar E (1998) On the variability of fracture surfaces in unsaturated chalk. Water Resour Res 34:1881–1887

Weisbrod N, Nativ R, Adar E, Ronen D (1999) Impact of intermittent rainwater and wastewater flow on coated and uncoated fractures in chalk. Water Resour Res 35:3211–3222

Weisbrod N, Nativ R, Adar E, Ronen D (2000a) Salt accumulation and flushing in unsaturated fractures in an arid invironment. Ground Water 38:452–461

Weisbrod N, Nativ R, Adar E, Ronen D, Ben-Nun A (2000b) Impact of coating and weathering on the properties of chalk fracture. Geophys Res Lett 105:27853–27864

Weisbrod N, Dahan O, Adar EM (2002) Particle transport in unsaturated fractured chalk under arid conditions. J Contam Hydrol 56:117–136

Westergaard HM (1939) Bearing pressure and cracks. Trans Am Soc Mech Eng 61A:49–53

White WS (1946) Rock-bursts in the granite quarries at Bare, Vermont. U.S. Geological Survey, Circular 13

Whittaker A, Green GW (1983) Geology of the country around Weston-super-Mare. Memoir of the Geological Survey of Great Britain. Sheet 279 and parts of 263 and 295

Wibberley CAJ, Petit JP, Rives T (2000) Micromechanics of shear rupture and the control of normal stress. J Struct Geol 22:411–427

Widom B (1966) Random sequential addition of hard spheres to a volume. J Chem Phys 44:3888

Wiederhorn SM (1967) Influence of water vapor on crack propagation in soda-lime glass. J Am Ceram Soc 50:407–414

Wiederhorn SM (1978) Mechanisms of subcritical crack growth in glass. In: Bradt RC, Hasselman DPH, Lange FF (eds) Fracture mechanics of ceramics, vol 4. Plenum, New York, pp 549–580

Wiederhorn SM, Bolz LH (1970) Stress corrosion and static fatigue of glass. J Am Ceram Soc 53:543–548

Wiederhorn SM, Johnson H (1973) Effect of electrolyte PH on crack propagation in glass. J Am Ceram Soc 56:192–197

Wiederhorn SM, Johnson H, Diness AM, Heuer AH (1974) Fracture of glass in vacuum. J Am Ceram Soc 57:336–341

Wiederhorn SM, Dretzke A, Rodel J (2002) Crack growth in soda-lime-silicate glass near the static fatigue limit. J Am Ceram Soc 85:2287–2292

Wilkins SJ, Michael RG, Wacker M, Eyal Y, Engelder T (2001) Faulted joints: Kinematics, displacement-length scaling relations and criteria for their identification. J Struct Geol 23:315–327

Willemse EJM, Pollard DD (1998) On the orientation and patterns of wing cracks and solution surfaces at the tips of a sliding flaw or fault. J Geophys Res 103B:2427–2438

Williams ML (1957) On the stress distribution at the base of a stationary crack. J Appl Mech-T ASME 24:109–114

Wise DU, McCrory TA (1982) A new method of fracture analysis: Azimuth versus traverse distance plots. Geol Soc Am Bull 93:889–897

Witherspoon PA, Newman SP (1967) Evaluating a slightly permeable caprock in aquifer gas storage. I. Caprock of infinite thickness. J Petrol Technol 19:949

Wong RHC, Chau KT (1998) Peak strength of replicated and real rocks containing cracks. Key Eng Mat 145–149:953–958

Wong RHC, Chau KT, Wang P (1996) Microcracking and grain size effect in Yuen Kong marbles. Int J Rock Mech Min 33:479–485

Woodworth JB (1895) Some features of joints. Science 2:903–904

Woodworth JB (1896) On the fracture system of joints, with remarks on certain great fractures. Boston Society Natural History Proceedings 27:63–184

Wu H, Pollard DD (1992) Propagation of a set of opening-mode fractures in layered brittle materials under uniaxial strain cycles. J Geophys Res 97:3381–3396

Wu H, Pollard DD (1995) An experimental study of relationship between joint spacing and layer thickness. J Struct Geol 17:887–905
Wyss M (2001) Why is earthquake prediction research not progressing faster? Tectonophysics 338: 217–223
Xu XP, Needleman A (1994) Numerical simulations of fast crack growth in brittle solids. J Mech Phys Solids 42:1397–1434
Yamada I, Masuda K, Mizutani H (1989) Electromagnetic and acoustic emission associated with rock fracture. Phys Earth Planet In 57:157–168
Yang B, Ravi-Chandar K (2001) Crack path instabilities in a quenched glass plate. J Mech Phys Solids 49:91–130
Yatomi C (1981) The effect of surface-energy on temperature rise around a fast running crack. Eng Fract Mech 14:759–762
Yepez E, Angulobrown F, Peralta J (1995) Electric field pastterns as seismic precursors. Geophys Res Lett 22(22):3087–3090
Yoffe EH (1951) The moving Griffith crack. Phil Mag 42:739–750
Yokobori T, Uozomi M, Ichikawa M (1971) Interaction between non-coplanar parallel straggered elastic cracks. Rep Res Inst Strength and Fract of Mater 7:25–47
Yoon KJ, Wiederhorn SM, Luecke WE (2000) Comparison of tensile and compressive creep behavior in silicon nitride. J Am Ceram Soc 85:2017–2022
Yoshino T, Tomizawa I (1989) Observation of low-frequency electromagnetic emissions as precursors to the volcanic eruption at Mt. Mihara during November, 1986. Phys Earth Planet In 57:32–39
Yoshino T, Tomizawa I, Sugimoto T (1993) Results of statistical analysis of low-frequency seismogenic EM emissions as precursors to earthquakes and volcanic eruptions. Phys Earth Planet In 77:21–31
Younes AL (1996) Tectonic implications from joints and fold history in the Appalachian Plateau, central New York. Geological Society of America Abstracts with Programs, vol 28, no 3, A-111
Younes AI, Engelder T (1995) Paleostress rotation as indicated by joints within the Appalachian Plateau: Clockwise, counterclockwise, or both? American Geological Society, Abstracts of the Annual Meeting, Session 135
Younes AI, Engelder T (1999) Fringe cracks: Key structures for the interpretation of the progressive Alleghanian deformation of the Appalachian Plateau. Geol Soc Am Bull 111:219–239
Yukawa S, Timo DP, Rubio A (1969) Fracture design practices for rotating equipment. In: Liebowitz H (ed) Fracture. Academic Press, New York, pp 65–157
Yuse A, Sano M (1993) Transition between crack patterns in quenched glass plates. Nature 362:329–331
Yuse A, Sano M (1997) Instabilities of quasi-static crack patterns in quenched glass plates. Physica D 108:365–378
Zawislanski P (1994) Surface fracture distribution at Raymond field site, Raymond, California, Lawrence Berkeley National Laboratory Report, LBNL-42675
Zhao M, Jacobi RD (1997) Formation of regional cross-fold joints in the northern Appalachian Plateau. J Struct Geol 19:817–834
Zhao YX, Gao Q, Wang JN (2000) The evolution of short fatigue crack lengths and crack density: Two approaches. Fatigue Fract Eng M 23:929–941
Zhurkov SN (1965) Kinetic concept of the strength of solids. Int J Fracture Mech 1:311–323
Zoback ML (1989) Global patterns of tectonic stress. Nature 34:291–298
Zoback ML (1992) First- and second-order patterns of stress in the lithosphere: The world stress map project. J Geophys Res 97:11703–11728
Zoback MD, Healy JH (1984) Friction, faulting, and in situ stress. Ann Geophys 2:689–698
Zurkov SN, Kuksenko VS, Slutsker AI (1969) Formation of submicroscopic cracks in polimers under load. Sov Phys Sol-State 11(2):238–245

Subject Index

F

G

K

L

M

Q

R

S

T

Ecological Studies, Vol. 134

Analysis and Synthesis

Ecological Studies

Volumes published since 1992 are listed at the end of this book.

Springer
New York
Berlin
Heidelberg
Barcelona
Hong Kong
London
Milan
Paris
Singapore
Tokyo

Paul R. Miller Joe R. McBride
Editors

Oxidant Air Pollution Impacts in the Montane Forests of Southern California

A Case Study of the San Bernardino Mountains

Foreword by O. Clifton Taylor

With 102 illustrations, 6 in color

Springer

Paul R. Miller
USDA Forest Service
Pacific Southwest Station
Riverside, CA 92507
USA

Joe R. McBride
Department of Environmental
Science, Policy, and Management
University of California
Berkeley, CA 94720
USA

Library of Congress Cataloging-in-Publication Data
Oxidant air pollution impacts in the montane forests of Southern
California: a case study of the San Bernardino Mountains/[edited
by] Paul Miller, Joe R. McBride.
p. cm. — (Ecological studies; 134)
Includes bibliographical references and index.
ISBN 0-387-98493-3 (hardcover: alk. paper)
1. Trees—Effect of air pollution on—California—San Bernardino
Mountains Region—Case studies. 2. Trees—Wounds and injuries—
California—San Bernardino Mountain Region—Case studies.
3. Conifers—Effect of air pollution on—California—San Bernardino
Mountains Region—Case studies. 4. Conifers—Wounds and injuries—
California—San Bernardino Mountain Region—Case studies.
5. Forest ecology—California—San Bernardino Mountains Region—Case
studies. I. Miller, Paul R., 1905– . II. McBride, Joe.
III. Series: Ecological studies; v. 134.
SB745.095 1998
577.3′27—DC21 98-11498

Printed on acid-free paper.

Production coordinated by Princeton Editorial Associates, Inc., and managed by Francine McNeill; manufacturing supervised by Jeffrey Taub.
Typeset by Princeton Editorial Associates, Inc., Roosevelt, NJ, and Scottsdale, AZ.
Printed and bound by Braun-Brumfield, Inc., Ann Arbor, MI.
Printed in the United States of America.

9 8 7 6 5 4 3 2 1

ISSN 0070-8356
ISBN 0-387-98493-3 Springer-Verlag New York Berlin Heidelberg SPIN 10670166

Foreword

The geographical location of the San Bernardino Mountains and the regional climate combine to create a potential for air pollution injury to midelevation conifer forests. The mountains are located downwind from one of the largest metropolitan regions in the United States, where accumulation of photochemical oxidant air pollution is enabled by the marine inversion layer that has a persistent influence during the summer months. The mixed conifer forest's most important species, ponderosa pine, was the first to show injury symptoms on needles in the mid-1950s. The condition was termed *x-disease* because the cause was unknown. Later, the term *chlorotic decline of ponderosa pine* was used interchangeably. At about the same time, experiments were being performed on citrus and grapes located in the Upland-Ontario area just a few miles upwind and in the urbanized valley west (upwind) of the San Bernardino Mountains. Whole trees or vines were enclosed in greenhouses and provided with carbon-filtered air or smoggy ambient air. The key results of this work included the recognition that leaf injury and declining yield of fruit were the result of exposure to smoggy air or conversely that carbon-filtered air protected these crops against injury and loss of yield. Furthermore, the term *oxidant stipple of grape*, due primarily to ozone, was used to describe the typical field symptoms on grape leaves.

In the meantime, the possible role of fluoride or sulfur pollutants from a nearby steel plant in causing injury to cultivated orchards, vineyards, and native vegetation, particularly ponderosa pine, was being investigated. These pollutants were dismissed as major causal agents. Observations of ornamental pine species in urban locations identified the same kind of chlorotic mottle symptoms on needles as were observed on native ponderosa pines. Several fumigation experiments with ponderosa pine seedlings or branches of field trees culminated as early as 1963 in publications that described how ozone could induce chlorotic mottle symptoms on ponderosa pine. Eventually, injury attributable to ozone was described on other native conifers, including Jeffrey pine, black oak, and white fir. Symptoms were harder to detect, or absent, on incense cedar, coulter pine, sugar pine, and several evergreen oaks.

More attention was drawn to ponderosa pine in the early 1970s because weakened trees were being killed by pine bark beetles, and sanitation-salvage logging operations were used to remove weakened trees before infestation. At about this time, multidisciplinary studies began for the purpose of studying the broader implications of ozone injury to the mixed conifer forest type (see Introduction). A feature of this work, which remains important today, was the study of pollutant effects on a west-to-east gradient of decreasing pollutant exposure. In this volume, we draw on the individual experiences of about 30 scientists who have contributed to our collective knowledge of pollutant effects as a result of their research over a period of up to 30 to 35 years.

The knowledge attained over the years has provided the first ecosystem-level interpretation of chronic ozone exposure effects on a forest ecosystem in the U.S. EPA Criteria Documents, used in support of the evaluation of National Air Quality Standards for ozone. The extension of these observations into the future is clearly an advantage because few forest ecosystems in the West have quality-assured data sets with this length of record. These results could be integrated into broader programs such as the National Environmental Monitoring Initiative. Ironically, the area has been regarded as somewhat of a pariah because at the extreme it represents a higher level of ozone exposure than has been observed anywhere in the United States. This is correct; however, observations have always been made at points on a gradient of decreasing pollutant exposure. Injury levels from other forested regions, particularly the Sierra Nevada and other southern California mountain ranges, have been carefully compared with the entire gradient of exposure/injury in the San Bernardino Mountains, and greater meaning can be attached to these observations. In addition, an important cause of the air pollution in the San Bernardino Mountains is the area's proximity to heavily urbanized areas—a condition becoming increasingly important for other forested areas in California and the West.

Air monitoring records for ozone from the South Coast Air Basin of California have shown steadily improving conditions. As reported in this volume, tree damage from ozone is also on the decline, but it has not reached levels below the threshold for chronic injury. At the same time, we have become aware of the increasing direct and indirect effects of nitrogen deposition in several forms. We do not yet fully understand how "nitrogen saturation" and ozone may interact. In light of the population increases predicted for the future, it is uncertain if current pollution control measures will continue to be adequate. The forests of the San Bernardino Mountains will continue to be an important monitor of progress in the protection of our air environment.

Riverside, California O. Clifton Taylor

Acknowledgments

A number of individuals are deserving of recognition because of their pioneering efforts during the earliest phases of work to investigate the cause of x-disease of ponderosa pine. These individuals include James Asher, Robert Bega, Eugene Cardiff, Fields Cobb, Donald Dahlsten, Tom Neff, Richard Parmeter, Jr., B.L. Richards, and O. Clifton Taylor. During the first multidisciplinary study between 1974 and 1979, the research group was managed and guided by O. Clifton Taylor and Joe R. McBride. Co-investigators were Rod Arkely, Johan Brun, Fields Cobb, Donald Dahlsten, Robert L. James, Ronald N. Kickert, Richard Laven, Robert Luck, Clif Ohmart, D.L. Rowney, Marshall White, Wayne Wilcox, and others from the University of California, Berkeley. Ronald Kickert deserves particular recognition for his early application of simulation modeling techniques as a way of integrating EPA-sponsored research and for completing the organization of the computerized project database, culminating in about 1979.

Throughout the past 30 years, we are grateful for the research support provided, at different times, by the U.S. Environmental Protection Agency, especially the Corvallis Environmental Research Laboratory, the California Air Resources Board, Research Division, and appropriated funds of the USDA Forest Service, Pacific Southwest Station. We also appreciate the long-standing support of the Southern California Edison Company, especially for underwriting the major share of the publication cost of the color illustrations of this volume. Facilities and services were provided by the Statewide Air Pollution Research Center at the University of California, Riverside.

We are particularly grateful to chapter reviewers, namely, Michael Barbour, John Battles, Allan Bell, Tom Cahill, John Carroll, Susan Conard, Paul Dunn, Mark Francis, Jay Garner, Alan Goldstein, David Grahm, David Grantz, William Hogsett, Jud Isebrands, Theodor Leininger, Gong-yu Lin, Carol Lovatt, John McColl, William Malm, David Nowak, Dennis Ojima, David Olszyk, David Peterson, Gregory Reams, Roger Rosentreter, Phillip Rundel, Michael Ryan, Andrew Sanders, Richard Smith, Ann Sutherland, James Weber, and Arthur Winer.

Riverside, California — Paul R. Miller
Berkeley, California — Joe R. McBride

Contents

Contributors

M.J. Arbaugh	USDA Forest Service, Riverside, CA 92507, USA
A. Bytnerowicz	USDA Forest Service, Riverside, CA 92507, USA
S.-C. Chen	Scripps Institution of Oceanography, La Jolla, CA 92093, USA
Judith C. Chow	Desert Research Institute, University of Nevada, Reno, NV 89506, USA
S. Copeland	USDA Forest Service, Fort Collins, CO 80526-2098, USA
D. Dahlsten	Department of Environmental Science, Policy, and Management, University of California, Berkeley, CA 94720, USA
M.E. Fenn	USDA Forest Service, Riverside, CA 92507, USA
Clifton A. Frazier	Desert Research Institute, University of Nevada, Reno, NV 89506, USA

F.M. Fujioka	USDA Forest Service, Riverside, CA 92507, USA
Mark Green	Desert Research Institute, University of Nevada, Reno, NV 89506, USA
N.E. Grulke	USDA Forest Service, Riverside, CA 92507, USA
Barbara Hinsvark	Desert Research Institute, University of Nevada, Reno, NV 89506, USA
J. Hom	USDA Forest Service, Radnor, PA 19087, USA
D.W. Johnson	Desert Research Institute, University of Nevada, Reno, NV 89506, USA
M. Krywult	W. Szafer Institute of Botany, Polish Academy of Sciences, Krakow 31-512, Poland
Richard D. Laven	Department of Forestry and Wood Science, Colorado State University, Fort Collins, CO 80521, USA
J.R. McBride	Department of Environmental Science, Policy, and Management, University of California, Berkeley, CA 94720, USA
L. Merrill	USDA Forest Service, San Bernardino, CA 92408-3430, USA
P.R. Miller	USDA Forest Service, Riverside, CA 92507, USA
Richard A. Minnich	Department of Earth Sciences/ Geography, University of California, Riverside, CA 92521, USA
T.H. Nash III	Department of Plant Biology, Arizona State University, Tempe, AZ 85287-6606, USA
P. Padgett	USDA Forest Service, Riverside, CA 92507, USA

K. Percy	Natural Resources Canada Forest Service, Fredericton, New Brunswick E38 5P7, Canada
D.L. Peterson	USGS, Biological Resources Division, Seattle, WA 98195, USA
M.A. Poth	USDA Forest Service, Riverside, CA 92507, USA
J. Pronos	USDA Forest Service, Sonora, CA 95370, USA
W.M. Pulliam	Natural Resource Ecology Laboratory, Colorado State University, Fort Collins, CO 80523, USA
J. Rechel	USDA Forest Service, Riverside, CA 92507, USA
G. Riechers	Agricultural Operations, University of California, Riverside, CA 92521, USA
J.O. Roads	Scripps Institute of Oceanography, La Jolla, CA 92093, USA
L.L. Sigal	112 Parma Road, Oak Ridge, TN 37830, USA
Patrick J. Temple	USDA Forest Service, Riverside, CA 92507, USA
John G. Watson	Desert Research Institute, University of Nevada, Reno, NV 89506, USA
P.L. Winter	USDA Forest Service, Riverside, CA 92507, USA
P. Wohlgemuth	USDA Forest Service, Riverside, CA 92507, USA

Introduction

P.R. Miller and J.R. McBride

Since the mid-1950s when native pines in the San Bernardino Mountains began to show unexplainable symptoms of decline, the area has been used as a natural laboratory to study the effects of oxidant air pollution on a mixed conifer forest ecosystem. Initial studies in 1963 confirmed that ozone, a major plant-damaging gas in photochemical oxidant air pollution, was responsible for the injury symptoms observed on pine (*Pinus ponderosa*) foliage and the decline in health of sensitive trees (Miller et al. 1963). Weakened trees were found to be vulnerable to being killed by bark beetles (Coleoptera) (Stark et al. 1968). The increasing number of weakened trees in the forests of the San Bernardino Mountains resulted in extensive sanitation salvage logging during the 1960s and 1970s (Hall 1958).

Beginning in 1974, an extensive multidisciplinary study was begun by University of California and Forest Service scientists to examine stand-level and ecosystem-level changes brought about by chronic ozone exposure (Taylor 1973; Miller et al. 1982). Publications resulting from this work and an international symposium held at the University of California, Riverside, in 1980 brought this area to prominence as an example of a forest ecosystem changed by long-term oxidant air pollution exposure (Miller 1980). This reputation is shared with the locality around Mexico City and some other regions worldwide that have a Mediterranean climate. A follow-up study of tree condition in permanent San Bernardino Mountain plots in 1989 (Miller et al. 1989), as part of a national research initiative on acidic pollution effects, indicated less damage attributed to ozone compared with initial measurements in 1974. During the period from 1991 to the present, another multidiscipinary study was completed under sponsorship of the California Air Resources Board (Miller et al. 1996a,b). This study was part of a larger program to examine the role of wet and dry deposited acidic pollutants on human health and natural ecosystems in California. In this case, the combined influences of acidic pollutants and ozone were emphasized.

The purpose of this book is to examine the body of research on oxidant and acidic air pollution effects on the mixed conifer forest ecosystem of the San Bernardino Mountains, beginning with the early 1950s and including the results of intensive work during the past 5 years. In its totality, this work provides much information about the response of a mixed conifer forest ecosystem to both natural and man-caused stresses. It also considers the sociological aspects of air pollution impacts on forest health and the impairment of visibility in relation to recreational use.

Chapters 1 and 2 explain background information concerning the geography and soils and the climate's relationship to regional air pollution. Chapter 3 describes the natural montane vegetation of the San Bernardino Mountains and discusses the historical changes in forest extent and composition, including the influence of fire.

Chapter 4 characterizes the chemistry of photochemical oxidant air pollution in relation to seasonal and local meteorology at Barton Flats, an intensively studied site in the San Bernardino Mountains. Chapter 5 examines the physics and chemistry of visibility impairment by the photochemical aerosol as determined since 1988 at an IMPROVE (*I*nteragency *M*onitoring of *PRO*tected *V*isual *E*nvironments) site, the San Gorgonio Wilderness Area.

Chapters 6–8 describe the physiological mechanism of ozone injury to sensitive pine species, the changes in crown injury due to ozone since 1974, and the dendroecological evidence for ozone and drought effects on stem growth. The effects of ozone pollution on understory plants are included in Chapter 9, which provides some results of a recent survey and evaluation. Chapter 10 considers the changes in extent and abundance of epiphytic lichens in the San Bernardino Mountains and several other mountain ranges in southern California, where the influence of oxidant air pollution exposure is present at both low and high levels.

Chapters 11–13 focus on the effects of nitrogen-containing pollutants. These chapters examine of the effect of the forest canopy on dry and wet deposition of gases and particles; Chapter 12 explores new hypotheses regarding the direct effects of nitric acid on foliage and the effects of nitrogen saturation on forest nutrient cycling. Because tree death is ultimately associated with insect and disease complexes, Chapter 14 describes the role of natural enemies in relation to pollution- and drought-stressed trees. Chapter 15 uses long-term changes in species frequency and size composition at permanent research plots as one approach for examining the nature of forest succession along a gradient of changing climate and pollution exposure. Chapter 16 appraises the application of simulation models for tree growth, stand succession, and nutrient cycling as tools for exploring questions about possible future conditions in forest stands of the San Bernardino Mountains.

Chapter 17 discusses recreational use in the San Bernardino Mountains and the possible relationships to regional air pollution. Chapter 18 discusses the various findings of preceding chapters in relation to other factors that present a risk to the sustainability of the forest ecosystem; Chapter 19 presents a summary of findings and recommendations for the future and considers the needs for additional research.

In summary, this book is the product of the collective knowledge of researchers who have worked together for up to 30 years in an effort to gain an improved understanding of the structure and function of a mixed conifer forest ecosystem and its responses to climate and long-term exposure to photochemical oxidant air pollution. Although the work is focused on the San Bernardino Mountain area of southern California, this synthesis of information will also apply to similar forest stands in the Sierra Nevada of central California where chronic oxidant pollutant injury to forests has reached levels similar to southern California.

References

Hall, R.C. Sanitation-salvage controls bark beetles in southern California recreation area. J. For. 56:9–11; 1958.

Miller, P.R., tech. coord. Proceedings of the symposium on the effects of air pollutants on Mediterranean and temperate forest ecosystems. Gen. Tech. Rep. PSW 43. Berkeley, CA: USDA Forest Service, Pacific Southwest Station; 1980.

Miller, P.R.; Parmeter, J.R.; Taylor, O.C.; Cardiff, E.A. Ozone injury to the foliage of *Pinus ponderosa*. Phytopathology 53:1072–1076; 1963.

Miller, P.R.; Taylor, O.C.; Wilhour, R.G. Oxidant air pollution effects on a western coniferous forest ecosystem. Research brief. EPA-600/D-82-276. Corvallis, OR: Environmental Protection Agency; 1982.

Miller, P.R.; McBride, J.R.; Schilling, S.L.; Gomez, A.P. Trend of ozone damage to conifer forests between 1974 and 1988 in the San Bernardino Mountains of southern California. In: Olson, R.K.; Lefohn, A.S., eds. Transactions Air and Waste Management Association symposium, Effects of air pollution on western forests. Pittsburgh, PA: Air and Waste Management Association; 1989:309–323.

Miller, P.; Chow, J.; Watson, J. Assessment of acidic deposition and ozone effects on conifer forests in the San Bernardino Mountains. Final report, Contract A032-180. NTIS PB97-115-612. Sacramento: California Environmental Protection Agency, Air Resources Board, Research Division; 1996a.

Miller P.; Poth, M.; Bytnerowicz, A.; Fenn, M.; Temple, P.; Chow, J.; Watson, J.; Frazier, C.; Green, M.; Johnson, D. Ecosystem level alterations in soil nutrient cycling: an integrated measure of cumulative effects of acidic deposition on a mixed conifer forest in southern California. Final report, Contract 92-335. NTIS PB97-106-223 Sacramento: California Environmental Protection Agency, Air Resources Board, Research Division; 1996b.

Stark, R.W.; Miller, P.R.; Cobb, F.W., Jr.; Wood, D.L.; Parmeter, J.R., Jr. Photochemical oxidant injury and bark beetle (*Coleoptera: Scolytidae*) infestation of ponderosa pine. I. Incidence of bark beetle infestation in injured trees. Hilgardia 39:121–126; 1968.

Taylor, O.C. Oxidant air pollutant effects on a western coniferous forest ecosystem. In: Task B report, Historical background and proposed systems study of the San Bernardino Mountain area. Riverside, CA: Air Pollution Research Center, University of California; 1973:A1–A36.

1. Description of Soils, Climate, Natural Resources, and Their Modification by Anthropogenic Influences

1. Geography, Geology, Geomorphology, and Forest Soils

M.A. Poth and P. Wohlgemuth

Introduction

This chapter describes the characteristics of the San Bernardino Mountains that form the substrate of the forests found there. The important geographical and geological features of the San Bernardino Mountains and the forces that have created them are summarized. The nature of forest soils produced as a result of the range of conditions within the San Bernardino Mountains and the current status and basis for assessing changes in the soils and their ability to supply water are discussed.

Geography

The San Bernardino Mountains are located on the western margin of the North American continent, centered on 34° 15′ north latitude and 117′ west longitude and northeast of Los Angeles in southern California (Fig. 1.1). Situated approximately 90 km from the Pacific Ocean, the San Bernardino Mountains form the apex of an arcuate series of mountain ranges across southern California that separate the coastal plains to the west and south from the interior deserts to the north and east (see Fig. 1.1). In plan view, the San Bernardino Mountains describe an east-west–trending wedge-shaped range, about 90 km long by 30 km wide and covering about 2,600 km^2. Elevations average 2,000 m across the San Bernardino

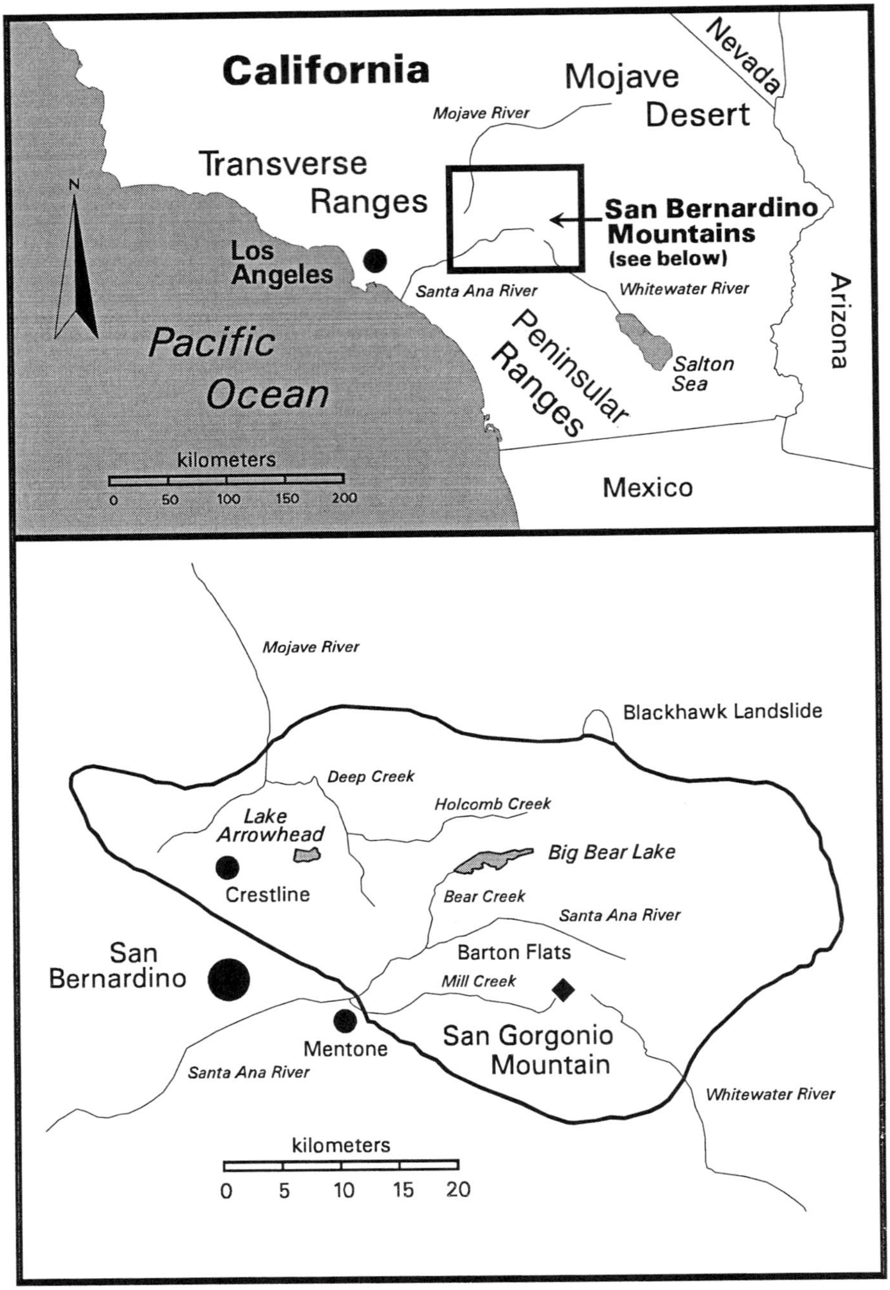

Figure 1.1. The position of the San Bernardino Mountains within southern California and detail of major watersheds and key place references for this chapter.

Mountains and culminate at 3,505 m, atop San Gorgonio Mountain, the highest point in southern California. Local relief varies from a gently rolling upland plateau to deeply dissected canyons and precipitous facets at the margins of the mountain block. The San Bernardino Mountains are relatively small in total area when compared with other mountain ranges in the western United States such as the Sierra Nevada or Cascade ranges. However, within a relatively small area the San Bernardino Mountains incorporate a large range of features.

Three major river systems head in the San Bernardino Mountains (Fig. 1.1). The Mojave River, including Deep Creek and Holcomb Creek, drains water from about 600 km^2 of the northwestern part of the range and flows north to become the master stream through the Mojave Desert. Flowing southeast from San Gorgonio Mountain, the Whitewater River drains about 105 km^2 of the San Bernardinos, sending water to the Coachella Valley and beyond to the Salton Sea. The Santa Ana River, including Bear Creek, has a drainage area that covers 550 km^2 of the central mountain block, flowing southwest and eventually reaching the Pacific Ocean. In addition, Mill Creek, which joins the Santa Ana River downstream in the city of Mentone, drains about 110 km^2 of the San Bernardinos off the southern flanks of San Gorgonio Mountain. Dams impound some of the runoff in the San Bernardino Mountains, creating two well-known recreational lakes: Lake Arrowhead and Big Bear Lake (Fig. 1.1).

Geology

The San Bernardino Mountains are composed of a complex assemblage of crystalline and sedimentary rocks ranging from ancient Precambrian basement terranes to modern Quaternary deposits. The regional chronology and landscape evolution primarily reflect the fault history of a tectonic plate boundary. Minor ore concentrations and carbonate deposits are of local economic importance (Bortugno and Spittler 1986).

Plate Tectonics

The lithospheric crust is believed to consist of several large rigid plates with brittle margins that form, move, and are destroyed by as yet poorly understood convection currents in the earth's interior (Condie 1989). In southern California, the regional geology is dominated by the San Andreas Fault zone, a northwest-southeast trending transform fault that forms the boundary between the North America and Pacific tectonic plates. Situated just north of the San Andreas Fault on the active plate margin, the San Bernardinos are a direct consequence of these plate interactions. Local compression and uplift caused by a realignment of the San Andreas Fault—associated with the opening of the Gulf of California some 4–5 million years ago (Ma) (Atwater 1970)—created the San Bernardino Mountains that we see today.

Rock Types

Lithologies in the San Bernardino Mountains are dominated by granitic plutons and remnants of the countryrock into which they were intruded. Isolated sections of nonmarine sedimentary rocks, small volcanic flows, and localized Quaternary deposits complete the geology (Fig. 1.2[1]).

The oldest rocks in the San Bernardino Mountains consist of Precambrian gneiss and late Precambrian through Paleozoic metasedimentary rocks—quartzite, marble, schist, and gneiss (Dibblee 1982; Matti and Morton 1993) and are located primarily along the northern and southern margins of the range and in a few interior islands. The northern and central exposures occur as roof pendants of the original host rocks within the younger granitics, whereas the southern outcrops are sections of exotic terranes sheared off at the tectonic plate margin and now juxtaposed along various strands of the San Andreas Fault zone (Dibblee 1982; Matti and Morton 1993).

The majority of the central San Bernardino Mountains is made up of plutonic igneous rocks of Mesozoic age (late Triassic through Cretaceous). These granitoid rocks, mostly quartz monzonite and granodiorite, are similar to and have affinities with those of the Mojave Desert to the north (Dibblee 1982; Matti and Morton 1993). For an overview of the mineralogy and petrography of these various granitic units, see Dibblee (1982).

The crystalline basement complex of the San Bernardino Mountains is overlain by small sections of late Cenozoic sedimentary rocks and volcanic flows. The sedimentary units accumulated in east-west–trending basins unrelated to the present topography and derived from nonlocal source material (Dibblee 1982; Sadler 1993). These nonmarine rocks are predominantly sandstone with some siltstone, conglomerate, and lacustrine clays. These units range in age from late Miocene to late Pliocene, based on diagnostic vertebrate fossils (Dibblee 1982; Trent 1990). Associated with several of the sandstone units are thin flows of vesicular olivine basalt. These basalts outcrop more extensively on the eastern edge of the San Bernardinos and have been radiometrically dated at 6.2 Ma (Sadler and Trent 1990).

Quaternary deposits include Pleistocene and modern alluvial fans, Pleistocene valley fill and modern stream alluvium, lake sediments, glacial till, and landslides. Together, these deposits cover only about 30% of the range. Alluvial fans occur along the mountain flanks as well as in several of the larger interior valleys, where fan material grades into valley fill and stream alluvium. Lake sediments cover the eastern portion of Big Bear Valley (Stout 1976), whereas glacial deposits occur on the northern slopes of San Gorgonio Mountain (Sharp et al. 1959). Landslides are pervasive throughout the San Bernardinos (Sadler and Morton 1989).

[1]Figure 1.2 will be found in the color insert following page 78.

Structure and Faulting

The San Bernardino Mountains have been broken into a number of structural blocks by tectonic activity, as the gently folded geological patterns of the pre-Cenozoic basement complex gave way to the sharply defined features of late-Cenozoic faulting. Several of these structural blocks are asymmetrically tilted to the north or east (Dibblee 1982). The range is completely bounded by faults, including the San Andreas Fault zone to the west and southwest, the Pinto Mountain Fault to the southeast, and a discontinuous fault system to the north (Dibblee 1982; Miller 1987). Four discrete strands of the San Andreas Fault have been identified, accounting for the sharply truncated rock units and the puzzling terrane juxtapositions on the southern edge of the San Bernardino Mountains (Matti and Morton 1993). In addition, several major faults have been mapped in the interior of the mountain range, and more are suggested by the topography and geomorphology. Most of these faults are highly active, as would be expected at an active tectonic plate margin and as witnessed by the 6.2-magnitude Big Bear earthquake on June 28, 1992 (Jones et al. 1992).

Regional Chronology

The San Bernardino Mountains were created 3 million years ago. Before this, the region was part of the Mojave Desert, an erosion surface of low relief (Dibblee 1982). The area to the south was continually changing as new terranes were constantly exposed by lateral slip along the San Andreas Fault. At one point in the Miocene, the San Gabriel Mountain block, now many kilometers to the west, was juxtaposed to the south of the incipient San Bernardino Mountains and contributed distinctive rock clasts to the Tertiary basins (Sadler 1993). A change to local source rocks in these sedimentary deposits at about 2.5 Ma marked the beginning of the elevation of the range (Sadler and Trent 1990), corresponding to compression along the restraining bend in the newly realigned San Andreas Fault (Atwater 1970). This resulted in differential uplift and structural dismemberment of the range throughout Quaternary time and indicates that the San Bernardinos are the youngest as well as the tallest mountains in southern California (Dibblee 1982; Trent 1990).

Mining and Mineral Resources

The mineral resource of major economic value in the San Bernardino Mountains is the carbonate rocks formed prior to the granitic intrusions. Limestone and marble from the north central flank of the range are mined for the manufacture of portland cement and were previously quarried as flux stone for a nearby steel mill (Dibblee 1982). In one area, an extremely pure grade of calcium carbonate is extracted for pharmaceutical products (Sadler and Trent 1990). Minor amounts of quartzite, schist, and sandstone have also been quarried for dimension stone (Dibblee 1982).

Gold production in the San Bernardinos was marginal compared with other parts of California, and abandoned mines still dot the landscape (Sadler and Trent 1990). Gold was discovered in placer gravels of Holcomb Valley in 1860 and was eventually traced to quartz veins associated with the granitic core of the San Bernardino Mountains. Although the placer deposits were soon exhausted, lode mining continued until the turn of the century (Dibblee 1982). Silver, lead, copper, and tungsten were also mined in small quantities (Dibblee 1982).

Geomorphology

The geomorphology of the San Bernardino Mountains is dominated by the fluvial dissection of an uplifted crustal block. Local relief varies from a gently rolling upland plateau to deep canyons and precipitous slopes at the margins of the mountain mass. The structural framework, reflecting the tectonic environment, has created a number of discrete geomorphic subunits across the range. Glaciation in the highest peaks and large landslides account for some of the distinctive landscape features in the San Bernardino Mountains.

Transverse Ranges

The San Bernardino Mountains are located at the east end of a series of elongated, parallel, east-west–trending mountain chains that make up the Transverse Ranges geomorphic province (Thornbury 1965). The Transverse Ranges are named for their anomalous orientation to the general northwest-southeast structural grain of the state of California. The San Bernardino Mountains are unique among the Transverse Ranges in that depositional flats and the remnants of a gentle, pre-orogenic erosion surface constitute a significant portion of the upland area (Sadler and Morton 1989). Although it is destined to be obliterated by fluvial downcutting, that this surface is preserved at all is testimony to the recent uplift and the youth of the range.

Geomorphic Subunits

The structural and topographical escarpments within the San Bernardino Mountains delineate five geomorphic subunits, excluding the front country around most of the perimeter of the range (Fig. 1.3): the Northern Plateau, the Eastern Ramp, the San Gorgonio Massif, Yucaipa Ridge, and the Santa Ana Basin (Sadler and Morton 1989).

The Northern Plateau consists of extensive remnants of the preorogenic erosion surface. Bounded by a thrust fault on the southern margin (Sadler 1993) and by steep reverse faults to the north (Miller 1987), the block has been elevated as a unit during the recent north-south regional episode of tectonic compression. The entire block is tilted sightly to the north (Dibblee 1982). The plateau surface is currently being deeply dissected by streams that were presumably inherited from

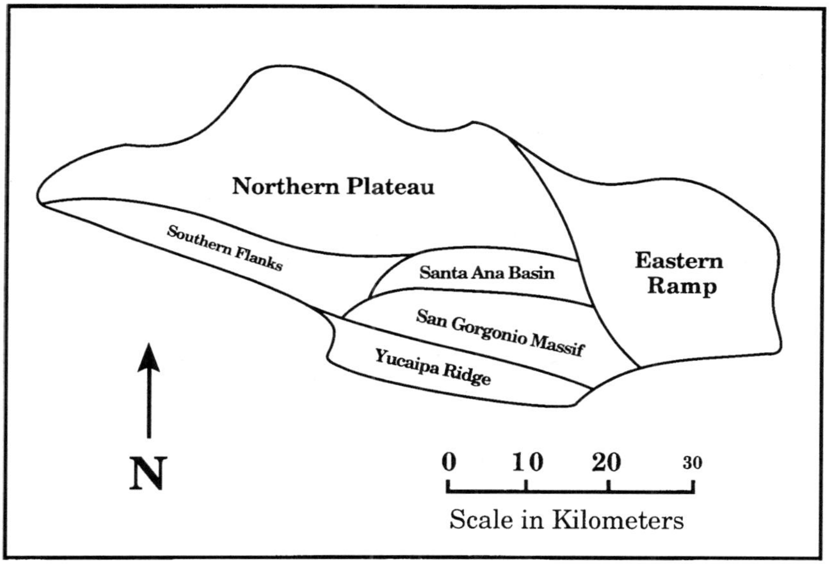

Figure 1.3. Geomorphic subunits within the San Bernardino Mountains.

the ancestral drainage network (Sadler and Morton 1989). The Northern Plateau supports most of the population and cultural activity within the San Bernardino Mountains.

The Eastern Ramp is similar to the Northern Plateau. It also contains remnants of the old erosion surface and is being dissected by a stream system inherited from the previous landscape (Sadler and Morton 1989). However, the structural block is tilted to the east, with the hinge line just to the east of Big Bear Valley, so that the land slopes gradually eastward to join the Mojave Desert (Dibblee 1982).

The San Gorgonio Massif consists of a structural knot in the south central portion of the San Bernardino Mountains that contains the highest elevations in the range. This northward-tilted block is bounded on the north by a normal fault with strong lateral slip components (Sadler 1993) and on the south by a strand of the San Andreas Fault zone (Matti and Morton 1993). The massif assumed its present elevation by the late Pleistocene and supported several small glaciers on its northern slopes. The large quantities of glacial and colluvial material from the San Gorgonio Massif continue to serve as a source of sediment for the downstream valleys (Sadler and Morton 1989).

Yucaipa Ridge is a composite geological feature located within the San Andreas Fault zone on the southern margins of the range. It has a synclinal structure, it has moved appreciably relative to the rest of the San Bernardino Mountains, and it has been uplifted between strands of the San Andreas Fault (Sadler and Morton 1989). Topography is steep and rugged on both sides of the ridge, yet the unit is traversed by the deep canyon of Mill Creek. Sadler and Morton (1989) speculate that an incised stream on the south side of the ridge breached the feature and captured the drainage behind it that formerly flowed west into the Santa Ana River before it debouched from the mountain front.

The Santa Ana Basin is an elongate intramontane basin located between the Northern Plateau and the San Gorgonio Massif in the central San Bernardino Mountains. Bounded by faults on both the north and south margins, this east-west–trending unit nearly coincides with one of the Tertiary sedimentary basins of the preorogenic landscape. However, the area has sustained considerable crustal shortening under the current compressional tectonic regime (Sadler 1993). The small steep drainages surrounding the basin have delivered significant quantities of alluvial, fluvial, and fluvioglacial sediments throughout the Quaternary. During the Pleistocene, this delivery of valley fill exceeded the uplift rate, creating an alluvial fan complex (Sadler and Morton 1989). The more prolific fluvioglacial debris from the San Gorgonio Massif displaced the Santa Ana River to the north. Subsequent climatic changes in the Holocene reduced the rate of sediment delivery to the basin, allowing uplift to become dominant and initiating stream channel incision (Sadler and Morton 1989).

Glaciation

Seven small glaciers once existed on San Gorgonio Mountain, five in the Santa Ana River drainage basin and two in the headwaters of the Whitewater River (Sharp et al. 1959). Care was necessary to distinguish glacial debris from mass wasting deposits. These glaciers all headed at elevations of 3,150–3,450 m, and none descended below 2,650 m. These small features varied in length from 0.8 to 2.7 km and ranged in area from 0.26 to 2.18 km^2 (Sharp et al. 1959). The steep valley gradients on San Gorgonio Mountain prevented the ice from attaining thicknesses of more than 100 m. Glacial landforms preserved in the San Bernardino Mountains include U-shaped valleys, sharp crested lateral moraines, nested end moraines that form natural lake basins, and cirques (Sadler and Trent 1990). Sharp et al. (1959) recognized two distinct episodes of glaciation but believed both were Wisconsin in age. Given the recent uplift history of the San Bernardinos, any prior glaciations are unlikely.

Landsliding

Landslides are nearly widespread across the San Bernardino Mountains and have significantly altered the local geomorphology. They are prevalent on both sides of Yucaipa Ridge and the San Gorgonio Massif, as well as along the margins of the Santa Ana Basin and the Northern Plateau, but are absent on the less precipitous slopes of the Eastern Ramp. Within the San Bernardino Mountains, there are several slide complexes that exceed 1 km^2 in area and many smaller ones (Sadler and Morton 1989). Most of these features probably date from the Pleistocene, when wetter climatic conditions induced mass movements. However, some of the landslide events are modern. For example, Sadler and Morton (1989) cite the case of Slide Lake, located on Bear Creek just upstream from its confluence with the

Santa Ana River; no lake is shown on a map based on an 1898 survey and no lake is present today, yet a small crescent-shaped lake can be seen on aerial photographs and topographical maps of the 1950s. Personal accounts of local residents suggest that the lake was formed in the 1930s by landslide debris off aptly named Slide Peak to the west, impounding the waters of Bear Creek. The lake was then breached and drained in the late 1970s (Sadler and Morton 1989).

Stout (1982) suggests that the Barton Flats area of the Santa Ana Basin is actually a single large composite landslide. Based on the morphology of the deposits and the deranged drainage of the Santa Ana River, this would have involved a slow-moving massive failure, as there are no lake features preserved upstream. More detailed fieldwork and analysis by Sadler and Morton (1989) indicate that the Barton Flats landform is probably a dissected alluvial fan.

The largest and most famous mass movement feature in the San Bernardino Mountains is the Blackhawk landslide. Located on the northern flank of the Northern Plateau and extending out into the Mojave Desert, the Blackhawk slide is 7.2 km long and covers an area of 14.2 km^2 (Shreve 1968). Triggered by a huge rockfall sometime in the Pleistocene, the deposit consists of more than 300 million m^3 of debris. Just how this massive feature was able to travel for such a long distance over relatively gentle slope gradients is a mystery. The most popular explanation is that the slide was able to generate enough momentum to launch itself over a low ridge and ride on a cushion of air trapped beneath the debris. Based on the geometry of the mountain front and the characteristics of the deposit, the velocity of the landslide must have been greater than 380 km/h (Shreve 1968).

Forest Soils

Soil-Forming Factors in the San Bernardino Mountains

Soils are the product of a complex web of interacting factors. The conceptual framework of soil-forming factors includes site geology, topography, climate, and vegetation (Jenny 1980). A key part of the concept of soil-forming factors is the element of time and the implicit nature of soils to change through time by the actions of the soil-forming factors. Many processes associated with soil development are slow, and so soil development is typically measured over a long time scale of hundreds to thousands of years. This means that soils are relatively stable and change slowly, unless there is a dramatic change in one of the soil-forming factors.

The atmospheric deposition of materials can be a part of a particular site or area climate. This part of the climate in southern California has undergone dramatic change in this century. Rainfall is naturally acidic because water in equilibrium with atmospheric carbon dioxide produces dissolved carbonic acid. Hence, rainfall in equilibrium with atmospheric carbon dioxide (at 340 ppm v/v) will have an acidic pH of 5.5. Reactions of geological parent materials with water and dissolved acids are important in the weathering of soil primary minerals into clays and other products. Even under pristine conditions, acid nitrogen compounds formed by lightning would be deposited with rain (Dise and Wright 1995). The amounts of material generated and deposited anthropogenically to forests in California are much higher than pristine conditions (Binkley et al. 1989; Bytnerowicz and Fenn 1995), and the highest deposition rates to forests are in the mountains that surround the greater Los Angeles Air Basin. If atmospheric deposition is defined as part of the climate, then there has been a dramatic change in one of the soil-forming factors for the San Bernardino Mountains.

The critical question for the future is whether these changes will pose a threat to ecosystem stability and important ecosystem outputs and benefits such as water, wildlife, and amenity values. Clues to the potential change in ecosystem stability and ecosystem outputs can be found in the current nature and status of forest soils and comparisons with past measurements. In one exceptional case, quantitative comparisons have been made between archived soil samples and large lysimeters in the San Gabriel Mountains (Patric 1961; Ulery et al. 1995). Unfortunately, soil descriptions traditionally have not included enough statistical detail to allow for quantitative determinations of change. This was due to the general consensus that soils would weather and change slowly, precluding some major physical disturbance leading to erosion. So, the available soil data in the literature for the San Bernardino Mountains are primarily descriptive rather that quantitative. Archived soil samples would also be valuable in the future, but there is not a current program to archive soil samples in a systematic way.

Available Soils Database for the San Bernardino Mountains

The first soils data for the San Bernardino Mountains were in reports filed in conjunction with operations conducted by the USDA Forest Service in managing federal lands within the San Bernardino Mountains begining in the 1920s. These reports were fragmentary and largely descriptive, as was required by the activity. For better long-term management of the forest resources in the San Bernardino Mountains, more comprehensive information was required, so in 1981 the San Bernardino Mountains soil survey and the associated soil names and descriptions were approved (USDA Forest Service et al. 1981). This third-order survey covered the entire San Bernardino Mountains. As a third-order survey, it relied heavily on aerial photography to define areas of particular soils series according to vegetation type coupled with field verification. This survey report included detailed chemistry and descriptions from individual soil pits that are representative of the particular soil series (Table 1.1).

The soil survey shows a picture of the range of soils present in the San Bernardino Mountains, reflecting the combinations of the soil-forming factors. Table 1.1 presents physical and chemical data for soils characteristic of major vegetation types found in the San Bernardino Mountains. The soils of the Osito series (Table 1.1) form on the steep slopes of 30–50% at lower elevations (550–1,300 m) with relatively little rain (380–635 mm), and the typical vegetation is chaparral. This combination of topography, climate, and vegetation soil-forming factors has re-

Table 1.1. Physical and Chemical Analyses of Selected Soils[a–c]

Soil series horizons	Depth (cm)	Clay (%)	Silt (%)	Sand (%)	C (g/kg)	N (g/kg)	Extractable cations ($cmol_c$/kg)				CEC	Base saturation (%)
							Ca	Mg	Na	K		
Brader												
A1	0–3	5.6	17.3	77.1	39.0	0.9	5.9	0.6	–	0.5	11.6	60
B2	3–16	4.9	21.1	74.0	11.5	0.4	4.1	0.5	–	0.4	7.2	69
C1	16–19	4.9	17.9	77.2	3.1	–	2.9	0.5	–	0.4	5.0	76
Hecker												
A1	0–6	9.2	29.3	61.5	18.2	1.05	8.8	1.7	–	0.7	10.6	100
B1t	6–20	9.8	30.4	59.8	4.8	0.3	4.4	1.1	–	0.4	6.9	86
B2t	20–37	13.8	29.8	56.4	2.1	–	6.4	1.9	–	0.4	9.9	88
B3	37–50	12.2	27.3	60.5	1.4	–	6.8	2.3	–	0.3	10.3	91
Modesto												
A1	o–5	11.7	28.6	59.7	16.3	1.06	9.9	1.7	–	0.5	12.0	100
B11	5–8	15.7	30.0	54.3	7.8	0.61	8.4	1.8	–	0.3	11.4	92
B12	8–15	16.4	33.3	50.3	4.6	0.41	9.1	2.2	–	0.2	12.4	93
B2t	15–28	19.0	30.3	50.7	2.4	–	12.8	4.2	0.4	0.1	13.5	100
B3	28–50	15.6	26.6	57.8	1.5	–	12.1	4.9	0.2	0.1	12.7	100
Osito												
A1	0–5	7.4	19.8	72.8	6.4	0.49	6.1	1.1	—	0.2	8.3	89
B2	5–13	10.2	15.5	74.3	1.8	0.17	10.5	1.9	—	0.1	11.7	100
Preston												
A1	0–4	3.4	14.8	81.8	33.0	1.04	7.9	0.5	—	0.4	11.7	75
C1	4–19	2.0	15.9	82.1	7.0	0.25	3.1	0.2	—	0.3	4.8	75
C2	19–28	3.2	15.2	81.6	3.2	—	2.4	0.3	—	0.3	3.7	81
Wind River												
A1	0–5	6.3	19.6	74.1	22.0	0.90	3.8	0.8	—	0.6	8.4	62
A3	5–19	8.6	18.8	72.6	7.2	0.43	3.8	0.4	—	0.4	6.3	73
B2	19–34	8.1	31.1	60.8	3.9	0.26	3.8	0.3	—	0.3	5.5	80
C1	34–45	9.3	20.88	69.9	2.3	—	3.6	0.5	—	0.2	5.2	82

[a]From USDA Forest Service et al. 1981.
[b]The characteristic vegetation of each soil series is Brader, pinyon-juniper; Hecker, mixed conifer and sugar pine; Modesto, canyon live oak; Osito, chaparral; Preston, coulter pine; Wind River, mixed conifer.
[c]Missing values are below detection limits.

sulted in a very shallow soil with little incorporation of C and N into the surface horizons. Adjacent to these soils, on rounded ridges with north-to-east aspects and at lower positions on the mountain slopes, are the Modesto series soils. The stability provided by topographical positions at the top or bottom of the mountainside, and the consequent presence of canyon live oak (*Quercus agrifolia*), has resulted in greater soil development in the Modesto series than the Osito series (Table 1.1). The Brader and Preston series support pinyon-coulter pine and coulter-Jeffrey pine, respectively. These somewhat excessively drained soils have low clay and silt contents in comparison with the Hecker and Wind River series. Mixed conifer stands, with a substantial amount of ponderosa pine and Jeffrey pine, are predominant on Hecker and Wind River soils. The Hecker and Wind River soil series form on gentle slopes (2–15 and 15–30%, respectively) at higher elevations (1,520–2,370 and 1,370–1,825 m, respectively) with a correspondingly higher annual rainfall of 435–890 mm.

Detailed soil analysis could not be performed at all survey sites because of limitations in the resources available to do the work. Consequently, the lack of a statistical foundation precludes the use of statistical methods to determine future changes. Nevertheless, the information is valuable in describing the general conditions found at the particular time.

In the 1970s, soils data were collected from a large number of mixed conifer sites as part of an extensive evaluation of the effects of ozone air pollution on the forests in the San Bernardino Mountains (Miller et al. 1977). Alhough forest conditions at these plots have been studied over time (Miller et al. 1977; Miller and Rechel, this volume), there has not been concomitant examination of the soils.

The study plots established in the 1970s extended from the more polluted western end of the San Bernardino Mountains near Crestline to the eastern end of the range beyond Barton Flats (Fig. 1.1). As part of the site characterization, soils data were collected that would aid in interpretation of the response of western yellow pine to ozone air pollution. Nitrogen is typically the most limiting nutrient for yellow pine in the western United States (Powers 1980), and water availability controls the growth of mixed conifer forests in the San Bernardino Mountains (Arkley 1981), as it does elsewhere. To evaluate soil N status and because the amount of soil organic carbon greatly influences water-holding capacity, total C and N determinations were made for all plots from the top 25 cm of the surface mineral soil horizon (Table 1.2). The table lists the sites in three groups, covering a wide range of soil conditions. The first group of sites are the western San Bernardino Mountains, the second group are the central, and the last group are the eastern sites. Soil C:N ratios are generally in the 20s, but the highest and lowest rations are both in the more polluted western end of the San Bernardino Mountains (Dogwood 3 and Camp O-Ongo, respectively; Table 1.2). Soil bulk density data from the western and central San Bernardino Mountain sites show a wide range as well (Table 1.3). The data in these two tables can help define the range of a particular characteristic but are not suited for evaluating future change because they are statistically incomplete. (What were the standard deviations of the determinations reported in Tables 1.1 and 1.2? The sample size?) Without knowing

Table 1.2. Organic Carbon (C) and Nitrogen (N) in the Surface Soils to a Depth of 25 cm[a]

Plot	Site	C (g/kg)	N (g/kg)	C/N
Dogwood	1	25.6	0.94	27.3
Dogwood	2	18.8	0.69	27.2
Dogwood	3	34.6	1.05	33.0
NE 13		27.0	0.97	27.9
S 22		17.0	0.71	23.9
Sky Forest R.S.		20.3	0.76	26.8
U.C. Conference Center		11.0	0.40	27.5
Camp Paivika		17.8	0.81	22.0
Breezy Point		20.8	1.02	20.4
Tunnel-2		16.3	0.51	32.0
Camp O-Ongo		17.2	1.22	14.1
Green Valley Creek		9.3	0.36	25.8
N.E. Green Valley		14.7	0.61	24.1
Snow Valley		8.3	0.41	20.4
Holcomb Valley		11.4	0.54	21.1
Bluff Lake		34.2	1.18	29.0
Sand Canyon	1	6.5	0.31	21.0
Sand Canyon	2	17.8	0.84	21.1
Schneider Creek		21.5	0.92	23.5
Barton Flat		29.9	1.19	25.1
Camp Osceola		11.4	0.43	26.8
Heart Bar		8.7	0.34	25.6

[a]From Miller et al. 1977.

this, it will be impossible to test for differences in the future when we may want to determine if small changes are statistically significant. These types of monitoring measurements would be an important adjunct to pedogenesis studies of climate effects that have identified the types and rates of soil change that might be expected (Dahlgren et al. 1997).

Detailed Chemistry and Physical Characteristics of Forest Soil at Barton Flats

To characterize variability and provide statistical measures of soil properties for the soils at Barton Flats, a series of 15 soil pits were dug, five adjacent to each of three primary long-term plot locations in the Barton Flats area (plot 1: 34° 10 9.5″ N, 116° 52′ 55.1″ W, 1,951-m elevation; plot 2: 34° 9′ 32.4″ N, 116° 52′ 52.8″ W, 2,142-m elevation; plot 3: 34° 9′ 22.5″ N, 116° 52′ 13.0″ W, 2,173-m elevation). Soil survey staff (Staff 1975) taxonomic procedures were followed. Specific soil pit locations and forest composition descriptions are available (Miller et al. 1996; Poth and Glaubig 1996).

At each pit, the average depth of visible soil horizons was measured. These typically included surface litter (the Oi horizon), decomposing organic materials

Table 1.3. Bulk Density (D_b) of Main Mineral Soil Horizons in Major Study Plots of Miller et al. 1977

		A Horizons		B Horizons		C Horizons	
Plot[a]	Site	Depth (cm)	D_b (mg/m^3)	Depth (cm)	D_b (mg/m^3)	Depth (cm)	D_b (mg/m^3)
DW	1	0–46	1.09	41–157	1.46	—	—
DW	2	0–43	1.13	43–147	1.64	147–267	1.57
DW	3	0–51	1.13	51–157	1.94	—	—
NE 13		0–50	1.07	50–189	1.61	189–272	1.53
S 22		0–20	1.20	20–145	1.53	145–290	1.70
SF		0–61	1.05	61–160	1.53	160–277	1.47
UCC		0–25	1.40	25–152	1.69	152–250	1.68
CP		0–27	1.12	27–80	1.65	—	—
BP		0–29	0.74	29–147	1.17	147–284	1.57
TUNE		0–28	1.18	28–90	1.67	—	—
COO		0–51	1.15	51–149	1.68	—	—
GVC		0–29	1.11	29–156	1.85	150–246	2.10
NEGV		0–22	1.31	22–150	2.15	—	—
SV		0–14	1.24	14–55	1.56	—	—
HV		0–13	1.46	13–117	1.77	—	—
BL		0–55	1.15	55–147	1.67	147–255	1.74
SC	1	0–25	1.30	25–110	1.80	—	—
SCR		0–20	1.19	20–145	1.75	145–234	1.73

[a]See Chapter 7 for plot locations.

(Oe), and the mineral horizons (A and B). For the mineral horizons, bulk density and percentage stones and gravel greater than 2 mm of subsamples were measured. Particle size distribution on the greater than 2-mm fraction was determined by the hydrometer method (Day 1965). Soil pH was determined with a 1:1 mixture of soil with 0.01 M $CaCl_2$. The Munsell system was used to define soil color. Soil samples from each mineral horizon in each pit were taken and were extracted with $SrCl_2$ in the laboratory. Extracts were analyzed for Ca and Mg with atomic adsorption spectroscopy and K and Na by emission. Cation exchange capacity (CEC) was also measured (Poth and Glaubig 1996). Base saturation was calculated as the sum of the extractable bases divided by the CEC times 100%. Total C and N were determined for each horizon by coupled combustion and gas chromatographical analysis by using a Carlo Erba 1500 system.

The results from the 15 soil pits at Barton Flats are summarized in Table 1.4. In general, these soils are well drained with coarse texture, because the soils are in a mountain range that is actively uplifting and producing the alluvial deposits that are the basis for soil formation at Barton Flats. These soils are relatively young as evidenced by the absence of clay movement, with no argillic horizons. Mineralogy appears mixed with some evidence of 2:1 layer clay formation (Bailey and Jahns 1954). This is direct evidence of the active weathering of primary minerals in the parent material at this site.

Table 1.4. Summary Soil Horizon Characteristics[a]

	Horizon			
	Oi	Ie	A	B
Depth	2.13 (0.64, 15)	3.00 (2.11, 14)	12.5 (9.61, 15)	33.1 (10.1, 15)
Bulk density (mg/m^3)	N/A[b]	N/A	1.23 (0.24, 15)	1.40 (0.19, 14)
>2 mm (%)	N/A	N/A	23.8 (5.14, 15)	25.5 (6.11, 14)
Sand (%)	N/A	N/A	68.1 (7.49, 15)	67.9 (5.02, 15)
Silt (%)	N/A	N/A	23.8 (5.15, 15)	24.2 (3.23, 15)
Clay (%)	N/A	N/A	8.09 (2.73, 15)	7.95 (2.14, 15)
pH	N/A	N/A	5.66 (0.37, 15)	5.40 (0.37, 15)
CEC (cmol$_c$/kg)	N/A	N/A	13.4 (5.51, 15)	7.25 (2.16, 15)
Extractable Ca (cmol$_c$/kg)	N/A	N/A	12.4 (5.74, 15)	5.94 (2.07, 15)
Extractable Mg (cmol$_c$/kg)	N/A	N/A	1.68 (0.62, 15)	1.20 (0.42, 15)
Extractable K (cmol$_c$/kg)	N/A	N/A	0.69 (0.23, 15)	0.43 (0.14, 15)
Extractable Na (cmol$_c$/kg)	N/A	N/A	0.02 (0.01, 15)	0.03 (0.01, 15)
Base saturation (%)	N/A	N/A	109 (8.41, 15)	104 (10.8, 15)
Total C (g/kg)	478 (37.6, 15)	359 61.3, 14)	27.8 (13.9, 15)	7.8 (3.9, 15)
Total N (g/kg)	11.3 (2.1, 15)	1.3 (3.3, 14)	1.2 (0.6, 15)	0.3 (0.1, 15)

[a]Values presented are the averages (standard deviation, number of soil pits) of the determinations from soil pits at Barton Flats in the San Bernardino Mountains.
[b]Not analyzed.

Soil base saturations are high in the mineral soil, where they are always near 100%. This again reflects the young nature of the these soils and the active weathering of primary soil parent materials to release base cations. The high base saturation of the soil exchange complex is offset by the relatively low CEC of the soils, which is a result of low organic matter (low total C) and clay contents. These soils are coarse mixed Typic Xerochrepts.

The soils at Barton Flats exhibit a considerable amount of spatial heterogeneity. The variability is a result of the heterogeneous nature of the vegetation distribution and the small-scale differences in topography and deposited alluvial parent materials. The variability observed is a key property of the forest soils in the area studied and is typical of many forested areas in the San Bernardino Mountains.

To characterize the spatial variability of the soils at Barton Flats, composite soil samples were collected from near each pine tree in all three plots. Samples were taken in 1991, with a 2.5-cm-diameter soil corer, from under midcanopy at the cardinal points around each mature yellow pine under study within the plot and composited by soil horizon. A horizon (a total of 150 individual samples) was analyzed for extractable bases, as described previously.

Exchangeable soil Ca and Mg are not normally distributed (Figure 1.4), but exchangeable K is normally distributed, as evidenced by the relatively straight line the data form on a normal probability plot. The lack of normality in soil characteristics must be considered if change is to be detected by future measurements. Subsequent sampling, measurements, and statistical interpretations of soil chemical characteristics should be designed with this in mind. Lognormal distributions are common in soils (Parkin et al. 1988). Sufficient numbers of samples must be made to allow an assessment of soil conditions because lognormally distributed data are characterized by small numbers of very high values. As a result, the arithmetic mean (dominated by the high values) may be much higher that the median. For example, the mean and median soil exchangeable Mg are 1.80 and 1.56 mEq/100 g, respectively, a difference of 15%.

Forest Soil Physical Properties and Water Availability

The growth of the dominant forest tree species is limited by water availability each year. Water availability is a function of the amount and timing of precipitation and the ability of soils, and associated sub strata, to store water in such a way that it can be used by forest trees and plants. Surface soils are able to store and supply only a limited amount of water in the San Bernardino Mountains. This is a result of the combined effects of the relatively coarse texture and shallow depth of the developed soil and the Mediterranean climate in the San Bernardino Mountains. The Mediterranean climate produces a long summer dry season just when the demand for water by trees is high, producing a situation in which the water demands of the trees cannot be met by surface forest soils (Arkley 1981; Yaalon 1997).

Recent studies in western forests are demonstrating that weathered bedrock may store significant amounts of water that would be available to trees (Jones and

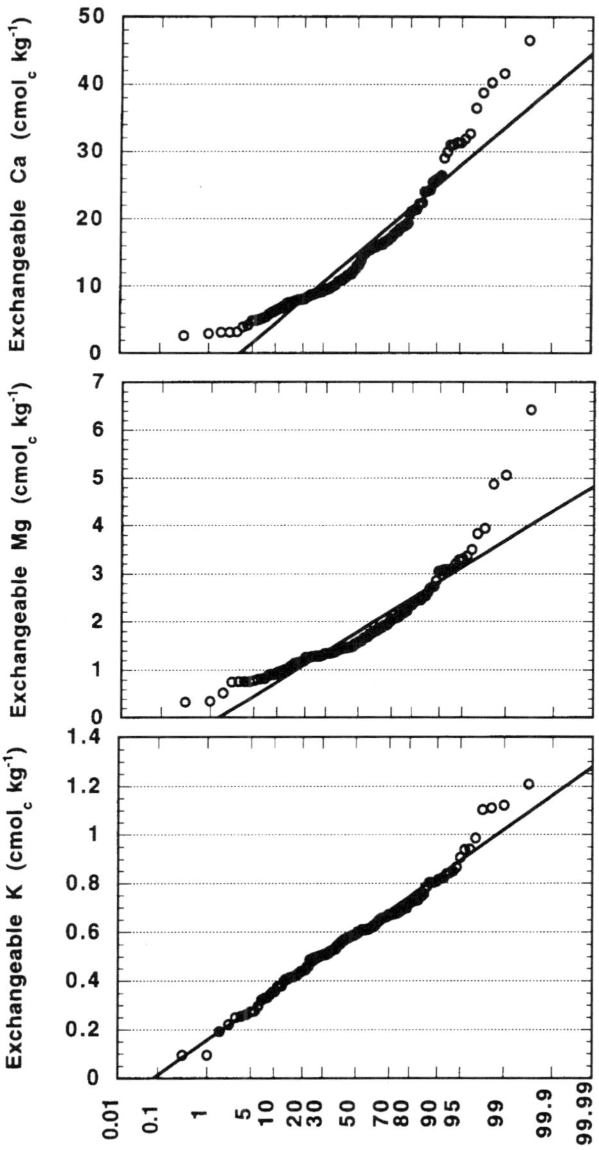

Figure 1.4. Normal probability plots of exchangeable bases from the A horizon soils under pines at Barton Flats in the San Bernardino Mountains ($N = 150$). Normally distributed data will form a straight line.

Graham 1993; Zweiniecki and Newton 1994). Weathered bedrock is the material that lies between the developed surface soil and the underlying unaltered bedrock. This material is typically fractured as the result of ongoing geological and tectonic processes. These fractures form natural channels for water movement and so are associated with weathering. The weathering processes lead to the formation of various clays depending on the nature of the bedrock involved. Clay formation and continued weathering greatly alter the water-holding capacity of the weathered bedrock from its original state. In water-limited ecosystems, plants have adapted to take advantage of this water source. Tree roots tend to follow fractures in weathered bedrock in the search for water. Root morphology is plastic enough to take advantage of the water availability within weathered bedrock. Roots become very thin and flattened in exploring the crevices within the weathered bedrock (Zweiniecki and Newton 1995). The quantity of water available is a function of the degree of weathering of the bedrock, with the most weathered materials storing the most water (Jones and Graham 1993). The total amount of water stored in the weathered bedrock is greater than the soil storage capacity in many forested areas (Jones and Graham 1993; Anderson et al. 1995).

There is evidence that trees must reach beyond the pole stage of growth to be mature enough to be able to tap this source of water. Ponderosa pine in western Oregon had not tapped this source after 14 years (trees 4–8 m tall) (Zweiniecki and Newton 1996). However, in the Sierra Nevada near Sequoia National Park mature Jeffrey pine were apparently using this resource (Anderson et al. 1995), as were ponderosa pine in the San Bernardino Mountains (Arkley 1981).

Summary

The San Bernardino Mountains, located in southern California between the deserts inland and the coastal plains to the southwest, cover about 2,600 km^2, lie adjacent to the San Andreas Fault, and were created by local compression and uplift caused by the realignment of the fault associated with the opening of the Gulf of California 4–5 million years ago. The San Bernardino Mountains support conifer forests and a range of other ecosystems on the edge of the greater Los Angeles metropolitan area.

Combinations of geology, topography, climate, and vegetation have produced a wide range of soils in the San Bernardino Mountains. These include relatively shallow soils at lower elevations associated with chaparral to productive soils associated with the ponderosa pine– and Jeffrey pine–dominated mixed conifer forests at higher elevations. The soils associated with the coniferous forests are representative of the well-drained soils associated with these types of forest throughout much of the western United States.

There are soil descriptions for the San Bernardino Mountains in the literature but little quantitative data with a statistical foundation. Future data should be collected at specifically identified sites (preferably with a geographical positioning system) with enough replication of sampling to provide good statistically based characterizations of soils. Without adopting these types of measures, it will not be possible to quantitatively monitor and detect future changes at specific sites and relate these to forest condition and growth. The development of a program to archive soils samples for future reference would also be valuable.

The Mediterranean climate of the region, with long dry summers and wet winters, make water a limiting factor for forest growth. The weathered bedrock of the soil sub stratum is apparently able to hold enough water to support mixed conifer forests in the San Bernardino Mountains.

References

Anderson, M.; Grahan, R.; Alyanakian, G.; Martynn, D. Late summer water status of soil and weathered bedrock in a giant sequoia grove. Soil Sci. 160:415–422; 1995.

Arkley, R.J. Soil moisture use by a mixed conifer forest in a summer-dry climate. Soil Sci. Soc. Am. J. 45:423–427; 1981.

Atwater, T. Implications of plate tectonics for the Cenozoic tectonic evolution of western North America. Geol. Soc. Am. Bull. 81:3513–3536; 1970.

Bailey, T.L.; Jahns, R.H. Geology of the Transverse Range province, southern California. In: Jahns, R.H., ed. Geology of southern California. Sacramento: California Division of Mines; 1954:83–106.

Binkley, D.; Driscoll, C.T.; Allen, H.L.; Schoeneberger, P.; McAvoy, D. Acid deposition and forest soils: context and case studies from the southeastern United States. Ecological studies 72. New York: Springer-Verlag; 1989.

Bortugno, E.J.; Spittler, T.E., compilers. Geologic map of the San Bernardino quadrangle, California. Regional Geologic Map Series, Map 3-A. California Division of Mines and Geology; Sacramento, CA; 1986.

Bytnerowicz, A.; Fenn, M.E. Nitrogen deposition in California forests: a review. Environ. Pollut. 92:127–146; 1995.

Condie, K.C. Plate tectonics and crustal evolution. Oxford: Pergamon; 1989.

Dahlgren, R.A.; Boettinger, J.L.; Huntington, G.L.; Amundson, R.G. Soil development along an elevational transect in the western Sierra Nevada, California. Geoderma 78:207–236; 1997.

Day, P.R. Particle fractionation and particle-size analysis. In: Black, C.A. et al., eds. Methods of soil analysis, Part I. Madison, WI: Soil Science Society of America; 1965:545–566.

Dibblee, T.W., Jr. Geology of the San Bernardino Mountains, southern California. In: Fife, D.L.; Minch, J.A., eds. Geology and mineral wealth of the California Transverse Ranges. Symposium and guidebook 10. Santa Ana, CA: South Coast Geology Society; 1982:149–169.

Dise, N.B.; Wright, R.F. Nitrogen leaching from European forests in relation to nitrogen deposition. For. Ecol. Manage. 71:153–161; 1995.

Jenny, H. The soil resource. Ecological studies. New York: Springer Verlag; 1980.

Jones, D.J.; Graham, R.C. Water-holding characteristics of weathered granitic rock in chaparral and forest ecosystems. Soil Sci. Soc. Am. J. 57:256–261; 1993.

Jones, L.; Mori, J.; Hauksson, E. The Landers earthquake: preliminary instrumental results. Earthquakes Volcanoes. 23:20–208; 1992.

Matti, J.C.; Morton, D.M. Paleogeographic evolution of the San Andreas Fault in southern California: a reconstruction based on a new cross-fault correlation. In: Powell, R.E.; Weldon, R.J., II; Matti, J.C., eds. The San Andreas Fault system: displacement, palinspastic reconstruction, and geologic evolution. Memoir 178. Boulder, CO: Geological Society of America; 1993:107–159.

Miller, F.K. Reverse-fault system bounding the north side of the San Bernardino Mountains. In: Morton, D.M.; Yerkes, R.F., eds. Recent reverse faulting in the Transverse Ranges, California. Prof. paper 1339. Washington, DC: U.S. Geological Survey; 1987:83–95.

Miller, P.R.; Kickert, R.N.; Taylor, O.C.; Arkley, R.J.; Cobb, F.W., Jr.; Dahlsten, D.L.; Gersper, P.J.; Luck, R.F.; McBride, J.R.; Parmeter, J.R., Jr.; Wenz, J.M.; White, M.; Wilcox, W.W. Photochemical oxidant air pollution effects on a mixed conifer forest ecosystem—a progress report. EPA-600/3-77-104. Corvallis, OR: U.S. Environmental Protection Agency; 1977.

Miller, P.R.; Chow, J.; Watson, J. Assessment of acidic deposition and ozone effects on conifer forests in the San Bernardino Mountains. Final report, Contract A032-180. NTIS PB97-115-612. Sacramento, CA: California EPA, Air Resources Board, Research Division; 1996.

Parkin, T.B.; Meisinger, J.J.; Chester, S.T.; Starr, J.L.; Robinson, J.A. Evaluation of statistical methods for lognormally distributed variables. Soil Sci. Soc. Am. J. 52:323–329; 1988.

Patric, J.H. The San Dimas large lysimeters. J. Soil Water Conserv. 16:13–17; 1961.

Poth, M.; Glaubig, R. Baseline soil morphology and chemistry at Barton Flats. In: Miller, P.R.; Chow, J.; Watson, J.G., eds. Assessment of acidic deposition and ozone effects on the conifer forests in the San Bernardino Mountains. Final report, Contract A032-180. Sacramento, CA: California Environmental Protection Agency, Air Resources Board; 1996:9-1–9-7.

Powers, R.F. Mineralizable soil nitrogen as an index of nitrogen availability to forest trees. Soil Sci. Soc. Am. J. 44:1314–1320; 1980.

Sadler, P.M. The Santa Ana basin of the central San Bernardino Mountains: evidence of the timing of uplift and strike slip relative to the San Gabriel Mountains. In: Powell, R.E.; Weldon, R.J., II; Matti, J.C., eds. The San Andreas Fault system: displacement, palinspastic reconstruction, and geologic evolution. Memoir 178. Boulder, CO: Geological Society of America; 307–321; 1993.

Sadler, P.M.; Morton, D.M. Landslides of the uppermost Santa Ana River basin and the adjacent San Bernardino Mountains of southern California. In: Sadler, P.M.; Morton, D.M., eds. Landslides in a semi-arid Environment with emphasis on the inland valleys of Southern California. Pub. Inland Geol. Soc. 1989:2:356–386.

Sadler, P.M.; Trent, D.D. Geologic transect of the central San Bernardino Mountains, southern California. In: Trent, D.D., coord. Geology of the central San Bernardino Mountains, southern California guidebook. Bellingham, WA: National Association of Geology Teachers Far West Section Spring Meeting; 1990:36–68.

Sharp, R.P.; Allen, C.R.; Meier, M.F. Pleistocene glaciers on southern California mountains. Am. J. Sci. 257:81–94; 1959.

Shreve, R.L. The Blackhawk landslide. Spec. paper 108. Boulder, CO: Geological Society of America; 1968.

Staff, Soil Survey. Soil taxonomy. Agriculture handbook 436. Washington, DC: USDA Soil Conservation Service; 1975.

Stout, M.L. Pleistocene and Holocene geology of Big Bear Valley, San Bernardino Mountains, California. In: Stout, M.L., ed. Geologic guide to the San Bernardino Mountains, southern California. A field trip. Los Angeles: Association of Engineering Geologists, South California Section; 70–72; 1976.

Stout, M.L. Barton Flats landslide, San Bernardino Mountains, southern California. In: Fife, D.L.; Minch, J.A., eds. Geology and mineral wealth of the California Transverse Ranges. Symposium and guidebook 10. Santa Ana, CA: South Coast Geological Society; 1982:634–636.

Thornbury, W.D. Regional geomorphology of the United States. New York: John Wiley; 1965.

Trent, D.D. An overview of the geology of the San Bernardino Mountains, southern

California. In: Trent, D.D. coordinator. Geology of the central San Bernardino Mountains, southern California guidebook. Bellingham, WA: National Association of Geology Teachers Far West Section Spring Meeting; 1–7; 1990.

Ulery, A.L.; Graham, R.C.; Chadwick, O.A.; Wood, H.B. Decade-scale changes of soil carbon, nitrogen and exchangable cations under chaparral and pine. Geoderma 65:121–134; 1995.

U.S. Department of Agriculture Forest Service, Soil Conservation Service, Regents of the University of California. Soil survey of San Bernardino National Forest area, California. National Cooperative Soil Survey; Berkeley, CA: Regents of the University of California; 1981.

Yaalon, D.H. Soils of the Mediterranean region: what makes them different? Catena 28:157–169; 1997.

Zwieniecki, M.; Newton, M. Root distribution of 12-year-old forests at rocky sites in southwestern Oregon: effects of rock physical properties. Can. J. For. Res. 24:1791–1796; 1994.

Zwieniecki, M.; Newton, M. Roots growing in rock fissures: their morphological adaptation. Plant Soil 172:181–187; 1995.

Zwieniecki, M.; Newton, M. Seasonal pattern of water depletion from soil-rock profiles in a Mediterranean climate of southwestern Oregon. Can. J. For. Res. 26:1346–1352; 1996.

2. Climatology

F.M. Fujioka, J.O. Roads, and S.-C. Chen

Introduction

In this chapter, we discuss the climatology of the South Coast Air Basin, with an emphasis on factors that affect air pollution. The relevant atmospheric processes span the spectrum from global to local scales, as typified by the subtropical anticyclone belt and the sea breeze, respectively. We also describe the El Niño phenomenon, which was developing as one of the strongest on record in summer 1997 and which has had profound but unpredictable consequences for the region.

The California coast ranges, like the California coastline, trend roughly north-northwest to south-southeast, except in the southern end of the state, where the Pacific Plate collides with the North American Plate. There, mountain building favors an east-west orientation, as evidenced by the Transverse Range. The Transverse Range defines a natural northern boundary for the Los Angeles Basin, particularly at the San Gabriel and San Bernardino mountains. This chapter is about the climatology of primarily the San Bernardino Mountains on the eastern end of the Transverse Range and, to a lesser extent, the South Coast Air Basin, which the California Air Resources Board identifies as the area encompassing those portions of Los Angeles and San Bernardino counties south of the San Gabriel and San Bernardino mountains, western Riverside County, and Orange County.

Early in the century, most authors discussed the healthful benefits of southern California climate, which was likened to the Mediterranean climates of Europe.

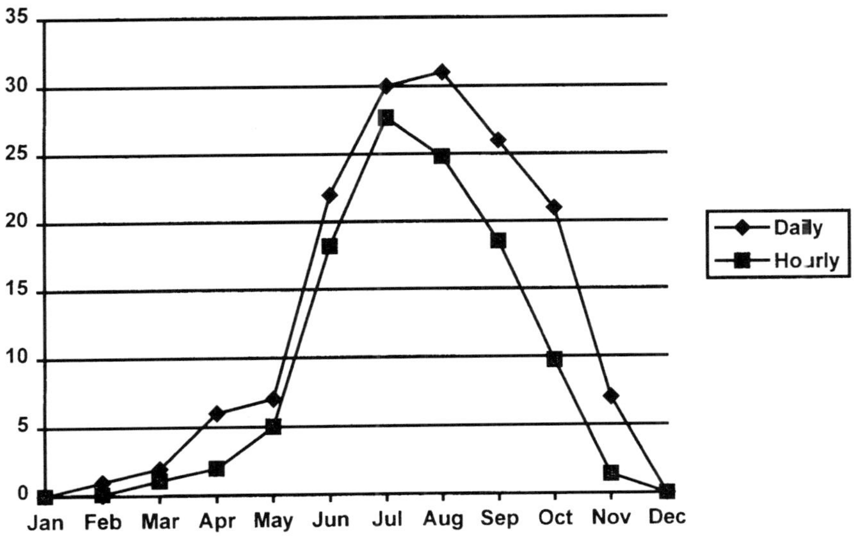

Figure 2.1. Number of ozone standards exceedances by month for 1995, South Coast Air Basin.

There are many similarities to the climate of the Mediterranean, where most of the rain occurs during the winter and the summers are warm and dry. During winter, storm systems from the extratropical eastern Pacific Ocean push into northern California, sometimes dipping as far south as Mexico. During summer, the semi-permanent Pacific subtropical anticyclone moves northward, displacing the storm belt within the midlatitude westerlies. Summer is also the season of the southerly monsoon, when thunderstorms form from the heat of the Southwest deserts and moisture from the Gulf of California or the Gulf of Mexico. Occasionally, the monsoon moisture extends westward into southern California, resulting in isolated thunderstorms, particularly over mountain areas.

As the population of southern California grew, air pollution problems emerged, adding an unwelcome characteristic to its climate. In the summer, the sun, a persistent temperature inversion, and emissions from automobiles and other sources elevate air pollution levels in the South Coast Air Basin. In 1995, the total number of days that ozone concentration in the basin exceeded the national standard was 153. Figure 2.1 shows the number of daily and hourly adverse pollution episodes (divided by 10) by month in 1995. Most of those in 1995 occurred in June through October: 130 days, compared with zero in January and December, and one in February (California EPA 1996).

The National Air Quality and Emissions Trends Report 1995 (EPA 1996) indicates two noteworthy ozone trends from one end of the South Coast Air Basin to the other (Fig. 2.2). The report summarized the annual number of days that the Pollution Standards Index values exceeded 100 for the Los Angeles–Long Beach

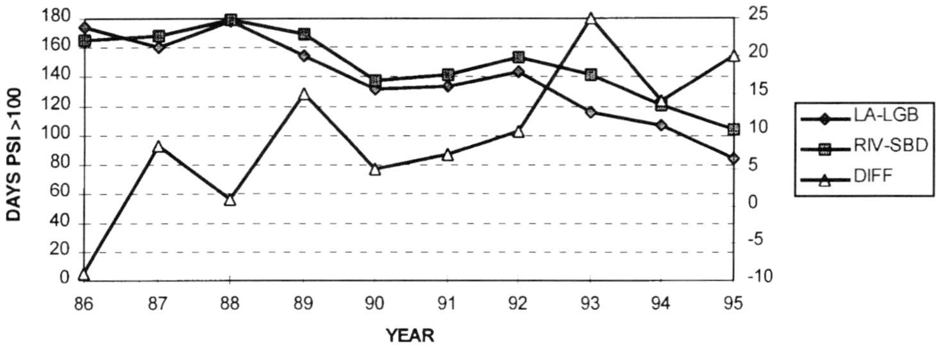

Figure 2.2. Number of days Pollution Standards Index (PSI) exceeded 100 in the western (LA-LGB) and eastern (RIV-SBD) parts of the Los Angeles Basin. The difference between the regions (western-eastern) has increased over time (difference axis on the right).

and Riverside–San Bernardino metropolitan statistical areas for the period 1986–1995. First, the annual number of adverse pollution episodes in both areas has declined over the 10-year period. Second, the difference in the number of days between the west and east sides has increased recently, because the decline in the Los Angeles–Long Beach area has occurred faster than in Riverside–San Bernardino.

Local Climatic Factors

Trewartha (1954) lists eight major factors that control climate: "(a) latitude or sun, (b) distribution of land and water, (c) the great semipermanent high- and low-pressure cells, (d) winds, (e) altitude, (f) mountain barriers, (g) ocean currents, [and] (h) storms of various kinds." These are clearly not independent of one another, as the sun's differential energy distribution over the earth gives rise to and controls the movement of the semipermanent pressure cells, which influence winds, which influence ocean currents and storms. In southern California, the regional topography, the proximity of the Pacific Ocean, and the semipermanent eastern Pacific high-pressure cell are dominant climatic factors. We consider briefly the effects of each.

Topographical Effect

The give and take between geological and climatic forces in southern California have sculpted a landscape marked by topographical complexity, which in turn affects climatic variations across the region. In the warm months, the Transverse and Peninsular ranges are barriers to the cool winds flowing inland from the

Pacific Ocean. The Peninsular Ranges run mainly northwest to southeast and include the San Jacinto Mountains at the north end and the Santa Rosa and Palomar mountains to the southeast and south; together, they separate the South Coast Air Basin from the Salton Trough. The Santa Ana Mountains to the west are also part of the Peninsular Ranges; they block the direct flow of sea air to the Perris-Hemet Plain. However, the sea breeze flows easier toward Riverside and San Bernardino following the Santa Ana River. Bailey (1966) cited the 915 m (3,000 ft) topographical contour as the normal upper limit of sea breeze penetration in southern California. Much of the Transverse Range lies above that limit, including the Cajon Pass, which dips to approximately 1,300 m at the Interstate 15 summit point but otherwise is a significant break connecting the South Coast Air Basin and the Mojave Desert. East of Riverside and San Bernardino, the Transverse Range rises to its highest point at the top of San Gorgonio Mountain, 3,501 m above mean sea level. Approximately 34 km to the south-southeast, the highest point of the Peninsular Range is found at the summit of Mt. San Jacinto, 3,293 m above mean sea level. Between them, the San Gorgonio Pass dips to an elevation of 797 m, allowing the westerly flow of surface winds from the direction of Riverside and the Pacific Ocean.

Marine Effect

The Pacific Ocean moderates temperature in the coastal regions of California, particularly in the warm season. Bailey (1966) constructed a curve showing mean maximum July temperatures of locations in southern California as a function of distance from the coast (Fig. 2.3). The curve rises relatively sharply in the first 50 km, then tapers monotonically. The prevailing northwest winds off the California coast drive the California current, which transports the cool waters of the North Pacific southward. Coastal cooling is further enhanced by upwelling, the process by which surface water forced seaward by the Coriolis effect is replaced by cooler water from below (Trewartha and Horn 1980). Upwelling can reduce the already cool surface temperatures of the California Current by another 2–3°C. The air in contact with the water for a sufficient period will likewise be cooled, not infrequently to the point that the water vapor within the air reaches saturation, resulting in fog and low clouds. When this so-called marine layer is thick enough, drizzle can be observed. Although the humidity levels tend to decrease away from the coast, it does so more slowly in the direction that the marine air is allowed to flow inland, hence topography is again a factor.

Another effect of the cool ocean temperature is the contrast it provides against the warmer inland, which sets up a sea breeze circulation during the day. A similar effect occurs due to differential heating and cooling on mountain slopes, which also forces a local circulation. The San Bernardino Mountain areas are subject to both mechanisms. At the now-defunct Norton Air Force Base in San Bernardino, for example, weather forecasters were taught to expect sea breezes consistently during the late spring, summer, and early fall. There, the sea breeze commences

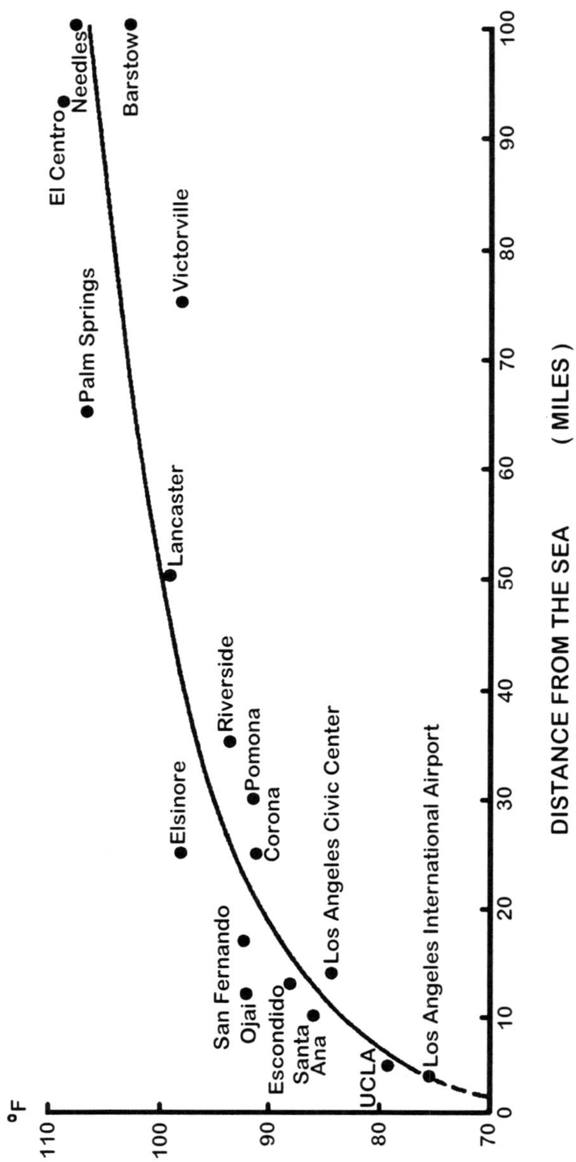

Figure 2.3. Summer afternoon temperatures (based on mean maximum temperature data for July) tend to increase with distance from the coast (Bailey 1966).

around noon, peaks at midafternoon, and diminishes after sunset.[1] By continuity, we expect a similar pattern within the nearby San Bernardino Mountains. However, as was previously pointed out, Bailey determined that the sea breeze was capped at approximately 915 m above mean sea level. This is due to the influence of the surface high-pressure cell in the eastern Pacific Ocean.

Pacific High Effect

The Pacific high-pressure cell is part of the belt of large subtropical anticyclones that girdles the Northern Hemisphere (Trewartha and Horn 1980). The Southern Hemisphere has a similar belt in approximately the same latitude zone south of the equator. Both are prominent features of the planetary circulation. In fact, the subtropical anticyclones are so persistent, they are described as semipermanent, changing somewhat in intensity and location from day to day and seasonally marching in phase with the solar zenith.

The Pacific high affects pollution transport and photochemical production in the South Coast Air Basin primarily through the subsidence inversion, which caps the mixing layer, and promotes cloud-free conditions above the inversion. In the absence of clouds, the sun causes temperatures to rise faster over land than over the ocean, which gives rise to a sea breeze directed from the ocean to the land. When the sea breeze forms rapidly, its progression inland can sometimes be observed visually, its leading edge marked by pollution concentrations that sharply decrease visibility. Behind this so-called sea breeze front, temperatures are also lower, and dew points and relative humidities are higher. Above the inversion, cloudless skies make more sunlight available for photochemical activity.

The subsidence inversion is tied directly to the eastern Pacific high and is characteristic of the eastern portion of oceanic subtropical anticyclones (Trewartha and Horn 1980). In this part of the high, the winds start to turn from an equatorward direction to an easterly direction, into the trade wind zone. As they do, the average wind speed increases, hence there is both speed and directional divergence of the streamlines, which is compensated for by strong subsidence. The subsiding air is heated adiabatically, while the surface layer over the ocean is cooled, thus producing an inversion layer in between the subsidence and the ocean surface, typically based at 500 m. The base of the subsidence inversion is usually the height of mixing, unless surface heating is sufficient to penetrate and break through the warm air layer above. Pollutants generated at the surface thus are confined to a thin boundary layer, increasing their concentrations to unhealthful levels for vulnerable humans and plants in southern California. The inversion base varies seasonally and diurnally. A lower inversion base is characteristic of a stronger inversion, which promotes higher and more persistent pollution concentrations.

---

[1] Terminal Forecast Reference Notebook, Detachment 14, 7th Weather Wing, Norton AFB, California, May 15, 1978.

Near-Surface Wind Flow Patterns

The effect of inversions is subject to local mechanisms that can lift, displace, penetrate, or break the inversion, hence affecting pollution dispersion.

Sea Breeze

Aloft and offshore over the ocean where surface friction is minimal, northwesterly winds prevail almost year-round. These winds are in geostrophic balance with the subtropical anticyclone, which moves air clockwise around the subtropical high. Off the coast, the California current, which is part of the North Pacific current system, mirrors the atmospheric circulation system. That is, the currents come from the north and move southward along the coast and then westward.

Inland, the atmospheric winds near the surface are affected by the surface friction. Considerable wind deflection therefore occurs near the coast and inland surface winds blow from high pressure to low pressure. Thus the overall climatic character of southern California is a contrast between subtropical maritime and subtropical continental. Near the shoreline, thermals rise over the land, causing the air to move from the ocean to the land. This air forms a sea breeze front, which moves inland during the day. Aloft, the circulation returns to the ocean. At night, the opposite occurs although not as strong, due to the relative coolness of the land in comparison with the ocean. Sailors take advantage of this diurnal variation to sail out from the coast in the early morning and then return in the afternoon. Seasonally, the inland movement is strongest during the summer, when the subtropical anticyclone is at its peak. Another feature affecting the summertime onshore flow pattern is the shallow heat low over the inland desert regions. During the winter, the offshore land breeze becomes more prevalent.

Santa Anas

Infrequently, a wind known as a Santa Ana blows in from the east. The Santa Ana is a foehn wind that emanates from a surface high-pressure cell, flows from high-elevation terrain to lower elevations, and undergoes compression heating and drying in the process. Winds similar to southern California's Santa Ana exist all over the world. In Colorado, these winds are called chinooks, or snoweaters. In Carson City, Nevada, they are called Washoe Zephyrs. In the Andes, they have the Zonda and Peluch.

The chief characteristics of these foehn winds are high temperatures and low relative humidities. Foehn winds are also noted for their violence and gustiness, but the magnitude depends on where the location is with respect to the topography. Santa Anas are usually associated with the worst fire disasters present in southern California, because the humidity is so low and the wind speed is so high. Synoptic characteristics include a high pressure at the surface over the Great Basin and a high-pressure ridge over the California coast (Fig. 2.4[2]). Air flows out

[2]Figure 2.4 will be found in the color insert following page 78.

of the Great Basin high southward and streams up and over the San Gabriel and San Bernardino mountains, where it descends into the Los Angeles Basin.

Catalina Eddy

The Catalina eddy is a circulation so named because it forms in the vicinity of Catalina Island during the summer. The eddy's diameter is about 150 km. As discussed above, the predominant surface flows in summer over the coastal areas and waters of southern California are northwesterly and westerly. These flows help air pollutants build up in the interior plains and valleys of the South Coast Air Basin. The Catalina eddy alters these airflow patterns by producing southerly flow along the coast north of San Diego. This altered circulation tends to thicken the marine layer, increasing low-level cloudiness, reducing daytime temperatures, and diluting air pollution.

Airflow Simulations

Meteorological model simulations can provide valuable insight into the processes that distribute air pollution. Early efforts at modeling weather in the South Coast Air Basin and the San Bernardino Mountains were related to studies of fire danger. Sommers (1976) examined the accuracy of interpolating limited-area fine mesh (LFM) forecasts to gridpoints over southern California, using b cubic splines (Rivlin 1969). He determined from case studies that the LFM model was insufficient in describing wind variations under Santa Ana conditions and even under the more quiescent sea breeze regime, because the model could not account for localized mechanisms. Ryan (1977) developed an empirical wind model incorporating local processes such as valley-mountain winds, slope winds, sea and land breezes, and what he termed the sheltering and diverting effect of topography. He found good agreement in the mean between the model and observations in the San Bernardino Mountains, but the empirical nature of the model poses problems of a data requirement nature, and it does not satisfy conservation of mass, which is essential in modeling air transport.

Lu and Turco (1995) conducted a study of air pollutant transport in the South Coast Air Basin, using a three-dimensional meteorological model. They simulated the airflow under stagnant conditions typical of summer, using passive tracers emitted from the western side of the basin, between 118.5° W and 117.7° W and 33.65° N and 34.2° N.

The model used by Lu and Turco (1995) was a hydrostatic primitive equation model, which incorporated equations for conservation of momentum and mass, thermodynamics, and hydrostatic balance. The model simulated vertical turbulent diffusion, horizontal subgrid diffusion, water vapor condensation and precipitation, radiative transfer and heating, soil moisture and heat transfer, and ground surface energy balance. The model domain consisted of 20 horizontal layers, each a grid of 85 by 55 points. The grid spacing was 4.6 km in the east-west direction and 5 km in the north-south direction. The model domain included that part of southern California between 119.25° W and 116.50° W and between 32.25° N and 35.00° N. This area covers San Diego to the south, Barstow to the north, Santa Barbara to the west, and Palm Springs to the east (Fig. 2.5).

The simulations started with horizontally uniform, no-wind conditions at 0500 PST. The vertical temperature and moisture distribution represented the July 1991 average sounding measured at the University of California, Los Angeles, characterized by a subsidence inversion with a base at approximately 500 m and a top at 1,500 m. Within the inversion layer, the temperature increased from 14°C at the base to 20°C at the top. The initial conditions were chosen to focus on the roles of mountains and sea breezes in distributing pollutants from the western part of the basin.

Not surprisingly, the mountains and sea breeze have significant roles in transporting pollutants around the South Coast Air Basin. By 1100 PST, a well-developed onshore flow crosses the coastline and proceeds upslope on the southern flanks of the San Gabriel and San Bernardino mountains. The airflow envelops the Santa Ana Mountains, but a weak upslope flow is apparent on the Elsinore side, and even stronger upslope components are found on the northern flanks of the San Gabriel and San Bernardino mountains. The Mojave Desert winds blow from the north but are not nearly as strong as the winds south of the ridgeline (Fig. 2.6A).

By 1700 PST, the simulated winds have strengthened over the entire basin, flowing into the southern portion of the high desert through the Newhall Pass, the Cajon Pass, the San Gorgonio Pass, and the Morongo Valley east of San Gorgonio Mountain. The winds turn from southerly to northerly, increasing in velocity from the morning upslope flow on the north side of the San Gabriel and San Bernardino mountains. The flow patterns are attributed to the coupling between the sea breeze and the upslope flow induced by heating of the mountain slopes. In fact, when the simulations were rerun first with no mountains, then with no ocean, the results showed that the sea breeze accelerates the flow within approximately 40 km of the coast, but the mountains provide the impetus further inland by thermal effects and channeling of flow (Fig. 2.6B).

Lu and Turco (1995) simulated the transport of a passive tracer across the South Coast Air Basin, from Santa Monica Bay to San Gorgonio Mountain. At 1700 PST, the sea breeze had penetrated 180 km inland to the slopes of the San Bernardino Mountains. The typically cool maritime air had become somewhat diffuse by that time, so that there was no clearly distinguishable sea breeze front. The layer of onshore flow was approximately 300 m thick, narrowing on the mountain slopes as it was squeezed between the ground and the inversion layer. Above, a return flow of air was apparent to approximately 2,500 m. Hence, tracer material was concentrated on the mountain slope below the inversion. But the return flow also spread the tracer back toward the coast, forming an elevated polluted layer concentrated over the eastern side of the basin. The approach of night stabilizes the boundary layer and kills the mixing process. Similar wind and tracer profiles occurred along a transect from the Santa Monica Bay over the San Gabriel Mountains to the Mojave Desert.

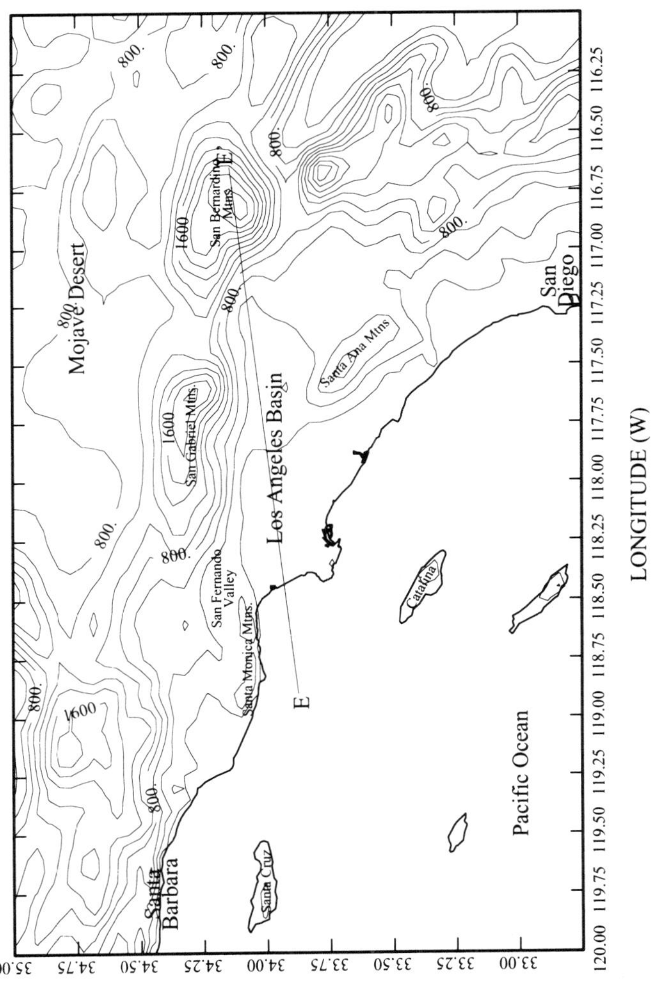

Figure 2.5. Lu and Turco (1995) model domain for airflow simulations under stagnant synoptic conditions. The vertical profiles of simulated winds and mass transport along the transect EE′ give a good cross section from the coast to the San Bernardino Mountains. Reprinted from *Atmospheric Environment*, Volume 29, Lu and Turco, "Air pollutant transport in a coastal environment—II: Three-dimensional simulations over Los Angeles Basin," 1499–1518, copyright 1995, with permission from Elsevier Science.

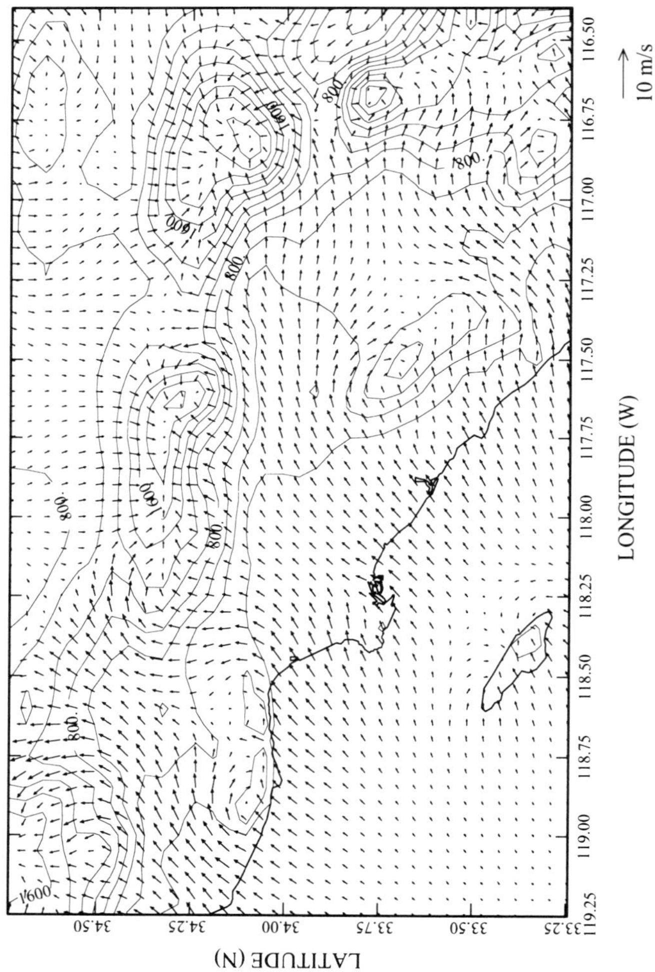
LATITUDE (N)
34.50
34.25
34.00
33.75
33.50
33.25
LONGITUDE (W)
119.25
119.00
118.75
118.50
118.25
118.00
117.75
117.50
117.25
117.00
116.75
116.50
10 m/s
800
1600

A

B

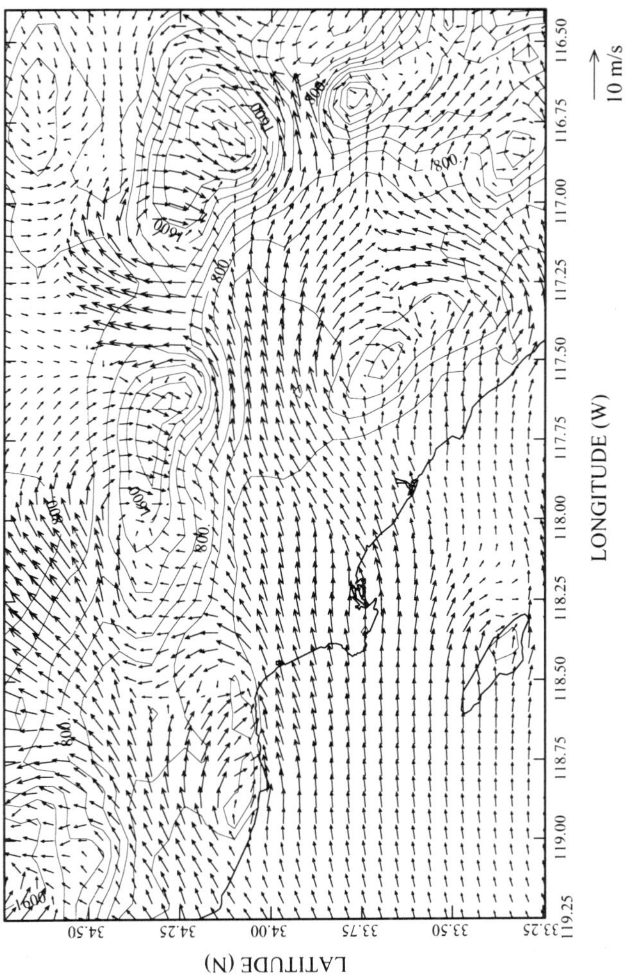

Figure 2.6. Lu and Turco (1995) simulations of surface wind flow in the South Coast Air Basin at (A) 1100 PST and (B) 1700 PST under synoptically stagnant July conditions. Reprinted from *Atmospheric Environment*, with permission from Elsevier Science.

The simulation results are in general agreement with the observations of Edinger et al. (1972), who sampled pollutants by ground and aircraft measurements from Santa Monica to Redlands and north over the crest of the San Bernardino Mountains. They found, in June 1970, conditions similar to the simulations, an increase in oxidant concentration in the marine layer from west to east. They also found high oxidant concentrations above the inversion base adjacent to the southern slopes of the San Bernardino and San Gabriel mountains; at 1600 PST, maximum oxidant concentrations occurred on the slopes between 900 and 1,200 m. Elevated aerosol layers were also observed in the evening over the eastern part of the South Coast Air Basin by Wakimoto and McElroy (1986).

Temperature and Precipitation

The spatial distribution of annual precipitation in southern California follows the orography. Over the mountainous regions of San Bernardino, more than 1,000 mm of precipitation falls in an average year, whereas over the coastal areas of Los Angeles, annual precipitation is only 400 mm. There is also a moderate northward increase from San Diego, which averages about 250 mm precipitation annually. The mountains have a distinct rainshadow, within which the lee areas have precipitation amounts approaching the average of the desert regions farther inland. The relationship of average total precipitation with distance from the Pacific Ocean and elevation above sea level for 24 southern California locations is shown in Table 2.1. Almost all the southern California precipitation falls during the winter, although some tropical storms, related to the storms of the Mexican monsoon, sometimes penetrate as far north as San Bernardino.

Contours of temperature also follow the topography, although the distance from the coast is also an important determinant (Table 2.2). It can be shown, for example, that although the temperature at San Bernardino is cooler than the coastal stations, it is actually warmer than the air temperature at the same elevation over the coastal regions. It is also much drier.

El Niño–Southern Oscillation

Interannual climate variations over southern California are almost random. However, one potential predictable climate signal is related to El Niño–southern oscillation (ENSO). The term *southern oscillation* was coined by Sir Gilbert Walker, head of the British Meteorological Service in New Delhi, to describe large-scale Southern Hemisphere sea-level pressure variations (Philander 1990). He named this irregular fluctuation the southern oscillation, to distinguish it from the northern oscillation found in the Northern Atlantic. This atmospheric pattern is related to seasonal to interannual variations in a southward countercurrent along the coast of Peru. Like California's cold upwelled currents from the north, cold upwelled currents from the south are the climatic normal along the South Ameri-

Table 2.1. Relationship of Average Total Precipitation with Distance from the Pacific Ocean and Elevation above Sea Level for 24 Southern California Locations[a]

Station name	Distance from Pacific Ocean (mi)	Elevation (m)	January average total precipitation (in.)	July average total precipitation (in.)	Annual average total precipitation (in.)
Elsinore	24	391.7	2.65	0.07	11.03
Corona	24	214.0	2.51	0.03	11.49
Santa Ana	10	41.1	3.20	0.01	12.93
Long Beach	15	7.6	2.81	0.06	12.08
LA International Airport	3	30.5	2.79	0.02	11.94
LA Civic Center	14	82.3	3.47	0.01	14.49
Pomona	29	225.6	3.74	0.01	17.41
Redlands	48	401.7	2.62	0.08	13.43
Riverside	34	256.0	2.07	0.04	10.29
San Bernardino	67	342.9	3.29	0.04	16.51
San Gabriel	25	137.1	3.90	0.02	17.53
San Pedro	0	3.0	2.09	0.00	10.83
Sta Monica Pier	0	4.3	3.01	0.02	12.39
Torrance	2	33.5	3.50	0.01	13.31
UCLA	5	129.0	4.09	0.02	17.00
San Fernando	17	296.0	3.48	0.01	16.20
Barstow	99	659.0	0.74	0.28	4.41
Big Bear Lake	120	2069.6	4.58	0.77	22.81
Escondido	13	201.0	2.75	0.02	15.59
El Centro	92	–9.1	0.51	0.09	2.63
Lancaster	50	712.9	1.73	0.09	7.67
Palm Springs	65	129.5	1.19	0.22	5.82
Needles	99	278.6	0.66	0.36	4.57
Victorville	74	871.1	1.08	0.13	5.39

[a]From National Weather Service 1988, 1993.

can western coast. Then, around Christmas, a warm southward flowing countercurrent occurs and was named El Niño, for the Christ child, by the local Peruvian sailors. Every few years, the current becomes more intense than usual and is accompanied by heavy rains.

However, it was not until the early 1970s that a UCLA meteorologist really connected the southern oscillation and El Niño variations. Bjerknes (1972) explained how dry air sinks over the cold water of the eastern tropical Pacific and flows westward along the equator as part of the trade winds. The air is warmed and moistened as it moves over the progressively warmer water until it reaches the western tropical Pacific, where it rises in towering rain clouds. Return flow in the upper troposphere then closes this loop, which Bjerknes named the *Walker circulation,* in honor of Sir Gilbert Walker. Bjerknes also proposed that the periodic warming of the eastern tropical Pacific Ocean weakens the Walker circulation and causes the convective zone of heavy rainfall to move eastward from the western

Table 2.2. Relationship of Average Temperatures with Distance from the Pacific Ocean and Elevation above Sea Level for 24 Southern California Locations[a]

Station name	Distance from Pacific Ocean (mi)	Elevation (m)	January average temperature (°F)	July average temperature (°F)	Annual average temperature (°F)
Elsinore	24	391.7	50.9	79.0	64.2
Corona	24	214.0	53.2	75.2	63.6
Santa Ana	10	41.1	56.3	72.5	64.5
Long Beach	15	7.6	56.1	73.1	64.5
LA International Airport	3	30.5	56.1	69.0	62.8
LA Civic Center	14	82.3	57.7	74.1	65.9
Pomona	29	225.6	51.9	74.1	62.6
Redlands	48	401.7	51.9	77.6	63.8
Riverside	34	256.0	52.9	77.0	64.3
San Bernardino	67	342.9	52.8	78.4	64.7
San Gabriel	25	137.1	55.0	74.7	64.7
San Pedro	0	3.0	54.9	67.7	61.6
Sta Monica Pier	0	4.3	56.8	66.0	61.4
Torrance	2	33.5	55.5	69.4	62.6
UCLA	5	129.0	57.5	68.7	63.2
San Fernando	17	296.0	54.1	74.6	63.8
Barstow	99	659.0	45.4	84.7	64.0
Big Bear Lake	120	2,069.6	33.4	63.6	46.7
Escondido	13	201.0	51.7	73.2	62.1
El Centro	92	–9.1	54.6	91.4	72.4
Lancaster	50	712.9	44.2	80.8	61.2
Palm Springs	65	129.5	55.4	91.4	72.6
Needles	99	278.6	52.4	95.8	73.3
Victorville	74	871.1	43.9	79.2	60.6

[a]From National Weather Service 1988, 1993.

Pacific into the central and eastern tropical Pacific and away from the monsoon regions.

It has been more difficult to pin down all the relationships between tropical region and remote region climate variations. However, Horel and Wallace (1981) showed that there were some clear physical teleconnections to middle latitudes. Basically, it is thought that equatorial thunderstorms cause two high-pressure regions at high elevations to straddle the equator, leading then to a lower-than-normal pressure in the region of the Aleutian low, a higher-than-normal high over the Canadian Rockies, and a lower-than-normal pressure over Florida. This climate variations pattern, which is known as the Pacific North American pattern, appears in one form or another from one El Niño to another and is related to a warming of the cold northern states and a moistening of the southern states, including southern California. Southern California seems to be most affected by the strongest ENSOs, and during the strongest one (up to now) during 1982 and 1983, large-scale precipitation and flooding occurred in California.

References

Bailey, H.P. Weather of southern California. Berkeley: University of California Press; 1966.

Bjerknes, J. Large-scale atmospheric response to the 1964–65 Pacific equatorial warming. J. Phys. Oceanogr. 2:212–217; 1972.

California Environmental Protection Agency, Air Resources Board. California air quality data, summary of 1995 air quality data. Vol. 27. Sacramento: California EPA; 1996.

Edinger, G.; McCutchan, M.H.; Miller, P.R.; Ryan, B.C.; Schroeder, M.J.; Behar, J.V. Penetration and duration of oxidant air pollution in the South Coast Air Basin of California. J. Air Pollut. Control Assoc. 22:882–886, 1972.

Environmental Protection Agency. National air quality and emissions trends report, 1995. EPA 454/R-96-005. Research Triangle Park, NC; Office of Air Quality Planning and Standards; 1996.

Horel, J.; Wallace, J.M. Planetary scale atmospheric phenomena associated with the southern oscillation. Monthly Weather Rev. 109:813–829; 1981.

Juang, H-M.H.; Kanamitsu, M. The NMC nested regional spectral model. Monthly Weather Rev. 122:3–26; 1994.

Lu, R.; Turco, R.P. Air pollutant transport in a coastal environment—II. Three-dimensional simulations over Los Angeles Basin. Atmos. Environ. 29:1499–1518; 1995.

National Weather Service. Surface observations, federal meteorological handbook 1 (FCM-H1-1988). NOAA-NWS. Silver Spring, MD: National Weather Service; 1988.

National Weather Service. National Weather Service observing handbook 7. Surface observations, NOAA-NWS. Silver Spring, MD: National Weather Service; 1993.

Philander, S.G. El Niño, La Niña, and the southern oscillation. Int. Geophysics Ser. 46; 1990.

Rivlin, T.J. An introduction to the approximation of functions. New York: Dover Publications, Inc.; 1969.

Ryan, B.C. A mathematical model for diagnosis and prediction of surface winds in mountainous terrain. J. Appl. Meteor. 16:571–584; 1977.

Sommers, W.T. On the accuracy of interpolated LFM forecasts and their use in predicting surface conditions in mountainous terrain. Monthly Weather Rev. 104:1573–1582; 1976.

Trewartha, G.T. An introduction to climate. New York: McGraw-Hill Book Co., Inc.; 1954.

Trewartha, G.T.; Horn, L.H. An introduction to climate. New York: McGraw-Hill; 1980.

Wakimoto, R.M.; McElroy, J.L. Lidar observations of elevated pollution layers over Los Angeles. J. Clim. Appl. Meteor. 25:1583–1599; 1986.

3. Vegetation, Fire Regimes, and Forest Dynamics

Richard A. Minnich

Introduction

The San Bernardino Mountains (SBM) have remarkably diverse shrubland and conifer forest communities in association with climatic gradients produced by large elevational relief. Wildland fire also plays a vital role affecting species composition, vegetation structure, and biogeography. In evaluating the effects of air pollution in SBM, it is important to consider fire disturbance as a mediator in stand dynamics and postfire successions and, in particular, how structural changes of some ecosystems due to a century of fire exclusion may have selectively altered the fitness and competitiveness of species, thereby affecting their resistance to air pollution. This chapter serves two goals: (1) to describe the vegetation of SBM, and (2) to evaluate the fire ecology of major ecosystems.

Present Vegetation

The vegetation of SBM (Fig. 3.1[1]) was mapped from color infrared and color aerial positive transparencies and prints obtained in 1969–1975 and 1984 (scale: 1:20,000 to 1:40,000). The photographs were interpreted using a roll-film stereo-

---

[1]Figure 3.1 will be found in the color insert following page 78.

scope with 3 and 8X magnification. Boundary data were then transferred onto 1:24,000 topographical sheets by using a zoom transfer scope. Plant species were identified on the basis of crown perimeter and apex configurations, vertical structure, shadows, and color (methods in Minnich 1987a).

To appraise the fuel characteristics in SBM, the vegetation was classified into three map coverages by lifeform: (1) conifer forest, (2) hardwood forest, and (3) an understory layer of chaparral, subshrub cover, and meadow vegetation. Types within each lifeform class were distinguished by dominant species. Map data were entered into the Arc-Info geographical information system (GIS) installed on a SUN workstation. Lifeform classes were composited to show vertical relationships of fuels in communities having more than one physiognomic layer. Conifer forest and arboreal hardwood map boundaries are defined where cover (as a function of canopy area) is 10% or greater and shrub layer types are 20% or greater.

Plant Communities

The vegetation of SBM exhibits broad zonation with altitude and with windward/leeward slope precipitation gradients associated with physiography, similar to other southern California coastal ranges and the Sierra Nevada (Hanes 1988; Rundel et al. 1988; Thorne 1988). Chaparral growing at lower elevations is replaced by extensive conifer forests above 1,400 m. Pinyon woodlands cover the leeward northern and eastern flanks of the range. The local patterning of communities is modified by slope, aspect, and substrate. In the following summary of SBM vegetation, the area of vegetation types is given in Table 3.1, data on shrub cover and species composition obtained from quadrat data taken by the California vegetation-type map survey (VTM) are given in Table 3.2, and forest density diameter class frequency distributions are given in Table 3.3.

Chaparral, an evergreen community of sclerophyllous shrubs in carpetlike stands 1–5 m tall, grows extensively on the southern escarpment and desert drainages opposite low passes—where rainshadows are diminished—including the West Fork of the Mojave River and the Whitewater River. Chamise chaparral, dominated by the linear, fasciculate-leaved shrub *Adenostoma fasciculatum*, covers the lower altitudes of the chaparral belt. This community is replaced by mixed chaparral, dominated by broadleaf sclerophyllous shrubs, above 700–900 m on north-facing slopes and 1,200–1,400 m on south-facing slopes. VTM data show that *Adenostoma fasciculatum* is the primary species from 300 to 1,200 m in association with members of the *Ceanothus* subg. *Cerastes*, mostly *C. crassifolius*. Other common shrubs in this type include *Ceanothus cuneatus*, *Rhus ovata*, *Rhamnus crocea*, *Cercocarpus betuloides*, and *Prunus ilicifolia*. *Quercus berberidifolia* dominates lower north-facing slopes, whereas stands at higher elevations consist of *Quercus berberidifolia*, *Q. wislizenii*, *Ceanothus* subg. *Ceanothus (C. oliganthus, C. leucodermis)*, *Arctostaphylos glandulosa*, and *A. glauca*. Common shrubs include *Heteromeles arbutifolia*, *Rhus ovata*, *Garrya flavescens*, and *Fraxinus dipetala*. *Cercocarpus betuloides* is abundant on cliff

Table 3.1. Spatial Extent of Vegetation Types in the San Bernardino Mountains

Vegetation type	Area (ha)[a]
Conifer forest	
Knobcone pine	985
Coulter pine	10,241
Bigcone Douglas-fir	8,431
Jeffrey pine	4,885
Mixed Jeffrey pine	17,721
Mixed ponderosa pine	11,328
Mixed white fir	12,507
Mixed western juniper	9,865
Lodgepole pine	6,198
Hardwood forest	
Canyon live oak	39,368
Black oak	13,023
Shrub and herbaceous communities	
Chamise chaparral	55,789
Mixed chaparral	21,618
Desert chaparral	24,443
Desert oak scrub	17,313
Deerbrush scrub	1,221
Timberland chaparral	8,424
Mountain mahogany scrub	17,907
Great Basin sage scrub	9,397

[a]Calculated from the GIS.

faces. Both chamise and mixed chaparral communities have average cover of greater than 90%.

On leeward slopes of SBM, mixed chaparral gives way to open stands of desert (petron) chaparral (cover, 50–75%), dominated by *Q. john-tuckeri, Q. cornelius-mulleri, C. betuloides, Arctostaphylos glandulosa, A. glauca,* and *Ceanothus greggii* var. *vestitus.* Desert chaparral nearly encircles the northern and eastern escarpments of SBM, with stands being most extensive in Deep Creek, middle slopes of Morongo Creek, and the Whitewater River, as well as in the rainshadows of upper Santa Ana River Canyon. More open stands of desert oak scrub, dominated by *Q. john-tuckeri* and lacking obligate postfire seeding shrubs such as *Ceanothus greggii* and *Fremontodendron californicum,* occur along the northern and eastern margins of SBM from Lucerne Valley to Pipes Wash. Desert chaparral and desert oak scrub are frequently mixed with shrubs, subshrubs, and succulents common at higher elevations of the Mojave Desert including *Prunus fasciculata, Ephedra viridus, E. nevadensis, Ericameria cooperi, Yucca schidigera,* and *Nolina parryi.*

Forests occur locally in the chaparral belt and increase in spatial extent with altitude. Streams are invariably lined with deciduous riparian forests consisting of *Alnus rhombifolia, Populus fremontii, Salix* spp., *Platanus racemosa,* and *Acer macrophyllum*. The evergreen *Umbellularia californica* is a frequent riparian tree, especially in western SBM. *Quercus chrysolepis* forest grows in deep canyons and steep convex north-facing slopes but forms shrubby thickets on smooth concave slopes in association with mixed chaparral. A few colonies of *Q. agrifolia* occur south of Oak Glen.

The closed-cone conifer *Pinus attenuata* forms small dense groves of even-aged trees along the southern escarpment between 800 and 1,400 m from City Creek eastward to Keller Peak, with stands concentrated on concave ridges and summits. *Pinus attenuata* is found entirely on granitic substrate and has an abundance of *Arctostaphylos (glandulosa),* similar to stands elsewhere in California (Vogl et al. 1988).

The chaparral belt contains patches of mixed evergreen forest, which consists of evergreen conifer forests with evergreen sclerophyllous understory of hardwoods and shrubs (Sawyer et al. 1988). *Pinus coulteri* occurs in disjunct stands on steep, smooth, concave slopes between 1,100 and 2,200 m along the southern escarpment east of Waterman Canyon. This community also forms a continuous belt on desert drainages of northwestern SBM from the West Fork of the Mojave River to Shay Mountain. On the southern escarpment below 1,300 m, it occurs in even-aged stands of pole-size and young canopy trees (diameter-breast-height [dbh] <60 cm) in association with dense mixed chaparral and thickets of *Quercus chrysolepis*. On the northern plateau, it occurs in mixed-aged stands in association with multilayered understories of *Q. chrysolepis* and *Q. kelloggii* woodlands and open cover of chaparral. Stands are mostly open with average stem densities of 137/ha. *Pinus coulteri* grows entirely on granitic substrate, except along Mill Creek Ridge and near Kitching Peak, where it is found on Precambrian metamorphics.

Pseudotsuga macrocarpa, a southern California endemic (Minnich 1982), occurs in compact discrete stands (mostly <10 ha) within mixed evergreen forests on precipitous sheltered convex canyons, steep slopes, and cliffs from 600 m to 2,200 m. It grows along the entire southern escarpment and extends onto the desert watersheds of the West Fork of the Mojave River and the Whitewater River. Although hundreds of colonies may be found in a single drainage, stands tend to increase in size and continuity with altitude. The largest stands (>320 ha) inhabit metamorphic rock units having high erosion rates along active faults, including Silverwood Reservoir, Santa Ana Canyon near Camp Angeles, and the northern flank of Mill Creek Ridge. *Pseudotsuga macrocarpa* is a long-lived tree (most stems >60 cm dbh) that grows in mixed-aged stands with arboreal *Quercus chrysolepis*. Stem densities average only 103/ha, but forests invariably form solid cover due to large spread of the pendulous *P. macrocarpa* canopies and dense cover of arboreal *Q. chrysolepis*. Shrub cover on the forest floor (*Q. chrysolepis, Ceanothus integerrimus*) is scarce due to high canopy cover, even though stands are frequently surrounded by chaparral.

Californian mixed conifer forest forms a continuous belt between 1,300 and 2,700 m from near Crestline in western SBM to Big Bear and southward into upper Santa Ana River Canyon and San Gorgonio Mountain, then eastward along

Table 3.2. Summary of Evergreen Sclerophyll Shrub Cover by Species (%) in the San Bernardino Mountains during the 1929–1934 California Vegetation Type Map Survey[a]

Vegetation type	Altitude (m)	Exp[c]	Species[b] Af	Ag	Agl	Ccr	Cl	Qb	Qw	Cb	Cg	Fc	Qjt	Ci	Ap	Cs	Cco	Cle	Total
Chaparral	<900	S	54			15	4	5		1									92
	900–1200	S	46	5	6	12	5	2	1	1	1								88
	1,200–1,500	S	33	7	7	7	14	3	3	2	1								89
	>1,500	S	21	5	20	2	22	3	5	2	2								96
	<900	N	13	4	4	11	6	32	12	5									98
	1,200–1,500	N	14	5	3	8	12	20	15	4			1	2					92
	>1,200	N	16	4	8	1	18	11	16	3	2			1					92
Desert chaparral		A	3	5	1					20	13	10							63
		N	2	2			1			21	12	4	2						71
Desert oak scrub			4	2						2	5	1	32						55
Knobcone pine			23		42	3		2											88
Mixed evergreen forest																			
Coulter pine–chaparral			11	9	7		3	6		12	13								84
Coulter pine–canyon oak				1					1	1				1					22
Coulter pine–black oak			4		1		2		2	3	5								34

Bigcone Douglas-fir							2					13
Mixed conifer forest												
Ponderosa–black oak							1			3		6
Mixed Jeffrey pine							2	1	1			7
Jeffrey pine–canyon oak							4	2		4	2	27
Jeffrey pine–timber chaparral								10	12	16		41
White fir							1	3	1			7
White fir-timber chaparral								1	5	18		34
Jeffrey pine–mountain mahogany											11	11
Lodgepole pine forest								1	9			11
Pinyon-juniper woodland												
Pinyon–mountain mahogany					4	1					13	25
Pinyon–canyon oak		1	1	1	5						4	43
Pinyon-Great Basin sage		1		1		1						30
Pinyon-desert chaparral	1		1	15	7	3						54

[a]From Minnich 1988.
[b]Af, *Adenostoma fasciculatum;* Ag, *Arctostaphylos glauca;* Agl, *Arctostaphylos glandulosa;* Ccr, *Ceanothus crassifolius;* Cl, *Ceanothus leucodermis;* Qb, *Quercus berberidifolia;* Qw, *Quercus wislizenii;* Cb, *Cercocarpus betuloides;* Cg, *Ceanothus greggii;* Fc, *Fremontodendron californicum;* Qjt, *Quercus john-tuckeri;* Ci, *Ceanothus integerrimus;* Ap, *Arctostaphylos patula;* Cs, *Chrysolepis sempervirons;* Cco, *Ceanothus cordulatus;* Cle, *Cercocarpus ledifolius.*
[c]Slope exposure divided hemispherically centered on north and south.

Table 3.3. Conifer Forest Stand Densities in 1932 and 1992 (per hectare)[a]

	VTM survey (1929–1934)						Replicate (1992)		Sierra San Pedro Mártir (1991)		Sierra Nevada (1969)
	Diameter class (in.)										
Forest type	4–12	12–24	24–36	>36	Total	SD	Total	SD	Total	SD	Total
Coulter pine	60	51	22	4	137	92	—		62		—
Bigcone Douglas-fir	16	45	29	13	103	62	—		—		—
Mixed ponderosa pine (logged)	154	52	14	10	232	250	378	232	—		—
Mixed ponderosa pine	14	42	54	34	144	131	250	60	—	48	549
Mixed Jeffrey pine	19	32	19	21	91	53	167	78	78	22	200
Jeffrey pine	14	12	30	24	79	26	169	105	66	20	—
Mixed white fir	54	47	41	35	174	75	246	110	143	52	519
Lodgepole pine	44	100	44	1	189	62	—	—	115	40	—
Singleleaf pinyon	136	42	4	1	220	113	253	90	—	—	—

[a]Original data in English units. 36 in. = 1.0 m. Sources: San Bernardino Mountains, 1929–1934, California Forest and Range Experiment Station; San Bernardino Mountain replicates (Minnich et al. 1995; Wangler and Minnich 1996); Sierra San Pedro Mártir, Baja California, estimated from aerial photographs flown in 1991 (Minnich et al., in press); Sierra Nevada (Vankat and Major 1978).

Mill Creek Ridge to Kitching Peak. The lower margin of this community seldom overlaps with chamise or mixed chaparral, the two ecosystems often forming sharp ecotones both in lifeform dominance and species composition. The most striking boundary occurs along the crest of the Rim of the World from Monument Peak to south of Big Bear, with forests occupying the north-facing plateaus and chaparral or mixed evergreen forest occupying the southern escarpment.

The composition of mixed conifer forest shows a relationship to topography and precipitation gradients. Forests along the crest of SBM at 1,300–2,000 m, including the western Rim of the World, Barton Flats, Camp Angeles, and Mill Creek Canyon, are dominated by *Pinus ponderosa*. Forests at higher altitudes, mostly in the eastern half of SBM, are dominated by *P. jeffreyi*. Both types have *Abies concolor* and *P. lambertiana* as associates. *Calocedrus decurrens* is common in ponderosa pine forest but is scarce above 2,100 m in mixed Jeffrey pine forest. Steep, north-facing slopes are covered by mixed white fir forest, dominated by *Abies concolor* and locally by *P. lambertiana*. White fir forests are most extensive on the north flanks of Butler Peak, Sugarloaf Mountain, San Gorgonio Mountain, and Mill Creek Ridge. Forests on the semiarid northeastern plateaus of SBM are dominated by *Pinus jeffreyi* mixed with *Juniperus occidentalis*. Monotypic stands of *P. jeffreyi* grow on the floors of Little Pine Flat, Big Pine Flat, Holcomb Valley, eastern Big Bear Basin, and Arrastre Flat. Although *Pinus lambertiana* and *Calocedrus decurrens* are virtually absent from these areas, a thin cover of *Abies concolor* extends to the upper reaches of the desert escarpment overlooking the Mojave Desert.

All mixed conifer types are characterized by dense mixed aged stands with an abundance of young understory trees less than 30 cm dbh (Minnich et al. 1995). Modern stand densities decrease along a precipitation gradient and range from 378 stems/ha in logged ponderosa pine forests in western SBM to 167 stems/ha in Jeffrey pine forest.

Shrub understory in mixed conifer forest is mostly open or absent (cover <20%), depending on altitude and climatic gradients. Deer brush scrub (*Ceanothus integerrimus, Rhamnus californica*) occurs in openings of ponderosa pine forest (1,400–1,900 m), with *Arctostaphylos pringlei* and *A. pungens* locally abundant in mature forests. Opens stands of timberland chaparral (*Arctostaphylos patula, Ceanothus cordulatus, Chrysolepis sempervirens*) are found above 2,200 m in association with mixed Jeffrey pine and mixed white fir forests (average shrub cover, 41%). Contiguous stands occur on a few exposed summits such as Butler Peak and the south face of San Gorgonio Mountain. Most semiarid Jeffrey pine forests on the northern and eastern flanks of SBM have open subcanopy of *Cercocarpus ledifolius*. This shrub also forms discrete stands on cliff faces of the mesic southern escarpment. Great Basin sage scrub (*Artemisia tridentata, Chrysothamnus nauseosus, Purshia tridentata*) covers high basin floors including Holcomb Valley, eastern Big Bear Basin, and upper Santa Ana River Canyon.

Two broadleaf tree species are common throughout mixed conifer forests. Open to closed stands of canyon live oak (*Quercus chrysolepis*) grow on steep slopes to as high as 1,900 m on northern exposures and 2,600 m on southern exposures.

Quercus kelloggii is common on gentle slopes and basins below 2,200 m in mesic forests along the Rim of the World to Big Bear Basin, as well as in Barton Flats, Mill Creek Canyon, and Raywood Flats.

Subalpine forests of *Pinus contorta* and *P. flexilis* cover the highest summits above 2,400–2,700 m, including the San Gorgonio Mountain block, Mill Creek Ridge, Sugarloaf Mountain, and Butler Peak. Forests on south-facing slopes tend to form open mixed-aged stands with discontinuous cover of timberland chaparral. Stands on north-facing slopes are frequently dense, lack shrub understory, and occasionally form even-aged patches (average stem densities, 189 stems/ha; shrub cover, 11%). Subalpine forests growing above 3,200 m on San Gorgonio Mountain are prostrate with trees flagged by winter snow and wind damage (Minnich 1984).

Wet meadows dominated by *Juncus* and *Carex* occur in the mixed conifer forest and subalpine forest zones, with some as large as 100 ha at Holcomb Valley and Big Bear Basin. Annual and perennial herbs are scarce throughout SBM, owing to summer drought. The common exotic annual, cheat grass (*Bromus tectorum*), forms scattered cover, especially along roadsides, forest openings, and edges of meadows. There is dense cover of ripgut brome (*B. diandrus*) in open forests near Lake Gregory.

Pinyon-juniper woodland, which consists mostly of mixed-aged monotypic forests of *P. monophylla*, covers the semiarid northern and eastern flanks of SBM between 1,200 and 2,700 m (Wangler and Minnich 1996). *Juniperus californica* is an occasional component of pinyon-juniper woodlands below 1,600 m, whereas *J. occidentalis* is common above 2,200 m. More mesic stands adjacent to mixed conifer forest attain densities greater than 300 stems/ha. Forests then become increasingly open (<100 stems/ha) approaching the Mojave Desert; average stand densities are 253 stems/ha. Pinyon-juniper woodlands possess extensive understory of drought-deciduous subshrubs and evergreen sclerophyllous shrubs. Low-elevation understories (<2,000 m) consist of a mixture of Great Basin sage scrub and high Mojave Desert scrub (*Artemisia californica, Chrysothamnus nauseosus, Purshia tridentata, Ephedra viridis, E. nevadescens, Prunus fasciculata, Lycium cooperi, Yucca brevifolia, Y. schidigera*) and open stands of desert chaparral (*Quercus john-tuckeri, Cercocarpus betuloides, Arctostaphylos glandulosa, Ceanothus greggii* var. *vestitus, Fremontodendron californicum*). Pinyon-juniper woodlands growing in basin floors contain abundant Great Basin sage scrub (*Artemisia, C. nauseosus, P. tridentata*). *Cercocarpus ledifolius* is abundant above 2,100 m in those areas where *J. occidentalis* is an associate. *Quercus chrysolepis* is common on steep north-facing slopes above 2,000 m.

Fire Regimes

Recurrent fire is an integral component of SBM ecosystems because the Mediterranean climate of winter rain and dry summers results in inefficient decomposition, rapid fuel buildup, and high fire hazard (Bruhn 1980; Stohlgren 1988; Hart et al. 1992). The establishment of fire exclusion on virtually all state and federal land in California by 1925 (Pyne 1982) and in the SBM by about 1900 (Minnich 1988) has altered fire regimes (fire size, frequency recurrence interval, intensity, forest damage) as well as ecosystem structure, making it increasingly difficult to deduce past fire regimes that shaped ecosystems. Models of presuppression fire regimes are usually developed from indirect site-specific evidences, including fire scar dating, reconstructions of stand age profiles, fragmentary historical records, and present stand structure. The extrapolation of landscape dynamics from point studies may not capture the regional processes that shaped forest ecosystems (McIntosh 1981).

Two hypotheses have been proposed concerning the broad-scale effects of suppression: (1) Fire control has lengthened fire intervals and contributed to excessive regional fuel buildup and greater fire intensities (e.g., Kilgore and Taylor 1979; Swetnam 1993; SNEP Science Team 1996). Evidence includes the sudden decline in the rate of fire scarring in pine forests throughout the western United States (e.g., Kilgore and Taylor 1979; Swetnam 1993). (2) Fire occurrence is time dependent, with minimum fire intervals being relatively long, regardless of management, due to the gradual cumulative fuel buildup (Loucks 1970; Heinselman 1981a,b; Minnich 1983; Minnich and Chou 1997; Minnich et al., in press). According to this hypothesis, fire regimes in the chaparral and conifer forests of the Pacific coast of North America are constrained in time and space by the rate of fuel accumulation and by previous fire history. At the scale of the landscape and over times of decades to centuries, the rate at which these ecosystems burn is nearly constant. The time threshold for burning is a property set by the rate of fuel accumulation inherent to the vegetation, with stands younger than that threshold constraining the progress of burns. Thus, the evolution of patch mosaics created by fire is a nonrandom and self-organizing process, because the occurrence of one event is affected by past events and, in turn, affects future events.

Changes in the fire regime in SBM since 1900 suggest that both models apply, depending on the vegetation type.

Postsuppression Changes

Accounts of SBM during the late 19th century—as recorded in books, manuscripts, newspapers, forest reports, diaries, and old photographs (summarized in Minnich 1988)—reveal that fire was most frequent in the chaparral, mixed evergreen forests, and mixed conifer forests on the mesic southern escarpment and at middle altitudes along the crest of the range. Fires were infrequent in pinyon forests on leeward slopes facing the Mojave Desert, as well as in lodgepole pine forests on the highest summits subject to cold winters and short growing seasons. Fire perimeter data show a similar pattern during suppression as return intervals have ranged from 40 to 60 years in chaparral to 450 years in pinyon-juniper woodland (Minnich 1978; Wangler and Minnich 1996). Fires recur at intervals of centuries in lodgepole pine forest (Minnich 1988). The trend for increasing fire intervals from the Pacific slope to the desert reflects a time dependence of fire occurrence due to heterogeneity in productivity and fuel buildup rates with mean annual precipitation. In woody ecosystems, where fire hazard is dependent on cumulative buildup of fuels, fires burned most frequently in areas with highest productivity (cf. Knight 1987; Veblen et al. 1991). The largest effect on postsuppression fire regimes, beginning about 1900 and taking hold by 1920 (Minnich 1987b; Minnich and Chou 1997), is that suppression forces have kept fires from penetrating mixed conifer forest from the chaparral belt where fire control has been less effective.

There have also been changes in fire behavior with the imposition of suppression. Anecdotal reports of smoke in newspapers and detailed reconstruction of three burns in the late 19th century reveal that individual fires persisted for long periods, with some lasting for months until extinguished by autumn storms (Minnich 1987b, 1988). Depending on weather and fuel conditions, flames spread actively for days, with long intervening periods when flames persisted in logs, standing dead trees, and other large fuels. New flame lines would later emerge from these sources. The relatively small size of burns with respect to the number of days when fires were actively moving implies low rates of fire spread and low flame line intensities before fire control. For example, a fire that persisted from July to November 1898 on Mt. Wilson in the San Gabriel Mountains consumed only 2,200 ha. Evidence of low intensities is seen in early photographs and written descriptions of braided reticulate configurations of denuded land and countless islands of unburned cover produced apparently by discontinuous flame fronts (Minnich 1987b). Newpaper reports of smoke were mostly in summer. Hence, fires were mostly pushed upslope by prevailing light-to-moderate southwesterly anabatic flows (5–10/ms) and afternoon relative humidity of 20–40% (Minnich 1987b, 1988). Similar fire weather and fire behavior occur presently without fire control in northern Baja California, Mexico (Minnich and Chou 1997).

By contrast, 20th century burns have been brief and extensive (>10,000 ha), with the largest burns invariably racing downslope across thousands of hectares per day with Santa Ana winds from the desert (wind, 10–20 m/sec, relative humidity, 5–20%) (U.S. Department of Agriculture 1982; Minnich and Chou 1997). Continuous flame fronts, sometimes 10–30 m tall, invariably denude chaparral inside perimeters. Fires seldom lasted more than a few days because "mopping up" efforts by fire crews prevented secondary recurrences once flame lines ceased spreading with improvement in the weather or fuels. The shift toward large intense fires in SBM, as well as in other parts of southern California, has been attributed to efficient "initial attack" suppression, in which the selective elimination of countless small fires in modal probability distribution of weather selects for nonrandom occurrence of large fires during the extreme weather (Minnich and Chou 1997).

The vegetation of SBM is well adapted to periodic fire, and ecosystems should be expected to persist in some steady state despite continuous local change arising from disturbance. This is supported by historical accounts of vegetation that appear to match up with present-day distributions of species and ecosystems (Minnich 1988). Perhaps the most striking example is the consistent accounts of the chaparral-mixed conifer forest ecotone along the crest of the Rim of the World, as described by Francisco Garcés in 1776, the Pacific Railroad survey in 1855, and shown on land survey maps in 1857, 1866, 1870, and 1893 that depict properties taken from public domain for the purpose of logging (Minnich 1988). The chaparral-forest boundary was also mapped along the ridge by Kinney (1887), Leiberg (1899, 1900), Grinnell (1908), and the VTM survey (Weislander 1929–1934).

The spatial extent, return interval, and vegetation damage from disturbances events tied to climate, productivity, and fuel accumulation rates may exert profound selection in the distribution of ecosystems (Veblen et al. 1991; Christensen 1993; Johnson and Gutsell 1994). Hence, the role of fire along environmental gradients can be viewed as a spatially changing balance between the forces of combustion in removing biomass and of the biological processes of establishment and growth, especially in ecosystems with contrasting physical structure, fire behavior, and survivorship. In SBM, conifer forest communities are associated with discrete ranges of understory assemblages and combustible biomass. Hence, the distributions of shrub and tree species possessing different life history attributes are shaped in part by gradients in fire intensity and recurrence interval between stand-replacement burns. Terrain, a spatially fixed variable, also contributes to species distributions by affecting fire damage through slope–wind interactions. This is seen in fire damage in SBM conifer forests, as interpreted from pre- and postburn aerial photographs matched site specifically to scale on a transfer scope (Table 3.4; for methods, see Minnich and Bahre 1995).

Allogenic factors such as weather, ignitions, and precipitation variability may engender variable fire outcomes and idiosyncratic responses by vegetal cover that promotes spatial variation in stand structure as a consequence of fortuitous events (Vale 1989; Sprugel 1991; Parker 1993). However, the occurrence of burns that comprise most of the total regional burn area was probably random in space and time before suppression, so that fire outcomes reflect modal probability distributions of allogenic factors (Minnich and Chou 1997). Over long time scales, the variance in disturbances would be expected to produce some expression of a steady state at the landscape scale.

The high-intensity conflagrations that have occurred in SBM over the past century have altered the vegetation in many areas since the first aerial photographs were first taken of the region in 1938. A critical question is whether recent changes reside within the natural variability of successions arising from disturbance or reflect long-term directional change. This question can be addressed if the long-term dynamics of ecosystems are understood.

The remainder of the chapter summarizes the fire regimes in chaparral and coniferous ecosystems of SBM, with discussion of disturbance pattern, postfire succession, and possible long-term change due to suppression, as summarized from Minnich (1978, 1988).

Table 3.4. Spatial Extent of Stand-Replacement Burns (% of Stand Area) of Conifer Forest Types in the San Bernardino Mountains[a]

		Slope aspect			Slope class (°)				
Forest type	Fire	North	South	Total	0–10	10–20	20–30	30–40	>40
Coulter pine–chaparral	Bear	87	90	89	—	87	87	94	100
Coulter pine–canyon oak	Bear	86	91	90	—	67	92	91	100
Coulter pine–black oak	Bear	—	30	30	—	43	0	—	0
Bigcone Douglas fir–canyon oak	Kitch	5	50	21	—	—	14	27	—
	Meyer	38	36	37	—	90	45	28	10
	Bear	27	25	26	—	32	42	13	2
	Village	25	30	25	—	—	49	9	5
	Baldy	16	18	18	—	—	33	14	6
Mixed conifer	Meyer	10	54	24	—	—	33	27	0
	Bear	16	25	20	—	0	20	18	17
	Baldy	19	21	20	—	47	17	21	—
Mixed conifer–black oak	Meyer	33	61	54	42	60	0	—	—
	Bear	16	25	20	15	20	30	33	—
Mixed conifer–canyon oak	Meyer	53	58	57	—	—	100	62	35
	Bear	71	76	75	—	70	80	79	45
	Baldy	—	26	26	—	—	27	25	—
Mixed conifer–timber chaparral	Meyer	0	35	33	—	—	—	33	—
	Baldy	—	48	48	—	14	52	38	—
Mixed conifer–mountain mahogany	Meyer	80	70	72	—	—	—	50	85
	Baldy	23	—	23	—	—	7	24	—
White fir	Meyer	52	50	51	—	—	50	52	—
	Bear	20	—	20	0	0	15	35	0
	Baldy	6	13	13	—	—	31	10	—
White fir–timber chaparral	Bear	22	41	33	0	29	43	43	—
Lodgepole pine	Baldy	47	—	47	—	58	21	7	—

[a]Interpreted from color and color infrared aerial photographs (1:20,000) obtained before and immediately after fires by site-specific matching of photographs, using a zoom transfer scope (Minnich and Chou 1997). Postdisturbance vegetation classified as burned (canopy defoliated), scorched (canopy dried but not burned), and unburned (some green canopy usually as canopy apices). With the exception of *Pseudotsuga macrocarpa*, which is capable of resprouting, stands burned or scorched were classified as stand-replacement burns. Stands with postburn green canopy are classified as understory burns (tree survivorship). The area of fire-killed and surviving stands was computed on the GIS and correlated with vegetation type determined from preburn photography. Slope aspect divided hemispherically centered to 0° and 180°. Years of fire: Bear, 1970; Meyer, 1970; Kitching, 1974; Village, 1975; Baldy, 1980.

Chaparral

The comparison of recent fire history in southern California (fire control) and northern Baja California (no fire control) suggests that fire intervals had not changed significantly (50–70 years) with the establishment of suppression, as would be expected in a time-dependent fire regime (Minnich and Chou 1997). Hence, presuppression chaparral fire intervals in SBM may have been similar to postsuppression intervals of 40–60 years.

Long fire intervals are related to low fire probabilities during the first decades of succession due to low biomass, open stand cover, low dead fuel buildup, and presence of a large volume of plant water essential to the evergreen habit of chaparral. The denudation of chaparral by canopy burns produced by high fuel continuity and large leaf/stem surface-to-volume ratios (Countryman and Philpot 1970) also discourages short-interval fire recurrences because carryover fuels through burn cycles are limited. During postfire succession, good stomatal controls of evergreen shrubs reduce transpiration under high evaporative demand, resulting in low photosynthesis rates and gradual cumulative fuel buildup (summarized in Hanes 1988; Keeley and Keeley 1989). The ratio of dead-to-live fuel is widely reported to gradually increase with time since fire, with dead fuel buildup in older stands related to the reduced ability of shrubs to meet concurrent demands of photosynthate storage and growth during spring. Dead fuel buildup may not correlate with stand age, especially in broadleaf types dominated by *Arctostaphylos, Quercus,* and *Ceanothus* spp. (Paysen and Cohen 1990). Riggan et al. (1988) propose that the expanding foliage area of dominant shrubs in maturing stands advances seasonal soil drying, hastening the onset of seasonal drought stress.

Written accounts of SBM before fire control, especially those by Leiberg (1899, 1900), indicate a fine-grained patch structure in chaparral, similar to that observed in other ranges of southern California during the period and in present-day Baja California (Minnich 1987b, 1988; Minnich and Chou 1997). For example, in SBM the 1903–1904 L.C. Miller Silviculture Survey party saw evidence of fire at ten localities between Waterman Canyon and Barton Flats. Leiberg (1900) made the most explicit account of patch mosaics in the San Jacinto Mountains, 50 km SE of SBM, describing the chaparral as "a growth which varies from extremely dense to thin or open, but rarely forms large uninterrupted patches. The dense portions are commonly separated by narrow lanes [burns], which are either wholly free of brush, or bear a scattered growth so thin as to offer no serious obstacle to travel" Leiberg (1899) also states that "recent fires" were "scattered throughout the reserve in small tracts."

Although fragmentary historical evidences in SBM do not permit a reconstruction of the size properties and recurrence intervals of burns at that time, the presuppression patch mosaic may have been similar to present-day Baja California, Mexico. Differences in fire history of southern California and northern Baja California indicate a time dependence in fire occurrence that is expressed in a negative feedback between fire size and frequency. Fires in Baja California are numerous (6.7 kha^{-1} 50 yr^{-1}) and small (<2,300 ha), and this has resulted in a highly fragmented patch mosaic. Patchiness was further enhanced by the erratic reticulate spread of summer-long fires. By contrast, 100 years of fire suppression in California have resulted in extensive stands of mature chaparral as the result of relatively few large fires (0.9 kha^{-1} 50 yr^{-1}) (Minnich 1983; Chou et al. 1993; Minnich and Chou 1997). Patch enlargement provoked by efficient containment of fire starts began early in the 20th century, as evidenced by an abrupt change in patch frequency and patch size distribution along the international boundary on aerial photographs taken as early as 1938 (Minnich 1987b; Chou et al. 1993; Minnich and Chou 1997). In SBM, the southern escarpment traversed by the 1903–1904 L.C. Miller party is now dominated by only three patches (burns in 1956, 1970, 1980).

High fire frequencies (events/area) observed in Baja California can be accomplished by lightning exclusively. Electromagnetic lightning detection data show that an area of 1.0 kha sustains a minimum of one strike every 1–2 years. A rate of 4% of discharges leading to fires would produce fire frequencies seen now in northern Baja California. At present, 2–11% of discharges in SBM produce fires that require suppression action (Minnich et al. 1993). Hence, natural ignition sources alone can readily produced patchiness comparable with that in Baja California.

Chaparral appears to be stable in SBM, despite fire enlargement, because shrubs either resprout or recruit from in situ dispersed seed caches, permitting efficient stand establishment under variable fire intervals and sizes (Keeley 1989; Minnich and Bahre 1995). Replication of 1929–1934 VTM brush plots in 1974 shows little change in species composition, except for the disappearance of short-lived *Ceanothus leucodermis* at sites unburned between the two surveys (Minnich 1978). Short-interval burns (<10 years) capable of degrading chaparral (Keeley 1989) occur rarely in SBM, except in lower southern exposures of Waterman Canyon and City Creek, where the invasion of flashy exotic annuals *Bromus rubens, B. diandrus, Brassica geniculata,* and *Avena barbata* (introduced in 1890; Minnich and Dezzani, in press) has contributed to short-interval recurrences and displacement of chaparral by exotic grassland, beginning in 1970 (cf. Zedler et al. 1983).

Increasing fire intervals toward the desert are paralleled by the decline in the presence of obligate seeders common to the moist southern escarpment, leaving long-lived postfire sprouters that establish by long-range dispersal of fleshy fruits or wind-dispersed seed, including *Quercus john-tuckeri, Cercocarpus betuloides,* and *Arctostaphylos glauca. A. glauca* adapts to desert conditions by layering and vegetative propagation (Vasek and Clovis 1976). Apparently, long fire intervals at the scale of centuries on the desert escarpment selectively eliminate the seedbanks of many obligate and nonobligate seeding shrubs.

Mixed Evergreen Forest

Pre- and postsuppression fire intervals in disjunct mixed evergreen forests were undoubtedly similar to the chaparral that surrounds this community (40–60 years). However, the structural diversity provided by chaparral and broadleaf evergreen hardwoods in mixed evergreen forests results in complex, disparate fire regimes and successions.

Studies in California indicate that *Pinus coulteri* and *P. attenuata,* characterized by serotinous or partially serotinous cone habits, respond to severe fires with high colonizing ability and reproductive effort that is concentrated in postburn growing seasons (Vale 1979; Talley and Griffin 1980; Griffin 1982; Borchart 1985; Vogl et al. 1988). In SBM, both species mostly experience stand-replacement burns (*P. attenuata* forests, 100% of stands; *P. coulteri* forests, 90% of stands) owing to their occurrence in dense chaparral and on steep, smooth, concave slopes subject to fire intensities, with denudation rates increasing with slope (Table 3.4). Mortality rates in *P. coulteri* tend to decline with increasing altitude as understory fuels shift from chaparral to *Quercus kelloggii* woodlands (Minnich 1980). A trend for decreasing serotiny in *Pinus coulteri* over a similar vegetation transition in Santa Barbara County suggests that strong selection may exist for the natural variation in cone habit (Borchart 1985). Presuppression accounts in SBM also describe stand-replacement fires in these types (Minnich 1988).

During postfire succession, seedlings germinate mostly during the first growing season from seed dispersed from serotinous cones persistent at the ends of branches or along the boles of fire-killed trees. Even-aged stands become pole size within 20–40 years. Cones are produced within 10–20 years, ensuring recruitment after short fire recurrences. The partial serotiny of *P. coulteri* permits recruitment between fire cycles, but establishment rates seem to be limited, perhaps in competition with dense shrub cover. Establishment by long-range dispersal from living distant stands is unlikely because seed is dispersed locally.

The large size of chaparral burns relative to the size of serotinous pine forests groves tends to synchronize stand age distributions across entire stream drainages, resulting in local temporal nonequilibria in their spatial extent at local scales. Even if fire size were smaller before fire control, postsuppression changes in fire intensity may not have altered distributions, because both *P. coulteri* and *P. attenuata* sustain stand-replacement burns with or without fire control. However, failures in postfire seedling establishment—due perhaps to drought or destruction of seed—would cause local extinctions because both the adult population and seed are eliminated. Such failures could possibly explain their erratic and fragmented distributions.

Pseudotsuga macrocarpa best illustrates how altered fire regimes may cause directional change in the structure and distribution of a forest ecosystem. Studies indicate that *P. macrocarpa* survives fire in convex canyons and cliffs and establishes offspring infrequently but continuously (Bolton and Vogl 1969; Minnich 1980). Nevertheless, presuppression accounts by Leiberg and others suggest that *P. macrocarpa* forests sustained severe fire damage and occasional defoliation

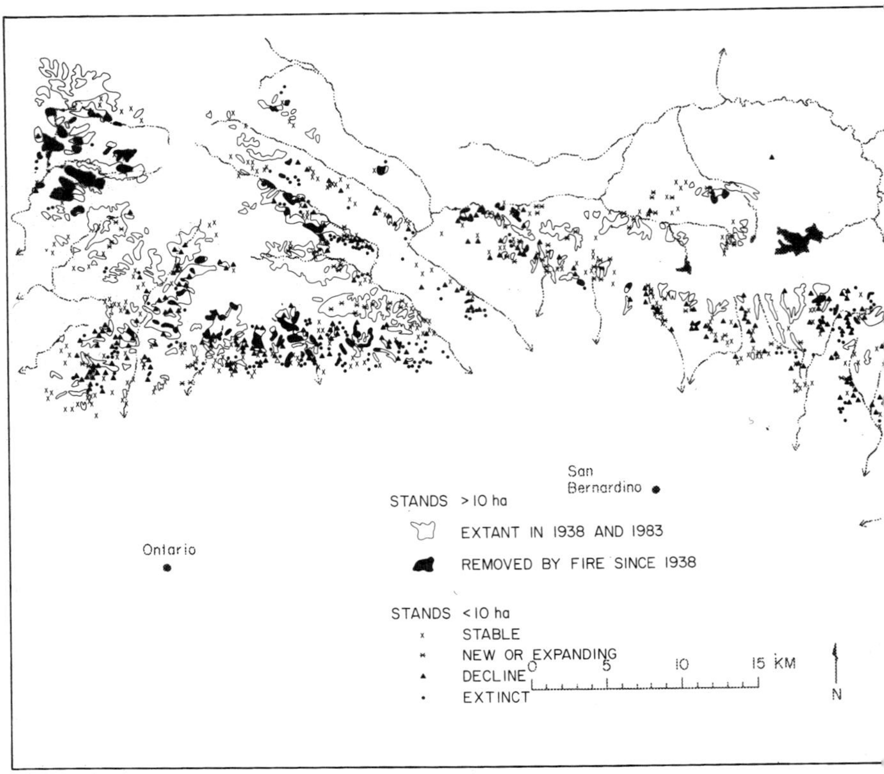

Figure 3.2. Distribution of *Pseudotsuga macrocarpa*. Stands present in 1938 and 1983 are shown by open polygons and (x) symbols (stands <10 ha). Stand-replacement fires since 1938 are shown by dark polygons and (·) symbols (stands <10 ha). Asterisks are new stands less than 10 ha establishing since 1938.

from chaparral fires. In fact, early settlers and naturalists thought that the tree was in decline based on observations of dead standing trees in most stands and the unsubstantiated deduction that fragmented forests were remnants of former continuous forests (Minnich 1988). However, *P. macrocarpa*'s ability to recover by sprouting after intense fires was never recognized nor was the abundance of standing dead trees understood in terms of *P. macrocarpa*'s resistance to windfall for 20–50 years.

Santa Ana wind–driven fires have produced large changes in the distribution of mature *Pseudotsuga macrocarpa* forest, as reconstructed from repeat aerial photographs since 1938 (Fig. 3.2). Nearly 50% of stands were defoliated in burns since 1938. About half of the defoliated stands resprouted from epicormic buds on large branches and along the bole (Bolton and Vogl 1969), resulting in an average of 18–37% of stands experiencing stand-replacement burns (Table 3.4). The spatial extent of stand-replacement burns was unrelated to slope exposure but inversely related with slope gradient. Trees along the perimeter of individual

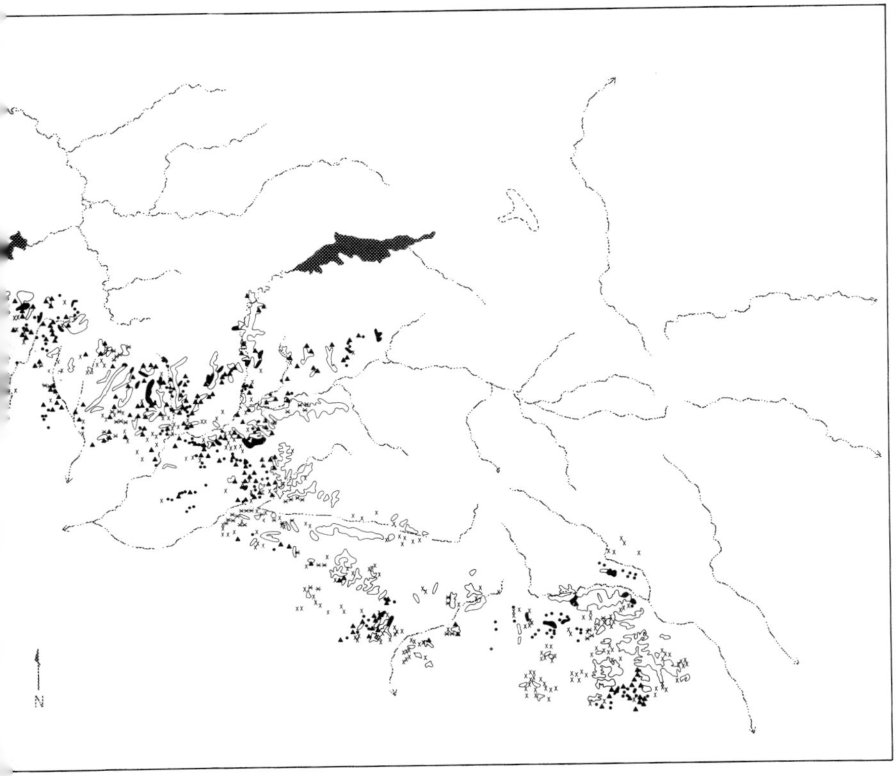

Table 3.5. Chronosequence of Postfire Shrub Cover (%) in Burned *Pseudotsuga macrocarpa* stands[a]

Time since fire (yr)	n	Species[b] Af	Ag	Cl	Ci	Cb	Aq	Qc	Total
4	12	2.1 (3.9)	3.1 (4.8)	6.1 (6.6)	17.1 (20.7)	0.4 (1.4)	7.0 (9.6)	27.7 (8.1)	62.5 (17.9)
10–12	4	0.3 (0.4)	3.5 (6.1)	0	24.0 (22.1)	4.5 (5.0)	2.5 (4.3)	37.5 (4.5)	71.0 (12.8)
18–19	5	5.2 (4.9)	4.8 (6.1)	3.8 (5.1)	12.7 (11.5)	9.5 (13.9)	5.7 (2.8)	27.1 (7.6)	81.7 (14.6)
21–25	6	0.2 (0.3)	0	0.7 (0.9)	40.0 (14.9)	0.5 (0.8)	5.0 (8.7)	29.3 (11.7)	84.5 (8.0)
38	1	0	0	0	2.0	0	0	63.0	65.0
56	1	0	0	0	1.0	0	0	90.0	91.0
64	1	0	0	0	3.0	0	0	69.0	87.0

[a]Total cover from 100-m line intercepts. Standard deviations given in parentheses.
[b]Af, *Adenostoma fasciculatum;* Ag, *Arctostaphylos glandulosa;* Cl, *Ceanothus leucodermis;* Ci, *C. integerrimus;* Cb, *Cercocarpus betuloides;* Qw, *Quercus wislizenii;* Qc, *Q. chrysolepis.*

groves were exposed to intense chaparral flame lines, whereas those in the interior were scorched or undamaged by understory fires due to limited chaparral fuels and the presence of fire-resistant *Quercus chrysolepis*.

Chronosequences show that *P. macrocarpa* forests experience divergent successional outcomes associated with fire intensity (Table 3.5; Fig. 3.3). Burned forests were covered by resprouts of *Quercus chrysolepis, Q. wislizenii,* and *Cercocarpus betuloides,* as well as dense thickets of *Ceanothus integerrimus, C. leucodermis,* and *C. oliganthus* establishing from seedbanks (Table 3.5). *Q. chrysolepis* was reduced to shrubby resprouts because its thin, dry, flaky bark does not protect arboreal stems from cambium damage in high-intensity surface fires (Plumb 1980). After 30 years, *Q. chrysolepis* developed canopy closure with contemporaneous dieback of *Ceanothus* spp. *P. macrocarpa* recruitment was less than 10/ha in stands as old as 40 years because juveniles appear to be poorly adapted to heat and drought conditions until the development of *Q. chrysolepis* overstory (Bolton and Vogl 1969). Establishment may also be delayed by inefficient seed dispersal from distant living stands. Sapling growth appears to be stalled by shade stress caused by the overstory (Gause 1966; Bolton and Vogl 1969; McDonald and Littrell 1976). Growth accelerates only after trees push through the *Q. chrysolepis* canopy at 50–70 years (i.e., it is tolerant of full sun only after it is well established) (McDonald and Littrell 1976).

Forests sustaining little canopy damage from surface fires retained the characteristics of mature stands, regardless of time since fire (i.e., mature forests, arboreal *Q. chrysolepis* understory, and scarce chaparral), even during postfire succession. Scorched forests experienced partial denudation of the canopy layer and moderate chaparral establishment in early succession.

Stand-replacement fires have produced large changes in *P. macrocarpa* forest distribution, with 2,746 ha being converted to *Quercus chrysolepis* woodlands. An average of 26% of stands within fire perimeters sustained stand-replacement fires (Fig. 3.4). The smallest groves having the highest perimeter/area ratios suffered the greatest losses, with 241 groves becoming extinct, nearly all less than 5 ha. During the same period, new stands established at 118 sites, covering only 239 ha, mostly in areas contiguous with extant groves *P. macrocarpa* forest. Sixty-two groves established at sites disjunct from extant forests. Post-1938 stand-replacement burns have resulted in a net decline of 2,507 ha or 18% of the 1938 forest area. The cumulative area of *P. macrocarpa* forest declined in most size classes. For each size class, the gains resulting from the fragmentation of larger stands were exceeded by even greater spatial fragmentation from stand-replacement burns (Fig. 3.4).

Current losses may be a natural perturbation due to variability in the fire process. Because mean chaparral fire return periods of 40–60 years coincide with initial *P. macrocarpa* establishment, stands persist primarily by the direct survival of old-growth trees (200–600 years) (Bolton and Vogl 1969). The small size of stand-replacement burns favors the establishment of mixed-aged groves resulting from both continuous recruitment and residual survivorship from multiple fire cycles spanning centuries (Bolton and Vogl 1969; McDonald and Littrell 1976).

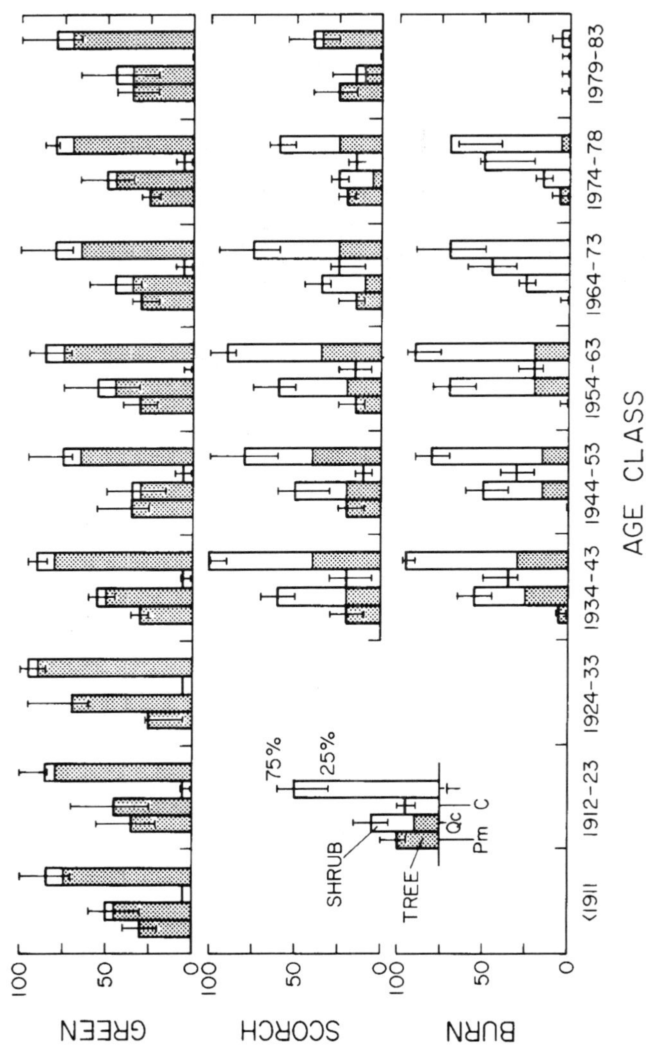

Figure 3.3. Chronosequences of *Pseudotsuga macrocarpa* forest successions with fire intensity. Fire damage: burn (canopy defoliated), scorch (canopy cured but not burned), and green (partial green canopy).

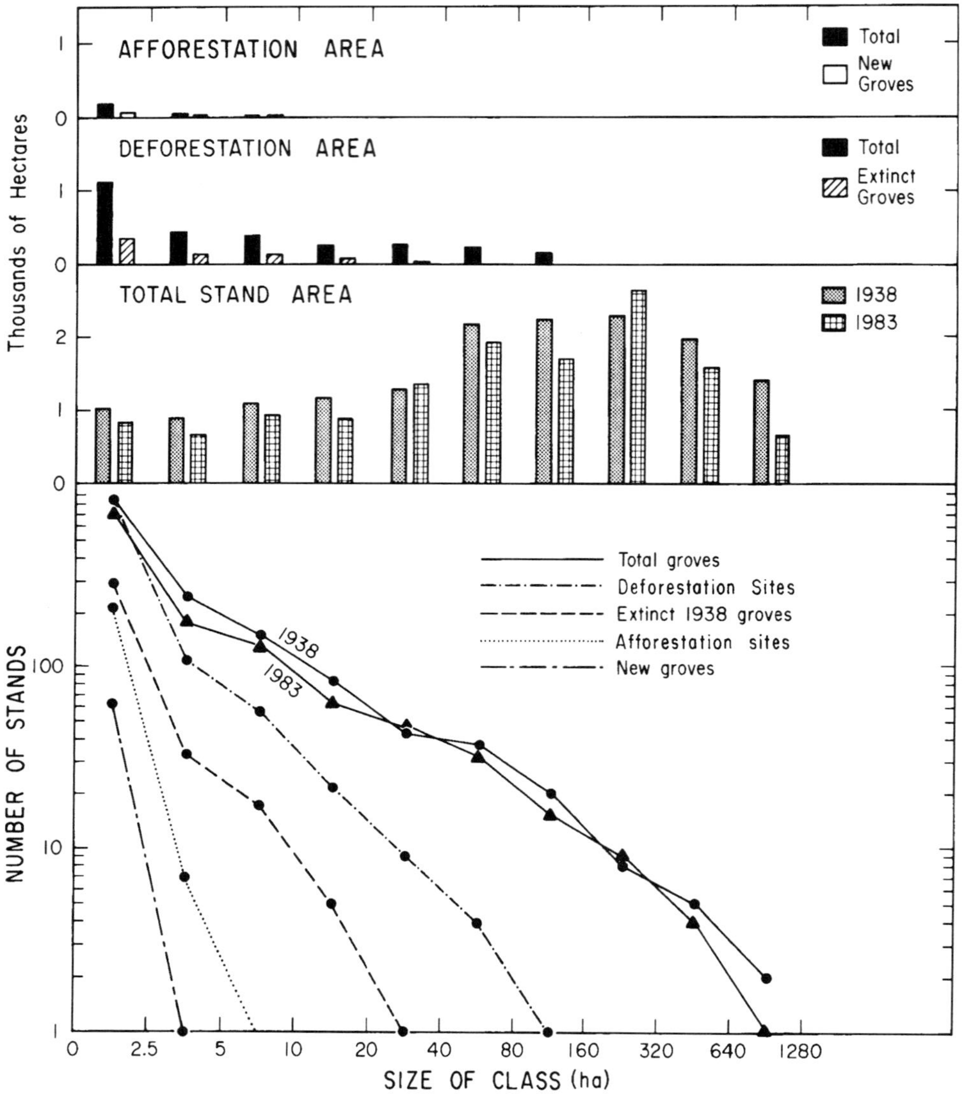

Figure 3.4. Size frequency distributions of afforestation (new stands), deforestation (stands removed), and total stands in *Pseudotsuga macrocarpa* forests of the San Bernardino and eastern San Gabriel mountains.

For this reason, the large size of chaparral burns relative to *P. macrocarpa* forest should not act to synchronize events in forest age class distribution nor produce temporal nonequilbria in the spatial extent of stands, at least to the degree seen for serotinous pines. Chronosequences provide no evidence of mass recruitment in postfire succession (cf. Bolton and Vogl 1969).

Climate does not appear to be a factor in changing *P. macrocarpa* distributions. Correlations between growth rings with precipitation for several tree species in central California reveal little evidence for changes in precipitation over the past 400 years (Michaelsen et al. 1987; Haston et al. 1988). Regressions of records dating to the 19th century at Los Angeles and San Diego reveal no trend in annual precipitation (Kahrl et al. 1978).

Current losses in *Pseudotsuga macrocarpa* may also be related to suppression-related effects on fire weather. In the late 19th century, low fire intensities associated with summer-long burns may have permitted sufficiently long intervals between stand-replacement burns so that slow-growing saplings could mature at comparable rates to the mortality of canopy trees. Excess fuel loadings may not have contributed to recent losses because chaparral fire intervals were affected by suppression (Minnich and Chou 1997). Present stand-replacement rates of 26% of stands from Santa Ana wind–blown fires appear to be excessive compared with the spatial extent of stand establishment. The continuation of fire control may eventually establish a new steady state in which *P. macrocarpa* forests become restricted to the steepest terrain that is resistant even to intense chaparral fires.

Cumulative air pollution effects may be limited in knobcone pine and mixed evergreen coulter pine forests where the lifespan of trees is mostly limited to single fire cycles. Even-aged stands of both species have recently regenerated after fire in many areas of the southern escarpment and some entirely within the period of air pollution beginning in the late 1950s. Perhaps the short generation times of these conifers may foster rapid selection of ecotypes resistant to ozone and other gases. The smog-sensitive species *Pseudotsuga macrocarpa,* which also grows mostly in the polluted southern escarpment, has sustained large losses in distribution from chaparral fires since 1938. Forests that have not burned since the arrival of air pollution in the 1950s appear to be stable or exhibit modest stand enlargement.

Mixed Conifer Forest

Mixed conifer forests of SBM illustrate how lengthening fire intervals with suppression can lead to excessive fuel load, alteration of vegetation structure, and species changes. A large body of historical writings, old photographs, and timely surveys of SBM, including the Forest Reserve survey by Leiberg (1899, 1900), reveals that recurrent surface fires in mixed conifer forest selectively eliminated shrubs and young trees, producing open forests of mature trees (Minnich 1988), similar to presuppression mixed conifer forests elsewhere in California (summaries in Rundel et al. 1988; McKelvey and Johnston 1992; SNEP Science Team 1996). VTM quadrats taken in SBM in 1930 (Minnich et al. 1995) show that old-growth forests had a flat multiple-age structure but some have a classic climax-type inverse J-shaped distribution, with maximum stem density in 70–100-cm dbh. Stand densities ranged from 70 trees/ha in *P. jeffreyi* forest to 170 trees/ha in *A. concolor* forest (Table 3.3). *Pinus jeffreyi* and *P. ponderosa* dominated south-facing slopes and flats, whereas *Abies concolor* dominated north-facing slopes.

Mixed ponderosa pine forest in western SBM was logged during the late 19th century, owing to the accessibility of these stands to the Los Angeles market (Minnich 1988). Cutting began west of Crestline (1857), then proceeded slowly eastward to Lake Arrowhead (1870), Running Springs (1900), and Green Valley Lake (1908). According to Leiberg (1900), the "cutting was never very clean," owing to the primitive steam sawmill technology of the period. Ground disturbance and the piecemeal elimination of old-growth forest resulted in the conversion of stands into hazardous dense stands of second growth, as VTM plots recorded densities as high as 470 stems/ha by 1932 (Minnich et al. 1995). Crown fires consumed several stands including Lake Gregory (1919), Heaps Peak (1922, 1956), and Jobs Peak (1936). Commercial logging on private lands ceased about 1910 with the establishment of the national forest. The forest service has since conducted widely scattered timber sales until 1970.

Livestock grazing was also practiced widely for 50 years in SBM beginning about 1860 (Minnich 1988). Interviews conducted with early settlers by T.P. Lukens in 1900 suggest that sawmill families in western SBM pastured livestock—mostly cattle—for the operation of the mills, as well as for subsistence. Farther east, cattle grazing was conducted in the major meadows at Big Bear Basin, Baldwin Lake, Holcomb Valley, Little Pine Flat, and the upper Santa Ana River. Sheep were grazed in open range, especially in the northeastern quarter of the range.

Consensus was lacking concerning the long-term effect of grazing during the period because the presence of livestock on the forest reserves was a political issue (e.g., Lockmann 1981). For example, during interviews with longtime settlers during the summer of 1900, T.P. Lukens—an early naturalist and confidant of John Muir—concluded that the openness of forests was due to livestock, but the effects of livestock were not differentiated from the effects of understory fires (Minnich 1988). By contrast, Leiberg (1899, 1900) wrote that grazing had little effect in the forests owing to the scarcity of herbaceous cover from dry summers.

The past effect of livestock on vegetation may be seen by examining a modern grazing regime in mixed conifer forest of the Sierra San Pedro Mártir of Baja California (Minnich et al., in press). This entire range is exposed to some grazing because stray cattle wander from the primary forage areas in the meadows. However, data obtained in small areas fenced to exclude cattle suggest that cattle grazing in forests has no measurable effect on herbaceous and shrub cover. Neither did cattle affect the recruitment of conifers. Most shrub species in the chaparral and mixed conifer forest were unpalatable to livestock (Minnich and Bahre 1995). Grazing annually removed about 50% of the biomass of meadows (*Juncus* spp., *Carex* spp., *Aster occidentalis, Eleocharis pauciflora, Muhlenbergia richardsonis, Trifolium wigginsii, Potentilla wheeleri*). Yet, interannual fluctations in meadow biomass due to precipitation variability greatly exceeded the effects of grazing pressure owing to the resilience of meadow species.

Anthropogenic control of fire has led to an increase in fire intervals, greater stand densities, and species changes during the 20th century (Vankat 1977; Vankat and Major 1978; Kilgore and Taylor 1979; McKelvey and Johnston 1992; Swetnam 1993). Minnich et al. (1995) replicated 72 field quadrats of the VTM taken in 1929–1934 in the SBM of southern California. Only six plots of 72 had burned during the intervening 60 years, giving an estimated fire rotation period of 360 years. They found that 60 years of suppression has caused stem density increases of 100–200 stems/ha (dbh >10 cm), with increases in density proportional to mean annual precipitation. Densities locally increased as much as 300 stems/ha in forests with mean annual precipitation of 100 cm, whereas forests with 40 cm experienced virtually no change. Species changes varied with forest type. White fir and monotypic Jeffrey pine experienced no change in composition, whereas recruitment in mesic mixed conifer Jeffrey and ponderosa pine forests was dominated by *Calocedrus decurrens* and *Abies concolor*. Tree densities on the west slope of the Sierra Nevada (mean annual precipitation, 100–150 cm) have climbed to 500 stems/ha and 200/ha in drier "eastside" *Pinus jeffreyi* forest (mean annual precipitation, 50–100 cm) (Table 3.3) (Vankat 1977; Vankat and Major 1978; McKelvey and Johnston 1992). Peterson and Arbaugh (1992) report stand densities of 175–2,770 stems/ha. Stand-thickening in the Sierra Nevada was accompanied by age-specific trends in which juvenile pole-size cohorts are dominated by *A. concolor* and *C. decurrens*, and mature overstory cohorts are dominated by *P. ponderosa* and *P. jeffreyi* (Vankat 1977; Vankat and Major 1978; Parsons and DeBenedetti 1979; McNeil and Zobel 1980; Bonnickson and Stone 1981, 1982; Kercher and Axelrod 1984; Rundel et al. 1988; MeKelvey and Johnston 1992).

Stand-thickening has been accompanied by an increasing density of standing dead trees (snags). Savage (1994), working in a dense mixed conifer forest in the San Jacinto Mountains, 40 km SE of SBM, found dead tree densities of 365/ha in severe mortality plots. Mortality was several times higher for *P. ponderosa, P. jeffreyi*, and *P. coulteri* than for *P. lambertiana, Abies concolor*, and *C. decurrens*. It was concluded that high mortality was due to an overabundance of trees competing for moisture and nutrients, with the shift in species dominance further heightened by differential response to drought, bark beetle infestation, and air pollution.

Evidence concerning the spatial extent and recurrence interval of presuppression fires that maintained the open forest structure of SBM is unclear. Fire scar dating (FSD) in SBM (McBride and Laven 1976) shows average fire return intervals of 15–29 years, with intervals increasing from the mesic Pacific escarpment on semiarid leeward slopes. SBM fire intervals are longer than in the Sierra Nevada, where reconstructions synthesized from FSD studies argue that fires recurred within a span of 4–20 years (Kilgore and Taylor 1979; Bonnickson and Stone 1981, 1982; Kercher and Axelrod 1984; Parsons and Swetnam 1989 Swetnam 1991, 1993). Short intervals are attributed to the survival of canopy trees and rapid deposition of litter. In white fir forests, longer fire-free periods led to more intense fires (Kilgore and Taylor 1979; Kercher and Axelrod 1984).

Site-specific fire scar sampling arrays seldom have sufficient resolution to distinguish the size of burns, nor can they distinguish landscape burns from spot fires that may have limited spatial extent or ecological impact (Johnson and

Gutsell 1994). A synoptic landscape-scale study of unmanaged fire in the Sierra San Pedro Mártir, using repeat aerial photographs, shows that forests there are open and parklike, dominated by trees greater than 60 cm dbh at densities of 50–150 stems/ha, well within the range of SBM densities in the VTM survey of 1929–1934 (Minnich et al., in press). Fires were also moderately large (500–6,000 ha), had fire rotation periods of 52 years, and had sufficient intensities to scorch canopy foliage 10–20 m aboveground. There was heavy mortality to both sapling and pole-size stems, as well as a few trees in the canopy layer. Long recurrence intervals are attributed to a gradual cumulative development of fire hazard related to low rates of production of fine fuels and buildup of understory canopy of shrubs, conifer saplings, and pole-size stems. Conifer needles have a limited capacity for supporting flash fires because of packing of the litter bed. Climatic data for locations in mixed conifer forests in the Sierra San Pedro Mártir, southern California mountains, and the Sierra Nevada suggest that environmental differences are not large enough to explain the differences in fire intervals.

Minnich et al. (in press) hypothesized that presuppression fire intervals in California using FSD methods have been underestimated due to problems of scale related to assumptions made with respect to site-based (point) sampling of fire as a spatial process (Johnson and Gutsell 1994). Under long-term lightning discharge densities of 3.0 kha^{-1} yr^{-1} (Minnich et al. 1993), the density of small fires may be sufficiently high that even small stands experience high frequencies of both local events and landscape burns. Evidence is seen in the occurrence of hundreds of lightning-initiated spot fires annually that died out on their own without suppression (Minnich et al., in press). Hence, FSD estimates include both landscape burns and and local events that collectively use up little fuel, cover little area, have little demographic impact on forests, thereby leading to underestimates of true spatial fire intervals.

The disparity in estimated fire intervals suggests that California forests may have experienced greater fuel buildup, fire intensities, and forest damage than previously envisioned. Kilgore and Taylor (1979), who correlated fire scar records with stand age data in the central Sierra Nevada, concluded that fires were small (1–800 ha), patchy, and irregularly shaped. However, the small size of the study area (2,300 ha) may have resulted in an underestimation of fire size. From FSD studies, it is hypothesized that recurring fires mostly killed saplings, resulting in the production of uneven-age aggregation (cohort) mosaics of mostly less than 2 ha (Bonnickson and Stone 1981, 1982; Kercher and Axelrod 1984). Small gaps created by local intense fires were colonized by seedling thickets that were thinned out by subsequent low-intensity fires, ultimately producing an old-growth cohort. Alternatively, data from the Sierra San Pedro Mártir suggest that mature forests result from broad-scale intense fires that uniformly eliminate most recruits to as large as pole-size status. Stands are mixed age in which relatively few pole-size trees join the canopy layer during successive fire cycles and relatively few canopy trees die of fire or other factors. The rate of entry into the overstory class is balanced by low mortality rates in the overstory class.

Estimated 15–29-year fire intervals in SBM (McBride and Laven 1976) may include both landscape and spot burns. No direct evidence exists on spatial recurrence intervals, which may be longer than predicted from FSD methods. Evidence of intense presuppression disturbances in SBM is seen in the form of numerous small forest gaps caused by presuppression canopy fires recorded on 1938 aerial photographs (Minnich 1988). Most occurred on steep smooth slopes and exposed ridges covered with greater fuel loads along chaparral ecotones or with understory of timberland chaparral and *Quercus chrysolepis* (Minnich 1988). In the Sierra San Pedro Mártir stand-replacement burns within perimeters of two burns in 1989 were also localized (<5 ha), with patches cumulatively accounting for 16.2% of forests within perimeters (Minnich et al., in press).

Postsuppression increases in fuel loadings (Kilgore 1973; van Wagtendonk 1974; Kilgore and Taylor 1979) have converted the fire regime from a patchy mosaic continuum to a sustained cyclic stand-replacement fire regime (Kilgore 1973, 1981; Parsons and DeBenedetti 1979; Parker and Peet 1984). Examples of large canopy fires may be found at Yosemite National Park and Stanislaus National Forest (McKelvey and Johnston 1992; Weatherspoon et al., 1992). In southern California, recent fires produced stand-replacement burns in 20–75% of forests (Table 3.4), with highest mortality rates occurring on steep slopes subject to fire "runs" and among stands with understory of *Quercus chrysolepis* or timberland chaparral. Stands on gentle surfaces with no shrub understory suffered the lowest incidence of stand-replacement burns. The spatial extent of stand-replacement burns was largely unrelated to slope exposure, indicating that fire intensities were more related to successional fuel loads than to microclimatic differences with terrain. During the 1970 Bear fire, nearly one-third of stands (970 ha) were fire-killed (Minnich 1980).

Chronosequences show that postfire succession in stand-replacement burns was dominated by timberland chaparral, *Ceanothus integerrimus, Quercus chrysolepis,* or Great Basin sage scrub *Artemisia californica* and *Chrysothamnus nauseosus* (Minnich 1978). Similar successions were documented in the Sierra Nevada, where shrubs and hardwoods either sprouted (*Quercus kelloggii, Q. chrysolepis*) or established from soils seedbanks stimulated by high fire intensities (*Ceanothus, Arctostaphylos*) (summarized in Kauffman and Martin 1985; Kauffman and Martin 1990, 1991). Some shrubs establish from seed dispersed long distances by wind.

Fire appears to influence the distribution of mixed conifer forests. This type mostly occurs in areas with low combustible chaparral biomass—either at higher altitudes above the chaparral belt or in gentle terrain at lower altitudes lacking significant shrub cover. Stands also descend into protected convex canyons and cliffs, which also lack shrub fuels. Mixed conifer forests are usually not found on steep undissected slopes covered with contiguous chaparral, *Quercus chrysolepis* scrublands, and mixed evergreen *Pinus coulteri* forests subject to greater fuel loads and stand-replacement fires. Apparently, the interval between stand-replacement burns in the shrub-dominated ecosystems selectively eliminate mixed conifer tree species. Alternatively, the persistence of mixed conifers in a regime of ground fires may selectively eliminate shade-intolerant chaparral species and *P. coulteri*. *Pseudotsuga macrocarpa* is seldom found with mixed conifer forest, even though both communities comprise long-lived species that survive understory fires. However, *P. macrocarpa* thrives on a specialized habitat of steep slopes subject to high erosion rates, and the dense shade of *Quercus chrysolepis* provides poor habitat for the sun-tolerant mixed conifer dominants *Pinus ponderosa* and *P. jeffreyi*. Shade-tolerant *Abies concolor* and *P. lambertiana* are also removed by infrequent stand-replacement burns, whereas *P. macrocarpa* survives by resprouting. In fact, the sprouting habit of *P. macrocarpa* may be the primary adaptation that permits it to occupy fire-resistant habitats to as low as 800 m in SBM.

Although mixed conifer forests growing on the moist southern escarpment and crest of SBM have the highest productivity, fuel accumulation rates, and potential fire hazard, these areas are also subject to the highest air pollution concentrations in summer due to the advection of marine air from the Los Angeles Basin as a consequence of onshore flow and anabatic circulations (Edinger et al. 1972). The trend for broad-scale stand-thickening in unburned mixed conifer forest over the past 60 years suggests that air pollution is not interfering with recruitment. Indeed, stand-thickening rates are highest in mesic forests most prone to high air pollution concentrations, such as Rim of the World. At the same time, stand-thickening may be synergistically contributing to enhanced damage and mortality from air pollution because of decreasing fitness of tree species as a consequence of increased competition for moisture and nutrients. The recent decline in the recruitment of *Pinus ponderosa* and *P. jeffreyi* (Minnich et al. 1995) may reflect the high sensitivity of these species to air pollution. However, the displacement of pines by *Abies concolor* and *Calocedrus decurrens*, which are tolerant to ozone and other gases, is also widely reported in less polluted forests of California, including the Sierra Nevada (SNEP Science Team 1996). Open semiarid mixed conifer forests adjacent to the Mojave Desert have experienced limited stand-thickening, but these stands are exposed to lower air pollution concentrations.

Pinyon Woodlands

Pinyon-juniper woodlands in SBM (*Pinus monophylla, Juniperus californica, J. occidentalis*) have experienced long-period stand-replacement fires both before and during suppression (fire rotation periods, 450 years) (Wangler and Minnich 1996), similar to findings in pinyon-juniper woodlands in the Southwest (Erdman 1970; Koniak 1985). The scarcity of fires in the past was indicated by Leiberg (1900) who states, "There are no areas outside the pinyon belt . . . which did not show evidence of having been burned over at various times in the past," although burns within pinyon forest were still observed "here and there."

Long fire intervals are due to low productivity and fuel buildup rates in this semiarid ecosystem. In SBM, denuded forests are colonized by a dense shrub layer dominated by species established by long-range seed dispersal (e.g., *Artemisia tridentata, Chrysothamnus* spp.), resprouting (*Purshia tridentata, Quercus john-tuckeri*), and scarification of soil seedbanks (e.g., *Ceanothus greggii, Fremontodendron californicum*). Although the seeds of pinyon pines are continually cached by birds, especially jays and nutcrackers (Vander Wall and Balda 1977), initial establishment of *P. monophylla* appears to be delayed 20–30 years by sun scald and/or freeze/thaw soil heaving until the establishment of the shrub layer, which acts as nurse plants (Wangler and Minnich 1996). The first pinyon recruits establish within shrub canopies, often near root axes. The growth of pine canopy eventually reduces freeze/thaw processes after 75 years, setting off a chain reaction of spatially random recruitment throughout burns. Pinyons develop canopy closure after 100–150 years, which is accompanied by a decline in the early successional shrub layer, due apparently to shrub senescence and shade stress. Pre- and postsuppression burns were most frequent in dense forests at highest elevations (Deep Creek to Baldwin Lake) and least frequent in open stands near the Mojave Desert, in response to changes productivity/fuel accumulation gradients with decreasing annual precipitation toward the desert. Short-period recurrences in mesic woodlands are precluded by the destruction of the tree layer as a source of litter, development of green fuels (high fuel moisture) during succession, and a limited presence of cured herbaceous cover from summer drought. Suppression appears to have had little impact on this ecosystem due to low primary productivity rates, as replicates of 1929–1934 VTM survey plots reveal no significant changes in forest densities. Without fire control in northern Baja California, return intervals in pinyon-juniper woodlands in the Sierra Juárez ranged from 100 to 225 years (Minnich and Chou 1997).

Pinyon-juniper woodlands typically form sharp ecotones with mixed conifer forest due to intense disturbance gradients related to fire behavior and stem mortality. The thin bark of *Pinus monophylla* virtually precludes the survivorship of these trees from surface fires of mixed conifer forest. At the same time, the removal of canopy by stand-replacement fires in pinyon forest arrests the rapid development of surface fuels characteristic of mixed conifer forest. In addition, shrub successions appear to lack sufficient biomass to support short-period burns. Thus, before fire suppression a discontinuity in fire return intervals may have existed along the ecotones between mixed conifer forests and pinyon woodlands in which predominantly understory fires shift to long-period stand-replacement burns in response to differences in stand structure, fire behavior, and survivorship.

Because pinyon-juniper woodlands largely grow on leeward slopes of SBM, stands are exposed to low concentrations of air pollution due to mixing associated with the dissipation of the marine inversion. Most stands have not experienced fire since the advent of air pollution beginning in the 1950s, and long-term dynamics of mature stands may permit the differentiation between air pollution and fire effects. The replication of VTM survey quadrats taken 60 years ago throughout SBM show little change in stand densities and diameter class distribution and species composition (Wangler and Minnich 1996). Nevertheless, this ecosystem may still be altered by the cumulative effects of air pollution because of the longevity of *Pinus monophylla, Juniperus occidentalis,* and other species in a stand-replacement fire regime.

Subalpine Forest

Fire is also infrequent in subalpine forests of SBM due to limited productivity and fuel accumulation in association with short growing seasons and heavy snow cover. Studies elsewhere in California and Oregon suggest that needle accumulation rates are lower than for mixed conifer forest, perhaps because of more rapid decomposition under deep snowpacks (Agee 1993; Parker 1993). Sheppard et al. (1988) found that the fire scars in *Pinus contorta/P. flexilis* forest on the summit of Mt. San Jacinto, 40 km SE of SBM, were caused by single-tree burns established by lightning strikes. In SBM, dense stands of subalpine forest show evidence of even-aged patches, mostly less than 40 ha, generated by stand-replacement fires (Minnich 1988). Several were both described by Leiberg (1900) and reported in local newspapers during the 1860s and 1890s. Under fire control, a 1,000-ha crown fire burned through subalpine forests in the upper Whitewater River in 1950, after 3 years of extreme drought, when the early snow melt permitted the drying of vegetation by late spring or early summer. In the San Gabriel Mountains, a Santa Ana wind–blown fire in 1980 destroyed 47% of *Pinus contorta* forests near Mt. Baldy (Table 3.4). Chronosequences of postfire succession show the establishment of timberland chaparral dominated by resprouting of *Chrysolepis sempervirens* and establishment of seedlings of *Arctostaphylos patula* and *Ceanothus cordulatus,* apparently from seed caches (Minnich 1978). *Pinus contorta* saplings were abundant by 20 years. *Pinus contorta* in SBM does not possess cone serotiny seen in Rocky Mountain populations (Critchfield 1980). Hence, saplings apparently may have established from wind-dispersed seed from distant unburned stands. Similar successions after presuppression fires were recorded on VTM vegetation maps.

Presuppression fires were not observed at close range for fire behavior. Perhaps they were similar to those in the Sierra Nevada, where lodgepole pine forests undergo variable fire patterns including low-intensity creeping, ground fires, moderately intense understory fires capable of scorching pole-size individuals, with canopy fires occurring in extreme weather (Kilgore and Briggs 1972; Kilgore 1981; Parker 1986; Agee 1993). Smoldering fires of long duration can still be fatal because of root and root collar heating (Gara et al. 1986). Stands in these areas tend to develop unimodal age and diameter structures after stand-destroying disturbances (Agee 1993; Parker 1993), similar to the patchiness seen in forests covering the north-facing slope of San Gorgonio Mountain, the San Bernardino Ridge, and Sugarloaf Mountain in SBM (Minnich 1988). Although presuppression fire intervals are not documented, the combination of low productivity, heavy snow cover, and a stand-replacement fire regime would indicate that present forests may be well within presuppression fire-free periods.

Subalpine forests of lodgepole and limber pines in SBM are exposed to low concentrations of air pollution because most stands cover high summits exposed to synoptic-scale ambient circulations above the tropospheric boundary layer. During the summer air pollution season, forests may still experience pollution brought to higher altitudes by anabatic slope winds, but long-term concentrations are reduced by mixing with ambient flows, as well as downdrafts caused by afternoon thunderstorms. Although fire recurrence intervals are long in subalpine forest, evidence that air pollution or other factors may be altering the dynamics of this ecosystem is equivocal. For example, on San Gorgonio Mountain, the site-specific scale matching of recent aerial photographs with those taken in 1938 reveals that stands are nearly static, with the same trees in individual stands existing in a fixed nearest-neighbor configuration; recruitment and mortality rates are very low (Minnich 1984). However, this ecosystem may still be subject to long-term cumulative effects of air pollution because of the longevity of lodgepole and limber pine in a stand-replacement fire regime.

Summary

The SBM have remarkably diverse shrubland and conifer forest communities in which recurrent fire plays a vital role affecting species composition, vegetation structure, and biogeography. The long-term effect of air pollution varies in SBM, depending on the intrinsic sensitivity of species to pollutants, spatial variation of pollutants, and spatial variation in fire regimes as a mediator in the dynamics of ecosystems along environmental gradients, including the alteration of fire's role over the past century by suppression. Because fire occurrence has time dependence in brushland and conifer forest ecosystems—in response to gradual cumulative fuel buildup—fire intervals are shortest in areas with highest productivity. Suppression forces have kept fires from penetrating mixed conifer forest during the 20th century. The most productive ecosystems on the southern escarpment and crest of SBM are subject to the highest air pollution concentrations in summer due to the advection of marine air from the Los Angeles Basin as a consequence of onshore flow and anabatic circulations.

Conifer forest communities are associated with discrete ranges of understory assemblages and combustible biomass. Terrain, a spatially fixed variable, also contributes to species distributions by affecting fire damage through slope–wind interactions. Hence, the distributions of shrub and tree species possessing different life history attributes are shaped in part by gradients in fire intensity and recurrence interval between stand-replacement burns. *Pinus coulteri* and *P. attenuata,* characterized by serotinous or partially serotinous cone habits, respond to frequent severe fires with high colonizing ability and reproductive effort that is concentrated in postburn growing seasons. *Pseudotsuga macrocarpa* survives in less disturbed convex canyons and cliffs and establishes offspring infrequently but continuously. Current losses in *Pseudotsuga macrocarpa* may also be related to suppression-related effects on the fire weather of large burns. Cumulative air pollution effects may be limited in knobcone pine and mixed evergreen coulter pine forests where the lifespan of trees is mostly limited to single fire cycles.

Although the smog-sensitive species *Pseudotsuga macrocarpa* has sustained large losses in distribution since 1938 due largely from intense chaparral fires, forests that have not burned since the arrival of air pollution in the 1950s appear to be stable or exhibit modest stand enlargement.

Anthropogenic control of fire in mixed conifer forest has led to an increase in fire intervals and greater stand densities during the 20th century. Stand-thickening was accompanied by age-specific trends in which juvenile pole-size cohorts are dominated by *A. concolor* and *C. decurrens* and mature overstory cohorts are dominated by *P. ponderosa* and *P. jeffreyi*. Although fire scar data suggest that presuppression intervals were about 15–29 years, presuppression fire intervals have been underestimated due to problems of scale related to assumptions made with respect to site-based (point) sampling of fire as a spatial process. The trend for broad-scale stand-thickening in mixed conifer forest over the past 60 years suggests that air pollution is not interfering with recruitment. Still, the increase in tree densities may be synergistically contributing to enhanced damage and mortality from air pollution because of decreasing fitness of tree species as a consequence of increased competition for moisture and nutrients.

Pinyon-juniper woodlands and subalpine forests both have experienced long-period stand-replacement fires both before and during suppression. Stands are exposed to low concentrations of air pollution owing to their distributions on leeward slopes or high mountain summits due to mixing associated with the dissipation of the marine inversion. Both ecosystems may still be altered by the cumulative effects of air pollution because of the longevity of dominant tree species in a stand-replacement fire regime.

References

Agee, J.K. Fire ecology of Pacific northwest forests. Washington, DC: Island Press; 1993.

Bolton, R.B., Jr.; Vogl, R.J. Ecological requirements of *Pseudotsuga macrocarpa* (Vasey) Mayr in the Santa Ana Mountains, California. J. For. 67:112–116; 1969.

Bonnicksen, T.M.; Stone, E.C. The giant sequoia-mixed conifer forest community characterized through pattern analysis as a mosaic of aggregations. For. Ecol. Manage. 3:307–328; 1981.

Bonnickson, T.M.; Stone, E.C. Reconstruction of a presettlement giant sequoia-mixed conifer forest community using the aggregation approach. Ecology 63:1134–1148; 1982.

Borchart, M. Serotiny and cone-habit variation in populations of *Pinus coulteri* (Pinaceae) in the southern coast ranges of California. Madroño 32:29–49; 1985.

Bruhn, J.N. Effects of oxidant air pollution on ponderosa and Jeffrey pine foliage decomposition. PhD Thesis, Univ. of California, Berkeley; 1980.

Chou, Y.H.; Minnich, R.A.; Dezzani, R.J. Do fire sizes differ between southern California and northern Baja California? For. Sci. 39:835–844; 1993.

Christensen, N.L. Fire regimes and ecosystem dynamics. In: Crutzen, P.J.; Goldammer, J.G., eds. Fire in the environment: the ecological, atmospheric, and climatic importance of vegetation fires. New York: John Wiley & Sons; 1993:233–244.

Countryman, C.M.; Philpot, C.W. Physical characteristics of chamise as a wildland fuel. Research paper PSW-66. Berkeley, CA: USDA Forest Service, Pacific Southwest Station; 1970.

Critchfield, W.B. Genetics of lodgepole pine. USDA Forest Service. Washington, DC: Government Printing Office; 1980.

Edinger, J.G.; McCutchan, M.H.; Miller, P.R.; Ryan, B.C.; Schroeder, M.J.; Behar, J.V. Penetration and duration of oxidant air pollution in the South Coast Air Basin of California. J. Air Pollut. Control Assoc. 22:882–886; 1972.

Erdman, J.A. Pinyon-juniper succession after natural fires on residual soils of Mesa Verde, Colorado. Brigham Young Univ. Sci. Bull. Biol. Ser. 11(2); 1970.

Gara, R.I.; Agee, J.K.; Littke, W.R.; Geizler, D.R. Fire wounds and beetle scars: distinguishing between the two can help reconstruct past disturbances. J. For. 84:47–50; 1986.

Gause, G.W. Silvicultural characteristics of bigcone Douglas-fir (*Pseudotsuga macrocarpa*). Research paper PSW-39. Berkeley, CA: USDA Forest Service, Pacific Southwest Station; 1966.

Griffin, J.R. Pine seedlings, native ground cover, and *Lolium multiflorum* on the marble cone burn, Santa Lucia Range, California. Madroño 29:177–188; 1982.

Grinnell, J. The biota of the San Bernardino Mountains. Univ. Calif. Pub. Bot. 5:1–170; 1908.

Hanes, T.L. Chaparral. In: Barbour, M.G.; Major, J., eds. Terrestrial vegetation of California. Special Publication. Davis, CA: California Botanical Society; 1988:417–469

Hart, S.C.; Firestone, M.K.; Paul, E.A. Decomposition and nutrient dynamics of ponderosa pine needles in a Mediterranean-type climate. Can. J. For. Res. 22:306–314; 1992.

Haston, L.L.; Davis, F.W.; Michaelson, J. Climatic response functions for bigcone spruce. Physical Geogr. 9:81–97; 1988.

Heinselman, M.L. Fire and succession in the conifer forests of northern North America. In: West, D.C.; Shugart, H.H.; Botkin, D.B., eds. Forest succession: concepts and application. New York: Springer-Verlag; 1981a:374–405.

Heinselman, M.L. Fire intensity and frequency as factors in the distribution and structure of northern ecosystems. In: Mooney, H.A.; Bonnicksen, T.M.; Christensen, N.L.; Lotan, J.E.; Reiners, W.A., tech. coords. Fire regimes and ecosystem properties. Gen. Tech. Rep. WO-26. Washington, DC: USDA Forest Service; 1981b:7–57.

Johnson, E.A.; Gutsell, S.L. Fire frequency models, methods and interpretations. Adv. Ecol. Res. 15:239–287; 1994.

Kahrl, W.L., et al. The California water atlas. Sacramento, CA: California Dept. Water Resources; 1978.

Kauffman, J.B.; Martin, R.E. Shrub and hardwood response to prescribed burning with varying season, weather, and fuel moisture. Fire For. Meteor. Conf. 8:279–286; 1985.

Kauffman, J.B.; Martin, R.E. Sprouting shrub response to different seasons and fuel consumption levels of prescribed fire in Sierra Nevada mixed conifer ecosystems. For Sci. 36:748–764; 1990.

Kauffman, J.B.; Martin, R.E. Factors influencing the scarification and germination of three montane Sierra Nevada shrubs. Northwest Sci. 65:180–187; 1991.

Keeley, J.E. Seed germination and life history syndromes in the California chaparral. Bot. Rev. 57:81–117; 1989.

Keeley, J.E.; Keeley, S.C. Chaparral. In: Barbour, M.G.; Billings, W.D., eds. North American terrestrial vegetation. Cambridge: Cambridge Univ. Press; 1989:166–207.

Kercher, J.R.; Axelrod, M.C. A process model of fire ecology and succession in a mixed-conifer forest. Ecology 65:1735–1742; 1984.

Kilgore, B.M. The ecological role of fire in Sierran conifer forests: its application to national park management. Quat. Res. 3:496–513; 1973.

Kilgore, B.M. Fire in ecosystem distribution and structure: western forests and scrublands. In: Mooney, H.A. et al., tech. coords. Fire regimes and ecosystem properties. Proceedings of the Conference. Gen. Tech. Rep. WO-26. USDA Forest Service; 1981:58–89.

Kilgore, B.M.; Briggs, G.S. Restoring fire to high elevation forests in California. J. For. 70:266–271; 1972.

Kilgore, B.M.; Taylor, D. Fire history in a sequoia-mixed conifer forest. Ecology 60:129–142; 1979.

Kinney, A. Report on the forests of the counties of Los Angeles, San Bernardino and San Diego, California. 1st Biennial report, 1885–86. Sacramento, CA: California State Board of Forestry; 1887.

Knight, D.H. Parasites, lightning, and the vegetation mosaics in wilderness landscapes. In: Turner, M.G., ed. Landscape heterogeneity and disturbance. Ecological studies 64. New York: Springer-Verlag; 1987:59–83.

Koniak, S. Succession in pinyon-juniper woodlands following wildfire in the Great Basin. Great Basin Nat. 45:81–116; 1985.

Leiberg, J.B. San Gabriel, San Bernardino, and San Jacinto forest reserves. In: Gannett, H., ed. Nineteenth annual report of the U.S. Geological Survey to the Secretary of Agriculture. Part 5, Forest reserves. Washington, DC: Government Printing Office; 1899:359–370.

Leiberg, J.B. San Gabriel, San Bernardino, and San Jacinto forest reserves. In: Twentieth annual report of the U.S. Geological Survey to the Secretary of Agriculture. Part 5, Forest reserves. Washington, DC; Government Printing Office; 1900:411–479.

Lockmann, R.F. Guarding the forest of southern California: evolving attitudes toward conservation of watershed, woodlands, and wilderness. Glendale, CA: Arthur H. Clark Co.; 1981.

Loucks, O.L. Evolution of diversity, efficiency, and community stability. Am. Zool. 10:17–25; 1970.

McBride, J.R.; Laven, R.D. Fire scars as an indicator of fire frequency in the San Bernardino Mountains, California. J. For. 74:439–442; 1976.

McDonald, P.M.; Littrell, E.E. The bigcone Douglas fir-canyon live oak community in southern California. Madroño 23:310–320; 1976.

McIntosh, R.P. Succession and ecological theory. In: West, C.; Shugart, H.H.; Botkin, D.B., eds. Forest succession: concepts and application. New York: Springer-Verlag; 1981:10–23.

McKelvey, K.S.; Johnston, J.D. Historical perspective on forests of the Sierra Nevada and the Transverse Ranges of southern California: forest conditions at the turn of the century. In: Verner, J.; McKelvey, K.S.; Noon, B.R.; Gutiérrez, R.J.; Gould, G.I.; Beck, T.W. tech. coords. The California spotted owl: a technical assessment of its current status. Gen. Tech. Rep. PSW-GTR-133. Albany, CA: USDA Forest Service; 1992:225–246.

McNeil, R.C.; Zobel, D.B. Vegetation and fire history of a ponderosa pine-white fir forest in Crater Lake National Park. Northwest Sci. 54:30–46; 1980.

Michaelsen, J.; Haston, L.; Davis, F.W. 400 Years of central California precipitation variability reconstructed from new tree ring chronologies. Water Resour. Bull. 23:809–818; 1987.

Minnich, R.A. The geography of fire and conifer forests in the eastern Transverse Ranges, California. PhD Thesis, Univ. of California, Los Angeles; 1978.

Minnich, R.A. Wildfire and the geographic relationships between canyon live oak, coulter pine, and bigcone Douglas fir forests. In: Plumb, T.R., tech. coord. Proceedings of the symposium on the ecology, management, and utilization of California oaks. Gen. Tech. Rep. PSW-44. Berkeley, CA: USDA Forest Service, Pacific Southwest Station; 1980:55–61.

Minnich, R.A. *Pseudotsuga macrocarpa* in Baja California? Madroño 29:22–31; 1982.

Minnich, R.A. Fire mosaics in southern California and northern Baja California. Science 219:1287–1294; 1983.

Minnich, R.A. Snow drifting and timberline dynamics on Mount San Gorgonio, California, USA. Arct. Alpine Res. 16:395–412; 1984.

Minnich, R.A. The distribution of forest trees in northern Baja California. Madroño 34:98–127; 1987a.

Minnich, R.A. Fire behavior in southern California chaparral before fire control: the Mount Wilson burns at the turn of the century. Ann. Assoc. Am. Geogr. 77:599–618; 1987b.

Minnich, R.A. The biogeography of fire in the San Bernardino Mountains of California: a historical study. Univ. Calif. Pub. Geogr. 27:1–121; 1988.

Minnich, R.A.; Bahre, C.J. Wildland fire and chaparral succession along the California-Baja California boundary. Int. J. Wildland Fire 5:13–24; 1995.

Minnich, R.A.; Chou, Y.H. Wildland fire patch dynamics in the Californian chaparral of southern California and northern Baja California. Int. J. Wildland Fire 7:221–248; 1997.

Minnich, R.A.; Franco-Vizcaíno, E.; Sosa-Ramírez, J.; Chou, Y.H. Lightning detection rates and wildland fire occurrence in the mountains of northern Baja California. Atmosfera 6:235–253; 1993.

Minnich, R.A.; Barbour, M.G.; Burk, J.H.; Fernau, R.F. Sixty years of change in conifer forests of the San Bernardino Mountains: reconstruction of Californian mixed conifer forests prior to fire suppression. Conserv. Biol. 9:902–914; 1993.

Minnich, R.A.; Dezzani, R.J. Historical decline of coastal sage scrub in the Riverside-Perris Plain. Western Birds (in press).

Minnich, R.A.; Barbour, M.G.; Burk, J.H.; Sosa-Ramírez, J. Californian conifer forests under unmanaged fire regimes in the Sierra San Pedro Mártir, Baja California, Mexico. J. Biogeogr. (in press).

Parker, A.J. Persistence of lodgepole pine forests in the central Sierra Nevada. Ecology 67:1560–1567; 1986.

Parker, A.J. Structural variation and dynamics of lodgepole pine forests in Lassen Volcanic National Park, California. Ann. Assoc. Am. Geogr. 83:613–629; 1993.

Parker, A.J.; Peet, R.K. Size and age structure of conifer forests. Ecology 65:1685–1689; 1984.

Parsons, D.J.; DeBenedetti, S.H. Impact of fire suppression on a mixed-conifer forest. For. Ecol. Manage. 2:21–33; 1979.

Parsons, D.J.; Swetnam, T.W. Restoring natural fire to the sequoia-mixed conifer forest: should intense fire play a role? Tall Timbers Fire Ecol. Conf. 20:20–30; 1989.

Paysen, T.E.; Cohen, J.D. Chamise chaparral dead fuel fraction is not reliably predicted by age. West. J. For. 5:127–131; 1990.

Peterson, D.L.; Arbaugh, M.J. Mixed conifer forests of the Sierra Nevada. In: Olson, R.K.; Binkley, D.; Bohm, M., eds. The response of western forests to air pollution. Ecological studies 97. New York: Springer-Verlag; 1992:433–459.

Plumb, T.R. Responses of oaks to fire. In: Plumb, T.R., tech. coord. Symposium on the ecology, management and utilization of California oaks. Gen. Tech. Rep. PSW-44. Berkeley, CA: USDA Forest Service; 1980:202–215.

Pyne, S.J. Fire in America: a cultural history of wildland and rural fire. Princeton, NJ: Princeton Univ. Press; 1982.

Riggan, P.J.; Goode, S.; Jacks, P.M.; Lockwood, R.N. Interaction of fire and community development in chaparral of southern California. Ecol. Monogr. 58:155–176; 1988.

Rundel, P.W.; Gordon, D.T.; Parsons, D.J. Montane and subalpine vegetation in the Sierra Nevada and Cascade ranges. In: Barbour, M.G.; Major, J., eds. Terrestrial vegetation of California. Special publication. Davis, CA: California Botanical Society; 1988:559–600.

Savage, M. Anthropogenic and natural disturbance and patterns of mortality in a mixed conifer forest stand in California. Can. J. For. Res. 24:1149–1159; 1994.

Sawyer, J.O.; Thornburgh, D.A.; Griffin, J.R. Mixed evergreen forest. In: Barbour, M.G.; Major, J., eds. Terrestrial vegetation of California. Special publication. Davis, CA: California Botanical Society; 1988:359–381.

Sheppard, P.R.; Means, J.E.; Lassoie, J.P. Cross-dating cores as a nondestructive method for dating living, scarred trees. For. Sci. 34:781–789; 1988.

SNEP Science Team, eds. Status of the Sierra Nevada, final report to Congress of the Sierra

Nevada Ecosystem Project. 3 vols. Wildland Resources Center Rep. 36. Davis, CA: Univ. of California; 1996.
Sprugel, D.G. Disturbance, equilibrium and environmental variability: what is "natural" vegetation in a changing environment? Conserv. Biol. 58:1–18; 1991.
Stohlgren, T.J. Litter dynamics in two Sierran mixed conifer forests. I: Litterfall and decomposition rates. Can. J. For. Res. 18:1127–1135; 1988.
Swetnam, T.W. Fire history and climate in the southwestern United States. In: Krammes, J.S.; Swolinski, M.J.; Covington, W.W., eds. Effects of fire in management of southwestern natural resources. Gen. Tech. Rep. INT-191. Missoula, MT: USDA Forest Service; 1991:6–17.
Swetnam, T.W. Fire history and climatic change in giant sequoia groves. Science 262:885–890; 1993.
Talley, S.N.; Griffin, J.R. Fire ecology of montane pine forest, Junipero Serra Peak, California. Madroño 27:49–60; 1980.
Thorne, R.F. Montane and subalpine forests of the Transverse and Peninsular ranges. In: Barbour, M.G.; Major, J., eds. Terrestrial vegetation of California. Special publication. Davis, CA: California Botanical Society; 1988:537–557.
U.S. Department of Agriculture. Estimating fire potential in California: atlas and guide for fire management planning. Vol. 10, sec. 9. Riverside, CA: USDA Forest Service, Pacific Southwest Station; 1982.
Vale, T.R. *Pinus coulteri* and wildfire on Mount Diablo, California. Madroño 26:135–139; 1979.
Vale, T.R. Vegetation management and nature protection. In: Malanson, G.P., ed. Natural areas facing climatic change. The Hague: SPB Academic Publishing; 1989:75–86.
Vander Wall, S.B.; Balda, R.P. Coadaptation of the Clark's nutcracker and the pinyon pine for efficient seed harvest and dispersal. Ecol. Monogr. 47:89–111; 1977.
Vankat, J.L. Fire and man in Sequoia National Park. Ann. Assoc. Am. Geogr. 67:17–27; 1977.
Vankat, J.L.; Major, J. Vegetation changes in Sequoia National Park, California. J. Biogeogr. 5:377–402; 1978.
van Wagtendonk, J.W. Refined burning prescriptions for Yosemite National Park. Occas. pap. 2. Washington, DC: U.S. Department of Interior, National Park Service; 1974.
Vasek, F.C.; Clovis, J.F. Growth forms in *Arctostaphylos glauca.* Madroño 26:135–139; 1976.
Veblen, T.T.; Hadley, K.S.; Reid, M.S. Disturbance and stand development of a Colorado subalpine forest. J. Biogeogr. 18:707–716; 1991.
Vogl, R.J.; Armstrong, W.P.; White, K.L.; Cole, K.L. The closed-cone pines and cypresses. In: Barbour, M.G.; Major, J., eds. Terrestrial vegetation of California. Special publication. Davis, CA: California Botanical Society; 1988:295–357.
Wangler, M.J.; Minnich, R.A. Fire and succession in pinyon-juniper woodlands of the San Bernardino Mountains, California. Madroño 43:493–514; 1996.
Weatherspoon, C.P.; Husari, S.J.; van Wagtendonk, J.W. Fire and fuels management in relation to owl habitat in forests of the Sierra Nevada and southern California. In: Verner, J.; McKelvey, K.S.; Noon, B.R.; Gutiérrez, R.J.; Gould, G.I.; Beck, T.W., tech. coords. The California spotted owl: a technical assessment of its current status. Gen. Tech. Rep. PSW-GTR-133. Albany, CA: USDA Forest Service; 1992.
Weislander, A.E. Vegetation types of California: San Bernardino, Redlands, Hesperia, Deep Creek quadrangles (1:62,500), San Gorgonio Quadrangle (1:125,000). Berkeley, CA: USDA Forest Service, California Forest and Range Experiment Station; 1929–1934.
Zedler, P.H.; Gautier, C.R.; MacMaster, G.S. Vegetation change in response to extreme events: the effect of a short interval between fires in California chaparral and coastal scrub. Ecology 64:809–818; 1983.

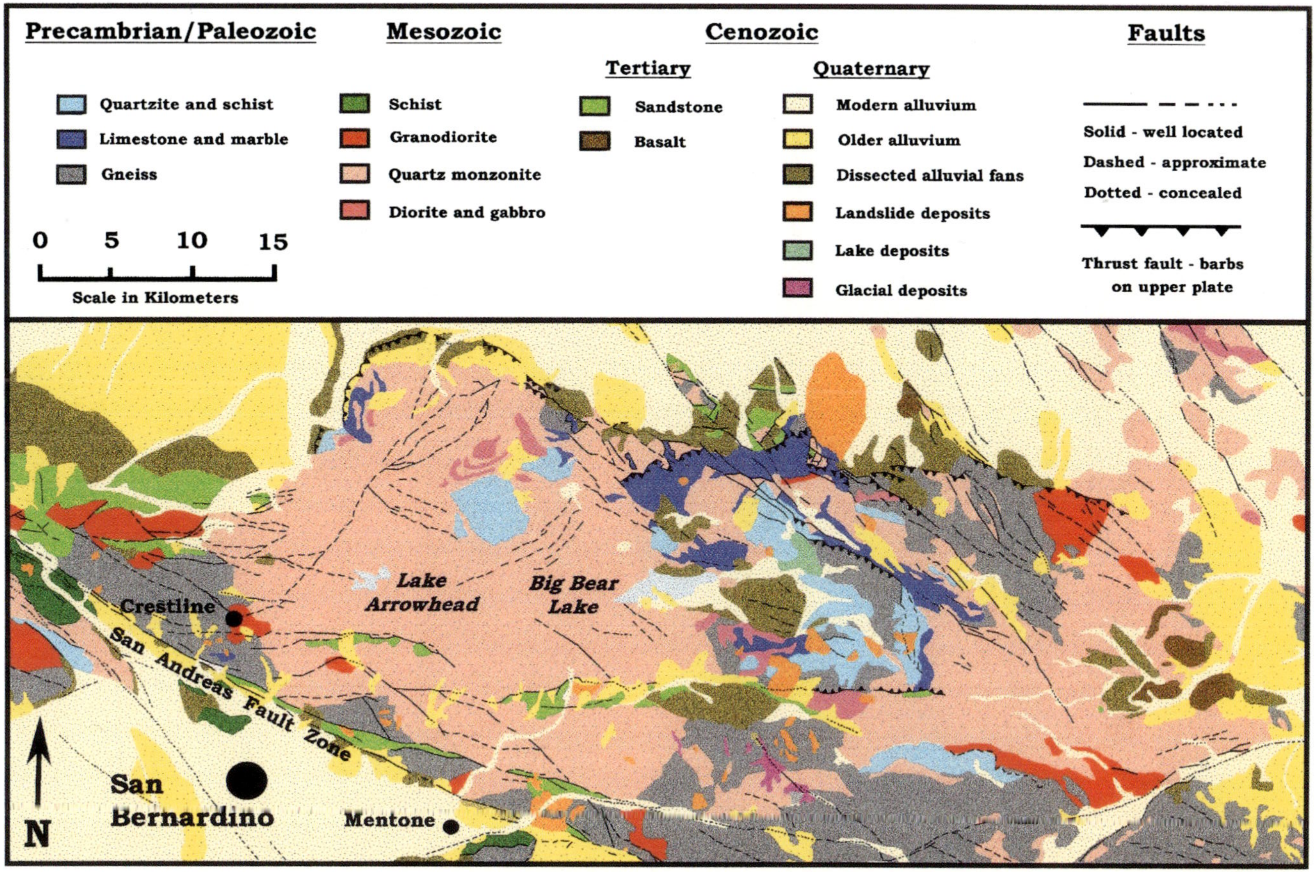

Figure 1.2. Rock types of the San Bernardino Mountains (Bortugno and Spittler 1986).

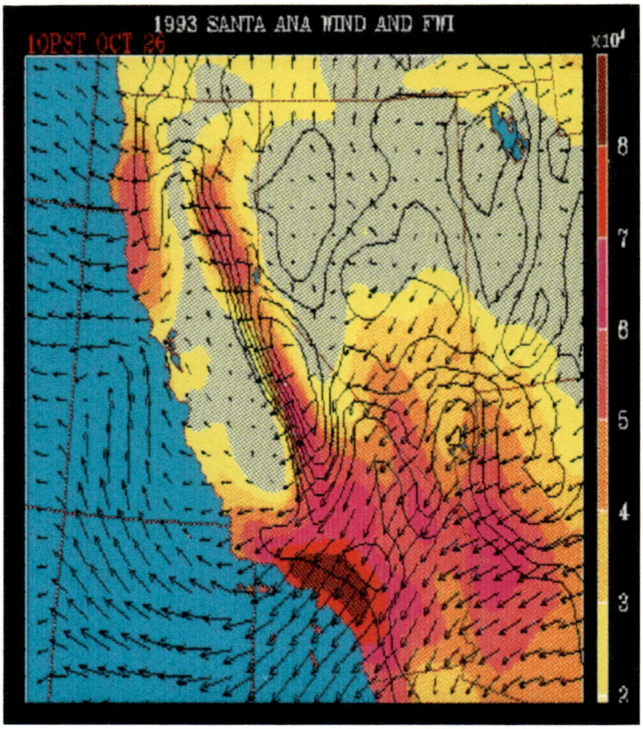

Figure 2.4. Santa Ana wind and fire weather index simulation generated from the Regional Spectral Model (Juang and Kanamitsu 1994). The fire weather index (shaded patterns with corresponding scale on the right) indicates fire potential and increases with increasing wind speed and decreasing relative humidity.

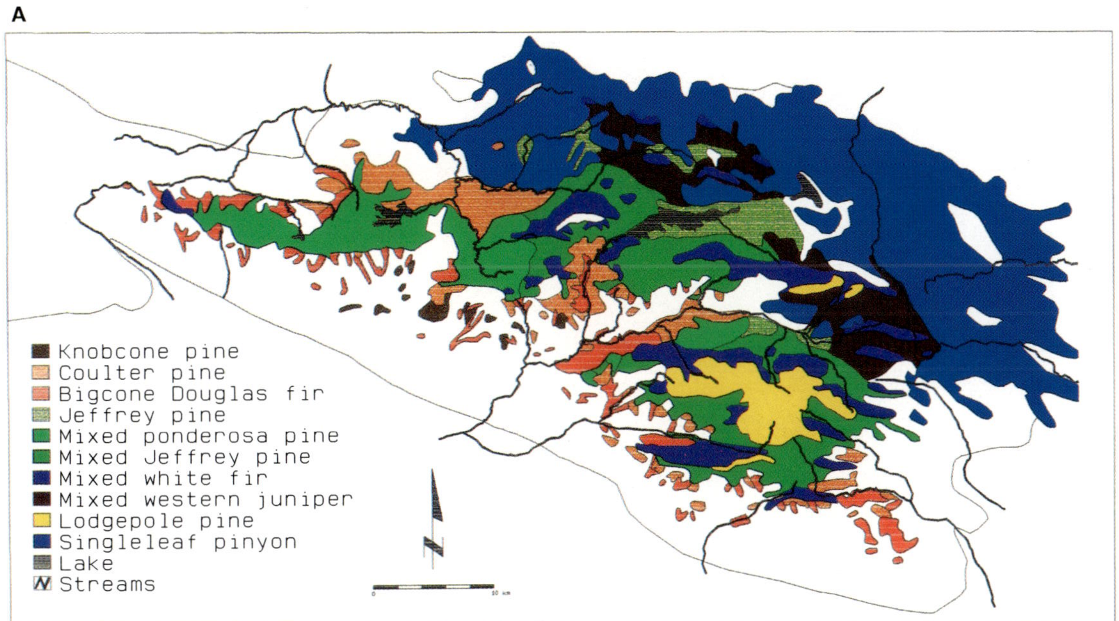

Figure 3.1. Vegetation of the San Bernardino Mountains: (A) conifer forest, (B) broadleaf forest, and (C) shrubland communities.

B

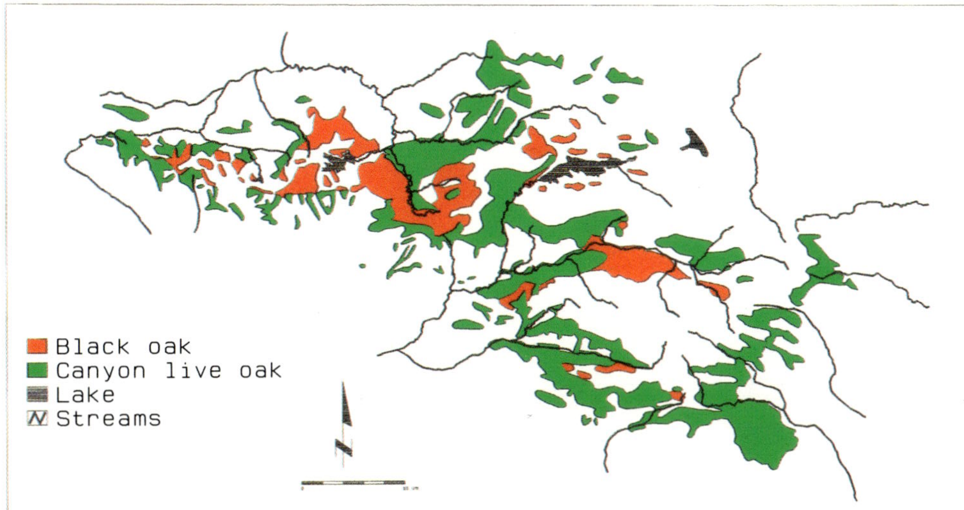

Figure 3.1. (*continued*)

c

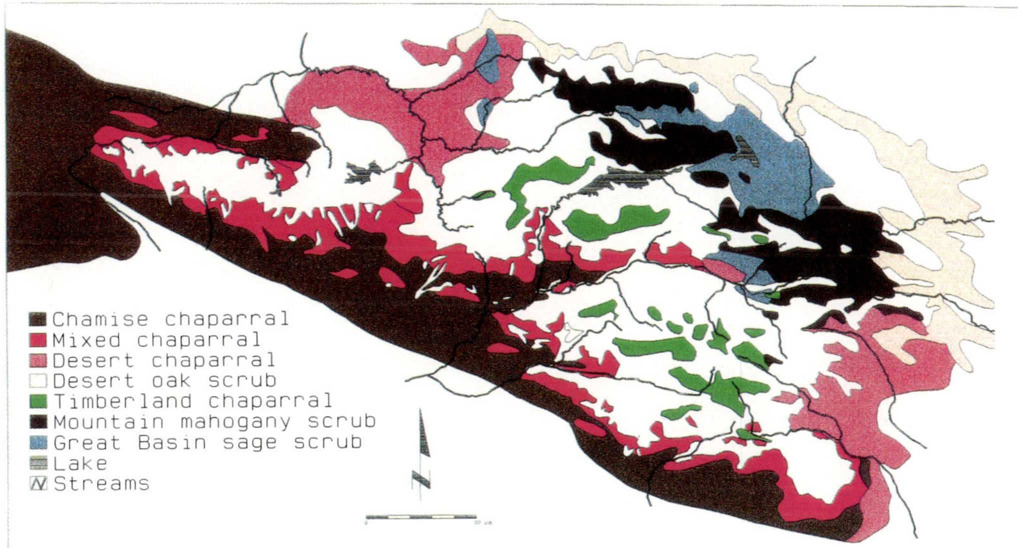

Figure 3.1. (*continued*)

San Gorgonio Wilderness
b_{ext}: 11–10/Mm Haziness: 1–0 dv Visual Range: 350–391 km

San Gorgonio Wilderness Area
b_{ext}: 23–20/Mm Haziness: 8–7 dv Visual Range: 170–200 km

San Gorgonio Wilderness Area
b_{ext}: 49–43/Mm Haziness: 16–15 dv Visual Range: 80–90 km

San Gorgonio Wilderness Area
b_{ext}: >98/Mm Haziness: >23 dv Visual Range: <40 km

Figure 5.3. Range of visibility conditions observed at San Gorgonio.

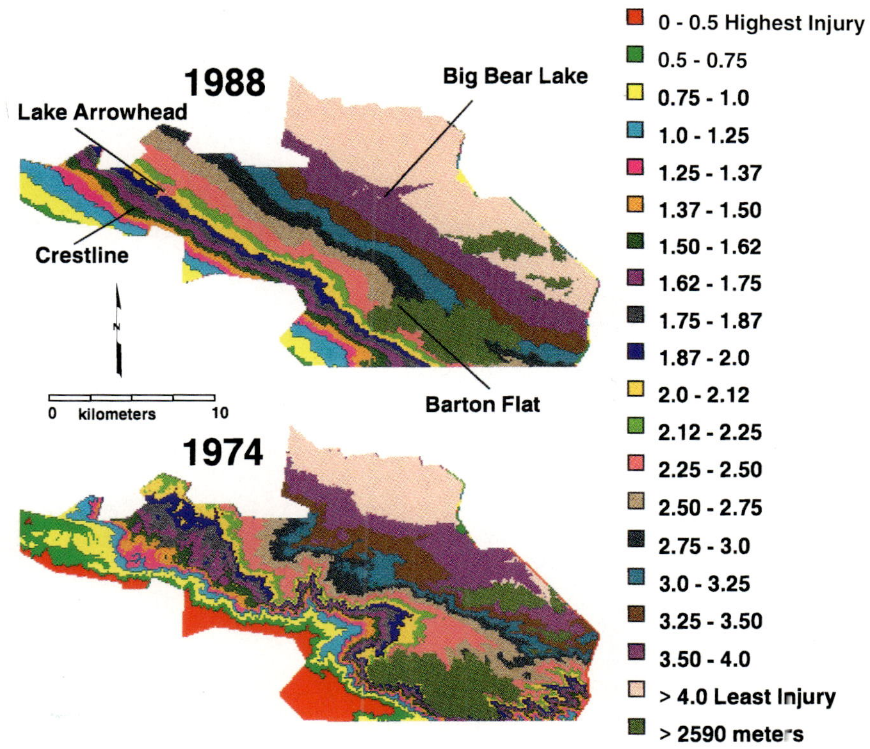

Figure 7.3. Spatial distribution of forest pest management (FPM) ozone injury indexes for 1974 and 1988 across the San Bernardino Mountains calculated with a simple GIS model based on ponderosa and Jeffrey pine crown injury (FPM index), the spatial coordinates, and elevation of tree plots.

Figure 9.1. (A) Ozone injury symptoms on leaflets of *Sambucus mexicana,* collected at Camp Paivika, San Bernardino Mountains, July 15, 1996. Leaflets on the left show under- and upper surfaces of young leaves with faint incipient chlorotic mottle on upper surfaces only. Leaflets on the right, taken from older leaves, show tan necrotic mottle on the upper surface and faint bifacial necrosis on the lower surface resulting from acute ozone exposure. (B) Ozone injury symptoms on the leaves of *Artemesia douglasiana,* collected at Camp Paivika, San Bernardino Mountains, July 15, 1996. Severity of injury decreases from severe chlorosis and necrosis of oldest leaves near the base of the stem, to chlorotic mottle of middle-aged leaves, to little or no injury on youngest leaves.

2. Effects of Ozone and Other Air Pollutants on Vegetation and Soils in the San Bernardino Mountains

4. Ambient Air Quality at Barton Flats and Other California Forests

John G. Watson, Judith C. Chow, Clifton A. Frazier, Barbara Hinsvark, and Mark Green

Introduction

This chapter documents ozone, nitric and sulfuric acid, and particle concentrations measured in the San Bernardino Mountains and compares these levels with those found in California's South Coast Air Basin (SoCAB) and other California forests. It identifies the meteorological conditions that favor and discourage transport between urban and forested portions of the SoCAB. The measured concentrations are related to observed and potential effects on vegetation in other chapters.

Many reactive chemicals are injected into the atmosphere in the SoCAB, which includes nearly 15 million people, hundreds of millions of miles per day of vehicle miles traveled, and some of the highest particle and ozone concentrations in the world. During late spring, summer, and early fall, ozone is formed from photochemical reactions of primarily light hydrocarbons (those with molecules containing two to ten carbon atoms) and oxides of nitrogen, much of which originate from SoCAB vehicle and fueling emissions (Harley et al. 1992; Fujita et al. 1994). The number and magnitude of the highest ozone concentrations in the SoCAB have decreased substantially during the 1970s and 1980s, owing to the implementation of emissions reduction strategies (Davidson 1993).

Fujioka et al. (this volume) show substantial year-to-year variability in the frequency of excessive SoCAB pollution levels from the mid-1980s through the mid-1990s. These years experienced large meteorological differences, with a prolonged drought from 1987 through 1992. Similar long-term measurements are insufficient to evaluate pollution trends in forests near the Los Angeles metropolitan area.

Photochemical processes also create nitric, sulfuric, and organic acids. Nitric acid results from atmospheric reactions of oxides of nitrogen emitted by fossil fuel combustion. It can combine with ammonia (emitted by animal husbandry, wastewater treatment, and human respiration) to create particles less than 2.5 μm in aerodynamic diameter ($PM_{2.5}$) (Pilinus and Seinfeld 1987). $PM_{2.5}$ is the size fraction that has the largest effect on visibility degradation (Watson and Chow 1994) and is the subject of recent National Ambient Air Quality Standards (U.S. EPA 1997; Vedal 1997) to protect public health.

Although nitric acid and ammonia gases rapidly deposit to the earth's surface, $PM_{2.5}$ remains in the atmosphere five to ten times longer and can be transported for hundreds of miles. Ammonium nitrate particles are unstable, remaining in equilibrium with ammonia and nitric acid concentrations in the atmosphere as a function of temperature and relative humidity (Watson et al. 1994). As temperature increases, relative humidity drops, nitric acid and ammonia gases deposit and disperse, and ammonium nitrate particles disassociate into ammonia and nitric acid gases (Russell et al. 1983). Ammonium nitrate particles transported from the SoCAB into the San Bernardino Mountains, therefore, can be effective delivery mechanisms for nitric acid to forested areas in conjunction with direct nitric acid transport.

Sulfuric acid is produced through gas- and aqueous-phase transformation pathways from sulfur dioxide emitted into the SoCAB atmosphere (Pandis et al. 1992). SoCAB sulfate concentrations are much lower than they once were (Cass 1979) owing to sulfur reductions in gasoline and diesel fuel and the use of natural gas instead of residual oil in SoCAB electrical generating stations. Sulfur dioxide converts to sulfate very slowly in dry air, but in the presence of fogs or clouds conversion rates exceed 10% per hour (Seinfeld 1986). Sulfuric acid in the $PM_{2.5}$ size fraction appears to be completely neutralized to ammonium sulfate by the time it reaches the eastern SoCAB for potential transport into the San Bernardino Mountains (Chow et al. 1992, 1994). In contrast to the transient properties of particulate ammonium nitrate, ammonium sulfate remains in the particle phase. Organic acids, specifically formate and acetate, are secondary acidic products that remain largely in the vapor phase and also have rapid deposition velocities.

In addition to secondary sulfates and nitrates, $PM_{2.5}$ from the SoCAB contains geological, organic carbon, and elemental carbon particles that may exist independently or mixed with ammonium sulfate and ammonium nitrate particles. These are not usually considered to be reactive species. Particles in the $PM_{2.5}$ size fraction have low gravitational settling rates and are usually transported with ozone, sulfur dioxide, nitric acid, nitrogen dioxide, and many volatile organic compounds.

Air quality measurements were taken at Barton Flats to better understand the concentrations of these interacting chemical components, how they vary by time of day and time of year, and how they arrive from the SoCAB to a high-elevation forest. The Barton Flats is located in the San Bernardino Mountains at an elevation of about 2,100 m above mean sea level (MSL). It is located in a canyon that opens to the west and curves to the east and southeast, with slopes rising to more than 3,000 m above MSL. By using measurement methods identical to those applied at other sites in the California Acid Deposition Monitoring Program (CADMP) (Watson et al. 1991), it is also possible to compare the concentrations of reactive species at Barton Flats with those from the eastern SoCAB and forested areas in or near Sequoia, Yosemite, and Redwoods national parks.

Ambient Air Quality Measurements

Aerosol, gas, and meteorological measurements were taken at the Barton Flats site from October 30, 1991, through September 30, 1993, and from June 1 to August 31, 1994 (Miller et al., 1996). Ozone, wind speed, wind direction, temperature, relative humidity, solar radiation, and time of wetness were measured every hour with continuous in situ monitors throughout these periods. Measurement methods applied in this study were identical to those applied in the CADMP (Watson et al. 1991; Chow et al. 1993; Fitz and Hering 1996).

Twelve-hour-average daytime (0600–1800 PST) and night-time (1800–0600 PST) filter samples were acquired every sixth day, corresponding to the U.S. Environmental Protection Agency's (EPA) particle sampling schedule, for $PM_{2.5}$ mass, water-soluble nitrate, sulfate, chloride, and ammonium. Impregnated filters and nitric acid denuders were used to collect sulfur dioxide, nitrogen oxide, ammonia, and nitric acid gases. An intensive experiment acquired filter measurements every day from July 18 through July 31, 1993, during morning (0600–1200 PST), afternoon (1200–1800 PST), and night-time (1800–0600 PST) to better understand diurnal variations and multiday accumulations of pollutants.

Filters were cleaned, weighed, impregnated, and loaded into filter holders in a laboratory and kept under refrigeration before and after field sampling. A sequential sampling system (Chow et al. 1993) drew air through a Teflon-coated $PM_{2.5}$ size-selective inlet and through the filters at constant flow rates. Mass concentrations were determined by gravimetric analysis. Filters were extracted in deionized distilled water and analyzed for chloride, nitrate, and sulfate by ion chromatography (Chow and Watson 1997) and for ammonium by automated colorimetry. Wet deposition samples were also taken weekly and were analyzed for ionic content, and these results are summarized by Miller et al. (1996).

Table 4.1 summarizes the number of valid values exceeding lower quantifiable limits (LQL) for the 12-hour samples acquired during the 2-year and one summer monitoring period. The LQLs are equal to or larger than the minimum detectable limits (MDL) because they include the standard deviation of the field blank and flow rate precision (Watson et al., 1995). $PM_{2.5}$ mass, sulfate, nitrate, and ammonium were detected on more than 90% of the samples. $PM_{2.5}$ chloride concentrations were generally low, especially during winter, and were detected on

Table 4.1. Analytical Specification of 12-Hour Filter and Hourly Ozone Measurements at the Barton Flats Station, California

	Analysis method[a]	MDL[b] [$\mu g/m^3$]	LQL[c] [$\mu g/m^3$]	No.[d] >LQL
$PM_{2.5}$ mass	Gravimetry	0.4859	1.4832	234
Chloride (Cl^-)	IC	0.0035	0.0425	43
Nitrate (NO_3^-)	IC	0.0035	0.0504	278
Sulfate (SO_4^{2-})	IC	0.0035	0.0421	291
Ammonium (NH_4^+)	AC	0.0035	0.0422	271
Ammonia (NH_3)	AC	0.0035	0.1188	267
Sulfur dioxide (SO_2)	IC	0.0035	0.2030	206
Nitrogen dioxide (NO_2)	AC	0.0347	0.1258	250
Total particulate nitrate (NO_3^-)	IC	0.0035	0.0484	263
Nitric acid (HNO_3)	IC	0.0060	0.0926	222
Ozone	UVA	1 ppbv	1 ppbv	26,363

[a]IC, ion chromatography; AC, automated colorimetry; UVA, ultraviolet absorption.
[b]Minimum detectable limits (MDL) are the concentrations at which instrument response equals three times the standard deviation of the response to a known concentration of zero. Typical sample volumes are 1.44 m^3 for NO_2 sampling port and 14.4 m^3 for the other sampling ports. Ozone MDL is determined by standard deviation of baseline signal.
[c]LQL, or lower quantifiable limit, is equal to two times the uncertainty of the field blank. The LQL is expressed here in terms of mass per cubic meter, after dividing by 14.4 m^3 (the nominal volume of a 12-hour sample at 20 L/min) for every sampling port but NO_2 in the dry deposition sampler and dividing by 1.44 m^3 (the nominal volume of a 12-hour sample at 2 L/min) for the NO_2 sampling port. Not applicable to ozone measurements.
[d]Number of nonvoided values reported.

15% of the samples. Sulfur dioxide, nitrogen dioxide, nitric acid, and ammonia gases were detected on more than 70% of the samples. The measurement process was sufficient to quantify measured species at levels well below those that might result in adverse effects on forests. Ozone was always detected at concentrations exceeding the ultraviolet absorption monitor's LQL of 1 ppb.

Monthly performance tests were performed on sequential sampler flow rates, and most were within ±5% of the nominal flow rates (Miller et al. 1996). Flow rates that deviated by more than ±10% from nominal values were re-adjusted. Interlaboratory comparisons of filter samples and U.S. EPA spiked filters were submitted for ion chromatographic analysis, and results were within ±10% of the standard values.

The sum of $PM_{2.5}$ Cl^-, NO_4^{2-}, and NH_4^- must be less than or equal to mass, within propagated precisions, and this is the case for all valid samples (Miller et al. 1996). These ionic species constituted less than half of the $PM_{2.5}$ mass for many of the samples, with the remainder presumably constituted by carbonaceous and geological material. All but a few of the measured $PM_{2.5}$ mass concentrations at Barton Flats were less than 15 $\mu g/m^3$, below the 65 $\mu g/m^3$ 24-hour average and 15 $\mu g/m^3$ annual average associated with National Ambient Air Quality Standard (NAAQS) (U.S. EPA 1997).

Figure 4.1 shows that ammonium nitrate volatilization was significant during all seasons, especially during summer when nearly 90% of the nitrate particles collected on quartz fiber filters turned to gas during sampling. Volatilization during winter is less certain, as many of the concentrations were substantially lower than those acquired during other seasons, but it exceeds 50% for several samples. Recent experiments (Chow and Egami 1997) in California's San Joaquin Valley show wintertime nitrate losses of less than 10%, but these are under conditions of much higher ammonia concentrations than those found at Barton Flats. Volatilized nitrate is not part of the measured $PM_{2.5}$ mass, so this loss does not show up in the sum of ionic species comparisons.

Sulfate and nitrate are most commonly found in the atmosphere as sulfuric acid (H_2SO_4), ammonium sulfate ($(NH_4)_2SO_4$), ammonium bisulfate (NH_4HSO_4), ammonium nitrate (NH_4NO_3), sodium nitrate ($NaNO_3$), sodium sulfate (Na_2SO_4), sodium chloride (NaCl), or potassium chloride (KCl), with NH_4NO_3 and $(NH_4)_2SO_4$ often being the most abundant. Miller et al. (1996) compare measured ammonium concentrations with those calculated from nitrate and sulfate concentrations (quartz fiber filter) assuming NH_4NO_3 plus either $(NH_4)_2SO_4$ or NH_4HSO_4 as the only ammonium-containing compounds. The $(NH_4)_2SO_4$ assumption provides the best explanation of measured ammonium, indicating that sulfuric acid and bisulfate were neutralized by reaction with available ammonia gas when they arrived at Barton Flats. Sulfuric acid is, therefore, not one of the reactive species that is likely to affect vegetation in the San Bernardino Mountains. This is consistent with sulfate neutralization observed by Chow et al. (1994) at inland SoCAB measurement sites. The consistency of this comparison also demonstrates the accuracy and precision of the ionic measurements, as random and systematic analysis errors would result in greater scatter among the calculated and measured ammonium levels.

Figure 4.2 compares hourly ozone concentrations from the long-term Barton Flats monitor with those from a solar-powered ozone monitor at the plot 2 site from July 16 through September 24, 1993, and from August 15 through August 21, 1994, and at the Heart Bar site from August 30 through September 5, 1994, and from September 12 through September 20, 1994. These comparisons show that the Barton Flats site generally represents ozone concentrations within the forest, but there are substantial spatial variations between individual values. These variations occur both because of the different site elevations, different air mass arrival times, and different effects of vegetation on ozone surrounding the measurement locations.

Results

Results of air quality monitoring are examined with respect to annual and seasonal averages and maximum concentrations at Barton Flats and in other California forests, conditions under the highest particle concentrations, changes in the

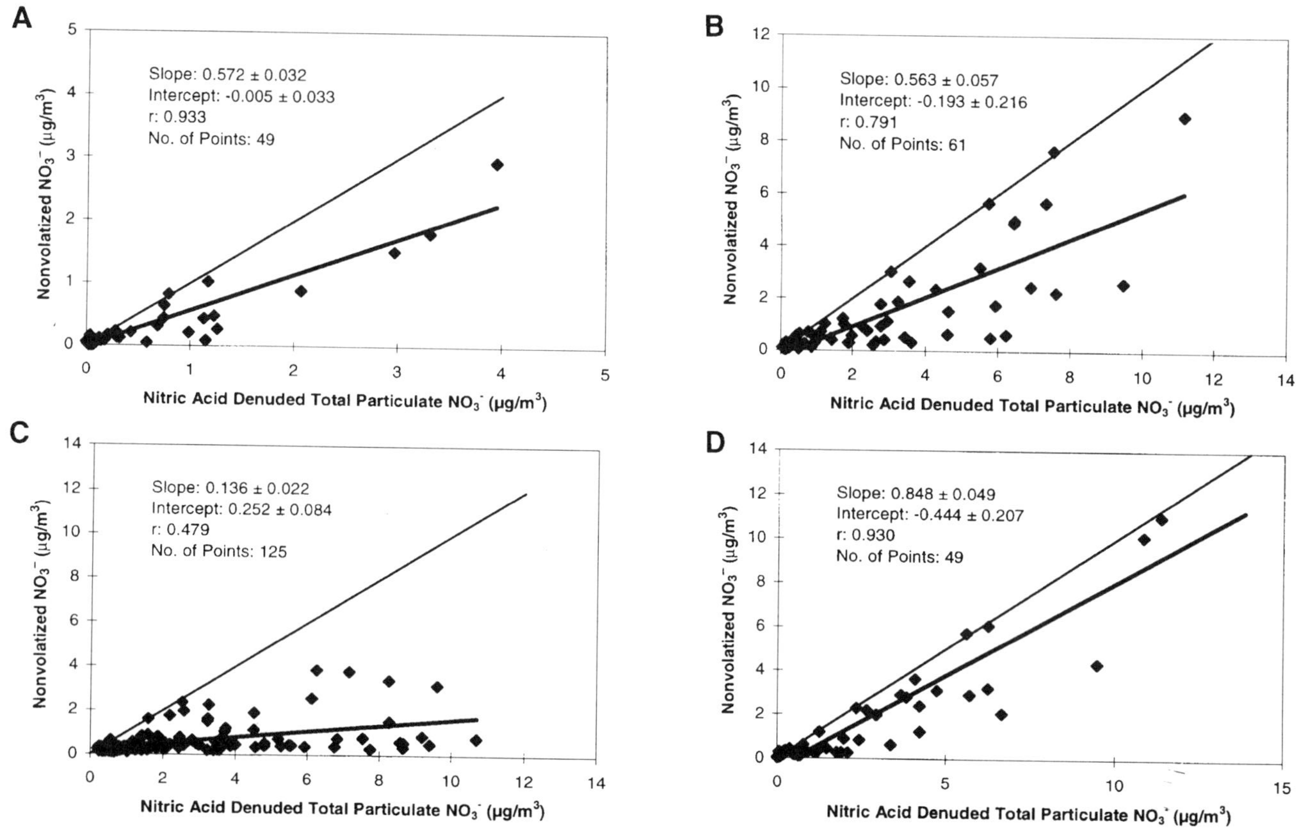

Figure 4.1. Scatter plots of $PM_{2.5}$ nonvolatilized nitrate versus nitric acid denuded nitrate at the Barton Flats Station, CA, during (A) winter (December 1991; January, February, and December 1992; and January and February 1993), (B) spring (March, April, and May 1992 and 1993), (C) summer (June, July, and August 1992, 1993, and 1994), and (D) fall (November 1991; September, October, and November 1992; and September 1993).

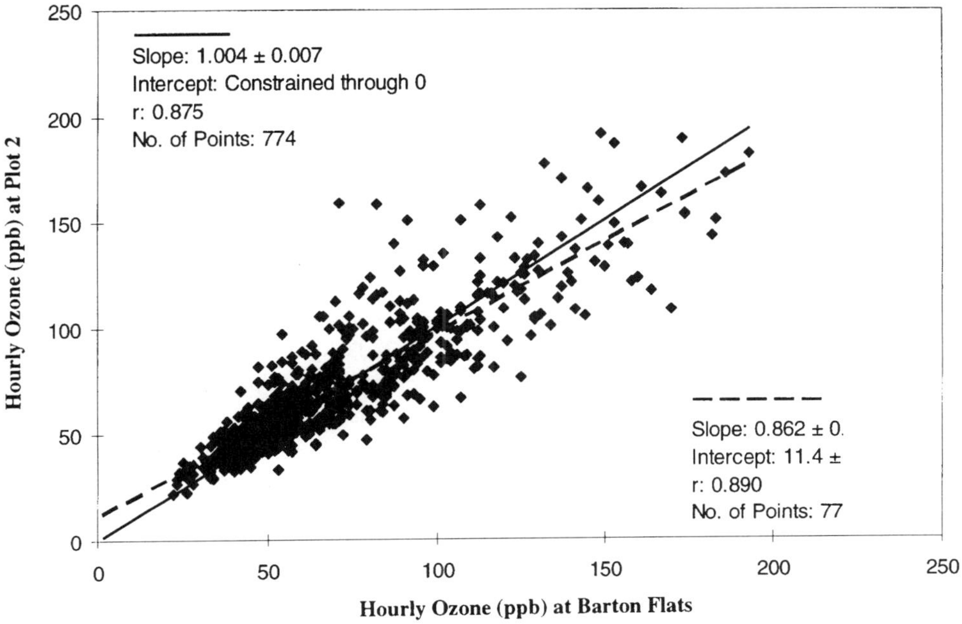

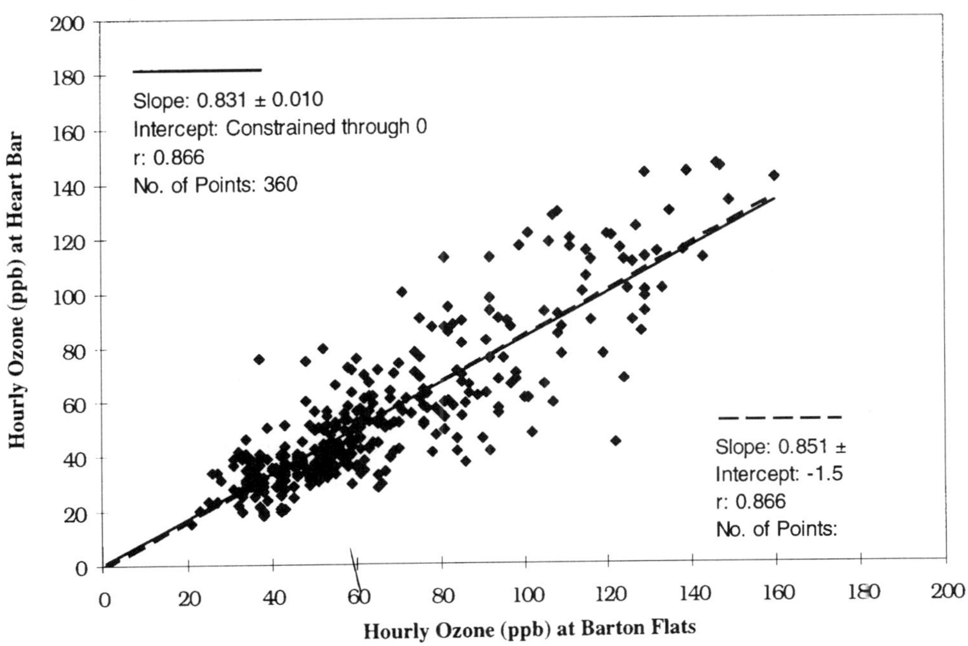

Figure 4.2. Scatter plot comparisons of ozone concentrations at the Barton Flats station with those acquired at the plot 2 and Heart Bar stations.

diurnal behavior of ozone concentrations for different seasons, and pollutant transport during the July 1993 intensive monitoring period.

Annual Average and Maximum Concentrations

Table 4.2 compares 2-year average and maximum concentrations for 12-hour averages of reactive particle and gas concentrations at Barton Flats with those from other forested sites from the CADMP network in Sequoia, Yosemite, and Redwoods national parks (Watson et al. 1991) for the same time period. Measurements from the CADMP Azusa site in the eastern SoCAB are also included for comparison with urban concentrations. Table 4.2 statistics are derived from samples taken over 12-hour periods, measured every 6 days (except for ozone, which was measured every day). Maxima are, therefore, lower than the 1-hour and 6-hour average concentrations presented later in this chapter.

Table 4.2 shows that 2-year average concentrations at Barton Flats were higher than those at other forested sites in California, but they were much lower (by factors of 3–5) than concentrations measured at Azusa. The exception was 2-year average ozone, for which daytime and night-time concentrations at Barton Flats exceeded those at any of the other sites, including urban Azusa.

Concentrations for most reactive components decreased from southern to northern forests in California, consistent with the proximity of measurement locations to major population centers and their corresponding emissions. Barton Flats $PM_{2.5}$ concentrations were similar to those measured at Sequoia and about 50% higher than those measured at Yosemite and Redwoods national parks. Daytime and night-time $PM_{2.5}$ maxima were highest at Redwoods, owing to sporadic fires in the region. The largest $PM_{2.5}$ mass concentrations at forested sites are sometimes affected by localized fires, as noted by Chow et al. (1996) at Sequoia and Yosemite during the summer of 1990. The Azusa site registered average and maximum $PM_{2.5}$ concentrations four to eight times larger than those found at any of the forested sites.

Barton Flats particle nitrate and nitric acid concentrations were more than twice those measured at Sequoia and nearly ten times those found at Redwoods. Average particle nitrate was about 20% of that measured at Azusa, and nitric acid was one-fourth of the nearby urban average. Barton Flats ammonia concentrations were slightly higher than those at Sequoia but two to five times those measured at Yosemite and Redwoods. Azusa ammonia concentrations were three to four times those at Barton Flats. Barton Flats nitrogen dioxide concentrations were lower than those measured in the other forests, probably owing to the isolation of Barton Flats from well-traveled roads and the proximity of such roads to the other sampling sites. The Redwoods site in Gasquet is at a civil airfield near a state highway, so its nitrogen dioxide levels are probably higher than those at other forested sites, owing to nearby engine exhaust emissions. Azusa nitrogen dioxide concentrations are affected by heavy traffic, and they were more than 30 times those at Barton Flats during the 2-year monitoring period.

Table 4.2. Average and Maximum Concentrations of Reactive Pollutants at Five California Sites from November 2, 1991, Through September 28, 1993

		Concentrations[a] at five sampling sites									
		Barton Flats		Sequoia		Yosemite		Redwoods		Azusa	
Species	Time[b]	Avg	Max	Avg	Max	Avg	Max	Avg	Max	Avg	Max
$PM_{2.5}$ mass	Day	6.4	27.5	6.6	21.0	4.4	16.0	3.9	30.0	30.1	109.0
	Night	5.5	19.6	7.3	24.0	4.3	16.0	5.8	45.0	21.1	90.0
Particle nitrate	Day	2.91	13.85	1.13	8.7	0.47	6.1	0.32	0.93	13.09	51
	Night	1.61	10.84	0.86	8.9	0.43	1.9	0.29	1.1	6.10	29
Nonvol nitrate[c]	Day	1.21	14.09	0.39	7.1	0.18	3.9	0.13	0.5	4.08	42
	Night	0.93	10.11	0.44	8	0.13	0.8	0.14	0.9	3.89	24
Sulfate	Day	1.04	3.53	0.96	3.2	0.72	2.3	0.52	2.1	3.80	12
	Night	0.89	2.55	1.06	3.8	0.72	2.3	0.57	2.2	2.85	9.5
Ammonium	Day	0.69	5.42	0.32	2.6	0.17	1.4	0.10	0.77	1.92	14
	Night	0.56	3.6	0.37	2.5	0.18	0.77	0.13	0.74	1.86	8
Chloride	Day	0.03	0.1	0.03	0.07	0.04	0.45	0.09	0.49	0.05	0.4
	Night	0.02	0.23	0.04	0.45	0.03	0.13	0.08	0.95	0.12	3.7
Ammonia	Day	2.0	7.0	1.5	4.8	1.0	13.0	0.4	3.4	6.8	13.0
	Night	0.9	4.9	1.0	4.0	0.5	1.7	0.3	3.3	4.2	13.0
Nitric acid	Day	1.13	4.33	0.43	1.93	0.45	1.93	0.10	0.45	4.67	12.4
	Night	0.20	1.16	0.17	0.92	0.20	0.85	0.09	0.71	0.71	12.4
Nitrogen dioxide	Day	2.2	10.2	3.5	9.9	2.3	6.1	5.5	13.0	74.1	219.0
	Night	1.8	7.4	3.7	11.0	2.1	5.0	5.6	16.0	70.9	195.0
Sulfur dioxide	Day	0.59	1.77	0.62	2.2	0.37	1.3	0.13	0.9	3.34	11
	Night	0.34	1.49	0.45	2.1	0.31	1.6	0.09	1.3	1.62	11
Ozone	Day	58	113	47	93	41	79	26	55	44	148
	Night	45	86	39	93	37	91	17	50	12	57

[a]Concentrations in $\mu g/m^3$ at station temperatures and pressures, except ozone in ppbv.
[b]Day corresponds to 0600–1800 PST and night corresponds to 1800–0600 PST. Samples taken every sixth day, except for ozone, which was monitored every day.
[c]Nonvolatilized nitrate is obtained from the front Teflon filter. Particle nitrate includes nonvolatilized nitrate and volatilized nitrate collected on the nylon backup filter.

Average sulfate concentrations were low, less than 1 μg/m^3, at all the forested sites. Barton Flats sulfate was comparable with that at Sequoia, about 50% higher than sulfate at Yosemite, and twice that at Redwoods. Even the maximum 12-hour sulfate concentration of 3.5 μg/m^3 was not excessive, especially when compared with sulfate concentrations in the eastern United States that average about 7 μg/m^3 during summer months (Eldred, 1997). Average and maximum sulfate concentrations at Azusa were nearly four times the levels at Barton Flats.

As noted above, ozone was the most abundant reactive component at Barton Flats, achieving a maximum 12-hour average of 113 ppbv, higher than that at other forested sites but lower than the corresponding 12-hour maximum 148-ppbv concentration measured at Azusa. Sequoia and Yosemite were also recipients of urban air from the San Joaquin Valley. Average daytime average ozone levels were comparable with daytime values at Azusa and nearly twice those encountered at Redwoods. Night-time average ozone decreased substantially at the Azusa site, owing to a lack of night-time production and titration with nitrogen oxide emitted by nearby traffic. Night-time average ozone was only about 25% less than daytime averages at the forested sites.

Daytime and night-time differences were small at the forested sites for most of the particulate species and precursor gases. Nitrate, nitric acid, and ammonia showed the largest differences, with night-time averages at Barton Flats and Azusa being about 50% of their daytime concentrations. Daytime/night-time differences at the other forested sites were not as large, although night-time levels were typically lower than daytime averages. The highest forest exposure to reactive components was during the daytime, corresponding to the period when plant photosynthesis is most active.

Seasonal Average and Maximum Concentrations

Table 4.3 examines the seasonal differences in average concentration for summer (June–August), winter (December–February), fall (September–November), and spring (March–May) seasons. Winter concentrations of all species were much lower than levels during other seasons at Barton Flats and other forested sites. The Redwoods site registered the highest wintertime $PM_{2.5}$ mass concentrations, but the measured constituent concentrations were lower. Although average $PM_{2.5}$ at the Azusa site was only three times that at Barton Flats in the summer, it was ten times higher during the winter. Nitric acid concentrations were negligible during the winter at the forested sites. Particle nitrate was largest at the Asuza site during fall, but it was comparable during all nonwinter seasons at Barton Flats.

Ozone again stands out as the most abundant of the reactive species at Barton Flats, with 12-hour averages higher than those at the other sites and averaging 76 ppbv during the summers of 1992 and 1993. The remaining reactive gases, however, showed substantial variability between daytime and night-time samples. The most noticeable of these is nitric acid, with Barton Flats average concentrations ranging from below quantifiable limits (winter night-time) to 4.3 μg/m^3 (summer daytime). Average nitric acid concentrations decreased by a factor of 2 from daytime to night-time during winter and by a factor of 7–8 from daytime to night-time during the other seasons.

Average sulfur dioxide concentrations were low, ranging from below quantifiable limits (winter night-time) to only 1.8 μg/m³ (summer daytime). The daytime average sulfur dioxide concentrations were 1.5–2.2 times the night-time averages at the Barton Flats station. Average ammonia concentrations ranged from below quantifiable limits (winter night-time) to 7 μg/m³ (summer and fall daytime). Daytime average ammonia concentrations were also more than twice the night-time average concentrations. Average nitrogen dioxide concentrations were highest during the summer, with equivalent daytime and night-time concentrations averaging 3.10 μg/m³ during daytime and 3.19 μg/m³ during night-time. These concentrations were variable during the other seasons and were 20–80% higher during the daytime as compared with the night-time averages.

Maximum precursor gas concentrations at Barton Flats occurred during daytime, with the maximum nitrogen dioxide concentration (10.21 ± 1.19 μg/m³) reported on October 27, 1992; maximum nitric acid concentration (4.33 ± 0.73 μg/m³) reported on July 30, 1993; maximum sulfur dioxide concentration (1.76 ± 0.14 μg/m³) reported on August 5, 1993; and maximum ammonia concentration (7.04 ± 0.32 μg/m³) reported on September 10, 1993.

Miller et al. (1996) summarize separately the averages and maximum concentrations for the June 1 through August 31, 1994, monitoring period. Average and maximum daytime and night-time concentrations were comparable with those of the previous summers of 1992 and 1993 for all species except the daytime nitric acid concentrations. The summer daytime average of 4.5 μg/m³ and the maximum 12-hour average of 12.2 ± 0.60 μg/m³ were found during the daytime on June 13, 1994. These acid concentrations were substantially higher than those measured during the previous two summers.

Highest 12-Hour Particle Concentrations

For the November 2, 1991, through September 28, 1993, period, the highest daytime average $PM_{2.5}$ concentration of 27.5 μg/m³ was measured at Barton Flats on Octber 21, 1992. This sample also corresponded to the highest particle nitrate concentration of 13.9 ± 0.08 μg/m³ and the highest particle sulfate concentration of 3.5 ± 0.3 μg/m³ measured at Barton Flats throughout the entire study. For this sample, the ratio of the sum of ionic species (i.e., chloride, nitrate, sulfate, and ammonium) to mass was 0.84, or 35% higher than the typical ratio of 0.50 (Miller et al. 1996). Ammonia (1.71 ± 0.08 μg/m³), sulfur dioxide (0.62 ± 0.09 μg/m³), and nitric acid (1.25 ± 1.06 μg/m³) concentrations were comparable with or less than the fall daytime averages, although the nitrogen dioxide concentration (6.32 ± 1.12 μg/m³) was twice the seasonal average. Daytime concentrations at Azusa on October 21, 1992, were not much larger than those at Barton Flats, with the CADMP network reporting 47 μg/m³ for $PM_{2.5}$, 13 μg/m³ for particle nitrate, 6.3 μg/m³ for sulfate, and only 2 μg/m³ for nitric acid.

Table 4.3. Seasonal Average Concentrations[a] of Reactive Pollutants at Five California Sites from November 2, 1991, Through September 28, 1993

		Summer		Winter		Fall		Spring	
Species	Site	Day	Night	Day	Night	Day	Night	Day	Night
$PM_{2.5}$ mass	Barton Flats	9.7	8.8	1.6	2.0	7.7	5.1	6.7	5.8
	Sequoia	11.8	12.4	1.8	1.6	7.7	8.0	5.3	6.8
	Yosemite	5.7	64	1.7	1.4	6.3	4.7	3.9	4.1
	Redwoods	3.5	3.4	4.0	7.4	5.4	7.0	3.1	5.7
	Azusa	29.9	18.3	13.1	16.9	35.9	26.7	30.3	21.8
Particle nitrate	Barton Flats	3.82	1.84	0.62	0.43	3.62	1.82	3.34	2.15
	Sequoia	1.23	0.92	0.33	0.28	1.85	1.43	1.13	0.79
	Yosemite	0.46	0.55	0.31	0.15	0.77	0.51	0.39	0.48
	Redwoods	0.40	0.37	0.24	0.23	0.28	0.26	0.35	0.29
	Azusa	13.50	4.00	9.43	7.57	15.63	7.80	13.62	5.31
Sulfate	Barton Flats	1.58	1.35	0.37	0.39	1.15	0.82	1.07	0.95
	Sequoia	1.66	1.77	0.26	0.26	0.96	1.01	0.97	1.12
	Yosemite	1.03	1.00	0.29	0.26	0.86	0.78	0.70	0.76
	Redwoods	0.69	0.75	0.28	0.32	0.49	0.47	0.57	0.72
	Azusa	5.36	3.75	1.21	1.03	3.98	3.39	4.25	2.78
Ammonium	Barton Flats	0.83	0.72	0.17	0.13	1.03	0.68	0.78	0.70
	Sequoia	0.39	0.57	0.07	0.07	0.45	0.39	0.37	0.41
	Yosemite	0.23	0.25	0.07	0.07	0.22	0.17	0.16	0.21
	Redwoods	0.11	0.14	0.05	0.07	0.09	0.10	0.13	0.19
	Azusa	1.39	1.44	1.83	1.71	2.40	2.61	2.11	1.74
Chloride	Barton Flats	0.03	0.02	0.03	0.02	0.02	0.02	0.03	0.03
	Sequoia	0.03	0.03	0.03	0.03	0.03	0.05	0.03	0.03
	Yosemite	0.03	0.03	0.03	0.03	0.03	0.04	0.05	0.04
	Redwoods	0.13	0.18	0.05	0.04	0.06	0.04	0.09	0.04
	Azusa	0.05	0.08	0.07	0.07	0.05	0.12	0.04	0.20

Ammonia	Barton Flats	3.90	1.81	0.34	0.16	2.08	0.78	1.74	0.88
	Sequoia	2.29	1.53	0.60	0.40	1.87	1.24	1.25	0.75
	Yosemite	0.91	0.55	0.52	0.24	1.86	0.66	0.72	0.52
	Redwoods	0.58	0.46	0.28	0.16	0.33	0.21	0.53	0.24
	Azusa	7.06	3.94	6.36	4.88	6.53	3.75	7.33	4.30
Nitric acid	Barton Flats	2.44	0.37	0.15	0.07	1.15	0.19	0.69	0.15
	Sequoia	0.99	0.28	0.11	0.11	0.49	0.21	0.15	0.07
	Yosemite	0.91	0.34	0.07	0.11	0.63	0.32	0.25	0.08
	Redwoods	0.10	0.06	0.10	0.11	0.10	0.13	0.11	0.09
	Azusa	6.79	0.75	1.25	0.33	5.67	1.42	4.12	0.43
Nitrogen dioxide	Barton Flats	3.06	3.19	0.79	0.44	3.21	1.81	1.99	1.66
	Sequoia	4.21	4.99	3.22	2.99	3.91	3.79	2.90	3.04
	Yosemite	2.65	2.31	2.05	1.66	2.69	2.46	2.08	1.90
	Redwoods	5.06	5.80	6.56	5.13	5.95	6.25	4.58	5.26
	Azusa	78.42	86.70	58.81	49.56	99.26	90.55	64.73	60.13
Sulfur dioxide	Barton Flats	0.82	0.54	0.24	0.19	0.69	0.37	0.59	0.27
	Sequoia	1.48	0.98	0.17	0.06	0.61	0.55	0.38	0.23
	Yosemite	0.70	0.66	0.10	0.08	0.42	0.32	0.30	0.22
	Redwoods	0.14	0.08	0.11	0.11	0.15	0.10	0.10	0.06
	Azusa	4.41	2.07	1.67	0.81	4.70	2.06	2.68	1.51
Ozone	Barton Flats	76	54	39	37	55	44	60	45
	Sequoia	70	60	29	25	51	42	38	30
	Yosemite	55	47	32	30	37	31	43	37
	Redwoods	27	17	22	18	26	16	28	15
	Azusa	67	11	17	12	43	9	46	15

[a] Average concentrations in $\mu g/m^3$ (ppbv for ozone).

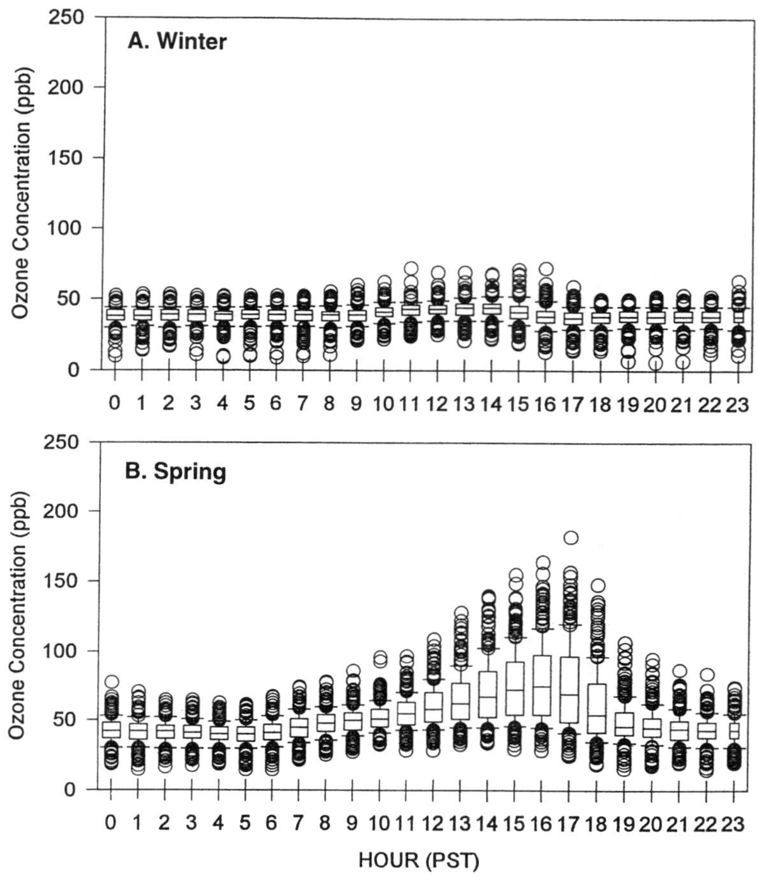

Figure 4.3. Diurnal box plots of hourly average ozone concentrations at the Barton Flats Station, CA, during (A) winter (December 1991; January, February, and December 1992 and 1993; and January and February 1994), (B) spring (March, April, and May 1992, 1993, and 1994) (C) summer (June, July, and August 1992, 1993, and 1994), and (D) fall (November 1991; September, October, and November 1992, 1993, and 1994).

On October 21, 1992, a weak surface front passed through the SoCAB and a weak upper-level low-pressure system hovered over the area. Approximately 5 mm of rain was recorded at Los Angeles International Airport. Under these conditions, transport of emissions from the SoCAB into the San Bernardino Mountains would be expected for many hours of the day, with abundant moisture available for aqueous-phase formation of secondary particles. Wind speeds at Barton Flats and in the SoCAB on October 21, 1992, were low (<3 m/sec) with predominant westerly to northwesterly wind flow during the daytime. The temperature was low (<12° C) during the daytime, with high relative humidities (80%) developed in the afternoon. Photochemical reaction on this day was minimal, with average daytime ozone of 60 ppbv, due to low (<300 watts/m^2) solar radiation in the SoCAB and surrounding areas. This is consistent with the finding

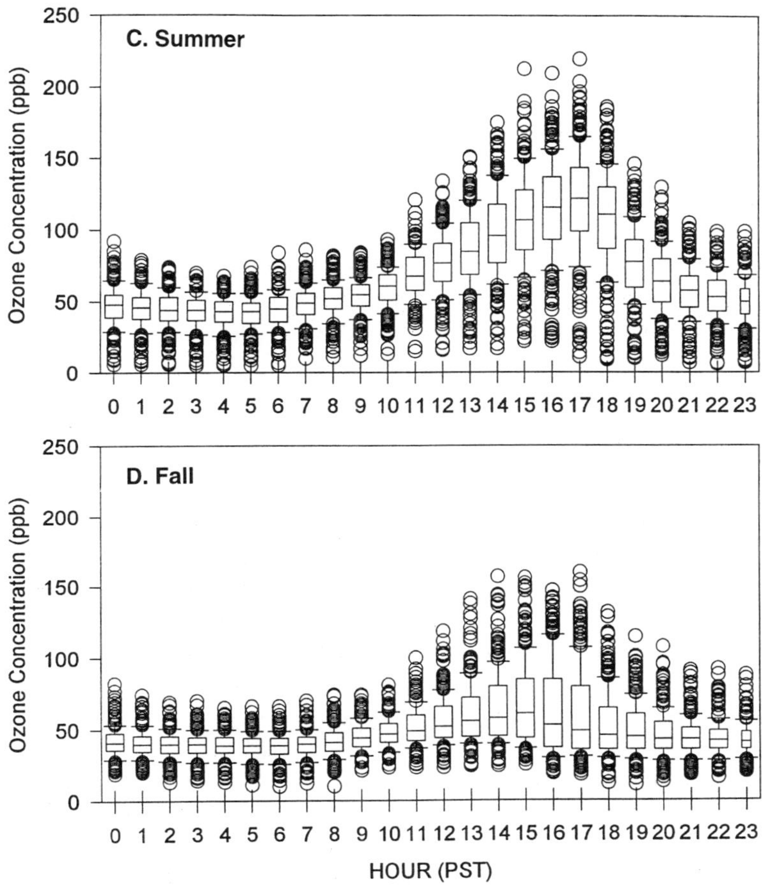

Figure 4.3. (*continued*)

of Russell et al. (1983) that when there is sufficient ammonia, nitrate will reside in the ammonium nitrate particle phase, rather than in the nitric acid gas phase, at temperatures lower than 15° C. It is apparent that low dispersion, low temperature, and high relative humidity on October 21, 1992, favored the formation of secondary sulfates and nitrates in the SoCAB that were transported to forests in the San Bernardino Mountains.

Diurnal Variation in Ozone Concentrations

High hourly average ozone concentrations were found frequently at the Barton Flats station. Hourly ozone concentrations exceeded the NAAQS of 120 ppb on 235 days (22% of all sampling days) between November 2, 1991, and September 30, 1994. These adverse pollution episodes occurred from late April through early October each year, between 1300–1800 PST. Peak hourly ozone concentrations often occurred in the late afternoon and lasted 2–3 hours. Maximum hourly ozone concentrations exceeded 200 ppb on July 18, 1992 (220 ppb at 1700 PST), on June 25, 1994 (204 ppb at 1700 PST), and on June 30, 1994 (213 ppb at 1500 PST and 210 ppb at 1600 PST).

Figure 4.3 shows how hourly ozone concentrations changed through the day during each season. During summer, both intense sunlight and transport of ozone with hydrocarbon and oxides of nitrogen precursors from source areas occur. During spring and fall, the sun angle and resulting solar radiation are lower than during summer. Flows from polluted areas are not as consistent or intense as during the summer, resulting in lower and less frequent hourly peak concentrations. During winter, there are more clouds and even lower sun angles, resulting in less solar radiation. Lower mixed layers in the SoCAB are also common during winter and mitigate against transport to higher elevations.

Afternoon hourly ozone concentrations were higher than night-time ozone concentrations by less than 10% during winter, by 40–60% during spring and fall, and by 80–90% during summer. Figure 4.3 shows that peak ozone concentrations occurred at about 1400 PST during winter, 1500 PST during fall, 1300 PST during spring, and 1700 PST during summer. Ozone usually peaked at the Azusa site in the SoCAB between 1200 and 1300 PST, regardless of season. Mornings experienced lower ozone concentrations than afternoon due to minimal transport from polluted areas and less time for the photochemical reactions.

Transport Dynamics

Green et al. (1992a,b) describe southern California wind flow patterns as influenced by three main features: (1) large-scale (synoptic) positions of high- and low-pressure systems; (2) differential heating of inland areas versus coastal areas (sea breezes and land breezes); and (3) terrain-driven circulation (upslope and downslope flows). Local winds and the transport of air pollutants are determined from the combined effects of these features.

Flow patterns in the summer are often influenced by the sea breeze and upslope/downslope flows. When a synoptic-scale high-pressure system develops east of California (typically over the Great Basin in Nevada and Utah), the strength of onshore flow is diminished, causing shallow mixing depths and low dispersion. When a low-pressure system develops over the Great Basin, onshore flow is more intense, mixed layers are deeper, and emissions are rapidly diluted by dispersion.

Winter shows more variability than summer. The Santa Ana regime results from high pressure to the northeast of the SoCAB and low pressure near the coast. Dry winds bring air into the SoCAB from the Mojave Desert. A second wintertime regime is characterized by weak synoptic scale pressure gradients with flows induced by local terrain and pressure and temperature gradients. Weak sea breeze/upslope flows occur during the day, and weak downslope/offshore flows occur at night. During a third pattern, low pressure over the Great Basin or Four Corners area results in continuous onshore flow and flow over the mountains into the Mojave Desert.

Figure 4.4 shows summertime and wintertime Barton Flats wind roses for daytime and night-time hours during this experiment. Summertime daytime winds showed upslope flow (from the west to northwest); the night-time winds show downslope flow (from the southeast). During winter, upslope and downslope flows were still dominant, but a significant amount of flow from the east was also observed. These easterly winds occurred during the stronger Santa Ana conditions.

During summer, afternoon mixed layers in the SoCAB can extend as high as Barton Flats, or the strong upslope velocities may be sufficient to carry pollutants above a lower mixed layer. During winter, the shallower layers and weaker winds result in lower likelihood of transport from the urban to forested area.

The July 18–31, 1993, intensive monitoring period illustrates the relationships between summertime meteorology and pollutant concentrations. Figure 4.5 shows the $PM_{2.5}$ concentrations with measured and unmeasured chemical components for the morning, afternoon, and night-time samples taken each day during the period. Figure 4.6 shows the corresponding precursor concentrations, and Figure 4.7 shows hourly ozone and wind directions. The maximum concentrations in these figures differ from those in Tables 4.2 and 4.3, owing to the every day (as opposed to every sixth day) measurements and shorter (6-hour) sample durations.

Figures 4.5 and 4.6 show that the highest mass, ion, and reactive species concentrations all occurred during the afternoon periods. By contrast, the lowest concentrations occurred during the morning period, with the exception of $PM_{2.5}$ chloride and gaseous ammonia, sulfur dioxide, and nitric acid concentrations. This could be the reason for the small differences between average daytime and night-time concentrations in Tables 4.1 and 4.2. Average daytime and night-time concentrations during the intensive period are comparable with those in Table 4.2, even though only three of the sixth-day samples from the intensive period were included in the Table 4.2 averages.

Maximum afternoon concentrations for mass, ions, and gaseous species during the intensive sampling period were lower than those observed for routine monitoring periods on an every-sixth-day sampling schedule between November 2, 1991, and September 28, 1993, with the exception of nitric acid and ammonia. Maximum nitric acid concentration (6.73 ± 0.53 $\mu g/m^3$) occurred during the afternoon of July 27, 1993, whereas maximum ammonia concentration (6.7 ± 0.4 $\mu g/m^3$) occurred during the afternoon of July 30, 1993. This is also the day (July 30, 1993) that maximum 12-hour nitric acid concentrations were found on the every-sixth-day sampling schedule between November 2, 1991, and September 28, 1993.

Figure 4.7 shows the repeatability of wind directions during different parts of the summer day and how these are likely to affect ozone concentrations. Miller et al. (1996) contains detailed maps of areawide changes in winds throughout the SoCAB for July 24, 1993. Light onshore flow occurred in the morning in the SoCAB, but this did not affect winds at Barton Flats that remained downslope until after sunrise. Upslope flow began by about 0700 PST at Barton Flats, but flows were still weak and northerly in the SoCAB. By 1200 PST, however, most SoCAB flows turned westerly, and this situation continued until after sunset, when at 2200 PST they returned to northerly or variable in the SoCAB. There was

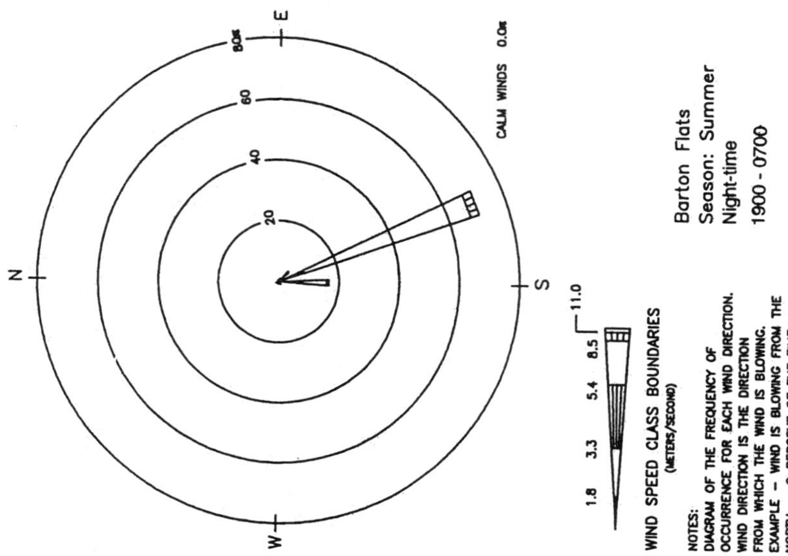
N
E
S
W
20
40
60
80%
CALM WINDS 0.0%
1.8 3.3 5.4 8.5 11.0
WIND SPEED CLASS BOUNDARIES
(METERS/SECOND)
NOTES:
DIAGRAM OF THE FREQUENCY OF
OCCURRENCE FOR EACH WIND DIRECTION.
WIND DIRECTION IS THE DIRECTION
FROM WHICH THE WIND IS BLOWING.
EXAMPLE – WIND IS BLOWING FROM THE
NORTH .0 PERCENT OF THE TIME.
Barton Flats
Season: Summer
Night-time
1900 - 0700

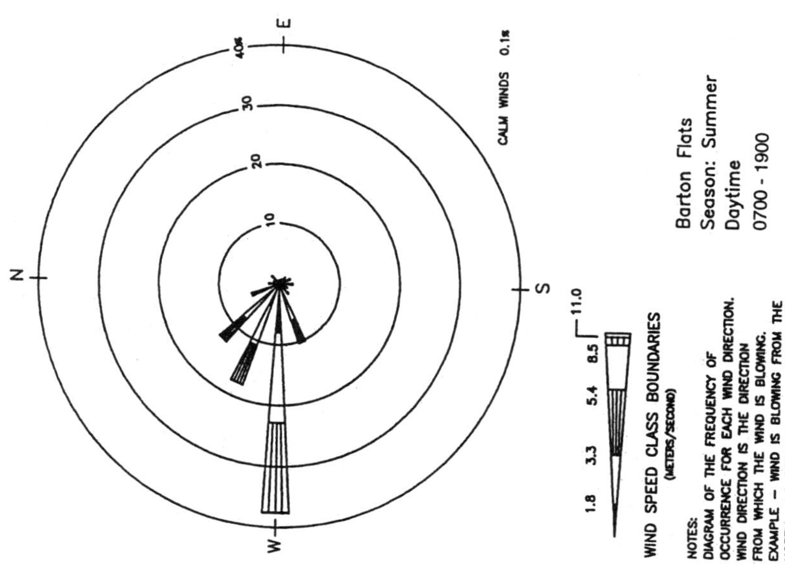
N
E
S
W
10
20
30
40%
CALM WINDS 0.1%
1.8 3.3 5.4 8.5 11.0
WIND SPEED CLASS BOUNDARIES
(METERS/SECOND)
NOTES:
DIAGRAM OF THE FREQUENCY OF
OCCURRENCE FOR EACH WIND DIRECTION.
WIND DIRECTION IS THE DIRECTION
FROM WHICH THE WIND IS BLOWING.
EXAMPLE – WIND IS BLOWING FROM THE
NORTH 1.8 PERCENT OF THE TIME.
Barton Flats
Season: Summer
Daytime
0700 - 1900

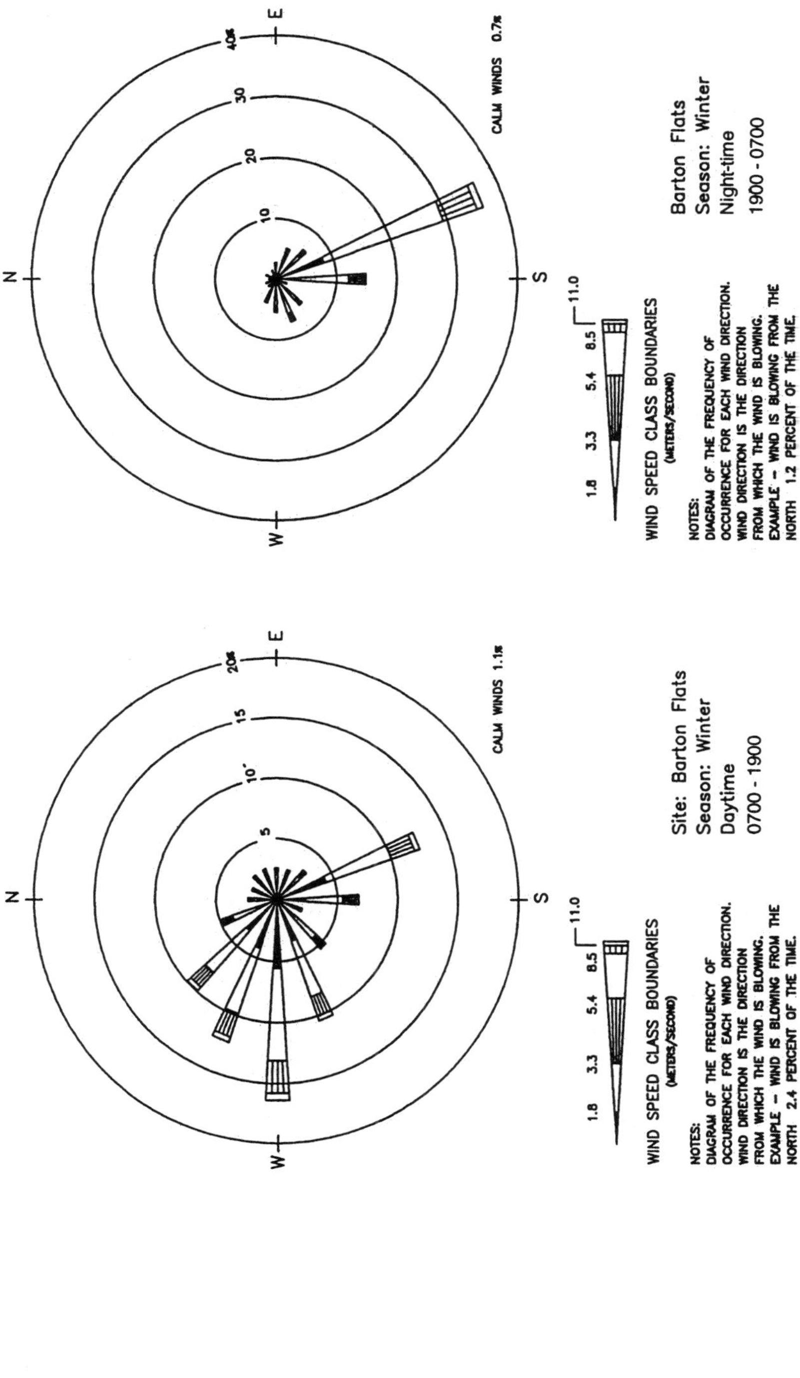

Figure 4.4. Summertime and wintertime Barton Flats wind roses for daytime and night-time hours.

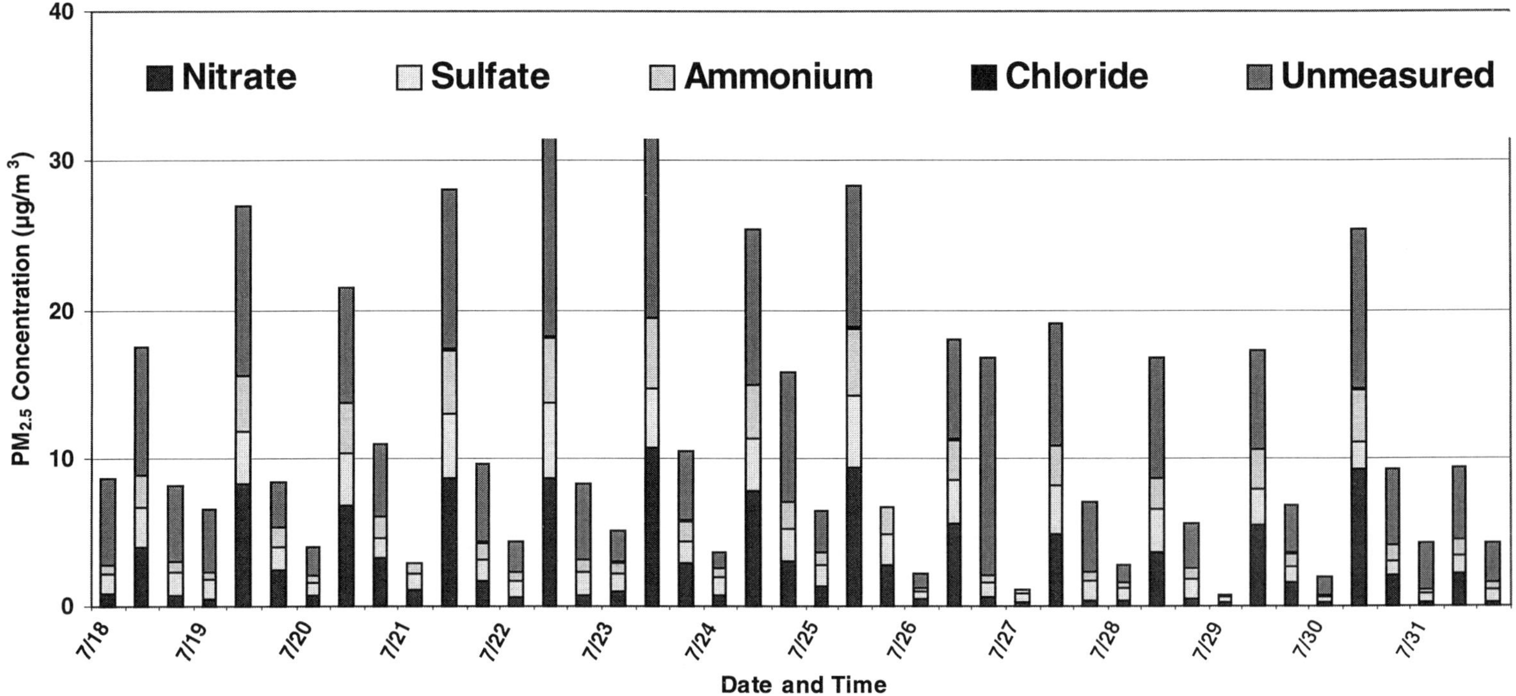

Figure 4.5. $PM_{2.5}$ mass and chemical concentrations for morning (0600–1200 PST), afternoon (1200–1800 PST), and night-time (1800–0600 PST) samples from July 8, 1992, through July 31, 1992. Particle mass, ammonium, and nitrate have been adjusted to include the effects of volatalized nitrate. The date symbol is below the 0600–1200 PST measurement.

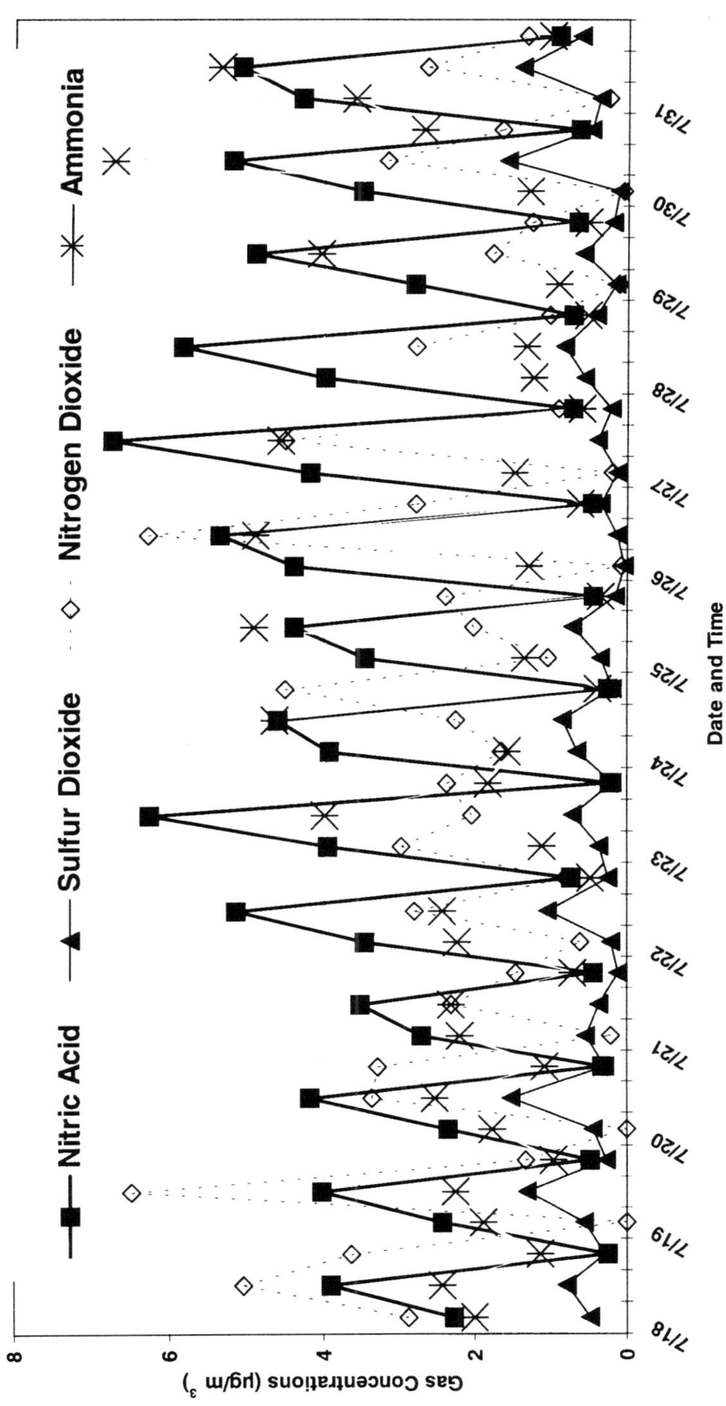

Figure 4.6. Six and twelve hour average nitric acid, nitrogen dioxide, sulfur dioxide, and ammonia concentrations from July 18, 1992, through July 31, 1992. The date symbol is below the 0600–1200 PST sample.

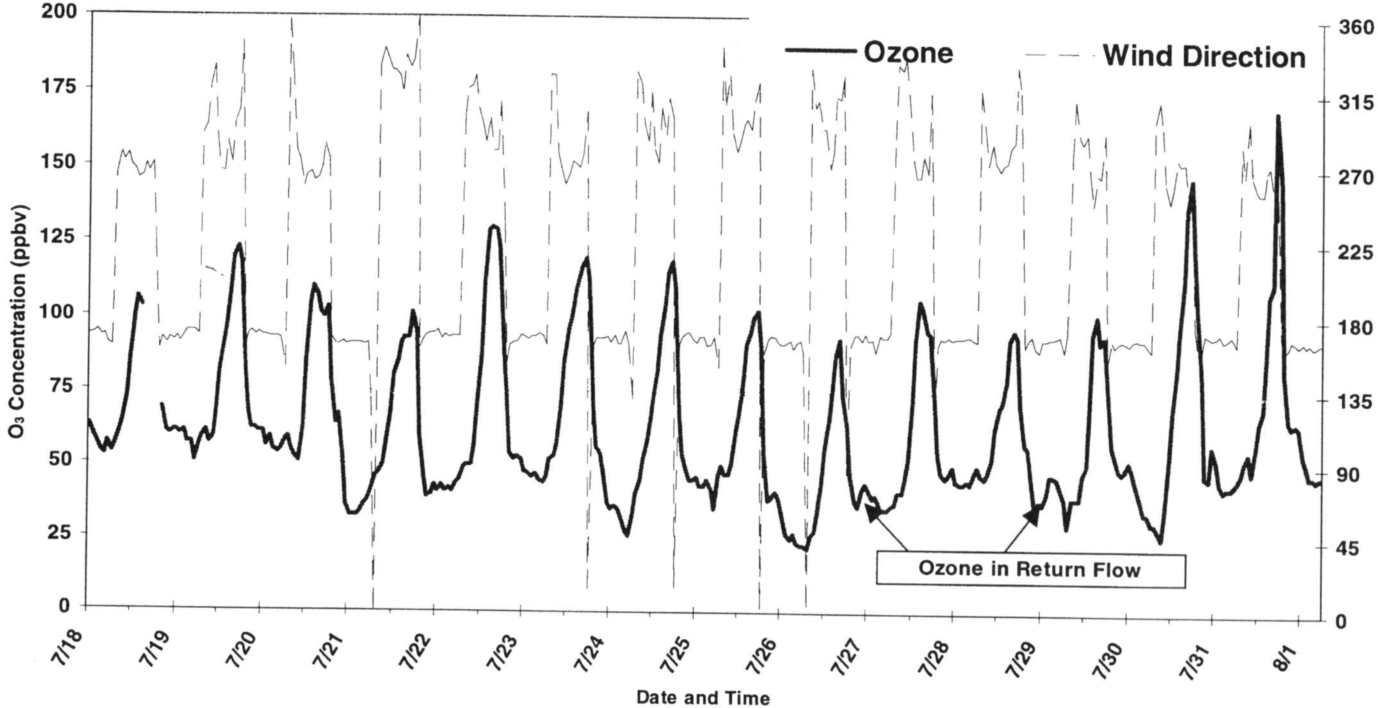

Figure 4.7. Hourly ozone concentrations and wind directions at Barton Flats from July 18, 1992, through July 31, 1992. Note the secondary ozone peak, usually early in the morning of the following day, as a result of downslope flows bringing back ozone transported upslope during the afternoon.

an abrupt shift in wind direction at Barton Flats between 1700 and 1900 PST to downslope flow, as shown in Figure 4.7. Owing to its location in a canyon, the sun drops from view earlier at Barton Flats than at more open sites, casting the slopes into shadows that cool them and the air above them. As Figure 4.7 shows, this pattern repeated itself day after day.

Figure 4.7 also provides another explanation for the high night-time ozone levels at Barton Flats. Nearly every ozone trace shows a peak near 1700 PST, but most show a shoulder or a secondary peak that occurs from 2 to 8 hours later. One of the most pronounced occurred on July 29, 1992, between 0100 and 0400 PST. This shows the regular and significant effect of ozone transported upslope into the San Bernardino Mountains being returned after flows shift to downslope. The forest at and near Barton Flats receives a double dose of the same reactive materials owing to the diurnal shifts in transport direction.

Conclusions

Aerosol, gas, and meteorological measurements were taken at the Barton Flats site from October 30, 1991, through September 30, 1993, and from June 1 to August 31, 1994. Ozone, wind speed, wind direction, temperature, relative humidity, solar radiation, and time of wetness were measured every hour with continuous in situ monitors throughout these periods.

Ambient air quality samples for $PM_{2.5}$ mass, nitrate, sulfate, and ammonium were found above the lower quantifiable limits (LQLs) for more than 80% of the samples. Precursor gas (i.e., sulfur dioxide, nitrogen dioxide, nitric acid, ammonia) concentrations were quantified in more than 70% of the samples. $PM_{2.5}$ chloride concentrations were often below LQLs, especially for the winter periods. Large fractions of $PM_{2.5}$ ammonium nitrate volatilized during sampling, and absorbing filters following denuders were needed to obtain accurate estimates of the particle nitrate and nitric acid concentrations.

Measured ammonium was completely explained by the presence of ammonium sulfate and ammonium nitrate in most cases. Sulfate was not in a strong acidic state by the time it reached Barton Flats. Daytime average concentrations for all the particulate species except nitrate were similar to the night-time averages at all seasons. Season-to-season variations were significant for $PM_{2.5}$ mass and ions, with the lowest values reported during winter and reaching maximum concentrations during summer. The highest reactive gas concentrations occurred during the daytime.

During summer, gas and particle concentrations were lowest during the morning (i.e., 0600–1200 PST) and highest during the afternoon (i.e., 1200–1800 PST). Particles emitted and formed in the Los Angeles Basin were transported into the Barton Flats Station area during late spring, summer, and early fall. This transport corresponded to the afternoon upslope flows. Downslope flows occurred after sunset and returned some of the pollutants to the Barton Flats Station site.

Ozone was the most frequently observed and most abundant reactive air pollutant measured at Barton Flats.. Hourly ozone concentrations exceeded NAAQS of

120 ppb on more than 20% of the sampling days. Peak ozone concentrations exceeding 200 ppb were found. Sulfuric acid concentrations were negligible. Twelve-hour average nitric acid concentrations did not exceed 4.5 $\mu g/m^3$ during the summers of 1992 and 1993 but achieved a maximum of 12.2 $\mu g/m^3$ during the summer of 1994. Daytime nitric acid concentrations averaged 2.4 $\mu g/m^3$ during the summers of 1992 and 1993 and 4.5 $\mu g/m^3$ during the summer of 1994.

References

Cass, G.R. On the relationship between sulfate, air quality, and visibility, with examples in Los Angeles. Atmos. Environ. 13:1069–1084; 1979.

Chow, J.C.; Egami, R.T. San Joaquin Valley 1995 integrated monitoring study: documentation, evaluation, and descriptive data analysis of PM_{10}, $PM_{2.5}$, and precursor gas measurements—technical support studies No. 4 and No. 8—final report. Prepared for the California Regional Particulate Air Quality Study, California Air Resources Board, Sacramento, CA, by the Desert Research Institute, Reno, NV, March 7, 1997.

Chow, J.C.; Watson, J.G. Ion chromatography. In: Landsberger, S.; Creatchman, M., eds. Elemental analysis of airborne particles. Newark, NJ: Gordon and Breach Publishers; 1997.

Chow, J.C.; Liu, C.S.; Cassmassi, J.; Watson, J.G.; Lu, Z.; Pritchett, L.C. A neighborhood-scale study of PM_{10} source contributions in Rubidoux, California. Atmos. Environ. 26A:693–706; 1992.

Chow, J.C.; Watson, J.G.; Bowen, J.L.; Gertler, A.W.; Frazier, C.A.; Fung, K.K.; Ashbaugh, L. A sampling system for reactive species in the western U.S. In: Winegar, E.; Keith, L.H., eds. Sampling and analysis of airborne pollutants. Washington, DC: American Chemical Society; 1993:209–228.

Chow, J.C.; Watson, J.G.; Fujita, E.M.; Lu, Z.; Lawson, D.R.; Ashbaugh, L.L. Temporal and spatial variations of $PM_{2.5}$ and PM_{10} aerosol in the southern California air quality study. Atmos. Environ. 28:2061–2080; 1994.

Chow, J.C.; Watson, J.G.; Lowenthal, D.H.; Countess, R.J. Sources and chemistry of PM_{10} aerosol in Santa Barbara County, CA. Atmos. Environ. 30(9):1489–1499; 1996.

Davidson, A. Update on ozone trends in California's Southern Coast Air Basin. JAWMA 43:226–227; 1993.

Eldred, R.A. Comparison of selenium and sulfur at remote sites throughout the United States. JAWMA 47:204–211; 1997.

Fitz, D.; Hering, S. Study to evaluate the CADMP sampler. Final report. Prepared under Contract 93–333 for the California Air Resources Board, Air Research Division, Sacramento, CA, by the University of California, Center for Environmental Research and Technology, College of Engineering, Riverside, CA, and Aerosol Dynamics Inc., Berkeley, CA, May 1996.

Fujita, E.M.; Watson, J.G.; Chow, J.C.; Lu, Z. Validation of the chemical mass balance receptor model applied to hydrocarbon source apportionment in the southern California air quality study. Environ. Sci. Technol. 28(9):1633–1649; 1994.

Green, M.C.; Flocchini, R.G.; Myrup, L.O. The relationship of the extinction coefficient distribution to wind field patterns in southern California. Atmos. Environ. 26A:827–840; 1992a.

Green, M.C.; Myrup, L.O.; Flocchini, R.G. A method for classification of wind field patterns and its application to southern California. Int. J. Climatol. 12:111–135; 1992b.

Harley, R.A.; Hannigan, M.P.; Cass, G.R. Respeciation of organic gas emissions and the detection of excess unburned gasoline in the atmosphere. Environ. Sci. Technol. 26(12):2395–2408; 1992.

Miller, P.R.; Watson, J.G.; Bytnerowicz, A.; Fenn, M.; Poth, M.; Chow, J.; Taylor, G. Assessment of acidic deposition and ozone effects on conifer forests in the San Bernardino Mountains, draft final report. Prepared for the California Air Resources Board, Sacramento, CA, by the USDA Forest Service, Pacific Southwest Research Station, Riverside, CA, and the Desert Research Institute, Reno, NV, 1996.

Pandis, S.N.; Harley, R.A.; Cass, G.R.; Seinfeld, J.H. Secondary organic aerosol formation and transport. Atmos. Environ. 26A:2269–2282; 1992.

Pilinus, C.; Seinfeld, J.H. Continued development of a general equilibrium model for inorganic multicomponent atmospheric aerosols. Atmos. Environ. 21:2453–2466; 1987.

Russell, A.G.; McRae, G.J.; Cass, G.R. Mathematical modeling of the formation and transport of ammonium nitrate aerosol. Atmos. Environ. 17(5):949–964; 1983.

Seinfeld, J.H. Air pollution: physical and chemical fundamentals. 2nd ed. New York: McGraw-Hill; 1986.

U.S. EPA. National ambient air quality standards for particulate matter—final rule. 40 CFR part 50. Fed. Reg. 62(138):38651–38760; July 18, 1997.

Vedal, S. Critical review—ambient particles and health: lines that divide. JAWMA 47:551–581; 1997.

Watson, J.G.; Chow, J.C. Clear sky visibility as a challenge for society. Annu. Rev. Energy Environ. 19:241–266; 1994.

Watson, J.G.; Chow, J.C.; Egami, R.T.; Bowen, J.L.; Frazier, C.A.; Gertler, A.W.; Lowenthal, D.H.; Fung, K.K. Measurements of dry deposition parameters for the California acid deposition monitoring program, final report. DRI Document 8063.2F1, Contract A6-076-32, prepared for California Air Resources Board, Sacramento, CA, by the Desert Research Institute, Reno, NV, May 31, 1991.

Watson, J.G.; Chow, J.C.; Lurmann, F.; Musarra, S. Ammonium nitrate, nitric acid, and ammonia equilibrium in wintertime Phoenix, AZ. J. Air Waste Assoc. 44:261–268; 1994.

Watson, J.G.; Lioy, P.J.; Mueller, P.K. The measurement process: precision, accuracy, and validity. In: Hering, S.V., ed. Air sampling instruments for evaluation of atmospheric contaminants, 8th ed. Cincinnati, OH: American Conference of Governmental Industrial Hygienists; 1995:187–194.

5. Visibility Impairment in the San Bernardino Mountains

S. Copeland

Introduction

IMPROVE Program

The data presented in this chapter were collected through the *I*nteragency *M*onitoring of *PRO*tected *V*isual *E*nvironments (IMPROVE) program. IMPROVE is a partnership of federal land management agencies, the Environmental Protection Agency, and other state organizations. The IMPROVE network began sampling at 30 sites nationwide (including San Gorgonio) in 1988. As of March 1997, there are about 65 sites in the network, which is largely unchanged from Figure 5.1.

The primary objectives of IMPROVE are the following:

1. To establish current background visibility levels in EPA class I areas
2. To identify chemical species and emission sources responsible for existing man-made visibility impairment
3. To document long-term trends for assessing progress toward the national visibility goal

The routine IMPROVE monitoring approach involves aerosol, optical, and view monitoring. Aerosol monitoring measures the mass concentration (in micrograms per cubic meter) and the elemental composition of the particles. Optical monitoring measures the light extinction coefficient (b_{ext}) by using a transmissometer or the light scattering coefficient (b_{scat}) by using a nephelometer. View monitoring

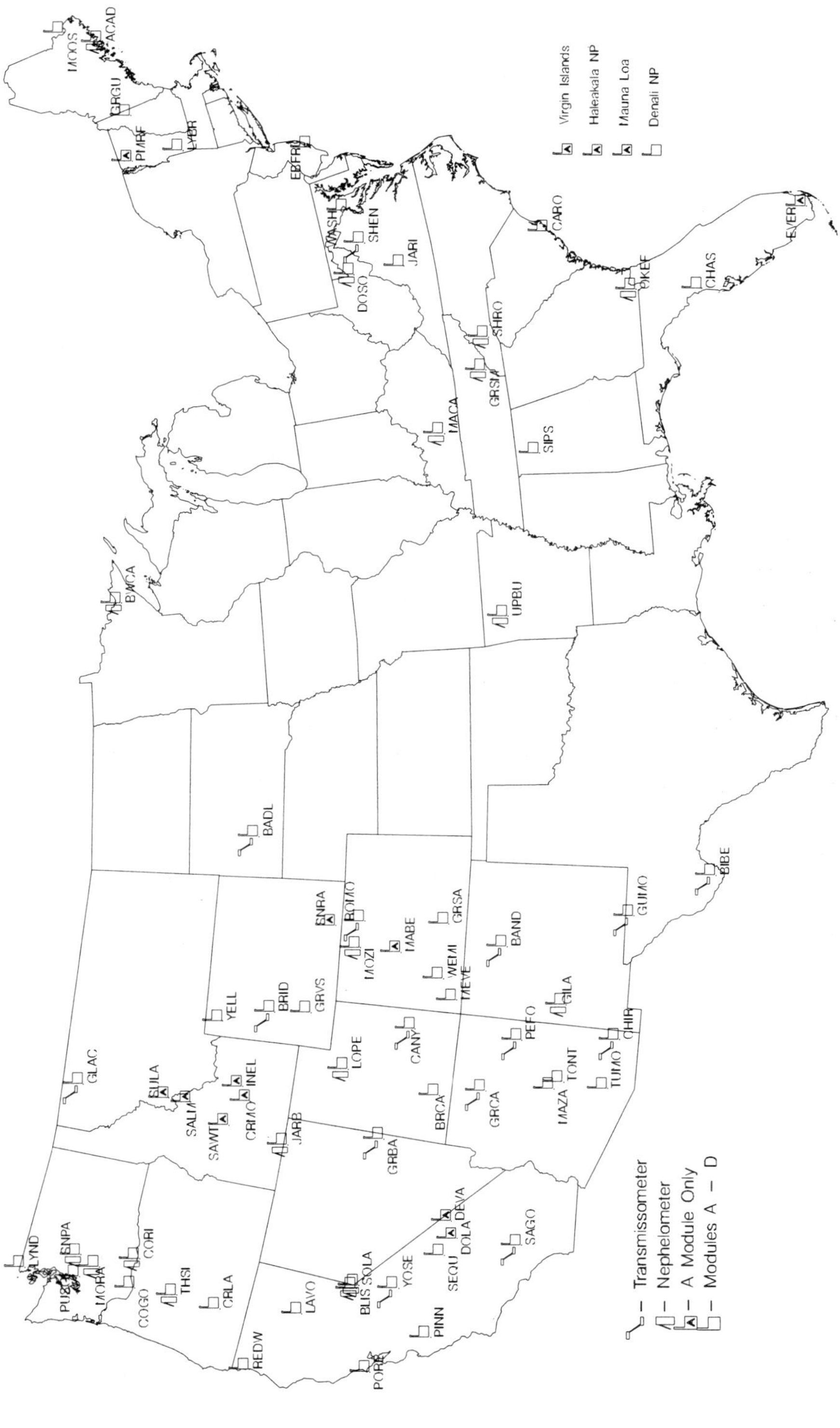

Figure 5.1. IMPROVE monitoring network 1996.

Table 5.1. Longitude, Latitude, and Elevations of San Gorgonio Samplers

Sampler	Latitude	Longitude	Elevation (m)
Aerosol	34° 11′ 36″	116° 54′ 48″	1,710
Transmitter	34° 09′ 45″	116° 56′ 07″	1,731
Receiver	34° 11′ 36″	116° 54′ 48″	1,710
Automatic camera	34° 13′ 55″	117° 14′ 02″	1,890

documents the appearance of the scene by automated photography using color slide film.

This chapter presents an overview of the aerosol, optical, and view monitoring data gathered at the San Gorgonio Wilderness Area site from March 1988 through May 1996.

Site Information

The aerosol sampler is located at the Converse Fire Station, at an elevation of 1,712 m. This location is approximately 3 km north of the wilderness boundary and about 200 m below the lower elevations of the wilderness area.

The transmissometer has a 4.1-km site path, which begins at 1,731 m and bears 31° to a receiver at 1,710 m near the Converse Fire Station. The site path is well above the ground along the entire path.

The automatic camera system was located on a ridge approximately 25 km west and slightly north of the wilderness area. The camera site was at an elevation of 1,890 m. Three photographs were taken each day at 0900, 1200, and 1500 local standard time.

The camera was removed in 1995, after collecting roughly 6.5 years of photographs, whereas the aerosol sampler and transmissometer are still operational.

Latitude, longitude, and elevation are listed below for each sampler in Table 5.1. Locations of all the samplers and the western wilderness boundary are shown in Figure 5.2.

Background

The sections titled "Instruments" and "Analytical Methods" are adapted directly from Sisler (1996). More detail is available in the original reference.

Instruments

Transmissometer

Transmissometers are calibrated to measure the irradiance, at a wavelength of 550 nm, of a light source after the light has traveled over a finite atmospheric path. The

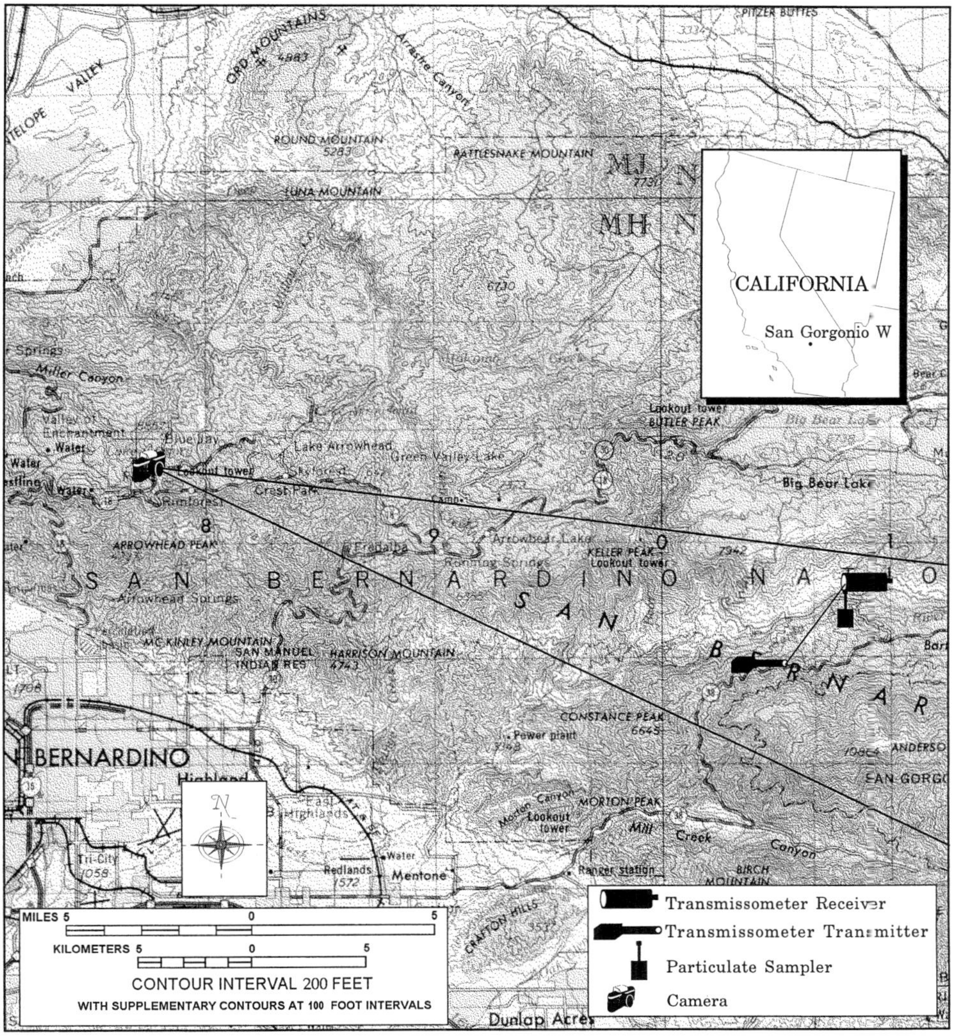

Figure 5.2. Sampler locations.

transmittance of the path is calculated by dividing the measured irradiance at the end of the path by the calibrated initial intensity of the light source. Bouger's law is applied to calculate the extinction. Because of the relatively clean atmospheres found in the western United States, path lengths of a few kilometers are required to achieve the necessary sensitivity to resolve extinctions near the Rayleigh limit.

The transmissometers used in this study are the OPTEC, Inc., LPV-2 instruments, which have been in use since 1986. Their use in remote locations such as national parks is discussed by Molenar et al. (1989), and their use in urban settings is presented by Dietrich et al. (1989). Hourly relative humidity and ambient temperature data are collected along with extinction measurements 365 days/year.

Data processing algorithms that incorporate corrections for interferences are thoroughly discussed by Molenar and Malm (1992). Basically, there are five checks the data must pass to be incorporated into a validated data set. They are:

1. Relative humidity must be less than 90%.
2. Maximum extinction cannot exceed a threshold value based on photometer sensitivity and path length.
3. Variability in extinction readings taken over a period of 1 hour cannot exceed a threshold value.
4. Rate of change of hourly average extinction measurements cannot exceed a threshold value.
5. By definition, any hourly average data point passing the above four criteria but falling in between two hourly average data points that have failed the criteria is referred to as "isolated." It is conservatively assumed that it has also been affected by interferences.

Molenar et al. (1989) discuss the inherent uncertainties associated with the measurement. The accuracy of the transmission measurement, as determined by field and laboratory calibrations, is better than 1%. However, the accuracy of the derived extinction is dependent on the accuracy of the transmission measurement in field conditions. The transmission calculation is determined from an absolute (as opposed to relative) measurement of irradiance of a light source of known intensity that is located some known distance from the receiver. The measurement is made through optics that are exposed to the ambient atmosphere but are assumed to be free of dust or other films, which tend to build up on the optical surfaces. The uncertainties associated with these parameters contribute to the overall uncertainty of the measurement. For a typical 5-km path length, the estimated uncertainty is about 4/Mm.

Aerosol Sampling System

The standard IMPROVE sampling module consists of (1) a size selective inlet, (2) a cyclone to provide a particle size cutoff based on the flow rate, (3) collection substrates, (4) a critical orifice that provides the proper flow rate for the desired particle size cutoff, and (5) a vacuum pump that produces the flow. The system is described in some detail by Malm et al. (1994) and Eldred et al. (1988) and is only briefly described here.

Two 24-hour aerosol samples are collected each week from midnight to midnight on Wednesdays and Saturdays. The sampling system consists of four independent sampling modules. Three modules (denoted A, B, and C) use a cyclone with a flow rate of 22.7 L/min that allows for collection of fine particles less than 2.5 μm in diameter. The fourth module (D) is a PM_{10} sampler with a wind-insensitive size selective inlet that collects particles less than 10 μm in diameter.

Table 5.2. Filter Media and Analysis Techniques Used to Determine Concentrations of Particulate Matter Species from IMPROVE Sampler Modules

Module	Filter media	Analyses[a]
A	Teflon	Gravimetric analysis for mass <2.5-μm diameter
		LIPM for optical absorption
		PIXE for elements Na to Mn
		PESA for H[c]
		XRF for elements Fe to Pb
B	Nylon (denuded)	Ion chromatography for NO_3 and SO_4
C	Quartz	TOR for organic and light-absorbing C
D	Teflon	Gravimetric analysis for mass <10-μm diameter

[a]LIPM, laser integrating plate method; PIXE, particle induced x-ray emission; PESA, proton elastic scattering; XRF, x-ray fluorescence; TOR, thermal optical reflectance.

Table 5.2 summarizes the substrates used and aerosol species measured on each filter.

Analytical Methods

Gravimetric mass (channel A fine mass, channel D PM_{10} mass) is measured as the difference between the weight of the substrates before and after sampling, using a microelectrobalance. The channel A Teflon substrates are analyzed for sulfur and other elements by particle induced x-ray emission and x-ray fluorescence and for hydrogen by proton elastic scattering analysis (Cahill et al. 1986).

The coefficient of light absorption for fine particles, b_{abs}, is also determined from the channel A Teflon substrates by using a laser integrating plate method (LIPM) (Cahill et al. 1986). This involves direct measurement of the absorption of a laser beam by a sample over the area of the sample.

Extract from the channel B nylon substrates are analyzed by ion chromatography for sulfate and nitrate ions from which the sulfate and nitrate compounds can be estimated (Cahill et al. 1986; Malm et al. 1994). The inlet on channel B has a sodium carbonate coated annual denuder, which is theoretically more than 99% efficient at removing gas-phase nitrate (nitric acid) from the sampling stream. This is significant because there is known to be a large amount of nitric acid in the Los Angeles Basin.

The channel C quartz substrates are analyzed by thermal optical reflectance combustion for organic and elemental carbon (Chow et al. 1993).

Characteristic Photographs

Photographs can be very useful when relating obtuse visibility data to nontechnical individuals. Although it is not reasonable to assign a single visibility parameter (e.g., "this photograph represents an SVR of 200 km") to any one photograph, it is practical to estimate bounds for any given visibility parameter for photos that

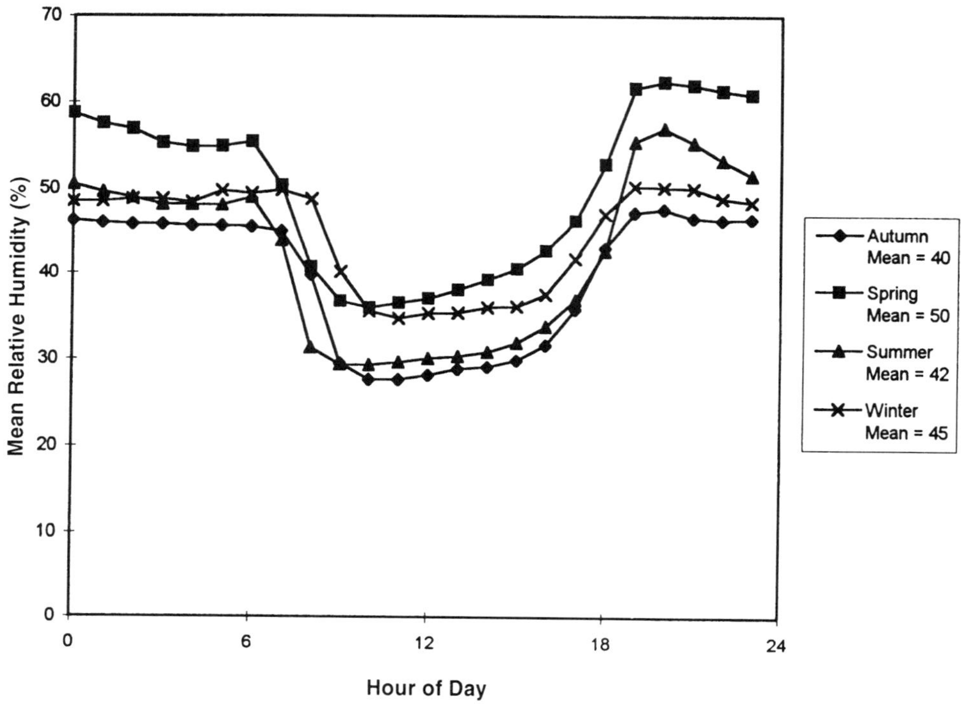

Figure 5.4. Diurnal pattern in mean relative humidity at San Gorgonio by season, 1988–1995.

meet certain criteria. Necessary criteria include presence of a uniform haze and a cloud-free sky. The following four photographs (Fig. 5.3[1]) demonstrate the range of visibility conditions that have been observed at San Gorgonio Wilderness Area and provide a range of corresponding visibility indices.

Transmissometer Data Interpretation

Figures 5.4 and 5.5 show the patterns in mean hourly relative humidity (RH) and mean hourly b_{ext} with weather-affected data removed and then sorted by season.

RH begins to decrease at sunrise (by definition, as air warms with a fixed moisture content, the RH decreases), reaching a minimum at about 10 AM. From about 10 AM to about 5 PM, the RH stays near its minimum and begins to climb rapidly near sunset, reaching a maximum at about 8 PM. RH is mainly constant

[1]Figure 5.3 will be found in the color insert following page 78.

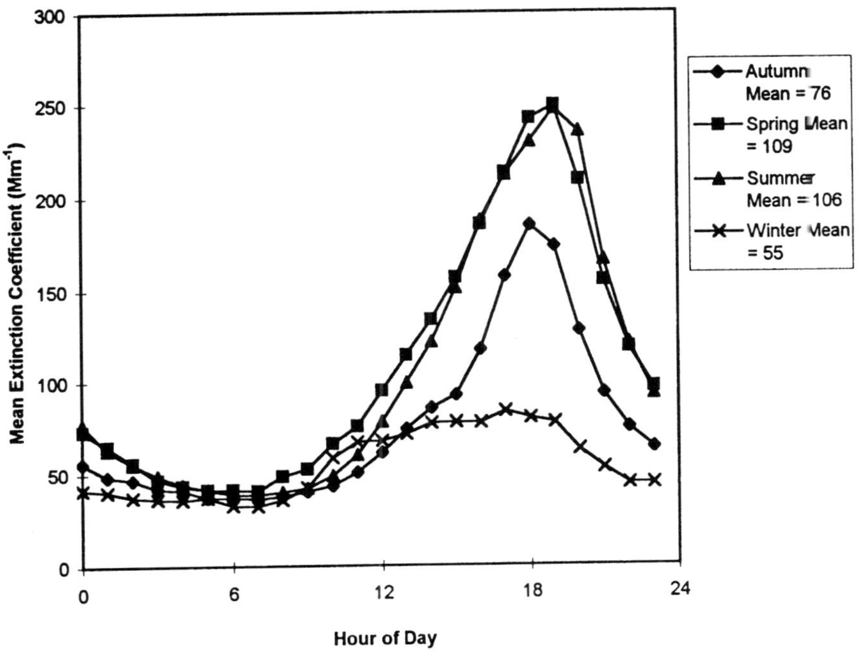

Figure 5.5. Diurnal pattern in mean b_{ext} at San Gorgonio by season, 1988–1995.

throughout the night except in the summer, when it decreases steadily through the night before dropping off at sunrise.

Two patterns are commonly seen in b_{ext} data from IMPROVE sites. The first pattern has RH and b_{ext} tracking each other throughout the day. When RH drops in the morning, b_{ext} drops simultaneously, both reaching minimums near noon and maximums near midnight. This can be explained by the hygroscopic behavior of some aerosols. Hygroscopic aerosols (notably ammonium nitrate and ammonium sulfate) grow and shrink in size as RH increases and decreases, respectively. When the aerosols grow, the size distribution generally becomes more efficient at scattering light, and conversely when the aerosols shrink, the distribution becomes less efficient. Because of this shift in aerosol extinction efficiency, one might expect b_{ext} to track RH. This pattern is seen at most IMPROVE sites, including Pinnacles National Monument and Yellowstone National Park, where the sampler is in a relatively flat region or in a low valley.

The second pattern is exactly the opposite. When RH drops in the morning, b_{ext} climbs, reaching a maximum at about noon and a minimum at about midnight. This effect can be described as the inversion or upslope wind effect. After sunrise, polluted air that may be trapped beneath an inversion layer heats, expands, and rises above the sampler's elevation, producing hazier conditions. At night, the polluted air cools and sinks down below the sampler's elevation, producing clear conditions. This pattern is seen at sites such as Shenandoah National Park and Mesa Verde National Park, which are elevated relative to their surroundings.

The pattern seen at San Gorgonio is unique among the IMPROVE sites. The peak b_{ext} occurs at 6 or 7 PM with a minimum at 6 or 7 AM, about 6 hours out of phase with respect to the RH pattern. Plotting 10th, 25th, 50th, 75th, and 90th percentiles in hourly b_{ext} reveals the same pattern in each variable. This suggests that the mean values plotted here are representative of the typical conditions, not statistical aberrations. It is possible that this pattern is due to pollutant transport that is driven by coastal influences, such that the pollutant load reaches the sampler at the same time that RH peaks in the evening.

Aerosol Data Interpretation

The following section is adapted from Sisler (1996). See this reference for more detail.

Determination of Aerosol Types

The fine aerosol species at most continental sites can be classified into five major types: sulfates, nitrates, organics, light-absorbing carbon, and soil. Methods for apportionment of measured mass to the various aerosol species are detailed in Malm et al. (1994), and only a summary is presented here. The major aerosol types are composites of the elements and ions measured in IMPROVE samplers, and their concentrations or masses are calculated from the masses of the measured elements and ions according to their presumed or probable composition and are summarized by Table 5.3. The convention used here to denote the mass concentration of a measured element, ion, or species is to enclose its symbol in brackets ([ ]).

The self-consistency and overall quality of the aerosol measurements are ensured by redundancy and intercomparisons between independently measured species. A detailed description of validation and quality assurance procedures is available in Malm et al. (1994), Sisler et al. (1993), and Eldred et al. (1988).

Mass Budgets

The aerosol sampler yields a large amount of information about trace elements that can, in some circumstances, be used for source apportionment. However, the pieces of information most directly related to quantifying visibility impairment are the reconstructed fine mass budget and the reconstructed extinction budget. These "budgets" apportion the mass and extinction into the categories described above. For purposes of mass budgets, "soot" is taken to be equal to elemental carbon.

"Reconstructed" mass means that the principal fine mass species shown in Table 5.3 were added together to reconstruct a fine mass. This reconstructed mass

Table 5.3. Formulas and Assumptions Applied to IMPROVE Sampler Measurements to Derive the Principal Fine Aerosol Species, Reconstructed Fine Mass, and Coarse Mass[a]

Species	Formula	Assumptions
Sulfate	4.125[S]	All elemental S is from sulfate. All sulfate is from ammonium sulfate.
Nitrate	1.29[NO_3]	Denuder efficiency is close to 100%. All nitrate is from ammonium nitrate.
Elemental carbon (EC)	[ECLT] + [ECHT]	All high-temperature carbon is elemental.
Organic mass from carbon (OMC)	1.4{[OCLT] + [OCHT]}	Average organic molecule is 70% carbon.
Soil (fine soil)	2.2[Al] + 2.49[Si] + 1.63[Ca] + 2.42[Fe] + 1.94[Ti]	[Soil K] = 0.6[Fe]. FeO and Fe_2O_3 are equally abundant. A factor of 1.16 is used for MgO, Na_2O, H_2O, CO_2.
Reconstructed fine mass (RCFM)	[Sulfate] + [Nitrate] + [LAC] + [OMC] + [Soil]	Represents dry ambient fine aerosol mass for continental sites.
Coarse mass (CM)	[PM_{10}] − [$PM_{2.5}$]	Consists only of insoluble soil particles.

[a]The brackets indicate the mass concentration of the aerosol species or element.

correlates well with gravimetrically determined fine mass but is not the same number.

Extinction Budgets

The approach used here to estimate scattering assumes externally mixed aerosols; that is, it is assumed that the fine mass species are not mixed within individual particles. The light-scattering coefficient can then be calculated (or reconstructed) from aerosol concentrations by taking Equation (1) and describing the light scattering contributed by aerosol component (i) as the product of the aerosol component's concentration (C_i) and its light scattering efficiency (β_i). Thus, the total light-scattering coefficient is simply the sum of the light extinctions of each aerosol component:

$$b_{\text{ext}} = b_{\text{Ray}} + \Sigma\beta_i C_i \tag{1}$$

Equation (1) can be cast into the following form for the aerosol components measured as part of the IMPROVE program:

$$b_{\text{ext}} = b_{\text{Ray}} + \beta_{\text{Sulfate}}[\text{Sulfate}] + \beta_{\text{Nitrate}}[\text{Nitrate}] + \beta_{\text{OCM}}[\text{OCM}] + \beta_{\text{Soil}}[\text{Soil}] + \beta_{\text{CM}}[\text{CM}] + b_{\text{abs}} \tag{2}$$

where b_{ext} is the total light extinction coefficient (in Mm^{-1}), b_{Ray} is the Rayleigh scattering coefficient (10/Mm), the β's are the light extinction efficiencies for each component (in m^2/g), and the parameters in brackets ([ ]) are the concentrations of the aerosol components (in $\mu\text{g/m}^3$). OCM is organic mass from carbon, and CM is coarse mass. To complete the equation for estimating extinction the channel A determination of absorption, b_{abs}, based on LIPM is used.

The values of light-scattering efficiency (in m^2/g) used in this report are as follows: sulfates and nitrates, 3 f_T(RH); organic carbon, 4; fine soil, 1; and coarse particles, 0.6.

In this chapter, it is assumed that coarse particles and fine soil particles are from a single natural source, wind-blown dust. Thus, the scattering calculated for these two components is combined into a single category and is reported as soil.

The function f_T(RH) is a correction factor to account for the liquid water that may be part of the hygroscopic aerosol components. Extinctions in this chapter are calculated by using an average f_T(RH) that represents mean conditions at San Gorgonio for each season. The effects of using this average in generating these budgets is minimal because there is no significant correlation ($r^2 = 0.01$) between daily average RH and daily average fine mass at this site. Hourly RH values of 98% or less are used in calculating this average factor. The specific seasonal factors used to reconstruct extinction at San Gorgonio are shown in Table 5.4. More information about the f_T(RH) curve used can be found in Sisler (1996).

For purposes of describing the aerosol data, it is convenient to sort the data into three groups: clear days, median days, and hazy days. Clear days are those observations (20%) with the lowest fine masses. Median and hazy days are those

Table 5.4. Seasonal f_T(RH) for San Gorgonio

Season	f_T(RH)
Spring	1.61
Summer	2.39
Autumn	1.54
Winter	2.29

days with the median 20% and the highest 20% of fine masses, respectively. Seasonal and annual reconstructed fine mass and total extinction budgets for clear, median, and hazy days are shown in Figures 5.6 and 5.7.

Mass Budget Interpretation

Ammonium nitrate is the largest single contributor to fine mass through all four seasons on median and hazy days and is roughly equal to elemental carbon annually on clear days. Carbon species (organic and elemental carbon) account for about the same mass fraction as nitrates on median and hazy days and exceed nitrates on clear days. Fine soil accounts for a small fraction of fine mass, and sulfates are approximately the same mass fraction as organic mass. The lowest fine mass days in the summer are nearly as high as the haziest days measured in the winter.

Extinction Budget Interpretation

Extinction budgets are, because of the assumptions used to derive them, very similar to fine mass budgets, except that sulfates and nitrates have an increased fraction of the budgets.

The clearest days at San Gorgonio occur in the winter, when extinction is relatively evenly split between the five principal aerosol species. On these days, standard visual ranges are about 211 km. The haziest days occur during the spring, when nitrates account for more than half of all extinction. Standard visual ranges on these days are about 23 km. Overall, the clearest to haziest seasons are winter, autumn, summer, and spring, respectively. On an annual average basis, the sum of both carbon species account for roughly the same fraction of extinction as nitrates.

Trends and Comparison

Long-Term Trends

As can be seen in the plots of reconstructed mass and extinction, the range of conditions within each season and the season-to-season variations are large. Additionally, individual years may have unique occurrences such as droughts or fires that dramatically affect the air quality. Combined with the fact that only 8 years of

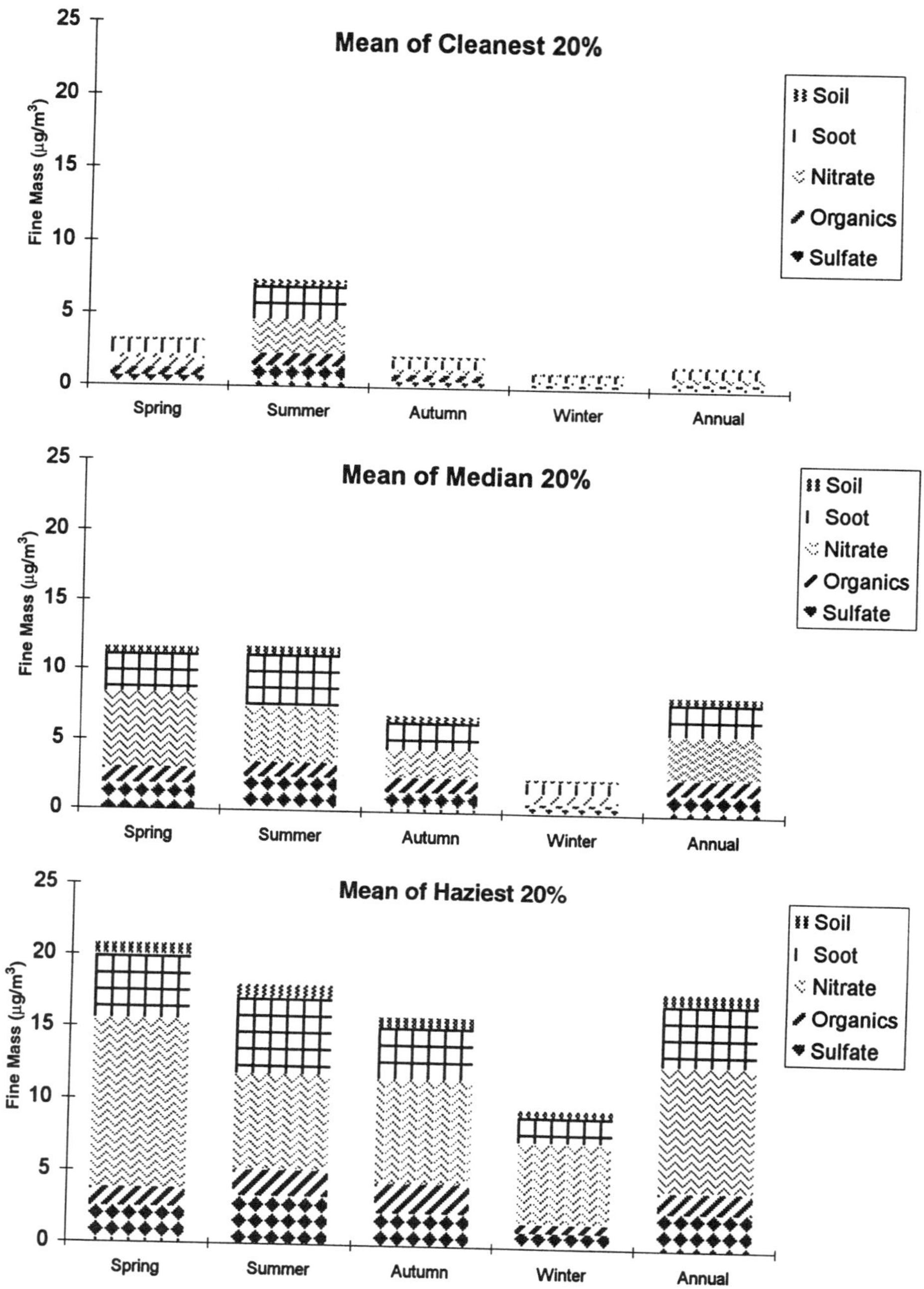

Figure 5.6. Seasonal fine mass budgets for clear, median, and hazy days at San Gorgonio Wilderness Area, March 1988–May 1996.

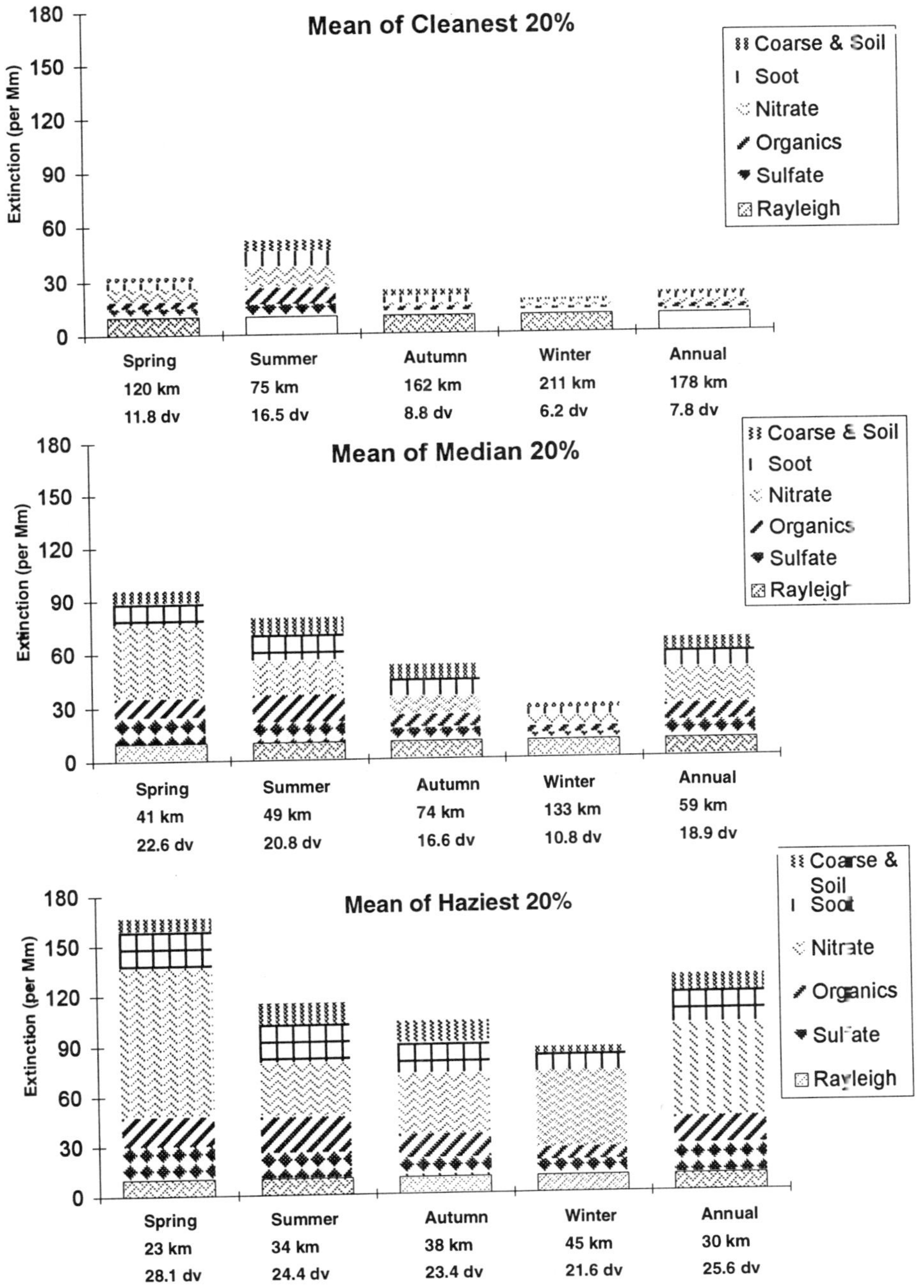

Figure 5.7. Seasonal extinction budgets for clear, median, and hazy days at San Gorgonio Wilderness Area, March 1988–May 1996.

data are available, these factors make long-term trends impossible to discern at this time.

Data Intercomparison

One of the reasons for measuring visibility parameters with both aerosol and optical monitors is to allow a cross-check between the methods. The assumption is that if both samplers yield the same result, then the likelihood of the result being incorrect is lower than for the case with only one sampler. Sixteen sites in the IMPROVE network have collocated transmissometers and aerosol samplers. For western U.S. sites, the 24-hour average b_{ext} values agree fairly well between the two samplers. Generally, a scatter plot of the two b_{ext} values yields a grouping of points about the 1:1 line, with scatter consistent with the uncertainties in the measurements. For every site, there is a group of points with significantly higher transmissometer b_{ext} than aerosol reconstructed b_{ext}, although the reverse (significantly higher aerosol reconstructed b_{ext}) tends to occur only on clear days. This is consistent with biases in the samplers.

For example, consider a rain event. Rain scavenges particulates from the atmosphere. The same rain event would also likely cause coincident transmissometer data to be flagged as invalid due to either too high an RH or b_{ext} above the threshold value. So, on a day with a rain event, the aerosol sampler filters rain-cleaned air, while the transmissometer collects invalidated data. Thus, rain events will generally bias results such that the transmissometer b_{ext} will be higher than the aerosol. In this case, neither is necessarily more or less accurate because both instruments are measuring what they are designed to measure.

Errors associated with the aerosol sampler tend to be random—associated with measurement uncertainties in volume calculation and species concentration, although there are questions regarding artifacts for some of the species, particularly the carbon species. Additionally, the aerosol sampler does not sample $NO_{2(g)}$ (which causes some light extinction) because of a denuder in the inlet stack. The transmissometer, however, could be characterized as having relatively low measurement uncertainty but an inherent tendency to overestimate b_{ext}. Anything that obstructs the sight path—water spots on the window, birds, fog layers that do not trip an invalidation code—will cause higher than desired b_{ext} measurements. Additionally, any alignment problem that is small enough to be undetected will cause higher b_{ext} measurements.

At most western U.S. sites, these errors produce the results described above: scatter about the 1:1 line with perhaps 10% of points having significantly higher transmissometer measured b_{ext}. At San Gorgonio, as shown in Figure 5.8, the same plot shows scatter about a line with a slope of roughly 1.4 (obtained by a regression forced through origin), with the same approximate 10% of relatively high transmissometer values.

For purposes of calculating a reconstructed b_{ext} from aerosol data, it is customary to assume that the aerosol concentration during the sample period remained constant. From Figures 5.4 and 5.5, it can reasonably be inferred that this is not the

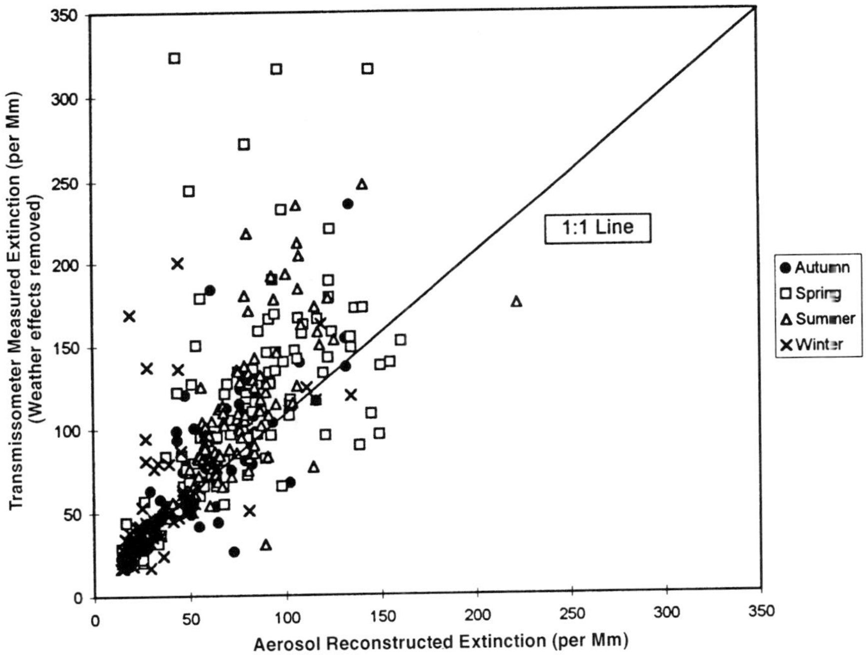

Figure 5.8. A comparison of aerosol reconstructed extinction with transmissometer-measured extinction.

case at San Gorgonio. Because the peaks in b_{ext} and RH are roughly coincident (i.e., they are in phase), the assumption of constant concentration causes an inherent underestimation of the effects of RH on extinction. Correcting for this bias by assuming that aerosol concentrations follow the same diurnal pattern as measured b_{ext} produced only an approximate 4% shift in the regression line (from $m = 1.45$ to $m = 1.41$). The correlation associated with the regression improved from $r^2 = 0.74$ to $r^2 = 0.78$.

Spatial Trends

There are seven other sites in California at which IMPROVE protocol aerosol samplers are located near EPA class I areas. By plotting extinction budget pie charts on a map, several interesting patterns can be seen. Figures 5.9–5.11 show clear, median, and hazy day extinction budgets for each of the eight California IMPROVE sites. The four-letter site codes are identified in Table 5.5.

Clear days can be split into two fairly distinct groups: high and low elevation. The higher-elevation sites (SAGO, 5,600 ft; YOSE, 5,300 ft; BLIS, 6,700 ft; LAVO, 5,900 ft) all have clear day conditions characterized by reconstructed extinctions between 16 and 22/Mm^{-1}. The four lower-elevation sites (PORE, 125 ft;

Table 5.5. Four-Letter Site Codes

Site code	Name
BLIS	D.L. Bliss State Park
LAVO	Lassen Volcanic National Park
PINN	Pinnacles National Monument
PORE	Point Reyes National Seashore
REDW	Redwood National Park
SAGO	San Gorgonio Wilderness Area
SEQU	Sequoia National Park
YOSE	Yosemite National Park

PINN, 1,040 ft; SEQU, 1,800 ft; REDW, 760 ft) have reconstructed extinctions ranging from 26 to 37/Mm. It can be inferred from this that elevation is more important than location (latitude or longitude) in determining clear day conditions but that elevation is not the only determining factor.

Median day conditions show a similar correlation with elevation, with nitrates becoming a more significant fraction of the extinction budget at many of the sites.

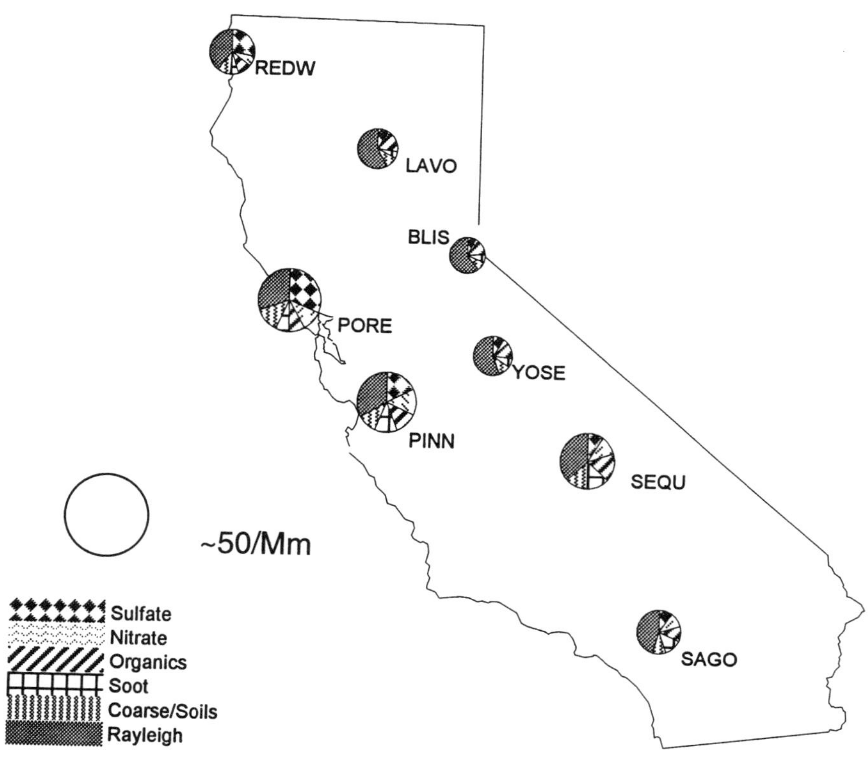

Figure 5.9. Clear day extinction budgets at California IMPROVE sites.

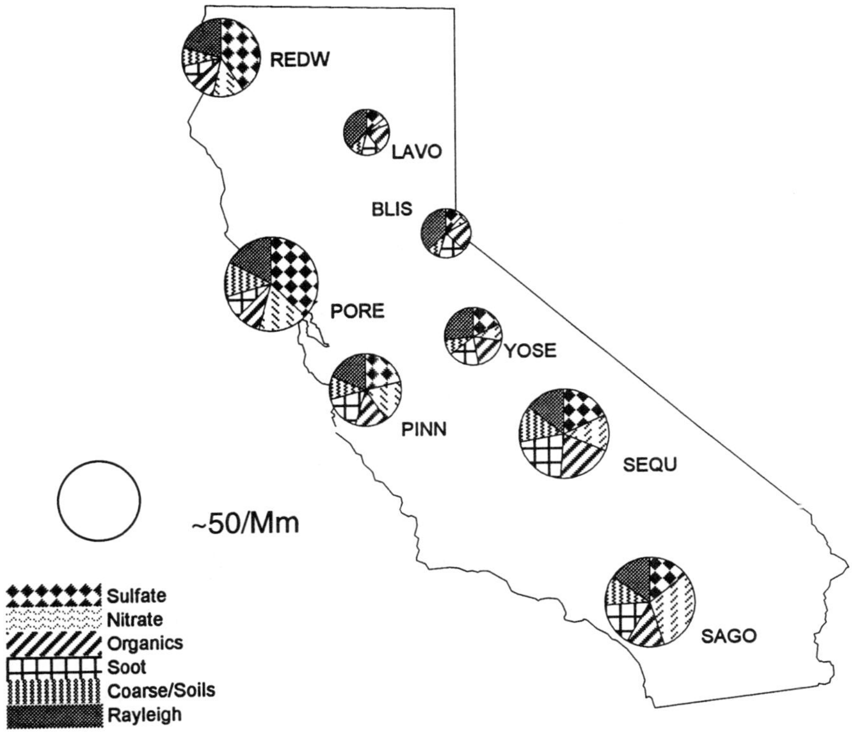

Figure 5.10. Median day extinction budgets at California IMPROVE sites.

Hazy days are grouped by high-nitrate or low-nitrate fraction. San Gorgonio and Point Reyes, which are near large urban areas, have unusually large nitrate extinction fractions on hazy days. Hazy days at Sequoia, with a reconstructed b_{ext} of about $150/Mm^{-1}$, would have a standard visual range of 26 km.

Conclusions

Visibility in the San Gorgonio Wilderness Area is characterized by a wide range of conditions, ranging from pristine conditions such as those seen in remote high-elevation wilderness areas to very hazy conditions such as those seen in and near urban areas. These conditions vary with respect to season, with winter experiencing the clearest conditions and spring experiencing the haziest. Visibility at San Gorgonio also has a strong diurnal component, with most mornings being relatively clear and most evenings being hazy. Aerosol measurements indicate that nitrates account for the majority of fine mass under most circumstances, and consequently nitrates account for the largest fraction of total extinction under most circumstances. Organic carbon and elemental carbon also account for a

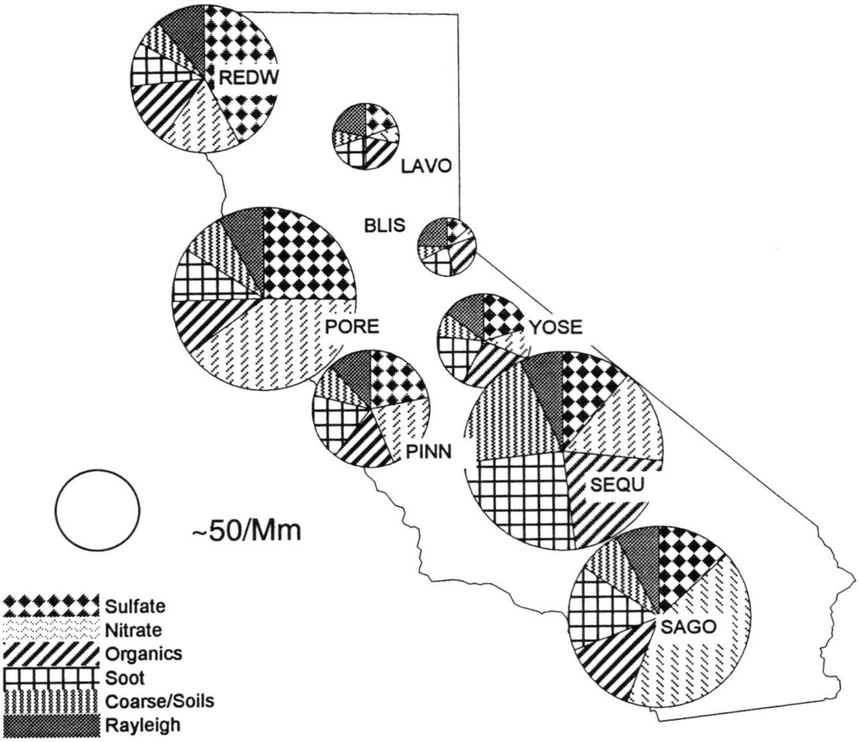

Figure 5.11. Hazy day extinction budgets at California IMPROVE sites.

significant fraction of fine mass. Combined, the two carbon species have a similar impact to nitrates. Sulfates and fine soil are less significant. Spatially, visibility conditions at San Gorgonio are similar to other high-elevation California sites on clear days and similar to lower-elevation sites in proximity to urban areas on hazy days.

The data set associated with San Gorgonio Wilderness Area is among the most complete in the IMPROVE network, with more than 900 aerosol samples and about 75,000 hourly b_{ext} and RH observations. This data set is a useful tool for measuring the effects of pollution control strategies and new sources that could influence the wilderness area. Although the two primary instruments do not agree perfectly, perhaps, in time, our understanding of visibility will enable us to further reconcile the data.

References

Cahill, T.A.; Eldred, R.A.; Feeney, P.J. Particulate monitoring and data analysis for the National Park Service 1982–1985. Davis, CA: Univ. of California; 1986.

Chow, J.C.; Watson, J.G.; Pritchett, L.C.; Pierson, W.R.; Frazier, C.A.; Purcell, R.G. The DRI thermal/optical reflectance carbon analysis system: description, evaluation, and applications in U.S. air quality studies. Atmos. Environ. 27(A), (8):1185–1201; 1993.

Dietrich, D.L.; Molenar, J.D.; Faust, J.F.; Watson, J.G. Transmissometer extinction measurements in an urban environment. In: Mathai, C.V., ed. Visibility and fine particles. Pittsburgh, PA: AWMA; 1989:374–383.

Eldred, R.A.; Cahill, T.A.; Pitchford, M.; Malm, W.C. IMPROVE—a new remote area particulate monitoring system for visibility studies. Proc. APCA Ann. Mtg. 81: –16; 1988.

Malm, W.C.; Sisler, J.F.; Huffman, D.; Eldred, R.A.; Cahill, T.A. Spatial and seasonal trends in particle concentration and optical extinction in the United States. J. Geophys. Res. 99(D1):1347–1370; 1994.

Molenar, J.F.; Malm, W.C. Ambient optical monitoring techniques. Presented at Conference on Visibility and Fine Particles, Vienna, Austria; 1992.

Molenar, J.F.; Dietrich, D.L.; Tree, R.M. Application of a long range transmissometer to measure the ambient atmospheric extinction coefficient in remote pristine environments. In: Mathai, C.V., ed. Visibility and fine particles, Pittsburgh, PA: AWMA; 1989:305–317.

Sisler, J.F. Spatial and seasonal patterns and long term variability of the composition of the haze in the United States: an analysis of data from the IMPROVE Network. Ft. Collins, CO: Colorado State University, Cooperative Institute for Research in the Atmosphere; 1996.

6. Physiological Responses of Ponderosa Pine to Gradients of Environmental Stressors

N.E. Grulke

Introduction

The general physiological responses of forest trees to oxidant pollution are well documented and summarized (McLaughin 1985; Darrall 1989; Bytnerowicz and Grulke 1992; Matyssek et al. 1994; Gower et al. 1995; Heath and Taylor 1997). Most conifers have a higher threshold of tolerance to air pollutants relative to broad-leaved deciduous species (Reich 1987; Darrall 1989). Resource (carbon, water, nutrients) acquisition, allocation, and partitioning are affected by exposure to oxidant pollution. In this chapter, resource acquisition refers to plant processes that gain carbon, nutrients, or water. Allocation is the redistribution of a resource to another plant part separated spatially from point of acquisition to its use (Dickson 1989). Resource partitioning is the chemical repackaging of carbon, nutrients, or water for metabolism, detoxification, reparation, or respiratory functions and may include compounds that are produced and stored until use within a plant part. The total pool of resources obtained, the cost of tissue maintenance, and the total pool of resources retained with environmental stressors ultimately determines the perceived stress and fate of the individual (Fig. 6.1).

The known effects of oxidant pollution on resource acquisition in conifers include lower photosynthetic capacity (Coyne and Bingham 1981; Bytnerowicz et al. 1991; Grulke et al., in press), lower rubisco carboxylation efficiency (Sasek and Richardson 1989), reduced apparent quantum efficiency (Grulke et al. 1989;

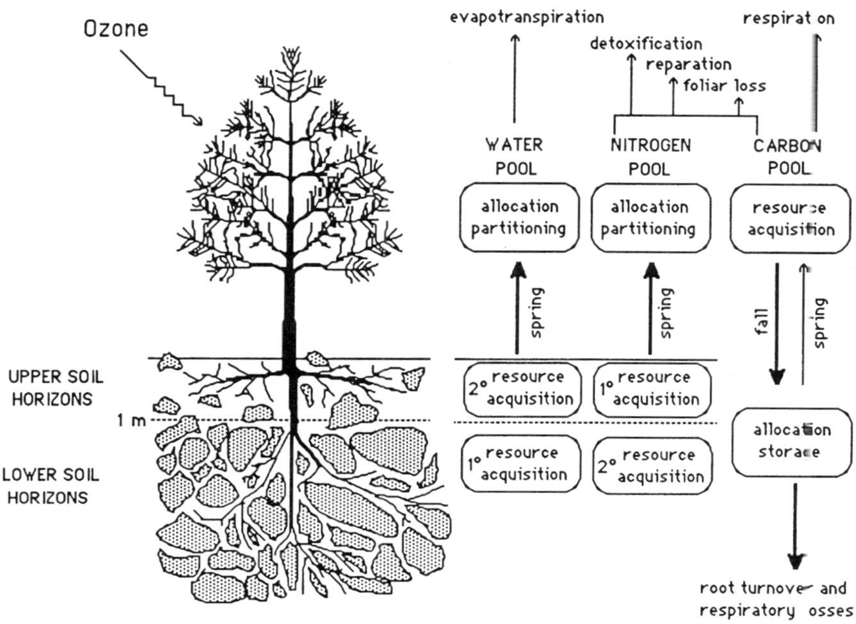

Figure 6.1. Illustration of spatial separation and timing of resource acquisition, allocation, and partitioning of water, nitrogen, and carbon pools within an idealized ponderosa pine tree. Definitions of acquisition, allocation, and partitioning follow that of Dickson (1989). Arrows indicate direction, strengths, and timing of flow of resource allocation. Respiratory loss of carbon, evapotranspirational losses of water, and both nitrogen and carbon losses from foliage and root turnover enter infinite pools in the environment. Some nitrogen and carbon resources may be recoverable from detoxification and reparation processes within the foliage. Resource acquisition by fine roots within the upper soil horizons is limited to periods when temperature and drought stress are not limiting.

Sasek and Richardson 1989), and impairment of the photo-electron transport system (Bytnerowicz et al. 1991; Patterson and Rundel 1995). Acquisitioned resources may be directly lost through higher respiration (Grulke et al. 1989) or indirectly after partitioning, such as loss of foliage (Miller et al. 1996) or chlorophyll or chloroplast degradation (Bytnerowicz et al. 1991; Grulke et al. 1996; Evans and Miller 1972). Resource allocation that is known to be affected by oxidant pollution includes reduced wood production (Miller et al. 1989; Peterson et al. 1991), lower root mass and carbohydrates (Andersen et al. 1991; Grulke et al., in review), and lower foliar nutrient content (Tingey et al. 1986; Bytnerowicz et al. 1990; Edwards et al. 1991). The reader is referred to other publications for general reviews of plant response to oxidant air pollution (Smith 1974; Heath 1975; Guderian 1985; Wellburn 1988).

Ponderosa pine is common throughout western midelevation forests (Oliver and Ryker 1994) within the Sierran mixed conifer zone (sensu Barbour 1988). It is one of the most sensitive trees to oxidant air pollution transported into these forests (Miller et al. 1983). We know more about this conifer's response to tropospheric ozone than any other western species, but most of the work has been on seedling response within exposure chambers (Barnes 1972; Aitken et al. 1984; Bytnerowicz et al. 1989; Beyers et al. 1992; Takemoto et al. 1997; Grulke et al., in press). In addition to original data presented here, three field studies in the central to southern Sierra Nevada and in the Transverse Range (north and east of Los Angeles, California) describe the response of reproductively mature ponderosa pine in natural stands under anthropogenic and natural stressors (Coyne and Bingham 1981, 1982; Patterson and Rundel 1989; Temple and Miller, in press).

The focus of this chapter is the physiological response to temporal variability (diurnal, seasonal, interannual) and spatial gradients (west to east across the San Bernardino Mountains as well as an atmospherically clean site near Lassen Volcanic National Park, California). The west-to-east transect of the San Bernardino Mountains represents gradients in oxidant exposure, nitrogen deposition, and moisture (relatively moist to drier variants in the Sierran mixed conifer type). Changes in the functioning of stomatal conductance across these temporal scales and spatial gradients are described. Components of carbon acquisition (net assimilation, or CO_2 exchange rate [CER], carboxylation efficiency, maximum photosynthetic capacity under saturating CO_2 in air [P_{max}]), and foliar respiratory losses across the spatial gradient are described. Much of the data presented in this chapter is original data from a study conducted 1992–1997 to specifically address the temporal variability of environmental stressors on growth, leaf chemistry, and gas exchange characteristics of ponderosa pine at sites distributed across the San Bernardino Mountains.

A stand-level study conducted by Miller and McBride over 14 years (Miller et al. 1989; see Miller and Rechel, this volume) established a strong gradient in ozone exposure as well as a demographic response of conifers at 15 sites across the San Bernardino Mountains. Three sites at comparable elevation along this transect were representative of the range of environmental stressors: Camp Paivika on the western end of the range with high oxidant exposure, high N deposition, and a relatively moist variant of the Sierran mixed conifer forest type; Strawberry Peak in the middle of the range with moderately high oxidant exposure, lower N deposition, and a relatively moist variant of the forest type; and Camp Osceola on the eastern end of the range with moderate oxidant exposure, low N deposition, and a drier variant of the forest type. The latter site was 3 km east of the intensive field studies described in other chapters focused at Barton Flats. Because of the aberrant physiological responses that were obtained in the elevated pollution environment, data for ponderosa pine from an atmospherically clean site near Lassen Volcanic National Park in northern California have also been included for comparison. Lassen has low oxidant exposure, no appreciable N deposition, and a drier variant of the forest type. Examples from other field studies at Camp Paivika (Coyne and Bingham 1981, 1982), Barton Flats (Temple and Miller, in press) for ponderosa and Jeffrey pine, and for Jeffrey pine in Sequoia National Park (Patterson and Rundel 1989) are also presented to illustrate important points.

Seasonal Timing of Oxidants and Environmental Stressors

The timing of oxidant exposure and other environmental stressors is critical to the level of stress ultimately experienced by the plant. The principal environmental stressors that influence ponderosa pine in the San Bernardino Mountains are the high oxidant pollution in the western and central parts of the range, seasonal (mid- to late summer) and extended multiyear periods of drought (1896–1902, 1947–1951, 1981–1991; summarized in Miller 1992), and the associated high nitrogen inputs on the western edge of the range. There are also subtle differences in the above- and belowground temperature regimes across the San Bernardino Mountains (Fig. 6.2). Other significant stressors include pests and pathogens (Miller 1993), other pollutants (Bytnerowicz and Grulke 1992), and human use (see Bytnerowicz et al., this volume; Pronos et al., this volume), but these are beyond the scope of this chapter. The reader is referred to other chapters for more thorough coverage of edaphic and atmospheric components in the San Bernardino Mountains (Poth and Wohl, this volume; Fujioka et al., this volume). The Mediterranean climate that dominates the San Bernardino Mountains (relative to more continental climates) allows us to predict the sequence of environmental events, their interaction with tropospheric ozone exposure, and potential plant uptake of ozone.

Tropospheric Ozone

The cumulative ozone exposure at the three sites distributed across the San Bernardino Mountains was calculated for the period April 15–October 15 in 1993 and 1994 based on 24-hour measures by using Grulke (1994) as a protocol for measurement. Camp Paivika experienced the highest ozone exposure (average of 80 ppb/h in both 1993 and 1994) and Camp Osceola the lowest (average of 62 ppb/h in 1993 and 64 ppb/h in 1994). Strawberry Peak was intermediate between the two but was more similar to Camp Paivika in 1993, the more mesic year (76 ppb/h), and to Camp Osceola in 1994, the more xeric year (69 ppb/h). Lassen was cleaner atmospherically, with an hourly average of 38 ppm/h in the more mesic year and 45 ppm/h in the more xeric year. Deposition of nitrate, sulfate, and ammonium across the same gradient of sites in southern California showed a twofold difference in all particulates across the gradient and exponentially covaried with ozone concentrations (Bytnerowicz and Fenn 1996; see Bytnerowicz et al., this volume).

Temperature Differences

The highest soil and air temperatures were found at Camp Paivika throughout the year (Fig. 6.2). Temperatures at Strawberry Peak were similar to Camp Paivika early in the growing season but were intermediate between the Camp Paivika and Camp Osceola later in the growing season. Lassen experienced lower early-season temperatures than the southern California sites but had a similar temperature regime to Camp Osceola in the latter half of the summer. Although air and

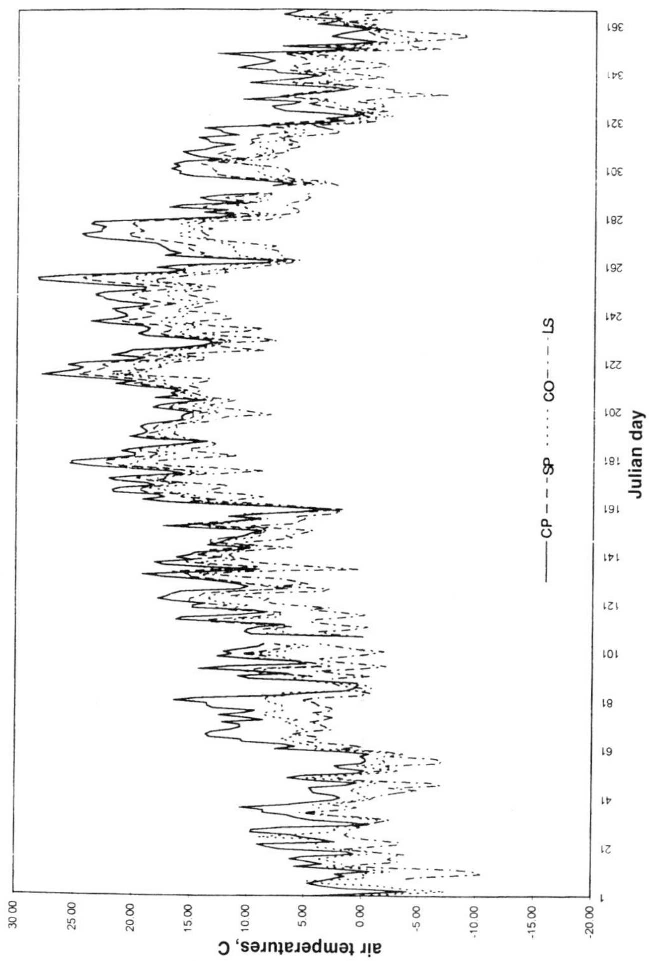
air temperatures, C
Julian day
CP
SP
CO
LS
30.00
25.00
20.00
15.00
10.00
5.00
0.00
-5.00
-10.00
-15.00
-20.00
1
21
41
61
81
101
121
141
161
181
201
221
241
261
281
301
321
341
361

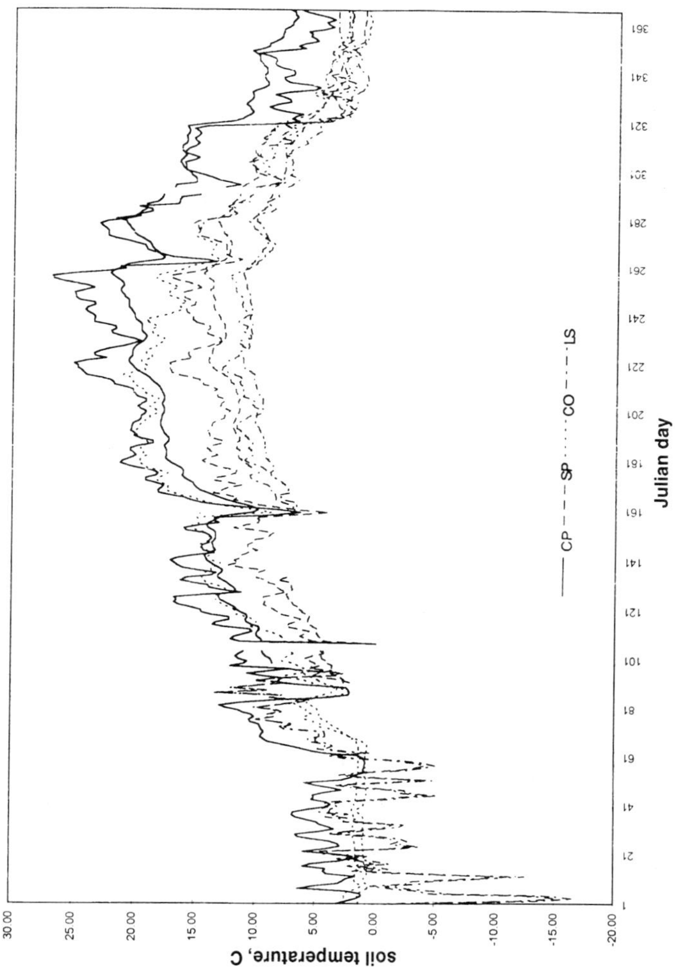

Figure 6.2. Seasonal course of daily mean air temperature at 1.5 m (upper graph) and soil temperatures at −10 and −50 cm (lower graph) for all four sites (solid line, CP: Camp Paivika; dashed line, SP: Strawberry Peak; dotted line, CO: Camp Osceola; dotted and dashed line, LS: Lassen Volcanic National Park). The two soil depths measured are represented by upper and lower lines, respectively. Data were taken in 1993.

Table 6.1. Summary of Needle Predawn Xylem Water Potential in September 1993, 1994, and 1995 at Each of the Four Sites[a]

Year	Lassen	Camp Osceola	Strawberry Peak	Camp Paivika	F	P
1993	–1.09 ± 0.04a	–1.14 ± 0.04a	–0.77 ± 0.05b	–0.99 ± 0.08b	8.746	.002
1994	–1.07 ± 0.26a	–1.70 ± 0.07b	–1.64 ± 0.08bc	–1.47 ± 0.09ac	3.64	.064
1995	–0.86 ± 0.09a	–1.12 ± 0.05b	–0.91 ± 0.05a	–0.88 ± 0.03a	4.17	.02

[a]Mean values ± 1 S.E. in MPa are given for six trees in the 21–60-year age class.

soil temperatures were greatest at Camp Paivika and least at Camp Osceola, ponderosa pine bud break varied by only 4 days across the San Bernardino Mountain gradient (Grulke et al., in review). Bud break was nearly 1 month later at Lassen, but maximum branch and needle elongation growth was obtained at the same time (by early August) at all four sites (Grulke et al., in review).

Drought

Winter snow is the primary source of moisture for the year in the San Bernardino Mountains, and snow release generally occurs in early to mid-May. Soil temperatures at 50 cm drop below 5°C for only 2–4 months from west to east across the San Bernardino Mountains (Grulke, unpublished data). Occasional summer thunderstorms occur in July and August, but precipitation is rarely great enough to moisten mineral horizons.

Accurate measures of soil moisture availability are difficult to obtain in rocky soils where roots extend well into fractured bedrock (Jones and Graham 1993). Two of the sites (Lassen and Camp Osceola) have shallow soils (80–100-cm rooting zone above fractured bedrock), and two of the sites (Strawberry Peak and Camp Paivika) have relatively deep soils (8–26 cm A_1 horizon, B_2 horizon to 100 cm) (Grulke et al., in review). Predawn xylem potential was used as a proxy (not a physiological response) for available soil moisture in the rooting zone, which perhaps better represented and integrated the level of drought stress experienced by trees at each site. At Camp Osceola, soil moisture content in the upper 50 cm dropped exponentially from June to mid-July without a concomitant response in predawn xylem potential or leaf conductance (Temple and Miller, in press).

Typical predawn needle xylem potential measured with a pressure chamber ranged between –0.3 and –0.8 MPa prior to the midsummer drought stress (July 19, Julian day 200) and from –0.8 to –1.2 MPa in 1993 and –1.3 to –1.5 MPa in 1994 for all trees older than 10 years. In September 1993, 1994, and 1995, trees at Camp Osceola experienced the lowest predawn xylem water potential (Table 6.1). In 1993, trees at Lassen and Camp Osceola had similar low predawn water potential at the peak of the drought, and predawn water potential at Strawberry Peak and Camp Paivika was significantly higher. In 1994, trees at Lassen and Camp Paivika had significantly higher predawn water potential, with lower potentials developing at Camp Osceola and Strawberry Peak. In 1995, trees at Camp

Osceola experienced the lowest predawn xylem water potentials, but they were similar across the other three sites. This is the best proxy available for estimating soil moisture in the entire rooting zones available to the tree.

Seasonal Timing of Physiological Activity Across the Pollution Gradient

The sequence of tissue growth (root growth, bud break, shoot growth, cambial growth) is described by theory (Denne 1979; Larson 1994; Luxmoore et al. 1995) but supported by few data in ponderosa pine. From studies of other temperate woody plants, the minimum temperature for root growth is 5°C (Kozlowski and Pallardy 1997). At the warmest site (Camp Paivika), soil temperatures at –10 cm fluctuate around 5°C for most of the winter, but there is a rapid increase in soil temperature at –50 cm for the western two sites in mid-April (Fig. 6.2). Root production is believed to be initiated by temperatures greater than 5°C in the rooting zone in the spring. Low soil water potential also limits root growth (Teskey and Hinckley 1981). The duration of spring root growth is not known but, based on field observations, probably ceases when soil water potential becomes limiting in mid- to late July over much of its range. Root mass was relatively high in early July and lower in early September across the San Bernardino Mountains (Grulke et al., in review).

If the theory of limiting internal resources holds (Waring and Schlesinger 1985; Luxmoore et al. 1995), root growth declines when aboveground growth is initiated. Coarse root carbohydrate stores dropped abruptly during a 4-day period in May when lateral branch growth was initiated at Camp Osceola (Grulke et al., in review). The initiation of shoot growth varies by 4 days across the southern Californian sites and varies more between years than between sites. The period from bud break to the onset of drought stress is the period of the highest gas exchange rates and presumably the greatest air pollutant uptake (Temple and Miller, in press). Branch and needle elongation growth declines with a significant drop in predawn water potential, regardless of the intensity of the drought stress. Control of elongation growth, thus, may be either photoperiodic control or a significant drop in water potential alone alters hormonal balance sufficiently as a signal (or some combination of the two).

Changes in Stomatal Control

The primary mechanism to limit air pollutant uptake is stomatal control (Tingey and Hogsett 1985; Reich 1987; Runeckles and Chevone 1992). The pathway by which ozone enters the leaf (Bytnerowicz and Grulke 1992) and the mechanisms by which atmospheric ozone induces stomatal closure have been well described for herbaceous species and are most likely generally applicable to conifers as well (Heath 1975; Mudd and Kozlowski 1975; Heath and Taylor 1997). For turgor to be maintained in the guard cells of the stomata, high intracellular concentrations of potassium ions must be maintained. Ozone alkalizes the cell membrane, increasing cell membrane permeability and loss of intracellular potassium ions, resulting in stomatal closure. In carefully controlled laboratory studies, this result is obtained in agronomic species that have been previously drought-stressed (Unsworth et al. 1972). In experimental treatments with ponderosa pine seedlings, some studies reported lower stomatal conductance after ozone exposure (Bytnerowicz and Takemoto 1989; Grulke et al., in press) and some reported greater stomatal conductance (Beyers et al. 1992). The responses obtained from multiple stressor experiments combining varying levels of ozone, acidic deposition, and drought stress may be the only way to explain opposing field observations (Temple et al. 1992), although meta-analysis has been suggested by others (Heath and Taylor 1997).

Most studies suggest that the greater the ozone uptake, the greater the expression of injury. However, it is difficult to separate stomatal response to ozone and endogenous (genetic) differences in stomatal conductance, leading to greater uptake and resulting in greater expression of symptoms. For example (Fig. 6.3), some individuals appear to have endogenously higher conductance relative to the population as a whole, and some have lower conductance (Taylor et al. 1991). If trees have lower conductance and are asymptomatic, they may be considered an "avoider." If trees have lower conductance with symptoms, they are more sensitive than trees with greater ozone uptake. In general, however, symptomatic trees may have endogenously greater conductance than asymptomatic trees (ponderosa pine, Coyne and Bingham 1981; Jeffrey pine, Patterson and Rundel 1989).

Sensitive trees, by either pathway, have a collection of characteristics that may predispose them to the stress of ozone uptake: the total pool of resources acquisitioned may be lower, resource allocation may be less effective (or blocked by exposure to the pollutant itself; see Heath and Taylor 1997), and/or resource partitioning may be less effective (e.g., insufficient or poorly timed available antioxidant) than tolerant trees. Any one of these characteristics in the extreme but most likely the collection of several physiological responses that are poorly timed or ineffective could contribute to the level of sensitivity and the expression of ozone injury symptoms (Grulke and Lee 1997). Conversely, tolerant trees have a collection of characteristics that allow either lower ozone uptake, greater resource acquisition, or more effective resource allocation or partitioning, which allows the individual to defend or repair tissues once elevated ozone levels are experienced in the substomatal cavity.

Environmental factors that decrease stomatal conductance also decrease ozone uptake and the expression of injury. Several studies have described the effects of environmental factors (temperature, light, humidity, available soil moisture) specifically on the stomatal response to ozone (Davis and Wood 1973; Jensen and Roberts 1986; Freer-Smith et al. 1989; Dobson et al. 1990; Wieser and Havranek 1993). In general, exposure of well-watered trees to elevated ozone results in a decrease in conductance. Exposure of drought-stressed trees to elevated ozone has been shown to increase stomatal conductance (Pearson and Mansfield 1993).

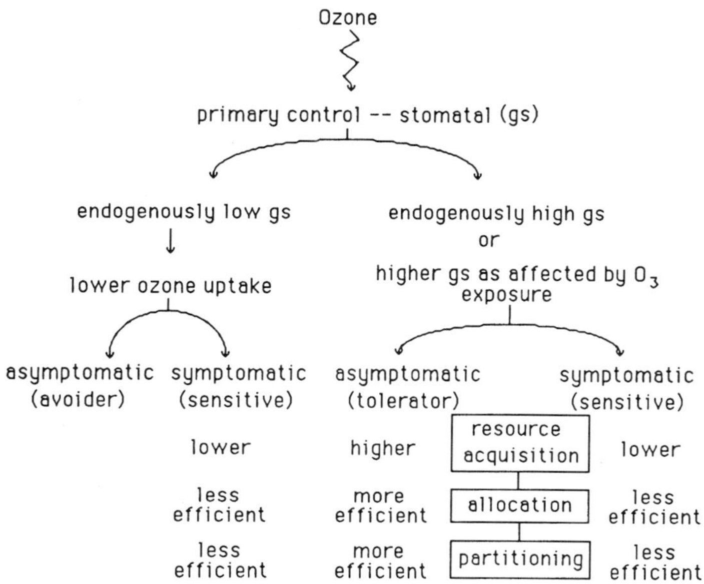

Figure 6.3. Word model of plant control of atmospheric ozone uptake and role of resource acquisition, allocation, and partitioning as collective characteristics that may pre-dispose plants to sensitivity to ozone (or other pollutant) uptake. Examples of resource acquisition include photosynthetic carbon uptake or root nitrogen uptake. Examples of resource allocation are shifts in the pool size of carbohydrates allocated to roots in the fall or the timing of allocation of nitrogen uptake from roots to foliage. Examples of resource partitioning include changes in amount or timing of antioxidants or maintenance of foliar chlorophyll content. The word *efficient* refers to an adaptive versus deleterious trait.

Spatial and Temporal Trends in Stomatal Conductance

The seasonal courses of stomatal conductance of different tree age classes of ponderosa pine across the San Bernardino Mountains and the Lassen Volcanic National Park site illustrate the range of responses to environmental stressors (Fig. 6.4). Gas exchange was measured monthly on the same six trees in each tree age class (except three each for mature trees) with a CO_2 depletion photosynthetic system in 1993 (Grulke 1994). The tree age classes were 3–8 years, 10–20 years, 21–40 years, 41–60 years, and 80 years or older. This data set, and comparable ones for 1994 and 1995, is used in subsequent examples given for the gradient study in the San Bernardino Mountains.

At the site with the lowest ozone exposure (Lassen), stomatal conductance of the different tree age classes had a similar magnitude and pattern, except for the 3–8-year-old seedlings. The seedlings experienced seasonal drought stress through a decline in stomatal conductance and accompanying net photosynthetic rates approximately 1 month sooner in both 1993 and 1994 because the upper soil

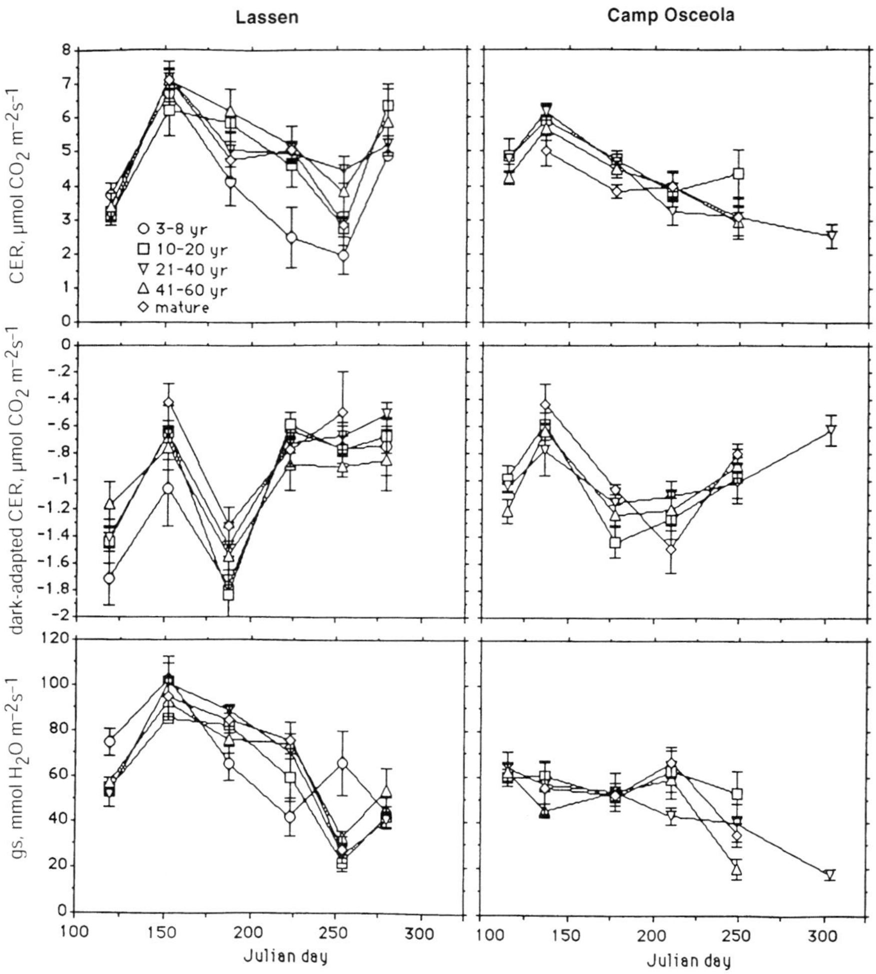

Figure 6.4. Seasonal course of CO_2 exchange rate (CER), respiration, and stomatal conductance (gs) of different tree age classes in 1993. Three sites are represented across the San Bernardino Mountains from west to east, from most polluted to less polluted: Camp Paivika, Strawberry Peak, and Camp Osceola. In addition, an atmospherically clean site in northern California is also presented (Lassen). Symbols represent mean of six trees in each tree age class except for the mature trees (>80-year-old trees) where only three measures were taken per sample date ($\pm$ 1 SE).

horizons dry out first (the only horizons accessible to the seedlings). There were no statistical differences between any of the other tree age classes at any sampling point in the growing season within a site. This implies that trees at first reproductive age ($\geq$ 20 years old in ponderosa pine) can be used as proxies for older tree age classes in assessing response to environmental stressors.

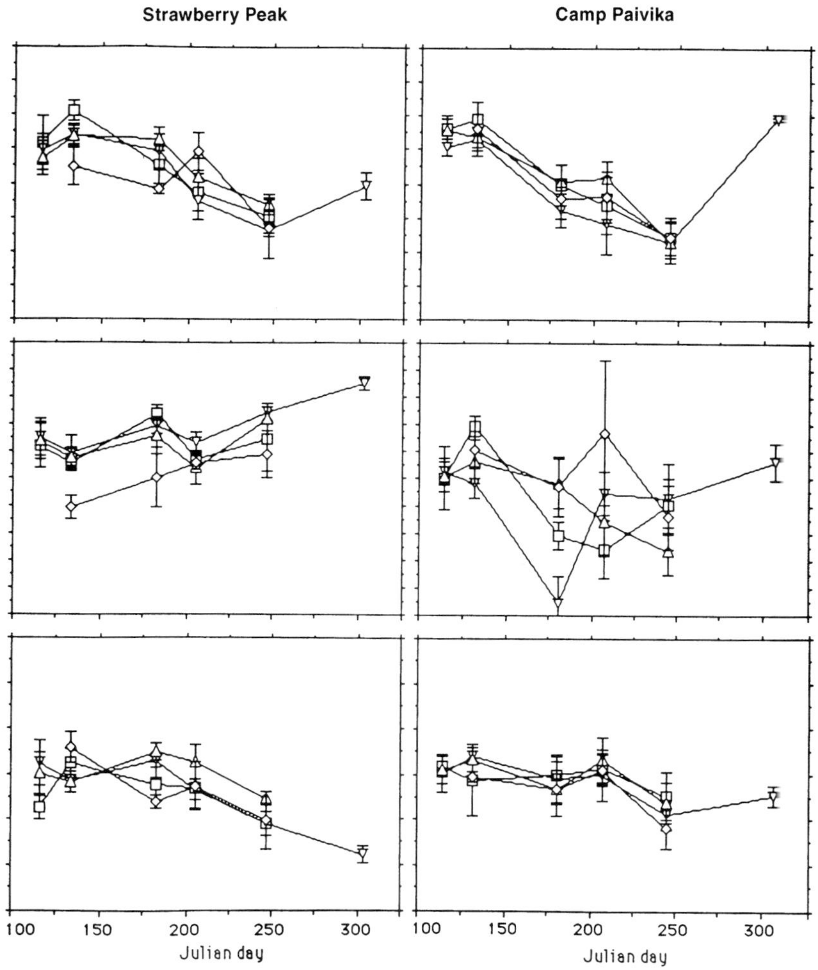

Figure 6.4. (*continued*)

At the end of October at Lassen and in mid-November at Camp Paivika, there was an increase in stomatal conductance. At Lassen, the increase was due to fall precipitation. At Camp Paivika, there had been no precipitation. Leaf temperature or vapor pressure deficit (VPD) differences between the sites did not explain the increase in gas exchange rates. Soil moisture may wick up into the rooting zone as air temperatures and evaporative demand drops. Where the end-of-season "rebound" occurred, gas exchange values were nearly as great as peak season values. Moderate winter photosynthetic rates have also been reported for other species where mild winter temperatures are experienced (slash pine, Cropper and Gholz 1993; Douglas-fir, Emmingham, unpublished data cited in Waring and Franklin 1979; loblolly pine, Boltz et al. 1986). Two exposure chamber studies

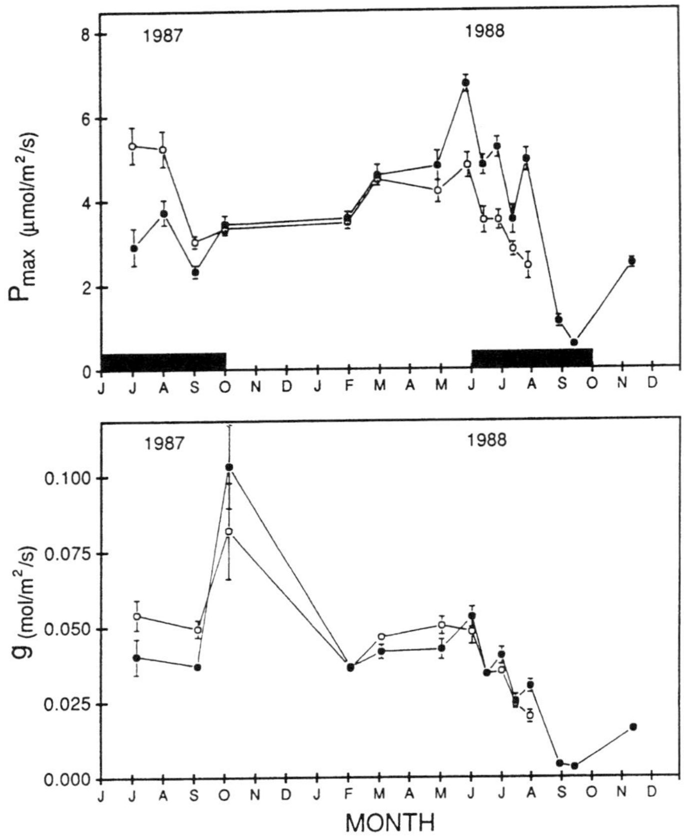

Figure 6.5. Seasonal course of mean stomatal conductance (g) and maximum photosynthetic rate (P_{max}) for sensitive or symptomatic trees (open circles) and tolerant or asymptomatic trees (closed circles) of Jeffrey pine in Sequoia National Park, California. Bars represent ± 1 SE for 9–35 trees per sample date. A "sluggish" stomatal response is illustrated: lower conductance when it should be at a maximum to accommodate conditions favorable to carbon acquisition, and higher conductance when it should be at a minimum to conserve water loss. (From Patterson and Rundel 1989.)

with ponderosa pine also exhibited an end-of-season rebound of stomatal conductance (Helms et al. 1994; Takemoto 1997). Across the three sites in the San Bernardino Mountains, there was greater drought stress in 1994 by mid-August, and the rebound of CER in mid-November did not occur at any site (data not presented).

Seasonal variability in needle conductance in the lower one-third of the canopy was measured for sensitive (symptomatic) and tolerant Jeffrey pine (less chlorotic mottle) (Fig. 6.5; Patterson and Rundel 1989). When conditions were favorable for gas exchange, conductance was lower for sensitive versus tolerant individuals.

When environmental conditions dictated lower conductance or closure, needles of sensitive individuals had greater conductance, indicating poorer or a more "sluggish" control (term first introduced by Keller and Hasler 1987). This trend held for diurnal measures of conductance as well (Fig. 6.6; Patterson and Rundel 1989). Stomatal control was not as good in sensitive trees in late August (greater conductance relative to less symptomatic individuals). In February, sensitive trees had a similar diurnal trend, but a lower morning maxima was achieved. A "sluggish" stomatal response (e.g., lower conductance when it should be at a maximum to accommodate conditions favorable to carbon acquisition, and higher conductance when it should be at a minimum to conserve water loss) may help to explain some of the discrepancies in the literature.

Ponderosa pine exposed to high oxidant concentrations (with 28% chlorotic mottle on average of 1-year-old needles) showed poorer stomatal control relative to cleaner sites (with 3% chlorotic mottle). Using the data presented for the ponderosa pine gradient study in Figure 6.4 and adding 1994 data as well, the relationship between CER and stomatal conductance was plotted for the predrought period (prior to July 19, ≤ Julian day 200) and the postdrought period for all four sites. The slope of the regression lines indicates relative differences in water use efficiency between trees in the 21–60-year tree age class between sites. During the predrought period, water use efficiency was low and constant across all four sites, regardless of pollution or other environmental stress history (Fig. 6.7). During the period of drought stress (after mid-July), water use efficiency increased at both the cleanest (Lassen) and next cleanest (Camp Osceola) sites, but stomata were not as responsive at the two most polluted sites (Strawberry Peak and Camp Paivika).

Another example of the sluggishness of stomata with increasing pollution is shown in the response of stomatal conductance to VPD (Fig. 6.8). Two sites along the pollution gradient had similar precipitation and predawn xylem water potentials with maximum drought (Camp Paivika and Strawberry Peak; Table 6.1) but different total ozone exposure (2-year average of 72 and 80 ppb/24 h, May 15–October 15). Stomatal conductance was measured on 1-year needles through the growing season in 1995.[1] The data were separated for early-season, predrought (prior to July 19, ≤ Julian day 200), and late-season postdrought responses (≥ Julian day 200). Both early-season and mid- to late-season conductance was lower at the most polluted site (Fig. 6.8). Although conductance was low at low VPD in the higher ozone environment, there was no appreciable decrease with increasing VPD. Stomatal closure as a function of VPD occurred around 4.8 kPa for ponderosa pine near an atmospherically clean site, Black Butte, Oregon (Hubbard et al., in press).

Stomatal conductance of ponderosa pine was measured from May through October of a relatively dry year (1977) at the most polluted site (Fig. 6.9; Camp Paivika) (Coyne and Bingham 1982). Current-year foliage of the least symptom-

---

[1]The standard technique for measuring gas exchange was modified for the LICOR 6200 to optimize for responsiveness to humidity changes within the cuvette (Grulke 1994).

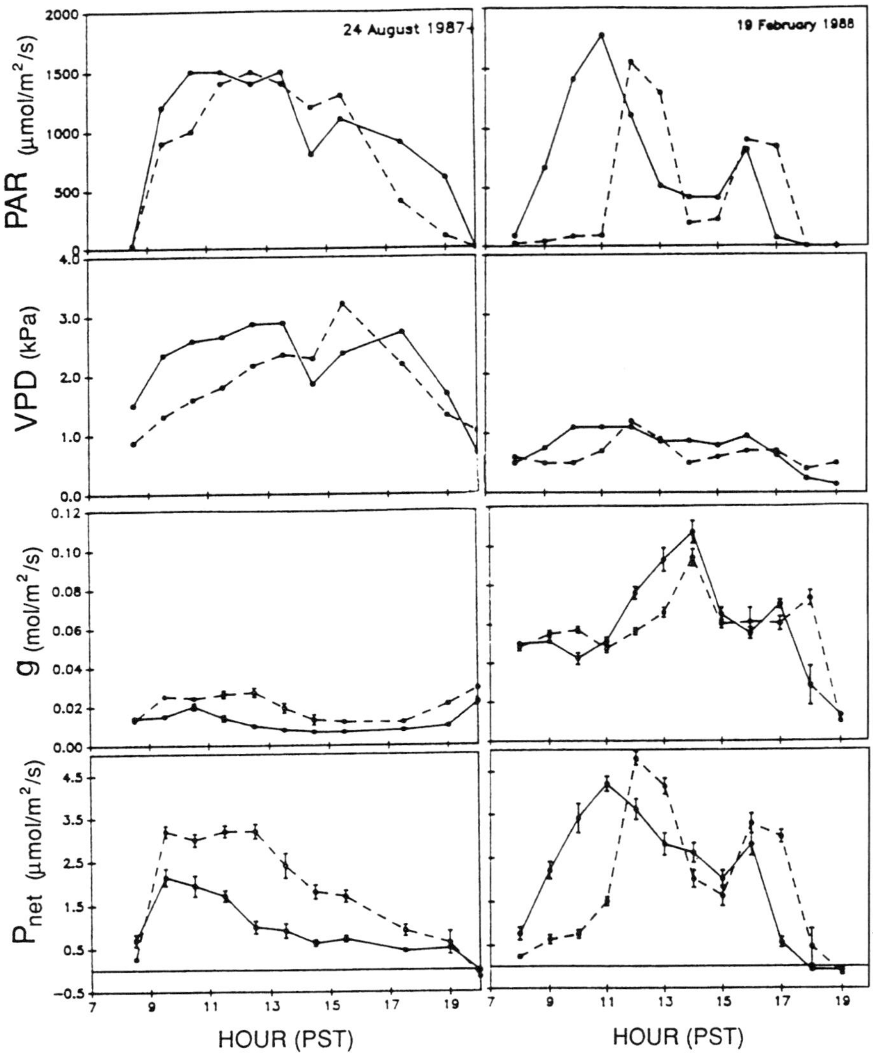

Figure 6.6. Diurnal course of mean stomatal conductance (g) and net photosynthesis (P_{net}) for sensitive (dashed lines) and tolerant (solid lines) of Jeffrey pine in Sequoia National Park, California. Bars represent ± 1 SE of ten measurements from representative trees. One day representing peak drought stress in August is compared with the winter state in February. (From Patterson and Rundel 1989.)

atic trees exhibited the best stomatal control, and the most symptomatic had the greatest conductance, similar to the results found for Jeffrey pine (Patterson and Rundel 1989). However, 2- and 3-year-old needle conductance of ponderosa pine was greatest for least symptomatic trees, in contrast with the Jeffrey pine data. One explanation may be that poorer stomatal control of more symptomatic trees

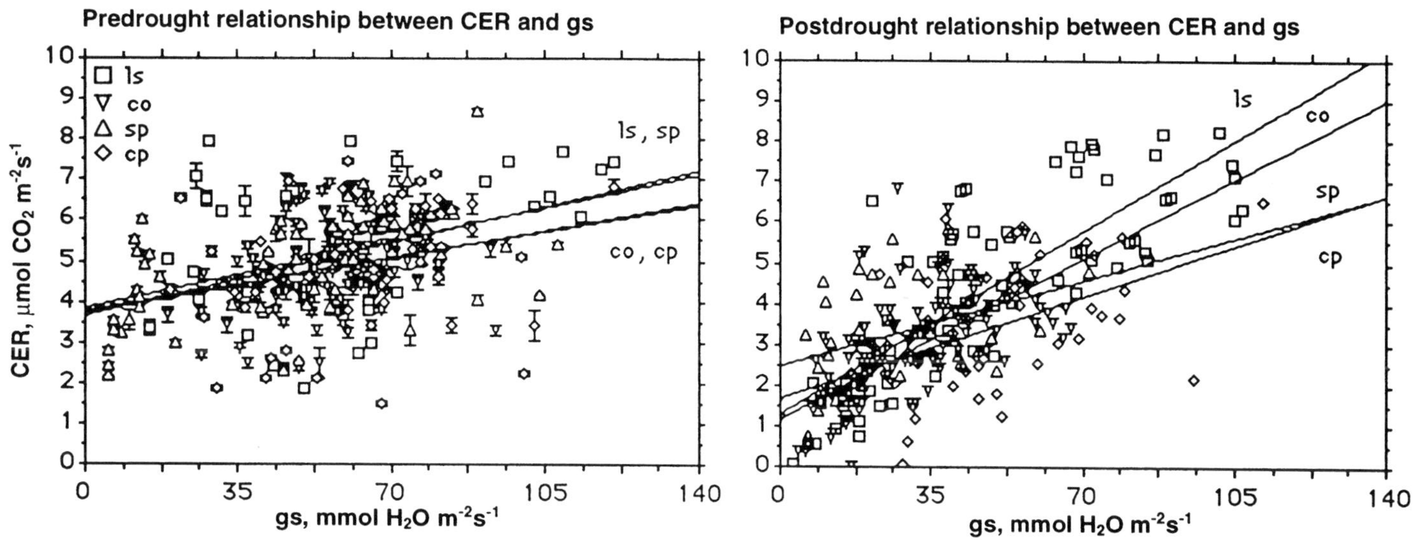

Figure 6.7. Relationship between CO_2 exchange rate (CER) and stomatal conductance (gs) prior to seasonal drought stress and after significant drought stress (after mid-July). Data have been pooled from 1993 and 1994. Symbols represent individual tree means (point in time) and bars indicate ± 1 SE. Open square indicates the atmospherically clean site, Lassen (LS), upside-down triangle indicates Camp Osceola (CO), right-side-up triangle indicates Strawberry Peak (SP), and the diamond indicates Camp Paivika (CP). Regression lines for the relationship is similar for LS and SP (upper lines) and CO and CP (lower lines) under predrought conditions. For the droughty period, the order of the regression lines is LS, CO, SP, and CP, in order from least to most pollution exposure.

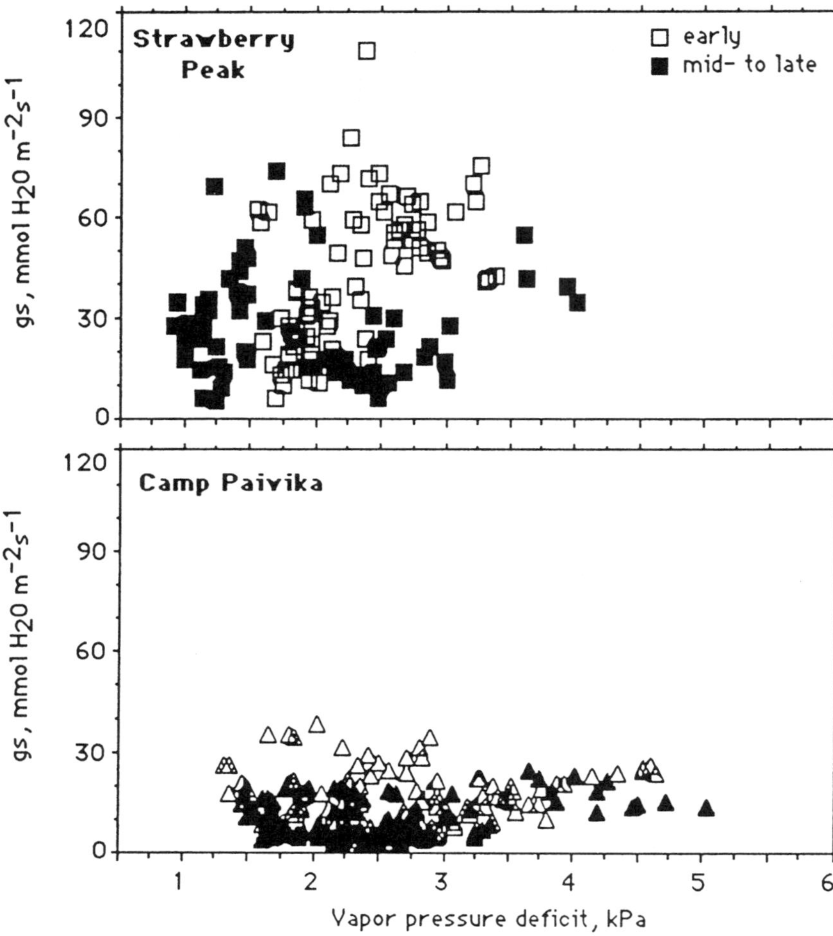

Figure 6.8. Relationship between stomatal conductance (gs) and vapor pressure deficit for pre- and postdrought periods (as in Fig. 6.7) for two sites with similar moisture regimes but different pollution loads. Strawberry Peak had moderately high ozone exposure and low N deposition, and Camp Paivika experienced high ozone exposure and high N deposition. Symbols represent individual tree mean stomatal conductances in 1995. In all cases, $\pm$ 1 SE was within symbol shape. Data were taken with a closed photosynthetic system with a modified technique to optimize humidity sensor responsivity to water vapor (Grulke 1994).

leads to greater drought stress, inducing stomatal closure in older needle age classes, but Coyne and Bingham (1982) state that there was no appreciable differences in needle water status among tree injury classes throughout the growing season. Another explanation may be that older foliage was responding indirectly to ozone through elevated ethylene or abscissic acid and exhibited premature aging (Heath and Taylor 1997). It is worthwhile to note that water use

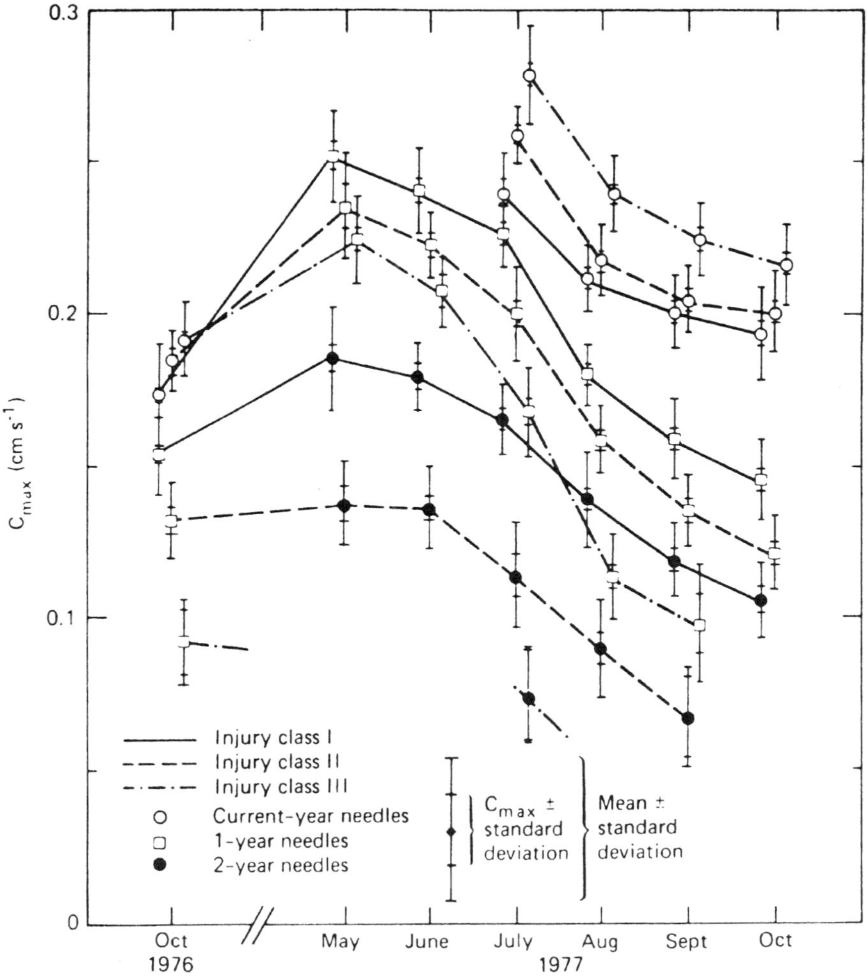

Figure 6.9. Seasonal course of maximum stomatal conductance of trees in three classes of foliar ozone injury (I: low [tolerant]; II: moderate [intermediate]; and III: high [sensitive]) at the most polluted site on the western edge of the San Bernardino Mountains (Camp Paivika). Symbols and bars represent the mean and ± 1 SE of three trees in each foliar ozone injury class. Stomatal conductance of current-year foliage is lowest for the low injury trees; for older needle age classes, the low injury trees had the greatest conductance. (From Coyne and Bingham 1982.)

efficiency of the current-year foliage of the least symptomatic trees was high relative to more symptomatic trees (e.g., despite lower conductance of the least symptomatic trees, they had the greatest maximum photosynthetic capacity) (Coyne and Bingham 1982). Increasing exposure to very high ozone concentrations typical of this site may cause increasingly lower conductance overall (see

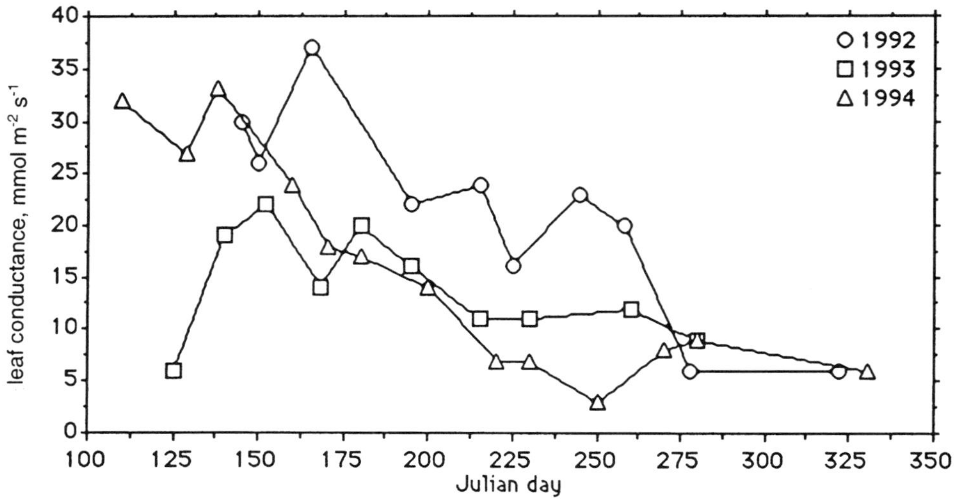

Figure 6.10. Interannual variation in stomatal conductance of yellow pine at Barton Flats, at the eastern (cleaner) end of the San Bernardino Mountains. Symbols represent the mean of two Jeffrey and one ponderosa pine measured at midcanopy (approximately 25 m). Leaf conductance in mid-July and August 1992 responded to light precipitation events that did not occur in 1993 and 1994. (Redrawn from Temple and Miller, in press).

Fig. 6.9, this study; data on 1-year-old foliage taken 1 km south of Coyne and Bingham's study site).

The effect of differing levels of drought stress on conductance between growing seasons is illustrated by Temple and Miller's (in press) study of yellow pine[2] at Camp Osceola at the eastern end of the pollution gradient (Fig. 6.10). Conductance was measured during the growing season in 1992, 1993, and 1994 at midcanopy of mature trees. All three years showed a seasonal maximum early in the growing season that declined with the progression of drought stress. Although percentage soil moisture was measured in the upper 15 cm concurrently with stomatal conductance, it was positively correlated[3] with soil moisture only in the driest year, during the predrought period (prior to July 19, ≤ Julian day 200). Precipitation during this warm dry year prolonged a period of greater stomatal conductance. At all other times, the scant precipitation after snow melt detectable at 15 cm was unrelated to stomatal conductance. Although yellow pine is deep rooted, apparently the fine root mass in the upper soil horizons contribute not only to the nutrient status of the tree but also significantly to the water status in early summer.

---

[2]Responses from two Jeffrey pines and one ponderosa pine were averaged.

[3]$y = 0.977x + 2.783$, adj. $r^2 = 0.754$, $F = 19.375$, $P = .007$, where y = stomatal conductance in mmol H_20 m^{-2} s^{-1} and x = percent soil moisture at 15 cm; calculated from Temple and Miller (in press).

Similarities in Response of Conductance to Ozone in Both Field and Chamber Studies

Field and chamber studies contain individual trees that appear to have higher stomatal conductance (relative to the population as a whole), and some have lower stomatal conductance. Seedlings of ponderosa pine from genetic lines known to be sensitive to ozone exposure have been shown to have greater stomatal conductance than relatively tolerant genetic lines in chamber studies with seedlings (Beyers et al. 1992; Temple et al. 1992; Temple and Miller 1994) and in field studies (Patterson and Rundel 1989). This finding is consistent for 1-year and older foliage in field studies of both ponderosa pine (Coyne and Bingham 1981) and Jeffrey pine (Patterson and Rundel 1989). Although greater leaf conductance has been attributable to increased erosion of the cuticle with oxidant exposure (*Pinus sylvestris;* Cape and Fowler 1981), cuticular uptake is negligible relative to stomatal uptake (Kerstiens and Lendzian 1989).

Increased transpiration with ozone fumigation was described for Norway spruce and Scots pine, presumably due to a decrease in the ability of the stomata to close (Keller and Hasler 1987; Skarby et al. 1987; Barnes et al. 1990). Conversely, a moderately high, constant ozone exposure (90 ppb) reduced stomatal conductance in Norway spruce without changes in any other field environmental condition (Wieser and Havranek 1989). Reduced stomatal conductance as well as possible changes in the hydraulic properties of the roots and stems in elevated ozone treatments explained leaf water potential differences.

Carbon Acquisition Changes Across the Pollution Gradient

Once ozone enters the foliage, individuals sensitive to ozone may have one or more attributes of carbon acquisition that lowers the total carbon pool and may predispose them to the stress of ozone uptake. In this section, changes in the seasonal course of CER, components of carboxylation efficiency (rubisco carboxylation efficiency, maximum photosynthetic capacity), and light handling capability (apparent quantum efficiency, chlorophyll content) are discussed.

Seasonal Changes in CER Across the Pollution Gradient

Across the pollution gradient, the data for the 21–60-year-old trees were pooled within each sampling date at each site and plotted versus Julian day and cumulative ozone exposure in the normal (1993) and droughty year (1994) (Fig. 6.11). The relatively short growing season at the northern, atmospherically clean site (Lassen) is apparent from the sharp increase in CER. Once the seasonal maxima was achieved, CER was greater at this site than any of the other sites in 1993. In the droughty year during August and September, CER was lowest for the least polluted sites (Lassen and Camp Osceola). The depression of CER from the end of

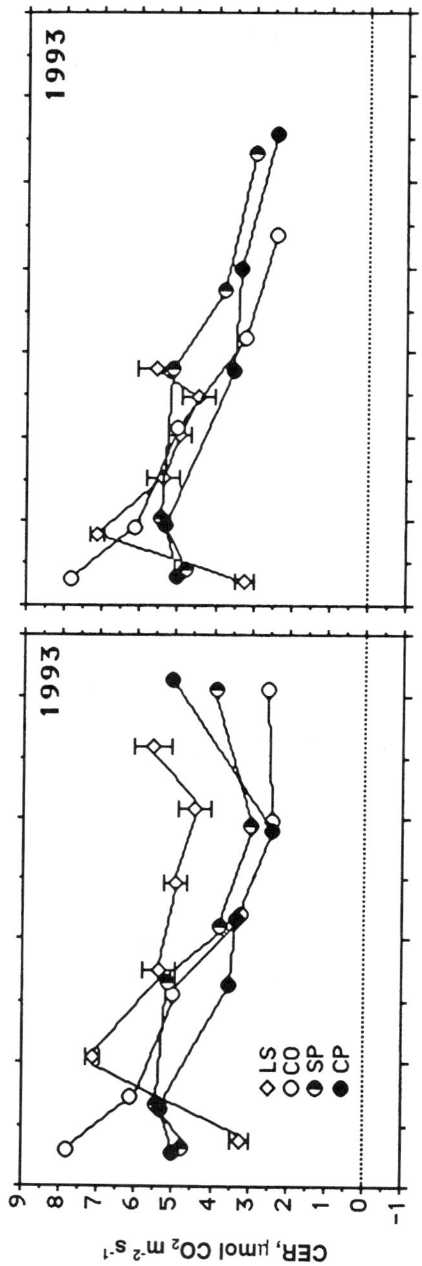
1993
1993
LS
CO
SP
CP
CER, µmol CO2 m-2 s-1
9
8
7
6
5
4
3
2
1
0
-1

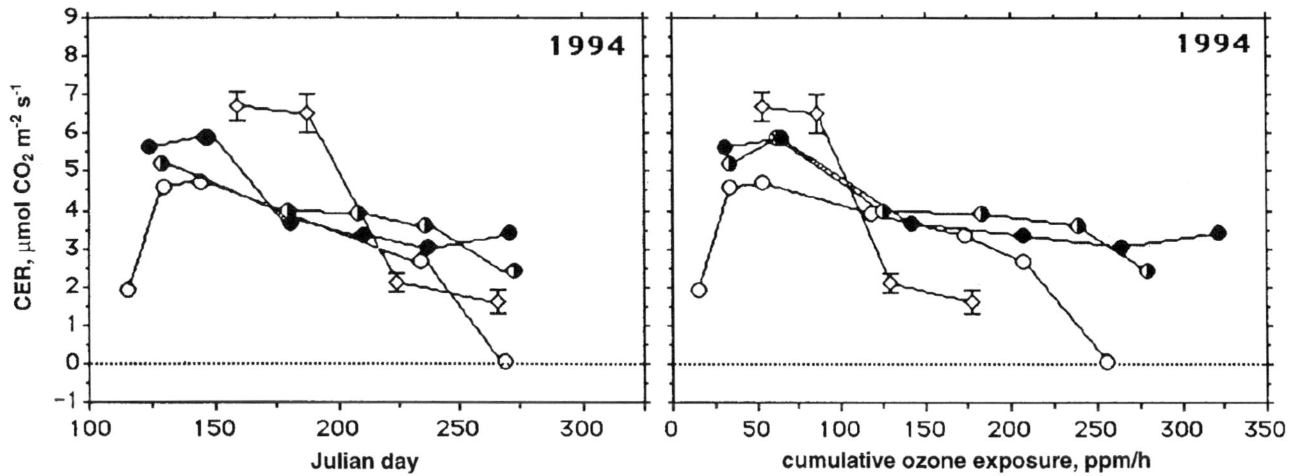

Figure 6.11. Comparison of the CO_2 exchange rate (CER) of the 21–60-year tree age class across the San Bernardino Mountains and at Lassen Volcanic National Park. Symbols and bars represent the mean and ± 1 SE of 12 trees at each site at each sampling date. The data were taken in a year of normal precipitation (1993) and in a droughty year (1994). If the data from the northernmost site is not included in the droughty year, the photosynthetic response of ponderosa pine to cumulative ozone exposure is similar across the four sites with widely ranging ozone exposure (hourly average of 38–80 ppb, averaged over 24 hours, for 6 months from April 15 to October 15). LS, Lassen; CO, Camp Osceola; SP, Strawberry Peak; CP Camp Paivika.

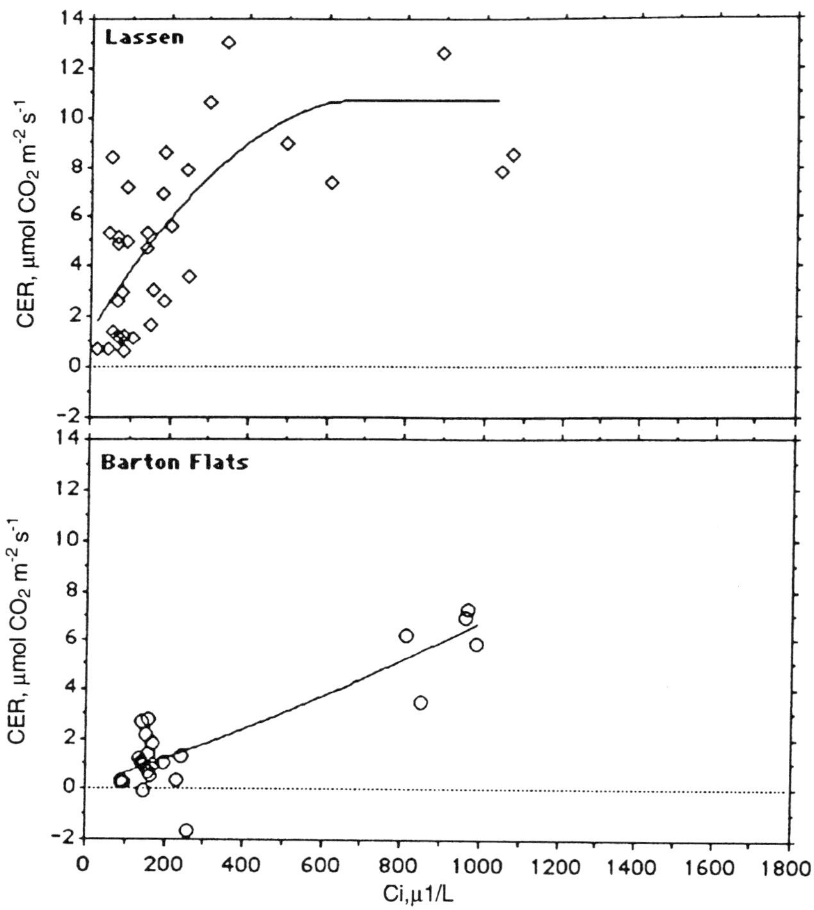

Figure 6.12. CO_2 exchange rate (CER, net assimilation) as a function of substomatal CO_2 concentration (Ci) for the four sites. These data were taken with a CO_2-, light-, and temperature-controlled cuvette during the first week of September 1995 for six trees in the 21–60-year tree age class. Standard error (± 1 SE) around the mean of individual measures fall within the symbol. Rubsico carboxylation efficiency (the initial slope) was similar for the six trees at each of Lassen, Camp Osceola, and Strawberry Peak (0.025) but was severely reduced at Camp Paivika (0.004). Maximum photosynthetic capacity under saturating CO_2 in air was greatest at Lassen and declined with increasing pollution exposure.

season drought is supported by low predawn leaf water potential at these two sites (Table 6.1).

When seasonal CER was plotted versus cumulative ozone exposure, one regression line described response of all sites if the early-season delays due to latitudinal differences and the late-season rebound in gas exchange at Lassen are omitted. Applying the same restrictions to the relationship between CER and cumulative ozone exposure in the droughty year, there was no significant differ-

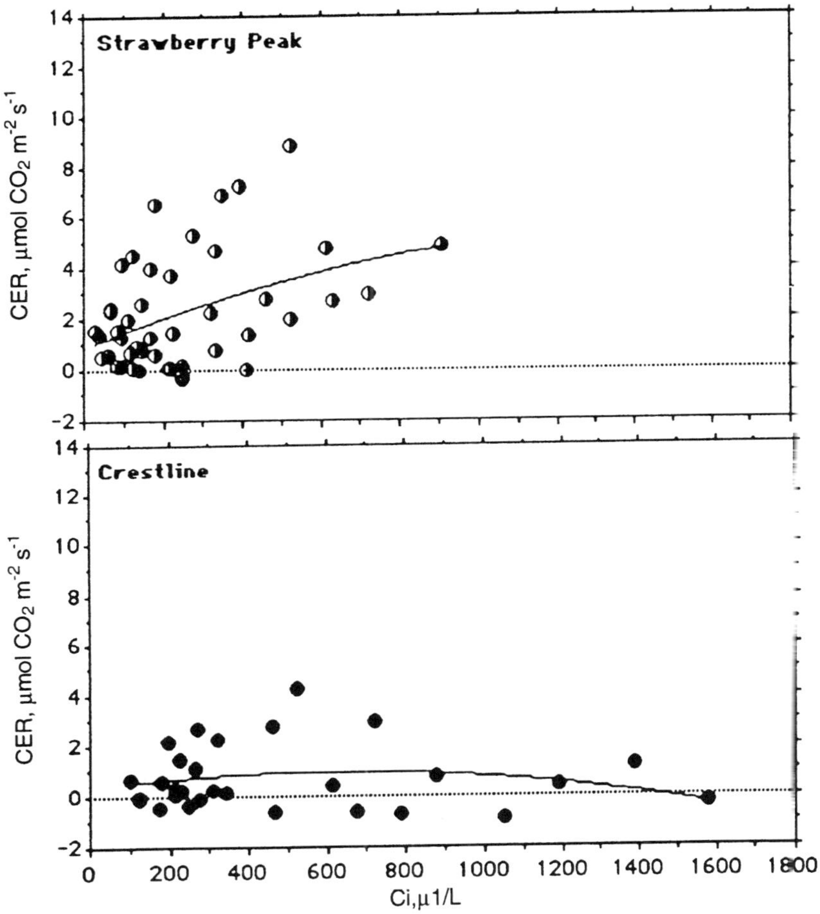

Figure 6.12. (*continued*)

ence in the regression lines between 1993 and 1994 if the Lassen data are excluded from the analysis.[4] The midsummer sharp decline in CER at the least polluted site (Lassen) in 1994 is a response to drought stress. Even the three southern Californian sites encompass a large range in ozone exposure, thus a similarity in response of CO_2 uptake to ozone exposure is in contrast to differing rates of decline in CO_2 uptake at differing ozone doses for pine described by Reich (1987). The single regression line describing ponderosa pine response to cumulative ozone exposure supports the assertion that there is not a unique response to ozone exposure: it accelerates senescence (Coyne and Bingham 1982; Reich 1983; Matyssek et al. 1994; Heath and Taylor 1997).

[4]1993: $y = -0.016x + 6.706$, adj. $r^2 = 0.774$, $F = 55.861$, $P = .0001$; 1994: $y = -0.01x + 5.513$, adj. $r^2 = 0.686$, $F = 35.934$, $P = .0001$, where y = CER in μmol CO_2 m^{-2} s^{-1} and x = cumulative ozone exposure in ppm/h.

When the data were analyzed on an early versus mid-to-late growing season time period with respect to ozone exposure using July 19 (Julian day 200) as a division, only the early-season CER declines sharply with ozone exposure (analysis with this data set not presented). Temple and Miller (in press) initially suggested that most of the uptake of pollution occurs prior to seasonal drought, occurring most years by mid- to late July.

Comparison of Field and Chamber Studies of Carbon Acquisition

Ozone exposure chamber studies with ponderosa pine showed decreases in CO_2 uptake (Weber et al. 1993; Takemoto et al. 1997; Grulke et al., in press), or needle biomass (Hogsett et al. 1989) with elevated ozone exposure. An ozone exposure chamber study with Jeffrey pine showed decreases in needle, stem, and root growth at moderately low ozone concentrations (0.2 ppm, three times per week for 6 weeks) (Temple 1988). Many examples of reduced carbon acquisition with elevated ozone exposure for other species can be found in general reviews on conifer response (Reich 1987; Darrrall 1989).

Several studies of ponderosa pine conducted in exposure chambers have described small increases rather than a decrease in CO_2 uptake with increasing ozone exposure (Beyers et al. 1992) with ponderosa pine, as well as other conifers (Freer-Smith et al. 1989; Eamus and Murray 1991). Perhaps the best explanation for increases in CO_2 uptake with elevated ozone exposure is reallocation of nitrogen to retained foliage (from greater leaf mortality of older tissue) or the increased sink for photosynthate within damaged leaves or to older tissue due to ozone injury. Others (Wellburn 1990; Bytnerowicz et al. 1995) have suggested that if air rather than oxygen is used to generate ozone, inadvertent production of nitrogen oxides can act as a fertilizer, enhancing photosynthetic rates.

Rubisco Activity and Maximum Photosynthetic Rate

Carboxylation efficiency, determined from the initial slope of the response of assimilation to internal stomatal CO_2 concentration, indicates the level of enzymatic activity and quantity (Caemmerer and Farquhar 1981; Dann and Pell 1989). Carboxylation efficiency was significantly greater at the least polluted site relative to the more polluted sites in southern California. The moderately high pollution site, Strawberry Peak, had the lowest carboxylation efficiency of the three southern Californian sites (Fig. 6.12). Although repeated exposure of drought stress has been shown to exacerbate the effect of oxidant pollution on carboxylation activity (Sasek and Richardson 1989; Wallin and Skarby 1992), both the droughty site (Camp Osceola) and the most polluted site (Camp Paivika) had similar carboxylation efficiency.

Matyssek et al. (1994) described the mechanism of conifer response to ozone via decreased photosynthetic rate (both carboxylation efficiency and maximum photosynthetic rate). A decline in carboxylation efficiency led to an increase in internal stomatal CO_2 concentration, which triggered a decline in stomatal conductance. Others have supported an alternative mechanism, such that ozone directly reduces stomatal aperture, decreases internal stomatal CO_2 concentration, and results in a photosynthetic decline (Weber et al. 1993). In a droughty year (1977) at the most polluted site (Camp Paivika), Coyne and Bingham (1981) found that photosynthesis declined at a rate greater than conductance, suggesting impairment to the photosynthetic machinery versus stomatal functioning in sensitive versus tolerant individuals. Sasek and Richardson (1989) found decreased carboxylation efficiency in elevated ozone levels relative to controls with loblolly pine in a chamber exposure study. In addition to the studies above, the best evidence that both damage to the photosynthetic machinery and stomatal sluggishness contributing to reduction in plant carbon gain is the diurnal response of sensitive Jeffrey pines: a temporal delay but equally high or higher maximum CER, and the greater leaf conductance during unfavorable and lower conductance during favorable microenvironmental conditions (Patterson and Rundel 1989). Further discussion of this topic occurs in a later section in this chapter.

The maximum photosynthetic rate at saturating CO_2 concentrations in air decreased with increasing pollution levels (Fig. 6.12). The response at the most polluted site was profoundly altered relative to the least polluted site. Data from six trees at each site were combined for each response curve. The flattening of the response of assimilation to internal stomatal CO_2 concentration was similar to that reported for a reduction in soil water potential from –0.9 to –2.0 MPa for ponderosa pine (DeLucia and Heckathorn 1989), although predawn needle xylem water potential was –0.9 ± 0.03 MPa for these trees at the time of measurement and was statistically significantly lower only for trees at Camp Osceola (see Table 6.1). This was not low relative to other growing seasons (1995 versus 1994 and 1993). The reduction in enzyme function was most likely due to atmospheric oxidant exposure because N deposition would act to increase photosynthesis. When sensitive (20% chlorotic mottle on needle surfaces) versus resistant Jeffrey pine individuals were compared, P_{max} was lower by 30% (Patterson and Rundel 1989). Biochemical explanations consistent for these field responses of ponderosa pine have been described for spruce (Hampp et al. 1990).

Light Response

Photosynthetic response to light was measured with an open gas exchange system (Grulke 1994) on three trees in the 21–60-year tree age class at each site across the pollution gradient as well as for Lassen. Measurements were made in the lower one-third of the canopy during the second and third weeks of July 1994, prior to sunrise. A measurement of dark respiration was made, then an artificial light (Grulke et al. 1989) was used to drive light response at 50, 100, 900, and 1,800 μmol quanta m^{-2} s^{-1}. A simple model[5] (Thornley 1976) was applied to predict photosynthetic response from photosynthetic photon flux density (PPFD) for a

---

[5]CER = P_{max} (1 – e(AQE * I / P_{max})), where P_{max} is maximum photosynthesis under saturating light, AQE is apparent quantum efficiency in the PPFD range of 0–100 μmol quanta m^{-2} s^{-1}, and I is incident PPFD.

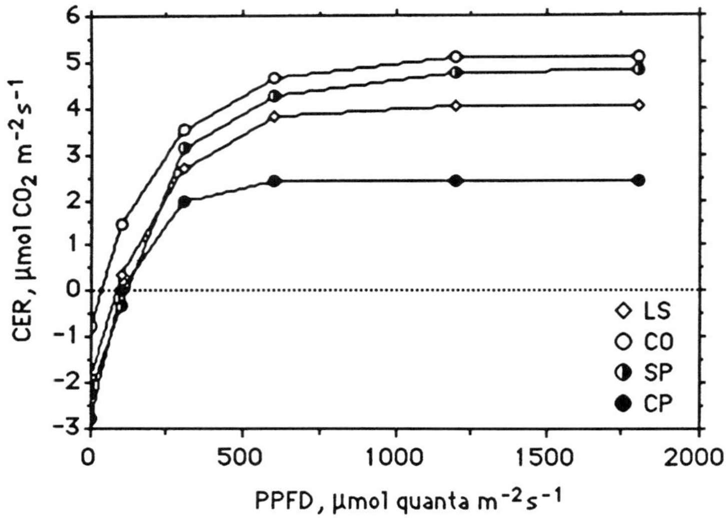

Figure 6.13. Summary regression of photosynthetic response to light for three trees at each of the four sites (LS: Lassen; CO: Camp Osceola; SP: Strawberry Peak; CP: Camp Paivika, ordered from least to most polluted) taken predawn in mid-July, 1994 with a temperature- and light-controlled cuvette. Within the southern Californian pollution gradient, dark respiration and maximum photosynthetic rate under saturating light decreased with increasing pollution exposure. Dark respiration at Lassen, with relatively fast growing individuals in clean air (see Fig. 6.16), was intermediate between SP and CO, which also probably contributed to the lower maximum photosynthetic rate under saturating light at this site relative to the moderate and moderately high ozone sites. Apparent quantum efficiency did not follow the rank order of pollution exposure.

limited number of points from the pooled data set for each site (Fig. 6.13). The maximum photosynthetic rate under saturating PPFD was greatest at Camp Osceola, followed by Strawberry Peak, Lassen, and Camp Paivika (5.13 ± 0.07, 4.86 ± 0.21, 4.03 ± 0.14, 2.41 ± 0.19 μmol CO_2 m^{-2} s^{-1}, respectively). Dark respiration was greatest at Camp Paivika, followed by Strawberry Peak, Lassen, and Camp Osceola (–2.77 ± 0.37, –2.22 ± 0.09, –1.90 ± 0.24, –0.75 ± 0.07 μmol CO_2 $m^{-2}s^{-1}$, respectively). The initial slope of the response of CER to PPFD, apparent quantum efficiency, was greatest at Camp Paivika, followed by Camp Osceola, Lassen, and Strawberry Peak (0.026, 0.020, 0.019, 0.017 μmol CO_2 per quanta of light m^{-2} s^{-1}, respectively). The individual tree variability in initial photosynthetic response to light was most variable at Lassen and was probably driven by prior light environment of the fascicles chosen. Across the gradient from least to greatest pollution, there was a progression of increasing dark respiration and decreasing apparent quantum efficiency and maximum photosynthetic rate under saturating light.

In a chamber exposure experiment with ponderosa pine, there was a significant difference in P_{max}, but no difference in apparent quantum efficiency or light compensation point between seedlings exposed to charcoal-filtered air or elevated ozone (Grulke et al., in press). However, this experiment was conducted in late season (early October), when the photosynthetic rates and the ability to detect treatment differences are low. Other published data defining pine response to light under controlled conditions were not found (Matyssek et al. 1994).

Photosynthetic Pigment Concentrations Across the Gradient

Photosynthetic pigments are an example of resource partitioning (nutrients and carbon) to the leaf. Trebst (1994) and Bjorkman and Demmig-Adams (1994) have recently reviewed photo-oxidation of pigments under both high light environments common in Mediterranean summers as well as likely effects and interactions with high oxidant exposure. Given this, one would expect that pigment concentration to be significantly lower at the high-pollution sites. Foliage samples for total chlorophyll was collected at regular intervals from the same trees measured for gas exchange (Fig. 6.4). Chlorophyll was extracted and analyzed by using dimethyl sulfoxide using standard techniques (Arnon 1949; Hiscox and Israelstam 1979).

The two sites with relatively little pollution (Lassen and Camp Osceola) had similar seasonal trends, and foliar chlorophyll content was low early season, increased until mid-July, then was held constant through the end of September (Fig. 6.14). Total chlorophyll content was highest at the two most polluted sites (Strawberry Peak and Camp Paivika) at each sampling date relative to the two sites with less pollution in 1994 (a year with significant drought stress). The seasonal trends were similar in 1995 (more mesic year) except total foliar chlorophyll content at Strawberry Peak was more similar to Camp Osceola and Lassen. Foliar nitrogen collected on the same sampling dates mirrors the total chlorophyll content (Grulke, unpublished data). Long-term high nitrogen deposition at the western two sites (Fenn et al. 1996) contributes to greater nitrogen availability in the environment and greater plant acquisition and partitioning to the photosynthetic machinery retained despite the stressors experienced.

Integration of CO_2 Exchange Rate and Stomatal Conductance Responses

A single characteristic that integrates stomatal control of pollutant uptake and reduction of carbon acquisition is carbon isotopic discrimination (Fig. 6.15). Stomatal closure causes decreased CO_2 concentration within the leaf, decreased discrimination of rubisco for the heavier carbon, and a less negative $\delta^{13}C$ content of the resulting tissue (Ehleringer and Rundel 1988). Damage to the photosynthetic machinery causes the opposite effect: increased discrimination of rubisco for

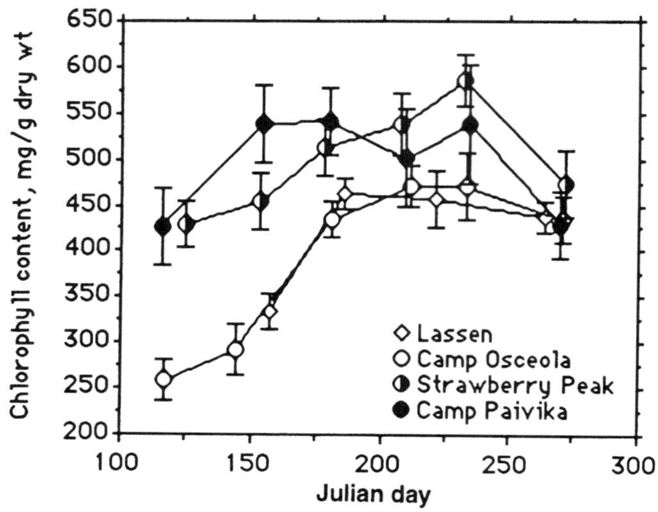

Figure 6.14. Seasonal course of chlorophyll content taken in 1994 for the previous year foliage. Foliar chlorophyll is greater at the sites with the greatest soil moisture content and atmospheric N deposition. Foliar N (not presented) followed similar seasonal trends. Mean and ±1 SE of 12 trees in the 21–60-year tree age class are represented by each point.

the heavier carbon and a more negative δ^{13} C content of the tissue (Greitner and Winner 1988). Analysis of carbon isotopic discrimination has also been used to elucidate differences in water use efficiency between species (Ehleringer and Rundel 1988; Farquhar et al. 1989).

Needle cellulose was extracted for subsequent carbon isotopic analysis (see Yoder et al. 1994 for methodology). Cellulose is found in cell walls, so that environmental conditions experienced during cell elongation (needle elongation growth 90% complete by late July; Grulke et al., in review) influenced the stomatal control of pollutant uptake and rubisco discrimination against the heavier isotope. Response in clean air in one plant may be mimicked by combined drought stress and ozone exposure in another plant (Greitner and Winner 1988). Therefore, carbon isotopic discrimination should reflect population differences in stomatal conductance (predrought) and cumulative ozone exposure until the onset of drought stress (Fig. 6.15). Water use efficiency was not significantly different between sites for this tree age class during the predrought period (as calculated from this data set).

Despite very similar cumulative ozone exposure in 1993 by July 20 at Camp Paivika and Strawberry Peak, δ^{13} C is more negative at Strawberry Peak relative to Camp Paivika. The pattern held true also in 1994, with greater ozone exposure at Camp Paivika. The very low stomatal conductance of the Camp Paivika trees (Fig. 6.15) resulted in greater rubisco discrimination against the heavier isotope

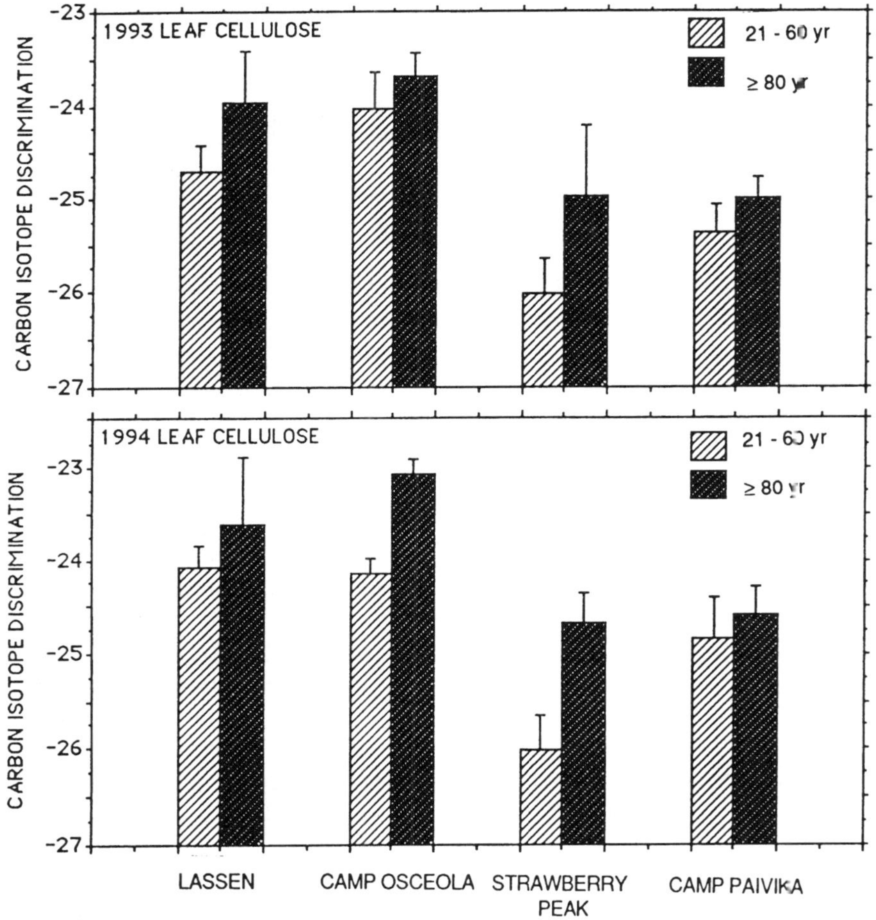

Figure 6.15. Integration of stomatal control and carbon acquisition using carbon isotopic discrimination. Bars represent the mean of six trees in the 21–60-year tree age class and three mature trees (≥ 80 years), ± 1 SE. Because leaf cellulose was extracted for carbon isotope discrimination, values represent environmental conditions over the period of needle growth (mid-May to mid-July).

and a less negative δ^{13} C. Reduced radial growth of Douglas-fir was associated with less negative δ^{13} C of cellulose purified from wood formed during those years (Martin and Sutherland 1990). However, leaf and bole cellulose δ^{13} C was not significantly correlated with foliar ozone injury level within a population exposed to high pollution (Camp Paivika) (Grulke and Lee 1997).

At each site, δ^{13} C of the mature tree foliage cellulose was less negative than that of the 30-year-old trees (average of six from 21–40-year-old trees). This is consistent with the change in δ^{13} C with increasing height as reported by Yoder et al. (1994). There were no differences in δ^{13} C of mature trees between Strawberry Peak and Camp Paivika which suggests that the stomatal conductance of the mature trees was similar between the two sites. For the younger tree age class, rubisco carboxylation efficiency was significantly reduced at Camp Paivika, but not at Strawberry Peak (Fig. 6.15). This suggests that stomatal limitations overrode carboxylation limitations in carbon isotope discrimination.

Changes in Foliar Respiration Across the Pollution Gradient

Respiration can account for 20–90% of the carbon acquisitioned in pine, depending on the time of the year (Kinerson 1975; Landsberg 1986). An increase in maintenance respiration has been reported for pathogens, diseases (Kozlowski 1990), and oxidant pollution (Barnes 1972; Coyne and Bingham 1981; Skarby et al. 1987; Grulke et al. 1989; Adams et al. 1990). Maximum seasonal respiration was greatest at the time of bud break (Fig. 6.4), which is a common phenomenon (Kozlowski 1990). In each of the four populations, there were individuals with high endogenous respiration rates, but the most variable population was Camp Paivika, the most polluted site. Sensitive individuals had greater respiration than tolerant trees at the most polluted site (Camp Paivika) (Coyne and Bingham 1981). If ozone exposure is associated with increased respiration, individuals with greater endogenous rates may be predisposed to ozone sensitivity. This response would not be anticipated in response to NO or NO_2 (Wellburn 1990).

An example of the effect of oxidant pollution on resource partitioning is illustrated by the relationship between foliar temperature and respiration across the pollution gradient for the 21–60-year tree age class (Fig. 6.16). At a given foliar temperature, respiration was greatest at the most polluted site and was similar at the moderately high and moderately polluted sites in the San Bernardino Mountains. The response of foliar respiration to temperature was much greater during the period of maximum needle and branchlet growth (early season, prior to mid-season drought stress) than after elongation growth slowed, presumably due to the high cost of growth respiration (the progressive order of the regression lines is the same as that of the symbols).

The ratio of net assimilation to foliar respiration has been linked to tree vigor at stressed sites (McLaughlin et al. 1990). During the period of active growth, CER was not significantly correlated with dark-adapted foliar respiration at any site. After the period of active growth, the correlation coefficient of the relationship between CER and dark respiration was low but significantly correlated at the three least polluted sites (Lassen, Camp Osceola, and Strawberry Peak). Perhaps more than 40 years of selection against individuals with greater respiration rate has eliminated the most sensitive at the most polluted site (Camp Paivika). In general, conifer response to environmental stressors is highly variable (Mitton 1995), specifically for ponderosa pine (Temple et al. 1992). Increased respiration is one of many factors contributing to ozone sensitivity.

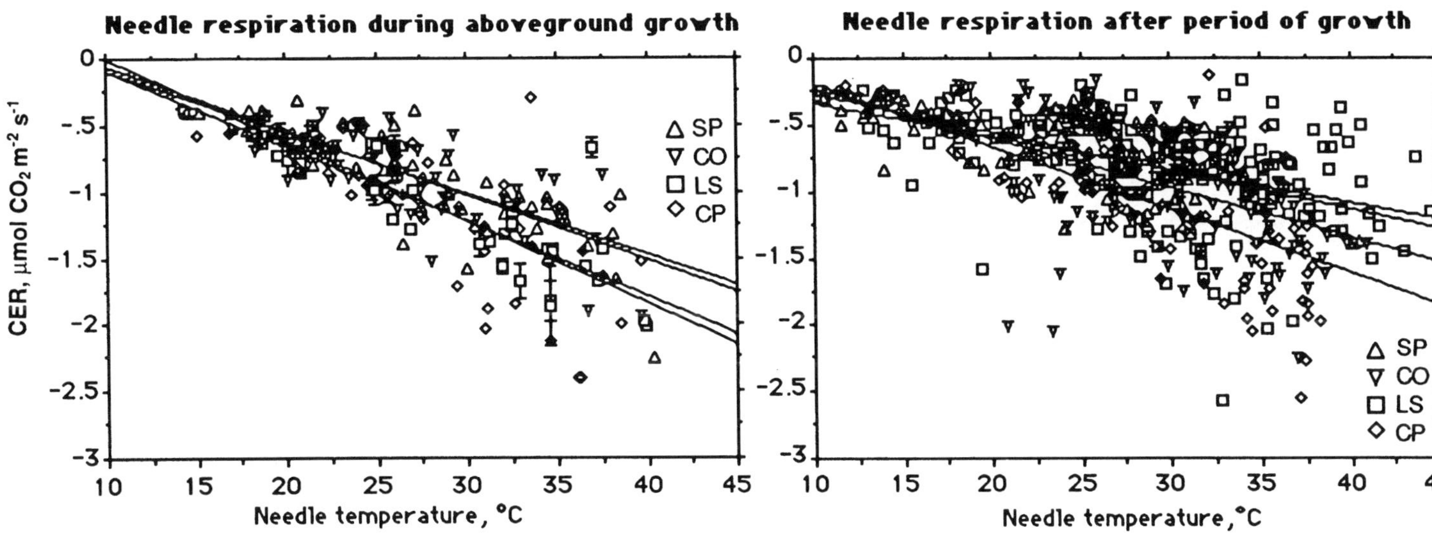

Figure 6.16. Needle respiration measured after 15-minute dark adaptation during the day. Individual tree values are plotted versus foliar temperature at time of measurement (± 1 SE) during predrought (prior to mid-July) and postdrought (after mid-July) periods compiled for 1993 and 1994. Predrought respiration was greater than postdrought at any one foliar temperature due to rapid growth. Similarly, foliar respiration was relatively high at Lassen, the atmospherically clean site, due to greater growth rates. Site acronyms are consistent with previous graphs.

Conclusions

In general, seasonal patterns of gas exchange were similar across the San Bernardino Mountains in a year of normal precipitation, but in a drier year, tree response within a site became more asynchronous with respect to other environmental triggers. Individual differences within a population (population variance) were greater within increasing pollution exposure. Stomatal conductance decreased with increasing pollution levels. Stomatal conductance was relatively nonresponsive to environmental factors during the period of drought stress. Normal stomatal function was lost at the most polluted site (Camp Paivika); stomata appeared to be partially open but not responsive to VPD or CO_2 concentration in the gradient study. CER decreased with increasing pollution, despite nitrogen loading accompanying high ozone levels. There was a profound effect of pollution on rubisco carboxylation and, combined with drought stress, on the maximum photosynthetic rate under saturating CO_2 concentrations in air. Predawn respiration and respiration measured after dark adaptation in the light were greatest at Camp Paivika. Population response to pollution was made more complex by differences in the timing of the onset of drought stress between the sites and between years in the same site. Despite the high oxidant pollution exposure at Camp Paivika, the most polluted forested site in the United States, the accompanying high nitrogen deposition ameliorated some of the expected responses in photosynthetic pigment concentration in both a xeric and mesic year. There is no unique response of plants to oxidant stress (Heath and Taylor 1997). It is the collection of characteristics (e.g., greater endogenous photosynthetic rate, lower respiration, greater ability to acquisition, allocate, and partition resources to photosynthetic machinery) that determines individual plant success or failure in the face of multiple stressors.

Acknowledgments

Although the research described in this article has been funded wholly by U.S. Environmental Protection Agency agreement DW 12934530, it has not been subjected to the agency's review and does not necessarily reflect the views of the agency, and no official endorsement should be inferred. The initial work of Dr. Paul Miller in this field of research and at these sites was instrumental to the success of this study. Special thanks are due to Drs. Bill Hogsett, Jim Weber, and Chris Andersen, who made many suggestions on the direction of this research, and to Drs. Jud Isebrands, Jim Weber, and Ruth Yanai, who carefully edited the manuscript. The field research effort was supported by Lisa Balduman, Janet Kirsch, Patti Haggerty, Diane Fisher, Lynn Morrison, Phil Dawson, Rob Lennox, Adam Cosner, Luis Murillo, and David Jones. Many thanks to the support staff, Jeannette Roberts, Debbie McKee, and Linda McDonough at the Corvallis Forestry Laboratory and Grady Nealy at the Environmental Protection Agency.

References

Adams, M.B.; Edwards, N.T.; Taylor, G.E.; Skaggs, B.L. Whole-plant carbon-14 photosynthate allocation in *Pinus taeda,* seasonal patterns at ambient and elevated ozone levels. Can. J. For. Res. 20:152–158; 1990.

Aitken, W.M.; Jacobi, W.R.; Staley, J.M. Ozone effects on seedlings of Rocky Mountain ponderosa pine. Plant Dis. 68:398–400; 1984.

Andersen, C.P.; Hogsett, W.E.; Wessling, R.; Plocher, M. Ozone decreases spring root growth and root carbohydrate content in ponderosa pine the year following exposure. Can. J. For. Res. 21:1288–1291; 1991.

Arnon, D.I. Copper enzymes in isolated chloroplasts: polyphenol-oxidase in *Beta vulgaris.* Plant Phys. 24:1–15; 1949.

Barbour, M.G. Californian upland forests and woodlands. In: Barbour, M.G.; Billings, W.D., eds. North American terrestrial vegetation. New York: Cambridge University Press; 1988:131–164.

Barnes, J.D.; Eamus, D.; Davison, A.W.; Ro-Poulsen, H.; Mortensen, L. Persistent effects of ozone on needle water loss and wettability in Norway spruce. Environ. Pollut. 63:345–363; 1990.

Barnes, R.L. Effects of chronic exposure to ozone on photosynthesis and respiration of pines. Environ. Pollut. 3:133–138; 1972.

Beyers, J.L.; Riechers, G.H.; Temple, P.J. Photosynthetic capacity of ponderosa pine seedlings exposed to different levels of atmospheric ozone and drought stress. N. Phyt. 122:81–90; 1992.

Bjorkman, O.; Demmig-Adams, B. Regulation of photosynthetic light energy capture, conversion, and dissipation in leaves. In: Schulze, E-D.; Caldwell, M.M., eds. Ecophysiology of photosynthesis. New York: Springer-Verlag; 1994:17–48.

Boltz, B.A.; Bongarten, B.C.; Teskey, R.O. Seasonal patterns of net photosynthesis of loblolly pine from diverse origins. Can. J. For. Res. 16:1063–1068; 1986.

Bytnerowicz, A.; Fenn, M.E. Nitrogen deposition in California forests: a review. Environ. Pollut. 92(2):127–146; 1996.

Bytnerowicz, A.; Grulke, N.E. Physiological effects of air pollutants on western forests. In: Olson, R.; Binkley, D.; Bohm, M., eds. The response of western forests to air pollution. Ecological studies 97. New York: Springer-Verlag; 1992:183–233.

Bytnerowicz, A.; Takemoto, B.K. Effects of photochemical smog on the growth and physiology of ponderosa pine seedlings grown under nitrogen and magnesium deficiencies. In: Olson, R.K.; Lefohn, A.S. Effects of Air Pollution on Western Forests. Pittsburgh, PA: Air Waste Manage. Assoc.; 1989:455–468.

Bytnerowicz, A.; Olszyk, D.M.; Huttunen, S.; Takemoto, B.K. Effects of photochemical smog on growth, injury, and gas exchange of pine seedlings. Can. J. Bot. 67:2175–2181; 1989.

Bytnerowicz, A.; Poth, M.A.; Takemoto, B.K. Effects of photochemical smog and mineral nutrition on ponderosa pine seedlings. Environ. Pollut. 67:233–248; 1990.

Bytnerowicz, A.; Dawson, P.J.; Morrison, C.L. Physiological and growth responses of ponderosa pine seedlings to ambient and elevated ozone concentrations in southern Sierra Nevada. Proceedings, 84th Annual Meeting, Vancouver, British Columbia Pittsburgh, PA: Air Waste Manage. Assoc.; 1991.

Bytnerowicz, A.; Tran, M.; Anderson, P. Effects of charcoal air filtration and ozone generation on concentrations of some N and S compounds in open-top field chambers. Atmos. Environ. 29(12):1355–1358; 1995.

Caemmerer, S. von; Farquhar, G.D. Some relationships between the biochemistry of photosynthesis and the gas exchange of leaves. Planta 153:376–387; 1981.

Cape, J.N.; Fowler, D. Changes in epicuticular was of *Pinus sylvestris* exposed to polluted air. Silva Fenn. 15:457–458; 1981.

Coyne, P.; Bingham, G.E. Comparative ozone dose response of gas exchange in a ponderosa pine stand exposed to long-term fumigations. J. Air Pollut. Control Assoc. 31:38–41; 1981.

Coyne, P.I.; Bingham, G.E. Variation in photosynthesis and stomatal conductance in an ozone-stressed ponderosa pine stand: light response. For. Sci. 28:257–273; 1982.

Cropper, W.P., Jr.; Gholz, H.L. Simulation of the carbon dynamics of a Florida slash pine plantation. Ecol. Model. 66:231–249; 1993.

Dann, M.S.; Pell, E.J. Decline of activity and quantity of ribulose bisphosphate carboxylase/oxygenase and net photosynthesis in ozone-treated potato foliage. Plant Phys. 91:427–432; 1989.

Darrall, N.M. The effect of air pollutants on physiological processes in plants. Plant Cell Environ. 12:1–30; 1989.

Davis, D.D.; Wood, W.A. The influence of environmental factors on the sensitivity of Virginia pine to ozone. Phytopathology 63:371–376; 1973.

DeLucia, E.H.; Heckathorn, S.A. The effect of soil drought on water-use efficiency in a contrasting Great Basin desert and Sierran montane species. Plant Cell Environ. 12:935–940; 1989.

Denne, M.P. Wood structure and production within the trunk and branches of *Picea sitchensis* in relation to canopy formation. Can. J. For. Res. 9:406–427; 1979.

Dickson, R.E. Carbon and nitrogen allocation in trees. In: Dreyer, E.; Aussenac, G.; Bonnet-Masimbert, M.; Dizengremel, P.; Favre, J.M.; Garrec, J.P.; Le Tacon, F.; Martin, F., eds. Forest tree physiology. Proceedings of the international symposium. Annales des Sciences Forestieres, vol. 46; Nancy, France: Laboratoire de Recherches en Sciences Forestieres; 1989:631–647.

Dobson, M.C.; Taylor, G.; Freer-Smith, P.H. The control of ozone uptake by *Picea abies* (L.) Karst. and *P. sitchensis* (Bong.) Carr. during drought and interacting effects on shoot water relations. N. Phyt. 116:465–474; 1990.

Eamus, D.; Murray, M. Photosynthetic and stomatal conductance responses of Norway spruce and beech to ozone, acid mist and frost—a conceptual model. Environ. Pollut. 72:23–44; 1991.

Edwards, G.S.; Edwards, N.T.; Kelly, J.M.; Mays, P.A. Ozone, acidic precipitation, and soil Mg effects on growth and nutrition of loblolly pine seedlings. Environ. Exp. Bot. 31(1):67–78; 1991.

Ehleringer, J.R.; Rundel, P.W. Stable isotopes: history, units, and instrumentation. In: Rundel, P.W.; Ehleringer, J.R.; Nagy, K.A., eds. Stable isotopes in ecological research. New York: Springer-Verlag; 1988:1–16.

Evans, L.S.; Miller, P.R. Ozone damage to ponderosa pine: a histological and histochemical appraisal. Am. J. Bot. 50:297–304; 1972.

Farquhar, G.D.; Ehleringer, J.R.; Hubick, K.T. Carbon isotope discrimination and photosynthesis. Annu. Rev. Plant Phys. Plant Mol. Biol. 40:503–537; 1989.

Fenn, M.E., Poth, M.A.; Johnson, D.W. Evidence for nitrogen saturation in the San Bernardino Mountains in southern California. For. Ecol. Manage. 82:211–230; 1996.

Freer-Smith, P.H.; Dobson, M.; Taylor, G. Factors controlling the rates of O_3 uptake by spruce and beech. In: Bucher, J.B.; Bucher-Wallin, I., eds. Air pollution and forest decline. Proceedings of the 14th international meeting for specialists in air pollution effects on forest ecosystems, IUFRO P2.05, Interlaken, Switzerland, 2–8 October 1988. EAFV Birmensdorf; 1989:407–409.

Gower, S.T.; Isebrands, J.G.; Sheriff, D.W. Carbon allocation and accumulation in conifers. In: Smith, W.K.; Hinckley, T.H., eds. Resource physiology of conifers—acquisition, allocation, and utilization. New York: Academic Press; 1995:217–254.

Greitner, C.S.; Winner, W.E. Increases in δ^{13} C of radish and soybean plants caused by ozone. N. Phyt. 108:489–494; 1988.

Grulke, N.E. Quality assurance/quality control document for physiological measurements.

Cooperative agreement DW 12934530. Available at: Corvallis Environmental Effects Laboratory, 200 SW 35th Street, Corvallis, OR 97333; 1994.
Grulke, N.E.; Lee, E.H. Assessing visible ozone-induced foliar injury in ponderosa pine. Can. J. For. Res. 27:1658–1668; 1997.
Grulke, N.E.; Miller, P.R.; Wilborn, R.D.; Hahn, S. Photosynthetic response of giant sequoia seedlings and rooted branchlets of mature foliage to ozone fumigation. In: Olson, R.K.; Lefohn, A.S., eds. Effects of air pollution on western forests. Pittsburgh, PA: Air Waste Manage. Assoc.; 1989:429–442.
Grulke, N.E.; Miller, P.R.; Scioli, D. Response of giant sequoia canopy foliage to experimentally manipulated levels of atmospheric ozone. Tree Phys. 16:575–581; 1996.
Grulke, N.E.; Miller, P.R.; Leininger, T.D. Effect of ozone exposure on seasonal gas exchange of five western conifers. Gen. Tech. Rep., PSW-167. Albany, CA: USDA Forest Service, Pacific Southwest Station; (in press).
Grulke, N.E.; Fenn, M.E.; Andersen, C.P.; Miller, P.R. Ozone exposure and nitrogen deposition reduced root standing biomass of ponderosa pine in the San Bernardino Mountains, California. Environ. Pollut. (in review).
Guderian, R., ed. Air pollution by photochemical oxidants. Ecological studies 52. Berlin: Springer-Verlag; 1985.
Hampp, R.; Einig, W.; Egger, B. Energy and redox status, and carbon allocation in one- to three-year-old spruce needles. Environ. Pollut. 68:305–318; 1990.
Heath, R.L. Ozone. In: Mudd, J.B.; Kozlowski, T.T., eds. Responses of plants to air pollution. New York: Academic Press; 1975:23–55.
Heath, R.L.; Taylor, G.E., Jr. Physiological processes and plant responses to ozone exposure. In: Sandermann, H.; Wellburn, A.R.; Heath, R.L., eds. Forest decline and ozone: a comparison of controlled chamber and field experiments. Ecological studies 127. New York: Springer-Verlag; 1997:310–368.
Helms, J.A.; Anderson, P.D.; Houpis, J.L.J. Gas exchange by *Pinus ponderosa* in relation to atmospheric pollutants. Final report, submitted to California Air Resources Board, Contract A132-101. Available from CARB, PO Box 2815, Sacramento, CA 98512-2812; 1994.
Hiscox, J.D.; Israelstam, G.F. A method for the extraction of chlorophyll from leaf tissue without maceration. Can. J. Bot. 57:1332–1334; 1979.
Hogsett, W.E.; Tingey, D.T.; Hendricks, C.; Rossi, D. Sensitivity of western conifers to SO_2 and seasonal interaction of acid fog and ozone. In: Olson, R.K.; Lefohn, A.S., eds. Effects of air pollution on western forests. Pittsburgh, PA: Air Waste Manage. Assoc.; 1989:469–492.
Hubbard, R.M.; Yoder, B.J.; Ryan, M.G. Evidence that hydraulic conductance limits photosynthesis in old *Pinus ponderosa* trees. Tree Phys. (in press).
Jensen, K.F.; Roberts, B.R. Changes in yellow poplar stomatal resistance with SO_2 and O_3 fumigation. Environ. Pollut. 41:235–245; 1986.
Jones, D.; Graham, R.C. Water-holding characteristics of weathered granitic rock in chaparral and forest ecosystems. Soil Sci. Am. J. 57(1):256–261; 1993.
Keller, T.; Hasler, R. The influence of a fall fumigation with ozone on the stomatal behavior of spruce and fir. Oecologia 64:284–286; 1987.
Kerstiens, G.; Lendzian, K.J. Interactions between ozone and plant cuticles. I. Ozone deposition and permeability. N. Phyt. 112:13–19; 1989.
Kinerson, R.S. Relationships between plant surface area and respiration in loblolly pine. J. Appl. Ecol. 12:965–971; 1975.
Kozlowski, T.T. Carbohydrate sources and sinks in woody plants. Bot. Rev. 58(2):107–223; 1990.
Kozlowski, T.T.; Pallardy, S.G. Physiology of woody plants. New York: Academic Press; 1997.
Landsberg, J.J. Physiological ecology of forest production. London: Academic Press; 1986.

Larson, P.R. The vascular cambium: development and structure. Berlin: Springer-Verlag; 1994.

Luxmoore, R.J.; Oren, R.; Sheriff, D.W.; Thomas, R.B. Source-sink-storage relationships of conifers. In: Smith, W.K.; Hinckley, T.M., eds. Resource physiology of conifers: acquisition, allocation, and utilization. New York: Academic Press; 1995:179–216.

Martin, B.; Sutherland, E.K. Air pollution in the past recorded in width and stable carbon isotope composition of annual growth rings of Douglas-fir. Plant Cell Environ. 13:839–844; 1990.

Matyssek, R.; Reich, P.; Oren, R.; Winner, W. Response mechanisms of conifers to air pollutants. In: Smith, W.K.; Hinckley, T.M., eds. Ecophysiology of coniferous forests. San Diego, CA: Academic Press; 1994:255–308.

McLaughlin, S.B. Effects of air pollution on forests: a critical review. J. Air Pollut. Control Assoc. 35:512–534; 1985.

McLaughlin, S.B.; Andersen, C.P.; Edwards, N.T.; Roy, W.K.; Layton, P.A. Seasonal patterns of photosynthesis and respiration of red spruce saplings from two elevations in declining southern Appalachian stands. Can. J. For. Res. 20:485–495; 1990.

Miller, P.R. Mixed conifer forests of the San Bernardino Mountains, California. In: Olson, R.K.; Binkley, D.; Bohm, M., eds. The response of western forests to air pollution. Ecological studies 97. New York: Springer-Verlag; 1992:461–497.

Miller, P.R. Abiotic diseases. In: Scharpf, R.F., ed. Diseases of Pacific coast conifers. Agricultural handbook 521. Washington, DC: USDA Forest Service; 1983:1–32.

Miller, P.R.; Longbotham, G.J.; Longbotham, C.R. Sensitivity of selected western conifers to ozone. Plant Dis. 67:1113–1115; 1983.

Miller, P.R.; McBride, J.R.; Schilling, S.L.; Gomez, A.P. Trend of ozone damage to conifer forests between 1974 and 1988 in the San Bernardino Mountains of southern California. In: Olson, R.K.; Lefohn, A.S., eds. Effects of air pollution on western forests. Pittsburgh, PA: Air Waste Manage. Assoc.; 1989:309–323.

Miller, P.R.; Stolte, K.W.; Duriscoe, E.; Pronos, J., tech. coords. Monitoring ozone air pollution effects on western pine forests. Gen. Tech. Rep. 155. Albany, CA: USDA Forest Service; 1996.

Mitton, J.B. Genetics and the physiological ecology of conifers. In: Smith, W.K.; Hinckley, T.M., eds. Ecophysiology of coniferous forests. New York: Academic Press; 1995:1–36.

Mudd, J.B.; Kozlowski, T.T., eds. Responses of plants to air pollution. New York: Academic Press; 1975.

Oliver, W.W.; Ryker, R.A. *Pinus ponderosa* Dougl. ex Laws. In: Burns, R.M.; Honkala, B.H., eds. Silvics of North America; vol. 1, conifers. Agricultural handbook 654. Washington, DC: USDA Forest Service; 1994:413–424.

Patterson, M.T.; Rundel, P.W. Seasonal physiological responses of ozone stressed Jeffrey pine in Sequoia National Park, California. In: Olson, R.K.; Lefohn, A.S., eds. Effects of air pollution on western forests. Pittsburgh, PA: Air Waste Manage. Assoc.; 1989:419–428.

Patterson, M.T.; Rundel, P.W. Stand characteristics of ozone-stressed populations of Pinus Jeffreyi (*Pinaceae*): extent, development, and physiological consequences of visible injury. Am. J. Bot. 82(2):150–158; 1995.

Pearson, M.; Mansfield, T.A. Interacting effects of ozone and water stress on the stomatal resistance of beech (*Fagus sylvatica* L.). N. Phyt. 123:351–358; 1993.

Peterson, D.L.; Arbaugh, M.J.; Robinson, L.J. Regional growth changes in ozone-stressed ponderosa pine (*Pinus ponderosa*) in the Sierra Nevada, California, USA. Holocene 1:50–61; 1991.

Reich, P.B. Effects of low concentrations of ozone on net photosynthesis, dark respiration, and chlorophyll contents in aging hybrid poplar leaves. Plant Phys. 73:291–296; 1983.

Reich, P.B. Quantifying plant response to ozone: a unifying theory. Tree Phys. 3:63–91; 1987.

Runeckles, V.C.; Chevone, B.I. Crop responses to O_3. In: Lefohn, A., ed. Surface-level O_3 exposure and their effects on vegetation. Chelsea, UK: Lewis Publisher; 1992:189–270.
Sasek, T.W.; Richardson, C.J. Effects of chronic doses of ozone on loblolly pine: photosynthetic characteristics in the third growing season. For. Sci. 35(3):745–755; 1989.
Skarby, L.; Troeng, E.; Bostrom, C-A. Ozone uptake and effects on transpiration, net photosynthesis, and dark respiration in Scots pine. For. Sci. 33:801–808; 1987.
Smith, W.H. Forest and air quality. J. For. 83:82–92; 1974.
Takemoto, B.K.; Bytnerowicz, A.; Dawson, P.J.; Morrison, C.L.; Temple, P.J. Effects of ozone on *Pinus ponderosa* seedlings: comparison of response in the first and second growing seasons of exposure. Can. J. For. Res. 27:23–30; 1997.
Taylor, G.E.; Pitelka, L.F.; Clegg, M.T., eds. Ecological genetics and air pollution. New York: Springer-Verlag; 1991.
Temple, P.J. Injury and growth of Jeffrey pine and giant sequoia in response to ozone and acidic mist. Environ. Exp. Bot. 28(4):323–333; 1988.
Temple, P.J.; Miller, P.R. Foliar injury and radial growth of ponderosa pine. Can. J. For. Res. 24:1877–1882; 1994.
Temple, P.J.; Miller, P.R. Seasonal influences on ozone uptake and foliar injury to ponderosa and Jeffrey pine at a southern California site. Gen. Tech. Rep. PSW-167. Albany, CA: USDA Forest Service, Pacific Southwest Station; (in press).
Temple, P.J.; Riechers, G.H.; Miller, P.R. Foliar injury responses of ponderosa pine seedlings to ozone, wet and dry deposition, and drought. Environ. Exp. Bot. 32(2):101–113; 1992.
Teskey, R.O.; Hinckley, T.M. Influence of temperature and water potential on root growth of white oak. Plant. Phys. 52:363–369; 1981.
Thornley, J.H.M. Mathematical models in plant physiology. London: Academic Press; 1976.
Tingey, D.T.; Hogsett, W.E. Water stress reduces ozone injury via stomatal mechanism. Plant Phys. 77:944–947; 1985.
Tingey, D.T.; Rodecap, K.D.; Lee, E.H.; Moser, T.J.; Hogsett, W.E. Ozone alters the concentrations of nutrients in bean tissue. Angew. Bot. 60:481–493; 1986.
Trebst, A. Dynamics in photosystem II structure and function. In: Schulze, E-D.; Caldwell, M.M., eds. Ecophysiology of photosynthesis. New York: Springer-Verlag; 1994:3–16.
Unsworth, M.H.; Bisce, P.V.; Pinckney, H.R. Stomatal responses to SO_2. Nature 239:458; 1972.
Wallin, G.; Skarby, L. The influence of ozone on the stomatal and non-stomatal limitation of photosynthesis in Norway spruce, *Picea abies* (L.) Karst, exposed to soil moisture deficit. Trees 6:128–136; 1992.
Waring, R.H.; Franklin, J.F. Evergreen forests of the Pacific Northwest. Science 204 1380–1386; 1979.
Waring, R.H.; Schlesinger, W.H. Forest ecosystems. New York: Academic Press; 1985.
Weber, J.A.; Clark, C.S.; Hogsett, W.E. Analysis of the relationships among O_3 uptake, conductance, and photosynthesis in needles of *Pinus ponderosa.* Tree Phys. 13:157–172; 1993.
Weiser, G.; Havranek, W.M. Ozone uptake in the sun and shade crown of spruce: quantifying the physiological effects of ozone exposure. Trees 7:227–232; 1989.
Wellburn, A.R. Air pollution and acid rain: the biological impact. Burnt Mill, UK: Longman Sci. Tech. Publ.; 1988.
Wellburn, A.R. Tansley review no. 24. Why are atmospheric oxides of nitrogen usually phytotoxic and not alternative fertilizers? N. Phyt. 115:395–429; 1990.
Yoder, B.J.; Ryan, M.G.; Waring, R.H.; Schoettle, A.W.; Kaufmann, M.R. Evidence of reduced photosynthetic rates in old trees. For. Sci. 40(3):513–527; 1994.

7. Temporal Changes in Crown Condition Indices, Needle Litterfall, and Collateral Needle Injuries of Ponderosa and Jeffrey Pines

P.R. Miller and J. Rechel

Introduction

Evaluating Ozone Injury to Pines

The green leafy canopy of a pine is generally described as the crown of the tree. The health of tree crowns is essentially an estimate of the amount of foliage actually present in relation to the potential amount that could be present under ideal conditions. The second major feature is the presence or absence of injury symptoms on each age increment of foliage present. Still another important feature is the length of the live crown compared with the total length of the crown (including both live and dead branches), sometimes expressed as percentage live crown. These general features of crown health have been interpreted separately or combined into indices to describe crown health changes with time. The evolution of procedures that have been used to describe crown injury to ponderosa and Jeffrey pines was described by Miller et al. (1996b). The three indices of crown injury that have been used are referred to as the forest pest management (FPM) index (Pronos et al. 1978), the ozone injury score (OIS) (Miller 1973), and the ozone injury index (OII; formerly known as the Eridanus Injury Index) (Durisoe 1988). The FPM method has one major feature in common with the other procedures, namely, a determination of the youngest age cohort (whorl) of needles that have chlorotic mottle symptoms, the definitive symptom of ozone injury.

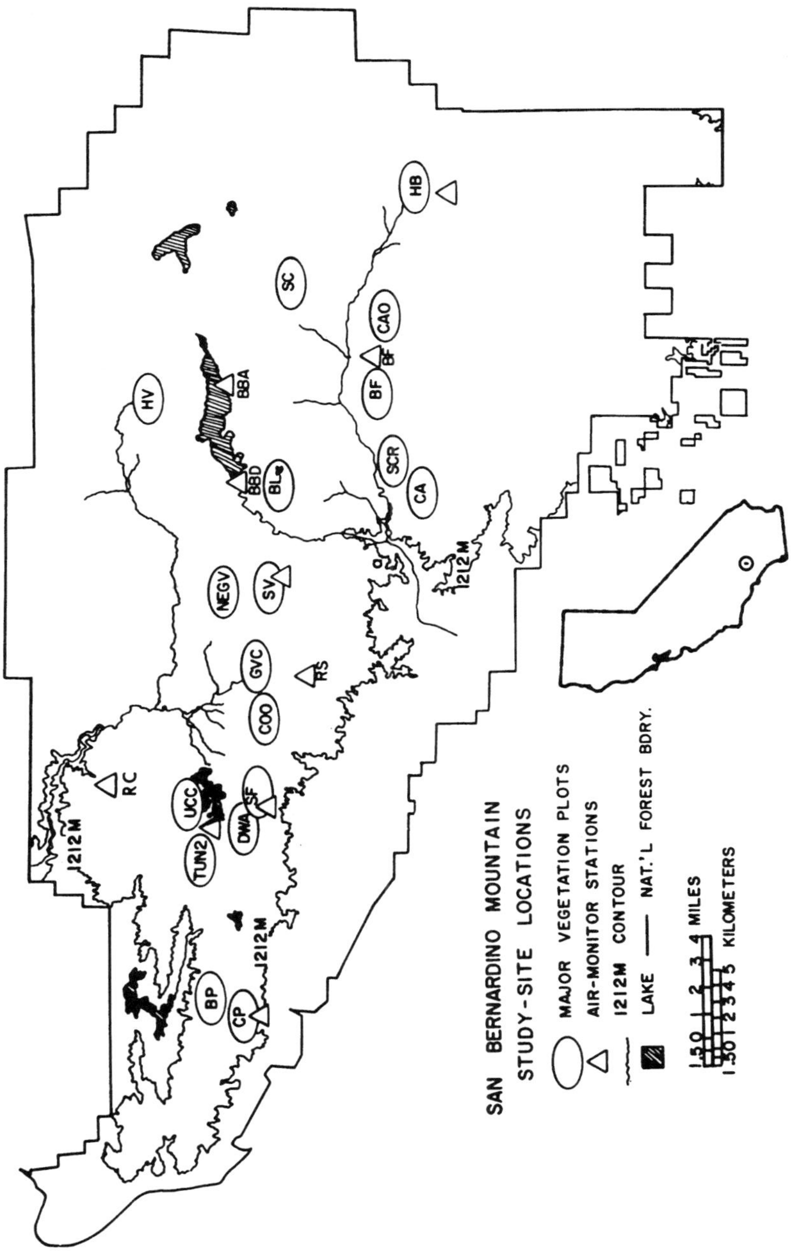

Figure 7.1. Map of the San Bernardino National Forest showing locations of forest plots for monitoring changes in ozone injury to ponderosa and Jeffrey pines, indicated by two- or three-letter abbreviations enclosed inside an oval symbol. Other details about plot names and location are shown in Table 7.2.

Table 7.1. Plot Average of the Number of Annual Needle Whorls Retained as Determined by Annual Observations

	Average of needle whorls for each year							
Plot name (abbreviation)	1973	1974	1975	1976	1977	1978	Mean	SD
Camp Paivika (CP)	2.4	2.3	1.7	2.4	1.8	1.5	2	0.4
Breezy Point (BP)	2.4	2.3	1.6	2.1	1.9	1.7	2	0.3
Dogwood A (D.A.)	2.8	2.6	1.9	2.6	1.9	1.8	2.2	0.4
Sky Forest (SF)	2.4	2.4	2	2.5	2.3	1.8	2.2	0.2
Camp O'Ongo (COO)	2.4	2.2	1.4	2.2	1.9	1.4	1.9	0.4
University Conf. Cen. (UCC)	2.5	2.5	1.9	2.3	1.9	1.9	2.2	0.3
Camp Angeles (CA)	3.4	2.3	2.1	2.4	1.8	2.3	2.4	0.5
Deerlick (DL)	na[a]	3	2.8	2.8	2.8	2.5	2.8	0.1
Tunnel 2 (T2)	2.8	2.8	2.2	2.7	2.5	2.2	2.5	0.3
Schneider Creek (SC)	2.2	1.9	1.9	2.2	2.2	2.1	2.1	0.1
Barton Flats (BF)	2.9	2.4	2.6	2.7	2.6	2.4	2.6	0.2
Green Valley Creek (GVC)	3.2	3	2.8	2.9	2.8	2.5	2.9	0.2
Northeast Green Valley (NEGV)	3.8	3.8	4.2	3.8	3.9	3.6	3.9	0.2
Camp Osceola (CAO)	3.2	3.1	3.3	3.7	2.8	2.8	3.1	0.3
Bluff Lake (BL)	4.1	4.1	3.8	3.8	4.3	4.1	4	0.2
Heart Bar (HB)	4.8	4.5	4.3	4.3	4.4	4.1	4.4	0.2
Holcomb Valley (HV)	5.5	5.4	5.6	5.5	6.4	5.2	5.6	0.4
Sand Canyon (SC)	4.4	4.9	4.8	4.9	5.3	4.9	4.9	0.3
Mean	3.1	3.1	2.8	3.1	3	2.7		
Standard Deviation	1.2	1	1.2	1	1.3	1.1		

[a]na, No data available.

Accordingly, the worst injury is rated as 0 and the least was set at 5, assuming that five youngest whorls are more productive and valuable to retain.

Regressions of the indices from sample populations for FPM and OIS and for FPM and OII resulted in r^2 values ≥ 0.95 (Arbaugh et al., in press). This lends great confidence to the interchangeable use of these indices.

Periodic assessments of crown condition have been useful for interpreting the trend of ozone injury to pines. In the San Bernardino Mountains, the sampling sites (tree plots) were established in the pine and mixed conifer forest types along a west-to-east gradient of decreasing ozone exposure (Fig. 7.1). Plot names (abbreviations) and location are also indicated in Tables 7.1 and 7.2.

Ozone Injury to Companion Tree Species

At the beginning of the monitoring work in San Bernardino Mountain plots, we determined that we should try to evaluate ozone injury symptoms to all tree species in each plot to determine the most sensitive indicators of year-to-year changes in ozone injury. The general features of the OIS were applied to the

Table 7.2. Changes in the FPM Scores at 17 San Bernardino Mountain Plots Between 1974 and 1988

Plot name (abbreviation)	FPM 1974	FPM 1988	Elevation (M)	Latitude	Longitude
Camp Paivika (CP)	0.63	0.32	1,600	34 14 05	117 19 25
Breezy Point, (BP)	0.6	0.7	1,525	34 16 20	117 18 30
Dogwood A (DA)	0.64	1.06	1,725	34 14 15	117 13 00
Sky Forest (SF)	0.39	0.99	1,705	34 14 10	117 10 55
Camp O'Ongo (COO)	0.28	1.15	1,920	34 14 00	117 08 20
University Conf. Cen. (UCC)	0.64	1.31	1,610	34 16 15	117 10 20
Camp Angeles (CA)	0.94	1.39	1,830	34 08 30	117 58 40
Deerlick (DL)	0.89	1.63	2,090	34 12 20	117 04 25
Tunnel 2 (T2)	1	1.67	1,640	34 16 10	117 13 15
Schneider Creek (SC)	0.19	1.67	1,830	34 08 55	116 56 50
Barton Flats (BF)	0.98	1.74	1,900	34 10 10	116 54 25
Green Valley Creek (GVC)	1.02	1.96	1,950	34 14 00	117 05 45
NE Green Valley (NEGV)	2.13	2.19	2,328	34 14 55	117 03 15
Camp Osceola (CAO)	1.02	2.39	2,135	34 09 40	116 51 20
Bluff Lake (BL)	1.88	2.65	2,320	34 13 30	116 58 10
Heart Bar (HB)	2.63	3.52	2,160	34 09 55	116 45 30
Holcomb Valley (HV)	3.57	3.99	2,225	34 17 45	117 54 55

description of ozone injury to sugar pine, white fir, and incense cedar present in the same plots with pines along the ozone gradient (Miller et al. 1977). White fir appeared in 15 of 19 plots, sugar pine in six of 19 plots, and incense cedar in five of 19 plots. The white fir scores for 1973–1975 were the lowest at Sky Forest and Dogwood (more injury) and highest at Northeast Green Valley and Holcomb Valley (very slight or no injury). Thus white fir did show a gradient of foliar injury similar to pines, but because it was only moderately sensitive compared with pine and was not present at every location along the gradient, its use as a bioindicator was de-emphasized. Incense cedar and sugar pine did not show any injury trends at the fewer locations where they were present, and they were much less injured than ponderosa pine and Jeffrey pine or white fir. California black oak was present at 15 of 19 plots. A visual rating scale was used to rate injury to leaves of the upper and lower crown, and the two estimates for each tree were added to produce an injury rating that ranged from 0 to 8 (Miller et al. 1980). Oaks also showed a gradient of ozone injury that ranged from 4.8 at one of the highest ozone exposure plots to 8.0 at the lowest-exposure plots. These ratings were an average of annual scores from 1975 to 1978 (Miller et al. 1980). The oaks were less sensitive to ozone than ponderosa pine and Jeffrey pine, and the deciduous leaves were useful mostly to single out particular years or particular places with higher ozone exposure. These initial observations of these companion tree species led to the decision to focus on ponderosa and Jeffrey pines as the most ubiquitous and most ozone-sensitive tree species.

Changes in Ponderosa and Jeffrey Pine Crown Condition Since the Early 1970s

Use of the Ozone Injury Score

The OIS method was used was between 1973 and 1988. Because a combination of hands-on inspection and fields of view with binoculars are used to count needle whorl number and estimate needle condition and needle length, we were concerned about the reliability of these repeated subjective measurements. Consecutive year determinations of tree condition at individual plots are shown as the plot average ($n \geq 50$ trees) for the total number of annual needle whorls retained (Table 7.1). The year-to-year variation at the same plot is relatively small compared with the expected wider variation between plots experiencing widely differing ozone exposures. During the 1973–1978 period, average needle retention did not change much from year to year. It appears that one of the most important components of the OIS, namely, needle whorl retention, could be estimated reliably, and thus OIS was a suitable method for tracking injury.

Application of the Forest Pest Management Score

Between 1973 and 1988, the FPM score (0 = worst injury) shows a distinct improvement of crown health at all but one sample plot (Camp Paivika) which has the greatest ozone exposure (Table 7.2). The ozone injury score also showed this improvement between 1974 and 1988 (Miller et al. 1989), which is coincident with a decrease in ozone exposure (Miller and McBride 1989).

Spatial Distribution of FPM Injury Classes Between 1974 and 1988

The FPM injury index was used to develop two spatial models of ozone injury trends in the San Bernardino Mountains for 1974 and 1988. The method allowed the visualization of ozone injury trends that are otherwise difficult to determine at the individual sites.

A geographical information system (GIS) was used to develop a spatial model of injury trends by using a slope map created from U.S. Geological Survey (USGS) 7.5-minute quadrangles. The values used in the analysis to estimate injury trends were derived from a regression equation using elevation, latitude, and longitude mapped variables. The mapped area included in the analysis covered the extent of conifer vegetation types displayed on a three-dimensional map of the San Bernardino Mountains (Fig. 7.2).

Injury trends, expressed as the 1974 and 1988 FPM indexes, for ponderosa pine and Jeffrey pine are shown in Figure 7.3.[1] In 1974, the highest injury (values ≤ 2.25) occurred in the lower to midelevations, generally between 1,000 and 1,800 m. In 1988, the highest ozone injury occurred at elevations between 1,000 and 1,400 m.

---

[1]Figure 7.3 will be found in the color insert following page 78.

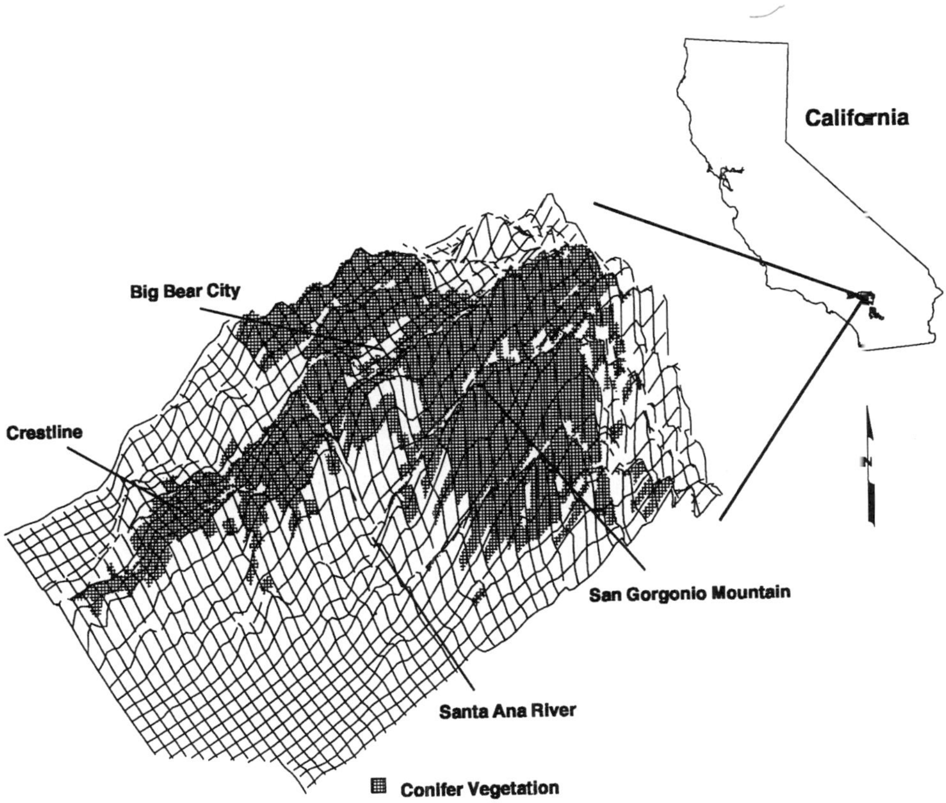

Figure 7.2. Three-dimensional representation of the terrain of the San Bernardino Mountains with the shaded area representing the extent of the pine and mixed conifer forest types.

The southwest-to-northeast orientation of the decreasing pine injury is consistent with the general direction of the wind fields caused by the onshore flow of air across the South Coast Air Basin and the San Bernardino Mountains (see Watson et al., this volume). Pines at elevations greater than 2,600 m are considered to have much lower ozone exposure because vertical mixing has diluted ozone concentrations (see Bytnerowicz et al., this volume, Fig. 11.1), and besides, the principal conifer species between there and treeline, namely, lodgepole pine (*Pinus contorta* Dougl. ex. Loud) and limber pine (*Pinus flexilis* James), are not known to be ozone-sensitive.

The overall trend is a decrease in ozone injury in the pine and mixed conifer forest types. Other than measured decreases in ozone injury over these 14 years, other possible reasons for the differences in patterns may be that terrain complexity, and not just elevation, is also an important factor determining ozone injury trends and that this factor may not be evident when ozone injury values are low. Clearly, these maps (Fig. 7.3) are exploratory representations, and additional analysis with GIS methodology will be helpful.

Crown Condition (FPM) Comparisons from 1973 to 1994/1995 at a Subset of Plots

The OII was applied to a subset of three of the plots listed in Table 7.1, and the FPM score was used as the common denominator to link earlier OIS evaluations to the most recent OII determinations. Camp Paivika (CP) was evaluated in 1995; Camp Angeles (CA) and Heart Bar (HB) were done in 1994. CP and CA each represent higher levels of chronic ozone exposure, whereas HB represents slight or very low chronic ozone exposure. The changes in the FPM score for these three locations are shown in Figure 7.4. These data represent a random sample of 50 trees from the original populations at CP and HB and 39 of the original 50 trees at CA. It shows that the mean of the age of the youngest whorl presenting a detectable level of chlorotic mottle at CA and HB generally increased between 1978 and 1988 and declined slightly between 1988 and 1994. The trees at CP have changed little over time. The magnitudes of difference among the three plots have remained about the same over the 1974–1994 time span.

Evidence of increased stem growth at sites with higher crown injury obtained from tree ring studies during this period (see Arbaugh et al., this volume) may mean that even though foliar injury is still obvious, there are other physiological mechanisms or selection processes that may be compensating for ozone injury. At one study site in the western section of the San Bernardino Mountains with high ozone exposure, there have actually been some increases of photosynthetic surface area because a significant number of trees retain the same or more whorls in recent years compared with the 1970s. This hypothesis is supported by needle whorl retention data acquired in 1992 from 219 ponderosa pines distributed in nine plots located in the heavily exposed area in the vicinity of Dogwood Campground Blue Jay (see DWA location on Fig. 7.1). Since 1973, 40.7 percent of the trees showed no change in numbers of whorls retained, 42.7 percent had more whorls, and 16.6 percent had fewer whorls (Fig. 7.5).

Annual Changes Determined with the OII

The usefulness of the OII for tracking year-to-year changes was tested in annual evaluations from 1991 to 1994 of 130 ponderosa or Jeffrey pines distributed in three plots in the Barton Flats area (Miller et al. 1996a). The OII ranges from 0 to 100, with 0 meaning the least injury. Over the course of the 1991–1994 study, ozone continued to cause injury each year. The frequency distributions of the 130 sampled ponderosa or Jeffrey pines for each year of measurement shows that a shift toward higher numbers in higher injury ranges began in 1992 and remained the same in 1993 and 1994 (Fig. 7.6). The corresponding OII means for all sample trees were 1991, 38.3; 1992, 47.5; 1993, 47.5; and 1994, 49.2. The only significant difference between years ($P < .05$), detected with a paired t-test, was between 1991 and 1992. This change was related to a higher level of ozone flux to foliage in 1992, the first year with favorable soil moisture following several preceding years of drought (Temple and Miller, in press). Readily available soil moisture

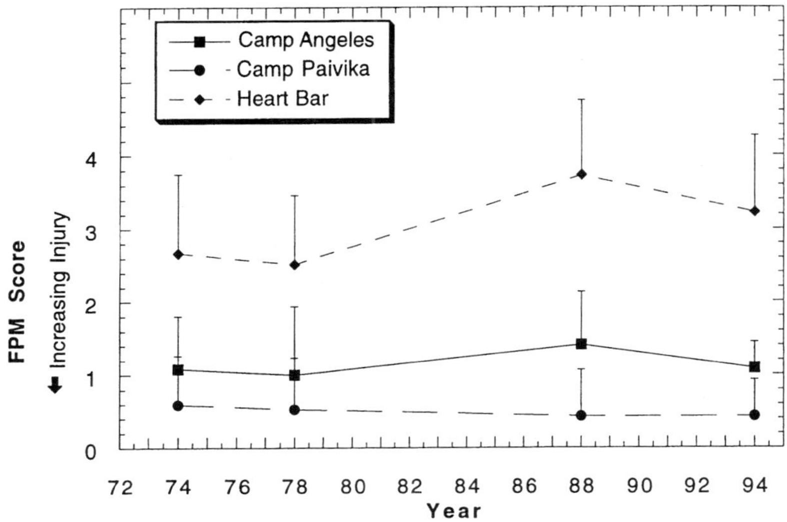

Figure 7.4. Changes in the forest pest management score for ozone injury between 1974 and 1994 to populations of ponderosa and Jeffrey pines at three of the plot locations shown in Figure 7.1. The trees at individual plots did not change significantly over time, and the difference between the slightly affected population at Heart Bar and the two more injured populations at Camp Angeles and Camp Paivika remained the same over the 1974–1994 period.

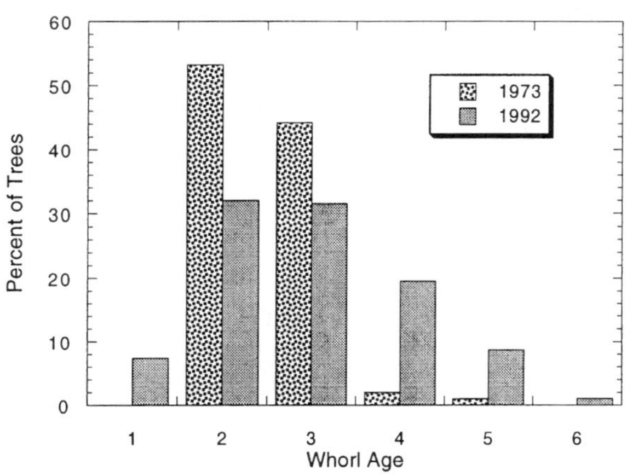

Figure 7.5. Changes from 1973 to 1992 in the average number of annual needle whorls retained by a sample of 219 ponderosa pines distributed in nine plots located in an area of severe ozone exposure (see DWA location on Figure 7.1). More whorls, or an improvement of crown health, is represented by 42.7% of the sample, whereas 40.7% showed no change and 16.6% had fewer whorls.

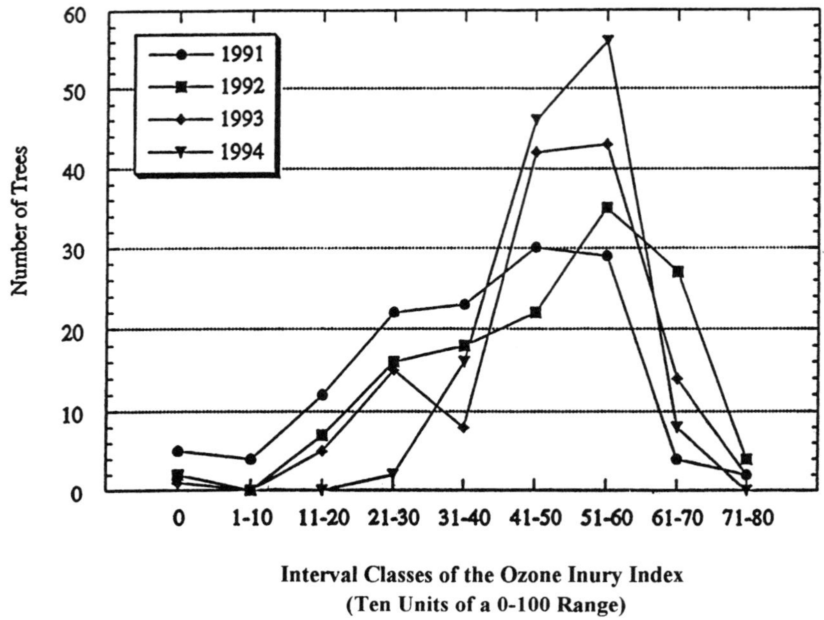

Figure 7.6. Annual changes of the distribution in ten increment intervals of the ozone injury index (OII) of 130 sample ponderosa and Jeffrey pines near Barton Flats (see BF location on Figure 7.1) in 1991–1994. Beginning in 1992, there was a shift to larger numbers of trees in higher injury intervals due to a larger flux of ozone to foliage in the early summer of 1992.

increases stomatal conductance results in greater gas exchange (i.e., ozone uptake by foliage).

These results from an area midway between CA and HB (Fig. 7.1), an area with moderate ozone exposure, confirm that ozone injury has remained fairly constant from year to year in recent years, with the exception of 1992, which was higher.

Amounts and Seasonal Pattern of Needle Litterfall

Seasonal Accumulation of Litter

The leaf litter collectors were 0.5-m × 0.5-m wooden frames with a fine-mesh nylon screen bottom. They were fixed in place and made level with dowels at each corner. The dowels were driven into the ground, and clamps at each corner were tightened around the dowel after the collector frame was level. The collectors were placed at a 1.0–1.2-m distance from the trunk of the pine being sampled. Because leaf litterfall is often coincident with windy periods, we could not guarantee that litter arriving in the collector originated only from the crown of the tree under which it was placed. At each of three forest plots at Barton Flats, there were ten litter collectors placed under pines with OII scores ranging from 10 to 80 (approximate, see Fig. 7.6). Litter was collected four times between Julian dates 180 and 325 in 1992 and five times in 1993 and 1994 between Julian dates 180 and 340. Bark and twig debris and needles or leaves of other species were excluded, and the ponderosa or Jeffrey pine needles were weighed air dry (Miller et al. 1996a).

We collected leaf litter in 90 collectors in 1992, 1993, and 1994 at Barton Flats plots under a subsample of 130 ponderosa or Jeffrey pines, which had been evaluated using the OII method (Miller et al. 1996a). Litterfall amount was the highest in 1992 at 31.2 g/collector. In Figure 7.4, there was a transition to higher levels of crown injury from 1991 to 1992. In 1993, it was 20.3 g/collector, and in 1994 it was 18.8 g. This result suggests that annual leaf litterfall amounts increase in years of higher ozone exposure and uptake (1992) but seem to stabilize at lower levels between these exceptional years. Certainly, other variables influence litterfall (e.g., the magnitude and timing of precipitation and freezing temperatures, or the presence or absence of cone crops); however, the increased flux of ozone in 1991 seemed to be a feature unique to that year.

Regardless of year, the first measurable litter amount (4–5 g) did not appear until Julian dates 250–260. The accumulation was most rapid between Julian dates 260 and 330–340. In 1994, when the total amount of litter was the least, the rate of accumulation was lower between Julian dates 300 and 340.

We tested the data to see if there was a relationship between the OII of trees under which litter collectors were placed and the amount of litterfall (Miller et al. 1996a). The OII values ranged between 10 and 80, and the litter amounts ranged from 5 to 60 g; however, the slopes were near zero and we could not demonstrate any relationship to OII with these data. Arkley and Glauser (1980) found higher litterfall under trees with moderate crown injury, less under more healthy trees, and the least under severely injured ponderosa pines. We believe that our collectors were gathering litter from more than one tree crown during windy periods, so the influence of the single crown directly above the collector was reduced. Further studies of stomtal conductance as it influences ozone uptake, photosynthesis, and visible needle injury are presented in Grulke (this volume).

Crown Health and Collateral Needle Injuries

Other abiotic agents besides ozone and biotic agents also influence needle condition and possibly retention. For example, necrotic flecking occurs following the first overwinter period of a new needle whorl and intensifies with each passing winter (Miller et al. 1996a). The histological evidence of winter fleck is extensive necrosis of mesophyll tissue immediately under the necrotic area visible at the epidermis. The injury extends as deep as the endodermis but does not show clear involvement of other cell types, whereas ozone injury affects mesophyll cells randomly and the visual evidence at the epidermis is not necrosis of epidermal

Table 7.3. 1992–1995 Evaluation of Winter Fleck on the Needle Whorl Formed in the Previous Year and Subjected to One Overwinter Period

	1992	1993	1994	1995
Percentage of sampled branches with				
≤ 24.9% winter fleck	62.3	72.0	22.8	30.5
≥ 25% winter fleck	1.9	22.0	0.8	0.2
Precipitation (cm)				
Sept–Dec (previous year)	6.5	16.7	7.7	7.3
Jan–April (current year)	36.1	77.4	28.7	64.5
Temperature °C, previous year				
Sept–Dec minimum	–7.3	–8.1	–5.2	–8.3
Sept–Dec maximum	19.8	25.5	27.7	24.0
Temperatue °C, current year				
Jan–April minimum	–5.9	–10.1	–7.7	–7.0
Jan–April maximum	21.5	20.4	20.7	18.2

cells but only a chlorotic area associated with the injury to underlying mesophyll cells (Miller and Evans 1974). There is no evidence to indicate the possible effect that winter fleck may have on needle photosynthesis; however, it would seem to decrease needle photosynthetic surface and hence reduce carbon fixation.

The upper surface needle fleck symptom (winter fleck) has been studied with relation to possible abiotic causal factors. The 1992–1995 evaluation of winter fleck on the needle whorl formed in the previous and subjected to one overwinter period (Table 7.3) shows that the needles evaluated in 1993 had the most winter fleck. An examination of precipitation and temperature (max/min) for the overwinter periods does not show any obvious connection between these weather variables and incidence of winter fleck except for the lowest minimum temperature and highest precipitation amount for the preceding January–April period (Miller et al. 1996a). The conditions contributing to winter fleck may be several and their interactions complex.

Localized collapse of needle tissue about one-third of the distance from the needle tip in the newest needle whorl is a symptom (Miller et al. 1996a) that seems to have no definite abiotic or biotic causal agent associated with it. All the needle tissue distal to that point becomes necrotic and drops away. The incidence of this symptom is always highest on the current-year whorl, suggesting that it originates only on elongating needles and that there are lower amounts in older whorls because some of the remaining broken needles abscise over time. At Barton Flats in 1995, the percentage of the remaining whorls with any needles with broken tips was 1995, 26.3%; 1994, 15.0%; 1993, 13.2%; 1992, 12.0%; 1991, 8.1%; and 1990, 0%.

Chlorotic bands not associated with broken needles represent a third symptom presumed to have an abiotic cause, but the incidence of whorls with any needles having chlorotic bands never exceeded 5% at Barton Flats (Miller et al. 1996a).

The separate evaluation of the presence of trace amounts of either scale (*Nuculaspis* sp.) or weevil (*Dichelonyx* sp.) injury (the detection threshold was

Table 7.4. Prevalence of Insect-Caused Needle Injury Attributed to Scale (*Nuculaspis* sp.) or Weevil (*Dichelonyx* sp.) Infestation[a]

Insect	Year 1991	1992	1993	1994	1995
			Percentage		
Scale (*Nuculaspis* sp.)	2.3	2.7	2.4	2.3	0.3
Weevil (*Dichelonyx* sp.)	20.3	25.5	26.2	18.3	2.5

[a]The incidence of sampled whorls with these injuries (≤24.9% of needles/whorl affected) is indicated for needle whorls still remaining on trees at three Barton Flats plots in 1995 (Miller et al. 1996a).

incidence ≤24.9% of needles per whorl) (Table 7.4) shows that scale infestation is not frequent and that weevil injury occurs in the first and second years and remains between 20 and 30% incidence in subsequent years (Miller et al. 1996a).

Summary and Conclusions

The need to assess the status of forest health has led to the development and refinement of field methods that allow for the reliable repeated observations of tree crown changes. During the 1980s, in particular, there was intensive development, testing, and application of methods for the evaluation of forest decline in Europe and the United States (Innes 1993a). Methods associated with the assessment of trees in forests have relied heavily on crown transparency (density) and crown discoloration (Innes 1993b); unfortunately, these are nonspecific indicators when it comes to distinguishing the effects of air pollutants and other environmental stresses. In addition, these measures are subjective and susceptible to serious error. It was also recognized that sample design (i.e., density of tree plots and numbers of trees per plot) could seriously limit place-to-place comparisons (Innes and Boswell 1990; Kohl et al. 1994). These shortcomings have prompted research on improved strategies for sampling in forest monitoring programs (Ghosh and Innes 1996) and better operational methods leading to error estimates for crown condition estimates (Ghosh and Innes 1995). The nonspecific nature of crown condition descriptors (density and discoloration) contributed importantly to the confusion surrounding the causal nature of "waldsterben" or "new forest decline" (Kandler and Innes 1993; Skelly and Innes 1994). Fortunately, none of these problems with crown evaluation and sampling presented themselves in the San Bernardino Mountain studies.

Although there are certainly multiple stresses involved, we learned from early fumigation studies that ozone was the only cause of the chlorotic mottle symptom and the subsequent abscission of older needles (Miller et al. 1963). Advanced defoliation without the presence of chlorotic mottle disqualified ozone as the causal agent. Each of the three crown evaluation methods relied on the presence of chlorotic mottle and loss of older needle whorls as determined by hand-held branch samples or close-up inspection of branches in the lower crowns of trees, sometimes aided by wide-field binoculars. Some of the possible variation of this visual inspection was limited by using a single observer from 1974 to the present. The sampling strategy was suitable because no less than 50 trees per plot were measured in plots that were located along a known gradient of decreasing ozone exposure (Miller et al. 1977). Also, the work was confined to a single mountain range, so comparison between distant points (e.g., parts of a region such as central Europe) were not required.

Three crown evaluation methods have been used in California to estimate chronic ozone injury to pines. In the San Bernardino Mountains, only two of these methods have been used, originally the OIS and recently the OII. The main advantage of the index is that all injury estimates are obtained from hands-on inspection of five branches collected from each tree. The third method, the FPM, has been used mainly in the Sierra Nevada. It also uses hands-on inspection of branches.

The FPM method has been used in this chapter to relate early results using the OIS to more recent estimate of injury to the same trees using the OII. The common feature that the FPM method uses to relate these estimates is the youngest needle whorl that has chlorotic mottle symptoms caused by ozone. By this measure, it is clear that there was less crown injury in 1988 compared with 1974 at all but one of 17 monitoring plots in the San Bernardino Mountains. This change was given better spatial representation by a simple GIS analysis. The one tree plot (CP) that did not decrease is located at the western portion of the ozone exposure gradient where ozone exposure is the highest. Using the FPM to represent the change in crown condition at three plots representing both ends and the approximate center of the gradient between 1974 and 1994 shows that the relationship between plots has remained about the same in 1974, 1988, and 1994; and over the entire time span the differences between times were small for individual plots. The OII values obtained for three plots at Barton Flats between 1991 and 1994 showed a very gradual increase in the injury index. Most of this increase happened in 1992 when ozone flux to foliage was exceptionally high. The first year of increased injury may have been associated with a larger amount of leaf litterfall; however, other factors that could influence litterfall are timing of precipitation or magnitude and timing of freezing temperatures.

Although a substantial range of ozone injury to white fir and California black oak can be observed along a gradient of decreasing ozone exposure in the San Bernardino Mountains (Miller et al. 1977), we settled on the use of ponderosa and Jeffrey pines as the most sensitive and convenient bioindicators of ozone effects to crowns.

Several other needle injuries to pine, including winter fleck, breaking of current-year needles during elongation, chlorotic bands, and minor damage due to scale insects or weevils, were also evaluated between 1991 and 1995. None of these injuries is as important singly or collectively as ozone injury, which results in complete abscission of needles in exposed whorls. The most ubiquitous of these injuries is winter fleck, and although the amount was particularly large in one year when winter precipitation was the highest and minimum temperatures the lowest, it could not be proved conclusively that a causal relationship with precipitation amount or temperature minimums was present during the first overwinter period.

Finally, the severity of ozone exposure in the San Bernardino Mountains is certainly less in recent years than in the 1970s, but it still seems to be sufficient, along with periodic droughts, to maintain continuing stress on ponderosa and Jeffrey pines. Therefore, we recommend that the trees in plots be sampled every 3–5 years by using the OII method (Milller et al. 1996b) to obtain quality-ensured assessments of the status of chronic ozone injury to ponderosa and Jeffrey pines. This measurement will help assess the sustainability of the forest ecosystem and evaluate the effectiveness of air pollution control measures in the South Coast Air Basin.

References

Arbaugh, M.J.; Miller, P.R.; Carroll, J.; Takemoto, B.; Procter, T. Relationship of ambient ozone with injury to pines in the Sierra Nevada and San Bernardino Mountains of California, USA. Environ. Pollut. (in press).

Arkley, R.J.; Glauser, R. Effects of oxidant air pollutants on pine litter-fall and the forest floor. In: Miller, P.R., tech. coord. Effects of air pollutants on Mediterranean and temperate forest ecosystems. Gen. tech. rep. PSW-43. Berkeley, CA: Pacific Southwest Station; 1980:225.

Duriscoe, D.M. Methods of sampling for *Pinus ponderosa* and *Pinus jeffreyi* for the evaluation of oxidant induced foliar injury. Final report. Three Rivers, CA: Eridanus Research Associates; 1988.

Ghosh, S.; Innes, J.L. Combining field and control team assessments to obtain error estimates for surveys of crown condition. Scand. J. Forest Res. 10:264–270; 1995.

Ghosh, S.; Innes, J.L. Comparing sampling strategies in forest monitoring programs. Forest Ecol. Manage. 82:231–238; 1996.

Innes, J.L. Forest health: its assessment and status. Wallingsford, Oxon, UK: CAB International; 1993a.

Innes, J.L. Methods to estimate forest health. Silva Fennica 27:145–157; 1993b.

Innes, J.L.; Boswell, R.C. Reliability presentation and relationships among data from inventories of forest condition. Can. J. Forest Res. 20:790–799; 1990.

Kandler, O.; Innes, J.L. Air pollution and forest decline in central Europe. Environ. Pollut. 90:171–180; 1993.

Kohl, M.; Innes, J.L.; Kaufmann, E. Reliability of differing densities of sample grids used for the monitoring of forest conditions in Europe. Environ. Monit. Assess. 29:201–220; 1994.

Miller, P.R. Oxidant-induced community change in a mixed conifer forest. In: Air pollution damage to vegetation. Adv. Chem. Ser. 122:101–117; 1973.

Miller, P.R.; Evans, L.S. Histopathology of oxidant injury and winter fleck injury on needles of western pine. Phytopathology 64(6):801–806; 1974.

Miller, P.R.; Parmeter, J.R.; Taylor, O.C.; Cardiff, E.A. Ozone injury to the foliage of *Pinus ponderosa*. Phytopathology 53:1072–1076; 1963.

Miller, P.R.; McBride, J.R. Trends of ozone damage to conifer forests in the western United States, particularly southern California. In: Bucher, J.B.; Bucher-Wallin, I., eds. Proceedings of the 14th international meeting for specialists in air pollution effects on forest ecosystems. IUFRO P2.05, Interlaken, Switzerland, October 2–8, 1988. Birmensdorf, Switzerland: Eidengenössische Anstalt für das forstliche Versuchwiesen (EAFV); 1989; 61–68.

Miller, P.R.; Kickert, R.N.; Taylor, O.C.; Arkely, R.J.; Cobb, F.W., Jr.; Dahlsten, D.L.; Gersper, P.J.; Luck, J.R.; McBride, J.R.; Parmeter, J.R., Jr.; Wenz, J.M.; White, M.;

Wilcox, W.W. Photochemical oxidant air pollutant effects on a mixed conifer forest ecosystem—a progress report, 1975–1976. EPA 600/3-77-104. Corvallis, OR: U.S. Environmental Protection Agency; 1977.

Miller, P.R.; Longbotham, G.J.; Van Doren, R.E.; Thomas, M.A. Effect of chronic oxidant air pollution exposure on California black oak in the San Bernardino Mountains. In: Proceedings of the symposium on ecology management and utilization of California oaks. Claremont, CA: Pacific Southwest Station; Gen. tech. rep. PSW 44. 1980:220–229.

Miller, P.R., McBride, J.R.; Schilling, S.L.; Gomez, A.P. Trend of ozone damage to conifer forests between 1974 and 1988 in the San Bernardino Mountains of southern California. In: Olson, R.K.; Lefohn, A.S., eds. Transactions Air and Waste Management Association symposium, Effects of air pollution on western forests. Pittsburgh, PA: Air and Waste Management Association; 1989:309–323.

Miller, P.R.; Chow, J.; Watson, J. Assessment of acidic deposition and ozone effects on conifer forests in the San Bernardino Mountains. Final report, contract A032-180. NTIS No. PB97-115-612. Sacramento: California Environmental Protection Agency, Air Resources Board, Research Division; 1996a.

Miller, P.R.; Stolte, K.W.; Duriscoe, D.; Pronos, J., tech. coords. Monitoring ozone air pollution effects on western pine forests. Gen. tech. rep. 155. Albany, CA: Pacific Southwest Research Station, Forest Service, U.S. Department of Agriculture; 1996b.

Pronos, J.; Vogler, D.R.; Smith, R.S. An evaluation of ozone injury to pines in the Sierra Nevada. Forest Pest Management Report 78-1. San Francisco, CA: Pacific Southwest Region, USDA Forest Service; 1978.

Skelly, J.M.; Innes, J.L. Waldsterben in the forests of central Europe and eastern North America: fantasy or reality? Plant Dis. 78:1021–1032; 1994.

Temple, P.J.; Miller, P.R. Seasonal influences of ozone uptake and foliar injury to ponderosa and Jeffrey pines at a southern california site. In: Bytnerowicz, A.; Arbaugh, M.; Schilling, S. tech. coords. Proceedings of the international symposium on air pollution and climate change effects on forest ecosystems, February 5–9, 1996, Riverside, CA, GTR-164 (in press).

8. Air Pollution Effects on Growth of Ponderosa Pine, Jeffrey Pine, and Bigcone Douglas-Fir

M.J. Arbaugh, D.L. Peterson, and P.R. Miller

Introduction

Air pollution from the Los Angeles/Orange County Metropolitan area travels eastward, entering the San Bernardino Mountains along the southern and western edges of the range. This inflow pattern results in elevated ambient levels of pollutants in most forest areas below 2,600 m and a northeast-to-southwest gradient of increasing pollution across the mountain range (Bytnerowicz et al., this volume). The highest air pollution levels in the San Bernardinos are at the lower ecotone of mixed conifer forest. Below the mixed conifer forest, bigcone Douglas-fir (*Pseudotsuga macrocarpa*) grow in canyon areas dominated by chaparral shrubland. This area has slightly lower amounts of ozone air pollution but has larger amounts of nitrogenous pollution, higher air temperatures, and lower amounts of precipitation.

Impacts of increased ambient ozone on crown condition of mature trees are well documented (Miller et al. 1963; Richards et al. 1968). Ozone was identified as a causal agent of crown damage to ponderosa and Jeffrey pine in the 1960s. Dead and weakened trees were removed between 1951 and 1984 by crown health classification, which corresponds to injury from ozone (Rosecrans 1958; Hall 1957).

The impact of ozone on radial growth of trees is less understood, although there is nearly as much long-term data as for crown condition. This is partially due to the lack of a unique injury symptomology for stem growth, such as is present for ozone damage to crowns. Efforts are underway to develop a biochemical marker (Higashi et al., in press) and to associate physiological and physical changes with radial growth changes (Grulke and Lee 1997), but definitive markers of ozone damage are not yet available.

The integrative function of radial stem growth also contributes to the lack of a clear ozone growth signal. Woody tissue accumulation is influenced by many different endogenous and exogenous factors associated with tree growth and development, including carbohydrate production in foliage, biomechanical support of the crown, and water and nutrient transport between roots and foliage. Some factors such as soil, aspect, and slope can differ greatly between sites. Other factors such as fluctuating weather and gradual and sudden changes in competitive relationships with other trees change through time. The combination of these influences make it difficult to separate individual factors influencing radial growth in the natural stands of the San Bernardino Mountains.

Limiting Factors to Growth (Stress Complex)

Atmospheric Environment

The most important climatic factor for tree growth in the San Bernardino Mountains is the timing and amount of annual precipitation. The Mediterranean climate is characterized by long dry seasons lasting 6–9 months each year and irregular precipitation amounts and patterns. One of the longest and most complete precipitation records is from Big Bear Dam at 2,077-m elevation (see Fig. 11.1, Bytnerowicz et al., this volume). This record indicates that the longest recorded dry period was the 7 years from 1895 to 1901, and the second longest was from 1946 to 1950. Precipitation between 1946 and 1976 was about 20% below the average of the entire length of record. More recently, 1978–1983 had higher-than-average winter precipitation and 1984–1987 had lower-than-average winter precipitation (Miller et al. 1982, 1989).

More than 90% of the precipitation between October and April comes in the form of rain or snow as a result of large-scale storms that originate in the Gulf of Alaska. Small amounts of summer precipitation result from thundershower activity in August and is concentrated mainly in the eastern one-third of the mountain area.

Arkeley (1981) found winter precipitation was almost always sufficient to recharge moisture content of the upper 270 cm of soils at experimental plots. Available water was virtually gone from this layer by the end of September. Because predawn xylem water potentials remained moderate even in September (Miller and Elderman 1977), it was concluded that trees were able to access water from sources in fractured rock deeper than 270 cm. However, studies by Temple and Miller (in press) found stomatal conductance at Barton Flats declined with diminishing soil water in early summer, with little physiological activity following that time (Grulke, this volume). Thus water availability may be site- or even tree-specific in many areas of the San Bernardinos.

Patterns of Ozone Impacts on Forests

The presence of crown injury to ponderosa and Jeffrey pines in the San Bernardinos has been documented since the early 1950s. Miller et al. (1963) and Richards et al. (1968) identified ozone as the cause of chlorotic mottle symptoms and early abscission of affected needles. A recent re-evaluation of 1974 crown damage patterns indicates that a general pattern of decreasing crown damage from the southwest to the northeast (see Fig. 7.3, Miller and Rechel, this volume) exists in the San Bernardinos. This gradient depends on both airflow and elevation, the maximum ozone levels occurring at 1,750 m at the southwestern end of the mountains. This elevation generally corresponds to the ecotone between the mixed conifer forest and chaparral ecosystems.

Interacting Stresses

Periods of prolonged drought (e.g., 1895–1901, 1946–1951) or lower-than-average precipitation (1946–1971, 1983–1988) have led to extensive mortality of stressed conifers from bark beetle infestation. Many pole-sized to mature trees on hill slopes may be killed in a 1–2-year period (USDA 1986). Sanitation logging of forest stands in specific areas in the San Bernardino Mountains has been used to reduce bark beetle infestations (La Fuze 1971). Stark et al. (1968) observed that bark beetle attack is almost always the ultimate cause of mortality of conifers (also see Pronos et al., this volume). Annual counts of bark beetle killed trees from the Lake Arrowhead area (Taylor 1973) show that high mortality was associated with dry years and low mortality with moist years through 1945. After 1945, counts of killed trees remained high through the end of the record in 1971. This sustained high level of mortality coincided with the beginning and continuation of tree injury due to ozone and a prolonged period of reduced precipitation.

Disturbance (Fire, Recreation, Urbanization)

One of the first accurately recorded descriptions of unharvested forest vegetation in the San Bernardino Mountains (Leiberg 1900) emphasized that the forest was comprised mainly of large-diameter yellow pines in very open stands with little or no understory growth. Euro-American settlement altered forest structure and composition, beginning with logging operations in the 1850s in the western section of the San Bernardinos that extended eastward to the Running Springs–Green Valley area by the late 1890s (La Fuze 1971). By 1900, forest regrowth was characterized as "dense young thickets 3–5 m high coming up after lumbering of 20 to 30 years ago" (Craig 1904). Some of the lands clear-cut by the Brookings Lumber and Box Company in the Running Springs area had burned repeatedly, and the vegetation cover is characterized today as timberland chaparral (Horton 1960). Logging was extensive, but complete removal of overstory trees occurred only in limited areas due to competition from northern forests (Rosecrans 1958; Hall 1957).

McBride and Laven (1976) determined that the interval between fires more than doubled since fire suppression started in 1905 (Pronos et al., this volume). The combination of partial overstory harvesting and fire suppression has resulted in extensive multiple age structured forests and increased buildup of live and dead biomass (Wilson and Dell 1971; Dodge 1972) in the San Bernardinos. Forest changes are most dramatic in western part of the mountain, where formerly open-canopy ponderosa pine stands have developed into closed-canopy stands dominated by white fir and incense cedar over the latter part of this century (see Minnich, this volume).

Tourism gradually increased in popularity, and permanent year-round communities began to appear in about 1925. Today, privately owned lands, which were formerly timber or mining claims, are intermixed with national forestland in the Lake Gregory–Lake Arrowhead area and enclosed by national forestland in the Big Bear Lake area. Urban development is extensive on private lands and by 1980 had already exceeded predictions for the year 2000. In addition, forests are within 2 hours' driving time of 14.8 million residents of the Los Angeles–Orange–Inland counties metropolitan area. These conditions have resulted in forests being managed primarily for the recreational use of residents and visitors for more than 40 years (Rosecrans 1958; USDA Forest Service 1986).

Types of Information Available for Detecting Tree Growth Trends

Two general types of information are available in the San Bernardino Mountains to assess changes in mature tree growth. The first is diameter measurements gathered sequentially through time at the same trees and sites. If the time period between surveys is long, then these data characterize changes in forest structure and composition (Minnich 1988). Shorter intervals, such as the 4–8 years used by Miller et al. (1989), are useful for understanding changes in growth due to stand dynamics, weather, air pollution, and interacting stressors such as pathogens and physical damage.

Unfortunately, long-term study plots are often established during or after critical forest changes have occurred. Understanding tree growth before stressors have affected growth is important to establish baseline growth rates for trees. One method of determining this is by using retrospective studies of annual growth rings. By using a known date of core extraction, carefully trained investigators can often associate each ring with a single year's radial growth (Stokes and Smiley 1968; Cook and Holmes 1985). Interannular differences in ring width and density can be associated with changes in the atmospheric and soil environment of the tree

(Fritts 1976). Long-term patterns not explained by physical or atmospheric changes in the bole have been associated in past studies with the presence of acute or regional air pollution exposure in California (Peterson et al. 1987, 1992, 1995; Peterson and Arbaugh 1992).

Dendrochronological data, although easily gathered, are difficult to interpret. Radial growth of a tree is influenced by many factors, few of which are measurable after the growth has occurred. Most cores are available only in surviving trees, thus extrapolation of tree core information to the stand level is difficult, especially when irregular mortality of trees has occurred during the period of interest.

In this chapter, both periodic survey data and retrospective dendrochronological data are combined to understand the extent and severity of the impact of ozone and other stressors on the growth and mortality of three species of mature conifers in the San Bernardino Mountains: ponderosa pine (*Pinus ponderosa*), Jeffrey pine (*Pinus jeffreyi*), and bigcone Douglas-fir (*Pseudotsuga macrocarpa*). These species represent the potentially most impacted conifers growing in the western United States.

Bigcone Douglas-Fir Growth Along an Air Pollution Gradient

Introduction

Bigcone Douglas-fir (*Pseudotsuga macrocarpa*) occupies a unique ecological niche in the San Bernardino Mountains. This coniferous species, which is found only in the mountains of southern California and Baja California, Mexico, occurs at elevations of 700–2,200 m in canyons and on dry slopes (Griffin and Critchfield 1972; McDonald and Littrel 1976; Minnich 1982). It is found in steep canyons in chaparral and lower-elevation mixed conifer forest communities and occasionally on ridges at higher elevations. It is typically associated with canyon live oak (*Quercus chrysolepis*), California bay (*Umbellularia californica*), scrub oak (*Q. dumosa*), and poison oak (*Rhus diversiloba*).

Perhaps the most interesting feature of bigcone Douglas-fir in this region is that it occurs primarily as old growth (up to 500 years old) throughout its range, in areas otherwise disturbed by human impacts and frequent fire. Large, old trees survive fires in canyon locations, which are less accessible to humans and provide moist refuges from fire (Bolton and Vogl 1969). Furthermore, this is one of the few coniferous species that sprouts epicormically after fire, allowing it to survive crown fires. The old age of most bigcone Douglas-fir means that long time series of growth are recorded in its tree rings. It is distributed from west to east along the entire range of the San Bernardino Mountains, making it an ideal species for studying long-term impacts of air pollution and climate on growth across the ozone and precipitation gradients in this area (Watson et al., this volume) of the San Bernardino Mountains.

Methods

Air Quality Data

Because there are few data available from ozone analyzers near the specific study sites, ozone exposure was interpolated by using ozone concentrations measured at foothill and valley locations adjacent to the San Bernardino Mountains. Hourly ozone data for 1976–1987 from monitoring sites in the South Coast Air Basin and Southeast Desert Basin were used to characterize ozone exposure. Hourly ozone data from the San Bernardino Mountains were also available for 1976–1978 (Miller and Elderman 1977). These data provided a sufficient time series and spatial spread to adequately describe ozone patterns in the study area.

Exposure statistics were calculated for April–October of each year for each site with at least 75% data capture at daily and quarterly levels. Ozone exposure was calculated as a 7-month (April–October), 12-hour (0800–1900) mean to express the extended daily exposure period in the less-populated mountainous region where bigcone Douglas-fir are found. The 12-hour seasonal mean is highly correlated with the number of peak hourly concentrations greater than 100 ppbv for the Los Angeles Basin (Lefohn and Benedict 1985). Both long-term low concentrations and short-term higher concentrations are potentially injurious to vegetation (Lee et al. 1988; Lefohn et al. 1987b; Grulke, this volume).

Ordinary kriging was used to determine spatial patterns of mean ozone concentrations across the San Bernardino Mountains for 1976–1987 (Biles 1995; Peterson et al. 1992). This technique has been used previously to estimate regional ozone exposure based on a limited number of monitoring stations (Lefohn et al. 1987a; Knudsen and Lefohn 1988). For each year, a linear variogram was found to best describe the sample variogram. Initial variograms were run through a kriging cross-validation process by using the "leave-one-out" technique to assess performance. The fit of the model was determined by regressing the chosen variogram against the sample variogram values. Examination of ozone concentration data across the study area indicates the possible presence of drift, suggesting the application of universal kriging. To check this, universal kriging was compared with the ordinary kriging results. Estimates were similar in all cases. Variogram models were used to develop maps of the 12-hour seasonal mean ozone concentration, with estimated values for each site for each year in the data set. Mean seasonal ozone estimates were calculated for the 1976–1987 period.

Climatic Data

Monthly precipitation data were compiled for weather stations in the San Bernardino Mountains and adjacent lower-elevation locations (data from the U.S. National Weather Service, California Department of Water Resources, and USDA Forest Service). Sufficient data were available to compile a high-quality summary for 1920–1988. Monthly precipitation at specific study sites was estimated by obtaining values from the nearest station with a complete record and using a regressed distance weighted least-squares approach to estimate precipitation at each study site. Monthly precipitation data were aggregated to calculate the final estimates of annual precipitation for each study site (Peterson et al. 1995).

Data Collection

Bigcone Douglas-fir is distributed generally from northwest to southeast across the San Bernardino Mountains. Eighteen sites were identified such that they were located throughout the range of the species (Fig. 8.1). Each site has bigcone Douglas-fir as the dominant overstory tree, the majority of trees more than 100 years old, no evidence of human disturbance (e.g., tree harvest), and no evidence of fire in the past 50 years.

Twenty trees (with dominant or codominant crowns) were sampled along transects at each site. Diameter at breast height (1.4 m) was measured on each tree. Three branches were pruned from the lower crown of sample trees (30% of the trees could not be reached with a pole pruner), and the number of needle whorls retained on each branch was counted to obtain an average for each tree. Branches were examined for chlorosis, which might indicate a visual manifestation of ozone injury. Two cores were extracted from the cross-slope sides of each tree at 1.4-m height, and sapwood length of cores was measured in the field. Cores were mounted in wooden blocks and sanded until individual tracheids were visible under the microscope.

Data Analysis

Ring widths were measured to the nearest 0.01 mm with an incremental measuring machine. Cores were included in the database only if they could be confidently cross-dated (Stokes and Smiley 1968). Time series of ring widths were converted to time series of basal area increments (BAI) for subsequent analysis. Only data for those trees with an estimated diameter greater than 20 cm in 1913 were analyzed to reduce the impact of juvenile growth on the analysis.

It was necessary to subjectively define time periods with and without high ozone concentrations. The year 1950 was chosen as a point in time at which ozone pollution had reached potentially phytotoxic levels. Therefore, tree growth records were divided into the equal periods 1951–1988 (period of potential ozone impacts) and 1913–1950 (reference period with relatively low pollution).

The analysis emphasizes variation in growth within and among sites, so most data are summarized by distribution of individual trees in various categories rather than by the mean growth of trees aggregated within sites. The percentage growth difference between time periods is expressed for individual trees as

$$[(\mathrm{BAI}_{1951-1988} - \mathrm{BAI}_{1913-1950}) / \mathrm{BAI}_{1913-1950}] \times 100 \quad (1)$$

where $\mathrm{BAI}_{1951-1988}$ is the mean BAI for 1951–1988 and $\mathrm{BAI}_{1913-1950}$ is the mean BAI for 1913–1950.

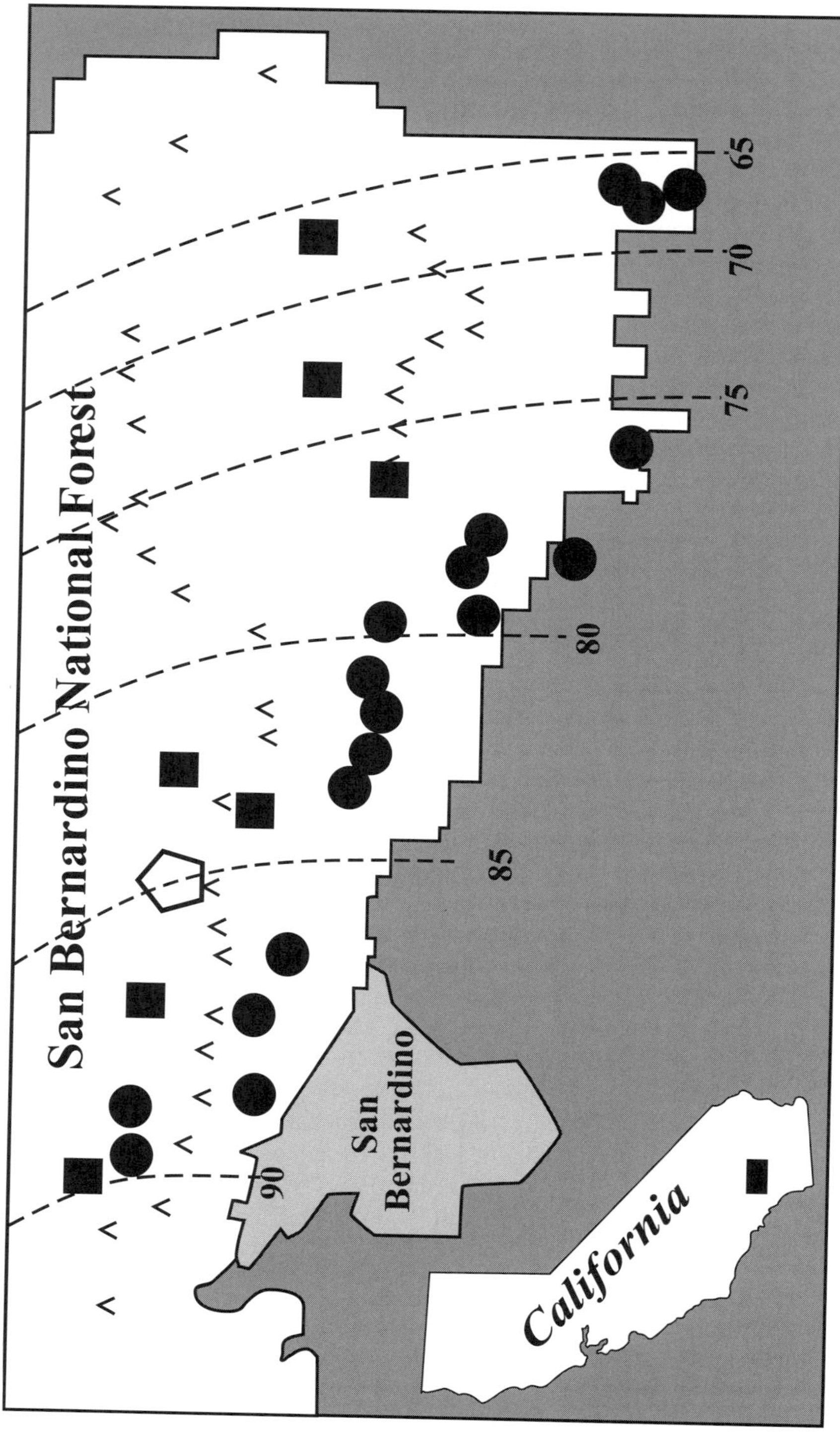

Figure 8.1. Location of study sites used in this chapter. Hatched lines are krigged estimates of 12-hour average seasonal mean ozone concentrations for 1976–1987 in ppb. Circles are bigcone Douglas-fir sites, squares are long-term mixed conifer sites, and the polygon is the location of the 11 high-exposure mixed conifer sites at Blue Jay, California.

The potential impact of precipitation was evaluated by calculating an index of response to short-term soil moisture deficit. Short-term percentage growth difference was calculated as

$$[(BAI_{post} - BAI_{pre}) / BAI_{pre}] \times 100 \qquad (2)$$

where BAI_{post} is the mean BAI for 3 years after the worst growth year and BAI_{pre} is the mean BAI for 3 years before the worst growth year.

The relationship of tree growth to precipitation was also examined. Growth index was calculated by fitting a cubic-spline function to the time series of BAI growth for each tree (Cook and Peters 1981). The BAI values for each year were divided by the spline function value for that year to calculate a time series of growth index values. These values were averaged for all trees at a site to develop a mean chronology. Correlations were calculated between growth index and total annual precipitation (October–September) for 1913–1950 and 1951–1988. Significant differences in correlations between time periods and among ozone exposure groups were tested with an analysis of variance *F*-test after determining that variances were homogeneous.

Interpretation of Growth Trends

General Patterns and Growth Characteristics

Estimates of ozone exposure calculated for this study are similar to those described in other analyses of the San Bernardino Mountains (see Watson et al , this volume). Values for 12-hour seasonal mean ozone concentrations range from 60 (east) to 95 ppbv (west) across the study area, with a clear gradient of ozone exposure from west (high) to east (low). Mean annual precipitation ranges from 46 to 70 cm, with highest values in the far west. The 1951–1988 period was wetter than 1913–1950 (77 cm/yr versus 70 cm/yr for the western San Bernardinos); the more recent period also had more drought periods (six versus two for the western San Bernardinos, where drought is defined as <60% of mean annual precipitation). There is a strong positive correlation ($r = 0.92$) between ozone concentration and annual precipitation across the study area.

For purpose of analysis, the study sites were aggregated into three groups (high, moderate, and low ozone) from west to east across the San Bernardinos. These groups also correspond to high, moderate, and low precipitation. Although these groups were subjectively chosen, it can be seen that they are quite discrete when displayed graphically (Fig. 8.2).

There is wide variation in growth trends for bigcone Douglas-fir at all study sites, a phenomenon typical of most forest ecosystems. When differences in BAI between 1913–1950 and 1951–1988 were examined, we found that more than 80% of all trees have reduced recent growth. Trees at sites with high-ozone exposure have larger growth decreases than at sites with low ozone. Approximately 30% of high-ozone exposure trees have growth reductions greater than 50%, and 60% have growth reductions greater than 25% (Fig. 8.3). Fewer than

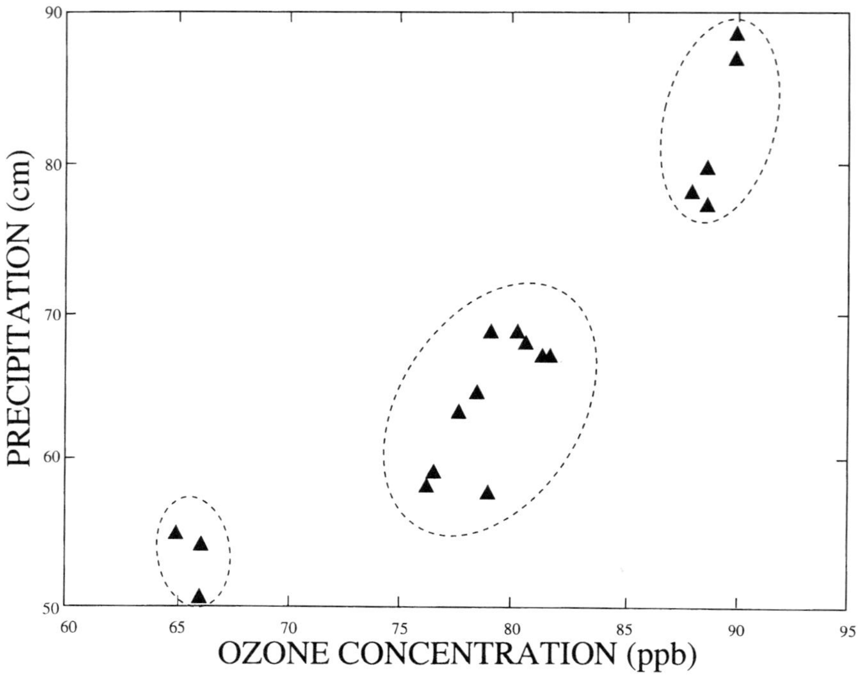

Figure 8.2. Study sites used in bigcone Douglas-fir study are plotted with respect to average 12-hour seasonal mean ozone concentration for 1976–1987 and average annual precipitation for 1913–1988. Sites are subjectively aggregated into high-, moderate-, and low-ozone exposure/precipitation groups. (From Peterson et al. 1995.)

20% of moderate-ozone trees have growth reductions greater than 50%, and fewer than 10% of trees in any ozone group have growth increases greater than 25%.

To assess this pattern of recent growth reductions, we need to consider the normal growth patterns of bigcone Douglas-fir, particularly because many of the trees in the study are more than 300 years old and might be expected to have a recent growth decrease. An assessment by tree-size class shows that severe growth reductions occur in all size classes, although they are less common in the smallest-diameter class (41–60 cm) and more common in larger trees (>100 cm) (Fig. 8.4). This analysis in itself does not prove that growth decreases are ozone-induced but corroborates previous studies of Jeffrey pine and ponderosa pine that show growth decreases tend to be larger in older trees (Peterson et al. 1987; Peterson and Arbaugh 1992).

Foliar Characteristics

Analysis of the foliage of bigcone Douglas-fir did not reveal any observable chlorosis that might be diagnosed as ozone injury. However, needle retention

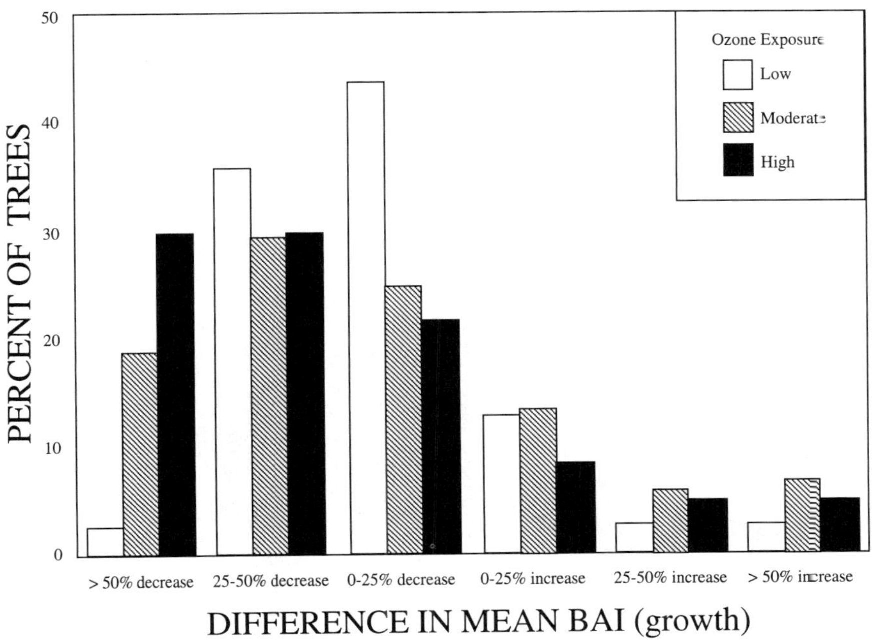

Figure 8.3. Summary of the difference in bigcone Douglas-fir mean BAI between 1913–1950 and 1951–1988 for sites with different levels of ozone exposure. (From Peterson et al. 1995.)

ranges from 4 to 9 years in the study area, with high variation even within study sites. Analysis of data by needle-retention classes (4–5, 6, 7, and 8–9 years) reveals that needle retention affects some aspects of growth (Table 8.1). For example, sapwood thickness is lower in the lowest needle-retention class than in the highest class. This is reasonable because sapwood is positively correlated with leaf area in many coniferous species (Waring et al. 1980). Trees with low sapwood and leaf area may indicate (ozone-induced or other) stress and susceptibility to further stress (Miller et al. 1983; Waring et al. 1981; Peterson and Arbaugh 1992). Mean annual BAI is similar for all needle-retention classes, although trees with 4–5 years of needles have a larger recent growth reduction (44%) than trees in all other needle-retention classes (Table 8.1). In addition, 47% of the trees in the 4–5-year needle-retention class have a recent growth reduction greater than 50%. There are no apparent differences in growth among the other needle-retention classes. This suggests that the trees with the least foliage retention (and probably lowest foliar biomass) have the greatest reductions in growth. Of course, current foliage characteristics reflect only the recent condition of trees.

There appears to be at least some connection between chronic ozone exposure and needle retention, because needle retention is lower for trees at high- and

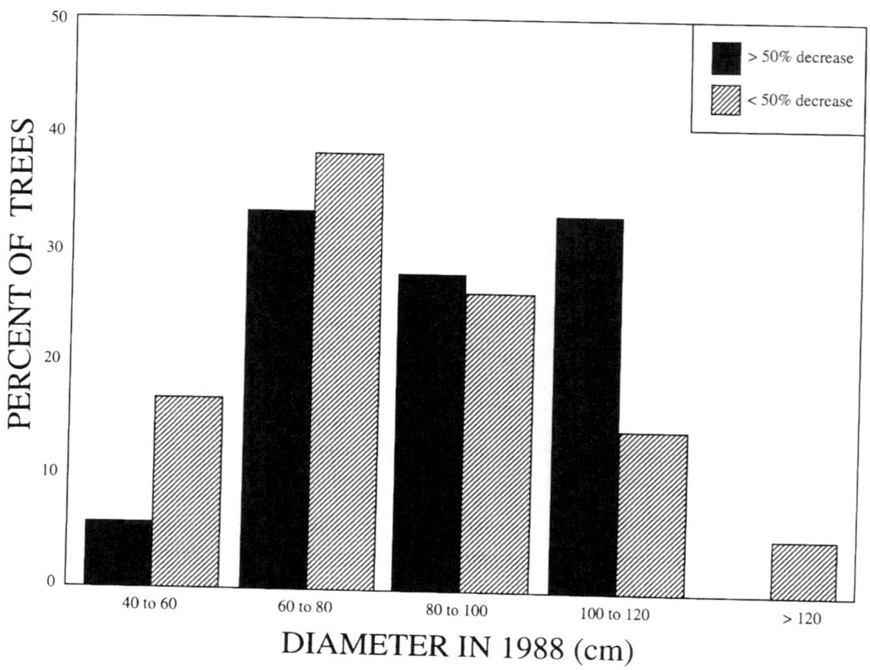

Figure 8.4. Diameter distribution of bigcone Douglas-fir at high-ozone sites, comparing trees with and without greater than 50% reduction in mean BAI between 1913–1950 and 1951–1988. (From Peterson et al. 1995.)

Table 8.1. Characteristics of Bigcone Douglas-Fir (±1 SD) with Different Needle Retention[a]

	Needle retention (years)			
	4–5	6	7	8–9
Number of trees	17	43	59	23
Diameter (cm)	74.0 (23.1)	70.5 (17.3)	75.2 (20.5)	79.1 (18.3)
Sapwood thickness (cm)	3.6 (2.1)	4.9 (2.1)	5.1 (1.9)	5.0 (2.3)
Mean annual BAI	13.5 (9.5)	12.6 (6.7)	14.5 (8.5)	12.2 (6.9)
Difference in BAI between 1913–1950 and 1951–1988 (%)	−44 (40)	−8 (45)	−15 (35)	−22 (29)
Trees with >50% reduction in BAI between 1913–1950 and 1951–1988 (%)	47	14	14	13

[a]Only trees with diameter >20 cm in 1913 are included. (From Peterson et al. 1995.)

Table 8.2. Characteristics of Bigcone Douglas-Fir (±1 SD) from Sites with Different Levels of Ozone Exposure[a]

	Current ozone exposure		
	High	Moderate	Low
Number of trees	60	132	39
Diameter (cm)	83.1 (20.4)	71.9 (21.4)	80.6 (18.7)
Sapwood thickness (cm)	5.9 (2.3)	4.5 (1.8)	5.1 (2.0)
Needle retention (years)	6.4 (1.0)	6.5 (.09)	7.5 (0.7)
Trees with <6 years needles (%)	24.3	9.9	0
Mean annual BAI	17.5 (9.6)	11.9 (6.0)	12.6 (5.8)
Difference in BAI between 1913–1950 and 1951–1988 (%)	–28 (35)	–18 (39)	–16 (29)
Trees with >50% reduction in BAI between 1913–1950 and 1951–1988 (%)	30	19	3

[a]Only trees with diameter >20 cm in 1913 are included. (From Peterson et al. 1995.)

moderate-ozone sites than at low-ozone sites (Table 8.2). Although mean needle retention is slightly (and not significantly) less at high- and moderate-ozone sites, 24% of high-ozone trees have less than 6 years of needles and all low-ozone trees have more than 6 years. In the absence of pollution stress, one might expect that the trees with highest precipitation would have highest needle retention. However, the opposite pattern occurs in this case, which suggests that ozone exposure is indeed having some impact on the foliage of bigcone Douglas-fir, even in the absence of chlorosis normally associated with ozone injury. Lower foliar biomass can be expected to be associated with lower growth rates in ozone-stressed trees (Peterson et al. 1987, 1992; Peterson and Arbaugh 1992; Temple et al. 1992; Grulke, this volume).

Growth Reductions: Ozone Stress and Climate

Data for trees from high-ozone sites provide insight into how ozone stress may be affecting the long-term growth of bigcone Douglas-fir. The high-ozone group consistently has more and larger growth reductions than the other groups (Table 8.2). For example, 50% of the trees had a mean growth reduction of 64%. In addition, 44% of these trees have less than 6 years of needles. This again suggests that foliar biomass has some impact on tree growth, particularly at sites with high ozone concentrations.

Climate interacts with ozone to affect long-term patterns of tree growth The majority of precipitation falls during winter in the Mediterranean climate of southern California (Fujioka et al., this volume). This precipitation is critical for sustaining soil moisture and plant growth throughout the rest of the year. In this study, positive correlations between annual precipitation and tree growth are significantly higher during 1951–1988 than 1913–1950. Correlations are lower at high-ozone sites (where precipitation is higher). The more recent time period has less precipitation, greater variance in precipitation, and more drought years. An

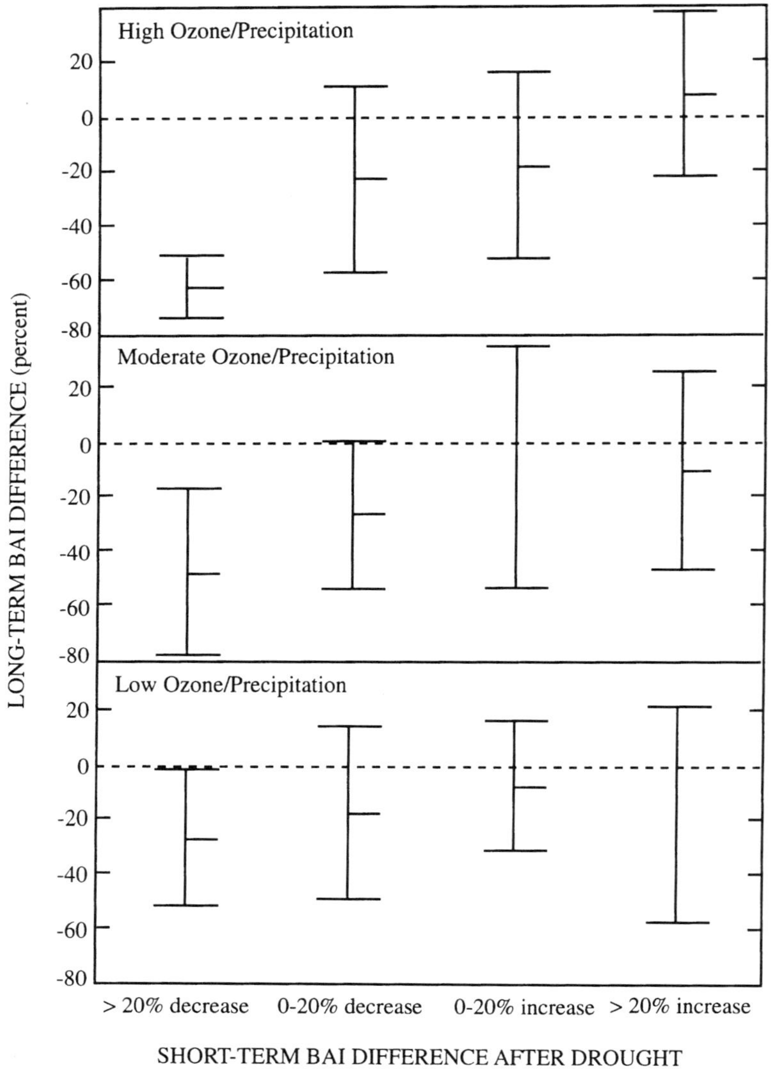

Figure 8.5. Relationship between short-term BAI change after drought and long–term difference in bigcone Douglas-fir mean BAI between 1913–1950 and 1951–1988 for sites with different ozone exposure levels. Mean ± SD is indicated on vertical bars. (From Peterson et al. 1995.)

examination of growth-index values for individual trees indicates that most trees have low growth in drought years.

The impact of drought was further examined by comparing long-term (1) and short-term (2) radial growth patterns, following the four most severe droughts in the precipitation record. High-ozone trees with the largest postdrought growth reductions also have the largest long-term mean growth reductions (Fig. 8.5).

Trees with greater than 20% postdrought growth reduction have a mean long-term growth reduction of greater than 60%. Trees with the largest postdrought increases have little change in overall growth. Low-ozone exposed trees have no apparent relationship between long-term growth and postdrought response.

What causes this disparate response of bigcone Douglas-fir to climate in different areas of the San Bernardino Mountains? Physiological stress at high-ozone sites may reduce the ability of trees to tolerate additional stress from low soil moisture, although the exact nature of climate–growth relationships varies by species. In addition, mature trees, which may already have considerable stress due to competition and increased physiological maintenance costs, may be more sensitive to drought periods (Manion 1981). This has been demonstrated for Jeffrey pine, in which older trees with ozone injury have larger postdrought growth reductions than younger trees (Peterson et al. 1987). Ozone stress can also reduce the resistance of conifers to additional stresses such as insects and fungi (Fenn et al. 1990; Miller et al. 1982; Pronos et al., this volume). Genetic differences associated with bigcone Douglas-fir across the precipitation gradient of the San Bernardinos may also play some role in response to both climate and ozone.

Precipitation is clearly a limiting factor to tree growth in the dry climate of southern California. However, chronic exposure to ozone in the Los Angeles Basin appears to be an important stressor of bigcone Douglas-fir. Although the relationship between ozone stress and growth in this species is based primarily on correlations without the support of experimental data (e.g., controlled fumigation studies), the overall evidence is compelling if not causal in nature. It is likely that pervasive ozone stress, particularly in the western San Bernardinos, predisposes bigcone Douglas-fir to additional stress during periods of exceptionally low precipitation. It is also clear that the growth of this species must be evaluated in the context of spatial and temporal variation of both climate and ozone.

Ponderosa and Jeffrey Pine Growth Along an Air Pollution Gradient

Description of the Long-Term Transect

Eighteen forest vegetation plots were established in 1974 along a west-to-east air pollution concentration gradient across the mountain area at sites with uneven age stands of ponderosa and/or Jeffrey pines and associated species (Fig. 8.1). Each plot had a minimum of 50 ponderosa and/or Jeffrey pines with diameters at breast height (dbh) greater than 29.9 cm. Plots were always 30 m wide but varied in length up to 400 m, depending on how much length was required to obtain 50 trees. Companion trees of all species were included in the plot if their dbh was greater than 10 cm (in 1974). The location of each tree was recorded as distance from the beginning of the plot and distance left and right of the center line (Miller and Elderman 1977). Tree diameters were measured with a steel tape to the nearest 0.1 cm in 1974, 1978, 1983, and 1988. Stem basal areas were computed from the measured dbh.

Results of the growth changes for ponderosa and Jeffrey pines for each of the three measurements have been reported in relation to rainfall and ozone injury (Miller et al. 1989).

During the winter of 1997, seven plots were remeasured to determine long-term mortality and basal area growth rates by species. Because of unmeasured ingrowth at the stands, sample trees no longer could be used to determine stand-level basal area. Instead average tree basal area changes were calculated for surviving ponderosa pine, Jeffrey pine, incense cedar, and white fir where there were ten or more trees of the species present.

Ponderosa and Jeffrey Pine Basal Area Changes over 22 Years

Basal area growth of ponderosa and Jeffrey pine was greatest at the western side of the mountains and lowest at the northeast side (Fig. 8.6). Growth increased from Camp Paivika at the southwestern edge of the mountains for about 20 km to Camp O'Ongo. Growth rates declined at more easterly sites, declining slightly toward the eastern end of the mountains. These growth patterns go in the opposite direction of the ozone (Copeland, this volume) and crown damage gradients to trees (see Fig. 7.3, Miller and Rechel, this volume).

Two other gradients occur along the transects that may contribute to the pattern of growth rates observed at the sites. As previously discussed, precipitation declines along a southwest-to-northeast gradient in these mountains (Fujioka et al., this volume). In addition, nitrogenous dry deposition decreases along the same gradient as ozone and precipitation, but much faster than either (Fenn and Bytnerowicz 1993). Although ozone declines by approximately 55% along the gradient, total N deposition may decline by 80% or more (Fenn and Bytnerowicz 1993). The long-term accumulation of N in soil and plant tissue may increase the effect of this difference between western and eastern mountain sites, especially since at least one western site is N saturated, whereas eastern sites are not (Fenn and Bytnerowicz 1993).

Trends of Radial Growth at Dogwood Campground: Blue Jay, CA

Study Description

The general gradient of 18 plots established by Miller et al. (1989) is useful for examining large-scale differences in tree response to ozone but is not useful for determining topographical influences on tree response to air pollution. This is an important consideration in the San Bernardino Mountains because of the highly variable terrain features. In these geologically young mountains, more than half of the slopes are greater than 50%. The effect of slope and aspect on tree growth is magnified by Mediterranean climate and the difference between moister onshore airflows arriving from the southwest and dry desert conditions to the north and east.

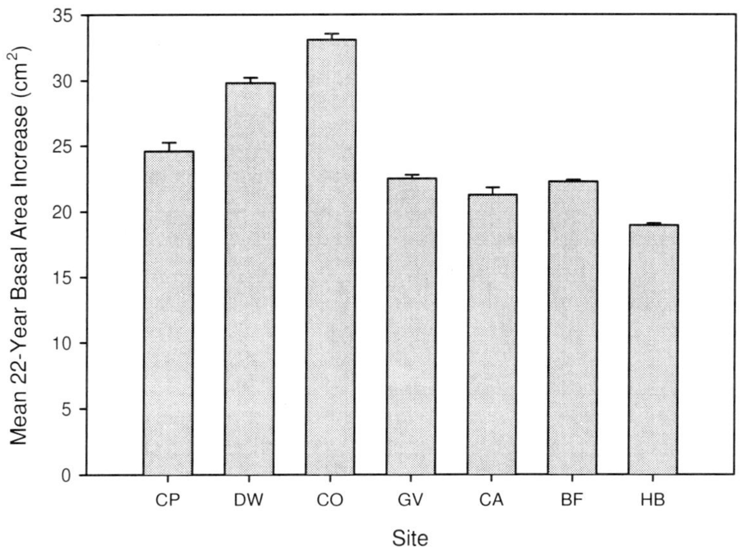

Figure 8.6. Mean basal area increase of surviving ponderosa pine at seven long-term monitoring plots located across an ozone gradient. Plots are listed from west to east (high to low ozone). CP, Camp Paivika, DW, Dogwood A; CO, Camp O'Ongo; GV, Green Valley Creek; CA, Camp Angeles; BF, Barton Flats; HB, Heart Bar.

The dramatic changes of large-scale weather in the mountains combined with frequent changes in slope and aspect may reduce our ability to extrapolate information from the 18 ozone damage sites to other locations that differ in slope and aspect. To examine the effect of topographical differences on crown condition and tree growth, Miller and McBride established 22 plots in a high-pollution area near Dogwood Campground at Blue Jay, California, in 1974. These plots were all located within 4 km^2 in a high-ozone exposure area along the southern rim area of the San Bernardino Mountains. Six different types of topography were included in the design: south, north, northeast, and northwest facing slopes, and ridge top and drainages.

In 1992, 11 plots were relocated and remeasured. Plots were chosen to be representative of each topography type. A combination of mensurational and dendrochronological information was collected to assess the effect of age, slope position, and proximity to the southern edge of the mountain (rim) on growth rates of ponderosa pine, white fir, and incense cedar. Crown condition (Miller et al. 1989), mortality, and dbh were remeasured.

In addition, two cores were extracted from sample trees at 1.4 m with an increment borer. Only trees without visible bole damage, disease, or extensive crown deformation were included in the dendrochronological sample. If there were insufficient tagged trees, then additional trees were sampled adjacent to the sample plot. A minimum of 15 trees (maximum of 20) per plot was sampled at each plot. Ring widths were measured to the nearest 0.01 mm with an incremental measuring machine equipped with digital encoder. Cores were included in the database only if they could be confidently cross-dated by accurately associating the rings of each core with a specific year (Stokes and Smiley 1968). The last ring measured was 1991, because cores were collected during the growing season of 1992. Following ring measurement, cross-dating was verified with the program COFECHA (Cook and Holmes 1985), which calculates correlation coefficients between the time series of ring widths for a single tree and for an entire collection of trees.

Analysis

Individual tree-ring widths were transformed into BAI for subsequent analysis. Trees were found to consist of three general age groups, probably as a result of successive harvesting operations between 1850 and 1925. Basal area growth was divided into three periods: 1925–1949, 1950–1974, and 1975–1991. These periods are chosen to be of generally equal length and allow comparison of the dendrochronological and plot remeasurement results.

Both absolute and relative growth rates were analyzed for the three time periods. Relative growth was expressed similarly to Equation (1) for individual trees as

$$\text{BAI difference 1} = [(\text{BAI}_{1950\text{–}1974} - \text{BAI}_{1925\text{–}1949}) / \text{BAI}_{1925\text{–}1949}] \times 100$$

$$\text{BAI difference 2} = [(\text{BAI}_{1975\text{–}1991} - \text{BAI}_{1950\text{–}1974}) / \text{BAI}_{1950\text{–}1974}] \times 100$$

$$\text{BAI difference 3} = [(\text{BAI}_{1975\text{–}1991} - \text{BAI}_{1925\text{–}1949}) / \text{BAI}_{1925\text{–}1949}] \times 100$$

Long-Term Growth Trends

Average long-term growth patterns of BAIs at Blue Jay are characterized by three distinct time periods (Fig. 8.7). The first period (1905–1945) is characterized by increasing BAIs that reached a maximum about 1945. BAIs decreased during the second period through 1972, with average BAIs declining by more than 40% during this time. Basal area growth rates recovered during the last period (1973–1991) to levels similar to those prior to 1945.

Relationship of Age and Growth

Sample trees were divided into three age classes to examine the relationship between tree age and BAI patterns through time. The oldest class was trees more than 100 years old. Ages within the group varied considerably, with some trees being 200–300 years old. The second age group originated between 1895 and 1910 and probably represents secondary forest establishment that followed the early logging operations in the mountains. The youngest group of trees dates between 1920 and 1930 and represents establishment that followed more recent logging operations in the area or establishment subsequent to fire suppression. Trees of each age group were found in all stands.

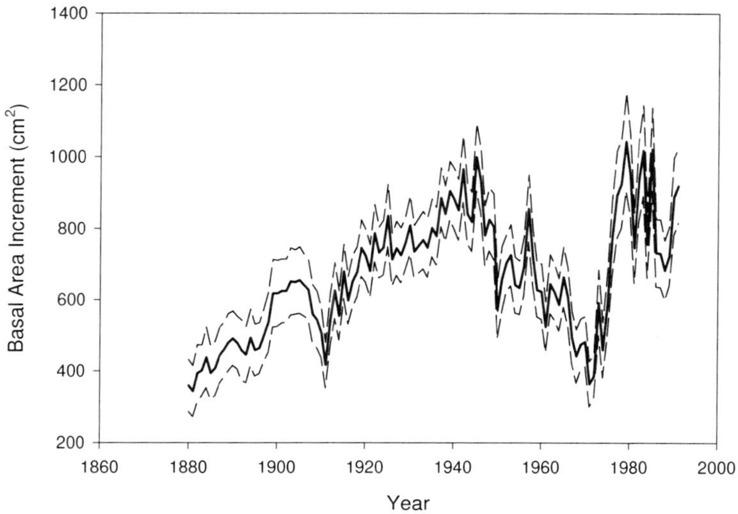

Figure 8.7. Average annual basal area increments for all ponderosa pine sampled at Blue Jay, California. Dashed lines are approximate ± 95% confidence intervals. Sample size varied through time (n = 40 at 1880, n = 166 at 1991).

Long-term patterns of growth for individual age groups were similar to that of the average growth for the entire study (Fig. 8.8). Mean BAI was not significantly different between age groups for the prepollution period (Table 8.3). Basal area growth of older trees was lowest during the 1949–1974 period. Growth for the youngest group of trees declined only slightly during the second period, while relative growth increased between periods 1 and 2.

The oldest group also had significantly lower mean BAIs and reduced growth (although not significant) during the 1974–1991 period compared with the prepollution period. This may indicate that older trees are both more affected by environmental stresses and have less ability to recover relative to younger trees.

Relationship of Climate and Ozone to Radial Growth

Annual precipitation was calculated for the three periods by using weather previously estimated for the bigcone Douglas-fir (see above). Precipitation at Blue Jay was similar to both regional weather and Big Bear Dam data (Miller et al. 1989). The 1925–1949 period was the wettest (131.5 cm/yr) and 1950–1974 the driest (98.6 cm/yr), although the 1975–1991 period was nearly as wet as the first period (122 cm/yr).

The role of ozone in the extended growth reduction that began in 1950 is difficult to evaluate directly. A group of 24 trees lacking the prolonged growth reduction (Fig. 8.9) indicates that much of the reduced growth between 1951 and 1969 may be due to many trees failing to recover from the severe drought occur-

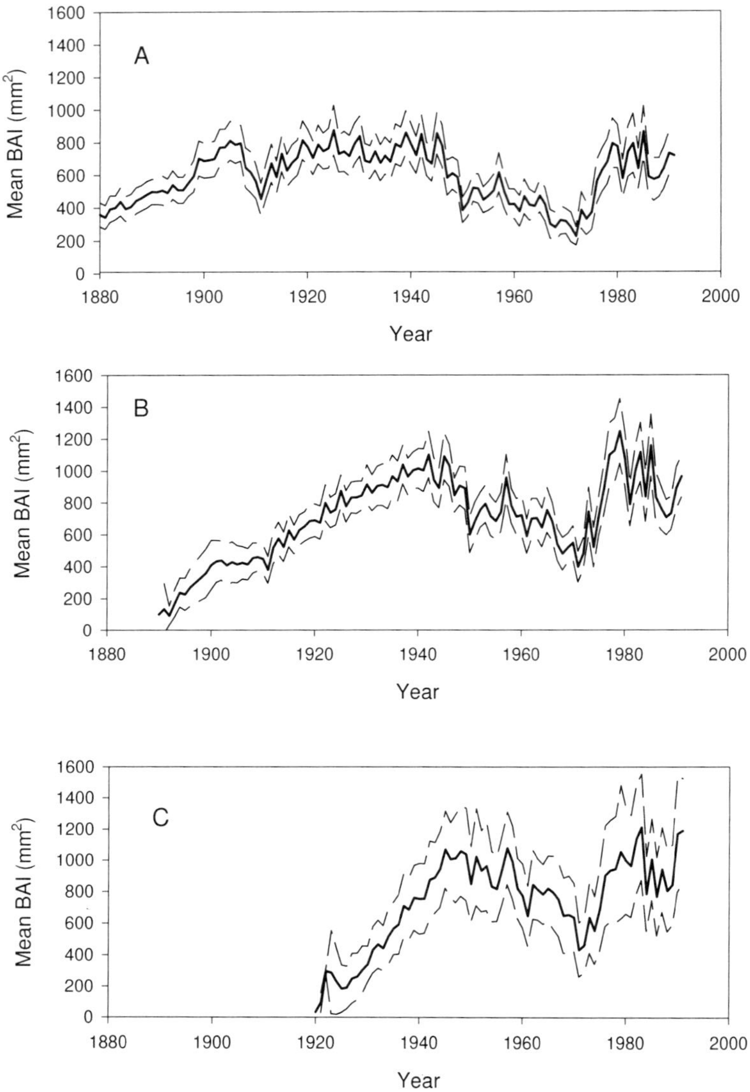

Figure 8.8. Average annual basal area increments for three age groups of ponderosa pine sampled at Blue Jay, California. Group A ages are more than 100 years, group B between 75 and 100 years, and group C is between 50 and 75 years old. Dashed lines are approximate ±95% confidence intervals. Sample size varied through time (sample sizes in Table 8.3).

ring during 1946–1950. The lack of a prolonged growth reduction was also found by Ohmart and Williams (1979) at three low-impact sites in the San Bernardino Mountains and in tree cores gathered on the northern side of the mountains (Arbaugh, in press) at sites with little crown injury.

Table 8.3. Average Basal Area Increments and Changes in Average Basal Area Increments for Three Age Groups of Trees Located at Blue Jay, California[a]

Time period	>100 Years Mean BAI (se) ($n = 55$)	70–100 Years Mean BAI (se) ($n = 67$)	50–70 Years Mean BAI (se) ($n = 31$)
1 (1925–1949)	7.4 (1.2)	9.3 (1.2)	7.2 (2.5)
2 (1950–1974)	4.1 (1.0)	6.6 (1.3)	7.7 (2.3)[b]
3 (1975–1991)	6.6 (1.7)[c,d]	9.4 (2.0)	9.6 (3.8)
BAI difference 1 and 2 (%)	–38 (5)*	–22 (5)*	60 (22)*
BAI difference 1 and 3 (%)	–14 (8)	13 (10)	107 (41)*
BAI difference 2 and 3 (%)	22 (5)*	25 (9)*	42 (21)*

*Significant at $P \leq .05$.
[a]Age groups were chosen arbitrarily and reflect two periods of ingrowth.
[b]Significantly different from trees >100 years at $P \leq .05$.
[c]Significantly different from trees 70–100 years at $P \leq .05$.
[d]Significantly different from trees 50–70 years at $P \leq .05$.

The increased growth rates of trees at this site after 1970 do not correspond to changes in ambient ozone levels. Data from Sky Forest (approximately 2 km from the study site locations) do not indicate decreases in ambient pollution levels between 1968 and 1974, and more recent work (Biles 1995) indicates that most of the decline in ambient ozone between 1976 and 1989 occurred during 1980–1981.

Relationship of Slope Position with Growth

The effect of slope position location on long-term growth patterns of trees exposed to chronic air pollution was examined by averaging BAI and BAI changes by slope position relative to the direction of the pollution transport. Sites directly exposed to the urban pollution are the southern slope sites and the ridge-top sites. Sites on the reverse slope from the pollution are the drainage, north, northeast and northwest aspect sites. Southern sites experience strong upslope winds during storms and the high transpiration rates associated with southern exposure but had some of the highest growth rates and the lowest growth decline during period 2 (Table 8.4). All locations had growth declines of 7.8–23.2% between the first and second periods. Growth declines during period 2 were significant, however, only for north, northwest, and ridge sites (t-test, sign test $P \leq .05$). The period 3 (1975–1991) average BAIs were significantly greater than period 2 (1950–1974) for all sites but were not significantly different from the prepollution period 1 at any slope positions.

Both south and ridge sites are located at the rim area of the mountain and receive unimpeded flow of air from the valley. Other sites are located on opposing interior slopes, so these might be more sheltered from the direct effects of ambient

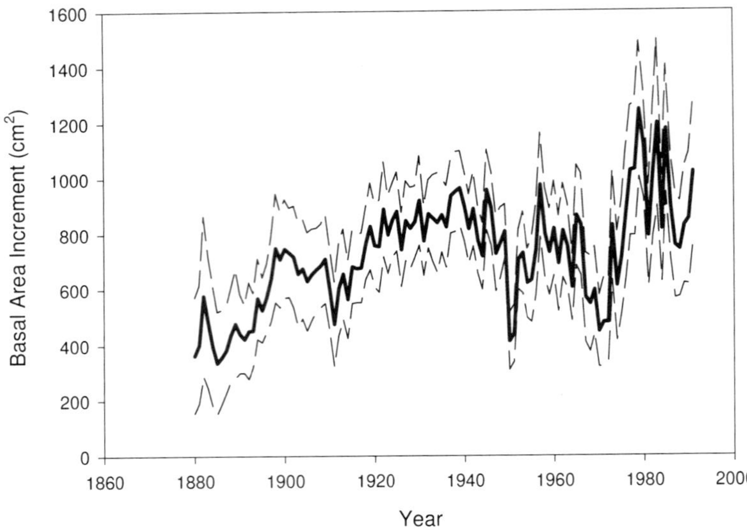

Figure 8.9. Average annual basal area increments for a subset of 24 ponderosa pine sampled at Blue Jay, California, that did not show the prolonged decline in growth during the 1950–1975 period. Dashed lines are approximate ±95% confidence intervals. Sample size varied through time (n = 15 at 1880, n = 24 at 1991).

air pollution. No significant differences between BAI changes (t-test on transformed data, $P \leq .05$) were found between the two groups.

Comparison of Dendrochronological and Long-Term Remeasurement Data

Average BAI differed greatly between remeasured and dendrochronological sample trees (Table 8.4). Remeasured tree basal area was 32–77% less than basal areas calculated from tree ring series. Relative growth between sites also differed between the two measures.

These differences are likely due to differences in sampling criteria used in the two studies. In remeasurement plots, all trees more than 10 cm in diameter at 1.4 m were measured and tagged, regardless of damage or injury in 1973–1974. In the dendrochronological sample, only trees with no apparent disease or physical injury were included in the sample set. Thus the BAI differences measured in the dendrochronological sample reflect the most healthy portion of the population at each site, whereas remeasured tree diameters integrate tree growth for all trees at each site, regardless of the factors contributing to growth changes.

Because of historical logging, fire, recreational use, and urbanization of the mountain area, many of the trees tagged in 1973–1974 had mechanical or pathological damage. Thus absolute growth rates derived from dendrochronological

Table 8.4. Basal Area (BA) Increments and Changes in California Basal Area Increments for Six Slope Positions for Ponderosa Pine Located at Blue Jay, California

	Drainage	North	Northeast	Northwest	Ridge	South
	Dendrochronological data					
dbh (se) (cm)	63.2 (3.0)	39.5 (4.1)	37.9 (1.1)	42.3 (1.3)	43.5 (4.1)	61.0 (4.0)
BA (1975–1991) (cm^2/yr)	4.7 (0.8)	8.0 (0.9)	8.1 (1.2)	9.6 (1.9)	9.5 (1.2)	9.8 (1.4)
BA difference between periods 1 and 2 (%)	–15.0 (9.4)	–23.0 (8.1)*	–13.5 (12.0)	–23.2 (12.3)*	–17.0 (6.9)	–7.8 (18.3)*
BA difference between periods 1 and 3 (%)	32.9 (19.6)	30.3 (18.6)	20.4 (20.4)	30.2 (28.7)	18.2 (11.1)	54.2 (43.1)
BA difference between periods 2 and 3 (%)	64.3 (27.2)*	68.1 (11.3)*	54.0 (21.6)*	54.0 (15.5)*	46.2 (10.3)*	65.1 (19.5)*
	Remeasure data					
dbh (se) (cm)	69 (4.3)	50 (6.0)	33 (2.9)	34 (2.5)	46.5 (2.7)	65.7 (4.4)
BA (1974–1991) (cm^2/yr)	3.1 (0.9)	3.8 (0.7)	2.4 (0.4)	2.7 (0.4)	4.1 (0.8)	6.0 (1.4)

*Significant at $P \leq .05$.

[a]Study sites were established in 1973–1974.

studies should be cautiously applied to the general forest, and growth changes may reflect conservative estimates of the general growth changes experienced by trees but are likely better estimators of the impact of air pollution on tree growth.

Tree Mortality in the San Bernardino Mountains

The effect of ozone on tree mortality has been largely indirect in the San Bernardino Mountains. Ozone increases stress by reducing carbon fixation, making trees vulnerable to insects, competition, and sanitation cuts implemented by forest managers.

Historically, the combination of these factors resulted in large amounts of harvesting between 1954 and 1984. The annual timber harvest averaged 11,162.4 (thousand board feet) with a high of 27,489 in 1963. Guidelines for tree selection was a four-category risk rating system based on crown condition.

The impetus for this management action was the persistent loss of trees to bark beetles, which became more pervasive when trees were weakened by other environmental stresses. Periods of prolonged drought (e.g., 1895–1901, 1946–1950) or lower-than-average precipitation (1983 to the present) led to extensive killing of stressed conifers by bark beetles.

In 1973–1974, ponderosa and Jeffrey pine mortality was 2–8% per year at sites with severe crown damage (Miller et al. 1982). Resurveyed plots from Dogwood indicate that 25% of the ponderosa pines died between 1973 and 1992 (Fig. 8.10). Southern sites had the highest mortality and the drainage site had the lowest, indicating that water stress may be an important factor associated with mortality.

In this study, mortality was not associated with either crown injury or ozone exposure across the gradient. Both crown surveys of 1974 and 1988 indicate that crown damage varies from a high at Camp Paivika to a low at Heart Bar, in the same order as in Figure 8.11. No particular pattern is apparent across the gradient, except sites that have the highest damage (Camp Paivika, Dogwood) also have higher mortality than the site with lowest crown damage (Heart Bar). The very high mortality at Camp Angeles is due to the combination of a bark beetle outbreak and a sanitation cut during the 1980s.

The lack of any pattern for ponderosa/Jeffrey pine mortality may be due to factors other than ambient ozone determining mortality rates, especially at intermediate and low crown damage sites. Stand density, nitrogen status, competition, and ingrowth can affect mortality rates. However, the low number of available sites surveyed makes it difficult to generalize causes of historical mortality patterns across these mountains. Further discussion of mortality factors and patterns across the San Bernardino Mountains can be found in Pronos et al. (this volume).

Discussion: Role of Ozone in Decreased Mature Tree Growth

Radial tree growth has varied greatly over the past century in the San Bernardino Mountains. Favorable weather in the early and middle parts of the century resulted

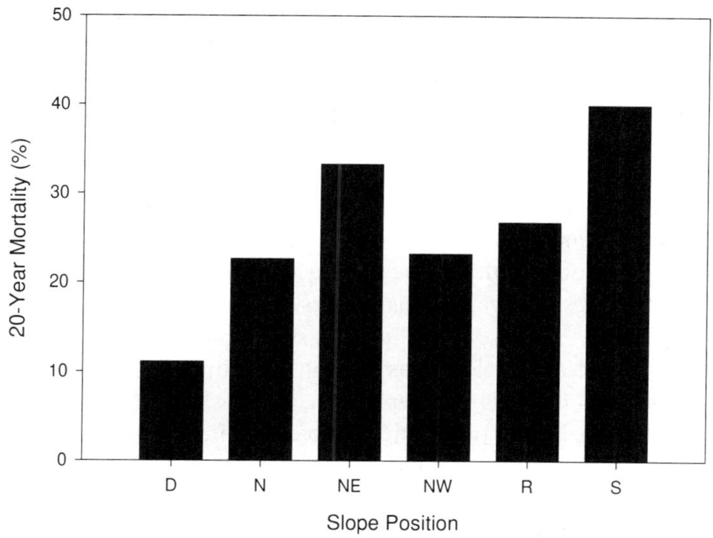

Figure 8.10. Cumulative mortality at 11 high-exposure plots at Blue Jay, California, by slope topography. Topographical features include drainage (D), north (N), northeast (NE), northwest (NW), ridge (R), and south (S) aspects.

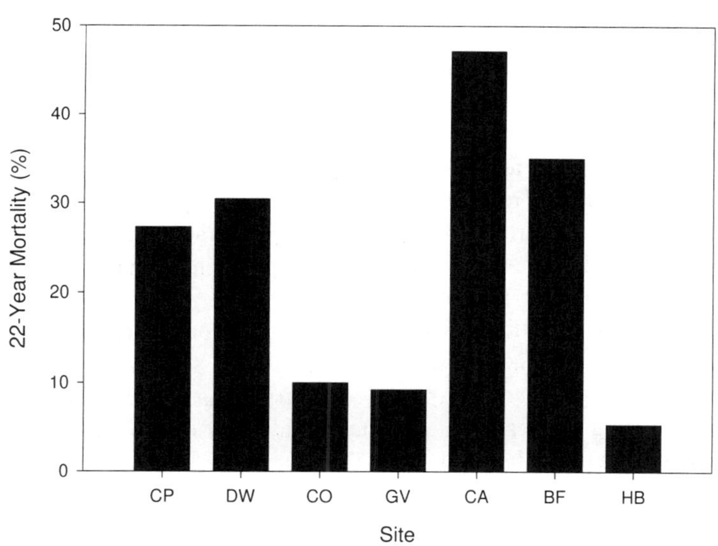

Figure 8.11. Cumulative mortality across seven long-term monitoring plots in the mixed conifer forest between 1974 and 1996.

in high rates of radial growth for all species. The 1946–1950 drought reduced BAIs, which remained low through the early 1970s at high-ozone exposure locations for many ponderosa pine, Jeffrey pine, and bigcone Douglas-fir.

The combination of drought and high ambient ozone may have contributed to decreased radial basal area growth rates of ponderosa and Jeffrey pine of 25–45% and bigcone Douglas-fir of 28% during the 1950–1975 period. At locations with lower ambient ozone, radial growth declined only 16% for bigcone Douglas-fir, even though precipitation was 36% less than at high ambient ozone sites.

The lower level of growth decline for bigcone Douglas-fir should not be inferred to indicate less danger to this species. Bigcone Douglas-fir is valuable because of its rarity, because its old-growth stands preserve the only long-term record of tree growth at low elevations in southern California, and as a nesting site for the southern spotted owl (J. Stephenson, personal communication). It also occurs primarily in a habitat type subject to frequent fire events and has a long reestablishment interval. Existing stands may take as much as 100 years to get reestablished (see Minnich, this volume).

The largest growth declines during this period occurred in older trees. Both older bigcone Douglas-fir and ponderosa pine were more affected by drought episodes and recovered from droughts less readily than younger trees. Increased maintenance respiration or other factors associated with tree aging may make older trees more vulnerable to multiple environmental stresses, including stress from ozone.

Between 1974 and 1984, average annual precipitation increased by almost 50% and ambient ozone decreased by 30%. Growth rates for ponderosa pine and Jeffrey pine increased by 25–30% at most locations, and at Dogwood, younger ponderosa pine radial growth rates returned to 1925–1945 levels. The great magnitude and abruptness of the recovery suggests that key environmental conditions switched rapidly across the mountain. This recovery is likely partially due to precipitation increases but does not appear to be linked with declines in seasonal ambient ozone.

Slope position appears to have little relationship with radial growth at high ambient ozone locations. Trees more directly exposed to upslope airflows did not have larger reductions in growth during the pollution period compared with more protected sites, nor did trees at a particular slope position appear to have greater growth declines than those at other positions.

Anthropogenic air pollution and climate changes are not the only factors affecting forest growth in the San Bernardinos. Fire suppression has resulted in increased stand density, especially in mesic western areas of the mountains. The effects of competing species, such as white fir and incense cedar, may contribute to the basal area declines observed in this study. It is more difficult to explain the sudden recovery of basal area growth after 1974 as a result of stand development. There is no evidence of a thinning operation conducted at Dogwood during the 1970s, which have been monitored since 1974.

Conclusions

The concurrence of drought, long-term reduction in precipitation, and high ozone have contributed to a period of growth decline in the San Bernardino Mountains for ponderosa pine, Jeffrey pine, and bigcone Douglas-fir. Other factors such as increased nitrogenous deposition (see Bytnerowicz et al., this volume) and stand development changes due to fire suppression (see Minnich, this volume) also contribute to this decline. Separation of the relative magnitude of the factors affecting growth is difficult because gradients co-occur across the mountains, thresholds of effects of factors are largely unknown, and interactions between factors have not been studied.

Older trees are an important component of the forest in the San Bernardino Mountains and are likely the most vulnerable to air pollution stress in the San Bernardinos. The additional stress due to high ambient ozone may make these trees more vulnerable to bark beetle infestations. Only scattered dominant trees remain in much of the mixed conifer forest but dominate bigcone Douglas-fir stands. In both forest types, older trees have important value for wildlife, seed production, and recreation.

The mid-1970s growth recovery may not persist indefinitely. General global circulation models indicate that rising levels of greenhouse gases have the potential to increase global temperatures and reduce precipitation. Results from nested global and mesoscale models (Giorgi et al. 1994) indicate that precipitation could decline as much as 20%, and mean warm season temperatures increase by 3–4°C in the southwestern United States with a doubling of CO_2. This would result in worse environmental conditions than the 1946–1972 period, which had a 20% reduction in precipitation but no increase in mean warm season temperature. In addition, the population of southern California is forecasted to increase during the next 20 years, which may reduce the effectiveness of present gains in air quality.

References

Arkeley, R.J. Soil moisture use by mixed conifer forest in a summer-dry climate. Soil Sci. Soc. Am. J. 45:423–427; 1981.

Biles, F.E. Estimates of two seasonal ozone indices across the San Bernardino Mountains using ordinary kriging. Master's thesis, University of Georgia, Athens; 1995.

Bolton, R.B.; Vogl, R.J. Ecological requirements of *Pseudotsuga macrocarpa* in the Santa Ana Mountains, California. J. Forest. 67:112–116; 1969.

Cook, E.R.; Peters, K. The smoothing spline: a new approach to standardizing forest interior tree-ring series. Tree-Ring Bull. 29:15; 1981.

Cook, E.R.; Holmes, R.L. User's manual for program ARSTAN. Tucson, AZ: Laboratory of Tree-Ring Research, University of Arizona; 1985.

Craig, R.D. Lumbering in the San Bernardino Mountains. Typescript. San Bernardino, CA: San Bernardino National Forest, Supervisors Office; 1904.

Dodge, M. Forest fuel accumulation—a growing problem. Science 177:139–142; 1972.

Fenn, M.E.; Bytnerowicz, A. Dry deposition of nitrogen and sulfur to ponderosa and Jeffrey pine in the San Bernardino National Forest in southern California. Environ. Pollut. 81: 277–285; 1993.

Fenn, M.E.; Dunn, P.H.; Wilborn, R. Black stain root disease in ozone-stressed ponderosa pine. Plant Dis. 74:426–430; 1990.

Fritts, H.C. Tree rings and climate. New York: Academic Press; 1976.

Giorgi, F.; Brodeur, C.S.; Bates, G.T. Regional climate change scenarios over the United States produced with a nested regional climate model. J. Clim. 7:375–399; 1994.

Griffin, J.R.; Critchfield, W.E. The distribution of forest trees in California. Gen. tech. rep. PSW-82. Berkeley, CA: USDA, Forest Service, Pacific Southwest Station; 1972.

Grulke, N.E.; Lee, E.H. Assessing visible ozone-induced foliar injury in ponderosa pine. Can. J. Forest Res. 27:1658–1668; 1997.

Hall, R.C. Sanitation-salvage control bark beetles in southern California recreation area. J. Forest. 56:9–11; 1957.

Higashi, R.M.; Arbaugh, M.J.; Fan, T.W-M. Cumulative biochemical markers of air pollutant effects in natural-grown pines. In: Proceedings of critical methodologies for the study of ecosystem health, Davis, CA, September 10–11, 1996 (in press).

Horton, J.S. Vegetation types of the San Bernardino Mountains. Forest Serv. tech. pap. 44. Berkeley, CA: USDA, Pacific Southwest Station; 1960.

Knudsen, P.H.; Lefohn, A.S. The use of geostatistics to characterize regional ozone exposures. In: Heck, W.W.; Taylor, O.C.; Tingey, D.T., eds. Assessment of crop loss from air pollution. London: Elsevier Applied Science; 1988:91–105.

La Fuze, P.B. Saga of the San Bernardino Mountains. Redlands, CA: San Bernardino Co. Museum Assoc.; 1971.

Lee, E.H.; Hogsett, W.E.; Tingey, D.T. Evaluation of ozone exposure indices in exposure-response modeling. Environ. Pollut. 53:43–62; 1988.

Lefohn, A.S.; Benedict, H.M. Exposure consideration associated with characterizing ozone ambient air quality monitoring data. In: Lee, S.D., ed. Evaluation of the scientific basis for ozone/oxidants standards. Pittsburgh, PA: Air Pollution Control Association; 1985:17–31.

Lefohn, A.S.; Knudsen, J.A.; Logan, J.A.; Simpson, J.; Bhumralkar, C. An evaluation of the kriging method to predict 7-h seasonal mean ozone concentrations for estimating crop losses. J. Air Pollut. Control Assoc. 37:595–602; 1987a.

Lefohn, A.S.; Laurence, J.A.; Kohut, R.J. A comparison of indices that describe the relationship between exposure to ozone and reduction in the yield of agricultural crops. Atmos. Environ. 22:1242–1243; 1987b.

Leiberg, J.B. San Gabriel, San Bernardino, and San Jacinto forest reserves. In: Gannett, H., ed. 20th Annu. Rep. U.S. Geological Survey to Sec. Agric. Part 5, Forest reserves. Washington, D.C.: Government Printing Office; 1900:411–479.

Manion, P.D. Tree disease concepts. Englewood Cliffs, NJ: Prentice-Hall; 1981.

McBride, J.R.; Laven, R.D. Fire scars as an indicator of fire frequency in the San Bernardino Mountains, California. J. Forest. 74:439–442; 1976.

McDonald, P.M.; Littrel, E.E. The big-cone Douglas-fir-canyon live oak community in southern California. Madroña 23:310–320; 1976.

Miller, P.R.; Elderman, M.H., eds. Photochemical oxidant air pollution effects on a mixed conifer forest ecosystem. EPA 600/3-77-104. Corvallis, OR: U.S. EPA; 1977.

Miller, P.R.; Parmeter, J.R.; Taylor, O.C.; Cardiff, E.A. Ozone injury to the foliage of *Pinus ponderosa*. Phytopathology 53:1072–1076; 1963.

Miller, P.R.; Taylor, O.C.; Wilhour, R.G. Oxidant air pollution effects on a western coniferous forest ecosystem. Research brief EPA 600/D-82-276. Corvallis, OR: U.S. EPA; 1982.

Miller, P.R.; Longbotham, G.J.; Longbotham, C.R. Sensitivity of selected western conifers to ozone. Plant Dis. 67:1113–1115; 1983.

Miller, P.R.; McBride, J.R.; Schilling, S.L.; Gomez, A.P. Trend of ozone damage to conifer forests between 1974 and 1988 in the San Bernardino Mountains of southern California. In: Proceedings of air pollution effects on western forests, 32nd annual meeting of Air and Waste Management Association, Anaheim, CA. 1989:309–323.

Minnich, R.A. *Pseudodotsuga macrocarpa* in Baja California. Madroña 29:22–31; 1982.
Minnich, R.A. The biogeography of fire in the San Bernardino mountains of California—a historical study. Geography 28. Berkeley, CA: Univ. of California Press; 1988.
Ohmart, C.P.; Williams, C.B., Jr. The effects of photochemica oxidants on radial growth increment for five species of conifers on the San Bernardino National Forest. Plant Dis. 63:1038–1042; 1979.
Peterson, D.L.; Silsbee, D.G., Poth, M.A., Arbaugh, M.J., Biles, F.E. Growth response of bigcone Douglas-fir (*Pseudotsuga macrocarpa*) to long-term ozone exposure in southern California. J. Air Waste Manage. Assoc. 45:36–45; 1995.
Peterson, D.L.; Arbaugh, M.J. Mixed conifer forests of the Sierra Nevada. In: Olson, R K.; Binkley, D.; Böhm, M. eds. The response of western forests to air pollution. New York: Springer-Verlag; 1992:433–459.
Peterson, D.L.; Arbaugh, M.J.; Wakefield, V.A.; Miller, P.R. Evidence of growth reduction in ozone-stressed Jeffrey pine (*Pinus jeffrey* Grev. and Balf.) in Sequoia and Kings Canyon national parks. J. Air Waste Manage. Assoc. 38:906–912; 1987.
Peterson, D.L.; Arbaugh, M.J.; Robinson, L.J. Regional growth changes in ozone-stressed ponderosa pine (*Pinus ponderosa*) in the Sierra Nevada, California, USA. Holocene 1:50–61; 1992.
Richards, B.L., Sr.; Taylor, O.C.; Edmunds, G.F., Jr. Ozone needle mottle of pine in southern California. J. Air Waste Manage. Assoc. 18:73–77; 1968.
Rosecrans, W.S. Logging in recreational forests. Am. Forests 63:20–22; 1958.
Spurr, S.H.; Barnes, B.V. Forest ecology. New York: John Wiley & Sons; 1980.
Stark, R.W.; Miller, P.R.; Cobb, F.W., Jr.; Wood, D.L.; Parmeter, J.R., Jr. Photochemical oxidant injury and bark beetle (Coleoptera: Scolytidae) infestation of ponderosa pine. I. Incidence of beetle infestation in injured trees. Hilgardia 39:121–126; 1968.
Stokes, M.A.; Smiley, T.L. An introduction to tree-ring dating. Chicago: Univ. of Chicago Press; 1968.
Taylor, O.C. Oxidant air pollutant effects on a western coniferous forest ecosystem. Task B report. Historical background and proposed systems study of the San Bernardino Mountain area. Riverside, CA: Statewide Air Pollution Research Center, University of California; 1973.
Temple, P.J.; Miller, P.R. Seasonal influences of ozone uptake and foliar injury to ponderosa and Jeffrey pines at a southern California site. In: Bytnerowicz, A.; Arbaugh, M.J.; Schilling, S. tech. coords. Proceedings of the international symposium on air pollution and climate change effects on forest ecosystems, February 5–9, 1996, Riverside, CA, GO-164 (in press). Internet: <http://www.rfl.psw.fs.fed.us/pubs/psw-gtr-164/index.html
Temple, P.J.; Reichers, G.H.; Miller, P.R. Foliar injury responses of ponderosa pine seedlings to ozone, wet and dry acidic deposition, and drought. Environ. Exp. Bot. 32:101–113; 1992.
USDA Forest Service. Draft forest plan. San Bernardino, CA: USDA Forest Service, San Bernardino National Forest, Supervisor's Office; 1986.
Waring, R.H.; Thies, W.G.; Muscato, D. Stem growth per leaf unit area: a measure of tree vigor. Forest Sci. 26:112–117; 1980.
Waring, R.H.; Newman, K.; Bell, J. Efficiency of tree crowns and stemwood production at different leaf canopy densities. Forestry 54:15–23; 1981.
Wilson, C.C.; Dell, J.D. The fuels buildup in American forests. J. Forest. 69:471–475; 1971.

9. Effects of Ozone on Understory Vegetation in the Mixed Conifer Forest

Patrick J. Temple

Introduction

Importance of Understory Plants in the Forest Ecosystem

The mixed conifer forest ecosystem of the San Bernardino Mountains includes all the producers (plants and algae), consumers (animals and other heterotrophs), and decomposers (bacteria, fungi, and other single-celled organisms) found in the forest, all of which contribute to the complex processes of energy flow, storage, and biogeochemical cycling. Although not the dominant vegetation, the herbaceous and woody plants of the mixed conifer forest understory are a significant part of the forest ecosystem. Although the effects of ozone on the dominant overstory trees, particularly ponderosa (*Pinus ponderosa* Laws.) and Jeffrey (*P. jeffreyi* Grev. & Balf.) pines, have been extensively documented, relatively little is known of the effects of ozone on the shrubs and herbaceous plants of the understory.

The flora of the San Bernardino Mountains is particularly rich and diverse. Although this mountain range covers only about 2% of the surface area of the state, nearly 25% of all the plant species in California grow here (A. Sanders, personal communication). Numerous annual and perennial plants are endemic to these mountains, and a variety of rare, threatened, and endangered species is also found in the San Bernardino National Forest (Hickman 1993). These understory plant species contribute to the structural complexity of the forest and provide food and shelter for a diverse list of vertebrate and invertebrate animals. Understory plants are an integral part of the complex web of food chains in the forest, through interactions with aboveground and belowground herbivores and fungi. Understory plants are part of the nutrient cycling component of the ecosystem, through litter decomposition and mycorrhizal associations (Kozlowski et al. 1991). Understory plants also interact directly with the dominant forest trees, through competition for water, light, and nutrients, which can limit the germination and establishment of tree seedlings. Other examples of important interactions among understory and overstory species include branch or root parasitism, such as mistletoes (*Arceuthobium* spp., *Phoradendron* spp.), and alternate hosts for serious tree diseases, such as white pine blister rust on currants (*Ribes* spp.). The effects of ozone in altering species diversity of understory plants of the mixed conifer forest or in altering food webs, competitive interactions, or biogeochemical cycling in the forest are largely unknown.

Understory Plants as Biomonitors of Ozone

The relatively thin leaves and fast growth rates of most annuals, herbaceous perennials, and deciduous shrubs suggest that these plants may be more susceptible to ozone than most coniferous or deciduous forest trees. Rates of gas exchange in annuals and perennials are generally an order of magnitude greater than those of coniferous tree species (Nobel 1991), and relative susceptibility to ozone among plant groups has been correlated with rates of leaf conductance to gases (Reich 1987). Understory plants have been used extensively as biomonitors of ozone effects in forests of eastern North America (Lewis and Conkling 1996), and surveys of ozone injury have begun on plants in forests of eastern Europe (Bytnerowicz et al. 1993). Examples of plant monitors include common milkweed (*Asclepias syriaca* L.), blackberry (*Rubus allegheniensis* Porter), and black cherry (*Prunus serotina* Ehrh.), all of which have been used in systematic surveys of ozone impacts on forested areas of the eastern United States as part of the forest health monitoring (FHM) program (Lewis and Conkling 1996).

Understory plants have not been used as biomonitors of ozone effects in coniferous forests of the West, in part because of an actual or perceived lack of suitable biomonitoring species (Smith 1995). Relatively few reports have been published of ambient ozone effects on western understory plants in the field. Laboratory or field exposures of understory plants of western forests to known concentrations of ozone are also sparse (Treshow and Stewart 1973; Mavity et al. 1995). Sensitive ozone biomonitors have generally not been needed to assess the impact of ozone on forest vegetation, particularly in southern California, because of the readily observed severe foliar injury symptoms on dominant tree species, including ponderosa and Jeffrey pines and California black oak (*Quercus kellog-gii* Newb.) (Miller et al. 1996). Beginning in the early 1970s, permanent pine observation plots were established in the San Bernardino Mountains and more recently in the Sierra Nevada to monitor for annual and spatial variations in foliar responses to ozone (Miller et al. 1996). However, development of suitable ozone biomonitoring plants using understory species would provide additional information on ozone effects, particularly early in the growing season when annual and perennial herbaceous plants are most active. Ozone injury on the pine plots is usually measured in September, at the end of the growing season for understory plants (Miller et al. 1996).

Effects of Ozone on Understory Plants

Foliar Injury Symptoms

The direct foliar effects of ozone on deciduous plants are characterized by upper leaf surface chlorotic or anthocyanotic lesions that coalesce on older leaves, producing a general chlorosis that leads to rapid leaf senescence. These prematurely senescent leaves frequently abscise, and the overall reduction in photosynthetic tissues may lead to reduced carbon assimilation, lower plant growth, and reduced reproductive potential. Often leaves with even small amounts (>30%) of chlorotic injury senesce and abscise rapidly. Unless deciduous plants are subjected to chronic exposures to toxic concentrations of ozone over their entire developmental cycle, plants in the field can appear to be uninjured, except for chlorotic or missing lower leaves. Thus, the "window of opportunity" to observe characteristic symptoms of upper surface chlorotic lesions on plants in the field can be short, depending on timing of ozone episodes and the phenology of foliar development. Because ozone-induced foliar senescence may be indistinguishable from normal leaf senescence, except for its more rapid development, and because timing of normal leaf senescence is seasonally variable, depending primarily on soil water availability, foliar senescence is often not a reliable indicator of ozone injury in the field.

Unlike understory plants in more humid forests, annual and perennial plants in the mixed conifer forest of the San Bernardino Mountains are dependent on available water stored in the upper soil horizons from winter precipitation and snowmelt. Most understory herbaceous plants are active for short periods from April to June, when ozone concentrations are lower than later in the summer. Understory shrubs active in the summer are predominantly drought-tolerant, such as the many species of *Arctostaphylos* and *Ceanothus* common in the mixed conifer forest. The leaves of these shrubs possess morphological and physiological adaptations to drought, such as thick cuticles, low stomatal density, and stomatal crypts. These adaptations to drought also confer resistance to ozone, except perhaps at the young seedling stage of development (Stolte 1982). In addition, daily peak ozone concentrations in the San Bernardino Forest normally occur between 1400 and 1800, when temperatures and vapor pressure deficits are greatest and rates of gas exchange are at a minimum (NRC 1977). Because careful assessments of ozone injury on pines and other dominant forest trees have been conducted late in the growing season, injury to herbaceous annuals and perennials active earlier in the year may have been underestimated. Thus, the timing of ozone episodes relative to the periods of active growth of potentially ozone-sensitive understory plants may account, in part, for the absence of long lists of ozone-injured plant species from the San Bernardino mixed conifer forest.

Another factor that may account for the relatively small number of reports of foliar injury on understory plants in the San Bernardino Mountains is the possibility that susceptible genotypes of plant species growing in areas exposed to high concentrations of ozone had already been eliminated by the time systematic studies of ozone effects on the San Bernardino Forest ecosystem had begun (Miller 1992). Daily summer maximum hourly ozone concentrations of 250–300 ppb or higher were recorded in the Crestline–Lake Gregory area of the San Bernardino Mountains in the early 1960s, with brief excursions to as high as 600 ppb (Miller 1992). This level of ozone may have been sufficient to eliminate susceptible genotypes from the forest community. Peak ozone concentrations in the San Bernardino Mountains have declined significantly in recent years (Miller et al. 1997) and at present may be below levels necessary to induce injury in the remaining plant populations. Despite considerable speculation on the possible effects of ozone on plant populations (Taylor et al. 1991; Taylor and Pitelka 1992), there are few reports of ozone as a selective force in altering the genetic structure of native plant populations (Berrang et al. 1991; Reiling and Davison 1992). Clearly, with greater than 25% mortality of ponderosa pines in the heavily damaged forest on the ridges facing Los Angeles (Miller and McBride 1975), many ozone-susceptible pine genotypes have been removed from this population of trees. However, the only report of the effects of ozone on populations of nonwoody plants in southern California is that of Dunn (1959), who found that susceptible genotypes of the annual lupine *Lupinus bicolor* Lindl. had been eliminated from high "smog" areas in the Los Angeles Basin. Unfortunately, there have been no recent studies of ozone as a selective pressure in high-ozone areas of the San Bernardino Mountains, so the hypothesis that susceptible elements of understory plant populations have been eliminated from these areas remains unproved.

Early Observations of Foliar Injury in the San Bernardino Forest

The few reports of ozone injury symptoms on plants in the San Bernardino Forest other than the dominant conifers and black oak begin in the late 1950s and early 1960s with the mention of injury symptoms on squawbush (*Rhus trilobata* Nutt.) (Richards et al. 1968), a shrub more common in the chaparral foothills than in the mixed conifer forest. The susceptibility of this species to ozone was later confirmed in greenhouse exposure studies (Temple 1989). Ozone injury was also noted on woody species common in canyon bottoms and riparian areas, such as western sycamore (*Platanus racemosa* Nutt.), bigleaf maple (*Acer macrophyllum* Pursh), wild grape (*Vitis girdiana* Muns.), and blue elderberry (*Sambucus mexicana* Presl.). Possible ozone injury symptoms were later reported in the 1970s on false indigo (*Amorpha californica* Nutt.). Symptoms on this common understory shrub included severe chlorotic mottle of leaflets and premature defoliation (NRC 1977). Other understory species with suspected ozone injury included blue ryegrass (*Elymus glaucus* Buckl.) and sweet-cicely (*Osmorhiza chilensis* H. & A.)

(Apiaceae). In some cases, following extremely high-ozone episodes early in the summer (e.g., a 1-hour average ozone concentration of 600 ppb on June 28, 1974), nearly every annual or perennial plant in the area had severe acute ozone injury (P. Miller, personal communication). Observations on the current status of ozone injury on understory plants in the mixed conifer forest are given in a later section.

Experimental Exposure Studies

Understory plants from the San Bernardino Forest have rarely been used in experimental controlled exposures to ozone. Because of the obvious severe damage to overstory conifers and oaks and because of the economic and ecological importance of these trees to the forest, the emphasis of experimental work has always been on trees. In addition, many of the annual and perennial native plants of the forest that have shown injury symptoms have proved to be difficult to propagate under standard greenhouse conditions. An exception to this pattern is *Elymus glaucus,* which has been used in experimental exposures to ozone (Yoshida 1994; P. Miller, unpublished data). Foliar injury symptoms on *E. glaucus* exposed to elevated ozone did not resemble those observed in the field, although the plants had accelerated senescence and reduced production of tillers in the high-ozone treatment (P. Miller, unpublished data). Shrubs native to the chaparral and coastal sage communities of the San Bernardino Mountains' foothills have also been used in experimental ozone exposures. In general, these drought-adapted species of *Artemisia, Adenostoma, Arctostaphylos, Ceanothus, Eriogonum, Encelia, Quercus,* and *Salvia* were highly resistant to ozone, except for recently germinated seedlings (Stolte 1982; Preston 1986; Temple, unpublished data). Numerous annual and perennial plants of the adjacent Mojave Desert have been screened for susceptibility to ozone (NRC 1977; Thompson et al. 1984). A few species [e.g., *Camissonia claviformis* (T. & F.) Raven], appeared to be susceptible to ozone; however, none of these species grows in the mixed conifer forest.

Plants native to the trembling aspen (*Populus tremuloides* Michx.) forest community of the Wasatch Mountains in Utah were exposed to ozone both in the greenhouse and in the field (Price 1973; Treshow and Stewart 1973). Several of the species used in these experiments also grow in the San Bernardino mixed conifer forest. Among these, the grasses *Bromus tectorum* L. and *B. carinatus* H. & A. appeared to be susceptible to ozone. Two species common in the dry mixed conifer zone, *Chenopodium fremontii* Wats. and *Collomia linearis* Nutt., and one found in moist mountain meadows, *Geranium richardsonii* Fisch. & Trautv., were also susceptible, showing injury at less than 0.25 ppm for 2 hours. Trembling aspen was the most susceptible tree, developing injury at an exposure of 0.15 ppm for 2 hours (Treshow and Stewart 1973). Trembling aspen is not common in the San Bernardino Mountains and is found only in isolated north-facing ravines far from areas of elevated ozone. No ozone injury has been reported on these plants from ambient exposures in the San Bernardino Mountains.

A recent series of controlled environment exposures of shrubs common in western forest communities was designed to screen understory plants for possible use as ozone bioindicator species (Mavity et al. 1995). Seven shrub species were selected, based on widespread distribution in western forests and on their relatively thin leaf morphology, which suggested that these species could be more susceptible to ozone than shrubs with sclerophyllous leaves. The greenhouse-grown plants were exposed to relatively high concentrations of ozone, designed to elicit symptoms of acute ozone injury. Three species, ninebark [*Physocarpus capitatus* (Pursh) Kunze], chokecherry [*Prunus virginiana* L. var. *demissa* (Nutt.) Torr.], and huckleberry (*Vaccinium* sp.), developed "classic" upper leaf surface purple stippling in response to ozone and were classified as showing good promise as ozone indicators. The other shrubs, including serviceberry (*Amelanchier alnifolia* Nutt.), sagebrush (*Artemisia* sp.), thimbleberry (*Rubus parviflorus* Nutt.), and elderberry (*Sambucus racemosa* L.) developed little visible injury or symptoms were not distinctive enough to be recognized as ozone injury in the field. Most of these species are either not found in the mixed conifer forest or are rare (Munz 1974; Hickman 1993). The sagebrush used in this study was not identified to species, so its status in the San Bernardino Mountains cannot be confirmed. Both *P. virginiana* var. *demissa* and *R. parviflorus* are found throughout the San Bernardino Forest, particularly in the more humid western portions. The extent of ambient ozone injury on these species is discussed in a later section.

Effects of Ozone on Plant Communities

Anthropogenic influences have profoundly altered the original structure of the montane forests of the San Bernardino Mountains. The original pine forests were shaped by frequent low-intensity wildfires that removed competing grasses and shrubs but did not damage canopies of larger trees. This fire regime produced a landscape dominated by large-diameter mature pines in open, irregularly distributed stands, with relatively little understory growth (Leiberg 1900). After 150 years of logging, intense fires followed by strict fire suppression policies, ozone damage, bark beetle attacks, and salvage logging, the open pine forest has been transformed into a closed-canopy mixed forest with thick understory growth and litter layer (Minnich, this volume). The increased understory growth is most apparent in the more mesic western portions of the forest, where ozone concentrations are highest and where the effects of ozone on pines have been most apparent (see Miller and Rechel, this volume). Ozone has worked in concert with fire suppression policies to reduce the domination of pines and increase understory cover of more ozone-resistant, but less fire-tolerant species, particularly incense cedar [*Calocedrus decurrens* (Torr.) Florin.] and white fir [*Abies concolor* (G. & G.) Lindl.] (Minnich et al. 1995). Within the closed-canopy forest, reduced light levels and rates of evapotranspiration and increased litter depth have significantly altered growing conditions for plants.

Ozone is not the only air pollutant to be deposited in the San Bernardino Forest. A strong gradient of nitrogen deposition exists in the mixed conifer zone from the western slopes and ridges facing Los Angeles to cleaner sites in the north and eastern parts of the forest (Fenn 1991; see Fenn and Poth, this volume). Rates of litter decomposition are greater in areas with higher deposition of nitrogen (Fenn and Dunn 1989). However, the litter layer is also thicker in these more highly polluted areas because of accelerated foliar abscission induced by ozone (Arkley and Glauser 1980). These complex interactions among pollutant effects, fire suppression, drought cycles, and the increased presence of weeds and garden escapes in the forest preclude defining a precise role for ozone in altering the structure of the understory plant community in the San Bernardino Forest.

Despite these complexities, attempts have been made to correlate exposure to ozone with changes in plant community structure and composition. The best example of this approach in southern California was that of Westman (1979), who measured the frequency and density of shrubs at 67 sites throughout the Coastal Sage Scrub ecosystem. A suite of climatic and edaphic variables was also measured at each site. Data on ambient ozone concentrations for each site were derived by extrapolation from the closest available air quality monitoring stations. Using multivariate analysis, he found the highest correlation between shrub density and environmental variables was with ozone concentration, with lowest percentage cover and lowest species diversity in plots located in areas with highest ambient ozone (e.g., Riverside). The inference was that ozone had played a significant role in altering species composition in the Coastal Sage Scrub community. The difficulty with this statistical approach is that causal relationships between variables can be inferred but not proved. Later fumigation studies showed that the shrub species with lowest densities in the high-ozone plots were generally resistant to ozone (Stolte 1982; Preston 1986; Temple, unpublished data). In addition, other factors that could account for lower shrub cover in the high-ozone plots were not assessed, such as direct competition between annual grasses and shrub seedlings for soil water; frequency and intensity of grazing, fire, and other disturbances; and high nitrogen deposition, which could favor the growth of annual grasses and other weeds (D'Antonio and Vitousek 1992). However, in addition to high ambient ozone exposure, plots with the lowest species richness were also located in areas with the highest summer temperatures and lowest annual precipitation (Westman 1981). This suggests that ozone could have acted as an additional stress on plants growing at the edge of their range on sites not optimal for their growth or reproduction.

In the San Bernardino Forest, frequency, density, and percentage cover of shrub and herbaceous layers were measured in quadrants and transects on pine observation plots located along the air pollution gradient from west to east (Taylor 1974). One of the objectives of this study was to determine the possible influence of ozone on plant succession in the mixed conifer forest. Because other environmental factors, such as precipitation, elevation, and disturbance, also vary along the air pollution gradient, as discussed above, it was not possible to assess the impact of ozone alone in altering the plant understory community. However, these data will provide an important baseline to assess changes in understory plant community composition and structure over time.

The effects of ozone on the lichen community of the San Bernardino Forest is discussed in Nash and Sigal (this volume).

Current Status of Ozone Injury on Understory Plants

Survey Methods

Observations of foliar ozone injury on understory vegetation were made at 14 locations in the San Bernardino Forest on, or in the vicinity of, the permanent pine measurement plots that were established in the 1970s to monitor changes in crown condition and growth of pines across the ozone gradient (see Miller and Rechel, this volume, for locations and site descriptions). Beginning in May 1996, plots in high-ozone areas were visited biweekly and other plots monthly until September 1996, when summer drought usually brings the growing season of understory plants to an end. At each site, plant species of known or potential usefulness as ozone biomonitors were located, and the degree and extent of foliar ozone injury symptoms were recorded on at least ten individuals of each species or however many were found at that location. The species selected for observation included those previously identified as potential ozone bioindicators in western forests (Mavity et al. 1995) and those for which ozone injury had been recorded in the San Bernardino Forest in the past. All other species with observed ozone injury symptoms in the field were also recorded. Special note was made of endemic, rare, threatened, or endangered species.

Trees and Shrubs

Characteristic ozone injury symptoms were first observed on *Sambucus mexicana* in the Camp Paivika (CP) area west of Crestline on July 8, 1996 (see Miller and Rechel, this volume, for plot location maps). Injury consisted of predominantly interveinal upper surface purple stipple and discoloration on older leaves of branches that had developed early in the growing season (Fig. 9.1A[1]). New leaves on older branches and all leaves on new branches showed little or no ozone injury. These acute ozone injury symptoms on *S. mexicana* apparently developed in response to a 2–3-day episode of elevated ozone that had occurred in early July. The majority (>70%) of *S. mexicana* shrubs showed ozone injury symptoms in this area. CP is located at 1,600 m on the southwestern slopes of the San Bernardino Mountains overlooking the Los Angeles Basin, and the highest ambient ozone concentrations and greatest amounts of ozone injury on pines have historically been recorded in this area (Miller et al. 1989). Similar but lesser amounts of ozone injury were observed on *S. mexicana* at Breezy Point (BP), 3 km NE, an area with lower exposure to ozone. By mid-August, most *S. mexicana* in an area from CP to BP to Lake Arrowhead University Conference Center (UCC) in the west and Mill Creek to Camp Angeles in the east had developed chronic ozone

[1]Figure 9.1 will be found in the color insert following page 78.

injury symptoms. These consisted of general upper surface to bifacial chlorosis of older leaves, followed by premature senescence and abscission of injured leaves. As with the acute injury symptoms, the most severe chronic injury on *S. mexicana* was observed at CP and injury decreased to the north and east of Crestline.

Ozone injury was first reported on *S. mexicana* (as *S. glauca,* Richards et al. 1968) in the San Bernardino Mountains in the late 1950s. Populations of *S. mexicana* growing on west-facing slopes and ridges of the mountains from 900 to 1,500 m have thus been exposed for the past 40 years to some of the highest ambient ozone concentrations recorded in North America. Observations of recent severe foliar ozone injury on *S. mexicana* in the same area in response to considerably lower ambient ozone concentrations than prevailed from 1955 to 1980 suggest that ozone-susceptible genotypes have not been eliminated from these *S. mexicana* populations.

Although ozone injury has been reported on *Amorpha californica* in the past, no symptoms characteristic of acute or chronic foliar ozone injury were observed on this species in 1996. The oldest leaves of some populations of *A. californica* underwent relatively rapid senescence and abscission during the third week in September. This coincided with similar rapid chlorosis and senescence of the oldest age class of Jeffrey pine needles. However, the role of ozone in inducing the rapid senescence of *A. californica* foliage could not be determined. No ozone injury was found on western dogwood (*Cornus nuttallii* Aud.) nor on two species of cherry, *Prunus emarginata* (Hook.) Walp. and *P. virginiana* L. var. *demissa* (Nutt.) Torr., even in areas where severe injury was observed on *S. mexicana* and other species. Species of *Prunus* in eastern forests, particularly *P. serotina* Ehrh. and *P. virginiana,* are highly susceptible to ozone and are used as bioindicator species for phytotoxic effects of ozone. No definitive ozone injury was observed on species of *Acer, Fraxinus,* or *Platanus* in riparian zones or on *Rubus parviflorus, Ribes* spp., or other understory woody plants.

Perennial Plants

Distinctive ozone injury symptoms were observed on *Artemisia douglasiana* Bess. (mugwort) on west-facing ridges at 1,500–1,600 m in the San Bernardino Forest beginning in early July. Injury consisted of dark brown to bronze upper surface discoloration and leaf curl on lower leaves, with symptoms gradually lessening to upper surface irregular chlorotic mottling on upper (younger) leaves (Fig. 9.1B). A population of *A. douglasiana* at CP observed on July 11, 1996, showed a wide range of susceptibility to ozone, from a few plants with no injury symptoms to many with severe upper surface irregular discoloration and bronzing on oldest seven or eight leaves on stems with 12 leaf nodes. Lowest, most severely injured leaves later abscised, but many leaves with moderate-to-severe injury persisted on the stem through August. Similar injury symptoms, but less severe in intensity, were observed in the Crestline area north to BP and east to Sky Forest, where injury was trace to light in early July. By mid- to late August, foliar injury on *A. douglasiana* at BP had become generally moderate, but some plants had no injury and a few had severe discoloration and leaf curl on the oldest leaves. Injury on *A. douglasiana* remained trace to light in other areas to the east, such as at Lake Arrowhead and UCC. *A. douglasiana* is widely distributed in the San Bernardino Mountains, particularly in the mesic western part of the range below 2,000 m. It is common along roadsides, in drainage ditches and damp ravines, and in semi-shaded disturbed areas. The combination of its susceptibility to ozone, its wide distribution and accessibility, and its persistence in damp soils where it can remain physiologically active through much of the summer makes *A. douglasiana* a good indicator of ozone effects on forest understory plants.

A second species of perennial Artemisia, *A. dracunculus* L.(tarragon), is also widespread in the San Bernardino Mountains and also showed symptoms of ozone injury. The foliar injury consisted of general or mottled upper surface chlorosis on the oldest leaves, followed by rapid senescence and abscission of injured leaves. *A. dracunculus* is common on dry disturbed soils, such as dry roadsides and construction areas. *A. dracunculus* may be as susceptible to ozone as *A. douglasiana.* However, because of the rapid abscission of injured leaves and their replacement by axillary leaves and because *A. dracunculus* is subject to drought stress on its dry sites, this species may not be as good an indicator of ozone injury as *A. douglasiana,* particularly in southern California.

Severe ozone injury was also observed on rosette and stem leaves of the biennial evening primrose [*Oenothera elata* Kunth var. *hookeri* (T.& G.) Dietr. & Wagn.] at scattered locations in the San Bernardino Mountains. Injury consisted of general chlorosis and mottled bronzy upper surface necrosis on the oldest leaves. Leaves that had moderate-to-severe injury by late July had abscised by late August. Although *Oe. elata* appeared to be highly susceptible to ozone injury and the injury symptoms were distinctive, this species is confined to damp sunny areas and is not widespread in the San Bernardino Mountains.

Milkweed (*Asclepias* spp.) has been used as a bioindicator of ozone effects in the eastern United States. However, two species of milkweed common in the San Bernardino Mountains (*A. californica* Greene and *A. fascicularis* Decne.) had no ozone injury, and both appeared to be resistant to current levels of ambient ozone. Blue ryegrass (*Elymus glaucus*) is another species that has been considered as susceptible to ozone, but foliar injury symptoms on *E. glaucus,* even in high-ozone sites, did not appear to be characteristic of ozone. Severe ozone injury symptoms were observed on *Tanacetum (Chrysanthemum) parthenium* (L.) Schultz-Bip. (feverfew) in the Crestline area as early as June 14, 1996, and this species appeared to be highly susceptible to ozone. However, *T. parthenium* is a garden plant that has become naturalized primarily along roadsides in the more developed areas around Crestline. It has too limited a distribution to be useful as an ozone indicator plant.

Annual Plants

Annual plants with characteristic foliar symptoms of ozone injury were observed at CP on June 14, 1996, after 9 days in which maximum hourly ozone exceeded 170 ppb. *Lepidium virginicum* L. var. *pubescens* (Greene) Thell., a weedy member of the mustard family (Brassicaceae), showed upper surface to bifacial bronzing and tissue collapse on the lowest leaves, with upper surface interveinal chlorosis on younger leaves. Most leaves with any amount of ozone injury had abscised by early July. *Gayophytum diffusum* T.& G. (Onagraceae) and *Cordylanthus rigidus* (Benth.) Jeps. (Scrophulariaceae) had upper surface to bifacial interveinal chlorosis and leaf curl on older leaves, with progressively lesser amounts of injury on younger leaves. As older injured leaves abscised over the next few weeks, symptoms became progressively more severe on remaining leaves. The distribution of foliar injury on these annual plants followed the same pattern as that found on *S. mexicana,* with greatest amounts of injury in the Crestline area, and lesser amounts to the north and east. No ozone injury symptoms were observed on *L. virginicum* or other annual plants growing at Heart Bar, 55 km east of Crestline. Other annual species with thin membranous leaves, such as *Clarkia rhomboidea* Doug. (Onagraceae) and *Collomia grandiflora* Lindl. (Polemoniaceae), growing next to ozone-injured species showed no ozone injury symptoms, indicating that not all annual species were equally susceptible to ozone.

Both *Gayophytum diffusum* (as *G. nuttallii*) and *Cordylanthus rigidus* (as *C. fulcratus*) were reported as present on pine plots in the high-ozone areas of CP and BP in the early 1970s (Taylor 1974; McBride and Semion 1976), although no observations of ozone injury symptoms on these species were documented at the time. These annual species have thus been exposed to high concentrations of ambient ozone for at least 25 years, yet many individuals in these plant populations continue to show moderate-to-severe ozone injury symptoms, even at ambient ozone levels considerably lower than those prevalent in the 1970s. Ozone has been shown to be a strong selective force in the production of susceptible and resistant clover clones (Heagle et al. 1994). In a controlled-environment exposure study, wild radish (*Raphanus sativus* L.) developed divergent susceptible and resistant lines after a single generation of exposure to ozone (Gillespie and Winner 1989). Apparent shifts in ozone resistance of populations of *Plantago major* L. in response to changing concentrations of ambient ozone have also been reported (Davison and Reiling 1995). Yet, susceptible genotypes of *G. diffusum* and *C. rigidus* have apparently not been eliminated from high-ozone areas in the San Bernardino Forest. The presence of ozone-susceptible genotypes of *S. mexicanus* at the same locations was noted earlier. The persistence of these ozone-susceptible genotypes of various plant species in areas of prevailingly high ambient ozone concentrations suggests that severe foliar ozone injury per se is not a sufficient selective pressure to alter significantly the genetic structure of these plant populations in the number of generations from the 1950s until the present.

Endangered Species

The center of distribution for most of the endemic rare, threatened, or endangered plant species in the San Bernardino Mountains is in Bear Valley, to the north and east of areas with significant impacts of ozone. However, at least 23 species or subspecies of rare plants in the San Bernardino Mountains grow in western areas of the forest with the potential for ozone injury (Skinner and Pavlik 1994). Only two of these were examined during this survey. No ozone injury was found on *Juglans californica* Wats., a riparian tree species found occasionally along shaded streams in the western parts of the mixed conifer zone. Another endangered species, *Horkelia wilderae* Parish (Rosaceae), grows in Barton Flats, an area in which ozone injury has been documented on ponderosa and Jeffrey pines. No ozone injury was observed in a small colony of *H. wilderae* during the summer of 1996, although moderate amounts of foliar ozone injury were seen on *Potentilla glandulosa* Lindl., a related member of the Rosaceae, growing in more exposed sites in Barton Flats. These scattered observations are insufficient to draw any conclusions regarding the actual or potential effects of ozone on rare or endangered plant species in the San Bernardino Forest.

Summary and Conclusions

Relatively little is known of the effects of ozone on annual, perennial, and woody understory plants of the mixed conifer forest in the San Bernardino Mountains. Because of the ecological and economic importance of ponderosa and Jeffrey pines to the forest and because of the severe impact of ozone on pine growth and mortality, the focus of both field research and experimental exposures has been on the dominant tree species in the mixed conifer forest. In addition, the perception has been that understory plants of the forest may not be particularly susceptible to ozone because of the hot dry climate or because susceptible genotypes have long since been extirpated by the extremely high ozone levels prevalent in the 1950s to 1970s. However, severe foliar ozone injury has been documented on understory species since the late 1950s, and despite significant reductions in ambient ozone concentrations, ozone injury continues to be observed on a number of plant species. The species most susceptible to ozone injury, based on a field survey conducted in 1996, include the shrub *Sambucus mexicana,* the perennials *Artemisia douglasiana, A. dracunculus,* and *Oenothera elata,* and several small annual plants, particularly *Lepidium virginicum* var. *pubescens, Gayophytum diffusum,* and *Cordylanthus rigidus.* The presence of severe foliar ozone injury on these plants, after 40 years or more of exposure to high ambient ozone, suggests that selection for ozone-resistant genotypes may not have occurred in these species. Both *S. mexicana* and *A. douglasiana* appear to be useful as bioindicator plants for ozone effects on forest vegetation because of their susceptibility to foliar ozone injury, their widespread distribution in California and other western states, and their relatively long growing seasons. Only a few observations were made of foliar injury symptoms on rare, threatened, or endangered plant species in the San Bernardino Forest, so no conclusions can be drawn regarding the actual or potential threat of ozone to the continued survival of these species. However, the presence of severe foliar ozone injury on a number of common plant species in the mixed conifer forest indicates that ozone pollution continues to be a significant stress on understory vegetation in the San Bernardino Mountains.

Acknowledgments

Thanks to J.R. McBride, University of California Berkeley, for numerous suggestions and to Tony Gomez, U.S. Forest Service, for the photography. Special thanks to A.C. Sanders, Department of Botany and Plant Sciences, University of California Riverside, for his careful review of the manuscript and for many valuable comments. A portion of this work was supported by a contract with the U.S. Forest Service, Pacific Southwest Station.

References

Arkley, R.J.; Glauser, R. Effects of oxidant air pollutants on pine litter fall and the forest floor. In: Miller, P.R., ed. Proc. symp. effects of air pollutants on Mediterranean and temperate forest ecosystems. Gen. tech. rep. PSW-43. Berkeley, CA: USDA For. Serv.; 1980.

Berrang, P.; Karnosky, D.F.; Bennett, J.P. Natural selection for ozone tolerance in *Populus tremuloides:* an evaluation of nationwide trends. Can. J. Forest Res. 21:1091–1097; 1991.

Bytnerowicz, A.; Manning, W.J.; Grosjean, D.; Chmielewski, W. Detecting ozone and demonstrating its phytotoxicity in forested areas of Poland—a pilot study. Environ. Pollut. 80:301–305; 1993.

D'Antonio, C.M.; Vitousek, P.M. Biological invasions by exotic grasses, the grass/fire cycle, and global change. Annu. Rev. Ecol. System. 23:63–87; 1992.

Davison, A.W.; Reiling, K. A rapid change in ozone resistance of *Plantago major* L. after summers with high ozone concentrations. N. Phytol. 131:337–344; 1995.

Dunn, D.B. Some effects of air pollution on Lupinus in the Los Angeles area. Ecology 40:621–625; 1959.

Fenn, M.E. Increased site fertility and litter decomposition rate in high pollution sites in the San Bernardino Mountains. Forest Sci. 37:1163–1181; 1991.

Fenn, M.E.; Dunn, P.H. Litter decomposition across an air pollution gradient in the San Bernardino Mountains. Soil Sci. Soc. Amer. J. 53:1560–1567; 1989.

Gillespie, C.T.; Winner, W.E. Development of lines of radish differing in resistance to O_3 and SO_2. N. Phytol. 112:353–361; 1989.

Heagle, A.S.; Miller, J.E.; Sherrill, D.E. A white clover system to estimate effects of trophospheric ozone on plants. J. Environ. Qual. 23:613–621; 1994.

Hickman, J.C., ed. The Jepson manual: higher plants of California. Berkeley, CA: Univ. of California Press; 1993.

Kozlowski, T.T.; Kramer, P.J.; Pallardy, S.G. The physiological ecology of woody plants. San Diego, CA: Academic Press; 1991

Leiberg, J.B. San Gabriel, San Bernardino, and San Jacinto forest reserves. In: Gannett, H., ed. 20th annual report, U.S. Geol. Survey to the Secretary of Agriculture. Part 5, Forest Reserves. Washington, DC: Government Printing Office; 1900:411–479.

Lewis, T.E.; Conkling, B.L., eds. Forest health monitoring. Indicators of forest health: indicator development research. Washington, DC: EMAP Center, U.S. EPA; 1996.

Mavity, E.; Stratton, D.; Barrang, P. Effects of ozone on several species of plants which are native to the western United States. Dry Branch, GA: USDA Forest Service Center for Forest Environmental Studies; 1995.

McBride, J.R.; Semion, V. Vegetation sub-committee progress report, June 1975–June 1976. In: Oxidant air pollution effects on a western coniferous forest ecosystem. EPA contract 68-03-0273. Berkeley, CA: Dept. of Forestry and Conservation, University of California; 1976.

Miller, P.R. Mixed conifer forests of the San Bernardino Mountains, California. In: Olson, R.K.; Binkley, D.; Bohm, M., eds. The response of western forests to air pollution. Ecol. studies 97. New York: Springer-Verlag; 1992:461–497.
Miller, P.R.; McBride, J.R. Effects of air pollutants on forests. In: Mudd, J.B.; Kozlowski, T.T., eds. Responses of plants to air pollution. New York: Academic Press; 1975:195–235.
Miller, P.R.; McBride, J.R.; Schilling, S.L; Gomez, A.P. Trends of ozone damage to conifer forests between 1974 and 1988 in the San Bernardino Mountains of southern California. In: Olson, R.K.; Lefohn, A.S., eds. Effects of air pollution on western forests. TR-16. Pittsburgh, PA: Air Waste Manage. Assoc., 1989:309–323.
Miller, P.R.; Stolte, K.W.; Duriscoe, D.M.; Pronos, J. Evaluating ozone air pollution effects on pines in the western United States. PSW-GTR-155. Albany, CA: U.S. Forest Service; 1996.
Miller, P.R.; Arbaugh, M.J.; Temple, P.J. Ozone and its known and potential effects on forests in western United States. In: Sandermann, H.; Wellburn, A.E.; Heath, R.L., eds. Forest decline and ozone. Ecol. studies 127. Berlin: Springer-Verlag; 1997:39–67.
Minnich, R.A.; Barbour, M.G.; Burk, J.H.; Fernau, R.F. Sixty years of change in Californian conifer forests of the San Bernardino Mountains. Conserv. Biol. 9:902–914; 1995.
Munz, P.A. A flora of southern California. Berkeley: Univ. of California Press; 1974:625.
National Research Council. Ozone and other photochemical oxidants. Washington DC: Committee on medical and biological effects of environmental pollutants, National Research Council; 1977.
Nobel, P.S. Physicochemical and environmental plant physiology. San Diego, CA: Academic Press; 1991.
Preston, K.P. Ozone and sulfur dioxide effects on the growth of California coastal sage scrub species. Ph.D. dissertation, Dept. of Geography, UCLA, Los Angeles; 1986.
Price, H.E. Effects of ozone on growth and reproduction of grasses. Ph.D. dissertation. Dept. of Biology, University of Utah, Salt Lake City; 1973.
Reich, P.B. Quantifying the response of plants to ozone: a unified explanation. Tree Physiol. 3:63–91; 1987.
Reiling, K.; Davison, A.W. Spatial variation in ozone resistance of British populations of *Plantago major* L. N. Phytol. 122:699–708; 1992.
Richards, B.L., Sr.; Taylor, O.C.; Edmunds, F.G., Jr. Ozone needle mottle of pines in southern California. J. Air Pollut. Control Assoc. 18:73–77; 1968.
Skinner, M.W.; Pavlik, B.M. California Native Plant Society's inventory of rare and endangered vascular plants of California. CNPS Special Publ. 1. Sacramento, CA: CNPS; 1994.
Smith, G. Forest health monitoring second ozone bioindicator workshop summary of proceedings. Amherst, MA: Dept. Forestry Wildlife Management, Univ, of Massachusetts; 1995.
Stolte, K.W. The effects of ozone on chaparral plants in the California South Coast Air Basin. M.S. thesis, Dept. of Botany and Plant Science, Univ. of California, Riverside; 1982.
Taylor, O.C., ed. Oxidant air pollution effects on a western coniferous forest ecosystem, Task D. Riverside, CA: Statewide Air Pollution Research Center, University of California; 1974.
Taylor, G.E., Jr.; Pitelka, L.F. Genetic diversity of plant populations and the role of air pollutants. In: Barker, J.R.; Tingey, D.T., eds. Air pollution effects on biodiversity. New York: Van Nostrand Reinholt; 1992.
Taylor, G.E., Jr.; Pitelka, L.F.; Clegg, M.T., eds. Ecological genetics and air pollution. New York: Springer-Verlag; 1991.
Temple, P.J. Oxidant air pollution effects on plants of Joshua Tree National Monument. Environ. Pollut. 57:35–47; 1989.
Thompson, C.R.; Olszyk, D.M.; Kats, G.; Bytnerowicz, A.; Dawson, P.J.; Wolf, J.W.

Effects of ozone and sulfur dioxide on annual plants of the Mojave Desert. J. Air Pollut. Control. Assoc. 34:1017–1022; 1984.
Treshow, M.; Stewart, D. Ozone sensitivity of plants in natural communities. Biol. Conserv. 5:209–214; 1973.
Westman, W.E. Oxidant effects on Californian coastal sage scrub. Science 205:1001–1003; 1979.
Westman, W.E. Factors influencing the distribution of species of Californian coastal sage scrub. Ecology 62:439–455; 1981.
Yoshida, L.C. Does ozone exposure alter growth and carbon allocation of Elymus glaucus, a mycorrhizal plant? M.S. thesis, Dept. of Biology and Microbiology, California State Univ., Los Angeles; 1994.

10. Epiphytic Lichens in the San Bernardino Mountains in Relation to Oxidant Gradients

T.H. Nash III and L.L. Sigal

Introduction

The use of lichens as indicators of air pollution, particularly in regions with high levels of SO_2, is well established (Nieboer et al. 1976; Richardson and Nieboer 1983; Nash and Wirth 1988; Galun and Ronen 1990; Gries 1996). Near sources of SO_2, lichen species show marked impoverishment, with species richness frequently reduced to less than 5% of normal levels. In Europe, such species patterns along SO_2 gradients can be directly correlated with long-term (winter months or annual) ambient SO_2 measurements (Hawksworth and Rose 1970, 1976). The apparent differential sensitivity to SO_2 among various lichen species based on these field studies is confirmed by similar patterns in SO_2 sensitivity based on short-term (hourly) fumigations with subsets of these species (Nash 1988). Further confirmation is provided by the fact that, following reductions in SO_2 emissions in the United Kingdom (and other areas of Europe) in recent years, several lichen species are migrating back into areas from which they had previously disappeared (Henderson-Sellers and Seaward 1979; Rose and Hawksworth 1981; Hawksworth and McManus 1989).

This chapter reviews investigations we have conducted on lichen communities in southern California, with an emphasis on investigations in the San Bernardino Mountains. These field data, together with ambient oxidant monitoring by P. Miller's research group, provided the first evidence that lichens may also respond to oxidant air pollutants. Experimental investigations with ozone and other oxidants provide some supporting evidence that the field patterns may actually reflect sensitivity to oxidants, but in general, lichens have not proved to be as sensitive as vascular plants to ozone. Consequently, it is not surprising that lichens appear to respond more to SO_2 than O_3 in regions such as Indiana, where both SO_2 gradients and oxidant gradients are present (McCune 1988). Even in the San Bernardino Mountains, pollutants other than oxidants may be partially responsible for the patterns observed in the epiphytic lichen community. Thus, the final section of this chapter discusses pollutants that may also contribute to lichen decline in the San Bernardino Mountains.

Lichen Biology and Use as Biomonitors of Air Pollution

A lichen is essentially an obligate, stable, self-supporting symbiosis between a fungus (mycobiont) and an alga and/or cyanobacterium (photobiont). The names of lichens refer strictly to the fungal component, and approximately 98% of all lichen-forming fungi are ascomycetes (Honegger 1996). Lichens are frequently useful as long-term integrators of pollutant response (Gries 1996). Because almost all lichens are slow-growing perennials, whose inherent morphological form does not change markedly over years, decades, and in some cases centuries (Nash 1996a), and because they accumulate atmospheric nutrients or contaminants (Nash 1996b), they are useful as monitors of atmospheric deposition. Their ability to accumulate some pollutants may lead to toxicity becoming manifest. Inherent differences between lichens and other plants may account for the apparent sensitivity of many lichens to air pollutants. For example, in contrast to vascular plants, lichens do not possess stomata and thus have no regulatory mechanism over gas exchange. Potentially then, the lichen may be exposed to a higher pollutant dose. The lack of cuticles in lichens may also result in enhanced pollutant uptake.

Overall Lichen Patterns in Southern California

Lichens, on the basis of extensive field studies (Nash and Sigal 1980; Ross 1982; Sigal and Nash 1983), are well known to exhibit differential sensitivity to the Los Angeles environment in general and the San Bernardino Mountains in particular. Field data for the region as a whole demonstrate that upward of 50% of the lichen species that occurred at the turn of the century have disappeared (Sigal and Nash 1983). A disproportionate number of these are cyanolichens (i.e., those lichens in which cyanobacteria are the photobionts), such as many *Collema, Peltigera,* and *Pseudocyphellaria* species. Because ozone is an unquestioned phytotoxic air pollutant to vascular plants in this area (e.g., Miller et al. 1997), it is reasonable to infer that lichens may also be responding to ozone. This inference is strengthened by the fact that gradients in oxidant levels do, in fact, reflect similar gradients in species richness and abundance of lichens occurring on both oak and conifer species (Nash and Sigal 1980; Ross 1982; Sigal and Nash 1983).

Table 10.1. Inferred Sensitivity of Selected Lichens Species to Air Pollution in Southern California[a]

Very sensitive	Sensitive	Moderately tolerant	Tolerant
Nodobryoria abbreviata	*Collema nigrescens*	*Hypogymnia imhaugii*	*Letharia vulpina*
Bryoria fremontii	*Leptogium californicum*	*Melanelia glabra*	*Physcia biziana*
Vulpicida canadensis	*Parmelia sulcata*	*Melanelia elegantula*	*Physcia tenella*
Evernia prunastri	*Parmelina quercina*	*Melanelia subolivacea*	*Physconia detersa*
Peltigera canina	*Peltigera rufescens*	*Xanthoria polycarpa*	*Xanthoria fallax*
Peltigera collina	*Phaeophyscia ciliata*		
Peltigera didactyla	*Phaeophyscia orbicularis*		
Phaeophyscia sciastra	*Leptochidium albociliatum*		
Platismatia glauca	*Tuckermannopsis merrillii*		
Pseudocyphellaria anthraspis	*Usnea* spp.		
Ramalina farinacea			
Ramalina menziesii			
Xanthoria candelaria			

[a]Based on Sigal and Nash 1983 and Ross 1982. The nomenclature is revised relative to Esslinger and Egan 1995.

Sigal and Nash (1983) developed an elaborate classification involving more than 20 species and their inferred differential sensitivity to oxidants (Table 10.1): very sensitive species had disappeared entirely from mountainous regions around Los Angeles; sensitive species, such as *Usnea* spp., could only be found in vestigal amounts compared with the turn of the century; tolerant species, such as *Hypogymnia imshaugii* (*H. enteromorpha* in older literature reports), were still commonly present but exhibited distinct patterns of morphological alteration: and the few very pollution tolerant species, such as *Letharia vulpina* and *Melanelia subolivacea,* occurred commonly and exhibited no morphological change compared with control (historically in time or geographically in regions remote from Los Angeles) material. Peterson et al. (1992) used these field data to propose that, for species potentially occurring in the San Bernardino Mountains, very sensitive species respond to long-term (7-hour mean for May–October) ozone concentrations of less than 20 ppb, whereas very tolerant species can tolerate more than 70 ppb (Table 10.2).

Comparison of Lichens Between the San Bernardino Mountains and Cuyamaca Rancho State Park

Probably there is no region in southern California where elevated oxidant levels do not occur relative to conditions in the early part of this century. However, differences in the degree of urbanization and its associated vehicular traffic load together with patterns of ventilation of the urban air into the surrounding mountainous regions results in considerable variation in pollutant exposure among different mountain regions. As part of a large-scale epiphytic lichen community

Table 10.2. Approximate Correspondence of Lichen Sensitivity Classes (Table 10.1) to Measured Pollutant Exposures in the San Bernardino Mountains

Pollutant	Sensitivity class			
	Very sensitive	Sensitive	Tolerant	Very tolerant
Ozone (ppb)[a]	< 20	21–40	41–70	> 70
S (kg ha^{-1} yr^{-1})	< 1.5	1.5–2.5	2.6–3.5	> 3.5
N (kg ha^{-1} yr^{-1})	< 2.5	2.6–5.0	5.1–7.0	> 7.0

[a]Ozone concentrations are based on 7-hour means from May to October. (Based on Peterson et al. 1992.)

survey conducted over 3 years in the late 1970s within five southern California mountain ranges (Sigal 1979), the Cuyamaca Rancho State Park in San Diego County was found to be the best "control" region for comparison with mideleva-tion plots of conifer and oak stands in the San Bernardino Mountains. Elevations (1,500–2,100 m), forest composition (*Abies concolor, Pinus jeffreyi, Pinus pon-derosa,* and *Quercus kelloggii),* annual precipitation (predominantly winter) amount (700–900 mm for most sites), and distribution and occurrence of summer fogs are comparable (Sigal 1979). However, one potentially important variable that was not comparable is the greater intensity of logging that has occurred in the San Bernardino Mountains this century (McBride, personal communication). The discussion of epiphytic lichens below emphasizes differences between these two regions as well as along the oxidant gradients within the San Bernardino Mountains.

Altogether 15 sites within the San Bernardino Mountains and 11 sites within Cuyamaca Rancho State Park were sampled in forests with tree densities of 200–375/ha. At each site, 100-m transects were established, and lichen cover by species at 1.5 m from the ground for ten trees (>38 cm dbh for the conifers and >14 cm dbh for the understory oak) per transect was estimated by using a line intercept method on conifers and small quadrats (5 × 16 cm) placed at the four cardinal directions for *Q. kelloggii.* Sites within the San Bernardino Mountains included the permanent plots established by Taylor (1974) reflecting an oxidant gradient. For conifers, at least 50 trees each were included for *Pinus jeffreyi, P. ponderosa,* and *Abies concolor* in each area. Further details of sampling pro-cedures are given in Sigal (1979) and Sigal and Nash (1983).

Lichens Occurring on Conifers

Altogether 15 lichen species were sampled in Cuyamaca Rancho State Park and seven lichen species in the San Bernardino Mountains (Table 10.3). Frequency of occurrence was consistently higher in the Cuyamaca Rancho State Park. On the basis of herbarium records and literature reports (Hasse 1913), all the lichens sampled in the Cuyamaca Rancho State Park were known to have occurred in the San Bernardino Mountains at the turn of the century. Hence it is inferred that the lichen flora has become impoverished in the latter area.

Table 10.3. Percentage Frequency (Number of Trees on Which Species Occurred, Divided by the Total Number of Trees Sample Times 100) of Lichen Species Occurring on Conifers in Cuyamaca Rancho State Park (CRSP) and the San Bernardino Mountains (SBM)[a]

Lichen species	CRSP	SBM
Hypogymnia imshaugii	76	33
Letharia vulpina	72	49
Nodobryoria abbreviata	49	–
Tuckermannopsis merrillii	34	16
Melanelia subolivacea	27	1
Platismatia glauca	16	–
Physcia tenella	12	–
Bryoria fremontii	9	–
Melanelia elegantula	7	1
Parmelia sulcata	5	+
Xanthoria fallax	4	1
Usnea spp.	1	–
Vulpicida canadensis	1	–
Parmelina quercina	1	–
Physconia detersa	1	–

[a]Species not found are indicated by –; those not sampled but observed in adjacent areas are indicated by +.

For the two most frequent lichens, *Hypogymnia imshaugii* and *Letharia vulpina,* average cover (approximately 12% each) was greatest on *Abies concolor* at Cuyamaca Rancho State Park; among the pines, their cover was less than 3% except that 7% cover for *Letharia vulpina* was found on *Pinus jeffreyi.* By contrast, lower cover values of both species were found in the San Bernardino Mountains (<4% on *Abies concolor* and 0–1% on the pines). For the eight San Bernardino sites at which oxidant and ozone data were available, the cover of the two species combined was less than 0.5% for the six sites with oxidant dose estimates of greater than 580 ppm/h/m^3 (Sigal and Nash 1983). At lower oxidant doses, lichen cover was similar to that found in the Cuyamaca Rancho State Park.

Within populations of *Hypogymnia imshaugii,* marked morphological changes were evident in randomly selected samples from the San Bernardino Mountains (n = 66) compared with the Cuyamaca Rancho State Park (n = 32) (Sigal and Nash 1983). Each sample was compared against standards representing four degrees of bleaching and convolution of the lobes with the following results: San Bernardino Mountains (0% unbleached, 30% slightly bleached, 27% moderately bleached, and 43% completely bleached; 0% unconvoluted, 3% slightly convoluted, 53% moderately convoluted, and 44% convoluted); Cuyamaca Rancho State Park (59% unbleached, 41% slightly bleached, 0% moderately or completely bleached; 37% unconvoluted, 47% slightly convoluted, 16% moderately convoluted, and 0% convoluted). In addition, fertility of specimens averaged only 33% in the San Bernardino Mountains compared with 75% in Cuyamaca Rancho State Park.

Table 10.4. Number of Lichen Species Found in Frequency Classes in the San Bernardino Mountains and Cuyamaca Rancho State Park (CRSP)[a]

Location	Abundant (> 74%)	Frequent (73–53%)	Occasional (52–28%)	Rare (> 27%)	Present (not sampled)	Total
SBM	6	3	2	14	5	30
CRSP	8	5	9	8	9	39

[a]Based on Sigal 1979.

Lichens Occurring on Hardwoods (*Quercus kelloggii*)

Although reduction in lichen frequency on black oak in the San Bernardino Mountains compared with Cuyamaca Rancho State Park was evident (Table 10.4), results were more complex and not as stark as those on the conifers. Altogether, 30 epiphytic species were found in sampling plots in Cuyamaca Rancho State Park and 25 species in the San Bernardino Mountains (Table 10.4). The most common eight species (*Collema nigrescens, Lecanora pacifica, Lecidella elaeochroma, Melanelia subolivacea, Physcia tenella, Physconia detersa, Rinodina laevigata,* and *Xanthoria fallax*) occurred on all transects and 75% or more of the trees sampled in both regions except for *Collema nigrescens,* which did not occur in any of the sample plots in the San Bernardino Mountains (Sigal 1979). Nine species were sampled in the Cuyamaca Rancho State Park but not in the San Bernardino Mountains, and these ranged in abundance from those occurring on 63–100% of the transects and 13–59% of the trees (*Caloplaca chrysophthalma, Ochrolechia mexicana, Pachyspora mutabilis, Phaeophyscia ciliata,* and *Physconia pulverulenta*) to a number occurring on less than 30% of the transects and on less than 5% of the trees (*Aspicilia* sp., *Caloplaca californica, Pertusaria* sp., and *Xanthoria candelaris*). In addition, *Buellia triphragmioides, Peltigera collina,* and *Physcia biziana* were collected on black oak outside the sampling plots in the Cuyamaca Rancho State Park but were not found in the San Bernardino Mountains. By contrast, *Lecanora* sp. and *Physcia callosa* occurred in low frequency in the sampling plots in the San Bernardino Mountains but were not found in Cuyamaca Rancho State Park; *Melanelia elegantula, Parmelina quercina,* and *Physcia stellaris,* also occurred infrequently in the sampling plots in the San Bernardinos but were only found outside the sampling plots in Cuyamaca Rancho State Park. *Hypocenomyce scalaris,* a species frequently colonizing burned trunks, was observed outside the sampling plots in the San Bernardino Mountains but was not found in the Cuyamaca Rancho State Park. Three species (*Leptogium californicum, Leptochidium albociliatum,* and *Peltigera rufescens*) were observed in both mountain regions but were restricted to tree bases outside of the sampling plots. The remaining 13 species (*Buellia punctata, Caloplaca* cf. *pinicola, Candelaria concolor, Candelariella vitellina, Lecanora hageni, Lecidea* sp., *Melanelia glabra, Parmelia sulcata, Phaeophyscia orbicularis, Physcia aipolia, Rinodina pyrina, Strangospora ochrophora,* and *Xanthoria polycarpa*) were sampled in both mountain regions, and ten of these occurred in approximately equal or greater frequency in the Cuyamaca Rancho State Park.

The most common five species (*L. pacifica, M. subolivacea, P. tenella P. detersa,* and *X. fallax*) averaged 2–5% cover each in the two mountain regions, and these species did not exhibit significant overall differences in cover between the two regions (Sigal 1979). *Collema nigrescens* also had high cover in Cuyamaca Rancho State Park but was not sampled (although it did occur outside the plots) in the San Bernardino Mountains. Although one might infer that *C. nigrescens* was the only oxidant-sensitive species among the six, further analysis revealed a correlation between lichen cover and oxidant dose within the San Bernardino Mountains.

For cover values of the dominant five lichens, a principal component analysis (PCA) ordination of the ten sites most homogeneous with respect to elevation and topographical position was run (Nash and Sigal 1980). Eighty percent of the stand variation was explained by the first axis of the PCA ordination, and this was primarily related to the total cover of these five lichens and especially the cover of the most common lichen, *M. subolivacea.* The first axis of the ordination was also strongly related to oxidant injury score estimates for pines at the same sites (Miller and McBride 1989) and inversely related ($P = .007$ for r_s) to oxidant dose estimates for the sites (Nash and Sigal 1980). Thus, oxidants appear to control the distribution of the most common black oak lichen, *M. subolivacea,* within the San Bernardino Mountains. By contrast, there was no significant correlation between oxidant dose and cover of the other four lichen species, and therefore, these species are inferred to be tolerant of oxidants.

Experimental Investigations with Oxidants

The degree to which the patterns observed in the field actually reflect the inferred differential sensitivity to oxidants requires verification by experimental fumigations with ozone and other oxidants, such as peroxyacetylnitrate (PAN). Unfortunately, in contrast to the vast literature on experimental studies with ozone and vascular plants, the literature dealing with fumigations of lichens with ozone is limited to about 15 species (Fields 1988; Scheidegger and Schroeter 1995; Gries 1996). Some differential effects among southern California species corresponding to the oxidant hypothesis have been demonstrated, but far more experimentation is necessary. For example, short-term fumigations with ozone (0.5 ppm) resulted in significantly greater decline of gross photosynthesis of *Parmelia sulcata,* a sensitive species (Table 10.1) than in *Hypogymnia imshaugii,* a moderately tolerant species (Nash and Sigal 1979). In addition, Scheidegger and Schroeter (1995) conducted long-term (80 days) ozone fumigations with much lower concentrations (90 ppb day; 40 ppb night) with seven lichen species, two of which occur in southern California. Significant reductions in the chlorophyll fluorescence parameter (F_v/F_m) and increases in the percentage of collapsed photobiont cells occurred in *Collema nigrescens* and *Evernia prunastri.* By contrast, *Ramalina men-*

ziesii, a low-elevation species known to be very sensitive to the Los Angeles ambient environment (Table 10.1) (Ross 1982; Boonpragob and Nash 1991), exhibited no significant decline in photosynthesis to a range of O_3 fumigations up to 0.8 ppm for as long as 12 hours (Ross and Nash 1983). Furthermore, some other lichen species also appear relatively tolerant of ozone, such as the terricolous *Cladina arbuscula* (Rosentreter and Ahmadjian 1977) and the epiphytic *Lobaria pulmonaria* (Sigal and Johnston 1986). Thus, at least some of the species listed in Table 10.1 probably reflect sensitivity to other factors.

In the case of PAN, Sigal and Taylor (1979) demonstrated that *Parmelia sulcata* was differentially more sensitive to 4-hour fumigations with 50 or 100 ppb PAN than *Hypogymnia imshaugii,* a result that parallels the ozone results discussed above. However, *Collema nigrescens* was relatively unaffected by PAN fumigations (Sigal and Taylor 1979).

Alternative Pollutants That May Contribute to Lichen Decline in the San Bernardino Mountains

It is reasonable to postulate that these southern California lichens may actually be responding to one or more alternative pollutants that also occur in the Los Angeles region. A multitude of potentially injurious pollutants can be suggested, including a wide range of organopollutants, such as aldehydes and peroxides, various metals, and SO_2 and HNO_3 gas. The paucity of information on organopollutant effects on lichens leaves us no basis to evaluate the plausibility of one or more of them being phytotoxic. The only investigation of which we are aware is one dealing with formaldehyde mists (Muir and Shirazi 1996) in which only a slight reduction in acetylene reduction activity (as an assay for nitrogen fixation) in *Lobaria pulmonaria* was found.

Many metal aerosols are known to be present in the Los Angeles Basin (Lyons 1993) and deposition gradients within the San Bernardino Mountains are known (P. Miller, personal communication). An investigation by Boonpragob and Nash (1990) on the lower slopes of the San Gabriel Mountains (immediately west of the San Bernardino Mountains) has shown that accumulation of Ba, Cd, Cu, Fe, K, Mg, Mn, Pb, Sr, Ti, and Zn occurred over a 10-week period. However, the magnitude of accumulation only increased by a factor of 1.3 to 3.7, and therefore toxicity was unlikely (Nash 1989), although long-term accumulation effects cannot be entirely dismissed.

The occurrence of high levels of oxidants in the San Bernardino Mountains and the ability of mimic field symptomology with controlled ozone fumigations of pines and other species has long lead to the assumption that oxidants were the major phytotoxic gases in the mountains (e.g., Miller et al. 1997). Nevertheless, sulfur dioxide is present albeit at lower concentrations than those found in many northern regions or around some SO_2 point sources. For example, during the time when field investigations of the lichen communities were conducted, hourly concentrations of SO_2 as high as 235 $\mu g\ SO_2/m^3$ were measured within the San

Bernardino Mountains (California Air Resources Board 1978, 1979). Significant reductions in lichen photosynthesis have been measured at concentrations as low as 260 μg SO_2/m^3 over 2–4 hours (Huebert et al. 1985; Sanz et al. 1992). One of the species used in those studies was *Evernia prunastri,* which is listed as being very sensitive to the Los Angeles environment (Table 10.1). It is therefore plausible that its decline is related to SO_2 and not oxidants. This inference is further supported by the fact that an S deposition gradient is also found within the San Bernardino Mountains (Table 10.2). Further experimentation with lichen species occurring in southern California is required to evaluate the degree to which SO_2 may be important in the decline of other species.

Another prime candidate for investigation is HNO_3 as a gas, which is a major constituent of acid deposition in this region and in the San Bernardino Mountains in particular (e.g., Grosjean and Bytnerowicz 1993; Bytnerowicz and Fenn 1996; Fenn et al. 1996). The suggestion is strengthened by ongoing and recent transplant studies with lichens involving two *Ramalina* species from relatively clean air habitats (respectively Mount Palomar and San Nicolas Island) to analogous habitats within the Los Angeles Basin and repeatedly observing that they have died over a few weeks during summer periods (Boonpragob and Nash 1991, Zambrano and Nash, unpublished data). Associated with this deterioration has been a massive accumulation of H^+ and NO_3^- (up to 45 μmol H^+/g and up to 170 μmol NO_3^-/g over 10 weeks) by the lichen thalli (Boonpragob et al. 1989; Zambrano et al., unpublished data). Although other sources of H^+ and NO_3^- ions may well exist, it is reasonable to postulate that most of these ions have originated from HNO_3 deposition, particularly in the light of the very high deposition velocity for this gas (Bytnerowicz et al. 1986; Bytnerowicz and Fenn 1996). Nitrate itself is, of course, a nutrient, and no detrimental effect on lichens has been demonstrated by short-term elevated treatments (Blum et al. 1989), although stimulation of one symbiont (the photobiont?) over the other and the subsequent breakdown of the symbiosis cannot be eliminated as a possible response. However, acidic responses associated with H^+ accumulation may well parallel responses to SO_2 incorporation where H_2SO_3 and H_2SO_4 are rapidly formed in the aqueous phase. Because we are aware of no literature dealing with the effects of HNO_3 gas on lichens, fumigations with HNO_3 are needed.

References

Blum, O.B.; Nash, T.H., III; Gebauer, R. The effect of nitrate on CO_2 exchange in the epiphytic lichens *Ramalina menziesii* Tayl. and *Pseudocyphellaria anthraspis* (Ach.) Magn. from central California. In: Noble, R.D.; Martin, J.L.; Jensen, K.F., eds. Air pollution effects on vegetation including forest ecosystems. Broomall, PA: USDA Forest Service, Northeastern Experiment Station; 1989:181–185.

Boonpragob, K.; Nash, T.H., III. Seasonal variation of elemental status in the lichen *Ramalina menziesii* Tayl. from two sites in southern California: evidence for dry deposition accumulation. Environ. Exp. Bot. 30:415–428; 1990.

Boonpragob, K.; Nash, T.H., III. Physiological responses of the lichen *Ramalina menziesii* Tayl. to the Los Angeles urban environment. Environ. Exp. Bot. 31:229–238; 1991.

Boonpragob, K.; Nash, T.H., III.; Fox, C.A. Seasonal deposition patterns of acidic ions and ammonium to the lichen *Ramalina menziesii* Tayl. in southern California. Environ. Exp. Bot. 29:187–197; 1989.

Bytnerowicz, A.; Fenn, M.E. Nitrogen deposition in California forests: a review. Environ. Pollut. 92:127–146; 1996.

Bytnerowicz, A.; Miller, P.R.; Olszyk, D.M.; Dawson, P.J.; Fox, C.A. Gaseous and particulate air pollution in the San Gabriel Mountains of southern California. Atmos. Environ. 21:1805–1814; 1986.

California Air Resources Board. California air quality data. Summary of 1978 air quality data. Sacramento, CA: CARB; 1978.

California Air Resources Board. California air quality data. Summary of 1979 air quality data. Sacramento, CA: CARB; 1979.

Esslinger, T.L.; Egan, R.S. A sixth checklist of the lichen-forming, lichenicolous, and allied fungi of the continental United States and Canada. Bryologist 98:467–549; 1995.

Fenn, M.E.; Poth, M.A.; Johnson, D.W. Evidence for nitrogen saturation in the San Bernardino Mountains in southern California. For. Ecol. Manage. 82:211–230; 1996.

Fields, R.F. Physiological responses of lichens to air pollutant fumigations. In: Nash, T.H., III; Wirth, V., eds. Lichens, bryophytes and air quality. Berlin and Stuttgart: J. Cramer; 1988:175–200.

Galun, M.; Ronen, R. Interaction of lichens and pollutants. In: Galun, M., ed. Handbook of lichenology, vol. III. Boca Raton, FL: CRC Press; 1990:55–72.

Gries, C. Lichens as indicators of air pollution. In: Nash, T.H., III, ed. Lichen biology. Cambridge, UK: Cambridge University Press; 1996:240–254.

Grosjean, D.; Bytnerowicz, A. Nitrogenous air pollutants at a southern California mountain forest smog receptor site. Atmos. Environ. 27A:483–492; 1993.

Hasse, H.E. The lichen flora of southern California. Contrib. U.S. Natl. Herb. 17:1–132; 1913.

Hawksworth, D.L.; McManus, P.M. Lichen recolonization in London under conditions of rapidly falling sulphur dioxide levels, and the concept of zone skipping. Bot. J. Linn. Soc. 100:99–109; 1989.

Hawksworth, D.L.; Rose, F. Qualitative scale for stimulating sulphur dioxide in air pollution in England and Wales using epiphytic lichens. Nature 227:145–148; 1970.

Hawksworth, D.L.; Rose, F. Lichens as air pollution monitors. Studies in biology 66. London: Edward Arnold; 1976.

Henderson-Sellers, A.; Seaward, M.R.D. Monitoring lichen reinvasion of ameliorating environments. Environ. Pollut. 19:207–213; 1979.

Honegger, R. Mycobionts. In: Nash, T.H., III, ed. Lichen biology. Cambridge, UK: Cambridge University Press; 1996:24–36.

Huebert, D.B.; L'Hirondelle, S.J.; Addison, P.A. The effects of sulfur dioxide on net CO_2 assimilation in the lichen *Evernia mesomorpha* Nyl. N. Phytol. 100:643–651; 1985.

Lyons, J.M. Size distribution of trace metals in the Los Angeles atmosphere. Atmos. Environ. 27:237–249; 1993.

McCune, B. Lichen communities along O_3 and SO_2 gradients in Indianapolis. Bryologist 91:223–228; 1988.

Miller, P.R.; McBride, J.R. Trends of ozone damage to conifer forests in the western United States, particularly southern California. In: Bucher, J.B.; Bucher-Wallin, I., eds. Proceedings of the 14th meeting for specialists in air pollution effects on forest ecosystems. Birmensdorf: IUFRO; 1989:61–68.

Miller, P.R.; Arbaugh, M.J.; Temple, P.J. Ozone and its known and potential effects on forests in western United States. In: Sandermann, H.; Wellburn, A.R.; Heath, R., eds. Forest decline and ozone. Ecological studies 127. Berlin-Heidelberg: Springer-Verlag; 1997:39–67.

Muir, P.S.; Shirazi, A.M. Effects of formaldehyde-enriched mists on *Pseudotsuga menziesii* (Mirbel) Franco and *Lobaria pulmonaria* (L.) Hoffm. Environ. Pollut.; 94:227–234; 1996.

Nash, T.H., III. Correlating fumigation studies with field effects. In: Nash, T.H., III; Wirth, V., eds. Lichens, bryophytes and air quality. Berlin-Stuttgart: J. Cramer; 1988:201–216.

Nash, T.H., III. Metal tolerance in lichens. In: Shaw, A.J., ed. Heavy metal tolerance in plants: evolutionary aspects. Boca Raton, FL: CRC Press; 1989:119–131.

Nash, T.H., III. Introduction. In: Nash, T.H. III, ed. Lichen biology. Cambridge, UK: Cambridge University Press; 1996a:1–7.

Nash, T.H., III. Nutrients, elemental accumulation and mineral cycling. In: Nash, T.H., III, ed. Lichen biology. Cambridge, UK: Cambridge University Press; 1996b:136–153.

Nash, T.H., III; Sigal, L.L. Gross photosynthetic response of lichens to short-term ozone fumigations. Bryologist 82:280–285; 1979.

Nash, T.H., III, Sigal, L.L. (1980) Sensitivity of lichens to air pollution with an emphasis on oxidant air pollutants. In: Miller, P.R., tech. coord. Proceedings of the symposium on effects of air pollutants on Mediterranean and temperate forest ecosystems. Gen. tech. rep. PSW-43. Berkeley, CA: USDA; 1980:117–124.

Nash, T.H., III; Wirth, V., eds. Lichens, bryophytes and air quality. Berlin-Stuttgart: J. Cramer; 1988.

Nieboer, E.; Richardson, D.H.S.; Puckett, K.J.; Tomassini, F.D. The phytotoxicity of sulphur dioxide in relation to measurable responses in lichens. In: Mansfield, T.A., ed. Effects of air pollutants on plants. Cambridge, UK: Cambridge University Press; 1976:61–85.

Peterson, D.L.; Schmoldt, D.L.; Eilers, J.M.; Fisher, R.W.; Doty, R.D. Guidelines for evaluating air pollution impacts on class I wilderness areas in California. Gen. tech. rep. PSW-GTR-136. Berkeley, CA: USDA; 1992.

Richardson, D.H.S.; Nieboer, E. Ecophysiological responses of lichens to sulphur dioxide. J. Hattori Bot. Lab. 54:331–351; 1983.

Rose, F.; Hawksworth, D.L. Lichen recolonization in London's cleaner air. Nature 289:289–292; 1981.

Rosentreter, R.; Ahmadjian, V. Effect of ozone on the lichen *Cladonia arbuscula* and the *Trebouxia* phycobiont of *Cladina stellaris*. Bryologist 80:600–605; 1977.

Ross, L.J. Lichens of coastal live oak in relation to ozone. M.S. thesis, Arizona State Univ., Tempe; 1982.

Ross, L.J.; Nash, T.H., III. Effect of ozone on gross photosynthesis of lichens. Environ. Exp. Bot. 23:71–77; 1983.

Sanz, M.J.; Gries, C., Nash, T.H., III. Dose-response relationships for SO_2 fumigations in the lichens *Evernia prunastri* (L.) Ach. and *Ramalina fraxinea* (L.) Ach. N. Phytol. 122:313–319; 1992.

Scheidegger, C.; Schroeter, B. Effects of ozone fumigation on epiphytic macrolichens: ultrastructure, CO_2 gas exchange and chlorophyll fluorescence. Environ. Pollut. 88:345–354; 1995.

Sigal, L.L. I. Lichen communities of southern California mountains: an ecological survey relative to oxidant air pollution. II. Gross photosynthetic response of lichens to short-term ozone fumigations. III. Preliminary studies of the gross photosynthetic response of lichens to peroxyacetyl nitrate fumigations. Ph.D. dissertation, Arizona State Univ., Tempe; 1979.

Sigal, L.L.; Johnston, J.W., Jr. Effects of acidic rain and ozone on nitrogen fixation and photosynthesis in the lichen *Lobaria pulmonaria* (L.) Hoffm. Environ. Exp. Bot. 26:59–64; 1986.

Sigal, L.L.; Nash, T.H., III. Lichen communities on conifers in southern California mountains: an ecological survey relative to oxidant air pollution. Ecology 64:1343–1354; 1983.

Sigal, L.L.; Taylor, O.C. Preliminary studies of the gross photosynthetic response of lichens to peroxyacetylnitrate fumigations. Bryologist 82:564–575; 1979.
Taylor, O.C., ed. Oxidant air pollutant effects on a western coniferous forest ecosystem. Annual progress report. Riverside, CA: University of California, Statewide Air Pollution Research Center; 1974.

11. Wet and Dry Pollutant Deposition to the Mixed Conifer Forest

A. Bytnerowicz, M.E. Fenn, P.R. Miller, and M.J. Arbaugh

Introduction

The Mediterranean climate in southern California regulates wet and dry deposition characteristics in the San Bernardino Mountains (SBM). Long dry periods in combination with the large air pollution emissions from mobile and stationary sources in the greater Los Angeles area lead to elevated concentrations of ozone and accumulation of gaseous and particulate N and S compounds within plant canopies and on the forest floor. Pollution plumes and transport of air from the nearby Mojave Desert also bring substantial amounts of basic cations to the SBM. Precipitation events occurring after extended dry periods result in concentrated pulses of nutrient transport from foliar surfaces within the canopy and onto the forest floor and underlying soil.

Wet deposition occurs as rain, snow, fog, and rime ice. Throughfall is precipitation that is intercepted by the canopy of understory or overstory vegetation before it reaches the forest floor. Depending on the ion (e.g., NH_4^+, SO_4^{2-}, NO_3^-, the throughfall concentration may be two to 18 times greater than in rain (Miller et al. 1996a). Ion concentrations in fog and rime ice are highly concentrated compared with rain or snow, especially in the highly polluted western region of the SBM (Berg et al. 1991; Fenn and Bytnerowicz 1997). However, precipitation volumes are much greater for rainfall and snowfall compared with fog or rime ice deposition.

Pollutant deposition to forest vegetation varies along both horizontal and vertical gradients in the SBM. Nitrogen concentrations and deposition decrease from west to east along the horizontal pollution gradient. Nitrogen deposition seems to decline more rapidly than ozone, probably as a result of the high deposition velocity of nitric acid (the main component of deposited N in the area) compared with ozone (Fenn and Bytnerowicz 1993). The horizontal or west-to-east pollution gradient has been discussed in Miller and Rechel (this volume) in terms of the gradient of ozone injury to the crowns or foliage of ponderosa and Jeffrey pines. In this chapter, we introduce two vertical scales that influence pollutant concentration and deposition. The first is the landscape scale (i.e., the influence of increasing elevation on pollutant concentration) and the second is the forest canopy scale (i.e., the effect of changing height within a mature tree canopy on the distribution of pollutant concentrations and deposition of airborne compounds) (see Figure 7.1, Miller and Rechel, this volume). We examine the evidence in the literature as well as some of our own unpublished studies regarding atmospheric concentrations and deposition of ozone, nitrogen, and sulfur compounds, and calcium, magnesium, potassium, and sodium.

Air Pollution Distribution in Relation to Elevation of Terrain

Ozone

Several recent studies have shown a consistent increase in ozone exposure with elevation. For example, monthly mean values of ozone measured between 1987 and 1994 at Wank Mountain, Bavaria (Stockwell et al. 1997), showed that concentrations were consistently higher at the mountain summit (1,782 m) and progressively lower at a midelevation station (1,175 m) and a valley station (735 m). When the nearby Zugspitze (2,962 m) is added to this comparison and all ozone exposure expressed as the sum of hourly concentrations equal to or greater than 40 ppb (AOT40) for 1987–1993, the Zugspitze had the highest exposure value (Stockwell et al. 1997).

A comparison of three Shenandoah National Park sites (Skelly et al. 1997), Big Meadows (1,067 m), Dickey Ridge (617 m), and Sawmill Run (457 m), usually showed higher exposure at the highest elevation when compared as the sum of concentrations greater than 60 ppb (SUM 60). That is, Big Meadows had the highest exposure of the three stations. Lefohn (1992) used several cumulative indices to test for differences between these stations during 1983–1987 and showed that in 2 of the 5 years, Big Meadows did not have the highest exposure. These observations emphasize the value of comparing data for multiple years on elevational gradients and also introduces caution in generalizing about the relationship between elevation and ozone exposure.

Ozone exposure was measured from Mill Creek Canyon, SBM (1,192 m), to near the top of nearby San Bernardino Peak (2,797 m) during the late summer of 1995 with Ogawa passive ozone monitors (Miller et al., unpublished data). Six

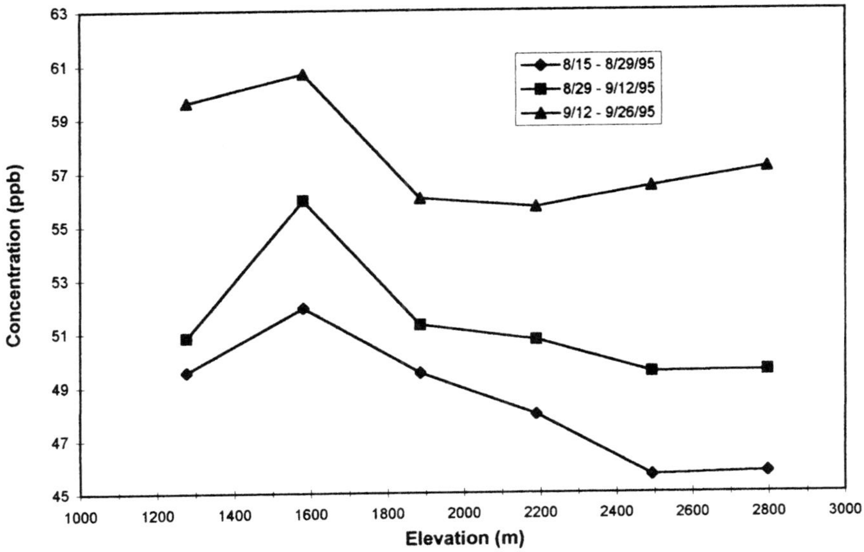

Figure 11.1. Gradient of ozone exposure during three 2-week exposures of Ogawa passive ozone monitors at six elevations from 4,200 to 9,200 ft (1,280–2,804 m) on the San Bernardino Peak trail indicates that maximum ozone exposure is at 5,200 ft (1,585 m), within the marine temperature inversion. The first 2-week interval showed the most ozone exposure and was also rainless compared with the other two periods.

passive monitors (one per elevation) were exposed at equal elevational intervals of about 1,000 m for three 2-week exposure periods. A real-time calibration was carried out at one of the sites during each exposure by comparing results obtained with a passive monitor with a conventional ultraviolet-type ozone monitoring instrument ($r^2 = 0.99$ for three exposures). During all three exposure periods, the highest cumulative exposure integrated by the passive monitors was at 1,585 m (Fig. 11.1). During two of three monitoring periods ozone concentrations consistently decreased with increased elevation between the 1,585- and 2,804-m portion of the transect.

This observation is consistent with an early study of the vertical distribution of ozone in the western SBM using a combination of ground stations and an aircraft equipped with a total oxidant (90% O_3) monitor (Miller et al. 1972). Sampling was conducted on three flights daily (0900, 1200, and 1600 hours). During all three flight times, the highest oxidant concentration was encountered between about 1,000 and 1,585 m. At 1200 and 1600 hours, the total oxidant concentration at 2,745 m was near 50 ppb (near background), but at 0900 hours the concentration at 2,745 m was frequently 80–90 ppb (Fig. 11.2). The higher level of total oxidant at that elevation at 0900 hours was interpreted as a remnant of the previous day's unmixed pollution that was displaced upward by the cool marine layer during the previous afternoon and evening. This remnant dispersed to higher elevations

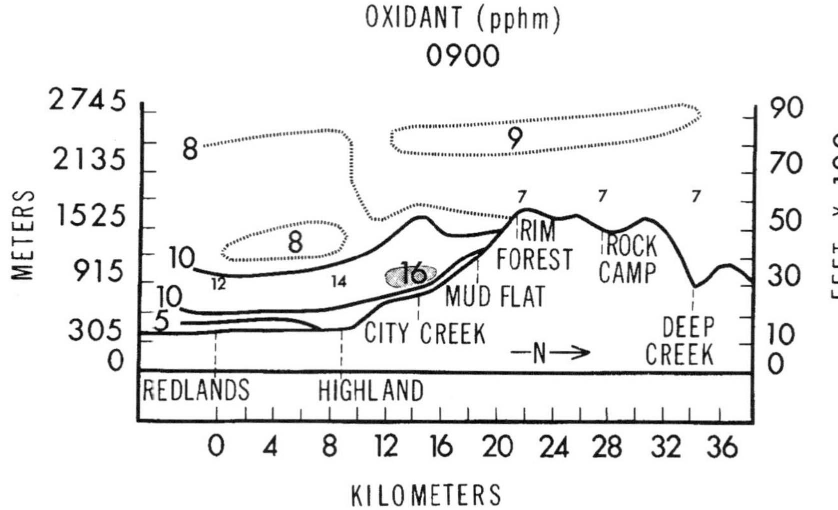

Figure 11.2. Aircraft measurements of ozone at 0900 hours from near the surface to 2,745 m above the western region of the San Bernardino Mountains show maximum concentrations at approximately 915 m at 0900 hours. Residual stable layers of ozone concentrations of 8–9 pphm above 2,100 m above sea level may have originated near the surface but were likely displaced upward by the late-afternoon influx of marine air on the previous afternoon/evening. By 1200 and 1600 hours, the mixing in the layers above 2,100 m had reduced ozone concentrations to about 4 pphm.

before 1200 hours as a result of mixing with increasing daytime temperatures. In summary, if integrated over entire time of measurements, concentrations between 1,000 and 1,585 m were consistently lower than at 2,745 m.

The peak ozone exposure level at 1,585 m coincided with the lower extent of the mixed conifer type, which may be a reflection of long-term exposures of these forests to high-ozone doses (Miller et al., unpublished data).

Nitrogen and Sulfur Deposition

Combined wet and dry deposition of N and S was determined along the elevational gradient from Mill Creek (1,192 m) to near the top of the San Bernardino Peak (2,797 m) by using collectors consisting of a cylinder of polyethylene-coated wire mesh nested in a funnel attached to a collection bottle (Bytnerowicz et al. 1998). After each 2-week exposure, the volume of precipitation was measured, and the samplers were rinsed and ionic concentrations in rinses and precipitation determined. In the first exposure period, rain increased from about 2 cm at the two lower sites (1,192 and 1,525 m) to 6 cm at 2,438 m. At the highest elevation (2,775 m), rain decreased to about 1 cm (Fig. 11.3A). Deposition of NO_3^-, SO_4^{2-}, and NH_4^+

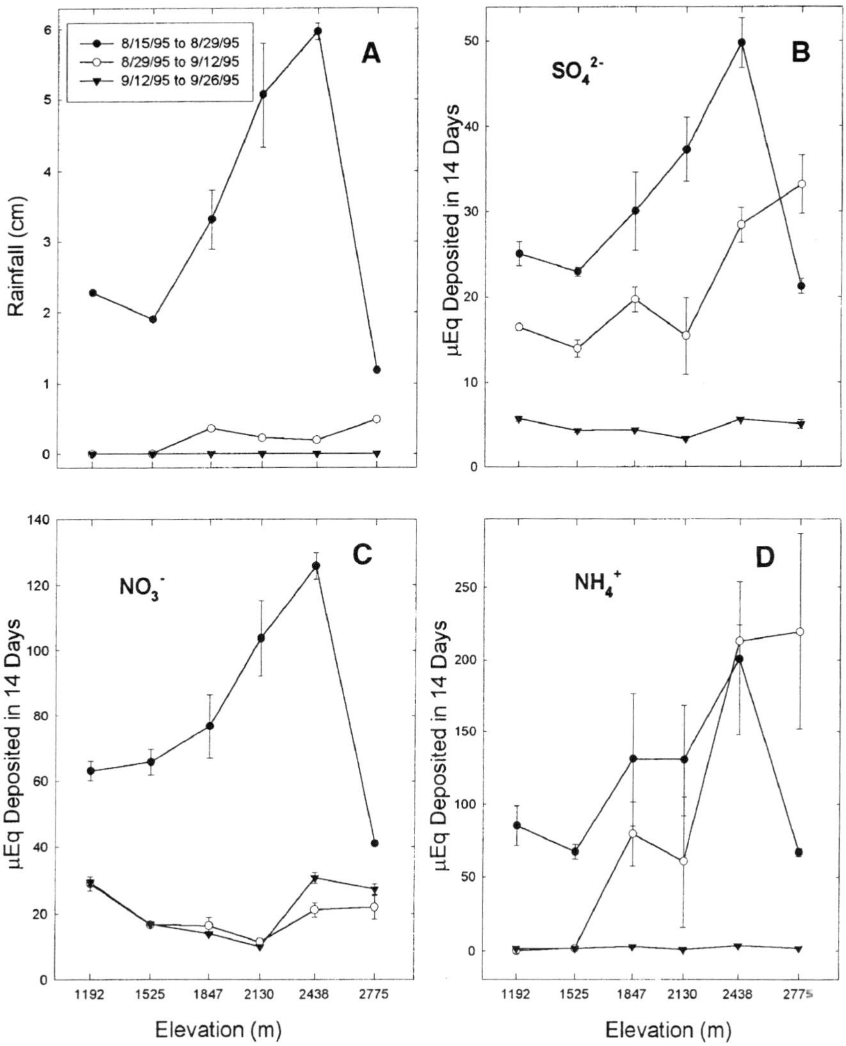

Figure 11.3. Rainfall (A) and wet-plus-dry deposition of sulfate (B), nitrate (C) and ammonium (D) at sites along an elevational gradient in the San Gorgonio Wilderness Area, San Bernardino Mountains.

followed a similar pattern, indicating that deposition was largely precipitation-driven. Rain was slight in the second exposure, and no rain occurred in the last exposure period. Deposition of SO_4^{2-}, and especially NO_3^-, was greatly reduced with lower rainfall. Ammonium deposition exhibited a more complex pattern, with very high NH_4^+ deposition at most of the higher elevations in the first two exposure periods but only trace amounts in the third rain-free period (Fig. 11.3B–D). Surprisingly, in the first two exposures, NH_4^+ deposition was generally much greater than NO_3^- deposition. Especially striking were the high NH_4^+ inputs in the second exposure period, when rainfall was negligible. These results suggest that NH_4^+ was scavenged from the atmosphere by rainfall much more efficiently than NO_3^-. Dry deposition of NH_4^+ to the collectors appears to be minimal, as shown by the data from the third rain-free exposure period. Nitrate deposition during the dry period was low—on average five times lower than during the first period when significant rainfall occurred.

Vertical Air Pollution Distribution in Mixed Conifer Canopy

Vertical distribution of pollutants in the forest canopy has been measured in two different locations in the mixed conifer forest of the Barton Flats area. Distribution of ozone concentrations was monitored at the "monitoring site," whereas the N and S pollutant concentrations and fluxes of airborne materials to branches were measured at "plot 2."

Distribution of Ozone Concentrations

The monitoring site occupied the point of a low ridge, and the terrain sloped downward on the south, west, and north. The forest composition was very different depending on slope direction. On the south, the canopy was comprised of black oak or canyon live oak with a shrub understory. On the north and west, there was a mixture of ponderosa pine and white fir with occasional black oaks. Compared with the typical overstory of pines in the general area, the stand on the north slope was of moderate stature. Its understory was mostly grasses, with the lower branches of pines reaching down to within 1 m of the ground. On the eastern side, the terrain rose gradually along the ridge crest, where mainly low-stature ponderosa pine, white fir, and black oak were present. The prevailing wind during the afternoon was from the west (upslope).

For continuous measurements of ozone, the air samples were taken above the canopy at 15 m on a tower and below the tree canopy in the understory at 1.5 m above the ground level. Airflow at the tower top was completely unobstructed because there were no nearby trees of equivalent height. The stand comprised of ponderosa pine, white fir, and black oak at a distance of 20 m north and slightly downslope from the tower.

The hourly ozone concentrations above and below tree canopy were compared during two multiple-day periods in 1992: July 27–30 (Fig. 11.4) and September 12–15 (Fig. 11.5). In July, both day and night concentrations were reduced slightly below the canopy (1.5 m) compared with above canopy (15 m) (Fig. 11.4). During September (Fig. 11.5) and October (not shown) sample periods, there were no differences in daytime concentrations above the canopy and below the canopy (upslope portions of the graph), but lower concentration below the canopy than above the canopy occurred in night-time (downslope portions of the graph).

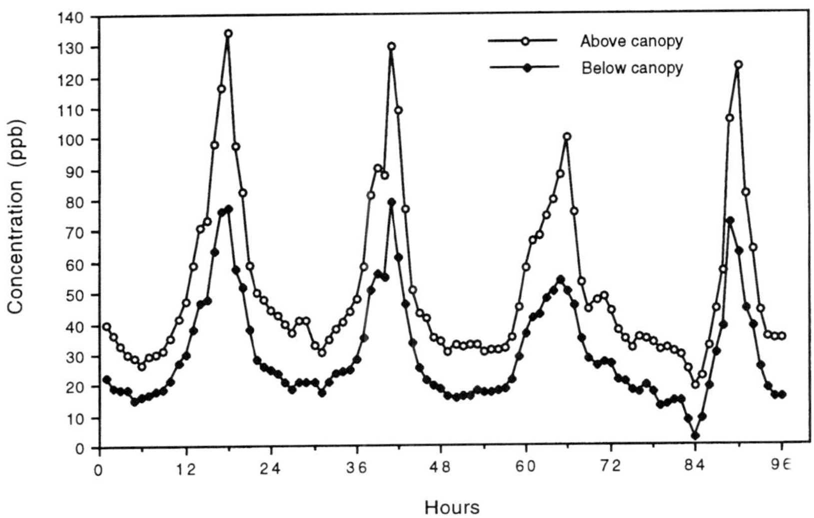

Figure 11.4. Day- and night-time ozone concentrations (ppm) above (15 m) and below the tree canopy (1.5 m) during July 27–30, 1992. Ozone concentrations above the canopy were consistently higher than below the canopy.

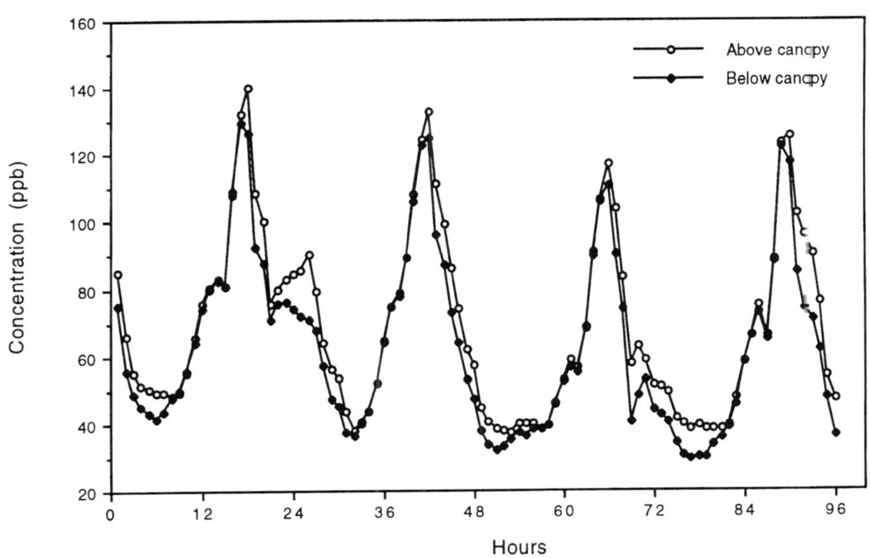

Figure 11.5. Day- and night-time ozone concentrations (ppm) above canopy (15 m) and within canopy (1.5 m) during September 12–15, 1992. The daytime ozone concentrations above and below the canopy were did not differ whereas the night-time ozone concentrations were higher above the canopy than below the canopy.

The maximum difference observed between above and below the canopy ozone concentrations reached about 50 ppb. A series of reactions could be involved in the depletion of ozone in the canopy and at the forest floor (Baldocchi et al. 1987). Differences in mixing processes or atmospheric turbulence between day and night are important. Daytime turbulence constantly replenishes canopy-level ozone, but with more stable night-time conditions, diffusion into the canopy is less rapid and so the sink effects predominate and canopy ozone is lower at night. Dew on leaf and other surfaces at night may increase the sink capacity of the canopy. Wetted foliar surfaces increased ozone deposition in a deciduous forest (Fuentes et al. 1992) and to grape leaves following dewfall (Grantz et al. 1995). During our study conducted in July, there were several small rains. The residual moisture stored in the litter layer could have contributed to more dew formation and wetness of the surfaces, which in turn increased night-time and early-morning deposition of ozone. Another effect of dew or higher residual moisture at night would be a stimulation of microbial activity in the litter layer with the resultant release of pulses of nitric oxide gas (Davidson and Shimel 1995), which would be available to react with ozone. Wetting the litter layer and upper soil has been reported to cause a 13-fold increase in NO emissions at Barton Flats (Fenn et al. 1996b).

During daylight hours, the major component of the total "ozone sink" in crop canopies is stomatal conductance (Leuning et al. 1979; Grantz et al. 1997). Adequate water availability in the soil had a substantial effect on ozone uptake and ozone injury to cotton (Temple 1988) and ponderosa pine seedlings in containers (Temple et al. 1992). During the July period of our study, the level of soil moisture availability was higher, stomatal conductance was higher, and greater ozone flux to leaf tissue may have reduced lower-canopy ozone concentrations compared with above-canopy concentrations. In September and October, there was no rain, soils were drier, stomatal conductance was lower, and ozone flux to leaf tissue was much less. Indeed, ozone flux calculations performed at Barton Flats in 1992–1994 (based on g_s and O_3 concentrations measurements) show that the early-summer flux rates are much higher than those of late summer (Temple and Miller 1998).

The complexity of interpreting the concentrations of ozone above canopy is illustrated by Enders et al. (1989). They measured ozone at three heights: 31 m, 41 m, and 51 m over a forest comprised mainly of spruce trees with an average height of 28 m. No measurements were included in the canopy. Inverse ozone vertical gradients (concentration decreasing with height) were observed during 80% of all days, especially during periods of high solar radiation. It was proposed that during the time when ozone concentrations were higher near the canopy, the formation of ozone was aided by biogenic hydrocarbons from the canopy. We did not have commensurate above-canopy measurements in our study.

The observed reduction of ozone concentrations above and below the tree canopy may have important significance. Reduced ozone concentrations would mean a lower dose of ozone and phytotoxic potential for the understory species compared with the mature trees. These kinds of reductions may also have importance from the human health point of view and should be considered for planning urban development and tourism activities.

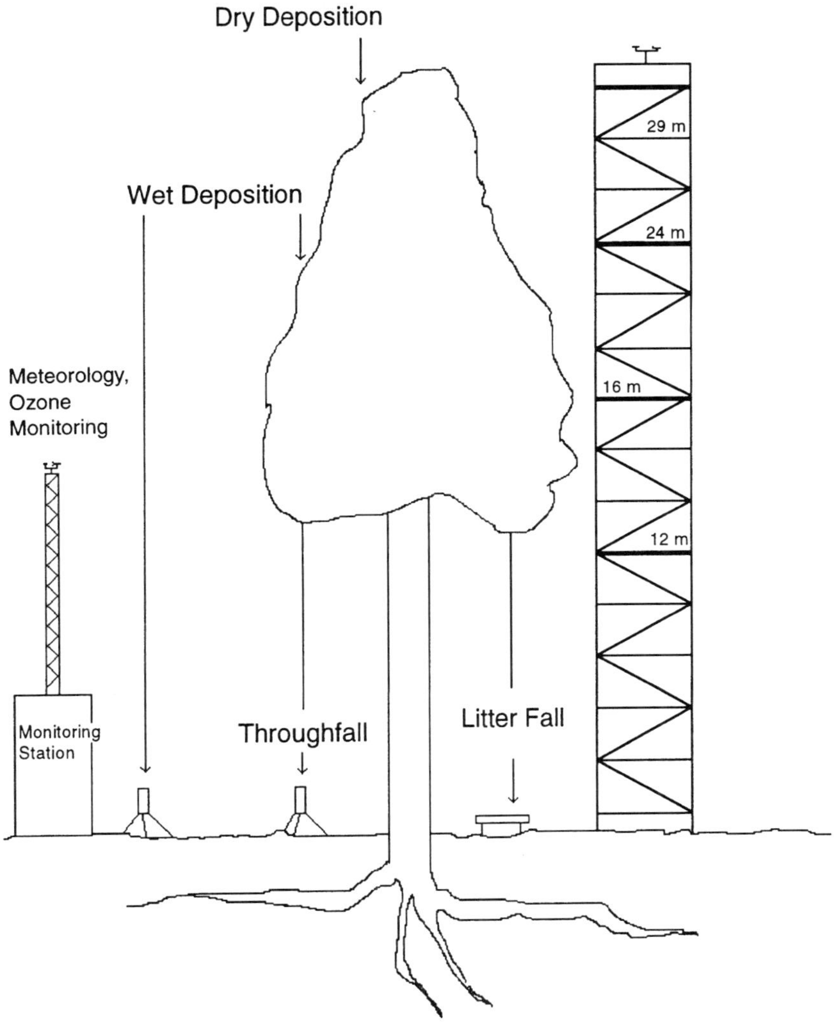

Figure 11.6. Schematic of tower used for concentration and dry deposition measurements in the mixed conifer canopy at plot 2, Barton Flats.

Distribution of Concentrations of N and S Pollutants and Deposition of Airborne Materials to Canopies

A scaffold tower was erected in plot 2 of the Barton Flats area at an elevation of 2,100 m (Fig. 11.6). The 30-m-tall tower was located in a mixed conifer forest stand of mature ponderosa/Jeffrey pines. The forest stand was open with about 67% canopy cover (Fenn and Bytnerowicz 1997). Among the trees 10.0 cm or greater (diameter at breast height), most were white fir (39%), followed by ponderosa pine (29%), black oak (20%), and Jeffrey pine (14%). Leaf area index (LAI) of the stand at Barton Flats was determined by a modification of the litterfall method (Marshall and Waring 1986). Conservative values for estimated LAI by species were 0.9 for white fir, 0.6 for black oak, and 0.4 for ponderosa/Jeffrey pine.

Distribution of Concentrations of N and S Pollutants

Single gas/particle collectors (annular denuders combined with filter packs) were placed at each of four different levels of the tower: 29 m (canopy top), 24 m and 16 m (canopy middle), and 12 m (canopy bottom) (Fig. 11.6). Average 24-hour concentrations of nitric acid (HNO_3) vapor, nitrous acid (HNO_2) vapor, ammonia (NH_3), sulfur dioxide (SO_2), fine particulate (particles <2.2 μm in diameter) nitrate (NO_3^-), ammonium (NH_4^+), and sulfate (SO_4^{2-}) were determined with an annular denuder system (Possanzini et al. 1983; Peake and Legge 1987; Grosjean and Bytnerowicz 1993). Concentrations of N pollutants were measured six times during the 1992 summer season, eight times during the 1993 summer season, and four times during the 1994 summer season.

In general, position within the canopy did not affect the concentrations of the studied gaseous and particulate N and S compounds; however, significant differences occurred between individual periods of measurements. Results of the 1993 season illustrate these findings (Fig. 11.7). A lack of significant effect of position within the canopy on concentrations of the sampled pollutants (with the exception of significantly higher concentrations of HNO_3 at 24 m than at 12 m during the 1993 season) was probably caused by the openness of the canopy upwind from the scaffold. In the westerly direction, from which pollutants were transported, absence of canopy allowed for vertical mixing of air masses. In more dense forest canopies, such as in the eastern United States, a vertical gradient of pollutants with diminishing concentrations from canopy top to bottom is more typical (Meyers et al. 1989).

Average summer concentrations of HNO_3 vapor (3.2–6.5 $\mu g/m^3$) at plot 2 were higher than the concentrations determined at the monitoring site by Watson et al. (this volume). The measured concentrations were moderately high compared with the summer average concentrations in forest locations in Tennessee (3.0 $\mu g/m^3$) and near Göttingen, Germany (5.8 $\mu g/m^3$) (Lindberg et al. 1990). At Tanbark Flat in the San Gabriel Mountains of southern California, a mountain site that is considered one of the most polluted in the United States (Bytnerowicz et al. 1987), the average 24-hour concentrations of HNO_3 during the 1988–1991 summer seasons ranged from 6.7 to 11.7 $\mu g/m^3$ (Grosjean and Bytnerowicz 1993). At various locations in the Sierra Nevada, the average 24-hour summer concentrations were less than 1 $\mu g/m^3$ (Ashbaugh et al. 1991).

Nitrous acid vapor concentrations (0.10–0.42 $\mu g/m^3$ in 1992, 0.00–0.35 $\mu g/m^3$ in 1993, and 0.00–0.12 $\mu g/m^3$ in 1994) (Miller et al. 1996b) were lower than the average summer concentrations at the Tanbark Flat site (0.55–0.65 $\mu g/m^3$) (Grosjean and Bytnerowicz 1993). However, they were higher than the annual mean

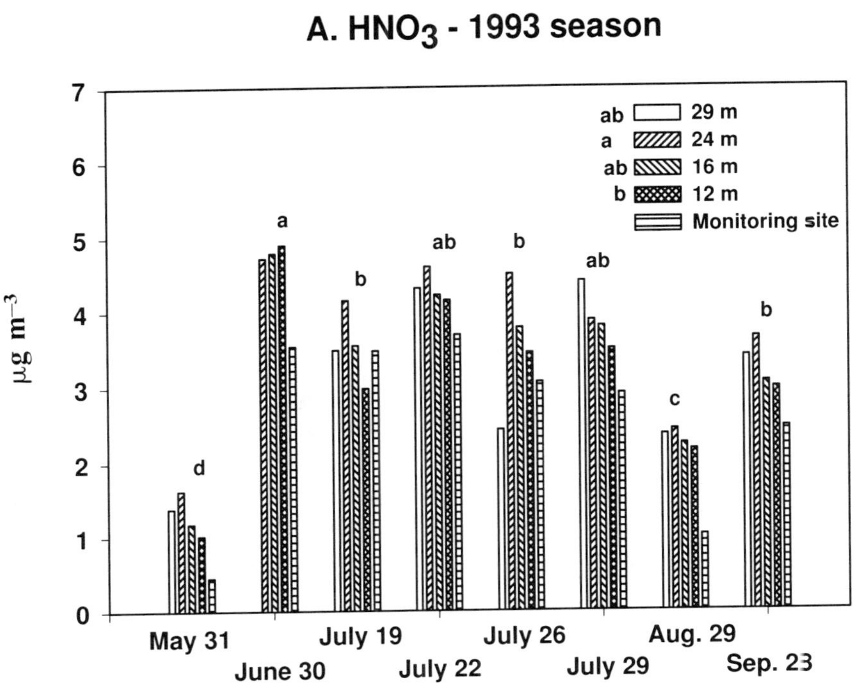

Figure 11.7. Concentrations of N and S gaseous and particulate pollutants at different heights of the mixed conifer canopy measured in 1993 polluted season: (A) nitric acid vapor (HNO_3), (B) ammonia (NH_3), (C) sulfur dioxide (SO_2), (D) nitrate in fine particles (NO_3^-), (E) ammonium in fine particles (NH_4^+), (F) sulfate in fine particles (SO_4^{2-}). Different letters indicate significant differences at the 0.05 probability level for location within the canopy or period of deposition.

concentration of 0.05 μg/m³ reported at a remote location in the Canadian Rockies (Legge and Krupa 1989).

Ammonia concentrations (1.40–1.48 μg/m³) were lower than the values determined at the monitoring site (Watson et al., this volume). The determined values were similar to the seasonal means (1.1–1.85 μg/m³) determined at Tanbark Flat (Grosjean and Bytnerowicz 1993) and lower than the estimated seasonal mean (3.0 μg/m³) near Göttingen, Germany (Lindberg et al. 1990). The measured concentrations were several times higher than the annual mean (0.26 μg/m³) measured in the Canadian Rockies (Legge and Krupa 1989) and the estimated seasonal mean (0.2 μg/m³) in Tennessee (Lindberg et al. 1990). However, concentrations of NH_3 in The Netherlands and Great Britain may be several times higher than at Barton Flats, reaching 15 μg/m³ (Keuken et al. 1988; Draaijers et al. 1989).

Concentrations of SO_2 were low (0.55–1.05 μg/m³), similar to the values determined at the monitoring site by Watson et al. (this volume), and at the levels

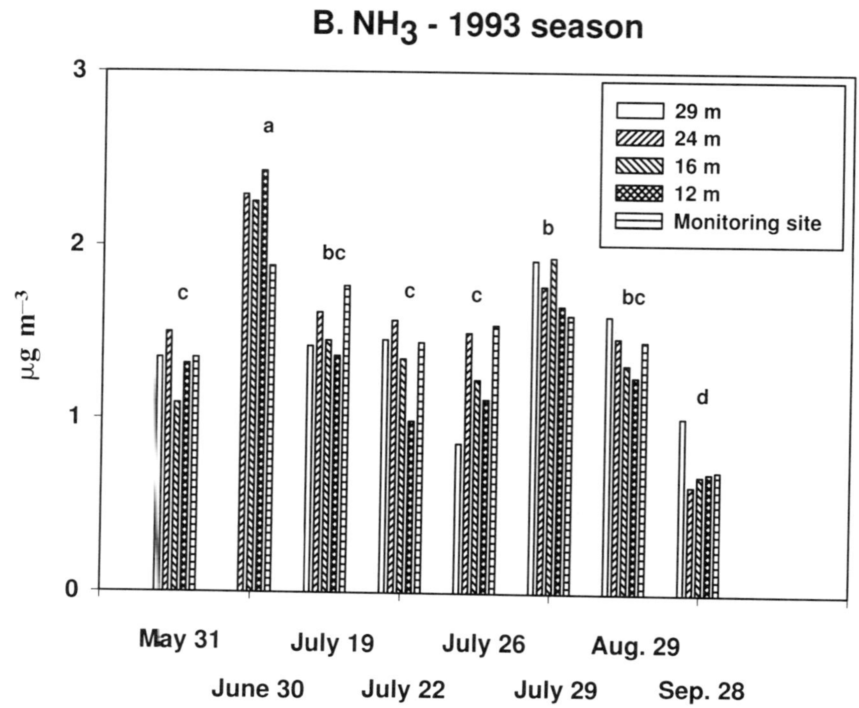

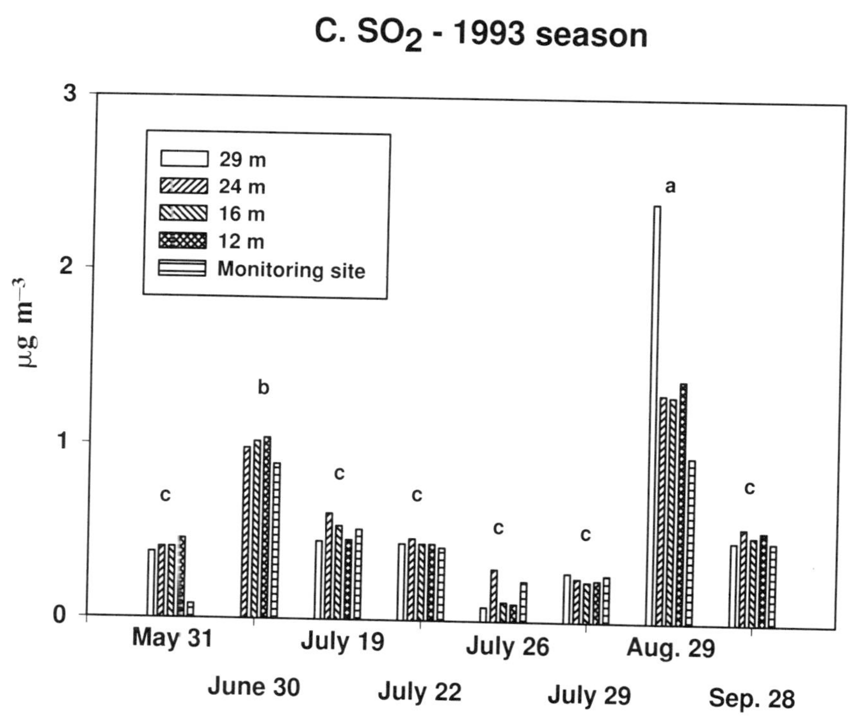

Figure 11.7. (*continued*)

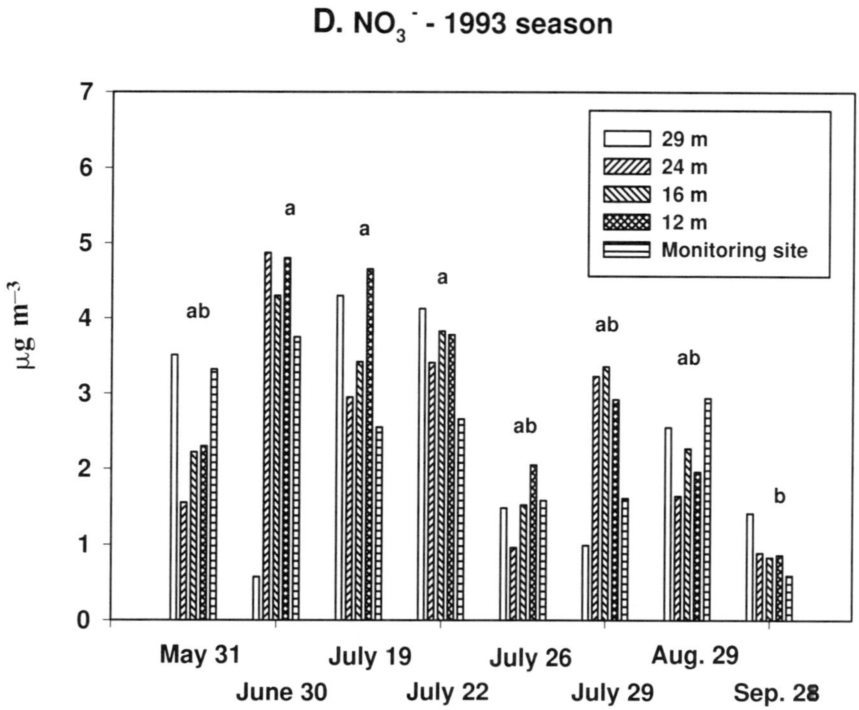

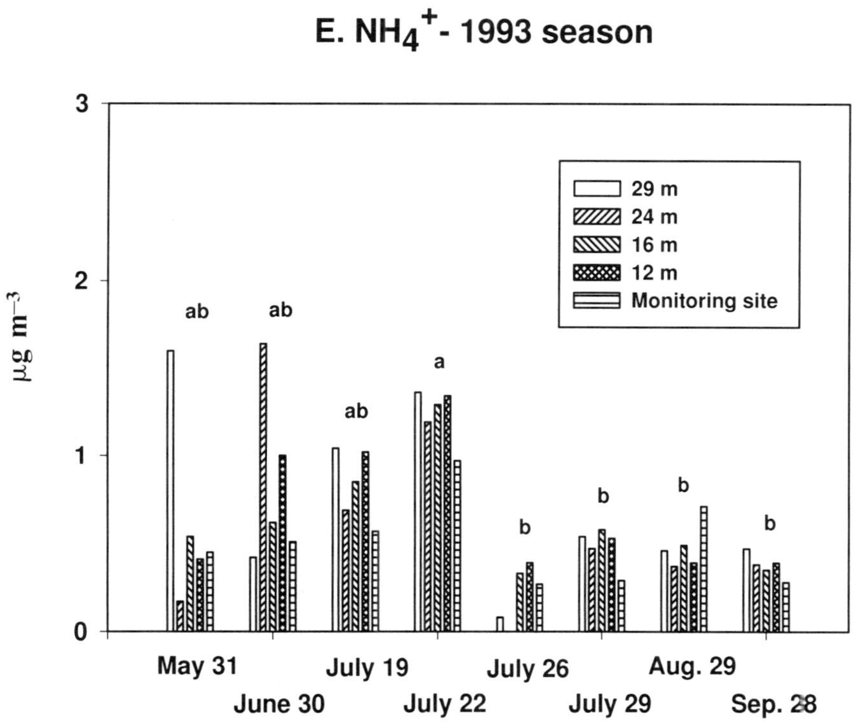

Figure 11.7. (*continued*)

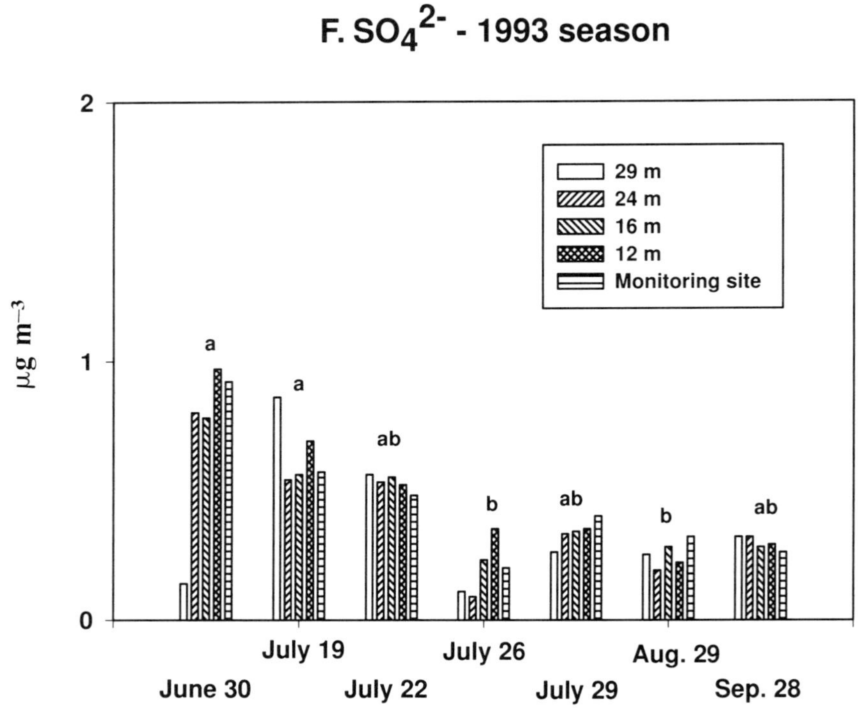

Figure 11.7. (*continued*)

normally occurring in southern California (Bytnerowicz et al. 1987). The measured concentrations were typical for western U.S. states and stayed well below phytotoxicity levels (Bohm 1992).

Concentrations of particulate NO_3^- were high (2.0–3.2 μg/m³) and similar to the values determined at the monitoring site (Watson et al., this volume), Tanbark Flat (seasonal means 1.75–2.9 μg/m³; Grosjean and Bytnerowicz 1993), in the SBM (24-hour means of 2.1–5.1 μg/m³; Fenn and Bytnerowicz 1993), and in the Los Angeles Basin (24-hour means of 3.3 μg/m³; Sickles et al. 1988). These values greatly exceeded the annual mean of 0.13 μg/m³ reported for the Canadian Rockies (Legge and Krupa 1989) and seasonal mean of 0.19 μg/m³ determined in Tennessee (Lindberg et al. 1990). The 24-hour mean concentration in Claremont, in the Los Angeles Basin (7.9 μg/m³; Pierson and Brachaczek 1988), was higher than the concentrations determined in this study.

Concentrations of particulate NH_4^+ (0.6–1.0 μg/m³) were similar to the values determined at the monitoring site (Watson et al., this volume), Tanbark Flat (seasonal means 0.65–0.85 μg/m³; Grosjean and Bytnerowicz 1993), other locations in the SBM (0.1–0.9 μg/m³; Fenn and Bytnerowicz 1993), and the mean concentration of 1.0 $\mu g/m^3$ determined in the urban Los Angeles area (Sickles et al. 1988). Ammonium concentrations reported for the forest locations in Tennessee and Germany (seasonal means of 2.4 and 5.5 $\mu g/m^3$, respectively; Lindberg et al. 1990) were higher than in this study.

Concentrations of SO_4^{2-} (0.4–1.1 $\mu g/m^3$) were slightly lower than the values determined at the monitoring site (Watson et al., this volume) but similar to the values determined at Tanbark Flat (Bytnerowicz et al. 1987), Camp Paivika (CP) (Fenn and Bytnerowicz 1993), and Fortress Mountain of the Canadian Rockies (Legge and Krupa 1989).

Deposition of Airborne Materials to Foliated Branches

A branch-washing technique (Lovett and Lindberg 1985) was used to determine deposition fluxes of airborne N and S compounds as well as some other elements on a vertical transect through the mature tree canopy. At the beginning of each collection period, seedling branches were thoroughly rinsed with double-deionized water. The branches were re-rinsed at the end of 1-week-long collection periods. Concentrations of NO_3^- in branch rinses were determined with ion chromatography and concentrations of NH_4^+ by colorimetry (Bytnerowicz et al. 1987). This relatively simple technique has been used extensively in our California studies, where lack of precipitation in the summer makes throughfall analysis unreliable as a method of measuring atmospheric deposition.

Fluxes of N and S Compounds. Deposition fluxes of NO_3^-, NH_4^+, and SO_4^{2-} to branches of ponderosa pine seedlings were determined on each of the four levels of the tower in the vicinity of the annular denuder systems used for N and S pollutants concentration determinations. Deposition fluxes were measured four times during the 1992 season, six times during the 1993 season, and five times during the 1994 photochemical smog season.

Deposition of NO_3^-, NH_4^+, and SO_4^{2-} to branches of the ponderosa pine seedlings was significantly affected by position within the canopy (Fig. 11.8). The highest deposition flux values for the studied ions were always determined at the canopy top (29 m). Deposition at 24 m was consistently lower than at the top but nearly always higher than at 16 m and the canopy bottom (12 m). Deposition fluxes of NO_3^-, NH_4^+, and SO_4^{2-} were similar among three years of the study, and the highest values were determined in July and August (Bytnerowicz et al. 1996).

Nitrate fluxes to branches (Fig. 11.8A) were elevated compared with other California locations (Bytnerowicz and Fenn 1996). The highest values were similar to the value of 180 $\mu g\ m^{-2}\ h^{-1}$ determined for mature ponderosa pine at CP in the SBM (Fenn and Bytnerowicz 1993). Higher NO_3^- deposition values were determined at Tanbark Flat, where deposition to hoaryleaf ceanothus *(Ceanothus crassifolius)* ranged from 88 to 455 $\mu g\ m^{-2}\ h^{-1}$ (Bytnerowicz et al. 1987). Deposition of nitrate to native California pines in other locations was much lower: about 4–6 $\mu g\ m^{-2}\ h^{-1}$ for Eastern Brook Lake in the eastern Sierra Nevada (Bytnerowicz et al. 1992); 27–30 $\mu g\ m^{-2}\ h^{-1}$ for Emerald Lake in the western Sierra

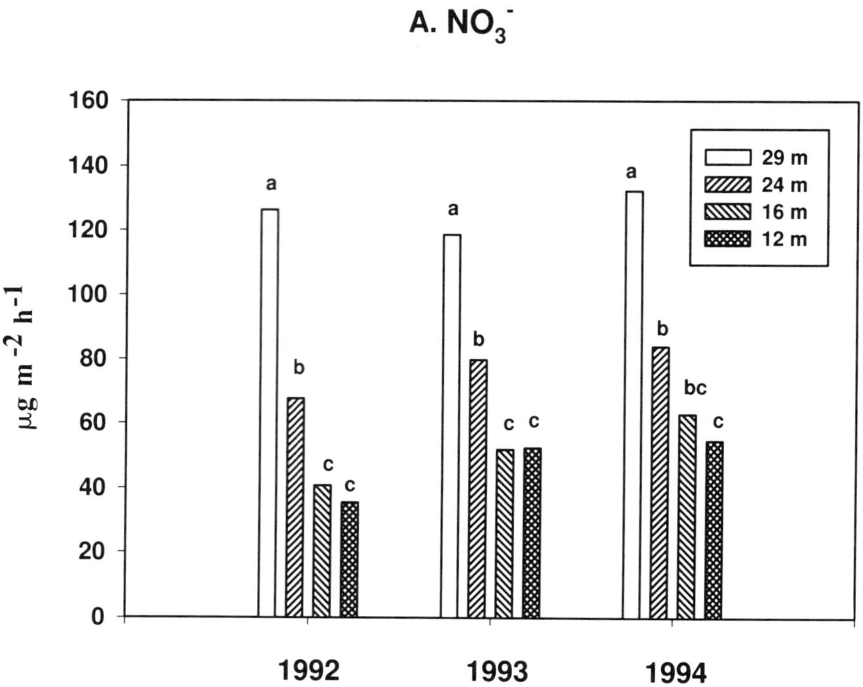

Figure 11.8. Deposition of ions to branches of ponderosa pine seedlings at different heights of the mixed conifer canopy: (A) nitrate (NO_3^-), (B) ammonium (NH_4^+), (C) sulfate (SO_4^{2-}). Different letters indicate significant differences at the 0.05 probability level for locations with the canopy.

Nevada (Bytnerowicz et al. 1991); and 6–31 $\mu g\ m^{-2}\ h^{-1}$ for Whitaker Forest in the western Sierra Nevada (Bytnerowicz and Riechers 1995).

Similar to the NO_3^- results, the deposition fluxes of NH_4^+ (Fig. 11.8B) were also elevated at Barton Flats—these values were similar to deposition of 21 $\mu g\ NH_4^+\ m^{-2}\ h^{-1}$ determined at CP, a high-pollution site in the western SBM (Fenn and Bytnerowicz 1993). Deposition of NH_4^+ to hoaryleaf ceanothus at Tanbark Flat (17–46 $\mu g\ m^{-2}\ h^{-1}$) was higher than at Barton Flats (Bytnerowicz et al. 1987). Deposition to native California pines at some other forest locations was lower than in the present study: at Eastern Brook Lake, NH_4^+ ranged from 0.8 to 1.6 $\mu g\ m^{-2}\ h^{-1}$ (Bytnerowicz et al. 1992); at Emerald Lake, it was 6.0–9.0 $\mu g\ m^{-2}\ h^{-1}$ (Bytnerowicz et al. 1991); and at Whitaker Forest, NH_4^+ was 1.0–1.5 $\mu g\ m^{-2}\ h^{-1}$ (Bytnerowicz and Riechers 1995).

Sulfate deposition values (Fig. 11.8C) were similar to the values determined for hoaryleaf ceanothus (7.7–43 $\mu g\ m^{-2}\ h^{-1}$) at Tanbark Flat (Bytnerowicz et al.

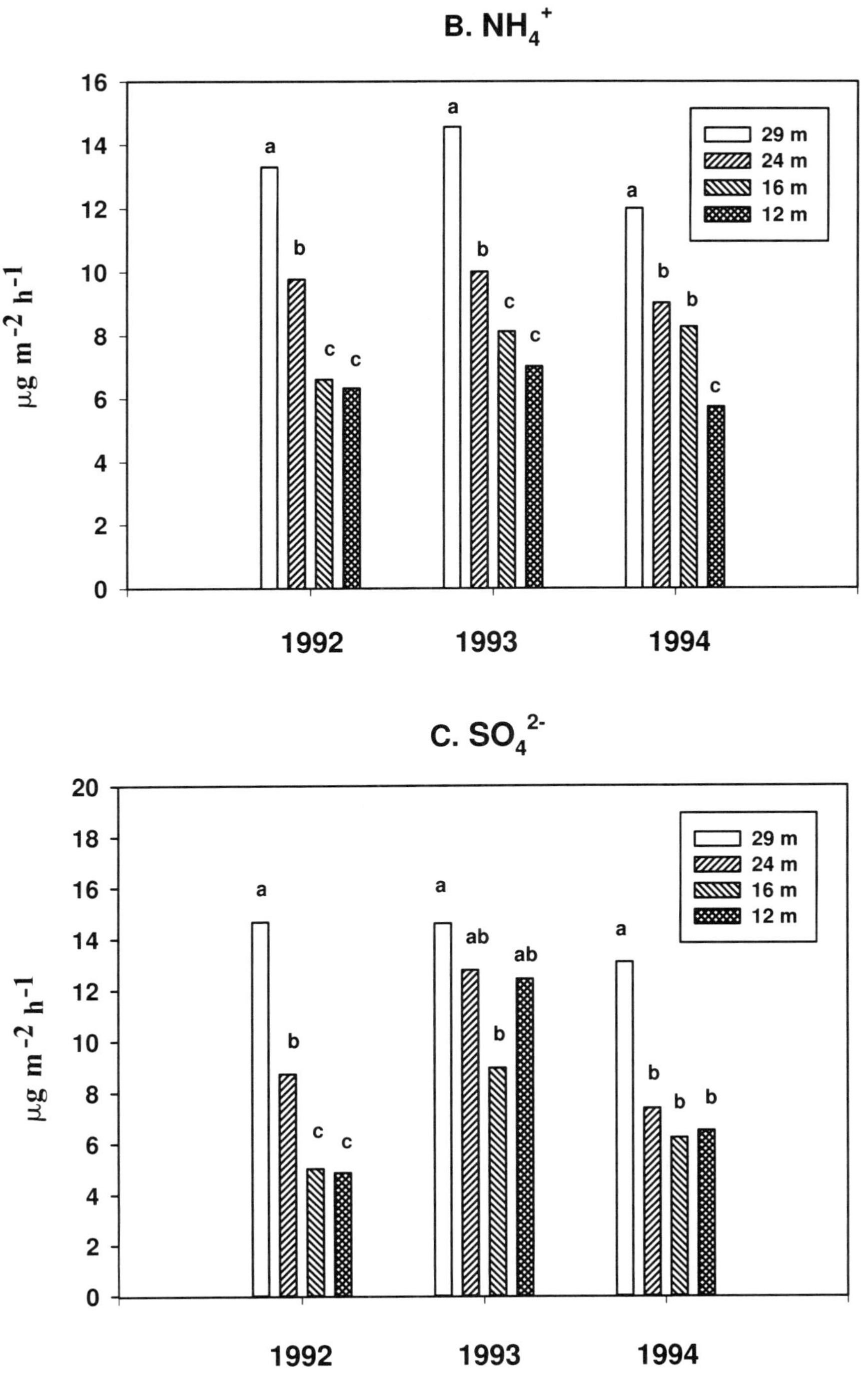

Figure 11.8. (*continued*)

1987). These average deposition values were higher than the average SO_4^{2-} deposition to ponderosa pine branches at CP (6.9 μg m^{-2} h^{-1}; Fenn and Bytnerowicz 1993), as well as native pines at Eastern Brook Lake (about 1.5 μg m^{-2} h^{-1}; Bytnerowicz et al. 1992) and at Emerald Lake (3.2–4.9 μg m^{-2} h^{-1}; Bytnerowicz et al. 1992).

The results of this study clearly indicate that NO_3^-, NH_4^+, and SO_4^{2-} deposition fluxes in a mixed coniferous forest strongly depend on the position of the branches within the canopy. Ion deposition increased with increasing branch height. Because concentrations of gaseous and particulate pollutants did not depend on canopy position, the observed distribution of deposition fluxes within the forest canopy could not be related to the air pollution concentration differences. Difference in deposition fluxes could be attributed to varying wind speed occurring along the vertical transect within the canopy because the highest deposition values occurred at the canopy top where wind speeds were highest (see later in chapter).

Fluxes of Basic Cations. Similar to nitrogen and sulfur, atmospheric dry deposition is an important source of physiologically important basic cations to forest ecosystems. Lindberg et al. (1986) determined that dry deposition represented more than 50% of the annual input of Ca^{2+} and K^+ in an oak forest in Tennessee. However, very little is known about deposition of basic cations to forests in the western United States, with the exception of a few studies in the subalpine ecosystem of the Sierra Nevada (Bytnerowicz et al. 1991, 1992; Brown and Lund 1994). No similar studies have been performed in the SBM.

Potassium, calcium, and magnesium are essential nutrients for higher plants (Salisbury and Ross 1978). Sodium is taken up by plants from the soil solution, and its role in plant physiology is not well established (Czerwinski 1978). It is believed that sodium may replace potassium in controlling cell turgor (Mengel and Kirkby 1982). With increased soil acidity (which could be caused by elevated deposition of nitrogen), potassium, calcium, and magnesium may be leached from soils (Falkengren-Grerup 1989; Duffy and Schreiber 1990; Huettl 1993). Acidic precipitation can also cause increased leaching of these metals from plant foliage (Barker and Ashenden 1992; Cape 1993). Therefore, forests with elevated acidic deposition may develop a deficiency of basic cations such as calcium or magnesium, as may be occurring in some forests in eastern North America (Likens et al. 1996). However, such cation deficiency may be compensated for by wet and dry deposition of these elements to forest trees and soils. The high base content of soils in the SBM indicates that base cation depletion is not a major threat to forest sustainability in the Barton Flats area, although in the highly polluted stands in the western end of the deposition gradient, percentage base saturation and soil pH are significantly lower than at Barton Flats (Fenn et al. 1996a,b).

Deposition fluxes of Ca^{2+} and Mg^{2+} to ponderosa pine branches were significantly affected by the position of branches in the canopy—deposition flux was consistently lower as the height decreased during the entire study. This pattern was similar for K^+ and Na^+ in the 1992 season; however, the results of the 1993 season were different (Table 11.1).

Table 11.1. Dry Deposition of Basic Cations to Branches of Ponderosa Pine Seedlings at Four Levels of the Mixed Conifer Forest Canopy at Plot 2, Barton Flats (μg m^{-2} h^{-1})

Position	K^+	Na^+	Ca^{2+}	Mg^{2+}
		1992		
29 m (top)	3.9 a[a]	13.2 a	27.6 a	4.1 a
24 m	2.5 b	6.2 b	15.9 b	2.1 b
16 m	2.9 ab	3.5 c	11.2 c	1.3 c
12 m (bottom)	2.2 b	2.9 c	8.8 c	1.1 c
P value	<.0001	<.0001	<.0001	<.0001
		1993		
29 m	6.9	10.6 a	25.3 a	3.7 a
24 m	6.1	7.6 b	19.6 b	2.9 b
16 m	8.8	5.5 b	13.1 c	2.0 c
12 m	10.7	10.7 b	13.6 c	2.5 bc
P value	.4773	.0002	<.0001	<.0001

[a]Means followed by different letters are statistically different.

Average values of potassium, sodium, and magnesium deposition to ponderosa pines at Barton Flats were similar to the values determined for western white pine and lodgepole pine in the western Sierra Nevada (Bytnerowicz et al. 1991; Brown and Lund 1994) and several times higher than the deposition to lodgepole and whitebark pines at the eastern Sierra Nevada site (Bytnerowicz et al. 1992). However, values of calcium deposition were about two times higher than values determined to native pines at the western Sierra Nevada site (Bytnerowicz et al. 1991; Brown and Lund 1994) and many times higher than in the eastern Sierra Nevada (Bytnerowicz et al. 1992). This difference may be related to the proximity of the Mojave Desert and long-range transport of desert dust containing high concentrations of calcium into the forests of the SBM. These relative differences between the Barton Flats site and other California locations indicate that dry deposition may be an important source of these metals in the mixed coniferous forests of the SBM.

Relationships of N Dry Deposition Flux Along a Vertical Canopy Gradient

The relationship between concentrations of pollutants along the vertical canopy gradient and the factors controlling the gradient is important to understand. For systems with uniform canopies, these relationships can be explored using inferential models developed over the past decade (Hicks and Meyers 1988; Meyers and Baldocchi 1988). These models estimate deposition flux (F_d) from air concentration measurements (C) and deposition conductance (K_1) using air concentrations, micrometeorological information, aerodynamic transport, boundary layer transport, and physicochemical surface interactions for model input.

The discontinuous canopy of the mixed conifer forest (Fig. 11.9) differs from the systems used to develop inferential models. There is little significant vertical

Figure 11.9. Aerial view of the mixed conifer forest stand (plot 2) and location of the monitoring tower (arrow).

air pollutant concentration gradients of N; however, deposition to leaf and branch surfaces does not occur equally along the vertical profile of the canopy. Thus neither inferential models nor a simple assumption of equal deposition to foliar surfaces is appropriate for this forest type.

A quasi-empirical data analysis approach was used to examine foliage rinse N flux relationships along a vertical canopy profile using foliage rinse data. The analysis was intended to suggest linkages between physical forces controlling deposition along the vertical gradient and to develop equations to reduce the measurements needed to estimate total canopy N flux in future studies.

Both the annular denuder and filter pack data show large significant correlations between particulate NO_3^- and particulate NH_4^+ concentrations ($r = 0.92$ and $r = 0.93$, respectively), indicating that the majority of particulates are in the form of NH_4NO_3. Based on this, we assumed that a dominant form of particulate NO_3^- is in the form of NH_4NO_3; however, we recognize that other particulate forms are also present.

By using the above assumptions and the relationship

$$NH_4NO_3 \Leftrightarrow HNO_3 + NH_3$$

flux (F) data can be expressed as

$$F_{\text{NO3 particulate}} = F_{\text{NH4 particulate}} \tag{1}$$

and

$$F_{\text{HNO3}} = F_{\text{NO3 total}} - F_{\text{NH4 particulate}} \tag{2}$$

Because the forest canopy is open and no vertical concentration gradient exists, it was assumed that the major factor controlling deposition rates to needle surfaces was wind speed. Using the standard equation, wind speed declines exponentially down a vertical canopy (Davidson and Wu 1990):

$$\text{Wind speed} = e^{[a1(1 - H_t/H_h)]} \tag{3}$$

a starting value of wind speed parameter $a_1 = -1.2$ was estimated from wind speed information gathered during 1994 at the tower location (Miller et al. 1996a).

A model of flux at lower canopy positions was then estimated by using Equations (2) and (3) for nitrate deposition as

$$F_{\text{NO3}h} = 0.76e[-1.4(1 - H_h/H_t)](F_{\text{HNO3}t} + F_{\text{NO3}t}) \qquad r^2 = 0.81$$

and by using Equations (1) and (3) for ammonium deposition as

$$F_{\text{NO3}h} = 0.85e[-1.0(1 - H_h/H_t)](F_{\text{NH4}t}) \qquad r^2 = 0.80$$

in which

h = vertical height along canopy (H)
$t = \max(h)$ = top of canopy
F_{Nt} = flux to leaves from either nitrate and nitric acid or ammonium

Models were fitted with a nonlinear regression procedure, NLIN (SAS 1990), by using the derivative-free multivariate secant method. A grid search was used to identify a starting value of a_0. Graphical and analytical analyses of residuals were conducted to determine if any model assumptions were violated. Additional details on model derivation, parameterization, and validation can be found in Arbaugh et al. (1998).

The high nonlinear r^2 (analogous but not equal to linear r^2) for both pollutants indicate that flux for both compounds at lower levels of the canopy is strongly linked to flux at the top of the canopy over a range of deposition flux rates. The decline in flux down the canopy is dominated by declines in wind speed from the top to the bottom of the canopy. The wind speed parameters a_1 (−1.4, −1.0) are lower than those reported for other systems (Davidson and Wu 1990), indicating that wind speed declines more linearly (the nonlinear relationships become linear when $a_1 = 1.0$) in this open forest relative to closed-canopy forests.

A single model was parameterized using NO_3^- and NH_4^+ expressed as mg N:

$$F_{Nh} = 0.79e[-1.2(1 - H_h/H_t)](F_{Nt}) \qquad r^2 = 0.69$$

where

F_{Nt} = canopy top flux to leaves from either nitrate and nitric acid or ammonium
F_{Nh} = flux to leaves at height h from either nitrate and nitric acid or ammonium

The single model, although explaining slightly less variation than the individual ion models, is a simpler representation of the system and is a better predictor at lower flux levels.

The scalar parameters a_0 (0.76, 0.85, 0.79) of the three models reflect an additional resistance to deposition at lower canopy positions not present at the canopy top. Although data are sparse, this resistance may be due to reduced turbulence occurring in the canopy interior that results in increased quasi-laminar layer depth and increased stability, which increases diffusive and interception resistance.

Throughfall, Fog, and Wet Deposition to Mixed Conifer Forests

In arid and semiarid climates, including mixed conifer forests of southern California, air pollutants are deposited mainly via dry deposition processes. However, during fog episodes or precipitation events, dry pollutants are scavenged or washed from the atmosphere and from plant surfaces and transported to the ground. Ions not retained within the plant/soil system are transported through the water chemistry profile (Kimmins 1987) consisting of ion transport in precipitation, throughfall, surface runoff, forest floor leachate, mineral soil leachate, groundwater, and streamwater. Thus high N inputs such as occur in the western end of the SBM result in elevated NO_3^- concentrations throughout the water chemistry profile, presumably because available N is in excess of biological demand and the N retention capacity of the plant/soil system.

Throughfall

Ionic concentrations in summer throughfall at Barton Flats (particularly NO_3^-) were as high or higher than values reported for forests in Europe and eastern North America, with much higher atmospheric deposition inputs. The high throughfall concentrations at Barton Flats were attributed to the washoff of accumulated dry pollutants from mature trees with large surface areas during infrequent low-volume rain events (Fenn and Bytnerowicz 1997). In more mesic systems in which precipitation is more evenly distributed throughout the year, ionic concentrations in throughfall are often lower than at Barton Flats during summer even when annual N and S deposition is greater (Lindberg et al. 1990; Probst et al. 1990; Stevens et al. 1990; Mitchell et al. 1992; Wesselink et al. 1995). White fir, the species with the greatest foliar surface area, had the highest throughfall concentrations and the lowest throughfall volumes in the study area. Canopy effects of ponderosa and Jeffrey pine on throughfall chemistry and volume were intermediate, whereas California black oak throughfall was less modified chemically than throughfall of the conifer species. Nitrogen deposition in throughfall was higher under fir than under pine or oak (Table 11.2).

Throughfall concentrations are much lower in winter than in summer in the mountains of southern California because of higher precipitation volumes and lower pollution deposition to tree canopies. Average annual NO_3^- concentrations in throughfall in a mixed conifer forest near Kratka Ridge in the San Gabriel Mountains were about five times lower than summer throughfall concentrations at Barton Flats, even though total N deposition was twice as high at Kratka Ridge (Fenn and Kiefer, unpublished data). Nearly all the throughfall deposition at Kratka Ridge occurred during the fall and winter wet season.

Table 11.2. Comparison of Dry Summertime Inorganic N ($NO_3^- + NH_4^+$) Deposition (kg/ha) at Plot 2, Barton Flats of the SBM, Based on Model Estimates, Branch Rinse, and Throughfall Data

	Branch rinse	Throughfall	Model estimate
Fir	0.99	1.10	0.87
Oak	1.70	0.45	1.24
Pine	0.64	0.51	0.44
Ground[a]	0.36	0.36	0.36
Total	3.69	2.06	2.55
Total without oak	1.99	1.61	0.95

[a]Estimates from dry deposition collections only and included in the other estimates for the purpose of comparison.

The marine influence was apparent at a forest site near Kratka Ridge, which is 65 km inland in the San Gabriel Mountains, but was not seen in the SBM further inland. Deposition of SO_4-S in three plots near Kratka Ridge (7.8–12.5 kg ha^{-1} yr^{-1}) occurred mainly as wet deposition, and total amounts were equivalent to total N deposition (6.0–11.5 kg ha^{-1} yr^{-1}). By contrast, N deposition was mostly as net throughfall (total throughfall deposition minus deposition in precipitation), which is an estimate of dry deposition inputs. The relatively high S deposition in the San Gabriel sites is attributed to deposition of seasalt sulfates. In more inland sites in the Los Angeles Air Basin, where seasalt deposition is negligible, S deposition is typically an order of magnitude lower than N deposition (Fenn and Bytnerowicz 1993; Bytnerowicz and Fenn 1996).

Fog and Winter Wet Deposition

Fog occurrence in the SBM occurs predominantly during winter and spring. Fog water ionic concentrations at CP were usually at least an order of magnitude greater than at Barton Flats. Ammonium and nitrate concentrations in fog at CP were similar except on March 19, 1997, when NH_4^+ concentrations were twice as high as NO_3^- concentrations (Fig. 11.10). Deposition of NH_4^+ and NO_3^- at CP in precipitation was highly similar (Fenn and Bytnerowicz 1997). Dry deposition of NH_4^+ is usually much lower than NO_3^- in polluted areas of California, but this does not seem to be true for fog and wet deposition. The high density of dairy farms in the Chino area to the west of CP is probably a major source of the unusually high levels of NH_4^+ deposited at that site.

Bulk deposition of N and S collected in canopy clearings during 5.5 months in the winter/early spring period of 1993–1994 at CP (11.9 and 2.7 kg ha^{-1} yr^{-1}) was greater than previously thought. Berg et al. (1991) also demonstrated that N and S concentrations in rime ice are significantly greater in the SBM than in montane sites in central and northern California. However, total inputs in rime ice may not be significant in the SBM because rime ice occurrence is not generally a persistent or widespread phenomenon in the SBM, especially in the lower elevation sites where N deposition is greatest.

Extrapolation of Experimental Results to Stand-Level Deposition of Airborne Materials

Deposition fluxes of N, S, K, Na, Ca, Mg, P, and Cl derived from branch washing experiments can be used to calculate the amount of those materials intercepted by the tree canopy. With some additional measurements and assumptions, deposition at the forest stand level can be calculated. Deposition estimates derived from branch wash data should be comparable with throughfall deposition data from the same forest stand. However, both branch rinsing and throughfall may underestimate deposition because of ion retention within the canopy or overestimate deposition if ions are leached from within the leaves by precipitation or by branch rinsing.

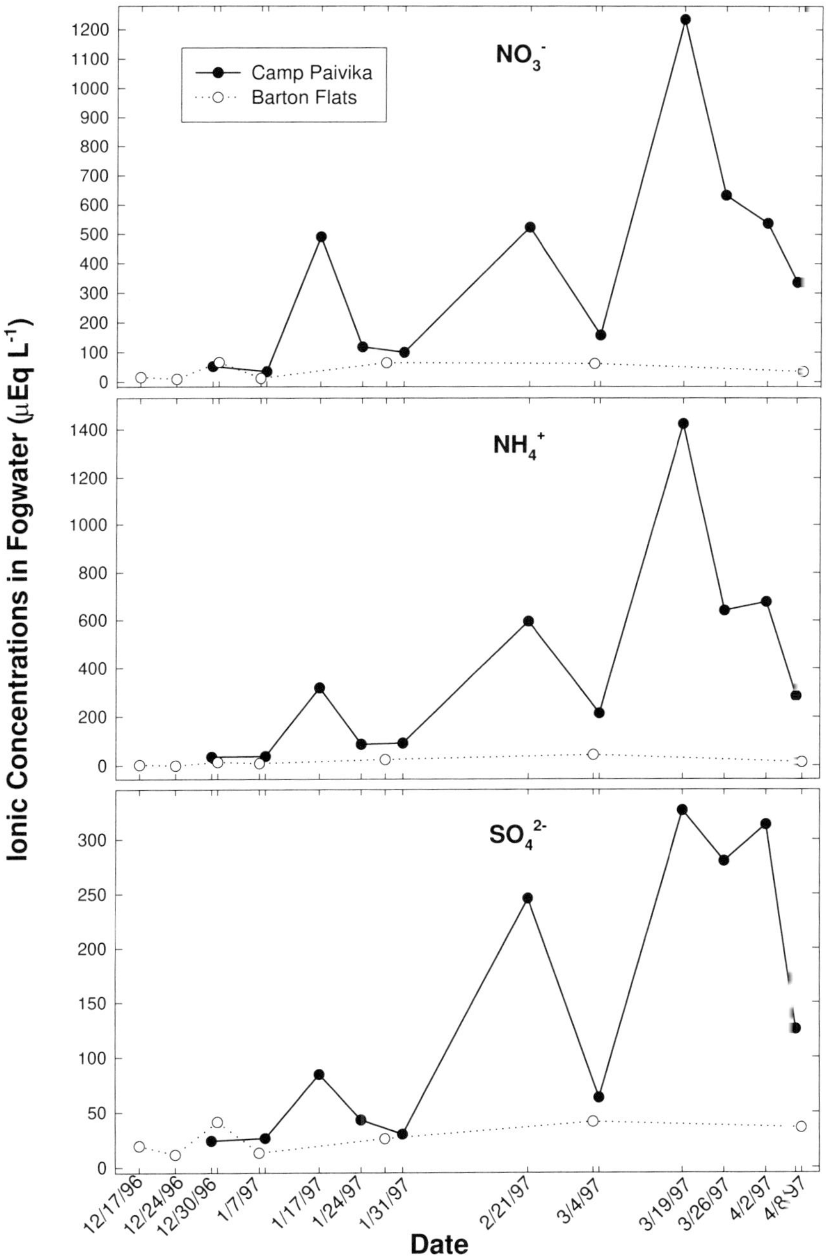

Figure 11.10. Ionic concentrations (NO_3^-, NH_4^+, and SO_4^{2-}) in fogwater at Camp Faivika and Barton Flats.

N Compounds

Surface and stomatal deposition of various N compounds to ponderosa pine, California black oak, and white fir during the polluted and clean seasons were calculated based on results of branch rinsing, ambient concentrations of pollutants, information on stomatal conductance, and LAI of these tree species (Bytnerowicz et al. 1996). Results of branch rinsing indicated pronounced differences in surface deposition of NO_3^- and NH_4^+ among the three species (Fig. 11.11). Differences in LAI values of the dominant tree species greatly affected the calculated stand deposition values. Apportionment of dry deposited N species for pine, fir, and oak is presented in Figure 11.12. Surface deposition of NO_3^- and NH_4^+ accounted for most of the deposited nitrogen, and only a small portion of the entire deposition could be attributed to the stomatal uptake of the gaseous N compounds. Big differences in deposition occurred between the polluted and clean seasons, and this was especially pronounced for the deciduous oak. The high capacity of California black oak to intercept gaseous HNO_3 and particulate NO_3 resulted in much greater N deposition to oak compared with the two coniferous species. Nitrogen deposition at the forest stand level was higher for fir than for pine, because of the higher foliar surface area (4,424 m^2 versus 1,827 m^2) of fir, even though the NO_3^- deposition flux to branches was similar for pine and fir and the NH_4^+ deposition flux was lower for fir.

Predictive Model of N Dry Deposition to Forest Stands

A simple model (see previous section) was developed to predict N flux to leaf surfaces at different canopy positions as a function of flux at the top of the tree canopy and tree height. This model was extended to predict surface N dry deposition for the entire forest canopy. It is assumed that the site consists of codominant trees of approximately equal height and that wind speed decreases for the whole forest are similar to decreases measured adjacent to an individual tree.

The modified model form is

$$F_{Nhi} = \alpha_0 \exp^{\alpha_1 P} F_{Nt}(\alpha_{2i} LA_{ih} + \alpha_{3i} BA_{ih}) + \epsilon_{ih}$$

where

F_{Nhi} = flux to the 5% of canopy centered at relative height h for tree species i

i = tree species

α_{2i} = constant parameters reflecting surface deposition differences between ponderosa pine foliage and foliage of other tree species

α_{3i} = constant parameters reflecting surface deposition differences between ponderosa pine foliage and wood surface area of tree species i

P = 5% relative vertical section of the forest; P = 0.025 corresponds to bottom 5% of canopy; and P = 0.975 corresponds to the top 5% of the forest canopy

LA_{ih}, BA_{ih} = leaf area and wood surface areas, respectively, for species i at relative height h

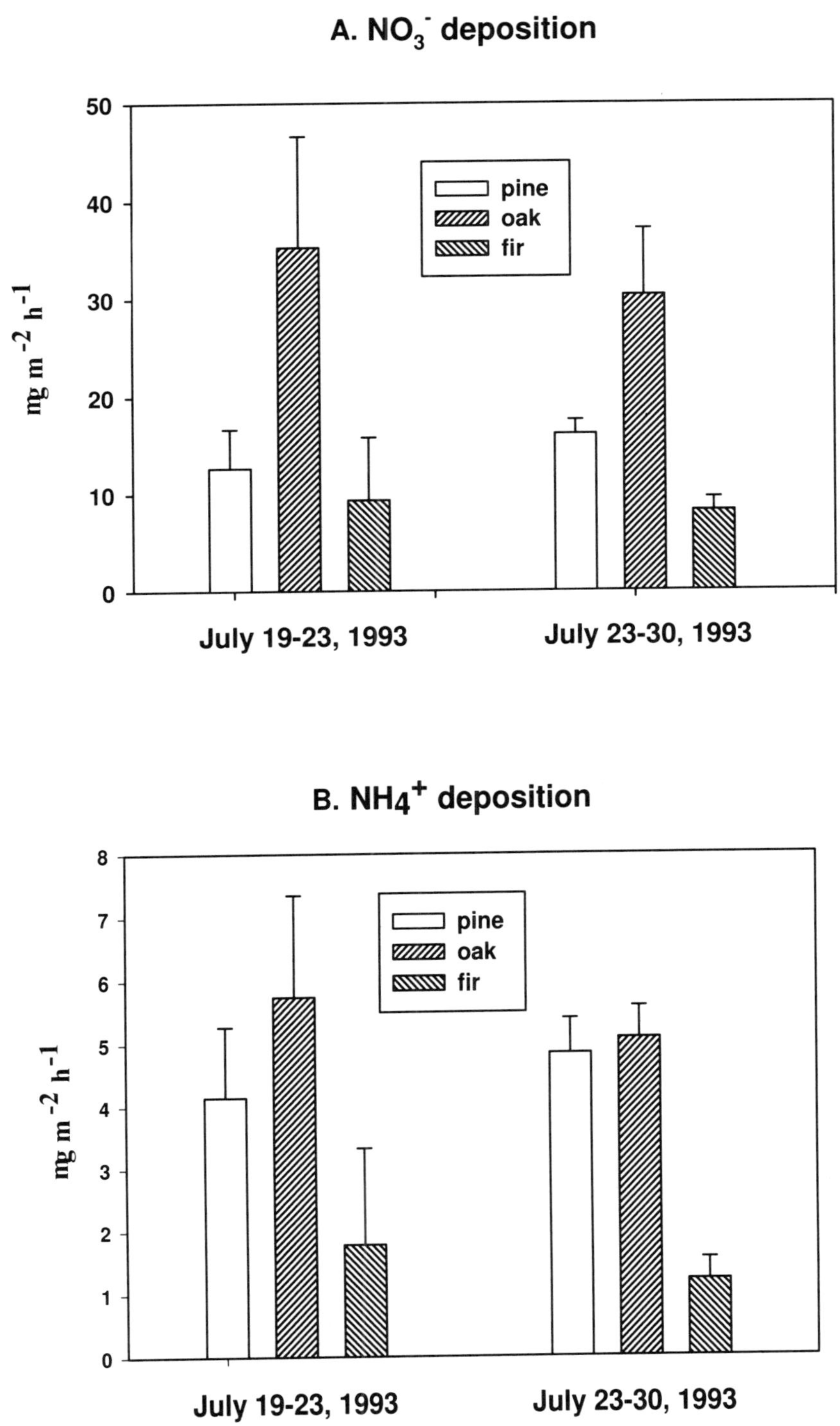

Figure 11.11. Deposition of (A) nitrate (NO_3^-) and (B) ammonium (NH_4^+) to branches of three dominant tree species at plot 2, Barton Flats, during two periods of measurements.

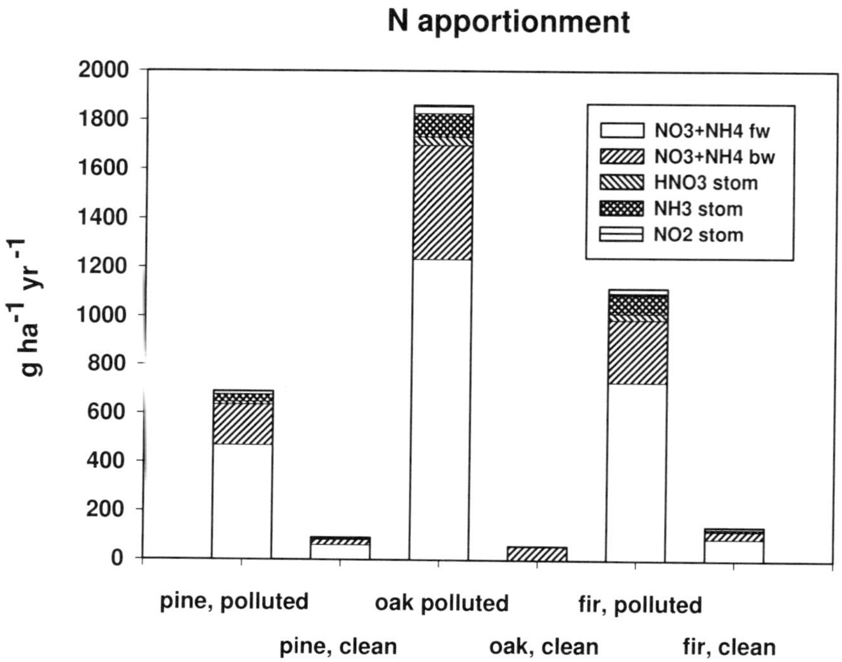

Figure 11.12. Apportionment of N dry deposition for the three dominant trees species at plot 2 during the polluted and clean seasons. Five different N forms are as follows: NO3+NH4 fw, foliar washable nitrate and ammonium; NO3+NH4 bw, branch (twig parts only) washable nitrate and ammonium; HNO3 stom, nitric acid taken up through stomata; NH3 stom, ammonia taken up through stomata; NO2 stom, nitrogen dioxide taken up through stomata.

The model was applied to predict deposition to leaf and branch surface at 5% vertical sections of the forest canopy. Parameterization information for fluxes and constants was developed from canopy deposition data used to determine empirical branch deposition (see previous section). Detailed description of this model can be found in Arbaugh et al. (1998).

For this study, separate models for F_{NO3} and F_{NH4} were used (Bytnerowicz et al. 1996) because the parameters α_{2i} and α_{3i} varied significantly between ions and tree species. For instance, NO_3^- surface deposition to oak foliage surfaces was found to be 1.99 times greater than to ponderosa pine, and NH_4^+ deposition to fir was found to be 0.49 times less than that of ponderosa pine. Other rates were not significantly different (Bytnerowicz et al. 1996). Surface deposition to branch wood was found to be approximately 7.9 times greater than deposition to foliage for pine and fir and 3.7 times greater for oak.

Deposition estimates without the oak data are very similar between the model, branch rinse, and throughfall estimates (Table 11.2). Similarities between branch rinse and throughfall estimates are expected because both are based on similar processes of dry deposition inputs and aqueous rinsing of the canopy surfaces.

Deposition estimated from branch wash data was highest for black oak, even though the leaf area was less than one-third of the entire stand. The high deposition for oak was primarily due to the large wood surface area of this species and high rates of deposition to the wood surface. Oak also had the greatest range of estimates. The throughfall estimate for black oak was just 25% of the estimate from branch rinses. The low throughfall in oak may have been due to stem flow not measured by throughfall collectors. For conifers, this is a minor component of deposition, but for deciduous forests as much as 30% of the water input to the forest may be from stem flow (Butler and Likens 1995). The irregular canopy of oak also makes measurement by throughfall more dependent on the position of collectors, which were often found to have ion concentrations similar to rainfall (Fenn and Bytnerowicz 1997), indicating that throughfall deposition to oaks may have been underestimated.

Estimated total annual N deposition for plot 2 of the Barton Flats area in the 1993 season was about 5 kg/ha, based on calculations using LAI values calculated from litterfall data (see previous section) for determining dry deposition inputs and by including estimated wet deposition. The predominant inputs were in dry-deposited materials to plant surfaces. Only a very small fraction of the total deposition could be attributed to stomatal uptake of gaseous N or to wet deposition. The estimated values of total N deposition indicate that the Barton Flats site experiences elevated N deposition compared with more pristine sites with little anthropogenic influence. For example, at Eastern Brook Lake in the eastern Sierras, total estimated N deposition was about 0.5 kg ha^{-1} yr^{-1} (Bytnerowicz et al. 1992). However, total N deposition at Barton Flats is low compared with CP in the western SBM (probably >30 kg ha^{-1} yr^{-1}; Fenn and Bytnerowicz 1997) or Tanbark Flat in the San Gabriel Mountains near Los Angeles (about 20–35 kg ha^{-1} yr^{-1}; Riggan et al. 1985; Bytnerowicz et al. 1987). Ongoing studies near Barton Flats and in the western more polluted region of the SBM focus on the ecological and environmental consequences of chronic atmospheric N inputs. Based on recent studies (Fenn et al. 1996b; Kiefer and Fenn 1997), the forest at Barton Flats is not yet N saturated, even though it has been affected by several decades of N deposition. However, the mixed conifer forest on the far western end of the deposition gradient in the SBM is N saturated (Aber et al. 1989), as evidenced by a number of plant and soil indicators and by high nitrate levels in streamwater (Fenn et al. 1996b).

Deposition of Sulfur, Phosphorus, Chloride, and Basic Cations

Calculations of sulfur and basic cation deposition to plot 2 site of the Barton Flats area during the "polluted" season of 1993 (i.e., April 15–October 15), based on results of branch rinsing, were analogous to the calculations of nitrate and ammonium deposition. Calculations were made for the three tree species as well as for the ground surface. A sum of all the above-mentioned deposition represents

Table 11.3. Calculated Dry Summertime Deposition of Sulfur Compounds, Basic Cations, and Chloride (g ha^{-1} polluted $season^{-1}$) to Trees at Plot 2, Barton Flats, during the 1993 Season Based on Branch Rinsing (Br) and Throughfall (Tf) Results

	S		K^+		Na^+		Ca^{2+}		Mg^{2+}		P-PO_4	Cl^-
Surface	Br	Tf	Br	Tf	Br	Tf	Br	Tf	Br	Tf	Tf	Tf
Pine	82	40	175	220	187	100	212	310	62	70	2	180
Oak	179	70	2,048	600	2,047	100	1,586	480	373	120	8	200
Fir	219	160	477	1,620	477	210	1,739	900	162	270	40	450
Ground	198	198	657	657	67	67	160	160	62	62	—	—
Total	678	468	3,357	3,097	2,778	477	3,697	1,850	659	522	50	830

total dry deposition to the forest stand. The assumptions used for the calculations and detailed calculations are presented elsewhere (Bytnerowicz et al. 1996). Surface washable sulfate was the major contributor in the total sulfur deposition. Stomatal uptake of SO_2 provided a very small amount of S to the forest stand, which is understandable knowing that SO_2 concentrations in southern California forests are near background levels (Bytnerowicz et al. 1996).

Comparison of results at the forest stand level for plot 2 obtained with the branch rinsing and throughfall techniques is presented in Table 11.3. Similar to the estimates of N deposition, the branch rinsing and throughfall techniques for ponderosa pine agreed well for dry deposition of sulfur and basic cations. For fir and oak, large differences in estimates of K^+, Na^+, and Ca^{2+} deposition occurred between the two methods. These differences may simply be due to the high degree of heterogeneity of the tree canopies and associated interception of dry deposition and rainfall, factors that contribute to variable throughfall deposition throughout the forest stand. Despite these uncertainties, our results indicate that deposition of K^+, Ca^{2+}, Mg^{2+}, and Na^+ was elevated and probably offsets much of the acidification associated with chronic deposition of N and S in the SBM.

Summary and Conclusions

At the landscape level, ozone concentrations in the SBM increased with elevation up to about 1,500 m and then gradually decreased with elevation. These findings do not support findings from Europe and the eastern United States indicating that ozone concentrations in the mountainous terrain continually increase with increasing altitude. It seems that ozone concentrations distribution patterns vary with climate and terrain features of individual mountain ranges. It was also determined that bulk deposition of NO_3^-, SO_4^{2-}, and NH_4^+ in late summer was proportional to the amount of rain precipitation and reached maximum at about 2,400 m.

Within the mixed conifer forest, concentrations of ozone below the tree canopy were lower than above the canopy. The observed differences may be attributed to various factors (e.g., increased moisture or scavenging of ozone by NO near the forest floor). Because of the reduced ozone concentrations below the tree canopy, the understory species would be exposed to a lower dose of ozone compared with the mature trees. Reduction of ozone concentrations under the canopy may also have importance from the human health point of view and for planning of urban development and tourism activities. Position within the canopy did not affect concentrations of gaseous HNO_3, HNO_2, NH_3, or SO_2 nor did it affect concentrations of particulate NO_3^- and NH_4^+. This is in contrast to the more dense forest stands in the eastern United States, where a vertical gradient of pollutant concentrations diminishing from top to bottom is typical.

Deposition fluxes of N and S compounds and basic cations strongly depend on position of branches within the forest canopy. Fluxes increased with increasing height. This gradient can be attributed to varying wind speeds and turbulence occurring on the vertical transect—the highest deposition values occurred at the canopy top where wind speeds were highest.

Along the west-to-east declining air pollution gradient in the SBM, N deposition decreases more rapidly than ozone exposure. Wet deposition of N at Barton Flats was low (1.5 kg N ha^{-1} yr^{-1}), whereas at CP on the western end of the gradient, wet deposition was 11.9 kg/ha during a 5.5-month period in winter and early spring in 1993–1994. Thus, N deposition during the relatively clean winter season is still quite high at CP and undoubtedly contributes to the N-saturated state of the forest and high NO_3^- export in streamwater from small catchments near CP (Fenn and Poth, unpublished data).

Ionic concentrations in summer throughfall were as high or higher than the values reported for eastern United States and Europe. These high concentrations can be attributed to the washoff of accumulated dry deposition from mature trees during infrequent low-volume summer rain events. Participation of NH_4^+ in rain and fog was higher than in dry deposition and greatly contributed to relatively high levels of winter/spring N deposition. Increased NH_4^+ deposition was probably caused by NH_3 emissions from the high-density dairy farms in the Chino area located about 60 km west of the SBM.

Estimates of summertime dry deposition of N to the mixed conifer forest at Barton Flats were very similar between the newly developed deposition model, branch rinse calculations, and throughfall estimates (2.1–3.7 kg ha^{-1} yr^{-1}). Most of the deposited N on plant surfaces was in the form of washable NO_3^- and NH_4^+, with only small participation of NO_2, NH_3, and gaseous HNO_3 taken up through stomata.

Estimated depositions of basic cations based on branch washing and throughfall analysis techniques were high, with a potential for offsetting much of the acidification associated with chronic deposition of N and S compounds.

References

Aber, J.D.; Nadelhofer, K.J.; Steudler, P.; Melillo, J.M. Nitrogen saturation in northern forests ecosystem. Bioscience 39:378–386; 1989.

Arbaugh, M.; Bytnerowicz, A.; Fenn, M.E. A quasi-empirical model to estimate N atmospheric deposition in mixed conifer forest stand of the San Bernardino Mountains. In: Bytnerowicz, A.; Arbaugh, M.; Schilling, S., eds. Proceedings of the international symposium air pollution and climate change effects on forest ecosystems, February 5–9, 1996, Riverside, CA. Gen. tech. rep. PSW-GTR-166. Albany, CA: Pacific Southwest Station, USDA Forest Service; 1998.

Ashbaugh, L.L.; Chow, J.C.; Watson, J.D. Atmospheric acidity data quality and measurement characteristics for California's acid deposition monitoring program. In: Proceedings of the 84th annual meeting and exhibition of the Air and Waste Management Association, Vancouver, BC; June 16–21, 1991.

Baldocchi, D.D.; Hicks, B.B.; Camara, P. A canopy stomatal resistance model for gaseous deposition to vegetated surfaces. Atmos. Environ. 21:97–101; 1987.

Barker, M.G.; Ashenden, T.W. Effects of acidic fog on cuticular permeability and cation leaching in holly *(Ilex aquifolium)*. Agric. Ecosystems Environ. 42:291–306; 1992.

Berg, N.; Dunn, P.; Fenn, M. Spatial and temporal variability of rime ice and snow chemistry at five sites in California. Atmos. Environ 25A:915–926; 1991.

Bohm, M. Air quality and deposition. In: Olson, R.K.; Binkley, D.; Bohm, M., eds. The responses of western forests to air pollution. New York: Springer-Verlag; 1992:63–152.

Brown, A.D.; Lund, L.J. Factors controlling throughfall characteristics at a high elevation Sierra Nevada site, California. J. Environ. Qual. 23:844–850; 1994.

Butler, J.T.; Likens, G.E. A direct comparison of throughfall plus stemflow to estimates of dry and total deposition for sulfur and nitrogen. Atmos. Environ. 29:1253–1265; 1995.

Bytnerowicz, A.; Riechers, G.H. Nitrogenous air pollutants in a mixed conifer stand of the western Sierra Nevada, California. Atmos. Environ. 29:1369–1377; 1995.

Bytnerowicz, A.; Miller, P.R.; Olszyk, D.M. Dry deposition of nitrate, ammonium and sulfate to a *Ceanothus crassifolius* canopy and surrogate surfaces. Atmos. Environ. 21:1749–1757; 1987.

Bytnerowicz, A.; Dawson, P.J.; Morrison, C.L.; Poe, M.P. Deposition of atmospheric ions to pine branches and surrogate surfaces in the vicinity of Emerald Lake watershed, Sequoia National Park. Atmos. Environ. 25A:2203–2210; 1991.

Bytnerowicz, A.; Dawson, P.J.; Morrison, C.L.; Poe, M.P. Atmospheric dry deposition on pines in the Eastern Brook Lake watershed, Sierra Nevada, California. Atmos. Environ. 26A:3195–3201; 1992.

Bytnerowicz, A.; Fenn, M.; Glaubig, R. Dry deposition of nitrogen and sulfur to forest canopies at three plots (calculated on the basis of foliage rinsing, internal uptake of gaseous pollutants, and estimates of deposition to ground), section 4. In: Miller, P.R.; Chow, J.; Watson, J.G., eds. Assessment of acidic deposition and ozone effects on conifer forests in the San Bernardino Mountains. Final report to the California Environmental Protection Agency, Air Resources Board, contract A032-180. Sacramento, CA: 1996:4-1–4-75.

Bytnerowicz, A.; Fenn, M. Nitrogen deposition in California forests: a review. Environ. Pollut. 92:127–146; 1996.

Bytnerowicz, A.; Carroll, J.; Takemoto, B.; Miller, P.; Fenn, M.; Musselman, R. Distribution and transport of air pollutants to vulnerable California ecosystems. In: Proceedings of the symposium on critical methodologies for the study of ecosystem health. Center for Ecological Health Research, University of California, Davis, CA, September 1996. Ann Arbor, MI: Ann Arbor Press; 1998.

Cape, J.N. Direct damage to vegetation caused by acidic rain and polluted cloud: definition of critical levels for forest trees. Environ. Pollut. 82:167–180; 1993.

Czerwinski, W. Fizjologia Roslin. Warszawa, Poland: PWN; 1978.

Davidson, C.I.; Wu, Y-I. Dry deposition of particles and vapors. In: Lindberg, A.S.; Page, E.L.; Norton, S.A., eds. Acidic precipitation, vol. 3. Sources, deposition and canopy interactions. New York: Springer-Verlag; 1990:103–216.

Davidson, E.A.; Shimel, J.P. Microbial processes of production and consumption of nitric oxide, nitrous oxide and methane. In: Matson, E.A.; Harriss, R.C., eds. Biogenic trace gases: measuring emissions from soil and water. Oxford: Blackwell Science; 1995:327–357.

Draaijers, G.P.J.; Ivens, W.P.M.F; Bos, M.M.; Blenten, W. The contribution of ammonia emissions from agriculture to the deposition of acidifying and eutrophying compounds onto forests. Environ. Pollut. 60:55–66; 1989.

Duffy, P.D.; Schreiber, J.D. Nutrient leaching of a loblolly pine forest floor by simulated rainfall. II. Environmental factors. Forest Sci. 36:777–789; 1990.

Enders, G.; Teichmann, U.; Kramm, G. Profiles of ozone and surface-layer parameters over a mature spruce stand. In: Georgii, H-W., ed. Mechanisms and effects of pollutant transfer into forests. Dordrecht: Kluwer; 1989:21–35.

Falkengren-Grerup, U. Soil acidification and its impact on ground vegetation. Ambio 18:179–183; 1989.

Fassnacht, K.S.; Gower, S.T.; Norman, J.M.; McMurtrie, R.E. A comparison of optical and direct methods for estimating foliage surface area index in forests. Agric. Forest Meteorol. 71:183–207; 1994.

Fenn, M.E.; Bytnerowicz, A. Dry deposition of nitrogen and sulfur to ponderosa and Jeffrey pine in the San Bernardino National Forest in southern California. Environ. Pollut. 81:277–285; 1993.

Fenn, M.E.; Bytnerowicz, A. Summer throughfall and winter deposition in the San Bernardino Mountains in southern California. Atmos. Environ. 31:673–683; 1997.

Fenn, M.E.; Glaubig, R.; Jones, D.; Schilling, S. Wet deposition and throughfall measurements in relation to the forest canopy at 3 plots. In: Miller, P.R., Chow, J., Watson, J., eds. Assessment of acidic deposition and ozone effects on conifer forests in the San Bernardino Mountains. Final report, Environmental Protection Agency, California Air Resources Board, contract A032-180. Sacramento, CA: 1996a:7-1–7-31.

Fenn, M.E.; Poth, M.A.; Johnson, D.W. Evidence for nitrogen saturation in the San Bernardino Mountains in southern California. Forest Ecol. Manage. 82:211–230; 1996b.

Fuentes, J.D.; Gillespie, T.J.; den Hartog, G.; Neumann, H.H. Ozone deposition onto a deciduous forest during dry and wet conditions Agric. Forest Meteorol. 62:1–18; 1992.

Grantz, D.A.; Zhang, X.J.; Massman, W.J.; den Hartog, G.; Neuman, H.H.; Pederson, J.R. Effects of stomatal conductance and surface wetness on ozone deposition in field-grown grape. Atmos. Environ. 29:3189–3198; 1995.

Grantz, D.A.; Zhang, X.J.; Massman, W.J.; Delany, A.; Pederson, J.R. Ozone deposition to cotton (*Gossypium hirsutum* L.) fields: stomatal and surface wetness effects during the California ozone deposition experiment. Agric. Forest Meteorol. 85:19–31; 1997.

Grosjean, D.; Bytnerowicz, A. Nitrogenous air pollutants at a southern California mountain forest smog receptor site. Atmos. Environ. 27A:483–492; 1993.

Hanson, P.J.; Garten, C.T., Jr. Deposition of $H^{15}NO_3$ vapour to white oak, red maple and loblolly pine foliage: experimental observations and a generalized model. N. Phytol. 122:329–337; 1992.

Hanson, P.J.; Lindberg, S.E. Dry deposition of reactive nitrogen compounds: a review of leaf, canopy and non-foliar measurements. Atmos. Environ. 25A:1615–1634; 1991.

Hicks, B.B.; Meyers, T.P. Measuring and modeling dry deposition in mountainous areas. In: Unsworth, M.H.; Fowler, B. eds. Acid deposition at high elevation sites. Amsterdam: Kluwer; 1988:541–552.

Huettl, R.F. Forest soil acidification. Angew. Bot. 67:66–75; 1993.

Keuken, M.P.; Schoonebek, C.A.M.; Van Wensween-Lonter, A.; Slanina, J. Simultaneous sampling of NH_3, HNO_3, HCl, SO_2 and H_2O_2 in ambient air by a wet annular denuder system. Atmos. Environ. 22:2541–2548; 1988.

Kiefer, J.W.; Fenn, M.E. Using vector analysis to assess nitrogen status of ponderosa and Jeffrey pine along deposition gradients in forests of southern California. Forest Ecol. Manage. 94:47–59; 1997.

Kimmins, J.P. Forest ecology. New York: Macmillan; 1987.

Lefohn, A.S. The characterization of ambient ozone exposures. In: Lefohn, A.S., ed.

Surface level ozone exposures and their effects on vegetation. Chelsea, MI: Lewis Publishers; 1992:31–92.

Legge, A.H.; Krupa, S.V. Air quality at a high elevation, remote site in western Canada. Proceedings of the 82nd annual meeting and exhibition of the Air and Waste Management Association, Anaheim, CA, June 29–30, 1989.

Leuning, R.; Neumann, H.H.; Thurtell, G.W. Ozone uptake by corn (*Zea mays* L.). A general approach. Agric. Meteorol. 20:115–135; 1979.

Likens, G.E.; Driscoll, C.T.; Buso, D.C. Long-term effects of acid rain: response and recovery of a forest ecosystem. Science 272:244–246; 1996.

Lindberg, S.E.; Lovett, G.M.; Richter, D.D.; Johnson, D.W. Atmospheric deposition and canopy interactions of major ions in a forest. Science 231:141–145; 1986.

Lindberg, S.E.; Bredemeier, M.; Schaefer, D.A.; Qi, L. Atmospheric concentrations and deposition of nitrogen and major ions in conifer forests in the United States and Federal Republic of Germany. Atmos. Environ. 24A:2207–2220; 1990.

Lovett, G.M.; Lindberg, S.E. Dry deposition and canopy exchange in a mixed oak forest as determined by analysis of throughfall. J. Appl. Ecol. 21:1013–1027; 1985.

Marshall, J.D.; Waring, R.H. Comparison of methods of estimating leaf-area index in old-growth Douglas-fir. Ecology 67:975–979; 1986.

Mengel, K.; Kirkby, E.A. Principles of plant nutrition. Bern, Switzerland: International Potash Institute; 1982.

Meyers, T.P.; Hulbert, B.J.; Hicks, B.B. HNO_3 deposition to a deciduous forest. Boundary-Layer Meteorol. 49:395–410; 1989.

Meyers, T.P.; Baldocchi, D.D. A comparison of models for deriving dry deposition fluxes of O_3 and SO_2 to a forest canopy. Tellus 40B:270–284; 1988.

Miller, P.R.; McCutchan, M.H.; Ryan, B.C. Influence of climate and topography on oxidant air pollution concentrations that damage conifer forests in southern California. Mitt. der Forstlichen Bundesversuchanstalt Wien. 97:585–607; 1972.

Miller, P.R.; Chow, J.; Watson, J.G.; Bytnerowicz, A.; Poth, M.; Taylor, G. Assessment of acidic deposition and ozone effects on conifer forests in the San Bernardino Mountains. Final report. Contract A032-180. Sacramento, CA: California Air Resources Board; 1996a.

Miller, P.R.; Bytnerowicz, A.; Jones, D.; Schilling, S. Characterization of ozone, particle chemistry and meteorology measurements in support of canopy deposition estimates. In: Assessment of acidic deposition and ozone effects on conifer forests in the San Bernardino Mountains. Final report. Contract A032–180. Sacramento, CA: California Air Resources Board; 1996b:3-1–3-42.

Mitchell, M.J.; Burke, M.K.; Shepard, J.P. Seasonal and spatial patterns of S, Ca, and N dynamics of a northern hardwood forest ecosystem. Biogeochemistry 17:165–189; 1992.

Nihlgard, B. Nitrogen pollutants seriously stressing Europe's forests. Swed. EPA Acid Magazine 9:22–24; 1990.

Peake, E.; Legge, A.H. Evaluation of methods used to collect air quality data in remote and rural sites in Alberta, Canada. EPA/APCA symposium on measurements of toxic and related air pollutants. Research Triangle Park, NC; May 3–6, 1987.

Pierce, L.L.; Running, S.W. Rapid estimation of coniferous forest leaf area index using a portable integrating radiometer. Ecology 69:1762–1767; 1988.

Pierson, W.R.; Brachaczek, W.W. Coarse and fine particulate atmospheric nitrate and HNO_3 (g) in Claremont, California, during the 1985 nitrogen species method comparison study. Atmos. Environ. 22:1665–1668; 1988.

Possanzini, M.; Febo, A.; Liberti, A. New design of a high-performance denuder for the sampling of atmospheric pollutants. Atmos. Environ. 17:2605–2612; 1983.

Probst, A.; Dambrine, E.; Viville, D.; Fritz, B. Influence of acid atmospheric inputs on surface water chemistry and mineral fluxes in a declining spruce stand within a small granitic catchment (Vosges Massif, France). J. Hydrol. 116:101–124; 1990.

Riggan, P.J.; Lockwood, R.N.; Lopez, E.N. Deposition and processing of airborne nitrogen pollutants in Mediterranean type ecosystem of southern California. Environ. Sci. Technol. 19:781–789; 1985.

Salisbury, F.B.; Ross, C.W. Plant physiology. Belmont, CA: Wadsworth; 1978.

SAS, Inc. SAS/STAT users guide, version 6, vol. 2, 4th ed. Cary, NC: SAS Institute; 1990.

Sickles, J.E., II; Perrino, C.; Allegrini, I.; Febo, A.; Possanzini, M. Sampling and analysis of ambient air near Los Angeles using an annular denuder system. Atmos. Environ. 22:1619–1625; 1988.

Skelly, J.M.; Chappelka, A.H.; Laurence, J.A.; Frederickson, T.S. Ozone and its known and potential effects on forests in the eastern United States. In: Sandermann, H.; Wellburn, A.R.; Heath, R.L., eds. Forest decline and ozone. Ecol. studies 127. Berlin-Heidelberg: Springer-Verlag; 1997:69–93.

Stevens, P.A.; Adamson, J.K.; Reynolds, B.; Hornung, M. Dissolved inorganic nitrogen concentrations and fluxes in three British Sitka spruce plantations. Plant Soil 128:103–108; 1990.

Stockwell, W.R.; Kramm, G.; Scheel, H-E.; Mohnen, V.A.; Seiler, W. Ozone formation, destruction and exposure in Europe and the United States. In: Sandermann, H.; Wellburn, A.R.; Heath, R.L., eds. Forest decline and ozone. Ecol. studies 127. Berlin-Heidelberg: Springer-Verlag; 1997:1–38.

Temple, P.J. Injury and yield responses of differentially irrigated cotton to ozone. Agron. J. 80:751–755; 1988.

Temple, P.J.; Miller, P.R. Seasonal influences of ozone uptake and foliar injury to ponderosa and Jeffrey pines at a southern California site. In: Bytnerowicz, A.; Arbaugh, M.; Schilling, S., eds. Proceedings of the international symposium on air pollution and climate change effects on forest ecosystems, February 5–9, 1996, Riverside, CA. GTR-164. Albany, CA: Forest Service, Pacific Southwest Station; 1998.

Temple, P.J.; Riechers, G.H.; Miller, P.R.; Lennox, R.W. Growth responses of ponderosa pine to long-term exposure to ozone, wet and dry acidic deposition, and drought. Can. J. For. Res. 23:59–66; 1992.

Wesselink, L.G.; Meiwes, K.J.; Matzner, E.; Stein, A. Long-term changes in water and soil chemistry in spruce and beech forests, Solling, Germany. Environ. Sci. Technol. 29:51–58; 1995.

12. Direct Effects of Nitric Acid on Forest Trees

A. Bytnerowicz, P. Padgett, K. Percy, M. Krywult,
G. Riechers, and J. Hom

Introduction

Concentrations of nitrogenous (N) pollutants have been steadily rising worldwide (Hameed and Dignon 1992). Anthropogenic activities are primarily responsible for these increases (Hanson and Lindberg 1991), and with the industrialization of the Third World nations, global N pollution will likely continue to increase over the next several decades (Vitousek 1995). In dense urban areas such as southern California, nitrogen oxides are a major component of photochemical smog. Although nitrogen dioxide (NO_2) is often the dominant N pollutant (Ollinger et al. 1993; Bytnerowicz and Fenn 1996), nitric acid vapor (HNO_3) may be more relevant to problems associated with N deposition to forests and other ecosystems because of its unusually high deposition velocity (Hanson and Lindberg 1991; Lovett 1994) and direct assimilation by plants and microorganisms. Anecdotal evidence suggests that in areas of high ambient concentrations, HNO_3 may also cause damage to the foliage of forest trees.

During the photochemical smog formation process, nitric oxide (NO) is rapidly oxidized to NO_2, which subsequently reacts with hydroxyl radicals, producing HNO_3 vapor (Crutzen 1979; Finlayson-Pitts and Pitts 1986). Concentrations of ozone (O_3) and HNO_3 in areas affected by photochemical smog are strongly correlated, and measurements performed in several locations of the Los Angeles Basin may serve as a good example of their co-occurrence (Cal Tech EQL 1986).

Nitric acid concentrations are the highest in urban plumes downwind from the pollution source areas (Tuazon et al. 1981; Bytnerowicz et al. 1987). In the mountains of southern California, the highest concentrations of HNO_3 typically occur in the afternoon. At Tanbark Flat in the San Gabriel Mountains, a highly contaminated location adjacent to the Los Angeles Basin, the daytime summer 12-hour average concentrations as high as 18.4 ppb (46.7 $\mu g/m^3$) have been recorded, whereas the average seasonal 12-hour daily concentrations were 4.4–7.9 ppb (11.3–20.3 $\mu g/m^3$) during 1989–1991 (Grosjean and Bytnerowicz 1993). In the moderately polluted Barton Flat site of the San Bernardino Mountains, the average 24-hour concentrations of HNO_3 ranged from 1.2 to 2.5 ppb (3.0–6.5 $\mu g/m^3$) during the 1993–1995 summer seasons (Bytnerowicz et al. 1996). By contrast, the Whitaker Forest of the western Sierra Nevadas receives significantly lower 12-hour daytime HNO_3 concentrations: on average 0.4–0.8 ppb (1.1–2.0 $\mu g/m^3$) (Bytnerowicz and Riechers 1995). And still further from population centers, eastern Sierra Nevada locations were much lower in atmospheric HNO_3 than the western Sierra Nevada sites, with average 24-hour concentrations of 0.15 ppb (0.36 $\mu g/m^3$) (Miller and Walsh 1991). Concentrations in the eastern Sierra Nevada were similar to other remote mountain locations of North America (Legge and Krupa 1989).

On a vertical transect (forest floor to treetop) through the ponderosa/Jeffrey pine canopy of the mixed coniferous forest of the San Bernardino Mountains during 1992–1994 summer seasons, the 24-hour average concentrations of HNO_3 were in the range of 0.4–3.7 ppb (1.0–9.6 $\mu g/m^3$). These values did not significantly differ along the vertical transect (Bytnerowicz et al., this volume). However, HNO_3 conductance or deposition velocity at a single branch level were significantly higher at the canopy top as compared with low branches (Fig. 12.1). Because only a fraction of the atmospheric load is detained by the leaf surfaces of an individual tree, deposition alone was not enough to reduce the atmospheric concentrations in the canopy. The differences in deposition velocity were, therefore, caused by higher wind speeds at the canopy top, thus higher rates of delivery. Wind speeds measured between June–October averaged 0.9 m/sec at 2 m above the ground and 3.0 m/sec at 29 m above the ground, resulting in an unequal distribution of deposited N throughout the canopy.

Controlled Studies on HNO_3 Effects on Forest Trees

A fumigation system has been developed for studying effects of HNO_3 vapor on tree seedlings (Krywult et al. 1996). The system, contained in growth chambers providing a constant temperature of 22°C, allowed for a short-term exposures of small seedlings or single branches to the pollutant at controlled levels.

Foliar Surface Interactions with Gaseous HNO_3

Strong evidence indicates that the stomatal uptake of O_3 and most oxides of nitrogen is the dominant pathway by which toxic pollutants enter the plant body

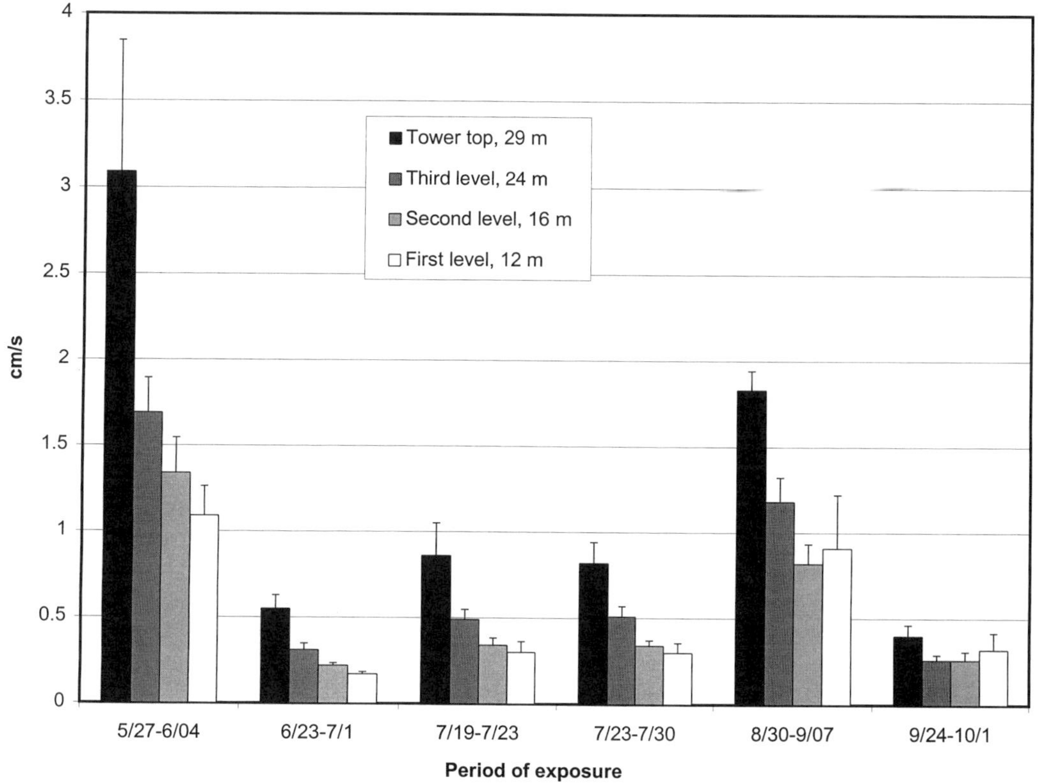

Figure 12.1. Estimated stomatal conductance (k_{HNO_3}) of nitric acid vapor to ponderosa pine foliage for different levels of canopy during the 1993 pollution season.

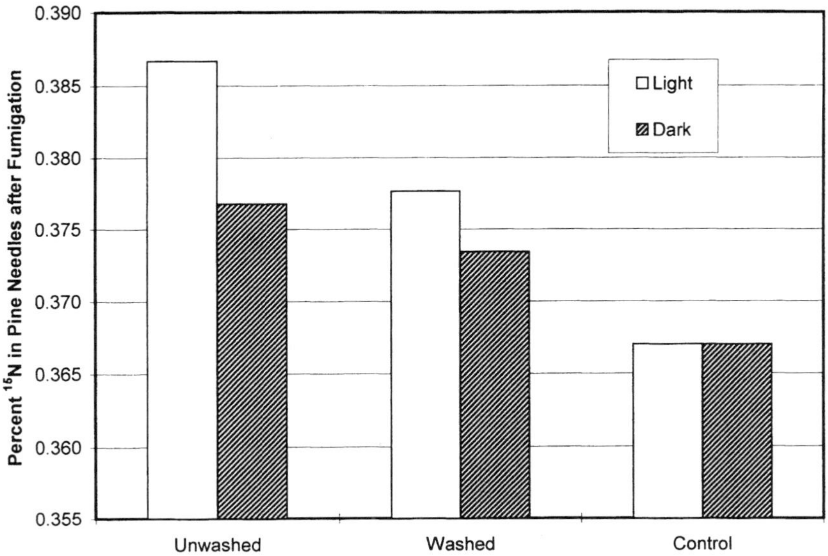

Figure 12.2. Deposition of $H^{15}NO_3$ to ponderosa pine seedlings under light and dark conditions.

(Darrall 1989; Wellburn 1990). Nitric acid, however, appears to be unique in that a significant quantity of deposited N enters the plant body via a transcuticular pathway (Marshall and Cadle 1989; Hanson and Garten 1992). Although this pathway has not been clearly elucidated, it has been fairly well accepted. The two mechanisms are not independent, however. The rate of deposition to the leaf surface is influenced by the stomatal conductance; increased stomatal conductance increases the apparent velocity of deposition by altering the leaf boundary layer conditions. As a consequence, greater surface accumulation due to changes in the leaf boundary layer conditions may enhance transcuticular uptake. In an effort to simplify the experimental system and to understand the transcuticular pathway for HNO_3 uptake, experiments were conducted in the dark when stomata are normally closed.

A comparison of deposition in the light versus in the dark indicated a significantly greater accumulation of $H^{15}NO_3$ on ponderosa pine (*Pinus ponderosa*) needles occurred in the light (Fig. 12.2). However, rinsing the needles with H_2O reduced ^{15}N by about 50% relative to the background. Accumulation of the ^{15}N label on dark fumigated seedlings remained significantly higher than the background controls, and the rinsing treatment resulted in an insignificant loss of label. Results of ^{15}N analysis indicated that after the dark fumigations, similar proportions of $H^{15}NO_3$ were retained on pine needle surfaces as in their interior. This supports previous reports (Norby et al. 1989; Krywult et al. 1996) that surface-deposited $H^{15}NO_3$ was incorporated into the leaf mesophyll by transcuticular transport. Much higher deposition of HNO_3 was determined for the California black oak (*Quercus kelloggii*) foliage compared with the pine needles (Fig. 12.3).

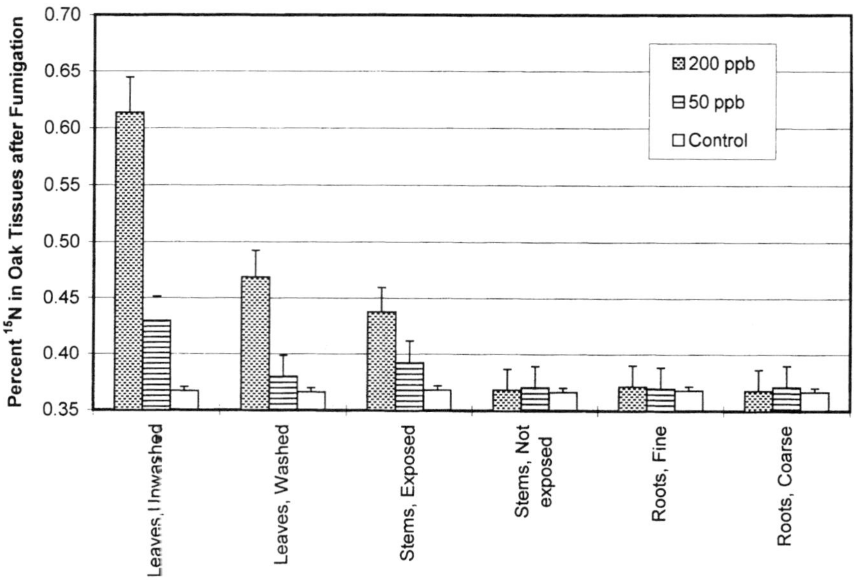

Figure 12.3. Deposition of $H^{15}NO_3$ to different parts of California black oak seedlings.

Portions of stems exposed to $H^{15}NO_3$ showed significantly more ^{15}N than the unexposed parts of stems and roots (Bytnerowicz et al. 1998). The exposure concentration had a significant effect on ^{15}N recovered following fumigation. Nearly 50% more $H^{15}NO_3$ was absorbed and retained in all plant parts by California black oak seedlings exposed to 200 ppb as compared with 50 ppb (Fig. 12.3). Results of these studies also suggested that only small portion of the pollutant was adsorbed to the epicuticular wax layer. The larger difference between the unrinsed and H_2O-rinsed oak foliage than for pine needles indicated that a larger proportion of $H^{15}NO_3$ was deposited on oak leaf surfaces and that a smaller proportion of $H^{15}NO_3$ was incorporated into oak leaf mesophyll. Results of the ^{15}N analysis were also confirmed by determinations of nitrate reductase (NR) activity (see below). These results were similar to those obtained by Marshall and Cadle (1989).

Modification of Foliar Surface Characteristics

The leaf surface is the first point of contact between air pollutants and forest trees. The outermost surfaces of all leaves and needles are covered by cuticular waxes, which are the most important functional components of higher plant aerial surfaces (Riederer and Markstadter 1996). These waxes are embedded within the underlying cuticular membranes and are, importantly, in direct contact with the atmosphere as epicuticular waxes. Among their many functions are control of water and solute diffusion across the cuticle (foliar uptake), "waterproofing," attenuation of damaging radiation, and protection against foliar insects and diseases.

Exposure of ponderosa pine seedlings to 50 ppb HNO_3 over 12 hours in the dark caused significant changes in wax structure similar to those observed following needle exposure to acid deposition applied either as simulated rain or fog (Percy and Baker 1990; Percy et al. 1990). When examined under the scanning electron microscope (SEM), needles of fumigated plants typically had highly occluded stomata and altered structure of crystalline wax tube aggregates between stomatal rows. The most common feature was the appearance of amorphous, platelike wax structures over, and a decrease in extent of tube crystallites within epistomatal chambers. Stomata of the exposed needles also had a sunken appearance in contrast to the control plants (Fig. 12.4A,B). Microscopic lesions on needles of the fumigated plants were observed as well under the SEM (Fig. 12.4C), pointing to an immediate interaction of the acidic vapor with needle surfaces leading to cuticle deterioration and epidermal cell collapse (Bytnerowicz et al., 1998).

Epicuticular wax structure is a species-dependent mixture of different crystalline geometries conferred by the chemical composition of individual homologues (carbon chain lengths) comprising it. Gas chromatographic analysis of wax recovered from ponderosa pine needles exposed to HNO_3 has shown that short-term fumigation can result in significant changes in composition. Proportions of fatty acids and alkyl esters were increased after seedlings were exposed at between 50 and 95 ppb HNO_3 for up to 48 hours (Krywult et al. 1996). In a later series of experiments, wax recovered from ponderosa pine needles exposed at 50 ppb over 12 hours had significantly decreased proportions of fatty acids and increased proportions of alkyl esters (Bytnerowicz et al. 1998). In both cases, proportions of alkane diols, alkanols, alkanes, and secondary alkanols were unaffected. There is little if any previous evidence for a direct short-term chemical interaction between inorganic air pollutants (O_3, acid deposition, SO_2) and epicuticular waxes in situ on developed needles. However, red spruce (*Picea rubens* Sarg.) wax recrystallized onto surrogate surfaces has been shown to be altered in a manner analogous with that just described for HNO_3 following brief immersion in strong acids at pH <3.0 (Percy et al. 1994).

By contrast, exposures to 50 ppb HNO_3 did not produce any similar microscopic (SEM) changes to the surfaces of the oak leaves, although some degree of epidermal cell collapse and trichome deformation was evident on oak leaves exposed at 200 ppb HNO_3 (Bytnerowicz et al. 1998). The epicuticular wax deposit, however, does not comprise a dense array of crystalline tube aggregates as was the case for ponderosa pine. Scattered small crystalline plates observed under the SEM were likely comprised of nonacosane, a C_{29} alkane and dominant homologue (ug/cm^2) present in wax recovered from this species. Gas chromatographic analysis revealed that the four main wax classes in order of decreasing relative proportion in wax recovered from leaf surfaces of control seedlings were alkanes, alkyl esters, fatty acids, and alkanols.

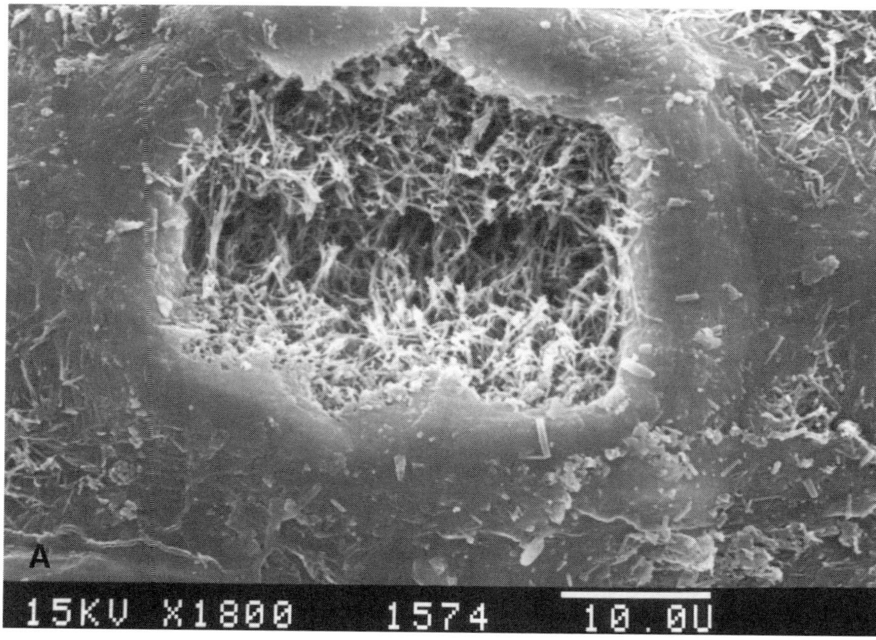

Figure 12.4. SEM evaluation of ponderosa pine needle surfaces: (A) close-up of the control plant stomata, (B) close-up of the stomata of 50 ppb $H^{15}NO_3$, 12-hour light fumigated plant, (C) close-up of the 50 ppb $H^{15}NO_3$, 12-hour light fumigated plant showing deep lesions on cuticle surfaces.

Cluster analysis was proved a useful tool for linking trees with pollution climate based on leaf wax chemical composition (Cape and Percy 1998). When oak leaf wax composition data were analyzed by using average-linked cluster analysis to sort effects on oak seedlings, more the 97% of the total variance in the data set was accounted for in the first two eigenvalues of the correlation matrix, and control and fumigated seedlings were successfully grouped by treatment (Bytnerowicz et al., unpublished data). This observation is noteworthy given the limited data set and is further confirmation for a subtle, yet direct interaction between HNO_3 and oak leaf surfaces.

These results have shown that short-term exposures to elevated levels of HNO_3 could cause damage of leaf surfaces of the two studied species. Ponderosa pine seems to be more sensitive to the pollutant than California black oak—very high levels of 200 ppb HNO_3 were needed to produce initial microscopic symptoms of damage for oak. It is unknown if field conditions (i.e., lower concentrations but longer exposures and the presence of other pollutants) could result in similar injury symptoms on ponderosa pine or California black oak. Longer exposures at more realistic concentrations will be required to determine chronic versus acute responses.

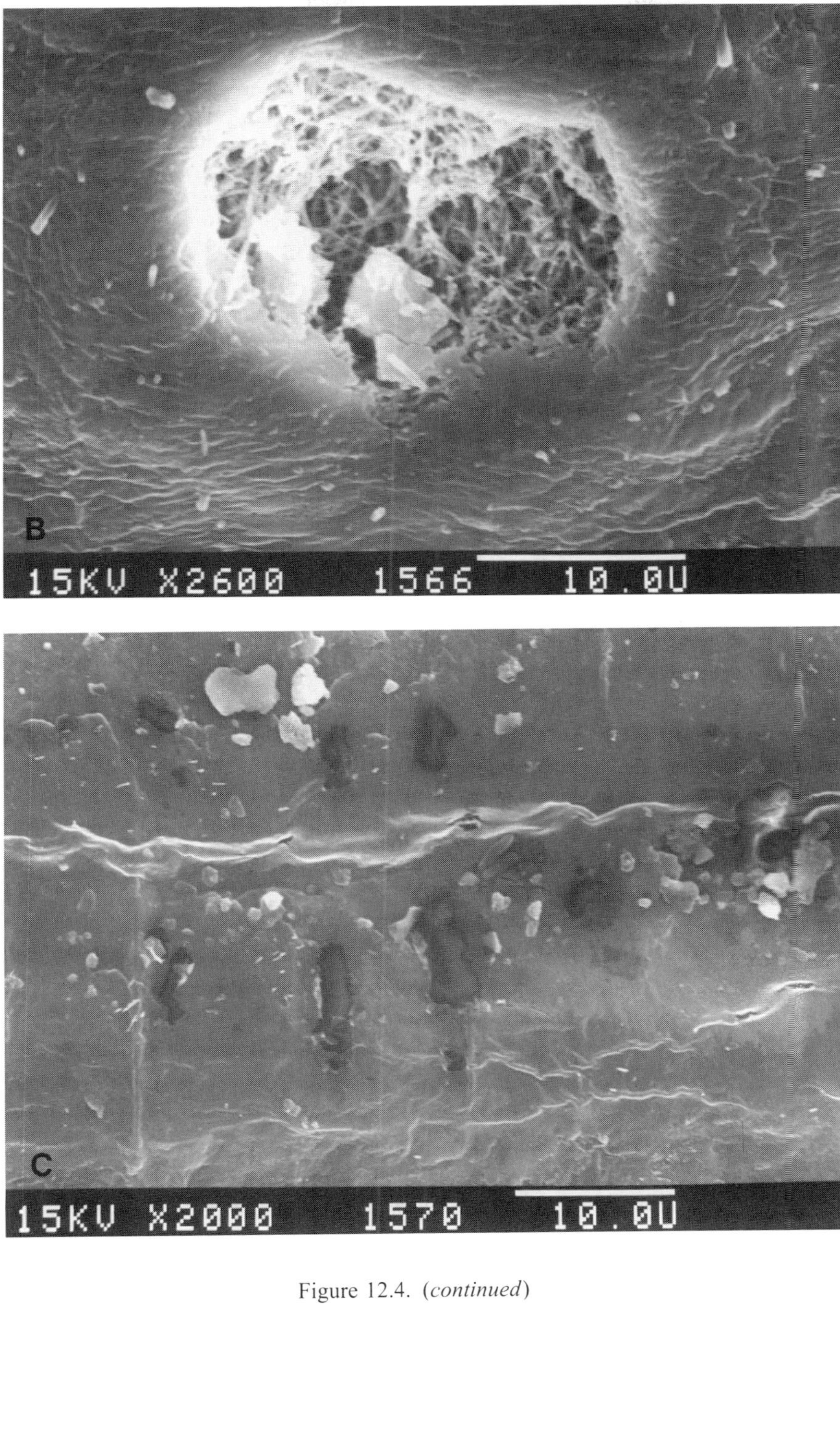

Figure 12.4. (*continued*)

Biochemical Fate of Absorbed HNO_3

Although the mechanism leading to transcuticular absorption is not fully understood, once HNO_3 or the disassociated NO_3^- ion enters the apoplastic space, the cellular response is assumed to be the same as if the ions were absorbed and translocated from roots. Ionic nitrate is relatively innocuous to plant cells and can therefore accumulate in the cytoplasm or be stored in vacuoles (Clarkson 1986). Nitrate absorbed by roots may be assimilated and used by root tissues or translocated via the transpiration stream to shoots either before or after reduction and assimilation. Whether foliar-absorbed HNO_3 is translocated as NO_3^- to roots is not known, but in general, nitrate concentrations in phloem sap are low or not detectable (Peuke et al. 1994).

The presence of NO_3^- either in the cell or at the cell membrane induces the synthesis of a transmembrane carrier (Clarkson 1986; Rosenberg et al. 1997), the enzymes nitrate reductase, nitrite reductase (Crawford and Campbell 1990; Aslam et al. 1993), and to a lesser extent, glutamine synthase and glutamate synthase (Miflin and Lea 1982; Lam et al. 1996). The first stable product of NO_3^- assimilation is the amino acid glutamine. Regulation of the various enzymes in N assimilation is complex, but it is largely a function of the substrate concentration, end-product demand, and feedback regulation (see Kleinhofs and Warner 1992; Imsande and Touraine 1994 for recent reviews). All higher plants investigated so far have the capacity to reduce nitrate in both leaf and root tissues, but dependence on one organ or the other is contingent on genetic predisposition and environmental conditions (Smirnoff et al. 1984; Andrews 1986). In general, it appears that conifers are mainly root reducers or have relatively low background levels of foliar nitrate reductase activity (NRA) (Smirnoff et al. 1984; Norby et al. 1989; Krywult et al. 1996). Our preliminary data suggest that California black oak is a shoot reducer or has significantly higher background levels of leaf NRA than that typically reported for pines.

Increased NRA, indicative of N assimilation, has been demonstrated in response to HNO_3 (Norby et al. 1989; Krywult et al. 1996). The results of fumigation experiments using ponderosa pine indicated that exposure to 50 ppb HNO_3 for 48 hours in the dark or during a 12-hour light/dark cycle increased NRA in leaves as much as fourfold as compared with control trees (Fig. 12.5). When half the needles from fumigated trees were washed with deionized water, the unwashed needles continued to increase in NRA while NRA remained constant in the washed needles (Fig. 12.6). These data imply that HNO_3 deposited on the leaf surface continues to be taken up after trees are removed from a HNO_3 vapor source. Such would be the case in nature as formation of HNO_3 occurs during daylight hours; after sunset, atmospheric concentrations drop significantly (Bytnerowicz and Fenn 1996). These results are similar to those reported by Norby et al. (1989), who showed that NRA levels remained high for several hours after removal of the HNO_3 source, but NRA induced in the presence of NO_2 declined immediately on removal from the NO_2 source. Nitrate reductase activity in roots of trees where only foliar surfaces were directly exposed to HNO_3 was largely unaffected (M. Krywult and P. Padgett, unpublished data).

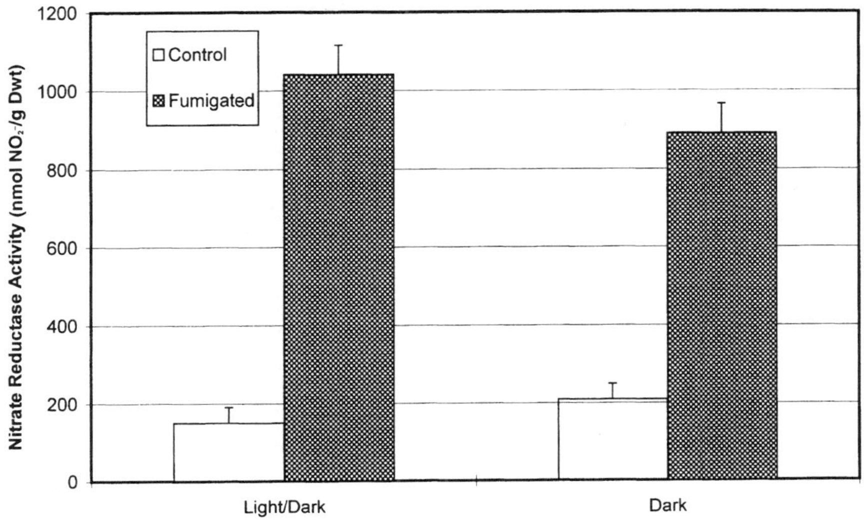

Figure 12.5. Nitrate reductase activity of ponderosa pine following 48-hour fumigation with HNO_3. Seedlings were exposed to HNO_3 either in complete darkness or during a 12-hour light/dark cycle.

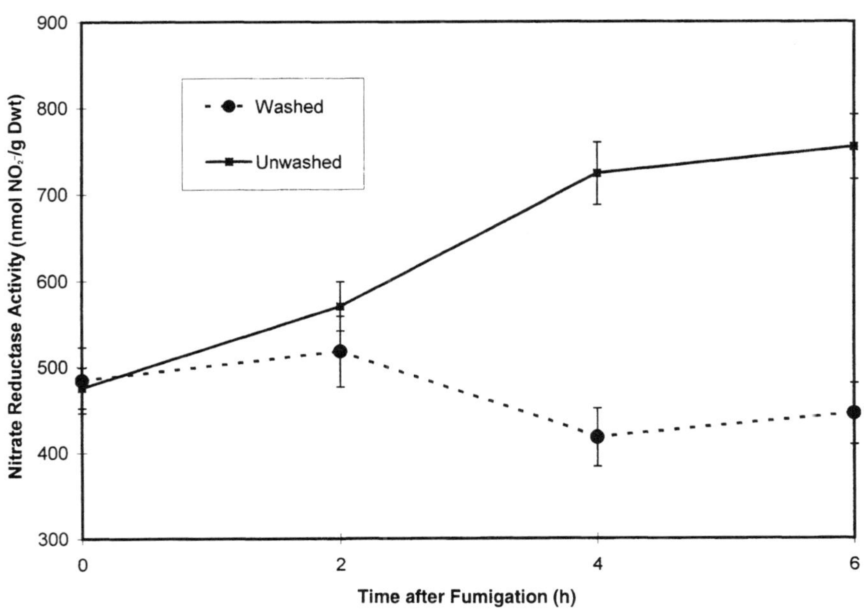

Figure 12.6. Nitrate reductase activity in ponderosa pine after 12-hour fumigations with 50 ppb HNO_3. Following fumigation, half the tree was rinsed to remove deposited material. Needles were assayed for nitrate reductase activity 2, 4, and 6 hours later.

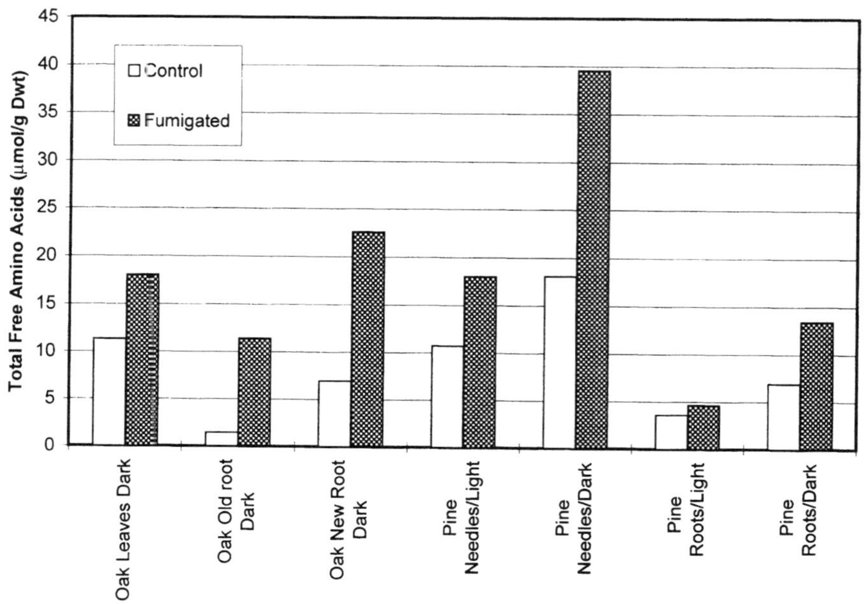

Figure 12.7. Free amino acid content of oak leaves, pine needles, and roots of both species following 12-hour fumigation at 50 ppb HNO_3.

Analysis of free amino acid content of roots and leaves/needles following fumigation under dark conditions suggests that atmospheric HNO_3 is not only incorporated in leaves but assimilation products are translocated to roots (Fig. 12.7). Changes in concentrations and distribution of amino acids after fumigation with NO_2 have also been reported (Nussbaum et al. 1993). Edfast et al. (1990) and Pietila et al. (1991) documented small changes in amino acid composition in Scots pine *(Pinus sylvestris)* that correlated with differences in N deposition (predominately NH_4^+) under field conditions.

Both NRA and nitrate transport appear to be regulated, in part, by free amino acid pools (Oaks et al. 1977; Cooper and Clarkson 1989; Muller and Tourraine 1992; Imsande and Tourraine 1994; Li et al. 1995; Padgett and Leonard 1996). Changes in free amino acids pools may be indicative of nitrate assimilation during foliar exposure to HNO_3. It is a reasonable expectation that changes in the pool sizes of amino acids would represent changes in the overall status of plant N and that foliar assimilation might have a profound effect on root uptake and function.

Effects on Plant Antioxidant System

Pine needles exposed to 50 ppb of HNO_3 for 16 hours in darkness showed substantial changes in antioxidant concentrations. Levels of glutathione and cysteine were markedly reduced, whereas the levels of oxidized glutathione and cystine increased in the HNO_3-exposed needles compared with the nonfumigated needles. At the same time, the treated plants showed slightly lower contents of ascorbate and a higher proportion of oxidized ascorbate (M. Tausz, unpublished data). More research on the effects of HNO_3 on antioxidative systems of pines and other species is needed to evaluate the ability of forest plants to cope with the pollutant. Such studies should also include investigations on effects of HNO_3 exposures on key enzymes controlling antioxidant system of plants (e.g., glutathione reductase activity).

HNO_3 Effects on Carbon Assimilation

The carbon and N metabolic pathways compete for carbon skeletons as well as reducing power and ATP (Goodwin and Mercer 1983; Stulen 1990; Huppe and Turpin 1994). The effect of N fertilization on carbon assimilation and storage as measured by yield components has been recognized for years (e.g., Brockley 1995). In general, abundant or excessive N results in stimulation in vegetative growth, delayed flower production and senescence, and delays and reduction in synthesis of storage products (Marschner 1995). The stimulation of growth, in particular, increases demand for water and other nutrients (Nilsson et al. 1995).

The results of 12-hour fumigations of *Quercus kelloggii* suggest that the normal mobilization of starch during in the dark hours is inhibited. This inhibition is accompanied by a reduction in soluble sugars in leaf tissue (Fig. 12.8). The

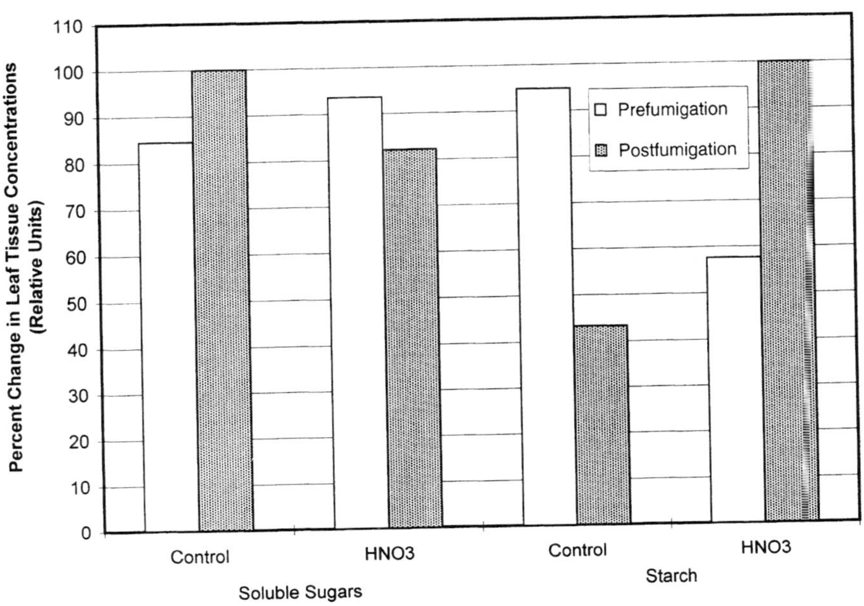

Figure 12.8. Soluble sugar and starch content of California black oak after 12-hour exposure to 50 ppb HNO_3.

reduction in sugars was across all classes of sugars including sucrose and a sugar specific to oaks, querbrachitol. A parallel increase in total amino acids, particularly glutamate, glutamine, and asparagine, hints that synthesis of amino acids may be at the expense of carbohydrate reserves. The mass balance and stoichiometry have not yet been determined and are from single 12-hour exposures, but the data suggest that chronic exposure to atmospheric HNO_3 may result in modifications to carbon assimilation and perhaps significant changes in the C/N ratio.

Ecological Relevance

HNO_3 as Plant Nutrient

It is not known whether the amount of HNO_3 air pollutants that are taken up by plant leaves are significant enough to approach the levels of those used in agricultural applications or foliar fertilization. A significant increase in N availability could, however, have a significant effect on function and structure of the forest. Vose and Swank (1990) addressed this issue by using whole branch chambers and $H^{15}NO_3$ fumigation of mature trees. Following mass spectrometry analysis, they calculated that 1% of the forest N requirement was met by foliar assimilation. In their discussion, they pointed out several inherent problems with these types of studies, particularly in reference to the assumptions required for the calculations. Our preliminary data following 12-hour fumigations of whole seedlings indicated that as much as one-fourth of the amino acids contained ^{15}N (data not shown). Neither of these studies provides enough information on which the contribution of foliar assimilation to whole plant nutrition can be calculated. But this is clearly an issue of importance to the long-term health of western forests.

Responses of Mature Forest Trees

In the mixed conifer forest stand of the San Bernardino National Forest at Barton Flats, the amount of deposited NO_3^- (most of which was of anthropogenic HNO_3 origin) from branches of California black oak was higher than deposition to ponderosa pine and white fir (*Abies concolor*) (Fig. 12.9). Deposition flux values from the forest stand were quite variable, especially for the oak seedlings. The data also indicated that for some species (i.e., ponderosa pine) differences in deposition between the branches of mature trees and seedlings were large. Therefore one must be cautious when extrapolating results of the seedling data to the mature forest stands.

Activity of nitrate reductase, a key enzyme in controlling HNO_3 assimilation by plants (Crawford et al. 1992), was well correlated with the amount of deposited HNO_3 vapor in controlled experiments (see above). Measurements of NR activity along the air pollution gradient in the San Bernardino Mountains also indicated that extensive N absorption in pine foliage was taking place and was correlated with the amount of airborne NO_3^- intercepted by the foliage. However, high variability within and between the experiment sites as well as between the individ-

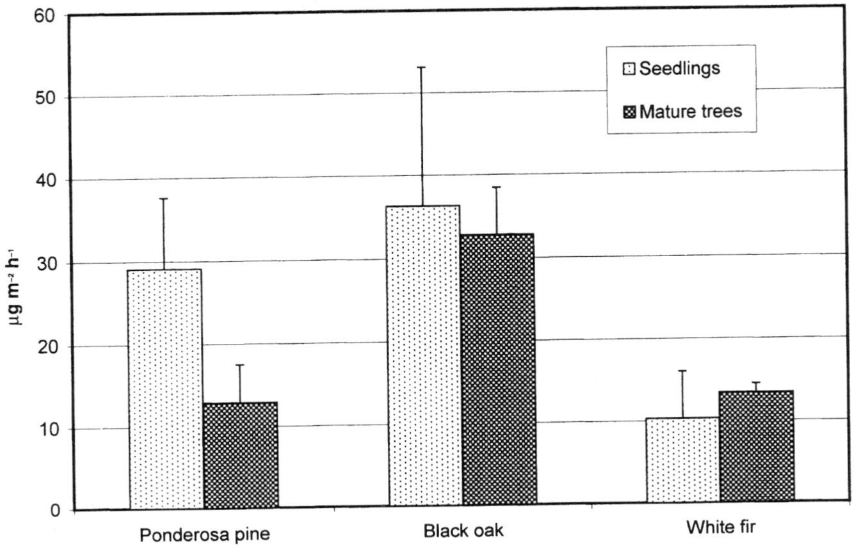

Figure 12.9. Comparison of nitrate deposition to foliage of mature trees versus seedlings of three tree species in the mixed conifer forest stand of the Barton Flats site of the San Bernardino Mountains.

ual collections interfered with the use of the assay as an indicator of N deposition (Krywult et al. 1996). The observed variability could be caused by differences in nutritional status of individual trees and other environmental factors. For example, elevated ozone concentrations, which co-occur with the elevated N deposition in the San Bernardino Mountains (Fenn and Bytnerowicz 1993), might also be responsible for the unpredictable changes in the activity of the enzyme. It has been long recognized that the general metabolic disruptions caused by ozone also interfere with NRA (Tingey et al. 1973; Guderian et al. 1989). The conflicting stresses of ozone inhibition and HNO_3^- induction could alter metabolism and lead to the prolonged presence of the pollutant in the intercellular spaces or the possible deleterious effects of stress-induced amides and amines (Wellburn 1988) This scenario could be similar to synergistic effects of O_3, NO_2, and SO_2 reported by Guderian et al. (1989).

Summary

Most of the results presented in this chapter are from the controlled fumigation system consisting of small Teflon cuvettes supplied with HNO_3 from permeation tubes. This system allowed for only short-term fumigations at unrealistically high concentrations (50–200 ppb). Although the highest concentrations used were about 11-fold higher than the maximum ambient levels recorded in the mountains of southern California (daytime average concentration of 18 ppb at Tanbark Flat, San Gabriel Mountains, in 1990), time of exposure was never longer than 48 hours. Consequently, during our experiments the plants were exposed to a HNO_3 dose of 1,200–9,600 ppb/h. For a comparison, the 1993 summertime HNO_3 dose at a moderately polluted site of Barton Flats was about 5,400 ppb/h (Bytnerowicz et al. 1996), whereas the 1989 summertime HNO_3 dose at the high-polluted site of Tanbark Flat of the San Gabriel Mountains was about 40,000 ppb/h (Grosjean and Bytnerowicz 1993). There is a need for studying responses of plants in more realistic scenarios of lower HNO_3 concentrations over longer-term (at least several months) exposures.

Nitric acid vapor is present in the air together with other components of photochemical smog. Therefore responses of plants to HNO_3 alone may be quite different from the responses in field conditions where plants are simultaneously exposed to high concentrations of ozone and other phytotoxic pollutants. Because of a high likelihood of synergistic interaction with ozone, exposures to mixtures of these pollutants seem to be a logical step after single-pollutant exposures. Such experiments should be designed based on a good understanding of seasonal and diurnal trends of both pollutants at the high-risk forest sites.

Responses of other forest species (including annuals) to HNO_3 vapor should be evaluated in forests of southern California mountains for a better understanding of expected effects of the pollutant on forest health and biological diversity. Due to its phytotoxic potential, as well as its importance as the N source, HNO_3 vapor has to be included in any future attempts to evaluate impacts of air pollution on forest ecosystems in the southern California mountains.

References

Andrews, M. The partitioning of nitrate assimilation between root and shoot of higher plants. Plant Cell Environ. 9:511–519; 1986.

Aslam, M.; Travis, R.T.; Huffaker, R.C. Comparative induction of nitrate and nitrite uptake and reduction systems by ambient nitrate and nitrite in intact roots of barley (*Hordeum vulgar* L.) seedlings. Plant Physiol. 102:811–819; 1993.

Brockley, T.P. Effects of nitrogen source and season of application on the nutrition and growth of lodgepole pine. Can. J. For. Res. 25:516–526; 1995.

Bytnerowicz, A.; Fenn, M.E. Nitrogen deposition in California forests: a review. Environ Pollut 92:127–146; 1996.

Bytnerowicz, A.; Riechers, G. Nitrogenous air pollutants in a mixed conifer stand of the western Sierra Nevada, California. Atmos. Environ. 29:1369–1377; 1995.

Bytnerowicz, A.; Miller, P.R.; Olszyk, D.M.; Dawson, P.J.; Fox, C.A. Gaseous and particulate air pollution in the San Gabriel Mountains of Southern California. Atmos. Environ. 21:1805–1814; 1987.

Bytnerowicz, A.; Fenn, M.; Glaubig, R. Dry deposition of nitrogen and sulfur to forests canopies at three plots. In: Miller, P.R., Chow, J., Watson, J., eds. Assessment of acidic deposition and ozone effects on conifer forests in the San Bernardino Mountains. Final report, California Environmental Protection Agency, Air Resources Board, contract A032-180. Sacramento, CA: 1996:4-1–4-75.

Bytnerowicz, A.; Percy, K.E.; Riechers, G.; Padgett, P.; Krywult, M. Nitric acid vapor effects on forest trees—deposition and cuticular changes. Chemosphere 36:697–702; 1998.

Cal Tech EQL. Acquisition of acid vapor and aerosol concentration data for use in dry deposition studies in the South Coast Air Basin. EQL report 25. Pasadena, CA: Environmental Quality Laboratory, California Institute of Technology; 1986.

Cape, J.N.; Percy, K.E. Use of needle epicuticular wax chemical composition in the early diagnosis of Norway spruce (*Picea abies* (L) Karst.) decline in Europe. Chemosphere 36:895–900; 1998.

Clarkson, D.T. Regulation of absorption and release of nitrate by plant cells: a review of current ideas and methodology. In: Lambers, H.; Neetson, J.J.; Stulen, I., eds. Fundamental, ecological and agricultural aspects of nitrogen metabolism in higher plants. Boston: Nijhoff Publishers; 1986.

Cooper, H.D.; Clarkson, D.T. Cycling of amino-nitrogen and other nutrients between shoots and roots in cereals—a plausible mechanism integrating shoot and root in the regulation of nutrient uptake. J. Exp. Bot. 40:753–762; 1989.

Crawford, N.M.; Campbell, W.H. Fertile fields. Plant Cell 2:829–835; 1990.

Crawford, N.M.; Wilkinson, J.Q.; LaBrie, S.T. Control of nitrate reduction in plants. Aust. J. Plant Physiol. 19:377–385; 1992.

Crutzen, P.J. The role of NO and NO_2 in chemistry of the troposphere and stratosphere. Annu. Rev. Earth Planet Sci. 7:443–472; 1979.

Darrall, N.M. The effects of air pollutants on physiological processes in plants. Plant Cell Environ. 12:1–30; 1989.

Edfast, A-B.; Nasholm, T.; Ericsson, A. Free amino acid concentrations in needles of Norway spruce and Scots pine trees on different sites in areas with two levels of nitrogen deposition. Can. J. For. Res. 20:1132–1136; 1990.

Fenn, M.E.; Bytnerowicz, A. Dry deposition of nitrogen and sulfur to ponderosa and Jeffrey pine in the San Bernardino National Forest in southern California. Environ. Pollut. 81:227–285; 1993.

Finlayson-Pitts, B.; Pitts, J.N., Jr. Atmospheric chemistry—fundamentals and experimental techniques. New York: John Wiley & Sons; 1986.

Goodwin, T.W.; Mercer, E.I. Introduction to plant biochemistry. New York: Pergamon Press; 1983.

Grosjean, D.; Bytnerowicz, A. Nitrogenous air pollutants at a southern California mountain forest smog receptor site. Atmos. Environ. 27:483–492; 1993.

Guderian, R.; Klumpp, G.; Klumpp, A. Effects of SO_2 O_3 and NO_2, singly and in combination, on forest species. In: Ozturk, M.A., ed. Plants and pollutants in developed and developing countries. Izmir, Turkey: Ege University; 1989:231–268.

Hameed, S.; Dignon, J. Global emissions of nitrogen and sulfur oxides in fossil fuel combustion 1970–1986. J. Air Waste Manage. 42:159–163; 1992.

Hanson, P.J.; Garten, C.T., Jr. Deposition of $H^{15}NO_3$ vapor to white oak, red maple and loblolly pine foliage: experimental observations and generalized model. N. Phytol. 22:329–337; 1992.

Hanson, P.J.; Lindberg, S.E. Dry deposition of reactive nitrogen compounds: a review of leaf, canopy and non-foliar measurements. Atmos. Environ. 25A:1615–1634; 1991.

Huppe, H.C.; Turpin, D.H. Integration of carbon and nitrogen metabolism in plant and algal cells. Annu. Rev. Plant Physiol. Plant Mol. Biol. 45:577–607; 1994.

Imsande, J.; Touraine, B. N demand and the regulation of nitrate uptake. Plant Physiol. 105:3–7; 1994.

Kleinhofs, A.; Warner, R.T. Molecular analysis of nitrate metabolism. In: P.R. Shewry, ed. Barley genetics, biochemistry, molecular biology, and biotechnology. Oxford, England: Alden Press; 1992:209–229.

Krywult, M.; Hom, J.; Bytnerowicz, A.; Percy, K.E. Deposition of gaseous nitric acid and its effects on foliage of ponderosa pine seedlings. In: Cox, R.M.; Percy, K.E ; Jensen K.F.; Simpson C.J., eds. Air pollution and multiple stresses. Proceedings of the 16th IUFRO international meeting for specialists in air pollution effects on forest ecosystems, September 7–9, 1994. Fredericton, Canada; 1996:45–51.

Lam, H-M.; Peng, S.S-Y.; Coruzzi, G.M. Metabolic regulation of the gene encoding glutamine-dependent asparagine synthetase in *Arabidopsis thaliana*. Plant Physiol. 106:1347–1357; 1996.

Legge, A.H.; Krupa, S.V. Air quality at a high elevation, remote site in western Canada. In: Olson, R.K.; Lefohn, A.S., eds. Effects of air pollution on western forests. Transactions ser. 16. Pittsburgh, PA: Air and Waste Management Association; 1989:193–206.

Li, X-Z.; Larson, D.E.; Glibetic, M.; Oaks, A. Effect of glutamine on the induction of nitrate reductase. Physiol. Plant. 93:740–744; 1995.

Lovett, G.M. Atmospheric deposition of nutrients and pollutants in North America: an ecological perspective. Ecol. Appl. 4:629–650; 1994.

Marschner, H. Mineral nutrition of higher plants. San Diego, CA: Academic Press; 1995.

Marshall, J.D.; Cadle, S.H. Evidence for trans-cuticular uptake of HNO_3 vapor by foliage of eastern white pine (*Pinus strobus*). Environ. Pollut. 60:15–28; 1989.

Miflin, B.J.; Lea, P.J. Ammonia assimilation and amino acid metabolism. In: Boutier, D.; Parthier, B., eds. Encyclopedia of plant physiology, vol. 14A. New York: Springer-Verlag; 1982:5–64.

Miller, D.F.; Walsh, P.A. Air quality and acidic deposition in the southeastern Sierra Nevada. Publication 91-59.3. 84th annual meeting and exhibition of the Air and Waste Management Association, Vancouver, British Columbia; 1991.

Muller, B.; Touraine, B. Inhibition of NO_3^- uptake by various phloem-translocated amino acids in soybean seedling. J. Exp. Bot. 43:617–623; 1992.

Nilsson, L-O.; Huttle, R.F.; Johansson, U.T.; Hockheim, H. Nutrient uptake and cycling in forest ecosystems—present status and future research directions. Plant Soil 168:5–13; 1995.

Norby, R.J.; Weerasuriya, Y.; Hanson, P.J. Induction of nitrate reductase activity in red spruce needles by NO_2 and HNO_3 vapor. Can. J. For. Res. 19:889–896; 1989.

Nussbaum, S.; von Ballmoos, P.; Gfeller, H.; Schlunegger, U.P.; Fuhrer, J.; Rhodes, D.; Brunold, C. Incorporation of atmospheric $^{15}NO_2$nitrogen into free amino acids by Norway spruce *Picea abies* (L.) Karst. Oecol. 94:408–414; 1993.

Oaks, A.; Aslam, M.; Boesel, I. Ammonium and amino acid as regulator of nitrate reductase in corn roots. Plant Physiol. 59:391–394; 1977.

Ollinger, S.V.; Aber, J.D.; Lovett, G.M.; Millham, S.E.; Lathrop, R.G.; Ellis, J.M. A spatial model of atmospheric deposition for the northeastern US. Ecol. Appl. 3:459–472; 1993.

Padgett, P.E.; Leonard, R.T. The interaction between cellular amino acid concentration and the regulation of nitrate transport in *Zea mays* cell suspension cultures. J. Exp. Bot. 47:871–883; 1996.

Percy, K.E.; Baker, E.A. Effects of simulated acid rain on epicuticular wax production, morphology, chemical composition and on cuticular membrane thickness in two clones of Sitka spruce *Picea sitchensis* (Bong.) Carr. N. Phytol. 116:79–87; 1990.

Percy, K.E.; Krause, C.R.; Jensen, K.F. Effects of ozone and acid fog on red spruce needle epicuticular wax ultrastructure. Can. J. For. Res. 20:117–120; 1990.

Percy, K.E.; McQuattie, C.J.; Rebbeck, J.A. Effects of air pollutants on epicuticular wax chemical composition. In: Percy, K.E.; Cape, J.N.; Jagels, R.; Simpson, C.J., eds. Air pollutants and the leaf cuticle. Heidelberg: Springer-Verlag; 1994:67–79.

Peuke, A.D.; Hartung, W.; Jeschke, W.D. The uptake and flow of C, N, and ions between roots and shoots in *Ricinus communis* L. II. Grown with low and high nitrate supply. J. Exp. Bot. 45:733–740; 1994.

Pietila, M.; Lahdesmaki, P.; Pietilainen, P.; Ferm, A.; Hytonen, J.; Patila, A. High nitrogen deposition causes changes in amino acid concentrations and protein spectra in needles of the scots pine (*Pinus sylvestris*). Environ. Pollut. 72:103–115; 1991.

Riederer, M.; Markstadter, C. Cuticular waxes: a critical assessment of current knowledge. In: Kerstiens, G., ed. Plant cuticles: an integrated functional approach. Oxford: Bios; 1996:189–200.

Rosenberg, L.A.; Padgett, P.E.; Assmann, S.M.; Walling, L.L.; Leonard, R.T. Identification of mRNAs and proteins in higher plants using probes from the band 3 anion transporter of mammals. J. Exp. Bot. 48:857–868; 1997.

Smirnoff, N.; Todd, P.; Stewart, G.R. The occurrence of nitrate reduction in the leaves of woody plants. Ann. Bot. 54:363–374; 1984.

Stulen, I. (1990) Interactions between carbon and nitrogen metabolism in relation to plant growth and productivity. In: Abrol, Y.P., ed. Nitrogen in higher plants. New York: John Wiley & Sons; 1990:297–312.

Tingey, D.T.; Fites, R.C.; Wickliff, C. Ozone alteration of nitrate reduction in soybean. Physiol. Plant 29:33–38; 1973.

Tuazon, E.C.; Winer, A.M.; Pitts, J.N., Jr. Trace pollutant concentrations in a multiday smog episode in the California South Coast Air Basin by long path Fourier transform infrared spectroscopy. Environ. Sci. Technol. 15:1232–1237; 1981.

Vitousek, P. Beyond global warming—ecology and global change. Ecology 75:1861–1876; 1995.

Vose, J.M.; Swank, W.T. Preliminary estimates of foliar absorption of ^{15}N-labeled nitric acid vapor (HNO_3) by mature eastern white pine (*Pinus strobus*). Can. J. For. Res. 20:857–860; 1990.

Wellburn, A. Air pollution and acid rain—the biological impacts. Burnt Hill, England: Longman; 1988.

Wellburn, A.R. Why are atmospheric oxides of nitrogen usually phytotoxic and not alternative fertilizers? N. Phytol. 115:395–429; 1990.

13. Nitrogen Deposition and Cycling in Mediterranean Forests: The New Paradigm of Nitrogen Excess

M.E. Fenn and M.A. Poth

Introduction

Although visible oxidant injury to ponderosa pine (*Pinus ponderosa* Dougl. ex Laws.) in the San Bernardino Mountains (SBM) has been documented for four decades (Miller 1992; Miller and Rechel, this volume), the ecosystem effects of chronic N deposition have only recently been investigated (Fenn and Dunn 1989; Fenn 1991; Fenn et al. 1996). Ozone exposure causes characteristic symptoms in ponderosa pine such as chlorotic mottle of foliage and reduced foliar retention (Miller et al. 1996a). The ecological effects of atmospheric N inputs in the SBM have not been studied until recently, largely because N deposition does not induce symptomatic visible injury. Furthermore, hypotheses suggesting that chronic N deposition causes serious perturbations of forest nutrient cycling and plant nutrient status were not seriously considered until the mid- to late 1980s (Nihlgård 1985; Aber et al. 1989).

Measurements of elevated N deposition in high-ozone sites in the SBM (Fenn and Bytnerowicz 1993) was not unexpected, considering the high emissions of N oxides in the Los Angeles Basin and because N oxides are known precursors of tropospheric ozone. Recent studies suggest that the productivity of some temperate forests may be negatively affected by excess N. In at least four long-term fertilization studies in eastern North America and Europe, low levels of N fertilization (30–150 kg ha^{-1} yr^{-1}) resulted in reduced wood production or greater mortality (Fenn et al. 1998). Allen et al. (1996) reported that high inorganic N concentrations in mineral soil in coastal sage ecosystems in the Los Angeles Air Basin with high N deposition may contribute to the observed decline of some species. The major impacts of excess N on forest sustainability and ecosystem function are likely to be chronic rather than acute. They will be subtle in nature and may involve complex interactions among ecosystem components (e.g., via effects of excess N on mycorrhizae, plant pathogens, or insect pests; shifts in plant species composition, soil acidification effects).

Similar to temperate forests in much of the Northern Hemisphere, N is the most commonly limiting nutrient for forest production in California (Powers et al. 1988). However, high temperatures and moisture limitation during the growing season in California interact with nutrient availability in controlling primary production. Notwithstanding the often severe water shortage during the growing season in forests of southern California, tree growth typically increases in response to added N (Kiefer and Fenn 1997). Nitrogen deposition is believed to stimulate forest productivity in areas of southern California receiving moderate N deposition. However, in years of successive drought such as occurred from 1987 to 1992 (Smith et al. 1994), tree responses to fertilization or N deposition are expected to be greatly decreased.

In this chapter, we do not review forest nutrient cycling, because this is covered in reviews elsewhere (Waring and Schlesinger 1985; Smith 1990; Johnson and Lindberg 1992; U.S. Environmental Protection Agency 1993; Perry 1994; Kimmins 1996). Rather, we emphasize the effects of chronic N deposition on N cycling in the summer-dry montane ecosystems of southern California. The phenomenon of N saturation (Aber et al. 1989) in chaparral and mixed conifer forests is described, and ecosystem responses to excess N peculiar to Mediterranean ecosystems are highlighted. The N cycle, with emphasis on N pools and N cycling processes affected by N deposition, is presented in Figure 13.1. Camp Paivika (CP), a highly N-saturated site on the western end of the SBM (Fig. 13.2), is used as a case study of ecosystem responses to long-term elevated N inputs in western xeric forests.

Perspectives on Nitrogen Cycling in Temperate Forests

Soil is the largest N pool in forest ecosystems with most of the N stored in organic forms that are largely unavailable to plants. This results in widespread N limitation in temperate forests. Nitrogen mineralization is a key process by which organic N is converted to inorganic ammonium, which is available for plant or microbial uptake. Ammonium can also be oxidized to nitrite and nitrate by nitrifier microbes. Nitrate can be used by microbes or plants, or it can be leached from the system (Fig. 13.1). Recent studies demonstrate that some plants and mycorrhizal roots can bypass the mineralization step by using N from simple organic compounds (Chapin et al. 1993). Internal N cycling or retranslocation of N within the plant is also an important process in perennial plants, especially evergreen species. Nitrogen is retranslocated from senescent foliage to younger foliage prior

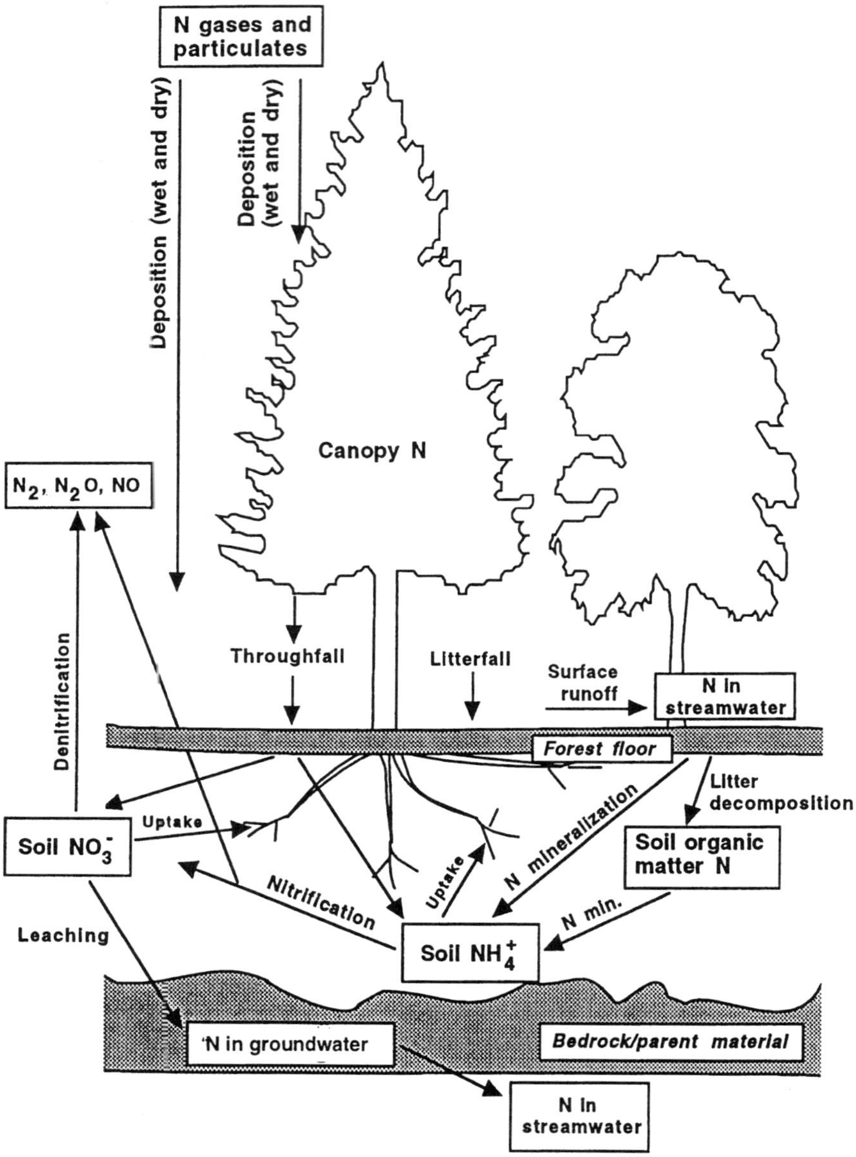

Figure 13.1. Diagrammatic representation of the N cycle with emphasis on N pools and processes affected by atmospheric N deposition.

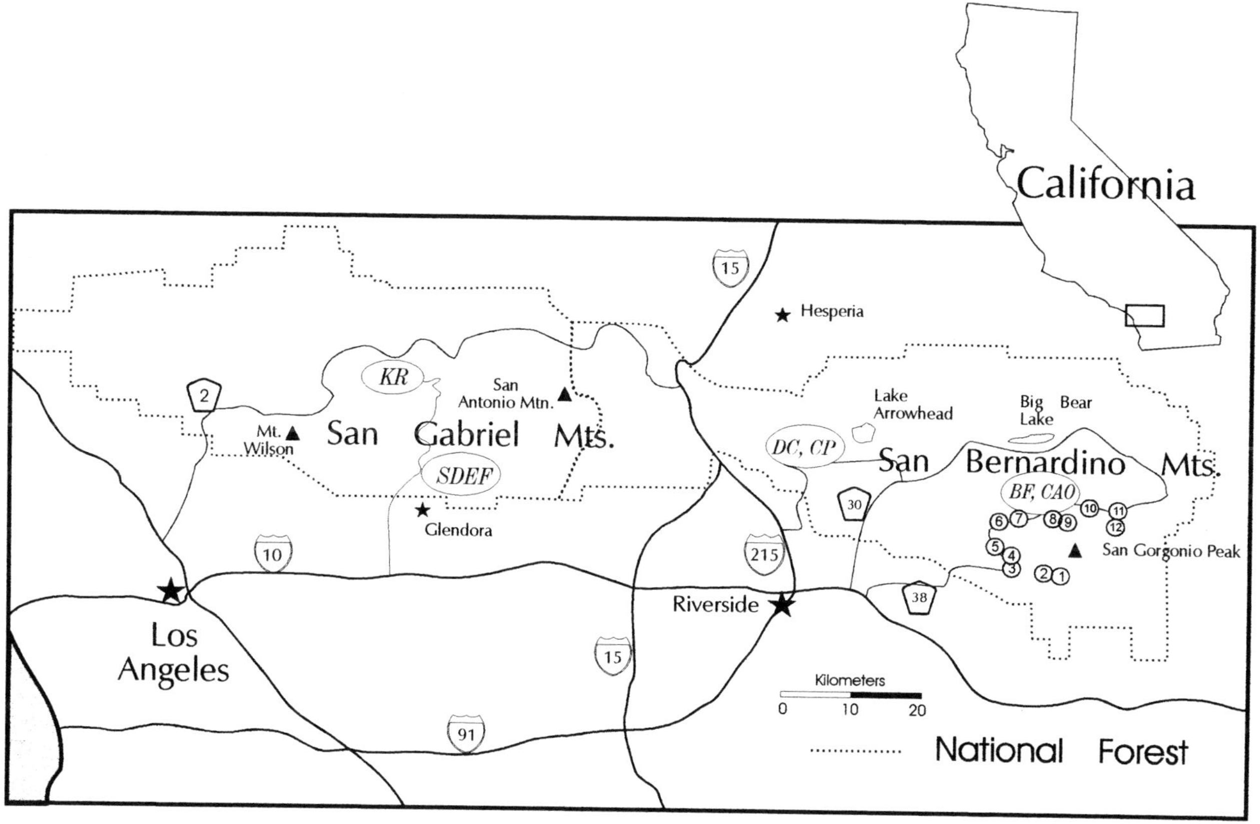

Figure 13.2. Location of research sites reviewed in this chapter. Site abbreviations are as follows. Kratka Ridge (KR), San Dimas Experimental Forest (SDEF), Devil Canyon (DC), Camp Paivika (CP), Barton Flats (BF), and Camp Osceola (CAO). The dotted line indicates the boundaries of the Angeles National Forest (San Gabriel Mountains) and the San Bernardino National Forest.

to foliar abscission as a way to conserve N. Many conifer species are well adapted to nutrient-poor sites, largely due to their ability to internally recycle N and other mobile nutrients (Chapin 1980). However, chronic N deposition can lead to abnormally high levels of available N in soil (Fenn et al. 1996), resulting in a N-rich environment for plants adapted to chronic N limitation.

The primary inputs of N to forest ecosystems include biological N_2 fixation (symbiotic or associative), atmospheric N deposition in both wet and dry forms, and fertilization. The most important pathways of N loss from forests are hydrological transport of organic and inorganic N (mainly as NO_3^-), N removal during harvest, and periodic losses during fires that volatilize N (Raison et al. 1985). Lesser amounts of N can be lost as gaseous emissions from soil during nitrification or denitrification (Bowden et al. 1990; Matson et al. 1991, 1992). Soil erosion can also result in N export from the forest. In ecosystems with elevated N inputs from biological fixation or from atmospheric deposition, increased N availability can overwhelm the capacity for abiotic and biotic N retention and result in a reversal of N limitation to the point of luxury consumption for many species. Under conditions of N excess, losses of N become abnormally high, mainly as NO_3^- export in groundwater or surface waters, and to a lesser degree, as gaseous emissions from soil. Forests exhibiting these symptoms are considered N-saturated according to the hypothesis or conceptual model of Aber et al. (1989, 1995).

Nitrogen Deposition Threshold and Geographic Extent of Nitrogen-Saturated Ecosystems

Studies along N deposition gradients in the San Gabriel Mountains (Kiefer and Fenn 1997) and the SBM (Fenn and Bytnerowicz 1993; Fenn et al. 1996)(Fig. 13.2) provide empirical indications of the levels of N inputs at which alterations in N cycling can be detected. In watersheds with high-smog exposure (N deposition of about 20 kg ha^{-1} yr^{-1} or greater for more than 40 years), nitrate concentrations in streamwater are commonly several orders of magnitude higher than in undisturbed watersheds with low anthropogenic N input (Riggan et al. 1985; Fig. 13.3). The mixed conifer forest at CP, near Crestline in the SBM (Fig. 13.2), is N-saturated based on a number of plant, soil, and hydrological indicators of N saturation (Fenn and Poth 1996; Fenn et al. 1996). Pine tree growth at CP appears to no longer be N-limited (Kiefer and Fenn 1997). Stands with relatively moderate N deposition (5–15 kg ha^{-1} yr^{-1}) are still N-limited, although moderate N deposition appears to lead to faster N cycling (Kiefer and Fenn 1997) and possibly greater plant growth. Extensive studies of the spatial patterns of N deposition in wildlands in southern California have not been done. However, available data suggest that only areas more directly exposed to incoming air pollution masses are affected to the degree that N limitation in overstory species is completely reversed. This contention is supported by the observation of a much steeper decline in the concentrations of nitrogenous air pollutants than of atmospheric ozone across an air pollution gradient in the SBM (Miller et al. 1986; Fenn and

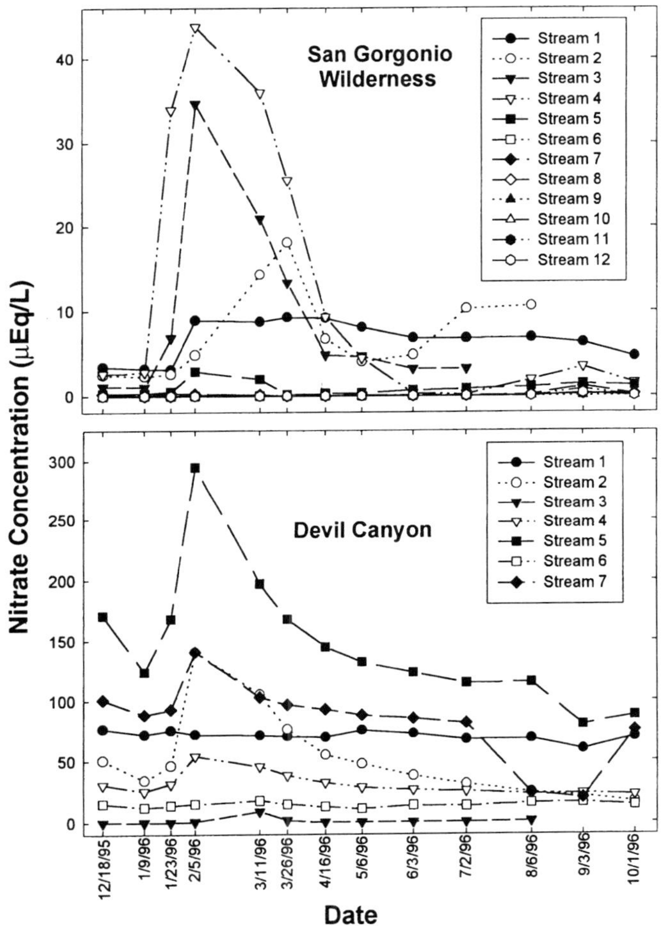

Figure 13.3. Nitrate concentrations in streamwater from streams in a high-pollution area in the western San Bernardino Mountains (Devil Canyon) and a relatively low-pollution area surrounding the San Gorgonio Wilderness Area in the eastern San Bernardino Mountains.

Bytnerowicz 1993). The steep N deposition gradient is thought to be a result of the high deposition velocity of nitric acid, a major component of nitrogenous, dry deposition. Forest canopies more directly exposed to incoming polluted air masses may rapidly reduce atmospheric concentrations of N, contributing to a steep N deposition gradient. Thus, the level of N deposition and the severity of perturbation of N cycling in wildlands within the South Coast Air Basin is probably quite heterogeneous, depending on distance from N emission sources, elevation, topography, atmospheric circulation patterns, stand characteristics, and weather conditions.

N deposition thresholds for development of N saturation are expected to vary, depending on the capacity of each ecosystem to retain and process available N. Vector analysis of fertilized Jeffrey pine (*Pinus jeffreyi* Grev. & Balf.) trees near Kratka Ridge in the San Gabriel Mountains (2,060-m elevation; Fig. 13.2) clearly demonstrated the N deficiency of trees in plots with annual throughfall inputs of 6–11 kg ha^{-1} yr^{-1} (Kiefer and Fenn 1997). Total N deposition is probably about 25% greater than throughfall inputs due to canopy retention of atmospheric N (see references in Fenn and Bytnerowicz 1997). However, chaparral watersheds in the San Dimas Experimental Forest (SDEF) in the San Gabriel Mountains (about 850-m elevation; Fig. 13.2) with throughfall inputs of 23 kg ha^{-1} yr^{-1} exported high levels of NO_3^- in streamwater and were clearly N saturated (Riggan et al. 1985, 1994). Nitrogen deposition at CP is estimated to be 25–35 kg ha^{-1} yr^{-1}, and the forest there is also N-saturated, whereas mixed conifer stands further east in the SBM at Barton Flats (BF) and Camp Osceola (CAO) (N deposition estimated at 5–10 kg ha^{-1} yr^{-1}) are N-limited. These findings from the San Gabriel Mountains and the SBM, although based on a limited number of sites, are in agreement with a survey of 65 forest plots and catchments in Europe, where N leaching occurred at all sites with N deposition greater than 25 kg ha^{-1} yr^{-1} (Dise and Wright 1995). Below a deposition threshold of 10 kg ha^{-1} yr^{-1}, no significant NO_3^- leaching occurred, whereas at intermediate levels of 10–25 kg ha^{-1} yr^{-1} leaching occurred at some sites. Further data on N deposition inputs and NO_3^- export from watersheds or NO_3^- leached beyond the soil rooting zone are needed to determine the areal extent of wildland ecosystems in southern California that exhibit symptoms of excess N.

Characteristics of N Deposition and Cycling in Montane Summer-Dry Ecosystems in Southern California

The distinct Mediterranean climate in much of California exerts a powerful influence on nutrient cycling and on the processing of atmospherically deposited N within the system. The atmospheric conditions associated with the extended dry "smog" season, numerous N emissions sources dominated by motor vehicles, and topography combine to transport large quantities of nitrogenous pollutants to the more exposed areas of the San Gabriel Mountains and the SBM. Much of this pollutant load is dry-deposited and accumulates on the forest canopy and the forest floor until the wet season begins, usually in October or November. The temporal dynamics and magnitude of hydrologic fluxes (e.g., precipitation, throughfall, snowmelt, percolation, runoff, subsurface flow) are major drivers of N transport within watersheds and exert major control on N-cycling processes at all scales. Water availability and hydrological transport of N through the plant/soil system is often a boom or bust phenomenon in California. Long periods of N accumulation in the canopy and forest floor are interrupted by precipitation, which transports concentrated pulses of N into the soil and hydrological system—thus providing nutrients and water needed by N-cycling microbes and vegetation.

Throughfall Dynamics

The dynamics of throughfall fluxes provide a good example of how the timing and characteristics of nutrient cycling in ecosystems with a Mediterranean climate differ from patterns in more mesic climates where precipitation is more evenly distributed or at least where extended droughts during the growing season are not the norm. Summer throughfall was studied for 3 years in three replicate stands at BF in the SBM. BF is a low-to-moderate deposition site near the eastern end on the mixed conifer zone. Annual N deposition at BF is estimated to be 6–10 kg ha^{-1} yr^{-1} (Fenn and Bytnerowicz 1997), which is low compared with CP (about 25–35 kg ha^{-1} yr^{-1}), some forests in eastern North America (Johnson and Lindberg 1992), or many forests in Europe where deposition inputs are similar to or greater than at CP (Van Breemen et al. 1987; Lamersdorf and Meyer 1993). Ionic concentrations in summer throughfall at BF, and particularly NO_3^- concentrations, were generally as high or higher than values reported for forests in Europe and eastern North America with much higher N deposition inputs. It was concluded that the high throughfall concentrations in the summer at BF were due to the washoff of accumulated dry pollutants from mature trees with large surface areas during the infrequent low-volume thunderstorms (Fenn and Bytnerowicz 1997). Thus, summer throughfall is characterized by relatively low-volume throughfall fluxes but with very high concentrations of nitrate and of other ions, even in sites such as BF with only moderate N deposition. In winter, throughfall volumes are much greater, but ionic concentrations are much lower than in summer. Canopy interactions with atmospherically deposited ions are also likely to be much different in summer and winter—because winter deposition is dominated by snow in the higher-elevation sites. Winter throughfall is a combination of snow that passes through the canopy during snowfall, followed by liquid throughfall from the melting of snow accumulated on the canopy.

Temporal Asynchrony in Hydrologic Fluxes and Biological Nutrient Demand

Temporal asynchrony between the major hydrological fluxes of NO_3^- through the soil profile (winter and early spring) and peak plant and microbial N demand (spring and early summer) limits plant and possibly microbial uptake of much of the chemically available N. This asynchrony, in combination with high nitrification rates in soil, leads to nonconservative "open N cycling" (Fenn et al. 1998) and elevated N export in California ecosystems with chronic N deposition. In the western SBM (near CP), reduced plant retention of N due to low temperatures and high water runoff volumes from the watershed are both factors that contribute to high levels of N export during the dormant season. The relative importance of these two factors will vary between sites based on elevation, temperature patterns, and hydrological fluxes. In the western SBM, the relatively low elevation (1,600–1,725 m) and generally mild winters may allow N uptake by some plant species and microbial activity to continue, although at reduced rates. Plant and microbial activity is expected to be temperature-restricted but not precluded during winter and moisture-limited during the extended dry summer. In fact, microbial N mineralization and nitrification in the moist N-enriched soils at CP may be a net source of nitrate in winter, thus contributing to the high levels of nitrate exported in streamwater (Fig. 13.3). In N-limited sites, microbial N immobilization activity under the snowpack or during snowmelt would likely increase the N retention capacity of the site and reduce N export from the watershed. According to the vernal dam hypothesis, microbial N retention is an important mechanism of N retention in spring before plant N uptake is fully active (Zak et al. 1990). Previous studies have also demonstrated that microbial activity can be important under snowpack (Brooks et al. 1996; Williams et al. 1996). We are currently testing these hypotheses in the SBM.

N retention capacity in the SBM is most apt to be overwhelmed during and after major precipitation events that follow long dry periods of N deposition (Bytnerowicz and Fenn 1996) and accumulation in the forest canopy, forest floor, and soil. Dry-season N mineralization and nitrification lead to soil nitrate loading. Nitrate concentrations in soil at CP reach very high levels during spring and summer (530–2,500 μmol/L; Fenn et al. 1996). A thick forest floor at CP enhances the conditions needed for microbial N cycling in soil by providing an effective mulch layer, reducing overheating of the mineral soil and soil moisture loss during summer and serving as an insulating cover in winter. Thus the forest floor extends the time of favorable soil conditions for nitrification and NO_3^- accumulation in soil. There may be a positive feedback mechanism in operation as elevated N deposition and nitrification rates cause greater litter production, reduced long-term litter decomposition rates (Fog 1988; Berg et al. 1995), and increased litter accumulation (Nohrstedt 1990; Kuyper 1994; Martikainen 1996).

The amount of water percolating through the system appears to be the controlling factor determining how much N leachate loss will occur (Riggan et al. 1985). The first precipitation events following periods of N accumulation result in washout of N pollutants from the atmosphere and cause periodic pulses of nutrients into the forest floor and soil as a result of canopy washoff (throughfall) and percolation through the forest floor and mineral soil. Nitrogen retention within the plant/soil system during these pulse events will depend on precipitation intensity and soil water flux rates, and the capacity for N retention by plants, microbes, and abiotic mechanisms within the soil and forest floor. Such nutrient pulse episodes are more common during infrequent low-volume summer thundershowers in some areas or at the beginning of the fall wet season.

The major periods of water use by overstory trees are during the spring and early summer and in the fall when rains begin. However, in montane regions of California most of the precipitation occurs as winter snowfall, most of which plants are not able to use. In watersheds with chronic N inputs, NO_3^- losses in streamwater during the off season are commonly ten to 40 times greater than in areas with low N inputs (Riggan et al. 1985; Peterjohn et al. 1996; Fenn et al. 1998). Nitrate export in streamwater in the SBM also peaks in winter, although in watersheds with high N deposition inputs, NO_3^- export is high year-round (Fig. 13.3).

There is another sometimes overlooked aspect to plant water use in western xeric forests. Overstory trees in semiarid forests of California, Oregon, and other parts of the West produce roots that access fissures and joint planes of rock layers and use water stored in these deep reservoirs (Jones and Graham 1993; Anderson et al. 1995; Beyers et al. 1995; Zwieniecki and Newton 1996). The N-cycling dynamics within these deep water stores is not known. It is expected that most of the inorganic N in deep water stores in N-enriched sites is inaccessible to plant roots because of sparse horizontal root coverage below the main rooting zone and is eventually leached into groundwater and streamwater.

Litter Decomposition

Litter Decomposition in Mediterranean Climatic Conditions

Moisture content is undoubtedly the major factor controlling biotic processes of litter decomposition in semiarid forests (Hart et al. 1992). The temporal separation of warm temperatures and moist conditions has profound effects on ecosystem processes such as litter decomposition. In Mediterranean climates, litter undergoes alternating cycles of decomposition dominated by biotic and abiotic processes. For example, intermittent low-volume summer rain events, such as occur in the SBM, probably result in bursts of microbial decomposition that last until moisture again limits litter decomposer organisms. Decomposer microorganisms are believed to respond rapidly to conditions favorable for decomposition (Gray and Schlesinger 1981). Trace gas emissions produced by soil nitrifiers responded dramatically within minutes after wetting dry soils in the San Gabriel Mountains and SBM (Anderson et al. 1988; Fenn et al. 1996), thus demonstrating the rapid response of soil microbes to moisture.

In summer-dry climates, litter buildup in fertile sites may be enhanced as water use by deep-rooted trees continues primary productivity during at least part of the summer drought (Arkley 1981; Anderson et al. 1995; Wang et al. 1995), while decomposition is negligible. Litter decomposition, especially the upper layers, typically becomes moisture-limited in early summer in the SBM and throughout much of California and western North America where xeric climates predominate. Greater moisture limitation of decomposition relative to litter production (Miller et al. 1996b) will result in net C gain in the forest floor. This scenario was also reported during a severe drought in the Harvard Forest in Massachusetts in 1995 (Goulden et al. 1996). The depletion of water near the soil surface reduced soil respiration by 30% while water deep in the soil column supported photosynthesis, resulting in above-average C storage in 1994–1995.

Higher site quality and fertility of a site can also contribute to litter accumulation in the forest floor as litter production is increased and a greater proportion of C is allocated aboveground (Haynes and Gower 1995). At CP, the growth rates, and presumably the rate of litter production, of many mature ponderosa pine trees have increased in recent years (P. Miller, personal communication). Carbon allocation to fine roots is also significantly lower at CP than at CAO, a low-pollution site in the SBM (Grulke et al., in press). Thus, drought stress of aboveground litter decomposer microbes, sustained litter production in trees capable of accessing deep water stores, greater C allocation aboveground, and N stimulation of litter production at the N-enriched CP site are all factors that likely contribute to long-term C storage in aboveground detritus at CP.

With increasing aridity, abiotic mechanisms of litter decomposition become more important as biotic decomposition decreases (Coûteaux et al. 1995). Physical processes involved in litter fragmentation include photochemical degradation of lignins (Moorhead and Reynolds 1989), ultraviolet radiation (Moorhead and Callaghan 1994), exposure to temperature extremes, and wetting and drying cycles (Gray and Schlesinger 1981). The open nature of ponderosa pine stands and the relatively high elevation of the mixed conifer zone (compared with low-elevation desert systems) also favor abiotic decomposition due to high radiation intensity (Moorhead and Callaghan 1994). Physical degradation of litter is likely to be important in the decomposition of surface litter directly exposed to solar radiation and the atmosphere during extended hot and dry summers.

Effects of Nitrogen Fertility on Litter Decomposition

During episodes of plentiful moisture, temperature and litter quality are the factors most likely to control decomposition rates. The decomposition rates of surface litter (L-layer) of ponderosa and Jeffrey pine, sugar pine (*Pinus lambertiana* Dougl.), and incense cedar (*Calocedrus decurrens* [Torr.] Florin.) in laboratory assays were higher for litter from western SBM high-deposition sites compared with eastern SBM low-deposition sites (Fenn and Dunn 1989; Fenn 1991). Decomposition rates of ponderosa pine litter were positively correlated with litter N concentration and negatively correlated with calcium concentrations in litter. The latter observation indicates that the more rapidly decomposing litter is from younger foliage, because foliar calcium concentration is largely a function of needle age (Fenn and Dunn 1989). The lower calcium content of pine needle litter in the high-pollution (ozone and N pollutants) plots can be attributed to the ozone-induced premature abscission of pine needles in these sites (Miller et al. 1986; Fenn and Dunn 1989). Nitrogen concentration of foliage is usually higher in younger foliage. However, litter from ozone-tolerant species (incense cedar and sugar pine) at the high-deposition sites was also higher in N and decomposed faster than litter from low-deposition sites. Chronic N deposition is believed to be the major factor resulting in greater N fertility and N concentration of litter from the western high-pollution plots (Fenn 1991).

Chronic N deposition, increased N availability in soil, and higher N concentrations in foliage and litter may not result in higher long-term litter decomposition rates, however. In fact, the unusually deep accumulation of litter in the forest floor at CP when compared with more easterly low-pollution sites such as BF supports the hypothesis that long-term decomposition rates are retarded by N-rich litter. Low calcium concentrations in litter at CP may also contribute to retarded decomposition rates of older litter, as reported for Scots pine in Sweden (Johansson 1994). At CP, the forest floor consists of a well-developed continuum of litter in various stages of decomposition, and the total litter depth is commonly 15–30 cm. By contrast, the forest floor at BF and other low-deposition sites is typically 1–4 cm in depth, but in many areas the litter is so sparse that bare mineral soil is visible at the surface of the forest floor. Decades of fire exclusion in the mixed conifer forest zone (Minnich et al. 1995) have allowed litter accumulation to occur, but greater N fertility as a result of elevated N deposition may enhance litter accumulation by reducing long-term decomposition rates. Many instances of litter accumulation have been reported in forests with high N inputs (Nohrstedt 1990; Kuyper 1994; Martikainen 1996).

In a synthesis paper, Berg et al. (1995) addressed the mechanism of litter and soil organic matter buildup in coniferous forests. They found a highly significant negative relationship between initial N concentration of litter and the degree to which litter eventually decomposed, even though the initial stages of litter decomposition may be stimulated by an ample N supply. Thus, according to the observations and hypothesis proposed by Berg et al. (1995) and others (Söderstrom et al. 1983; Fog 1988; Melillo et al. 1989), the buildup of the forest floor at CP may be accounted for by the excess available soil N and high N concentrations of litter (Fenn and Dunn 1989; Fenn et al. 1996), in addition to the factors outlined above. Decreased decomposition of N-rich litter may be due to the formation of stable nitrogenous compounds from lignin by-products (Berg et al. 1995). It is reasonable to hypothesize that CO_2 emissions per unit C in the forest floor will be lower at CP compared with N-limited sites such as BF, but CO_2 fluxes from the forest floor per unit land area will be higher due to greater accumulation of C in the forest floor. The dense litter accumulation at CP may have feedback effects on plant growth, because the thick forest floor functions as an effective mulch layer, reducing soil heating and drying and soil erosion from runoff. However, the thick forest floor probably also inhibits regeneration of ponderosa pine (Savage 1994).

Effects of Nitrogen Deposition on Soil Processes

Nitrogen Mineralization and Nitrification

N mineralization (the release of inorganic N from organic N substrates) rates are generally higher in N-saturated stands than in more N-limited sites (Aber et al. 1989; Van Miegroet et al. 1992). Nitrogen mineralization was the factor most highly correlated with NO_3^- leaching in the integrated forest study, which included 16 forests, mainly in North America (Johnson and Lindberg 1992). Nitrogen mineralization and NO_3^- leaching rates were highest in soils with total soil N content greater than 5,000 kg/ha in the top 60 cm of soil and a C:N ratio less than 20 (Cole et al. 1992). In the SBM, net N mineralization rates in soil in laboratory assays were greater at two high-pollution sites than at two low-pollution sites The soil C:N ratio at CP was 19 compared with 27 at CAO, a low-deposition site. Net nitrification rates were particularly high in mineral soil from CP, and conversion of NH_4^+ to NO_3^- was virtually complete (Fenn et al. 1996). Nitrogen deposition may increase nitrification rates by maintaining greater nitrifier populations in soil (Johnson 1992).

Nitrification is an active process in many if not most western forest soils. Several investigators have reported high nitrification rates or nitrification potential in western forests (Hart and Firestone 1989; Powers 1990; Frazer et al. 1990; Covington and Sackett 1992; Davidson et al. 1992), even with low N deposition rates. Riggan et al. (1985) reported rapid nitrification in soil and streams in a chaparral watershed with high N deposition in the San Gabriel Mountains following treatment with water or NH_4SO_4. Stark and Hart (1996) measured high gross nitrification rates in 11 undisturbed conifer forests in New Mexico and Oregon. Net nitrification rates poorly predicted gross rates because the soil microbial communities had the capacity to assimilate almost all the NO_3^- produced (Stark and Hart 1996). Evidence from the SBM suggests that net nitrification rates and NO_3^- concentrations in soil are much higher in stands with elevated N deposition.

Stable isotope analysis from four sites traversing the N deposition gradient in the SBM also suggest significant differences in N cycling at sites with high and low N deposition. Mean bulk N $\delta^{15}N$ values of soil and vegetation at CP were at least 1.7‰ more enriched than the other less polluted sites (Korontzi et al. 1996). Ammonium in soil extracts was more ^{15}N-enriched and NO_3^- was more ^{15}N-depleted at CP than in the less-polluted sites (Fig. 13.4). The high nitrification rates at CP are probably a major factor accounting for the ^{15}N depletion at CP, because nitrification causes isotopic fractionation, favoring the lighter N-14 isotope during NO_3^- production.

Mean foliar $\delta^{15}N$ values were positively correlated with net mineralization and nitrification potentials in soil in sites along an elevational and N deposition gradient in the Great Smoky Mountains in eastern Tennessee (Garten and Van Miegroet 1994). Stable isotope data from the SBM were very similar to those from the Smoky Mountains, with higher $\delta^{15}N$ values in foliage and higher net nitrification rates at CP compared with less fertile sites (Fenn et al. 1996; Korontzi et al. 1996). Garten and Van Miegroet (1994) concluded that foliar $\delta^{15}N$ values may be useful as a preliminary indicator of forest N status, especially because the correlation between foliar ^{15}N abundance and nitrification included a wide range of forest types, soils, and climatic environments. Our results from the SBM also confirm the potential usefulness of $\delta^{15}N$ as an indicator of forest N status. However, within-site variation in foliar $\delta^{15}N$ may limit this indicator to distinguishing between sites that exhibit rather large differences in soil N availability (Garten and Van Miegroet 1994).

It was postulated that proportionally greater NH_4^+ uptake and reduced NO_3^- uptake accounted for the observed foliar ^{15}N enrichment in the N-saturated stands in the Great Smoky Mountains (Garten and Van Miegroet 1994). We do not know to what extent greater relative NH_4^+ uptake at CP may contribute to higher ^{15}N levels in oak and pine foliage. However, Durka et al. (1994), using stable isotope data for N and oxygen, reported that in declining spruce plantations in Germany 60–100% of atmospherically deposited N leaches through the system into spring

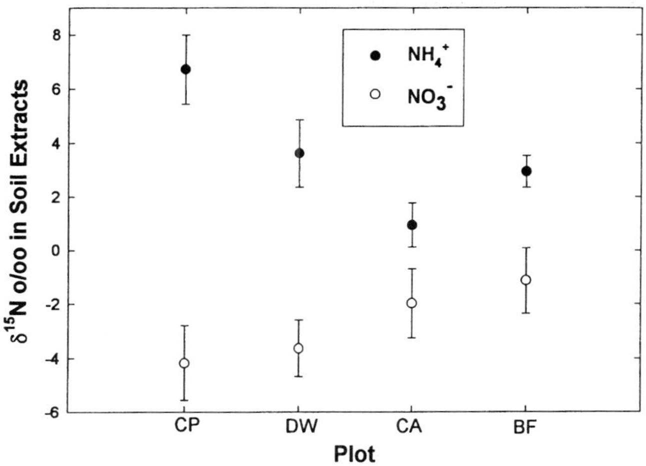

Figure 13.4. $\delta^{15}N$ isotope analysis of NO_3^- and NH_4^+ in soil extracts from four sites in the San Bernardino Mountains. Air pollution severity decreases from left to right on the *x*-axis as follows: Camp Paivika (CP) > Dogwood (DW) > Camp Angeles (CA) > Barton Flats (BF).

water without being biologically assimilated. The elevated NO_3^- levels in soil and streamwater in the western SBM demonstrate that a significant fraction of the NO_3^- from nitrification and possibly from atmospheric deposition also leaches from the plant/soil system and may account for the greater isotopic fractionation observed at CP (Fig. 13.4). Further studies are needed to determine if at CP NO_3^- is leached while NH_4^+ uptake becomes the preferred N form for plant uptake. However, the nitrate:ammonium ratio was 45:1 in saturation extracts and in the soil solution at CP. At more easterly sites with much lower N deposition (two BF plots, CAO and Heart Bar), nitrate:ammonium ratios ranged from 2:1 to 10:1 (Fenn et al. 1996). The low available NH_4^+ concentrations in soil at CP suggest that NO_3^- is probably still the major N ion taken up by the vegetation. However, the high $\delta^{15}N$ values for extractable NH_4^+ at CP (Fig. 13.4) suggest that even low-to-moderate NH_4^+ uptake could have significant effects on ^{15}N levels in foliage. High NO_3^- reductase activity in pine and oak foliage in the SBM suggests that significant NO_3^- uptake occurs (Krywult et al. 1996), although elevated atmospheric concentrations of nitric acid vapor, such as occur in the SBM, can also stimulate foliar NO_3^- reductase activity (see Bytnerowicz et al., this volume).

Soil Solution Nitrate

Concentrations of NO_3^- in the soil solution and in saturation extracts at CP ranged from 530 to 2,500 μEq/L during the growing season, similar to values obtained previously from computer simulations of the effects of N deposition (using the NuCM, Nutrient Cycling Model; Fenn et al. 1996). Nitrate concentrations in the soil solution and saturation extracts at BF and CAO (a relatively low-deposition site about 5 km east of BF) ranged from 25 to 900 μEq/L during the growing season. Concentrations of NH_4^+ and SO_4^{-2} were similar among the three plots. The high NO_3^- concentrations during summer, even at BF and CAO, are presumably due to the prolonged summer drought. Lack of soil water allows the highly mobile nitrate ion to accumulate in these soils. The extremely high NO_3^- concentrations in soil at CP compared with other ions, however, reflect the N-saturated condition of the forest at CP. The high NO_3^- content of soils at CP leads to high leachate losses from the site during the fall/winter wet season. Geographical patterns of stream-water NO_3^- concentrations across the air pollution gradient in the SBM (Fig. 13.3) parallel the results from N status indicators such as soil solution NO_3^-, soil C:N ratio, foliar N content, and foliar N:P ratios (Fenn et al. 1996).

Trace Gas Emissions

According to the Aber conceptual model of a N-saturated forest (Aber et al. 1989), emissions of nitrogenous trace gases also increase as N inputs accrue in the ecosystem. Studies in a number of forests in North America and Europe support this hypothesis. In many reports, nitrous oxide (N_2O) emissions increase with greater N deposition or site fertility. However, in many cases nitric oxide (NO) was not measured, because NO is highly reactive and is difficult to quantify. At CP (Fenn et al. 1996) and in a N-saturated chaparral stand in the San Gabriel Mountains (Anderson et al. 1988), NO was by far the dominant trace N gas. Davidson (1991, 1993) reported that NO emissions were much higher than emissions of N_2O in soils with less than 60% water-filled pore space and that chemo-autotrophic nitrification was the dominant source of both N_2O and NO. Denitrification is the major process of N_2O production in soils above field capacity water content (Davidson 1992). Elevated NO emissions in the highly polluted sites in southern California are presumably due to the coarse-textured and highly aerated conditions of the soil and the high concentrations of available N in soil (Fenn et al. 1996). High NO emissions from undisturbed soils in western forests may be indicative of a N-saturated condition. However, use of this indicator in a monitoring program will be limited due to the specialized equipment and expertise needed to measure soil fluxes of NO (Fenn and Poth 1996).

Soil Acidity and Base Cation Saturation

Forest soils tend to slowly acidify naturally as stands age (Waring and Schlesinger 1985). Acidic deposition, high soil nitrification rates, and base cation leaching from soil as counterbalancing ions for leached NO_3^- are processes that lead to accelerated soil acidification, base cation depletion, and increased risk of aluminum toxicity (Shortle and Smith 1988; Johnson et al. 1991). Soil acidification and associated consequences for plant health, forest sustainability, and aquatic effects are of major concern in mesic forests in Europe and eastern North America, where the soils are naturally highly leached and of low base cation saturation (Shortle and Smith 1988; Federer et al. 1989; Johnson et al. 1991). Soils in the Mediterranean climate of California are typically of high base saturation, not highly acidic, and with low aluminum levels in the soil solution. Nonetheless, spatial trends of soil acidity and comparisons of recent soil pH data with data from the 1970s in the San Gabriel Mountains and the SBM indicate that soils in areas with elevated atmospheric deposition have acidified at a greater rate and are of lower base saturation than in less-polluted areas (Wood et al. 1992; Fenn et al. 1996). Soil pH and percentage base saturation were highly correlated at sites across the N deposition gradient in the SBM (Fenn et al. 1996). However, even at CP where soil pH averaged 4.0, percentage base saturation ranged from 35 to 65%. In a mature N-saturated red spruce site in the Great Smoky Mountains in eastern Tennessee, base saturation was 4% (Johnson and Lindberg 1992), illustrating that even in the most-polluted site in the SBM, base saturation was manyfold greater than in eastern mesic forests.

Effects of N Deposition on Vegetation

Increased growth rates because of partial or complete alleviation of N limitation is probably the most widespread effect of N deposition over forested areas (Kauppi et al. 1992; Eriksson and Johansson 1993; Binkley and Hogberg 1997). Tree response to greater N fertility is expected to be more pronounced in years with greater available soil water during the growing season. Ponderosa pine trees at CP are no longer N-limited, based on vector analysis of foliar growth responses to N fertilization (Kiefer and Fenn 1997). Notwithstanding the high ozone concentrations at CP (Miller et al. 1986), increases in ponderosa pine basal area at CP from 1974 to 1988 were as great or greater than sites with low air pollution exposure (Miller et al. 1991). It is not known to what extent high N deposition may offset ozone effects on pine growth.

N concentrations in foliage and litter are often greater in forests with elevated N deposition (Fenn 1991; Aber et al. 1995; Fenn et al. 1996; McNulty et al. 1996). Ratios of N and other nutrients are effective indicators of increased N uptake and/or reduced base cation uptake in cation-depleted soils (Hüttl 1990). Nitrogen-to-phosphorus (Zinke 1980; Ericsson et al. 1993; Fenn et al. 1996), C:N, and cation:N (Hüttl 1990; McNulty et al. 1996) ratios in foliage have also been used as indicators of altered nutrient status in forests with chronic N deposition. In N-saturated forests, excess N may be stored in vegetation as amino acids (Näsholm et al. 1997) or as NO_3^-. Foliar NO_3^- concentrations were much higher at CP, a high-deposition site, than in more easterly low-deposition sites. Bracken fern has been shown to be an effective N accumulator (Gerloff et al. 1966). At CP and at a high-deposition site in The Netherlands, NO_3^- accumulation was much higher in bracken fern compared with other plant species (Stams and Schipholt 1990; Fenn et al. 1996). Bracken fern may be an excellent indicator species for monitoring a site's N status and for identifying forests at risk of N saturation.

N toxicity has been observed in ponderosa pine seedlings at N levels that were not toxic to loblolly pine (*Pinus taeda* L.) (Griffin et al. 1995; Johnson et al.

1995), suggesting that the high N concentrations in soils of N-saturated stands may be detrimental to growth and development of ponderosa pine. At present, we have no evidence from field studies to support this hypothesis, except for one fertilization experiment in which the dry weight of annual needle whorls were significantly lower in fertilized ponderosa pine trees compared with unfertilized trees at CP. We are just beginning a fertilization study in the SBM in which growth responses are being monitored with dendrometer bands. In another study in the Los Angeles Air Basin, potted seedlings of coastal sagebrush (*Artemisia californica*) grown in soil with 50 μg inorganic N per g soil senesced prematurely and died. Inorganic N levels in soil in polluted coastal sage sites in the Los Angeles Air Basin were as high as 86 μg/g soil prior to the fall rains, suggesting that high soil N levels may contribute to the observed shrub mortality (Allen et al. 1996).

When considering the impacts of accumulated atmospheric N inputs in forest ecosystems, and especially coniferous systems, it is insightful to remember that these ecosystems are best adapted to chronic N limitation (Vitousek and Howarth 1991). The magnitude of changes in N cycling in N-saturated forests suggests that long-term unpredictable consequences are likely to occur. It is not known how many years of elevated N deposition are required to induce observable symptoms of N saturation in the Mediterranean ecosystems of California, but the time period is relatively short when considered within the context of forest developmental history. To become N-saturated within a period of years represents unprecedented changes in nutrient status for these systems. Disturbances such as fire, storm damage, or timber harvesting may have similarly drastic effects on nutrient cycling, but in these instances significant acute N losses from the system occur (Vitousek and Melillo 1979; Riggan et al. 1994). In the case of chronic N deposition, the perturbation is a more gradual process of N accumulation until abnormally high chronic N losses begin to occur as the ecosystem reaches a N-saturated state.

Some of the most critical effects of high N fertility on terrestrial ecosystems can be exerted indirectly via interactions with other biotic and abiotic stressors. For example, excess N has the potential to exert significant effects on ecosystem function by altering plant biodiversity, mycorrhizal development (Termorshuizen 1993), biological N fixation, insect infestations, plant disease, drought severity, and ozone injury (Skeffington and Wilson 1988). Little is known of the interactions between these factors and N deposition in the ecosystems in southern California. However, enough is known of the effects of increased N fertility on these factors and ecosystem components in forests to predict that N does in all likelihood affect many of these processes in California forests (Skeffington and Wilson 1988).

Streamwater Nitrate Export

Streamwater NO_3^- export from forest watersheds is an important environmental concern, because water from montane watersheds is often depended on as a source of clean water. Streamwater ionic flux is also a useful integrator of overall ecosystem integrity and serves as a watershed-level indicator of ecosystem nutrient loss or conservation. N-saturated watersheds export above-normal amounts of NO_3^- once the capacity for N retention by the biota and soil chemical fixation mechanisms is exceeded (Stoddard 1994). Streamwater NO_3^- export is the best indicator of the N status of a watershed inasmuch as N export in streamwater is the net balance between N retention and release mechanisms within the entire watershed (except for gaseous losses of N, which are usually small compared with streamwater N export in N-enriched forests). A brief and relatively small NO_3^- loss from temperate forest watersheds is expected during spring snowmelt runoff. During this early spring season, water flows are greatest and soil temperatures are still low, and plant and microbial nutrient demand is typically insufficient to retain all the dissolved N. However, a state of N saturation is indicated when NO_3^- concentrations begin to increase above this "normal" background level and when losses are prolonged. When N excess in the watershed becomes severe, elevated streamwater NO_3^- concentrations are found year-round, with little seasonal variability (Fig. 13.3) (Stoddard 1994; Peterjohn et al. 1996).

Peak flows in N-enriched California ecosystems seem to occur during major winter storms or during subsequent snowmelt periods when more complete washout of the soil profile occurs. Annual discharge rates from two chaparral watersheds in the SDEF in the San Gabriel Mountains (Fig. 13.2) over 4 years range from 0.04 to 10.0 kg ha^{-1} yr^{-1}, and discharge rates for two grassland watersheds ranged from 0.11 to 19.4 kg ha^{-1} yr^{-1} (Riggan et al. 1985). The NO_3^- yield in these watersheds was limited by the amount of water percolating through the soil. Relatively low NO_3^- concentrations occurred during base flow in the summer. Riggan et al. (1985) attributed the low summertime concentrations to biological NO_3^- uptake within the aquatic system of these chaparral watersheds. Notably, groundwater NO_3^- pollution in the eastern San Gabriel Basin just below the SDEF is a serious problem, with NO_3^- concentrations often exceeding the federal drinking water standard (Riggan et al. 1985). Atmospheric deposition is thought to be a major source of the groundwater NO_3^-.

Nitrate concentrations in streams draining small catchments in Devil Canyon near CP in the western SBM are unusually high compared with forested watersheds elsewhere in North America (Stoddard 1994). Nitrate concentrations are frequently 75 μEq/L or greater and peak at about 300 μEq/L in streams within Devil Canyon, whereas streamwater NO_3^- from streams in the BF–San Gorgonio Wilderness Area (low N deposition) is typically an order of magnitude lower and frequently near or below detection limits (Fig. 13.3). Nitrate concentrations in some streams at Devil Canyon are high year-round, although concentrations decline in spring and summer and peak in winter (Fig. 13.3). The high base-level NO_3^- concentrations during summer in some streams indicate a highly N-saturated state of these catchments. A similar phenomenon was reported for the highly N-saturated watersheds in the Fernow Experimental Forest in West Virginia (Peterjohn et al. 1996). Consistent differences in NO_3^- concentrations among the seven streams at Devil Canyon suggest that N processing in the terrestrial catchments and/or in the riparian zones varies significantly from site to site.

Management of Nitrogen-Saturated Stands

Historically, forest management practices have not dealt with chronic or sustained NO_3^- export from undisturbed forested watersheds. Some rethinking may be in order before implementing management practices effective for N-limited forests in N-saturated forests—or in forests that have accumulated large stores of N, with high potential to export excess NO_3^- as the stand matures. Management of N-saturated watersheds with the objectives of reducing NO_3^- losses and detrimental effects on forest nutrition and water quality presents new challenges for land managers. Forests that are still N-limited may not require management action to offset the apparently beneficial effects of increased available N—unless stand history, stand conditions, and N inputs suggest that N saturation will worsen. In some cases, high NO_3^- levels in streamwater can be reduced by denitrification in water storage facilities (e.g., in reservoirs) if favorable conditions exist for denitrification, (sufficient available C and anaerobic conditions). A drawback of this strategy is that N_2O, a major product of denitrification is also an important greenhouse gas (Lammel and Graßl 1995).

Successful fire suppression efforts may have contributed to the development of N saturation in fire-adapted ecosystems in southern California by allowing N to accumulate in soil and the forest floor and by maintaining dense, over mature stands with reduced N demand (Fenn et al. 1996, Riggan et al. 1985). The use of prescribed fire has been proposed as a possible management tool when N deposition and streamwater NO_3^- concentrations are very high in chaparral ecosystems near Los Angeles (Riggan et al. 1994). Implementing prescribed fire in N-saturated fire-adapted ecosystems could serve a twofold purpose: to reduce N stores from the system, and to prepare a seedbed for regeneration of seedlings or to induce resprouting of shrubs that will later be much stronger N sinks than decadent vegetation Burning will initially cause a strong pulse of released N previously immobilized in vegetation, detritus, and soil organic matter. A fraction of the released N will be volatized, whereas NH_4^+ released will likely be nitrified after the fire and result in leached NO_3^- (Dunn et al. 1979; Riggan et al. 1994). However, the long-term benefits of fire should outweigh the consequences of short-term atmospheric emissions and episodes of high NO_3^- concentrations in streamwater. Prescribed burns can be done when air pollution concentrations are lowest to reduce impacts on air quality.

Silvicultural treatments that maintain vigorous forest growth are likely to aid in retaining N more effectively as N is incorporated into plant biomass, and as higher stand evapotranspiration reduces streamwater discharge and nutrient losses from the watershed (Swank 1988). Such treatments could include selective harvesting of decadent trees, replanting, and thinning of overcrowded stands or fertilizing with nutrients that have become limiting with N sufficiency or excess. In thinning or harvest operations, slash and debris should preferably be removed from the site, or possibly burned, because these materials are significant sources of leachable NO_3^- and dissolved organic N (Rosén and Lundmark-Thelin 1987). Nitrate losses following whole-tree harvesting are lower than in conventional harvesting, probably because of the N removed from the site in the slash material and because slash removal allows for more rapid regeneration of vegetation cover, resulting in greater vegetative N demand and N retention (Hendrickson et al. 1989; Stevens and Hornung 1990; Staaf and Olsson 1994). However, in many situations, such as in national parks or wilderness areas, the practices mentioned above will have limited application due to economic, policy, and regulatory restraints.

Population and vehicle miles continue to increase in California, and future trends in N emissions, N deposition, and the extent of N-saturated systems will largely depend on the success of political and social agreements in controlling NO_x emissions. However, with increased fossil fuel use, it seems unlikely that N emissions will be greatly reduced, at least in the near future. If N emissions increase, the effects of chronic N inputs observed in forests in the Los Angeles Air Basin may also develop in the southern and western Sierra Nevada. However, if reductions in NO_x emissions occur, the areal extent of N-saturated ecosystems in southern California may diminish slightly. However, the lag time from when emissions are reduced until some forests are no longer a net source of streamwater NO_3^- is unknown. In the so-called roof experiments in Europe (transparent roofs built over the forest canopy for manipulating throughfall N inputs), NO_3^- leaching through the soil profile was rapidly reduced when N inputs were diminished to background levels (Bredemeier et al. 1995; Wright et al. 1995). Biotic responses were slower to produce measurable effects of reduced N deposition (Boxman et al. 1994, 1995). We expect that similar responses may occur in southern California ecosystems, unless mineralization of large organic N stores in the forest floor and soil produces excess leachable NO_3^- as observed in other systems with moderate (about 8-10 kg ha^{-1} yr^{-1}) N deposition (Foster et al. 1989). However, without major outside N inputs such as N deposition or N fixation, excessive NO_3^- export is not expected to continue indefinitely.

Summary

Chronic N deposition to montane ecosystems in the South Coast Air Basin is causing major perturbations in biogeochemical cycling of N. Symptoms of N excess in these systems are analogous to overfertilization of arable land. Limited field studies indicate that chaparral and mixed conifer forests with N deposition of approximately 20–25 kg ha^{-1} yr^{-1} or greater for more than 40 years exhibit the symptoms of N saturation. Systems with N deposition less than 15 kg ha^{-1} yr^{-1} are still N-limited. Further research is needed to verify these tentative thresholds, but they are in accord with a survey of 65 forested watersheds in Europe (Dise and Wright 1995). Edaphic, vegetative, and hydrological indicators support the conclusion that chaparral and forest watersheds in southern California with high air pollution exposure are possibly the most severely saturated in North America (Fenn et al. 1996, 1998). Principal indicators demonstrating N saturation in these systems include increased foliar N and N:P ratios, NO_3^- storage in foliage, high

NO_3^- concentrations in soil and in soil solution, high N mineralization and nitrification rates, elevated concentrations of NO_3^- in streamwater year-round, and high emissions of nitrogenous trace gases from soil. Soil acidification and reduced base saturation have also been observed in N-saturated sites, although not yet to the extent expected to have detrimental effects on plant growth.

The long-term ecological effects of N excess in these systems is a critical unknown. Factors that are likely to influence the impacts of high N fertility include fire exclusion, interactions with ozone, stand succession favoring more ozone-tolerant and fire-sensitive species (white fir [*Abies concolor* Gord. & Glend.] and incense cedar), reduction in C allocation to roots, and drought. Indirect effects of N saturation on mycorrhizal fungi, insect infestations, and tree diseases have the potential to cause rapid and significant ecological effects that may drastically affect forest sustainability. Threatened and endangered plant species may also be further endangered as a result of the ecological effects of excess N. Increased tree growth in areas of moderate N deposition may also worsen the undesirable densification of forest stands as a result of long-term fire exclusion, thus increasing the risk of catastrophic fire (Savage 1994). Reduced water quality as a result of elevated N deposition and NO_3^- concentrations in streamwater is probably the most important environmental issue of immediate concern.

References

Aber, J.D.; Nadelhoffer, K.J.; Steudler, P.; Melillo, J.M. Nitrogen saturation in northern forest ecosystems. BioScience 39:378–386; 1989.

Aber, J.D.; Magill, A.; McNulty, S.G.; Boone, R.D.; Nadelhoffer, K.J.; Downs, M.; Hallett, R. Forest biogeochemistry and primary production altered by nitrogen saturation. Water Air Soil Pollut. 85:1665–1670; 1995.

Allen, E.B.; Padgett, P.E.; Bytnerowicz, A.; Minnich, R.A. Nitrogen deposition effects on Coastal Sage vegetation of southern California. In: Bytnerowicz, A., ed. Proceedings air pollution and climate change effects on forest ecosystems, February 5–9, 1996, Riverside, CA. Electronic access to proceedings: http://www.rfl.pswfs.gov/pubs/psw-gtr-164/index.html; Albany, CA: USDA, Forest Service, Pacific Southwest Station; 1996.

Anderson, I.C.; Levine, J.S.; Poth, M.A.; Riggan, P.J. Enhanced biogenic emissions of nitric oxide and nitrous oxide following surface biomass burning. J. Geophys. Res. 93:3893–3898; 1988.

Anderson, M.A.; Graham, R.C.; Alyanakian, G.J.; Martynn, D.Z. Late summer water status of soils and weathered bedrock in a giant Sequoia grove. Soil Sci. 160:415–422; 1995.

Arkley, R.J. Soil moisture use by mixed conifer forest in a summer-dry climate. Soil Sci. Soc. Am. J. 45:423–427; 1981.

Berg, B.; McClaugherty, C.; de Santo, A.V.; Johansson, M.J.; Gunnar, E. Decomposition of litter and soil organic matter—Can we distinguish a mechanism for soil organic matter buildup? Scand. J. For. Res. 10:108–119; 1995.

Beyers, J.L.; Sternberg, P.D.; Anderson, M.A.; Graham, R.C. Chaparral shrubs make use of water stored within weathered bedrock. Suppl. Bull. Ecol. Soc. Am. 76:302; 1995.

Binkley, D.; Hogberg, P. Does atmospheric deposition of nitrogen threaten Swedish forests? For. Ecol. Manage. 92:119–152; 1997.

Bowden, R.D.; Steudler, P.A.; Melillo, J.M. Annual nitrous oxide fluxes from temperate forest soils in the northeastern United States. J. Geophys. Res. 95:13997–14005; 1990.

Boxman, A.W.; van Dijk, H.F.G.; Roelofs, J.G.M. Soil and vegetation responses to decreased atmospheric nitrogen and sulphur inputs into a Scots pine stand in The Netherlands. For. Ecol. Manage. 68:39–45; 1994.
Boxman, A.W.; van Dam, D.; van Dijk, H.F.G.; Hogervorst, R.F.; Koopmans, C.J. Ecosystem responses to reduced nitrogen and sulphur inputs into two coniferous stands in The Netherlands. For. Ecol. Manage. 71:7–29; 1995.
Bredemeier, M.; Blanck, K.; Lamersdorf, N.; Wiedey, G.A. Response of soil water chemistry to experimental "clean rain" in the NITREX roof experiment at Solling, Germany. For. Ecol. Manage. 71:31–44; 1995.
Brooks, P.D.; Williams, M.W.; Schmidt, S.K. Microbial activity under alpine snowpacks, Niwot Ridge, Colorado. Biogeochemistry 32:93–113; 1996.
Bytnerowicz, A.; Fenn, M.E. Nitrogen deposition in California forests: a review. Environ. Pollut. 92:127–146; 1996.
Chapin, F.S., III. The mineral nutrition of wild plants. Annu. Rev. Ecol. Syst. 11:233–260; 1980.
Chapin, F.S., III; Moilanen, L.; Kielland, K. Preferential use of organic nitrogen for growth by a nonmycorrhizal arctic sedge. Nature 361:150–153; 1993.
Cole, D.W.; Van Miegroet, H.; Foster, N.W. Retention or loss of N in IFS sites and evaluation of relative importance of processes. In: Johnson, D.W.; Lindberg, S.E., eds. Atmospheric deposition and forest nutrient cycling. Ecological studies 91. New York: Springer-Verlag; 1992:196–199.
Coûteaux, M.M.; Bottner, P.; Berg, B. Litter decomposition, climate and litter quality. Trends Ecol. Evol. 10:63–66; 1995.
Covington, W.W.; Sackett, S.S. Soil mineral nitrogen changes following prescribed burning in ponderosa pine. For. Ecol. Manage. 54:175–191; 1992.
Davidson, E.A. Fluxes of nitrous oxide and nitric oxide from terrestrial ecosystems. In: Rogers, J.E.; Whitman, W.B., eds. Microbial production and consumption of greenhouse gases: methane, nitrogen oxides, and halomethanes. Washington, DC: Am Soc. Microbiol.; 1991:219–235.
Davidson, E.A. Sources of nitric oxide and nitrous oxide following wetting of dry soil. Soil Sci. Soc. Am. J. 56:95–102; 1992.
Davidson, E.A. Soil water content and the ratio of nitrous oxide to nitric oxide emitted from soil. In: Oremland, R.S., ed. Biogeochemistry of global change: radiatively active trace gases. New York: Chapman & Hall; 1993:369–386.
Davidson, E.A.; Hart, S.C.; Firestone, M.K. Internal cycling of nitrate in soils of a mature coniferous forest. Ecology 73:1148–1156; 1992.
Dise, N.B.; Wright, R.F. Nitrogen leaching from European forests in relation to nitrogen deposition. For. Ecol. Manage. 71:153–161; 1995.
Dunn, P.H.; De Bano, L.F.; Eberlein, G.E. Effects of burning chaparral soils: II. Soil microbes and nitrogen mineralization. Soil Sci. Soc. Am. J. 43:509–514; 1979.
Durka, W.; Schulze, E-D.; Gebauer, G.; Voerkelius, D. Effects of forest decline on uptake and leaching of deposited nitrate determined from ^{15}N and ^{18}O measurements. Nature 372:765–767; 1994.
Ericsson, A.; Nordén, L-G.; Näsholm, T.; Walheim, M. Mineral nutrient imbalances and arginine concentrations in needles of *Picea abies* (L.) Karst. from two areas with different levels of airborne deposition. Trees 8:67–74; 1993.
Eriksson, H.; Johansson, U. Yields of Norway spruce (*Picea abies* (L.) Karst.) in two consecutive rotations in southwestern Sweden. Plant Soil 154:239–247; 1993.
Federer, C.A.; Hornbeck, J.W.; Tritton, L.M.; Martin, C.W.; Pierce, R.S.; Smith, C.T. Long-term depletion of calcium and other nutrients in eastern U.S. forests. Environ. Manage. 13:593–601; 1989.
Fenn, M.E. Increased site fertility and litter decomposition rate in high-pollution sites in the San Bernardino Mountains. For. Sci. 37:1163–1181; 1991.

Fenn, M.E.; Bytnerowicz, A. Dry deposition of nitrogen and sulfur to ponderosa and Jeffrey pine in the San Bernardino National Forest in southern California. Environ. Pollut. 81:277–285; 1993.

Fenn, M.E.; Bytnerowicz, A. Summer throughfall and winter deposition in the San Bernardino Mountains in southern California. Atmos. Environ. 31:673–683; 1997.

Fenn, M.E.; Poth, M.A.; Aber, J.D.; Baron, J.S.; Bormann, B.T.; Johnson, D.W.; Lemly, A.D.; McNulty, S.G.; Ryan, D.F.; Stottlemyer, R. Nitrogen excess in North American ecosystems: predisposing factors, ecosystem responses, and management strategies. Ecol. Applic 8:706–733; 1998.

Fenn, M.E.; Dunn, P.H. Litter decomposition across an air-pollution gradient in the San Bernardino Mountains. Soil Sci. Soc. Am. J. 53:1560–1567; 1989.

Fenn, M.E.; Poth, M.A. Indicators of nitrogen status in California forests. Gen. tech. rep. PSW-164. Published electronically: http://www.rfl.pswfs.gov/pubs/psw-gtr-164/index.html. USDA, Forest Service, Pacific Southwest Station; Albany, CA: 1996.

Fenn, M.E.; Poth, M.A.; Johnson, D.W. Evidence for nitrogen saturation in the San Bernardino Mountains in southern California. For. Ecol. Manage. 82:211–230; 1996.

Fog, K. The effect of added nitrogen on the rate of decomposition of organic matter. Biol. Rev. 63:433–462; 1988.

Foster, N.W.; Hazlett, P.W.; Nicolson, J.A.; Morrison, I.K. Ion leaching from a sugar maple forest in response to acidic deposition and nitrification. Water Air Soil Pollut. 48:251–261; 1989.

Fraser, D.W.; McColl, J.G.; Powers, R.F. Soil nitrogen mineralization in a clearcutting chronosequence in a northern California conifer forest. Soil Sci. Soc. Am. J. 54:1145–1152; 1990.

Garten, C.T., Jr.; Van Miegroet, H. Relationships between soil nitrogen dynamics and natural ^{15}N abundance in plant foliage from Great Smoky Mountains National Park. Can. J. For. Res. 24:1636–1645; 1994.

Gerloff, G.C.; Moore, D.G.; Curtis, J.T. Selective absorption of mineral elements by native plants of Wisconsin. Plant Soil 25:393–405; 1966.

Goulden, M.L.; Munger, J.W.; Fan, S.M.; Daube, B.C.; Wofsy, S.C. Exchange of carbon dioxide by a deciduous forest: response to interannual climate variability. Science 271:1576–1578; 1996.

Gray, J.T.; Schlesinger, W.H. Nutrient cycling in Mediterranean type ecosystems. In: Miller, P.C., ed. Resource use by chaparral and matorral. Ecological studies 39. New York: Springer-Verlag; 1981:259–285.

Griffin, K.L.; Winner, W.E.; Strain, B.R. Growth and dry matter partitioning in loblolly and ponderosa pine seedlings in response to carbon and nitrogen availability. N. Phytol. 129:547–556; 1995.

Grulke, N.E.; Andersen, C.P.; Fenn, M.E.; Miller, P.R. Pollution reduces root standing biomass of ponderosa pine in the San Bernardino Mountains, California. Environ. Pollut. (in press).

Hart, S.C.; Firestone, M.K. Evaluation of three in situ soil nitrogen availability assays. Can. J. For. Res. 19:185–191; 1989.

Hart, S.C.; Firestone, M.K.; Paul, E.A. Decomposition and nutrient dynamics of ponderosa pine needles in a Mediterranean-type climate. Can. J. For. Res. 22:306–314; 1992.

Haynes, B.E.; Gower, S.T. Belowground carbon allocation in unfertilized and fertilized red pine plantations in northern Wisconsin. Tree Physiol. 15:317–325; 1995.

Hendrickson, O.Q.; Chatarpaul, L.; Burgess, D. Nutrient cycling following whole-tree and conventional harvest in northern mixed forest. Can. J. For. Res. 19:725–735; 1989.

Hüttl, R.F. Nutrient supply and fertilizer experiments in view of N saturation. Plant Soil 128:45–58; 1990.

Johansson, M-B. Decomposition rates of Scots pine needle litter related to site properties, litter quality, and climate. Can. J. For. Res. 24:1771–1781; 1994.

Johnson, D.W. Nitrogen retention in forest soils. J. Environ. Qual. 21:1–12; 1992.

Johnson, D.W.; Lindberg, S.E., eds. Atmospheric deposition and forest nutrient cycling: a synthesis of the integrated forest study. New York: Springer-Verlag; 1992.

Johnson, D.W.; Van Miegroet, H.; Lindberg, S.E.; Harrison, R.B.; Todd, D.E. Nutrient cycling in red spruce forests of the Great Smoky Mountains. Can. J. For. Res. 21:769–787; 1991.

Johnson, D.W.; Walker, R.F.; Ball, J.T. Lessons from lysimeters: soil N release from disturbance compromises controlled environment study. Ecol. Applic. 5:395–400; 1995.

Jones, D.P.; Graham, R.C. Water-holding characteristics of weathered granitic rock in chaparral and forest ecosystems. Soil Sci. Soc. Am. J. 57:256–261; 1993.

Kauppi, P.E.; Mielikäinen, K.; Kuusela, K. Biomass and carbon budget of European forests, 1971 to 1990. Science 256:70–74; 1992.

Kiefer, J.W.; Fenn, M.E. Using vector analysis to assess nitrogen status of ponderosa and Jeffrey pine along deposition gradients in forests of southern California. For. Ecol. Manage. 94:47–59; 1997.

Kimmins, J.P. Forest ecology, a foundation for sustainable management, 2nd ed. Upper Saddle River, NJ: Prentice Hall; 1996.

Korontzi, S.; Macko, S.A.; Anderson, I.C.; Poth, M.A. Use of stable isotopes to trace the effects of increased nitrogen deposition to San Bernardino National Forest. Eos Supplement 77:F108, American Geophysical Union, Transactions of the Fall Meeting, Dec. 15–19, 1996, San Francisco, CA.

Krywult, M.; Karolak, A.; Bytnerowicz, A. Nitrate reductase activity as an indicator of ponderosa pine response to atmospheric nitrogen deposition in the San Bernardino Mountains. Environ. Pollut. 93:141–146; 1996.

Kuyper, T.W. Fungal species diversity and forest ecosytem functioning in The Netherlands. In: Boyle, T.J.B.; Boyle, C.E.B., eds. NATO ASI series, vol. I, 20. Biodiversity, temperate ecosytems, and global change. Berlin: Springer-Verlag; 1994:99–122.

Lamersdorf, N.P.; Meyer, M. Nutrient cycling and acidification of a northwest German forest site with high atmospheric nitrogen deposition. For. Ecol. Manage. 62:323–354; 1993.

Lammel, G.; Graßl, H. Greenhouse effect of NO_x. Environ. Sci. Pollut. Res. 2:40–45; 1995.

Martikainen, P.J. Microbial processes in boreal forest soils as affected by forest management practices and atmospheric stress. In: Soil biochemistry, vol. 9. Stotzky, G.; Bollag, J.-M., eds. New York: Marcel Dekker, Inc.; 1996:196–232.

Matson, P.A.; Volkmann, C.; Coppinger, K.; Reiners, W.A. Annual nitrous oxide flux and soil nitrogen characteristics in sagebrush steppe ecosystems. Biogeochemistry 14:1–12; 1991.

Matson, P.A.; Gower, S.T.; Volkmann, C.; Billow, C.; Grier, C.C. Soil nitrogen cycling and nitrous oxide flux in a Rocky Mountain Douglas-fir forest: effects of fertilization, irrigation and carbon addition. Biogeochemistry 18:101–117; 1992.

McNulty, S.G.; Aber, J.D.; Newman, S.D. Nitrogen saturation in a high elevation New England spruce-fir stand. For. Ecol. Manage. 84:109–121; 1996.

Melillo, J.M.; Aber, J.D.; Linkins, A.E.; Ricca, A.; Fry, B.; Nadelhoffer, K.J. Carbon and nitrogen dynamics along the decay continuum: plant litter to soil organic matter. Plant Soil 115:189–198; 1989.

Miller, P.R. Mixed conifer forests of the San Bernardino Mountains, California. In: Olson, R.K.; Binkley, D.; Böhm, M., eds. The response of western forests to air pollution. Ecological studies 97. New York: Springer-Verlag; 1992:461–497.

Miller, P.R.; Taylor, O.C.; Poe, M.P. Spatial variation of summer ozone concentrations in the San Bernardino Mountains. In: Proceedings of air pollution control association annual meeting, vol. 3, Minneapolis, MN, June 22–27, 1986. Pittsburgh, PA: Air Pollut. Control Assoc.; 1986:86-39.2.

Miller, P.R.; McBride, J.R.; Schilling, S.L. Chronic ozone injury and associated stresses affect relative competitive capacity of species comprising the California mixed conifer forest type. In: Proceedings symposium on sustainable agriculture: an option for development without environmental degradation. Montecillo, Mexico, Dec. 9–10, 1991. Montecillo, Mexico: Postgraduate College and M.O.A. International; 1991:161–172.

Miller, P.R.; Stolte, K.W.; Duriscoe, D.M.; Pronos, J., tech. coords. Evaluating ozone air pollution effects on pines in the western United States. Gen. tech. rep. PSW-GTR-155. USDA Forest Service, Pacific Southwest Station; 1996a.

Miller, P.R.; Schilling, S.; Jones, D.; Guthrey, D. Annual changes in crown condition of ponderosa and Jeffrey pines between 1991 and 1994 at Barton Flats plots. In: Miller, P.R.; Watson, J., eds. Assessment of acidic deposition and ozone effects on conifer forests in the San Bernardino Mountains. Contract A032-180. Final report, Environmental Protection Agency, California Air Resources Board, contract A032-180. Sacramento, CA. 1996b:12/1–12/29.

Minnich, R.A.; Barbour, M.G.; Burk, J.H.; Fernau, R.F. Sixty years of change in California forests of the San Bernardino Mountains. Conserv. Biol. 9:902–914; 1995.

Moorhead, D.L.; Callaghan, T. Effects of increasing ultraviolet B radiation on decomposition and soil organic matter dynamics: a synthesis and modelling study. Biol. Fert. Soils 18:19–26; 1994.

Moorhead, D.L.; Reynolds, J.F. Mechanisms of surface litter mass loss in the northern Chihuahuan Desert: a re-interpretation. J. Arid. Environ. 16:157–163; 1989.

Näsholm, T.; Nordin, A.; Edfast, A-B.; Högberg, P. Identification of coniferous forests with incipient nitrogen saturation through analysis of arginine and nitrogen-15 abundance of trees. J. Environ. Qual. 26:302–309; 1997.

Nihlgård, B. The ammonium hypothesis—an additional explanation to the forest dieback in Europe. Ambio 14:2–8; 1985.

Nohrstedt, H-O. Effects of repeated nitrogen fertilization with different doses on soil properties in a *Pinus sylvestris* stand. Scand. J. For. Res. 5:3–15; 1990.

Olson, J.S. Energy storage and the balance of producers and decomposers in ecological systems. Ecology 44:322–331; 1963.

Perry, D.A. Forest ecosystems. Baltimore, MD: Johns Hopkins University Press; 1994.

Peterjohn, W.T.; Adams, M.B.; Gilliam, F.S. Symptoms of nitrogen saturation in two central Appalachian hardwood forest ecosystems. Biogeochemistry 35:507–522; 1996.

Powers, R.F. Nitrogen mineralization along an altitudinal gradient: interactions of soil temperature, moisture, and substrate quality. For. Ecol. Manage. 30:19–29; 1990.

Powers, R.F.; Webster, S.R.; Cochran, P.H. Estimating the response of ponderosa pine forests to fertilization. Gen. tech. rep. INT-243. USDA, Forest Service, Intermountain Research Station; Ogden, UT: 1988:219–225.

Raison, R.J.; Khanna, P.K.; Woods, P.V. Mechanisms of element transfer to the atmosphere during vegetation fires. Can. J. For. Res. 15:132–140; 1985.

Riggan, P.J.; Lockwood, R.N.; Lopez, E.N. Deposition and processing of airborne nitrogen pollutants in Mediterranean-type ecosystems of southern California. Environ. Sci. Technol. 19:781–789; 1985.

Riggan, P.J.; Lockwood, R.N.; Jacks, P.M.; Colver, C.G.; Weirich, F.; DeBano, L.F.; Brass, J.A. Effects of fire severity on nitrate mobilization in watersheds subject to chronic atmospheric deposition. Environ. Sci. Technol. 28:369–375; 1994.

Rosén, K.; Lundmark-Thelin, A. Increased nitrogen leaching under piles of slash—a consequence of modern forest harvesting techniques. Scand. J. For. Res. 2:21–29; 1987.

Savage, M. Anthropogenic and natural disturbance and patterns of mortality in a mixed conifer forest in California. Can. J. For. Res. 24:1149–1159; 1994.

Shortle, W.C.; Smith, K.T. Aluminum-induced calcium deficiency syndrome in declining red spruce. Science 240:239–240; 1988.

Skeffington, R.A.; Wilson, E.J. Excess nitrogen deposition: issues for consideration. Environ. Pollut. 54:159–184; 1988.

Smith, S.L.; Dale, J.; DeNitto, G.; Marshall, J.; Owen, D. California forest health, past and present. Rep. R5-FPM-PR-001. San Francisco: USDA, Forest Service, Pacific Southwest Station; 1994.

Smith, W.H. Air pollution and forests: interactions between air contaminants and forest ecosystems, 2nd ed. New York: Springer-Verlag; 1990.

Söderström, B.; Bääth, E.; Lundgren, B. Decrease in soil microbial activity and biomasses owing to nitrogen amendments. Can. J. Microbiol. 29:1500–1506; 1983.

Staaf, H.; Olsson, B.A. Effects of slash removal and stump harvesting on soil water chemistry in a clearcutting in SW Sweden. Scand. J. For. Res. 9:305–310; 1994.

Stams, A.J.M.; Schipholt, I.J.L. Nitrate accumulation in leaves of vegetation of a forested ecosystem receiving high amounts of atmospheric ammonium sulfate. Plant Soil 125:143–145; 1990.

Stark, J.M.; Hart, S.C. High rates of nitrification and nitrate turnover in undisturbed coniferous forests. Nature 385:61–64; 1996.

Stevens, P.A.; Hornung, M. Effect of harvest intensity and ground flora establishment on inorganic-N leaching from a Sitka spruce plantation in north Wales, UK. Biogeochemistry 10:53–65; 1990.

Stoddard, J.L. Long-term changes in watershed retention of nitrogen: its causes and aquatic consequences. In: Baker, L.A., ed. Environmental chemistry of lakes and reservoirs. Advances in Chemistry Series 237. Washington, DC: American Chemical Society; 1994:223–284.

Swank, W.T. Stream chemistry responses to disturbance. In: Swank, W.T.; Crossley, D.A., Jr., eds. Forest hydrology and ecology at Coweeta. Ecological studies 66. New York: Springer-Verlag; 1988.339–357.

Termorshuizen, A.J. The influence of nitrogen fertilisers on ectomycorrhizas and their fungal carpophores in young stands of *Pinus sylvestris*. For. Ecol. Manage. 57:179–189; 1993.

U.S. Environmental Protection Agency. Air quality criteria for oxides of nitrogen. U.S. EPA/600/8-91/049bF. Research Triangle Park, NC: Office of Health and Environmental Assessment, Environmental Criteria and Assessment Office; 1993.

Van Breeemen, N.; Mulder, J.; Van Grinsven, J.J.M. Impacts of acid atmospheric deposition on woodland soils in The Netherlands: II. Nitrogen transformations. Soil Sci. Soc. Am. J. 51:1634–1640; 1987.

Van Miegroet, H.; Lovett, G.M.; Cole, D.W. Nitrogen chemistry, deposition, and cycling in forests: summary and conclusions. In: Johnson, D.W.; Lindberg, S.E., eds. Atmospheric deposition and forest nutrient cycling. Ecological studies 91. New York: Springer-Verlag; 1992:202–207.

Vitousek, P.M.; Howarth, R.W. Nitrogen limitation on land and in the sea: how can it occur? Biogeochemistry 13:87–115; 1991.

Vitousek, P.M.; Melillo, J.M. Nitrate losses from disturbed forests: patterns and mechanisms. For. Sci. 25:605–619; 1979.

Wang, Z.Q.; Newton, M.; Tappeiner, J.C., II. Competitive relations between Douglas-fir and pacific madrone on shallow soils in a Mediterranean climate. For. Sci. 41:744–757; 1995.

Waring, R.H.; Schlesinger, W.H. Forest ecosystems: concepts and management. Orlando, FL: Academic Press; 1985.

Williams, M.W.; Brooks, P.D.; Mosier, A.; Tonnessen, K.A. Mineral nitrogen transformations in and under seasonal snow in a high-elevation catchment in the Rocky Mountains, United States. Water Res. 32:3161–3171; 1996.

Wood, H.B.; Olivier, K.L.; Ryan, T.M. Surface soil acidification in smog-polluted chaparral ecosystems in the San Gabriel Mountains, California (abstract). Bull. Ecol. Soc. Am. 73:392; 1992.

Wright, R.F.; Roelofs, J.G.M.; Bredemeier, M.; Blanck, K.; Boxman, A.W.; Emmett, B.A.; Gundersen, P.; Hultberg, H.; Kjønaas, O.J.; Moldan, F.; Tietema, A.; van Breemen, N.;

van Dijk, H.F.G. NITREX: responses of coniferous forest ecosystems to experimentally changed deposition of nitrogen. For. Ecol. Manage. 71:163–169; 1995.

Zak, D.R.; Groffman, P.M.; Pregitzer, K.S.; Christensen, S.; Tiedje, J.M. The vernal dam: plant-microbe competition for nitrogen in northern hardwood forests. Ecology 71:651–656; 1990.

Zinke, P.J. Influence of chronic air pollution on mineral cycling in forests. In: Miller, P.R. ed. Proceedings symposium on effects of air pollutants on Mediterranean and temperate forest ecosystems, Riverside, CA, June 22–27, 1980. Gen. tech. rep. PSW-43. Berkeley, CA: USDA Forest Service; 1980:88–99.

Zwieniecki, M.A.; Newton, M. Seasonal pattern of water depletion from soil-rock profiles in a Mediterranean climate in southwestern Oregon. Can. J. For. Res. 26:1346–1352; 1996.

3. Interactions of Physical, Chemical, and Biological Factors and Their Influences on Forest Health and Forest Use

14. Insects and Pathogens in a Pollution-Stressed Forest

J. Pronos, L. Merrill, and D. Dahlsten

Introduction

A healthy forest is comprised of trees and the plant, animal, and microbial communities with which they are normally associated. Native insect herbivores and pathogens of forest trees perform important functions in natural ecosystems, killing decadent trees, recycling nutrients, and creating gaps for regeneration. These insects and diseases are defined as pests when their activities interfere with management objectives for a stand. Such activities may cause tree mortality, injury, or debilitation.

Insects and tree pathogens ("pests"), their natural enemies, trees, and air pollution may be regarded as the corners of a square embedded in a matrix composed of the environment, the latter including climate, soils, topography, and other factors important to the functioning of particular species (Fig. 14.1). The environment affects pests, their hosts, and their natural enemies, as well as the movement and concentration of air pollutants at a particular site. Trees and their insects and pathogens affect each other's survival and reproduction, and natural enemies of insects and pathogens (including the whole array of biotic agents that are predators, parasites, or competitors of insect and plant pathogen pests) are influenced by their hosts as well as directly influencing their hosts and indirectly influencing the growth and survival of the trees. Thus pollution may have direct effects on the insect or disease organism, indirect effects on the insect or pathogen through changes to the host, and direct and indirect effects on natural enemies of the insect

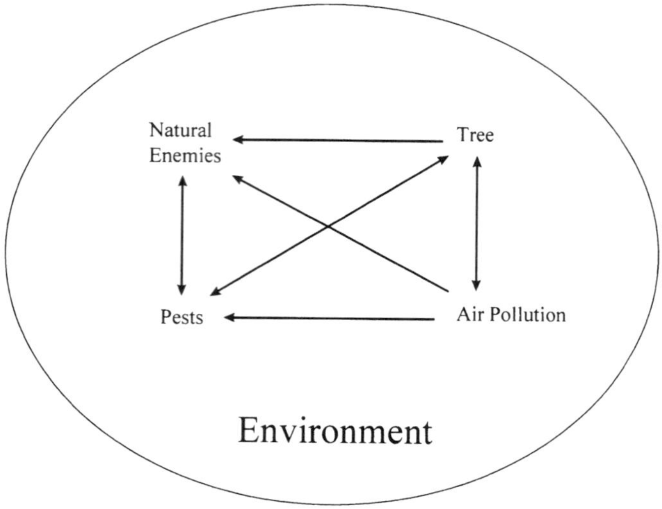

Figure 14.1. Schematic representation of relationships between trees, pests, natural enemies, air pollution, and the environment. Arrows indicate direct effects of one component on another.

or pathogen. Heagle (1973), Alstad et al. (1982), Riemer and Whittaker (1989), Whittaker and Warrington (1990), Heliovaara and Vaisanen (1993), and other authors have reviewed the research on these interactions.

The following discussion concerns what is known about pest–host–air pollution interactions in the San Bernardino Mountains. Specific information on the influences of pollution and interrelationships shown in Figure 14.1 are for the most part unknown for southern California mountain ecosystems. This chapter addresses the most important, in terms of numbers of trees killed or debilitated, pests and discusses research on specific pests. Most attention has been focused on economically important tree species and their pests.

Causes of Tree Mortality in the San Bernardino Mountains

In the late 1970s and early 1980s, two studies (pest damage inventories) were conducted to determine the most important causes of mortality to commercial timber species on the San Bernardino National Forest (including the San Bernardino, San Jacinto, and Santa Rosa mountains). The two studies measured mortality over 12-month periods beginning in May for the 3 years 1975–1978 (Smith and Roettgering 1982) and for the year beginning in May 1981 (DeNitto and Pierce 1983). Although the San Bernardino National Forest is managed for recreation, wildlife, watersheds, and other ecosystem and aesthetic values, not timber production, the same protocol was followed as on the timber forests of the Sierra Nevada.

Both studies used a stratified random sampling survey design. Three commercial forest types are in the San Bernardino National Forest: ponderosa and Jeffrey pine, mixed conifer, and coulter pine types. Pinyon-juniper and hardwood woodlands were not included in these studies, although there are small components of *Pinus monophylla* and hardwoods in the major forest types. Aerial photographs were used to randomly select a 5% minimum sample of the three types. In May of each year of the study, aerial photos were taken of each plot, and these were used to enumerate trees that had died over the previous 12 months. Each group of dead trees was visited within a few months, and the following data were collected: stand composition; basal area (a measure of stand density, the area occupied by live tree stems at 4.5 ft above the ground); number of dead trees; species, height, and diameter of each tree; year in which each tree died; and pest data for branches, stems, and roots. Because the trees were dead when visited, ozone damage, needle diseases, and defoliating insects could not be observed. The following species are common and serious pests, and their presence, along with mechanical injury, was noted in the surveys:

Heterobasidion annosum	Annosus root disease of conifers
Armillaria mellea	*Armillaria* root disease of pine
Arceuthobium spp.	Dwarf mistletoes, pines
Phoradendron pauciflorum	True mistletoe, fir
Dendroctonus jeffreyi	Jeffrey pine beetle, host-specific
D. brevicomis	Western pine beetle, wide host range but particularly common on coulter pine in southern California
D. ponderosae	Mountain pine beetle, attacks primarily sugar and ponderosa pines in southern California
Scolytus ventralis	Fir engraver, true fir
Melanophila californica	California flatheaded borer, attacks Jeffrey, ponderosa, coulter, and other pines
Tetropium abietis	Roundheaded fir borer, fir

The insects listed are all beetles that feed as larvae (last two species) or as larvae and adults in the phloem of their hosts. Dwarf mistletoes are water and nutrient parasites (they do not photosynthesize), whereas true mistletoes are primarily water parasites and thus are somewhat less damaging to their hosts. The root diseases may rot or girdle the roots and root crowns of trees, which may lead to mortality or windthrow.

Moisture availability is critical in the Mediterranean climate of southern California. Drought stress impairs the primary defense of conifers against successful bark beetle attacks, for example, and lack of moisture can kill trees even in the absence of biotic agents. Drought stress also aggravates the effects of diseases such as root disease and dwarf mistletoes. Thus it might be expected that more mortality would occur during and immediately after periods of low rainfall. As

Table 14.1. Mean Annual Tree Mortality (dbh ≥12) on the San Bernardino National Forest During the Survey Periods of 1975–1978 and 1981–1982

	Number of trees		Total volume (MBF)	
Tree species	1975–1978	1981–1982[a]	1975–1978	1981–1982
White fir	2,408	10,314	544.5	4,253.2
Jeffrey pine	3,112	4,300	2,683.3	3,438.8
Coulter pine	1,390	751	273.7	205.4
Total	6,910	15,907	3,501.5	8,148.4

[a]Totals for 1981–1982 include tree species not listed and thus exceed the sum of the figures given. (From DeNitto and Pierce 1983.)

discussed below, the pest damage inventories did not substantiate a correlation between tree mortality and total annual rainfall.

More than twice as many trees died on an annual basis in the course of the second study (Table 14.1), a dry year preceded by 3 years of above-average precipitation, as died annually between May 1975 and May 1978. Water years 1976 and 1977 were at the end of an 8-year drought (interrupted by 1 year of slightly above-average precipitation in 1973), and 1978 was a year of high precipitation. Numbers of dead trees per acre were as follows: 1975–1976, 0.12; 1976–1977, 0.05; 1977–1978, 0.05; and 1981–1982, 0.13. Although these levels of mortality appear to bear little relationship to precipitation totals for these years (Fig. 14.2), several factors may compound the interpretation of the data. First, the gross mortality figures lump several tree species, and these vary in their adaptation to drought. Second, the timing of annual precipitation may be as important as the absolute quantity; for example, several small rainfall events in late spring may be better absorbed and used than heavy storms in autumn or winter. Summer rain, which occurs sporadically in the San Bernardino Mountains, may be of critical importance in dry years. Third, the population ecology of biotic agents such as bark beetles may delay or prolong the exact time of tree mortality where drought is the fundamental cause. In conclusion, it may not always be possible to show a cause-and-effect relationship between soil moisture and tree mortality on a landscape scale. Finding a strong correlation would require more years of mortality data than are available.

Pest guilds involved in the mortality of the three major tree species are listed in Tables 14.2–14.4. Note that amount of mortality associated with certain causal agents such as the insects (*Scolytus ventralis* and *Tetropium abietis*) plus true mistletoe (*Phoradendron pauciflorum*) in white fir (Table 14.2) varied greatly between the two studies. This is not unexpected—populations of pest organisms fluctuate over time, as do numbers of trees killed (rather than just being infected) by them. Tree mortality, not pest population size, was measured in these studies. Also note that most mortality was caused by complexes of pests (e.g., insects attacking root-disease- and/or mistletoe-weakened trees). This is also important with respect to reducing mortality in a forest. For example, even if it were possible

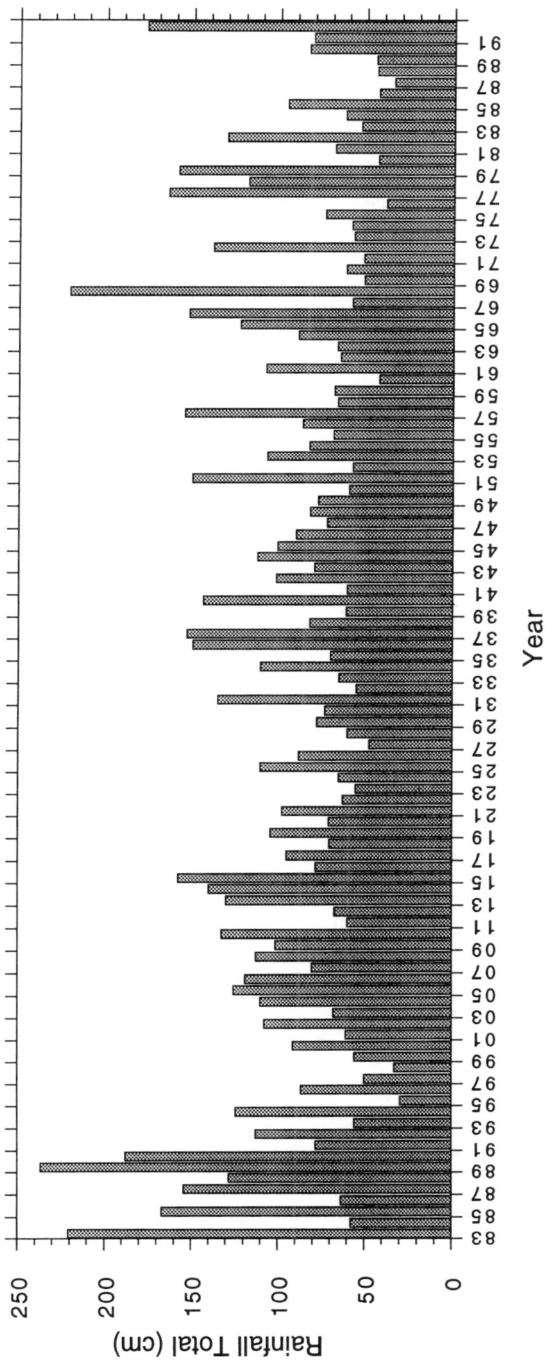

Figure 14.2. Precipitation record from 1884 to 1991 at Big Bear Dam. Totals shown are for water years from October 1 to September 30. For example, 1970 represents precipitation for October 1969 to September 1970. Mean annual precipitation at Big Bear Dam is 92.95 cm. (From Miller 1992.)

Table 14.2. Pest Guilds Involved in White Fir Mortality on the San Bernardino National Forest Expressed as Percent of the Total Number of Dead Trees for the Survey Period[a]

Pest guilds	1975–1978	1981–1982
Insects + root disease	58 ± 15%	36%
Insects + true mistletoe	4 ± 2%	27%
Insects + true mistletoe + root disease	16 ± 9%	—[b]
Insects alone	21 ± 11%	34%
Diseases alone	1 ± 1%	—[b]
Other	—[b]	3%

[a]Data from Smith and Roettgering 1982 and DeNitto and Pierce 1983.
[b]Category of data not reported in that study.

Table 14.3. Pest Guilds Involved in Jeffrey Pine Mortality on the San Bernardino National Forest, Expressed as Percentage of the Total Number of Dead Trees for the Survey Period[a]

Pest guilds	1975–1978	1981–1982
Insects + root disease	22 ± 9%	24%
Insects + dwarf mistletoe	15 ± 8%	8%
Insects + root disease + dwarf mistletoe	15 ± 8%	17%
Insects + other disease	16 ± 11%	—[b]
Insects + injury	8 ± 4%	—[b]
Insects alone	21 ± 7%	—[b]
Diseases alone	2 ± 2%	—[b]
No pest apparent	1 ± 1%	—[b]
Injury	—[b]	4%
Other	—[b]	11%

[a]Data from Smith and Roettgering 1982 and DeNitto and Pierce 1983.
[b]Category of data not reported in that study.

Table 14.4. Pest Guilds Involved in Coulter Pine Mortality on the San Bernardino National Forest, Expressed as Percent of the Total Number of Dead Trees for the Survey Period[a]

Pest guilds	1975–1978	1981–1982
Insects + root disease	10 ± 5%	—[b]
Insects + dwarf mistletoe	21 ± 11%	47%
Insects + root disease + dwarf mistletoe	—[b]	18%
Insects + injury	7 ± 7%	—[b]
Insects alone	61 ± 23%	35%
Diseases alone	1 ± 1%	—[b]

[a]Data from Smith and Roettgering 1982 and DeNitto and Pierce 1983.
[b]Category of data not reported in that study.

to remove Jeffrey pine beetles from the forest, Jeffrey pines would continue to be killed by annosus root disease, dwarf mistletoe infections, drought, ozone, and/or the California flatheaded borer.

Although the details of the patterns of mortality differed in these two studies, both point to stress as the factor that leads to death of a tree. Dwarf mistletoe reduces the amount of water and photosynthates available for tree survival and growth, root diseases may reduce water absorption (e.g., annosus root rot) or water transport (e.g., black stain root disease, an important mortality agent in pinyon woodlands in the San Bernardino Mountains), drought reduces moisture availability, and so forth. Although these debilitating agents or events can be lethal in the absence of other factors, diseased or drought-stressed conifers are usually killed by bark beetles or other phloem-feeding insects.

The data collected in these studies also document the identities of major mortality agents. However, as noted above, because the data were collected after tree death, the role of air pollutants was not assessed. In addition, these studies were not designed to document relationships between pollutants and pest impacts, thus it is not possible to use the data to relate mortality levels to the air pollution gradient in the San Bernardino Mountains.

Interactions Between Oxidant Air Pollutants and Bark Beetles (Coleoptera: Scolytidae)

The predisposition of ponderosa pine by photochemical oxidants to bark beetle attack and the eventual death of the tree is clearer with bark beetles than that with the pathogens discussed below. Numerous workers have observed that trees predisposed by a number of factors such as drought, overstocking, physiological stress, flooding, injury, and disease, particularly root disease, are susceptible to attack by bark beetles (Miller and Keen 1960; Stark et al. 1968; Furniss and Carolin 1992). It was in the San Bernardino Mountains that another bark beetle predisposing factor was discovered.

In the early 1960s, workers in the San Bernardino Mountains reported a decline in the health of ponderosa pines characterized by needle chlorosis (Miller et al. 1963). A study was done in which needles of ponderosa pine were exposed to ozone for 9–18 days under field conditions; chlorotic mottle, terminal dieback, and abscission similar to the needle symptoms of the chlorotic decline were observed. This evidence suggested that the ozone component of photochemical smog was one of the principal causes of the chlorotic decline observed in the San Bernardino Mountains.

Beginning in 1966, scientists from the University of California at Berkeley and the USDA Forest Service began studies to determine the relationships between oxidant injury and bark beetle infestation in ponderosa pines in the San Bernardino Mountains. The two important bark beetles attacking ponderosa pines are the mountain pine beetle, *Dendroctonus ponderosae,* and the western pine beetle, *D. brevicomis* (Fig. 14.3). Of about 1,000 trees examined, 55% had symptoms of

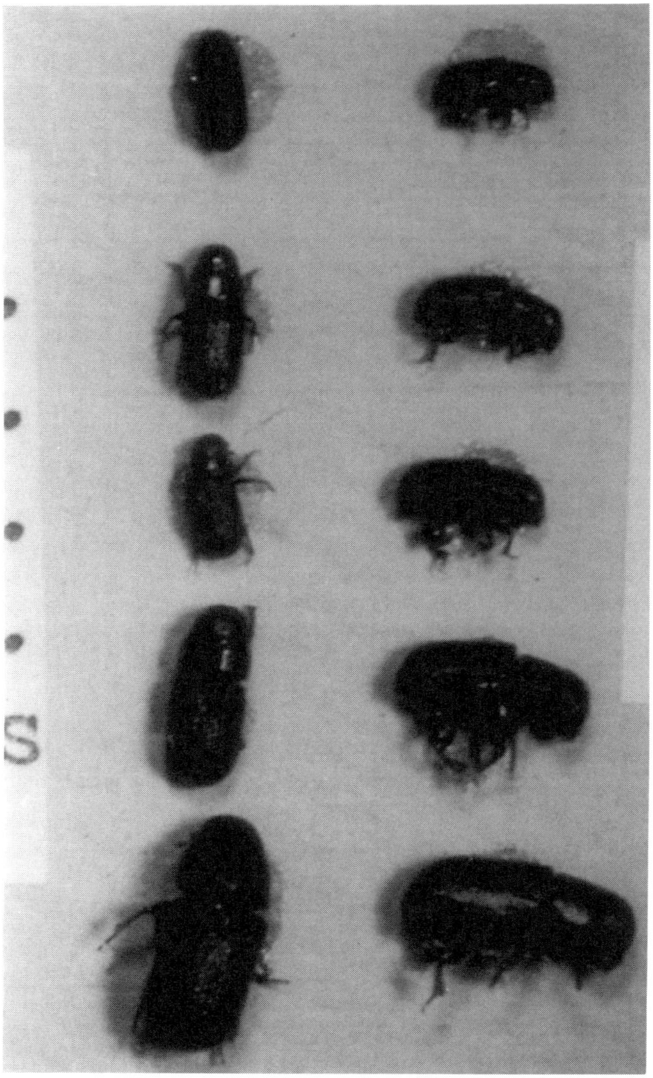

Figure 14.3. Four important *Dendroctonus* species that occur in the San Bernardino Mountains. From the top: western pine beetle (*D. bervicomis*), two specimens of the mountain pine beetle (*D. ponderosae*), Jeffrey pine beetle (*D. jeffreyi*), and the red turpentine beetle (*D. valens*) on the bottom.

oxidant injury, and 41% of those most severely affected had bark beetle attacks, whereas only 3.5% of the apparently healthy trees were attacked (Stark and Cobb 1969). These studies showed that oxidant injury results in reduced oleoresin yield, rate of flow, and exudation pressure; sapwood and phloem moisture content; and phloem thickness, all of which are important in the defense of the tree against beetle attack.

Bark Beetle Life Histories

The western pine beetle is thought to be more aggressive than the mountain pine beetle, and this was supported by Stark et al. (1968) in their study in the San Bernardino Mountains. The western pine beetle killed more trees in the healthy and intermediate categories of chlorotic decline, whereas *D. ponderosae* killed more trees with advanced decline. Fourteen of the 107 trees in all categories were victims of a combined attack of the two species (Stark et al. 1968).

The western pine beetle can have up to four generations per year in southern California. Flight and attacks by the beetle begin in spring or early summer and continue until the weather turns cold. Initial attacks are at midbole and then proceed to the upper bole. The egg galleries are a maze and cross one another repeatedly, giving the appearance of spaghetti. The adults are dark brown and from 3.0 to 5.0 mm in length. The attacks produce small pitch tubes on the surface of the bark, and red boring dust, which is an indication of successful attack, can also be seen. The eggs are laid singly in small frass-filled niches on the sides of the gallery. The larvae feed for a short distance (5–10 mm) and then turn into the outer bark, where they continue to feed and complete their development. Adult emergence holes look like small "bb" holes scattered over the bark.

The mountain pine beetle, *D. ponderosae,* differs from the western pine beetle in several ways. Although the western pine beetle attacks only ponderosa and coulter pine, the mountain pine beetle attacks several pine species in addition to ponderosa pine (e.g., *P. lambertiana, P. contorta, P. monticola*). The adults are larger, black, and rather stout and from 4 to 8 mm in length. The egg galleries are long (30–90 cm) and perpendicular and are through the inner bark. Both the bark and wood are engraved. At the bottom of the gallery, there is usually a 25–50-mm crook or bend. As with the western pine beetle, eggs are laid singly in small frass-covered niches, but they are placed in groups on alternate sides of the gallery. The larvae make galleries that vary in length at right angles to the egg gallery. The larvae of the mountain pine beetle remain in the inner bark, and they construct small pupal chambers at the end of the larval mines. Then mines and the pupal cells are exposed when the bark is removed.

Effects of Oxidants on Western Pine Beetle Populations

In the 1973–1976 period, we undertook a study of how oxidant air pollution, by damaging the ponderosa pine hosts of the western pine beetle, was affecting the populations of this beetle in the San Bernardino Mountains (Dahlsten and Rowney 1980).

Table 14.5. Mean and Standard Error of Productivity of Western Pine Beetle in Ponderosa Pines Sampled in the San Bernardino National Forest, 1973–1976

	Generation 1		Generation 2	
	Oxidant-damaged	Normal	Oxidant-damaged	Normal
Productivity: emerged brood per attacking female	5.03 (±1.36)	2.63 (±0.30)	2.53 (±1.06)	2.02 (±0.62)

The attack rates, brood survival, and emergence rates of the western pine beetle and incidence of entomophagus associates were compared between photochemical oxidant–damaged and apparently healthy ponderosa pines. Seventy-one attacked trees were sampled over the 4-year period. We found that effects of oxidant damage on fecundity varied with beetle generations. In the first generation, productivity of the attacking beetles was significantly higher in the oxidant damaged trees (i.e., fewer attacking parent adults were required to produce a similar number of emerging adults) (Table 14.5) (Dahlsten et al. 1997). In the second generation, both attacks and emergence were much lower, and effects were small, with similar numbers of attacking beetles producing similar broods.

The results suggest that oxidant-damaged trees attacked by western pine beetle produce about the same total brood with fewer initial attacks compared with healthier trees. This higher productivity trend is most evident in trees attacked by the first generation. Trees attacked by the second generation, both damaged and healthy, had significantly fewer western pine beetle broods than generation-one attacked trees with oxidant damage. Another significant observation was that predators and parasitoids had higher densities in the healthier trees. The implication of these results is that in stands with a higher proportion of oxidant-damaged trees, a given population of western pine beetle could kill more trees and increase at a greater rate than in a stand with a lower proportion of damaged trees.

Other Insect–Tree–Pollution Interactions

Ozone fumigation of plants has been shown to change carbohydrate metabolism, causing an increase in soluble sugars in aboveground parts of ponderosa pine seedlings and a decrease in starch and sugars in the roots. High concentrations of ozone cause an increase in free amino acid and total protein content of bean leaves, and ozone fumigation is also associated with higher concentrations of certain vitamins and with changes in secondary plant compounds. The biochemical changes that occur in pollutant-exposed plants are complex and have been reviewed by Riemer and Whittaker (1989).

Experiments from other systems have been contradictory with respect to feeding, growth, and survival of chewing and sapsucking insects on ozone-fumigated foliage (Whittaker and Warrington 1990). Several such species in the San Bernardino Mountains damage but usually do not kill forest trees, but for the most part, these pests have not been studied in relation to the air pollution gradient. An exception is a needleminer on Jeffrey pine, *Coleotechnites* sp. (Lepidoptera: Gelechiidae). Luck (1976) found no relationship between the distribution of this insect and oxidant air pollution. During the period in late summer when oxidant air pollution peaks, this *Coleotechnites* sp. is completing larval development (through mid-August)—an active feeding stage that should be influenced by foliar quality. Pupation, the adult stage, and egg laying occur from mid-July through mid-October, and there are no larvae feeding from mid-August to mid-September, when eggs begin to hatch. However, this needleminer is not escaping oxidant air pollution to its host, as larvae tend to mine older needles, which have the most accumulated pollutant damage. Whether needleminer populations are restricted by premature senescence of ozone-damaged needles has not been investigated.

A different situation exists with respect to the fruittree leafroller, *Archips argyrospila* (Lepidoptera: Tortricidae), on California black oak, *Quercus kelloggii*. The larvae of this insect feed on foliage in early summer before oxidant air pollution damage becomes apparent. When populations of leafrollers are high, as occurs every several years, trees are heavily or entirely defoliated. The trees produce a second flush of growth to compensate for the lost photosynthetic ability, and this new foliage is then damaged by oxidant air pollution. Thus the insects likely increase the pollution-associated growth loss of the trees, although definitive data are difficult to extract in natural systems (Miller et al. 1979). Miller et al. (1979) noted higher levels of defoliation in the western end of the San Bernardino Mountains, where oxidant air pollution is also most concentrated. These two factors may be unrelated, however, as other site factors may be responsible for the distribution of the fruittree leafroller.

Interactions Between Ozone and Forest Tree Pathogens

Ozone may influence any or all components of forest ecosystems, including the trees and the pathogens that affect trees. Although the exact mechanism of ozone injury may not be known at the cellular or biochemical level (Mudd 1996), the effects of oxidant air pollution on tree/stand growth and health in the San Bernardino Mountains have been well documented (Miller et al. 1989; Miller 1992; USEPA 1996). Much less information, however, is available about how forest pathogens interact with ozone-stressed trees. Current protocols for evaluating ozone effects on western pine forests recommend avoiding sites with obvious insect and disease problems that would "confound the evaluation of air pollution injury" (Miller et al. 1996). Physiological research on a variety of plant hosts has concluded that ozone stress causes reactions in plants similar to those induced by microbial pathogens (Heagle 1973; Sandermann 1996). This pollutant-induced stress changes the plant's predisposition to either increased or decreased susceptibility to a second pest agent. There are many examples of environmental stresses that lead to a decline in the health of forest trees (Manion and Lachance 1992). The question we ask here is, does the air pollution stress in the San Bernardino Mountains make forest trees more susceptible to common indigenous tree diseases?

Root diseases and dwarf mistletoes account for most of the pathogen-caused growth loss and tree mortality in the western United States, including the San Bernardino Mountains (Childs and Shea 1967; Smith and Roettgering 1982; USDA 1994). The following section looks at these categories of tree pests, discusses how they operate in the forest, and sees whether they might affect ozone-weakened trees differently than healthy ones. All these pests include ponderosa (*Pinus ponderosa*) and Jeffrey (*P. jeffreyi*) pines as hosts.

Root Diseases

The three most common root diseases in the San Bernardino Mountains are annosus root disease, caused by *Heterobasidion annosum; Armillaria* root disease, caused by *Armillaria mellea;* and black stain root disease, caused by *Leptographium wageneri.* The fungi that cause annosus and *Armillaria* root diseases are Basidiomycetes, which can kill living trees quickly if they become established in the cambium and can also decay wood. The fungus that incites black stain root disease is an Ascomycete with an imperfect stage (*Leptographium*). Black stain's sexual stage, which is rare in nature, is in the genus *Ophiostoma* and was formerly in the genus *Ceratocystis*. Fungi in these genera typically cause vascular wilts and kill their host by plugging water-conducting tissues; they are not capable of decaying wood. Closely related to *Leptographium wageneri* are the pathogens that incite Dutch elm disease and oak wilt.

Annosus Root Disease

Annosus root disease begins when microscopic airborne spores land on recently cut stump surfaces or occasionally on other bole wounds with exposed inner wood. The spores germinate, penetrate into the wood, and grow down into the root system. After colonizing part or all of the root system, the fungus can move from infected stump roots to adjacent healthy trees if there is root contact. Once *H. annosum* is established underground in tree roots, it can continue to spread as long as there are contacts between susceptible host roots large enough to sustain the fungus. After a suitable colonization period, usually several years, the fungus produces fruiting bodies, or conks, inside the stump (Fig. 14.4). These structures contain basidiospores that can be carried in the air and establish new infections.

Successful attack or infection by many forest pests is often determined by the vigor of its host. Pine bark beetles, discussed earlier, are excellent examples of organisms that are most damaging when trees are under stress. In California, *H. annosum* is not generally known to require a weakened host for infection. Research conducted primarily in the southeastern United States has identified many characteristics of sites prone to serious root disease, including high soil pH, low

Figure 14.4. Fruiting bodies (arrows), or conks, of *Heterobasidion annosum* inside a decaying conifer stump.

soil organic matter, and high sand content (Froelich et al. 1966); low field capacity (Alexander et al. 1975); low soil microbial activity (Veech and Boyce 1964); soil moisture stress (Towers and Stambaugh 1968); and crown suppression (Shain 1967). Other than field observations, little research has been done in California to identify similar relationships.

One notable exception to the lack of research in the western United States involving root diseases and abiotic stress agents is the work done in the San Bernardino Mountains by James and co-workers (James et al. 1980). They looked at the susceptibility to infection and rate of vertical colonization of stumps cut from ponderosa and Jeffrey pines in the Barton Flat area. Trees selected for cutting had crown decline symptoms rated as either (1) moderate to severe or (2) slight to no injury from ozone. After artificial inoculation with conidia, *H. annosum* colonized a larger proportion of stump surface area and grew faster down into the stumps of trees showing severe injury compared with those showing little or no injury. The conidia used in these trials were asexual spores produced by the fungus, whereas sexual spores (basidiospores) from conks are believed to account for most natural infection (Otrosina and Cobb 1989). Attempts to inoculate roots cut from pines exhibiting different amounts of foliar ozone injury showed no link between infection and the amount of injury expressed by the tree's crown (James et al. 1980). In the greenhouse, a higher percentage of fumigated seedlings was infected using *H. annosum*-colonized wood segments than nonfumigated. In the field, there was no correlation between natural basidiospore deposition rates and ambient ozone level (James 1977), but fumigation of fungal cultures resulted in slower growth and reduced conidia production (James et al. 1982).

A general conclusion of the research on annosus root disease and oxidant air pollution in the San Bernardino Mountains was that ozone may cause changes in pine hosts that result in increased amounts of root disease. Information from field observations and surveys do not support this conclusion. Heart Bar campground, on the far eastern end of the ozone gradient in the San Bernardino Mountains, has many pines with very little or no chlorotic mottle, which are infected with annosus root disease. A survey of the campground completed in 1979 revealed 36 confirmed or suspected root disease centers in an area of less than 30 acres (Freeman 1979, 1980). Other areas in California that have serious annosus root disease problems include Yosemite Valley in Yosemite National Park and most of the pine forests east of the Sierra Nevada and Cascade Mountain crest (called eastside pine) in central and northern California. Like the San Bernardino Mountains, these areas consist primarily of ponderosa and/or Jeffrey pines on well-drained soils, where annual precipitation occurs almost entirely during the winter months.

The amount of ozone injury to pines in Yosemite National Park monitoring plots has been described as "slight to moderate" and is less than that reported in the vicinity of Heart Bar (Takemoto 1996). Annosus root disease, which can cause large living trees to fall, has been studied in Yosemite Valley for more than 25 years. More than 120 disease centers have been documented in 17 developed sites (Parmeter et al. 1978). In eastside pine forests, surveys of stands managed for timber production commonly showed that 50% or more of the existing stumps, as indicated by the presence of conks, are colonized by *H. annosum* (Kliejunas 1989). Ambient ozone levels are generally quite low in these remote eastside pine forests, and as a result, most sites do not have trees with foliar injury, although they have not been formally surveyed.

Unfortunately, for specific forested sites in California, data may be available for either ozone injury (based on visible foliar symptoms) or root disease incidence, but not both. Field surveys to assess root disease frequency and occurrence, as have been done in many central and northern California forest stands, are lacking in the San Bernardino Mountains. Whether ambient ozone levels in the San Bernardino Mountains can cause an increase in the incidence of annosus root disease remains unclear.

The primary factor that determines the severity of annosus root disease in California pine stands has long been believed to be the number of stumps that have been created in a given area (Bega and Smith 1966). The treatment of freshly cut stumps with chemicals containing borate to control annosus root disease in California began in the early 1970s. Compounds such as borax (sodium tetraborate decahydrate), now called "Sporax," are toxic to *H. annosum* spores and can be 90% effective in preventing the fungus from invading new stumps (Smith 1970). In the late 1970s, borate treatment of stumps became standard procedure in many forested areas of California. It is currently USDA Forest Service policy to treat all conifer stumps cut in developed recreation sites on California national forests.

For decades preceding 1970, many trees were cut in locations such as Yosemite Valley (Yosemite National Park, CA) and the mountains of southern California. The reasons for tree removal included clearing to establish recreation and residence sites, salvage of dead trees, and direct control of bark beetle–infested pines. This was all done without knowing about the presence of *H. annosum,* how it spread, and how it could be controlled. Much of the root disease in the San Bernardino Mountains was established before the research on oxidant air pollution began.

Armillaria *Root Disease*

The genus *Armillaria* is worldwide in distribution and has been reported as a pathogen on hundreds of hosts, mainly woody species. Historically, there has been much confusion about the taxonomy of this fungus, which until the late 1970s was thought to be a single but extremely variable species (Tainter and Baker 1996). At least seven species are now recognized in North America. There has also been considerable disagreement in the literature about the pathogenic role that *Armillaria* plays in the forest ecosystem, much of which can now be explained by differences between species. Unfortunately, publications prior to the 1980s refer almost exclusively to *A. mellea* as the pathogen, whereas in the 1980s the generic term *Armillaria* spp. was used when the species was unknown. It has only been during the past decade that researchers have begun to sort out and identify which species are active in specific geographical locations.

Published reports repeatedly characterize *Armillaria* spp. as a secondary pathogen that usually requires its host to be under some type of stress for successful infection. Stresses that have been described as predisposing hosts to attack include insect defoliation (Houston 1981; Allen et al. 1992), drought (Houston 1987), suppression (Davidson and Rishbeth 1988), other diseases (Kulhavy et al. 1984), and air pollution (Wargo and Harrington 1991). The air pollutants reported included sulfur dioxide, hydrogen fluoride, or simply polluted air from industrial sites but did not include ozone.

Armillaria mellea is the most frequently found species in California (Jacobs et al. 1994; Harris 1995). Here, the fungus may be present in most forest types, has a preference for oaks (primarily California black oak, *Quercus kelloggii*), but is not aggressively pathogenic. Healthy oaks are normally resistant to *Armillaria.* Weakened oaks may be infected but tend to decline over a long period of time. The presence of the fungus may go unnoticed unless sudden tree mortality occurs or fruiting bodies are produced. The fruiting bodies are mushrooms that appear at the base of infected trees in the fall (Williams et al. 1986). Much of the tree injury in California forests from *A. mellea* occurs when California black oaks are cut and the site is then replanted to conifers. The fungus builds up in the oak stump and root system and can spread through the soil and infect adjacent conifers. Young conifers are quite susceptible to this pathogen and do not have to be under stress to be infected.

There has been no field research involving this root disease and ozone. One study in the San Bernardino Mountains looked at the effects of oxidant air pollution and other pests on the health of California black oak (Miller et al. 1979). Insect defoliation, canker dieback, true mistletoe infection, stem decay, and snow damage were documented during a 4-year period, but no root disease was mentioned. If there was a relationship between this root disease and ozone, it would most likely be manifested by the decline in health of California black oaks and/or mortality of conifers near declining oaks. Such information is lacking, and there is no evidence suggesting that *A. mellea* is more active, or damaging, in the San Bernardino Mountains than elsewhere in California.

Black Stain Root Disease

The pathogen *Leptographium wageneri* is the most recently recognized and least studied of the three major root disease organisms in the San Bernardino Mountains. Two features of the disease are that it can spread short distances directly through the soil and can also be carried long distances by root-inhabiting insects. Once established within a host, the fungus colonizes the water-conducting vascular tissues and interferes with water movement. Trees die from black stain because of inadequate moisture and not from the destruction of living tissues as with the previous two root diseases. Most reports of black stain conclude that its occurrence is strongly linked to site disturbances such as road building, tractor logging, and thinning (Cobb 1988). A reasonable scenario is that stressed trees are more attractive to certain insects that introduce the pathogen when they feed on the roots of these weakened trees. This strongly suggests that ozone stressed pines could also be more susceptible to black stain.

Leptographium wageneri has three variants that vary in their ability to infect different conifer hosts (Smith 1993). The most common black stain variant in the San Bernardino Mountains infects single leaf pinyon pine (*P. monophylla*), where several thousand acres of land have been identified as affected (Tainter and Baker 1996). This tree species does not grow in the pine and mixed conifer stands that characterize the high-use recreation sites of southern California. Rather, it is found in the arid forests on the eastern fringes of the San Bernardino Mountains that blend into the high deserts of southern California. Typically, ambient ozone levels and foliar injury are at a minimum on sites where pinyon pine grows. A second variant of black stain infects ponderosa and Jeffrey pines, which are the pine species that grow in the most heavily ozone-polluted portions of the San Bernardino Mountains. There has been only one report of a Jeffrey pine infected by black stain in this area.

A field study by Lackner and Alexander (1983) completed in the southeastern United States looked at the relationship between a root disease (caused by *Leptographium procera*) similar to black stain and its occurrence on ozone-sensitive eastern white pine (*P. strobus*). It found pines with severe ozone injury had a much higher root disease infection frequency than ozone-tolerant pines. To test this relationship under California conditions, Fenn and co-workers (1990) inoculated ozone-fumigated pine seedlings with the ponderosa pine variant of black stain. The fumigation profiles that they used simulated diurnal cycles representative of conditions that occur in northern and southern California forests. More fumigated seedlings were infected by black stain than nonfumigated, and the linear extent of characteristic root staining was greater in ozone-stressed trees. This is convincing evidence that ozone may predispose ponderosa pine to black stain root disease, but as the authors point out, the pathogen variant tested is rare in the heavily polluted pine and mixed conifer forests of the San Bernardino Mountains.

Mistletoes

Mistletoes are widespread, native, seed-producing plants that parasitize conifers and hardwoods throughout the western United States. Within this group are the true (leafy) mistletoes in the genus *Phoradendron,* which affect many hardwoods and a few conifers, and the dwarf mistletoes in the genus *Arceuthobium,* which attack only conifers. Unlike the microscopic fungal pathogens described so far, mistletoes are fairly large, conspicuous higher plants that are easy to recognize (Fig. 14.5). True mistletoes take water from their host and produce their own photosynthates, whereas the dwarf mistletoes take both water and nutrients from their host. Because of this, dwarf mistletoes cause much more damage to their host and have received far more attention from managers of forest ecosystems and from researchers. There is a tremendous amount of knowledge about the biology of dwarf mistletoes and considerable experience in controlling them. Once a dwarf mistletoe occupies one-half or more of its host's living crown, tree growth and vigor are reduced (Hawksworth et al. 1992). Parasitism by this pest also disrupts the normal growth patterns of conifers and causes the formation of "witches' brooms"—dense clusters of branches and foliage that give the tree an asymmetrical appearance. Over many years, the mistletoe can continue to spread within the crown and weaken the tree to the point where it is an easy target for bark beetles. Pest damage inventories, described earlier in this chapter, have concluded that during periods of drought, conifers heavily infected by dwarf mistletoes are often the first to die.

In the San Bernardino Mountains, oak (true) mistletoe, *Phoradendron villosum,* spp. *villosum,* is commonly found infecting California black oak. Western dwarf mistletoe, *Arceuthobium campylopodum,* is a frequent parasite on ponderosa, Jeffrey, and coulter pines. Both of these pests have been reported as present in sites used for ozone research in southern California (Miller 1973; Miller and White 1977). The relationship between ozone-sensitive trees and mistletoe, however, has not been studied; neither has the effect of ozone on mistletoe plants. What has been assumed, and probably correctly, is that the drain on tree vigor from dwarf mistletoe infection, plus the weakening effect of ozone, is additive at least, and results in more rapid or more frequent host mortality.

Figure 14.5. Female (A) and male (B) plants of western dwarf mistletoe, *Arceuthobium campylopodium,* on ponderosa pine branches.

The question presented at the beginning of this section asked whether ozone-sensitive trees are more susceptible to indigenous biotic disease agents. There is some good convincing evidence that this may be true, especially for annosus root disease and black stain. What has not been shown is that tree diseases are actually more common and more severe on sites with high ambient ozone levels. Mortality surveys (e.g., pest damage inventories) routinely show that trees are killed by a combination of environmental factors, insects, and diseases (pest complexes). Unfortunately, such surveys targeted dead trees for examination, and these trees do not have foliage that can be assessed for chlorotic mottle.

Conclusion

Trees, which may live for several hundred years, are exposed to numerous and repeated stresses. Most environmental and biotic stresses, such as drought and insect defoliation, are recurrent but on irregular cycles and are difficult to predict. Air pollution, however, has been intruding into the San Bernardino Mountains year after year for many decades. Sensitive plants unable to live with this stress have already died. Other less susceptible plants may have survived but are unthrifty and easy targets for the periodic, but unpredictable, environmental and biotic stresses. The documented impact of ozone, by itself, is enough to justify a continuing concern over the health of forest ecosystems in the San Bernardino Mountains. It also underscores the importance of maintaining air pollution standards that will help minimize future injury.

References

Alexander, S.A.; Skelly, J.M.; Morris, C.L. Edaphic factors associated with the incidence and severity of disease caused by *Fomes annosus* in loblolly pine plantations in Virginia. Phytopathology 65:585–591; 1975.

Allen, D.C.; Bauce, E.; Barnett, C.J. Sugar maple declines—causes, effects, and recommendations. In: Manion, P.D.; Lachance, D., eds. Forest decline concepts. St. Paul, MN: APS Press; 1992:123–136.

Alstad, D.N.; Edmunds, G.F.; Weinstein, L.H. Effects of air pollutants on insect populations. Annu. Rev. Entomol. 27:369–384; 1982.

Bega, R.V.; Smith, R.S., Jr. Distribution of *Fomes annosus* in natural forests of California. Plant Dis. Rep. 50:832–836; 1966.

Childs, T.W.; Shea, K.R. Annual losses from disease in Pacific northwest forests. Res. bull. PNW-20. Portland, OR: USDA Forest Service; 1967.

Cobb, F.W., Jr. *Leptographium wageneri,* cause of black-stain root disease: a review of its discovery, occurrence and biology with emphasis on pinyon and ponderosa pine. In: Harrington, T.C.; Cobb, F.W., Jr., eds. *Leptographium* root diseases on conifers. St. Paul, MN: American Phytopathological Society Press; 1988:41–62.

Dahlsten, D.L.; Rowney, D.L. Influence of air pollution on population dynamics of forest insects and on tree mortality. In: Miller, P.R., tech. coord. Proceedings of symposium on effects of air pollutants on Mediterranean and temperate forest ecosystems. Gen. tech. rep. PSW-43. Berkeley, CA: USDA Forest Service, Pacific Southwest Station; 1980:125–130.

Dahlsten, D.L.; Rowney, D.L.; Kickert, R.N. Effects of oxidant air pollutants on western pine beetle (Coleoptera: Scolytidae) populations in southern California. Environ. Pollut. 96:415–423; 1997.

Davidson, A.J.; Rishbeth, J. Effect of suppression and felling on infection of oak and Scots pine by *Armillaria.* Eur. J. For. Pathol. 18:161–168; 1988.

DeNitto, G.A.; Pierce, J. An evaluation of conifer mortality on the San Bernardino National Forest between May 1981 and May 1982. Forest pest management rep. 83-35. San Francisco, CA: USDA Forest Service, Pacific Southwest Station; 1983.

Fenn, M.E.; Dunn, P.H.; Wilborn, R. Black stain root disease in ozone-stressed ponderosa pine. Plant Dis. 74:426–430; 1990.

Freeman, W.L., Jr. Biological evaluation of Heart Bar campground. San Francisco: USDA Forest Service, Forest Insect and Disease Management Report. Pacific Southwest Region; 1979.

Freeman, W.L., Jr. *Fomes annosus* disease centers in Heart Bar campground. San Francisco: USDA Forest Service, Forest Insect and Disease Management Report. Pacific Southwest Region; 1980.

Froelich, R.C.; Dell, T.R.; Walkinshaw, C.H. Soil factors associated with *Fomes annosus* in the Gulf states. Forest Sci. 12:356–361; 1966.

Furniss, R.L.; Carolin, V.M. Western forest insects. Misc. publ. 1339. Washington, DC: USDA Forest Service; 1992.

Harris, J.L. Identification of *Armillaria* species in California Sierra Nevada forests and their strengths of associations with young conifer plantations. MS thesis, Utah State University, Logan; 1995.

Hawksworth, F.G.; Williams-Cipriani, J.C.; Eav, B.B.; Geils, B.W.; Johnson, R.R.; Marsden, M.A.; Beatty, J.S.; Shubert, G.D. Interim dwarf mistletoe impact modeling sys-

tem: users guide and reference manual. Rep. MAG-91-3. Ft. Collins, CO: USDA Forest Service, Forest Pest Management; 1992.
Heagle, A.S. Interactions between air pollutants and plant parasites. Annu. Rev. Phytopathol. 11 365–488; 1973.
Heliovaara, K.; Vaisanen, R. Insects and pollution. Boca Raton, FL: CRC Press; 1993.
Houston, D.R. Stress triggered tree diseases—the diebacks and declines. NE-INF-41-81. Radnor, PA: USDA Forest Service; 1981.
Houston, D.R. Forest tree declines past and present: current understanding. Can. J. Plant Pathol. 9:349–360; 1987.
Jacobs, K.A.; MacDonald, J.D.; Cobb, F.W., Jr.; Wells, K. Identification of *Armillaria* species in California. Mycologia 86:113–116; 1994.
James, R.L. The effects of photochemical air pollution on the epidemiology of *Fomes annosus*. Unpublished Ph.D. dissertation, University of California, Berkeley; 1977.
James, R.L.; Cobb, F.W.; Wilcox, W.W.; Rowney, D.L. Effects of photochemical oxidant injury of ponderosa and Jeffrey pines on susceptibility of sapwood and freshly-cut stumps to *Fomes annosus*. Phytopathology 70:704–708; 1980.
James, R.L.; Cobb, F.W., Jr.; Parmeter, J.R. Effects of ozone on sporulation, spore germination, and growth of *Fomes annosus*. Phytopathology 72:1205–1208; 1982.
Kliejunas, J.T. Borax stump treatment for control of annosus root disease in the eastside pine type forests of northeastern California. In: Otrosina, W.J.; Scharpf, R.F., eds. Proceedings of the symposium on research and management of annosus root disease (*Heterobasidion annosum*) in western North America. Gen. tech. rep. PSW-116. Berkeley, CA: USDA Forest Service; 1989:159–166.
Kulhavy, D.L.; Partridge, A.D.; Stark, R.W. Root diseases and blister rust associated with bark beetles (Coleoptera: Scolytidae) in western white pine in Idaho. Environ. Entomol. 13:813–817; 1984.
Lackner, A.L.; Alexander, S.A. Root disease and insect infestations on air-pollution-sensitive *Pinus strobus* and studies of pathogenicity of *Verticicladiella procera*. Plant Dis. 67:679–681; 1983.
Luck, R.F. Bionomics and parasites of a needle miner, *Coleotechnites* sp., infesting Jeffrey pine in southern California. Environ. Entomol. 5:937–942; 1976.
Manion, P.D.; Lachance, D. Forest decline concepts. St. Paul, MN: American Phytopathological Society Press; 1992.
Miller, J.M.; Keen, F.P. Biology and control of the western pine beetle. Misc. publ. 800. USDA Forest Service; 1960.
Miller, P.R. Oxidant-induced community change in a mixed conifer forest. In: Air pollution damage to vegetation. Advances in chemistry series 122. Washington, DC: American Chem. Soc. 1973:101–117.
Miller, P.R. Mixed conifer forests of the San Bernardino Mountains, California. In: Olsen, R.K.; Binkley, D.; Bohn, M., eds. The response of western forests to air pollution. New York: Springer-Verlag; 1992:461–497.
Miller, P.R.; Parmeter, J.R., Jr.; Taylor, O.C.; Cardiff, E.A. Ozone injury to foliage of *Pinus ponderosa*. Phytopathology 53:1072–1076; 1963.
Miller, P.R.; Longbotham, G.J.; Van Doren, R.E.; Thomas, M.A. Effect of chronic oxidant air pollution exposure on California black oak in the San Bernardino Mountains. In: Plumb, T.R., tech. coord. Ecology, management, and utilization of California oaks. Proceedings of the symposium. Gen. tech. rep. PSW-44. Berkeley, CA: USDA Forest Service; 1979:220–229.
Miller, P.R.; McBride, J.R.; Schilling, S.L.; Gomez, A.P. Trends of ozone damage to conifer forests between 1974 and 1988 in the San Bernardino Mountains of southern California. In: Olson, R.K., Lefohn, A.S., eds. Effects of air pollution on western forests. Transactions series 16. Pittsburgh, PA: Air and Waste Management Association; 1989:309–323.
Miller, P.R.; White, M. Ecosystems. In: Ozone and other photochemical oxidants. Washington, DC: National Research Council, Committee on Medical and Biological Effects of Environmental Pollutants, National Academy of Sciences; 1977:586–642.

Miller, P.R.; Stolte, K.W.; Duriscoe, D.M.; Pronos, J., tech coords. Evaluating ozone air pollution effects on pines in the western United States. Gen. tech. rep. PSW-GTR-155. Albany, CA: USDA Forest Service; 1996.

Mudd, J.B. Toxicity of ozone. In: Yunus, M.; Iqbal, M., eds. Plant response to air pollution. New York: John Wiley & Sons; 1966:267–283.

Otrosina, W.J.; Cobb, F.W., Jr. Biology, ecology, and epidemiology of *Heterobasidion annosum*. In: Otrosina, W.J.; Scharpf, R.F., eds. Proceedings of the symposium on research and management of annosus root disease (*Heterobasidion annosum*) in western North America. Gen. tech. rep. PSW-116. USDA Forest Service; 1989:26–33.

Parmeter, J.R., Jr.; MacGregor, N.J.; Smith, R.S., Jr. An evaluation of *Fomes annosus* in Yosemite National Park. Forest insect and disease management rep. 78-2. San Francisco: USDA Forest Service, Pacific Southwest Region; 1978.

Reimer, J.; Whittaker, J.B. Air pollution and insect herbivores: observed interactions and possible mechanisms. In: Bernays, E.A., ed. Insect–plant interactions, vol. 1. Boca Raton, FL: CRC Press; 1989:73–105.

Sandermann, H. Ozone and plant health. Annu. Rev. Phytopathol. 34:347–366; 1996.

Shain, L. Resistance of sapwood in stems of loblolly pine to infection by *Fomes annosus*. Phytopathology 57:1034–1045; 1967.

Smith, R.S., Jr. Borax to control *Fomes annosus* infection of white fir stumps. Plant Dis. Rep. 54:872–875; 1970.

Smith, R.S., Jr. Root diseases. In: Scharpf, R.F., tech coord. Diseases of Pacific coast conifers. Agriculture handbook 521. Washington, DC: USDA Forest Service; 1993.

Smith, R.S., Jr; Roettgering, B.R. A biological evaluation of three years of pest-caused tree mortality on the San Bernardino National Forest. Forest pest management report 82-4. San Francisco, CA: USDA Forest Service, Pacific Southwest Region; 1982.

Stark, R.W.; Cobb, F.W., Jr. Smog injury, root diseases and bark beetle damage in ponderosa pine. Calif. Agric. 23(9):13–15; 1969.

Stark, R.W.; Miller, P.R.; Cobb, F.W., Jr.; Wood, D.L.; Parmeter, J.R., Jr. Photochemical oxidant injury and bark beetle (Coleoptera: Scolytidae) infestation of ponderosa pine. I. Incidence of bark beetle infestation in injured trees. Hilgardia 39:121–126; 1968.

Tainter, F.H.; Baker, F.A. Principles of forest pathology. New York: John Wiley & Sons, Inc.; 1996.

Takemoto, B.K. Ozone air quality in California's mixed conifer forests: status of air monitoring and tree injury assessment activities in the Sierra Nevada and San Bernardino Mountains. Sacramento, CA: California Air Resources Board, Research Division; 1996.

Towers, B.; Stambaugh, J.W. The influence of induced moisture stress upon *Fomes annosus* root rot of loblolly pine. Phytopathology 58:269–272; 1968.

USDA. California forest health—past and present. Rep. R5-FPM-PR-001. San Francisco: USDA Forest Service, Pacific Southwest Region; 1994.

U.S. EPA. Air quality criteria for ozone and related photochemical oxidants. vol. II. EPA/600/p-93/004bF. Research Triangle Park, NC: National Center for Environmental Assessment, Office of Research and Development, U.S. Environmental Protection Agency; 1996.

Veech, J.A.; Boyce, J.S. Soil microorganisms in two Georgia slash pine plantations with annosus root rot. US Plant Dis. Rep. 48:873–874; 1964.

Wargo, P.M.; Harrington, T.C. Host stress and susceptibility. In: Shaw, C.G., III; Kile, G.A., eds. *Armillaria* root disease. Agriculture handbook 691. Washington, DC: USDA Forest Service; 1991.

Whittaker, J.B.; Warrington, S. Effects of atmospheric pollutants on interactions between insects and their food plants. In: Burdon, J.J.; Leather, S.R., eds. Pests, pathogens and plant communities. Oxford: Blackwell Scientific; 1990:97–110.

Williams, R.E.; Shaw, C.G.; Wargo, P.M.; Sites, W.H. *Armillaria* root disease. Forest insect & disease leaflet 78. Washington, DC: USDA Forest Service; 1986.

15. Impact of Oxidant Air Pollutants on Forest Succession in the Mixed Conifer Forests of the San Bernardino Mountains

Joe R. McBride and Richard D. Laven

Introduction

Chronic air pollution is an anthropogenic disturbance that may be influencing plant succession in the San Bernardino Mountains. Westman (1979, 1981) suggested that air pollution induced successional changes that were occurring in the coastal sage scrub communities of southern California. Air pollution differentially affects the survival, growth, and reproduction of the species making up a plant community. These impacts influence the competitive relationships among the species. Our ability to predict successional changes and eventual steady-state communities under conditions of chronic oxidant air pollution has been limited by the relatively short period that ecologists have had to observe this phenomenon. We are also limited in our ability to predict how patterns of plant succession will change in the future once chronic air pollution has been eliminated. It is important that we grapple with these questions to understand the potential impacts of the current episode of air pollution on the forests of southern California. An understanding of these impacts can inform forest managers and conservation biologists of the need to develop management strategies for the protection of vulnerable species and vegetation types. The purpose of this chapter is to review the results of a study initiated in the 1970s to evaluate the long-term impact of chronic air pollution on succession in the mixed conifer forests of the San Bernardino Mountains.

Studies of Forest Succession in the San Bernardino Mountains

Previous work concerning forest succession in the San Bernardino Mountains has been reported by Minnich (1978, 1988, and this volume) and Laven (1978, 1982). The work of these authors has focused on postfire succession and has not addressed the impact of oxidant air pollution on succession.

Minnich (1987, 1988, and this volume) reported that following stand-replacement burns the initial stage of secondary succession was dominated by montane chaparral species (*Ceanothus* spp., *Arctostaphylos* spp., and *Quercus* spp.) in the ponderosa pine (*Pinus ponderosa*) zone of the mixed conifer forest and by Great Basin sage brush (*Artemisia tridentata*) and rabbit brush (*Chrysothamnus nauseous*) in the portions of the mixed conifer forest dominated by Jeffrey pine (*Pinus jeffreyi*). The early brushfields that develop following stand-replacing fires often support seedlings of the mixed conifer forest dominants. These seedlings, or others that become established in the chaparral as its canopy begins to break apart with age, eventually succeed to the prefire mixed conifer forest. The pattern of postfire succession following stand-replacement burns is best described by Connell and Slatyer's (1977) tolerance pathway.

Laven (1978, 1982) examined postfire succession following surface fires in the mixed conifer forest. He identified four patterns of postfire regeneration based on topographical position, aspect, and dominant overstory species. Postfire succession following surface burns was characterized by a slight increase in cover of shrub species in the forest understory and the establishment of forest tree seedlings, especially ponderosa and Jeffrey pine. Surface fires were shown to favor the regeneration and maintenance of the pine component of the mixed conifer forest. Similar findings have been reported for the mixed conifer forest in the Sierra Nevada (Sugihara and McBride 1996).

Minnich et al. (1995) reviewed 60 years of change in the conifer forests of the San Bernardino Mountains in the light of the fire suppression activities of the U.S. Forest Service and the California Department of Forestry and Fire Protection. Their work indicates an increase in tree density and a shift toward greater dominance of white fir (*Abies concolor*) and incense cedar (*Calocedrus decurrens*). Similar successional trends have been reported for the mixed conifer forest type in the Sierra Nevada as a result of fire suppression (Vankat and Major 1978; Parson and DeBenedetti 1979; Bonnickson and Stone 1981; Christensen et al. 1987).

The role of chronic air pollution in forest succession in the San Bernardino Mountains was addressed in a series of reports by McBride (1976, 1977, 1978a, 1979), McBride and Miller (1977, 1987), and McBride et al. (1985). These reports were based on an interdisciplinary study of the impact of oxidant air pollutants on the mixed conifer forest conducted in the 1970s in the San Bernardino Mountains. This chapter brings together the findings of those reports and examines early conclusions in the context of the changing patterns of air pollution and forest conditions in the San Bernardino Mountains.

The San Bernardino Mountain Study

In 1972, the U.S. Environmental Protection Agency, in cooperation with the University of California and the U.S. Forest Service, initiated a study of the effects of oxidant air pollutants on the mixed conifer forest ecosystem in the San Bernardino Mountains (Taylor 1973). This interdisciplinary study focused on a series of plots established along air pollution gradients. The establishment of these plots and their levels of air pollution exposure have been described in this book (Watson et al., Miller and Rechel, and Arbaugh et al.) and in reports to the Environmental Protection Agency (Taylor 1974; Kickert 1976; Kickert et al. 1977, 1980).

The plots established to monitor air pollution injury to forest trees were 30 m wide and sufficiently long to contain 50 ponderosa or Jeffery pines more than 10 cm diameter at breast height (dbh). Other major tree species occurring on these plots were white fir, incense cedar, sugar pine (*Pinus lambertiana*), and black oak (*Quercus kelloggii*). Depending on tree density, these plots ranged from 70 to 140 m in length. Tree density decreased along gradients of decreasing air pollution and decreasing precipitation. The potential for plot bias in tree mortality due to density could not be eliminated due to the nature of the air pollution and precipitation gradients. The location of each tree greater than 10 cm dbh was mapped on each plot. The trees were measured at breast height (1.4 m), numbered, and tagged. A center line rectangular subplot 5 m wide was established to record the seedlings, saplings, and poles occurring in the plots. The diameter of each tree seedling (plants $<$0.25 m tall) at 2.5 cm above the ground was measured. Tree sapling (plants $>$0.25 m tall but $<$1.4 m tall dbh) diameters were measured at 10 cm above the ground. The diameter of each pole-sized tree (plants $>$1.4 m tall, but $<$10 cm DBH) was measured at breast height (1.4 m).

Increment cores were taken to determine the age of the poles and trees on the plots, unless the poles were too small for coring. In that case, their age was estimated on the basis of the number of annual lateral branch whorls. Seedlings and saplings occurring off the plots were destructively sampled to establish a correlation between age and diameter. Correlations were also developed between number of annual rings and number of lateral branch whorls for the pine and white fir saplings.

The plots were examined annually in the late summer or early fall for a period of 5 years to record tree mortality and air pollution injury. The air pollution injury of each tree was assessed according to a procedure developed by Miller (1973). Ten years after the establishment of the plots, the center line subplots were remeasured to record the establishment of new tree seedlings and the mortality of any seedlings, saplings, or poles. During these surveys, tree mortality on the larger plots was also recorded.

In 1985, a projection of future successional change was developed based observations of air pollution injury, mortality, and seedling establishment from 1974 to 1983 (McBride et al. 1985). The life table method (Deevey 1947; Rabotnov 1964; Silverton 1982) was used for this projection. This method involves projecting future population size on the basis of birth and death in a population. Rates of births and deaths are established by observations over a given period of time: in this case, the 10-year period from 1974 to 1983.

The projection of population dynamics in the mixed conifer forests of the San Bernardino Mountains was limited to eight plots that were dominated by ponderosa pine and had not been seriously disturbed by tree cutting or other activities at the time of remeasurement in 1983. The eight plots were grouped into two classes on the basis of foliar injury scores. Six of the plots had average injury scores for ponderosa pine, which ranged from 13 to 17 (Sky Forest = 13, Breezy Point = 16, Camp O'Ongo = 16, U.C. Conference Center = 16, Camp Paivika = 16, Dogwood = 17) on a scale of 1 (maximum) to 36 (minimum)(Miller and Ederman 1977; Miller and Rechel, this volume). This group of plots was considered to have severe foliar injury. These plots were located in areas of higher concentrations of oxidant air pollutants in the San Bernardino Mountains. The two remaining plots, of the eight plots used for the succession projection, had average air pollution injury scores of 22 (Tunnel Two) and 23 (Camp Angeles). Trees here were ranked as having slight foliar injury. These two plots were located at the low end of the air pollution gradient for mixed conifer forests dominated by ponderosa pine.

Tree Mortality

Data collected from the plots were used to construct an age distribution table for each plot by tallying the number of individuals (of each species) present by 10-year age classes in 1973 and 1984. An example of these data, using the Dogwood plot, is shown in Table 15.1. These tables were then used to determine the percentage mortality that had occurred in each 10-year age class over the 10-year period (Table 15.2).

Average percentage mortality by age class groups (seedlings = 0–9; saplings and poles = 10–49; young mature = 50–99; mature = 100+ years old) for the plots showing severe injury symptoms was compared with average percentage mortality for the plots showing slight injury (Table 15.3). A significant difference (0.01%; Student's *t*-test) occurred between ponderosa pine in the young mature age class on severe injury plots and slight injury plots. Approximately 33% of the young mature ponderosa pine trees died on the severe injury plots, whereas only 7% died on the slight injury plots. In the same age class group, 24% of the white fir trees died on the severe injury plots, whereas no mortality was observed in this age class group on the slight injury plots. Other species in other age class groups showed significantly more mortality on the severe injury plots.

The high mortality percentages observed in the seedling, saplings, and poles were not significantly different between the severe and slight injury plots for any species except sugar pine. Mortality in these age classes in the San Bernardino Mountains may have been more a function of competition for light and soil moisture than air pollution injury. The crowns of trees in these age classes are not in the upper portions of the forest canopy. As a result, they may be exposed to

Table 15.1. Number of Trees by Age Class on the Dogwood Plot in 1974 and 1983

		Age class (years)											
Species[a]	Year	0–9	10–19	20–29	30–39	40–49	50–59	60–69	70–79	80–89	90–99	100–109	110–119
PP	1974	7	11	24	21	19	19	12	7	4	0	1	0
	1983	6	7	10	12	16	15	18	11	5	3	0	1
WF	1974	2	2	10	2	0	1	0	0	0	0	0	0
	1983	10	2	2	10	2	0	1	0	0	0	0	0
IC	1974	0	15	9	7	11	3	2	0	0	0	0	0
	1983	12	0	15	9	7	9	3	2	0	0	0	0
SP	1974	0	0	2	1	1	0	0	0	0	0	0	0
	1983	2	0	0	2	1	1	0	0	0	0	0	0
BO	1974	14	10	5	11	7	1	1	1	0	0	0	0
	1983	12	14	9	4	11	6	1	1	0	0	0	0

[a] PP, *Pinus ponderosa;* WF, *Abies concolor;* IC, *Calocedrus decurrens;* SP, *Pinus lambertiana;* BO, *Quercus kelloggii.*

Table 15.2. Mortality Percentage from 1974 to 1983 of Trees on the Dogwood Plot in 1974 and 1983

Age class group (years)	Species[a]	Number of trees alive 1974	Number of trees alive 1983	Mortality (%)	Age class (years)	Species	Number of trees alive 1974	Number of trees alive 1983	Mortality (%)
0–9	PP	7	7	0	60–69	PP	12	11	8
	WF	2	2	0		WF	0	0	—
	IC	0	0	—		IC	2	2	0
	SP	0	0	—		SP	0	0	—
	BO	14	14	—		BO	1	1	0
10–19	PP	11	10	10	70–79	PP	7	5	29
	WF	2	2	0		WF	0	0	—
	IC	15	15	0		IC	0	0	—
	SP	0	0	—		SP	0	0	—
	BO	10	9	10		BO	0	0	—
20–29	PP	24	12	50	80–89	PP	4	3	25
	WF	10	10	0		WF	0	0	—
	IC	9	9	0		IC	0	0	—
	SP	2	2	0		SP	0	0	—
	BO	5	4	20		BO	0	0	—
30–39	PP	21	16	24	90–99	PP	0	0	—
	WF	2	2	0		WF	0	0	—
	IC	7	7	0		IC	0	0	—
	SP	1	1	0		SP	0	0	—
	BO	11	11	0		BO	0	0	—
40–49	PP	19	15	21	100–109	PP	1	1	0
	WF	0	0	—		WF	0	0	—
	IC	11	9	18		IC	0	0	—
	SP	1	1	0		SP	0	0	—
	BO	7	6	14		BO	0	0	—
50–59	PP	19	18	5					
	WF	1	1	0					
	IC	3	3	0					
	SP	0	0	—					
	BO	1	1	0					

[a]PP, *Pinus ponderosa;* WF, *Abies concolor;* IC, *Calocedrus decurrens;* SP, *Pinus lambertiana;* BO, *Quercus kelloggii.*

Table 15.3. Comparison of Percent Mortality by Age Class Group in Areas of Severe and Slight Foliar Ozone Injury, 1974–1984

		Foliar zone injury		
Age class group	Species[a]	Severe	Slight	Difference[b]
Seedling (0–9 yrs)	PP	54.5	41.2	
	WF	57.5	44.1	
	IC	41.1	—	
	SP	51.1	0	*
	BO	63.5	68.5	
Saplings and poles (10–49 yrs)	PP	67.8	85.7	
	WF	42.4	45.8	
	IC	19.4	—	
	SP	30.9	100	*
	BO	56.5	69.7	
Young mature (50–99 yrs)	PP	33.2	6.9	*
	WF	24.2	0	*
	IC	4.2	—	
	SP	35.6	20	
	BO	3.3	0	*
Mature (100 yrs)	PP	14.4	22.5	
	WF	0	50	*
	IC	—	—	
	SP	—	0	
	BO	0	50	*

[a]PP, *Pinus ponderosa;* WF, *Abies concolor;* IC, *Calocedrus decurrens;* Sp, *Pinus lambertiana;* BO, *Quercus kelloggii.*
[b]Difference significant at 0.01%.

lower concentrations of ozone. Trees in the young mature age class group are growing into the upper portions of the forest canopy and are exposed to larger doses of ozone. The larger percentage of mortality of young mature ponderosa pine, white fir, and black oak compared with mature trees of the same species on the severe injury plots may be the result of their exposure to more ozone as they grow into the forest canopy. The exposure of an individual needle to ozone per unit of time is increased with wind velocity. As wind velocity increases with the distance above the ground, the foliage of taller trees is experiencing a greater amount of ozone per unit of time. The more ozone-sensitive genotypes may have succumbed as young mature trees, whereas the more ozone-resistant genotypes survive to join the mature age class. The lower mortality rates exhibited by trees in the mature age class may be a reflection of greater genetic resistance to ozone injury. However, other factors must be effecting mortality of mature trees because greater mortality in this class was observed on the slight injury plots. One factor might have been the age of trees in the mature tree class on the slight injury plots. Trees on these plots were generally older, by as much as 100 years, and therefore closer to senescence than the trees on the severe injury plots.

Seedling Establishment

Initial seedling populations, represented by trees in the 0–9-year age class at the beginning of the study, were compared with seedling populations 10 years later to determine the change over time on each group of plots and the difference in establishment between the groups (Table 15.4). Ponderosa pine seedling establishment was not significantly different between the two time periods on each group of plots, but the number of seedlings established on the severe injury plots was significantly lower than the number established on the slight injury plots in both time periods. White fir seedling establishment showed a significant decrease on both plot groups from 1965–1974 to 1975–1984. A statistically greater number of white fir seedlings was established on the severe injury plots than on the slight injury plots in both periods. Black oak seedling populations significantly decreased in numbers only on the slight injury plots between 1965–1974 and 1975–1984. Like white fir, the number of black oak seedlings established on the severe injury plots was greater in both time periods than on the slight injury plots. Incense cedar seedlings were only observed on the severe injury plots. Their numbers significantly increased from 1965–1974 to 1975–1984. Sugar pine seedling populations were very small and showed no significant difference between the two time periods or the two foliar injury groups.

The variability in seedling establishment between the two periods might be attributed to differences in the abundance of seed production as conifer trees are known to show variability in annual cone crops. White fir is considered to produce heavy cone crops at 3–9-year intervals, black oak every 5–6 years, ponderosa pine every 8 years, incense cedar every 3–6 years, and sugar pine every 4 years (Fowells and Schubert 1965). Black oak produces heavy mast crops every 5–6 years (McDonald 1978). These frequencies of large seed crops suggest that the decrease in seedling establishment of white fir and black oak during the second 10-year period was not likely due to a normal infrequency of large seed crops. Drought conditions in the second time period (1975–1984) may account for reduced number of seedlings of all species except incense cedar. Drought would decrease the production of seeds and limited successful seedling establishment (Kozlowski and Pallardy 1997). The reduction in the numbers of white fir, incense cedar, and black oak seedlings established on the slight injury plots in comparison with the severe injury plots may be a function of the more open canopy conditions of the plots in the slight injury group. Both the Tunnel Two and the Camp Angeles plots have relatively open crown canopies, which allow considerable sunlight to reach the forest floor. This results in seedbed conditions that are hotter and drier and have higher levels of radiation. These conditions would favor the establishment of ponderosa pine seedlings but not white fir, incense cedar, or black oak (McBride 1978b). The lower number of ponderosa pine seedlings on the severe injury plots in comparison with the slight injury plots could be the result of decreased cone production under conditions of high levels of oxidant air pollutants. Luck (1977) found a significant decrease in cone production in ponderosa pine trees showing severe foliar injury symptoms on the study plots in the

Table 15.4. Comparison of 10-Year Seeding Establishment in Areas of Severe and Slight Foliar Ozone Injury

	Average number of seedlings established per hectare					
	Severe			Slight		
Species[a]	1965–1974	1975–1984	Diff.[b]	1965–1974	1975–1984	Diff.[b]
PP	92	94		180	131	
WF	1,321	226	*	172	94	*
IC	193	890	*	0	0	
SP	0	14	*	5	5	
BO	327	175		265	83	*

[a]PP, *Pinus ponderosa;* WF, *Abies concolor;* IC, *Calocedrus decurrens;* Sp, *Pinus lambertiana;* BO, *Quercus kelloggii.*
[b]Difference significant at 0.01%.

San Bernardino Mountains. Increased seedling establishment was observed for incense cedar on the severe injury plots but not on the slight injury plots from 1965–1974 to 1975–1984. This may be due to a greater number of incense cedar trees growing into seed-producing age during the 10-year period from 1974 to 1983 as well as the difference in canopy closure between the two groups of plots.

Projected Succession (Based on 1984 Data)

Succession was projected for the years 2024 and 2074 using the life table method. A separate life table was developed for each of the two groups of plots (severe injury, slight injury) so that any effects of chronic oxidant air pollutants on succession could be identified. Data from the two groups of plots were used to establish an average number of trees per hectare for each class at the beginning of the study (1973). Average mortality rates were calculated for each 10-year age class for each species. These mortality rate were used to project the number of individuals in each age class forward every 10 years until 2024 and 2074. The projected number of seedlings established by each species in each 10-year period from 1974 to 2024 and 2074 was held constant at averages based on the average number of seedlings established in the 10 years before and after 1974 for each species. This averaging procedure was used because of large variance in regeneration of tree species.

Results of the life table projection of the numbers of seedlings, saplings and poles, young trees, and mature trees are shown in Tables 15.5 and 15.6. The projection of succession shows a shift away from the dominance of ponderosa pine by the year 2074 on both the severe and slight foliar injury plots. Ponderosa pine's percentage composition in the mature tree age class will drop from 85.5% in 1974 to 65.8% in 2024 and to 13.5% in 2074 on severe injury plots. On the slight injury plots, the decline will be from 58.3% in 1974 to 16.7% in 2024 and

Table 15.5. Tree Species Composition in 1974 and Projected Species Composition in 2024 and 2074 on an Average Hectare in the Severe Foliar Ozone Injury Zone of the San Bernardino Mountains

	Species composition (%)											
	Seedlings			Saplings and poles			Young mature			Mature		
Species[a]	1974	2024	2074	1974	2024	2074	1974	2024	2074	1974	2024	2074
PP	4.8	5.5	5.5	16.8	3.6	3.6	65.9	2.0	0.6	85.8	65.8	13.5
WF	68.3	46.4	46.4	28.9	41.1	41.1	3.3	48.5	27.0	7.1	2.4	26.9
IC	10.0	32.6	32.6	13.3	36.7	36.7	9.3	24.6	52.9	0	11.6	27.2
SP	0	0.4	0.4	2.5	1.5	1.5	0.5	1.3	1.9	0	0.6	2.4
BO	16.9	15.1	15.1	38.5	17.1	17.1	21.0	23.4	17.6	7.1	19.6	30.0

[a]PP, *Pinus ponderosa;* WF, *Abies concolor;* IC, *Calocedrus decurrens;* Sp, *Pinus lambertiana;* BO, *Quercus kelloggii.*

Table 15.6. Tree Species Composition in 1974 and Projected Species Composition in 2024 and 2074 on an Average Hectare in the Slight Foliar Ozone Injury Zone of the San Bernardino Mountains

	Species composition (%)											
	Seedlings			Saplings and poles			Young mature			Mature		
Species[a]	1974	2024	2074	1974	2024	2074	1974	2024	2074	1974	2024	2074
PP	28.8	33.3	33.3	44.2	42.8	42.8	56.2	39.8	39.9	58.3	16.7	16.7
WF	28.0	28.4	28.4	43.8	29.7	29.7	10.0	40.6	33.7	8.3	16.7	53.6
IC	0	0	0	0	0	0	0	0	0	0	0	0
SP	0.8	1.1	1.1	1.0	2.9	2.9	7.5	3.7	2.5	8.3	9.5	6.2
BO	42.4	37.2	37.2	11.0	24.6	24.6	26.3	15.9	23.9	25.1	57.1	23.5

[a]PP, *Pinus ponderosa;* WF, *Abies concolor;* IC, *Calocedrus decurrens;* SP, *Pinus lambertiana;* BO, *Quercus kelloggii.*

will remain at 16.7% in 2074. Future recruitment of ponderosa pine into the mature age class on the severe injury plots will be very limited following the year 2074, considering that only 0.6% of the trees in the young mature class are ponderosa pine. The majority of trees in the young mature class on the severe injury plots will be incense cedar; however, white fir will be well represented in both the young mature (27%) and mature (26.9%) age classes. Plots showing slight injury will be dominated by white fir after the year 2074 but will also support ponderosa pine. Recruitment of ponderosa pine into the mature age class will continue because of shared dominance in the young mature class by ponderosa pine and white fir.

The contrast between the projected composition of the two groups of plots indicates the impact of continued chronic oxidant air pollutants may have on forest succession. Under higher levels of oxidant air pollution, the mixed conifer forest will succeed to a white fir-and-incense cedar–dominated type. This successional trend may in part be due to fire suppression as suggested by Minnich et al. (1995). However, the absence of selective thinning and the increase in stem density suggested by these authors cannot account for the increased mortality observed in ponderosa pine on severe foliar injury plots. The successional trend projected here by using the life table method is in large part due to the greater sensitivity of ponderosa pine to oxidant pollutants. A variety of processes that affect ponderosa pine's competitive ability is affected by exposure to high levels of oxidant air pollutants. Coyne and Bingham (1981) demonstrated a reduction in photosynthetic rate in naturally occurring ponderosa pine saplings and poles exposed to high levels of oxidant air pollutants. McBride et al. (1975), Petersen et al. (1989), and Miller et al. (1989) have shown that exposure to high levels of oxidant air pollutants under field conditions reduces growth in ponderosa pine. Luck (1977) demonstrated a correlation between foliar injury score and cone production in ponderosa pine in the San Bernardino Mountains. An additional factor contributing to the decline of ponderosa pine is the increased depth of litter reported under trees with high foliar injury scores by Arkley and Glauser (1980) and Miller and Rechel (this volume). Ponderosa pine does not exhibit high levels of seedling establishment where litter accumulations exceed 2 cm (Fowells 1965). As the capacity for photosynthesis, growth, and regeneration is diminished by air pollution, the capacity of ponderosa pine to succeed in competition with other more tolerant species is also diminished. Likewise, the increased susceptibility of ponderosa pine to attack by the western pine beetle (*Dendroctonus brevicomis*) has been shown to be correlated with foliar injury score (Byler et al. 1978; McBride et al. 1979; Pronos et al., this volume).

Projected Succession (Based on Post-1984 Data)

The life table model used to predict forest succession was not updated after 1984 because of disturbance on several plots by tree removal and ground-level activities that destroyed regeneration. Field observations during the 10 years since

1984 suggest increased seedling establishment and survival of saplings and poles on all plots. Surveys to assess ozone injury scores and tree mortality since 1984 show decreased levels of foliar injury and decreased rates of mortality in comparison with the period before 1984 (Miller et al. 1989, this volume; Arbaugh et al., this volume). The reduction in tree mortality and the apparent increase in seedling establishment would change the projection of succession, based on the life table method, toward a greater percentage of ponderosa pine in the future mixed conifer forest stands. It may be that the initial exposure of the mixed conifer forest in the San Bernardino Mountains to chronic levels of air pollution eliminated the more susceptible individuals. This could account for the greater rates of mortality observed in the 1960s and 1970s. The surviving individuals have since become the breeding populations that are producing seedlings with somewhat greater resistance to oxidant air pollutants. These more resistant genotypes will allow for the continued presence of some ponderosa pine in the mixed conifer forests of the San Bernardino Mountains, provided that levels of oxidant air pollutants do not increase.

References

Arkley, R.J.; Glauser, R. Effects of photochemical air pollution on pine litter fall and the forest floor. In: Miller, P.R., ed. Proceedings of the symposium on effects of air pollutants on Mediterranean and temperate forest ecosystems. Gen. Tech. Rep. PSW-43. Berkeley, CA: USDA Forest Service; 1980:225.

Bonnickson, T.M.; Stone, E.C. The giant sequoia-mixed conifer forest community characterized through pattern analysis as a mosaic of aggregations. For. Ecol. Manage. 3:307–328; 1981.

Burns, R.; Honkala, G. Silvics of forest trees of the United States, vol. 1. Conifers. Agric. handbook 271A. Washington, DC: USDA Forest Service; 1990:431–471.

Byler, J.; Cobb, F., Jr.; Dahlsten, D.; McBride, J.R. A forest-wide survey of pests and pest complexes associated with tree mortality on the San Bernardino National Forest. In: Workshop proceedings. Simulation modeling of oxidant air pollution effects on mixed conifer forests and the possible role of models in timber management planning for southern California national forests, March 20–21, 1978, University of California, Riverside. Berkeley, CA: Statewide Air Pollution Research Center and U.S. Forest Service; 1978.

Christensen, N.L.; Cotton, L.; Harvey, T.; Martin, R.; McBride, J.; Rundel, P.; Wakimoto, R. Review of fire management program for sequoia mixed conifer forests of Yosemite, Sequoia and Kings Canyon national parks. Occasional paper. Washington, DC: U.S. National Park Service; 1987.

Connell, J.H.; Slatyer, R.O. Mechanisms of succession in natural communities and their role in community stability and organization. Am. Nat. 111:1119–1144; 1977.

Coyne, P.; Bingham, G. Comparative dose response of gas exchange in ponderosa pine stands exposed to long term fumigations. JAPCA 31:38–41; 1981.

Deevey, E.S. Life tables for natural populations of animals. Q. Rev. Biol. 22:283–314; 1947.

Fowells, H.A.; Schubert, G.H. Seed crops of forest trees in the pine region of California. USDA tech. bull. 1150. Washington, DC: USDA; 1965.

Kickert, R., ed. Photochemical air pollutant impact on mixed conifer forest ecosystems. A progress report. Contract 68-03-0273 (EPA). Riverside, CA: Statewide Air Pollution Research Center, Univ. California; 1976:49–61.

Kickert, R.N.; Miller, P.R.; Taylor, D.C.; McBride, J.R.; Barbieri, J.; Arkley, R.; Cobb, F., Jr.; Dahlsten, D.; Wilcox, W.W.; Wenz, J.; Parmeter, J.R., Jr.; Luck, R.F.; White, M. Photochemical air pollutant effects on mixed conifer ecosystems. Progress rep. 1974-75. EPA-600/3-77-058. Corvallis, OR: U.S. Environmental Protection Agency, Corvallis Environmental Research Laboratory; 1977.

Kickert, R.N.; Miller, P.R.; Taylor, D.C.; McBride, J.R.; Arkley, R.J.; Cobb, F., Jr.; Dahlsten, D.; Wilcox, W.W.; Wenz, J.; Parmeter, J.R., Jr.; Luck, R.F. Photochemical oxidant air pollution effects on a mixed conifer forest ecosystem. Ecological series. Corvallis, OR: Research Laboratory; 1980.

Kozlowski, T.T.; Pallardy, S.G. Physiology of woody plants. San Diego, CA: Academic Press; 1997.

Laven, R.D. Application of numerical taxonomy to the study of post-fire succession in the Jeffrey pine region of the San Bernardino Mountains, California. Ph.D. dissertation, University of California, Berkeley; 1978.

Laven, R.D. Establishing homogeneity in studies of forest succession. For. Ecol. Manage 4:161–177; 1982.

Luck, R.F. Cone production in the mixed conifer forests of the San Bernardino Nountains in relation to air pollution. In: Kickert, R.N., ed. Photochemical air pollutant effects on mixed conifer ecosystems. Progress rep. 1974-75. EPA-600/3-77-058. Corvallis, OR: U.S. Environmental Protection Agency, Corvallis Environmental Research Laboratory; 1977.

MacDonald, P.M. Silviculture—ecology of three native California hardwoods on high sites in north central California. Ph.D. dissertation, Oregon State University, Corvallis; 1978.

McBride, J. Annual report of the vegetation sub-committee. In: Taylor, O.C.; Miller, P.R., eds. Photochemical oxidant air pollutant effects on a mixed conifer forest ecosystem. Riverside, CA: Statewide Air Pollution Research Center, Univ. California; 1977.

McBride, J.R. Long-term change in forest competition. In: Kickert, R., ed. Photochemical air pollutant impact on mixed conifer forest ecosystems. A progress report. Contract 68-03-0273 (EPA). Riverside, CA: Statewide Air Pollution Research Center, Univ. California; 1976:216–221.

McBride, J.R. Predictions of changes in forest species composition associated with the effects of oxidant injury, pest complexes, and fire. In: Workshop proceedings, simulation modeling of oxidant air pollution effects on mixed conifer forests and the possible role of models in timber management planning for southern California national forests, March 20–21, 1978, University of California, Riverside. Riverside, CA: Statewide Air Pollution Research Center and U.S. Forest Service; 1978a:12.

McBride, J.R. Occurrence of forest tree seedlings in relation to variation in forest overstory. In: Workshop proceedings, simulation modeling of oxidant air pollution effects on mixed conifer forests and the possible role of models in timber management planning for southern California national forests, March 20–21, 1978, University of California, Riverside. Riverside, CA: Statewide Air Pollution Research Center and U.S. Forest Service; 1978b:20–21.

McBride, J.R.; Miller, P.R. Tree population dynamics subsystem. In: Miller, P.R.; Elderman, M.J., eds. Photochemical oxidant air pollution effects on a mixed conifer forest ecosystem—A progress report, 1976. EPA 600/3-77-104. Corvallis, OR: U.S. Environmental Protection Agency, Corvallis Environmental Research Laboratory; 1977:106–121.

McBride, J.R.; Miller, P.R. Response of American forests to photochemical oxidants. In: Hutchinson, T.C., ed. Proceedings of the acid deposition workshop, Univ. Toronto, May 12–17, 1985, Toronto, CA. New York: Springer-Verlag: 1987:217–228.

McBride, J.R.; Semion, V.P.; Miller, P.R. Impact of air pollution on the growth of ponderosa pine. Calif. Agric. 29(12):8–9; 1975.

McBride, J.R.; Dahlsten, D.L.; Cobb, F.W. Cause and extent of tree mortality. In: Taylor, O.C., ed. Photochemical oxidant air pollution effects on a mixed conifer forest eco-

system—final report 1977. Riverside, CA: Statewide Air Pollution Research Center, University of California; 1979:132–135.
McBride, J.R.; Miller, P.R.; Laven, R. Effects of oxidant air pollutants on forest succession in the mixed conifer type of southern California. Proceedings of the air pollution effects on forest ecosystem symposium, May 8–9, 1985, St. Paul, MN. St Paul, MN: The Acid Rain Foundation; 1985:157–167.
Miller, P.R. Oxidant-induced community change in a mixed conifer forest. In: Naegele, J.A. Air pollution damage to vegetation. Advances in chemistry series 112. Washington, DC:Am. Chem. Soc. 1973:101–117.
Miller, P.R.; Elderman, M.J. Photochemical oxidant air pollution effects on a mixed conifer forest ecosystem: a progress report. EPA-600/3-77-104. Corvallis, OR: U.S. Environmental Protection Agency; 1977.
Miller, P.R.; McBride, J.R.; Schilling, S.L.; Gomez, A.P. Trends of ozone damage to conifer forests between 1974 and 1989 in the San Bernardino Mountains of southern California. In: Olson, R.K.; Lefohn, A.S., eds. Effects of air pollution on western forests. Pittsburgh, PA: Air and Waste Management Association; 1989:309–324.
Minnich, R.A. The geography of fire and conifer forests in the eastern transverse ranges. Ph.D. thesis, University of California, Los Angeles; 1978.
Minnich, R.A. The biogeography of fire in the San Bernardino Mountains of California: a historical study. Univ. Calif. Pub. Geogr. 27:1–121; 1988.
Minnich, R.A.; Barbour, M.G.; Burk, J.H.; Fernau, R.F. Sixty years of change in conifer forests of the San Bernardino Mountains: reconstruction of California mixed conifer forest prior to fire suppression. Conserv. Biol. 9:902–914; 1995.
Parsons, D.J.; DeBenedetti, S.H. Impact of fire suppression on a mixed-conifer forest. For. Ecol. Manage 2:21–33; 1979.
Peterson, D.L.; Arbaugh, M.J.; Robinson, L.J. Ozone injury and growth trends of ponderosa pine in the Sierra Nevada. In: Olson, R.K.; Lefohn, A.S., eds. Effects of air pollution on western forests. Pittsburgh, PA: Air and Waste Management Association; 1989:293–308.
Rabotnov, T.A. The biology of monocarp perennial meadow plants. Russian Translation Service 8739, British Library. Bull. Moscow Soc. Nature 69:57–66; 1964.
Silverton, J.W. Introduction to plant population ecology. London: Longman; 1982.
Sugihara, N.G.; McBride, J.R. Dynamics of sugar pine and associated species following non-stand-replacing fires in white fir-dominated mixed conifer forests. In: Kinloch, B.B., Jr.; Marosy, M.; Huddleston, M.E., eds. Sugar pine: status, value, and roles in ecosystems: proceedings of a symposium presented by the California Sugar Pine Management Committee. Publication 3362. Davis, CA: University of California, Division of Agriculture and Natural Resources; 1996:39–44.
Taylor, O.C. Oxidant air pollutant effects on a western coniferous forest ecosystem, Task C report, annual progress 1972–73. Study site selection and verification data on pollutants and species. EPA-600/3-78-052c, U.S. Environmental Protection Agency. NTIS No. PB 228 333/AS. Riverside, CA: Air Pollution Research Center, University of California; 1973.
Taylor, O.C. Oxidant air pollutant effects on a western coniferous forest ecosystem, Task D report, annual progress 1973–74. EPA-600/3-78-052d, U. S. Environmental Protection Agency. NTIS No. PB 281 858/AS. Riverside, CA: Air Pollution Research Center, University of California; 1974.
Vankat, J.L.; Major, J. Vegetation changes in Sequoia National Park, California. J. Biogeog. 5:377–402; 1978.
Westman, W.E. Oxidant effects on California coastal sage scrub. Science 205:1001–1003; 1979.
Westman, W.E. Factors influencing the distribution of species of California coastal sage scrub. Ecology 62:439–455; 1981.

16. Simulated Effects of N Deposition, Ozone Injury, and Climate Change on a Forest Stand in the San Bernardino Mountains

M.J. Arbaugh, D.W. Johnson, and W.M. Pulliam

Introduction

Euro-American settlement, fire suppression, harvest, and grazing have altered forest development and distribution of many western forests including the San Bernardino Mountains (Veblen and Lorenz 1991; Minnich, this volume). In addition to these influences, southern California forests also have been affected by high levels of air pollution over the past five decades. Changes in soil and water chemistry, biomass accumulation, and distribution of above- and belowground carbon due to air pollution have been identified (Miller et al. 1989; Fenn et al. 1996; Miller 1996).

The impact of these changes on forest health and sustainability is not clear. Interactions between environmental stressors are largely unknown and are not directly measurable. Sequential interactions between ozone, drought, and mountain pine beetle have been identified as causes of forest mortality and species composition changes in the San Bernardino Mountains (Miller et al. 1989), but the cumulative and interacting effects of ozone, nitrogen (N) deposition, and climate change have not been studied.

In this chapter, two models will be used to simulate the effects of N deposition, ozone, and climate change on biogeochemical processes of forest growth at a site in the eastern San Bernardino Mountains. Two models were chosen because the complexity of stressors (ozone, N deposition, and climate) is beyond the ability of any single simulation model presently available. The simulations are intended to show the logical implications of the present state of knowledge of carbon and nutrient flows and accumulation in the mixed conifer forest system of the San Bernardino Mountains. They are not intended as general parameterizations for this forest type.

Similarly, this is not intended as a model comparison study. The models differ greatly in philosophy and application. Models were chosen to be complementary, each extending the ability of the other. Some discussion of model input and output parameter is conducted to determine that the two models generally agree on the major pools of the system and that model results are not strongly dependent on different initial conditions.

Model Selection and Description

The models selected for this study are constructed at similar spatial scales but emphasize different biochemical aspects of ecosystem functioning. The Nutrient Cycling Model (NuCM) was developed primarily to explore the effects of atmospheric deposition on forest ecosystems, especially those within the integrated forest study (Johnson and Lindberg 1992). It has detailed soil chemistry and hydrology components found in few other models. By using mass balance and transport formulations, the model tracks 16 solution-phase components including the major cations and anions (analytical totals), acid-neutralizing capacity (ANC), an organic acid analog, and total aluminum (Liu et al. 1991). Nutrient pools associated with soil solution, the ion exchange complex, minerals, and soil organic matter are explicitly simulated (Fig. 16.1).

The model has been used to explore the effects of changes in atmospheric deposition of S, N, and base cations at a mixed deciduous forest site at Coweeta, North Carolina (Johnson et al. 1991), a loblolly pine site at Duke, North Carolina (Johnson et al. 1995a), and a red spruce site in the Smoky Mountains (Johnson et al. 1996), as well as the Barton Flats site (Fenn et al. 1996). In addition, NuCM has been used to simulate the effects of forest harvesting and species change at both the Coweeta and Duke sites (Johnson et al. 1995a) and to simulate the effects of lime application at Coweeta (Johnson et al. 1995b).

In contrast to the complex simulation of soil chemistry, nutrient flows, and hydrology, simulations of biomass accumulation and litter decomposition rates are less detailed in NuCM. Aboveground growth is separated into canopy (foliage) and woody tissue (bole, limbs, and branches). Maximum standing biomass and litter decay rates are set by the user and are reduced by nutrient limitations. This approach limits the ability of the model to examine the effects of ozone and interactions of ozone, nitrogen, and climate on the rates of biomass accumulation and decomposition. NuCM does not conserve carbon; above and below grown biomass are not explicitly linked and are calculated from user-defined growth rate functions.

For these reasons, Century (Parton et al. 1987, 1988) was chosen to simulate production, decomposition, and carbon (C) storage in these forests. Century is a

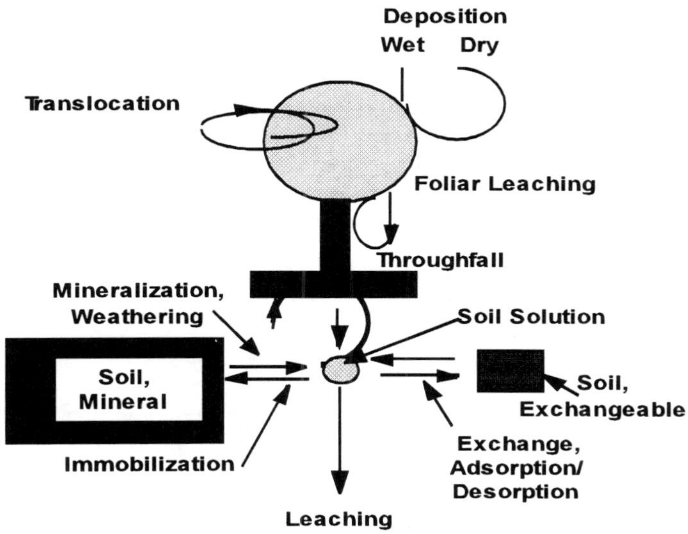

Figure 16.1. Major flows and pools of NuCM.

biogeochemical model that conserves mass balance between inputs, outputs, and changes in storage and uses simple but robust environmental controls on decomposition and primary production.

Century is a linked production-decomposition biogeochemical model with aggregated biomass pools and multiple soil organic matter (SOM) pools. The forest model (Fig. 16.2) introduces additional structural variables to describe the flow of carbon through a larger number of living tissue classes than is present in the NuCM model.

The forest version of Century was initially described by Sanford et al. (1991), who developed forest Century to investigate long-term impacts of recurrent hurricanes on production and SOM dynamics in a tropical forest in Puerto Rico. It contains five biomass pools: live leaves, fine roots, and three live wood biomass fractions. Each biomass pool has separate lignin, N, P, and S contents. Retranslocation of nutrients from senescent leaves is simulated explicitly. Prior to leafdrop, a specified fraction of each nutrient element is retranslocated from leaves and transferred to an internal storage pool, where it is available for future biomass production.

Net primary production (NPP) is calculated as a function of climatic, biotic, and nutrient limitations without an explicit photosynthesis simulation. Nutrient limitations on NPP are applied after climate-based potential is calculated. If one or more of the nutrient elements being simulated is not available in sufficient quantity to satisfy this demand, NPP is reduced in proportion to the most limiting nutrient. Allocation of new production among the five live biomass pools takes place by a

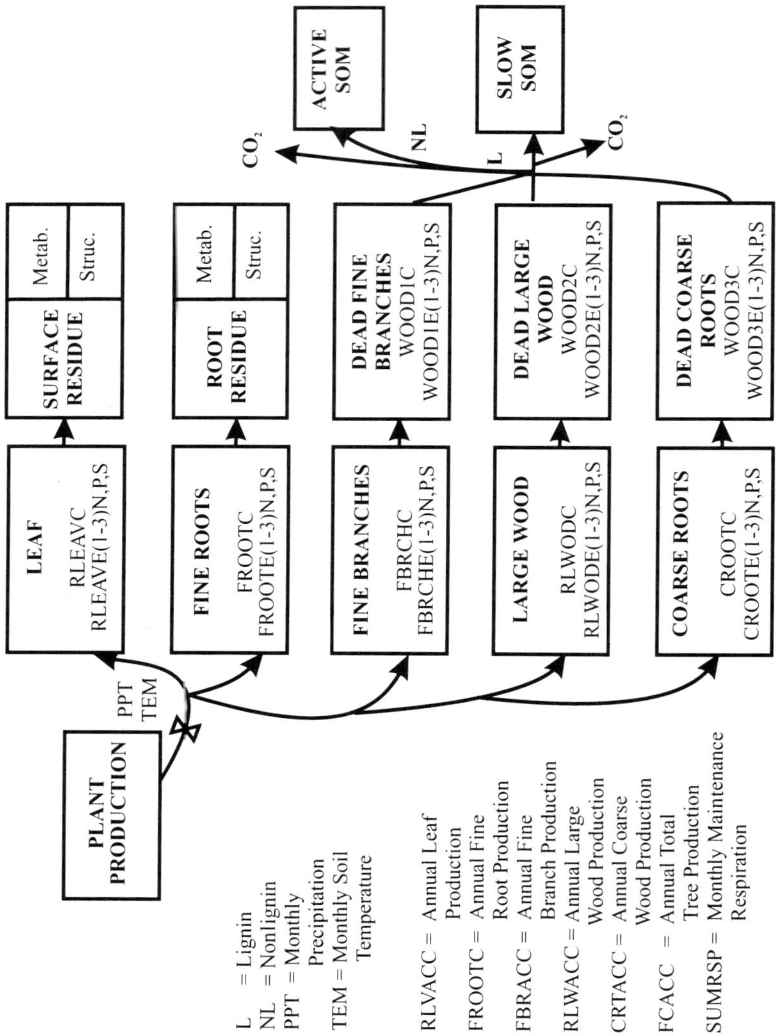

Figure 16.2. Major flows and pools of the forest submodel of Century.

set of fixed specified allocation fractions that are based on observed patterns of NPP allocation.

Loss of live biomass and generation of detritus takes place via both continuous turnover and mortality associated with discrete disturbance events. Continuous turnover represents senescence, self-pruning, baseline herbivory and wind damage, and other processes that can be considered ongoing on a time scale of several years. Disturbance events are particular incidents during which biomass is killed or removed, such as fires, windstorms, or harvests.

Ryan et al. (1996) presented comparisons of the ability of six forest ecosystem simulation models, including Century, to capture the dynamics of young pine plantations at sites in Sweden and Australia. This study included multiple irrigation and fertilization treatments at both sites, providing extensive tests of these models. Century was able to reproduce biomass and NPP of treated stands with a mean error of less than 20% after calibration against the control stands. However, Century (as most of the models examined) had less ability and agreement for simulating the allocation of N fertilizer amendments to the forest systems.

Parameterization Data

The site used for the simulation is located at Barton Flats and was the subject of an intensive 3-year study that examined many aspects of the forest, soils, and atmospheric environments (Miller et al. 1996; also see Watson et al., Miller and Rechel, and Bytnerowicz et al., this volume). Extensive data were gathered to describe the effect of wet and dry deposited acidic compounds and ozone on components and processes of the mixed conifer forest in the San Bernardino Mountains. Detailed studies of soil chemistry, air chemistry, dry and wet deposition, ambient ozone, and litter decomposition were conducted, as well as describing vegetation distribution and crown condition.

NuCM Simulations

Modeling Approach and Deposition Scenarios

Five N deposition scenarios were run for NuCM and Century: inputs of 1.9, 9.5, 28.7, 38.2, and 95.6 kg ha^{-1} yr^{-1}. These inputs were 0.1, 0.5, 2, and 5 times the baseline input value in the model of 19 kg ha^{-1} yr^{-1} for Barton Flats Each simulation was run from the same initial conditions, with no input variable changed except N deposition. Simulations were run for 40 years each.

NuCM was calibrated for the Barton Flats site by using data collected from the site for vegetation and soil nutrient concentrations, soil and litter mass, and soil solution concentrations. At the time of this exercise, no data for biomass from the Barton Flats site were available, so the data on a ponderosa pine stand from Klemmedson (1975) were used (Table 16.1).

Table 16.1. Initial Conditions for Major Carbon and Nitrogen Pools for NuCM and Century[a]

	NuCM			Century		
Organic pool	Initial biomass	Allocation (%)	Max. growth rate	Initial biomass	Allocation (%)	Turnover (%)
Carbon						
Foliage	6,350	25	600[b]	1,400	13	13
Branch	—	—	—	8,000	12	0.4
Large wood	—	—	—	70,000	38	0.12
Total wood	63,400	56	1,350	78,000	50	0.15
Fine root	—	—	—	5,000	25	2
Coarse root	—	—	—	15,000	12	0.2
Total root	12,800	19	465	20,000	37	0.65
Litter	6,486	—	600[b]	1,870	—	23
SOM	284,400	—	—	64,170	—	0.84
Nitrogen						
Foliage	122	7	9[b]	33	14	13
Branch	—	—	—	64	12	0.4
Large wood	—	—	—	280	38	0.12
Total wood	457	78	9.7	344	79	0.15
Fine root	—	—	—	100	25	2
Coarse root	—	—	—	150	12	0.2
Total root	262	14	9.5	250	20	0.65
Litter	274	—	9[b]	51	—	31
Soil exchange N	5,832	—	—	3,036	—	0.22

[a]Unlike Century, NuCM does not conserve either C or N; thus biomass processes are not linked and turnover rates are set independently during model parameterization (kg ha^{-1} and kg ha^{-1} yr^{-1}).
[b]Approximate turnover rates (kg ha^{-1} yr^{-1}).

Procedures outlined in the user's manual (Munson et al. 1992) were used in the calibration; details are described elsewhere (Liu et al. 1991; Johnson et al. 1993). Several model parameters that were of no great consequence to our particular ecosystem and for which no data were available (e.g., organic acid adsorption, phosphate adsorption, snowmelt characteristics, fractions of leachable nutrients in litter) were left as in the original model formulation (Liu et al. 1991).

NuCM Results

Changes in Nitrate Leaching and N Retention

As expected, increases in simulated N deposition resulted in increases in simulated NO_3^- leaching (Table 16.2). However, simulated NO_3^- leaching rates increased substantially only after N deposition rates exceeded the current rates of deposition. Cumulative net ecosystem N (vegetation, SOM, and soil exchangeable N) retention was actually lower at 0.1 × N (63% of input) than at 0.5 × N or 1 × N (87 and 89% of input, respectively), even though NO_3^- leaching rates increased progressively with N deposition throughout these scenarios (Table 16.2). Apparently, the NO_3^- leaching rate at 0.1 × N was near a baseline level. Simulated vegetation and forest floor N contents increased progressively with N deposition from 0.1 to 1 × N, accounting for the retention of N in the ecosystem (Table 16.3). Between 0.1 and 1 × N, vegetation N content increased by 86%, forest floor N increased by 22%, and the sum of vegetation, forest floor, and soil exchangeable N increased by 59% (Table 16.3). At N deposition levels of 2 × N and 5 × N, net ecosystem retention decreased to 55 and 25% of input, respectively (Table 16.2). There were some increases in vegetation (11%) and forest floor (4%) N contents from 1 to 2 × N but no further increases at 5 × N (Table 16.3). There was a large relative increase in exchangeable N from the 2 to the 5 × N scenarios (16,380%), but the absolute magnitude of this increase was too small to cause a large increase in ecosystem N retention (the sum of vegetation, forest floor, and exchangeable N increased by only 7%, all in the form of exchangeable N) (Table 16.3).

Forest Response to Changes in N

The patterns of simulated leaching and N retention closely corresponded to those in growth. Simulated growth increased substantially from the 0.1 to 0.5 and 1 × N scenarios, indicating N limited growth at less than 1 × N input (Fig. 16.3) The growth rate also increased slightly from the 1 to 2 × N scenarios, but there was no further growth increase from the 2 to the 5 × N scenario.

Changes in Soil Biogeochemistry

The N deposition scenarios caused large reductions in P leaching due mostly to accumulation of P in vegetation and forest floor (Table 16.2). However, P leaching rates were low in relation to ecosystem P pools, so that the changes in the P pool were very slight (<0.5% variation among scenarios) with increasing N deposition (Table 16.3).

Table 16.2. Simulated Net Ecosystem Balances of N, P, Ca, K, Mg, and S for Different N Deposition Scenarios (kg ha^{-1} 40 yr^{-1})[a]

Scenario	0.1 × N		0.5 × N		1 × N		2 × N		5 × N	
Nitrogen	NCM	CEN	NCM	CEN	NCM	CEN	NCM	CEN	NCM	CEN
Deposition	76	76	382	382	764	764	1,530	1,530	3,823	3,823
Leaching	28	240	49	252	79	266	688	276	2,848	860
% Retention	63%	—	87%	34%	89%	65%	55%	82%	25%	78%
Phosphorus										
Deposition	0.000		0.000		0.000		0.000		0.000	
Leaching	1.2		0.9		0.8		0.4		0.2	
% Retention	—		—		—		—		—	
Calcium										
Deposition	19		19		19		19		19	
Leaching	208		147		163		338		1,206	
% Retention	—		—		—		—		—	
Potassium										
Deposition	2		2		2		2		2	
Leaching	280		216		229		258		414	
% Retention	—		—		—		—		—	
Magnesium										
Deposition	2		2		2		2		2	
Leaching	86		61		68		138		406	
% Retention	—		—		—		—		—	
Sulfur										
Deposition	236		236		236		236		236	
Leaching	190		165		164		136		136	
% Retention	20%		30%		31%		42%		43%	

[a]Estimates are included for both NuCM (NCM) and Century (CEN) for nitrogen.

Table 16.3. Simulated Nutrient Distribution after 40 Years under Varying N Deposition Scenarios (kg ha^{-1} and kg ha^{-1} yr^{-1})[a]

Scenario	0.1 × N		0.5 × N		1 × N		2 × N		5 × N	
N Deposition	1.9		9.5		19		38.2		95.6	
Nitrogen	NCM	CEN	NCM	CEN	NCM	CEN	NCM	CEN	NCM	CEN
Vegetation	560	569	755	613	1,044	666	1,164	755	1,164	888
Forest floor	415	305	470	315	506	327	528	347	528	380
Soil exchange	<1		<1		<1		1		115	
Total	975		1,225		1,550		1,693		1,807	
Phosphorus										
Vegetation	91		122		137		176		176	
Forest floor	22		30		26		33		33	
Soil exchange	3,444		3,404		3,398		3,351		3,360	
Total	3,557		3,556		3,561		3,560		3,569	
Calcium										
Vegetation	717		867		958		1,227		1,273	
Forest floor	153		172		163		190		195	
Soil exchange	7,652		7,548		7,456		6,996		6,112	
Total	8,522		8,587		8,577		8,413		7,580	
Potassium										
Vegetation	416		527		582		737		759	
Forest floor	291		339		317		389		403	
Soil exchange	757		651		623		333		144	
Total	1,464		1,517		1,522		1,459		1,306	
Magnesium										
Vegetation	142		177		195		256		256	
Forest floor	57		64		60		71		71	
Soil exchange	1,055		1,046		1,031		916		691	
Total	1,254		1,287		1,286		1,243		1,018	
Sulfur										
Vegetation	54		70		76		94		94	
Forest floor	316		321		319		324		323	
Soil exchange	28		30		30		32		34	
Total	398		421		425		450		451	

[a]Estimates are included for both NuCM (NCM) and Century (CEN) for nitrogen.

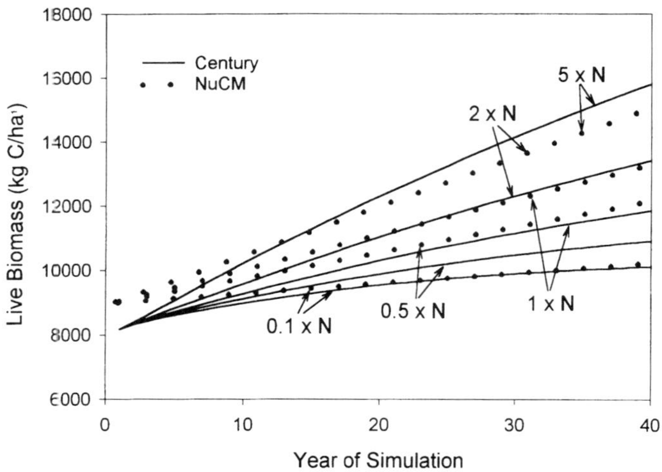

Figure 16.3. NuCM and Century simulated biomass at Barton Flats under various N deposition scenarios: 0.1 × = 1.9, 0.5 × = 9.5, 1 × = 28.7, 2 × = 38.2, and 5 × = 95.6 kg ha^{-1} yr^{-1}. NuCM estimates are represented by dots and Century estimates by solid lines.

Increases in N deposition from 0.1 to 1 × N caused unexpected reductions in simulated K^+, Ca^{2+}, Mg^{2+}, base cation, and ANC leaching rates (Table 16.2). This was due to a complicated set of chemical interactions that resulted in lower total anion concentrations with increasing N deposition from 0.1 to 1 × N during winter months (October–February) when most leaching occurred. Although K^+, Ca^{2+}, and Mg^{2+} leaching rates decreased from 0.1 to 1 × N, soil exchangeable K^+, Ca^{2+}, Mg^{2+} pools decreased continuously from 0.1 to 5 × N deposition (Table 16.3). The decreases from 0.1 to 1 × N were due to uptake and accumulation in vegetation, which offset the reductions in leaching. As N deposition rates increased from 1 to 5 × N, base cation leaching rates increased again, and at 5 × N, there were large decreases in soil exchangeable base cation pools (Tables 16.2 and 16.3). The decrease in exchangeable pools was much larger for K^+ (decreased by 77% from 1 to 5 × N) than for Ca^{2+} and Mg^{2+} (decreased by 18 and 23%, respectively, from 1 to 5 × N) (Table 16.3).

Discussion

The NuCM simulations for Barton Flats produced some results that were expected and some that were unexpected. The reductions in N retention and increases in NO_3^- leaching with increasing N deposition were certainly expected; the only question was at what levels of N deposition the effects would occur. There was a continuous increase in NO_3^- leaching with N deposition, but the leaching rates were relatively low until N deposition increased beyond 1 × N (19.2 kg ha^{-1} yr^{-1}).

At some level between 2 × N and 5 × N deposition (38.4 and 95.6 kg ha^{-1} yr^{-1}), growth responses to N ceased and N retention remained constant. Thus, N "saturation" occurred at some point above 38.4 kg ha^{-1} $yr.^{-1}$, if N saturation is defined as the inability to accumulate further N. Below 38.4 kg ha^{-1} yr^{-1}, a pronounced N deficiency occurred, as would also be expected as the norm in less-polluted semiarid forests of this type. The threshold at which N saturation occurs in a given simulation will vary with input values for vegetation N uptake (which in turn are determined by the maximum growth rates and C:N concentrations in foliage, woody, and root pools of trees) and the degree of soil N retention. In the absence of any further data for these parameters in varying N deposition regimes, we have left them constant.

Although growth response to N occurred up through 38.4 kg ha^{-1} yr^{-1}, soil solution NO_3^- concentrations peaked at relatively high midsummer values (>150 $\mu mol_c/L^1$) even at the lowest N deposition scenario (1.9 kg ha^{-1} yr^{-1}). Peak NO_3^- concentrations, which were later found in the field in spring sampling (Fenn et al. 1996), would pose no serious threat for ground- or surface water pollution because of the lack of water movement at that time.

The reductions in Ca^{2+}, Mg^{2+}, and K^+ leaching due to reduced N deposition rates were unexpected. These reductions were an indirect result of complicated interactions among soil solution pH, base saturation, NO_3^-, SO_4^{2-}, and HCO_3^- concentration, which could only have been explored through simulation modeling. The general direction of the responses in soil solution pH, SO_4^{2-}, and HCO_3^- concentration are consistent with the soil chemical theory built into the model, but the magnitudes of the responses and their interactions with NO_3^- and base cations were not intuitively predictable, very interesting, and worthy of testing either with field or laboratory column studies.

Century Modeling

Century Canopy Nitrogen Model

For the specific application of Century to the air pollution–affected San Bernardino forests, we implemented several additional phenomena. Pulliam et al. (1995) and Sievering et al. (1996) developed a canopy nitrogen interaction submodel within forest Century. This model is designed to simulate the dynamics and fates of atmospheric N deposition in more detail than the baseline model, for use at sites affected by large amounts of anthropogenic N input. In the canopy submodel, N deposition undergoes a series of fractionations based on specified constants The effect of the additional details of canopy N uptake and allocation was minor at Barton Flats but is similar in type than those reported previously by Pulliam et al. (1995) and Sievering et al. (1996) for sites in eastern North America. In general, the model predicts that the increased accuracy of representation of the canopy interaction with N deposition is less important at Barton Flats than in the eastern United States.

Parameterization of Century for Barton Flats

For application at the Barton Flats site, Century was parameterized to simulate a semideciduous forest (25% deciduous), simulating C and N dynamics. Initial values and parameters for the biomass production model are summarized in Table 16.1. These values are based on extensive data summarized in Miller et al. (1996).

However, even with the considerable data presented for this site, it was necessary to estimate some parameter values (most notably those for lignin as well as belowground biomass and production). Soil texture and initial soil organic matter pools were based on Miller et al. (1996). Bulk soil organic matter was apportioned into 3% active, 71% slow, and 26% passive, values that have produced stable initial model behavior for other temperate conifer ecosystems (Ryan et al. 1996). As the spilt between these pools is conceptual rather than physical, there is no direct means to measure this subdivision.

Little data were available for determining values for the canopy N interaction model. Default values from Sievering et al. (1996) were used, and the sensitivity of the model to this parameterization was evaluated.

N deposition rates were specified monthly, with 80% of the annual total allocated to the period of May–September. N deposition scenarios used in NuCM were also used in these simulations, thus $1 \times N$ was taken to be 1.9 g m^{-2} yr^{-1}, and additional scenarios of 0.1, 0.5, 2, and $5 \times N$ present-day deposition were also simulated.

One of the principal effects of ozone damage observed in the San Bernardino Mountains (Miller et al. 1989) is the decrease in needle retention. Ozone effects were simulated in Century by halving the rate of foliar turnover in the forest over the baseline condition (Table 16.4). This roughly corresponds to the effects of a halving of cumulative ambient ozone concentrations (Fenn et al. 1996), based on relative differences in the ozone gradient from the west to the eastside of the mountains. Other important effects of ozone is to reduce net photosynthetic rates by destruction of chloroplasts, altering photosynthate partitioning between sucrose and starch, and interference in carbohydrate transport to the stem and roots. These were not simulated by Century, their incorporation requiring extensive

Table 16.4. Century Simulated Effects of Ozone Damage on Barton Flats Site on Different Carbon Pools and Process Rates (kg ha^{-1} and kg ha^{-1} yr^{-1})

Process or pool	Unimpacted by ozone	Impacted by ozone
Foliar biomass	2,940	2,470
Foliar production	940	1,430
Foliar C/N	29	30
Total aboveground production	4,550	4,510
Total live biomass	238,070	203,420
Litter mass	3,150	4,970
Soil organic carbon	120,490	119,640
Soil organic matter C/N ratio	20.8	20.0
N mineralization	69	74

model alteration. Thus simulated ozone effects should be considered conservative and indicate the direction of trends rather than quantitative estimates of impacts of increased ozone on forest growth.

Two climate scenarios were used: present day and altered by greenhouse warming. The altered climate was created by increasing temperatures by 3.8°C and reducing precipitation by 21.5%. These estimates were derived from Giorgio et al. (1994) and represent long-term changes due to 2 $\times CO_2$ atmospheric increase derived from nested mesoscale-global scale model simulations of the continental United States. They are scenarios only and are not intended to represent predictions of actual future climate change, thus indicating the direction of the change between present-day and the 2 $\times CO_2$ scenarios.

Century Results and Discussion

Ozone Damage

Overall, simulated ozone damage resulted in faster production and turnover of foliar biomass and a shifting of bulk C storage from the forest canopy to the forest floor (Fig. 16.4). The simulated impact of a doubling of foliar turnover rates is a 15% decrease in foliar biomass along with a 50–60% increase in leaf production. The storage of recent litter on the forest floor increases by 50–60% as well. Along with this projected increase in foliar production is a decrease in woody production, so that the total aboveground productivity and biomass are only slightly affected. Soil organic matter and nitrogen mineralization are also only slightly affected by ozone (Table 16.4).

Interactions Between Ozone and Nitrogen Deposition

In general, Century estimates that N saturation will occur when N deposition rates approach five times present-day amounts. There is a strong positive effect of N on production, biomass, and soil organic matter standing stock across the range of scenarios simulated (Fig. 16.4). Foliar and SOM C/N ratios decline steadily with increasing N deposition. In concert with ozone impacts, high N deposition rates result in dramatic increases in the storage of forest floor litter. Present-day N deposition with ozone is predicted to yield more than three times the litter mass as 0.1 × present-day N deposition without ozone (conditions that are probably similar to the preindustrial era).

Climate Change Scenarios

For most N deposition scenarios, the simulated climate change (Fig. 16.5) results in modest changes in litter mass (±5–15%). At the intermediate N deposition rates, climate change yields small (3–7%) decreases in productivity; at the lowest and highest N deposition scenarios, it yields 10–15% increases in productivity. Total SOM is affected similarly. The most significant impact of simulated climate change is a sharp decrease in soil moisture, especially in May, June, and November. This indicates a lengthening of the summer drought season by, on average, about 2 months.

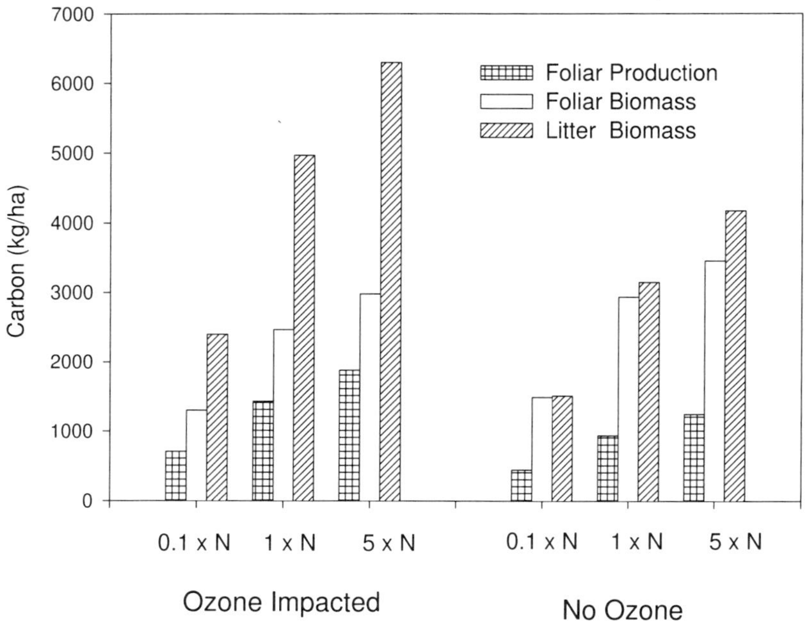

Figure 16.4. Century predictions of foliar production and biomass changes associated with doubled foliar turnover rates at different levels of N deposition. Total biomass and aboveground biomass patterns are similar. Notice that as N deposition increases the effect of ozone-N on litter biomass increases, indicating that the effect of the two pollutants together is interactive rather than additive.

Discussion of Comparative Model Results

The modeling emphasis differs substantially between models; Century conserves the mass of carbon and nutrients, whereas NuCM conserves nutrients but not carbon and emphasizes soil chemistry process rates. However, estimates of C and N for foliage, woody, and root biomass, litter accumulation, and SOM can be obtained from both models.

Initial conditions used in the models differ primarily for foliage, litter, and SOM. NuCM has about 4.5 × C and N more in these pools than does Century (Table 16.1). The lower initial pools of NuCM were due to using data from Arizona ponderosa pine stands (Klemmedson 1975) and the uncertainty of LAI estimates for this stand, which varied from 1.9 to 5.7 (Miller et al. 1996). Similarly, initial pools of Century were influenced by using the lower estimate of LAI (1.9) for model simulations.

Both models predicted increased nitrogen pools for vegetation, litter, and SOM (Table 16.3) with increasing amounts of N deposition. NuCM predicted the great-

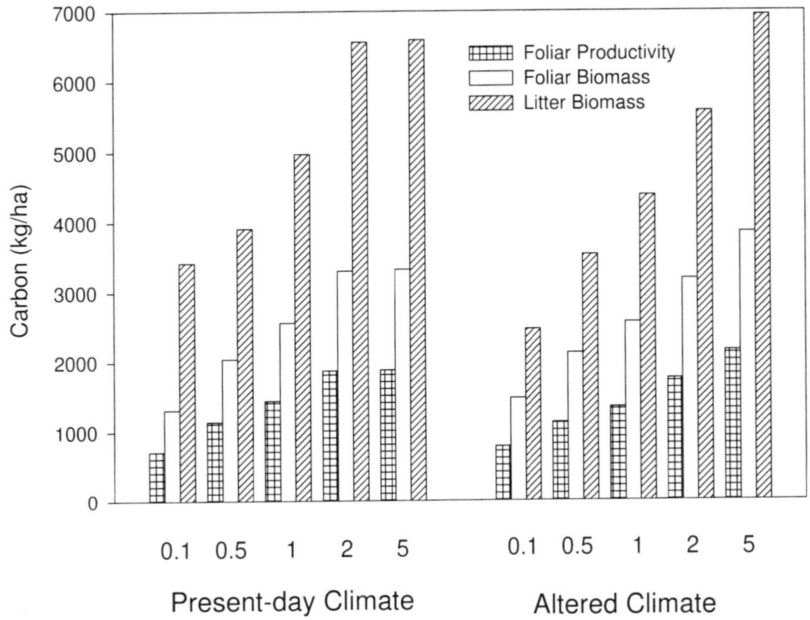

Figure 16.5. Century predictions of foliar production and biomass changes associated with altered climate (21.5% decrease in precipitation and 3.8° increase in temperature) for different levels of N deposition. Total biomass and aboveground biomass patterns are similar. Increased N deposition has a much greater effect on productivity than does altered climate.

est biomass response would be between the 1 × N and 2 × N scenarios, whereas Century predicts the highest growth response to be between the 2 × N and 5 × N scenarios. The pattern of N leaching was similar for the two models, although baseline amounts of N leached were different for the models (Table 16.2).

Despite the predicted foliar and N budget differences, biomass response to N deposition was comparable for the two models (Fig. 16.3). The greatest differences between models were at intermediate N deposition levels. These differences reflect the different mechanisms used by models for carbon gain, and longer simulation periods would lead to more divergent estimates.

Model Predictions Compared with Field Observations

Direct validation data is scarce because the Barton Flats study was not replicated, no experimental nutrient or water amendments were conducted, nor was the study long enough for predictive comparisons through time. Several studies have been conducted at Barton Flats, and long-term remeasure data on crown condition and basal area changes exist across the mountains. In addition, at the western side of the mountains, Camp Paivika, a site that has strong evidence of N saturation (Fenn et al. 1996), can be used as a crude spatial surrogate for higher ozone and N deposition scenarios.

Both models predict that high levels of N deposition will lead to N saturation in the San Bernardino Mountains. NuCM estimates that the amounts will be between 38.5 and 95 kg ha^{-1}/yr^{-1} ($2 \times N$ to $5 \times N$), whereas Century predicts that $5 \times N$ will be needed before saturation occurs. In addition, the Century model predicts that the high ozone-N deposition combination will result in increased litter biomass and reduced rates of woody biomass increase.

Current estimates of annual N deposition at Camp Paivika are between 35 and 45 kg ha^{-1}/yr^{-1}. (Fenn et al. 1996) and are believed to have been existing for 40–50 years, which support the NuCM results. Evidence for increased litter biomass is supported by observations at Camp Paivika, which has litter depths 15–50 cm, in contrast to Barton Flats, which has 1–4 cm of litter (M. Fenn, unpublished data).

Field data support decreased foliar biomass predicted by Century at high ambient ozone and N deposition locations. Long-term repeated measurements of crown injury indicates that foliage biomass is lower at most western sites than at Barton Flats (see Miller and Rechel, this volume). Century simulations also predict that woody biomass will be higher at sites with high ambient ozone and N deposition relative to cleaner sites. Field data supports indicate that basal area increases were greater over the past 20 years at high-injury sites relative to low-injury sites (see Arbaugh et al., this volume). Observed growth responses may be partially due to precipitation differences between sites or differences through time at the same site rather than to the effects of ozone or ozone and N deposition.

Additional data needed for true model validation include measures of litterfall and litter decomposition along the air pollution gradient. An N fertilization experiment is presently underway (M. Fenn and M. Poth, personal communication) that should provide much of the data needed for validating model predictions. Two sites (one near Barton Flats and Camp Paivika) are having two replicated levels of N fertilizer applied over 3 years. At the end of this time, sufficient data should be available to more thoroughly examine model predictions. Information from this study should also help understand discrepancies between model predictions of the N budget and baseline N leaching rates.

Synthesis and Conclusions

The results of NuCM simulations suggest that at the present levels of N deposition to the Barton Flats site, N is still growth-limiting, whereas a doubling in deposition will result in N saturation and increased groundwater pollution due to NO_3^-. There is some suggestion from these simulations that N saturation will lead to K limitations to growth because of large reductions in soil-available pools. However, the rates of K weathering from soil primary minerals are unknown, so this cannot be predicted from these simulations.

The simulations also suggest that reducing N deposition rates below 19.8 kg ha^{-1} yr^{-1} would cause increases in the rate of base cation leaching from soils due to increases in soil solution HCO_3^- and SO_4^{2-} concentrations. This interesting result was not intuitively predictable and is worthy of testing either in the field or with laboratory column studies. Despite the reductions in K^+, Ca^{2+}, and Mg^{2+} leaching, however, soil exchangeable K^+, Ca^{2+}, Mg^{2+} pools decreased continuously from 0.1 to 5 × N deposition due to uptake and accumulation in vegetation.

Century results indicate that the combination of high ozone and N deposition will lead to increased rates of biomass production, especially for litter accumulation. The accumulation of litter, already observed at western sites, is likely to increase the potential of fires eliminating mature trees. Increased numbers of fire-intolerant species and increased litter biomass may lead to conditions in which even low-to-moderate-intensity fires can result in widespread tree mortality through cambial death. Large tree mortality could then result in even more dangerous fire conditions in the future.

The predicted effects of long-term changes in mean temperature and total precipitation on the mixed conifer forest of San Bernardino are small (Century and ZELIG-TREGRO) compared with the effects of ozone and N deposition. This is largely due to most species being drought-resistant and already experiencing 6–10 months of continuous drought each year. The simulations do not include changes in the frequency of extreme single- and multiple-year weather events. Historically, these consist of years with high frequencies of Santa Anna wind events, or multiple year droughts, which are associated with increased mortality due to the mountain pine beetle. Simulations including the effects of these extreme weather events are needed before a complete assessment of the importance of climate change impacts can be made.

This study illustrates the difficulty of using a single model to simulate all aspects of pollution impacts on the forest systems of the San Bernardino Mountains. Air pollution directly affects physiological processes and soil chemistry and tree nutrition. These are of less interest to managers, however, than long-term changes in species composition and structure. At the present time, we lack a single model (or linked models) that can examine all consequences of ozone, N deposition, and climate change to long-term forest composition and structure.

Future Model Development Needs

Neither model presently has adequate scope to simulate all complexities of this ecosystem. The NuCM model was designed primarily to address atmospheric deposition-related issues and soil acidification and thus does not have explicit C budgets. When dealing with N and N saturation issues, as well as global C sequestering issues that have arisen since the development of NuCM, the addition of an explicit C budget component would be highly desirable.

NuCM also lacks an effective mechanism for including the effects of ozone on the system. Foliar loss and litter decomposition are set with single rates, limiting the ability to use NuCM to examine interactive effects of ozone and N deposition on litter decomposition rates and litter accumulation.

For the purposes of simulating short-term dynamics of N, microbial uptake of N needs to be explicitly simulated by NuCM. At present, soil organic N accumulation is simulated only as the residue of decomposing litter, which is perhaps adequate over the long term (>10 years, which is typical of NuCM simulations) but will not capture short-term pulses of N into the system such as happens during fertilization.

Century lacks the ability to simulate reductions in carbohydrate production due to chlorotic mottle independent of needle loss. This loss of carbohydrate is important for understanding the consequences of ozone effects on tree growth (see Grulke, this volume).

There were large absolute estimated differences in the N budget by the two models. Past studies of forest ecosystems indicate that many forest ecosystem models, including Century, may underestimate plant N uptake due to an incomplete understanding of the controls for these processes (Ryan et al. 1996). Further development of N-cycle processes would be valuable for both models, especially because of increased interest in the consequences of long-term N deposition and climate change on forest health.

Neither of the simulation models used in this study was able to determine the effect of N deposition, ozone, and climate change on small-scale changes associated with tree physiology or larger-scale changes associated with changes in species composition. A study currently in progress is developing a linked model system between a detailed physiological simulation model (TREGRO) and a stand development model (ZELIG) (Laurence 1996). TREGRO (Weinstein et al 1991; Weinstein and Yanai 1994) was used to examine tree response to ozone of two tree species (*Pinus ponderosa, Abies concolor*) in the mixed conifer forest at two locations (Barton Flats and Sky Forest, a western site). Three variables, total tree biomass, fine root biomass, and root-to-shoot ratio were developed for several scenarios of ozone and precipitation. The values were then used by ZELIG (Urban 1990) to simulate the effects of ozone and the effects of ozone and precipitation on forest structure and composition over a 100-year period.

A similar process may be advantageous for future NuCM and Century simulations of N deposition and climate change effects on forests of the western United States. Both models clearly have strengths (soil chemistry for NuCM, carbon budget for Century), but neither model can examine all aspects of the ecosystem directly and indirectly affecting the combination of elevated ambient ozone, N deposition, and climate change.

Acknowledgments

Research supported by the U.S. Forest Service, California Air Resources Board, the Nevada Agricultural Experiment Station, and U.S. Environmental Protection Agency. Many thanks are due Robin Peterson, Richard Susfalk, and Kastlie Beck for technical help. D.W. Johnson's research was also funded by the USDA—

Forest Service Pacific Global Change program for the integrated modeling project. Century modeling work was supported through a grant from U.S. Environmental Protection Agency. This chapter has not been subject to either Forest Service or EPA policy review and should not be construed to represent the policies of either agency.

References

Fenn, M.E.; Poth, M.A.; Johnson, D.W. Evidence for nitrogen saturation in the San Bernardino Mountains in southern California. For. Ecol. Manage. 82:211–230; 1996.

Giorgi, F.; Brodeur, C.S.; Bates, G. Regional climate change scenarios over the United States produced with a nested regional climate model. J. Climate 7:375–394; 1994.

Johnson, D.W.; Lindberg, S.E., eds. Atmospheric deposition and nutrient cycling in forest ecosystems. Ecological studies 91. New York: Springer-Verlag; 1992.

Johnson, D.W.; Van Miegroet, H.; Lindberg, S.E.; Harrison, R.B.; Todd, D.E. Nutrient cycling in red spruce forests of the Great Smoky Mountains. Can. J. For. Res. 21:769–787; 1991.

Johnson, D.W.; Swank, W.T.; Vose, J.M. Simulated effects of atmospheric sulfur deposition on nutrient cycling in a mixed deciduous forest. Biogeochemistry 23:169–196; 1993.

Johnson, D.W.; Binkley, D.; Conklin, P. Simulated effects of atmospheric deposition, harvesting, and species change on nutrient cycling in a loblolly pine forest. For. Ecol. Manage. 76:29–45; 1995a.

Johnson, D.W.; Swank, W.T.; Vose, J.M. Effects of liming on soils and streamwaters in a deciduous forest: comparison of field results and simulations. J. Environ. Qual. 24:1104–1117; 1995b.

Klemmedson, J.O. Nitrogen and carbon regimes in an ecosystem of young dense ponderosa pine in Arizona. For. Sci. 21:163–168; 1975.

Laurence, J.A. Validation check of Zelig using data from the San Bernardino Mountains. Final report to EPA. Corvallis, OR: 1996.

Liu, S.; Munson, R.; Johnson, D.W.; Gherini, S.; Summers, K.; Hudson, R.; Wilkinson, K.; Pitelka, L.F. The Nutrient Cycling Model (NuCM): overview and ppplication. In: Johnson, D.W.; Lindberg, S.E., eds. Atmospheric deposition and forest nutrient ecological series 91. New York: Springer-Verlag; 1991:583–609.

Miller, P.R.; McBride, J.R.; Schilling, S.L.; Gomez, A.P. Trend of ozone damage to conifer forests between 1974 and 1988 in the San Bernardino Mountains of southern California. Proceedings of symposium on effects on western forests, 32nd annual meeting of the Air and Waste Management Association. Anaheim, CA: Air and Waste Management Association; 1989:309–323.

Miller, P.R.; Chow, J.; Watson, J.G., eds. Assessment of acidic deposition and ozone effects on conifer forest in the San Bernardino Mountains. Final report to California Air Resources Board A032-180. Sacramento, CA: California environmental protection agency; 1996.

Munson, R.K.; Liu, S; Gherinin, S.A.; Johnson, D.W.; Wilkinson, K.J.; Hudson, R.J.M.; White, K.S.; Summers, K.V. NuCM code version 2.0: An IBM PC code for simulating nutrient cycling in forest ecosystems. Final report for Electric Power Research Institute. Palo Alto, CA: Tetra Tech, Inc.; 1992.

Parton, W.J.; Schimel, D.S.; Cole, C.V.; Ojima, D.S. Analysis of factors controlling soil organic levels of grasslands in the Great Plains. Soil Sci. Soc. Am. J. 51:1173–1179; 1987.

Parton, W.J.; Stewart, J.W.B.; Cole, C.B. Dynamics of C, N, P, and S in grassland soils: a model. Biogeochemistry 5:109–131; 1988.

Pulliam, W.M.; Sievering, S.; Payton, R.; Wilson, J.; Mangis, D. Modeling canopy uptake of atmospheric N deposition: modification of the Century model. Bull. Ecol. Soc. Am. 76:219; 1995.

Ryan, M.G.; McMurtie, R.E.; Agren, G.I.; Hunt, E.R., Jr.; Aber, J.D.; Friend, A.D.; Rastetler, E.B.; Pulliam, W.M. Comparing models of ecosystem function for temperate conifer forests. II. Simulations of the effect of climate change. In: Bremeyer, A.I.; Hall, D.O.; Melillo, J.M.; Agren, G.I., eds. Global change: effects on coniferous forest and grasslands. New York: John Wiley & Sons; 1996:363–387.

Sandford, R.L., Jr.; Parton, W.J.; Ojima, D.S.; Lodge, D.J. Hurricane effects on soil organic matter dynamics and forest production in the Luquillo Experimental Forest, Puerto Rico: results of simulation modelling. Biotropica 23(4a):364–372; 1991.

Sievering, H.; Rusch, D.; Marquen, L. Nitric acid, particulate nitrate and ammonium in the continental free troposphere: nitrogen deposition to an alpine tundra ecosystem. Atmos. Environ. 30:23–27; 1996.

Urban, D.L. A versatile model to simulate forest pattern: a user's guide to ZELIG version 1.0. Charlottesville, VA: Environmental Sciences Department, University of Virginia; 1990.

Veblen, T.T.; Lorenz, D.C. The Colorado front range: a century of ecological change. Salt Lake City, UT: University of Utah Press; 1991.

Weinstein, D.A.; Yanai, R.D. Integrating the effects of simultaneous multiple stresses on plants using the simulation model TREGRO. J. Environ. Qual. 23:418–428; 1994.

Weinstein, D.A.; Beloin, R.M.; Yanai, R.D. Modeling changes in red spruce carbon balance and allocation in response to interacting ozone and nutrient stresses. Tree Physiol. 9:127–146; 1991.

17. Human Aspects of Air Quality in the San Bernardino Mountains

P.L. Winter

Introduction

The preceding chapters of this book have provided information on the ecological characteristics of the San Bernardino Mountains, as well as the effects of ozone and other air pollutants on vegetation and soil in the San Bernardinos, and additional interactions with air pollution and forest health. This chapter focuses on the human aspects of air quality in the San Bernardino Mountains because of the desired connection between available scientific information, wider understanding and acceptance of that information, and movement toward policy and behavioral changes necessary to act on scientific knowledge. Much of the literature has drawn on results from studies conducted in locations other than the San Bernardino Mountains. The author makes the assumption that the majority of findings from differing, or larger geographical regions are applicable to the San Bernardinos, the focus of this book. Where applicability is tenuous, it is so noted. Examining human–environment interactions was referred to by Stern (1993) as a critical second environmental science, wherein the linkages between human activity and environmental protection can be explored. A connection to perceptions of air quality and recreation is explored in the light of the significant resource production represented by recreation in the San Bernardino National Forest (as demonstrated by the reporting of more than 6 million recreation visitor-days in 1995 alone for the forest; USDA Forest Service 1996). An early proposal to further examine social and economic issues related to the San Bernardino Mountains is found in Taylor (1973), wherein valuable historical background of the area is provided.

How much of a scientific concern is there regarding air quality and forests? There appears to be agreement among scientists studying air pollution in forests that ozone is the greatest threat to forest growth, with southern California conifer and Appalachian high-elevation spruce-fir ecosystems suffering the most (de Steiguer et al. 1990). Growth declines are expected to accelerate over the next 20 years, even if emissions remain constant. Threats of damage from ozone exposure extend into Class I wilderness areas, where air quality is of major protection concern. Based on their findings from New Hampshire and Vermont, Lefohn and Manning (1995) suggest that ozone presents a considerable risk to Class I areas nationally, necessitating continued monitoring efforts as well as continued research on cause-and-effect relationships between ozone and vegetative health and vigor.

A general discussion of the varied costs from air pollution damage to vegetation revealed economic losses (in terms of loss of crops and increased need to use pesticides when pollutants make the vegetation more susceptible to pests); losses to ecological structure and function, which can also have economic linkages, including impaired long-term sustainability of a system; loss or altering of genetic diversity of populations; and loss of cultural values based on the intrinsic worth of ecosystems (Tingey et al. 1990).

Agreement over degree of damage from air pollution and limits to growth in forests is not global, however. Kandler and Innes (1995) contend that much of the discussion of decline and peril of central European forests is based on poor information and that growth rates of trees and stands in central Europe are higher than at any other time of recorded growth. It is their assessment that improved forest management (e.g., regular thinning and forest pest management) may actually be more important than other environmental changes in examining forest growth. The main pollutants examined by these authors were sulfur dioxide, nitrogen dioxide, and ozone.

Such discussions make it difficult for laypersons to understand what scientists know thus far and how it affects them, if at all. These difficulties complicate any hopes of support for current or future policies designed to improve air quality and protect ecosystems. By focusing on the interactions between humans and the environment, however, there may be a natural linkage that the layperson can understand and more readily embrace.

Human Dimensions of Air Quality

The examination of the human dimensions of air quality reflects the varied impacts that air pollution has on quality of life.

Physical Impacts of Poor Air Quality

"A critical question in many urban environments is not whether the air is unhealthy, but, given that the air quality is poor, how severely is health affected?"

(Hall 1996, p. 743). "At low concentrations ozone causes coughing, nausea, irritation of the eyes, nose and throat, and headaches. At higher concentrations, 150–200 parts per billion, it damages lung function" (Godlee 1991, p. 1539). The damaged lung function is revealed through reduced forced vital capacity, forced expiratory volume, and peak expiratory flow rate. These findings are present for both individuals with asthma (considered at highest risk) as well as in healthy volunteers. Effects of ozone are worsened with exercise and prolonged exposure (Godlee 1991).

A comprehensive assessment of health effects of ozone and inhalable particulate exposure was conducted in the South Coast Air Basin of southern California. That assessment revealed nearly every resident being exposed to levels of ozone above federal air quality standards at least some time during the year. Thirty percent of San Bernardino County residents were exposed to ozone levels well above these standards (28 pphm), but only 7% of Orange County residents were exposed to such levels. School-aged children received higher exposures and dosages, as well as experiencing more health effects. Eye and throat irritation were the most frequent health effects identified from ozone exposure (Hall 1996). Similar findings are cited by Aunan (1996). A State of California Air Resources Board–sponsored study shows the 24-hour average ozone level in 1994 in the Lake Arrowhead community (in the San Bernardino Mountains) to be almost double the levels seen in 11 other comparison communities being monitored. This important study is examining the long-term exposure of air pollutants and its impact on children's health in southern California (1995).

An analysis of hospital emergency outpatient visits and admissions and measurements of air pollutants including smoke, carbon monoxide, and nitrogen dioxide were analyzed for the greater Athens, Greece, area during 1988 (Pantazopoulou et al. 1995). During the winter season, the effect of all air pollutants was statistically significantly related to both the increase in the number of cardiac admissions and emergency respiratory admissions. These were analyzed by using multiple linear regression and controlling for meteorological variables. An explanation for the lack of a significant relationship during summer months between air pollution and hospital emergency department visits was not offered. Ozone was not examined in this study because of its relatively low levels in Athens. The authors cite a study by Touloumi and others demonstrating statistically significant effects of air pollution on mortality in Athens using time series analysis of 1983–1991 data. Caution should be taken, however, in applying health effects results to other areas because of differences in pollution sources and levels, weather, time activity patterns, overall health status, age distribution of the population, and access to medical care (Aunan 1996; Lipfert 1997). Progress in studies of air quality and health linkages has been somewhat limited by scarce resources and few multidisciplinary studies (Dunn and Kingham 1996).

Fisher (1976) contends that the cost of air pollution to the urbanite is significant in terms of bronchitis, emphysema, other respiratory diseases, and even death from lung cancer. Structural damage to the lungs may result from chronic exposure to ozone (Godlee 1991). However, the offsetting benefit of greater access to medical treatment for urbanites is viewed as counteracting some of this cost (Fisher 1976), and there may be some physical, behavioral, and attitudinal adaptation to pollution exposure over time (Kates 1976; Evans and Jacobs 1982; Godlee 1991; Bell et al. 1996). Whether this adaptation should be viewed as positive or negative is addressed in the following section, examining some of the psychological and behavioral impacts of air pollution exposure.

Psychological and Behavioral Impacts of Air Pollution

Examinations of the psychological and behavioral impacts of chronic air pollution exposure have been limited and are characterized by a primary focus on short-term exposure to pollutants or toxic odors in a laboratory setting (Evans 1994). There are, however, exceptions to this short-term exposure focus. The research conducted by Rotton and colleagues examines social data during high-ozone periods to identify linkages between pollutant exposure and psychological and behavioral indices. For example, Rotton (1993) studied incidences of reported sex crimes and found a statistically significant relationship between photochemical oxidant levels and obscene phone calls. The statistical relationship held, even after the effects of temperature were partialled out. A similarly disturbing crime trend was reported by Rotton and Frey (1985), wherein archival data spanning 2 years revealed a significant relationship between ozone and reports of family disturbances. Both of these studies are indicative of the negative social cost of ozone, most probably linked to increases in negative affect and irritability (characteristic of the larger line of research on environmental stressors). Such affective disturbances were also indicated by Rotton and Frey's 1984 article, revealing an increase in calls to police to handle psychiatric emergencies on days when photochemical oxidants were high.

Evans et al. (1982) examined instrumental and palliative coping strategies used in response to air pollution. Instrumental strategies included active forms of coping, such as gathering additional information about air pollution or changing behaviors by limiting time outside on smoggy days or using mass transit. Palliative strategies included more passive forms of coping, such as shifting one's perception of the air quality in one's immediate neighborhood. Length of residency was positively related to tendency to use palliative strategies such that individuals who had lived in a high-smog area longer were more likely to use these forms of coping to address air quality. Individuals living less time in a high-smog area were more likely to use instrumental strategies, including observing health advisories. These differences were indicated by long-term residents believing that smog had a less negative effect on them than did new residents, less of a belief that use of mass transit or car pooling would reduce smog, and a shifting of the amount of smog necessary in a visual scene before detection. Beliefs about the potential negative health effects of smog are important, as they have been found to be the best predictor of compliance with health advisories (Evans et al. 1988). In addition, beliefs about the ability to reduce smog are important in understanding compliance with health advisories (Evans et al. 1988). Viewing these studies in combination raises a significant concern about the well-being of long-term residents in areas with high smog levels. If residents believe that their actions will not make a difference and that the personal risk is minimal, they will be unlikely to comply with health advisories or to take any advocacy role to address pollution (Evans 1994).

Economic Impacts of Air Pollution

Lee (1993) contends that "unless environmental problems are recognized as economic problems they will not be addressed sensibly, no matter how sophisticated our scientific knowledge" (p. 1980). Economic impacts of air pollution on health are viewed as well-established components of cost models using impact pathway approaches that capitalize on the best scientific data available (Holland 1995). Assessment is based on "willingness to pay for improvement" or "willingness to accept" further risk, rather than the absolute value of health or life. Pollution is also costly to organizations through employee illnesses (Fisher 1976). Economic costs have also been applied to the evidenced decrease in lifespan of buildings, materials, and cultural monuments (Kucera and Fritz 1995). Finally, air pollution controls have economic impacts in the form of pollution abatement, regulation, monitoring, and research and development (Vob and Schmid 1991; Linn et al. 1994). Although poor air quality has costs attached, improvements in air quality offer economic benefits. Analyses of economic impacts of air pollution and regulatory efforts should be offset by net increases due to spending changes and additional in-migration as air quality improves (Lieu and Treyz 1992), generally calculated in terms of wage-rate increase equivalents.

However, costs in terms of loss to natural resources is much less clear, as data are more limited (Vob and Schmid 1991; Holland 1995) and an agreed-on, systematic commodity structure for environmental resources is still lacking (Smith 1993). Nevertheless, it has been argued that applying economic values to public resources is possible, can be held constant across a variety of ideologies, and can be tied into intervention efforts when translated into individual economic consequences (Bell 1995).

In an attempt to counteract the varied costs of air pollution, several regulations have been enacted. The next section explores some of those regulatory responses and their effectiveness.

Regulatory Responses and Their Effectiveness

"Air pollution control in the United States and most other countries of the world has been a series of disjointed, uncoordinated efforts dealing with narrow issues and specific industries" (McIllvaine 1993). McIllvaine (1994) contends that older energy plants are not affected by the regulatory guidelines established in the United States (see also Schulze 1993), and some pollutants are heavily focused on through regulation whereas others are overlooked. Policies on air pollution and controls should be set on a global level (Vob and Schmid 1991; McIllvaine 1993) and focused on the variety of sources and industries causing pollution, instead of a select few (Anonymous 1993; McIllvaine 1993). Only marginal results can be expected out of national efforts without cooperation at the international level (Michel Mousel, President of Agence de l'environnement et de la maîtrise de l'énergie in France, paraphrased in Rector 1992). "The design and enforcement of environmental regulations are based almost exclusively on prior scientific evidence concerning the links between particular environmental factors and health outcomes rather than on postintervention evaluation research" (Stokols 1992, p. 17).

Shortcomings in demand-side management practices for energy production have led to an increase in air pollution in some instances, rather than a decrease (Hall et al. 1995). In part, this is explained by the focus on installation of emission control technology, without consideration of actual emissions or air quality impacts when operating systems. The cleanest power plants are not necessarily the ones that are dispatched first (Hall et al. 1995). Air quality improvement measures addressed through the 1991 Air Quality Management Plan in the Los Angeles Basin were examined by Bae (1993). It is her assessment, based on 1991 data, that the program measures focused on reducing vehicle miles traveled have only a modest impact on air pollution reduction. The level of travel behavior changes that are assumed in the Air Quality Management Plan is not viewed as feasible, given the multiple determinants of travel behavior, including location of jobs some distances from households (Bae 1993). She argues that more transit use, ridesharing, and telecommuting are not needed to achieve clean air objectives, whereas technological solutions to the automobile emissions problem are of great importance.

Although some countries have been able to improve their air quality, developing countries are now considered to be at great risk because many lack the necessary monitoring and evaluation systems currently in place for major air pollutants in megacities such as Los Angeles and Tokyo (Mage et al. 1996). Tolley (1979, cited in Shukla and Parikh 1992) suggests that pollutant concentrations caused by industrial establishments, household fuels, and motor vehicles all increase with city size. However, sustainable development offers the opportunity to seek economic prosperity while guarding natural resources (Ramon John Hnatyshyn, paraphrased in Rector 1992). In fact, examinations of the ability to use pollution controls accompanied by economic growth and adoption of pollution abatement measures paint a more positive picture of the future as nations develop (Shulka and Parikh 1992). Such measures must be accompanied by a reduction in per capita consumption and a decrease in environmental throughput, especially in the more affluent nations (Vlek 1995).

Although several assessments have led to the conclusion that regulatory policies, have been ineffective, Rinquist's (1995) review of data from the Environmental Protection Agency spanning the mid-1970s to the late 1980s reveals that states with stronger regulatory programs have been effective in reducing atmospheric concentrations of nitrogen dioxide and sulfur dioxide. Therefore, although criticisms have been aimed at current regulatory policies, they are, in fact, demonstrating effectiveness.

Attitudinal Assessments and Behavioral Responses to Air Quality and Regulation Efforts

"Although the role of human behavior in air pollution is considerable, research has emphasized the technical, economic, and physical health aspects of the problem. Human behaviors such as energy use or choice of transportation mode directly affect the production of air pollution. Less directly, our attitudes and political and economic decisions about environmental protection also affect pollution levels" (Evans and Jacobs 1982, p. 105). The heavy focus on technology and emission standards (see, e.g., Zundel et al. 1995) directs our energies away from the potential gain that could be appreciated from examining human attitudes and behaviors that might also be addressed in pollution control. In fact, social scientists have much to offer in the way of pollution control (Vlek 1995).

Assessments examining awareness of pollution have pinpointed typical cues individuals used to detect the presence of air pollution, including dust, odor, discoloration and damage to property, respiratory irritation, eye irritation, and poor visibility (Evans and Jacobs 1982; Bell et al. 1996). Adaptation, perceived control, and other individual factors explain important differences in perceptions of air quality (Kates 1976; Evans and Jacobs 1982). Although laypersons tend to define or describe air pollution in terms of particular effects, experts are most likely to focus on causal factors and health effects. Even though individuals may be aware of air pollution, according to Evans and Jacobs (1982), there is little evidence that individuals significantly alter their outdoor activities in response to pollution, and very few actually lodge complaints to appropriate agencies about perceived poor air quality. A denial of personal adverse effects of air pollution and of the disadvantages of living in an area with poor air quality compound the pattern of nonresponsiveness to air quality issues (Liu 1996).

Possible Innovations in Intervention

The behavioral intervention literature offered by environmental and social psychologists offers an excellent reference point for potential innovations in pollution intervention, given the complexities of regulation in the face of attitudes, values, and behaviors (Oskamp 1991). However, much of the behavioral intervention literature has yet to be translated into efforts at the community or regional scale. Dwyer and colleagues' (1993) taxonomy of intervention strategies provides an organizing structure for the exploration of behavioral interventions. Intervention strategies may be targeted at the individual or group level, may be active or passive, and may involve antecedent or consequence conditions. The following discussion focuses on individuals or groups rather than organizations. Stern (1995) contends that "most pollution is caused by organizational behavior, not individual behavior." He also points out that the amount of pollution decrease that would result from changes in individual actions, assuming they could be changed by "simple acts of will" is marginal. Ozone should be noted as the exception to Stern's contention. Use of more environmentally sensitive technologies, educating the next generation, and taking collective action to promote technical and social changes are far more important initiatives (Stern 1995). Nevertheless, the individual can play an important role in collective action and changing direction, maintaining the merit of examining findings on individual and small group interventions. There is yet one more important caveat to address before moving forward, however. It is sometimes attractive among those holding knowledge about the environment and its protection to believe that to change individual behavior they need to share their knowledge with others, thereby affecting the others' knowledge base and consequently their values (Eden 1993). This presumably would then lead to a change in behavior, making the person more environmentally sensitive and responsible. Yet, there is no credible evidence that proenvironmental values are sufficient to guarantee environmental protection (Hutton and Ahtola 1991; Oskamp 1991; Stern 1995). In fact, although environmental concern is high among the general population, inaction remains the norm. Factors external to the individual seem to overwhelm many of the proenvironmental behaviors that might take place (Oskamp 1991; Stern 1995). The real interest will lie in long-term behavioral changes and commitments to environmentalism (Dwyer et al. 1993), accompanied by situations that allow individuals to act on their commitment (Stern 1995).

Individual-passive-antecedent strategies would involve things such as exposing an individual to written or oral information providing knowledge that would encourage a target behavior (Dwyer et al. 1993). Examples of brochures and signs that can be used in wilderness education (USDA 1989) serve as a tangible reminder of agency investment in these strategies. "This kind of strategy is based on the assumption that the relevant behavior is 'reasoned,' and that influencing the cognitive processes underlying behavior choices will also affect those choices themselves" (Vlek 1995, p. 15). These can be information- or morally based (Gardner and Stern 1996), although morality serves as less of an effective appeal than other bases (Eden 1993). Controlled studies of written prompts, in the forms of handbills or signs, have shown some effectiveness in increasing environmentally responsible behaviors (Dwyer et al. 1993). These written messages are usually simple, and ones that are worded positively appear to have more behavioral impact than those worded negatively. Modeling desired behaviors is another example of an individual-passive-antecedent strategy, wherein others are exposed to a video or first-hand viewing of environmentally responsible acts. Social demonstrations of desired behavior tend to work best under circumstances wherein the desired action is unclear (Cialdini 1988). Modeling assumes that the behavior and its underlying cognitive structure are strongly determined by social factors such as norms and customs, social comparison processes, privacy versus collectiveness, and public examples set by sometimes prominent members of society (Vlek 1995). It is important that models are detected by target observers, however, if one hopes for any behavioral impact. Modeling was used as part of a forest environmental education effort in dispersed recreation settings in hopes of encouraging responsible action. Although program educators were observed modeling during the majority of visitor contacts, recreationists who were interviewed rarely reported noticing their actions (Absher and Winter 1997).

Active-individual-antecedent strategies have also been applied for behavioral interventions. Of particular interest are the studies examining commitment, either written or oral, of individuals to behave in an environmentally responsible fashion. Written commitments appear to have longer-lasting effects than verbal ones under experimental conditions (Dwyer et al. 1993), and commitments made in public appear to strengthen people's personal, private commitment (Gardner and Stern 1996). Group strategies would be used in similar fashion to the individual ones described above, although team or competitive strategies can be brought into the intervention plan.

Individual-consequent strategies would involve methods such as direct feedback, indirect feedback, rewards, or penalties (Dwyer et al. 1993). An example of feedback is the monthly usage comparisons for current and past year appearing on gas and electric bills. Feedback, even on an intermittent basis, has been found to have some effect on energy consumption (Dwyer et al. 1993). I currently receive a rate-break on my electricity bill because I have permitted the utility to install a cycling device, wherein they can reduce my usage during peak periods. Penalties are found in the higher costs attached to usage above baseline levels. Such strategies can be effective; however, tax-based rewards and penalties tend to be more favorable to upper-income consumers (Howard 1993; Stern 1993). This would be especially true in circumstances wherein a desirable conservation behavior requires some significant up-front investment to be made, to enjoy the longer-term benefit of reduced consumption (Howard 1993). Group-consequent strategies would apply the same principles just described to a set of persons rather than being individually focused. Apartment dwellings are an example of one setting wherein group feedback and incentives have been used with some success (Dwyer et al. 1993). At the community level, caution must be exercised because not all communities distribute resources equally and not all members are necessarily happy. Concerns about individual freedom and procedural justice remain (Gardner and Stern 1996).

Human behavior and environmentally destructive or harmful actions need to be analyzed in terms of limiting factors such as technology, attitudes, knowledge, money, convenience, attention, and trust, as all are needed for behavior change (Stern 1993; Gardner and Stern 1996). It is clear that interventions must have multiple features, and because limiting factors vary by individuals and can change over time, they must be dynamic and flexible (Stern 1993; Gardner and Stern 1996). The ability for interventions to maintain behavioral changes over the long term has been underexplored. There is a distinction between changing behavior and maintaining that behavior change. For example, antecedents and consequences that were removed after an experiment was over took the behavioral changes with them (Dwyer et al. 1993). It is important that other contingencies, such as social approval or acceptance, fill in once the initial intervention effort subsides. In many cases, environmentally responsible actors are still made to feel that they are "on the fringe," despite the high level of environmental concern expressed (Eden 1993). An increased sense of competence, as well as the motivators of frugality and participation, can be intrinsically satisfying and may, in fact, serve to sustain environmentally responsible acts when external contingencies are not relied on (De Young 1996). At an individual level, it is important to recognize that behaving in an ecologically responsible fashion represents a wise personal investment, whereas behaving irresponsibly becomes even more costly on a daily basis (Howard 1993). Environmental behaviors must be viewed as personal responsibilities, as efficacious, and as possible (Eden 1993). "A good argument can be made that air pollution reduction may be society's most difficult environmental issue. . . . For air pollution reduction, effects cannot be measured on an individual level, and benefits may not accrue at all if others do not participate" (Hutton and Ahtola 1991, p. 253). The complexity of intervention is compounded by attitudinal and behavioral strategies used to cope with chronic exposure to air pollution (Evans 1994).

This chapter now turns to the discussion of the linkage between recreationists and air quality. Recreationists represent a large constituency, and there is some evidence that avid recreationists are more likely to be environmentally responsible (Winter 1996a), although the explanation for this difference is not yet fully understood.

Perceived Air Quality and Recreation

Amendments to the Clean Air Act in 1977 provide for the protection of visibility in national parks and wilderness areas (Class I areas) (Bell et al. 1985). The amendments came about in part because of the increased concern that scenic views in national parks were deteriorating and that existing clean air provisions were not protecting visual air quality in parks and rural areas (Borie 1988). Inclusion of references to "humanly detectable" reductions in visibility in sections of the Clean Air Act amendments draw attention to the importance of human perception (Daniel 1990). Haze is one of the most heavily relied predictors in human estimates of air quality (Malm et al. 1981; Molenar et al. 1994; Bell et al. 1996), and ratings of visual air quality appear to be reliable and valid (Latimer et al. 1981; Malm et al. 1981). Part of the management difficulty with reduced visibility from regional haze is that it is very complex. The atmospheric processes involved, along with the transport of haze over hundreds of miles from the source, explain just part of the complexity (Borie 1988). One example of the long-distance transport was provided by Borie (1988), who stated that one-third of the sulfates in the Grand Canyon were found to originate in urban southern California. There are additional values to natural resources running far beyond how they might be enjoyed first-hand through recreation. Wilderness affords therapeutic opportunities not available in other settings (Kale 1995). "Many of us receive satisfaction simply from knowing wild lands exist" (Dickensen 1981, p. 3), and the reported importance of aesthetic appreciation and ecological protection are demonstrative of those values (Manning and Valliere 1996). There are a multitude of values attached to wilderness, and recreation may be declining in importance among those values (McCloskey 1990). Nevertheless, recreation remains a significant commodity offered by forests.

Recreationists' Perspectives on Air Quality

Air quality and the more general scenic quality of an area are important considerations in the recreational experience. Visibility and visual quality are an important component of scenic beauty measures (Daniel 1990), and vistas are highly preferred among choices of outdoor scenes (Nelson and Loewen 1993). Of 11 potential values of wilderness presented to respondents in a survey of visitors to the Breadloaf Wilderness in Vermont, "Wilderness is a place to enjoy the beauty of nature" received the highest overall rating (mean of 5.55, scale was 1 = not at all important to 6 = extremely important) (Manning and Valliere 1996). Visitors to the Indian Peaks backcountry area were surveyed regarding general trip characteristics, sociodemographics, and physical resource attributes affecting trip satisfaction. For the 286 respondents, the most important attribute contributing to their backcountry visit was "clean fresh air" (average rating of 1.124 based on a scale from 1 to 9, 1 = most strongly adds to satisfaction and 9 = most strongly detracts from satisfaction). Overall, the most positively rated items were those dealing with a pollution-free environment and scenic beauty (Brown et al. 1977). Respondents to a survey examining recreation in Bryce Canyon (USDI 1980) revealed the high importance assigned to scenic views (rated as important, very important, or extremely important by approximately 99% of respondents) and to clean air (rated as important, very important, or extremely important by approximately 94% of respondents).

Physical Costs of Recreating in Pollution

An additional significant adverse impact of poor air quality on recreation, especially that which involves physical exertion, is well documented. Brunekreef and others (1994) studied young, healthy males participating in a series of amateur cycling training sessions and competitive races. Low concentrations of ozone were significantly associated with declines in lung function, as well as shortness of breath and wheeze. Eye irritation and cough were not significant. Even when the highest ozone concentration days were excluded from analysis, the significant relationship between ozone exposure and decreased lung function during exercise remained. Concentrations of other pollutants were low during the life of the study, and adjustments for humidity and air temperature did not eliminate the significant findings. Brunekreef and colleagues note that the exercise their subjects were engaged in was vigorous and that such significant relationships between low ozone levels and decreased lung function were not found in less strenuous exercise situations.

A 10% reduction in aerobic capacity of subjects breathing ozone comparable with peak levels in smog-prone cities has been demonstrated. Long-term exposure has also been shown to have a cost, through obstruction in respiratory bronchioles, for example (Stamford 1990).

Gong and Krishnareddy (1995) summarize the effects of pollutants on exercise as well. High ozone levels are known to cause restrictive lung dysfunction, leading to coughing, chest tightness, and shortness of breath; high carbon monoxide levels can impair oxygen delivery; sulfur dioxide worsens nasal symptoms in people with allergies and causes bronchospasms in asthmatics; and airway irritation from fine particulates can lead to bronchospasms. Each of these is contrary to the positive goal of outdoor exercise, which is to increase and maintain physical health.

Recreationists are counseled to avoid the adverse effects of poor air quality. Castleman (1991) advises readers to jog in the early morning hours when air pollutants, especially auto-related ones, are at their lowest. He states that warm months are the worst time for ozone, a key component of smog. Ozone effects on the jogger can include tickling in the throat, raspiness, coughing, shortness of breath, tightness in the chest, and possibly headache and nausea. Castleman recommends avoiding physical exertion during those periods when air quality is the worst; Stamford (1990) provides similar advice for city dwellers interested in exercise. Health advisories, however, are typically ignored among adult urban dwellers, as previously discussed (Evans et al. 1982).

Diminished Enjoyment of the Recreational Experience

Scenery is very important in the appreciation and enjoyment of natural resource settings (Daniel 1990; Taylor and Winter 1995), especially viewing it through clean, fresh air (USDA 1990). Given this importance of scenery, and the potential physical effects that can detract from a visit to natural resource areas, it is clear that air quality plays an important role in preservation and existence of natural resource areas. Wilderness offers an opportunity to escape from the everyday environment of buildings and streets and from urban ugliness and environmental concerns (Kaplan and Talbot 1983). Especially true for urbanites, wilderness offers a reprieve from the intense daily filtering of our environment, which leads to inattention of details in a setting (Milgram 1976). Air quality is a component of the overall scenery, and "pleasant scenery is often the key to the quality of an outdoor recreation activity" (Cordell et al. 1990, p. 87). Evidence of pollution can clearly detract from the nature experience. A survey of Coloradoans revealed "viewing the scenery" as the most important experience in a wilderness visit out of 20 possible experiences presented (Walsh et al. 1981). The wilderness experience is supposed to offer contact with "unadulterated expressions of nature" (McCloskey 1990, p. 17). However, the degree to which the relationship between atmospheric pollutants and wilderness experiences goes beyond valuing clean air and good visibility is not well known (Watson and Williams 1995). Concerns about air quality affecting recreation extend into the San Bernardino Mountains specifically, as demonstrated in the following sections.

San Gorgonio Wilderness Case Study

The San Gorgonio Wilderness, located in the San Bernardino National Forest in southern California, encompasses 56,722 acres (Hendee et al. 1990). It was among the 54 wildernesses specifically designated in the Wilderness Act of 1964,

Public Law 88-577; and its original acreage of 33,898 was increased to the present size with the California Wilderness Act of 1984, Public Law 98-425 (Browning et al. 1988). Originally proposed as a downhill ski resort, hiking enthusiasts banded together as "Defenders of the San Gorgonio Wilderness" and successfully fought for wilderness designation (Wondelleck and Yaffee 1994). According to comparisons using recreation visits per acre from 1986, the San Gorgonio ranks as the second most-visited wilderness in California and falls within the top 12 in the nation (Hendee et al. 1990). The high use is managed through a mandatory permit system for both day and overnight use year-round. A volunteer association of approximately 120 members contributes thousands of person-hours annually (Wondelleck and Yaffee 1994), adding to the Forest Service's efforts for wilderness patrols, trail maintenance, and visitor education.

Visitors to the San Gorgonio Wilderness

Located within an hour's drive of more than 7.5 million people (Wondelleck and Yaffee 1994), the vast majority (98%) of San Gorgonio Wilderness visitors come from the five-county Los Angeles metropolitan area (Winter and Tocco 1995). The reported pattern of geographically proximate wilderness visitation is in keeping with wildernesses near other large urban centers (Hendee et al. 1990).

Fall Visitors' Air Quality Experiences

A wilderness recreation survey was conducted in Fall 1994 on the San Gorgonio Wilderness (Winter 1996b). Recreationists were contacted at a trailhead and asked to complete a mini-survey. Those expressing a willingness to participate in a longer mailed survey were queried regarding trip characteristics, reactions to potential wilderness management strategies, wilderness knowledge, and sociodemographics. Of the 65 Fall respondents to the mailed survey, approximately 75% were on a repeat San Gorgonio Wilderness visit in 1994. The average years reported for San Gorgonio Wilderness repeat visitors was 11 years.

Respondents were asked a series of questions regarding perceived air quality in the San Gorgonio Wilderness. Items queried are typical indicators reported in judgments of environmental quality (Evans and Jacobs 1982), and perceptions of areas are valuable in assessing the conditions of environments (Craik and Zube 1976). Almost 37% of San Gorgonio visitors reported an impaired view of the landscape due to poor air quality, whereas only 8% reported experiencing difficulty breathing and 2% reported "burning eyes due to air pollution." Only a few (6%) reported a lack of discomfort or impaired view. Therefore, the majority listed at least some adverse experience due to poor air quality. These questions were responded to based on recollections of the experience, through the mailed survey completed within 1–3 months maximum after the wilderness visit. Recalled experiences were of interest because of their longer-term potential role in wilderness attitudes and opinions. There may be some very distinct differences between what would have been reported in situ and what was recalled for the mailed survey (Stewart and Hull 1994). Assessments of impaired view would have been interesting to contrast, when made on site, with data from the IMPROVE program as presented in Copeland (this volume).

Wilderness Volunteers' Perceptions of San Gorgonio Air Quality

In 1995, the San Gorgonio Volunteer Association members were contacted by the author and invited to participate in survey querying their wilderness volunteer experiences. They were asked to provide information on their degree of volunteerism, perceived uses of the wilderness, perceptions of visitors, perceptions of wilderness quality, opinions regarding potential management interventions, benefits and drawback to wilderness volunteerism, and sociodemographics.

The 90 volunteer respondents averaged 3.5 years (median) of volunteer work on the San Gorgonio Wilderness. A significant monthly contribution of person-hours was reported, with 31 hours as the median. Because of their familiarity with the wilderness and its attributes, the San Gorgonio Volunteer Association was viewed as an ideal information source regarding air quality. Respondents were asked to indicate what wilderness visitors should expect to experience on their visits to the San Gorgonio. Almost half (49.5%) indicated that impaired view of the landscape due to air pollution should be expected, whereas 11% indicated that burning eyes due to air pollution should be expected, and about 9% reported that visitors should expect some difficulty breathing. Slightly less than half (46.2%) indicated that none of those effects should be expected, that the air quality during their wilderness visits had been fine. Again, querying recollections of air quality and its impact on the recreational visit to the San Gorgonio may have yielded very different results than responses in situ may have provided. However, it is clear from these data that approximately half the responding volunteers perceived some air quality problem within the San Gorgonio Wilderness that would affect recreation.

Recreationists' Willingness to Ensure Improved Air Quality

When a problem with air quality is perceived, as it was in the San Gorgonio Wilderness case study previously described, how willing are individuals to make some investment to change their experiences? Results from other geographical areas provide information useful toward solutions for the San Gorgonio.

Trade-Offs for Visibility in the Grand Canyon

An experiment by Bell and others (1985) demonstrates the willingness of individuals to make trade-offs to increase visibility of vistas in recreational settings. Experimental subjects were presented with three scenic vistas of the Grand Canyon with varying degrees of visibility (low, moderate, high), information on driving time, and information on waiting time to see an archeological site. Based on a visit/choice scenario, experimental subjects showed a clear preference for good visibility and short waiting and driving times. In fact, the average person was willing to spend almost an extra hour in driving time (55.8 minutes) for one incremental improvement in visibility of the San Francisco Peaks and about half an hour in waiting time (30.4 minutes). Other potential factors in the decision-making process were not considered such as sociodemographics, entrance fees, or crowds. Bell and colleagues contend that "one way of documenting how people value visibility is through their willingness to increase effort and time commitment (costs) for an increase in visibility (reward). Visibility should be considered a natural resource that contributes positive or reward outcomes to a recreation setting" (Bell et al. 1985, p. 473).

Willingness to Pay for Improved Air Quality Among Recreationists

Persons who had registered for races or had purchased sporting goods at a community sporting goods store were contacted in a questionnaire mailing in Louisiana (Farber and Rambaldi 1993). The researchers contend that outdoor exercise enthusiasts should be better informed about the relationships between air quality and health, due to their generally higher education levels. Respondents ($n = 790$, with a 61.8% response rate) were informed that the local area had 14 days of unhealthful air quality in the prior year. They were asked to select a range of money that represented what they would be willing to pay annually in taxes or product costs to guarantee that the local air was always in the healthful range. The median response fell in the \$91–\$100 interval, for annual willingness to pay, with a daily median of \$6.79 per person per day. Of greatest interest in this study is the indicated willingness to make a financial commitment to improved air quality, in keeping with other studies cited by Farber and Rambaldi (1993). Exact ranges of willingness to pay for cleaner air in the San Bernardinos are not known.

Maintaining Air Quality in Natural Areas Extends Beyond Recreational Concerns

Valuing natural areas' relationships to air quality was revealed by wilderness users and nonusers alike in a survey examining potential purposes of wilderness. Virden (1990) reported that 80.1% of wilderness users and 80.9% of wilderness nonusers listed the protection of air quality among the top five purposes of wilderness. Coloradoans were asked which, among 13 reasons provided, were the most important reasons for valuing existing or potential wilderness areas. Protecting air quality was the third highest reason provided for wilderness, with 77.2% indicating that is was either "very" or "extremely important" (Haas et al. 1986). Significant differences in these responses were not found between wilderness visitors and those who had not visited Colorado wildernesses. The clean air found in wilderness areas can also help protect the scenic integrity of areas adjacent to them (Cordell et al. 1990). However, as Lackey (1996) points out, although there is public support of the environment and an appreciation of it, publics vary in their views and intensities of opinions. The decisions made in environmental management occur in a larger political context rarely addressed in public opinion polls.

Implications for the San Bernardino National Forest

It has been argued that the value of wilderness to the present generation will be dwarfed by the values accrued to future generations (Haas et al. 1986). Several strategies are already available for relaying information to visitors at natural resource areas (USDA Forest Service 1989). Unfortunately, talks specifically focused on air quality in the San Bernardinos appear to be of little interest to recreationists (Chavez and Mainieri 1995). However, information about the natural resources in the area **is** of interest. The innovative interventions described earlier in this chapter may offer additional intervention methods, which when paired with those already in place, can serve as an effective mechanism in creating the awareness, agreement, and behavioral commitments necessary to improve air quality. More than 6 million recreation visitor-days were reported by the San Bernardino National Forest in 1995 (USDA Forest Service 1996). These visitor-days represent a critical constituency base for improving the understanding of air quality concerns in the San Bernardino Mountains and hopes for support for policy-based changes. In addition, residents of the Los Angeles Basin may have definite interests in the health and sustainability of the flora and fauna of the San Bernardinos, which go beyond recreation-related concerns. Residents in the San Bernardino mountain range, both year-round and seasonal, have a vested interest as well.

Conclusions

"Clean air cannot be achieved through study and research. Clean air can only be achieved through an appropriate amount of study followed by a major commitment to a plan of action" (Schulze 1993, p. 21). Better approaches to presenting options in environmental management and their implications need to be developed (Lackey 1996). We desperately need to present scientific findings in an informative and interesting manner to build public support (Wiggins 1997). Although knowledge about the San Bernardino Mountains and degree of damage to the forests is fairly well advanced, translations into understandable information for the layperson will very much depend on personally relevant and important information being relayed. Poor air quality affects the ecosystem and humans within it through reduced environmental experiential quality and health risks. Being made aware of these conditions is not enough (Hutton and Ahtola 1991). Air quality scientists also need to translate these conditions into personal impacts, as well as offering practical and possible solutions to reduce impacts (Hutton and Ahtola 1991; Eden 1993; Vlek 1995). They also have to ensure that the information provided is specific to the area, relaying and understanding of local conditions and their related hazards (Peltu 1991). Scientists must work hard at relaying the balances of varying environmental risks to publics whose perceptions of the greatest hazards may not match technical experts' assessments. Priorities and legislative agendas are more likely to be driven by public perceptions than by scientists' perceptions, increasing the importance of effective and credible communication of scientific knowledge (Slovic 1997). Although recreationists were this chapter's main focus for such intervention efforts, more broadly focused interventions also need to be explored.

References

Absher, J.; Winter, P.L. Evaluation of the 1995 eco-team program. Riverside, CA: USDA Forest Service, Pacific Southwest Research Station; 1997.

Anonymous. Seeing the parkland through the haze. Sci. News 143:95; 1993.

Aunan, K. Exposure-response functions for health effects of air pollutants based on epidemiological findings. Risk Anal. 16:693–709; 1996.

Bae, C.C. Air quality and travel behavior: untying the knot. J. Amer. Planning Assoc. 59:65–74; 1993.

Bell, P.A. It's the economy psychologist! Paper presented at the American Psychological Association Convention, August, New York. Ft. Collins, CO: Colorado State University; 1995.

Bell, P.A.; Greene, T.E.; Fisher, J.D.; Baum, A. Environmental psychology, 4th ed. Fort Worth, TX: Holt, Rinehart & Winston, Inc.; 1996.

Bell, P.A.; Malm, W.; Loomis, R.J.; McGlothin, G.E. Impact of impaired visibility on visitor enjoyment of the Grand Canyon: a test of an Ordered Logit Utility Model. Environ. Behav. 17:459–474; 1985.

Borie, L. A tale of sites unseen. Sierra 73:30–31, 33–36; 1988.

Brown, P.J.; Haas, G.E.; Manfredo, M.J. Identifying resource attributes providing opportunities for dispersed recreation. Final report. Ft. Collins, CO: College of Forestry and Natural Resources, Colorado State Univ.; 1977.

Browning, J.A.; Hendee, J.C.; Roggenbuck, J.W. 103 Wilderness laws: milestones and management direction in wilderness legislation, 1964–1987. Bull. 51. Moscow, ID: University of Idaho, College of Forestry, Wildlife and Range Sciences; 1988.

Brunekreef, B.; Hoek, G.; Breugelmans, O.; Leentvaar, M. Respiratory effects of low-level photochemical air pollution in amateur cyclists. Am. J. Respir. Crit. Care. Med. 150:962–966; 1994.

Castleman, M. A jog in the smog. Sierra 76:20; 1991.

Chavez, D.J.; Mainieri, T. Recreation day use series report 1: San Bernardino National Forest. Report submitted to the San Bernardino National Forest. Riverside, CA: USDA Forest Service, Southwest Pacific Research Station; 1995.

Cialdini, R.B. Influence: science and practice, 2nd ed. Glenview, IL: Scott Foresman, Little, Brown College Div.; 1988.

Cordell, H.K.; Bergstrom, J.C.; Hartmann, L.A.; English, B.K. An analysis of the outdoor recreation and wilderness situation in the United States: 1989–2040. A technical document supporting the 1989 USDA Forest Service RPA assessment. Gen. tech. rep. RM-189. Ft. Collins, CO: USDA Forest Service; 1990.

Craik, K.H.; Zube, E.H. Perceiving environmental quality, research and applications. New York: Plenum Press; 1976.

Daniel, T.C. Measuring the quality of the natural environment: a psychophysical approach. Am. Psychol. 45:633–637; 1990.

deSteiguer, J.E.; Pye, J.M.; Love, C.S. Air pollution damage to U.S. forests: a survey of perceptions and estimates by scientists. J. For. 88(8):17–22; 1990.

De Young, R. Some psychological aspects of reduced consumption behavior: the role of intrinsic satisfaction and competence motivation. Environ. Behav. 28:358–409; 1996.

Dickenson, R.E. Wilderness values in the national parks. Wilderness Resource Distinguished Lectureship. Moscow, ID: University of Idaho Wilderness Research Center; 1981.

Dunn, C.; Kingham, S. Establishing links between air quality and health: searching for the impossible? Soc. Sci. Med. 42:831–841; 1996.
Dwyer, W.O.; Leeming, F.C.; Cobern, M.K.; Porter, B.E.; Jackson, J.M. Critical review of behavioral interventions to preserve the environment: research since 1980. Environ. Behav. 25:275–321; 1993.
Eden, S.E. Individual environmental responsibility and its role in public environmentalism. Environ. Planning 25:1743–1758; 1993.
Evans, G.W. The psychological costs of chronic exposure to ambient air pollution. In: Isaacson, R.L.; Jensen, K.F., eds. The vulnerable brain and environmental risks. vol. 3. Toxins in air and water. New York: Plenum Press; 1994:167–182.
Evans, G.W.; Jacobs, S.V. Air pollution and human behavior. In: Evans, G.W., ed. Environmental stress. Cambridge: Cambridge University Press; 1982:105–132.
Evans, G.W.; Jacobs, S.V.; Frager, N.B. Behavioral responses to air pollution. In: Baum, A.; Singer, J.E., eds. Advances in environmental psychology. vol. 4. Environment and health. New York: Lawrence Erlbaum Associates; 1982:237–269.
Evans, G.W.; Colome, S.D.; Shearer, D.F. Psychological reactions to air pollution. Environ. Res. 45:1–15; 1988.
Farber, S.; Rambaldi, A. Willingness to pay for air quality: the case of outdoor exercise. Contemp. Policy Issues 11:19–30; 1993.
Fisher, C.S. The urban experience. New York: Harcourt Brace Jovanovich; 1976.
Gardner, G.T.; Stern, P.C. Environmental problems and human behavior. Needham Heights, MA: Allyn & Bacon; 1996.
Godlee, F. Air pollution: II-road traffic and modern industry. BMJ. 303:1539–1543; 1991.
Gong, H., Jr.; Krishnareddy, S. How pollution and airborne allergens affect exercise. Physician Sportsmed. 23:35–38, 39–42; 1995.
Haas, G.E.; Hermann, E.; Walsh, R. Wilderness values. Natural Areas J. 6:37–43; 1986.
Hall, D.C.; Win, M.S.; Hall, J.V. Air pollution impacts from demand–side management. Energy 20:27–35; 1995.
Hall, J.V. Assessing health effects of air pollution. Atmos. Environ. 30:743–746; 1996.
Hendee, J.C.; Stankey, G.H.; Lucas, R.C. Wilderness management. Golden CO: International Wilderness Leadership Foundation; 1990.
Holland, M.R. Assessment of the economic costs of damage caused by air-pollution. Water Air Soil Pollut. 85:2583–2588; 1995.
Howard, G.S. Thoughts on saving our planet: political, economic, cultural and bureaucratic impediments to ecological activism. Counseling Psychol. 21:597–617; 1993.
Hutton, R.B.; Ahtola, O.T. Consumer response to a five-year campaign to combat air pollution. J. Public Policy Marketing 10:242–256; 1991.
Kale, M. Wilderness and the human spirit. Am. For. 101:39–43; 1995.
Kandler, O.; Innes, J.L. Air pollution and forest decline in central Europe. Environ. Pollut. 90:171–180; 1995.
Kaplan, S.; Talbot, J.F. Psychological benefits of a wilderness experience. In: Altman, I.; Wohlwill, J.F., eds. Behavior and the natural environment. New York: Plenum Press; 1983:163–203.
Kates, R.W. Experiencing the environment as hazard. In: Proshansky, H.M.; Ittelson, W.H.; Rivlin, L.G., eds. Environmental psychology, 2nd ed. New York: Holt, Rinehart & Winston; 1976:401–418.
Kucera, V.; Fritz, S. Direct and indirect effects on materials including cultural monuments. Water Air Soil Pollut. 85:153–165; 1995
Lackey, R.T. Ecological risk assessment. In: Molak, V., ed. Fundamentals of risk analysis and risk management. New York: CRC Press/Lewis Publishers; 1996.
Latimer, D.A.; Hogo, H.; Daniel, T.C. The effects of atmospheric optical conditions on perceived scenic beauty. Atmos. Environ. 15:1865–1874; 1981.
Lee, D.R. An economist's perspective on air pollution. Environ. Sci. Technol. 27:1980–1982; 1993.

Lefohn, A.S.; Manning, W.J. Ozone exposures near class I wilderness areas in New Hampshire and Vermont. Atmos. Environ. 29:601–606; 1995.

Lieu, S.; Treyz, G.I. Estimating the economic and demographic effects of an air quality management plan: the case of southern California. Environ. Planning 24:1799–1811; 1992.

Linn, X.; Polenske, K.R.; Robinson, K. Economic impact analyses in U.S. state and local air pollution control agencies: a survey. J. Air Waste Manage. Assoc 44:134–140; 1994.

Lipfert, F.W. Air pollution and human health: perspectives for the '90s and beyond Risk Analysis 17:137–146; 1997.

Liu, F. Urban ozone plumes and population distribution by income and race: a case study of New York and Philadelphia. J. Air Waste Manage. Assoc. 46:207–215; 1996.

Mage, D.; Ozolins, G.; Peterson, P.; Webster, A.; Orthofer, R.; Vandeweerd, V.; Gwynne, M. Urban air pollution in megacities of the world. Atmos. Environ. 30:681–686; 1996.

Malm, W.; Kelley, K.; Molenar, J.; Daniel, T. Human perception of visual air quality (uniform haze). Atmos. Environ. 15:1875–1890; 1981.

Manning, R.E.; Valliere, W.A. Environmental values, environmental ethics, and wilderness management. Int. J. Wilderness 2:27–32; 1996.

McCloskey, M. Evolving perspectives on wilderness values: putting wilderness values in order. In: Preparing to manage wilderness in the 21st century: proceedings of the conference, State Botanical Gardens, April 4–6. Gen. tech. rep. SE-66. Athens, GA: USDA, Southeastern Experiment Station; 1990.

McIllvaine, R.W. Air pollution control and waste management in a new world. J. Air Waste Manage. Assoc. 43:293–297; 1993.

McIllvaine, R.W. Air pollution challenges and forecasts. J. Air Waste Manage. Assoc. 44:248–252; 1994.

Milgram, S. The experience of living in cities. In: Proshansky, H.M.; Ittelson, W.H.; Rivlin, L.G., eds. Environmental Psychology, 2nd ed.: people and their physical settings. New York: Holt, Rinehart and Winston; 1976:494–507.

Molenar, J.V.; Malm, W.C.; Johnson, C.E. Visual air quality simulation techniques. Atmos. Environ. 28:1055–1063; 1994

Nelson, T.M.; Loewen, L.J. Factors affecting perception of outdoor public environments. Perceptual Motor Skills 76:139–146; 1993.

Oskamp, S. Attitudes and opinions, 2nd ed. Englewood Cliffs, NJ: Prentice Hall; 1991.

Pantazopoulou, A.; Katsouyanni, K.; Kourea-Kremastinou, J.; Trichopoulos, D. Short-term effects of air pollution on hospital emergency outpatient visits and admissions in the greater Athens, Greece, area. Environ. Res. 69:31–36; 1995.

Peltu, M. Risk perception and the real world. N. Scientist 1782:4; 1991.

Rector, A.B. Towards the year 2000: critical issues in the global environment. J. Air Waste Manage. Assoc. 42:1573–1576; 1992.

Ringquist, E.J. Is "effective regulation" always oxymoronic?: the states and ambient air quality. Soc. Sci. Q. 76:69–87; 1995.

Rotton, J. (1993) Atmospheric and temporal correlates of sex crimes: endogenous factors do not explain seasonal differences in rape. Environ. Behav. 25:625–642; 1993.

Rotton, J.; Frey, J. Psychological costs of air pollution: atmospheric conditions, seasonal trends, and psychiatric emergencies. Pop. Environ. 7:3–16; 1984.

Rotton, J.; Frey, J. Air pollution, weather, and violent crimes: concomitant time-series analysis of archival data. J. Personality Soc. Psychol. 49:1207–1220; 1985.

Schulze, R.H. The 20-year history of the evolution of air pollution control legislation in the USA. Atmos. Environ. 27B:15–22; 1993.

Shukla, V.; Parikh, K. The environmental consequences of urban growth: cross-national perspectives on economic development, air pollution, and city size. Urban Geogr. 13:422–449; 1992.

Slovic, P. Public perception of risk. Environ. Health 59:22–23, 54; 1997.

Smith, V.K. Nonmarket valuation of environmental resources: an interpretive appraisal. Land Econ. 69:1–26; 1993.
Stamford, B. Exercise and air pollution. Physician Sportsmed. 18:153–154; 1990.
Stern, P.C. A second environmental science: human–environment interactions. Science 260:1897–1899; 1993.
Stern, P.C. Understanding and overcoming barriers to pollution prevention. Paper presented at the American Psychological Association Convention, August, New York. Washington, DC: National Research Council; 1995.
Stewart, W.P.; Hull, R.B. Capturing the present moments of leisure: concerns of in situ research. Paper presented at the NRPA leisure research symposium, October 12–16, Minneapolis, MN. College Station, TX: Texas A&M University, and Blacksburg, VA: Virginia Polytechnic Institute and State University; 1995.
Stokols, D. Establishing and maintaining healthy environments: toward a social ecology of health promotion. Am. Psychol. 47:6–22; 1992.
Taylor, D.; Winter, P. Environmental values, ethics and depreciative behavior. In: Chavez, D.J., tech. coord. Proceedings of the second symposium on social aspects and recreation research. Gen. tech. rep. PSW-156. Albany, CA: USDA, Pacific Southwest Research Station; 1995:59–66.
Taylor, O.C. Oxidant air pollutant effects on a western coniferous forest ecosystem: historical background and proposed systems study of the San Bernardino Mountain area. Riverside, CA: Statewide Air Pollution Research Center, University of California; 1973.
Tingey, D.T.; Hodsett, W.E.; Henderson, S. Definition of adverse effects for the purpose of establishing secondary national ambient air quality standards. J. Environ. Qual. 19:635–639; 1990.
USDA, Forest Service. Rocky Mountain region annual report. Ft. Collins: USDA; 1990.
USDA, Forest Service. San Bernardino National Forest recreation use report, 1995. San Bernardino, CA: USDA; 1996.
USDA, Forest Service Recreation Staff. Ideas for wilderness information and education. Washington, DC: USDA; 1989.
USDI. Descriptive analysis of the data from the Bryce Canyon National Park visitor survey. Washington, DC: Economic Staff, Office of Policy Analysis; 1980.
Virden, R.J. A comparison study of wilderness users and nonusers: implications for managers and policymakers. J. Park Recreation Admin. 8:13–24; 1990.
Vlek, C.A. Technical versus socio-behavioral solutions to environmental problems: psychology's unexploited potential. Paper presented at the American Psychological Association Convention, August, New York. The Netherlands: University of Groningen; 1995.
Vob, A.; Schmid, G. Cost-effectiveness of air-pollution control measures. Energy 16:1215–1224; 1991.
Walsh, R.G.; Gillman, R.A.; Loomis, J.B. Wilderness resource economics: recreation use and preservation values. Ft. Collins, CO: Department of Economics, Colorado State University; 1981.
Watson, A.E.; Williams, D.R. Priorities for human experience research in wilderness. Trends 32:14–18; 1995.
Wiggins, R.C. Scientists must clarify the societal relevance of research. Scientist 8(3):10; 1997.
Winter, P.L. Environmental concern and environmental action: how do recreationists fare? The sixth international symposium on society and resource management: social behavior, natural resources, and the environment, May 18–23, University Park, PA. University Park, PA: Pennsylvania State University; 1996a.
Winter, P.L. San Gorgonio Wilderness visitor survey. Technical report. Riverside, CA: USDA, Pacific Southwest Station; 1996b.
Winter, P.L.; Tocco, L. Capturing urban wilderness visitation on the San Gorgonio. Paper

presented at the 4th international outdoor recreation and tourism trends symposium, St. Paul, MN, May 14–17. Riverside, CA: USDA, Pacific Southwest Station; 1995.
Wondelleck, J.M.; Yaffee, S.L. Building bridges across agency boundaries: in search of excellence in the United States Forest Service. Ann Arbor, MI: School of Natural Resources and the Environment, The University of Michigan; 1994.
Zundel, T.; Rentz, O.; Dorn, R.; Jattke, A.; Wietschel, M. Control techniques and strategies for regional air pollution control from energy and industrial sectors. Water Air Soil Pollut. 85:213–224; 1995.

4. Synthesis and Conclusions

18. Assessment of Ecological Risks and Implications for Policy and Management in the San Bernardino Mountains

P.R. Miller and J.R. McBride

Introduction

Air pollution and its relationship to the forest ecosystems of the San Bernardino Mountains must be evaluated in relation to a complex web of resource values and management issues. A clear identification and assessment of these values and issues are essential for the formulation and application of future policies needed to sustain the productivity of the forest ecosystem. Future policy must be formulated by resource managers, land-use planners, and stakeholders from the community. The needs and expectations of future generations must be one of the important issues addressed by this collaborative stewardship process.

Ecological Risk Assessment as a Means of Identifying Important Issues

Ecological risk assessment evaluates the likelihood that adverse ecological effects are occurring or may occur as a result of exposure to one or more stressors (Lackey 1994, 1997). Ecological risk assessment has been used most effectively for situations in which a single pollutant affects specific measurable outcomes or end points (Lackey 1994). Some critics maintain that the ecological risk assessment process may become distorted because there is not general agreement about the appropriateness of particular end points or properties measured (Lackey

1995). A more value-neutral approach is ecological alternatives analysis, which tends to evaluate stressors effects on a wider spectrum of possible end points and is not likely to be limited to highly valued particular end points. A current challenge is to use ecological risk assessment to implement ecosystem management (Lackey 1997)

Myers (1997) has suggested that the three toughest issues facing society are population growth, the continuing adverse impact of technology, and increasing consumption. Each of these will act as sources of stressors on the forests of the San Bernardino Mountains through their relationship to air pollution. Increased population means more individuals will need to occupy the same land area, bringing with them the impacts of adverse technology, including air pollution. For example, more on-site pressure might be expected from off-road vehicles and foot traffic as well as the accumulating impact from regional air pollution. The increased consumption of fossil fuel by vehicles with higher fuel consumption, namely, sports utility vehicles, is one example of a popular trend that will exacerbate the air pollution problem. More complete compliance with existing air pollution control regulations and the possible adoption of new regulations may be required to maintain air quality at levels at least similar to the present.

Possible events related to air pollution in the San Bernardino Mountains are identified in Table 18.1. The events were identified in part in an informal workshop conducted by the U.S. Forest Service at the Riverside Fire Laboratory in July 1997 to introduce ecological risk assessment to land managers and specialists concerned with forest lands in southern California. Additional events were identified by the authors. The probability of occurrence of these adverse events during the next 30 years has been ranked as high, intermediate, or low. Many adverse events that are not always directly related or interactive with air pollution have been omitted from the list (e.g., major channel erosion from high rainfall years). We clearly recognize that the risk of large stand-replacing fires belongs in the highest likelihood category but because the focus of this book is on air pollution, the list of events in Table 18.1 would be superimposed on the effects of and recovery from a major fire disturbance. The list focuses on air pollution–related events to provide an ecological risk assessment that is relevant to the focus of this book.

Furthermore, the items in Table 18.1 are an initial attempt at the problem formulation phase of the ecological risk assessment method (U.S. EPA 1992) (Fig. 18.1). The analysis phase (Fig. 18.1) of the ecological risk assessment method can best be represented by a general model of ecosystem response (Miller et al. 1998) (Fig. 18.2), in which various stresses from Table 18.1 exert an influence on a process (arrows) or a state (box). The general model is intended to show that analysis and interpretation is partially accomplished in actuality through the use of specific process models such as NuCM and Century, which represent subunits of the general model. These process models are described in Arbaugh et al. (this volume).

Other scenarios that would have an indeterminate adverse influence that is difficult to express in the context of the general model are reversals of environ-

Table 18.1. Partial List of Scenarios Related to Air Pollution That May Occur Within the Next 30 Years in the San Bernardino Mountains

High likelihood of occurrence

1. Increased pressures from population growth in the Los Angeles Basin leading to increased air pollution exposure in the mountains.
2. Increased year-round human population in the San Bernardino Mountains leading to increased disturbance of stands as new space is occupied.
3. Increased forest density of ozone-tolerant species creates more live fuel for potential wildfires and puts additional competitive stress on the remaining overstory trees, making them more susceptible to air pollution injury.
4. Elevated N deposition negatively affects the competitive ability of some forest species.
5. Changes in soil chemistry from air pollution deposition.
6. Increased mortality rates largely from bark beetles, exacerbated by air pollution injury, is triggered by lengthy droughts.

Intermediate likelihood of occurrence

7. Global warming and attendant climate changes exacerbates air pollution exposure and stress on forest trees.
8. Decline in numbers of genotypes of trees, herbs, and lichens previously resistant to ozone injury, following long-term exposure.
9. Decline in frequency of good cone crops in conifer species due to increased air pollution and environmental stress causes successional shift toward chaparral dominance on dry sites.

Low likelihood of occurrence

10. New synergistic interactions between ozone and minor components of air pollution increase tree stress.
11. Air pollution injury exacerbated due to increasing levels of ultraviolet B radiation if stratospheric ozone reaches new levels of depletion.

mental protection and environmental policy regulations that diminish the choices that managers have available to prevent pollution or carry out mitigating strategies. Also, visual aspects of the recreational experience may be diminished due to decreasing air quality and depreciation of forest appearance.

The scenarios identified in Table 18.1 range across many issues that relate to the projected increase in human population in southern California, changes in the air pollution response of the forest ecosystem due to prolonged exposure to stress, changes in physiological stress on forest trees as a result of global climate change, and the potential adverse effect of a changing political climate. The possibilities for forest management to alleviate the effects of these events on the forest ecosystem may be diminished because of urbanization, regulation of other agencies, and greater stakeholder pressure.

A process or method for evaluating the influences of different scenarios and a characterization of the risk posed by these scenarios is represented in Figure 18.1. It provides the framework in which risk assessors and risk managers can discover possible remedies.

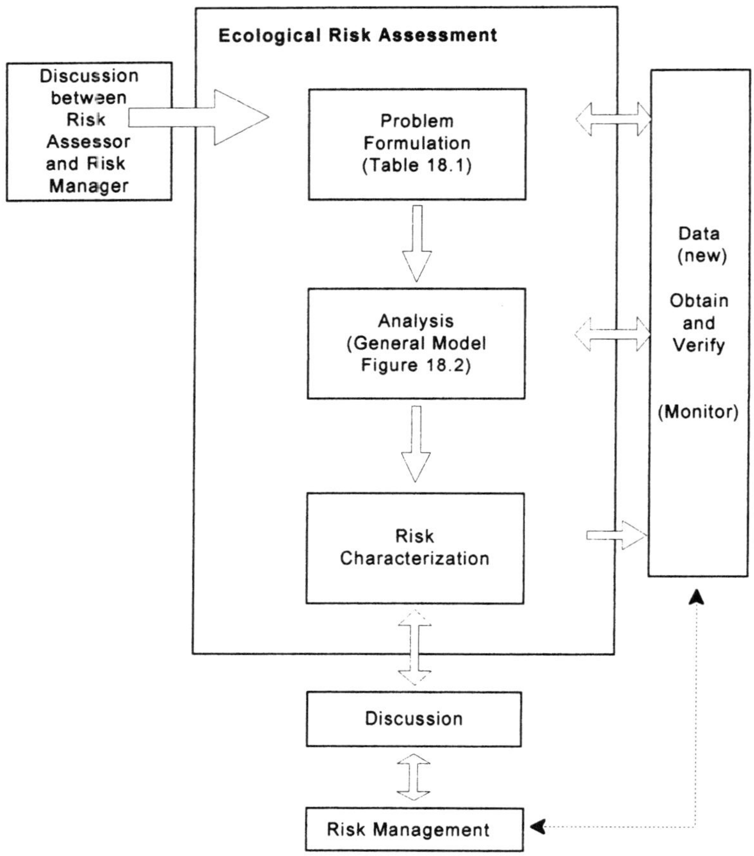

Figure 18.1. Paradigm for ecological risk assessment. (Modified from U.S. EPA 1992.)

Actions to Address the Ecological Risks Associated with Air Pollution in the San Bernardino Mountains

Sustainability of Ecosystems

The foremost management goal of federal and state agencies responsible for the management of public lands must be the sustainability of ecosystem structure and function. The U.S. Forest Service's San Bernardino Forest Management Plan de-emphasizes timber as a product from the forest; however, recreation and water have largely replaced timber as leading forest products. Timber harvesting in the past contributed to a reduction in fire hazard (Minnich, this volume). The Forest Service plan recognizes that the maintence of a healthy forest is essential for fire management and the protection of wildlife habitat and watershed stability. Very selective mechanical thinning is proposed to increase the vigor and survivability

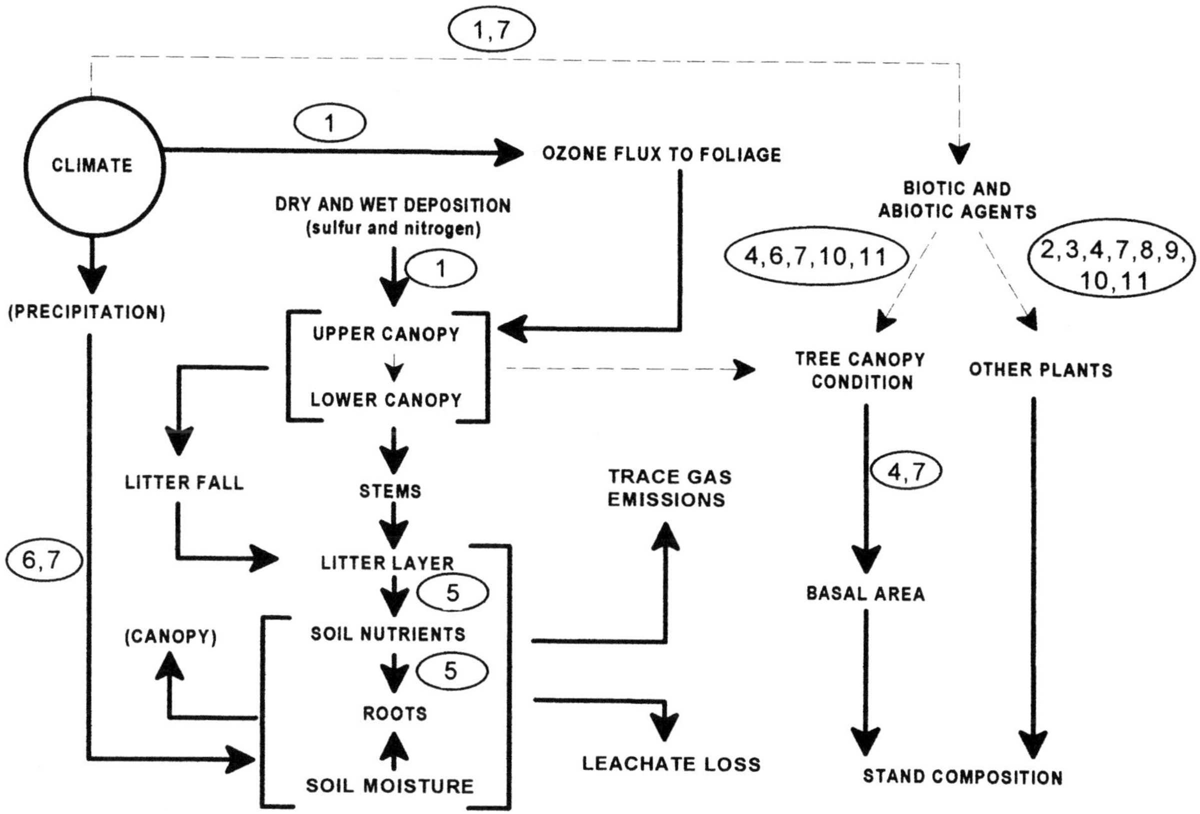

Figure 18.2. General model of forest ecosystem response to air pollution in which numbers placed on arrows represent the influences of scenarios listed in Table 18.1.

of the remaining trees. Heavy fuel removal by prescribed burning and other means will decrease the risk of losing the present overstory to catastrophic fires.

One of the most important actions to maintain forest health will be to minimize the production and delivery of air pollution into the San Bernardino Mountains. Progress was made in the control of ozone in the 1980s by reducing emissions from automobiles and industrial sources in the Los Angeles Basin (Miller et al. 1989). The growing population of Los Angeles and its surrounding cities will reverse this trend unless more stringent regulations are enacted and enforced. Technological innovation (e.g., electrical cars, low-pollution fuels) may offer the possibility of pollution reduction, but enforcement of existing and/or new regulations will be of paramount importance in minimizing air pollution.

The deposition of nitrogen and particulates remains a concern. Because they accumulate and can exert either primary effects or perhaps interactive effects with other pollutants, particularly ozone, there is a need to be cautious about these possible effects and to express a need for reduction in emissions.

Urbanization

Urbanization will intensify both on existing private lands inside national forest boundaries and at the periphery of federal land. The increase in urbanization outside of federal land will have a major impact on air quality in the forests of the San Bernardino Mountains because of the denser emissions from mobile sources (cars) coupled with the air transport characteristics of southern California (Fujioka et al., this volume). Within forest boundaries, urban areas must evaluate the needs to regulate the production of air pollution from the burning of fire wood and solid wastes. The influence of polluted stormwater runoff from surfaces of asphalt streets and parking lots within these urban areas may exacerbate the impact of airborne pollutants on the forest trees and directly affect aquatic habitats. Communities should be encouraged to set aside "natural areas" and consider the efficacy of air pollution–tolerant street and yard trees in reducing air pollution (Nowak 1994).

Global Climate Change

Global climate change is likely to exacerbate the impacts of air pollution on the forest ecosystem in the San Bernardino Mountains. Global models have predicted future increases in temperature and changes in several forest types in California (Botkin et al. 1991). However, one must try to interpret climate change effects as close to the landscape level as possible. The initial modeling experiment using the Century model in Arbaugh et al. (this volume) suggests that the San Bernardino Forest ecosystem is fairly robust (able to adjust) for the projected mean climate change predicted, but it is more susceptible to the combined effects of ozone and N deposition. Research to better understand the synergistic relationship between changing patterns of temperature, precipitation, air pollution, and ecosystem response is also critical. Modeling is one of the most useful research tools available to do this.

Assessment and Monitoring

The effort to achieve sustainable ecosystems can be maintained only if there is a database on the condition of particular resource components or end points that is constantly replenished with new information. Air pollution monitoring must be continued and extended in the forests of the San Bernardino Mountains. The permanent plots established in the Environmental Protection Agency study (Taylor 1974) should be revisited periodically, and the need should be considered to establish a new series of plots for the observation of air pollution injury. The National Science Foundation should consider the establishment of a long-term ecological research site in the San Bernardino Mountains for the continued observation of a western forest ecosystem under air pollution stress.

Collaborative Stewardship

The actions that will be needed to maintain a sustainable ecosystem will require closer collaboration among stakeholder groups. Some existing examples of collaborative resource management planning show the ways in which public awareness and participation in efforts to minimize air pollution effects may be encouraged. The work of the Quincy Library Group and the California Natural Communities Conservation Planning program (Wheeler 1993) are examples of bringing together resource managers, land-use planners, and local community members. Education and communication must be encouraged for all age groups. A model for future action in education in the San Bernardino National Forest is represented by the San Gorgonio Volunteer Association (SGVA), dedicated to protecting the San Bernardino National Forest and the San Gorgonio Wilderness Area. The activities of members include nature interpretation and other assistance to visitors, litter pickup, and trail maintenance. The SGVA web page provides a wealth of information including links to other web sites describing recreational uses of public lands. The Wilderness Society, the Izaak Walton League, and similar organizations promote the observance of National Lands Celebration Day. Many events across the nation celebrate the 623 million acres of parks, forests, and other natural areas and the benefits we derive from this natural heritage. The National Children's Forest, located near Running Springs, California, and the associated program encourage children to have a feeling of ownership and to develop an attitude of personal responsibility for resources and their management through a variety of educational projects.

Summary

The San Bernardino Mountains and their surrounding area are subject to many adverse influences that both directly and indirectly influence air pollution and the response of the forest ecosystem to air pollution. Oxidant air pollution effects from ozone may have peaked in the 1970s; however, predictions of an increasing human population (United Nations 1995) suggest increasing levels of ozone in the future. The continuing accumulation of nitrogen from air pollution will grow in importance as it affects soils and water quality. Many adverse events such as fire will occur and result in immediate disturbances. The recovery from these disturbances can be influenced by air quality. The ecological risk assessment procedure is proposed as a method for analyzing adverse impacts and identifying response strategies. Management actions at the local level that are intended to sustain the integrity and productivity of ecosystems will require increasing levels of cooperation between forest resource managers, land-use planners, and stakeholders. At the state, national, and international levels, a variety of new technologies and policies will need to be adopted to minimize background levels of air pollution and global climate change.

References

Botkin, D.I.; Nisbet, R.A.; Bicknell, S.; Woodhouse, C.; Bentley, B.; Ferren, W. Global climate change and California's natural ecosystems. In: Knox, J.B.; Foley Scheuring, A. Global climate change and California: potential impacts and responses. Berkeley, CA: University of California Press; 1991:123–149.

Lackey, R.T. Ecological risk assessment. Fisheries 19(9):14–18; 1994.

Lackey, R.T. The future of ecological risk assessment. Hum. Ecol. Risk Assess. 1(3):1–5; 1995.

Lackey, R.T. Challenges to the use of ecological risk assessment to implement ecosystem management. Water Resources Update 103:46–49; 1997.

Miller P., Bytnerowicz, A.; Fenn, M.; Poth, M.; Temple, P.; Schilling, S.; Jones, D.; Johnson, D.; Chow, J.; Watson, J. Multidisciplinary study of ozone, acidic deposition and climate effects on a mixed conifer forest in California, USA. In: Stress factors and air pollution, proceedings of the 17th international meeting for specialists in air pollution effects on forest ecosystems, Florence, Sept 14–19, 1996. Chemosphere 36:1001–1006; 1998.

Miller, P.R.; McBride, J.R.; Schilling, S.L.; Gomez, A.P. Trends in ozone damage to conifer forests between 1974 and 1989 in the San Bernardino Mountains of southern California. In: Olson, R.K.; Lefohn, A.S., eds. Effects of air pollution on western forests. Pittsburgh, PA: Air and Waste Management Association; 1989:309–323.

Myers, N. Consumption: challenge to sustainable development. Science 276:53–55; 1997.

Nowak, D.J. Air pollution removal by Chicago's urban forest. In: McPherson, E.G.; Nowak D.J.; Rowntree, R.A., eds. Chicago's urban forest ecosystem: results of the Chicago urban forest climate project. Gen. tech. rep. NE-186. Radnor, PA: U.S. Department of Agriculture, Forest Service, Northeastern Forest Experiment Station; 1994:63–82.

Taylor, O.C. Oxidant air pollutant effects on a western coniferous forest ecosystem. Task D report, annual progress 1973–74. EPA-600/3-78-052d, U.S. Environmental Protection Agency. NTIS No. PB 281 858/AS. Riverside, CA: Air Pollution Research Center, University of California; 1974.

United Nations. World population prospects: the 1994 revision. New York: United Nations Population Division; 1995.

U.S. EPA. Framework for ecological risk assessment. EPA/630/R-92/001. Washington, DC: U.S. Environmental Protection Agency; 1992.

Wheeler, D.P. Resource management in an era of biodiversity. In: Keeley, J.E., ed. Interface between ecology and land development in California. Los Angeles: Southern California Academy of Sciences; 1993:9–10.

19. Implications of Chronic Air Pollution in the San Bernardino Mountains for Forest Management and Future Research

J.R. McBride and P.R. Miller

Findings of Individual Chapters Concerning Air Pollution and the Mixed Conifer Forest Ecosystem

The research brought together in this book documents a history of chronic air pollution in the San Bernardino Mountains (SBM). The decline of ponderosa pine trees first noted in the 1950s was the result of exposure to ozone, a toxic component of photochemical smog (Miller and McBride, introduction to this volume). Tree mortality reached a peak during the drought years of the 1970s. Although air pollution levels dropped during the 1980s and early 1990s, foliar injury symptoms are still in evidence, and some tree mortality continues. A similar history is unfolding in other parts of the world where photochemical smog moves from urban and industrial areas into adjacent forests (De Lourdes De Bauer and Krupa 1990). Some areas, such as the pine and sacred fir forests in the mountains southwest of Mexico City, have already shown high levels of mortality similar to those in the SBM in the 1970s (McBride and Miller 1987). The objectives of this chapter are to review the information brought together in this book and the implications of that information for forest management and future research in the SBM.

The SBM are typical of other mountains in the Transverse and Peninsular Ranges of southern California (Poth and Wohlgemuth, this volume). These authors provide an historical account of the processes and events that have built and shaped these mountains, which are composed of five geomorphic subunits: the

Northern Plateau, the Eastern Ramp, the San Gorgonio Massif, Yucaipa Ridge, and the Santa Ana Basin. They discuss soil-forming factors and describe the characteristics of soils that support the present vegetation. The climate of southern California includes distinct patterns of airflow that vary with season (Fujioka et al. and Watson et al., this volume). Photochemical oxidants are generated from automobile exhaust and other sources of hydrocarbons in the South Coast Air Basin to the west of the SBM. These and other pollutants are often concentrated due to persistent temperature inversions. In addition to the transport of pollutants by summertime regional airflow patterns, there is an enhancement of pollutant drift upward into the middle to higher elevations of the mountains as a result of a chimney effect as the south- and southwest-facing slopes warm up in the afternoon (Miller et al. 1986). The complex interactions of the strength of the inversion, the velocity of offshore winds, and the difference in atmospheric pressure between the ocean and the desert influence the pulse of air pollutants moving into the mountains (Fujioka et al., this volume). Passive ozone monitors show that ozone concentrations tend to reach a maximum at about 1,525 m above sea level (a.s.l.) and decrease with increasing elevation to the highest point of observation at 2,797 m a.s.l. on San Bernardino Peak (Bytnerowicz et al., Chap. 11, this volume). The reversal of flow to downslope at night results in a return of pollution to midelevations locations such as Barton Flats (Watson et al., this volume).

In addition to ozone, other pollutants move into the mountains. Probably, most of these significant from the standpoint of the forest ecosystem in the SBM are the oxides of nitrogen in both gaseous and particulate phases (Watson et al., this volume). Nitrate aerosol is an important component of air pollution from the standpoint of the reduction of visibility and the impairment of the visual enjoyment of the mountain landscape (Copeland, this volume). These authors monitored visibility impairment in the vicinity of the San Gorgonio Wilderness Area from 1988 to 1995 as part of the IMPROVE (Interagency Monitoring of Protected Visual Environments) program. The various other chemical components of the aerosol besides nitrates were sulfate, organics, soil, and soot. The spring months had the highest levels of visibility impairment followed by summer months and autumn months. The winter months had the least visibility impairment.

The vegetation of the SBM is not unlike that of other higher-elevation ranges in southern California and northern Baja California (Minnich, this volume). The south and western slopes of these mountains support coastal sage scrub at lower elevations and chaparral at midelevations. These scrub formations are interrupted by riparian woodlands, coast live oak woodlands, knobcone pine woodlands, and isolated groves of bigcone Douglas-fir. The occurrence of these other vegetation types, within the broad belt of coastal sage scrub and chaparral, is due to local variations in soil parent material, alluvial deposits, and topography. On the lower east- and north-facing flanks of the mountains, one encounters scrub types dominated by desert species, which give way at midelevations to Great Basin sagebrush. These scrub and chaparral formations are also intersected by riparian woodlands and interrupted at midelevations by pinyon-juniper woodlands on rockier soils. Above the scrub and woodland formations on the west- and south-facing sides of the mountains occurs a mosaic of oak woodlands and pine forests. The oak woodlands are dominated by several evergreen oak species on drier east- and north-facing slopes. Woodlands of California black oak may occur on a variety of slopes where recent fires have destroyed the former codominant pines on these sites. The pine forests of these upper slope elevations are dominated by coulter pine on the more xeric sites and ponderosa pine on the more mesic sites. The pine forest belt blends into an extensive mixed conifer forest at higher elevation. Dominance of this forest was shared by ponderosa pine, white fir, incense cedar, and sugar pine in presettlement times. Jeffrey pine replaces ponderosa pine as the elevation increases with greater distance eastward. Selective logging and fire suppression have shifted the dominance of this mixed conifer forest toward white fir and incense cedar. On the slopes above the mixed conifer forests in the SBM, white fir and Jeffrey pine form stands that give way to stands of lodgepole pine or limber pine. A timberline occurs near the crest of San Gorgonio Mountain at the eastern edge of the SBM. Above this timberline is the alpine fell-field community.

The dynamics, composition, and structure of the vegetation types in the SBM have been influenced by a changing pattern of fire frequency, intensity, and extent of burning. Minnich (this volume) has characterized the fire history of the mountains as changing from frequent, low-intensity fire of small extent to less frequent, high-intensity fire of considerable extent. The consequence of less frequent fires has been an increase in the density of forest trees marked by an increase in less-fire-resistant species such as white fir and incense cedar in the mixed conifer forest. When fires do occur, they are more destructive of the existing vegetation. Stand-replacing fires in the latter half of the twentieth century have reduced some areas of the mixed conifer forest to chaparral.

The response of the dominant tree species, and particularly ponderosa and Jeffrey pines, to the air pollutants moving into the forest is a function of the physiological uptake of the ozone through the stoma and the impact of pollutants on leaf surfaces (Grulke, this volume; Bytnerowicz et al., Chap. 11, this volume). Nitric acid, resulting from both wet and dry deposition, is the principal pollutant, in addition to ozone, to which the trees are exposed (Bytnerowicz et al., Chap. 12, this volume). The physiological impact of ozone is the reduction of photosynthetic rate and the production of carbohydrates, however, the seasonal course of stomatal conductance, as regulated by water availability, is a key process controlling injury development each summer (Grulke, this volume). Leaf tissue responds to ozone exposure by developing a pattern of chlorotic mottling, which leads to necrosis of leaf tissue and the eventual death of the leaf (Miller and Rechel, this volume). The elements of crown injury including number of needle whorls, the presence or absence of chlorotic mottle on needles of each whorl, needle length, and percentage live crown (or lower branch mortality) have been used in indices that chronicle changes in crown condition of ponderosa and Jeffrey pines in plots located across the gradient of decreasing pollution. The simplest of these indices, called the FPM (forest pest management) index, shows that crown condition may have improved a little in the 1980s; however, there was no significant change in the 1990s. The difference in condition between high-exposure and low-exposure plots has remained the same between 1974 and 1994 (Miller and Rechel, this volume).

Reduced photosynthetic productivity limits the carbohydrates available for the growth of the tree. Ring width and the increment of cross-sectional growth are reduced where ponderosa and Jeffrey pines and bigcone Douglas-fir are exposed to chronic levels of oxidant air pollution (Arbaugh et al., Chap. 8, this volume). Decreased crown vigor and decreased tree growth predisposed pine trees to attack by insects, primarily the western bark beetle, which results in the death of the tree. Root diseases and dwarf mistletoe on pines are important pests that can act alone or sometimes act in conjunction with bark beetles to cause tree death, but the prevalence of these two diseases or their severity seems not to be acting in coincidence with air pollution exposure under present circumstances in the SBM (Pronos et al., this volume).

The work of Nash and Sigal (this volume) demonstrated that a major decrease in species richness and the percentage cover of remaining epiphytic lichens is evident in the SBM compared with earlier survey records and compared with a nearly pristine site in the Cuyamaca Mountains in southern California. Temple (this volume) has investigated a number of understory herbs and shrubs in the mixed conifer zone. In the areas of highest ozone exposure, he found that elderberry (*Sambucus mexicana*) and mugwort (*Artemisia douglasiana*) could be used as monitors of ozone injury because they occupy a fairly extensive habitat between them and because the symptoms were easy to distinguish. These studies demonstrated that the effects on the forest are not limited to the tree species.

The forest canopy is generally very discontinuous in uniformity of cover and variable in species composition. Both factors increase the difficulty of assessing canopy effects on gas and particle deposition rates; however, it was demonstrated that ozone concentration decreased within the canopy (presumably by both active and passive mechanisms) compared with above canopy. Also, particle deposition (e.g., nitrate) was higher in the upper canopy than in the lower canopy. This is primarily governed by the increased wind speed at the top of the canopy because air concentrations of particulate species were found to be uniform throughout the canopy (Bytnerowicz et al., Chap. 11, this volume).

Other consequences of air pollution in the mixed conifer forest ecosystem in the SBM involve the effects of deposition of nitrogen and other airborne materials to the canopy and the forest floor. Through both wet and dry deposition, nitrogen input into the system has increased (Bytnerowicz et al., Chap. 11, this volume). This increased input of nitrogen has affected the nitrogen cycle of the forest soils by increasing the levels of nitrate nitrogen in the cycle and generally elevating the nitrogen level in the forest soils (Fenn and Poth, this volume). The long-term effects of the combination of ozone and nitrogen deposition to the canopy of the dominant primary producer species, such as ponderosa pine in the SBM, and the accumulated effects on forest soil can be investigated by computer simulation models such as NuCm (Nutrient Cycling Model) and the forest version of Century, a carbon fixation model (Arbaugh et al., Chap. 16, this volume). NuCM results indicate that most areas of the SBM have a nitrogen deficiency at current deposition rates, and both the NuCM and Century model also indicated that increased N deposition rates could lead to N saturation at some sites. Century simulated ozone effects as a slower turnover of foliage in the system and predicted lower accumulation of woody biomass. The loss in primary production of species such as ponderosa pine has serious implications for long-term forest health. Reduced primary production leads to the reduction in cone and seed production, reduction in resin pressure, and the potential exacerbation of drought stress. These changes emanating from the loss of primary productivity set the stage for a number of negative feedbacks within the mixed conifer forest ecosystem. Reduced primary productivity of the forest dominants also has consequences for any future extraction of commercial forest products.

The reduction of cone and seed production in ponderosa pine noted by Luck (1977) weakens its regeneration potential. The projections of forest succession by McBride and Laven (this volume) suggest a shift in dominance toward white fir and incense cedar under conditions of continued chronic air pollution. Their work does not predict a crash of the mixed conifer ecosystem, but a change in the composition of dominant species. This shift in dominance away from ponderosa pine to white fir and incense cedar was noted by Minnich (this volume) in his reexamination of forest survey plots. He concluded that it was a response to the increased time intervals between fires. The longer intervals between fires, which have resulted from fire suppression in the SBM, have allowed fire-sensitive species such as white fir and incense cedar to become established in greater numbers. This phenomenon is, no doubt, augmented by the impacts of air pollution on the reproductive capacity of the pines. The implications of this change for wildlife habitat, fire management, and production of high-value forest products are significant. The white fir and incense cedar dominance of the mixed conifer forest will result in a shift away from the more open habitat of the ponderosa pine–dominated mixed conifer forest. The forest habitat will become darker, exhibit a more diverse vertical structure, and provide a different menu of plant food for wildlife species. These changes are expected to favor more interior, late successional species such as brown creeper, pygmy nuthatch, and Douglas squirrel over species such as the common flicker, olive-sided flycatcher, and the golden-mantled ground squirrel (Verner and Boss 1980). This change in biodiversity is not of itself a negative change, but it may involve species of special interest that depend on the habitat conditions of the ponderosa pine–dominated mixed conifer forest.

The decrease in resin pressure in conifers suffering from air pollution injury predisposes these trees to attack by bark beetles. The work of Pronos et al. (this volume) summarized observations over several years of the development of bark beetle mortality centers around trees showing severe injury symptoms. In contrast to the populations of bark beetle adults produced from trees not showing air pollution injury symptoms, populations arising from trees with severe foliar injury symptoms are smaller. The reduced populations of bark beetles emerging from air pollution–stressed trees are due to the reduced cambial thickness of trees with lower primary productivity. However, it took a smaller number of bark beetles to successfully attack weakened trees. The pattern of distribution of severely injured trees observed in the 1970s at the height of air pollution injury and the reduced populations of adult bark beetles emerging from these suggests that bark beetle–caused mortality should not lead to the collapse of the ponderosa pine–dominated mixed conifer forest ecosystem. Loss of individual trees and groups of trees in mortality centers would, however, contribute to the shift in dominance to white fir and incense cedar.

The reduction of primary productivity of the dominant species in the mixed conifer forest ecosystem due to air pollution stress would exacerbate the stress on trees during drought years (Arbaugh et al., Chap. 8, this volume). The climate of the region is characterized by periodic droughts (Fujioka et al., this volume). The mid-1970s were years of severe drought and high tree mortality in the SBM. Drought stress reduces the photosynthetic capacity of the trees and results in reduced levels of carbohydrates, which can be used for growth, tissue repair, and cone production. Drought stress also contributes to the predisposition of trees to attack by insects and pathogens, increased fire hazard, and failure of regeneration. The combined effects of drought stress and air pollution injury present the greatest threat to the mixed conifer forest ecosystem in the SBM. A possible scenario involving widespread fires during drought years characterized by high levels of air pollution, could lead to the replacement of the significant portions of the mixed conifer forest by stands of chaparral and oak woodland. This change, coupled with global warming and increasing levels of air pollution, could make the eventual recovery of the mixed conifer forest through natural succession problematic. However, the Century simulation (Arbaugh et al., Chap. 16, this volume) suggests that the projected global climate changes of the next millennium would not cause species of the mixed conifer forest to be unable to respond successfully, apparently because they already experience very large year-to-year differences.

Implications for Policy Formulation

The mixed conifer forests of the SBM and other mountain areas of southern California under the administration of the U.S. Forest Service are not currently being managed for timber products. Some forest harvesting does occur in these mountains on privately owned lands. Air pollution injury to the foliage of trees results in decreased levels of the growth (Arbaugh et al., Chap. 8, this volume) This reduction in productivity coupled with the change in tree species composition is bringing about a loss of value to those private lands managed for timber products and a potential loss in value of federal lands should the American people reverse their attitude concerning the management of Forest Service lands at some time in the future A shift to white fir and incense cedar dominance of the forests would mean a considerable shift in value. The average stumpage price (price paid for live trees to be harvested, measured in units of 1,000 board feet (12″ × 12″ × 1″) in 1997 of ponderosa pine in California was about three times that of white fir and incense cedar. The reduction in volume growth of 60% in ponderosa pine due to air pollution injury reported by McBride et al. (1975) would mean a considerable reduction in the sustainability of profits to forest landowners managing land for timber products in the SBM.

In addition to the impacts of air pollution on the components and processes in the mixed conifer forest ecosystem of the SBM, there is an important human dimension to consider (e.g., the San Bernardino National Forest reported more than 6 million visitor-days in 1995; USDA Forest Service 1996). It could be argued that human health may be affected by air quality inasmuch as the California Air Resources Board has included the Lake Arrowhead area in a southern California epidemiological study of the health of school children. The area was selected largely because ozone daily averages in the mountains can be higher than in nearby basin cities (Watson et al., this volume). The results of this study will not be known for several more years. Winter (this volume) points to surveys in which approximately 80% of both wilderness users and nonusers listed the protection of air quality among the top five purposes of wilderness. The impact to visibility and the visual characteristics of the forest landscapes were reported to be affected by Copeland (this volume). This reduced visibility did not, however, play a significant role in the decision process that recreationists used in deciding to go into the mountains (Winter, this volume); however, the majority of San Gorgonio visitors polled are residents of the Los Angeles Basin and may have developed a greater tolerance to impaired visibility and other degradative features of air pollution (photochemical smog). This attitude may largely be due to the failure of the various scientific disciplines represented by the chapters of this book to communicate their results to the general public.

Implications for Forest Management

Miller and McBride (this volume) identified a number of directions that should be taken to reduce the generation of air pollution outside of the SBM. These measures must be taken if a long-term solution is to be found for the problem. A chronic problem will continue in the SBM unless the emission of primary, smog-forming air pollutants can be controlled. Until that time, there are some forest management practices that could be used to address the problems associated with air pollution discussed above. These include forest thinning, fire hazard reduction, regeneration, and group selection management.

Forest thinning could be used to address the problem of tree stress, which exacerbates the effects of air pollution injury. Minnich (this volume) has documented the increase in tree density resulting from fire protection in the forests of the SBM. As tree density increases, competition for water and light increases in a forest. A thinning program would reduce competition for soil moisture and provide greater access to sunlight for the remaining trees. The literature of forestry documents numerous examples of the growth response of trees following thinning (Daniels et al. 1979; Nyland 1996; Smith et al. 1996). A growth response was noted by Arbaugh et al. (Chap. 8, this volume) during the above-average rainfall years in the SBM in the early 1980s. Thinning could not only improve tree growth and the primary productivity of the forest ecosystem, but it could be used to favor the retention of ponderosa pine trees in the mixed conifer forest ecosystem. A thinning program that would favor pine could help to sustain the potential of pine to remain a viable part of the forest ecosystem. A thinning program that favored pine would also help to sustain the long-term profitability of managing the mixed conifer forest for timber products.

Management for fire hazard reduction is a major obligation of resource management agencies in the SBM as well as other wildland areas in southern California. The accumulation of surface fuels, the development of ladder fuels, and the shift in forest species composition to less-fire-resistant species is a major problem threatening the safety of individuals living in the mountains and the sustainablilty of the natural ecosystems. Fire hazard reduction should start with the thinning of forest stands proposed above. Thinning will be required to prepare the stands for prescribed burning necessary to reduce the accumulations of surface fuels. There is no practical way at present to eliminate the accumulated surface fuels other than by prescribed burning. Implementation of a prescribed burning program on the scale needed in the SBM will require cooperation with local air pollution control boards, California Department of Fish and Game, and an education program for local residents and visitors to the forest. The new program being undertaken in the Tahoe Basin in central California may serve as a model for the type of program needed in the SBM. An effective program of fire hazard reduction will help to stem the shift in species composition toward white fir and incense cedar. It will, no doubt, have other benefits to the mixed conifer forest ecosystem, such as the stimulation of mineral cycling (release of nitrogen into available forms for postfire species) and improvement of runoff water quality where calcium released can precipitate the phosphorus before it enters streamwaters and lakes where it can cause eutrophication.

Forest management activities should address the issue of tree regeneration in the mixed conifer forests of the SBM. Two aspects of this issue are related to the problem of continuing chronic air pollution. The first is the decrease in cone and seed production of trees exhibiting air pollution injury. The second is the failure of forest regeneration in areas burned by wildfire during the twentieth century. The decrease in cone and seed production, primarily in ponderosa pine, is resulting in decreased regeneration of pine. This is exacerbated by increased tree density in the forest as a result of fire suppression and increasing litter depths. Litter depth has increased as a result of fire suppression and loss of needle retention caused by air pollution. The thinning proposed above will be necessary to reduce competition for sunlight to achieve successful regeneration of ponderosa pine. Prescribed burning will reduce the thick layers of needles that interfere with pine seedling establishment. These measures alone cannot ensure successful pine regeneration unless a cone and seed production occurs in the forest. An aggressive planting program may need to be developed by forest managers to ensure that ponderosa pine remains a component of the mixed conifer forest.

Areas where the mixed conifer forest has been eliminated by wildfires should be examined for their potential for restoration management. Restoring these areas through a planting program may be in the best long-term interest of fire safety, recreational use, timber production, and watershed management. Site preparation involving the removal of brush species and, on some sites, the thinning of oak trees will be required prior to planting. Control of competition following planting will be a key to successful regeneration of these areas. Any planting program must address the question of the seed source used for planting. Local seed sources must be used to maintain the genetic biodiversity of the area and to ensure the greatest long-term survival of the planted trees. Research by Temple et al. (1993) indicated that genetic resistance to air pollution injury exists in seedlings produced from seeds collected from trees in the Sierra Nevada that do not exhibit foliar injury symptoms. An effort should be made to identify trees without foliar injury symptoms to use as seed sources for regeneration projects.

In those portions of the SBM where the objective of the owners of forestland is to sustain profitability through timber management, forest harvesting will take place. Sanitation-salvage logging has been practiced in the past to address the problems of insects and disease in the forests of southern California. This management strategy attempts to reduce the impacts of insects and disease by removing dead trees and trees that are judged to be at risk. As a management practice, it does not always ensure the regeneration of the forest and it does not provide a very valuable forest planning approach to sustaining the profitability of forest management. A flexible group selection approach to the management of the mixed conifer forest would provide a better way of planning forest operation to maintain profitability, ensure the regeneration of all species, and provide some flexibility in dealing with tree mortality. Group selection involves harvesting small areas of trees (usually less than 3 acres). The opening created by group selection provides sufficient sunlight for the regeneration of ponderosa pine but is not usually offensive to the public opposed to clear-cutting. The shape of the openings can reflect the characteristics of local terrain and the pattern of age or size classes in the forest. The initial layout in a flexible group selection approach could address current problems associated with dead and dying trees. Subsequent planned entries could have the flexibility of "picking up" new mortality centers.

Implications for Research

Future research is needed to carry on the unfinished business of previously initiated research and to address emerging areas related to those potential events identified by Miller and McBride (this volume). Some chapter authors in this book have indicated areas where further research is needed to further understand the areas they were addressing. It is not the purpose of this summary to repeat those individual suggestions but to look at the broader implication of air pollution for research. Areas of major concern should be (1) the potential problems arising from global climate change, (2) the long-term response of the mixed conifer forest ecosystem to continued perturbation by air pollution, (3) the interaction of succession, fire risk, and air pollution, (4) potential synergistic or additive interactions of ozone and nitrogen pollutants, and (5) forest management activities to reduce the impacts of air pollution on the mixed conifer forest ecosystem. Research needed to address each of these concerns is suggested in the following paragraphs.

Potential Problems Arising from Global Climate Change

Global climate change will lead to increased temperature, decreased rainfall, and increased levels of ultraviolet B (UV-B) in the mountains of southern California. Research is needed to understand the potential interactions, depending on sequences of occurrence, between global warming, drought, and air pollution on the physiology and growth of the major forest dominants. Similarly, studies will be needed on the synergistic effects of increased UV-B on air pollution injury and forest health. The northwest-facing slope of San Gorgonio Mountain would provide an excellent location for such studies. It provides an elevation gradient along which both air pollution and UV-B vary. Growth chamber and open-topped chambers would also be amenable for the research needed to study global warming and decreased precipitation.

Long-Term Response to the Continued Ecosystem Perturbation by Air Pollution

Natural ecosystems have evolved over very long time periods in response to environmental gradients and the composition of the regional biota. The understanding of the mixed conifer forest ecosystem in the SBM brought together in this book is based on research covering only a few decades. A much longer observation period is needed to fully understand the long-term implications of air pollution for this ecosystem. Specific long-term studies should focus on the following.

Long-Term Effects of Air Pollution on Wildlife

Very little is known about the direct and indirect effects of air pollution on the wildlife. Long-term field observations are needed to understand the physiological and ecological impacts on wildlife species.

Long-Term Effects of Oxidant Air Pollution on the Breakdown of Organic Matter and Mineral Cycling

Studies of organic matter breakdown and mineral cycling in the northeastern United States (McDonell and Pickett 1990) have demonstrated an increase in the lignin content of leaves of hardwood species exposed to ozone. The increased lignin content interfered with the breakdown of leaf litter by soil organisms and slowed the rate of mineral cycling. Similar studies are called for in the SBM.

Long-Term Effects of Air Pollution on Characteristics of Forest Stands

Although modeling may provide some insights into the impacts of air pollutants on ecosystems, we do not currently have the data to appropriately model air pollution impacts at the forest stand level. Observations should be continued on the permanent plots established during the Environmental Protection Agency–sponsored study in the 1970s. Some of these plots have been heavily affected by human activity and are no longer suitable for continued observations. Other plots should be brought into a system of plots for long-term observation of forest stand level responses.

The need for long-term observations and study suggests that the National Science Foundation should be approached and asked to consider the establishment of a long-term ecological research site in the SBM. A designation of this type along with appropriate funding could help to sustain the long-term research that is needed.

Interaction of Succession, Fire Risk, and Air Pollution

Our understanding of the role of air pollution in the dynamics of the landscape in the SBM is far from complete. This is an area in which the newly emerging fields of landscape ecology and fire behavior modeling could be brought together to further our understanding. Specific research projects in this area should include the following.

Regional Landscape Succession Modeling

Research that brings together an understanding of the influence of air pollution on the dynamics of forest stands with the level of landscape information developed with geographical information system programs could be used to predict future landscape dynamics. Future mosaics of vegetation could be projected using different air pollution levels. Such an approach could provide resource managers with a better understanding of their future management problems and responsibilities.

Linking Air Pollution, Succession, and Fire Risk in the San Bernardino Mountains

Research could take the information generated by research on regional landscape succession one step farther by adding a component of fire behavior modeling. Recently developed fire behavior models, such as FARSITE, provide an opportunity to link stand conditions, topography, and climate to model fire behavioral characteristics. Application of this approach in the urban-wildland fringe in Oakland, California, has informed agencies responsible for vegetation management and fire suppression as to the major areas where the risk of fire is extreme. Research of this type that could bring in the air pollution component would be of great value in understanding the fire management problem in the SBM.

Potential Synergistic Interactions of Various Pollutants

More research is needed on the potential synergistic interactions of ozone and nitrogen air pollutants in the SBM. Nitrogen additions to plot with various levels of chronic ozone injury have just begun, and observation of results should be done for at least a decade.

Forest Management Research

Forest management practices can be used to alleviate the current level of injury to the mixed conifer forest of the SBM. What is needed are studies to compare different management treatments. Studies of the effects of forest thinning, manipulation of the forest litter through prescribed burning, and the identification and selection of genetically superior individuals (in terms of air pollution resistance) are called for. These studies could also guide forest practice in other areas where air pollution was not a problem.

References

Daniels, T.W., et al. Principles of silviculture. New York: McGraw Hill; 1979.

De Lourdes De Bauer, M.A.; Krupa, S.V. The valley of Mexico: summary of observational studies on its air quality and effects on vegetation. Environ. Pollut. 65:109–118; 1990.

Luck, R.F. Cone and seed production for dominant conifer tree reproduction. In: Miller, P.R.; Elderman, M.J., eds. Photochemical air pollutant effects on a mixed conifer ecosystem. Progress rep. 1974–75. EPA-600/3-77-058. Corvallis, OR: U.S. Environmental Protection Agency, Corvallis Environmental Research Laboratory; 1976:191–203.

McBride, J.R.; Miller, P.R. Response of American forests to photochemical oxidants. In: Hutchinson, T.C., ed. Proceedings of the acid deposition workshop, Univ. Toronto, May 12–17, 1985.,Toronto, Canada. New York: Springer-Verlag; 1987:217–228.

McBride, J.R.; Semion, V.P.; Miller, P.R. Impact of air pollution on the growth of ponderosa pine. Calif. Agric. 29(12):8–9; 1975.

McDonnell, M.J.; Pickett, S.T.A. Ecosystem structure and function along gradients of urbanization. Ecology 71:1231–1237; 1990.

Miller, P.R.; Taylor, O.C.; Poe, M.P. Spatial variation of summer ozone concentrations in the San Bernardino Mountains. Paper 86-39.2. Proceedings of the 79th annual APCA meetings, Minneapolis, MN. Pittsburgh, PA: Air Pollution Control Association; 1986.

Nyland, R.D. Silviculture: concepts and applications. New York: McGraw Hill; 1996.

Smith, D.M.; Larson, B.C.; Kelty, M.J.; Ashton, P. The practice of silviculture: applied forest ecology. New York: Wiley; 1996.

Temple, P.J.; Riechers, G.H.; Miller, P.R.; Lennox, R.W. Growth responses of ponderosa pine to long-term exposure to ozone, wet and dry acidic deposition, and drought. Can. J. For. Res. 23:59–66; 1993.

Verner, J.; Boss, A.S. California wildlife and their habitats: western Sierra Nevada. Gen. tech. rep. PSW-37. Berkeley, CA: Pacific Southwest Station, Forest Service, USDA; 1980.

Index

Ecological Studies

Volumes published since 1992

Volume 89
Plantago: A Multidisciplinary Study (1992)
P.J.C. Kuiper and M. Bos (Eds.)

Volume 90
Biogeochemistry of a Subalpine Ecosystem: Loch Vale Watershed (1992)
J. Baron (Ed.)

Volume 91
Atmospheric Deposition and Forest Nutrient Cycling (1992)
D.W. Johnson and S.E. Lindberg (Eds.)

Volume 92
Landscape Boundaries: Consequences for Biotic Diversity and Ecological Flows (1992)
A.J. Hansen and F. di Castri (Eds.)

Volume 93
Fire in South African Mountain Fynbos: Ecosystem, Community, and Species Response at Swartboskloof (1992)
B.W. van Wilgen et al. (Eds.)

Volume 94
The Ecology of Aquatic Hyphomycetes (1992)
F. Bärlocher (Ed.)

Volume 95
Palms in Forest-Ecosystems of Amazonia (1992)
F. Kahn and J.-J. DeGranville

Volume 96
Ecology and Decline of Red Spruce in the Eastern United States (1992)
C. Eagar and M.B. Adams (Eds.)

Volume 97
The Response of Western Forests to Air Pollution (1992)
R.K. Olson, D. Binkley, and M. Böhm (Eds.)

Volume 98
Plankton Regulation Dynamics (1993)
N. Walz (Ed.)

Volume 99
Biodiversity and Ecosystem Function (1993)
E.-D. Schulze and H.A. Mooney (Eds.)

Volume 100
Ecophysiology of Photosynthesis (1994)
E.-D. Schulze and M.M. Caldwell (Eds.)

Volume 101
Effects of Land-Use Change on Atmospheric CO_2 Concentrations: South and South East Asia as a Case Study (1993)
V.H. Dale (Ed.)

Volume 102
Coral Reef Ecology (1993)
Y.I. Sorokin (Ed.)

Volume 103
Rocky Shores: Exploitation in Chile and South Africa (1993)
W.R. Siegfried (Ed.)

Volume 104
Long-Term Experiments With Acid Rain in Norwegian Forest Ecosystems (1993)
G. Abrahamsen et al. (Eds.)

Volume 105
Microbial Ecology of Lake Plußsee (1993)
J. Overbeck and R.J. Chrost (Eds.)

Volume 106
Minimum Animal Populations (1994)
H. Remmert (Ed.)

Volume 107
The Role of Fire in Mediterranean-Type Ecosystems (1994)
J.M. Moreno and W.C. Oechel (Eds.)

Volume 108
Ecology and Biogeography of Mediterranean Ecosystems in Chile, California, and Australia (1994)
M.T.K. Arroyo, P.H. Zedler, and M.D. Fox (Eds.)

Volume 109
Mediterranean Type Ecosystems: The Function of Biodiversity (1994)
G.W. Davis and D.M. Richardson (Eds.)

Volume 110
Tropical Montane Cloud Forests (1994)
L.S. Hamilton, J.O. Juvik, and
F.N. Scatena (Eds.)

Volume 111
Peatland Forestry: Ecology and Principles (1995)
E. Paavilainen and J. Päivänen

Volume 112
Tropical Forests: Management and Ecology (1995)
A.E. Lugo and C. Lowe (Eds.)

Volume 113
Arctic and Alpine Biodiversity: Patterns, Causes and Ecosystem Consequences (1995)
F.S. Chapin III and C. Körner (Eds.)

Volume 114
Crassulacean Acid Metabolism: Biochemistry, Ecophysiology and Evolution (1995)
K. Winter and J.A.C. Smith (Eds.)

Volume 115
Islands: Biological Diversity and Ecosystem Function (1995)
P.M. Vitousek, H. Andersen, and
L. Loope (Eds.)

Volume 116
High-Latitude Rainforests and Associate Ecosystems of the West Coast of the Americas: Climate, Hydrology, Ecology and Conservation (1995)
R.G. Lawford, P.B. Alaback, and
E.R. Fuentes (Eds.)

Volume 117
Anticipated Effects of a Changing Global Environment on Mediterranean-Type Ecosystems (1995)
J.M. Moreno and W.C. Oechel (Eds.)

Volume 118
Impact of Air Pollutants on Southern Pine Forests (1995)
S. Fox and R.A. Mickler (Eds.)

Volume 119
Freshwaters of Alaska: Ecological Synthesis (1997)
A.M. Milner and M.W. Oswood (Eds.)

Volume 120
Landscape Function and Disturbance in Arctic Tundra (1996)
J.F. Reynolds and J.D. Tenhunen (Eds.)

Volume 121
Biodiversity and Savanna Ecosystem Processes: A Global Perspective (1996)
O.T. Solbrig, E. Medina, and J.F. Silva (Eds.)

Volume 122
Biodiversity and Ecosystem Processes in Tropical Forests (1996)
G.H. Orians, R. Dirzo, and J.H. Cushman (Eds.)

Volume 123
Marine Benthic Vegetation: Recent Changes and the Effects of Eutrophication (1996)
W. Schramm and P.H. Nienhuis (Eds.)

Volume 124
Global Change and Arctic Terrestrial Ecosystems (1996)
W.C. Oechel (Ed.)

Volume 125
Ecology and Conservation of Great Plains Vertebrates (1997)
F.L. Knopf and F.B. Samson (Eds.)

Volume 126
The Central Amazon Floodplain: Ecology of a Pulsing System (1997)
W.J. Junk (Ed.)

Volume 127
Forest Design and Ozone: A Comparison of Controlled Chamber and Field Experiments (1997)
H. Sanderman, A.R. Wellburn, and
R.L. Heath (Eds.)

Volume 128
The Productivity and Sustainability of Southern Forest Ecosystems in a Changing Environment (1998)
R.A. Mickler and S. Fox (Eds.)

Volume 129
Pelagic Nutrient Cycles (1997)
T. Andersen

Volume 130
Vertical Food Web Interactions (1997)
K. Dettner, G. Bauer, and W. Völkl (Eds.)

Volume 131
The Structuring Role of Submerged Macrophytes in Lakes (1998)
E. Jeppesen, M. Søndergaard,
M. Søndergaard, and K. Christoffersen (Eds.)

Volume 132
Vegetation of the Tropical Pacific Islands (1998)
D. Mueller-Dombois and F.R. Fosberg

Volume 133
Aquatic Humic Substances (1998)
D.O. Hessen and L.J. Tranvik

Volume 134
Oxidant Air Pollution Impacts in the Montane Forests of Southern California: A Case Study of the San Bernardino Mountains (1999)
P.R. Miller and J.R. McBride (Eds.)